1800   1900

OTHER CONTRIBUTORS
Pierre de Fermat (1601-1665)
Michel Rolle (1652-1719)
Brook Taylor (1685-1731)
Colin Maclaurin (1698-1746)
Thomas Simpson (1710-1761)
Pierre-Simon de Laplace (1749-1827)
George Green (1793–1841)
George Gabriel Stokes (1819-1903)

Lagrange

Gauss

Cauchy

Riemann

Lebesgue

99) ⎯⎯⎯⎯⎯⎯⎯⎯⎯

agrange (1736-1813) ⎯⎯⎯⎯⎯⎯⎯⎯

⎯C. Gauss (1777-1855) ⎯⎯⎯⎯⎯⎯⎯⎯⎯

⎯A. Cauchy (1789-1857) ⎯⎯⎯⎯⎯⎯

⎯K. Weierstrass (1815-1897) ⎯⎯⎯⎯⎯⎯⎯

⎯G. Riemann (1826-1866) ⎯⎯⎯⎯

⎯J. Gibbs (1839-1903) ⎯⎯⎯⎯⎯⎯⎯

⎯S. Kovalevsky (1850-1891)

⎯H. Lebesgue (1875-1941) ⎯⎯⎯⎯⎯

Agnesi

Weierstrass

Kovalevsky

Gibbs

1756   1799   1821   1854   1873   1902

Gauss proves Fundamental
Theorem of Algebra

Riemann integral

Lebesgue integral

Lagrange begins
Mécanique analytique

Precise notion of
limit (Cauchy)

e is transcendental
(Hermite)

FORMULAS FROM GEOMETRY

Triangle

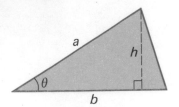

$\text{Area} = \frac{1}{2}bh$

$\text{Area} = \frac{1}{2}ab\sin\theta$

Parallelogram

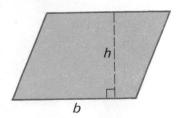

$\text{Area} = bh$

Trapezoid

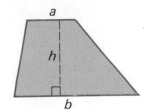

$\text{Area} = \frac{a+b}{2}h$

Circle

$\text{Circumference} = 2\pi r$

$\text{Area} = \pi r^2$

Sector of Circle

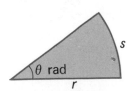

$\text{Arc length } s = r\theta$

$\text{Area} = \frac{1}{2}r^2\theta$

Polar Rectangle

$\text{Area} = \frac{R+r}{2}(R-r)\theta$

Right Circular Cylinder

$\text{Lateral area} = 2\pi rh$

$\text{Volume} = \pi r^2 h$

Ball

$\text{Area} = 4\pi r^2$

$\text{Volume} = \frac{4}{3}\pi r^3$

Right Circular Cone

$\text{Lateral area} = \pi rs$

$\text{Volume} = \frac{1}{3}\pi r^2 h$

Frustum of Right Circular Cone

$\text{Lateral area} = \pi s(r+R)$

$\text{Volume} =$

$\frac{1}{3}\pi(r^2 + rR + R^2)h$

General Cone

$\text{Volume} = \frac{1}{3}(\text{area } B)h$

Wedge

$\text{Area } A = (\text{area } B)\sec\theta$

CALCULUS
with Analytic Geometry

CALCULUS
with Analytic Geometry
FIFTH EDITION

Prentice-Hall, Inc., Englewood Cliffs, New Jersey 07632

Edwin J. Purcell
University of Arizona

Dale Varberg
Hamline University

Library of Congress Cataloging-in-Publication Data

Purcell, Edwin J. (Edwin Joseph), (date)
  Calculus with analytic geometry.

  Includes index.
    1. Calculus.   2. Geometry, Analytic.
  I. Varberg, Dale E.   II. Title.
  QA303.P99 1987   515'.15   86-17011
  ISBN 0-13-111105-1

Editorial/production supervision: Zita de Schauensee
Interior design: Jayne Conte
Illustrator: Eric G. Hieber
Manufacturing buyer: John Hall
Cover design: Lundgren Graphics, Ltd.
Cover photo: Megatek Corp.

Credit for chapter-opening illustrations. Chapters 1, 3, 4, 6, 7, 10, 12, 13, 14, 15, 17, 18: New York Public Library Picture Collection. Chapters 2, 8 , 11: The Bettmann Archive, Inc. Chapter 5: Culver Pictures, Inc. Chapters 9, 16: Bibliothèque Nationale.

Credit for chapter-opening quotations. Chapter 5: From Dirk J. Struik, *A Concise History of Mathematics,* vol. 2 (New York: Dover Publications, 1948), p. 232. Chapter 6: From Sir Thomas Heath, *A History of Greek Mathematics,* vol. 2 (Oxford at the Clarendon Press, 1921), p. 20. Chapter 13: From Howard Eves, *An Introduction to the History of Mathematics,* Fifth Edition (Philadelphia: Saunders College Publishing, 1983), p.265: © 1983 by CBS College Publishing. Chapter 18: From Lynn M. Osen, *Women in Mathematics* (Cambridge, MA: The MIT Press, 1974), p.117; copyright © 1974 by The Massachusetts Institute of Technology.

Credit for quotations in marginal boxes. Page 5: From G. H. Hardy, *A Mathematician's Apology* (New York: Cambridge University Press, 1941), p. 34. Page 129: From Galileo Galilei, *Saggiatore,* Opere VI, p. 232, as translated on the cover of George Polya, *Mathematical Methods in Science* (Mathematical Association of America; vol. 26 in the MAA New Mathematical Library Series). Page 331: From François Le Lionnais (ed.) *Great Currents of Mathematical Thought,* vol. 1 (New York: Dover Publications), pp. 68-69. Page 452: From Donald E. Knuth, "Computer Science and Its Relation to Mathematics" (*Mathematical Monthly* 81, 1974, p. 323).

Credit for quotation on front end papers: From the Foreword by Richard Courant to Carl B. Boyer, *The History of the Calculus and Its Conceptual Development* (New York: Dover Publications, 1949).

Calculus with Analytic Geometry, Fifth Edition
Edwin J. Purcell and Dale Varberg

Printed in the United States of America

10   9   8   7   6

ISBN 0-13-111105-1   01

Prentice-Hall International (UK) Limited, *London*
Prentice-Hall of Australia Pty. Limited, *Sydney*
Prentice-Hall Hispanoamericana, S.A., *Mexico*
Prentice-Hall Canada Inc., *Toronto*
Prentice-Hall of India Private Limited, *New Delhi*
Prentice-Hall of Japan, Inc., *Tokyo*
Prentice-Hall of Southeast Asia Pte. Ltd., *Singapore*
Editora Prentice-Hall do Brasil, Ltda., *Rio de Janeiro*

Contents

Appendix   823

Answers to Odd-Numbered Problems   842

Index   897

Preface

We believe that teachers and students who enjoyed earlier editions of this book will like this newest revision even more. We have sought to polish the rough edges, to clarify difficult ideas, and (dare we admit it) to correct a few errors. The organization and major features of the fourth edition have been retained. Perhaps the most notable change is the addition of **approximately 600 new problems**. The authors have searched the literature looking for good problems, primarily of two types: applied problems and challenge problems. Two to ten of these problems have been added to each problem set usually at the end (though in some cases we have reworked the whole problem set to give it more coherence). We are confident that our problem sets offer to all students an exceptional learning experience, beginning as they do with plenty of routine practice problems, following with numerous applications, and ending with spicy problems that may challenge even budding mathematicians.

For those who have not used our book before, we mention some of the features that have contributed to the success of earlier editions. First, this book is filled with solid mathematics presented in a straightforward, uncluttered way. It is a middle-of-the-road book, **not overly rigorous and yet with due attention to the logic of the subject**. We have kept in mind that the majority of users will be scientists and engineers but that the book should provide an appropriate foundation for mathematics majors.

Second, the book is **modern in spirit without being faddish**. For example, we assume that users have electronic calculators and mark with $\boxed{c}$ problems that are simplified by use of these machines. More significant is our decision to devote a whole chapter (Chapter 10) to **numerical calculus**, expounding topics suitable for computer calculation (but still workable with a calculator). We think our treatment of the derivative in n-dimensions (Chapter 15) is both clear and up to date. This forms the basis for a rather full introduction to **vector calculus** (Chapter 17) including the theorems of Green, Gauss, and Stokes. Our discussion of **maxima-minima theory** is notable for its completeness and for the coherence between the one dimensional case (Chapter 4) and the n-dimensional case (Chapter 15). We do include a section on Lagrange's method.

Third, we have aimed for **good pedagogy without pedantry**. As examples, we suggest prospective users look at our **introduction of limit** (Chapter 2), our consistent use of a technique (**slice, approximate, integrate**) in showing how integrals arise in applications (see, for example, pages 267, 296, 561, and 739), and our highlighting of the concept of **linearity** throughout the book. Many students will appreciate the historical **timeline chart** inside the front cover, the **biographies** that open each chapter, the **marginal boxes** (new in this edition) that point to tangential material, and the **summary chart** in Section A.3 of the appendix. Included in each book following the index is a **formula card** which should be cut out, folded, and kept handy at all times. Teachers will agree that the chief learning device of any calculus book is its **problem sets**. We have already alluded to our efforts to make those in this book especially outstanding. We mention also the **chapter reviews**. These consist of true-false questions designed to check the student's mastery of the theory of calculus and sample test problems of the kind a teacher might reasonably put on a chapter test.

It is a pleasure to acknowledge the constructive criticism and helpful advice of a host of reviewers. The junior author's colleagues at Hamline University (Gary Anderson, Kay Blair, Walter Fleming, Wojciech Komornicki, Nadine Myers) have been supportive. Prentice-Hall kindly arranged a two-day reviewer conference for a page by page analysis of the book. Participating were:

> Wilson Banks, Illinois State University
> Richard Grassl, University of New Mexico
> Louis Guillou, St. Mary's College
> Paul Liebnitz, University of Missouri, Kansas City
> Wayne Roberts, Macalester College
> Robert Sickles, Prentice-Hall, Inc

Louis Guillou offered innumerable suggestions, worked all the new problems, and prepared both the student's and instructor's manuals. Wayne Roberts made detailed comments on the whole manuscript; he and Kay Blair helped correct the page proofs. Others who read and commented on part or all of the manuscript were:

> Gene Brown, Northern Virginia Community College
> James H. Fife, The University of Richmond
> John Spellman, Southwest Texas State University
> Paul Weichsel, The University of Illinois
> Eugene Wingo, Thomas Nelson Community College

We thank Norton Starr, Amherst College, for supplying the computer graphics for Chapter 15 and C. H. Edwards, Jr. and David E. Penney, University of Georgia, for granting permission to use the table of integrals from their book, *Calculus and Analytic Geometry*, Prentice-Hall, 1982.

Finally, it is a pleasure to thank the staff of Prentice-Hall for their continued encouragement and expert production work. We mention for special recognition Robert Sickles (Executive Editor), David Ostrow (Mathematics Editor), Zita de Schauensee (Project Editor), and Jayne Conte (Designer).

Edwin J. Purcell

Dale Varberg

1 Preliminaries

[Coordinate geometry], far more than any of his metaphysical speculations, immortalized the name of Descartes, and constitutes the greatest single step ever made in the progress of the exact sciences.
John Stuart Mill

René Descartes
1596–1650

René Descartes is best known as the first great modern philosopher. He was also a founder of modern biology, a physicist, and a mathematician.

Born in Touraine, France, Descartes was the son of a moderately wealthy lawyer who sent him to a Jesuit school at the age of eight. Because of delicate health, Descartes was permitted to spend his mornings studying in bed, a practice he found so useful that he continued it throughout the rest of his life. At age 20, he obtained a law degree and thereafter lived the life of a gentleman, serving for a few years in the army and living at times in Paris, at others in the Netherlands. Invited to instruct Queen Christina, he went to Sweden, where he died of pneumonia in 1650.

Descartes searched for a general method of thinking that would give coherence to knowledge and lead to truth in the sciences. The search led him to mathematics, which he concluded was the means of establishing truth in all fields. His most influential mathematical work was *La Géométrie*, published in 1637. In it, he attempted a unification of the ancient and venerable *geometry* with the still infant *algebra*. Together with another Frenchman, Pierre Fermat (1601–1665), he is credited with the union that we today call analytic geometry, or coordinate geometry. The full development of calculus could not have occurred without it.

1.1
The Real Number System

Calculus is based on the real number system and its properties. But what are the real numbers and what are their properties? To answer, we start with some simpler number systems.

THE INTEGERS AND THE RATIONAL NUMBERS The simplest numbers of all are the **natural numbers**,

$$1, 2, 3, 4, 5, 6, \ldots$$

With them we can *count*: our books, our friends, and our money. If we adjoin their negatives and zero, we obtain the **integers**:

$$\ldots, -3, -2, -1, 0, 1, 2, 3, \ldots$$

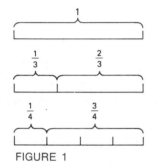

FIGURE 1

When we try to *measure* length, weight, or voltage, the integers are inadequate. They are spaced too far apart to give sufficient precision. We are led to consider quotients (ratios) of integers (Figure 1), numbers such as

$$\frac{3}{4}, \frac{-7}{8}, \frac{21}{5}, \frac{19}{-2}, \frac{16}{2}, \quad \text{and} \quad \frac{-17}{1}$$

Note that we included $\frac{16}{2}$ and $\frac{-17}{1}$, though we would normally write them as 8 and -17, since they are equal to the latter by the ordinary meaning of division. We did not include $\frac{5}{0}$ or $\frac{-9}{0}$, since it is impossible to make sense out of these symbols (see Problem 36). In fact, let us agree once and for all to banish division by zero from this book (Figure 2). Numbers which can be written in the form m/n, where m and n are integers with $n \neq 0$, are called **rational numbers**.

FIGURE 2

Do the rational numbers serve to measure all lengths? No. This surprising fact was discovered by the ancient Greeks several centuries before Christ. They showed that while $\sqrt{2}$ measures the hypotenuse of a right triangle with sides of length 1 (Figure 3), it cannot be written as a quotient of two integers (see Problem 43). Thus, $\sqrt{2}$ is an irrational (not rational) number. So are $\sqrt{3}$, $\sqrt{5}$, $\sqrt[3]{7}$, π, and a host of other numbers.

FIGURE 3

THE REAL NUMBERS Consider the set of all numbers (rational and irrational) that can measure lengths, together with their negatives and zero. We call these numbers the **real numbers**.

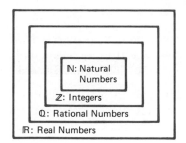

FIGURE 4

FIGURE 5

The real numbers may be viewed as labels for points along a horizontal line. There they measure the distance to the right or left (the **directed distance**) from a fixed point called the **origin** and labeled 0 (Figure 4). Though we cannot possibly show all the labels, each point does have a unique real number label. This number is called the **coordinate** of the point. And the resulting coordinate line is referred to as the **real line**.

There are standard symbols to identify the classes of numbers so far discussed. From now on, $\mathbb{N}$ will denote the set of natural numbers (positive integers), $\mathbb{Z}$ (from the German *Zahlen*) will denote the set of integers, $\mathbb{Q}$ (quotients of integers) the set of rational numbers, and $\mathbb{R}$ the set of real numbers. As suggested by Figure 5,

$$\mathbb{N} \subset \mathbb{Z} \subset \mathbb{Q} \subset \mathbb{R}$$

Here $\subset$ is the set inclusion symbol; it is read "is a subset of."

Most students will remember that the number system can be enlarged still more—to the so-called **complex numbers**. These are numbers of the form $a + b\sqrt{-1}$, where a and b are real numbers. Complex numbers will rarely be used in this book. In fact, if we say *number* without any qualifying adjective, you can assume we mean real number. The real numbers are the principal characters in calculus.

THE FOUR ARITHMETIC OPERATIONS   Given two real numbers x and y, we may add or multiply them to obtain two new real numbers $x + y$ and $x \cdot y$ (also written simply as xy). Addition and multiplication have the following familiar properties. We call them the field properties.

The Field Properties

1. **Commutative laws.** $x + y = y + x$ and $xy = yx$.
2. **Associative laws.** $x + (y + z) = (x + y) + z$ and $x(yz) = (xy)z$.
3. **Distributive law.** $x(y + z) = xy + xz$.
4. **Identity elements.** There are two distinct numbers 0 and 1 satisfying $x + 0 = x$ and $x \cdot 1 = x$.
5. **Inverses.** Each number x has an *additive inverse* (also called a *negative*), $-x$, satisfying $x + (-x) = 0$. Also, each number x except 0 has a *multiplicative inverse* (also called a *reciprocal*), x^{-1}, satisfying $x \cdot x^{-1} = 1$.

Subtraction and division are defined by

$$x - y = x + (-y)$$

and

$$\frac{x}{y} = x \cdot y^{-1}$$

From these basic facts, many others follow. In fact, most of algebra ultimately rests on the five field properties and the definition of subtraction and division.

ORDER ON THE REAL LINE

To say that $x < y$ means that x is to the left of y on the real line.

ORDER The nonzero real numbers separate nicely into two disjoint sets—the positive real numbers and the negative real numbers. This fact allows us to introduce the order relation $<$ (read "is less than") by

$$x < y \quad \Leftrightarrow \quad y - x \text{ is positive}$$

Here the double-arrow symbol $\Leftrightarrow$ is the conjunction of $\Rightarrow$ (implies) and $\Leftarrow$ (is implied by). Thus, $\Leftrightarrow$ may be read "is equivalent to" or as "if and only if." We agree that $x < y$ and $y > x$ shall mean the same thing. Thus $3 < 4$, $4 > 3$, $-3 < -2$, and $-2 > -3$. Note the geometric interpretation of $<$ shown in the box in the margin.

The Order Properties

1. **Trichotomy.** If x and y are numbers, exactly one of the following holds:

$$x < y \quad \text{or} \quad x = y \quad \text{or} \quad x > y$$

2. **Transitivity.** $x < y$ and $y < z \quad \Rightarrow \quad x < z$
3. **Addition.** $x < y \quad \Leftrightarrow \quad x + z < y + z$
4. **Multiplication.** When z is positive, $x < y \quad \Leftrightarrow \quad xz < yz$. When z is negative, $x < y \quad \Leftrightarrow \quad xz > yz$.

The order relation $\leq$ (read "is less than or equal to") is a first cousin of $<$. It is defined by

$$x \leq y \quad \Leftrightarrow \quad y - x \text{ is positive or zero}$$

Order properties 2, 3, and 4 hold with the symbols $<$ and $>$ replaced by $\leq$ and $\geq$.

A BIT OF LOGIC Important results in mathematics are called **theorems**; you will find many theorems in this book. The most important ones occur in the text with the label *Theorem* and are usually given names (for example, Pythagorean Theorem). Others occur in the problem sets and are introduced with the words *show that* or *prove that*. In contrast to axioms or definitions, which are taken for granted, theorems require proof.

A theorem is either stated in the form "If P, then Q" or can be restated so it has this form. We often abbreviate this by $P \Rightarrow Q$. We call P the hypothesis and Q the conclusion of the theorem. A proof consists in showing that P necessarily implies Q.

Beginning students (and some mature ones) may confuse $P \Rightarrow Q$ with its converse, $Q \Rightarrow P$. Clearly, these two statements are not equivalent. "If John is a Minnesotan, then he is an American" is a true statement, but its converse, "If John is an American, then he is a Minnesotan" is clearly false. On the other hand, $\sim Q \Rightarrow \sim P$, read "not Q implies not P" and called the

contrapositive, is equivalent to $P \Rightarrow Q$. In our example, it is true that "If John is not an American, then he is not a Minnesotan."

Because a statement and its contrapositive are equivalent, we often choose to prove a theorem stated in the latter form. This is the basis of what is called **proof by contradiction**. Thus, to prove $P \Rightarrow Q$, we may suppose $\sim Q$ and try to deduce $\sim P$ from it; that is, we try to contradict P. Here is a simple example.

Theorem

The sum of a rational number and an irrational number is irrational.

Proof   The theorem could be stated as follows: "If $x = m/n$, where m and n are integers, and if y is an irrational number, then $x + y$ is irrational." *We suppose $x + y$ is rational*, that is, we suppose that $x + y = p/q$, where p and q are integers. Then

$$y = \frac{p}{q} - x = \frac{p}{q} - \frac{m}{n} = \frac{np - mq}{qn}$$

This means that y is a rational number, *contrary to our hypothesis*. The theorem is proved. ∎

REDUCTIO AD ABSURDUM

Proof by contradiction also goes by the name *reductio ad absurdum*. Here is what the great mathematician G. H. Hardy had to say about it.

"Reductio ad absurdum, which Euclid loved so much, is one of a mathematician's finest weapons. It is a far finer gambit than any chess gambit; a chess player may offer the sacrifice of a pawn or even a piece, but a mathematician offers the game."

Another way to view proof by contradiction is through the *Law of the Excluded Middle*, which says: Either R or $\sim R$, not both. In the theorem above, let R be the statement: The sum of a rational number and an irrational number is irrational. Our proof shows that $\sim R$ is false (absurd). We conclude that R must be true.

Occasionally, we will need another type of proof called **Mathematical Induction**. It would take us too far afield to describe this now but we have given a complete discussion as Appendix A.1.

PROBLEM SET   1.1

We assume that you remember how to manipulate numbers, but it won't hurt to practice a little. In Problems 1–20, simplify as much as possible. Be sure to remove all parentheses and reduce all fractions.

1. $4 - 3(8 - 12) - 6$

2. $2[3 - 2(4 - 8)]$

3. $-4[3(-6 + 13) - 2(5 - 9)]$

4. $5[-1(7 + 12 - 16) + 4] + 2$

5. $\frac{5}{6} - (\frac{1}{4} + \frac{2}{3})$

6. $\frac{3}{4} - (\frac{7}{12} - \frac{2}{9})$

7. $\frac{1}{3}[\frac{1}{2}(\frac{1}{4} - \frac{1}{3}) + \frac{1}{6}]$

8. $-\frac{1}{3}[\frac{2}{5} - \frac{1}{2}(\frac{1}{3} - \frac{1}{5})]$

9. $\frac{14}{33}(\frac{2}{3} - \frac{1}{7})^2$

10. $(\frac{5}{7} + \frac{7}{9})/(1 + \frac{1}{2})$

11. $\dfrac{\frac{11}{49} - \frac{3}{7}}{\frac{11}{49} + \frac{3}{7}}$

12. $\dfrac{\frac{1}{2} - \frac{3}{4} + \frac{7}{8}}{\frac{1}{2} + \frac{3}{4} - \frac{7}{8}}$

13. $1 - \dfrac{2}{2 + \frac{3}{4}}$

14. $2 + \dfrac{3}{1 + \frac{5}{2}}$

15. $(\sqrt{2} + \sqrt{3})(\sqrt{2} - \sqrt{3})$

16. $(\sqrt{2} + \sqrt{3})^2$

17. $3\sqrt{2}(\sqrt{2} - \sqrt{8})$

18. $2\sqrt[3]{4}[\sqrt[3]{2} + \sqrt[3]{16}]$

19. $(\frac{5}{6} + \frac{1}{3})^{-2}$

20. $\left(\dfrac{1}{\sqrt{2}} - \dfrac{5}{2\sqrt{2}}\right)^{-2}$

A little algebraic practice is good for any calculus student. In Problems 21–34, perform the indicated operations and simplify.

21. $(2x - 3)(2x + 3)$

22. $(2x - 3)^2$

23. $(3x - 9)(2x + 1)$

24. $(3x + 11)(2x - 4)$

25. $(3t^2 - t + 1)^2$

26. $(2t - 1)^3$

27. $\dfrac{x^2 - 4}{x - 2}$

28. $\dfrac{x^2 - x - 6}{x - 3}$

29. $\dfrac{x^3 - 8}{2x - 4}$

30. $\dfrac{2x - 2x^2}{x^3 - 2x^2 + x}$

31. $\dfrac{18}{x^2 + 3x} - \dfrac{4}{x} + \dfrac{6}{x + 3}$

32. $\dfrac{2}{6y - 2} + \dfrac{y}{9y^2 - 1} - \dfrac{2y + 1}{1 - 3y}$

33. $\dfrac{x^2 + x - 6}{x^2 - 1} \cdot \dfrac{x^2 + x - 2}{x^2 + 5x + 6}$

34. $\dfrac{\dfrac{x}{x - 3} - \dfrac{2}{x^2 - 4x + 3}}{\dfrac{5}{x - 1} + \dfrac{5}{x - 3}}$

35. Find the value of each of the following; if undefined, say so.

(a) $0 \cdot 0$

(b) $\frac{0}{0}$

(c) $\frac{0}{8}$

(d) $\frac{8}{0}$

(e) 8^0

(f) 0^8

36. Show that division by 0 is meaningless as follows: Suppose $a \neq 0$. If $a/0 = b$, then $a = 0 \cdot b = 0$, which is a contradiction. Now find a reason why $0/0$ is also meaningless.

37. Tell whether each of the following is true or false.

(a) $-2 < -20$

(b) $1 > -39$

(c) $-3 < \frac{5}{9}$

(d) $-4 > -16$

(e) $\frac{6}{7} < \frac{34}{39}$

(f) $-\frac{5}{7} < -\frac{44}{59}$

38. Prove each if $a > 0$, $b > 0$.

(a) $a < b \iff a^2 < b^2$

(b) $a < b \iff \dfrac{1}{a} > \dfrac{1}{b}$

39. Prove that the average of two numbers is between the two numbers; that is, prove that

$$a < b \implies a < \frac{a + b}{2} < b$$

40. Which of the following are always correct if $a \leq b$?

(a) $a - 4 \leq b - 4$

(b) $-a \leq -b$

(c) $a^2 \leq ab$

(d) $a^3 \leq a^2 b$

41. A prime number is a natural number (positive integer) with exactly two natural number divisors, itself and 1. The first few primes are 2, 3, 5, 7, 11, 13, 17. According to the Fundamental Theorem of Arithmetic, every natural number (other than 1) can be written as the product of a unique set of primes. For example, $45 = 3 \cdot 3 \cdot 5$. Write each of the following as a product of primes. *Note*: The product is trivial if the number is prime—that is, it has only one factor.

(a) 240

(b) 310

(c) 119

(d) 5400

42. Use the Fundamental Theorem of Arithmetic (Problem 41) to show that the square of any natural number (other than 1) can be written as the product of a unique set of primes, with each of these primes occurring an *even* number of times. For example, $(45)^2 = 3 \cdot 3 \cdot 3 \cdot 3 \cdot 5 \cdot 5$.

43. Show that $\sqrt{2}$ is irrational. *Hint*: Try a proof by contradiction. Suppose $\sqrt{2} = p/q$, where p and q are natural numbers (necessarily different from 1). Then $2 = p^2/q^2$, and so $2q^2 = p^2$. Now use Problem 42 to get a contradiction.

44. Show that $\sqrt{3}$ is irrational (see Problem 43).

45. Show that the sum of two rational numbers is rational.

46. Show that the product of a rational number (other than 0) and an irrational number is irrational. *Hint*: Try proof by contradiction.

47. Which of the following are rational and which are irrational?

(a) $\sqrt{4}$

(b) 0.375

(c) $1 + \sqrt{2}$

(d) $(1 + \sqrt{3})^2$

(e) $(3\sqrt{2})(5\sqrt{2})$

(f) $5\sqrt{2}$

48. Is the sum of two irrational numbers necessarily irrational? Explain.

49. Show that if the natural number m is not a perfect square, then $\sqrt{m}$ is irrational.

50. Show that $\sqrt{6} + \sqrt{3}$ is irrational.

51. Show that $\sqrt{2} - \sqrt{3} + \sqrt{6}$ is irrational.

52. Show that $\log_{10} 5$ is irrational.

1.2
Decimals, Denseness, Calculators

Any rational number can be written as a decimal, since by definition it can always be expressed as the quotient of two integers; if we divide the denominator into the numerator, we obtain a decimal. For example (Figure 1),

$$\frac{1}{2} = 0.5 \qquad\qquad \frac{3}{8} = 0.375$$

$$\frac{13}{11} = 1.181818\ldots \qquad \frac{3}{7} = 0.428571428571428571\ldots$$

Irrational numbers, too, can be expressed as decimals. For instance,

$$\sqrt{2} = 1.4142135623\ldots, \qquad \sqrt{3} = 1.7320508075\ldots$$
$$\pi = 3.1415926535\ldots$$

```
      .375            1.181
  8 ⌐3.000       11 ⌐13.000
    2 4               11
     60               2 0
     56               1 1
     40                 90
     40                 88
                        20
  3/8 = .375            11
                         9

                 13/11 = 1.181818...
```

FIGURE 1

REPEATING AND NONREPEATING DECIMALS The decimal representation of a rational number either terminates (as in $\frac{3}{8} = 0.375$) or else repeats in regular cycles forever (as in $\frac{13}{11} = 1.181818\ldots$). A little experimenting with the long division process will show you why. (It is because there can be only a finite number of different remainders.) A terminating decimal can be regarded as a repeating decimal ending in all zeros. For instance,

$$\frac{3}{8} = 0.375 = 0.3750000\ldots$$

Thus, every rational number can be written as a repeating decimal. It is a remarkable fact that the converse is also true. Every repeating decimal represents a rational number. This is obvious in the case of a terminating decimal (for instance, $3.137 = 3137/1000$) and is easy to show in the general case.

EXAMPLE 1 **(Repeating decimals are rational).** Show that

$$x = 0.136136136\ldots \quad \text{and} \quad y = 0.27171717\ldots$$

represent rational numbers.

Solution We subtract x from $1000x$ and then solve for x.

$$1000x = 136.136136\ldots$$
$$\underline{x = 0.136136\ldots}$$
$$999x = 136$$

$$x = \frac{136}{999}$$

Similarly,

$$100y = 27.17171717\ldots$$

$$y = 0.27171717\ldots$$

$$99y = 26.9$$

$$y = \frac{26.9}{99} = \frac{269}{990}$$

In general, our first step is to multiply a repeating decimal z by 10^m if the decimal repeats in a cycle consisting of m digits. ∎

The decimal representations of irrational numbers do not repeat in cycles. Conversely, a nonrepeating decimal must represent an irrational number. Thus, for example,

$$0.101001000100001\ldots$$

must represent an irrational number. The diagram in Figure 2 summarizes what we have said.

The Real Numbers

Rational Numbers (the repeating decimals)	Irrational Numbers (the nonrepeating decimals)

FIGURE 2

DENSENESS Between any two different real numbers x and y, there is another real number. In particular, the number $z = (x + y)/2$ is a number midway between x and y (Figure 3). Since there is also a number s between x and z and another number t between z and y and since this argument can be repeated ad infinitum, we are forced to the amazing but correct conclusion that between any two different real numbers (no matter how close together), there are infinitely many other real numbers. This should destroy once and for all such notions as "the number just larger than 3." There is no such number.

Actually we can say more. Between any two distinct real numbers, there are both a rational number and an irrational number—and hence infinitely many of each variety.

EXAMPLE 2 Find a rational number and an irrational number between x and y if

$$x = 0.31234158\ldots$$

$$y = 0.31234200\ldots$$

Solution Let

$$z = 0.312341600000\ldots$$

$$w = 0.3123416010010001\ldots$$

Then z is rational (it ends in repeating 0s), whereas w is irrational (note the pattern of inserting more and more 0s between the 1s). It should be clear that $x < z < w < y$. ∎

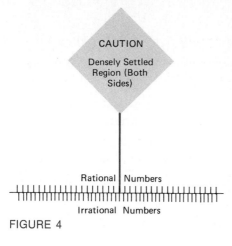

CAUTION

Densely Settled Region (Both Sides)

Rational | Numbers

Irrational Numbers

FIGURE 4

1

1.4

1.41

1.414

$\sqrt{2}$

FIGURE 5

One way that mathematicians describe the situation we have been discussing is to say that both the rational numbers and the irrational numbers are **dense** along the real line (Figure 4). Every number has both rational and irrational neighbors arbitrarily close to it. The two types of numbers are inseparably intertwined and relentlessly crowded together.

One manifestation of the density property just described is that any irrational number can be approximated as closely as we please by a rational number. Take $\sqrt{2}$ as an example. The sequence of rational numbers 1, 1.4, 1.41, 1.414, 1.4142, 1.41421, 1.414213,... marches steadily and inexorably toward $\sqrt{2}$ (Figure 5). By going far enough along in this sequence, we can get as near $\sqrt{2}$ as we wish.

CALCULATORS   There was a time when all budding scientists and engineers walked around campuses with slide rules attached to their belts. Today they have calculators in their pockets. If you do not own one of these electronic wizards, we urge you to buy one. Be sure to get a scientific model (with sines, cosines, and logarithms) and, if you can afford it, we recommend a programmable version. You will find plenty of use for a calculator in this book, especially in problems marked with a ⃞C.

One fact that should be immediately apparent is that we cannot enter an infinite decimal into a calculator. Calculators work exclusively with decimals of some prescribed length (for example, ten digits). In fact, then, calculators handle only rational numbers with decimal expansions that terminate quickly. Thus, we must often round a number to enter it into a calculator, and usually the answer given by the calculator will also be rounded. For example, a calculator can never use the true value of $\sqrt{2}$, but must be content with an approximation such as

$$\sqrt{2} \approx 1.414213562$$

Here we have used the symbol $\approx$ to abbreviate the phrase "is approximately equal to."

Our advice is this: Do calculations that can easily be done by hand without a calculator, especially if this allows an exact answer. For example, we generally prefer the exact answer $\sqrt{3}/2$ for the sine of 60° to the calculator value 0.8660254. However, in any complicated calculation we encourage use of a calculator. You will note that our answer key at the end of the book often gives both the exact answer and a decimal approximation obtained by use of a calculator.

PROBLEM SET   1.2

In Problems 1–6, change each rational number to a decimal by performing a long division.

1. $\frac{7}{8}$ 2. $\frac{3}{7}$

3. $\frac{3}{20}$ 4. $\frac{5}{13}$

5. $\frac{11}{3}$ 6. $\frac{11}{7}$

In Problems 7–12, change each repeating decimal to a ratio of two integers (see Example 1).

7. 0.123123123...   8. 0.217171717...

9. 2.56565656...   10. 3.929292...

11. 0.199999...   12. 0.399999...

13. Since $\quad 0.199999\ldots = 0.200000\ldots \quad$ and $0.399999\ldots = 0.400000\ldots$ (see Problems 11 and 12), we see that certain rational numbers have two different decimal expansions. Which rational numbers have this property?

14. Show that any rational number p/q, where the prime factorization of q consists entirely of 2s and 5s, has a terminating decimal expansion.

15. Find a positive rational number and a positive irrational number both smaller than 0.00001.

16. What is the smallest positive integer? The smallest positive rational number? The smallest positive irrational number?

17. Find an irrational number between 3.14159 and π (see Example 2 and note that $\pi = 3.141592\ldots$).

18. Is $\pi - \frac{22}{7}$ positive, negative, or zero?

19. Is there a number between 0.9999... (repeating 9s) and 1?

20. Find a rational number between $\frac{17}{37}$ and $\frac{52}{111}$.

21. Is $0.1234567891011121314\ldots$ rational or irrational? (You should see a pattern in the given sequence of digits.)

22. Find two irrational numbers whose sum is rational.

C In Problems 23–32, find the best decimal approximation that your calculator allows.

23. $(\sqrt{2} + 1)^3$

24. $(\sqrt{3} - \sqrt{2})^4$

25. $\sqrt[4]{1.215} - \sqrt[3]{1.015}$

26. $(3.617)^{-1/2}$

27. $\dfrac{\sqrt{130} - \sqrt{5}}{3^{1.2} - 3}$

28. $\dfrac{\sqrt{143.2} + \sqrt{36.1}}{(23.41)^4 - (11.2)^2}$

29. $\dfrac{(6.34 \times 10^7)(5.23 \times 10^6)}{4.21 \times 10^9}$

30. $\dfrac{(0.00121)(5.23 \times 10^{-3})}{6.16 \times 10^{-4}}$

31. $\sqrt{3\pi^2 + 2} + 4$

32. $\sqrt[4]{(6\pi^2 - 2)\pi}$

C **33.** Note that

$$2x^3 - 7x^2 + 11x - 2 = [(2x - 7)x + 11]x - 2$$

To calculate the expression on the right at $x = 3$, press the following keys on an algebraic logic calculator.

$$2 \;\boxed{\times}\; 3 \;\boxed{-}\; 7 \;\boxed{=}\; \boxed{\times}\; 3 \;\boxed{+}\; 11 \;\boxed{=}\; \boxed{\times}\; 3 \;\boxed{-}\; 2 \;\boxed{=}$$

Use this idea to calculate the given expression in each case.

(a) $x = \pi$ (b) $x = 2.15$
(c) $x = 11.19$

C **34.** Use the idea discussed in Problem 33 to evaluate $x^4 - 3x^3 + 5x^2 + 6x - 10$ at each value.

(a) $x = 1$ (b) $x = \pi$
(c) $x = 13.53$

35. A number b is called an **upper bound** for a set S of numbers if $x \le b$ for all x in S. For Example, 5, 6.5, and 13 are upper bounds for the set $S = \{1, 2, 3, 4, 5\}$. The number 5 is the **least upper bound** for S (the smallest of the upper bounds). Similarly, 1.6, 2, and 2.5 are upper bounds for the infinite set $T = \{1.4, 1.49, 1.499, 1.4999, \ldots\}$, whereas 1.5 is its least upper bound. Find the least upper bound of each of the following sets.

(a) $S = \{-10, -8, -6, -4, -2\}$
(b) $S = \{-2, -2.1, -2.11, -2.111, -2.1111, \ldots\}$
(c) $S = \{2.4, 2.44, 2.444, 2.4444, \ldots\}$
(d) $S = \{1 - \frac{1}{2}, 1 - \frac{1}{3}, 1 - \frac{1}{4}, 1 - \frac{1}{5}, \ldots\}$
(e) $S = \{x : x = (-1)^n + 1/n, \ n \text{ a positive integer}\}$; that is, S is the set of all numbers x that have the form $x = (-1)^n + 1/n$, where n is a positive integer.
(f) $S = \{x : x^2 < 2, \ x \text{ a rational number}\}$.

36. The **Axiom of Completeness** for the real numbers says: *Every set of real numbers that has an upper bound has a least upper bound that is a real number.*

(a) Show that the italicized statement is false if the word *real* is replaced by *rational*.
(b) Would the italicized statement be true or false if the word *real* were replaced by *natural*?

Note: The real numbers $\mathbb{R}$ are the only set of numbers simultaneously having the field properties, the order properties, and the completeness property.

1.3 Inequalities

Solving equations (for instance, $3x - 17 = 6$ or $x^2 - x - 6 = 0$) is one of the traditional tasks of mathematics; it will be important in this course and we assume you remember how to do it. But of almost equal significance in calculus is the notion of solving an inequality (for example, $3x - 17 < 6$ or $x^2 - x - 6 \ge 0$). To **solve** an inequality is to find the set of all real numbers

$(-1, 6) = \{x : -1 < x < 6\}$

FIGURE 1

$[-1, 5] = \{x : -1 \leq x \leq 5\}$

FIGURE 2

which make the inequality true. In contrast to an equation, whose solution set normally consists of one number or perhaps a finite set of numbers, the solution set of an inequality usually consists of a whole interval of numbers or, in some cases, a union of such intervals.

INTERVALS Several kinds of intervals will arise in our work and we introduce special terminology and notation for them. The double inequality $a < x < b$ describes the **open interval** consisting of all numbers between a and b, not including the endpoints a and b. We denote it by the symbol (a, b) (Figure 1). In contrast, the inequality $a \leq x \leq b$ describes the corresponding **closed interval**, which does include the endpoints a and b. It is denoted by $[a, b]$ (Figure 2). The table indicates the wide variety of possibilities and introduces our notation.

Set Notation	Interval Notation	Graph
$\{x : a < x < b\}$	(a, b)	
$\{x : a \leq x \leq b\}$	$[a, b]$	
$\{x : a \leq x < b\}$	$[a, b)$	
$\{x : a < x \leq b\}$	$(a, b]$	
$\{x : x \leq b\}$	$(-\infty, b]$	
$\{x : x < b\}$	$(-\infty, b)$	
$\{x : x \geq a\}$	$[a, \infty)$	
$\{x : x > a\}$	(a, ∞)	
$\mathbb{R}$	$(-\infty, \infty)$	

SOLVING INEQUALITIES Just as with equations, the procedure for solving an inequality consists in transforming the inequality a step at a time until the solution set is obvious. The chief tools are the order properties from Section 1.1. They imply that we may perform certain operations on an inequality without changing its solution set. In particular:

1. *we may add the same number to both sides of an inequality;*
2. *we may multiply both sides of an inequality by a positive number;*
3. *we may multiply both sides by a negative number, but then we must reverse the direction of the inequality sign.*

EXAMPLE 1 Solve the inequality $2x - 7 < 4x - 2$ and show the graph of its solution set.

Solution

$$2x - 7 < 4x - 2$$

$$2x < 4x + 5 \qquad \text{(add 7)}$$

$$-2x < 5 \qquad \text{(add } -4x)$$

$$x > -\tfrac{5}{2} \qquad \text{(multiply by } -\tfrac{1}{2})$$

$$\left(-\tfrac{5}{2}, \infty\right) = \left\{x : x > \tfrac{-5}{2}\right\}$$

FIGURE 3

The graph appears in Figure 3. ∎

EXAMPLE 2 Solve $-5 \le 2x + 6 < 4$.

Solution

$$-5 \le 2x + 6 < 4$$

$$-11 \le 2x < -2 \qquad \text{(add } -6)$$

$$-\tfrac{11}{2} \le x < -1 \qquad \text{(multiply by } \tfrac{1}{2})$$

$$\left[\tfrac{-11}{2}, -1\right) = \left\{x : \tfrac{-11}{2} \le x < -1\right\}$$

FIGURE 4

Figure 4 shows the corresponding graph. ∎

Before tackling a quadratic inequality, we point out that a linear factor of the form $x - a$ is positive for $x > a$ and negative for $x < a$. It follows that a product $(x - a)(x - b)$ can change from being positive to negative, or vice versa, only at a or b. These points, where a factor is zero, are called **split points**. They are the keys to determining the solution sets of quadratic and higher-degree inequalities.

EXAMPLE 3 Solve the quadratic inequality $x^2 - x < 6$.

Solution As with quadratic equations, we move all nonzero terms to one side and factor.

$$x^2 - x < 6$$

$$x^2 - x - 6 < 0 \qquad \text{(add } -6)$$

$$(x - 3)(x + 2) < 0 \qquad \text{(factor)}$$

We see that -2 and 3 are the split points; they divide the real line into the three intervals $(-\infty, -2)$, $(-2, 3)$, and $(3, \infty)$. On each of these intervals, $(x - 3)(x + 2)$ is of constant sign—that is, it is either always positive or always negative. To find this sign in each interval, we use the **test points** -3, 0, and 5 (any points on the three intervals would do). Our results are shown below.

Test Point	Value of $(x - 3)(x + 2)$	Sign
-3	6	$(+)$
0	-6	$(-)$
5	14	$(+)$

Split points

Test points

−2   3

(−2, 3)

FIGURE 5

−2/3   1

$\left(-\infty, -\tfrac{2}{3}\right) \cup \left(1, \infty\right)$

FIGURE 6

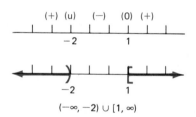

−2   1

−2   1

$(-\infty, -2) \cup [1, \infty)$

FIGURE 7

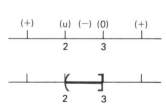

2   3

2   3

FIGURE 8

The information we have obtained is summarized in the top half of Figure 5. We conclude that the solution set for $(x - 3)(x + 2) < 0$ is the interval $(-2, 3)$. Its graph is shown in the bottom half of Figure 5. ∎

EXAMPLE 4   Solve $3x^2 - x - 2 > 0$.

Solution   Since

$$3x^2 - x - 2 = (x - 1)(3x + 2) = 3(x - 1)(x + \tfrac{2}{3})$$

the split points are $-\tfrac{2}{3}$ and 1. These points, together with the test points -2, 0, and 2, establish the information shown in Figure 6. We conclude that the solution set of the inequality consists of the points in either $(-\infty, -\tfrac{2}{3})$ or $(1, \infty)$. In set language, the solution set is the **union** (symbolized by $\cup$) of these two intervals; that is, it is $(-\infty, -\tfrac{2}{3}) \cup (1, \infty)$. ∎

EXAMPLE 5   Solve $\dfrac{x - 1}{x + 2} \geq 0$.

Solution   Our inclination to multiply both sides by $x + 2$ leads to an immediate dilemma, since $x + 2$ may be either positive or negative. Should we reverse the inequality sign or leave it alone? Rather than try to untangle this problem (which would require breaking it into two cases), we observe that the quotient $(x - 1)/(x + 2)$ can change sign only at the split points of the numerator and denominator, namely, at 1 and -2. The test points -3, 0, and 2 yield the information displayed in Figure 7. The symbol u indicates that the quotient is undefined at -2. We conclude that the solution set is $(-\infty, -2) \cup [1, \infty)$. Note that -2 is not in the solution set because the quotient is undefined there. On the other hand, 1 is included as the inequality is valid at 1. ∎

EXAMPLE 6   Solve $\dfrac{2x - 5}{x - 2} \leq 1$.

Solution   Rewrite the inequality successively as

$$\frac{2x - 5}{x - 2} - 1 \leq 0$$

$$\frac{2x - 5 - (x - 2)}{x - 2} \leq 0$$

$$\frac{x - 3}{x - 2} \leq 0$$

Then proceed as in Example 6. The summary shown in Figure 8 yields the solution set $(2, 3]$. ∎

EXAMPLE 7   Solve $x^3 - 5x^2 + 4x \leq 0$.

FIGURE 9

FIGURE 10

Solution The cubic $x^3 - 5x^2 + 4x$ factors as $x(x - 1)(x - 4)$, so there are 3 split points—0, 1, and 4—dividing the real line into four intervals. When we test these intervals, we obtain the information in Figure 9. The solution set is $(-\infty, 0] \cup [1, 4]$. ∎

EXAMPLE 8 Solve $(x + 1)(x - 1)^2(x - 3) \le 0$.

Solution The split points are -1, 1, and 3, which divide the real line into four intervals, as shown in Figure 10. After testing these intervals, we conclude that the solution set is $[-1, 1] \cup [1, 3]$; that is, it is the interval $[-1, 3]$. ∎

PROBLEM SET 1.3 _____

1. Show each of the following intervals on the real line.

(a) $(-4, 1)$ (b) $[-4, 1]$
(c) $(-4, 1]$ (d) $[-4, 1)$
(e) $[1, \infty)$ (f) $(-\infty, -4]$

2. Use the notation of Problem 1 to describe the intervals below.

(a)

(b)

(c)

(d)

In each of Problems 3–34, express the solution set of the given inequality in interval notation and sketch its graph.

3. $4x - 7 < 3x + 5$ **4.** $2x + 16 < x + 25$

5. $7x - 1 \le 10x + 4$ **6.** $6x - 10 \ge 5x - 16$

7. $10x + 1 > 8x + 5$ **8.** $3x + 5 > 7x + 17$

9. $-6 < 2x + 3 < -1$ **10.** $-3 < 4x - 9 < 11$

11. $-2 < 1 - 5x \le 3$ **12.** $4 < 5 - 3x < 7$

13. $2 + 3x < 5x + 1 < 16$

14. $2x - 4 \le 6 - 7x \le 3x + 6$

15. $x^2 + x - 12 < 0$ **16.** $x^2 - 5x + 6 > 0$

17. $3x^2 - 11x - 4 \le 0$ **18.** $2x^2 + 7x - 15 \ge 0$

19. $2x^2 + 5x - 3 > 0$ **20.** $4x^2 - 5x - 6 < 0$

21. $\dfrac{x + 5}{2x - 1} \le 0$ **22.** $\dfrac{2x - 3}{x + 1} > 0$

23. $\dfrac{1}{x} < 5$ **24.** $\dfrac{7}{2x} < 3$

25. $\dfrac{1}{3x - 2} \le 4$ **26.** $\dfrac{3}{x + 5} > 2$

27. $\dfrac{x - 2}{x + 4} < 2$ **28.** $\dfrac{2x - 1}{x - 3} > 1$

29. $(x + 2)(2x - 1)(3x + 7) \ge 0$

30. $(2x + 3)(3x - 1)(x - 2) < 0$

31. $(2x + 3)(3x - 1)^2(x - 5) < 0$

32. $(x + 5)(x + 2)^2(2x - 1) > 0$

33. $x^3 - 5x^2 - 6x < 0$

34. $x^3 - x^2 - x + 1 > 0$

35. Find all values of x that satisfy both inequalities simultaneously.

(a) $3x + 7 > 1$ and $2x + 1 < 3$
(b) $3x + 7 > 1$ and $2x + 1 > -4$
(c) $3x + 7 > 1$ and $2x + 1 < -4$

36. Find all values of x that satisfy at least one of the two inequalities.

(a) $3x + 7 > 1$ or $2x + 1 < -5$
(b) $3x + 7 \le 1$ or $2x + 1 < -8$
(c) $3x + 7 \le 1$ or $2x + 1 > -8$

37. Solve for x, expressing your answer in interval notation.

(a) $(x + 1)(x^2 + 2x - 7) \ge x^2 - 1$
(b) $x^4 - 2x^2 \ge 8$
(c) $(x^2 + 1)^2 - 7(x^2 + 1) + 10 < 0$

38. Solve $1 + x + x^2 + x^3 + \cdots + x^{99} \leq 0$.

39. The formula

$$\frac{1}{R} = \frac{1}{R_1} + \frac{1}{R_2} + \frac{1}{R_3}$$

gives the total resistance R in an electric circuit due to three resistances, R_1, R_2, and R_3, connected in parallel. If $10 \leq R_1 \leq 20$, $20 \leq R_2 \leq 30$, and $30 \leq R_3 \leq 40$, find the range of values for R.

1.4
Absolute Values,
Square Roots, Squares

FIGURE 1

The concept of absolute value is extremely useful in calculus and the reader should acquire skill in working with it. The **absolute value** of a real number x, denoted by $|x|$, is defined by

$$
\begin{array}{ll}
|x| = x & \text{if } x \geq 0 \\
|x| = -x & \text{if } x < 0
\end{array}
$$

For example, $|6| = 6$, $|0| = 0$, and $|-5| = -(-5) = 5$.

This two-pronged definition merits careful study. Note that it does not say that $|-x| = x$ (try $x = -5$ to see why). It is true that $|x|$ is always nonnegative; it is also true that $|-x| = |x|$.

One of the best ways to think of absolute value is as an (undirected) distance. In particular, $|x|$ is the distance between x and the origin. Similarly, $|x - a|$ is the distance between x and a (Figure 1).

PROPERTIES   Absolute values behave nicely under multiplication and division, but not so well under addition and subtraction.

Properties of Absolute Value

1. $|ab| = |a|\,|b|$

2. $\left|\dfrac{a}{b}\right| = \dfrac{|a|}{|b|}$

3. $|a + b| \leq |a| + |b|$ (Triangle Inequality)

4. $|a - b| \geq \big||a| - |b|\big|$

INEQUALITIES INVOLVING ABSOLUTE VALUES   If $|x| < 3$, then x must be simultaneously less than 3 and greater than -3; that is, $-3 < x < 3$. In contrast if $|x| > 3$, then either $x < -3$ or $x > 3$ (Figure 2). These are special cases of the following general statements.

FIGURE 2

$$
\begin{array}{lll}
|x| < a & \Leftrightarrow & -a < x < a \\
|x| > a & \Leftrightarrow & x < -a \ \text{ or } \ x > a
\end{array}
$$

We can use these facts to solve inequalities involving absolute values, since they provide a way of removing absolute value signs.

EXAMPLE 1 Solve the inequality $|x - 4| < 1.5$ and show the solution set on the real line.

Solution From the first boxed statement with x replaced by $x - 4$, we see that

$$|x - 4| < 1.5 \quad \Leftrightarrow \quad -1.5 < x - 4 < 1.5$$

When we add 4 to all three members of the latter inequality, we obtain $2.5 < x < 5.5$. The graph is shown in Figure 3.

There is another way to look at this problem, and it is of equal importance. The symbol $|x - 4|$ represents the distance between x and 4. Thus to say $|x - 4| < 1.5$ is equivalent to saying *the distance between x and 4 is less than* 1.5. The numbers x with this property are the numbers between 2.5 and 5.5, that is, $2.5 < x < 5.5$. ∎

$|x - 4| < 1.5$

FIGURE 3

The boxed statements displayed just before Example 1 are valid with $<$ and $>$ replaced by $\leq$ and $\geq$, respectively. We need the second statement in this form in our next example.

EXAMPLE 2 Solve the inequality $|3x - 5| \geq 1$ and show its solution set on the real line.

Solution The given inequality may be written successively as

$$3x - 5 \leq -1 \quad \text{or} \quad 3x - 5 \geq 1$$
$$3x \leq 4 \quad \text{or} \quad 3x \geq 6$$
$$x \leq \tfrac{4}{3} \quad \text{or} \quad x \geq 2$$

The solution set is the union of two intervals; it is the set $(-\infty, \tfrac{4}{3}] \cup [2, \infty)$ shown in Figure 4. ∎

$\left(-\infty, \tfrac{4}{3}\right] \cup [2, \infty)$

FIGURE 4

When we get to the "epsilon, delta" definition of limit in Chapter 2, we will need to make the kind of manipulations illustrated by the next two examples. Delta and epsilon are the fourth and fifth letters, respectively, of the Greek alphabet and are traditionally used to stand for small positive numbers.

EXAMPLE 3 Let ε (epsilon) be a positive number. Show that

$$|x - 2| < \frac{\varepsilon}{5} \quad \Leftrightarrow \quad |5x - 10| < \varepsilon$$

Solution

$$|x - 2| < \frac{\varepsilon}{5} \quad \Leftrightarrow \quad 5|x - 2| < \varepsilon \qquad \text{(multiply by 5)}$$
$$\Leftrightarrow \quad |5|\,|x - 2| < \varepsilon \qquad (|5| = 5)$$
$$\Leftrightarrow \quad |5(x - 2)| < \varepsilon \qquad (|a|\,|b| = |ab|)$$
$$\Leftrightarrow \quad |5x - 10| < \varepsilon$$
∎

EXAMPLE 4 Let ε be a positive number. Find a positive number δ (delta) such that

$$|x - 3| < \delta \quad \Rightarrow \quad |6x - 18| < \varepsilon$$

Solution

$$|6x - 18| < \varepsilon \quad \Leftrightarrow \quad |6(x - 3)| < \varepsilon$$

$$\Leftrightarrow \quad 6|x - 3| < \varepsilon \qquad (|ab| = |a|\,|b|)$$

$$\Leftrightarrow \quad |x - 3| < \frac{\varepsilon}{6} \qquad \left(\text{multiply by } \frac{1}{6}\right)$$

Therefore, we choose $\delta = \varepsilon/6$. Following the implications backwards, we see that

$$|x - 3| < \delta \quad \Rightarrow \quad |6x - 18| < \varepsilon \qquad\qquad \blacksquare$$

SQUARE ROOTS Every positive number has two square roots. For example, the two square roots of 9 are -3 and 3; the two square roots of 100 are -10 and 10. For $a \geq 0$, the symbol $\sqrt{a}$, called the **principal square root** of a, denotes the nonnegative square root of a. Thus, $\sqrt{9} = 3$ and $\sqrt{(-10)^2} = \sqrt{100} = 10$. The two square roots of 7 are $\pm\sqrt{7}$. It is incorrect to write $\sqrt{16} = \pm 4$; rather, $\sqrt{16} = 4$. Here is an important fact worth remembering.

$$\boxed{\sqrt{x^2} = |x|}$$

Most students will recall the **Quadratic Formula**. The solutions to $ax^2 + bx + c = 0$ are given by

$$\boxed{x = \frac{-b \pm \sqrt{b^2 - 4ac}}{2a}}$$

The number $d = b^2 - 4ac$ is called the **discriminant** of the quadratic equation $ax^2 + bx + c = 0$. This equation has two real solutions if $d > 0$, one real solution if $d = 0$, and no real solutions if $d < 0$.

With the Quadratic Formula, we can easily solve quadratic inequalities even if they do not factor by inspection.

EXAMPLE 5 Solve $x^2 - 2x - 4 \leq 0$.

Solution The two solutions of $x^2 - 2x - 4 = 0$ are

$$x_1 = \frac{2 - \sqrt{4 + 16}}{2} = 1 - \sqrt{5} \approx -1.24$$

and

$$x_2 = \frac{2 + \sqrt{4 + 16}}{2} = 1 + \sqrt{5} \approx 3.24$$

Thus,

$$x^2 - 2x - 4 = (x - x_1)(x - x_2) = (x - 1 + \sqrt{5})(x - 1 - \sqrt{5})$$

The split points $1 - \sqrt{5}$ and $1 + \sqrt{5}$ divide the real line into three intervals (Figure 5). When we test them with the test points $-2, 0,$ and 4, we conclude that the solution set for $x^2 - 2x - 4 \leq 0$ is $[1 - \sqrt{5}, 1 + \sqrt{5}]$. ∎

$$\left[1 - \sqrt{5}, 1 + \sqrt{5}\right]$$

FIGURE 5

We mention in passing that if n is even and $a \geq 0$, the symbol $\sqrt[n]{a}$ always denotes the nonnegative nth root of a. When n is odd, there is only one real nth root of a, denoted by the symbol $\sqrt[n]{a}$. Thus $\sqrt[4]{16} = 2$, $\sqrt[3]{27} = 3$, and $\sqrt[3]{-8} = -2$.

SQUARES Turning to squares, we notice that

$$\boxed{|x|^2 = x^2}$$

This follows from the property $|a||b| = |ab|$.

Does the squaring operation preserve inequalities? In general, the answer is no. For instance, $-3 < 2$, but $(-3)^2 > 2^2$. On the other hand, $2 < 3$ and $2^2 < 3^2$. If we are dealing with nonnegative numbers, then $a < b \iff a^2 < b^2$. A useful variant of this is

$$\boxed{|x| < |y| \iff x^2 < y^2}$$

For a proof of this fact, see Problem 29.

EXAMPLE 6 Solve the inequality $|3x + 1| < 2|x - 6|$.

Solution This inequality is more difficult to solve than our earlier examples, because there are two sets of absolute value signs. We can get rid of both of them by using the last boxed result.

$$
\begin{aligned}
|3x + 1| < 2|x - 6| \quad &\Leftrightarrow \quad |3x + 1| < |2x - 12| \\
&\Leftrightarrow \quad (3x + 1)^2 < (2x - 12)^2 \\
&\Leftrightarrow \quad 9x^2 + 6x + 1 < 4x^2 - 48x + 144 \\
&\Leftrightarrow \quad 5x^2 + 54x - 143 < 0 \\
&\Leftrightarrow \quad (5x - 11)(x + 13) < 0
\end{aligned}
$$

The split points for this quadratic inequality are -13 and $\frac{11}{5}$; they divide the real line into the three intervals $(-\infty, -13)$, $(-13, \frac{11}{5})$, and $(\frac{11}{5}, \infty)$. When we use the test points $-14, 0$, and 3, we discover that only the points in $(-13, \frac{11}{5})$ satisfy the inequality. ∎

PROBLEM SET 1.4

In Problems 1–12, find the solution sets of the given inequalities (see Examples 1 and 2).

1. $|x + 1| < 4$

2. $|x - 2| < 5$

3. $|3x + 4| < 8$

4. $|2x - 7| < 3$

5. $\left|\dfrac{x}{3} - 2\right| \le 6$

6. $\left|\dfrac{3x}{5} + 1\right| \le 4$

7. $|2x - 7| > 3$

8. $|5x - 6| > 1$

9. $|4x + 2| \ge 10$

10. $\left|\dfrac{x}{2} + 7\right| \ge 2$

11. $\left|2 + \dfrac{5}{x}\right| > 1$

12. $\left|\dfrac{1}{x} - 3\right| > 6$

In Problems 13–16, solve the given quadratic inequality using the quadratic formula (see Example 5).

13. $2x^2 - 5x - 4 \le 0$

14. $3x^2 + x - 1 > 0$

15. $4x^2 + x - 2 > 0$

16. $x^2 + 2x - 5 < 0$

In Problems 17–20, show that the indicated implication is true (see Example 3).

17. $|x - 3| < 0.5 \quad \Rightarrow \quad |5x - 15| < 2.5$

18. $|x + 2| < 0.3 \quad \Rightarrow \quad |4x + 8| < 1.2$

19. $|x - 2| < \dfrac{\varepsilon}{6} \quad \Rightarrow \quad |6x - 12| < \varepsilon$

20. $|x + 4| < \dfrac{\varepsilon}{2} \quad \Rightarrow \quad |2x + 8| < \varepsilon$

In Problems 21–24, find δ (depending on ε) so that the given implication is true (see Example 4).

21. $|x - 5| < \delta \quad \Rightarrow \quad |3x - 15| < \varepsilon$

22. $|x - 2| < \delta \quad \Rightarrow \quad |4x - 8| < \varepsilon$

23. $|x + 6| < \delta \quad \Rightarrow \quad |6x + 36| < \varepsilon$

24. $|x + 5| < \delta \quad \Rightarrow \quad |5x + 25| < \varepsilon$

In Problems 25–28, solve the inequalities (see Example 6).

25. $|x - 2| < 3|x + 7|$

26. $|2x - 5| < |x + 4|$

27. $2|2x - 3| < |x + 10|$

28. $|3x - 1| < 2|x + 6|$

29. Prove $|x| < |y| \quad \Leftrightarrow \quad x^2 < y^2$ by giving a reason for each step below.

$$|x| < |y| \quad \Rightarrow \quad |x|\,|x| \le |x|\,|y| \quad \text{and} \quad |x|\,|y| < |y|\,|y|$$
$$\Rightarrow \quad |x|^2 < |y|^2$$
$$\Rightarrow \quad x^2 < y^2$$

Conversely,

$$x^2 < y^2 \quad \Rightarrow \quad |x|^2 < |y|^2$$
$$\Rightarrow \quad |x|^2 - |y|^2 < 0$$
$$\Rightarrow \quad (|x| - |y|)(|x| + |y|) < 0$$
$$\Rightarrow \quad |x| - |y| < 0$$
$$\Rightarrow \quad |x| < |y|$$

30. Use the result of Problem 29 to show that

$$0 < a < b \quad \Rightarrow \quad \sqrt{a} < \sqrt{b}$$

31. Use the Triangle Inequality to show that each of the following is true.

(a) $|a - b| \le |a| + |b|$
(b) $|a - b| \ge |a| - |b|$
(c) $|a + b + c| \le |a| + |b| + |c|$

32. Use the Triangle Inequality and the fact that $0 < |a| < |b| \quad \Rightarrow \quad 1/|b| < 1/|a|$ to establish the following chain of inequalities.

$$\left|\dfrac{1}{x^2 + 3} - \dfrac{1}{|x| + 2}\right| \le \dfrac{1}{x^2 + 3} + \dfrac{1}{|x| + 2} \le \dfrac{1}{3} + \dfrac{1}{2}$$

33. Show that (see Problem 32)

$$\left|\dfrac{x - 2}{x^2 + 9}\right| \le \dfrac{|x| + 2}{9}$$

34. Show that

$$|x| \le 2 \quad \Rightarrow \quad \left|\dfrac{x^2 + 2x + 7}{x^2 + 1}\right| \le 15$$

35. Show that

$$|x| \le 1 \quad \Rightarrow \quad \left|x^4 + \tfrac{1}{2}x^3 + \tfrac{1}{4}x^2 + \tfrac{1}{8}x + \tfrac{1}{16}\right| < 2$$

36. Show each of the following.

(a) $x < x^2$ for $x < 0$ or $x > 1$
(b) $x^2 < x$ for $0 < x < 1$

37. Show $a \neq 0 \Rightarrow a^2 + 1/a^2 \geq 2$. *Hint*: Consider $(a - 1/a)^2$.

38. The number $\frac{1}{2}(a + b)$ is called the average, or **arithmetic mean**, of a and b. Show that the arithmetic mean of two numbers is between the two numbers; that is, prove that

$$a < b \Rightarrow a < \frac{a + b}{2} < b$$

39. The number $\sqrt{ab}$ is called the **geometric mean** of two positive numbers a and b. Prove that

$$0 < a < b \Rightarrow a < \sqrt{ab} < b$$

40. For two positive numbers a and b, prove that

$$\sqrt{ab} \leq \tfrac{1}{2}(a + b)$$

This is the simplest version of a famous inequality called the **geometric mean–arithmetic mean inequality**.

41. Show that, among all rectangles with given perimeter p, the square has largest area. *Hint*: If a and b denote the lengths of adjacent sides of a rectangle of perimeter p, then the area is ab, and for the square the area is $a^2 = [(a + b)/2]^2$. Now see Problem 40.

1.5
The Rectangular Coordinate System

Two Frenchmen deserve credit for the idea of a coordinate system. Pierre de Fermat was a lawyer who made mathematics his hobby. In 1629, he wrote a paper which, in effect, made use of coordinates to describe points and curves. René Descartes was a philosopher who thought mathematics could unlock the secrets of the universe. He published his *La Géométrie* in 1637. It is a famous book and though it does emphasize the role of algebra in solving geometric problems, one finds only a hint of coordinates there. By virtue of having the idea first and more explicitly, Fermat ought to get the major credit. History can be a fickle friend; coordinates are known as Cartesian coordinates, named after René Descartes.

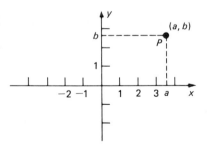

FIGURE 1

CARTESIAN COORDINATES In the plane, produce two copies of the real line, one horizontal and the other vertical, so that they intersect at the zero points of the two lines. The two lines are called **coordinate axes**; their intersection is labeled O and is called the **origin**. By convention, the horizontal line is called the **x-axis** and the vertical line is called the **y-axis**. The positive half of the x-axis is to the right; the positive half of the y-axis is upward. The coordinate axes divide the plane into four regions, called **quadrants**, labeled I, II, III, and IV, as shown in Figure 1.

Each point P in the plane can now be assigned a pair of numbers, called its **Cartesian coordinates**. If vertical and horizontal lines through P intersect the x- and y-axes at a and b, respectively, then P has coordinates (a, b) (see Figure 2). We call (a, b) an **ordered pair** of numbers because it makes a difference which number is first. The first number a is the **x-coordinate** (or abscissa); the second number b is the **y-coordinate** (or ordinate).

Conversely, take any ordered pair (a, b) of real numbers. The vertical line through a on the x-axis and the horizontal line through b on the y-axis meet in a point P, whose coordinates are (a, b).

Think of it this way: The coordinates of a point are the address of that point. If you have found a house (or a point), you can read its address. Conversely, if you know the address of a house (or a point), you can always locate it. In Figure 3, we have shown the coordinates of several points.

FIGURE 2

FIGURE 3

FIGURE 4

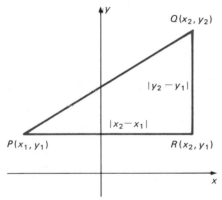

FIGURE 5

THE DISTANCE FORMULA   With coordinates in hand, we can introduce a simple formula for the distance between any two points in the plane. It is based on the **Pythagorean Theorem**, which says that if a and b measure the two legs of a right triangle and c measures its hypotenuse (Figure 4), then

$$a^2 + b^2 = c^2$$

Conversely, this relationship between the three sides of a triangle holds only for a right triangle.

Now consider any two points P and Q, with coordinates (x_1, y_1) and (x_2, y_2), respectively. Together with R—the point with coordinates (x_2, y_1)—P and Q are vertices of a right triangle (Figure 5). The lengths of PR and RQ are $|x_2 - x_1|$ and $|y_2 - y_1|$, respectively. When we apply the Pythagorean Theorem and take the principal square root of both sides, we obtain the following expression for $d(P, Q)$, the (undirected) distance between P and Q.

$$d(P, Q) = \sqrt{(x_2 - x_1)^2 + (y_2 - y_1)^2}$$

This is called the **Distance Formula**.

EXAMPLE 1   Find the distance between
   (a) $P(-2, 3)$ and $Q(4, -1)$
   (b) $P(\sqrt{2}, \sqrt{3})$ and $Q(\pi, \pi)$

Solution
   (a) $d(P, Q) = \sqrt{(4 - (-2))^2 + (-1 - 3)^2} = \sqrt{36 + 16} = \sqrt{52} \approx 7.21$
   (b) $d(P, Q) = \sqrt{(\pi - \sqrt{2})^2 + (\pi - \sqrt{3})^2} \approx \sqrt{4.971} \approx 2.23$ ∎

The formula holds even if the two points lie on the same horizontal line or the same vertical line. Thus, the distance between $P(-2, 2)$ and $Q(6, 2)$ is

$$\sqrt{(-2 - 6)^2 + (2 - 2)^2} = \sqrt{64} = 8$$

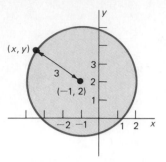

FIGURE 6

THE EQUATION OF A CIRCLE   It is a small step from the distance formula to the equation of a circle. A **circle** is the set of points which lie at a fixed distance (the *radius*) from a fixed point (the *center*). Consider, for example, the circle of radius 3 with center at $(-1, 2)$ (Figure 6). Let (x, y) denote any point on this circle. By the Distance Formula,

$$\sqrt{(x + 1)^2 + (y - 2)^2} = 3$$

When we square both sides, we obtain

$$(x + 1)^2 + (y - 2)^2 = 9$$

which we call the equation of this circle.

More generally, the circle of radius r and center (h, k) has the equation

$$\boxed{(x - h)^2 + (y - k)^2 = r^2}$$

We call this the **standard equation of a circle**.

EXAMPLE 2   Find the equation of a circle of radius 5 and center $(1, -5)$. Also find the y-coordinates of the two points on this circle with x-coordinate 2.

Solution   The desired equation is

$$(x - 1)^2 + (y + 5)^2 = 25$$

To accomplish the second task, we substitute $x = 2$ in the equation and solve for y.

$$(2 - 1)^2 + (y + 5)^2 = 25$$
$$(y + 5)^2 = 24$$
$$y + 5 = \pm\sqrt{24}$$
$$y = -5 \pm \sqrt{24} = -5 \pm 2\sqrt{6} \qquad \blacksquare$$

If we expanded the two squares in the boxed equation and combined the constants, the equation would take the form

$$x^2 + ax + y^2 + by = c$$

This suggests asking whether every equation of the latter form is the equation of a circle. The answer is yes (with some obvious exceptions), as we see in the following example.

In this example, we need to *complete the square*, a process important in many contexts. To complete the square of $x^2 \pm ax$, add $(a/2)^2$. Thus,

$$x^2 - 12x + 6^2 = (x - 6)^2$$
$$x^2 + \tfrac{2}{5}x + (\tfrac{1}{5})^2 = (x + \tfrac{1}{5})^2$$

EXAMPLE 3 Show that the equation

$$x^2 - 2x + y^2 + 6y = -6$$

represents a circle, and find its center and radius.

Solution We complete the squares for the expressions in both x and y by adding the same numbers to both sides of the equation.

$$(x^2 - 2x \quad) + (y^2 + 6y \quad) = -6$$
$$(x^2 - 2x + 1) + (y^2 + 6y + 9) = -6 + 1 + 9$$
$$(x - 1)^2 + (y + 3)^2 = 4$$

The last equation is in standard form. It is the equation of a circle with center $(1, -3)$ and radius 2. If, as a result of this process, we had come up with a negative number on the right side of the final equation, the equation would not have represented any curve. If we had come up with zero, the equation would have represented the single point $(1, -3)$. ∎

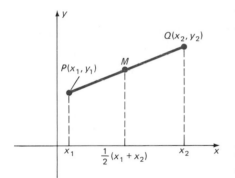

FIGURE 7

THE MIDPOINT FORMULA Consider two points $P(x_1, y_1)$ and $Q(x_2, y_2)$ with $x_1 \leq x_2$, as in Figure 7. Note that

$$x_1 + \tfrac{1}{2}(x_2 - x_1) = x_1 + \tfrac{1}{2}x_2 - \tfrac{1}{2}x_1$$
$$= \tfrac{1}{2}x_1 + \tfrac{1}{2}x_2$$
$$= \frac{x_1 + x_2}{2}$$

It follows that the point $(x_1 + x_2)/2$ is midway between x_1 and x_2 on the x-axis and, consequently, that the midpoint M of the segment PQ has $(x_1 + x_2)/2$ as its x-coordinate. In a completely similar manner, we can show that $(y_1 + y_2)/2$ is the y-coordinate of M. Thus, we have the following result.

> The midpoint of the line segment from $P(x_1, y_1)$ to $Q(x_2, y_2)$ is
> $$\left(\frac{x_1 + x_2}{2}, \frac{y_1 + y_2}{2} \right)$$

EXAMPLE 4 Find the equation of the circle having the segment from $(1, 3)$ to $(7, 11)$ as a diameter.

Solution The center of the circle is at the midpoint of the diameter; thus, the center has coordinates $(1 + 7)/2 = 4$ and $(3 + 11)/2 = 7$. The length of the diameter, obtained from the distance formula, is

$$[(7 - 1)^2 + (11 - 3)^2]^{1/2} = [36 + 64]^{1/2} = 10$$

and so the radius of the circle is 5. The equation of the circle is

$$(x - 4)^2 + (y - 7)^2 = 25$$ ∎

In Problems 1–6, plot the given points in the coordinate plane and then find the distance between them.

1. $(2, -1), (5, 3)$ **2.** $(-2, 1), (7, 13)$

3. $(4, 2), (2, 4)$ **4.** $(-1, 5), (6, 3)$

$\boxed{C}$ **5.** $(1.232, 4.153), (\pi, \sqrt{2})$

$\boxed{C}$ **6.** $(2.71, -3.42), (5.16, 4.33)$

7. Show that the triangle whose vertices are $(5, 3)$, $(-2, 4)$, and $(10, 8)$ is isosceles.

8. Show that the triangle whose vertices are $(2, -4)$, $(4, 0)$, and $(8, -2)$ is a right triangle.

9. The points $(3, -1)$ and $(3, 3)$ are two vertices of a square. Give three other pairs of possible vertices.

10. Find the point on the x-axis that is equidistant from $(3, 1)$ and $(6, 4)$.

11. Find the distance between $(-2, 3)$ and the midpoint of the segment joining $(-2, -2)$ and $(4, 3)$.

12. Find the length of the line segment joining the midpoints of the segments AB and CD, where $A = (1, 3)$, $B = (2, 6)$, $C = (4, 7)$, and $D = (3, 4)$.

In Problems 13–18, find the equation of the circle satisfying the given conditions.

13. Center $(1, -2)$, radius 6.

14. Center $(-3, 4)$, radius 8.

15. Center $(2, -1)$, goes through $(5, 3)$.

16. Center $(4, 3)$, goes through $(6, 2)$.

17. Diameter AB, where $A = (-1, 2)$ and $B = (3, 8)$.

18. Center $(3, 4)$ and tangent to x-axis.

19. Find the y-coordinates of the two points on the circle of Problem 13 with x-coordinate of 3 (see Example 2).

20. Find the x-coordinates of the two points on the circle of Problem 14 with y-coordinate of 8.

In Problems 21–26, find the center and radius of the circle with the given equation (see Example 3).

21. $x^2 + y^2 + 2x - 10y + 25 = 0$

22. $x^2 + y^2 - 6y = 16$

23. $x^2 + y^2 - 12x + 35 = 0$

24. $x^2 + y^2 - 10x + 10y = 0$

25. $4x^2 + 4y^2 + 4x - 12y + 1 = 0$

26. $3x^2 + 3y^2 - 2x + 4y = \frac{20}{3}$

27. The points $(2, 3)$, $(6, 3)$, $(6, -1)$, and $(2, -1)$ are corners of a square. Find the equations of the inscribed and circumscribed circles.

28. A belt fits tightly around the two circles, with equations $(x - 1)^2 + (y + 2)^2 = 16$ and $(x + 9)^2 + (y - 10)^2 = 16$. How long is this belt?

$\boxed{C}$ **29.** Cities at A, B, and C are vertices of a right triangle, with the right angle at vertex B. Also, AB and BC are roads of lengths 214 and 179 miles, respectively. An airplane can fly the route AC, which is not a road. It costs $3.71 per mile to ship a certain product by truck and $4.82 per mile by plane. Decide whether it is cheaper to ship the product from A to C by truck or plane and find the total cost by the cheaper method.

$\boxed{C}$ **30.** City B is 10 miles downstream from city A and on the opposite side of a river $\frac{1}{2}$ mile wide. Mary Crane will run from city A along the river for 6 miles, then swim diagonally to city B. If she runs at 8 miles per hour and swims at 3 miles per hour, how long will it take her to get from city A to city B? Assume the rate of the current is negligible.

31. Show that the midpoint of the hypotenuse of any right triangle is equidistant from the three vertices.

32. Find the equation of the circle circumscribed about the right triangle whose vertices are $(0, 0)$, $(8, 0)$, and $(0, 6)$.

33. Show that the two circles $x^2 + y^2 - 4x - 2y - 11 = 0$ and $x^2 + y^2 + 20x - 12y + 72 = 0$ do not intersect. *Hint*: Find the distance between their centers.

34. What relationship between a, b, and c must hold if $x^2 + ax + y^2 + by + c = 0$ is the equation of a circle?

$\boxed{C}$ **35.** Find the length of the crossed belt in Figure 8 that fits tightly around the circles $(x - 2)^2 + (y - 2)^2 = 9$ and $(x - 10)^2 + (y - 8)^2 = 9$. *Note*: A little trigonometry is needed to complete this problem.

FIGURE 8

36. Show that the set of points that are twice as far from $(3, 4)$ as from $(1, 1)$ form a circle. Find its center and radius.

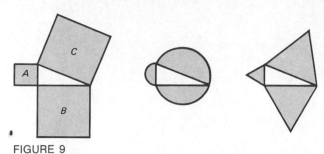

FIGURE 9

37. The Pythagorean Theorem says that the areas A, B, and C of the squares in Figure 9 satisfy $A + B = C$. Show that semicircles and equilateral triangles satisfy the same relation and then guess at a very general theorem.

38. Consider a circle C and a point P exterior to the circle. Let line segment PT be tangent to C at T and let the line through P and the center of C intersect C at M and N. Show that $(PM)(PN) = (PT)^2$.

1.6 The Straight Line

Of all curves, the straight line is in many ways the simplest. We assume that our readers have a good intuitive notion of this concept from looking at a taut string or sighting along the edge of a ruler. In any case, let us agree that two points—for example, $A(3, 2)$ and $B(8, 4)$ shown in Figure 1—determine a unique straight line through them. And from now on, we use the word *line* as a synonym for *straight line*.

A line is a geometric object. When it is placed in a coordinate plane, it ought to have an equation, just as a circle does. How do we find the equation of a line? To answer, we will need the notion of slope.

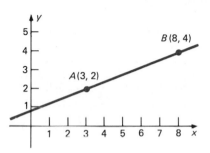

FIGURE 1

THE SLOPE OF A LINE   Consider the line in Figure 1. From point A to point B, there is a **rise** (vertical change) of 2 units and a **run** (horizontal change) of 5 units. We say that the line has a slope of $\frac{2}{5}$. In general (Figure 2), for a line through $A(x_1, y_1)$ and $B(x_2, y_2)$, where $x_1 \neq x_2$, we define the **slope** m of that line by

$$m = \frac{\text{rise}}{\text{run}} = \frac{y_2 - y_1}{x_2 - x_1}$$

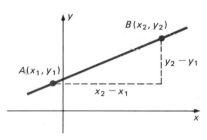

FIGURE 2

You should immediately raise a question. A line has many points. Does the value we get for the slope depend on which pair of points we use for A and B? The similar triangles in Figure 3 show us that

$$\frac{y_2' - y_1'}{x_2' - x_1'} = \frac{y_2 - y_1}{x_2 - x_1}$$

Thus, points A' and B' would do just as well as A and B. It does not even matter whether A is to the left or right of B, since

$$\frac{y_1 - y_2}{x_1 - x_2} = \frac{y_2 - y_1}{x_2 - x_1}$$

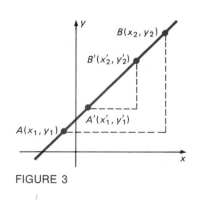

FIGURE 3

All that matters is that we subtract the coordinates in the same order in numerator and denominator.

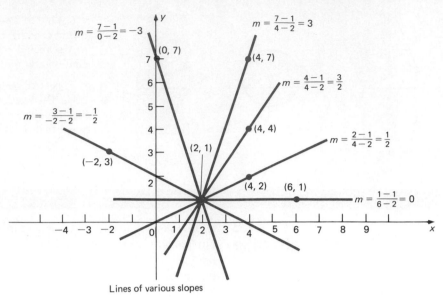

FIGURE 4

Lines of various slopes

The slope m is a measure of the steepness of a line, as Figure 4 illustrates. Notice that a horizontal line has zero slope, a line that rises to the right has positive slope, and a line that falls to the right has negative slope. The larger the magnitude of the slope, the steeper the line. The concept of slope for a vertical line makes no sense, since it would involve division by zero. Therefore, slope for a vertical line is left undefined.

THE POINT-SLOPE FORM Consider again the line of our opening discussion; it is reproduced in Figure 5. We know that this line:

1. passes through $(3, 2)$;
2. has slope $\frac{2}{5}$.

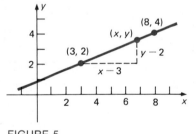

FIGURE 5

Take any other point on that line, such as one with coordinates (x, y). If we use this point and the point $(3, 2)$ to measure slope, we must get $\frac{2}{5}$—that is,

$$\frac{y - 2}{x - 3} = \frac{2}{5}$$

or, after multiplying by $x - 3$,

$$y - 2 = \tfrac{2}{5}(x - 3)$$

Notice that this last equation is satisfied by all points on the line, even by $(3, 2)$. Moreover, none of the points not on the line can satisfy this equation.

What we have just done in an example can be done in general. The line passing through the (fixed) point (x_1, y_1) with slope m has equation

$$\boxed{y - y_1 = m(x - x_1)}$$

We call this the **point-slope** form of the equation of a line.

Consider once more the line of our example. That line passes through $(8, 4)$ as well as $(3, 2)$. If we use $(8, 4)$ as (x_1, y_1), we get the equation

$$y - 4 = \tfrac{2}{5}(x - 8)$$

which looks quite different from

$$y - 2 = \tfrac{2}{5}(x - 3)$$

However, both can be simplified to $5y - 2x = 4$; they are equivalent.

EXAMPLE 1   Find an equation of the line through $(-4, 2)$ and $(6, -1)$.

Solution   The slope m is $(-1 - 2)/(6 + 4) = -\tfrac{3}{10}$. Thus, using $(-4, 2)$ as the fixed point, we obtain the equation

$$y - 2 = -\tfrac{3}{10}(x + 4) \qquad \blacksquare$$

THE SLOPE-INTERCEPT FORM   The equation of a line can be expressed in various forms. Suppose we are given the slope m for a line and the y-intercept b (that is, the line intersects the y-axis at $(0, b)$), as shown in Figure 6. Choosing $(0, b)$ as (x_1, y_1) and applying the point-slope form, we get

$$y - b - m(x - 0)$$

which we can rewrite as

$$\boxed{y = mx + b}$$

FIGURE 6

The latter is called the **slope-intercept** form.

Why get excited about that, you ask? Any time we see an equation written this way, we recognize it as a line and can immediately read its slope and y-intercept. For example, consider the equation

$$3x - 2y + 4 = 0$$

If we solve for y, we get

$$y = \tfrac{3}{2}x + 2$$

It is the equation of a line with slope $\tfrac{3}{2}$ and y-intercept 2.

EQUATION OF A VERTICAL LINE   Vertical lines do not fit within the discussion above since the concept of slope is not defined for them. But they do have equations, very simple ones. The line in Figure 7 has equation $x = \tfrac{5}{2}$, since a point is on the line if and only if it satisfies this equation. The equation of any vertical line can be put in the form

$$\boxed{x = k}$$

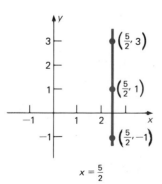

FIGURE 7

where k is a constant. It should be noted that the equation of a horizontal line can be written in the form $y = k$.

THE FORM $Ax + By + C = 0$ It would be nice to have a form that covered all lines, including vertical lines. Consider for example,

(1) $y - 2 = -4(x + 2)$

(2) $y = 5x - 3$

(3) $x = 5$

These can be rewritten (by taking everything to the left-hand side) as follows:

(1) $4x + y + 6 = 0$

(2) $-5x + y + 3 = 0$

(3) $x + 0y - 5 = 0$

All are of the form

$$Ax + By + C = 0, \qquad A \text{ and } B \text{ not both } 0$$

which we call the **general linear equation**. It takes only a moment's thought to see that the equation of any line can be put in this form. Conversely, the graph of the general linear equation is always a line (see Problem 43).

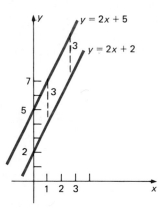

y = 2x + 5

y = 2x + 2

FIGURE 8

PARALLEL LINES   If two lines have the same slope, they are parallel. Thus, $y = 2x + 2$ and $y = 2x + 5$ represent parallel lines; both have slope of 2. The second line is 3 units above the first for every value of x (see Figure 8).

Similarly, the lines with equations $-2x + 3y + 12 = 0$ and $4x - 6y = 5$ are parallel. To see this, solve these equations for y (that is, find the slope-intercept form); you get $y = \frac{2}{3}x - 4$ and $y = \frac{2}{3}x - \frac{5}{6}$, respectively. Both have slope $\frac{2}{3}$; they are parallel.

We may summarize by stating that *two nonvertical lines are parallel if and only if they have the same slope.*

EXAMPLE 2   Find the equation of the line through $(6, 8)$ which is parallel to the line with equation $3x - 5y = 11$.

Solution   When we solve $3x - 5y = 11$ for y, we obtain

$$y = \tfrac{3}{5}x - \tfrac{11}{5}$$

from which we read the slope of the line to be $\frac{3}{5}$. The equation of the desired line is

$$y - 8 = \tfrac{3}{5}(x - 6)$$

or, equivalently, $3x - 5y + 22 = 0$.   ■

FIGURE 9

PERPENDICULAR LINES Is there a simple slope condition which characterizes perpendicular lines? Yes; *two nonvertical lines are perpendicular if and only if their slopes are negative reciprocals of each other.* To see why this is true, consider two nonvertical intersecting lines l_1 and l_2. Without loss of generality, we may suppose they intersect at the origin since if not, we may translate them so they do without changing their slopes. Let $P_1(x_1, y_1)$ be a point on l_1 and $P_2(x_2, y_2)$ be a point on l_2, as shown in Figure 9. By the Pythagorean theorem and its converse (Section 1.5), P_1OP_2 is a right angle if and only if

$$[d(P_1, O)]^2 + [d(P_2, O)]^2 = [d(P_1, P_2)]^2$$

that is, if and only if

$$(x_1^2 + y_1^2) + (x_2^2 + y_2^2) = (x_1 - x_2)^2 + (y_1 - y_2)^2$$

After expansion and simplification, this equation becomes $2x_1x_2 + 2y_1y_2 = 0$, or

$$\frac{y_1}{x_1} = -\frac{x_2}{y_2}$$

Now, y_1/x_1 is the slope of l_1, whereas y_2/x_2 is the slope of l_2. Thus, P_1OP_2 is a right angle if and only if the slopes of the two lines are negative reciprocals of each other.

The lines $y = \frac{3}{4}x$ and $y = -\frac{4}{3}x$ are perpendicular. So are $2x - 3y = 5$ and $3x + 2y = -4$, since after solving these equations for y, we see that the first line has slope $\frac{2}{3}$ and the second has slope $-\frac{3}{2}$.

EXAMPLE 3 Find the equation of the line through the point of intersection of the lines with equations $3x + 4y = 8$ and $6x - 10y = 7$ which is perpendicular to the first of these two lines.

Solution To find the point of intersection of the two lines, we multiply the first equation by -2 and add it to the second equation.

$$\begin{array}{r} -6x - 8y = -16 \\ \underline{6x - 10y = 7} \\ -18y = -9 \\ y = \tfrac{1}{2} \end{array}$$

Substituting $y = \frac{1}{2}$ in either of the original equations yields $x = 2$. The point of intersection is $(2, \frac{1}{2})$.

When we solve the first equation for y (to put it in slope-intercept form), we get $y = -\frac{3}{4}x + 2$. A line perpendicular to it has slope $\frac{4}{3}$. The equation of the required line is

$$y - \tfrac{1}{2} = \tfrac{4}{3}(x - 2) \qquad \blacksquare$$

In Problems 1–8, find the slope of the line containing the given two points.

    1. $(2, 3)$ and $(4, 8)$ **2.** $(4, 1)$ and $(8, 2)$

    3. $(-4, 2)$ and $(3, 0)$ **4.** $(2, -4)$ and $(0, -6)$

    5. $(3, 0)$ and $(0, 5)$ **6.** $(-6, 0)$ and $(0, 6)$

    ⊡   **7.** $(-1.732, 5.014)$ and $(4.315, 6.175)$

    ⊡   **8.** $(\pi, \sqrt{3})$ and $(1.642, \sqrt{2})$

In Problems 9–16, find an equation for each line. Then write your answer in the form $Ax + By + C = 0$.

    9. Through $(2, 3)$ with slope 4.

    10. Through $(3, -4)$ with slope -2.

    11. With y-intercept 4 and slope -2.

    12. With y-intercept 5 and slope 0.

    13. Through $(2, 3)$ and $(4, 8)$.

    14. Through $(4, 1)$ and $(8, 2)$.

    15. Through $(2, -3)$ and $(2, 5)$.

    16. Through $(-5, 0)$ and $(-5, 4)$.

In Problems 17–20, find the slope and y-intercept of each line.

    17. $3y = 2x - 4$ **18.** $2y = 5x + 2$

    19. $2x + 3y = 6$ **20.** $4x + 5y = -20$

    21. Write an equation for the line through $(3, -3)$ which is:

(a) parallel to the line $y = 2x + 5$;
(b) perpendicular to the line $y = 2x + 5$;
(c) parallel to the line $2x + 3y = 6$;
(d) perpendicular to the line $2x + 3y = 6$;
(e) parallel to the line through $(-1, 2)$ and $(3, -1)$;
(f) parallel to the line $x = 8$;
(g) perpendicular to the line $x = 8$.

    22. Find the value of k for which the line $4x + ky = 5$:

(a) passes through the point $(2, 1)$;
(b) is parallel to the y-axis;
(c) is parallel to the line $6x - 9y = 10$;
(d) has equal x- and y-intercepts;
(e) is perpendicular to the line $y - 2 = 2(x + 1)$.

    23. Write an equation for the line through $(0, -4)$ which is perpendicular to the line $y + 2 = -\frac{1}{2}(x - 1)$.

    24. Find the value of k such that the line $kx - 3y = 10$:

(a) is parallel to the line $y = 2x + 4$;

(b) is perpendicular to the line $y = 2x + 4$;
(c) is perpendicular to the line $2x + 3y = 6$.

    25. Does $(3, 9)$ lie above or below the line $y = 3x - 1$?

    26. Show that the equation of the line with x-intercept $a \neq 0$ and y-intercept $b \neq 0$ can be written as

$$\frac{x}{a} + \frac{y}{b} = 1$$

In Problems 27–30, find the coordinates of the point of intersection. Then write an equation for the line through that point perpendicular to the line given first (see Example 3).

    27. $2x + 3y = 4$ **28.** $4x - 5y = 8$
          $-3x + y = 5$ $2x + y = -10$

    29. $3x - 4y = 5$ **30.** $5x - 2y = 5$
         $2x + 3y = 9$ $2x + 3y = 6$

It can be shown that the distance d from the point (x_1, y_1) to the line $Ax + By + C = 0$ is

$$d = \frac{|Ax_1 + By_1 + C|}{\sqrt{A^2 + B^2}}$$

Use this result to find the distance from the given point to the given line.

    31. $(-3, 2)$; $3x + 4y = 6$

    32. $(4, -1)$; $2x - 2y + 4 = 0$

    33. $(-2, -1)$; $5y = 12x + 1$

    34. $(3, -1)$; $y = 2x - 5$

In Problems 35 and 36, find the (perpendicular) distance between the given parallel lines. *Hint*: First find a point on one of the lines.

    35. $3x + 4y = 6$, $3x + 4y = 12$

    36. $5x + 12y = 2$, $5x + 12y = 7$

    37. A bulldozer costs \$120,000 and each year it depreciates 8% of its original value. Find a formula for the value V of the bulldozer after t years.

    38. The graph of the answer to Problem 37 is a straight line. What is its slope, assuming the t-axis to be horizontal? Interpret the slope.

    39. Past experience indicates that egg production in Matlin County is growing linearly. In 1960 it was 700,000 cases, and in 1970 it was 820,000 cases. Write a formula for the number N of cases produced n years after 1960 and use it to predict egg production in the year 2000.

40. A piece of equipment purchased today for $80,000 will depreciate linearly to a scrap value of $2000 after 20 years. Write a formula for its value V after n years.

41. Suppose that the profit P that a company realizes in selling x items of a certain commodity is given by $P = 450x - 2000$ dollars.

(a) Interpret the value of P when $x = 0$.
(b) Find the slope of the graph of the above equation. This slope is called the **marginal profit**. What is its economic interpretation?

42. The cost C of producing x items of a certain commodity is given by $C = 0.75x + 200$ dollars. The slope of its graph is called the **marginal cost**. Find it and give an economic interpretation.

43. Show that the graph of $Ax + By + C = 0$ is always a line (provided A and B are not both 0). *Hint*: Consider two cases: (1) $B = 0$ and (2) $B \neq 0$.

44. Find an equation for the line through $(2, 3)$ which has equal x- and y-intercepts. *Hint*: Use Problem 26.

45. Show that for each value of k, the equation

$$2x - y + 4 + k(x + 3y - 6) = 0$$

represents a line through the intersection of the two lines $2x - y + 4 = 0$ and $x + 3y - 6 = 0$. *Hint*: It is not necessary to find the point of intersection (x_0, y_0).

46. Find an equation for the line which bisects the line segment from $(-2, 1)$ to $(4, -7)$ and is at right angles to this line segment.

47. The center of the circumscribed circle of a triangle lies on the perpendicular bisectors of the sides. Use this fact to find the center of the circle that circumscribes the triangle with vertices $(0, 4)$, $(2, 0)$, and $(4, 6)$.

48. Suppose that (a, b) is on the circle $x^2 + y^2 = r^2$. Show that the line $ax + by = r^2$ is tangent to the circle at (a, b).

49. Find the equations of the two tangent lines to the circle $x^2 + y^2 = 36$ that go through $(12, 0)$. *Hint*: See Problem 48.

50. Express the perpendicular distance between the parallel lines $y = mx + b$ and $y = mx + B$ in terms of m, b, and B. *Hint*: The required distance is the same as that between $y = mx$ and $y = mx + B - b$.

51. Show that the line through the midpoints of two sides of a triangle is parallel to the third side. *Hint*: You may assume the triangle has vertices at $(0, 0)$, $(a, 0)$, and (b, c).

52. Show that the line segments joining the midpoints of adjacent sides of any quadrilateral (four-sided polygon) form a parallelogram.

1.7
Graphs of Equations

The use of coordinates for points in the plane allows us to describe a curve (a geometric object) by an equation (an algebraic object). We saw how this was done for circles and lines in the previous sections. Now we want to consider the reverse process: graphing an equation. The **graph of an equation** in x and y consists of those points in the plane whose coordinates (x, y) satisfy the equation—that is, make it a true equality.

THE GRAPHING PROCEDURE To graph an equation—for example, $y = 2x^3 - x + 19$—we follow a simple three-step procedure:

Step 1 Obtain the coordinates of a few points that satisfy the equation.
Step 2 Plot those points in the plane.
Step 3 Connect the points with a smooth curve.

The best way to do Step 1 is to make a table of values. Assign values to one of the variables, such as x, and determine the corresponding values of the other variable, listing the results in tabular form.

EXAMPLE 1 Graph the equation $y = x^2 - 3$.

Solution   The three-step procedure is shown in Figure 1.

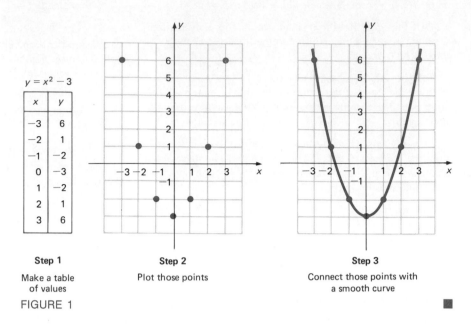

$y = x^2 - 3$	
x	y
-3	6
-2	1
-1	-2
0	-3
1	-2
2	1
3	6

Step 1
Make a table of values

Step 2
Plot those points

Step 3
Connect those points with a smooth curve

FIGURE 1

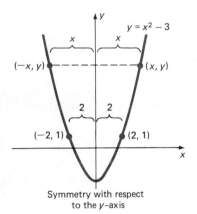

Symmetry with respect to the *y*-axis

FIGURE 2

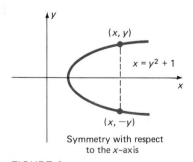

Symmetry with respect to the *x*-axis

FIGURE 3

Of course, you need to use common sense and even a little faith. When you connect the points you have plotted with a smooth curve, you are assuming that the curve behaves nicely between consecutive points, which is faith. That is why you should plot enough points so the outline of the curve seems very clear; the more points you plot, the less faith you will need. Also, you should recognize that you can seldom display the whole curve. In our example, the curve has infinitely long arms, opening wider and wider. But our graph does show the essential features. This is our goal in graphing: Show enough of the graph so the essential features are visible.

SYMMETRY OF A GRAPH   We can save ourselves some work and also draw more accurate graphs if we can recognize certain symmetries of the graph by examining the corresponding equation. Look at the graph of $y = x^2 - 3$, drawn above and again in Figure 2. If the coordinate plane were folded along the *y*-axis, the two branches would coincide. For example, $(3, 6)$ would coincide with $(-3, 6)$, $(2, 1)$ would coincide with $(-2, 1)$, and, more generally, (x, y) would coincide with $(-x, y)$. Algebraically, this corresponds to the fact that replacing x by $-x$ in the equation $y = x^2 - 3$ results in an equivalent equation.

Consider an arbitrary graph. It is **symmetric with respect to the *y*-axis** if whenever (x, y) is on the graph, $(-x, y)$ is also on the graph (Figure 2). Similarly, it is **symmetric with respect to the *x*-axis** if whenever (x, y) is on the graph, $(x, -y)$ is also on the graph (Figure 3). Finally, a graph is **symmetric with respect to the origin** if whenever (x, y) is on the graph, $(-x, -y)$ is also on the graph (see Example 2).

32

In terms of equations, we have three simple tests.

The graph of an equation is:

1. symmetric with respect to the y-axis if replacing x by $-x$ gives an equivalent equation (for example, $y = x^2$).
2. symmetric with respect to the x-axis if replacing y by $-y$ gives an equivalent equation (for example, $x = 1 + y^2$).
3. symmetric with respect to the origin if replacing x by $-x$ and y by $-y$ gives an equivalent equation ($y = x^3$ is a good example since $-y = (-x)^3$ is equivalent to $y = x^3$).

EXAMPLE 2 Sketch the graph of $y = x^3$.

Solution We note, as pointed out above, that the graph will be symmetric with respect to the origin. Thus we need only get a table of values for non-negative x's; we can find matching points by symmetry (Figure 4).

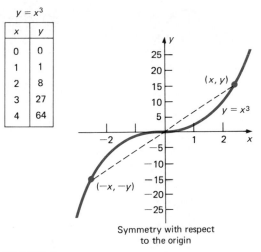

$y = x^3$

x	y
0	0
1	1
2	8
3	27
4	64

Symmetry with respect
to the origin

FIGURE 4

In graphing $y = x^3$, we used a smaller scale on the y-axis than on the x-axis. This made it possible to show a larger portion of the graph (it also distorted the graph by flattening it). We suggest that before putting scales on the two axes, you should examine your table of values. Choose scales so that all or most of your points can be plotted and still keep your graph of reasonable size.

INTERCEPTS The points where the graph of an equation crosses the two coordinate axes play a significant role in many problems. Consider, for example,

$$y = x^3 - 2x^2 - 5x + 6 = (x + 2)(x - 1)(x - 3)$$

Notice that $y = 0$ when $x = -2, 1, 3$. The numbers -2, 1, and 3 are called **x-intercepts**. Similarly, $x = 0$ when $y = 6$, and so 6 is called the **y-intercept**.

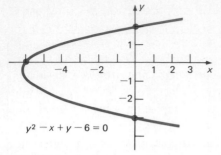

$y^2 - x + y - 6 = 0$

FIGURE 5

EXAMPLE 3 Sketch the graph of $y^2 - x + y - 6 = 0$, showing all intercepts clearly.

Solution Putting $y = 0$ in the given equation, we get $x = -6$, and so the x-intercept is -6. Putting $x = 0$ in the equation, we find $y^2 + y - 6 = 0$, or $(y + 3)(y - 2) = 0$; the y-intercepts are -3 and 2. A check on symmetries indicates that the graph has none of the three types discussed earlier. The graph is displayed in Figure 5. ∎

GENERAL QUADRATIC AND CUBIC EQUATIONS Since quadratic and cubic equations will often be used as examples in later work, we display their typical graphs in Figure 6.

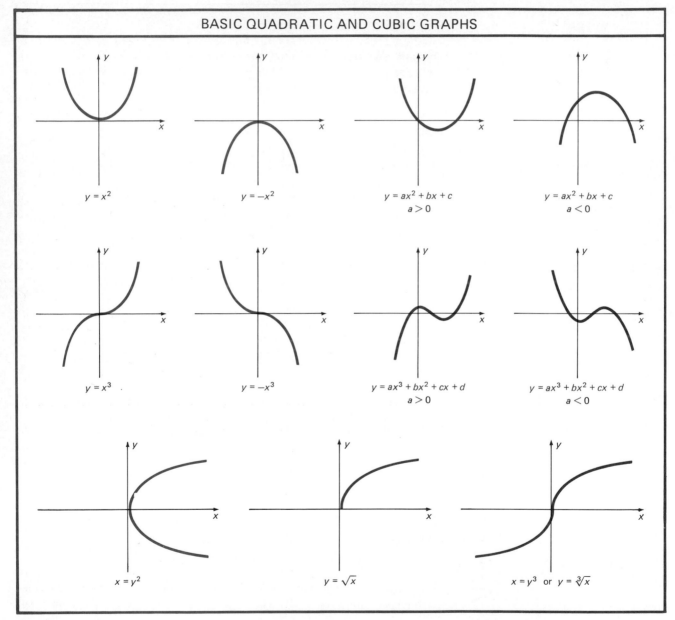

BASIC QUADRATIC AND CUBIC GRAPHS

$y = x^2$

$y = -x^2$

$y = ax^2 + bx + c$
$a > 0$

$y = ax^2 + bx + c$
$a < 0$

$y = x^3$

$y = -x^3$

$y = ax^3 + bx^2 + cx + d$
$a > 0$

$y = ax^3 + bx^2 + cx + d$
$a < 0$

$x = y^2$

$y = \sqrt{x}$

$x = y^3$ or $y = \sqrt[3]{x}$

FIGURE 6

The graphs of quadratic equations are cup-shaped curves called **parabolas**. If an equation has the form $y = ax^2 + bx + c$ or $x = ay^2 + by + c$ with $a \neq 0$, its graph will always be a parabola. In the first case, the graph opens up or down, according as $a > 0$ or $a < 0$. In the second case, the graph opens right or left, according as $a > 0$ or $a < 0$. Note that the equation of Example 3 can be put in the form $x = y^2 + y - 6$.

INTERSECTIONS OF GRAPHS Occasionally, we need to know the points of intersection of two graphs. These points are found by solving the two equations for the graphs simultaneously.

EXAMPLE 4 Find the points of intersection of the line $y = -2x + 2$ and the parabola $y = 2x^2 - 4x - 2$ and sketch both graphs on the same coordinate plane.

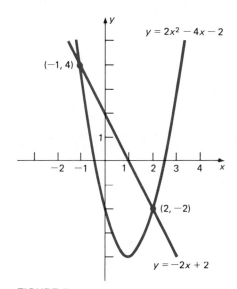

Solution We must solve the two equations simultaneously. This is easy to do by substituting the expressions for y from the first equation into the second equation and then solving the resulting equation for x.

$$-2x + 2 = 2x^2 - 4x - 2$$
$$0 = 2x^2 - 2x - 4$$
$$0 = 2(x - 2)(x + 1)$$
$$x = -1, \quad x = 2$$

By substitution, we find the corresponding values of y to be 4 and -2; the intersection points are therefore $(-1, 4)$ and $(2, -2)$. The two graphs are shown in Figure 7. ∎

FIGURE 7

PROBLEM SET 1.7

In Problems 1–18, sketch the graph of the given equation. Begin by checking for symmetries and be sure to find all x- and y-intercepts.

1. $y = -x^2 + 4$

2. $x = -y^2 + 4$

3. $3x^2 + 4y = 0$

4. $y = 2x^2 - x$

5. $x^2 + y^2 = 36$

6. $(x - 2)^2 + y^2 = 4$

7. $4x^2 + 9y^2 = 36$

8. $16x^2 + y^2 = 16$

9. $y = x^3 - 3x$

10. $y = x^3 + 1$

11. $y = \dfrac{1}{x^2 + 1}$

12. $y = \dfrac{x}{x^2 + 1}$

13. $x^3 - y^2 = 0$

14. $x^4 + y^4 = 16$

15. $y = (x - 2)(x + 1)(x + 3)$

16. $y = x(x - 3)(x - 5)$

17. $y = x^2(x - 2)$

18. $|x| + |y| = 4$

In Problems 19–26, sketch the graphs of both equations on the same coordinate plane. Be sure to find the points of intersection of the two graphs (see Example 4). You will need the quadratic formula in Problems 23–26.

19. $y = -x + 1$
    $y = x^2 + 2x + 1$

20. $y = -x + 4$
    $y = -x^2 + 2x + 4$

21. $y = -2x + 1$
    $y = -x^2 - x + 3$

22. $y = -3x + 15$
    $y = 3x^2 - 3x + 12$

$\boxed{\text{C}}$ **23.** $y = 1.5x + 3.2$
$y = x^2 - 2.9x$

$\boxed{\text{C}}$ **24.** $y = 2.1x - 6.4$
$y = -1.2x^2 + 4.3$

$\boxed{\text{C}}$ **25.** $y = 4x + 3$
$x^2 + y^2 = 4$

$\boxed{\text{C}}$ **26.** $y - 3x = 1$
$x^2 + 2x + y^2 = 15$

27. Find the distance between the two points on the graph of $y = 3x^4 - 2x + 1$ with x-coordinates -1 and 1.

$\boxed{\text{C}}$ **28.** Find the distance between the points on the curve $y = 3x^2 - 2x + 1$ corresponding to $x = 1$ and $x = \pi$, accurate to four decimal places.

29. Determine any symmetries and sketch the graph of $y = 2^x + 2^{-x}$.

30. Determine any symmetries and sketch the graph of $y = 2^x - 2^{-x}$.

$\boxed{\text{C}}$ **31.** Sketch the graph of $y = (1 + x^{3/2})/x$ for $0 < x \le 16$ by making an extensive table of values. *Note*: Be careful near $x = 0$.

32. Rewrite the equation of Problem 31 as $y = 1/x + \sqrt{x}$. Now sketch its graph by separately graphing $y = 1/x$ and $y = \sqrt{x}$ on the same coordinate plane and then adding ordinates (y-coordinates).

33. What information can we deduce about the graph of $y = ax^2 + bx + c$ from the discriminant $d = b^2 - 4ac$? *Hint*: Use the quadratic formula and consider the three cases $d > 0$, $d = 0$, and $d < 0$.

34. Use the process of completing the square to show that the vertex (high or low point) of the parabola $y = ax^2 + bx + c$ has x-coordinate $-b/2a$. Find the corresponding y-coordinate.

1.8 Chapter Review

TRUE-FALSE QUIZ

Respond with true or false to each of the following assertions. Be prepared to justify your answer.

1. Any number that can be written as a fraction p/q is rational.

2. The difference of two rational numbers is rational.

3. The difference of two irrational numbers is irrational.

4. Between two distinct irrational numbers there is always another irrational number.

5. $0.999\ldots$ (repeating 9s) is less than 1.

6. The operation of exponentiation is commutative; that is, $(a^m)^n = (a^n)^m$.

7. The operation $*$ defined by $m * n = m^n$ is associative.

8. The inequalities $x \le y$, $y \le z$, and $z \le x$ together imply that $x = y = z$.

9. If $|x| < \varepsilon$ for every positive number ε, then $x = 0$.

10. If x and y are real numbers, then $(x - y)(y - x) \le 0$.

11. If $a < b < 0$, then $1/a > 1/b$.

12. It is possible for two closed intervals to have exactly one point in common.

13. If two open intervals have a point in common, then they have infinitely many points in common.

14. If $x < 0$, then $\sqrt{x^2} = -x$.

15. If $|x| < |y|$, then $x < y$.

16. If $|x| < |y|$, then $x^4 < y^4$.

17. If x and y are both negative, then $|x + y| = |x| + |y|$.

18. It is possible to have an inequality whose solution set consists of exactly one number.

19. The equation $x^2 + y^2 + ax + y = 0$ represents a circle for every real number a.

20. If (a, b) is on a line with slope $\frac{3}{4}$, then $(a + 4, b + 3)$ is also on that line.

21. If $ab > 0$, then (a, b) lies in either the first or third quadrant.

22. If $ab = 0$, then (a, b) lies on either the x-axis or the y-axis.

23. If $\sqrt{(x_2 - x_1)^2 + (y_2 - y_1)^2} = |x_2 - x_1|$, then (x_1, y_1) and (x_2, y_2) lie on the same horizontal line.

24. The distance between $(a + b, a)$ and $(a - b, a)$ is $|2b|$.

25. The equation of any line can be written in point-slope form.

26. If two nonvertical lines are parallel, they have the same slope.

27. It is possible for two lines to have positive slopes and be mutually perpendicular.

28. If the x- and y-intercepts of a line are rational and nonzero, then the slope of the line is rational.

29. The lines $ax + y = c$ and $ax - y = c$ are perpendicular.

30. $(3x - 2y + 4) + m(2x + 6y - 2) = 0$ is the equation of a straight line for each real number m.

SAMPLE TEST PROBLEMS

1. Calculate each value for $n = 1, 2,$ and -2.

(a) $\left(n + \dfrac{1}{n}\right)^n$

(b) $(n^2 - n + 1)^2$

(c) $4^{3/n}$

2. Simplify.

(a) $\left(1 + \dfrac{1}{m} + \dfrac{1}{n}\right)\left(1 - \dfrac{1}{m} + \dfrac{1}{n}\right)^{-1}$

(b) $\dfrac{\dfrac{2}{x + 1} - \dfrac{x}{x^2 - x - 2}}{\dfrac{3}{x + 1} - \dfrac{2}{x - 2}}$

3. Show that the average of two rational numbers is a rational number.

4. Write the repeating decimal $4.1282828\ldots$ as a ratio of two integers.

5. Find an irrational number between $\frac{1}{2}$ and $\frac{13}{25}$.

$\boxed{\text{C}}$ **6.** Calculate $(\sqrt[3]{8.15 \times 10^4} - 1.32)^2/3.24$.

In Problems 7–14, find the solution set, graph this set on the real line, and express this set in interval notation.

7. $6x + 3 > 2x - 5$

8. $3 - 2x \le 4x + 1 \le 2x + 7$

9. $2x^2 + 5x - 3 < 0$

10. $\dfrac{2x - 1}{x - 2} > 0$

11. $(x + 4)(2x - 1)^2(x - 3) \le 0$

12. $|3x - 4| < 6$

13. $\dfrac{3}{1 - x} \le 2$

14. $|8 - 3x| \ge |2x|$

15. Suppose $|x| \le 2$. Use properties of absolute values to show

$$\left|\frac{2x^2 + 3x + 2}{x^2 + 2}\right| \le 8$$

16. Sketch the triangle with vertices $A(-2, 6)$, $B(1, 2)$, and $C(5, 5)$, and show that it is a right triangle.

17. Find the distance from $(3, -6)$ to the midpoint of the line segment from $A(1, 2)$ to $B(7, 8)$.

18. Find the equation of the circle with diameter AB if $A = (2, 0)$ and $B = (10, 4)$.

19. Find the center and radius of the circle with equation $x^2 + y^2 - 8x + 6y = 0$.

20. Find the distance between the centers of the circles with equations

$$x^2 - 2x + y^2 + 2y = 2 \quad \text{and} \quad x^2 + 6x + y^2 - 4y = -7$$

21. Write the equation of the line through $(-2, 1)$ which:

(a) goes through $(7, 3)$;
(b) is parallel to $3x - 2y = 5$;
(c) is perpendicular to $3x + 4y = 9$;
(d) is perpendicular to $y = 4$;
(e) has y-intercept 3.

22. Show that $(2, -1), (5, 3),$ and $(11, 11)$ are on the same line.

In Problems 23–26, sketch the graph of each equation.

23. $3y - 4x = 6$

24. $x^2 - 2x + y^2 = 3$

25. $y = \dfrac{2x}{x^2 + 2}$

26. $x = y^2 - 3$

27. Find the points of intersection of the graphs of $y = x^2 - 2x + 4$ and $y - x = 4$.

28. Among all lines perpendicular to $4x - y = 2$, find the equation of the one which—together with the positive x- and y-axes—forms a triangle of area 8.

2 Functions and Limits

The distinguished work [*Cauchy's Cours d'analyse*] *should be read by everyone who loves rigor in mathematical investigations.*
Niels Henrik Abel

Augustin-Louis Cauchy
1789–1857

Augustin-Louis Cauchy was born in Paris and educated at the École Polytechnique. Because of ill health, he was advised to concentrate on mathematics. During his career, he held professorships at the École Polytechnique, the Sorbonne, and the Collège de France. His mathematical contributions were both brilliant and staggering in quantity. His productivity was so enormous that the Paris Academy chose to restrict the size of papers in its journal in order to cope with Cauchy's output.

Cauchy was a devout Catholic and an ardent Royalist. Refusing to swear allegiance to the French government that took control in 1830, he exiled himself to Italy for some years and then taught in several religious institutions in Paris until oaths of allegiance were dropped after the revolution of 1848.

Cauchy was a man of very wide interests. He loved poetry and authored a work on Hebrew prosody. His religious convictions led him to sponsor charitable work for prisoners and other disadvantaged people.

While calculus was invented in the late seventeenth century, its foundations remained in a state of confusion and disarray until Cauchy and his contemporaries (Gauss, Abel, and Bolzano) imposed standards of rigor. To Cauchy we owe the idea of basing calculus on a clear definition of the limit concept. All modern textbooks follow, at least in essence, Cauchy's exposition of calculus.

2.1
Functions and
Their Graphs

Think of a function as a gun. It takes ammunition from one set called the *domain* and fires it at a target set. Each bullet hits a *single* target point, but it could happen that several bullets land on the same point. We can state the definition more formally and introduce some notation at the same time.

Definition

A **function** f is a rule of correspondence that associates with each object x in one set, called the **domain**, a unique value $f(x)$ from a second set. The set of values so obtained is called the **range** of the function. (See Figure 1.)

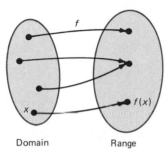

Domain   Range

FIGURE 1

The definition puts no restriction on the domain and range sets. The domain might consist of the set of people in your calculus class, the range the set of grades $\{A, B, C, D, F\}$ that will be given, and the rule of correspondence the procedure your teacher uses in assigning grades.

More relevant in calculus will be examples where both the domain and the range consist of sets of real numbers. For example, the function g might take a real number x and square it, thus producing the real number x^2. In this case, we have a formula that gives the rule of correspondence, namely, $g(x) = x^2$. A schematic diagram for this function is shown in Figure 2.

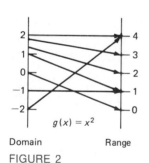

$g(x) = x^2$

Domain   Range

FIGURE 2

FUNCTIONAL NOTATION   A single letter like f (or g or F) is used to name a function. Then $f(x)$, read "f of x" or "f at x," denotes the value that f assigns to x. Thus, if $f(x) = x^3 - 4$,

$$f(2) = 2^3 - 4 = 4$$

$$f(-1) = (-1)^3 - 4 = -5$$

$$f(a) = a^3 - 4$$

$$f(a + h) = (a + h)^3 - 4 = a^3 + 3a^2h + 3ah^2 + h^3 - 4$$

A clear understanding of functional notation is critical in calculus. Study the following examples carefully. They will play an important role in the next chapter.

EXAMPLE 1   For $f(x) = x^2 - 2x$, find and simplify: (a) $f(4)$, (b) $f(4 + h)$, (c) $f(4 + h) - f(4)$, (d) $[f(4 + h) - f(4)]/h$

Solution

(a) $f(4) = 4^2 - 2 \cdot 4 = 8$

(b) $f(4 + h) = (4 + h)^2 - 2(4 + h) = 16 + 8h + h^2 - 8 - 2h$
$$= 8 + 6h + h^2$$

(c) $f(4 + h) - f(4) = 8 + 6h + h^2 - 8 = 6h + h^2$

(d) $\dfrac{f(4 + h) - f(4)}{h} = \dfrac{6h + h^2}{h} = \dfrac{h(6 + h)}{h} = 6 + h$ ∎

EXAMPLE 2 For $g(x) = 1/x$, find and simplify $[g(a + h) - g(a)]/h$.

Solution

$$\frac{g(a + h) - g(a)}{h} = \frac{\dfrac{1}{a + h} - \dfrac{1}{a}}{h} = \frac{\dfrac{a - (a + h)}{(a + h)a}}{h}$$

$$= \frac{-h}{(a + h)a} \cdot \frac{1}{h} = \frac{-1}{(a + h)a} = \frac{-1}{a^2 + ah}$$ ∎

$F(x) = x^2 + 1$

Domain   Range

FIGURE 3

DOMAIN AND RANGE The rule of correspondence is the heart of a function, but a function is not completely determined until its domain is given. Recall that the *domain* is the set of elements to which the function assigns values. The *range* is the set of values so obtained. For example, if F is the function with rule $F(x) = x^2 + 1$ and if the domain is specified as $\{-1, 0, 1, 2, 3\}$ (Figure 3), then the range is $\{1, 2, 5, 10\}$. The domain and the rule determine the range.

When no domain is specified for a function, we always assume that it is the largest set of real numbers for which the rule for the function makes sense and gives real number values. This is called the **natural domain**.

EXAMPLE 3 Find the natural domains for: (a) $f(x) = 1/(x - 3)$; (b) $g(t) = \sqrt{9 - t^2}$.

Solution

(a) The natural domain for f is $\{x \in \mathbb{R}: x \neq 3\}$. This is read "the set of x's in $\mathbb{R}$ (the real numbers) such that x is not equal to 3." We excluded 3 to avoid division by 0.

(b) Here we must restrict t so that $9 - t^2 \geq 0$ in order to avoid nonreal values for $\sqrt{9 - t^2}$. This is achieved by requiring that $|t| \leq 3$. Thus, the natural domain is $\{t \in \mathbb{R}: |t| \leq 3\}$. In interval notation, we could write the domain as $[-3, 3]$. ∎

When the rule for a function is given by an equation of the form $y = f(x)$ (for example, $y = x^3 + 3x - 6$), x is often called the **independent variable** and y the **dependent variable**. Any element of the domain may be chosen as a value of the independent variable x, but that choice completely determines the corresponding value of the dependent variable. Thus, the value of y is dependent on the chosen value of x.

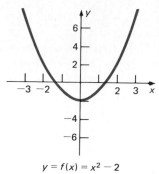

$y = f(x) = x^2 - 2$

FIGURE 4

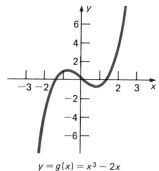

$y = g(x) = x^3 - 2x$

FIGURE 5

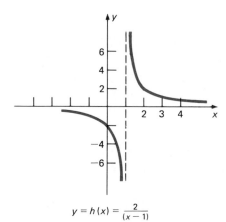

$y = h(x) = \dfrac{2}{(x-1)}$

FIGURE 6

GRAPHS OF FUNCTIONS   When both the domain and range of a function consist of real numbers, we can picture the function by drawing its graph on a coordinate plane. And the **graph of a function** f is simply the graph of the equation $y = f(x)$.

EXAMPLE 4   Sketch the graphs of: (a) $f(x) = x^2 - 2$; (b) $g(x) = x^3 - 2x$; (c) $h(x) = 2/(x-1)$.

Solution   We use the natural domains. In the case of f and g, this is the set $\mathbb{R}$ of all real numbers; for h, it is all of $\mathbb{R}$ except 1. Following the procedure described in Section 1.7 (make a table of values, plot the corresponding points, connect these points with a smooth curve), we obtain the three graphs shown in Figures 4 through 6.   ∎

Pay special attention to the graph of h; it points to an oversimplification we have made and now need to correct. When connecting the plotted points by a smooth curve, do not do so in a mechanical way that ignores special features that may be apparent from the formula for the function. In the case of $h(x) = 2/(x-1)$, it is clear that something dramatic must happen as x nears 1. In fact, the values of $|h(x)|$ increase without bound (for example, $h(0.99) = -200$ and $h(1.001) = 2000$). We have indicated this by drawing a vertical line, called an **asymptote**, at $x = 1$. As x approaches 1, the graph gets closer and closer to this line, though this line itself is not part of the graph. Rather, it is a guide line. Notice that the graph of h also has a horizontal asymptote, namely, the x-axis.

We might ask: What is the range for each of these three functions? Our answer, obtained by looking at the graphs, is given in the table below.

Function	Domain	Range
$f(x) = x^2 - 2$	$\mathbb{R}$	$\{y \in \mathbb{R} : y \geq -2\}$
$g(x) = x^3 - 2x$	$\mathbb{R}$	$\mathbb{R}$
$h(x) = \dfrac{2}{x-1}$	$\{x \in \mathbb{R} : x \neq 1\}$	$\{y \in \mathbb{R} : y \neq 0\}$

EVEN AND ODD FUNCTIONS   We can often predict the symmetries of the graph of a function by inspecting the formula for the function. If $f(-x) = f(x)$, then the graph is symmetric with respect to the y-axis. Such a function is called an **even function**, probably because a function that specifies $f(x)$ as a sum of even powers of x is even. The function $f(x) = x^2 - 2$ (graphed in Figure 4) is even; so are $f(x) = 3x^6 - 2x^4 + 11x^2 - 5$, $f(x) = x^2/(1 + x^4)$, and $f(x) = (x^3 - 2x)/3x$.

If $f(-x) = -f(x)$, the graph is symmetric with respect to the origin. We call such a function an **odd function**. A function that gives $f(x)$ as a sum of odd powers of x is odd. Thus, $g(x) = x^3 - 2x$ (graphed in Figure 5) is odd. Note that

$$g(-x) = (-x)^3 - 2(-x) = -x^3 + 2x = -(x^3 - 2x) = -g(x)$$

Consider the third function $h(x) = 2/(x - 1)$, which we graphed in Figure 6. It is neither even nor odd. To see this, observe that $h(-x) = 2/(-x - 1)$, which is not equal to either $h(x)$ or $-h(x)$. Note that its graph is neither symmetric with respect to the y-axis nor the origin. (The graph of h does have symmetry with respect to the point $(1, 0)$, a result that follows from the fact that $h(1 + x) = -h(1 - x)$.)

EXAMPLE 5   Is $f(x) = \dfrac{x^3 + 3x}{x^4 - 3x^2 + 4}$ even, odd, or neither?

Solution   Since

$$f(-x) = \frac{(-x)^3 + 3(-x)}{(-x)^4 - 3(-x)^2 + 4} = \frac{-(x^3 + 3x)}{x^4 - 3x^2 + 4} = -f(x)$$

f is an odd function.   ∎

For more on even and odd functions, see Problems 23 and 24 of Section 2.2.

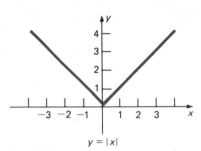

FIGURE 7

$y = |x|$

TWO SPECIAL FUNCTIONS   Among the functions that will often be used as examples are two very special ones: the absolute value function $|\ \ |$ and the greatest integer function $[\![\ \ ]\!]$. They are defined by

$$|x| = \begin{cases} x & \text{if } x \geq 0 \\ -x & \text{if } x < 0 \end{cases}$$

and

$$[\![x]\!] = \text{the greatest integer less than or equal to } x$$

Thus, $|-3.1| = |3.1| = 3.1$, while $[\![-3.1]\!] = -4$ and $[\![3.1]\!] = 3$. The first is an even function, since $|-x| = |x|$. Is $[\![\ \ ]\!]$ even, or odd, or neither? We show the graphs of these two functions in Figures 7 and 8.

We will often appeal to the following special features of these graphs. The graph of $|x|$ has a sharp corner at the origin, while the graph of $[\![x]\!]$ takes a jump at each integer.

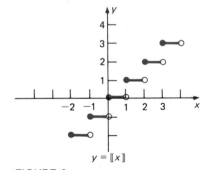

$y = [\![x]\!]$

FIGURE 8

PROBLEM SET 2.1

1. For $f(x) = x^2 - 1$, find each value.

(a) $f(1)$ (b) $f(-2)$ (c) $f(0)$
(d) $f(k)$ (e) $f(-6)$ (f) $f(\frac{1}{2})$

(g) $f(2t)$ (h) $f(3x)$ (i) $f\left(\dfrac{1}{x}\right)$

2. For $F(x) = 3x^3 + x$, find each value.

(a) $F(-6)$ (b) $F(\frac{1}{2})$ (c) $F(3.2)$

(d) $F(\sqrt{3})$ (e) $F(\pi)$ (f) $F\left(\dfrac{1}{x}\right)$

3. For $G(y) = 1/(y - 1)$, find each value.

(a) $G(0)$ (b) $G(0.999)$ (c) $G(1.01)$

(d) $G(y^2)$ (e) $G(-x)$ (f) $G\left(\dfrac{1}{x^2}\right)$

4. For $\phi(t) = \sqrt{t}/(1 + t^2)$, find each value.

(a) $\phi(0)$ (b) $\phi(\frac{1}{4})$ (c) $\phi(x^3)$

(d) $\phi(x + 2)$ (e) $\phi(-t)$ (f) $\phi\left(\dfrac{1}{z^4}\right)$

C 5. For

$$f(x) = x^4 + 3x^3 - 6x^2 + 2x + 1$$
$$= \{[(x + 3)x - 6]x + 2\}x + 1,$$

find each value.

(a) $f(0.32)$ (b) $f(\pi)$ (c) $f(3 + \sqrt{2})$

C 6. For

$$g(x) = x^5 - 5x^3 + 2x^2 - \pi x + \sqrt{3}$$
$$= \{[(x^2 - 5)x + 2]x - \pi\}x + \sqrt{3},$$

find each value.

(a) $g(-1.71)$ (b) $g(3.01)$ (c) $g(\sqrt{3})$

C 7. For $f(x) = \sqrt{x^2 + 9}/(x - \sqrt{3})$, find each value.

(a) $f(0.79)$ (b) $f(12.26)$ (c) $f(\sqrt{3})$

C 8. If $G(t) = t^{2/3} - 4\sqrt{t}$, find each value.

(a) $G(1.46)$ (b) $G(\pi)$ (c) $G(-1.23)$

9. Which of the following determine a function f with formula $y = f(x)$? For those that do, find $f(x)$. *Hint*: Solve for y in terms of x and note that the definition of function requires a unique y for each x.

(a) $x^2 + y^2 = 4$ (b) $xy + y + 3x = 4$

(c) $x = \sqrt{3y + 1}$ (d) $3x = \dfrac{y}{y + 1}$

10. Which of the graphs in Figure 9 are graphs of functions with formulas of the form $y = f(x)$?

FIGURE 9

This problem suggests a rule: *For a graph to be the graph of a function $y = f(x)$, each vertical line must meet the graph in at most one point.*

11. For $f(x) = 2x^2 - 1$, find and simplify $[f(a + h) - f(a)]/h$. (See Examples 1 and 2.)

12. For $F(t) = 4t^3$, find and simplify $[F(a + h) - F(a)]/h$.

13. For $g(u) = 3/(u - 2)$, find and simplify $[g(x + h) - g(x)]/h$.

14. For $G(t) = t/(t + 4)$, find and simplify $[G(a + h) - G(a)]/h$.

15. Find the natural domain for each of the following. (See Example 3.)

(a) $F(z) = \sqrt{2z + 3}$ (b) $g(v) = 1/(4v - 1)$
(c) $\psi(x) = \sqrt{x^2 - 9}$ (d) $H(y) = -\sqrt{625 - y^4}$

16. Find the natural domain in each case.

(a) $f(x) = \dfrac{4 - x^2}{x^2 - x - 6}$ (b) $G(y) = \sqrt{(y + 1)^{-1}}$

(c) $\phi(u) = |2u + 3|$ (d) $F(t) = t^{2/3} - 4$

In Problems 17–32, specify whether the given function is even, odd, or neither, and then sketch its graph. (See Examples 4 and 5.)

17. $f(x) = -4$ 18. $f(x) = 3x$

19. $F(x) = 2x + 1$ 20. $F(x) = 3x - \sqrt{2}$

21. $g(x) = 3x^2 + 2x - 1$ 22. $g(u) = \dfrac{u^3}{8}$

23. $g(x) = \dfrac{x}{x^2 - 1}$.   24. $\phi(z) = \dfrac{2z + 1}{z - 1}$

25. $f(w) = \sqrt{w - 1}$ 26. $h(x) = \sqrt{x^2 + 4}$

27. $f(x) = |2x|$ 28. $F(t) = -|t + 3|$

29. $g(x) = \left[\!\!\left[\dfrac{x}{2} \right]\!\!\right]$ 30. $G(x) = [\![2x - 1]\!]$

31. $g(t) = \begin{cases} 1 & \text{if } t \le 0 \\ t + 1 & \text{if } 0 < t < 2 \\ t^2 - 1 & \text{if } t \ge 2 \end{cases}$

32. $h(x) = \begin{cases} -x^2 + 4 & \text{if } x \le 1 \\ 3x & \text{if } x > 1 \end{cases}$

33. A plant has the capacity to produce from 0 to 100 refrigerators per day. The daily overhead for the plant is $2200 and the direct cost (labor and material) of producing one refrigerator is $151. Write a formula for $T(x)$, the total cost of producing x refrigerators in one day, and also the unit cost $u(x)$ (average cost per refrigerator). What are the domains for these functions?

C 34. It costs the ABC Company $400 + 5\sqrt{x(x - 4)}$ dollars to make x toy stoves that sell for $6 each.

(a) Find a formula for $P(x)$, the total profit in making x stoves.

(b) Evaluate $P(200)$ and $P(1000)$.

(c) How many stoves does ABC have to make to just break even?

C 35. Find a formula for the amount $E(x)$ by which a number x exceeds its cube. Draw a very accurate graph of $E(x)$ for $0 \leq x \leq 1$. Use the graph to estimate the positive number that exceeds its cube by the maximum amount.

36. Let p denote the perimeter of an equilateral triangle. Find a formula for $A(p)$, the area of such a triangle.

37. The Acme Car Rental Agency charges $24 a day for the rental of a car plus $0.40 per mile.

(a) Write a formula for the total rental expense $E(x)$ for one day where x is the number of miles driven.

(b) If you rent a car for one day, how many miles can you drive for $120?

38. A right circular cylinder of radius r is inscribed in a sphere of radius $2r$. Find a formula for $V(r)$, the volume of the cylinder, in terms of r.

39. A 1-mile track has parallel sides and equal semi-circular ends. Find a formula for the area enclosed by the track, $A(d)$, in terms of the diameter of the semicircles. What is the natural domain for this function?

40. Let f be a function with domain $\mathbb{N}$ satisfying $f(1) = 3$, $f(2) = 1$, $f(3) = 4$, $f(4) = 1$, $f(5) = 5$, and $f(6) = 9$. After finding a pattern, give a rule for $f(n)$. What is the range of this function?

41. What is the range of the function f if $f(n)$ is the nth digit in the decimal expansion of $\frac{3}{13}$?

42. Which of the following functions satisfies $f(x + y) = f(x) + f(y)$ for all x and y in $\mathbb{R}$?

(a) $f(t) = 2t$　　　　　　(b) $f(t) = t^2$

(c) $f(t) = 2t + 1$　　　　(d) $f(t) = -3t$

43. Let $f(x + y) = f(x) + f(y)$ for all x and y in $\mathbb{R}$. Prove that there is a number m such that $f(t) = mt$ for all rational numbers t. *Hint:* First decide what m has to be. Then proceed in steps, starting with $f(0) = 0$, $f(p) = mp$ for p in $\mathbb{N}$, $f(1/p) = m/p$, and so on.

2.2
Operations on Functions

Functions are not numbers. But just as two numbers a and b can be added to produce a new number $a + b$, so two functions f and g can be added to produce a new function $f + g$. This is just one of several operations on functions that we will describe in this section.

SUMS, DIFFERENCES, PRODUCTS, QUOTIENTS, POWERS　Consider functions f and g with formulas

$$f(x) = \frac{x - 3}{2}, \qquad g(x) = \sqrt{x}$$

We can make a new function $f + g$ by having it assign to x the value $(x - 3)/2 + \sqrt{x}$—that is,

$$(f + g)(x) = f(x) + g(x) = \frac{x - 3}{2} + \sqrt{x}$$

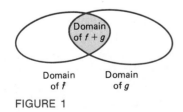

Domain of f　　Domain of g

FIGURE 1

Of course, we must be a little careful about domains. Clearly, x must be a number on which both f and g can work. In other words, the domain of $f + g$ is the intersection (common part) of the domains of f and g (Figure 1).

The functions $f - g$, $f \cdot g$, and f/g are introduced in a completely analogous way. Assuming that f and g have their natural domains, we have the following.

Formula	Domain
$(f + g)(x) = f(x) + g(x) = \dfrac{x-3}{2} + \sqrt{x}$	$[0, \infty)$
$(f - g)(x) = f(x) - g(x) = \dfrac{x-3}{2} - \sqrt{x}$	$[0, \infty)$
$(f \cdot g)(x) = f(x) \cdot g(x) = \dfrac{x-3}{2}\sqrt{x}$	$[0, \infty)$
$\left(\dfrac{f}{g}\right)(x) = \dfrac{f(x)}{g(x)} = \dfrac{x-3}{2\sqrt{x}}$	$(0, \infty)$

We had to exclude 0 from the domain of f/g to avoid division by 0.

We may also raise a function to a power. By f^n, we mean the function that assigns to x the value $[f(x)]^n$. Thus,

$$f^2(x) = [f(x)]^2 = \left[\frac{x-3}{2}\right]^2 = \frac{x^2 - 6x + 9}{4}$$

and

$$g^3(x) = [g(x)]^3 = (\sqrt{x})^3 = x^{3/2}$$

There is one exception to the above agreement, namely, $n = -1$. We reserve the symbol f^{-1} for the inverse function, which will be discussed in Section 7.2. Thus, f^{-1} does not mean $1/f$.

EXAMPLE 1 Let $F(x) = \sqrt[4]{x+1}$ and $G(x) = \sqrt{9 - x^2}$, with respective natural domains $[-1, \infty)$ and $[-3, 3]$. Find formulas for $F + G$, $F - G$, $F \cdot G$, F/G, and F^5 and give their natural domains.

Solution

Formula	Domain
$(F + G)(x) = F(x) + G(x) = \sqrt[4]{x+1} + \sqrt{9 - x^2}$	$[-1, 3]$
$(F - G)(x) = F(x) - G(x) = \sqrt[4]{x+1} - \sqrt{9 - x^2}$	$[-1, 3]$
$(F \cdot G)(x) = F(x) \cdot G(x) = \sqrt[4]{x+1}\,\sqrt{9 - x^2}$	$[-1, 3]$
$\left(\dfrac{F}{G}\right)(x) = \dfrac{F(x)}{G(x)} = \dfrac{\sqrt[4]{x+1}}{\sqrt{9 - x^2}}$	$[-1, 3)$
$F^5(x) = [F(x)]^5 = (\sqrt[4]{x+1})^5 = (x+1)^{5/4}$	$[-1, \infty)$

∎

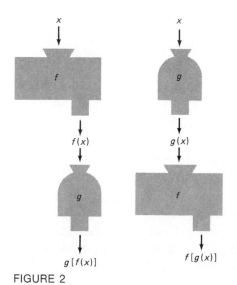

FIGURE 2

COMPOSITION OF FUNCTIONS Earlier, we asked you to think of a function as a gun. Now we want you to think of the function f as a machine. It accepts x as input, works on x, and produces $f(x)$ as output. Two machines may often be put together in tandem to make a more complicated machine; so may two functions f and g (Figure 2). If f works on x to produce $f(x)$ and g then works on $f(x)$ to produce $g(f(x))$, we say that we have *composed*

g with f. The resulting function, called the **composite** of g with f, is denoted by $g \circ f$. Thus,

$$(g \circ f)(x) = g(f(x))$$

Recall our earlier examples, $f(x) = (x - 3)/2$ and $g(x) = \sqrt{x}$. We may compose them in two ways.

$$(g \circ f)(x) = g(f(x)) = g\left(\frac{x - 3}{2}\right) = \sqrt{\frac{x - 3}{2}}$$

$$(f \circ g)(x) = f(g(x)) = f(\sqrt{x}) = \frac{\sqrt{x} - 3}{2}$$

We note one thing right away: Composition of functions is not commutative; $g \circ f$ and $f \circ g$ are usually different. You shouldn't be too surprised at this. If you remove your clothes and then take a bath, you are going to get a quite different result from that of doing these two operations in the opposite order.

We must also be careful in describing the domain of a composite function. The domain of $g \circ f$ is that part of the domain of f (that is, those values x) for which g can accept $f(x)$ as input. In our example, the domain of $g \circ f$ is $[3, \infty)$, since x must be greater than or equal to 3 in order to give a nonnegative number $(x - 3)/2$ for g to work on. The diagram in Figure 3 gives another view of these matters.

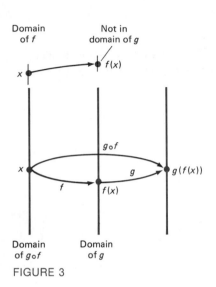

Domain of f — Not in domain of g — $f(x)$ — x

$g \circ f$ — x — g — $g(f(x))$ — f — $f(x)$

Domain of $g \circ f$ — Domain of g

FIGURE 3

EXAMPLE 2 Let $f(x) = 6x/(x^2 - 9)$ and $g(x) = \sqrt{3x}$. First, find $(f \circ g)(12)$; then find $(f \circ g)(x)$ and give its domain.

Solution

$$(f \circ g)(12) = f(g(12)) = f(\sqrt{36}) = f(6) = \frac{36}{27} = \frac{4}{3}$$

$$(f \circ g)(x) = f(g(x)) = f(\sqrt{3x})$$

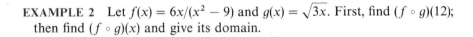

$$= \frac{6\sqrt{3x}}{(\sqrt{3x})^2 - 9} = \frac{6\sqrt{3x}}{3x - 9} = \frac{2\sqrt{3x}}{x - 3}$$

The domain of $f \circ g$ is $[0, 3) \cup (3, \infty)$. (Recall that $\cup$ denotes the set operation of union.) Note that 3 is excluded from the domain to avoid division by 0. ∎

In calculus, we shall often need to take a given function and decompose it—that is, break it into composite pieces. Usually, this can be done in several ways. Take $p(x) = \sqrt{x^2 + 4}$, for example. We may think of it as

$$p(x) = g(f(x)) \qquad \text{where } g(x) = \sqrt{x} \quad \text{and} \quad f(x) = x^2 + 4$$

or as

$$p(x) = g(f(x)) \quad \text{where } g(x) = \sqrt{x + 4} \quad \text{and} \quad f(x) = x^2$$

EXAMPLE 3   Write the function $p(x) = (x + 2)^5$ as a composite function $g \circ f$.

Solution   The most obvious way to do it is to write

$$p(x) = g(f(x)) \quad \text{where } g(x) = x^5 \quad \text{and} \quad f(x) = x + 2 \qquad \blacksquare$$

TRANSLATIONS   Observing how a function is built up from simpler ones can be a big aid in graphing. We may ask this question: How are the graphs of

$$y = f(x) \qquad y = f(x - 3) \qquad y = f(x) + 2 \qquad y = f(x - 3) + 2$$

related to each other? Consider $f(x) = |x|$ as an example. The corresponding four graphs are displayed in Figure 4.

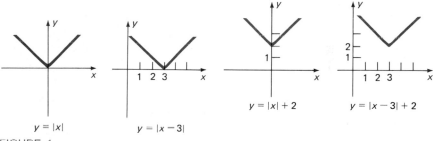

FIGURE 4

What happened with $f(x) = |x|$ is typical. Notice that all four graphs have the same shape; the last three are just translations of the first. Replacing x by $x - 3$ translates the graph 3 units to the right; adding 2 translates it upward by 2 units.

Figure 5 offers another illustration of these principles for the function $f(x) = x^3 + x^2$.

FIGURE 5

FIGURE 6

Exactly the same principles apply in the general situation. They are illustrated in Figure 6.

If $h < 0$, the translation is to the left; if $k < 0$, the translation is downward.

$y = f(x) = \sqrt{x}$

FIGURE 7

EXAMPLE 4 Sketch the graph of $g(x) = \sqrt{x + 3} + 1$ by first graphing $f(x) = \sqrt{x}$ and then making appropriate translations.

Solution The graph of g (Figure 8) is obtained by translating the graph of f (Figure 7) 3 units left and 1 unit up. ∎

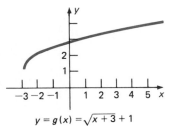

$y = g(x) = \sqrt{x + 3} + 1$

FIGURE 8

PARTIAL CATALOGUE OF FUNCTIONS A function of the form $f(x) = k$, where k is a constant (real number) is called a **constant function**. Its graph is a horizontal line (Figure 9). The function $f(x) = x$ is called the **identity function**. Its graph is a line through the origin of slope 1 (Figure 10). From these simple functions, we can build many of the important functions of calculus.

Any function that can be obtained from the constant functions and the identity function by use of the operations of addition, subtraction, and multiplication is called a **polynomial function**. This amounts to saying that f is a polynomial function if it is of the form

$$f(x) = a_n x^n + a_{n-1} x^{n-1} + \cdots + a_1 x + a_0$$

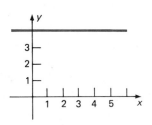

The constant
function $f(x) = 4$

FIGURE 9

where the a's are real numbers and n is a nonnegative integer. If $a_n \neq 0$, n is the degree of the polynomial function. In particular, $f(x) = ax + b$ is a first-degree function, or **linear function**, and $f(x) = ax^2 + bx + c$ is a second-degree function, or **quadratic function**.

Quotients of polynomial functions are called rational functions. Thus f is a **rational function** if it is of the form

$$f(x) = \frac{a_n x^n + a_{n-1} x^{n-1} + \cdots + a_1 x + a_0}{b_m x^m + b_{m-1} x^{m-1} + \cdots + b_1 x + b_0}$$

The identity
function $f(x) = x$

FIGURE 10

An **explicit algebraic function** is one that can be obtained from the constant functions and the identity function via the five operations of addition, subtraction, multiplication, division and root extraction. Examples are

$$f(x) = 3x^{2/5} = 3\sqrt[5]{x^2} \qquad g(x) = \frac{(x + 2)\sqrt{x}}{x^3 + \sqrt[3]{x^2 - 1}}$$

CHAPTER 2/Functions and Limits

The functions so far listed, together with the trigonometric, inverse trigonometric, exponential, and logarithmic functions (to be introduced later), are the basic raw material for calculus.

PROBLEM SET 2.2

1. For $f(x) = x/(x - 1)$ and $g(x) = \sqrt{1 + x^2}$, find each value (if possible).

(a) $(f + g)(2)$ (b) $(f \cdot g)(0)$ (c) $(g/f)(3)$
(d) $(f \circ g)(0)$ (e) $(f \circ g)(\sqrt{8})$ (f) $(g \circ f)(0)$

2. For $f(x) = x^2 + x$ and $g(x) = 2/(x + 3)$, find each value.

(a) $(f - g)(2)$ (b) $(f/g)(1)$ (c) $g^2(3)$
(d) $(f \circ g)(1)$ (e) $(g \circ f)(1)$ (f) $(g \circ g)(3)$

3. If $f(x) = x^3 + 2$ and $g(x) = 2/(x - 1)$, find a formula for each of the following and state its domain. (See Examples 1 and 2.)

(a) $(f + g)(x)$ (b) $(g/f)(x)$
(c) $(f \circ g)(x)$ (d) $(g \circ f)(x)$

4. If $f(x) = \sqrt{x^2 - 1}$ and $g(x) = 2/x$, find formulas for the following and state their domains.

(a) $(f \cdot g)(x)$ (b) $f^4(x) + g^4(x)$
(c) $(f \circ g)(x)$ (d) $(g \circ f)(x)$

5. If $f(x) = \sqrt{x - 4}$ and $g(x) = |x|$, find formulas for $(f \circ g)(x)$ and $(g \circ f)(x)$.

6. If $g(x) = x^2 + 1$, find formulas for $g^3(x)$ and $(g \circ g \circ g)(x)$.

$\boxed{C}$ **7.** Calculate $f(3.12)$ if $f(x) = \left(\dfrac{\sqrt{x^2 + x}}{1 + x^3}\right)^5$.

$\boxed{C}$ **8.** Calculate $g(2.03)$ if $g(x) = \dfrac{(\sqrt{x} - \sqrt[3]{x})^4}{1 - x + x^2}$.

$\boxed{C}$ **9.** Calculate $\sqrt{f^2(3.46) + 4f(3.46)}$ if $f(x) = 1/x$.

$\boxed{C}$ **10.** Calculate $[g^3(\pi) - g(\pi)]^{1/3}$ if $g(x) = 6x - 11$.

11. Find f and g so that $F = g \circ f$. (See Example 3.)

(a) $F(x) = \sqrt{x + 7}$ (b) $F(x) = (x^2 + x)^{15}$

12. Find f and g so that $p = f \circ g$.

(a) $p(x) = \dfrac{2}{(x^2 + x + 1)^3}$ (b) $p(x) = \log(x^3 + 3x)$

13. Write $p(x) = \log\sqrt{x^2 + 1}$ as a composite of three functions in two different ways.

14. Write $p(x) = \log\sqrt{x^2 + 1}$ as a composite of four functions.

15. Sketch the graph of $f(x) = \sqrt{x - 2} - 3$ by first sketching $g(x) = \sqrt{x}$. (See Example 4.)

16. Sketch the graph of $g(x) = |x + 3| - 4$ by first sketching $h(x) = |x|$ and then translating.

17. Sketch the graph of $f(x) = (x - 2)^2 - 4$ by making use of translations.

18. Sketch the graph of $g(x) = (x + 1)^3 - 3$ by making use of translations.

19. Sketch the graphs of $f(x) = (x - 3)/2$ and $g(x) = \sqrt{x}$ using the same coordinate axes. Then sketch $f + g$ by adding ordinates.

20. Follow the directions of Problem 19 for $f(x) = x$ and $g(x) = |x|$.

21. Sketch the graph of $F(t) = |t|/t$.

22. Sketch the graph of $G(t) = t - [\![t]\!]$.

23. State whether each of the following is an odd function, an even function, or neither. Prove your statements.

(a) The sum of two even functions.
(b) The sum of two odd functions.
(c) The product of two even functions.
(d) The product of two odd functions.
(e) The product of an even function and an odd function.

24. Let F be any function whose domain contains $-x$ whenever it contains x. Prove each of the following.

(a) $F(x) - F(-x)$ determines an odd function.
(b) $F(x) + F(-x)$ determines an even function.
(c) F can always be expressed as the sum of an odd and an even function.

25. Classify each of the following as a PF (polynomial function), RF (rational function but not a polynomial function), or AF (explicit algebraic function but neither PF nor RF).

(a) $f(x) = 3x^{1/2} + 1$ (b) $f(x) = 3$
(c) $f(x) = 3x^2 + 2x^{-1}$ (d) $f(x) = \pi x^3 - 3\pi$
(e) $f(x) = \dfrac{1}{x + 1}$ (f) $f(x) = \dfrac{x + 1}{\sqrt{x + 3}}$

26. Follow the directions of Problem 25 for each function.

(a) $g(x) = 1 - 3x + x^{52}$ (b) $g(x) = (1 + 5x)^{-2}$
(c) $g(x) = (1 + 5x)^{-1/2}$ (d) $g(x) = x^{-2} + 2x^{-1} + 3$

(e) $g(x) = x + \sqrt{x}$ (f) $g(x) = \left(\dfrac{1}{x + 2}\right)^{-2}$

C **27.** The relationship between the unit price P (in cents) for a certain product and the demand D (in thousands of units) appears to satisfy

$$P = \sqrt{29 - 3D + D^2}$$

On the other hand, the demand has risen over the t years since 1970 according to $D = 2 + \sqrt{t}$.

(a) Express P as a function of t.
(b) Evaluate P when $t = 15$.

28. After being in business x years, a manufacturer of tractors is making $100 + x + 2x^2$ units per year. The sales price (in dollars) per unit has risen according to the formula $P = 500 + 60x$. Write a formula for the manufacturer's yearly revenue $R(x)$ after x years.

29. Starting at noon, plane A flies due north at 400 miles per hour. Starting 1 hour later, plane B flies due east at 300 miles per hour. Neglecting the curvature of the earth and assuming they fly at the same altitude, find a formula for $D(t)$, the distance between the two planes t hours after noon. *Hint*: There will be two formulas for $D(t)$, one if $0 \leq t \leq 1$, the other if $t > 1$.

C **30.** Find the distance between the planes of Problem 29 at 2:30 P.M.

31. Let $f(x) = \dfrac{ax + b}{cx - a}$. Show that $f(f(x)) = x$, provided $a^2 + bc \neq 0$ and $x \neq a/c$.

32. Let $f(x) = \dfrac{x - 3}{x + 1}$. Show that $f(f(f(x))) = x$, provided $x \neq \pm 1$.

33. Let $f(x) = \dfrac{x}{x - 1}$. Find and simplify each value.

(a) $f(1/x)$ (b) $f(f(x))$ (c) $f(1/f(x))$

34. Let $f_1(x) = x$, $f_2(x) = 1/x$, $f_3(x) = 1 - x$, $f_4(x) = 1/(1 - x)$, $f_5(x) = (x - 1)/x$, and $f_6(x) = x/(x - 1)$. Note that $f_3(f_4(x)) = f_3(1/(1 - x)) = 1 - 1/(1 - x) = x/(x - 1) = f_6(x)$, that is, $f_3 \circ f_4 = f_6$. In fact, the composition of any two of these functions is another one in the list. Fill in the composition table in Figure 11.

$\circ$	f_1	f_2	f_3	f_4	f_5	f_6
f_1						
f_2						
f_3				f_6		
f_4						
f_5						
f_6						

FIGURE 11

2.3
The Trigonometric Functions

We assume that you have studied trigonometry and are familiar with the definitions of trigonometric functions based on angles and right triangles. We remind you of three of these definitions in Figure 1. Do not forget them. Here, however, we are more interested in the trigonometric functions as based on the unit circle. When considered in this way, their domains are sets of numbers rather than sets of angles.

Let C be the unit circle—that is, the circle $x^2 + y^2 = 1$ centered at the origin with radius 1 (Figure 2). Denote by A the point $(1, 0)$ and let t be any positive number. Then there is exactly one point $P(x, y)$ on C such that the length of arc $\overset{\frown}{AP}$, measured in the *counterclockwise* direction from A along the unit circle, is t. The circumference of C is 2π; so if $t > 2\pi$, it will take more than a complete circuit of the unit circle to trace the arc $\overset{\frown}{AP}$. If $t = 0$, $P = A$.

Similarly if $t < 0$, there is exactly one point $P(x, y)$ on the unit circle such that the length of the arc $\overset{\frown}{AP}$, measured *clockwise* on C, is $|t|$. Thus, with any real number t, we associate a unique point $P(x, y)$. This allows us to make the key definitions of sine and cosine (sin and cos).

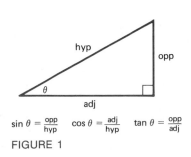

$$\sin \theta = \frac{\text{opp}}{\text{hyp}} \quad \cos \theta = \frac{\text{adj}}{\text{hyp}} \quad \tan \theta = \frac{\text{opp}}{\text{adj}}$$

FIGURE 1

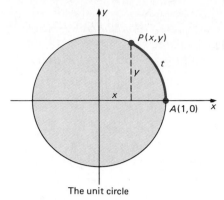

The unit circle

FIGURE 2

FIGURE 3

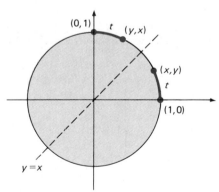

FIGURE 4

Definition

Let t determine the point $P(x, y)$ as indicated above. Then

$$\sin t = y \qquad \cos t = x$$

BASIC PROPERTIES OF SINE AND COSINE   Several facts are almost immediately apparent from the definitions just given. First, x and y vary between -1 and 1, so

$$|\sin t| \le 1 \qquad |\cos t| \le 1$$

Since t and $t + 2\pi$ determine the same point $P(x, y)$,

$$\sin(t + 2\pi) = \sin t \qquad \cos(t + 2\pi) = \cos t$$

We say that sine and cosine are periodic with period 2π. More generally, a function f is **periodic** if there is a positive number p such that $f(t + p) = f(t)$ for all t in the domain of f. And the smallest such p is called the **period** of f.

The points P corresponding to t and $-t$ are symmetric with respect to the x-axis (Figure 3), and thus their x-coordinates are equal and their y-coordinates differ in sign only. Consequently,

$$\sin(-t) = -\sin t \qquad \cos(-t) = \cos t$$

In other words, sine is an odd function while cosine is an even function.

The points P corresponding to t and $\pi/2 - t$ are symmetric with respect to the line $y = x$ (Figure 4), and thus have their coordinates interchanged. This means that

$$\sin\left(\frac{\pi}{2} - t\right) = \cos t \qquad \cos\left(\frac{\pi}{2} - t\right) = \sin t$$

Finally, we mention an important identity connecting the sine and cosine functions.

$$\sin^2 t + \cos^2 t = 1$$

This identity follows from the fact that on the unit circle, $y^2 + x^2 = 1$.

GRAPHS OF SINE AND COSINE   To graph $y = \sin t$ and $y = \cos t$, we follow our standard procedure (make a table of values, plot the corresponding points, and connect these points with a smooth curve). But how do we

t	$\sin t$	$\cos t$
0	0	1
$\pi/6$	1/2	$\sqrt{3}/2$
$\pi/4$	$\sqrt{2}/2$	$\sqrt{2}/2$
$\pi/3$	$\sqrt{3}/2$	1/2
$\pi/2$	1	0
$2\pi/3$	$\sqrt{3}/2$	$-1/2$
$3\pi/4$	$\sqrt{2}/2$	$-\sqrt{2}/2$
$5\pi/6$	1/2	$-\sqrt{3}/2$
π	0	-1

FIGURE 5

make a table of values? Fortunately, others have done the work for us. Extensive tables for the sine and cosine are available. One such table is Table II of the Appendix; a brief table for special numbers is shown in Figure 5. With the aid of these tables or from computations on a calculator (in radian mode), we can draw the graphs in Figure 6.

FIGURE 6

Even a casual observer might notice four things about these graphs:

1. Both $\sin t$ and $\cos t$ range from -1 to 1.
2. Both graphs repeat themselves on adjacent intervals of length 2π.
3. The graph of $y = \sin t$ is symmetric about the origin, $y = \cos t$ about the y-axis.
4. The graph of $y = \sin t$ is the same as that of $y = \cos t$ but shifted $\pi/2$ units to the right.

There are no surprises here. These are the graphical interpretations of the first four boxed formulas of the previous subsection.

FOUR OTHER TRIGONOMETRIC FUNCTIONS   We could get by with just the sine and cosine, but it is convenient to introduce four additional trigonometric functions: tangent, cotangent, secant, and cosecant.

$$\tan t = \frac{\sin t}{\cos t} \qquad \cot t = \frac{\cos t}{\sin t}$$

$$\sec t = \frac{1}{\cos t} \qquad \csc t = \frac{1}{\sin t}$$

What we know about sine and cosine will automatically give us knowledge about these four new functions.

EXAMPLE 1   Show that tangent is an odd function.

Solution

$$\tan(-t) = \frac{\sin(-t)}{\cos(-t)} = \frac{-\sin t}{\cos t} = -\tan t$$

■

EXAMPLE 2 Verify the following identities.

$$\boxed{1 + \tan^2 t = \sec^2 t \qquad 1 + \cot^2 t = \csc^2 t}$$

Solution

$$1 + \tan^2 t = 1 + \frac{\sin^2 t}{\cos^2 t} = \frac{\cos^2 t + \sin^2 t}{\cos^2 t} = \frac{1}{\cos^2 t} = \sec^2 t$$

$$1 + \cot^2 t = 1 + \frac{\cos^2 t}{\sin^2 t} = \frac{\sin^2 t + \cos^2 t}{\sin^2 t} = \frac{1}{\sin^2 t} = \csc^2 t \qquad ■$$

Next we graph the tangent function (Figure 7). Here we are in for two minor surprises.

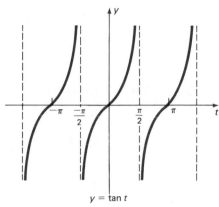

$y = \tan t$

FIGURE 7

First, we notice that there are vertical asymptotes at $-3\pi/2$, $-\pi/2$, $\pi/2$, $3\pi/2$, and so on. We should have anticipated this since $\cos t = 0$ at these values of t, which means that $(\sin t)/(\cos t)$ involves a division by zero. Second, it appears that the tangent is periodic (which we expected), but with period π (which we might not have expected). You will see the analytic reason for this in Problem 19.

RELATION TO ANGLE TRIGONOMETRY   Angles are commonly measured either in degrees or in radians. The angle corresponding to a complete revolution measures $360°$, but only 2π radians. Equivalently, a straight angle measures $180°$ or π radians, a fact worth remembering.

$$\boxed{180° = \pi \text{ radians} \approx 3.1415927 \text{ radians}}$$

This leads to the common conversions shown in Figure 8 and also to the following facts.

$$1 \text{ radian} \approx 57.29578° \qquad 1° \approx 0.0174533 \text{ radian}$$

The division of a revolution into 360 parts is quite arbitrary (due to the ancient Babylonians, who liked multiples of 60). The division into 2π

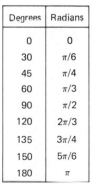

Degrees	Radians
0	0
30	$\pi/6$
45	$\pi/4$
60	$\pi/3$
90	$\pi/2$
120	$2\pi/3$
135	$3\pi/4$
150	$5\pi/6$
180	π

FIGURE 8

$s = rt$

FIGURE 9

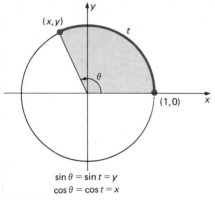

$\sin\theta = \sin t = y$
$\cos\theta = \cos t = x$

FIGURE 10

parts is more fundamental and lies behind the almost universal use of radian measure in calculus. Notice, in particular, that the length s of the arc cut off on a circle of radius r by a central angle of t radians satisfies (see Figure 9)

$$\frac{s}{2\pi r} = \frac{t}{2\pi}$$

That is

$$\boxed{s = rt}$$

When $r = 1$, this gives $s = t$. In words, *the length of arc on the unit circle cut off by a central angle of t radians is t.* This is correct even if t is negative, provided we interpret length to be negative when measured in the clockwise direction.

EXAMPLE 3   Find the distance traveled by a bicycle with wheels of radius 30 centimeters when the wheels turn through 100 revolutions.

Solution   We use the boxed formula, recognizing that 100 revolutions correspond to $100 \cdot (2\pi)$ radians.

$$s = (30)(100)(2\pi) = 6000\pi$$
$$\approx 18849.6 \text{ centimeters} \qquad \blacksquare$$

Now we can make the connection between angle trigonometry and unit circle trigonometry. If θ is an angle measuring t radians (Figure 10), then

$$\boxed{\sin\theta = \sin t \qquad \cos\theta = \cos t}$$

In calculus, when we meet an angle measured in degrees, we almost always change it to radians before doing any calculations. For example,

$$\sin 31.6° = \sin\left(31.6 \cdot \frac{\pi}{180} \text{ radian}\right) \approx \sin(0.552)$$

EXAMPLE 4   Find $\cos 51.8°$.

Solution   The simplest procedure is to press the right keys on a calculator. But if we want to use Table II of the appendix, we first change $51.8°$ to radians.

$$51.8° = 51.8\left(\frac{\pi}{180}\right) \approx 0.904 \text{ radian}$$

Thus,

$$\cos(51.8°) \approx \cos(0.904) \approx 0.6184 \qquad \blacksquare$$

ANOTHER VIEW

We have based our discussion of trigonometry on the unit circle. We could as well have used a circle of radius r.

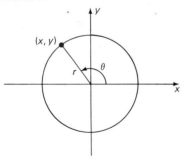

Then

$$\sin\theta = \frac{y}{r}$$

$$\cos\theta = \frac{x}{r}$$

54

LIST OF IMPORTANT IDENTITIES   We will not take space to verify all of the following identities. We simply assert their truth and suggest that most of them will be needed somewhere in this book.

Trigonometric Identities

Odd-even identities   *Cofunction identities*

$$\sin(-x) = -\sin x \qquad \sin\left(\frac{\pi}{2} - x\right) = \cos x$$

$$\cos(-x) = \cos x \qquad \cos\left(\frac{\pi}{2} - x\right) = \sin x$$

$$\tan(-x) = -\tan x \qquad \tan\left(\frac{\pi}{2} - x\right) = \cot x$$

Pythagorean identities

$$\sin^2 x + \cos^2 x = 1 \qquad 1 + \tan^2 x = \sec^2 x$$

$$1 + \cot^2 x = \csc^2 x$$

Addition identities

$$\sin(x + y) = \sin x \cos y + \cos x \sin y$$

$$\cos(x + y) = \cos x \cos y - \sin x \sin y$$

$$\tan(x + y) = \frac{\tan x + \tan y}{1 - \tan x \tan y}$$

Double-angle identities

$$\sin 2x = 2 \sin x \cos x$$

$$\cos 2x = \cos^2 x - \sin^2 x = 2 \cos^2 x - 1 = 1 - 2 \sin^2 x$$

Half-angle identities

$$\sin^2 x = \frac{1 - \cos 2x}{2} \qquad \cos^2 x = \frac{1 + \cos 2x}{2}$$

Sum identities

$$\sin x + \sin y = 2 \sin\left(\frac{x + y}{2}\right)\cos\left(\frac{x - y}{2}\right)$$

$$\cos x + \cos y = 2 \cos\left(\frac{x + y}{2}\right)\cos\left(\frac{x - y}{2}\right)$$

Product identities

$$\sin x \sin y = -\tfrac{1}{2}[\cos(x + y) - \cos(x - y)]$$

$$\cos x \cos y = \tfrac{1}{2}[\cos(x + y) + \cos(x - y)]$$

$$\sin x \cos y = \tfrac{1}{2}[\sin(x + y) + \sin(x - y)]$$

---

1. Convert the following to radians (leave π in your answer).

(a) $240°$ (b) $-60°$ (c) $-135°$
(d) $540°$ (e) $600°$ (f) $720°$
(g) $18°$ (h) $22.5°$ (i) $6°$

2. Convert the following radian measures to degrees.

(a) $\dfrac{7\pi}{6}$ (b) $\dfrac{-\pi}{3}$ (c) 8π

(d) $\dfrac{5\pi}{4}$ (e) $\dfrac{3\pi}{2}$ (f) $\dfrac{-11\pi}{12}$

(g) $\dfrac{\pi}{18}$ (h) $\dfrac{7\pi}{4}$ (i) $\dfrac{-\pi}{5}$

C **3.** Convert the following to radians $(1° = \pi/180$ radians).

(a) $33.3°$ (b) $471.5°$ (c) $-391.4°$
(d) $14.9°$ (e) $4.02°$ (f) $-1.52°$

C **4.** Convert the following radian measures to degrees (1 radian $= 180/\pi$ degrees).

(a) 1.51 (b) -3.1416 (c) 2.31
(d) 34.25 (e) -0.002 (f) 6.28

C **5.** Calculate (be sure your calculator is in radian mode).

(a) $\sin(0.452)$ (b) $\cos(0.452)$
(c) $\tan(0.452)$ (d) $\sin(-0.361)$
(e) $\cos(-0.361)$ (f) $\tan(-0.361)$

6. Use Table II of the Appendix to find each value.

(a) $\sin(1.23)$ (b) $\cos(0.63)$
(c) $\tan(1.55)$ (d) $\sin(-1.23)$
(e) $\cos(-0.63)$ (f) $\tan(-1.55)$

C **7.** Calculate.

(a) $\dfrac{234.1 \sin(1.56)}{\cos(0.34)}$ (b) $\sin^2(2.51) + \sqrt{\cos(0.51)}$

C **8.** Calculate.

(a) $\dfrac{56.3 \tan 34.2°}{\sin 56.1°}$ (b) $\left(\dfrac{\sin 35°}{\sin 26° + \cos 26°}\right)^3$

9. Evaluate without use of a calculator.

(a) $\tan\left(\dfrac{\pi}{6}\right)$ (b) $\sec(\pi)$ (c) $\sec\left(\dfrac{3\pi}{4}\right)$

(d) $\csc\left(\dfrac{\pi}{2}\right)$ (e) $\cot\left(\dfrac{\pi}{4}\right)$ (f) $\tan\left(-\dfrac{\pi}{4}\right)$

10. Evaluate without use of a calculator.

(a) $\tan\left(\dfrac{\pi}{3}\right)$ (b) $\sec\left(\dfrac{\pi}{3}\right)$ (c) $\cot\left(\dfrac{\pi}{3}\right)$

(d) $\csc\left(\dfrac{\pi}{4}\right)$ (e) $\tan\left(-\dfrac{\pi}{6}\right)$ (f) $\cos\left(-\dfrac{\pi}{3}\right)$

11. Verify that the following are identities (see Example 2).

(a) $(1 + \sin z)(1 - \sin z) = \dfrac{1}{\sec^2 z}$

(b) $(\sec t - 1)(\sec t + 1) = \tan^2 t$

(c) $\sec t - \sin t \tan t = \cos t$

(d) $\dfrac{\sec^2 t - 1}{\sec^2 t} = \sin^2 t$

(e) $\cos t\,(\tan t + \cot t) = \csc t$

12. Verify that the following are identities.

(a) $\dfrac{\sin u}{\csc u} + \dfrac{\cos u}{\sec u} = 1$

(b) $(1 - \cos^2 x)(1 + \cot^2 x) = 1$
(c) $\sin t\,(\csc t - \sin t) = \cos^2 t$

(d) $\dfrac{1 - \csc^2 t}{\csc^2 t} = \dfrac{-1}{\sec^2 t}$

(e) $\dfrac{1}{\sin t \cos t} - \dfrac{\cos t}{\sin t} = \tan t$

13. Sketch the graphs of the following on $[-\pi, 2\pi]$.

(a) $y = \sec t$ (b) $y = 3 \sin t$

(c) $y = \sin 2t$ (d) $y = \sin\left(t - \dfrac{\pi}{4}\right)$

14. Sketch the graphs of the following on $[-\pi, 2\pi]$.

(a) $y = \csc t$ (b) $y = 2 \cos t$

(c) $y = \cos 3t$ (d) $y = \cos\left(t + \dfrac{\pi}{3}\right)$

15. Find the quadrant in which the point $P(x, y)$ will lie for each t below, and thereby determine the sign of $\cos t$. *Hint*: See the unit circle definition of $\cos t$.

(a) $t = 5.97$ (b) $t = 9.34$ (c) $t = -16.1$

16. Follow the directions of Problem 15 to determine the sign of $\tan t$.

(a) $t = 4.34$ (b) $t = -15$ (c) $t = 21.9$

17. Which of the following are odd functions? Even functions? Neither?

(a) $\sec t$ (b) $\csc t$

(c) $t \sin t$ (d) $x \cos x$

(e) $\sin^2 x$ (f) $\sin x + \cos x$

18. Find identities analogous to the addition identities for each expression.

(a) $\sin(x - y)$ (b) $\cos(x - y)$ (c) $\tan(x - y)$

19. Use the addition identity for the tangent to show that $\tan(t + \pi) = \tan t$ for all t in the domain of $\tan t$.

20. Show that $\cos(x - \pi) = -\cos x$ for all x.

21. Find the length of arc on a circle of radius 2.5 centimeters cut off by each central angle (see Example 3).

(a) 6 radians   (b) $225°$

22. How far does a wheel of radius 2 feet roll along level ground in making 150 revolutions? (See Example 3.)

C **23.** Suppose that a tire on a truck has an outer radius of 2.5 feet. How many revolutions per minute does the tire make when the truck is traveling 60 miles per hour?

C **24.** A belt passes around two wheels, as shown in Figure 11. How many revolutions per second does the small wheel make when the large wheel makes 21 revolutions per second?

FIGURE 11

25. A 50-pound bag of corn is being dragged along the ground by a man whose arm makes an angle of t radians with the ground. The force F (in pounds) required is given by

$$F(t) = \frac{50\mu}{\mu \sin t + \cos t}$$

Here μ is a constant relating to the friction involved. Find F in each case.

(a) $t = \dfrac{\pi}{4}$ (b) $t = 0$

C (c) $t = 1$ (d) The angle is $90°$.

26. The **angle of inclination** α of a line is the smallest positive angle from the positive x-axis to the line ($\alpha = 0$ for a horizontal line). Show that the slope m of the line is equal to $\tan \alpha$.

27. Find the angle of inclination of the following lines (see Problem 26).

(a) $y = \sqrt{3}x - 7$ (b) $\sqrt{3}x + 3y = 6$

28. Let l_1 and l_2 be two nonvertical lines with slopes m_1 and m_2, respectively. If θ, the angle from l_1 to l_2, is not a right angle, then

$$\tan \theta = \frac{m_2 - m_1}{1 + m_1 m_2}$$

Show this using the fact that $\theta = \theta_2 - \theta_1$ in Figure 12.

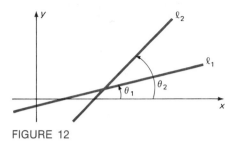

FIGURE 12

C **29.** Find the angle (in radians) from the first line to the second (see Problem 28).

(a) $y = 2x, \; y = 3x$ (b) $y = \dfrac{x}{2}, \; y = -x$

(c) $2x - 6y = 12, \; 2x + y = 0$

30. Derive the formula $A = \frac{1}{2}r^2\theta$ for the area of a sector of a circle. Here r is the radius and θ is the radian measure of the vertex angle (see Figure 13).

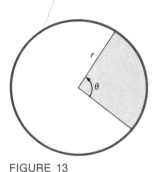

FIGURE 13

31. Find the area of the sector of a circle of radius 5 centimeters and vertex angle 2 radians (see Problem 30).

32. Suppose that the spoke of a mounted wheel of radius 2 feet rotates 10 times a second. Find the area swept out by the spoke in $\frac{1}{40}$ second; in 3 seconds.

33. A $33\frac{1}{3}$-rpm stereo record has a spiral groove starting 6 inches from the center and ending 3 inches from the center. If it took 18 minutes to play the record, approximately how long is the spiral groove?

34. A regular polygon of n sides is inscribed in a circle of radius r. Find formulas for the perimeter, P, and area, A, of the polygon in terms of n and r.

35. An isosceles triangle is topped by a semicircle, as shown in Figure 14. Find a formula for the area A of the whole figure in terms of the side length r and vertex angle t (radians).

FIGURE 14

36. From a product identity, we obtain

$$\cos \frac{x}{2} \cos \frac{x}{4} = \tfrac{1}{2}\left[\cos \tfrac{3}{4}x + \cos \tfrac{1}{4}x\right]$$

Find the corresponding sum of cosines for

$$\cos \frac{x}{2} \cos \frac{x}{4} \cos \frac{x}{8} \cos \frac{x}{16}$$

Do you see a generalization?

2.4
Introduction to Limits

The topics discussed so far are part of what is called *precalculus*. They provide the foundation for calculus, but they are not calculus. Now we are ready for an important new idea, the notion of *limit*. It is this idea that distinguishes calculus from other branches of mathematics. In fact, we might define calculus as *the study of limits*.

Of course, the word *limit* is used in everyday language as when one says, "I'm nearing the limit of my patience." That usage has something to do with calculus, but not very much.

AN INTUITIVE UNDERSTANDING   Consider the function determined by the formula

$$f(x) = \frac{x^3 - 1}{x - 1}$$

Note that it is not defined at $x = 1$ since at this point $f(x)$ has the form $\frac{0}{0}$, which is meaningless. We can, however, still ask what is happening to $f(x)$ as x approaches 1. More precisely, is $f(x)$ approaching some specific number as x approaches 1? To get at this question, we have done three things. We have calculated some values of $f(x)$ for x near 1, we have shown these values in a schematic diagram, and we have sketched the graph of $y = f(x)$ (Figure 1).

All the information we have assembled seems to point to the same conclusion: $f(x)$ approaches 3 as x approaches 1. In mathematical symbols, we write

$$\lim_{x \to 1} \frac{x^3 - 1}{x - 1} = 3$$

This is read "the limit as x approaches 1 of $(x^3 - 1)/(x - 1)$ is 3."

Being good algebraists (thus knowing how to factor the difference of cubes), we can provide more and better evidence.

x	$y = \dfrac{x^3 - 1}{x - 1}$
1.25	3.813
1.1	3.310
1.01	3.030
1.001	3.003
↓	↓
1.000	?
↑	↑
0.999	2.997
0.99	2.970
0.9	2.710
0.75	2.313

Table
of values

Schematic
diagram

Graph of $y = f(x) = \dfrac{x^3 - 1}{x - 1}$

FIGURE 1

$$\lim_{x \to 1} \frac{x^3 - 1}{x - 1} = \lim_{x \to 1} \frac{(x - 1)(x^2 + x + 1)}{x - 1}$$

$$= \lim_{x \to 1}(x^2 + x + 1) = 1^2 + 1 + 1 = 3$$

Note that $(x - 1)/(x - 1) = 1$ as long as $x \neq 1$. This justifies the second step.

To be sure we are on the right track, we need to have a clearly understood meaning for the word *limit*. Here is our first attempt at a definition.

Definition

(Intuitive meaning of limit). To say that $\lim_{x \to c} f(x) = L$ means that when x is near but different from c, then $f(x)$ is near L.

Notice that we do not require anything to be true right at c. The function f need not even be defined at c; it was not in the example $f(x) = (x^3 - 1)/(x - 1)$ just considered. The notion of limit is associated with the behavior of a function near c, not at c.

A cautious reader is sure to object to our use of the word *near*. What does *near* mean? How near is near? For precise answers, you will have to study the next section; however, some further examples will help to clarify the idea.

MORE EXAMPLES Our first example is almost trivial, but is nonetheless important.

EXAMPLE 1 Find $\lim_{x \to 3}(4x - 5)$.

Solution When x is near 3, $4x - 5$ is near $4 \cdot 3 - 5 = 7$. We write

$$\lim_{x \to 3}(4x - 5) = 7 \qquad \blacksquare$$

EXAMPLE 2 Find $\lim_{x \to 3} \dfrac{x^2 - x - 6}{x - 3}$.

Solution Note that $(x^2 - x - 6)/(x - 3)$ is not defined at $x = 3$, but that is all right. To get an idea of what is happening as x approaches 3, we could use a calculator to evaluate the given expression, for example, at 3.1, 3.01, 3.001, and so on. But it is much better to use a little algebra to simplify the problem.

$$\lim_{x \to 3} \frac{x^2 - x - 6}{x - 3} = \lim_{x \to 3} \frac{(x - 3)(x + 2)}{x - 3} = \lim_{x \to 3}(x + 2) = 3 + 2 = 5$$

The cancellation of $x - 3$ in the second step is legitimate because the definition of limit ignores the behavior right at $x = 3$. Thus, we did not divide by 0. $\qquad \blacksquare$

EXAMPLE 3 Find $\lim_{x \to 1} \dfrac{x - 1}{\sqrt{x} - 1}$.

Solution

$$\lim_{x \to 1} \frac{x - 1}{\sqrt{x} - 1} = \lim_{x \to 1} \frac{(\sqrt{x} - 1)(\sqrt{x} + 1)}{\sqrt{x} - 1} = \lim_{x \to 1}(\sqrt{x} + 1) = \sqrt{1} + 1 = 2 \qquad \blacksquare$$

EXAMPLE 4 Find $\lim_{x \to 0} \dfrac{\sin x}{x}$.

Solution No algebraic trick will simplify our task; certainly we cannot cancel the x's. A calculator will help us get an idea of the limit. Use your own calculator (radian mode) to check the values in the table of Figure 2. Our conclusion—though we admit it is a bit shaky—is that

$$\lim_{x \to 0} \frac{\sin x}{x} = 1$$

We will give a rigorous demonstration in Section 3.4. $\qquad \blacksquare$

SOME WARNING FLAGS Things are not quite as simple as they may appear. Calculators may mislead us; so may our own intuition. The examples that follow suggest some possible pitfalls.

x	$\frac{\sin x}{x}$
1.0	0.84147
0.5	0.95885
0.1	0.99833
0.01	0.99998
↓	↓
0	?
↑	↑
−0.01	0.99998
−0.1	0.99833
−0.5	0.95885
−1.0	0.84147

FIGURE 2

x	$x^2 - \dfrac{\cos x}{10000}$
±1	0.99995
±0.5	0.24991
±0.1	0.00990
±0.01	0.000000005
↓	↓
0	?

FIGURE 3

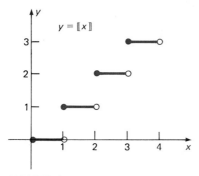

FIGURE 4

EXAMPLE 5 (Your calculator may fool you). Find $\lim\limits_{x\to 0}\left[x^2 - \dfrac{\cos x}{10{,}000}\right]$.

Solution Following the procedure used earlier, we construct the table of values shown in Figure 3. The conclusion it suggests is that the desired limit is 0. But that is wrong. If we recall the graph of $y = \cos x$, we realize that $\cos x$ approaches 1 as x approaches 0. Thus,

$$\lim_{x\to 0}\left[x^2 - \frac{\cos x}{10{,}000}\right] = 0^2 - \frac{1}{10{,}000} = -\frac{1}{10{,}000}\qquad \blacksquare$$

EXAMPLE 6 (No limit at a jump). Find $\lim\limits_{x\to 2}[\![x]\!]$

Solution Recall that $[\![x]\!]$ denotes the greatest integer in x (see Section 2.1). The graph of $y = [\![x]\!]$ is shown in Figure 4. For all numbers x less than 2 but near 2, $[\![x]\!] = 1$, but for all numbers x greater than 2 but near 2, $[\![x]\!] = 2$. Is $[\![x]\!]$ near to a single number L when x is near 2? No. No matter what number we propose for L, there will be x's arbitrarily close to 2 on one side or the other, where $[\![x]\!]$ differs from L by at least $\frac{1}{2}$. Our conclusion is that $\lim\limits_{x\to 2}[\![x]\!]$ does not exist. If you check back, you will see that we have not claimed that every limit we can write must exist. $\blacksquare$

EXAMPLE 7 (Too many wiggles). Find $\lim\limits_{x\to 0}\sin(1/x)$.

Solution This example poses the most subtle limit question asked yet. Since we do not want to make too big a story out of it, we ask you to do two things. First, pick a sequence of x-values approaching 0. Use your calculator to evaluate $\sin(1/x)$ at these x's. Unless you happen on some very lucky choices, your values will oscillate wildly.

Second, consider trying to graph $y = \sin(1/x)$. No one will ever do this very well, but the table of values in Figure 5 gives us a good clue about what is happening. In any neighborhood of the origin, the graph wiggles up and down between -1 and 1 infinitely many times (Figure 6). Clearly $\sin(1/x)$ is not near a unique number L when x is near 0. We conclude that $\lim\limits_{x\to 0}\sin(1/x)$ does not exist. $\blacksquare$

x	$\sin\dfrac{1}{x}$
$2/\pi$	1
$2/2\pi$	0
$2/3\pi$	−1
$2/4\pi$	0
$2/5\pi$	1
$2/6\pi$	0
$2/7\pi$	−1
$2/8\pi$	0
$2/9\pi$	1
$2/10\pi$	0
$2/11\pi$	−1
$2/12\pi$	0
↓	↓
0	?

FIGURE 5

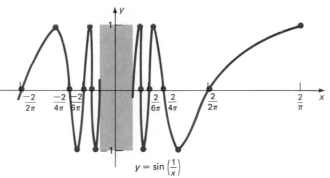

$$y = \sin\left(\frac{1}{x}\right)$$

FIGURE 6

SECTION 2.4/Introduction to Limits

ONE-SIDED LIMITS When a function takes a jump (as does $[\![x]\!]$ at each integer, in Example 6), then the limit does not exist at the jump points. For such functions, it is natural to introduce one-sided limits. Let the symbol $x \to c^+$ mean that x approaches c from the right, and let $x \to c^-$ mean that x approaches c from the left.

Definition

(Right- and left-hand limits). To say that $\lim\limits_{x \to c^+} f(x) = L$ means that when x is near but on the right of c, then $f(x)$ is near L. Similarly, to say that $\lim\limits_{x \to c^-} f(x) = L$ means that when x is near but on the left of c, then $f(x)$ is near L.

Thus, while $\lim\limits_{x \to 2}[\![x]\!]$ does not exist, it is correct to write (look at the graph in Example 6)

$$\lim_{x \to 2^-}[\![x]\!] = 1 \qquad \lim_{x \to 2^+}[\![x]\!] = 2$$

We believe that you will find the following theorem quite reasonable.

Theorem A

$\lim\limits_{x \to c} f(x) = L$ if and only if $\lim\limits_{x \to c^-} f(x) = L$ and $\lim\limits_{x \to c^+} f(x) = L$.

Figure 7 should give additional insight.

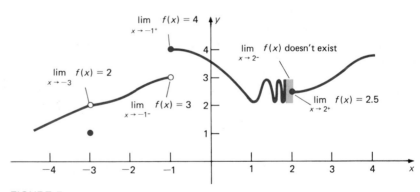

FIGURE 7

PROBLEM SET 2.4

In Problems 1–6, find the indicated limit by inspection.

1. $\lim\limits_{x \to 3}(2x - 8)$

2. $\lim\limits_{x \to 3}\left(\dfrac{2}{x} + 1\right)$

3. $\lim\limits_{x \to -2}(x^2 - 3x + 1)$

4. $\lim\limits_{x \to 4}\dfrac{\sqrt{9 + x^2}}{x - 3}$

5. $\lim\limits_{x \to \sqrt{3}}\dfrac{\sqrt{12 - x^2}}{x^4}$

6. $\lim\limits_{x \to 1}\dfrac{5x - x^2}{x^2 + 2x - 4}$

In Problems 7–16, find the indicated limit. In most cases, it will be wise to do some algebra first (see Examples 2 and 3).

7. $\lim_{x \to 1} \dfrac{x^2 + 3x - 4}{x - 1}$

8. $\lim_{x \to 3} \dfrac{x^2 - 2x - 3}{x - 3}$

9. $\lim_{x \to -3} \dfrac{2x^2 + 5x - 3}{x + 3}$

10. $\lim_{x \to 0} \dfrac{x^3 - 16x}{x^2 + 4x}$

11. $\lim_{x \to 9} \dfrac{x - 9}{\sqrt{x} - 3}$

12. $\lim_{x \to 2} \dfrac{x^3 - 8}{x - 2}$

13. $\lim_{x \to 2} \dfrac{x^2 + x - 6}{x + 2}$

14. $\lim_{x \to -3} \dfrac{x^2 + 4x - 4}{x - 3}$

15. $\lim_{t \to 2} \dfrac{t^2 - 5t + 6}{t^2 - t - 2}$

16. $\lim_{u \to 1} \dfrac{u^2 + 6u - 7}{u^2 - 1}$

In Problems 17–26, use a calculator (or Table II) to calculate values of the given function near the limit point c. Organize these values in a table (as in Examples 4, 5, and 7) and use the results to find the required limit or to conclude that it does not exist. Be sure to put your calculator in radian mode.

[C] 17. $\lim_{x \to 0} \dfrac{\tan x}{2x}$

[C] 18. $\lim_{x \to 0} \dfrac{1 - \cos x}{x}$

[C] 19. $\lim_{x \to 0} \dfrac{x - \sin x}{x^3}$

[C] 20. $\lim_{x \to 0} \dfrac{1 - \cos x}{x^2}$

[C] 21. $\lim_{t \to 0} \dfrac{\sin t}{t^2}$

[C] 22. $\lim_{t \to 0} \dfrac{\sin^2 t}{t}$

[C] 23. $\lim_{x \to \pi} \dfrac{1 + \cos x}{\sin 2x}$

[C] 24. $\lim_{x \to 0} \dfrac{\tan x - \sin x}{x^3}$

[C] 25. $\lim_{x \to 1} \dfrac{\sin 3x}{x - 1}$

[C] 26. $\lim_{u \to \pi/2} \dfrac{\tan u}{u}$

27. For the function f graphed in Figure 8, find the indicated limit or function value, or state that it does not exist.

(a) $\lim_{x \to -3} f(x)$

(b) $f(-3)$

(c) $f(-1)$

(d) $\lim_{x \to -1} f(x)$

(e) $f(1)$

(f) $\lim_{x \to 1} f(x)$

(g) $\lim_{x \to 1^-} f(x)$

(h) $\lim_{x \to 1^+} f(x)$

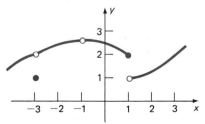

FIGURE 8

28. Follow the directions of Problem 27 for the function f graphed in Figure 9.

FIGURE 9

29. Sketch the graph of
$$f(x) = \begin{cases} x^2 & \text{if } x \le 0 \\ x & \text{if } 0 < x < 1 \\ 1 + x^2 & \text{if } x \ge 1 \end{cases}$$

Then find each of the following or state that it does not exist.

(a) $\lim_{x \to 0} f(x)$

(b) $f(1)$

(c) $\lim_{x \to 1} f(x)$

(d) $\lim_{x \to 1^-} f(x)$

30. Sketch the graph of
$$g(x) = \begin{cases} -x + 1 & \text{if } x < 1 \\ x - 1 & \text{if } 1 < x < 2 \\ 5 - x^2 & \text{if } x \ge 2 \end{cases}$$

Then find each of the following or state that it does not exist.

(a) $\lim_{x \to 1} g(x)$

(b) $g(1)$

(c) $\lim_{x \to 2} g(x)$

(d) $\lim_{x \to 2^+} g(x)$

31. Sketch the graph of $f(x) = x - [\![x]\!]$; then find each of the following or state that it does not exist.

(a) $f(0)$

(b) $\lim_{x \to 0} f(x)$

(c) $\lim_{x \to 0^-} f(x)$

(d) $\lim_{x \to 1/2} f(x)$

32. Follow the directions of Problem 31 for $f(x) = x/|x|$.

33. Find $\lim_{x \to 1}(x^2 - 1)/|x - 1|$ or state that it does not exist.

34. Evaluate $\lim_{x \to 0}(\sqrt{x + 3} - \sqrt{3})/x$. *Hint*: Rationalize the numerator.

35. Let
$$f(x) = \begin{cases} x & \text{if } x \text{ is rational} \\ -x & \text{if } x \text{ is irrational} \end{cases}$$

Find each value, if possible.

(a) $\lim\limits_{x \to 1} f(x)$

(b) $\lim\limits_{x \to 0} f(x)$

36. Sketch, as best you can, the graph of a function f that satisfies all the following conditions.

(a) Its domain is the interval $[0, 4]$.
(b) $f(0) = f(1) = f(2) = f(3) = f(4) = 1$

(c) $\lim\limits_{x \to 1} f(x) = 2$

(d) $\lim\limits_{x \to 2} f(x) = 1$

(e) $\lim\limits_{x \to 3^-} f(x) = 2$

(f) $\lim\limits_{x \to 3^+} f(x) = 1$

37. Let

$$f(x) = \begin{cases} x^2 & \text{if } x \text{ is rational} \\ x^4 & \text{if } x \text{ is irrational} \end{cases}$$

For what values of a does $\lim\limits_{x \to a} f(x)$ exist?

38. The function $f(x) = x^2$ had been carefully graphed, but during the night a mysterious visitor changed the values of f at a million different places. Did this affect the value of $\lim\limits_{x \to a} f(x)$ at any a? Explain.

39. Find each of the following limits or state that it does not exist.

(a) $\lim\limits_{x \to 1} \dfrac{|x - 1|}{x - 1}$

(b) $\lim\limits_{x \to 1^-} \dfrac{|x - 1|}{x - 1}$

(c) $\lim\limits_{x \to 1^-} \dfrac{x^2 - |x - 1| - 1}{|x - 1|}$

(d) $\lim\limits_{x \to 1^-} \left[\dfrac{1}{x - 1} - \dfrac{1}{|x - 1|} \right]$

40. Find each of the following limits or state that it does not exist.

(a) $\lim\limits_{x \to 1^+} \sqrt{x - [\![x]\!]}$

(b) $\lim\limits_{x \to 0^+} [\![1/x]\!]$

(c) $\lim\limits_{x \to 0^+} x(-1)^{[\![1/x]\!]}$

(d) $\lim\limits_{x \to 0^+} [\![x]\!](-1)^{[\![1/x]\!]}$

(e) $\lim\limits_{x \to 0^+} x[\![1/x]\!]$

(f) $\lim\limits_{x \to 0^+} x^2[\![1/x]\!]$

2.5
Rigorous Study of Limits

You should not believe everything you are told. It is prudent to be skeptical— not so skeptical that you will not believe anything, but skeptical enough to check a statement before you accept it. Mathematicians tend to be very skeptical people. Tell a mathematician that something is true and you will probably get the response: Prove it. But to be able to prove something requires that we be very clear about the meaning of the words we are using. This is especially true of the word *limit*, because all of calculus rests on the meaning of that word.

We gave an informal definition of limit in the previous section. Here is a slightly better, but still informal, rewording of that definition. *To say that $\lim\limits_{x \to c} f(x) = L$ means that the difference between $f(x)$ and L can be made arbitrarily small by requiring that x be sufficiently close to but different from c.* Now let us try to pin this down.

MAKING THE DEFINITION PRECISE   First, we follow a long tradition in using the Greek letters ε (epsilon) and δ (delta) to stand for arbitrary positive numbers. Think of ε and δ as small positive numbers.

To say that $f(x)$ differs from L by less than ε is to say that

$$|f(x) - L| < \varepsilon$$

or, equivalently,

$$L - \varepsilon < f(x) < L + \varepsilon$$

This means that $f(x)$ lies in the open interval $(L - \varepsilon, L + \varepsilon)$ shown on the graph in Figure 1.

$|f(x) - L| < \epsilon$

FIGURE 1

$$0 < |x - c| < \delta$$

FIGURE 2

Next, to say that x is sufficiently close to but different from c is to say that for some δ, x is in the open interval $(c - \delta, c + \delta)$ with c deleted. Perhaps the best way to say this is to write

$$0 < |x - c| < \delta$$

Note that $|x - c| < \delta$ would describe the interval $c - \delta < x < c + \delta$, while $0 < |x - c|$ requires that $x = c$ be excluded. The deleted interval we are describing is shown in Figure 2.

We are ready for what some people have called the most important definition in calculus.

Definition

(Precise meaning of limit). To say that $\lim_{x \to c} f(x) = L$ means that for each given $\varepsilon > 0$ (no matter how small), there is a corresponding $\delta > 0$ such that $|f(x) - L| < \varepsilon$ provided that $0 < |x - c| < \delta$; that is,

$$0 < |x - c| < \delta \quad \Rightarrow \quad |f(x) - L| < \varepsilon$$

The pictures in Figure 3 may help you absorb this definition.

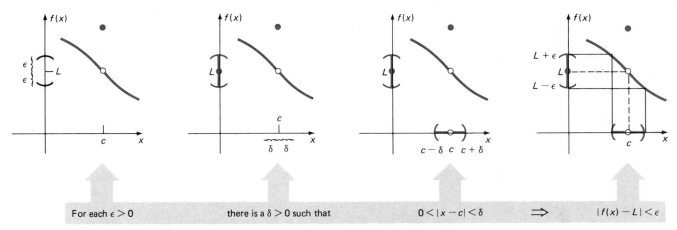

FIGURE 3

It is to be emphasized that the number ε is given *first*; the number δ is to be produced. Suppose Doris wishes to prove to Edward that $\lim_{x \to c} f(x) = L$. Edward can challenge Doris with any particular ε he chooses (for example, $\varepsilon = 0.00001$) and demand that Doris produce a corresponding δ.

Moreover, since the definition says "for each $\varepsilon > 0$" (not "for some $\varepsilon > 0$"), Doris will not have proved the limit statement if she produces a δ for one particular ε, nor even for several of them. She must be able to produce for each ε a corresponding δ. Of course, the δ she produces may depend on ε. In general, the smaller the ε Edward gives her, the smaller the δ Doris will need to return.

Is there any possible way that Doris can meet such a challenge? There is, but do not expect it to be easy.

SOME LIMIT PROOFS (OPTIONAL) In each of the examples below, we begin with what we call a preliminary analysis. This is not part of the proof. It is the kind of work you should do on scratch paper. We include it so our proofs will not appear to float down from heaven.

EXAMPLE 1 Prove that $\lim\limits_{x \to 4}(3x - 7) = 5$.

PRELIMINARY ANALYSIS Let ε be any positive number. We must produce a $\delta > 0$ such that

$$0 < |x - 4| < \delta \quad \Rightarrow \quad |(3x - 7) - 5| < \varepsilon$$

Consider the inequality on the right.

$$
\begin{aligned}
|(3x - 7) - 5| < \varepsilon \quad &\Leftrightarrow \quad |3x - 12| < \varepsilon \\
&\Leftrightarrow \quad |3(x - 4)| < \varepsilon \\
&\Leftrightarrow \quad |3||x - 4| < \varepsilon \\
&\Leftrightarrow \quad |x - 4| < \frac{\varepsilon}{3}
\end{aligned}
$$

Now we see how to choose δ, namely, $\delta = \varepsilon/3$. Of course, any smaller δ would work.

FORMAL PROOF Let $\varepsilon > 0$ be given. Choose $\delta = \varepsilon/3$. Then $0 < |x - 4| < \delta$ implies

$$|(3x - 7) - 5| = |3x - 12| = |3(x - 4)| = 3|x - 4| < 3\delta = \varepsilon$$

If you read this chain of equalities and an inequality from left to right and use the transitive laws, you see that

$$|(3x - 7) - 5| < \varepsilon$$

If Edward were to challenge Doris with $\varepsilon = 0.01$ in this example, Doris would respond with $\delta = 0.01/3 = 0.0033$. If Edward said $\varepsilon = 0.000003$, Doris would say $\delta = 0.000001$. If she gave an even smaller δ, that would be fine.

Of course, if you think about the graph of $y = 3x - 7$ (a line with slope 3, as in Figure 4), you know that to force $3x - 7$ to be close to 5, you had better make x even closer (closer by a factor of one-third) to 4. ∎

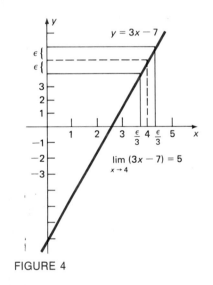

FIGURE 4

EXAMPLE 2 Prove that $\lim\limits_{x \to 2} \dfrac{2x^2 - 3x - 2}{x - 2} = 5$.

PRELIMINARY ANALYSIS We are looking for δ such that

$$0 < |x - 2| < \delta \quad \Rightarrow \quad \left| \frac{2x^2 - 3x - 2}{x - 2} - 5 \right| < \varepsilon$$

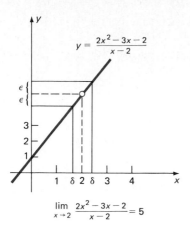

$$y = \frac{2x^2 - 3x - 2}{x - 2}$$

$$\lim_{x \to 2} \frac{2x^2 - 3x - 2}{x - 2} = 5$$

FIGURE 5

Now for $x \neq 2$,

$$\left| \frac{2x^2 - 3x - 2}{x - 2} - 5 \right| < \varepsilon \quad \Leftrightarrow \quad \left| \frac{(2x + 1)(x - 2)}{x - 2} - 5 \right| < \varepsilon$$

$$\Leftrightarrow \quad |(2x + 1) - 5| < \varepsilon$$

$$\Leftrightarrow \quad |2(x - 2)| < \varepsilon$$

$$\Leftrightarrow \quad |2||x - 2| < \varepsilon$$

$$\Leftrightarrow \quad |x - 2| < \frac{\varepsilon}{2}$$

This indicates that $\delta = \varepsilon/2$ will work (see Figure 5).

FORMAL PROOF   Let $\varepsilon > 0$ be given. Choose $\delta = \varepsilon/2$. Then $0 < |x - 2| < \delta$ implies

$$\left| \frac{2x^2 - 3x - 2}{x - 2} - 5 \right| = \left| \frac{(2x + 1)(x - 2)}{x - 2} - 5 \right| = |2x + 1 - 5|$$

$$= |2(x - 2)| = 2|x - 2| < 2\delta = \varepsilon$$

The cancellation of the factor $x - 2$ is legitimate because $0 < |x - 2|$ implies $x \neq 2$; thus, division by 0 is avoided. ∎

EXAMPLE 3   Prove that $\lim_{x \to c}(mx + b) = mc + b$.

PRELIMINARY ANALYSIS   We want to find δ such that

$$0 < |x - c| < \delta \quad \Rightarrow \quad |(mx + b) - (mc + b)| < \varepsilon$$

Now

$$|(mx + b) - (mc + b)| = |mx - mc| = |m(x - c)| = |m||x - c|$$

It appears that $\delta = \varepsilon/|m|$ should do.

FORMAL PROOF   Let $\varepsilon > 0$ be given. Choose $\delta = \varepsilon/|m|$. Then $0 < |x - c| < \delta$ implies

$$|(mx + b) - (mc + b)| = |mx - mc| = |m||x - c| < |m|\delta = \varepsilon$$

There is only one problem with our choice of δ. What if $m = 0$? In that case, any δ will do just fine since

$$|(0x + b) - (0c + b)| = |0| = 0$$

The latter is less than ε for all x. ∎

EXAMPLE 4   Prove that if $c > 0$, $\lim_{x \to c}\sqrt{x} = \sqrt{c}$.

PRELIMINARY ANALYSIS   We must find δ such that

$$0 < |x - c| < \delta \quad \Rightarrow \quad \left|\sqrt{x} - \sqrt{c}\right| < \varepsilon$$

Now

$$\left|\sqrt{x} - \sqrt{c}\right| = \left|\frac{(\sqrt{x} - \sqrt{c})(\sqrt{x} + \sqrt{c})}{\sqrt{x} + \sqrt{c}}\right| = \left|\frac{x - c}{\sqrt{x} + \sqrt{c}}\right|$$

$$= \frac{|x - c|}{\sqrt{x} + \sqrt{c}} \leq \frac{|x - c|}{\sqrt{c}}$$

To make the latter less than ε requires that we make $|x - c| < \varepsilon\sqrt{c}$ (Figure 6).

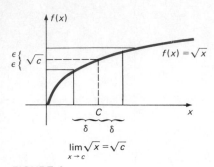

$$\lim_{x \to c} \sqrt{x} = \sqrt{c}$$

FIGURE 6

FORMAL PROOF   Let $\varepsilon > 0$ be given. Choose $\delta = \varepsilon\sqrt{c}$. Then $0 < |x - c| < \delta$ implies

$$\left|\sqrt{x} - \sqrt{c}\right| = \left|\frac{(\sqrt{x} - \sqrt{c})(\sqrt{x} + \sqrt{c})}{\sqrt{x} + \sqrt{c}}\right| = \left|\frac{x - c}{\sqrt{x} + \sqrt{c}}\right|$$

$$= \frac{|x - c|}{\sqrt{x} + \sqrt{c}} \leq \frac{|x - c|}{\sqrt{c}} < \frac{\delta}{\sqrt{c}} = \varepsilon$$

There is one further technical point. We should insist that $\delta \leq c$, for then $|x - c| < \delta$ implies $x > 0$ so that $\sqrt{x}$ is defined. Thus, for absolute rigor, choose δ to be the smaller of c and $\varepsilon\sqrt{c}$.   ∎

Our demonstration in Example 4 depended on rationalizing the numerator, a trick frequently useful in calculus.

EXAMPLE 5   Prove that $\lim_{x \to 3}(x^2 + x - 5) = 7$.

PRELIMINARY ANALYSIS   Our task is to find δ such that

$$0 < |x - 3| < \delta \quad \Rightarrow \quad \left|(x^2 + x - 5) - 7\right| < \varepsilon$$

Now

$$\left|(x^2 + x - 5) - 7\right| = \left|x^2 + x - 12\right| = |x + 4|\,|x - 3|$$

Since the second factor $|x - 3|$ can be made as small as we wish, it is enough to bound the factor $|x + 4|$. To do this, we first agree to make $\delta \leq 1$. Then $|x - 3| < \delta$ implies

$$|x + 4| = |x - 3 + 7|$$
$$\leq |x - 3| + |7| \qquad \text{(Triangle Inequality)}$$
$$< 1 + 7 = 8$$

$$|x - 3| < 1 \Rightarrow 2 < x < 4$$
$$\Rightarrow 6 < x + 4 < 8$$
$$\Rightarrow |x + 4| < 8$$

FIGURE 7

Figure 7 offers an alternative demonstration of this fact. If we also require $\delta \leq \varepsilon/8$, the product $|x + 4||x - 3|$ will be less than ε.

FORMAL PROOF Let $\varepsilon > 0$ be given. Choose $\delta = \min\{1, \varepsilon/8\}$; that is, choose δ to be the smaller of 1 and $\varepsilon/8$. Then $0 < |x - 3| < \delta$ implies

$$|(x^2 + x - 5) - 7| = |x^2 + x - 12| = |x + 4||x - 3| < 8 \cdot \frac{\varepsilon}{8} = \varepsilon \quad \blacksquare$$

EXAMPLE 6 Prove that $\lim\limits_{x \to c} x^2 = c^2$.

PROOF We mimic the proof in Example 5. Let $\varepsilon > 0$ be given. Choose $\delta = \min\{1, \varepsilon/(1 + 2|c|)\}$. Then $0 < |x - c| < \delta$ implies

$$|x^2 - c^2| = |x + c||x - c| = |x - c + 2c||x - c|$$

$$\leq (|x - c| + 2|c|)|x - c| \qquad \text{(Triangle Inequality)}$$

$$< \frac{(1 + 2|c|) \cdot \varepsilon}{1 + 2|c|} = \varepsilon \qquad \blacksquare$$

EXAMPLE 7 Prove that $\lim\limits_{x \to c} \dfrac{1}{x} = \dfrac{1}{c}$, $c \neq 0$.

PRELIMINARY ANALYSIS We must find δ such that

$$0 < |x - c| < \delta \quad \Rightarrow \quad \left| \frac{1}{x} - \frac{1}{c} \right| < \varepsilon$$

Now

$$\left| \frac{1}{x} - \frac{1}{c} \right| = \left| \frac{c - x}{xc} \right| = \frac{1}{|x|} \cdot \frac{1}{|c|} \cdot |x - c|$$

The factor $1/|x|$ is troublesome, especially if x is near 0. We can bound this factor if we can keep x away from 0. To that end, note that

$$|c| = |c - x + x| \leq |c - x| + |x|$$

so

$$|x| \geq |c| - |x - c|$$

Thus, if we choose $\delta \leq |c|/2$, we succeed in making $|x| \geq |c|/2$. Finally, if we also require $\delta \leq \varepsilon c^2/2$, then

$$\frac{1}{|x|} \cdot \frac{1}{|c|} \cdot |x - c| < \frac{1}{|c|/2} \cdot \frac{1}{|c|} \cdot \frac{\varepsilon c^2}{2} = \varepsilon$$

FORMAL PROOF   Let $\varepsilon > 0$ be given. Choose $\delta = \min\{|c|/2, \varepsilon c^2/2\}$. Then $0 < |x - c| < \delta$ implies

$$\left|\frac{1}{x} - \frac{1}{c}\right| = \left|\frac{c - x}{xc}\right| = \frac{1}{|x|} \cdot \frac{1}{|c|} \cdot |x - c| < \frac{1}{|c|/2} \cdot \frac{1}{|c|} \cdot \frac{\varepsilon c^2}{2} = \varepsilon \qquad \blacksquare$$

ONE-SIDED LIMITS   It does not take much imagination to give the ε, δ definitions of right- and left-hand limits.

Definition

To say $\lim\limits_{x \to c^+} f(x) = L$ means that for each $\varepsilon > 0$, there is a corresponding $\delta > 0$ such that

$$0 < x - c < \delta \quad \Rightarrow \quad |f(x) - L| < \varepsilon$$

We leave the corresponding ε, δ definition for the left-hand limit to the reader.

PROBLEM SET   2.5

In Problems 1-6, give the appropriate ε, δ definition of each statement.

1. $\lim\limits_{t \to a} f(t) = M$ **2.** $\lim\limits_{u \to b} g(u) = L$

3. $\lim\limits_{z \to d} h(z) = P$ **4.** $\lim\limits_{y \to e} \phi(y) = B$

5. $\lim\limits_{x \to c^-} f(x) = L$ **6.** $\lim\limits_{t \to a^+} g(t) = D$

In Problems 7–18, give an ε, δ proof of each limit fact (see Examples 1–5).

7. $\lim\limits_{x \to 3}(5x - 11) = 4$ **8.** $\lim\limits_{x \to -2}(2x - 4) = -8$

9. $\lim\limits_{x \to 5}\dfrac{x^2 - 25}{x - 5} = 10$ **10.** $\lim\limits_{x \to 1}\dfrac{x^2 + 5x - 6}{x - 1} = 7$

11. $\lim\limits_{x \to -1}\dfrac{2x^2 - x - 3}{x + 1} = -5$

12. $\lim\limits_{x \to 0}\dfrac{3x^2 - 4x}{x} = -4$

13. $\lim\limits_{x \to 2}\sqrt{2x} = 2$

14. $\lim\limits_{x \to 5}\sqrt{x - 1} = 2$

15. $\lim\limits_{x \to 1}\dfrac{2x^3 + 3x^2 - 2x - 3}{x^2 - 1} = 5$. *Hint:* $x^2 - 1$ is a factor of the numerator.

16. $\lim\limits_{x \to 2}(2x^2 - 4x + 3) = 3$

17. $\lim\limits_{x \to 3}(x^2 - 2x) = 3$

18. $\lim\limits_{x \to 0} x^4 = 0$

19. Prove that if $\lim\limits_{x \to c} f(x) = L$ and $\lim\limits_{x \to c} f(x) = M$, then $L = M$.

20. Let F and G be functions such that $0 \le F(x) \le G(x)$ for all x near c, except possibly at c. Prove that if $\lim\limits_{x \to c} G(x) = 0$, then $\lim\limits_{x \to c} F(x) = 0$.

21. Prove that $\lim\limits_{x \to 0} x^4 \sin^2(1/x) = 0$. *Hint:* Use Problems 18 and 20.

22. Prove that $\lim\limits_{x \to 0^+} \sqrt{x} = 0$.

23. By considering left- and right-hand limits, prove that $\lim\limits_{x \to 0} |x| = 0$.

24. Prove that if $|f(x)| < B$ for $|x - a| < 1$ and $\lim\limits_{x \to a} g(x) = 0$, then $\lim\limits_{x \to a} f(x)g(x) = 0$.

25. Suppose that $\lim\limits_{x \to a} f(x) = L$ and that $f(a)$ exists (though it may be different from L). Prove that f is bounded on some interval containing a; that is, show that there is an interval (c, d) with $c < a < d$ and a constant M such that $|f(x)| \le M$ for all x in (c, d).

26. Prove that if $f(x) \leq g(x)$ for all x in some deleted interval about a and if $\lim\limits_{x \to a} f(x) = L$ and $\lim\limits_{x \to a} g(x) = M$, then $L \leq M$.

27. Which of the following are equivalent to the definition of limit?

(a) For some $\varepsilon > 0$ and every $\delta > 0$, $0 < |x - c| < \delta \Rightarrow |f(x) - L| < \varepsilon$.

(b) For every $\delta > 0$, there is a corresponding $\varepsilon > 0$ such that
$$0 < |x - c| < \varepsilon \quad \Rightarrow \quad |f(x) - L| < \delta$$

(c) For every positive integer N, there is a corresponding positive integer M such that $0 < |x - c| < 1/M \Rightarrow |f(x) - L| < 1/N$.

(d) For every $\varepsilon > 0$, there is a corresponding $\delta > 0$ such that $0 < |x - c| < \delta$ and $|f(x) - L| < \varepsilon$ for some x.

28. State in ε, δ language what it means to say $\lim\limits_{x \to c} f(x) \neq L$.

2.6
Limit Theorems

Most readers will agree that proving the existence of limits using the ε, δ definition of the preceding section is both time-consuming and difficult. That is why the theorems of this section are so welcome. Our first theorem is the big one. With it, we can handle most limit problems that we will face for quite some time.

Theorem A

(Main Limit Theorem). Let n be a positive integer, k be a constant, and f and g be functions which have limits at c. Then

1. $\lim\limits_{x \to c} k = k$;

2. $\lim\limits_{x \to c} x = c$;

3. $\lim\limits_{x \to c} kf(x) = k \lim\limits_{x \to c} f(x)$;

4. $\lim\limits_{x \to c}[f(x) + g(x)] = \lim\limits_{x \to c} f(x) + \lim\limits_{x \to c} g(x)$;

5. $\lim\limits_{x \to c}[f(x) - g(x)] = \lim\limits_{x \to c} f(x) - \lim\limits_{x \to c} g(x)$;

6. $\lim\limits_{x \to c}[f(x) \cdot g(x)] = \lim\limits_{x \to c} f(x) \cdot \lim\limits_{x \to c} g(x)$;

7. $\lim\limits_{x \to c} \dfrac{f(x)}{g(x)} = \dfrac{\lim\limits_{x \to c} f(x)}{\lim\limits_{x \to c} g(x)}$, provided $\lim\limits_{x \to c} g(x) \neq 0$;

8. $\lim\limits_{x \to c}[f(x)]^n = \left[\lim\limits_{x \to c} f(x)\right]^n$;

9. $\lim\limits_{x \to c}\sqrt[n]{f(x)} = \sqrt[n]{\lim\limits_{x \to c} f(x)}$, provided $\lim\limits_{x \to c} f(x) > 0$ when n is even.

These important results are remembered best if learned in words. For example, Statement 4 translates as: *The limit of a sum is the sum of the limits.*

Of course, Theorem A needs to be proved. We postpone that job till the end of the section, choosing first to show you how this multipart theorem is used.

APPLICATIONS OF THE MAIN LIMIT THEOREM   In the next examples, the circled numbers refer to the numbered statements from the list given earlier. Each equality is justified by the indicated statement.

EXAMPLE 1   Find $\lim\limits_{x \to 3} 2x^4$.

Solution

$$\lim_{x \to 3} 2x^4 \overset{\text{③}}{=} 2 \lim_{x \to 3} x^4 \overset{\text{⑧}}{=} 2\left[\lim_{x \to 3} x\right]^4 \overset{\text{②}}{=} 2[3]^4 = 162 \qquad \blacksquare$$

EXAMPLE 2   Find $\lim\limits_{x \to 4}(3x^2 - 2x)$.

Solution

$$\lim_{x \to 4}(3x^2 - 2x) \overset{\text{⑤}}{=} \lim_{x \to 4} 3x^2 - \lim_{x \to 4} 2x \overset{\text{③}}{=} 3 \lim_{x \to 4} x^2 - 2 \lim_{x \to 4} x$$

$$\overset{\text{⑧}}{=} 3\left(\lim_{x \to 4} x\right)^2 - 2 \lim_{x \to 4} x \overset{\text{②}}{=} 3(4)^2 - 2(4)$$

$$= 40 \qquad \blacksquare$$

EXAMPLE 3   Find $\lim\limits_{x \to 4} \dfrac{\sqrt{x^2 + 9}}{x}$.

Solution

$$\lim_{x \to 4} \frac{\sqrt{x^2 + 9}}{x} \overset{\text{⑦}}{=} \frac{\lim\limits_{x \to 4} \sqrt{x^2 + 9}}{\lim\limits_{x \to 4} x} \overset{\text{⑨}}{\underset{\text{②}}{=}} \frac{\sqrt{\lim\limits_{x \to 4}(x^2 + 9)}}{4} \overset{\text{④}}{=} \frac{1}{4}\sqrt{\lim_{x \to 4} x^2 + \lim_{x \to 4} 9}$$

$$\overset{\text{⑧,1}}{=} \frac{1}{4}\sqrt{\left[\lim_{x \to 4} x\right]^2 + 9} \overset{\text{②}}{=} \frac{1}{4}\sqrt{4^2 + 9} = \frac{5}{4} \qquad \blacksquare$$

EXAMPLE 4   If $\lim\limits_{x \to 3} f(x) = 4$ and $\lim\limits_{x \to 3} g(x) = 8$, find

$$\lim_{x \to 3}\left[f^2(x) \cdot \sqrt[3]{g(x)}\right]$$

Solution

$$\lim_{x\to 3}\left[f^2(x)\cdot\sqrt[3]{g(x)}\right]\overset{\text{\textcircled{6}}}{=}\lim_{x\to 3}f^2(x)\cdot\lim_{x\to 3}\sqrt[3]{g(x)}$$

$$\overset{\text{\textcircled{8,9}}}{=}\left[\lim_{x\to 3}f(x)\right]^2\cdot\sqrt[3]{\lim_{x\to 3}g(x)}$$

$$=[4]^2\cdot\sqrt[3]{8}$$

$$=32 \qquad \blacksquare$$

Recall that a polynomial function f has the form

$$f(x)=a_nx^n+a_{n-1}x^{n-1}+\cdots+a_1x+a_0$$

whereas a rational function f is the quotient of two polynomial functions— that is,

$$f(x)=\frac{a_nx^n+a_{n-1}x^{n-1}+\cdots+a_1x+a_0}{b_mx^m+b_{m-1}x^{m-1}+\cdots+b_1x+b_0}$$

Theorem B

(Substitution Theorem). If f is a polynomial function or a rational function, then

$$\lim_{x\to c}f(x)=f(c)$$

provided in the case of a rational function that the value of the denominator at c is not zero.

The proof of Theorem B follows from repeated applications of Theorem A. Note that Theorem B allows us to find limits for polynomial and rational functions by simply substituting c for x throughout.

EXAMPLE 5 Find $\lim\limits_{x\to 2}\dfrac{7x^5-10x^4-13x+6}{3x^2-6x-8}$.

Solution

$$\lim_{x\to 2}\frac{7x^5-10x^4-13x+6}{3x^2-6x-8}=\frac{7(2)^5-10(2)^4-13(2)+6}{3(2)^2-6(2)-8}=-\frac{11}{2}\qquad\blacksquare$$

EXAMPLE 6 Find $\lim\limits_{x\to 1}\dfrac{x^3+3x+7}{x^2-2x+1}=\lim\limits_{x\to 1}\dfrac{x^3+3x+7}{(x-1)^2}$.

Solution Neither Theorem B nor Statement 7 of Theorem A applies, since the limit of the denominator is 0. However, since the limit of the numerator is 11, we see that as x nears 1, we are dividing a number near 11 by a positive number near 0. The result is a large positive number. In fact, the resulting number can be made as large as you like by letting x get close enough to 1. We say that the limit does not exist. (Later in the book—see Section 4.6—we will allow ourselves to say the limit is $+\infty$.) ∎

EXAMPLE 7 Find $\lim\limits_{t \to 2} \dfrac{t^2 + 3t - 10}{t^2 + t - 6}$.

Solution Again, Theorem B does not apply. But this time, the quotient takes the meaningless form 0/0 at $t = 2$. Whenever this happens one should look for an algebraic simplification (factorization) of the quotient before trying to take the limit.

$$\lim_{t \to 2} \frac{t^2 + 3t - 10}{t^2 + t - 6} = \lim_{t \to 2} \frac{(t-2)(t+5)}{(t-2)(t+3)} = \lim_{t \to 2} \frac{t+5}{t+3} = \frac{7}{5}$$ ∎

PROOF OF THEOREM A (OPTIONAL) You should not be too surprised when we say that the proofs of some parts of Theorem A are quite sophisticated. Because of this, we prove only the first five parts here, deferring the others to the Appendix (Section A.2, Theorem A). To get your feet wet, you might try Problems 33 and 34.

Proof of Statements 1 and 2 These statements result from $\lim\limits_{x \to c}(mx + b) = mc + b$ (Example 3 of Section 2.5) using first $m = 0$ and then $m = 1$, $b = 0$. ∎

Proof of Statement 3 If $k = 0$, the result is trivial, so we suppose $k \neq 0$. Let $\varepsilon > 0$ be given. By hypothesis, $\lim\limits_{x \to c} f(x)$ exists; call its value L. By definition · of limit, there is a number δ such that

$$0 < |x - c| < \delta \quad \Rightarrow \quad |f(x) - L| < \frac{\varepsilon}{|k|}$$

Someone is sure to complain that we put $\varepsilon/|k|$ rather than ε at the end of the inequality above. Well, isn't $\varepsilon/|k|$ a positive number? Yes. Doesn't the definition of limit require that for any positive number, there is a corresponding δ? Yes.

Now with δ determined, we may assert that $0 < |x - c| < \delta$ implies

$$|kf(x) - kL| = |k|\,|f(x) - L| < |k|\frac{\varepsilon}{|k|} = \varepsilon$$

This shows that

$$\lim_{x \to c} kf(x) = kL = k \lim_{x \to c} f(x)$$ ∎

CHAPTER 2/Functions and Limits

Proof of Statement 4   Let $\lim\limits_{x\to c} f(x) = L$ and $\lim\limits_{x\to c} g(x) = M$. If ε is any given positive number, then $\varepsilon/2$ is positive. Since $\lim\limits_{x\to c} f(x) = L$, there is a positive number δ_1 such that

$$0 < |x - c| < \delta_1 \quad \Rightarrow \quad |f(x) - L| < \frac{\varepsilon}{2}$$

Since $\lim\limits_{x\to c} g(x) = M$, there is a positive number δ_2 such that

$$0 < |x - c| < \delta_2 \quad \Rightarrow \quad |g(x) - M| < \frac{\varepsilon}{2}$$

Choose $\delta = \min\{\delta_1, \delta_2\}$; that is, choose δ to be the smaller of δ_1 and δ_2. Then $0 < |x - c| < \delta$ implies

$$
\begin{aligned}
|f(x) + g(x) - (L + M)| &= |[f(x) - L] + [g(x) - M]| \\
&\le |f(x) - L| + |g(x) - M| \\
&< \frac{\varepsilon}{2} + \frac{\varepsilon}{2} = \varepsilon
\end{aligned}
$$

In this chain, the first inequality is the triangle inequality (Section 1.4); the second results from the choice of δ. We have just shown that

$$0 < |x - c| < \delta \quad \Rightarrow \quad |f(x) + g(x) - (L + M)| < \varepsilon$$

Thus

$$\lim_{x\to c}[f(x) + g(x)] = L + M = \lim_{x\to c} f(x) + \lim_{x\to c} g(x) \qquad ■$$

Proof of Statement 5

$$
\begin{aligned}
\lim_{x\to c}[f(x) - g(x)] &= \lim_{x\to c}[f(x) + (-1)g(x)] \\
&= \lim_{x\to c} f(x) + \lim_{x\to c}(-1)g(x) \\
&= \lim_{x\to c} f(x) + (-1)\lim_{x\to c} g(x) \\
&= \lim_{x\to c} f(x) - \lim_{x\to c} g(x) \qquad ■
\end{aligned}
$$

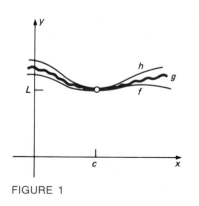

FIGURE 1

THE SQUEEZE THEOREM   Did you ever hear anyone say, "I was caught between a rock and a hard place?" This is what has happened to g in the following theorem (see Figure 1).

> **Theorem C**
>
> **(Squeeze Theorem).** Let f, g, and h be functions satisfying $f(x) \le g(x) \le h(x)$ for all x near c, except possibly at c. If $\lim_{x \to c} f(x) = \lim_{x \to c} h(x) = L$, then $\lim_{x \to c} g(x) = L$.

Proof (Optional)   Let $\varepsilon > 0$ be given. Choose δ_1 such that

$$0 < |x - c| < \delta_1 \quad \Rightarrow \quad L - \varepsilon < f(x) < L + \varepsilon$$

and δ_2 such that

$$0 < |x - c| < \delta_2 \quad \Rightarrow \quad L - \varepsilon < h(x) < L + \varepsilon$$

Choose δ_3 so that

$$0 < |x - c| < \delta_3 \quad \Rightarrow \quad f(x) \le g(x) \le h(x)$$

Let $\delta = \min\{\delta_1, \delta_2, \delta_3\}$. Then

$$0 < |x - c| < \delta \quad \Rightarrow \quad L - \varepsilon < f(x) \le g(x) \le h(x) < L + \varepsilon$$

We conclude that $\lim_{x \to c} g(x) = L$. ∎

EXAMPLE 8   Assume that we have proved $1 - x^2/6 \le (\sin x)/x \le 1$ for all x near but different from 0. What can we conclude from this?

Solution   Let $f(x) = 1 - x^2/6$, $g(x) = (\sin x)/x$, and $h(x) = 1$. It follows that $\lim_{x \to 0} f(x) = \lim_{x \to 0} h(x) = 1$, and so by Theorem C,

$$\lim_{x \to 0} \frac{\sin x}{x} = 1$$

∎

We conclude with an important remark. *All theorems of this section are valid for right- and left-hand limits.*

PROBLEM SET 2.6 _____

In Problems 1–12, use Theorem A to find each of the limits. Justify each step by appealing to a numbered statement, as in Examples 1–4.

1. $\lim_{x \to 3} (7x - 4)$

2. $\lim_{x \to -1} (2x^3 - 5x)$

3. $\lim_{x \to 2} [(x^2 + 1)(3x - 1)]$

4. $\lim_{x \to 0} [(4x^2 - 3)(7x^3 + 2x)]$

5. $\lim_{x \to 4} \dfrac{2x}{3x^3 - 16}$

6. $\lim_{x \to -2} \dfrac{3x^4 - 8}{x^3 + 24}$

7. $\lim_{x \to 3} \sqrt{3x - 5}$

8. $\lim_{x \to -3} \sqrt{5x^2 + 2x}$

9. $\lim_{t \to -2} (2t^3 + 15)^{13}$

10. $\lim_{w \to -2} \sqrt{-3w^3 + 7w^2}$

11. $\lim_{y \to 2} \left(\dfrac{4y^3 + 8y}{y + 4} \right)^{1/3}$

12. $\lim_{w \to 5} (2w^4 - 9w^3 + 19)^{-1/2}$

In Problems 13–22, find the indicated limit or state that it does not exist. In many cases, you will want to do some algebra before trying to evaluate the limit (see Examples 5–7).

13. $\displaystyle\lim_{x\to 3}\frac{x^4 - x^3 - 2x^2 + 1}{3x^2 - 5x + 7}$

14. $\displaystyle\lim_{x\to -1}\frac{x^{14} - 3x^{11} + 2x^3 - 6}{3x^9 + 2x + 1}$

15. $\displaystyle\lim_{x\to 4}\frac{x^2 + 2x - 24}{x - 4}$
16. $\displaystyle\lim_{x\to -2}\frac{x^2 + 7x + 10}{x + 2}$

17. $\displaystyle\lim_{x\to -1}\frac{x^2 + 7x + 6}{x^2 - 4x - 5}$
18. $\displaystyle\lim_{u\to 2}\frac{u^2 - 2u}{u^2 - 4}$

19. $\displaystyle\lim_{t\to -1}\frac{t^2 + 7t + 7}{t^2 - 4t - 5}$
20. $\displaystyle\lim_{u\to 2}\frac{u^2 - 2u + 1}{u^2 - 4}$

21. $\displaystyle\lim_{y\to 1}\frac{(y - 1)(y^2 + 2y - 3)}{y^2 - 2y + 1}$

22. $\displaystyle\lim_{w\to -2}\frac{(w + 2)(w^2 - w - 6)}{w^2 + 4w + 4}$

In Problems 23–28, find the limits if $\lim\limits_{x\to a} f(x) = 3$ and $\lim\limits_{x\to a} g(x) = -1$ (see Example 4).

23. $\displaystyle\lim_{x\to a}\sqrt{f^2(x) + g^2(x)}$
24. $\displaystyle\lim_{x\to a}\frac{2f(x) - 3g(x)}{f(x) + g(x)}$

25. $\displaystyle\lim_{x\to a}\sqrt[3]{g(x)}[f(x) + 3]$
26. $\displaystyle\lim_{x\to a}[f(x) - 3]^4$

27. $\displaystyle\lim_{t\to u}[f(t) + (t - a)g(t)]$
28. $\displaystyle\lim_{u\to a}[f(u) + 3g(u)]^3$

In Problems 29–32, find $\lim\limits_{x\to 2}[f(x) - f(2)]/(x - 2)$ for each given function f.

29. $f(x) = 5x^2$
30. $f(x) = 3x^2 - 5$

31. $f(x) = \dfrac{1}{x}$
32. $f(x) = \dfrac{3}{x^2}$

33. Prove Statement 6 of Theorem A. *Hint·*

$$|f(x)g(x) - LM| = |f(x)g(x) - Lg(x) + Lg(x) - LM|$$
$$= |g(x)[f(x) - L] + L[g(x) - M]|$$
$$\leq |g(x)||f(x) - L| + |L||g(x) - M|$$

Now show that if $\lim\limits_{x\to c} g(x) = M$, there is a number δ_1 such that

$$0 < |x - c| < \delta_1 \quad\Rightarrow\quad |g(x)| < |M| + 1$$

34. Prove Statement 7 of Theorem A by first giving an ε, δ proof that $\lim\limits_{x\to c}[1/g(x)] = 1\Big/\Big[\lim\limits_{x\to c} g(x)\Big]$ and then applying Statement 6.

35. Prove that $\lim\limits_{x\to c} f(x) = L \;\Leftrightarrow\; \lim\limits_{x\to c}[f(x) - L] = 0.$

36. Prove that $\lim\limits_{x\to c} f(x) = 0 \;\Leftrightarrow\; \lim\limits_{x\to c}|f(x)| = 0.$

37. Prove that if $\lim\limits_{x\to c} f(x) = L > 0$, then there is an interval $(c - \delta, c + \delta)$ such that $f(x) > 0$ for all x in $(c - \delta, c + \delta)$, $x \neq c$.

38. Prove that $\lim\limits_{x\to c}|x| = |c|.$

39. Find examples to show that:

(a) $\lim\limits_{x\to c}[f(x) + g(x)]$ exists does not imply that either $\lim\limits_{x\to c} f(x)$ or $\lim\limits_{x\to c} g(x)$ exists;

(b) $\lim\limits_{x\to c}[f(x) \cdot g(x)]$ exists does not imply that either $\lim\limits_{x\to c} f(x)$ or $\lim\limits_{x\to c} g(x)$ exists.

40. Prove that if $\lim\limits_{x\to c}[f(x) + g(x)]$ and $\lim\limits_{x\to c} g(x)$ both exist, then $\lim\limits_{x\to c} f(x)$ must exist.

In Problems 41–48, find each of the right-hand or left-hand limits or state that it does not exist.

41. $\displaystyle\lim_{x\to 1^+}\frac{\sqrt{x - 1}}{2x + 1}$
42. $\displaystyle\lim_{x\to 3^-}\sqrt{9 - x^2}$

43. $\displaystyle\lim_{x\to 3^+}\frac{x - 3}{\sqrt{x^2 - 9}}$
44. $\displaystyle\lim_{x\to 1^-}\frac{\sqrt{1 + x}}{4 + 4x}$

45. $\displaystyle\lim_{x\to 2^+}\frac{(x^2 + 1)[\![x]\!]}{(3x - 1)^2}$
46. $\displaystyle\lim_{x\to 3^-}(x - [\![x]\!])$

47. $\displaystyle\lim_{x\to 0^-}\frac{x}{|x|}$
48. $\displaystyle\lim_{x\to 3^+}[\![x^2 + 2x]\!]$

49. Suppose $f(x)g(x) = 1$ for all x and $\lim\limits_{x\to a} g(x) = 0$. Prove that $\lim\limits_{x\to a} f(x)$ does not exist.

50. Let R be the rectangle joining the midpoints of the sides of the quadrilateral Q having vertices $(\pm x, 0)$ and $(0, \pm 1)$. Calculate

$$\lim_{x\to 0^+}\frac{\text{perimeter of } R}{\text{perimeter of } Q}$$

51. Let $y = \sqrt{x}$ and consider the points M, N, O, and P with coordinates $(1, 0)$, $(0, 1)$, $(0, 0)$, and (x, y), respectively. Calculate:

(a) $\displaystyle\lim_{x\to 0^+}\frac{\text{perimeter of } \triangle NOP}{\text{perimeter of } \triangle MOP}$;

(b) $\displaystyle\lim_{x\to 0^+}\frac{\text{area of } \triangle NOP}{\text{area of } \triangle MOP}.$

In ordinary language, we use the word *continuous* to describe a process that goes on without abrupt changes. It is this notion as it pertains to functions that we now want to make precise. Consider the three graphs shown in Figure 1. Only the third graph exhibits continuity at c. Here is the formal definition.

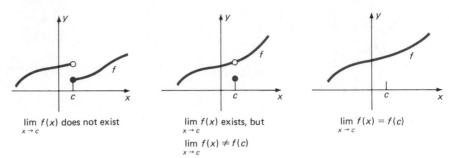

$\lim\limits_{x \to c} f(x)$ does not exist

$\lim\limits_{x \to c} f(x)$ exists, but

$\lim\limits_{x \to c} f(x) \neq f(c)$

$\lim\limits_{x \to c} f(x) = f(c)$

FIGURE 1

Definition

(Continuity at a point). Let f be defined on an open interval containing c. We say that f is **continuous** at c if

$$\lim_{x \to c} f(x) = f(c)$$

We mean by this definition to require three things: (1) $\lim\limits_{x \to c} f(x)$ exists, (2) $f(c)$ exists (that is, c is in the domain of f), and (3) $\lim\limits_{x \to c} f(x) = f(c)$. If any of these three fails, then f is **discontinuous** at c. Thus, the functions represented by the first and second graphs above are discontinuous at c. They are, however, continuous at other points of their domains.

EXAMPLE 1 Let $f(x) = \dfrac{x^2 - 4}{x - 2}$, $x \neq 2$. How should f be defined at $x = 2$ in order to make it continuous there?

Solution

$$\lim_{x \to 2} \frac{x^2 - 4}{x - 2} = \lim_{x \to 2} \frac{(x - 2)(x + 2)}{x - 2} = \lim_{x \to 2}(x + 2) = 4$$

Therefore, we define $f(2) = 4$. The graph of the resulting function is shown in Figure 2. In fact, we see that $f(x) = x + 2$ for all x. ■

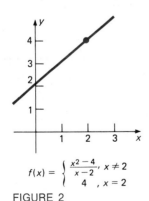

$f(x) = \begin{cases} \dfrac{x^2 - 4}{x - 2}, & x \neq 2 \\ 4, & x = 2 \end{cases}$

FIGURE 2

CONTINUITY OF FAMILIAR FUNCTIONS Most functions that we will meet in this book are either continuous everywhere or everywhere except at a few points. In particular, Theorem 2.6B implies the following result.

Theorem A

A polynomial function is continuous at every real number c. A rational function is continuous at every real number c in its domain, that is, except where its denominator is zero.

Recall the function $f(x) = |x|$; its graph is shown in Figure 3. For $x < 0$, $f(x) = -x$, a polynomial; for $x > 0$, $f(x) = x$, another polynomial. Thus, $|x|$ is continuous at all numbers different from 0 by Theorem A. But

$$\lim_{x \to 0} |x| = 0 = |0|$$

(see Problem 23 of Section 2.5). Therefore, $|x|$ is also continuous at 0; it is continuous everywhere.

By the Main Limit Theorem (Theorem 2.6A)

$$\lim_{x \to c} \sqrt[n]{x} = \sqrt[n]{\lim_{x \to c} x} = \sqrt[n]{c}$$

provided $c > 0$ when n is even. This means that $f(x) = \sqrt[n]{x}$ is continuous at each point where it makes sense to talk about continuity. In particular, $f(x) = \sqrt{x}$ is continuous at each real number $c > 0$ (Figure 4). We summarize.

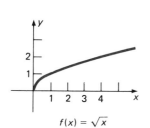

$f(x) = |x|$

FIGURE 3

$f(x) = \sqrt{x}$

FIGURE 4

Theorem B

The absolute value function is continuous at every real number c. If n is odd, the nth root function is continuous at every real number c; if n is even, it is continuous at every positive real number c.

CONTINUITY UNDER FUNCTIONAL OPERATIONS Do the standard functional operations preserve continuity? Yes, according to Theorem C. In it, f and g are functions, k is a constant, and n is a positive integer.

Theorem C

If f and g are continuous at c, then so are kf, $f + g$, $f - g$, $f \cdot g$, f/g (provided $g(c) \neq 0$), f^n, and $\sqrt[n]{f}$ (provided $f(c) > 0$ if n is even).

Proof All of these results are easy consequences of the corresponding facts for limits from Theorem 2.6A. For example, that theorem, combined with the fact that f and g are continuous at c, gives

$$\lim_{x \to c} f(x)g(x) = \lim_{x \to c} f(x) \cdot \lim_{x \to c} g(x) = f(c)g(c)$$

This is precisely what it means to say that $f \cdot g$ is continuous at c.

EXAMPLE 2 At what numbers is $F(x) = (3|x| - x^2)/(\sqrt{x} + \sqrt[3]{x})$ continuous?

Solution We need not even consider nonpositive numbers, since F is not defined at such numbers. For any positive number, the functions $\sqrt{x}$, $\sqrt[3]{x}$, $|x|$, and x^2 are all continuous (Theorems A and B). It follows from Theorem C that $3|x|$, $3|x| - x^2$, $\sqrt{x} + \sqrt[3]{x}$, and—finally

$$\frac{(3|x| - x^2)}{(\sqrt{x} + \sqrt[3]{x})}$$

are continuous at each positive number. ∎

There is another functional operation that will be very important in later work, namely, composition. It, too preserves continuity.

Theorem D

(Composite Limit Theorem). If $\lim\limits_{x \to c} g(x) = L$ and if f is continuous at L, then

$$\lim_{x \to c} f(g(x)) = f\left(\lim_{x \to c} g(x)\right) = f(L)$$

In particular, if g is continuous at c and f is continuous at $g(c)$, then the composite $f \circ g$ is continuous at c.

We defer the proof to the end of the section.

EXAMPLE 3 Show that $h(x) = |x^2 - 3x + 6|$ is continuous at each real number.

Solution Let $f(x) = |x|$ and $g(x) = x^2 - 3x + 6$. Both are continuous at each real number, and so their composite

$$h(x) = f(g(x)) = |x^2 - 3x + 6|$$

is also. ∎

EXAMPLE 4 It will be shown later that $f(x) = \sin x$ is continuous at each real number. Conclude that

$$h(x) = \sin\left(\frac{x^4 - 3x + 1}{x^2 - x - 6}\right)$$

is continuous except at 3 and -2.

Solution $x^2 - x - 6 = (x - 3)(x + 2)$. Thus, the rational function

$$g(x) = \frac{x^4 - 3x + 1}{x^2 - x - 6}$$

is continuous except at 3 and -2 (Theorem A). From Theorem D, we conclude that, since $h(x) = f(g(x))$, h is also continuous except at 3 and -2. ∎

CONTINUITY ON AN INTERVAL So far, we have been discussing continuity at a point. We wish to discuss continuity on an interval. Continuity on an interval ought to mean continuity at each point of that interval. That is exactly what it does mean for an open interval.

When we consider a closed interval $[a, b]$, we face a problem. It might be that f is not even defined to the left of a (for example, $f(x) = \sqrt{x}$ has this problem at $a = 0$), so that strictly speaking $\lim_{x \to a} f(x)$ does not exist. We choose to get around this problem by calling f continuous on $[a, b]$ if it is continuous at each point of (a, b) and if $\lim_{x \to a^+} f(x) = f(a)$ and $\lim_{x \to b^-} f(x) = f(b)$ (called, respectively, *right continuity* at a and *left continuity* at b). We summarize in a formal definition.

Definition

We say f is **continuous on an open interval** if it is continuous at each point of that interval. It is **continuous on the closed interval** $[a, b]$ if it is continuous on (a, b), right continuous at a, and left continuous at b.

For example, it is correct to say that $f(x) = 1/x$ is continuous on $(0, 1)$ and that $g(x) = \sqrt{x}$ is continuous on $[0, 1]$.

EXAMPLE 5 Using the definition above, describe the continuity properties of the function whose graph is sketched in Figure 5.

FIGURE 5

Solution The function is continuous on the open intervals $(-\infty, 0)$, $(0, 3)$, and $(5, \infty)$ and also on the closed interval $[3, 5]$. ∎

For f to be continuous on $[a, b]$ means that when x_1 and x_2 are near each other and both are in $[a, b]$, then $f(x_1)$ and $f(x_2)$ are near each other. There are no jumps or abrupt changes, and so we may "draw" the graph of f on $[a, b]$ without lifting our pencil from the paper. This is related to the fact that a continuous function must assume every value between any two of its values, a result we now state precisely.

Theorem E

(Intermediate Value Theorem). If f is continuous on $[a, b]$ and if W is a number between $f(a)$ and $f(b)$, then there is a number c between a and b such that $f(c) = W$.

W

a   b   x

Not continuous;
Intermediate Value
property fails.

FIGURE 6

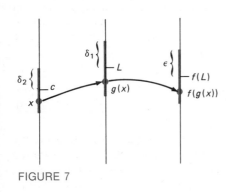

FIGURE 7

It is clear from the graph in Figure 6 that continuity is needed for this theorem. It also seems clear that continuity is enough, and yet a formal proof of this turns out to be difficult. We leave the proof to more advanced works.

PROOF OF THEOREM D (OPTIONAL)

Proof   Let $\varepsilon > 0$ be given. Since f is continuous at L, there is a corresponding $\delta_1 > 0$ such that

$$|t - L| < \delta_1 \quad \Rightarrow \quad |f(t) - f(L)| < \varepsilon$$

and so (see Figure 7)

$$|g(x) - L| < \delta_1 \quad \Rightarrow \quad |f(g(x)) - f(L)| < \varepsilon$$

But because $\lim_{x \to c} g(x) = L$, for a given $\delta_1 > 0$ there is a corresponding $\delta_2 > 0$ such that

$$0 < |x - c| < \delta_2 \quad \Rightarrow \quad |g(x) - L| < \delta_1$$

When we put these two facts together, we have

$$0 < |x - c| < \delta_2 \quad \Rightarrow \quad |f(g(x)) - f(L)| < \varepsilon$$

This shows that

$$\lim_{x \to c} f(g(x)) = f(L)$$

The second statement in Theorem D follows from the observation that if g is continuous at c, then $L = g(c)$.   ∎

PROBLEM SET   2.7

In Problems 1–14, state whether the indicated function is or is not continuous at 2; if not, tell why not.

1. $f(x) = 4x^2 - 2x + 12$ **2.** $f(x) = \dfrac{8}{x - 2}$

3. $g(x) = \dfrac{3x^2}{x - 2}$ **4.** $g(x) = \sqrt{x - 1}$

5. $h(x) = \sqrt{x - 3}$ **6.** $h(x) = |3 - 5x^2|$

7. $f(t) = [\![t]\!]$ **8.** $f(t) = [\![t - \tfrac{1}{2}]\!]$

9. $g(t) = \dfrac{t^3 - 8}{t - 2}$ **10.** $g(t) = \dfrac{4t - 8}{t - 2}$

11. $h(t) = \begin{cases} \dfrac{t^3 - 8}{t - 2} & \text{if } t \neq 2 \\ 12 & \text{if } t = 2 \end{cases}$

12. $h(t) = \begin{cases} \dfrac{4t - 8}{t - 2} & \text{if } t \neq 2 \\ 2 & \text{if } t = 2 \end{cases}$

13. $f(x) = \begin{cases} x + 3 & \text{if } x < 2 \\ x^2 + 1 & \text{if } x \geq 2 \end{cases}$

14. $f(x) = \begin{cases} -3x + 4 & \text{if } x \leq 2 \\ -2 & \text{if } x > 2 \end{cases}$

In Problems 15–20, the given function is not defined at a certain point. How should it be defined there in order to make it continuous at this point? (See Example 1.)

15. $f(x) = \dfrac{x^2 - 9}{x - 3}$ **16.** $g(x) = \dfrac{9x^2 - 4}{3x + 2}$

17. $f(t) = \dfrac{t-1}{\sqrt{t}-1}$ **18.** $g(t) = \dfrac{\sin t}{t}$

19. $\phi(x) = \dfrac{x^4 + 2x^2 - 3}{x+1}$ **20.** $F(x) = \sin\left(\dfrac{x^2 - 1}{x+1}\right)$

In Problems 21–32, at what points, if any, are the functions discontinuous?

21. $f(x) = \dfrac{2x+3}{x^2 - x - 6}$ **22.** $g(x) = \dfrac{x}{2x^2 - x - 1}$

23. $f(t) = |t^2 - 2t + 5|$ **24.** $g(t) = \dfrac{1}{\sqrt{t-1}}$

25. $f(u) = \dfrac{2u+7}{\sqrt{u+5}}$ **26.** $g(u) = \dfrac{u^2 + |u-1|}{\sqrt[3]{u+1}}$

27. $F(x) = \dfrac{1}{\sqrt{4+x^2}}$ **28.** $G(x) = \dfrac{1}{\sqrt{4-x^2}}$

29. $f(x) = \begin{cases} x & \text{if } x < 0 \\ x^2 & \text{if } 0 \le x \le 1 \\ 2 - x & \text{if } x > 1 \end{cases}$

30. $g(x) = \begin{cases} x^2 & \text{if } x < 0 \\ -x & \text{if } 0 \le x \le 1 \\ x & \text{if } x > 1 \end{cases}$

31. $f(t) = [\![t]\!]$

32. $g(t) = [\![t + \frac{1}{2}]\!]$

33. Sketch the graph of a function f that satisfies all the following conditions.

(a) Its domain is $[-2, 2]$.
(b) $f(-2) = f(-1) = f(1) = f(2) = 1$.
(c) It is discontinuous at -1 and 1.
(d) It is right continuous at -1 and left continuous at 1.

34. Let
$$f(x) = \begin{cases} x & \text{if } x \text{ is rational} \\ -x & \text{if } x \text{ is irrational} \end{cases}$$

Sketch the graph of this function as best you can and decide where it is continuous.

35. Use the Intermediate Value Theorem to prove that $x^3 + 3x - 2 = 0$ has a real solution between 0 and 1.

36. Show that the equation $x^5 + 4x^3 - 7x + 14 = 0$ has at least one real solution. *Hint*: Intermediate Value Theorem.

37. Prove that f is continuous at c if and only if $\lim_{t \to 0} f(t+c) = f(c)$.

38. Prove that if f is continuous at c and $f(c) > 0$, then there is an interval $(c - \delta, c + \delta)$ such that $f(x) > 0$ on this interval.

39. Prove that if f is continuous on $[0, 1]$ and satisfies $0 \le f(x) \le 1$ there, then f has a *fixed point*—that is, there is a number c in $[0, 1]$ such that $f(c) = c$. *Hint*: Apply the Intermediate Value Theorem to $g(x) = x - f(x)$.

40. Find the values of a and b so that the following function is continuous everywhere.

$$f(x) = \begin{cases} x + 1 & \text{if } x < 1 \\ ax + b & \text{if } 1 \le x < 2 \\ 3x & \text{if } x \ge 2 \end{cases}$$

41. A stretched elastic string covers the interval $[0, 1]$. The ends are let go and the string contracts so that it covers the interval $[a, b]$, $a \ge 0$, $b \le 1$. Prove that this results in one point of the string (actually exactly one point) being where it was originally. See Problem 39.

42. Let f be continuous on $[0, 1]$ with $f(0) = f(1) = 0$. Prove that the graph of f has a chord (line segment with both ends on the graph) of length W, where W is any number between 0 and 1.

43. Use the Intermediate Value Theorem to show that there are always two points on a circular wire ring with the same temperature. *Hint*: Put the center at the origin and let θ be the angle that a diameter makes with the x-axis. Define $f(\theta)$ appropriately.

44. Starting at 4 A.M., a monk slowly climbed to the top of a mountain, arriving at noon. The next day, he returned along the same path, starting at 5 A.M. and getting to the bottom at 11 A.M. Show that at some point along the path, his watch showed the same time on both days.

45. Prove that any bounded region can be circumscribed by a square. Equivalently, let D be a bounded but otherwise arbitrary region in the first quadrant. Given an angle θ, $0 \le \theta \le \pi/2$, D can be circumscribed by a rectangle whose base makes angle θ with the x-axis as shown in Figure 8. Prove that at some angle, this rectangle is a square.

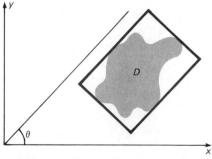

FIGURE 8

46. Show that if we know the values of a continuous function on the rational numbers, we know its values everywhere. Equivalently, prove that if f and g are continuous functions and $f(x) = g(x)$ for all x in $\mathbb{Q}$ (the rationals), then $f(x) = g(x)$ for all x in $\mathbb{R}$.

47. Let $f(x + y) = f(x) + f(y)$ for all x and y in $\mathbb{R}$ and suppose that f is continuous at $x = 0$.

(a) Prove that f is continuous everywhere.

(b) Prove that there is a constant m such that $f(t) = mt$ for all t in $\mathbb{R}$ (see Problem 43 of Section 2.1).

48. (Famous Problem). Let $f(x) = 0$ if x is irrational and let $f(x) = 1/q$ if x is the rational number p/q in reduced form ($q > 0$).

(a) Sketch (as best you can) the graph of f on $(0, 1)$.

(b) Show that f is continuous at each irrational number in $(0, 1)$ but is discontinuous at each rational number in $(0, 1)$.

2.8 Chapter Review

TRUE-FALSE QUIZ

Respond with true or false to each of the following assertions. Be prepared to justify your answer.

1. The equation $xy + x^2 = 3y$ determines a function with formula of the form $y = f(x)$.
2. The equation $xy^2 + x^2 = 3x$ determines a function with formula of the form $y = f(x)$.
3. The natural domain of

$$f(x) = \sqrt{\frac{x}{4 - x}}$$

   is the interval $[0, 4)$.
4. The range of $f(x) = x^2 - 6$ is the interval $[-6, \infty)$.
5. The sum of two even functions (with the same domain) is an even function.
6. The product of two odd functions (with the same domain) is an odd function.
7. The function $f(x) = (2x^3 + x)/(x^2 + 1)$ is odd.
8. If the range of a function consists of just one number, then its domain also consists of just one number.
9. If the domain of a function contains at least two numbers, then the range also contains at least two numbers.
10. If $g(x) = \llbracket x/2 \rrbracket$, then $g(-1.8) = -1$.
11. If $f(x) = x^2$ and $g(x) = x^3$, then $f \circ g = g \circ f$.
12. If f and g have the same domain, then f/g also has that domain.
13. If the graph of $y = f(x)$ has an x-intercept at $x = a$, then the graph of $y = f(x + h)$ has an x-intercept at $x = a - h$.
14. The cotangent is an odd function.
15. The natural domain of the tangent function is the set of all real numbers.
16. If $\cos s = \cos t$, then $s = t$.
17. If $\lim\limits_{x \to c} f(x) = L$, then $f(c) = L$.
18. If $f(c)$ is not defined, then $\lim\limits_{x \to c} f(x)$ does not exist.
19. The coordinates of the hole in the graph of $y = \dfrac{x^2 - 25}{x - 5}$ are $(5, 10)$.
20. If $p(x)$ is a polynomial, then $\lim\limits_{x \to c} p(x) = p(c)$.
21. If $\lim\limits_{x \to c^+} f(x) = \lim\limits_{x \to c^-} f(x)$, then f is continuous at $x = c$.
22. The function $f(x) = \llbracket x/2 \rrbracket$ is continuous at $x = 2.3$.
23. If $\lim\limits_{x \to 2} f(x) = f(2) > 0$, then $f(x) < 1.001 f(2)$ for all x in some interval containing 2.

24. If $\lim_{x \to c} [f(x) + g(x)]$ exists, then $\lim_{x \to c} f(x)$ and $\lim_{x \to c} g(x)$ both exist.

25. If $0 \le f(x) \le 3x^2 + 2x^4$ for all x, then $\lim_{x \to 0} f(x) = 0$.

26. If $\lim_{x \to a} f(x) = L$ and $\lim_{x \to a} f(x) = M$, then $L = M$.

27. If $f(x) \ne g(x)$ for all x, then $\lim_{x \to c} f(x) \ne \lim_{x \to c} g(x)$.

28. If $f(x) < 10$ for all x and $\lim_{x \to 2} f(x)$ exists, then $\lim_{x \to 2} f(x) < 10$.

29. If $\lim_{x \to a} f(x) = b$, then $\lim_{x \to a} |f(x)| = |b|$.

30. If f is continuous and positive on $[a, b]$, then $1/f$ must assume every value between $1/f(a)$ and $1/f(b)$.

SAMPLE TEST PROBLEMS

1. For $f(x) = 1/(x + 1) - 1/x$, find each value (if possible).

(a) $f(1)$ (b) $f(-\frac{1}{2})$ (c) $f(-1)$

(d) $f(t - 1)$ (e) $f\left(\dfrac{1}{t}\right)$

2. For $g(x) = (x + 1)/x$, find and simplify.

(a) $g(2)$ (b) $g(\frac{1}{2})$

(c) $g(\frac{1}{10})$ (d) $\dfrac{g(2 + h) - g(2)}{h}$

3. Describe the natural domains of each function.

(a) $f(x) = \dfrac{x}{x^2 - 1}$ (b) $g(x) = \sqrt{4 - x^2}$

(c) $h(x) = \dfrac{\sqrt{1 + x^2}}{|2x + 3|}$

4. Which of the following functions are odd? Even? Neither even nor odd?

(a) $f(x) = \dfrac{3x}{x^2 + 1}$ (b) $g(x) = |\sin x| + \cos x$

(c) $h(x) = x^3 + \sin x$ (d) $k(x) = \dfrac{x^2 + 1}{|x| + x^4}$

5. Sketch the graphs of each of the following functions.

(a) $f(x) = x^2 - 1$ (b) $g(x) = \dfrac{x}{x^2 + 1}$

(c) $h(x) = \begin{cases} x^2 & \text{if } 0 \le x \le 2 \\ 6 - x & \text{if } x > 2 \end{cases}$

6. Suppose that f is an even function satisfying $f(x) = -1 + \sqrt{x}$ for $x \ge 0$. Sketch the graph of f for $-4 \le x \le 4$.

7. An open box is made by cutting squares of side x inches from the four corners of a sheet of cardboard 24 inches by 32 inches and then turning up the sides. Express the volume $V(x)$ in terms of x. What is the domain for this function?

8. Let $f(x) = x - 1/x$ and $g(x) = x^2 + 1$. Find each value.

(a) $(f + g)(2)$ (b) $(f \cdot g)(2)$
(c) $(f \circ g)(2)$ (d) $(g \circ f)(2)$
(e) $f^3(-1)$ (f) $f^2(2) + g^2(2)$

9. Sketch the graphs of each of the following, making use of translations.

(a) $y = \frac{1}{4}x^2$ (b) $y = \frac{1}{4}(x + 2)^2$
(c) $y = -1 + \frac{1}{4}(x + 2)^2$

10. Let $f(x) = \sqrt{16 - x}$ and $g(x) = x^4$. What is the domain of each of the following?

(a) f (b) $f \circ g$ (c) $g \circ f$

11. Write $F(x) = \sqrt{1 + \sin^2 x}$ as the composite of four functions, $f \circ g \circ h \circ k$.

12. Calculate each of the following without using a calculator or tables.

(a) $\sin(570°)$ (b) $\cos\left(\dfrac{9\pi}{2}\right)$

(c) $\sin^2(5) + \cos^2(5)$ (d) $\cos\left(\dfrac{-13\pi}{6}\right)$

13. If $\sin t = 0.8$ and $\cos t < 0$, find each value.

(a) $\sin(-t)$ (b) $\cos t$
(c) $\sin 2t$ (d) $\tan t$

(e) $\cos\left(\dfrac{\pi}{2} - t\right)$ (f) $\sin(\pi + t)$

14. Write $\sin 3t$ in terms of $\sin t$. *Hint*: $3t = 2t + t$.

15. A fly sits on the rim of a wheel spinning at the rate of 20 revolutions per minute. If the radius of the wheel is 9 inches, how far does the fly travel in 1 second?

In Problems 16–27, find the indicated limit or state that it doesn't exist.

16. $\lim\limits_{u \to 1} \dfrac{u^2 - 1}{u + 1}$

17. $\lim\limits_{u \to 1} \dfrac{u^2 - 1}{u - 1}$

18. $\lim\limits_{u \to 1} \dfrac{u + 1}{u^2 - 1}$

19. $\lim\limits_{x \to 2} \dfrac{1 - 2/x}{x^2 - 4}$

20. $\lim\limits_{z \to 2} \dfrac{z^2 - 4}{z^2 + z - 6}$

21. $\lim\limits_{x \to 0} \dfrac{\tan x}{\sin 2x}$

22. $\lim\limits_{y \to 1} \dfrac{y^3 - 1}{y^2 - 1}$

23. $\lim\limits_{x \to 4} \dfrac{x - 4}{\sqrt{x} - 2}$

24. $\lim\limits_{x \to 0} \dfrac{\cos x}{x}$

25. $\lim\limits_{x \to 0^-} \dfrac{|x|}{x}$

26. $\lim\limits_{x \to 1/2^+} [\![4x]\!]$

27. $\lim\limits_{t \to 2^-} ([\![t]\!] - t)$

28. Let $f(x) = \begin{cases} x^3 & \text{if } x < -1 \\ x & \text{if } -1 < x < 1 \\ 1 - x & \text{if } x \geq 1 \end{cases}$

Find each value.

(a) $f(1)$

(b) $\lim\limits_{x \to 1^+} f(x)$

(c) $\lim\limits_{x \to 1^-} f(x)$

(d) $\lim\limits_{x \to -1} f(x)$

29. Refer to f of Problem 28. (a) What are the values of x at which f is discontinuous? (b) How should f be defined at $x = -1$ to make it continuous there?

30. Give the ε, δ definition in each case.

(a) $\lim\limits_{u \to a} g(u) = M$

(b) $\lim\limits_{x \to a^-} f(x) = L$

31. If $\lim\limits_{x \to 3} f(x) = 3$ and $\lim\limits_{x \to 3} g(x) = -2$ and if g is continuous at $x = 3$, find each value.

(a) $\lim\limits_{x \to 3} [2f(x) - 4g(x)]$

(b) $\lim\limits_{x \to 3} g(x) \dfrac{x^2 - 9}{x - 3}$

(c) $g(3)$

(d) $\lim\limits_{x \to 3} g(f(x))$

(e) $\lim\limits_{x \to 3} \sqrt{f^2(x) - 8g(x)}$

(f) $\lim\limits_{x \to 3} \dfrac{|g(x) - g(3)|}{f(x)}$

32. Sketch the graph of a function f that satisfies all of the following conditions.

(a) Its domain is $[0, 6]$.
(b) $f(0) = f(2) = f(4) = f(6) = 2$.
(c) f is continuous except at $x = 2$.
(d) $\lim\limits_{x \to 2^-} f(x) = 1$ and $\lim\limits_{x \to 5^+} f(x) = 3$.

33. Let $f(x) = \begin{cases} -1 & \text{if } x \leq 0 \\ ax + b & \text{if } 0 < x < 1 \\ 1 & \text{if } x \geq 1 \end{cases}$

Determine a and b so that f is continuous everywhere.

34. Use the Intermediate Value Theorem to prove that the equation $x^5 - 4x^3 - 3x + 1 = 0$ has at least one solution between $x = 2$ and $x = 3$.

3 The Derivative

Leibniz's discoveries lay in the direction in which all modern progress in science lies, in establishing order, symmetry, and harmony, i.e., comprehensiveness and perspicuity—rather than in dealing with single problems, in the solution of which followers soon attained greater dexterity than himself.
J. T. Merz

Gottfried Wilhelm Leibniz
1646–1716

Gottfried Wilhelm Leibniz was a universal genius, an expert in law, religion, philosophy, literature, politics, geology, history, and mathematics. Born in Leipzig, Germany, he enrolled at the University of Leipzig and obtained a doctorate from the University of Altdorf. Like Descartes, whose work he studied, Leibniz searched for a universal method by which he could obtain knowledge and understand its essential unity. One of his passions was to reconcile the Catholic and Protestant faiths.

With Isaac Newton, he shares credit for the discovery of calculus. The question of priority caused endless controversy between the followers of these two great men, one English, the other German. History has judged that Newton had the main ideas first (1665–66) but that Leibniz discovered them independently during the years 1673–76. Despite his greatness, Leibniz did not receive the honors showered on Newton. He died a lonely man with one mourner at his funeral, his secretary.

Leibniz was perhaps the greatest inventor of mathematical symbols. To him we owe the names *differential calculus* and *integral calculus*, as well as the standard symbols dy/dx and $\int$ for the derivative and the integral. The term *function* and the consistent use of $=$ for equality are other contributions. Partly because of the superiority of his symbolism, calculus developed much more rapidly on the European continent than it did in England.

3.1
Two Problems with One Theme

Our first problem is very old; it dates back to the great Greek scientist Archimedes (287–212 B.C.). We refer to the problem of the *tangent line*.

Our second problem is newer. It grew out of attempts by Kepler (1571–1630), Galileo (1564–1642), Newton (1642–1727), and others to describe the speed of a moving body. It is the problem of *instantaneous velocity*.

The two problems, one geometric and the other mechanical, appear to be quite unrelated. In this case, appearances are deceptive. The two problems are identical twins.

THE TANGENT LINE   Euclid's notion of a tangent as a line touching a curve at one point is all right for circles (Figure 1) but completely unsatisfactory for most other curves (Figure 2). The idea of a tangent to a curve at P as that line which best approximates the curve near P is better, but is still too vague for mathematical precision. The concept of limit provides a way of getting the best description.

Let P be a fixed point on a curve and let Q be a nearby movable point on that curve. Consider the line through P and Q, called a **secant line**. The **tangent line** at P is the limiting position (if it exists) of the secant line as Q moves toward P along the curve (Figure 3).

Suppose that the curve is the graph of the equation $y = f(x)$. Then P has coordinates $(c, f(c))$, a nearby point Q has coordinates $(c + h, f(c + h))$, and the secant line through P and Q has slope m_{sec} given by (Figure 4)

$$m_{\text{sec}} = \frac{f(c + h) - f(c)}{h}$$

Tangent line at P

FIGURE 1

Tangent line at P

FIGURE 2

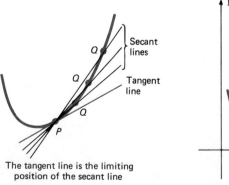

The tangent line is the limiting position of the secant line

FIGURE 3

FIGURE 4

$$m_{\text{tan}} = \lim_{h \to 0} m_{\text{sec}}$$

CHAPTER 3/The Derivative

Consequently, the tangent line—if not vertical—is that line through P with slope m_{tan} satisfying

$$m_{\text{tan}} = \lim_{h \to 0} m_{\text{sec}} = \lim_{h \to 0} \frac{f(c + h) - f(c)}{h}$$

EXAMPLE 1   Find the slope of the tangent line to the curve $y = f(x) = x^2$ at the point $(2, 4)$.

Solution   The line whose slope we are seeking is shown in Figure 5. Clearly it has a large positive slope.

$$
\begin{aligned}
m_{\text{tan}} &= \lim_{h \to 0} \frac{f(2 + h) - f(2)}{h} \\
&= \lim_{h \to 0} \frac{(2 + h)^2 - 2^2}{h} \\
&= \lim_{h \to 0} \frac{4 + 4h + h^2 - 4}{h} \\
&= \lim_{h \to 0} \frac{h(4 + h)}{h} \\
&= 4
\end{aligned}
$$
∎

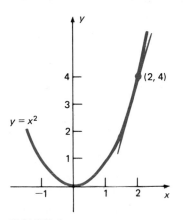

FIGURE 5

EXAMPLE 2   Find the slope of the tangent line to the curve $y = f(x) = -x^2 + 2x + 2$ at the points with x-coordinates $-1, \frac{1}{2}, 2$, and 3.

Solution   Rather than make four separate calculations, it seems wise to calculate the slope at the point with x-coordinate c and then obtain the four desired answers by substitution.

$$
\begin{aligned}
m_{\text{tan}} &= \lim_{h \to 0} \frac{f(c + h) - f(c)}{h} \\
&= \lim_{h \to 0} \frac{-(c + h)^2 + 2(c + h) + 2 - (-c^2 + 2c + 2)}{h} \\
&= \lim_{h \to 0} \frac{-c^2 - 2ch - h^2 + 2c + 2h + 2 + c^2 - 2c - 2}{h} \\
&= \lim_{h \to 0} \frac{h(-2c - h + 2)}{h} \\
&= -2c + 2
\end{aligned}
$$

The four desired slopes (obtained by letting $c = -1, \frac{1}{2}, 2, 3$) are $4, 1, -2$, and -4. These answers do appear to be consistent with the graph in Figure 6. ∎

FIGURE 6

FIGURE 7

EXAMPLE 3 Find the equation of the tangent line to the curve $y = 1/(2x)$ at $(\frac{1}{2}, 1)$ (see Figure 7).

Solution Let $f(x) = 1/(2x)$.

$$m_{\tan} = \lim_{h \to 0} \frac{f(\frac{1}{2} + h) - f(\frac{1}{2})}{h}$$

$$= \lim_{h \to 0} \frac{\dfrac{1}{2(\frac{1}{2} + h)} - \dfrac{1}{2(\frac{1}{2})}}{h}$$

$$= \lim_{h \to 0} \frac{\dfrac{1}{1 + 2h} - \dfrac{1}{1}}{h}$$

$$= \lim_{h \to 0} \frac{1 - (1 + 2h)}{h(1 + 2h)}$$

$$= \lim_{h \to 0} \frac{-2h}{h(1 + 2h)}$$

$$= \lim_{h \to 0} \frac{-2}{1 + 2h} = -2$$

Knowing the slope of the line ($m = -2$) and the point $(\frac{1}{2}, 1)$ on it, we can easily write its equation by using the point-slope form $y - y_0 = m(x - x_0)$. The result is $y - 1 = -2(x - \frac{1}{2})$. ∎

INSTANTANEOUS VELOCITY   If we drive an automobile from one town to another 80 miles away in 2 hours, our average velocity is 40 miles per hour. That is, *average velocity* is the distance from the first position to the second position divided by the elapsed time.

But during our trip the speedometer reading was often different from 40. At the start, it registered 0; at times it rose as high as 57; at the end it fell back to 0 again. Just what does the speedometer measure? Certainly it does not indicate average velocity.

Consider the more precise example of an object P falling in a vacuum. Experiment shows that if it starts from rest, P falls $16t^2$ feet in t seconds. Thus it falls 16 feet in the first second and 64 feet during the first 2 seconds (Figure 8); clearly it falls faster and faster as time goes on.

During the second second (that is, in the time interval from $t = 1$ to $t = 2$), P fell $(64 - 16)$ feet. Its average velocity was

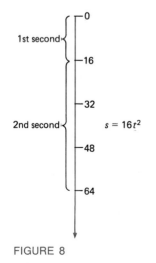

FIGURE 8

$$v_{\text{ave}} = \frac{64 - 16}{2 - 1} = 48 \text{ feet per second}$$

During the time interval from $t = 1$ to $t = 1.5$, it fell $16(1.5)^2 - 16 = 20$ feet. Its average velocity was

$$v_{\text{ave}} = \frac{16(1.5)^2 - 16}{1.5 - 1} = \frac{20}{0.5} = 40 \text{ feet per second}$$

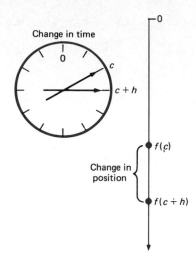

Change in time

Change in position

FIGURE 9

Similarly, on the time intervals $t = 1$ to $t = 1.1$ and $t = 1$ to $t = 1.01$, we calculate the respective average velocities to be

$$v_{ave} = \frac{16(1.1)^2 - 16}{1.1 - 1} = \frac{3.36}{0.1} = 33.6 \text{ feet per second}$$

$$v_{ave} = \frac{16(1.01)^2 - 16}{1.01 - 1} = \frac{0.3216}{0.01} = 32.16 \text{ feet per second}$$

What we have done is to calculate the average velocity over shorter and shorter time intervals, each starting at $t = 1$. The shorter the time interval, the better we should approximate the "true" velocity at the instant $t = 1$. Looking at the numbers 48, 40, 33.6, and 32.16, you might guess 32 to be the instantaneous velocity.

But let us be more precise. Suppose that an object P moves along a coordinate line so that its position at time t is given by $s = f(t)$. At time c the object is at $f(c)$; at the nearby time $c + h$, it is at $f(c + h)$ (see Figure 9). Thus, the average velocity on this interval is

$$v_{ave} = \frac{f(c + h) - f(c)}{h}$$

And now we define the **instantaneous velocity** v at c by

$$\boxed{v = \lim_{h \to 0} v_{ave} = \lim_{h \to 0} \frac{f(c + h) - f(c)}{h}}$$

In the case where $f(t) = 16t^2$, the instantaneous velocity at $t = 1$ is

$$v = \lim_{h \to 0} \frac{f(1 + h) - f(1)}{h}$$

$$= \lim_{h \to 0} \frac{16(1 + h)^2 - 16}{h}$$

$$= \lim_{h \to 0} \frac{16 + 32h + 16h^2 - 16}{h}$$

$$= \lim_{h \to 0} (32 + 16h) = 32$$

This confirms our earlier guess.

Now you can see why we called the *slope of the tangent line* and *instantaneous velocity* identical twins. Look at the two boxed formulas in this section. They give different names to the same concept.

EXAMPLE 4 Find the instantaneous velocity of a falling object, starting from rest, at $t = 3.8$ seconds and at $t = 5.4$ seconds.

Solution We calculate the instantaneous velocity at $t = c$ seconds.

$$v = \lim_{h \to 0} \frac{f(c + h) - f(c)}{h}$$

$$= \lim_{h \to 0} \frac{16(c + h)^2 - 16c^2}{h}$$

$$= \lim_{h \to 0} \frac{16c^2 + 32ch + 16h^2 - 16c^2}{h}$$

$$= \lim_{h \to 0} (32c + 16h) = 32c$$

Thus, the instantaneous velocity at 3.8 seconds is $32(3.8) = 121.6$ feet per second; at 5.4 seconds, it is $32(5.4) = 172.8$ feet per second. ∎

EXAMPLE 5 How long will it take the falling object of Example 4 to reach an instantaneous velocity of 112 feet per second?

Solution We learned in Example 4 that the instantaneous velocity after c seconds is $32c$. Thus we must solve the equation $32c = 112$. The solution is $c = \frac{112}{32} = 3.5$ seconds. ∎

EXAMPLE 6 A particle moves along a coordinate line and s, its directed distance in centimeters from the origin at the end of t seconds, is given by $s = \sqrt{5t + 1}$. Find the instantaneous velocity of the particle at the end of 3 seconds.

Solution

$$v = \lim_{h \to 0} \frac{f(3 + h) - f(3)}{h}$$

$$= \lim_{h \to 0} \frac{\sqrt{5(3 + h) + 1} - \sqrt{5(3) + 1}}{h}$$

$$= \lim_{h \to 0} \frac{\sqrt{16 + 5h} - 4}{h}$$

To evaluate this limit, we rationalize the numerator (by multiplying numerator and denominator by $\sqrt{16 + 5h} + 4$). We obtain

$$v = \lim_{h \to 0} \left(\frac{\sqrt{16 + 5h} - 4}{h} \cdot \frac{\sqrt{16 + 5h} + 4}{\sqrt{16 + 5h} + 4} \right)$$

$$= \lim_{h \to 0} \frac{16 + 5h - 16}{h(\sqrt{16 + 5h} + 4)}$$

$$= \lim_{h \to 0} \frac{5}{\sqrt{16 + 5h} + 4} = \frac{5}{8}$$

We conclude that the instantaneous velocity at the end of 3 seconds is $\frac{5}{8}$ centimeter per second. ∎

VELOCITY OR SPEED

For the time being, we will use the terms *velocity* and *speed* interchangeably. Later, in Section 3.7, we will distinguish between these two words.

RATES OF CHANGE Velocity is only one of many rates of change that will be important in this course; it is the rate of change of distance with respect to time. Other rates of change that will interest us are density for a wire (the rate of change of mass with respect to distance), marginal revenue (the rate of change of revenue with respect to the number of items produced), and current (the rate of change of electrical charge with respect to time). These rates and many more are discussed in the problem set. In each case, we may discuss an *average* rate of change or an *instantaneous* rate of change. The phrase *rate of change* without an adjective will mean instantaneous rate of change.

PROBLEM SET 3.1

In Problems 1–2, a tangent line to a curve is drawn. Estimate its slope (slope = rise/run). Be careful to note the difference in scales on the two axes.

1.

2.

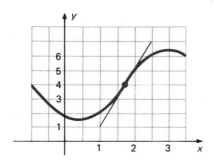

In Problems 3–6, draw the tangent line to the curve through the indicated point and estimate its slope.

3.

4.

5.

6.

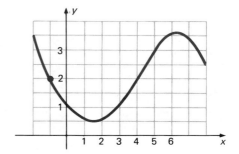

7. Consider $y = 4 - x^2$.

(a) Sketch its graph as carefully as you can.

(b) Draw the tangent line at $(3, -5)$.

(c) Estimate the slope of this tangent line.

C (d) Calculate the slope of the secant line through $(3, -5)$ and $(3.01, 4 - 3.01^2)$.

(e) Find by the limit process (see Example 1) the exact slope of the tangent at $(3, -5)$.

8. Consider $y = x^3 + 1$.

(a) Sketch its graph.

(b) Draw the tangent line at $(2, 9)$.

(c) Estimate the slope of this tangent line.

C (d) Calculate the slope of the secant line through $(2, 9)$ and $(1.999, 1.999^3 + 1)$.

(e) Use the limit process to find the exact slope of the tangent at $(2, 9)$.

9. Find the slope of the tangent line to the curve $y = x^2 - 3x + 2$ at the points where $x = -2, 1.5, 2, 5$ (see Example 2).

10. Find the slope of the tangent line to the curve $y = x^3 - 2x$ at the points where $x = -3, -1.5, 0, 3$.

11. Sketch the graph of $y = 1/(x + 1)$ and then find the equation of the tangent line at $(1, \frac{1}{2})$ (see Example 3).

12. Find the equation of the tangent line to $y = 2/(x - 2)$ at $(0, -1)$.

13. Assume that a falling body will fall $16t^2$ feet in t seconds.

(a) How far will it fall between $t = 3$ and $t = 4$?

(b) What is its average velocity on the interval $3 \leq t \leq 4$?

C (c) What is its average velocity on the interval $3 \leq t \leq 3.02$?

(d) Find its instantaneous velocity at $t = 3$ (see Example 4).

14. An object travels along a line so that its position s is $s = 2t^2 + 2$ meters after t seconds.

(a) What is its average velocity on the interval $2 \leq t \leq 3$?

C (b) What is its average velocity on the interval $2 \leq t \leq 2.001$?

(c) What is its average velocity on the interval $2 \leq t \leq 2 + h$?

(d) Find its instantaneous velocity at $t = 2$.

15. Suppose an object moves along a line $\sqrt{t}$ feet in t seconds.

(a) Find its instantaneous velocity at $t = c$, $c > 0$.

(b) When will it reach a velocity of $\frac{1}{6}$ foot per second? (See Example 5.)

16. If a particle moves along a coordinate line so that its directed distance from the origin after t seconds is $(-t^2 + 4t)$ feet, when did the particle come to a momentary stop (that is, when did its instantaneous velocity become zero)?

17. A certain bacterial culture is growing so that it has a mass of $\frac{1}{2}t^2 + 1$ grams after t hours.

C (a) How much did it grow during the interval $2 \leq t \leq 2.01$?

(b) What was its average growth rate during the interval $2 \leq t \leq 2.01$?

(c) What was its instantaneous growth rate at $t = 2$?

18. A business is prospering in such a way that its total (accumulated) profit after t years is $1000t^2$ dollars.

(a) How much did the business make during the third year (that is, between $t = 2$ and $t = 3$)?

(b) What was its average rate of profit during the first half of the third year (that is, between $t = 2$ and $t = 2.5$)?

(c) What was its instantaneous rate of profit at $t = 2$?

19. A wire of length 8 centimeters is such that the mass between its left end and a point x centimeters to the right is x^3 grams (Figure 10).

Mass is x^3 g

FIGURE 10

(a) What is the average density of the middle 2-centimeter segment of this wire? *Note*: Average density equals mass/length.

(b) What is the actual density at the point 3 centimeters from the left end?

20. Suppose that the revenue in dollars from producing x pounds of a product is given by $R(x) = 0.5x - 0.002x^2$. Find the instantaneous rate of change of revenue when $x = 10$; when $x = 100$. (The instantaneous rate of change of revenue with respect to the amount produced is called the *marginal revenue*.)

21. The weight in grams of a malignant tumor at time t is $W(t) = 0.2t^2 - 0.09t$, where t is measured in weeks. Find the rate of growth of the tumor when $t = 10$.

22. A city is hit by an Asian flu epidemic. Officials estimate that t days after the beginning of the epidemic, the number of persons sick with the flu is given by $p(t) = 120t^2 - 2t^3$, provided that $0 \leq t \leq 40$. At what rate is the flu spreading at time $t = 10$; $t = 20$; $t = 40$?

23. The graph in Figure 11 shows the amount of water in a city water tank during one 24-hour day when no water was pumped into the tank. What was the average rate of water usage during the day? How fast was water being used at 8 o'clock?

FIGURE 11

24. The rate of change of electric charge with respect to time is called **current**. Suppose $\frac{1}{3}t^3 + t$ coulombs of charge flow through a wire in t seconds. Find the current in amperes (coulombs per second) after 3 seconds. When will a 20-amp fuse in the line blow?

25. The radius of a circular oil spill is growing at a constant rate of 2 kilometers per day. At what rate is the area of the spill growing 3 days after it began?

26. Find the rate of change of the area of a circle with respect to its circumference when the circumference is 6 centimeters.

3.2
The Derivative

We have seen that *slope of tangent line* and *instantaneous velocity* are manifestations of the same basic idea. Rate of growth of an organism (biology), marginal profit (economics), density of a wire (physics), and dissolution rates (chemistry) are other versions of the same basic concept. Good mathematical sense suggests that we study this concept independently of these specialized vocabularies and diverse applications. We choose the neutral name *derivative*. Add it to *function* and *limit* as one of the key words in calculus.

Definition

The **derivative** of a function f is another function f' (read "eff prime") whose value at any number c is

$$f'(c) = \lim_{h \to 0} \frac{f(c + h) - f(c)}{h}$$

provided that this limit exists.

If this limit does exist, we say that f is **differentiable** at c. Finding a derivative is called **differentiation**; the part of calculus associated with the derivative is called **differential calculus**.

ILLUSTRATIVE EXAMPLES

EXAMPLE 1   Let $f(x) = 13x - 6$. Find $f'(4)$.

Solution

$$f'(4) = \lim_{h \to 0} \frac{f(4 + h) - f(4)}{h} = \lim_{h \to 0} \frac{[13(4 + h) - 6] - [13(4) - 6]}{h}$$

$$= \lim_{h \to 0} \frac{13h}{h} = \lim_{h \to 0} 13 = 13$$

■

EXAMPLE 2 If $f(x) = x^3 + 7x$, find $f'(c)$.

Solution

$$f'(c) = \lim_{h \to 0} \frac{f(c + h) - f(c)}{h}$$

$$= \lim_{h \to 0} \frac{[(c + h)^3 + 7(c + h)] - [c^3 + 7c]}{h}$$

$$= \lim_{h \to 0} \frac{3c^2 h + 3ch^2 + h^3 + 7h}{h}$$

$$= \lim_{h \to 0} (3c^2 + 3ch + h^2 + 7)$$

$$= 3c^2 + 7 \qquad \blacksquare$$

EXAMPLE 3 If $f(x) = 1/x$, find $f'(x)$.

Solution Note a subtle change in the way this example is worded. So far we have used the letter c to denote a fixed number at which a derivative is to be evaluated. Accordingly, we have calculated $f'(c)$. To calculate $f'(x)$, we simply think of x as a fixed, but arbitrary, number and proceed as before.

$$f'(x) = \lim_{h \to 0} \frac{f(x + h) - f(x)}{h} = \lim_{h \to 0} \frac{\dfrac{1}{x + h} - \dfrac{1}{x}}{h}$$

$$= \lim_{h \to 0} \left[\frac{x - (x + h)}{(x + h)x} \cdot \frac{1}{h} \right] = \lim_{h \to 0} \left[\frac{-h}{(x + h)x} \cdot \frac{1}{h} \right]$$

$$= \lim_{h \to 0} \frac{-1}{(x + h)x} = \frac{-1}{x^2}$$

Thus f' is the function given by $f'(x) = -1/x^2$. Its domain is all real numbers except $x = 0$. $\qquad \blacksquare$

EXAMPLE 4 Find the derivative of F if $F(x) = \sqrt{x}, x > 0$.

Solution

$$F'(x) = \lim_{h \to 0} \frac{F(x + h) - F(x)}{h}$$

$$= \lim_{h \to 0} \frac{\sqrt{x + h} - \sqrt{x}}{h}$$

By this time you will have noticed that finding a derivative always involves taking the limit of a quotient where both numerator and denominator are

---

FORM IS ALL THAT MATTERS

We use x in place of c in Example 3, F in place of f in Example 4, and $x - c$ in place of h in Example 5. The symbols we use are matters of taste only.

"Mathematicians do not study objects, but relations between objects; thus, they are free to replace some objects by others so long as the relations remain unchanged. Content to them is irrelevant: they are interested in form only."

Henri Poincaré

---

approaching zero. Our task is to simplify this quotient so that we can cancel a factor h from numerator and denominator, thus allowing us to evaluate the limit. In the present example, this can be accomplished by rationalizing the numerator.

$$F'(x) = \lim_{h \to 0}\left[\frac{\sqrt{x+h} - \sqrt{x}}{h} \cdot \frac{\sqrt{x+h} + \sqrt{x}}{\sqrt{x+h} + \sqrt{x}}\right]$$

$$= \lim_{h \to 0} \frac{x + h - x}{h(\sqrt{x+h} + \sqrt{x})}$$

$$= \lim_{h \to 0} \frac{h}{h(\sqrt{x+h} + \sqrt{x})}$$

$$= \lim_{h \to 0} \frac{1}{\sqrt{x+h} + \sqrt{x}}$$

$$= \frac{1}{\sqrt{x} + \sqrt{x}} = \frac{1}{2\sqrt{x}}$$

Thus, F'— the derivative of F—is given by $F'(x) = 1/2\sqrt{x}$. Its domain is $(0, \infty)$. ∎

FIGURE 1

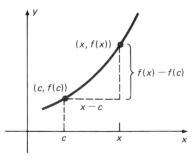

FIGURE 2

EQUIVALENT FORMS FOR THE DERIVATIVE There is nothing sacred about use of the letter h in defining $f'(c)$. Notice, for example, that

$$f'(c) = \lim_{h \to 0} \frac{f(c + h) - f(c)}{h}$$

$$= \lim_{p \to 0} \frac{f(c + p) - f(c)}{p}$$

$$= \lim_{s \to 0} \frac{f(c + s) - f(c)}{s}$$

A more radical change, but still just a change of notation, may be understood by comparing Figure 1 and Figure 2. Note how x takes the place of $c + h$, and so $x - c$ replaces h. Thus,

$$f'(c) = \lim_{x \to c} \frac{f(x) - f(c)}{x - c}$$

In a similar vein, we may write

$$f'(x) = \lim_{t \to x} \frac{f(t) - f(x)}{t - x} = \lim_{p \to x} \frac{f(p) - f(x)}{p - x}$$

Note that in all cases, the number at which f' is evaluated is held fixed during the limit operation.

EXAMPLE 5   Use the boxed result above to find $g'(c)$ if $g(x) = 2/(x + 3)$.

Solution

$$g'(c) = \lim_{x \to c} \frac{g(x) - g(c)}{x - c} = \lim_{x \to c} \frac{\dfrac{2}{x + 3} - \dfrac{2}{c + 3}}{x - c}$$

$$= \lim_{x \to c} \left[\frac{2(c + 3) - 2(x + 3)}{(x + 3)(c + 3)} \cdot \frac{1}{x - c} \right]$$

$$= \lim_{x \to c} \left[\frac{-2(x - c)}{(x + 3)(c + 3)} \cdot \frac{1}{x - c} \right]$$

$$= \lim_{x \to c} \frac{-2}{(x + 3)(c + 3)} = \frac{-2}{(c + 3)^2}$$

Here we manipulated the quotient until we could cancel a factor of $x - c$ from numerator and denominator. Then we could evaluate the limit.   ■

EXAMPLE 6   Each of the following is a derivative, but of what function and at what point?

(a) $\displaystyle \lim_{h \to 0} \frac{(4 + h)^2 - 16}{h}$ (b) $\displaystyle \lim_{x \to 3} \frac{\dfrac{2}{x} - \dfrac{2}{3}}{x - 3}$

Solution
 (a) This is the derivative of $f(x) = x^2$ at $x = 4$.
 (b) This is the derivative of $f(x) = 2/x$ at $x = 3$.   ■

DIFFERENTIABILITY IMPLIES CONTINUITY   If a curve has a tangent line at a point, then that curve cannot take a jump or wiggle badly at the point. The precise formulation of this fact is an important theorem.

Theorem A

If $f'(c)$ exists, then f is continuous at c.

Proof   We need to show that $\lim_{x \to c} f(x) = f(c)$. Now

$$f(x) = f(c) + \frac{f(x) - f(c)}{x - c} \cdot (x - c), \qquad x \neq c$$

Therefore,

$$\lim_{x \to c} f(x) = \lim_{x \to c} \left[f(c) + \frac{f(x) - f(c)}{x - c} \cdot (x - c) \right]$$

$$= \lim_{x \to c} f(c) + \lim_{x \to c} \frac{f(x) - f(c)}{x - c} \cdot \lim_{x \to c} (x - c)$$

$$= f(c) + f'(c) \cdot 0$$

$$= f(c) \qquad \blacksquare$$

The converse of this theorem is false. If a function f is continuous at c, it does not follow that f has a derivative at c. This is easily seen by considering $f(x) = |x|$ at the origin (Figure 3). This function is certainly continuous at zero, but it does not have a derivative there. To see the latter, observe that

$$\lim_{x \to 0} \frac{f(x) - f(0)}{x - 0} = \lim_{x \to 0} \frac{|x| - |0|}{x} = \lim_{x \to 0} \frac{|x|}{x}$$

$f(x) = |x|$

FIGURE 3

This limit does not exist since

$$\lim_{x \to 0^+} \frac{|x|}{x} = \lim_{x \to 0^+} \frac{x}{x} = 1$$

while

$$\lim_{x \to 0^-} \frac{|x|}{x} = \lim_{x \to 0^-} \frac{-x}{x} = -1$$

The argument just presented shows that at any point where the graph of a function has a sharp corner, it is continuous but not differentiable. The graph in Figure 4 indicates a number of possibilities.

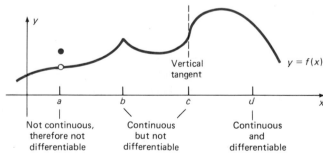

$y = f(x)$

Not continuous, therefore not differentiable

Continuous but not differentiable

Vertical tangent

Continuous and differentiable

FIGURE 4

We claim in Figure 4 that the derivative does not exist at the point c where the tangent line is vertical. This is because

$$\frac{f(c + h) - f(c)}{h}$$

increases without bound as $h \to 0$. It corresponds to the fact that the slope of a vertical line is undefined.

PROBLEM SET 3.2

In Problems 1–4, use the definition

$$f'(c) = \lim_{h \to 0} \frac{f(c + h) - f(c)}{h}$$

to find the indicated derivative (see Examples 1 and 2).

1. $f'(3)$ if $f(x) = x^2 - x$

2. $f'(-2)$ if $f(x) = x^3$

3. $f'(-1)$ if $f(x) = x^3 + 2x^2$

4. $f'(4)$ if $f(x) = \dfrac{3}{x + 1}$

In Problems 5–22, use $f'(x) = \lim_{h \to 0}[f(x + h) - f(x)]/h$ to find the derivative at x (see Examples 3 and 4).

5. $f(x) = 5x - 4$

6. $f(x) = ax + b$

7. $f(x) = 8x^2 - 1$

8. $f(x) = x^2 + 3x + 4$

9. $f(x) = ax^2 + bx + c$

10. $f(x) = 2x^3$

11. $f(x) = x^3 - 2x$

12. $f(x) = x^4$

13. $g(x) = \dfrac{3}{5x}$

14. $g(x) = \dfrac{2}{x + 6}$

15. $F(x) = \dfrac{6}{x^2 + 1}$

16. $F(x) = \dfrac{x - 1}{x + 1}$

17. $G(x) = \dfrac{2x - 1}{x - 4}$

18. $G(x) = \dfrac{2x}{x^2 - x}$

19. $g(x) = \sqrt{3x}$

20. $g(x) = \dfrac{1}{\sqrt{3x}}$

21. $H(x) = \dfrac{3}{\sqrt{x - 2}}$

22. $H(x) = \sqrt{x^2 + 4}$

In Problems 23–26, use $f'(x) = \lim_{t \to x}[f(t) - f(x)]/[t - x]$ to find $f'(x)$ (see Example 5).

23. $f(x) = x^2 - 3x$

24. $f(x) = x^3 + 5x$

25. $f(x) = \dfrac{x}{x - 5}$

26. $f(x) = \dfrac{x + 3}{x}$

In Problems 27–36, the given limit is a derivative, but of what function and at what point? (See Example 6.)

27. $\lim_{h \to 0} \dfrac{2(5 + h)^3 - 2(5)^3}{h}$

28. $\lim_{h \to 0} \dfrac{(3 + h)^2 + 2(3 + h) - 15}{h}$

29. $\lim_{x \to 2} \dfrac{x^2 - 4}{x - 2}$

30. $\lim_{x \to 3} \dfrac{x^3 + x - 30}{x - 3}$

31. $\lim_{t \to x} \dfrac{t^2 - x^2}{t - x}$

32. $\lim_{p \to x} \dfrac{p^3 - x^3}{p - x}$

33. $\lim_{x \to t} \dfrac{\dfrac{2}{x} - \dfrac{2}{t}}{x - t}$

34. $\lim_{x \to y} \dfrac{\sin x - \sin y}{x - y}$

35. $\lim_{h \to 0} \dfrac{\cos(x + h) - \cos x}{h}$

36. $\lim_{h \to 0} \dfrac{\tan(t + h) - \tan t}{h}$

37. From Figure 5, estimate $f'(0)$, $f'(2)$, $f'(5)$, and $f'(7)$.

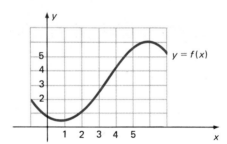

FIGURE 5

38. From Figure 6, estimate $g'(-1)$, $g'(1)$, $g'(4)$, and $g'(6)$.

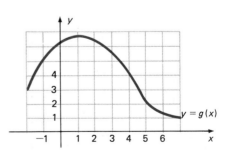

FIGURE 6

39. Sketch the graph of $y = f'(x)$ on $-1 < x < 7$ for the function f of Problem 37.

40. Sketch the graph of $y = g'(x)$ on $-1 < x < 7$ for the function g of Problem 38.

41. Consider the function $y = f(x)$, whose graph is sketched in Figure 7.

FIGURE 7

(a) Estimate $f(2)$, $f'(2)$, $f(0.5)$, and $f'(0.5)$.
(b) Estimate the average rate of change in f on the interval $0.5 \le x \le 2.5$.
(c) Where on the interval $-1 < x < 7$ does $\lim_{u \to x} f(u)$ fail to exist?
(d) Where on the interval $-1 < x < 7$ does f fail to be continuous?

(e) Where on the interval $-1 < x < 7$ does f fail to have a derivative?
(f) Where on the interval $-1 < x < 7$ is $f'(x) = 0$?
(g) Where on the interval $-1 < x < 7$ is $f'(x) = 1$?

42. Suppose that $f(x + y) = f(x)f(y)$ for all x and y. Show that if $f'(0)$ exists, then $f'(a)$ exists and $f'(a) = f(a)f'(0)$.

43. Let $f(x) = \begin{cases} mx + b & \text{if } x < 2 \\ x^2 & \text{if } x \ge 2 \end{cases}$

Determine m and b so that f is differentiable everywhere.

44. The **symmetric derivative** $f_s(x)$ is defined by

$$f_s(x) = \lim_{h \to 0} \frac{f(x + h) - f(x - h)}{2h}$$

Show that if $f'(x)$ exists, then $f_s(x)$ exists but that the converse is false.

45. Let f be differentiable and let $f'(x_0) = m$. Find $f'(-x_0)$ if (a) f is an odd function; (b) f is an even function.

46. Prove that the derivative of an odd function is an even function and that the derivative of an even function is an odd function.

3.3
Rules for Finding Derivatives

The process of finding the derivative of a function directly from the definition of the derivative, that is, by setting up the difference quotient

$$\frac{f(x + h) - f(x)}{h}$$

and evaluating its limit, can be time-consuming and tedious. We are going to develop tools that will allow us to shortcut this lengthy process—that will, in fact, allow us to find derivatives of the most complicated looking functions almost instantly.

Recall that the derivative of a function f is another function f'. For example, if $f(x) = x^2$ is the formula for f, then $f'(x) = 2x$ is the formula for f'. Taking the derivative of f (differentiating f) is operating on f to produce f'. We often use the letter D to indicate this operation. Thus we write $Df = f'$, $Df(x) = f'(x)$, or (in the example mentioned above) $D(x^2) = 2x$. All the theorems below are stated both in functional notation and in the operator D notation.

FIGURE 1

THE CONSTANT AND POWER RULES The graph of the constant function $f(x) = k$ is a horizontal line (Figure 1) which, therefore, has slope zero everywhere. This is one way to understand our first theorem.

Theorem A

(Constant Function Rule). If $f(x) = k$, where k is a constant, then for any x, $f'(x) = 0$—that is,

$$D(k) = 0$$

Proof

$$f'(x) = \lim_{h \to 0} \frac{f(x + h) - f(x)}{h} = \lim_{h \to 0} \frac{k - k}{h} = \lim_{h \to 0} 0 = 0 \quad \blacksquare$$

The graph of $f(x) = x$ is a line through the origin with slope 1 (Figure 2); so we should expect the derivative of this function to be 1 for all x.

Theorem B

(Identity Function Rule). If $f(x) = x$, then $f'(x) = 1$—that is,

$$D(x) = 1$$

Proof

$$f'(x) = \lim_{h \to 0} \frac{f(x + h) - f(x)}{h} = \lim_{h \to 0} \frac{x + h - x}{h} = \lim_{h \to 0} \frac{h}{h} = 1 \quad \blacksquare$$

Before stating our next theorem, we recall something from algebra: how to raise a binomial to a power.

$$(a + b)^2 = a^2 + 2ab + b^2$$
$$(a + b)^3 = a^3 + 3a^2b + 3ab^2 + b^3$$
$$(a + b)^4 = a^4 + 4a^3b + 6a^2b^2 + 4ab^3 + b^4$$
$$\vdots$$
$$(a + b)^n = a^n + na^{n-1}b + \frac{n(n - 1)}{2} a^{n-2}b^2 + \cdots + nab^{n-1} + b^n$$

Theorem C

(Power Rule). If $f(x) = x^n$, where n is a positive integer, then $f'(x) = nx^{n-1}$—that is,

$$D(x^n) = nx^{n-1}$$

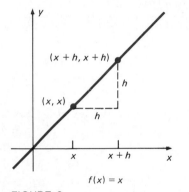

FIGURE 2

The graph labels: $(x + h, x + h)$, (x, x), h, x, $x + h$, $f(x) = x$

Proof

$$f'(x) = \lim_{h \to 0} \frac{f(x + h) - f(x)}{h} = \lim_{h \to 0} \frac{(x + h)^n - x^n}{h}$$

$$= \lim_{h \to 0} \frac{x^n + nx^{n-1}h + \dfrac{n(n-1)}{2} x^{n-2}h^2 + \cdots + nxh^{n-1} + h^n - x^n}{h}$$

$$= \lim_{h \to 0} \frac{\cancel{h}\left[nx^{n-1} + \dfrac{n(n-1)}{2} x^{n-2}h + \cdots + nxh^{n-2} + h^{n-1} \right]}{\cancel{h}}$$

Within the brackets, all terms except the first have h as a factor, and so each of these terms has limit zero as h approaches zero. Thus

$$f'(x) = nx^{n-1}$$ ∎

As illustrations of Theorem C, note that

$$D(x^3) = 3x^2, \qquad D(x^9) = 9x^8, \qquad D(x^{100}) = 100x^{99}$$

D IS A LINEAR OPERATOR   The operator D behaves very well when applied to constant multiples of functions or to sums of functions.

Theorem D

(Constant Multiple Rule). If k is a constant and f is a differentiable function, then $(kf)'(x) = k \cdot f'(x)$—that is,

$$D[k \cdot f(x)] = k \cdot Df(x)$$

In words, this says that *a constant multiplier k can be passed across the operator D.*

Proof   Let $F(x) = k \cdot f(x)$. Then

$$F'(x) = \lim_{h \to 0} \frac{F(x + h) - F(x)}{h} = \lim_{h \to 0} \frac{k \cdot f(x + h) - k \cdot f(x)}{h}$$

$$= \lim_{h \to 0} k \cdot \frac{f(x + h) - f(x)}{h} = k \cdot \lim_{h \to 0} \frac{f(x + h) - f(x)}{h}$$

$$= k \cdot f'(x)$$

The next-to-the-last step was the critical one. We could shift k past the limit sign because of the Main Limit Theorem (Part 3). ∎

Examples that illustrate this result are

$$D(-7x^3) = -7D(x^3) = -7 \cdot 3x^2 = -21x^2$$

and

$$D(\tfrac{4}{3}x^9) = \tfrac{4}{3}D(x^9) = \tfrac{4}{3} \cdot 9x^8 = 12x^8$$

Theorem E

(Sum Rule). If f and g are differentiable functions, then $(f + g)'(x) = f'(x) + g'(x)$—that is,

$$D[f(x) + g(x)] = Df(x) + Dg(x)$$

In words, this says that *the derivative of a sum is the sum of the derivatives.*

Proof Let $F(x) = f(x) + g(x)$. Then

$$F'(x) = \lim_{h \to 0} \frac{[f(x + h) + g(x + h)] - [f(x) + g(x)]}{h}$$

$$= \lim_{h \to 0} \left[\frac{f(x + h) - f(x)}{h} + \frac{g(x + h) - g(x)}{h} \right]$$

$$= \lim_{h \to 0} \frac{f(x + h) - f(x)}{h} + \lim_{h \to 0} \frac{g(x + h) - g(x)}{h}$$

$$= f'(x) + g'(x)$$

Again, the next-to-the-last step was the critical one. It is justified by appealing to the Main Limit Theorem (Part 4). ∎

Any operator L with the properties stated in Theorems D and E is called *linear*; that is, L is **linear** if:

1. $L(kf) = kL(f)$, k a constant;
2. $L(f + g) = L(f) + L(g)$.

Linear operators will appear again and again in this book; D is a particularly important example. A linear operator always satisfies the difference rule $L(f - g) = L(f) - L(g)$ stated next for D.

Theorem F

(Difference Rule). If f and g are differentiable functions, then $(f - g)'(x) = f'(x) - g'(x)$—that is,

$$D[f(x) - g(x)] = Df(x) - Dg(x)$$

LINEARITY

The fundamental meaning of the word *linear* as used in mathematics is that given in this section. An operator L is linear if it satisfies the two key conditions:

- $L(ku) = kL(u)$
- $L(u + v) = L(u) + L(v)$

Linear operators play the central role in the *linear algebra* course that many readers of this book will eventually take.

Unfortunately, functions of the form $f(x) = mx + b$ are called linear functions (because of their connection with lines) even though they are not linear in the operator sense. To see this, note that

$$f(kx) = mkx + b$$

whereas

$$kf(x) = k(mx + b)$$

Thus, $f(kx) \neq kf(x)$ unless b happens to be zero.

Proof

$$D[f(x) - g(x)] = D[f(x) + (-1)g(x)]$$
$$= Df(x) + D[(-1)g(x)] \qquad \text{(Theorem E)}$$
$$= Df(x) + (-1)Dg(x) \qquad \text{(Theorem D)}$$
$$= Df(x) - Dg(x)$$

EXAMPLE 1 Find the derivatives of $5x^2 + 7x - 6$ and $4x^6 - 3x^5 - 10x^2 + 5x + 16$.

Solution

$$D(5x^2 + 7x - 6) = D(5x^2 + 7x) - D(6) \qquad \text{(Theorem F)}$$
$$= D(5x^2) + D(7x) - D(6) \qquad \text{(Theorem E)}$$
$$= 5D(x^2) + 7D(x) - D(6) \qquad \text{(Theorem D)}$$
$$= 5 \cdot 2x + 7 \cdot 1 + 0 \qquad \text{(Theorems C, B, A)}$$
$$= 10x + 7$$

To find the next derivative, we note that the theorems on sums and differences extend to any finite number of terms. Thus,

$$D(4x^6 - 3x^5 - 10x^2 + 5x + 16)$$
$$= D(4x^6) - D(3x^5) - D(10x^2) + D(5x) + D(16)$$
$$= 4D(x^6) - 3D(x^5) - 10D(x^2) + 5D(x) + D(16)$$
$$= 4(6x^5) - 3(5x^4) - 10(2x) + 5(1) + 0$$
$$= 24x^5 - 15x^4 - 20x + 5 \qquad \blacksquare$$

The method of Example 1 allows us to find the derivative of any polynomial. If you know the Power Rule and do what comes naturally, you are almost sure to get the right result. If you can write the answer without any intermediate steps, that is fine.

PRODUCT AND QUOTIENT RULES Now we are in for a surprise. The derivative of a product of functions is *not* equal to the product of the derivatives of the functions.

Theorem G

(Product Rule). If f and g are differentiable functions, then $(f \cdot g)'(x) = f(x)g'(x) + g(x)f'(x)$—that is,

$$D[f(x)g(x)] = f(x)Dg(x) + g(x)Df(x)$$

This should be memorized in words as follows: *The derivative of a product of two functions is the first times the derivative of the second plus the second times the derivative of the first.*

Proof Let $F(x) = f(x)g(x)$. Then

$$F'(x) = \lim_{h \to 0} \frac{F(x + h) - F(x)}{h}$$

$$= \lim_{h \to 0} \frac{f(x + h)g(x + h) - f(x)g(x)}{h}$$

$$= \lim_{h \to 0} \frac{f(x + h)g(x + h) - f(x + h)g(x) + f(x + h)g(x) - f(x)g(x)}{h}$$

$$= \lim_{h \to 0} \left[f(x + h) \cdot \frac{g(x + h) - g(x)}{h} + g(x) \cdot \frac{f(x + h) - f(x)}{h} \right]$$

$$= \lim_{h \to 0} f(x + h) \cdot \lim_{h \to 0} \frac{g(x + h) - g(x)}{h} + g(x) \cdot \lim_{h \to 0} \frac{f(x + h) - f(x)}{h}$$

$$= f(x)g'(x) + g(x)f'(x)$$

The derivation just given relied first on the trick of adding and subtracting the same thing, namely, $f(x + h)g(x)$. Second, at the very end, we used the fact that

$$\lim_{h \to 0} f(x + h) = f(x)$$

This is just an application of Theorem 3.2A, which says that differentiability at a point implies continuity there. ∎

EXAMPLE 2 Find the derivative of $(3x^2 - 5)(2x^4 - x)$ by use of the Product Rule. Check the answer by doing the problem a different way.

Solution

$$D[(3x^2 - 5)(2x^4 - x)] = (3x^2 - 5)D(2x^4 - x) + (2x^4 - x)D(3x^2 - 5)$$

$$= (3x^2 - 5)(8x^3 - 1) + (2x^4 - x)(6x)$$

$$= 24x^5 - 3x^2 - 40x^3 + 5 + 12x^5 - 6x^2$$

$$= 36x^5 - 40x^3 - 9x^2 + 5$$

To check, we first multiply and then take the derivative.

$$(3x^2 - 5)(2x^4 - x) = 6x^6 - 10x^4 - 3x^3 + 5x$$

Thus,

$$D[(3x^2 - 5)(2x^4 - x)] = D(6x^6) - D(10x^4) - D(3x^3) + D(5x)$$
$$= 36x^5 - 40x^3 - 9x^2 + 5 \qquad \blacksquare$$

Theorem H

(Quotient Rule). Let f and g be differentiable functions with $g(x) \neq 0$. Then

$$\left(\frac{f}{g}\right)'(x) = \frac{g(x)f'(x) - f(x)g'(x)}{g^2(x)}$$

That is,

$$D\,\frac{f(x)}{g(x)} = \frac{g(x)Df(x) - f(x)Dg(x)}{g^2(x)}$$

We strongly urge you to memorize this in words, as follows: *The derivative of a quotient is equal to the denominator times the derivative of the numerator minus the numerator times the derivative of the denominator, all divided by the square of the denominator.*

Proof   Let $F(x) = f(x)/g(x)$. Then

$$\cdot \; F'(x) = \lim_{h \to 0} \frac{F(x + h) - F(x)}{h}$$

$$= \lim_{h \to 0} \frac{\dfrac{f(x + h)}{g(x + h)} - \dfrac{f(x)}{g(x)}}{h}$$

$$= \lim_{h \to 0} \frac{g(x)f(x + h) - f(x)g(x + h)}{h} \cdot \frac{1}{g(x)g(x + h)}$$

$$= \lim_{h \to 0} \left[\frac{g(x)f(x + h) - g(x)f(x) + f(x)g(x) - f(x)g(x + h)}{h} \cdot \frac{1}{g(x)g(x + h)} \right]$$

$$= \lim_{h \to 0} \left\{ \left[g(x)\frac{f(x + h) - f(x)}{h} - f(x)\frac{g(x + h) - g(x)}{h} \right] \frac{1}{g(x)g(x + h)} \right\}$$

$$= \left[g(x)f'(x) - f(x)g'(x) \right] \frac{1}{g(x)g(x)} \qquad \blacksquare$$

EXAMPLE 3   Find the derivative of $\dfrac{(3x - 5)}{(x^2 + 7)}$.

Solution

$$D\left[\frac{3x - 5}{x^2 + 7}\right] = \frac{(x^2 + 7)D(3x - 5) - (3x - 5)D(x^2 + 7)}{(x^2 + 7)^2}$$

$$= \frac{(x^2 + 7)(3) - (3x - 5)(2x)}{(x^2 + 7)^2}$$

$$= \frac{-3x^2 + 10x + 21}{(x^2 + 7)^2} \qquad \blacksquare$$

EXAMPLE 4   Find Dy if $y = \dfrac{2}{x^4 + 1} + \dfrac{3}{x}$.

Solution

$$Dy = D\left(\frac{2}{x^4 + 1}\right) + D\left(\frac{3}{x}\right)$$

$$= \frac{(x^4 + 1)D(2) - 2D(x^4 + 1)}{(x^4 + 1)^2} + \frac{xD(3) - 3D(x)}{x^2}$$

$$= \frac{(x^4 + 1)(0) - (2)(4x^3)}{(x^4 + 1)^2} + \frac{(x)(0) - (3)(1)}{x^2}$$

$$= \frac{-8x^3}{(x^4 + 1)^2} - \frac{3}{x^2} \qquad \blacksquare$$

EXAMPLE 5   Show that the Power Rule holds for negative integral exponents; that is,

$$\boxed{D(x^{-n}) = -nx^{-n-1}}$$

Solution

$$D(x^{-n}) = D\left(\frac{1}{x^n}\right) = \frac{x^n \cdot 0 - 1 \cdot nx^{n-1}}{x^{2n}} = \frac{-nx^{n-1}}{x^{2n}} = -nx^{-n-1} \qquad \blacksquare$$

We saw as part of Example 4 that $D(3/x) = -3/x^2$. Now we have another way to see the same thing.

$$D\left(\frac{3}{x}\right) = D(3x^{-1}) = 3D(x^{-1}) = 3(-1)x^{-2} = -\frac{3}{x^2}$$

In Problems 1–44, find Dy using the rules of this section.

1. $y = 2x^3$

2. $y = 3x^4$

3. $y = \pi x^2$

4. $y = \sqrt{2}x^5$

5. $y = -3x^{-3}$

6. $y = 4x^{-2}$

7. $y = \dfrac{-2}{x^4}$

8. $y = \dfrac{-8}{x^{10}}$

9. $y = \dfrac{3}{5x^5}$

10. $y = \dfrac{2}{3x^6}$

11. $y = -x^3 + 2x$

12. $y = 2x^4 - 3x$

13. $y = -x^4 + 3x^2 - 6x + 1$

14. $y = 11x^4 - 3x + 19$

15. $y = 5x^6 - 3x^5 + 11x - 9$

16. $y = 3x^7 - 9x^2 + 21$

17. $y = 3x^{-5} + 2x^{-3}$

18. $y = 2x^{-6} + x^{-1}$

19. $y = \dfrac{2}{x} - \dfrac{1}{x^2}$

20. $y = \dfrac{3}{x^3} - \dfrac{1}{x^4}$

21. $y = \dfrac{1}{2x} + 2x$

22. $y = \dfrac{2}{3x} - \dfrac{2}{3}$

23. $y = x(x^2 + 1)$

24. $y = 3x(x^3 - 1)$

25. $y = (2x + 1)^2$

26. $y = (-3x + 2)^2$

27. $y = (x^2 + 2)(x^3 + 1)$

28. $y = (x^4 - 1)(x^2 + 1)$

29. $y = (x^2 + 17)(x^3 - 3x + 1)$

30. $y = (x^4 + 2x)(x^3 + 2x^2 + 1)$

31. $y = (5x^2 - 7)(3x^2 - 2x + 1)$

32. $y = (3x^2 + 2x)(x^4 - 3x + 1)$

33. $y = \dfrac{1}{3x^2 + 1}$

34. $y = \dfrac{2}{5x^2 - 1}$

35. $y = \dfrac{1}{4x^2 - 3x + 9}$

36. $y = \dfrac{4}{2x^3 - 3x}$

37. $y = \dfrac{x - 1}{x + 1}$

38. $y = \dfrac{2x - 1}{x - 1}$

39. $y = \dfrac{2x^2 - 1}{3x + 5}$

40. $y = \dfrac{5x - 4}{3x^2 + 1}$

41. $y = \dfrac{2x^2 - 3x + 1}{2x + 1}$

42. $y = \dfrac{5x^2 + 2x - 6}{3x - 1}$

43. $y = \dfrac{x^2 - x + 1}{x^2 + 1}$

44. $y = \dfrac{x^2 - 2x + 5}{x^2 + 2x - 3}$

45. If $f(0) = 4$, $f'(0) = -1$, $g(0) = -3$, and $g'(0) = 5$, find (a) $(f \cdot g)'(0)$; (b) $(f + g)'(0)$; (c) $(f/g)'(0)$.

46. If $f(3) = 7$, $f'(3) = 2$, $g(3) = 6$, and $g'(3) = -10$, find (a) $(f - g)'(3)$; (b) $(f \cdot g)'(3)$; (c) $(g/f)'(3)$.

47. Use the Product Rule to show that $D[f(x)]^2 = 2 \cdot f(x) \cdot f'(x)$.

48. Develop a rule for $D[f(x)g(x)h(x)]$.

49. Find the equation of the tangent line to $y = 3x^2 - 6x + 1$ at $(1, -2)$.

50. Find the equation of the tangent line to $y = 1/(x^2 + 1)$ at $(1, \frac{1}{2})$.

51. Find all points on the graph of $y = x^3 - x^2$ where the tangent line is horizontal.

52. Find all points on the graph of $y = \frac{1}{3}x^3 + x^2 - x$ where the tangent line has slope 1.

53. The height s in feet of a ball above the ground at t seconds is given by $s = -16t^2 + 40t + 100$.
(a) What is its instantaneous velocity at $t = 2$?
(b) When is its instantaneous velocity 0?

54. A ball rolls down a long inclined plane so that its distance s from its starting point after t seconds is $s = 4.5t^2 + 2t$ feet. When will its instantaneous velocity be 30 feet per second?

55. There are two tangent lines to the curve $y = 4x - x^2$ that go through $(2, 5)$. Find the equations of both of them. *Hint*: Let (x_0, y_0) be the point of tangency. Find two conditions that (x_0, y_0) must satisfy.

56. A space traveler is moving from left to right along the curve $y = x^2$. When she shuts off the engines, she will go off along the tangent line at the point where she is at that time. At what point should she shut off the engines in order to reach the point $(4, 15)$?

57. A fly is crawling from left to right along the top of the curve $y = 7 - x^2$. A spider waits at the point $(4, 0)$. Find the distance between the two insects when they first see each other.

58. Let $P(a, b)$ be a point on the first quadrant portion of the curve $y = 1/x$ and let the tangent line at P intersect the x-axis at A. Show that triangle AOP is isosceles and determine its area.

59. The radius of a spherical watermelon is growing at a constant rate of 2 centimeters per week. The thickness of the rind is always one-tenth of the radius. How fast is the volume of the rind growing at the end of the fifth week?

3.4
Derivatives of Sines and Cosines

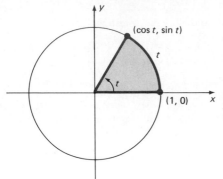

FIGURE 1

Our modern world runs on wheels. Questions about rotating wheels and velocities of points on them lead inevitably to the study of sines and cosines and their derivatives. To prepare for this study it would be well to review Section 2.3. Figure 1 reminds us of the definition of the sine and cosine functions. In what follows, t should be thought of as a number measuring the length of an arc on the unit circle or, equivalently, as the number of radians in the corresponding angle. Thus, $f(t) = \sin t$ and $g(t) = \cos t$ are functions for which both domain and range are sets of real numbers. We may consider the problem of finding their derivatives.

THE DERIVATIVE FORMULAS   We choose to use x rather than t as our basic variable. To find $D(\sin x)$, we appeal to the definition of derivative and use the addition identity for $\sin(x + h)$.

$$D(\sin x) = \lim_{h \to 0} \frac{\sin(x + h) - \sin x}{h}$$

$$= \lim_{h \to 0} \frac{\sin x \cos h + \cos x \sin h - \sin x}{h}$$

$$= \lim_{h \to 0} \left(-\sin x \frac{1 - \cos h}{h} + \cos x \frac{\sin h}{h} \right)$$

$$= (-\sin x) \left[\lim_{h \to 0} \frac{1 - \cos h}{h} \right] + (\cos x) \left[\lim_{h \to 0} \frac{\sin h}{h} \right]$$

To finish our derivation, we have two limits to evaluate. A calculator provided the table in Figure 2. It suggests and, later in this section, we will prove that

$$\lim_{h \to 0} \frac{1 - \cos h}{h} = 0 \qquad \lim_{h \to 0} \frac{\sin h}{h} = 1$$

Thus,

$$D(\sin x) = (-\sin x) \cdot 0 + (\cos x) \cdot 1 = \cos x$$

Similarly,

$$D(\cos x) = \lim_{h \to 0} \frac{\cos(x + h) - \cos x}{h}$$

$$= \lim_{h \to 0} \frac{\cos x \cos h - \sin x \sin h - \cos x}{h}$$

$$= \lim_{h \to 0} \left(-\cos x \frac{1 - \cos h}{h} - \sin x \frac{\sin h}{h} \right)$$

$$= (-\cos x) \cdot 0 - (\sin x) \cdot 1$$

$$= -\sin x$$

h	$\dfrac{1 - \cos h}{h}$	$\dfrac{\sin h}{h}$
1.0	0.45970	0.84147
0.5	0.24483	0.95885
0.1	0.04996	0.99833
0.01	0.00500	0.99998
↓	↓	↓
0	?	?
↑	↑	↑
−0.01	−0.00500	0.99998
−0.1	−0.04996	0.99833
−0.5	−0.24483	0.95885
−1.0	−0.45970	0.84147

FIGURE 2

We summarize these results in an important theorem.

Theorem A

The functions $f(x) = \sin x$ and $g(x) = \cos x$ are both differentiable. In fact,

$$D(\sin x) = \cos x \qquad D(\cos x) = -\sin x$$

EXAMPLE 1 Find $D(3 \sin x - 2 \cos x)$.

Solution

$$D(3 \sin x - 2 \cos x) = 3\, D(\sin x) - 2\, D(\cos x)$$
$$= 3 \cos x + 2 \sin x \qquad \blacksquare$$

EXAMPLE 2 Find $D(\tan x)$.

Solution

$$D(\tan x) = D\left(\frac{\sin x}{\cos x}\right)$$
$$= \frac{\cos x\, D(\sin x) - \sin x\, D(\cos x)}{\cos^2 x}$$
$$= \frac{\cos x \cos x + \sin x \sin x}{\cos^2 x}$$
$$= \frac{1}{\cos^2 x} = \sec^2 x \qquad \blacksquare$$

EXAMPLE 3 Find the equation of the tangent line to the graph of $y = 3 \sin 2x$ at the point $(\pi/2, 0)$ (see Figure 3).

Solution We need the derivative of $\sin 2x$; unfortunately, we know how to find only the derivative of $\sin x$. However, $\sin 2x = 2 \sin x \cos x$. Thus,

$$D(3 \sin 2x) = D(6 \sin x \cos x)$$
$$= 6\, D(\sin x \cos x)$$
$$= 6[\sin x\, D(\cos x) + \cos x\, D(\sin x)]$$
$$= 6[(\sin x)(-\sin x) + \cos x \cos x]$$
$$= 6[\cos^2 x - \sin^2 x]$$
$$= 6 \cos 2x$$

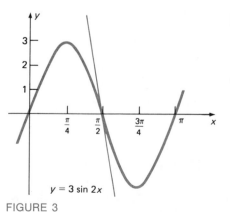

FIGURE 3

$y = 3 \sin 2x$

At $x = \pi/2$, this derivative has the value -6, which is therefore the slope of the desired tangent line. The equation of this line is

$$y - 0 = -6\left(x - \frac{\pi}{2}\right)$$ ∎

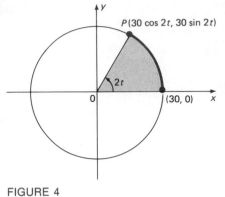

FIGURE 4

EXAMPLE 4   Consider a ferris wheel of radius 30 feet, which is rotating counterclockwise with an angular velocity of 2 radians per second. How fast is a seat on the rim rising (in the vertical direction) when it is 15 feet above the horizontal line through the center of the wheel?

Solution   We may suppose that the wheel is centered at the origin and that the seat P was at $(30,0)$ at time $t = 0$ (Figure 4). Thus at time t, P has moved through an angle of $2t$ radians and so has coordinates $(30 \cos 2t, 30 \sin 2t)$. The rate at which P is rising is just the derivative of the vertical coordinate $30 \sin 2t$ measured at an appropriate t-value. By Example 3,

$$D(30 \sin 2t) = 60 \cos 2t$$

The appropriate t at which to evaluate this derivative is $t = \pi/12$, since $30 \sin(2 \cdot \pi/12) = 15$. We conclude that at $t = \pi/12$, the seat P is rising at

$$60 \cos\left(2 \cdot \frac{\pi}{12}\right) = 60\sqrt{3}/2 \approx 51.96 \text{ feet per second}$$ ∎

PROOF OF TWO LIMIT STATEMENTS   Everything we have done in this section depends on the two limit statements,

$$\lim_{t \to 0} \frac{\sin t}{t} = 1 \qquad \lim_{t \to 0} \frac{1 - \cos t}{t} = 0$$

They require proofs.

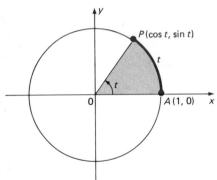

FIGURE 5

Proof   Consider the (by now familiar) diagram in Figure 5. Note that as $t \to 0$, the point P moves toward $(1,0)$ and, therefore, that

$$\lim_{t \to 0} \cos t = 1 \qquad \lim_{t \to 0} \sin t = 0$$

Next, for $-\pi/2 < t < \pi/2$, $t \neq 0$, draw the vertical line segment BP and the circular arc BC, as shown in Figure 6. (If $t < 0$, the shaded region will be reflected across the x-axis.) Clearly,

$$\text{Area(sector } OBC) \leq \text{area}(\triangle OBP) \leq \text{area(sector } OAP)$$

From the formulas $\frac{1}{2}bh$ (area of a triangle) and $\frac{1}{2}r^2|t|$ (area of a circular sector; see Problem 30 of Section 2.3), we obtain

$$\tfrac{1}{2}(\cos t)^2|t| \leq \tfrac{1}{2} \cos t|\sin t| \leq \tfrac{1}{2}(1)^2|t|$$

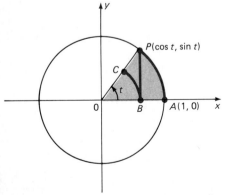

FIGURE 6

112   CHAPTER 3/The Derivative

or, after multiplying by 2 and dividing by the positive number $|t| \cos t$ and noting that $(\sin t)/t$ is positive,

$$\cos t \leq \frac{\sin t}{t} \leq \frac{1}{\cos t}$$

This double inequality simply begs for the Squeeze Theorem. When we apply it, we get

$$\lim_{t \to 0} \frac{\sin t}{t} = 1$$

the first of our claimed results.

The second result follows easily from the first.

$$\lim_{t \to 0} \frac{1 - \cos t}{t} = \lim_{t \to 0} \frac{1 - \cos t}{t} \frac{1 + \cos t}{1 + \cos t} = \lim_{t \to 0} \frac{1 - \cos^2 t}{t(1 + \cos t)}$$

$$= \lim_{t \to 0} \frac{\sin^2 t}{t(1 + \cos t)} = \lim_{t \to 0} \frac{\sin t}{t} \cdot \frac{\lim_{t \to 0} \sin t}{\lim_{t \to 0}(1 + \cos t)}$$

$$= 1 \cdot \frac{0}{2} = 0 \qquad \blacksquare$$

While the main use of these two limit statements was to prove the derivative formulas, we also can use them to evaluate other limits.

EXAMPLE 5   Find the following limits.

(a) $\displaystyle \lim_{t \to 0} \frac{1 - \cos t}{\sin t}$ (b) $\displaystyle \lim_{x \to 0} \frac{\sin 5x}{3x}$

Solution

(a) $\displaystyle \lim_{t \to 0} \frac{1 - \cos t}{\sin t} = \lim_{t \to 0} \frac{\dfrac{1 - \cos t}{t}}{\dfrac{\sin t}{t}} = \frac{0}{1} = 0$

(b) $\displaystyle \lim_{x \to 0} \frac{\sin 5x}{3x} = \lim_{x \to 0} \frac{5}{3} \cdot \frac{\sin 5x}{5x} = \frac{5}{3} \lim_{x \to 0} \frac{\sin 5x}{5x} = \frac{5}{3} \cdot 1 = \frac{5}{3}$

At the next-to-the-last step, we used the fact that when $x \to 0$, $5x \to 0$. Thus,

$$\lim_{x \to 0} \frac{\sin 5x}{5x} = \lim_{5x \to 0} \frac{\sin 5x}{5x} = \lim_{t \to 0} \frac{\sin t}{t} = 1 \qquad \blacksquare$$

In Problems 1–12, find Dy.

1. $y = 3 \sin x - 5 \cos x$

2. $y = \sin x \cos x$

3. $y = \cot x = \dfrac{\cos x}{\sin x}$

4. $y = \sec x = \dfrac{1}{\cos x}$

5. $y = \csc x = \dfrac{1}{\sin x}$

6. $y = \sin^2 x$

7. $y = \dfrac{\sin x}{\sin x + \cos x}$

8. $y = \dfrac{\tan x}{\sin x - \cos x}$

9. $y = \sin^2 x + \cos^2 x$

10. $y = x^2 \sin x$

11. $y = \dfrac{\cos x}{x}$

12. $y = \dfrac{x^2 + 1}{x \sin x}$

C 13. Find the equation of the tangent line to $y = \sin x$ at $x = 1$.

14. Find the equation of the tangent line to $y = \tan x$ at $x = \pi/4$.

15. Consider the ferris wheel of Example 4. At what rate is the seat on the rim moving horizontally when $t = \pi/4$ seconds (that is, when the set reaches the very top of the wheel)?

16. A ferris wheel of radius 20 feet is rotating counterclockwise at an angular velocity of 1 radian per second. One seat on the rim is at $(20,0)$ at $t = 0$.
(a) What are its coordinates at $t = \pi/6$?
(b) How fast is it rising (vertically) at $t = \pi/6$?
(c) How fast is it rising (vertically) when it is rising at the fastest rate?

In Problems 17–22, follow the procedure of Example 5 to find each of the limits.

17. $\lim\limits_{x \to 0} \dfrac{\sin 2x}{3x}$

18. $\lim\limits_{x \to 0} \dfrac{\sin(\frac{1}{2}x)}{3x}$

19. $\lim\limits_{x \to 0} \dfrac{\sin 2x}{\tan x}$

20. $\lim\limits_{x \to 0} \dfrac{\cot x}{\csc x - 1}$

21. $\lim\limits_{x \to 0} \dfrac{\tan x - \sin x}{x \cos x}$

22. $\lim\limits_{x \to 0} \dfrac{\sin 2x}{\tan 3x}$

23. Show that the curves $y = \sqrt{2} \sin x$ and $y = \sqrt{2} \cos x$ intersect at right angles at a certain point with $0 < x < \pi/2$.

24. At time t seconds, the center of a bobbing cork is $2 \sin t$ centimeters above (or below) water level. What is the velocity of the cork at $t = 0, \pi/2, \pi$?

25. Use the definition of derivative to show that $D(\sin x^2) = 2x \cos x^2$.

26. Use the definition of derivative to show that $D(\sin 5x) = 5 \cos 5x$.

27. Let x_0 be the smallest positive value of x at which the curves $y = \sin x$ and $y = \sin 2x$ intersect. Find x_0 and also the acute angle at which the two curves intersect at x_0 (see Problem 28 of Section 2.3).

28. From area$(OBP) \le$ area$(OAP) \le$ area$(OBP) +$ area$(ABPQ)$ in Figure 7, show that

$$\cos t \le \frac{t}{\sin t} \le 2 - \cos t$$

and thus obtain another proof that $\lim\limits_{t \to 0}(\sin t)/t = 1$.

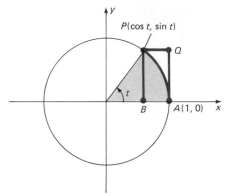

FIGURE 7

29. In Figure 8, let D be the area of triangle ABP and E the area of the shaded region.
(a) Guess the value of $\lim\limits_{t \to 0^+}(D/E)$ by looking at the figure.
(b) Find a formula for D/E in terms of t.
C (c) Use a calculator to get an accurate estimate of $\lim\limits_{t \to 0^+}(D/E)$.

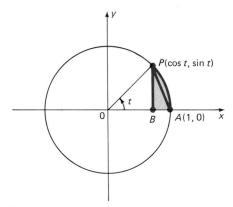

FIGURE 8

CHAPTER 3/The Derivative

Note: The exact answer to (c) will be found in Problem 27 of Section 9.1.

30. An isosceles triangle is topped by a semicircle, as shown in Figure 9. Let D be the area of triangle AOB and E be the area of the shaded region. Find a formula for D/E in terms of t and then calculate

$$\lim_{t \to 0^+} \frac{D}{E}$$

FIGURE 9

3.5
The Chain Rule

Imagine trying to find the derivative of

$$F(x) = (2x^2 - 4x + 1)^{60}$$

You would first have to multiply together the 60 quadratic factors of $2x^2 - 4x + 1$ and then differentiate the resulting polynomial of degree 120.

Fortunately, there is a better way. After you have learned the *Chain Rule*, you will be able to write the answer

$$F'(x) = 60(2x^2 - 4x + 1)^{59}(4x - 4)$$

as fast as you can move your pencil. In fact, the *Chain Rule* is so important that you will seldom again differentiate any function without using it. But in order to state the rule properly, we need to introduce a slight refinement of our *D*-notation.

THE D_x-NOTATION   If a problem involves more than one variable, it will help to have a notational device for indicating which variable is under consideration at a particular instant. Thus if $y = s^2 x^3$ and we wish to treat x as the independent variable and s as a constant, we write $D_x y$ and obtain

$$D_x y = D_x(s^2 x^3) = s^2 D_x(x^3) = s^2 \cdot 3x^2$$

The symbol $D_x y$ is read *the derivative of y with respect to x*.

More important is the following example. Suppose $y = u^{60}$ and $u = 2x^2 - 4x + 1$. Then $D_u y = 60u^{59}$ and $D_x u = 4x - 4$. But notice that when we substitute $u = 2x^2 - 4x + 1$ in $y = u^{60}$, we obtain

$$y = (2x^2 - 4x + 1)^{60}$$

so it makes sense to ask for $D_x y$. What is $D_x y$ and how is it related to $D_u y$ and $D_x u$? More generally, how do you differentiate a composite function?

DIFFERENTIATING A COMPOSITE FUNCTION   If David can type twice as fast as Mary and Mary can type three times as fast as Jack, then David can type $2 \cdot 3 = 6$ times as fast as Jack. The two rates are multiplied.

Suppose that

$$y = f(u) \quad \text{and} \quad u = g(x)$$

determine the composite function $y = f(g(x))$. Since a derivative indicates a rate of change, we can say that

y changes $D_u y$ times as fast as u

u changes $D_x u$ times as fast as x

It seems reasonable to conclude that

y changes $D_u y \cdot D_x u$ times as fast as x

This is in fact true and we will give a formal proof in the next section. The result is called the **Chain Rule**.

Theorem A

(Chain Rule). Let $y = f(u)$ and $u = g(x)$ determine the composite function $y = f(g(x)) = (f \circ g)(x)$. If g is differentiable at x and f is differentiable at $u = g(x)$, then $f \circ g$ is differentiable at x and

$$(f \circ g)'(x) = f'(g(x))g'(x)$$

That is,

$$D_x y = D_u y D_x u$$

Maybe it will help you to remember it this way.

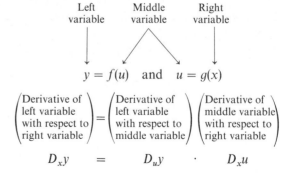

Thought of this way, you should have no trouble seeing that if

$$w = f(s) \quad \text{and} \quad s = g(t)$$

then

$$D_t w = D_s w\, D_t s$$

APPLICATIONS OF THE CHAIN RULE We begin with the example $(2x^2 - 4x + 1)^{60}$ introduced at the beginning of this section.

EXAMPLE 1 If $y = (2x^2 - 4x + 1)^{60}$, find $D_x y$.

Solution We think of this as

$$y = u^{60} \quad \text{and} \quad u = 2x^2 - 4x + 1$$

Thus,

$$D_x y = D_u y \cdot D_x u$$
$$= (60u^{59})(4x - 4)$$
$$= 60(2x^2 - 4x + 1)^{59}(4x - 4) \qquad \blacksquare$$

EXAMPLE 2 If $y = 1/(2x^5 - 7)^3$, find $D_x y$.

Solution Think of it this way.

$$y = \frac{1}{u^3} = u^{-3} \quad \text{and} \quad u = 2x^5 - 7$$

Thus,

$$D_x y = D_u y \cdot D_x u$$
$$= (-3u^{-4})(10x^4)$$
$$= \frac{-3}{u^4} \cdot 10x^4$$
$$= \frac{-30x^4}{(2x^5 - 7)^4} \qquad \blacksquare$$

EXAMPLE 3 If $y = \sin(x^3 - 3x)$, find $D_x y$.

Solution

$$y = \sin u \quad \text{and} \quad u = x^3 - 3x$$

Hence,

$$D_x y = D_u y \cdot D_x u$$
$$= (\cos u) \cdot (3x^2 - 3)$$
$$= [\cos(x^3 - 3x)] \cdot (3x^2 - 3) \qquad \blacksquare$$

EXAMPLE 4 Find $D_t \left(\dfrac{t^3 - 2t + 1}{t^4 + 3} \right)^{13}$.

Solution Think of this as finding $D_t y$, where

$$y = u^{13} \quad \text{and} \quad u = \frac{t^3 - 2t + 1}{t^4 + 3}$$

Then,

$$D_t y = D_u y \cdot D_t u$$

$$= 13 u^{12} \frac{(t^4 + 3)(3t^2 - 2) - (t^3 - 2t + 1)(4t^3)}{(t^4 + 3)^2}$$

$$= 13 \left(\frac{t^3 - 2t + 1}{t^4 + 3} \right)^{12} \cdot \frac{-t^6 + 6t^4 - 4t^3 + 9t^2 - 6}{(t^4 + 3)^2} \qquad \blacksquare$$

Soon you will learn to make a mental introduction of the middle variable without actually writing it. Thus, an expert immediately writes:

$$D_x(\cos 3x) = (-\sin 3x) \cdot 3 = -3 \sin 3x$$

$$D_x(x^3 + \sin x)^6 = 6(x^3 + \sin x)^5(3x^2 + \cos x)$$

$$D_t \left(\frac{t}{\cos 3t} \right)^4 = 4 \left(\frac{t}{\cos 3t} \right)^3 \frac{\cos 3t - t(-\sin 3t)3}{\cos^2 3t}$$

$$= \frac{4t^3(\cos 3t + 3t \sin 3t)}{\cos^5 3t}$$

THE COMPOUND CHAIN RULE   Suppose

$$y = f(u) \quad \text{and} \quad u = g(v) \quad \text{and} \quad v = h(x)$$

Then

$$D_x y = D_u y D_v u D_x v$$

EXAMPLE 5   Find $D_x[\sin^3(4x)]$.

Solution Think of this as finding $D_x y$, where

$$y = u^3 \quad \text{and} \quad u = \sin v \quad \text{and} \quad v = 4x$$

Then,

$$D_x y = D_u y \cdot D_v u \cdot D_x v$$

$$= 3u^2 \cdot \cos v \cdot 4$$

$$= 3 \sin^2(4x) \cdot \cos(4x) \cdot 4$$

$$= 12 \sin^2(4x)\cos(4x) \qquad \blacksquare$$

Here, too, you will soon make these substitutions in your head and write the answer immediately. It may help if you notice that, in differentiating compound composite functions, you work from the outer parentheses inward, like peeling an onion.

Let's do Example 5 again, making explicit what we have just said.

$$D_x[\sin(4x)]^3 = 3\sin^2(4x)D_x\sin(4x)$$
$$= 3\sin^2(4x)\cos(4x)D_x(4x)$$
$$= 3\sin^2(4x)\cos(4x) \cdot 4$$
$$= 12\sin^2(4x)\cos(4x)$$

EXAMPLE 6 Find $D_x\{\sin[\cos(x^2)]\}$.

Solution

$$D_x\{\sin[\cos(x^2)]\} = \cos[\cos(x^2)] \cdot [-\sin(x^2)] \cdot 2x \qquad ■$$

PROBLEM SET 3.5

In Problems 1–26, find D_xy.

1. $y = (2 - 9x)^{15}$

2. $y = (4x + 7)^{23}$

3. $y = (5x^2 + 2x - 8)^5$

4. $y = (3x^3 - 11x)^7$

5. $y = (x^3 - 3x^2 + 11x)^9$

6. $y = (2x^4 - 12x^2 + 11x - 9)^{10}$

7. $y = (3x^4 + x - 8)^{-3}$

8. $y = (4x^3 - 3x^2 + 11x - 1)^{-5}$

9. $y = \dfrac{1}{(3x^4 + x - 8)^9}$

10. $y = \dfrac{3}{(4x^3 + 11x)^7}$

11. $y = \sin(3x^2 + 11x)$

12. $y = \cos(4x^5 - 11x)$

13. $y = \sin^3 x$

14. $y = \cos^5 x$

15. $y = \left(\dfrac{x^2 - 1}{x + 4}\right)^4$

16. $y = \left(\dfrac{3x - 1}{2x + 5}\right)^6$

17. $y = \sin\left(\dfrac{3x - 1}{2x + 5}\right)$

18. $y = \cos\left(\dfrac{x^2 - 1}{x + 4}\right)$

19. $y = (4x - 7)^2(2x + 3)$

20. $y = (5x + 6)^2(x - 13)^3$

21. $y = (2x - 1)^3(x^2 - 3)^2$

22. $y = (3x^2 + 5)^2(x^3 - 11)^4$

23. $y = \dfrac{(x + 1)^2}{3x - 4}$

24. $y = \dfrac{2x - 3}{(x^2 + 4)^2}$

25. $y = \dfrac{(3x^2 + 2)^2}{2x^2 - 5}$

26. $y = \dfrac{(x^2 - 1)^3}{(4x^3 - 5)^2}$

In Problems 27–34, find the indicated derivative.

27. $D_t\left(\dfrac{3t - 2}{t + 5}\right)^3$

28. $D_s\left(\dfrac{s^2 - 9}{s + 4}\right)$

29. $D_t\left(\dfrac{(3t - 2)^3}{t + 5}\right)$

30. $D_s\left(\dfrac{s^2 - 9}{s + 4}\right)^3$

31. $D_\theta(\sin^3\theta)$

32. $D_\theta(\cos^4\theta)$

33. $D_x\left(\dfrac{\sin x}{\cos 2x}\right)^3$

34. $D_t[\sin t \tan(t^2 + 1)]$

In Problems 35–38, evaluate the indicated derivative.

35. $f'(3)$ if $f(x) = \left(\dfrac{x^2 + 1}{x + 2}\right)^3$

36. $G'(1)$ if $G(t) = (t^2 + 9)^3(t^2 - 2)^4$

37. $F'(1)$ if $F(t) = \sin(t^2 + 3t + 1)$

38. $g'(\tfrac{1}{2})$ if $g(s) = \cos \pi s \sin^2 \pi s$

In Problems 39–46, use the Compound Chain Rule (Example 5) to find the indicated derivative.

39. $D_x[\sin^4(x^2 + 3x)]$

40. $D_t[\cos^5(4t - 19)]$

41. $D_t[\sin^3(\cos t)]$

42. $D_u\left[\cos^4\left(\dfrac{u + 1}{u - 1}\right)\right]$

43. $D_\theta[\cos^4(\sin \theta^2)]$

44. $D_x[x \sin^2(2x)]$

45. $D_x\{\sin[\cos(\sin 2x)]\}$

46. $D_t\{\cos^2[\cos(\cos t)]\}$

47. Find the equation of the tangent line to $y = (x^2 + 1)^3(x^4 + 1)^2$ at $(1, 32)$.

48. A point P is moving in the plane so that its coordinates after t seconds are $(4 \cos 2t, 7 \sin 2t)$, measured in feet.

(a) Show that P is following an elliptical path. *Hint*: Show $(x/4)^2 + (y/7)^2 = 1$, which is an equation of an ellipse.
(b) Obtain an expression for L, its distance from the origin at time t.
(c) How fast is P moving away from the origin at $t = \pi/8$? You will need the fact that $D_u(\sqrt{u}) = 1/(2\sqrt{u})$ (see Example 4 of Section 3.2).

49. A wheel centered at the origin and of radius 10 centimeters is rotating counterclockwise at a rate of 4 revolutions per second. A point P on the rim is at $(10,0)$ at $t = 0$.
(a) What are the coordinates of P at time t seconds?
(b) At what rate is P rising (or falling) at time $t = 1$?

50. Consider the wheel-piston device in Figure 1. The wheel has radius 1 foot and rotates counterclockwise at 2 radians per second. The connecting rod is 5 feet long. The point P is at $(1,0)$ at time $t = 0$.

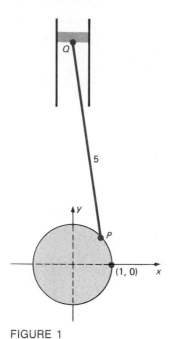

FIGURE 1

(a) Find the coordinates of P at time t.
(b) Find the y-coordinate of Q at time t (the x-coordinate is always zero).
(c) Find the velocity of Q at time t. You will need the fact that $D_u(\sqrt{u}) = 1/(2\sqrt{u})$.

51. Do Problem 50 assuming the wheel is rotating at 60 revolutions per minute.

52. Show that $D_x|x| = |x|/x$, $x \neq 0$. *Hint*: Write $|x| = \sqrt{x^2}$ and use the Chain Rule with $u = x^2$.

53. Apply the result in Problem 52 to find each derivative.
(a) $D_x|x^2 - 1|$
(b) $D_x|\sin x|$

54. Later in the book (Chapter 7), we will study a function L satisfying $L'(x) = 1/x$. Find each of the following derivatives.
(a) $D_x L\left(\dfrac{x}{x + 1}\right)$
(b) $D_x L(\cos^4 x)$

55. In each of the following, find $f'(x)$ and write your answer in simplest form in terms of $\sin 2x$.
(a) $f(x) = -\cos 2x + \frac{1}{3}\cos^3 2x$
(b) $f(x) = \frac{3}{8}x - \frac{3}{32}\sin 4x - \frac{1}{8}\sin^3 2x \cos 2x$

56. Show that if a polynomial $p(x)$ is divisible by $(ax + b)^2$, then $p'(x)$ is divisible by $ax + b$.

57. Let $f(0) = 0$ and $f'(0) = 2$. Find the derivative of $f(f(f(f(x))))$ at $x = 0$.

58. Use the Chain Rule to show that the derivative of an odd function is even and the derivative of an even function is odd.

3.6
Leibniz Notation

Gottfried Wilhelm Leibniz was one of the two principal founders of the calculus (the other was Isaac Newton). His notation for the derivative is still widely used, especially in applied fields such as physics, chemistry, and economics. Its attraction lies in its form, a form that often suggests true results and sometimes suggests how to prove them. After we have mastered the Leibniz notation, we will use it to restate the Chain Rule, and then actually prove that rule .

INCREMENTS If the value of a variable x changes from x_1 to x_2, then $x_2 - x_1$, the change in x, is called an **increment** of x and is commonly denoted by Δx (read "delta x"). Note immediately that Δx does *not* mean Δ times x.

If $x_1 = 4.1$ and $x_2 = 5.7$, then

$$\Delta x = x_2 - x_1 = 5.7 - 4.1 = 1.6$$

If $x_1 = c$ and $x_2 = c + h$, then

$$\Delta x = x_2 - x_1 = c + h - c = h$$

Suppose next that $y = f(x)$ determines a function. If x changes from x_1 to x_2, then y changes from $y_1 = f(x_1)$ to $y_2 = f(x_2)$. Thus, corresponding to the increment $\Delta x = x_2 - x_1$ in x, there is an increment in y given by

$$\Delta y = y_2 - y_1 = f(x_2) - f(x_1)$$

EXAMPLE 1   Let $y = f(x) = 2 - x^2$. Find Δy when x changes from 0.4 to 1.3 (see Figure 1).

Solution

$$\Delta y = f(1.3) - f(0.4) = \left[2 - (1.3)^2\right] - \left[2 - (0.4)^2\right]$$
$$= 0.31 - 1.84 = -1.53 \qquad \blacksquare$$

FIGURE 1

THE *dy/dx* SYMBOL FOR THE DERIVATIVE   Suppose now that the independent variable changes from x to $x + \Delta x$. The corresponding change in the dependent variable, y, will be

$$\Delta y = f(x + \Delta x) - f(x)$$

and the ratio

$$\frac{\Delta y}{\Delta x} = \frac{f(x + \Delta x) - f(x)}{\Delta x}$$

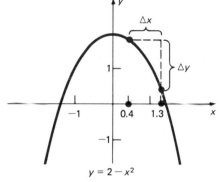

FIGURE 2

represents the slope of a secant line through $(x, f(x))$, as shown in Figure 2. As $\Delta x \to 0$, the slope of this secant line approaches that of the tangent line, and for this latter slope Leibniz used the symbol dy/dx. Thus,

$$\frac{dy}{dx} = \lim_{\Delta x \to 0} \frac{\Delta y}{\Delta x} = \lim_{\Delta x \to 0} \frac{f(x + \Delta x) - f(x)}{\Delta x} = f'(x)$$

Leibniz called dy/dx a quotient of two infinitesimals. The meaning of the word *infinitesimal* is vague, and we will not use it. However, dy/dx is a standard symbol for the derivative; we will use it frequently from now on. For the present, think of d/dx as an operator symbol with the same meaning as D_x, and read it "the derivative with respect to x."

EXAMPLE 2   Find dy/dx if $y = x^3 - 3x^2 + 7x$.

Solution

$$\frac{dy}{dx} = \frac{d}{dx}(x^3 - 3x^2 + 7x)$$

$$= \frac{d(x^3)}{dx} - 3\frac{d(x^2)}{dx} + 7\frac{d(x)}{dx}$$

$$= 3x^2 - 3(2x) + 7(1)$$

$$= 3x^2 - 6x + 7 \qquad\blacksquare$$

EXAMPLE 3   Find $\dfrac{d}{dt}\left(\dfrac{3t}{t^2 + 1}\right)$.

Solution   By the Quotient Rule,

$$\frac{d}{dt}\left(\frac{3t}{t^2 + 1}\right) = \frac{(t^2 + 1)(3) - (3t)(2t)}{(t^2 + 1)^2} = \frac{-3t^2 + 3}{(t^2 + 1)^2} \qquad\blacksquare$$

THE CHAIN RULE AGAIN   Suppose that $y = f(u)$ and $u = g(x)$. In Leibniz notation, the Chain Rule takes the particularly elegant form

$$\boxed{\frac{dy}{dx} = \frac{dy}{du}\frac{du}{dx}}$$

It is elegant because it is easy to remember. Just cancel the du's on the right side and you have the left side. Do not try to make mathematical sense out of this cancellation, but use it as a memory aid if it helps.

EXAMPLE 4   Find dy/dx if $y = (x^3 - 2x)^{12}$.

Solution   Think of $x^3 - 2x$ as u. Then $y = u^{12}$ and

$$\frac{dy}{dx} = \frac{dy}{du}\frac{du}{dx}$$

$$= (12u^{11})(3x^2 - 2)$$

$$= 12(x^3 - 2x)^{11}(3x^2 - 2) \qquad\blacksquare$$

If $y = f(u)$, $u = g(v)$, and $v = h(x)$, then

$$\frac{dy}{dx} = \frac{dy}{du}\frac{du}{dv}\frac{dv}{dx}$$

EXAMPLE 5 Find dy/dx if $y = \cos^3(x^2 + 1)$.

Solution We can think of this as $y = u^3$, $u = \cos v$, and $v = x^2 + 1$.

$$\begin{aligned}
\frac{dy}{dx} &= \frac{dy}{du}\frac{du}{dv}\frac{dv}{dx} \\
&= (3u^2)(-\sin v)(2x) \\
&= (3\cos^2 v)[-\sin(x^2 + 1)](2x) \\
&= -6x\cos^2(x^2 + 1)\sin(x^2 + 1) \quad\blacksquare
\end{aligned}$$

PARTIAL PROOF OF THE CHAIN RULE

Proof We suppose that $y = f(u)$ and $u = g(x)$, that g is differentiable at x, and that f is differentiable at $u = g(x)$. When x is given an increment Δx, there are corresponding increments in u and y given by

$$\begin{aligned}
\Lambda u &= g(x + \Delta x) - g(x) \\
\Delta y &= f(g(x + \Delta x)) - f(g(x)) \\
&= f(u + \Delta u) - f(u)
\end{aligned}$$

Thus,

$$\begin{aligned}
\frac{dy}{dx} &= \lim_{\Delta x \to 0}\frac{\Delta y}{\Delta x} = \lim_{\Delta x \to 0}\frac{\Delta y}{\Delta u}\frac{\Delta u}{\Delta x} \\
&= \lim_{\Delta x \to 0}\frac{\Delta y}{\Delta u} \cdot \lim_{\Delta x \to 0}\frac{\Delta u}{\Delta x}
\end{aligned}$$

Since g is differentiable at x, it is continuous there (Theorem 3.2A), and so $\Delta x \to 0$ forces $\Delta u \to 0$. Hence,

$$\frac{dy}{dx} = \lim_{\Delta u \to 0}\frac{\Delta y}{\Delta u} \cdot \lim_{\Delta x \to 0}\frac{\Delta u}{\Delta x} = \frac{dy}{du} \cdot \frac{du}{dx}$$

This proof was very slick, but unfortunately it contains a subtle flaw. There are functions $u = g(x)$ that have the property that $\Delta u = 0$ for some points in every neighborhood of x (the constant function $g(x) = k$ is a good example). This means the division by Δu at our first step might not be legal. There is no simple way to get around this difficulty, though the Chain Rule is valid even in this case. We give a complete proof of the Chain Rule in the appendix (Section A.2, Theorem B). $\blacksquare$

In Problems 1–4, find Δy for the given values of x_1 and x_2 (see Example 1).

1. $y = x^2 - 2x + 4$, $x_1 = 1$, $x_2 = 1.5$

2. $y = 2x + \dfrac{1}{x}$, $x_1 = 2$, $x_2 = 2.2$

[C] **3.** $y = \dfrac{3}{x+1}$, $x_1 = 2.34$, $x_2 = 2.31$

[C] **4.** $y = \cos 2x$, $x_1 = 0.571$, $x_2 = 0.573$

In Problems 5–8, first find and simplify

$$\frac{\Delta y}{\Delta x} = \frac{f(x + \Delta x) - f(x)}{\Delta x}$$

Then find dy/dx by taking the limit of your answer as $\Delta x \to 0$.

5. $y = x^2 - 3x$ **6.** $y = x^3$

7. $y = \dfrac{x}{x+1}$ **8.** $y = \dfrac{x^2 + 1}{x}$

In Problems 9–20, use the Chain Rule to find dy/dx.

9. $y = u^3$ and $u = x^2 + 3x$

10. $y = \dfrac{1}{u^2} = u^{-2}$ and $u = \sin x$

11. $y = \sin(x^2)$ **12.** $y = \sin^2 x$

13. $y = \left(\dfrac{x^2 + 1}{\cos x}\right)^4$ **14.** $y = [(x^2 + 1)\sin x]^3$

15. $y = \cos(x^2)\sin^2 x$ **16.** $y = \dfrac{(x^3 + 2x)^4}{x^4 + 1}$

17. $y = \sin^4(x^2 + 3)$ (See Example 5.)

18. $y = \sin[(x^2 + 3)^4]$

19. $y = \cos^2\left(\dfrac{x^2 + 2}{x^2 - 2}\right)$

20. $y = \sin^2[\cos^2(x^2)]$

In Problems 21–26, find the indicated derivatives.

21. $\dfrac{d}{dt}(\sin^3 t + \cos^3 t)$

22. $\dfrac{d}{ds}[(s^2 + 3)^3 - (s^2 + 3)^{-3}]$

23. $D_r[\pi(r + 3)^2 - 3\pi r(r + 2)^2]$

24. $D_t[u^3 + 3u]$ if $u = t^2$

25. $f'(2)$ if $f(x) = \left(x + \dfrac{1}{x}\right)^4$

26. $F'(0)$ if $F(t) = \cos(t^2)\sin 3t$

27. Suppose that $f(3) = 2$, $f'(3) = -1$, $g(3) = 3$, and $g'(3) = -4$. Calculate each value.
(a) $(f + g)'(3)$ (b) $(f \cdot g)'(3)$
(c) $(f/g)'(3)$ (d) $(f \circ g)'(3)$

28. If $f(2) = 4$, $f'(4) = 6$, and $f'(2) = -2$, calculate each value.
(a) $\dfrac{d}{dx}[f(x)]^3$ at $x = 2$ (b) $\dfrac{d}{dx}\left[\dfrac{3}{f(x)}\right]$ at $x = 2$

(c) $(f \circ f)'(2)$

Problems 29 and 30 refer to the graphs in Figures 3 and 4.

FIGURE 3

FIGURE 4

29. Find each value approximately.
(a) $(f + g)'(4)$ (b) $(f \circ g)'(6)$

30. Find each value approximately.
(a) $(f/g)'(2)$ (b) $(g \circ f)'(3)$

31. The edge of a cube is increasing at the constant rate of 16 centimeters per minute.
(a) Find the rate at which the volume of the cube is increasing at the instant when the edge is 20 centimeters.
(b) Find the rate at which the total surface area of the cube is increasing at the instant when the edge is 15 centimeters.

32. Ships A and B start from the origin at the same time. Ship A travels due east at a rate of 20 miles per hour and ship B travels due north at the rate of 12 miles per hour. How fast are they separating after 3 hours? After 6 hours?

33. Where does the tangent line to the curve $y = x^2 \cos^2(x^2)$ at $x = \sqrt{\pi}$ intersect the x-axis?

34. The dial of a standard clock has a 10-centimeter radius. The two ends of an elastic string are attached to the rim at 12 and to the tip of the 10-centimeter minute hand. At what rate is the string stretching at 12:15 (assuming the clock is not slowed down by this stretching)?

35. Suppose f is differentiable and that there are points x_1 and x_2 such that $f(x_1) = x_2$ and $f(x_2) = x_1$. Let $g(x) = f(f(f(f(x))))$. Show that $g'(x_1) = g'(x_2)$.

36. Let $f(x) = \begin{cases} x^2 \sin \dfrac{1}{x} & \text{if } x \neq 0 \\ 0 & \text{if } x = 0 \end{cases}$

(a) Find $f'(x)$ for $x \neq 0$ by our rules.
(b) Find $f'(0)$ from the definition of the derivative.
(c) Show that $f'(x)$ is discontinuous at $x = 0$.

3.7
Higher-Order Derivatives

The operation of differentiation takes a function f and produces a new function f'. If we now differentiate f', we produce still another function, denoted by f'' (read "eff double prime") and called the **second derivative** of f. It, in turn, may be differentiated, thereby producing f''', which is called the **third derivative** of f, and so on. Thus, for example, let

$$f(x) = 2x^3 - 4x^2 + 7x - 8$$

Then

$$f'(x) = 6x^2 - 8x + 7$$
$$f''(x) = 12x - 8$$
$$f'''(x) = 12$$
$$f''''(x) = 0$$

Since the derivative of the zero function is zero, all higher-order derivatives will be zero.

We have introduced three notations for the derivative (now also called the *first derivative*) of $y = f(x)$. They are

$$f'(x) \qquad D_x y \qquad \frac{dy}{dx}$$

called, respectively, the *prime notation*, the *dee notation*, and the *Leibniz notation*. There is a variation of the prime notation—namely, y'—that we will also use occasionally. All of these notations have extensions for higher-order derivatives, as shown in the chart on the next page. Note especially the Leibniz notation, which—though complicated—seemed most appropriate to Leibniz. What, thought he, is more natural than to write

$$\frac{d}{dx}\left(\frac{dy}{dx}\right) \quad \text{as} \quad \frac{d^2 y}{dx^2}$$

Notations for Derivatives of $y = f(x)$

Derivative	f' Notation	y' Notation	D Notation	Leibniz Notation
First	$f'(x)$	y'	$D_x y$	$\dfrac{dy}{dx}$
Second	$f''(x)$	y''	$D_x^2 y$	$\dfrac{d^2 y}{dx^2}$
Third	$f'''(x)$	y'''	$D_x^3 y$	$\dfrac{d^3 y}{dx^3}$
Fourth	$f''''(x)$	y''''	$D_x^4 y$	$\dfrac{d^4 y}{dx^4}$
Fifth	$f^{(5)}(x)$	$y^{(5)}$	$D_x^5 y$	$\dfrac{d^5 y}{dx^5}$
Sixth	$f^{(6)}(x)$	$y^{(6)}$	$D_x^6 y$	$\dfrac{d^6 y}{dx^6}$
$\vdots$	$\vdots$	$\vdots$	$\vdots$	$\vdots$
nth	$f^{(n)}(x)$	$y^{(n)}$	$D_x^n y$	$\dfrac{d^n y}{dx^n}$

EXAMPLE 1 If $y = \sin 2x$, find $d^3 y/dx^3$, $d^4 y/dx^4$, and $d^{12} y/dx^{12}$.

Solution

$$\frac{dy}{dx} = 2 \cos 2x$$

$$\frac{d^2 y}{dx^2} = -2^2 \sin 2x$$

$$\frac{d^3 y}{dx^3} = -2^3 \cos 2x$$

$$\frac{d^4 y}{dx^4} = 2^4 \sin 2x$$

$$\frac{d^5 y}{dx^5} = 2^5 \cos 2x$$

$$\vdots$$

$$\frac{d^{12} y}{dx^{12}} = 2^{12} \sin 2x$$

∎

VELOCITY AND ACCELERATION   In Section 3.1, we used the notion of instantaneous velocity to motivate the definition of the derivative. Let's review this notion by means of an example. Also, from now on, we will use the single word *velocity* in place of the more cumbersome phrase *instantaneous velocity*.

EXAMPLE 2 An object moves along a coordinate line so that its position s satisfies $s = 2t^2 - 12t + 8$, where s is measured in centimeters and t in seconds. Determine the velocity of the object when $t = 1$ and when $t = 6$. When is the velocity 0? When is it positive?

Solution If we use the symbol $v(t)$ for the velocity at time t, then

$$v(t) = \frac{ds}{dt} = 4t - 12$$

Thus,

$$v(1) = 4(1) - 12 = -8 \text{ centimeters per second}$$

$$v(6) = 4(6) - 12 = 12 \text{ centimeters per second}$$

The velocity is 0 when $4t - 12 = 0$, that is, when $t = 3$. The velocity is positive when $4t - 12 > 0$, or when $t > 3$. All of this is shown schematically in Figure 1.

FIGURE 1

The object is, of course, moving along the s-axis, not on the colored path above it. But the colored path shows what is happening to the object. Between $t = 0$ and $t = 3$, the velocity is negative; the object is moving to the left (backing up). By the time $t = 3$, it has "slowed" to a zero velocity, and then starts moving to the right as its velocity becomes positive. Thus, negative velocity corresponds to moving in the direction of decreasing s; positive velocity corresponds to moving in the direction of increasing s. A rigorous discussion of these points will be given in Section 4.8. ■

There is a technical distinction between the words *velocity* and *speed*. Velocity has a sign associated with it; it may be positive or negative. **Speed** is defined to be the absolute value of the velocity. Thus, in the example above, the speed at $t = 1$ is $|-8| = 8$ centimeters per second. The meter in most cars is a speedometer; it always gives nonnegative values.

Now we want to give a physical interpretation of the second derivative d^2s/dt^2. It is, of course, just the first derivative of the velocity. Thus, it measures the rate of change of velocity with respect to time, which has the name **acceleration**. If it is denoted by a, then

$$\boxed{a = \frac{dv}{dt} = \frac{d^2s}{dt^2}}$$

In Example 2, $s = 2t^2 - 12t + 8$. Thus,

$$v = \frac{ds}{dt} = 4t - 12$$

$$a = \frac{d^2s}{dt^2} = 4$$

This means that the velocity is increasing at a constant rate of 4 centimeters per second every second, which we write as 4 centimeters per second per second.

EXAMPLE 3   A point moves along a horizontal coordinate line in such a way that its position at time t is specified by

$$s = t^3 - 12t^2 + 36t - 30$$

Here s is measured in feet and t in seconds.
(a) When is the velocity 0?
(b) When is the velocity positive?
(c) When is the point moving backward (that is, to the left)?
(d) When is the acceleration positive?

FIGURE 2

Solution
(a) $v = ds/dt = 3t^2 - 24t + 36 = 3(t - 2)(t - 6)$. Thus, $v = 0$ at $t = 2$ and $t = 6$.
(b) $v > 0$ when $(t - 2)(t - 6) > 0$. We learned how to solve quadratic inequalities in Section 1.3. The solution is $\{t : t < 2 \text{ or } t > 6\}$ or, in interval notation, $(-\infty, 2) \cup (6, \infty)$; see Figure 2.
(c) The point is moving left when $v < 0$—that is, when $(t - 2)(t - 6) < 0$. This inequality has as its solution the interval $(2, 6)$.
(d) $a = dv/dt = 6t - 24 = 6(t - 4)$. Thus, $a > 0$ when $t > 4$.
     The motion of the point is shown schematically in Figure 3.

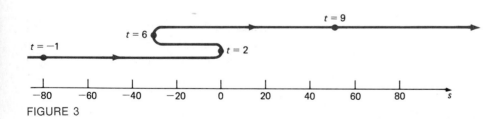

FIGURE 3

FALLING-BODY PROBLEMS   If an object is thrown straight up (or down) from an initial height of s_0 feet with an initial velocity of v_0 feet per second and if s is its height above the ground in feet after t seconds, then

$$s = -16t^2 + v_0 t + s_0$$

$v = v_0$ at $t = 0$

s_0

Ground level

FIGURE 4

This assumes that the experiment takes place near sea level and that air resistance can be neglected. The diagram in Figure 4 portrays the situation we have in mind.

EXAMPLE 4 Suppose that a ball was thrown upward from the top of a building 160 feet high with an initial velocity of 64 feet per second.
(a) When did it reach maximum height?
(b) What was its maximum height?
(c) When did it hit the ground?
(d) With what speed did it hit the ground?
(e) What was its acceleration at $t = 2$?

Solution Here $s_0 = 160$ and $v_0 = 64$, so

$$s = -16t^2 + 64t + 160$$

$$v = \frac{ds}{dt} = -32t + 64$$

$$a = \frac{dv}{dt} = -32$$

(a) The ball reached its maximum height at the time its velocity was 0—that is, when $-32t + 64 = 0$, or when $t = 2$ seconds.
(b) At $t = 2$, $s = -16(2)^2 + 64(2) + 160 = 224$ feet.
(c) The ball hit the ground when $s = 0$—that is, when

$$-16t^2 + 64t + 160 = 0$$

If we divide by -16 and use the quadratic formula, we obtain

$$t^2 - 4t - 10 = 0$$

$$t = \frac{4 \pm \sqrt{16 + 40}}{2} = \frac{4 \pm 2\sqrt{14}}{2} = 2 \pm \sqrt{14}$$

Only the positive answer makes sense. Thus, the ball hit the ground at $t = 2 + \sqrt{14} \approx 5.74$ seconds.
(d) At $t = 2 + \sqrt{14}$, $v = -32(2 + \sqrt{14}) + 64 \approx -119.73$. Thus, the ball hit the ground at a speed of 119.73 feet per second.
(e) The acceleration is always -32 feet per second per second. This is the acceleration of gravity near sea level. ∎

MATHEMATICAL MODELING Galileo may have been right in claiming that the book of nature is written in mathematical language. Certainly, the scientific enterprise seems largely an effort to prove him correct. The task of taking a physical phenomenon and representing it in mathematical symbols is called **mathematical modeling**. One of its basic elements is translating word descriptions into mathematical language. Doing this, especially in connection with rates of change, will become increasingly important as we go on. Here are some simple illustrations.

THE BOOK OF NATURE

"The great book of Nature lies ever open before our eyes and the true philosophy is written in it. . . . But we cannot read it unless we have first learned the language and the characters in which it is written. . . . It is written in mathematical language and the characters are triangles, circles, and other geometrical figures."

Galileo Galilei

Word Description		Mathematical Model

Water is leaking from a cylindrical tank at a rate proportional to the depth of the water.

If V denotes the volume of the water at time t, then $\dfrac{dV}{dt} = -kh$.

A wheel is spinning at a constant rate of 6 revolutions per minute.

$\dfrac{d\theta}{dt} = 6(2\pi)$

The density (in grams per centimeter) of a wire at a point is twice its distance from the left end.

If m denotes the mass of the left x centimeters of the wire, then $\dfrac{dm}{dx} = 2x$.

The height of a tree continues to increase but at a slower and slower rate.

$\dfrac{dh}{dt} > 0, \qquad \dfrac{d^2h}{dt^2} < 0$

FIGURE 5

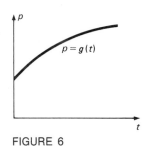

FIGURE 6

The use of mathematical language is not limited to the physical sciences; it is also appropriate in the social sciences, especially in economics.

EXAMPLE 5 KTS News reported in May, 1980, that unemployment continued to increase at an increasing rate. On the other hand, the price of food was increasing, but at a slower rate than before. Interpret these statements in the language of calculus.

Solution Let $u = f(t)$ denote the number of people unemployed at time t. Though u actually jumps by unit amounts, we follow standard practice in representing u by a nice smooth curve, as in Figure 5. To say unemployment is increasing is to say $du/dt > 0$; to say that it is increasing at an increasing rate is to say $d^2u/dt^2 > 0$.

Similarly if $p = g(t)$ represents the price of food (for instance, the typical cost of one day's groceries for one person) at time t, then $dp/dt > 0$ but $d^2p/dt^2 < 0$; see Figure 6. ∎

PROBLEM SET 3.7

In Problems 1–8, find d^3y/dx^3.

1. $y = x^3 + 3x^2 - 2x - 8$

2. $y = 2x^5 - x^4$

3. $y = (2x + 5)^4$

4. $y = (3x - 2)^5$

5. $y = \sin(3x)$

6. $y = \cos(x^2)$

7. $y = \dfrac{1}{x - 3}$

8. $y = \dfrac{x}{2x + 1}$

In Problems 9–16, find $f''(2)$.

9. $f(x) = 2x^3 - 7$

10. $f(x) = 5x^3 + 1$

11. $f(t) = \dfrac{1}{t}$

12. $f(u) = \dfrac{1}{2u - 5}$

13. $f(x) = x(x^2 + 1)^3$

14. $f(x) = \dfrac{2x + 1}{x^2 + 1}$

15. $f(x) = \sin^2(\pi x)$

16. $f(x) = x \cos(\pi x)$

17. Let $n! = n(n - 1)(n - 2) \cdots 3 \cdot 2 \cdot 1$. Thus, $4! = 4 \cdot 3 \cdot 2 \cdot 1 = 24$ and $5! = 5 \cdot 4 \cdot 3 \cdot 2 \cdot 1$. We give $n!$ the name *n* **factorial**. Show that $D_x^n(x^n) = n!$.

18. Using the factorial symbol of Problem 17, find a formula for

$$D_x^n(a_n x^n + a_{n-1} x^{n-1} + \cdots + a_1 x + a_0)$$

19. Without doing any calculating, find each derivative.

(a) $D_x^4(3x^3 + 2x - 19)$

(b) $D_x^{12}(100x^{11} - 79x^{10})$

(c) $D_x^{11}(x^2 - 3)^5$

20. Find a formula for $D_x^n(1/x)$.

21. If $f(x) = x^3 + 3x^2 - 45x - 6$, find the value of f'' at each zero of f'—that is, at each point c where $f'(c) = 0$.

22. Suppose $g(t) = at^2 + bt + c$ and $g(1) = 5$, $g'(1) = 3$, and $g''(1) = -4$. Find a, b, and c.

In Problems 23–28, an object is moving along a horizontal coordinate line according to the formula $s = f(t)$, where s, the directed distance from the origin, is in feet and t is in seconds. In each case, answer the following questions (see Examples 2 and 3).

(a) What are $v(t)$ and $a(t)$, the velocity and acceleration at time t?

(b) When is the object moving right?

(c) When is it moving left?

(d) When is the acceleration negative?

(e) Draw a schematic diagram, showing the motion of the object.

23. $s = 12t - 2t^2$ **24.** $s = t^3 - 6t^2$

25. $s = t^3 - 9t^2 + 24t$ **26.** $s = 2t^3 - 6t + 5$

27. $s = t^2 + \dfrac{16}{t}, t > 0$ **28.** $s = t + \dfrac{4}{t}, t > 0$

29. If $s = \frac{1}{2}t^4 - 5t^3 + 12t^2$, find the velocity of the moving object when its acceleration is zero.

30. If $s = \frac{1}{10}(t^4 - 14t^3 + 60t^2)$, find the velocity of the moving object when its acceleration is zero.

31. Two particles move along a coordinate line. At the end of t seconds their directed distances from the origin, in feet, are given by $s_1 = 4t - 3t^2$ and $s_2 = t^2 - 2t$, respectively.

(a) When do they have the same velocity?

(b) When do they have the same speed? (The speed of a particle is the absolute value of its velocity.)

(c) When do they have the same position?

32. The positions of two particles, P_1 and P_2, on a coordinate line at the end of t seconds are given by $s_1 = 3t^3 - 12t^2 + 18t + 5$ and $s_2 = -t^3 + 9t^2 - 12t$, respectively. When do the two particles have the same velocity?

33. An object thrown directly upward is at a height $s = -16t^2 + 48t + 256$ feet after t seconds (see Example 4).

(a) What was its initial velocity?

(b) When did it reach maximum height?

(c) What was its maximum height?

[C] (d) When did it hit the ground?

[C] (e) With what speed did it hit the ground?

34. An object thrown directly upward from ground level with an initial velocity of 48 feet per second is approximately $s = 48t - 16t^2$ feet high at the end of t seconds.

(a) What is the maximum height attained?

(b) How fast is it moving, and in which direction, at the end of 1 second?

(c) How long does it take to return to its original position?

[C] **35.** A projectile is fired directly upward from the ground with an initial velocity of v_0 feet per second. Its height in t seconds is given by $s = v_0 t - 16t^2$ feet. What must its initial velocity be for the projectile to reach a maximum height of 1 mile?

36. An object thrown directly downward from the top of a cliff with an initial velocity of v_0 feet per second falls approximately $s = v_0 t + 16t^2$ feet in t seconds. If it strikes the ocean below in 3 seconds with a velocity of 140 feet per second, how high is the cliff?

37. A point moves along a horizontal coordinate line in such a way that its position at time t is specified by $s = t^3 - 3t^2 - 24t - 6$. Here s is measured in centimeters and t in seconds. When is the point slowing down, that is, when is its *speed* decreasing?

38. Convince yourself that a point moving along a line is slowing down when its velocity and acceleration have opposite signs (see Problem 37).

39. Translate each of the following into the language of first, second, and third derivatives of distance with respect to time.

(a) The speed of that car is proportional to the distance it has traveled.

(b) That car is speeding up.

(c) I didn't say that car was slowing down; I said its rate of increase in speed was slowing down.

(d) That car's velocity is increasing 10 miles per hour every minute.

(e) That car is slowing very gently to a stop.

(f) That car always travels the same distance in equal time intervals.

40. Translate each of the following into the language of derivatives.

(a) Water is evaporating from that tank at a constant rate.

(b) Water is being poured into that tank at 3 gallons per minute but is also leaking out at $\frac{1}{2}$ gallon per minute.

(c) Since water is being poured into that conical tank at a constant rate, the water level is rising at a slower and slower rate.

(d) Inflation held steady this year but is expected to rise more and more rapidly in the years ahead.

(e) At present the price of oil is dropping but this trend is expected to slow and then reverse directions in 2 years.

(f) David's temperature is still rising but the penicillin seems to be taking effect.

41. Translate each of the following statements into mathematical language as in Example 5.

(a) The cost of a car continues to increase and at a faster and faster rate.

(b) During the last 2 years, the United States has continued to cut its consumption of oil, but at a slower and slower rate.

(c) World population continues to grow, but at a slower and slower rate.

(d) That car is going faster and faster at a constant rate.

(e) The angle the Leaning Tower of Pisa makes with the vertical is increasing more rapidly.

(f) Upper Midwest firm's profit growth slows.

(g) The XYZ Company has been losing money but will soon turn this situation around.

42. Translate each statement from the following newspaper column into a statement about derivatives.

(a) In the United States, the ratio R of government debt to national income remained unchanged at around 28% up to 1981 but (b) then it began to increase more and more sharply, reaching 36% during 1983. (c) The IMF released a table showing that the speed of increase of R was greater in the United States than in Japan.

43. Leibniz obtained a general formula for $D_x^n(uv)$, where u and v are both functions of x. See if you can find it. *Hint:* Begin by considering the cases $n = 1$, $n = 2$, and $n = 3$.

3.8
Implicit
Differentiation

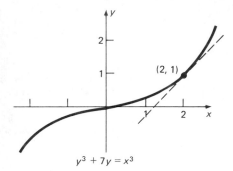

$y^3 + 7y = x^3$

FIGURE 1

With a little work, most students will be able to check that the graph of

$$y^3 + 7y = x^3$$

looks something like what is shown in Figure 1. Certainly the point $(2, 1)$ is on the graph, and there appears to be a well-defined tangent line at that point. How shall we find the slope of this tangent? Easy, you may answer; just calculate dy/dx at this point. But that is the rub; we do not know how to find dy/dx in this situation.

The new element in this problem is that we are given an equation which is not explicitly solved for y. And try as we will, we cannot seem to solve it for y. Is it possible to find dy/dx in circumstances like these? Yes. Differentiate both sides of the equation

$$y^3 + 7y = x^3$$

with respect to x and equate the results. In doing this, we assume that the given equation does determine y as a function of x (it is just that we do not know how to find it explicitly). Thus, after using the Chain Rule on the first term, we get

$$3y^2 \cdot \frac{dy}{dx} + 7\frac{dy}{dx} = 3x^2$$

The latter can be solved for dy/dx as follows.

$$\frac{dy}{dx}(3y^2 + 7) = 3x^2$$

$$\frac{dy}{dx} = \frac{3x^2}{3y^2 + 7}$$

Note that our expression for dy/dx involves both x and y, a fact that is often a nuisance. But if we wish only to find a slope at a point where we know both coordinates, no difficulty exists. At $(2, 1)$,

$$\frac{dy}{dx} = \frac{3(2)^2}{3(1)^2 + 7} = \frac{12}{10} = \frac{6}{5}$$

The slope is $\frac{6}{5}$.

The method just illustrated for finding dy/dx without first solving the given equation for y explicitly in terms of x is called **implicit differentiation**. But is the method legitimate—does it give the right answer?

AN EXAMPLE THAT CAN BE CHECKED To give some evidence for the correctness of the method, consider the following example, which can be worked two ways.

EXAMPLE 1 Find dy/dx if $4x^2y - 3y = x^3 - 1$.

Solution

METHOD 1 We can solve the given equation explicitly for y as follows.

$$y(4x^2 - 3) = x^3 - 1$$

$$y = \frac{x^3 - 1}{4x^2 - 3}$$

Thus,

$$\frac{dy}{dx} = \frac{(4x^2 - 3)(3x^2) - (x^3 - 1)(8x)}{(4x^2 - 3)^2} = \frac{4x^4 - 9x^2 + 8x}{(4x^2 - 3)^2}$$

METHOD 2 **(Implicit differentiation).** We equate the derivatives of the two sides of

$$4x^2y - 3y = x^3 - 1$$

We obtain, after using the Product Rule on the first term,

$$4x^2 \cdot \frac{dy}{dx} + y \cdot 8x - 3\frac{dy}{dx} = 3x^2$$

$$\frac{dy}{dx}(4x^2 - 3) = 3x^2 - 8xy$$

$$\frac{dy}{dx} = \frac{3x^2 - 8xy}{4x^2 - 3}$$

Though this answer looks different from the one obtained earlier, they are equivalent. To see this, substitute $y = (x^3 - 1)/(4x^2 - 3)$ in the expression for dy/dx just obtained.

$$\frac{dy}{dx} = \frac{3x^2 - 8xy}{4x^2 - 3} = \frac{3x^2 - 8x\dfrac{x^3 - 1}{4x^2 - 3}}{4x^2 - 3}$$

$$= \frac{12x^4 - 9x^2 - 8x^4 + 8x}{(4x^2 - 3)^2} = \frac{4x^4 - 9x^2 + 8x}{(4x^2 - 3)^2} \quad \blacksquare$$

SOME SUBTLE DIFFICULTIES If an equation in x and y determines a function $y = f(x)$ and if this function is differentiable, then the method of implicit differentiation will yield a correct expression for dy/dx. There are two big *ifs* in this statement.

Consider first the equation

$$x^2 + y^2 = -1$$

It has no solutions and therefore does not determine a function.

On the other hand,

$$x^2 + y^2 = 25$$

determines both the function $y = f(x) = \sqrt{25 - x^2}$ and the function $y = g(x) = -\sqrt{25 - x^2}$. Their graphs are shown in Figure 2.

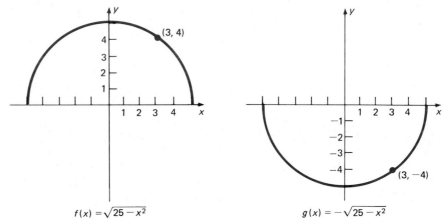

FIGURE 2

Happily, both of these functions are differentiable on $(-5, 5)$. Consider f first. It satisfies

$$x^2 + [f(x)]^2 = 25$$

When we differentiate implicitly and solve for $f'(x)$, we obtain

$$2x + 2f(x)f'(x) = 0$$

$$f'(x) = -\frac{x}{f(x)} = -\frac{x}{\sqrt{25 - x^2}}$$

A completely similar treatment of $g(x)$ yields

$$g'(x) = -\frac{x}{g(x)} = \frac{x}{\sqrt{25 - x^2}}$$

For practical purposes, we can obtain both of these results simultaneously by implicit differentiation of $x^2 + y^2 = 25$. This gives

$$2x + 2y\frac{dy}{dx} = 0$$

$$\frac{dy}{dx} = -\frac{x}{y} = \begin{cases} \dfrac{-x}{\sqrt{25 - x^2}} & \text{if } y = f(x) \\[3mm] \dfrac{-x}{-\sqrt{25 - x^2}} & \text{if } y = g(x) \end{cases}$$

Naturally, the results are identical with those obtained above.

Note that it is enough to know $dy/dx = -x/y$ in order to apply our results. Suppose we want to know the slope of the tangent line to the circle $x^2 + y^2 = 25$ when $x = 3$. The corresponding y-values are 4 and -4. The slopes at $(3, 4)$ and $(3, -4)$, obtained by substituting in $-x/y$, are $-\frac{3}{4}$ and $\frac{3}{4}$, respectively (see Figure 2).

To complicate matters, we point out that

$$x^2 + y^2 = 25$$

determines many other functions. Consider the function h defined by

$$h(x) = \begin{cases} \sqrt{25 - x^2} & \text{if } -5 \le x \le 3 \\ -\sqrt{25 - x^2} & \text{if } 3 < x \le 5 \end{cases}$$

It too satisfies $x^2 + y^2 = 25$, since $x^2 + [h(x)]^2 = 25$. But it is not even continuous at $x = 3$, so it certainly does not have a derivative there (see Figure 3).

While the subject of implicit functions leads to difficult technical questions (treated in advanced calculus), the problems we study have straightforward solutions.

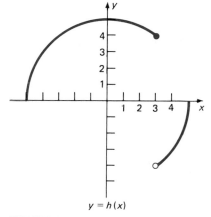

$y = h(x)$

FIGURE 3

MORE EXAMPLES In the examples that follow, we assume that the given equation determines one or more differentiable functions whose derivatives can be found by implicit differentiation.

EXAMPLE 2 Find dy/dx if $x^2 + 5y^3 = x + 9$.

Solution

$$\frac{d}{dx}(x^2 + 5y^3) = \frac{d}{dx}(x + 9)$$

$$2x + 15y^2\frac{dy}{dx} = 1$$

$$\frac{dy}{dx} = \frac{1 - 2x}{15y^2}$$

EXAMPLE 3 Find $D_t y$ if $t^3 + t^2 y - 10y^4 = 0$.

Solution

$$D_t(t^3 + t^2 y - 10y^4) = D_t(0)$$

$$3t^2 + t^2 D_t y + y(2t) - 40y^3 D_t y = 0$$

$$D_t y(t^2 - 40y^3) = -3t^2 - 2ty$$

$$D_t y = \frac{3t^2 + 2ty}{40y^3 - t^2}$$ ∎

EXAMPLE 4 Find the equation of the tangent line to the curve

$$y^3 - xy^2 + \cos xy = 2$$

at the point $(0, 1)$.

Solution For simplicity, let us use the notation y' for dy/dx. When we differentiate both sides and equate the results, we obtain

$$3y^2 y' - x(2yy') - y^2 - (\sin xy)(xy' + y) = 0$$

$$y'(3y^2 - 2xy - x \sin xy) = y^2 + y \sin xy$$

$$y' = \frac{y^2 + y \sin xy}{3y^2 - 2xy - x \sin xy}$$

At $(0, 1)$, $y' = \frac{1}{3}$. Thus, the equation of the tangent line at $(0, 1)$ is

$$y - 1 = \tfrac{1}{3}(x - 0)$$ ∎

THE POWER RULE AGAIN   We have learned that $D_x(x^n) = nx^{n-1}$, where n is any integer. We now extend this to the case where n is any rational number.

Theorem A

(Power Rule). Let r be any rational number. Then

$$D_x(x^r) = rx^{r-1}$$

Proof Since r is rational, r may be written as p/q, where p and q are integers with $q > 0$. Let

$$y = x^r = x^{p/q}$$

Then

$$y^q = x^p$$

and, by implicit differentiation,

$$qy^{q-1}D_xy = px^{p-1}$$

Thus,

$$D_xy = \frac{px^{p-1}}{qy^{q-1}} = \frac{p}{q}\frac{x^{p-1}}{(x^{p/q})^{q-1}} = \frac{p}{q}\frac{x^{p-1}}{x^{p-p/q}}$$

$$= \frac{p}{q}x^{p-1-p+p/q} = \frac{p}{q}x^{p/q-1} = rx^{r-1}$$

We have obtained the desired result, but—to be honest—we must point out a flaw in our argument. In the implicit differentiation step, we assumed that D_xy exists—that is, that $y = x^{p/q}$ is differentiable. We can fill this gap but since it is hard work, we relegate the complete proof to the appendix (Section A.2, Theorem C). ∎

EXAMPLE 5 Find D_xy if $y = 2x^{11/3} + 4x^{3/4} - 6/x^{2/3}$.

Solution First we write

$$D_xy = 2D_x(x^{11/3}) + 4D_x(x^{3/4}) - 6D_x(x^{-2/3})$$

Then, by the rule just proved,

$$D_xy = 2 \cdot \tfrac{11}{3}x^{8/3} + 4 \cdot \tfrac{3}{4}x^{-1/4} - 6(-\tfrac{2}{3})x^{-5/3}$$

$$= \tfrac{22}{3}x^{8/3} + 3x^{-1/4} + 4x^{-5/3}$$ ∎

EXAMPLE 6 If $y = \sqrt{t^4 - 3t + 17}$, find dy/dt.

Solution Think of this as

$$y = u^{1/2} \quad \text{and} \quad u = t^4 - 3t + 17$$

and apply the Chain Rule.

$$\frac{dy}{dt} = \frac{dy}{du}\frac{du}{dt}$$

$$= \left(\frac{1}{2}u^{-1/2}\right)(4t^3 - 3)$$

$$= \left(\frac{1}{2\sqrt{u}}\right)(4t^3 - 3)$$

$$= \frac{4t^3 - 3}{2\sqrt{t^4 - 3t + 17}}$$ ∎

Assuming that each equation in Problems 1–12 defines a differentiable function of x, find $D_x y$ by implicit differentiation.

1. $x^2 - y^2 = 9$

2. $4x^2 + 9y^2 = 36$

3. $xy = 4$

4. $b^2 x^2 + a^2 y^2 = a^2 b^2$, where a, b are constants.

5. $xy^2 - x + 16 = 0$

6. $x^3 - 3x^2 y + 19xy = 0$

7. $4x^3 + 11xy^2 - 2y^3 = 0$

8. $\sqrt{xy} + 3y = 10x$

9. $6x - \sqrt{2xy} + xy^3 = y^2$

10. $\dfrac{y^2}{x^3} - 1 = y^{3/2}$

11. $xy + \sin y = x^2$

12. $\cos(xy) = y^2 + 2x$

In Problems 13–18, find the equation of the tangent line at the indicated point (see Example 4).

13. $x^3 y + y^3 x = 10$; $(1, 2)$

14. $x^2 y^2 + 3xy = 10y$; $(2, 1)$

15. $\sin(xy) = y$; $(\pi/2, 1)$

16. $y + \cos(xy^2) + 3x^2 = 4$; $(1, 0)$

17. $x^{2/3} - y^{2/3} - 2y = 2$; $(1, -1)$

18. $\sqrt{y} + xy^2 = 5$; $(4, 1)$

In Problems 19–32, find dy/dx (see Examples 5 and 6).

19. $y = 3x^{5/3} + \sqrt{x}$

20. $y = \sqrt[3]{x} - 2x^{7/2}$

21. $y = \sqrt[3]{x} + \dfrac{1}{\sqrt[3]{x}}$

22. $y = \sqrt[4]{2x + 1}$

23. $y = \sqrt[4]{3x^2 - 4x}$

24. $y = (x^3 - 2x)^{1/3}$

25. $y = \dfrac{1}{(x^3 + 2x)^{2/3}}$

26. $y = (3x - 9)^{-5/3}$

27. $y = \sqrt{x^2 + \sin x}$

28. $y = \sqrt{x^2 \cos x}$

29. $y = \dfrac{1}{\sqrt[3]{x^2 \sin x}}$

30. $y = \sqrt[4]{1 + \sin 5x}$

31. $y = \sqrt[4]{1 + \cos(x^2 + 2x)}$

32. $y = \sqrt{\tan^2 x + \sin^2 x}$

33. If $s^2 t + t^3 = 1$, find ds/dt and dt/ds.

34. If $y = \sin(x^2) + 2x^3$, find dx/dy.

35. Sketch the graph of the circle

$$x^2 - 4x + y^2 + 3 = 0$$

and then find equations for the two tangent lines that pass through the origin.

36. Find the equation of the **normal line** (line perpendicular to the tangent line) to the curve $8(x^2 + y^2)^2 = 100(x^2 - y^2)$ at $(3, 1)$.

37. Suppose $xy + y^3 = 2$. Then implicit differentiation twice with respect to x yields in turn:
(a) $xy' + y + 3y^2 y' = 0$;
(b) $xy'' + y' + y' + 3y^2 y'' + 6y(y')^2 = 0$.

Solve (a) for y' and substitute in (b), and then solve for y''.

38. Find y'' if $x^3 - 4y^2 + 3 = 0$ (see Problem 37).

39. Find y'' at $(2, 1)$ if $2x^2 y - 4y^3 = 4$ (see Problem 37).

40. Use implicit differentiation twice to find y'' at $(3, 4)$ if $x^2 + y^2 = 25$.

41. Show that the normal line to $x^3 + y^3 = 3xy$ at $(\frac{3}{2}, \frac{3}{2})$ passes through the origin.

42. Show that the hyperbolas $xy = 1$ and $x^2 - y^2 = 1$ intersect at right angles.

43. Show that the graphs of $2x^2 + y^2 = 6$ and $y^2 = 4x$ intersect at right angles.

44. Suppose curves C_1 and C_2 intersect at (x_0, y_0) with slopes m_1 and m_2, respectively. Then (see Problem 28 of Section 2.3) the positive angle θ from C_1 (that is, from the tangent line to C_1 at (x_0, y_0)) to C_2 satisfies

$$\tan \theta = \frac{m_2 - m_1}{1 + m_1 m_2}$$

Find the angles from the circle $x^2 + y^2 = 1$ to the circle $(x - 1)^2 + y^2 = 1$ at the two points of intersection.

45. Find the angle from the line $y = 2x$ to the curve $x^2 - xy + 2y^2 = 28$ at their point of intersection in the first quadrant (see Problem 44).

46. A particle of mass m moves along the x-axis so that its position x and velocity $v = dx/dt$ satisfy

$$m(v^2 - v_0^2) = k(x_0^2 - x^2)$$

where v_0, x_0, and k are constants. Show by implicit differentiation that

$$m \frac{dv}{dt} = -kx$$

whenever $v \neq 0$.

47. The curve $x^2 - xy + y^2 = 16$ is an ellipse centered at the origin and with the line $y = x$ as its major axis. Find the equations of the tangent lines at the two points where the ellipse intersects the x-axis.

48. Find any points on the curve $x^2y - xy^2 = 2$ where the tangent line is vertical, that is, where $dx/dy = 0$.

3.9
Related Rates

If a variable y depends on time t, then its derivative dy/dt is called a **time rate of change**. Of course, if y measures distance, then this time rate of change is also called velocity. We are interested in a wide variety of time rates: the rate at which water is flowing into a bucket, the rate at which the area of an oil spill is growing, the rate at which the value of a piece of real estate is increasing, and so on. If y is given explicitly in terms of t, the problem is simple; we just differentiate and then evaluate the derivative at the required time.

It may be that, in place of knowing y explicitly in terms of t, we know a relationship that connects y and another variable x and that we also know something about dx/dt. We may still be able to find dy/dt, since dy/dt and dx/dt are **related rates**. This will usually require implicit differentiation.

TWO SIMPLE EXAMPLES　In preparation for outlining a systematic procedure for solving related rate problems, we discuss two examples.

EXAMPLE 1　A small balloon is released at a point 150 feet away from an observer, who is on level ground. If the balloon goes straight up at a rate of 8 feet per second, how fast is the distance from the observer to the balloon increasing when the balloon is 50 feet high?

FIGURE 1

Solution　Let t denote the number of seconds after the balloon is released. Let h denote the height of the balloon and s its distance from the observer (see Figure 1). Both h and s are variables that depend on t; however, the base of the triangle (the distance from the observer to the point of release) remains unchanged as t increases. We stress that the figure we have drawn is valid for all $t > 0$.

Next we ask (and answer) two basic questions about h and s.

(a) What is given? *Answer*: $dh/dt = 8$.

(b) What do we want to know? *Answer*: We want to know ds/dt at the instant when $h = 50$.

The variables s and h change with time (they are implicit functions of t), but they are always related by the Pythagorean equation

$$s^2 = h^2 + (150)^2$$

If we differentiate implicitly with respect to t and use the Chain Rule, we obtain

$$2s \frac{ds}{dt} = 2h \frac{dh}{dt}$$

or

$$s\frac{ds}{dt} = h\frac{dh}{dt}$$

This relationship also holds for all $t > 0$.

Now, and *not before now*, we turn to the situation when $h = 50$. From the Pythagorean Theorem, we see that, when $h = 50$,

$$s = \sqrt{(50)^2 + (150)^2} = 50\sqrt{10}$$

Substituting in $s(ds/dt) = h(dh/dt)$ yields

$$50\sqrt{10}\,\frac{ds}{dt} = 50(8)$$

or

$$\frac{ds}{dt} = \frac{8}{\sqrt{10}} \approx 2.53$$

At the instant when $h = 50$, the distance between the balloon and the observer is increasing at 2.53 feet per second. ∎

FIGURE 2

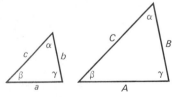
EXAMPLE 2 Water is pouring into a conical cistern at the rate of 8 cubic feet per minute. If the height of the cistern is 12 feet and the radius of its circular opening is 6 feet, how fast is the water level rising when the water is 4 feet deep?

Solution Denote the depth of the water in the cistern at any time t by h and let r be the corresponding radius of the surface of the water (see Figure 2).

We are *given* that the volume, V, of water in the cistern is increasing at the rate of 8 cubic feet per minute; that is, $dV/dt = 8$. We *want to know* how fast the water is rising—that is, dh/dt—at the instant when $h = 4$.

We need to find an equation relating V and h. The formula for the volume of water in the cistern, $V = \frac{1}{3}\pi r^2 h$, contains the unwanted variable r, unwanted because we do not know its rate dr/dt. However, by similar triangles (see the marginal box), we have $r/h = 6/12$, so $r = h/2$. Substituting this in $V = \frac{1}{3}\pi r^2 h$ gives

$$V = \frac{\pi h^3}{12}$$

a relation holding for all $t > 0$.

Now we differentiate implicitly, keeping in mind that h depends on t. We obtain

$$\frac{dV}{dt} = \frac{3\pi h^2}{12}\frac{dh}{dt}$$

or

$$\frac{dV}{dt} = \frac{\pi h^2}{4} \frac{dh}{dt}$$

At this point, and not earlier, we consider the situation when $h = 4$. Substituting $h = 4$ and $dV/dt = 8$, we obtain

$$8 = \frac{\pi(4)^2}{4} \frac{dh}{dt}$$

from which

$$\frac{dh}{dt} = \frac{2}{\pi} \approx 0.637$$

When the depth of the water is 4 feet, the water level is rising at 0.637 foot per minute.

If you think about it for a moment, you realize that the water level will rise more and more slowly as time goes on. For example, when $h = 10$

$$8 = \frac{\pi(10)^2}{4} \frac{dh}{dt}$$

so $dh/dt = 32/100\pi \approx 0.102$ foot per minute.

What we are really saying is that the acceleration d^2h/dt^2 is negative. We can calculate an expression for it. At any time t,

$$8 = \frac{\pi h^2}{4} \frac{dh}{dt}$$

so

$$\frac{32}{\pi} = h^2 \frac{dh}{dt}$$

If we differentiate implicitly again, we get

$$0 = h^2 \frac{d^2h}{dt^2} + \frac{dh}{dt} \left(2h \frac{dh}{dt} \right)$$

from which

$$\frac{d^2h}{dt^2} = \frac{-2\left(\dfrac{dh}{dt}\right)^2}{h}$$

This is clearly negative. ■

A SYSTEMATIC PROCEDURE　　Examples 1 and 2 suggest the following method for solving a related rates problem.

Step 1 Let t denote the elapsed time. Draw a diagram that is valid for all $t > 0$. Label those quantities whose values do not change as t increases with their given constant values. Assign letters to the quantities that vary with t, and label the appropriate parts of the figure with these variables.

Step 2 State what is given about the variables and what information is wanted about them. This information will be in the form of derivatives with respect to t.

Step 3 Write an equation relating the variables that is valid at all times $t > 0$, not just at some particular instant.

Step 4 Differentiate the equation found in Step 3 implicitly with respect to t. The resulting equation, containing derivatives with respect to t, is true for all $t > 0$.

Step 5 Substitute in the equation found in Step 4 all data that are valid *at the particular instant* for which the answer to the problem is required. Solve for the desired derivative.

EXAMPLE 3 A plane flying north at 640 miles per hour passes over a certain town at noon, and a second plane going east at 600 miles per hour is directly over the same town 15 minutes later. If the planes are flying at the same altitude, how fast will they be separating at 1:15 P.M.?

Solution

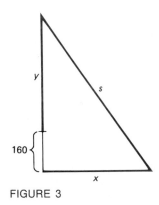

FIGURE 3

Step 1 Let t denote the number of hours after 12:15 P.M. Figure 3 shows the situation for all $t > 0$. The distance in miles from the town to the northbound plane when $t = 0$ (12:15 P.M.) is labeled with the constant $\frac{640}{4} = 160$. For any $t > 0$, we let y denote the distance in miles flown by the northbound plane (after 12:15 P.M.), x the distance flown by the eastbound plane, and s the distance between the planes.

Step 2 We are given that for all $t > 0$, $dy/dt = 640$ and $dx/dt = 600$. We want to know ds/dt at $t = 1$, that is, at 1:15 P.M.

Step 3 By the Pythagorean Theorem,

$$s^2 = x^2 + (y + 160)^2$$

Step 4 Differentiating implicitly with respect to t and using the Chain Rule, we have

$$2s \frac{ds}{dt} = 2x \frac{dx}{dt} + 2(y + 160) \frac{dy}{dt}$$

or

$$s \frac{ds}{dt} = x \frac{dx}{dt} + (y + 160) \frac{dy}{dt}$$

Step 5 For all $t > 0$, $dx/dt = 600$ and $dy/dt = 640$, while at the particular instant $t = 1$, $x = 600$, $y = 640$, and $s = \sqrt{(600)^2 + (640 + 160)^2} = 1000$. When we substitute these data in the equation of Step 4, we obtain

$$1000 \frac{ds}{dt} = (600)(600) + (640 + 160)(640)$$

from which

$$\frac{ds}{dt} = 872$$

At 1:15 P.M., the planes are separating at 872 miles per hour. ∎

CHAPTER 3/The Derivative

EXAMPLE 4 A woman standing on a cliff is watching a motor boat through a telescope as the boat approaches the shoreline directly below her. If the telescope is 250 feet above the water level and if the boat is approaching at 20 feet per second, at what rate is the angle of the telescope changing when the boat is 250 feet from shore?

Solution

Step 1 We draw a figure (Figure 4) and introduce variables x and θ, as shown.

Step 2 We are given that $dx/dt = -20$; the sign is negative because x is decreasing with time. We want to known $d\theta/dt$ at the instant when $x = 250$.

Step 3 From trigonometry,

$$\tan \theta = \frac{x}{250}$$

Step 4 We differentiate implicitly using the fact that $D_\theta \tan \theta = \sec^2 \theta$ (Example 2 of Section 3.4). We get

$$\sec^2 \theta \frac{d\theta}{dt} = \frac{1}{250} \frac{dx}{dt}$$

Step 5 At the instant when $x = 250$, θ is $\pi/4$ radians and $\sec^2 \theta = \sec^2(\pi/4) = 2$. Thus,

$$2\frac{d\theta}{dt} = \frac{1}{250}(-20)$$

or

$$\frac{d\theta}{dt} = \frac{-1}{25} = -0.04$$

The angle is changing at -0.04 radians per second. The sign is negative because θ is decreasing with time. ∎

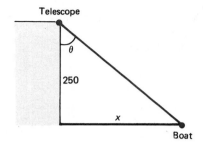

Telescope

θ

250

x

Boat

FIGURE 4

A GRAPHICAL RELATED RATE PROBLEM Often in a real-life situation, we do not know a formula for a certain function, but rather have an empirically determined graph for it. We may still be able to answer questions about rates.

EXAMPLE 5 Webster City monitors the height of the water in its cylindrical water tank with an automatic recording device. Water is constantly pumped into the tank at a rate of 2400 cubic feet per hour, as shown in Figure 5. During a certain 12-hour period (beginning at midnight), the water level rose and fell according to the graph in Figure 6. If the radius of the tank is 20 feet, at what rate was water being used at 7:00 A.M.?

2400 ft³/hr

h

20

$2400 - \dfrac{dV}{dt}$

FIGURE 5

h(ft)

18
15
12
9
6
3

1 2 3 4 5 6 7 8 9 10 11 t(hr)

FIGURE 6

Solution Let t denote the number of hours past midnight, h the height of the water in the tank at time t, and V the volume of water in the tank at that time (see Figure 5). Then $2400 - dV/dt$ is the rate at which water is being used at any time t. Since the slope of the tangent line at $t = 7$ is approximately -3 (Figure 6), we conclude that $dh/dt \approx -3$ at that time.

For a cylinder, $V = \pi r^2 h$, and so

$$V = \pi(20)^2 h$$

from which

$$\frac{dV}{dt} = 400\pi \frac{dh}{dt}$$

At $t = 7$,

$$\frac{dV}{dt} \approx 400\pi(-3) \approx -3770$$

Thus Webster City residents were using water at the rate of $2400 + 3770 = 6170$ cubic feet per hour at 7:00 A.M. ∎

PROBLEM SET 3.9

1. Each edge of a variable cube is increasing at the rate of 3 inches per second. How fast is the volume of the cube increasing when an edge is 10 inches long?

2. Assuming that a soap bubble retains its spherical shape as it expands, how fast is its radius increasing when its radius is 2 inches, if air is blown into it at the rate of 4 cubic inches a second?

3. An airplane, flying horizontally at an altitude of 1 mile, passes directly over an observer. If the constant speed of the plane is 240 miles per hour, how fast is its distance from the observer increasing 30 seconds later? *Hint*: Use Figure 7 and note that in 30 seconds ($\frac{1}{120}$ hour), the plane goes 2 miles.

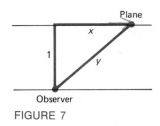

Plane

FIGURE 7

4. A student is using a straw to drink from a conical paper cup, whose axis is vertical, at the rate of 3 cubic centimeters a second. If the height of the cup is 10 centimeters and the diameter of its opening is 6 centimeters, how fast is the level of the liquid falling when the depth of the liquid is 5 centimeters?

5. An airplane, flying west at 400 miles per hour, goes over a certain town at 11:30 A.M., and a second plane at the same altitude, flying south at 500 miles per hour, goes over

the town at noon. How fast are they separating at 1:00 P.M.? *Hint*: See Example 3.

6. A man on a dock is pulling in a rope fastened to the bow of a small boat. If the man's hands are 12 feet higher than the point where the rope is attached to the boat and if he is retrieving the rope at the rate of 3 feet per second, how fast is the boat approaching the dock when 20 feet of rope are still out?

7. A 20-foot ladder is leaning against a wall. If the bottom of the ladder is pulled along the level pavement directly away from the wall at 2 feet per second, how fast is the top of the ladder moving down the wall when the foot of the ladder is 4 feet from the wall?

8. Oil from a ruptured tanker spreads in a circular pattern. If the radius of the circle increases at the constant rate of 1.5 feet per second, how fast is the enclosed area increasing at the end of 2 hours?

9. Sand is pouring from a pipe at the rate of 16 cubic feet per second. If the falling sand forms a conical pile on the ground whose altitude is always $\frac{1}{4}$ the diameter of the base, how fast is the altitude increasing when the pile is 4 feet high? *Hint*: Refer to Figure 8 and use the fact that $V = \frac{1}{3}\pi r^2 h$.

FIGURE 8

10. A child is flying a kite. If the kite is 90 feet above the child's hand level and the wind is blowing it on a horizontal course at 5 feet per second, how fast is the child paying out cord when 150 feet of cord is out? (Assume the cord forms a line—actually an unrealistic assumption.)

11. A swimming pool is 40 feet long, 20 feet wide, 8 feet deep at the deep end, and 3 feet deep at the shallow end; the bottom is rectangular (see Figure 9). If the pool is filled by pumping water into it at the rate of 40 cubic feet per minute, how fast is the water level rising when it is 3 feet deep at the deep end?

FIGURE 9

12. A particle P is moving along the graph of $y = \sqrt{x^2 - 4}$, $x \geq 2$, so that the x-coordinate of P is increasing at the rate of 5 units per second. How fast is the y-coordinate of P increasing when $x = 3$?

13. A metal disk expands during heating. If its radius increases at the rate of 0.02 inch per second, how fast is the area of one of its faces increasing when its radius is 8.1 inches?

14. Two ships sail from the same island port, one going north at 24 knots (24 nautical miles per hour) and the other east at 30 knots. The northbound ship departed at 9:00 A.M. and the eastbound ship left at 11:00 A.M. How fast is the distance between them increasing at 2:00 P.M.? *Hint*: Let $t = 0$ at 11:00 A.M.

15. A light in a lighthouse 1 kilometer offshore from a straight shoreline is rotating at 2 revolutions per minute. How fast is the beam moving along the shoreline when it passes the point $\frac{1}{2}$ kilometer from the point opposite the lighthouse?

C **16.** An aircraft spotter observes a plane flying toward her at an altitude of 4000 feet. She notes that when the angle of elevation is $\frac{1}{2}$ radian, it is increasing at a rate of $\frac{1}{10}$ radian per second. What is the ground speed of the airplane?

17. Andy, who is 6 feet tall, is walking away from a street light pole 30 feet high at a rate of 2 feet per second.
(a) How fast is his shadow increasing in length when Andy is 24 feet from the pole? 30 feet?
(b) How fast is the tip of his shadow moving?
(c) To follow the tip of his shadow, at what angular rate must he lift his head when his shadow is 6 feet long?

18. The vertex angle opposite the base of an isosceles triangle with equal sides of length 100 centimeters is increasing at $\frac{1}{10}$ radian per minute. How fast is the area of the triangle increasing when the vertex angle measures $\pi/6$ radians? *Hint*: $A = \frac{1}{2}ab \sin \gamma$.

19. A long, level highway bridge passes over a railroad track that is 100 feet below it and at right angles to it. If an automobile traveling 45 miles per hour (66 feet per second) is directly above a train going 60 miles per hour (88 feet per second), how fast will they be separating 10 seconds later?

20. Water is pumped at a uniform rate of 2 liters (1 liter = 1000 cubic centimeters) per minute into a tank shaped like a frustum of a right circular cone. The tank has altitude 80 centimeters and lower and upper radii of 20 and 40 centimeters, respectively (Figure 10). How fast is the water level rising when the depth of the water is 30 centimeters? *Note*: The volume, V, of a frustum of a right circular cone of altitude h and lower and upper radii a and b is $V = \frac{1}{3}\pi h \cdot (a^2 + ab + b^2)$.

FIGURE 10

21. Water is leaking out the bottom of a hemispherical tank of radius 8 feet at a rate of 2 cubic feet per hour. The tank was full at a certain time. How fast is the water level changing when its height h is 3 feet? *Note*: The volume of a segment of height h in a sphere of radius r is $\pi h^2[r - (h/3)]$. (See Figure 11.)

FIGURE 11

22. The hands on a clock are of length 5 inches (minute hand) and 4 inches (hour hand). How fast is the distance between the tips of the hands changing at 3:00?

23. A right circular cylinder with a piston at one end is filled with gas. Its volume is continually changing because of the movement of the piston. If the temperature of the gas is kept constant, then—by Boyle's Law—$PV = k$, where P is the pressure (pounds per square inch), V is the volume (cubic inches), and k is a constant. The pressure was monitored by a recording device over one 10-minute period. The results are shown in Figure 12. Approximately how fast was the volume changing at $t = 6.5$ if its volume was 300 cubic inches at that instant? (See Example 5.)

FIGURE 12

24. Work Example 5 in the text assuming the water tank is a sphere of radius 20 feet. (See Problem 21 for the volume of a spherical segment.)

25. An 18-foot ladder leans against a 12-foot vertical wall, its top extending over the wall. The bottom end of the ladder is pulled along the ground away from the wall at 2 feet per second.

(a) Find the vertical velocity of the top end when the ladder makes an angle of 60° with the ground.

(b) Find the vertical acceleration at the same instant.

26. A spherical steel ball rests at the bottom of the tank of Problem 21. Answer the question posed there if the ball has radius (a) 6 inches and (b) 2 feet.

27. A snowball melts at a rate proportional to its surface area.

(a) Show that its radius shrinks at a constant rate.

(b) If it melts to $\frac{8}{27}$ its original volume in one hour, how long will it take to melt completely?

28. A steel ball will drop $16t^2$ feet in t seconds. Such a ball is dropped from a height of 64 feet at a horizontal distance 10 feet from a 48-foot street light. How fast is the ball's shadow moving when the ball hits the ground?

3.10
Differentials and Approximations

We have been using the Leibniz notation dy/dx for the derivative of y with respect to x. Up to now, we have treated dy/dx as a single symbol and have not tried to give separate meanings to dy and dx. That is what we propose to do now.

To motivate our definition, let $P(x_0, y_0)$ be a fixed point on the graph of $y = f(x)$, as shown in Figure 1. With P as the origin, introduce new co-ordinate axes (the dx- and dy-axes) parallel to the old x- and y-axes. In this new coordinate system, the tangent line at P has a particularly simple equation, namely, $dy = m\,dx$, where m is its slope. But the slope m in the new coordinate system is the same as it was in the old xy-system. Thus $m = f'(x_0)$, and so the equation of the tangent line may be written

$$dy = f'(x_0)\,dx$$

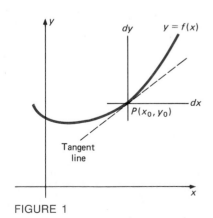

FIGURE 1

The usefulness of this notion rests on the fundamental fact that the tangent line is very close to the curve $y = f(x)$ in the vicinity of $P(x_0, y_0)$ (see Figure 2). Thus if x is given a small increment $\Delta x = dx$, the corresponding increment in y on the curve is $\Delta y = f(x_0 + \Delta x) - f(x_0)$, while on the tangent line it is $dy = f'(x_0)\Delta x$. But dy is a good approximation to Δy and, being just a constant times Δx, it is normally much easier to calculate. We explore the ramifications of this fact later in this section.

FIGURE 2

DIFFERENTIALS DEFINED   Here is the formal definition of *differentials*.

Definition

(Differentials). Let $y = f(x)$ be differentiable at x and suppose that dx, the differential of the independent variable x, denotes an arbitrary increment of x. The corresponding differential dy of the dependent variable y is defined by

$$dy = f'(x) \, dx$$

EXAMPLE 1   Find dy if (a) $y = x^3 - 3x + 1$, (b) $y = \sqrt{x^2 + 3x}$, (c) $y = \sin(x^4 - 3x^2 + 11)$.

Solution   If we know how to calculate derivatives, we know how to calculate differentials. We simply calculate the derivative and multiply it by dx.

(a) $dy = (3x^2 - 3) \, dx$

(b) $dy = \frac{1}{2}(x^2 + 3x)^{-1/2}(2x + 3) \, dx = \dfrac{2x + 3}{2\sqrt{x^2 + 3x}} \, dx$

(c) $dy = \cos(x^4 - 3x^2 + 11) \cdot (4x^3 - 6x) \, dx$ ∎

Now we ask you to note several things. First, since $dy = f'(x) \, dx$, division of both sides by dx yields

$$f'(x) = \frac{dy}{dx}$$

and we can, if we wish, interpret the derivative as a quotient of two differentials.

Second, corresponding to every derivative rule, there is a differential rule obtained from the former by multiplying through by dx. We illustrate the major rules in the table below.

Derivative Rule	Differential Rule
1. $\dfrac{dk}{dx} = 0$	**1.** $dk = 0$
2. $\dfrac{d(ku)}{dx} = k\dfrac{du}{dx}$	**2.** $d(ku) = k \, du$
3. $\dfrac{d(u + v)}{dx} = \dfrac{du}{dx} + \dfrac{dv}{dx}$	**3.** $d(u + v) = du + dv$
4. $\dfrac{d(uv)}{dx} = u\dfrac{dv}{dx} + v\dfrac{du}{dx}$	**4.** $d(uv) = u \, dv + v \, du$
5. $\dfrac{d(u/v)}{dx} = \dfrac{v(du/dx) - u(dv/dx)}{v^2}$	**5.** $d\left(\dfrac{u}{v}\right) = \dfrac{v \, du - u \, dv}{v^2}$
6. $\dfrac{d(u^n)}{dx} = nu^{n-1}\dfrac{du}{dx}$	**6.** $d(u^n) = nu^{n-1} \, du$

Third, while the definition of dy assumes that x is an independent variable, that assumption is not important. Suppose $y = f(x)$, where $x = g(t)$. Then t is the independent variable and both x and y depend on it. Now

$$dx = g'(t)\,dt$$

and since

$$y = f(g(t))$$
$$dy = f'(g(t))g'(t)\,dt$$
$$= f'(x)\,dx$$

Note that dy turns out to be $f'(x)\,dx$, as it would have if x had been the independent variable.

Finally, we interject a warning. *Be careful to distinguish between derivatives and differentials.* They are not the same. When you write $D_x y$ or dy/dx, you are using a symbol for the derivative; when you write dy, you are denoting a differential. Do not be sloppy and write dy when you mean to label a derivative. That would lead to boundless confusion.

APPROXIMATIONS Differentials will play several roles in this book, but for now their chief use is in providing approximations. We hinted at this earlier.

Suppose $y = f(x)$, as shown in Figure 3. When x is given an increment Δx, y receives a corresponding increment Δy, which can be approximated by dy. Thus, $f(x + \Delta x)$ is approximated by

$$\boxed{f(x + \Delta x) \approx f(x) + dy = f(x) + f'(x)\,\Delta x}$$

This is the basis for all the examples that follow.

EXAMPLE 2 Suppose you need good approximations to $\sqrt{4.6}$ and $\sqrt{8.2}$, but your calculator has died. What might you do?

Solution Consider the graph of $y = \sqrt{x}$ sketched in Figure 4. When x changes from 4 to 4.6, $\sqrt{x}$ changes from $\sqrt{4} = 2$ to (approximately) $\sqrt{4} + dy$. Now

$$dy = \frac{1}{2}x^{-1/2}\,dx = \frac{1}{2\sqrt{x}}\,dx$$

which, at $x = 4$ and $dx = 0.6$, has the value

$$dy = \frac{1}{2\sqrt{4}}(0.6) = \frac{0.6}{4} = 0.15$$

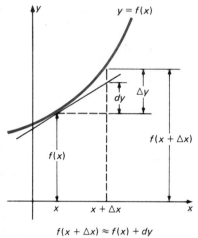

$$f(x + \Delta x) \approx f(x) + dy$$

FIGURE 3

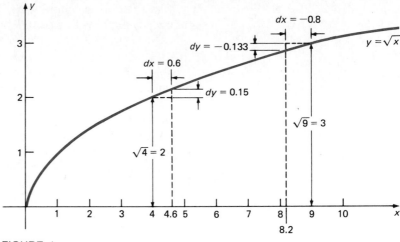

FIGURE 4

Thus,

$$\sqrt{4.6} \approx \sqrt{4} + dy = 2 + 0.15 = 2.15$$

Similarly, at $x = 9$ and $dx = -0.8$,

$$dy = \frac{1}{2\sqrt{9}}(-0.8) = \frac{-0.8}{6} = -0.133$$

Hence

$$\sqrt{8.2} \approx \sqrt{9} + dy = 3 - 0.133 = 2.867$$

Note that both dx and dy were negative in this case.

The approximate values 2.15 and 2.867 may be compared to the true values (to four decimal places) of 2.1448 and 2.8636. ∎

EXAMPLE 3 Use differentials to approximate the increase in the area of a soap bubble when its radius increases from 3 inches to 3.025 inches.

Solution The area of a spherical soap bubble is given by $A = 4\pi r^2$. We may approximate the exact change, ΔA, by the differential dA, where

$$dA = 8\pi r \, dr$$

At $r = 3$ and $dr = \Delta r = 0.025$,

$$dA = 8\pi(3)(0.025) \approx 1.885 \text{ square inches} \quad ∎$$

ESTIMATING ERRORS Here is a typical problem in science. A researcher measures a certain variable x to have a value x_0 with a possible error of size $\pm\Delta x$. The value x_0 is then used to calculate a value y_0 for y that depends on x. The value y_0 is contaminated by the error in x, but how badly? The standard procedure is to estimate this error by means of differentials.

EXAMPLE 4   The side of a cube is measured as 11.4 centimeters with a possible error of ± 0.05 centimeter. Evaluate the volume of the cube and give an estimate for the error in this value.

Solution   The volume V of a cube of side x is $V = x^3$. Thus $dV = 3x^2\,dx$. If $x = 11.4$ and $dx = 0.05$, then $V = (11.4)^3 \approx 1482$ and

$$dV = 3(11.4)^2(0.05) \approx 19.$$

Thus, we might report the volume of the cube as 1482 ± 19 cubic centimeters. ∎

EXAMPLE 5   It is known that $y = 3 \sin 2t + 4 \cos^2 t$. If t is measured as 1.13 ± 0.005, calculate y and give an estimate for the error.

Solution

$$y = 3 \sin(2)(1.13) + 4 \cos^2(1.13) \approx 3.043$$

$$dy = \left[(3)(2)\cos 2t - (4)(2)\cos t \sin t\right] dt$$

$$= \left[6 \cos(2)(1.13) - 8 \cos(1.13)\sin(1.13)\right](0.005)$$

$$\approx -0.035$$

Thus, $y = 3.043 \pm 0.035$. ∎

PROBLEM SET 3.10

In Problems 1–6, find dy.

1. $y = 2x^2 - 3x + 5$ **2.** $y = 7x^3 - 3x^2 + 4$

3. $y = (3 + 2x^3)^{-4}$ **4.** $y = \dfrac{13x}{5x^2 + 2}$

5. $y = \sqrt{4x^5 + 2x^4 - 5}$

6. $y = (6x^8 - 11x^5 + x^2)^{-2/3}$

7. If $s = \sqrt[5]{(t^2 - 3)^2}$, find ds.

8. If $F(x) = (5x^2 + 1)^2(x - 7)^5$, find dF.

9. Let $y = f(x) = x^3$. Find the value of dy in each case.
(a) $x = 0.5, dx = 1$ (b) $x = -1, dx = 0.75$

Make a careful drawing of the graph of f for $-1.5 \le x \le 1.5$ and the tangents to the curve at $x = 0.5$ and $x = -1$; on this drawing label dy and dx for each of the given sets of data in (a) and (b).

10. Let $y = 1/x$. Find the value of dy in each case.
(a) $x = 1, dx = 0.5$ (b) $x = -2, dx = 0.75$

Make a large scale drawing, as in Problem 9, for $-3 \le x < 0$ and $0 < x \le 3$.

[C] **11.** For the data of Problem 9, find the actual changes in y, namely, Δy.

[C] **12.** For the data of Problem 10, find the changes in y, namely, Δy.

13. If $y = x^2 - 3$, find the values of Δy and dy in each case.
(a) $x = 2$ and $dx = \Delta x = 0.5$
[C] (b) $x = 3$ and $dx = \Delta x = -0.12$

14. If $y = x^4 + 2x$, find the values of Δy and dy in each case.
(a) $x = 2$ and $dx = \Delta x = 1$
[C] (b) $x = 2$ and $dx = \Delta x = 0.005$

In Problems 15–18, use differentials to approximate the given number (see Example 2). Compare with calculator values.

15. $\sqrt{402}$ **16.** $\sqrt{35.9}$

17. $\sqrt[3]{26.91}$ **18.** $\sqrt[6]{64.05}$

C **19.** Approximate the volume of material in a spherical shell of inner radius 5 centimeters and outer radius 5.125 centimeters (see Example 3).

C **20.** All six sides of a cubical metal box are 0.25 inch thick, and the volume of the interior of the box is 40 cubic inches. Use differentials to find the approximate volume of metal used to make the box.

21. The outside diameter of a thin spherical shell is 12 feet. If the shell is 0.3 inch thick, use differentials to approximate the volume of the region interior to the shell.

22. The interior of an open cylindrical tank is 12 feet in diameter and 8 feet deep. The bottom is copper and the sides are steel. Use differentials to find approximately how many gallons of waterproofing paint are needed to apply a 0.05-inch coat to the steel part of the inside of the tank (1 gallon $\approx$ 231 cubic inches).

23. Assuming that the equator is a circle whose radius is approximately 4000 miles, how much longer than the equator would a concentric, coplanar circle be if each point on it were 2 feet above the equator? Use differentials.

24. The period of a simple pendulum of length L feet is given by $T = 2\pi\sqrt{L/g}$ seconds. We assume that g, the acceleration due to gravity on (or very near) the surface of the earth, is 32 feet per second per second. If the pendulum is that of a clock that keeps good time when $L = 4$ feet, how much time will the clock gain in 24 hours if the length of the pendulum is decreased to 3.97 feet?

25. The diameter of a sphere is measured as 20 ± 0.1 centimeters. Calculate the volume with an estimate for the error (see Examples 4 and 5).

26. A cylindrical roller is exactly 12 inches long and its diameter is measured as 6 ± 0.005 inches. Calculate its volume with an estimate for the error.

C **27.** The angle between the two equal sides of an isosceles triangle measures 0.53 ± 0.005 radians. The two equal sides are exactly 151 centimeters long. Calculate the length of the third side with an estimate for the error.

C **28.** Calculate the area of the triangle of Problem 27 with an estimate for the error. *Hint*: $A = \frac{1}{2}ab \sin \gamma$.

29. It can be shown that if $|d^2y/dx^2| \le M$ on a closed interval with c and $c + \Delta x$ as endpoints, then

$$|\Delta y - dy| \le \tfrac{1}{2}M(\Delta x)^2$$

Find, using differentials, the change in $y = 3x^2 - 2x + 11$ when x increases from 2 to 2.001 and then give a bound for the error you have made by using differentials.

30. Find $f(0.02)$ approximately if

$$f(x) = \sin[\sin(\sin 2x)]/\cos(\sin x).$$

31. A tank has the shape of a cylinder with hemispherical ends. If the cylindrical part is 100 centimeters long and has a radius of 10 centimeters, about how much paint is required to coat the outside of a tank to a thickness of 1 millimeter?

32. A conical cup, 10 centimeters high and 8 centimeters wide at the top, is filled with water to a depth of 9 centimeters. An ice cube 3 centimeters on a side is about to be dropped in. Use differentials to decide whether the cup will overflow.

3.11 Chapter Review

TRUE-FALSE QUIZ

Respond with true or false to each of the following assertions. Be prepared to defend your answer.

1. The tangent line to a curve at a point cannot cross the curve at that point.

2. The slope of the tangent line to the curve $y = x^4$ is different at every point of the curve.

3. It is possible for the velocity of an object to be increasing while its speed is decreasing.

4. If the tangent line to the graph of $y = f(x)$ is horizontal at $x = c$, then $f'(c) = 0$.

5. If $f'(x) = g'(x)$ for all x, then $f(x) = g(x)$ for all x.

6. If $y = \pi^5$, then $D_x y = 5\pi^4$.

7. If $f'(c)$ exists, then f is continuous at c.

8. The graph of $y = \sqrt[3]{x}$ has a tangent line at $x = 0$ and yet $D_x y$ does not exist there.

9. The derivative of a product is the product of the derivatives.

10. If the acceleration of an object is negative, then its velocity is decreasing.

11. If x^3 is a factor of the differentiable function $f(x)$, then x^2 is a factor of its derivative.

12. The equation of the line tangent to the graph of $y = x^3$ at $(1, 1)$ is $y - 1 = 3x^2(x - 1)$.

13. If $y = f(x)g(x)$, then $D_x^2 y = f(x)g''(x) + g(x)f''(x)$.

14. If $y = (x^3 + x)^8$, then $D_x^{25} y = 0$.

15. The derivative of a polynomial is a polynomial.

16. The derivative of a rational function is a rational function.

17. If $f'(c) = g'(c) = 0$ and $h(x) = f(x)g(x)$, then $h'(c) = 0$.

18. The expression

$$\lim_{x \to \pi/2} \frac{\sin x - 1}{x - \pi/2}$$

is the derivative of $f(x) = \sin x$ at $x = \pi/2$.

19. The operator D^2 is linear.

20. If $h(x) = f(g(x))$ where both f and g are differentiable, then $g'(c) = 0$ implies $h'(c) = 0$.

21. If $f'(2) = g'(2) = g(2) = 2$, then $(f \circ g)'(2) = 4$.

22. If f is differentiable and increasing and if $dx = \Delta x > 0$, then $\Delta y > dy$.

23. If the radius of a sphere is increasing at 3 feet per second, then its volume is increasing at 27 cubic feet per second.

24. $D_x^{n+4}(\sin x) = D_x^n(\sin x)$ for every positive integer n.

25. $\displaystyle\lim_{x \to 0} \frac{\tan x}{3x} = \frac{1}{3}$.

26. If $s = 5t^3 + 6t - 300$ gives the position of an object on a horizontal coordinate line at time t, then that object is always moving to the right (the direction of increasing s).

27. If air is being pumped into a spherical rubber balloon at a constant rate of 3 cubic inches per second, then the radius will increase but at a slower and slower rate.

28. If water is being pumped into a spherical tank of fixed radius at a rate of 3 gallons per second, the height of the water in the tank will increase more and more rapidly as the tank nears being full.

29. If an error Δr is made in measuring the radius of a sphere, the corresponding error in the calculated volume will be approximately $S \cdot \Delta r$ where S is the surface area of the sphere.

30. If $y = x^5$, then $dy \geq 0$.

SAMPLE TEST PROBLEMS

1. Use $f'(x) = \lim_{h \to 0} [f(x + h) - f(x)]/h$ to find the derivative of each of the following.

(a) $f(x) = x^2 - 5x$ (b) $f(x) = \dfrac{1}{x - 3}$

(c) $f(x) = \sqrt{9 - x}$

2. Use $g'(x) = \lim_{t \to x} \dfrac{g(t) - g(x)}{t - x}$ to find $g'(x)$ in each case.

(a) $g(x) = x^3$ (b) $g(x) = \sqrt{x}$

3. The given limit is a derivative, but of what function f and at what point?

(a) $\displaystyle\lim_{h \to 0} \frac{3(2 + h)^2 - 3(2)^2}{h}$ (b) $\displaystyle\lim_{\Delta x \to 0} \frac{\tan(\pi/4 + \Delta x) - 1}{\Delta x}$

(c) $\displaystyle\lim_{p \to x} \frac{3/p - 3/x}{p - x}$

4.

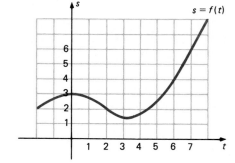

FIGURE 1

Use the sketch in the Figure 1 to approximate each of the following.

(a) $f'(2)$ (b) $f'(6)$

(c) v_{ave} on $[3, 7]$ (d) $\dfrac{d}{dt} f(t^2)$ at $t = 2$

(e) $\dfrac{d}{dt} [f^2(t)]$ at $t = 2$

In Problems 5–14, find each derivative by using the rules we have developed.

5. $D_x(x^3 - 3x^2 + x^{-2})$.

6. $D_x\left(\dfrac{3x - 5}{x^2 + 1}\right)$

7. $D_x^2(3x + 2)^{2/3}$

8. $D_t(t\sqrt{2t + 6})$ 9. $\dfrac{d}{dx}\left(\dfrac{1}{\sqrt{x^2 + 4}}\right)$

10. $\dfrac{d}{dt}\left[\sin(t^2) - \sin^2(t)\right]$

11. $\dfrac{d}{dx}(\cos^3 5x)$

12. $\dfrac{d}{dt}\left[\sin^2(\cos 4t)\right]$

13. $f'(2)$ if $f(x) = (x^2 - 1)^2(3x^3 - 4x)$

14. $g''(0)$ if $g(x) = \sin 3x + \sin^2 3x$

15. Find the coordinates of the point on the curve $y = (x - 2)^2$ at which the tangent line is perpendicular to the line $2x - y + 2 = 0$.

16. A spherical balloon is expanding from the sun's heat. Find the rate of change of the volume of the balloon with respect to its radius when the radius is 5 meters.

17. Use differentials to approximate the change in volume of the balloon of Problem 16 when its radius increased from 5 to 5.1 meters.

18. If the volume of the balloon of Problem 16 is increasing at a constant rate of 10 cubic meters per hour, how fast is its radius increasing when the radius is 5 meters?

19. A trough 12 feet long has a cross section in the form of an isosceles triangle 4 feet deep and 6 feet across at the top. If water is filling the trough at the rate of 9 cubic feet per minute, how fast is the water level rising when the water is 3 feet deep?

20. An object is projected directly upward from the ground with an initial velocity of 128 feet per second. Its height s at the end of t seconds is approximately $s = 128t - 16t^2$ feet.

(a) When does it reach its maximum height and what is this height?

(b) When does it hit the ground and with what velocity?

21. An object moves on a horizontal coordinate line. Its directed distance s from the origin at the end of t seconds is $s = t^3 - 6t^2 + 9t$ feet.

(a) When is the object moving to the left?

(b) What is its acceleration when its velocity is zero?

(c) When is its acceleration positive?

22. Find $D_x^{20}y$ in each case.

(a) $y = 13x^{19} - 2x^{12} - 6x^5 + 18$

(b) $y = \dfrac{1}{x}$

23. Find dy/dx in each case.

(a) $x^3 + y^3 = x^3y^3$

(b) $x \sin(xy) = x^2 + 1$

24. Show that the tangents to the curves $y^2 = 4x^3$ and $2x^2 + 3y^2 = 14$ at $(1, 2)$ are perpendicular to each other. *Hint*: Use implicit differentiation.

25. Let $y = \sin(\pi x) + x^2$. If x changes from 2 to 2.01, approximately how much does y change?

26. Suppose $f(2) = 3$, $f'(2) = 4$, $f''(2) = -1$, $g(2) = 2$, and $g'(2) = 5$. Find each value.

(a) $\dfrac{d}{dx}\left[f^2(x) + g^3(x)\right]$ at $x = 2$

(b) $\dfrac{d}{dx}\left[f(x)g(x)\right]$ at $x = 2$

(c) $\dfrac{d}{dx}\left[f(g(x))\right]$ at $x = 2$

(d) $D_x^2[f^2(x)]$ at $x = 2$

27. A 13-foot ladder is leaning against a vertical wall. If the bottom of the ladder is being pulled along the ground at a constant rate of 2 feet per second, how fast is the top end of the ladder moving down the wall when it is 5 feet above the ground?

28. An airplane is climbing at a $15°$ angle to the horizontal. How fast is it gaining altitude if its speed is 400 miles per hour?

29. Given that $D_x|x| = |x|/x$, $x \neq 0$, find a formula for $D_x|\sin x|$.

4 Applications of the Derivative

I do not know what I may appear to the world; but to myself I seem to have been only like a boy playing on the seashore, and diverting myself now and then finding a smoother pebble or a prettier shell than ordinary, whilst the great ocean of truth lay all undiscovered before me.
Isaac Newton

Isaac Newton
1642–1727

Born to an English farm family on Christmas day, 1642, Isaac Newton showed little academic promise as a young boy. He was bored with school, preferring to build kites, water wheels, clocks, and other gadgets. An uncle was the first to recognize the lad's unusual talents; he persuaded Newton's mother to send Newton off to Trinity College of Cambridge University. There he fell under the influence of Isaac Barrow, a theologian and professor of mathematics. Barrow saw in Newton a greater ability than his own and yielded his professorship to Newton when Newton was only 26.

Before that, just after graduation from Trinity, Newton had gone home to escape the great bubonic plague of 1664–65. For 18 months, beginning in January, 1665, he meditated on the outstanding problems of mathematics and science. There is no comparable burst of genius in the history of science.

In that short time, Newton discovered the general binomial theorem, the elements of both the differential and integral calculus, the theory of colors, and the universal law of gravitation. Lagrange has remarked that Newton was the greatest genius that ever lived and the most fortunate, since only once can the system of the universe be established.

Like so many of his contemporary scientists, Newton was devoutly religious and is said to have given as much time to study of the Bible as to mathematics. He died an honored man at age 85 and was buried with his nation's great in Westminster Abbey.

4.1
Maxima and Minima

FIGURE 1

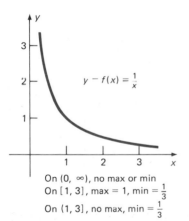

On $(0, \infty)$, no max or min
On $[1, 3]$, max = 1, min = $\frac{1}{3}$
On $(1, 3]$, no max, min = $\frac{1}{3}$

FIGURE 2

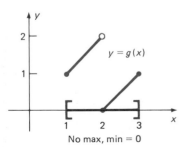

No max, min = 0

FIGURE 3

Often in life, we are faced with the problem of finding the *best* way to do something. For example, a farmer wants to choose the mix of crops that is likely to produce the largest profit. A doctor wishes to select the smallest dosage of a drug that will cure a certain disease. A manufacturer would like to minimize the cost of distributing its products. Sometimes a problem of this type can be formulated so it involves maximizing or minimizing a function over a specified set. If so, the methods of calculus provide a powerful tool for solving the problem.

Suppose then that we are given a function f and a domain S as in Figure 1. Our first job is to decide whether f even has a maximum value or a minimum value on S. Assuming that such values exist, we want to know where on S they are attained. Finally, we wish to determine the maximum and the minimum values. Analyzing these three tasks is the principal goal of this section.

We begin by introducing a precise vocabulary.

Definition

Let S, the domain of f, contain the point c. We say that:

(i) $f(c)$ is the **maximum value** of f on S if $f(c) \geq f(x)$ for all x in S;
(ii) $f(c)$ is the **minimum value** of f on S if $f(c) \leq f(x)$ for all x in S;
(iii) $f(c)$ is an **extreme value** of f on S if it is either the maximum value or the minimum value.

THE EXISTENCE QUESTION *Does f have a maximum (or minimum) value on S?* The answer depends first of all on the set S. Consider $f(x) = 1/x$ on $S = (0, \infty)$; it has neither a maximum value nor a minimum value (Figure 2). On the other hand, the same function on $S = [1, 3]$ has the maximum value of $f(1) = 1$ and the minimum value of $f(3) = \frac{1}{3}$. On $S = (1, 3]$, f has no maximum value and the minimum value of $f(3) = \frac{1}{3}$.

The answer also depends on the type of function. Consider the discontinuous function g (Figure 3) defined by

$$g(x) = \begin{cases} x & \text{if } 1 \leq x < 2 \\ x - 2 & \text{if } 2 \leq x \leq 3 \end{cases}$$

On $S = [1, 3]$, g has no maximum value (it gets arbitrarily close to 2 but never attains it). However, g has the minimum value $g(2) = 0$.

There is a nice theorem which answers the existence question for some of the problems that come up in practice. Though it is intuitively obvious, a rigorous proof is quite difficult; we leave that for more advanced textbooks.

Theorem A

(Max-Min Existence Theorem). If f is continuous on a closed interval $[a, b]$, then f attains both a maximum value and a minimum value there.

Note the key words; f is required to be *continuous* and the set S is required to be a *closed interval*.

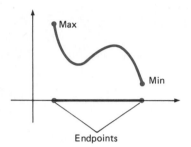

FIGURE 4

WHERE DO EXTREME VALUES OCCUR?   Usually a function that we want to maximize or minimize will have an interval I as its domain. But this interval may be any of the nine types discussed in Section 1.3. Some of them contain their **endpoints**; some do not. For instance, $I = [a, b]$ contains both its endpoints; $[a, b)$ contains only its left endpoint; (a, b) contains neither endpoint. Extreme values of functions defined on closed intervals often occur at endpoints (see Figure 4).

If c is a point at which $f'(c) = 0$, we call c a **stationary point**. The name derives from the fact that at a stationary point, the graph of f levels off, since the tangent line is horizontal. Extreme values often occur at stationary points (see Figure 5).

Finally, if c is an interior point of I where f' fails to exist, we call c a **singular point**. It is a point where the graph of f has a sharp corner, a vertical tangent, or perhaps takes a jump (or near which it wiggles very badly). Extreme values can occur at singular points (Figure 6), though in practical problems this is quite rare.

These three kinds of points (endpoints, stationary points, and singular points) are the key points of max-min theory. Any point in the domain of a function f of one of these three types is called a **critical point** of f.

FIGURE 5

EXAMPLE 1   Find the critical points of $f(x) = -2x^3 + 3x^2$ on $[-\frac{1}{2}, 2]$.

Solution   The endpoints are $-\frac{1}{2}$ and 2. To find the stationary points, we solve $f'(x) = -6x^2 + 6x = 0$ for x, obtaining 0 and 1. There are no singular points. Thus, the critical points are $-\frac{1}{2}$, 0, 1, 2.   ∎

FIGURE 6

Theorem B

(Critical Point Theorem). Let f be defined on an interval I containing the point c. If $f(c)$ is an extreme value, then c must be a critical point; that is, either c is:

(i)   an endpoint of I;
(ii)   a stationary point of f ($f'(c) = 0$);   or
(iii)   a singular point of f ($f'(c)$ does not exist).

Proof Consider first the case where $f(c)$ is the maximum value of f on I and suppose that c is neither an endpoint nor a singular point. It will be enough to show that c is then a stationary point.

Now since $f(c)$ is the maximum value, $f(x) \leq f(c)$ for all x in I; that is,

$$f(x) - f(c) \leq 0$$

Thus if $x < c$, so that $x - c < 0$, then

(1)
$$\frac{f(x) - f(c)}{x - c} \geq 0$$

whereas if $x > c$, then

(2)
$$\frac{f(x) - f(c)}{x - c} \leq 0$$

But $f'(c)$ exists, since c is not a singular point. Consequently, when we let $x \to c^-$ in (1) and $x \to c^+$ in (2), we obtain, respectively, $f'(c) \geq 0$ and $f'(c) \leq 0$. We conclude that $f'(c) = 0$, as desired.

The case where $f(c)$ is the minimum value is handled similarly. ∎

In the proof just given, we used the fact that the inequality $\leq$ is preserved under the operation of taking limits (see Problem 26 of Section 2.5).

WHAT ARE THE EXTREME VALUES? In view of Theorems A and B, we can now state a very simple procedure for finding the maximum value or minimum value of a continuous function f on a *closed interval I*.

Step 1 Find the critical points of f on I.
Step 2 Evaluate f at each of these critical points. The largest of these values is the maximum value; the smallest is the minimum value.

EXAMPLE 2 Find the maximum and minimum values of

$$f(x) = -2x^3 + 3x^2$$

on $\left[-\frac{1}{2}, 2\right]$.

Solution In Example 1, we identified $-\frac{1}{2}$, 0, 1, 2 as the critical points. Now $f(-\frac{1}{2}) = 1$, $f(0) = 0$, $f(1) = 1$, and $f(2) = -4$. Thus the maximum value is 1 (attained at both $-\frac{1}{2}$ and 1) and the minimum value is -4 (attained at 2). The graph of f is shown in Figure 7. ∎

EXAMPLE 3 The function $F(x) = x^{2/3}$ is continuous everywhere. Find its maximum and minimum values on $[-1, 2]$.

Solution $F'(x) = \frac{2}{3}x^{-1/3}$, which is never 0. However, $F'(0)$ does not exist, so 0 is a critical point, as are the endpoints -1 and 2. Now $F(-1) = 1$, $F(0) = 0$, and $F(2) = \sqrt[3]{4} \approx 1.59$. Thus the maximum value is $\sqrt[3]{4}$; the minimum value is 0. The graph is shown in Figure 8. ∎

FIGURE 7

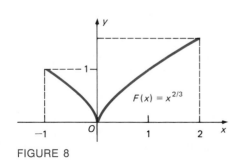

FIGURE 8

PRACTICAL PROBLEMS By a practical problem, we mean a problem that might arise in everyday life. Such problems rarely have singular points; in fact, for them, maximum and minimum values usually occur at stationary points, though endpoints should be checked. Here are two typical examples.

FIGURE 9

EXAMPLE 4 A rectangular box is to be made from a piece of cardboard 24 inches long and 9 inches wide by cutting out identical squares from the four corners and turning up the sides, as in Figure 9. Find the dimensions of the box of maximum volume. What is this volume?

Solution Let x be the width of the square to be cut out and V the volume of the resulting box. Then

$$V = x(9 - 2x)(24 - 2x) = 216x - 66x^2 + 4x^3$$

Now x cannot be less than 0 nor more than 4.5. Thus, our problem is to maximize V on $[0, 4.5]$. The stationary points are found by setting dV/dx equal to 0 and solving the resulting equation:

$$\frac{dV}{dx} = 216 - 132x + 12x^2 = 12(18 - 11x + x^2) = 12(9 - x)(2 - x) = 0$$

This gives $x = 2$ or $x = 9$, but 9 is not in the interval $[0, 4.5]$. We see that there are only three critical points, namely, 0, 2, and 4.5. At the endpoints 0 and 4.5, $V = 0$; at 2, $V = 200$. We conclude that the box has a maximum volume of 200 cubic inches if $x = 2$—that is, if the box is 20 inches long, 5 inches wide, and 2 inches deep. ∎

EXAMPLE 5 Farmer Brown has 100 meters of wire fence with which he plans to build two identical adjacent pens, as shown in Figure 10. What are the dimensions of the total enclosure for which its area is a maximum?

Solution Let x be the width and y the length of the total enclosure, both in meters. Because there are 100 meters of fence, $3x + 2y = 100$—that is,

$$y = 50 - \tfrac{3}{2}x$$

FIGURE 10

The total area A is given by

$$A = xy = 50x - \tfrac{3}{2}x^2$$

Since there must be three sides of length x, we see that $0 \le x \le \frac{100}{3}$. Thus, our problem is to maximize A on $[0, \frac{100}{3}]$. Now

$$\frac{dA}{dx} = 50 - 3x$$

When we set $50 - 3x$ equal to 0 and solve, we get $x = \frac{50}{3}$. Thus, there are three critical points: 0, $\frac{50}{3}$, and $\frac{100}{3}$. The two endpoints 0 and $\frac{100}{3}$ give

$A = 0$, while $x = \frac{50}{3}$ yields $A = 416.67$. The desired dimensions are $x = \frac{50}{3}$ meters and $y = 50 - \frac{3}{2}(\frac{50}{3}) = 25$ meters. ∎

Our last example illustrates a problem faced by a firm that delivers its products by truck. As the speed at which a truck is driven increases, the operating cost (gasoline, oil, and so) increases, whereas the driver cost goes down. What is the most economical speed at which a truck should be driven?

EXAMPLE 6 The operating cost for a certain truck is estimated to be $(30 + v/2)$ cents per mile when it is driven at a speed of v miles per hour. The driver is paid \$14 per hour. What speed will minimize the cost of making a delivery to a city k miles away? Assume that the law restricts the speed to $40 \leq v \leq 60$.

Solution Let C be the total cost in cents of driving the truck k miles. Then,

$$C = \text{driver cost} + \text{operating cost}$$

$$= \frac{k}{v}(1400) + k\left(30 + \frac{v}{2}\right) = 1400kv^{-1} + \left(\frac{k}{2}\right)v + 30k$$

Thus,

$$\frac{dC}{dv} = -1400kv^{-2} + \frac{k}{2} + 0$$

Setting dC/dv equal to 0 gives

$$\frac{1400k}{v^2} = \frac{k}{2}$$

$$v^2 = 2800$$

$$v \approx 53$$

A speed of 53 miles per hour would appear to be optimum, but we must evaluate C at the three critical points 40, 53, and 60 to be sure.

$$\text{At } v = 40, \quad C = k\left(\frac{1400}{40}\right) + k(30 + 20) = 85k$$

$$\text{At } v = 53, \quad C = k\left(\frac{1400}{53}\right) + k\left(30 + \frac{53}{2}\right) = 82.9k$$

$$\text{At } v = 60, \quad C = k\left(\frac{1400}{60}\right) + k(30 + 30) = 83.3k$$

We conclude that a speed of 53 miles per hour is best. ∎

We will have a good deal more to say about applied maxima and minima problems in Section 4.4.

In Problems 1–16, identify the critical points and find the maximum value and minimum value (see Examples 1, 2, and 3).

1. $f(x) = -x^2 + 4x - 1;\ I = [0, 3]$

2. $f(x) = x^2 + 3x;\ I = [-2, 1]$

3. $G(x) = \frac{1}{5}(2x^3 + 3x^2 - 12x);\ I = [-3, 3]$

4. $h(t) = 4t^3 + 3t^2 - 6t + 1;\ I = [-2, 1]$

5. $f(x) = x^3 - 3x + 1;\ I = (-\frac{3}{2}, 3)$ *Hint*: Sketch the graph.

6. $f(x) = x^3 - 3x + 1;\ I = [-\frac{3}{2}, 3]$

7. $g(x) = \dfrac{1}{1 + x^2};\ I = [-2, 1]$

8. $g(x) = \dfrac{1}{1 + x^2};\ I = (-\infty, \infty)$ *Hint*: Sketch the graph.

9. $f(x) = \dfrac{x}{x^2 + 2};\ I = [-1, 4]$

10. $F(t) = \sin t - \cos t;\ I = [0, \pi]$

11. $f(x) = |x - 2|;\ I = [1, 5]$

12. $f(x) = |5 - 3x|;\ I = [0, 3]$

13. $g(x) = x^{2/5};\ I = [-1, 32]$

14. $f(x) = [\![5x]\!];\ I = [-\frac{3}{2}, 4]$

15. $h(x) = x^{2/5};\ I = (-1, 32)$

16. $f(t) = t - \tan t;\ I = [-\pi/4, \pi/4]$

17. Find two nonnegative numbers whose sum is 10 and whose product is a maximum. *Hint*: If x is one number, $10 - x$ is the other.

18. What number exceeds its square by the maximum amount? Begin by convincing yourself that this number is on the interval $[0, 1]$.

19. John has 200 feet of fence with which he plans to enclose a rectangular yard for his dog. If he wishes the area to be a maximum, what should the dimensions be?

20. Show that for a rectangle of given perimeter K, the one of maximum area is a square.

21. Find the volume of the largest open box that can be made from a piece of cardboard 24 inches square by cutting equal squares from the corners and turning up the sides (see Example 4).

22. A piece of wire 16 inches long is cut into two pieces; one piece is bent to form a square and the other is bent to form a circle. Where should the cut be made in order that the sum of the areas of the square and the circle be a minimum? A maximum? (Allow the possibility of no cut.)

23. Farmer Brown has 80 feet of fence with which he plans to enclose a rectangular pen along one side of his 100-foot barn, as shown in Figure 11 (the side along the barn needs no fence). What are the dimensions of the pen that has maximum area?

FIGURE 11

24. Farmer Brown of Problem 23 decides to make three identical pens with his 80 feet of fence, as shown in Figure 12. What dimensions for the total enclosure make the area of the pens as large as possible?

FIGURE 12

25. Suppose that Farmer Brown of Problem 23 has 180 feet of fence and wants the pen to adjoin to the whole side of the barn, as shown in Figure 13. What should the dimensions be for maximum area? Note that $0 \le x \le 40$ in this case.

FIGURE 13

26. Suppose that Farmer Brown of Problem 23 decides to use his 80 feet of fence to make a rectangular pen to fit a 20-foot by 40-foot corner, as shown in Figure 14 (all the corner must be used and does not require fence). What dimensions give the pen a maximum area? *Hint*: Begin by deciding on the allowable values for x.

FIGURE 14

FIGURE 18

27. A rectangle has two corners on the *x*-axis and the other two on the parabola $y = 12 - x^2$, with $y \geq 0$. What are the dimensions of the rectangle of this type (Figure 15) with maximum area?

FIGURE 15

28. A rectangle is to be inscribed in a semicircle of radius *r*, as shown in Figure 16. What are the dimensions of the rectangle if its area is to be maximized?

FIGURE 16

29. A metal rain gutter is to have 3-inch sides and a 3-inch horizontal bottom, the sides making an equal angle θ with the bottom (Figure 17). What should θ be in order to maximize the carrying capacity of the gutter? *Note*: $0 \leq \theta \leq \pi/2$.

FIGURE 17

30. A huge conical tank is to be made from a circular piece of sheet metal of radius 10 meters by cutting out a sector with vertex angle θ and then welding the straight edges of the remaining piece together (Figure 18). Find θ so that the resulting cone has the largest possible volume.

31. The operating cost of a certain truck is $25 + x/4$ cents per mile if the truck travels at *x* miles per hour. In addition, the driver gets $12 per hour. What is the most economical speed at which to operate the truck on a 400-mile run if the highway speed is required to be between 40 and 55 miles per hour?

32. Redo Problem 31 assuming the operating cost is $40 + 0.05x^{3/2}$ cents per mile.

33. Find the points *P* and *Q* on the curve $y = x^2/4$, $0 \leq x \leq 2\sqrt{3}$, which are closest to and farthest from the point $(0, 4)$. *Hint*: The algebra is simpler if you consider the square of the required distance rather than the distance itself.

34. A humidifier uses a rotating disk of radius *r*, which is partially submerged in water. The most evaporation occurs when the exposed wetted region (shown shaded in Figure 19) is maximized. Show that this happens when *h* (the distance from the center to the water) is equal to $r/\sqrt{1 + \pi^2}$.

FIGURE 19

35. A covered box is to be made from a rectangular sheet of cardboard measuring 5 feet by 8 feet. This is done by cutting out the shaded regions of Figure 20 and then folding on the dotted lines. What are the dimensions *x*, *y*, and *z* that maximize the volume?

FIGURE 20

Consider the graph in Figure 1. No one will be surprised when we say that f is decreasing to the left of c and increasing to the right of c. But to make sure that we agree on terminology, we give precise definitions.

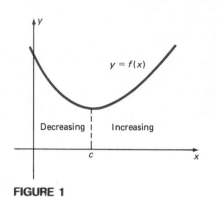

FIGURE 1

Definition

Let f be defined on an interval I (open, closed, or neither). We say that:

(i) f is **increasing** on I if for every pair of numbers x_1 and x_2 in I,

$$x_1 < x_2 \quad \Rightarrow \quad f(x_1) < f(x_2)$$

(ii) f is **decreasing** on I if for every pair of numbers x_1 and x_2 in I,

$$x_1 < x_2 \quad \Rightarrow \quad f(x_1) > f(x_2)$$

(iii) f is **strictly monotonic** on I if it is either increasing on I or decreasing on I.

How shall we decide where a function is increasing? Someone might suggest that we draw its graph and look at it. But a graph is usually drawn by plotting a few points and connecting those points with a smooth curve. Who can be sure that the graph does not wiggle between the plotted points? We need a better procedure.

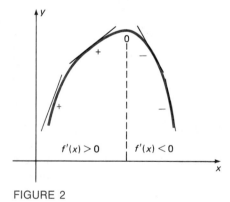

FIGURE 2

THE FIRST DERIVATIVE AND MONOTONICITY   Recall that the first derivative $f'(x)$ gives us the slope of the tangent line to the graph of f at the point x. Then, if $f'(x) > 0$, the tangent line is rising to the right (see Figure 2). Similarly, if $f'(x) < 0$, the tangent line is falling to the right. These facts make the following theorem intuitively clear. We postpone a rigorous proof until Section 4.8.

Theorem A

(Monotonicity Theorem). Let f be continuous on an interval I and differentiable at every interior point of I.

(i) If $f'(x) > 0$ for all x interior to I, then f is increasing on I.
(ii) If $f'(x) < 0$ for all x interior to I, then f is decreasing on I.

This theorem usually allows us to determine precisely where a differentiable function increases and where it decreases. It is a matter of solving two inequalities.

EXAMPLE 1   If $f(x) = 2x^3 - 3x^2 - 12x + 7$, find where f is increasing and where it is decreasing.

Values of f'

FIGURE 3

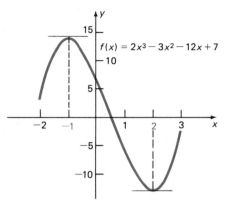

$f(x) = 2x^3 - 3x^2 - 12x + 7$

FIGURE 4

Solution We begin by finding the derivative of f.

$$f'(x) = 6x^2 - 6x - 12 = 6(x + 1)(x - 2)$$

We need to determine where

$$(x + 1)(x - 2) > 0$$

and also where

$$(x + 1)(x - 2) < 0.$$

This problem was discussed in great detail in Section 1.3, a section worth reviewing now. The split points are -1 and 2; they split the x-axis into three intervals: $(-\infty, -1)$, $(-1, 2)$, and $(2, \infty)$. Using the test points -2, 0, and 3, we conclude that $f'(x) > 0$ on the first and last of these intervals and that $f'(x) < 0$ on the middle interval (Figure 3). Thus, by Theorem A, f is increasing on $(-\infty, -1]$ and $[2, \infty)$; it is decreasing on $[-1, 2]$. Note that the theorem allows us to include the endpoints of these intervals, even though $f'(x) = 0$ at those points. The graph of f is shown in Figure 4. ∎

EXAMPLE 2 Determine where $g(x) = x/(1 + x^2)$ is increasing and where it is decreasing.

Solution

$$g'(x) = \frac{(1 + x^2) - x(2x)}{(1 + x^2)^2} = \frac{1 - x^2}{(1 + x^2)^2} = \frac{(1 - x)(1 + x)}{(1 + x^2)^2}$$

Since the denominator is always positive, $g'(x)$ has the same sign as $(1 - x)(1 + x)$. The split points, -1 and 1, determine the three intervals $(-\infty, -1)$, $(-1, 1)$, and $(1, \infty)$. When we test them, we find that $g'(x) < 0$ on the first and last of these intervals and that $g'(x) > 0$ on the middle one (Figure 5). We conclude from Theorem A that g is decreasing on $(-\infty, -1]$ and $[1, \infty)$, that it is increasing on $[-1, 1]$. We postpone graphing g until later, but if you want to see the graph, turn to Example 4. ∎

Values of g'

FIGURE 5

THE SECOND DERIVATIVE AND CONCAVITY A function may be increasing and still have a very wiggly graph (Figure 6). To analyze wiggles, we need to study how the tangent line turns as we move from left to right along the graph. If the tangent line turns steadily in the counterclockwise direction, we say that the graph is *concave up*; if the tangent turns in the clockwise direction, the graph is *concave down*. Both definitions are better stated in terms of functions and their derivatives.

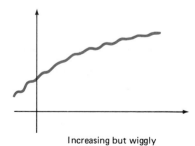

Increasing but wiggly

FIGURE 6

Definition

Let f be differentiable on an open interval I. If f' is increasing on I, f (and its graph) is **concave up** there; if f' is decreasing on I, f is **concave down** on I.

The diagrams in Figure 7 will help to clarify these notions. Note that a curve that is concave *up* is shaped like a *cup*.

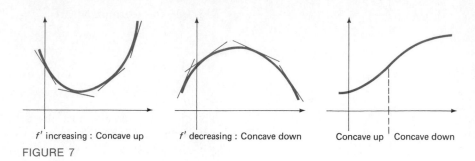

f' increasing : Concave up f' decreasing : Concave down   Concave up   Concave down

FIGURE 7

In view of Theorem A, we have a simple criterion for deciding where a curve is concave up and where it is concave down. We simply keep in mind that the second derivative of f is the first derivative of f'. Thus f' is increasing if f'' is positive; it is decreasing if f'' is negative.

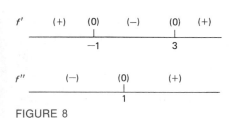

FIGURE 8

Theorem B

(Concavity Theorem). Let f be twice differentiable on the open interval I.

(i) If $f''(x) > 0$ for all x in I, then f is concave up on I.
(ii) If $f''(x) < 0$ for all x in I, then f is concave down on I.

For most functions, this theorem reduces the problem of determining concavity to the problem of solving inequalities. We are experts at this.

EXAMPLE 3 Where is $f(x) = \frac{1}{3}x^3 - x^2 - 3x + 4$ increasing, decreasing, concave up, and concave down?

Solution

$$f'(x) = x^2 - 2x - 3 = (x + 1)(x - 3)$$
$$f''(x) = 2x - 2 = 2(x - 1)$$

By solving the inequality $(x + 1)(x - 3) > 0$ and its opposite we conclude that f is increasing on $(-\infty, -1]$ and $[3, \infty)$ and decreasing on $[-1, 3]$ (Figure 8). Similarly, solving $2(x - 1) > 0$ and $2(x - 1) < 0$ shows that f is concave up on $(1, \infty)$, concave down on $(-\infty, 1)$. The graph of f is shown in Figure 9. ∎

EXAMPLE 4 Where is $g(x) = x/(1 + x^2)$ concave up and where is it concave down? Sketch the graph of g.

$y = f(x) = \frac{1}{3}x^3 - x^2 - 3x + 4$

FIGURE 9

Solution We began our study of this function in Example 2. There we learned that g is decreasing on $(-\infty, -1]$ and $[1, \infty)$ and increasing on $[-1, 1]$. To analyze concavity, we calculate g''.

$$g'(x) = \frac{1 - x^2}{(1 + x^2)^2}$$

$$g''(x) = \frac{(1 + x^2)^2(-2x) - (1 - x^2)(2)(1 + x^2)(2x)}{(1 + x^2)^4}$$

$$= \frac{(1 + x^2)[(1 + x^2)(-2x) - (1 - x^2)(4x)]}{(1 + x^2)^4}$$

$$= \frac{2x^3 - 6x}{(1 + x^2)^3} = \frac{2x(x^2 - 3)}{(1 + x^2)^3}$$

g' $(-)$ (0) $(+)$ (0) $(-)$

-1 1

g'' $(-)$ (0) $(+)$ (0) $(-)$ (0) $(+)$

$-\sqrt{3}$ 0 $\sqrt{3}$

FIGURE 10

Since the denominator is always positive, we need only solve $x(x^2 - 3) > 0$ and its opposite. The split points are $-\sqrt{3}$, 0, and $\sqrt{3}$. These three split points determine four intervals. After testing them (Figure 10), we conclude that g is concave up on $(-\sqrt{3}, 0)$ and $(\sqrt{3}, \infty)$ and that it is concave down on $(-\infty, -\sqrt{3})$ and $(0, \sqrt{3})$.

To sketch the graph of g we make use of all the information obtained so far, plus the fact that g is an odd function whose graph is symmetric with respect to the origin (Figure 11). ∎

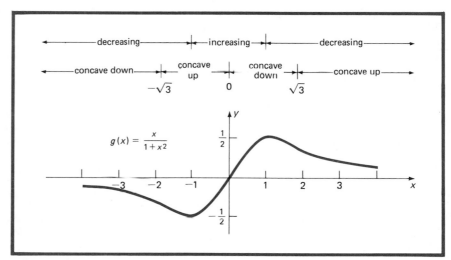

FIGURE 11

INFLECTION POINTS Let f be continuous at c. We call $(c, f(c))$ an **inflection point** of the graph of f if f is concave up on one side of c and concave down on the other side. The graph in Figure 12 indicates a number of possibilities.

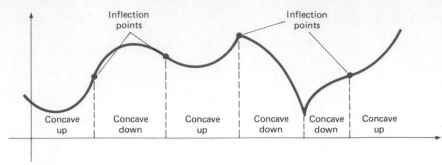

FIGURE 12

As you might guess, points where $f''(x) = 0$ or where $f''(x)$ does not exist are the candidates for points of inflection. We use the word *candidate* deliberately. Just as a candidate for political office may fail to be elected, so—for example—may a point where $f''(x) = 0$ fail to be a point of inflection. Consider $f(x) = x^4$, which has the graph shown in Figure 13. It is true that $f''(0) = 0$; yet the origin is not a point of inflection. However, in searching for inflection points, we begin by identifying those points where $f''(x) = 0$ (and where $f''(x)$ does not exist). Then we check to see if they really are inflection points.

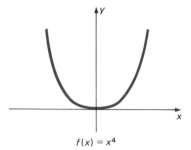

$f(x) = x^4$

FIGURE 13

Look back at the graph in Example 4. You will see that it has three inflection points. They are $(-\sqrt{3}, -\sqrt{3}/4)$, $(0, 0)$, and $(\sqrt{3}, \sqrt{3}/4)$.

EXAMPLE 5   Find all inflection points of the graph of $f(x) = \frac{1}{6}x^3 - 2x$.

Solution

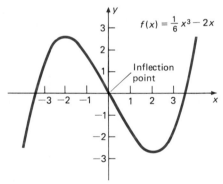

$f(x) = \frac{1}{6}x^3 - 2x$

Inflection point

FIGURE 14

$$f'(x) = \tfrac{1}{2}x^2 - 2$$
$$f''(x) = x$$

There is only one candidate for an inflection point, namely, the point where $f''(x) = 0$. This occurs at the origin, $(0, 0)$. That $(0, 0)$ is an inflection point follows from the fact that $f''(x) < 0$ for $x < 0$ and $f''(x) > 0$ for $x > 0$. Thus, the concavity changes direction at $(0, 0)$. The graph is shown in Figure 14.   ■

EXAMPLE 6   Find all points of inflection for $F(x) = x^{1/3} + 2$.

Solution

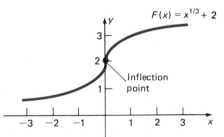

$F(x) = x^{1/3} + 2$

Inflection point

FIGURE 15

$$F'(x) = \frac{1}{3x^{2/3}} \qquad F''(x) = \frac{-2}{9x^{5/3}}$$

The second derivative, $F''(x)$, is never 0; however, it fails to exist at $x = 0$. The point $(0, 2)$ is an inflection point since $F''(x) > 0$ for $x < 0$ and $F''(x) < 0$ for $x > 0$. The graph is sketched in Figure 15.   ■

In Problems 1–10, use the Monotonicity Theorem to find where the given function is increasing and where it is decreasing.

1. $f(x) = x^2 - 4x + 2$

2. $f(x) = 2x - x^2$

3. $F(x) = x^3 - 1$

4. $F(x) = 2x^3 + 9x^2 - 13$

5. $g(t) = t^4 + 4t$

6. $g(t) = \dfrac{t^4}{2} - \dfrac{4t^3}{3}$

7. $f(x) = 2x^5 - 15x^4 + 30x^3 - 6$

8. $f(x) = \dfrac{2 - x}{x^2}$

9. $H(t) = \sin^2 2t,\ 0 \le t \le \pi$

10. $H(t) = \cos t + \sin t,\ 0 \le t \le 2\pi$

In Problems 11–18, use the Concavity Theorem to determine where the given function is concave up and where it is concave down. Also find all inflection points.

11. $f(x) = (x - 3)^2$

12. $f(x) = 4 - x^2$

13. $F(x) = x^3 - 12x$

14. $F(x) = (x - 3)^3 + 4$

15. $g(x) = 3x^2 - \dfrac{1}{x^2}$

16. $g(x) = x^4 - 6x^3 - 24x^2 + x + 2$

17. $g(x) = 2x^6 + 15x^4 + 90x^2 + 120x - 4$

18. $g(x) = 2x^2 + \cos^2 x$

In Problems 19–28, determine where the graph of the given function is increasing, decreasing, concave up, and concave down. Then sketch the graph (see Example 4).

19. $f(x) = x^3 - 3x - 1$

20. $g(x) = x^3 - 2x^2 + x + 1$

21. $g(x) = 3x^4 - 4x^3 + 2$

22. $F(x) = x^6 - 3x^4$

23. $G(x) = 3x^5 - 5x^3 + 1$

24. $H(x) = \dfrac{x^2}{x^2 + 1}$

25. $f(x) = \sqrt{\sin x}$ on $[0, \pi]$

26. $g(x) = x\sqrt{x - 2}$

27. $f(x) = x^{2/3}(1 - x)$

28. $g(x) = 8x^{1/3} + x^{4/3}$

In Problems 29–32, on $[0, 6]$ sketch the graph of a continuous function f that satisfies all of the stated conditions.

29. $f(0) = 3;\ f(3) = 0;\ f(6) = 4;$
$f'(x) < 0$ on $(0, 3);\ f'(x) > 0$ on $(3, 6);$
$f''(x) > 0$ on $(0, 5);\ f''(x) < 0$ on $(5, 6).$

30. $f(0) = 3;\ f(2) = 2;\ f(6) = 0;$
$f'(x) < 0$ on $(0, 2) \cup (2, 6);\ f'(2) = 0;$
$f''(x) < 0$ on $(0, 1) \cup (2, 6);\ f''(x) > 0$ on $(1, 2).$

31. $f(0) = f(4) = 1;\ f(2) = 2;\ f(6) = 0;$
$f'(x) > 0$ on $(0, 2);\ f'(x) < 0$ on $(2, 4) \cup (4, 6);$
$f'(2) = f'(4) = 0;\ f''(x) > 0$ on $(0, 1) \cup (3, 4);$
$f''(x) < 0$ on $(1, 3) \cup (4, 6).$

32. $f(0) = f(3) = 3;\ f(2) = 4;\ f(4) = 2;\ f(6) = 0;$
$f'(x) > 0$ on $(0, 2);\ f'(x) < 0$ on $(2, 4) \cup (4, 5);$
$f'(2) = f'(4) = 0;\ f'(x) = -1$ on $(5, 6);$
$f''(x) < 0$ on $(0, 3) \cup (4, 5);\ f''(x) > 0$ on $(3, 4).$

33. Prove that a quadratic function has no points of inflection.

34. Prove that a cubic function has exactly one point of inflection.

35. Prove that if $f'(x)$ exists and is continuous on an interval I and if $f'(x) \ne 0$ at all interior points of I, then either f is increasing throughout I or decreasing throughout I. *Hint*: Use the Intermediate Value Theorem to show that there cannot be two points x_1 and x_2 of I where f' has opposite signs.

36. Suppose f is a function whose derivative is $f'(x) = (x^2 - x + 1)/(x^2 + 1)$. Use Problem 35 to prove that f is increasing everywhere.

37. Use the Monotonicity Theorem to prove each statement if $0 < x < y$.

(a) $x^2 < y^2$ (b) $\sqrt{x} < \sqrt{y}$ (c) $\dfrac{1}{x} > \dfrac{1}{y}$

38. What conditions on a, b, and c will make $f(x) = ax^3 + bx^2 + cx + d$ always increasing?

39. Determine a and b so that $f(x) = a\sqrt{x} + b/\sqrt{x}$ has the point $(4, 13)$ as an inflection point.

40. The general cubic function $f(x)$ has three zeros, r_1, r_2, and r_3. Show that its inflection point has x-coordinate $(r_1 + r_2 + r_3)/3$. *Hint*: $f(x) = a(x - r_1)(x - r_2)(x - r_3)$.

41. Suppose that $f'(x) > 0$ and $g'(x) > 0$ for all x. What simple additional conditions (if any) are needed to guarantee that:

(a) $f(x) + g(x)$ is increasing for all x;
(b) $f(x) \cdot g(x)$ is increasing for all x;
(c) $f(g(x))$ is increasing for all x?

42. Suppose that $f''(x) > 0$ and $g''(x) > 0$ for all x. What simple additional conditions (if any) are needed to guarantee that:

(a) $f(x) + g(x)$ is concave up for all x;
(b) $f(x) \cdot g(x)$ is concave up for all x;
(c) $f(g(x))$ is concave up for all x?

4.3
Local Maxima and Minima

We recall from Section 4.1 that the maximum value (if it exists) of a function f on a set S is the largest value f attains on the whole set S. It is sometimes referred to as the **global maximum value**, or the *absolute maximum value* of f. Thus for the function f with domain $S = [a, b]$ whose graph is sketched in Figure 1, $f(a)$ is the global maximum value. But what about $f(c)$? It may not be king of the country, but at least it is chief of its own locality. We call it a **local maximum value**, or a *relative maximum value*. Of course a global maximum value is automatically a local maximum value. Figure 2 illustrates a number of possibilities. Note that the global maximum value (if it exists) is simply the largest of the local maximum values. Similarly, the global minimum value is the smallest of the local minimum values.

FIGURE 1

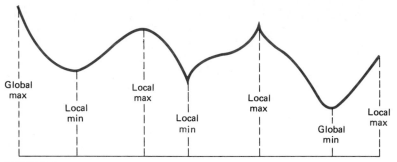

FIGURE 2

Here is the formal definition of local maxima and local minima. Recall that the symbol $\cap$ denotes the intersection (common part) of two sets.

Definition

Let S, the domain of f, contain the point c. We say that:

(i) $f(c)$ is a **local maximum value** of f if there is an interval (a, b) containing c such that $f(c)$ is the maximum value of f on $(a, b) \cap S$;

(ii) $f(c)$ is a **local minimum value** of f if there is an interval (a, b) containing c such that $f(c)$ is the minimum value of f on $(a, b) \cap S$;

(iii) $f(c)$ is a **local extreme value** of f if it is either a local maximum or a local minimum value.

WHERE DO LOCAL EXTREME VALUES OCCUR? The Critical Point Theorem (Theorem 4.1B) holds as stated, with the phrase *extreme value* replaced by *local extreme value*; the proof is essentially the same. Thus, the critical points (endpoints, stationary points, and singular points) are the candidates for points where local extrema may occur. We say *candidates* because we are not claiming that there must be a local extremum at every critical point. The left graph in Figure 3 makes this clear. However, if the derivative is positive on one side of the critical point and negative on the other, then we have a local extremum.

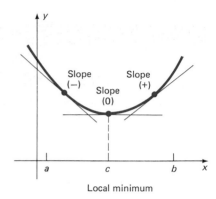

FIGURE 3

No local extreme value (left)

Local maximum (middle)

Local minimum (right)

Theorem A

(First Derivative Test for Local Extrema). Let f be continuous on an open interval (a, b) that contains a critical point c.

(i) If $f'(x) > 0$ for all x in (a, c) and $f'(x) < 0$ for all x in (c, b), then $f(c)$ is a local maximum value of f.

(ii) If $f'(x) < 0$ for all x in (a, c) and $f'(x) > 0$ for all x in (c, b), then $f(c)$ is a local minimum value of f.

(iii) If $f'(x)$ has the same sign on both sides of c, then $f(c)$ is not a local extreme value of f.

Proof of (i) Since $f'(x) > 0$ for all x in (a, c), f is increasing on $(a, c]$ by the Monotonicity Theorem. Again, since $f'(x) < 0$ for all x in (c, b), f is decreasing on $[c, b)$. Thus, $f(x) < f(c)$ for all x in (a, b), except of course at $x = c$. We conclude that $f(c)$ is a local maximum.

The proofs of (ii) and (iii) are similar. ∎

EXAMPLE 1 Find the local extreme values of the function $f(x) = x^2 - 6x + 5$ on $(-\infty, \infty)$.

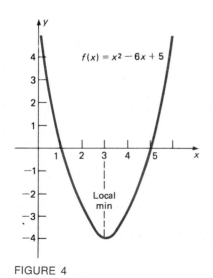

FIGURE 4

Solution The polynomial function f is continuous everywhere, and its derivative, $f'(x) = 2x - 6$, exists for all x. Thus, the only critical point for f is the single solution of $f'(x) = 0$, namely, $x = 3$.

Since $f'(x) = 2(x - 3) < 0$ for $x < 3$, f is decreasing on $(-\infty, 3]$; and because $2(x - 3) > 0$ for $x > 3$, f is increasing on $[3, \infty)$. Therefore, by the First Derivative Test, $f(3) = -4$ is a local minimum value of f. Since 3 is the only critical number, there are no other extreme values. The graph of f is shown in Figure 4. Note that $f(3)$ is actually the (global) minimum value in this case. ∎

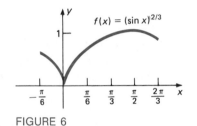

FIGURE 5

$$f(x) = \frac{1}{3}x^3 - x^2 - 3x + 4$$

FIGURE 6

EXAMPLE 2 Find the local extreme values of $f(x) = \frac{1}{3}x^3 - x^2 - 3x + 4$ on $(-\infty, \infty)$.

Solution Since $f'(x) = x^2 - 2x - 3 = (x + 1)(x - 3)$, the only critical points of f are -1 and 3. When we use the test points -2, 0, and 4, we learn that $(x + 1)(x - 3) > 0$ on $(-\infty, -1)$ and $(3, \infty)$ and $(x + 1)(x - 3) < 0$ on $(-1, 3)$. By the First Derivative Test, we conclude that $f(-1) = \frac{17}{3}$ is a local maximum value and that $f(3) = -5$ is a local minimum value (Figure 5). ∎

EXAMPLE 3 Find the local extreme values of $f(x) = (\sin x)^{2/3}$ on $(-\pi/6, 2\pi/3)$.

Solution

$$f'(x) = \frac{2 \cos x}{3(\sin x)^{1/3}}, \qquad x \neq 0$$

The points 0 and $\pi/2$ are critical points, since $f'(0)$ does not exist and $f'(\pi/2) = 0$. Now $f'(x) < 0$ on $(-\pi/6, 0)$ and on $(\pi/2, 2\pi/3)$, while $f'(x) > 0$ on $(0, \pi/2)$. By the First Derivative Test, we conclude that $f(0) = 0$ is a local minimum value and that $f(\pi/2) = 1$ is a local maximum value. The graph of f is shown in Figure 6. ∎

THE SECOND DERIVATIVE TEST There is another test for local maxima and minima that is sometimes easier to apply than the First Derivative Test. It involves evaluating the second derivative at the stationary points. It does not apply to singular points.

Theorem B

(Second Derivative Test for Local Extrema). Let f' and f'' exist at every point in an open interval (a, b) containing c, and suppose $f'(c) = 0$.

(i) If $f''(c) < 0$, $f(c)$ is a local maximum value of f.
(ii) If $f''(c) > 0$, $f(c)$ is a local minimum value of f.

Proof of **(i)** It is tempting to say that since $f''(c) < 0$, f is concave downward near c and to claim that this proves (i). However, to be sure that f is concave downward in a neighborhood of c, we need $f''(x) < 0$ in that neighborhood (not just at c), and nothing in our hypothesis guarantees that. We take a different tack.

By definition and hypothesis,

$$f''(c) = \lim_{x \to c} \frac{f'(x) - f'(c)}{x - c} = \lim_{x \to c} \frac{f'(x) - 0}{x - c} < 0$$

so we can conclude that there is a (possibly small) interval (α, β) around c where

$$\frac{f'(x)}{x - c} < 0, \qquad x \neq c$$

But this inequality implies that $f'(x) > 0$ for $\alpha < x < c$ and $f'(x) < 0$ for $c < x < \beta$. Thus, by the First Derivative Test, $f(c)$ is a local maximum value. The proof of (ii) is similar. ∎

EXAMPLE 4 For $f(x) = x^2 - 6x + 5$, use the Second Derivative Test to identify local extrema.

Solution This is the function of Example 1. Note that

$$f'(x) = 2x - 6 = 2(x - 3)$$
$$f''(x) = 2$$

Thus $f'(3) = 0$ and $f''(3) > 0$. Therefore, by the Second Derivative Test, $f(3)$ is a local minimum value. ∎

EXAMPLE 5 For $f(x) = \frac{1}{3}x^3 - x^2 - 3x + 4$, use the Second Derivative Test to identify local extrema.

Solution This is the function of Example 2.

$$f'(x) = x^2 - 2x - 3 = (x + 1)(x - 3)$$
$$f''(x) = 2x - 2$$

The critical points are -1 and 3 ($f'(-1) = f'(3) = 0$). Since $f''(-1) = -4$ and $f''(3) = 4$, we conclude—by the Second Derivative Test—that $f(-1)$ is a local maximum value and that $f(3)$ is a local minimum value. ∎

$f(x) = x^3$

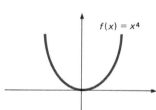

$f(x) = x^4$

FIGURE 7

Unfortunately, the Second Derivative Test sometimes fails, since $f''(x)$ may be 0 at a stationary point. For both $f(x) = x^3$ and $f(x) = x^4$, $f'(0) = 0$ and $f''(0) = 0$ (see Figure 7). The first does not have a local maximum or minimum value at 0; the second has a local minimum there. This shows that if $f''(x) = 0$ at a stationary point, we are unable to draw a conclusion about maxima or minima without more information.

PROBLEM SET 4.3

In Problems 1–6, identify the critical points. Then use (a) the First Derivative Test, and (if possible) (b) the Second Derivative Test to decide which of the critical points gives a local maximum value and which gives a local minimum value.

1. $f(x) = x^3 - 3x^2 + 2$

2. $f(x) = x^3 - 3x + 4$

3. $f(x) = \frac{1}{2}x - \sin x$, $0 < x < 2\pi$

4. $f(x) = \cos^2 x$, $-\pi/2 < x < 3\pi/2$

5. $g(x) = \frac{1}{4}x^4 + 1$

6. $g(x) = 3x^4 - 4x^3$

In Problems 7–16, find the critical points and use the test you prefer to decide which give a local maximum value and which give a local minimum value. What are these local maximum and minimum values?

7. $f(x) = \frac{1}{4}x^3 - 3x - 1$

8. $g(x) = x^4 - 2x^2 + 3$

9. $h(x) = x^4 + 2x^3$

10. $f(x) = (x - 1)^5$

11. $g(t) = 2 - (t - 1)^{2/3}$

12. $h(t) = 2t + t^{2/3}$

13. $f(x) = x + \dfrac{1}{x}$, $x \neq 0$

14. $g(x) = \dfrac{x^2}{\sqrt{x^2 + 1}}$

15. $f(t) = \dfrac{\sin t}{2 + \cos t}$, $0 < t < 2\pi$

16. $g(t) = |\cos t|$, $0 < t < 2\pi$

17. Find the (global) maximum and minimum values of $F(x) = 6\sqrt{x} - 3x$ on $[0, 9]$.

18. Do Problem 17 on the interval $[0, \infty)$.

19. Find (if possible) the maximum and minimum values of

$$f(x) = \frac{64}{\sin x} + \frac{27}{\cos x}$$

on $(0, \pi/2)$.

20. Find (if possible) the maximum and minimum values of

$$f(x) = x^2 + \frac{1}{x^2}$$

on $(0, \infty)$.

$\boxed{C}$ **21.** Find the minimum value of

$$g(x) = x^2 + \frac{16x^2}{(8-x)^2}, \qquad x > 8$$

22. Consider $f(x) = Ax^2 + Bx + C$ with $A > 0$. Show that $f(x) \geq 0$ for all x if and only if $B^2 - 4AC \leq 0$.

23. If $f'(x) = 2(x+2)(x+1)^2(x-2)^4(x-3)^3$, what values of x make $f(x)$ a local maximum? A local minimum?

24. What conclusions can you draw about f from the information that $f'(c) = f''(c) = 0$ and $f'''(c) > 0$?

25. Let f be a continuous function and let f' have the graph shown in Figure 8. Try to sketch a graph for f and answer the following questions.
(a) Where is f increasing? Decreasing?
(b) Where is f concave up? Concave down?
(c) Where does f attain a local maximum? A local minimum?
(d) Where are there inflection points for f?

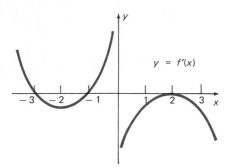

FIGURE 8

26. Suppose that both f and g are positive functions with local minima at c. Show that $f \cdot g$ must have a local minimum at c.

4.4
More Max-Min Problems

The problems we studied in Section 4.1 usually assumed that the set on which we wanted to maximize or minimize a function was a *closed* interval. However, the intervals that arise in practice are not always closed; they are sometimes open or even half open, half closed. We can still handle these problems if we correctly apply the theory developed in Section 4.3. Keep in mind that maximum (minimum) with no qualifying adjective means global maximum (minimum).

EXTREMA ON OPEN INTERVALS We give two examples to illustrate appropriate procedures for intervals that are open or half open.

EXAMPLE 1 Find (if possible) the minimum and maximum values of $f(x) = x^4 - 4x$ on $(-\infty, \infty)$.

Solution

$$f'(x) = 4x^3 - 4 = 4(x^3 - 1) = 4(x-1)(x^2 + x + 1)$$

Since $x^2 + x + 1 = 0$ has no real solutions (quadratic formula), there is only one critical point, namely, $x = 1$. For $x < 1$, $f'(x) < 0$, whereas for $x > 1$, $f'(x) > 0$. We conclude that $f(1) = -3$ is a local minimum value for f; and since f is decreasing on the left of 1 and increasing on the right of 1, it must actually be the minimum value of f.

The facts stated above imply that f cannot have a maximum value. The graph of f is shown in Figure 1. ∎

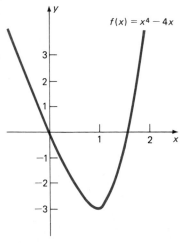

FIGURE 1

$f(x) = x^4 - 4x$

EXAMPLE 2 Find (if possible) the maximum and minimum values of $g(x) = x/(x^3 + 2)$ on $[0, \infty)$.

Solution

$$g'(x) = \frac{x^3 + 2 - x(3x^2)}{(x^3 + 2)^2} = \frac{2 - 2x^3}{(x^3 + 2)^2} = \frac{2(1 - x)(1 + x + x^2)}{(x^3 + 2)^2}$$

On $[0, \infty)$, there are two critical points: the endpoint 0 and the stationary point 1. For $0 < x < 1$, $g'(x) > 0$, while for $x > 1$, $g'(x) < 0$. Thus, $g(1) = \frac{1}{3}$ is the maximum value of g on $[0, \infty)$.

If g has a minimum value, it must occur at the other critical point, namely, $x = 0$. Now $g(0) = 0$ and $g(x) > 0$ for $x > 0$, so $g(0) = 0$ is the minimum value of g on $[0, \infty)$. The graph is shown in Figure 2. ∎

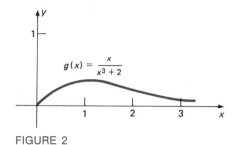

FIGURE 2

PRACTICAL PROBLEMS Each of the examples below is different; yet there are common elements in the procedures that we use to solve them. At the end of the section, we will suggest a set of steps to use in solving any max-min problem.

EXAMPLE 3 A handbill is to contain 50 square inches of printed matter, with 4-inch margins at top and bottom and 2-inch margins on each side. What dimensions for the handbill would use the least paper?

Solution Let x be the width and y the height of the handbill (Figure 3). Its area is

$$A = xy$$

We wish to minimize A.

As it stands, A is expressed in terms of two variables, a situation we do not know how to handle. However, we will find an equation connecting x and y so that one of these variables can be eliminated in the expression for A. The dimensions of the printed matter are $x - 4$ and $y - 8$ and its area is 50 square inches; so $(x - 4)(y - 8) = 50$. When we solve this equation for y, we obtain

$$y = \frac{50}{x - 4} + 8$$

FIGURE 3

Substituting this expression for y in $A = xy$ gives

$$A = \frac{50x}{x - 4} + 8x$$

The allowable values for x are $4 < x < \infty$; we want to minimize A on the open interval $(4, \infty)$.

Now

$$\frac{dA}{dx} = \frac{(x - 4)50 - 50x}{(x - 4)^2} + 8 = \frac{8x^2 - 64x - 72}{(x - 4)^2} = \frac{8(x + 1)(x - 9)}{(x - 4)^2}$$

The only critical points are obtained by solving $dA/dx = 0$; this yields $x = 9$ and $x = -1$. We reject $x = -1$ because it is not in the interval $(4, \infty)$. Since $dA/dx < 0$ for x in $(4, 9)$ and $dA/dx > 0$ for x in $(9, \infty)$, we conclude that A attains its minimum value at $x = 9$. This value of x makes $y = 18$ (found by substituting in the equation connecting x and y). So the dimensions for the handbill that will use the least amount of paper are 9 inches by 18 inches. ∎

EXAMPLE 4 A rectangular beam is to be cut from a log with circular cross section. If the strength of the beam is proportional to the product of its width and the square of its depth, find the dimensions of the cross section that give the strongest beam.

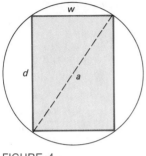

FIGURE 4

Solution Denote the diameter of the log by a (constant) and the width and depth of the beam by w and d, respectively (Figure 4). We wish to maximize S, the strength of the beam.

From the condition of the problem,

$$S = kwd^2$$

where k is a constant of proportionality. The strength S depends on the two variables w and d, but there is a simple relationship between them,

$$d^2 + w^2 = a^2$$

When we solve this equation for d^2 and substitute in the formula for S, we obtain

$$S = kw(a^2 - w^2) = ka^2w - kw^3$$

We consider the allowable values for w to be $0 < w < a$, an open interval.

To find the critical points, we calculate dS/dw, set it equal to 0, and solve for w.

$$\frac{dS}{dw} = ka^2 - 3kw^2 = k(a^2 - 3w^2)$$

$$k(a^2 - 3w^2) = 0$$

$$w = \frac{a}{\sqrt{3}}$$

Since $a/\sqrt{3}$ is the only critical point in $(0, a)$, it is likely that it gives the maximum S. A check on the sign of dS/dw to the left and right of $a/\sqrt{3}$ will confirm this.

When we substitute $w = a/\sqrt{3}$ in $d^2 + w^2 = a^2$, we learn that $d = \sqrt{2}a/\sqrt{3}$. The desired dimensions are $w = a/\sqrt{3}$ and $d = \sqrt{2}a/\sqrt{3}$. Note that $d = \sqrt{2}w$. ∎

EXAMPLE 5 Henry, who is in a rowboat 2 miles from the nearest point B on a straight shoreline, notices smoke billowing from his house, which is 6 miles down the shoreline from B. He figures he can row at 6 miles per hour and run at 10 miles per hour. How should he proceed in order to get to his house in least time?

Solution We interpret the problem to mean that we are to determine the x in Figure 5 that will make Henry's traveling time a minimum. It is clear that we should restrict x to the closed interval $[0, 6]$.

The distance AD is $\sqrt{x^2 + 4}$ miles and the time to row it is $\sqrt{x^2 + 4}/6$ hours. The distance DC is $6 - x$ miles and the time to run it is $(6 - x)/10$ hours. Thus the total time T in hours is

$$T = \frac{\sqrt{x^2 + 4}}{6} + \frac{6 - x}{10}$$

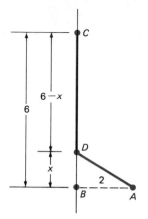

FIGURE 5

We wish to minimize T on $[0, 6]$.

This time there are three critical points, the endpoints 0 and 6 and a stationary point obtained by setting dT/dx equal to 0.

$$\frac{dT}{dx} = \frac{1}{6} \cdot \frac{1}{2}(x^2 + 4)^{-1/2}(2x) - \frac{1}{10}$$

$$= \frac{x}{6\sqrt{x^2 + 4}} - \frac{1}{10}$$

$$= \frac{10x - 6\sqrt{x^2 + 4}}{60\sqrt{x^2 + 4}}$$

When we set this equal to 0 and solve, we obtain, in successive steps,

$$10x - 6\sqrt{x^2 + 4} = 0$$

$$5x = 3\sqrt{x^2 + 4}$$

$$25x^2 = 9(x^2 + 4)$$

$$16x^2 = 36$$

$$x^2 = \tfrac{36}{16}$$

$$x = \tfrac{3}{2}$$

x	T
0	.93
1.5	.87
6	1.05

FIGURE 6

Since the domain for T is a closed interval, T has a minimum (Theorem 4.1A), and this minimum must occur at a critical point (Theorem 4.1B). After consulting the table in Figure 6, we conclude that Henry should row to a point 1.5 miles down the shoreline and then run the rest of the way. It will take him about 0.87 hour, or 52 minutes. For a similar problem in which one of the endpoints produces the minimum time, see Problem 15. ∎

EXAMPLE 6 Find the dimensions of the right circular cylinder of greatest volume that can be inscribed in a given right circular cone.

FIGURE 7

Solution Let a be the altitude and b the radius of the base of the given cone (both constants). Denote by h, r, and V the altitude, radius, and volume, respectively, of an inscribed cylinder (see Figure 7).

The volume of the cylinder is

$$V = \pi r^2 h$$

From similar triangles,

$$\frac{a - h}{r} = \frac{a}{b}$$

which gives

$$h = a - \frac{a}{b}r$$

When we substitute this expression for h in the formula for V, we obtain

$$V = \pi r^2 \left(a - \frac{a}{b}r \right) = \pi a r^2 - \pi \frac{a}{b}r^3$$

We wish to maximize V for r in the interval $[0, b]$. (Someone is sure to argue—and with good reason—that the appropriate interval is $(0, b)$. Actually, the answer is the same either way, though we have to use the First Derivative Test if we do the problem using $(0, b)$ as the domain.)
Now

$$\frac{dV}{dr} = 2\pi a r - 3\pi \frac{a}{b}r^2 = \pi a r \left(2 - \frac{3}{b}r \right)$$

This yields the stationary point $r = 2b/3$, giving us three critical points to consider: 0, $2b/3$, and b. One look at the picture shows that $r = 0$ and $r = b$ both give a volume of 0. Thus, $r = 2b/3$ has to give the maximum volume. When we substitute this value for r in the equation connecting r and h, we find that $h = a/3$. ∎

A SUMMARY OF THE METHOD   Based on the examples above, we suggest a step-by-step method to use in applied max-min problems. Do not follow it slavishly; common sense may sometimes suggest an alternative approach or omission of some steps.

Step 1   Draw a picture for the problem and assign appropriate variables to key quantities.
Step 2   Write a formula for the quantity Q to be maximized (minimized) in terms of those variables.
Step 3   Use the conditions of the problem to eliminate all but one of these variables and thereby express Q as a function of a single variable, such as x.
Step 4   Determine the set of possible values for x, usually an interval.
Step 5   Find the critical points (endpoints, stationary points, singular points). Most often, the key critical points are the stationary points where $dQ/dx = 0$.
Step 6   Use the theory of this chapter to decide which critical point gives the maximum (minimum).

1. Find two numbers whose product is -12 and the sum of whose squares is a minimum.

2. For what number does the principal fourth root exceed twice the number by the largest amount?

3. Find the points on the parabola $x = 2y^2$ that are closest to the point $(10, 0)$. *Hint:* Minimize the square of the distance between (x, y) and $(10, 0)$.

4. Find the points on the hyperbola $x^2/4 - y^2 = 1$ that are closest to the point $(5, 0)$.

5. A farmer wishes to fence off two identical adjoining rectangular pens, each with 900 square feet of area, as shown in Figure 8. What are x and y so that the least amount of fence is required?

FIGURE 8

6. Suppose that the outer boundary of the pens of Problem 5 requires heavy fence that costs $2 per foot, but that the middle partition requires fence costing only $1 per foot. What dimensions x and y will produce the least expensive fence?

7. Suppose the farmer of Problem 5 chose to make three adjoining pens each of 900 square feet, as shown in Figure 9. What are x and y so that the least amount of fence is required?

FIGURE 9

8. Study the solutions to Problems 5 and 7 to see if you can make a conjecture about the ratio of the amount of fence required in the x-direction to the amount required in the y-direction in all problems of this type. Then, if you are ambitious, try to prove your conjecture.

9. A cistern with a square base is to be constructed to hold 12,000 cubic feet of water. If the metal top costs twice as much per square foot as the concrete sides and base, what are the most economical dimensions for the cistern?

10. An open box with a capacity of 36,000 cubic inches is needed. If the box must be twice as long as it is wide, what dimensions would require the least amount of material?

11. If the strength of a rectangular beam is proportional to the product of its width and the square of its depth, find the dimensions of the strongest beam that can be cut from a log whose cross section has the form of the ellipse $9x^2 + 8y^2 = 72$. See Example 4.

12. The illumination at a point is inversely proportional to the square of the distance of the point from the light source and directly proportional to the intensity of the light source. If two light sources are s feet apart and their intensities are I_1 and I_2, respectively, at what point between them will the sum of their illuminations be a minimum?

13. A small island is 2 miles from the nearest point P on the straight shoreline of a large lake. If a man on the island can row his boat 3 miles per hour and can walk 4 miles per hour, where should he land his boat in order to arrive at a town 10 miles down the shore from P in the shortest time? See Example 5.

14. In Problem 13, suppose the man will be picked up by a car that will average 50 miles per hour when he gets to the shore. Then where should he land?

15. In Problem 13, suppose the man uses his motorboat that goes 20 miles per hour. Then where should he land?

16. A powerhouse is located on one bank of a straight river that is w feet wide. A factory is situated on the opposite bank of the river, L feet downstream from the point A directly opposite the powerhouse. What is the most economical path for a cable connecting the powerhouse to the factory if it costs a dollars per foot to lay the cable under water and b dollars on land $(a > b)$?

17. At 7:00 A.M. one ship was 60 miles due east from a second ship. If the first ship sailed west at 20 miles per hour and the second ship sailed southeast at 30 miles per hour, when were they closest together?

18. Find the equation of the line that is tangent to the ellipse $b^2x^2 + a^2y^2 = a^2b^2$ in the first quadrant and that forms with the coordinate axes the triangle with smallest possible area (a and b are positive constants).

19. Find the greatest volume that a right circular cylinder can have if it is inscribed in a sphere of radius r.

20. Show that the rectangle with maximum perimeter that can be inscribed in a circle is a square.

21. What are the relative dimensions of the right circular cylinder with greatest curved surface area that can be inscribed in a given sphere?

22. A right circular cone is to be inscribed in another right circular cone of given volume, with the same axis and with the vertex of the inner cone touching the base of the outer cone. What must be the ratio of their altitudes for the inscribed cone to have maximum volume?

23. A wire of length 100 centimeters is cut into two pieces; one is bent to form a square, the other an equilateral triangle. Where should the cut be made if (a) the sum of the two areas is to be a minimum; (b) a maximum? (Allow the possibility of no cut.)

24. A closed box in the form of a rectangular parallelepiped with a square base is to have a given volume. If the material used in the bottom costs 20% more per square inch than the material in the sides and the top costs 50% more per square inch than the sides, find the most economical proportions for the box.

25. An observatory is to be in the form of a right circular cylinder surmounted by a hemispherical dome. If the hemispherical dome costs twice as much per square foot as the cylindrical wall, what are the most economical proportions for a given volume?

26. A weight connected to a spring moves along the x-axis so that its x-coordinate at time t is

$$x = \sin 2t + \sqrt{3} \cos 2t$$

What is the farthest the weight gets from the origin?

27. A flower bed will be in the shape of a sector of a circle (a pie-shaped region) of radius r and vertex angle θ. Find r and θ if its area is a constant A and the perimeter is a minimum.

28. A fence h feet high runs parallel to a tall building and w feet from it (Figure 10). Find the length of the shortest ladder that will reach from the ground across the top of the fence to the wall of the building.

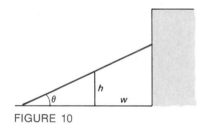

FIGURE 10

29. (Snell's Law). Fermat's Principle in optics says that light travels from point A to point B along the path that requires least time. Suppose that light travels in one medium at velocity c_1 and in a second medium at velocity c_2. If A is in medium 1 and B in medium 2 and the x-axis separates the two media, as shown in Figure 11, show that

$$\frac{\sin \theta_1}{c_1} = \frac{\sin \theta_2}{c_2}$$

FIGURE 11

30. Light from A is reflected to B by a plane mirror. Use Fermat's Principle (Problem 29) to show that the angle of incidence is equal to the angle of reflection.

31. One end of a 27-foot ladder rests on the ground and the other end rests on the top of an 8-foot wall. As the bottom end is pushed along the ground, the top end extends beyond the wall. Find the maximum horizontal overhang of the top end.

32. I have enough pure silver to coat 1 square meter of surface area. I plan to coat a sphere and a cube. What dimensions should they be if the total volume of the silvered solids is to be a maximum? A minimum? (Allow the possibility of all the silver going onto one solid.)

33. One corner of a long narrow strip of paper is folded over so it just touches the opposite side, as shown in Figure 12. With parts labeled as indicated, determine x in order to:

(a) maximize the area of triangle A;
(b) minimize the area of triangle B;
(c) minimize the length z.

FIGURE 12

34. Determine θ so that the area of the symmetric cross shown in Figure 13 is maximized. Then find this maximum area.

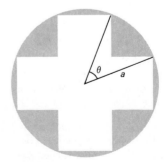

FIGURE 13

4.5
Economic Applications

FIGURE 1

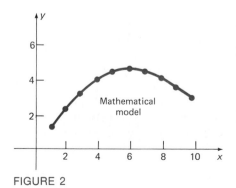

FIGURE 2

Each discipline has its own language. This is certainly true of economics, which has a highly developed special vocabulary. Once we learn this vocabulary, we will discover that many of the problems of economics are just ordinary calculus problems dressed in new clothes.

Consider a typical company, the ABC Company. For simplicity, assume that ABC produces and markets a single product; it might be television sets, car batteries, or bars of soap. If it sells x units of the product in a fixed period of time (for example, a year), it will be able to charge a **price**, $p(x)$, for each unit. We indicate that p depends on x because if ABC increases its output, it will probably need to reduce the price per unit in order to sell its total output. The **total revenue** that ABC can expect is given by $R(x) = xp(x)$, the number of units times the price per unit.

To produce and market x units, ABC will have a total cost, $C(x)$. This is normally the sum of a **fixed cost** (office utilities, real estate taxes, and so on) plus a **variable cost**, which depends directly on the number of units produced.

The key concept for a company is the **total profit**, $P(x)$. It is just the difference between revenue and cost—that is,

$$P(x) = R(x) - C(x) = R(x) - xp(x)$$

Generally, a company seeks to maximize its total profit.

Note a feature that tends to distinguish problems in economics from those in the physical sciences. In most cases, the product will be in discrete units (you can't make or sell 0.23 televisions sets or π car batteries). Thus, the functions $R(x)$, $C(x)$, and $P(x)$ are usually defined only for $x = 0, 1, 2, \ldots$ and consequently, their graphs consist of discrete points (Figure 1). In order to make the tools of calculus available, we connect these points with a smooth curve (Figure 2), thereby pretending that R, C, and P are nice differentiable functions. This illustrates an aspect of *mathematical modeling* that is almost always necessary, especially in economics. To model a real world problem, we must make simplifying assumptions. This means that the answers we get only approximate the answers we seek—one of the reasons economics is a less than perfect science.

A related problem for an economist is how to obtain formulas for the functions $C(x)$ and $p(x)$. In a simple case, $C(x)$ might have the form

$$C(x) = 10,000 + 50x$$

If so, $10,000 is the fixed cost, and there is a $50 direct cost for each unit produced. Perhaps a more typical situation is

$$\hat{C}(x) = 10,000 + 45x + 100\sqrt{x}$$

Note that in this case the average variable cost per unit is

$$\frac{45x + 100\sqrt{x}}{x} = 45 + \frac{100}{\sqrt{x}}$$

FIGURE 3

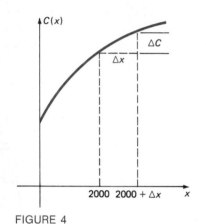

FIGURE 4

an amount that decreases as x increases (efficiency of size). The cost functions $C(x)$ and $\hat{C}(x)$ are graphed together in Figure 3.

Selecting appropriate cost and price functions is a difficult task. Occasionally, they can be inferred from basic assumptions. In other cases, a careful study of the history of the firm will suggest reasonable choices. Sometimes, we must simply make intelligent guesses.

USE OF THE WORD *MARGINAL*   Suppose that ABC knows its cost function $C(x)$ and that it has tentatively planned to produce 2000 units this year. President Hornblower would like to determine the additional cost per unit if ABC increased production slightly. Would it, for example, be less than the additional revenue per unit? If so, it would make good economic sense to increase production.

If the cost function is the one shown in Figure 4, President Hornblower is asking for the value of $\Delta C / \Delta x$ when $\Delta x = 1$. But we expect that this will be very close to the value of

$$\lim_{\Delta x \to 0} \frac{\Delta C}{\Delta x}$$

when $x = 2000$. This is called the **marginal cost**. We mathematicians recognize it as dC/dx, the derivative of C with respect to x.

In a similar vein, we define **marginal price** as dp/dx, **marginal revenue** as dR/dx, and **marginal profit** as dP/dx.

EXAMPLES   We now illustrate how to solve a wide variety of economic problems.

EXAMPLE 1   Suppose $C(x) = 8300 + 3.25x + 40\sqrt[3]{x}$ dollars. Find the average cost per unit and the marginal cost and then evaluate them when $x = 1000$.

Solution

$$\text{Average cost:} \quad \frac{C(x)}{x} = \frac{8300 + 3.25x + 40x^{1/3}}{x}$$

$$\text{Marginal cost:} \quad \frac{dC}{dx} = 3.25 + \frac{40}{3}x^{-2/3}$$

At $x = 1000$, these have the values 11.95 and 3.38, respectively. This means that it costs, on the average, $11.95 per unit to produce the first 1000 units; to produce one additional unit beyond 1000 costs only about $3.38. ∎

EXAMPLE 2   A company estimates that it can sell 1000 units per week if it sets the unit price at $3.00, but that its weekly sales will rise by 100 units for each 10¢ decrease in price. If x is the number of units sold each week ($x \geq 1000$), find:
(a) the price function, $p(x)$;

CHAPTER 4/Applications of the Derivative

(b) the number of units and the corresponding price that will maximize weekly revenue;

(c) the maximum weekly revenue.

Solution

(a) We are given that

$$x = 1000 + \frac{3.00 - p(x)}{0.10}(100)$$

or, equivalently,

$$p(x) = 3.00 - (0.10)\frac{(x - 1000)}{100} = 4 - 0.001x$$

(b)
$$R(x) = xp(x) = 4x - 0.001x^2$$

$$\frac{dR}{dx} = 4 - 0.002x$$

The only critical points are the endpoint 1000 and the stationary point 2000, obtained by setting $dR/dx = 0$. The First Derivative Test ($R'(x) > 0$ for $1000 \le x < 2000$ and $R'(x) < 0$ for $x > 2000$) shows that $x = 2000$ gives the maximum revenue. This corresponds to a unit price $p(2000) = \$2.00$.

(c) The maximum weekly revenue is $R(2000) = \$4000$. ∎

EXAMPLE 3   In manufacturing and selling x units of a certain commodity, the price function p and the cost function C (in dollars) are given by

$$p(x) = 5.00 - 0.002x$$

$$C(x) = 3.00 + 1.10x$$

Find expressions for the marginal revenue, marginal cost, and marginal profit; determine the production level that will produce the maximum total profit.

Solution

$$R(x) = xp(x) = 5.00x - 0.002x^2$$

$$P(x) = R(x) - C(x) = -3.00 + 3.90x - 0.002x^2$$

Thus, we have the following derivatives.

Marginal revenue: $\dfrac{dR}{dx} = 5 - 0.004x$

Marginal cost: $\dfrac{dC}{dx} = 1.1$

Marginal profit: $\dfrac{dP}{dx} = \dfrac{dR}{dx} - \dfrac{dC}{dx} = 3.9 - 0.004x$

To maximize profit, we set $dP/dx = 0$ and solve. This gives $x = 975$ as the only critical number to consider. It does provide a maximum, as may be checked by the First Derivative Test. The maximum profit is $P(975) = \$1898.25$. ∎

Note that at $x = 975$, both the marginal revenue and the marginal cost are \$1.10. In general, a company should expect to be at a maximum profit level when the cost of producing an additional unit exactly equals the revenue from that unit.

The statement just made assumes that the cost function and the revenue function are nice, differentiable functions and that endpoints are not significant. In some situations, the cost function may take large jumps, as when a new employee or a new piece of equipment is added; also, a manufacturing plant may have a maximum capacity, thereby introducing an important endpoint. We illustrate these possibilities in the next two examples.

EXAMPLE 4 The XYZ Company manufactures wicker chairs. With its present machines, it has a maximum yearly output of 500 units. If it makes x chairs, it can set a price of $p(x) = 200 - 0.15x$ dollars each and will have a total yearly cost of $C(x) = 4000 + 6x - (0.001)x^2$ dollars. What production level maximizes the total yearly profit?

Solution

$$R(x) = xp(x) = x(200 - 0.15x) = 200x - 0.15x^2$$

and so

$$P(x) = 200x - 0.15x^2 - (4000 + 6x - 0.001x^2)$$
$$= -4000 + 194x - 0.149x^2$$

Thus,

$$\frac{dP}{dx} = 194 - 0.298x$$

which yields the stationary point 651. However 651 is not in the interval $[0, 500]$, so the only critical points to check are the two endpoints, 0 and 500. If the maximum is at 0, the company better go out of business fast. It is not. The maximum occurs at 500, and the maximum profit is $P(500) = \$55,750$. ∎

EXAMPLE 5 With the addition of a new machine, the XYZ Company of Example 4 could boost its yearly production of chairs to 750. However, its cost function $C(x)$ will then take the form

$$C(x) = \begin{cases} 4000 + 6x - (0.001)x^2 & \text{if } 0 \le x \le 500 \\ 6000 + 6x - (0.003)x^2 & \text{if } 500 < x \le 750 \end{cases}$$

What production level maximizes total yearly profit under these circumstances?

Solution The new cost function results in the new profit function

$$P(x) = \begin{cases} -4000 + 194x - 0.149x^2 & \text{if } 0 \le x \le 500 \\ -6000 + 194x - 0.147x^2 & \text{if } 500 < x \le 750 \end{cases}$$

On the interval $500 < x < 750$,

$$\frac{dP}{dx} = 194 - 0.294x$$

FIGURE 5

which gives the stationary point 660. There are four critical points, namely, 0, 500, 660, and 750. The corresponding values of P are -4000, 55,750, 58,007, and 56,813. We conclude that a production level of 660 units gives the maximum profit. The graph in Figure 5 clarifies this example. ∎

PROBLEM SET 4.5

1. The fixed monthly cost for operating a manufacturing plant that makes zeebars is $8000 and there are direct costs of $110 for each unit produced. Write an expression for $C(x)$, the total cost of making x zeebars in a month.

2. The manufacturer of Problem 1 estimates that 100 units per month can be sold if the unit price is $250, and that sales will increase by 20 units for each $10 decrease in price. Write an expression for the price $p(x)$ and the revenue $R(x)$ if x units are sold in 1 month, $x \ge 100$ (see Example 2).

3. Use the information in Problems 1 and 2 to write an expression for the total monthly profit $P(x)$, $x \ge 100$.

4. Sketch the graph of $P(x)$ of Problem 3, and from it estimate the value of x that maximizes P. Then find this x exactly by the methods of calculus.

[C] **5.** The total cost of producing and selling x units of a certain commodity per month is $C(x) = 1200 + (3.25)x - (0.0002)x^2$. If the production level is 1800 units per month, find the average cost, $C(x)/x$, of each unit and the marginal cost.

6. The total cost of producing and selling x units of a certain commodity per week is $C(x) = 1100 + x^2/1200$. Find the average cost, $C(x)/x$, of each unit and the marginal cost at a production level of 900 units per week.

7. The total cost of producing and marketing x units of a certain commodity is given by

$$C(x) = \frac{80{,}000x - 400x^2 + x^3}{40{,}000}$$

For what number x is the average cost a minimum?

8. The total cost of producing and selling $100x$ units of a particular commodity per week is

$$C(x) = 1000 + 33x - 9x^2 + x^3$$

Find (a) the level of production at which the marginal cost is a minimum, and (b) the minimum marginal cost.

9. A price function, p, is defined by

$$p(x) = 20 + 4x - \frac{x^2}{3}$$

where $x \ge 0$ is the number of units.
(a) Find the total revenue function and the marginal revenue function.
(b) On what interval is the total revenue increasing?
(c) For what number x is the marginal revenue a maximum?

[C] **10.** For the price function defined by $p(x) = (182 - x/36)^{1/2}$, find the number of units x_1 that makes the total revenue a maximum and state the maximum possible revenue. What is the marginal revenue when the optimum number of units, x_1, are sold?

11. For the price function given by

$$p(x) = 800/(x + 3) - 3,$$

find the number of units x_1 that makes the total revenue a maximum and state the maximum possible revenue. What is the marginal revenue when the optimum number of units, x_1, is sold?

12. A river boat company offers a Fourth of July excursion to a fraternal organization with the understanding that there will be at least 400 passengers. The price of each ticket will be $12.00, and the company agrees to refund $0.20 to every passenger for each 10 passengers in excess of 400. Write an expression for the price function $p(x)$ and find the number x_1 of passengers that makes the total revenue a maximum.

13. A merchant finds that he can sell 4000 yards of a particular fabric each month if he prices it at $6.00 per yard, and that his monthly sales will increase by 250 yards for each $0.15 reduction in the price per yard. Write an expression for $p(x)$ and find the price per yard that would bring maximum revenue.

14. A manufacturer estimates that she can sell 500 articles per week if her unit price is $20.00, and that her weekly sales will rise by 50 with each $0.50 reduction in price. The cost of producing and selling x articles a week is $C(x) = 4200 + 5.10x + 0.0001x^2$. Find each of the following.

(a) The price function.
(b) The level of weekly production for maximum profit.
(c) The price per article at the optimum level of production.
(d) The marginal price at that level of production.

15. The monthly overhead of a manufacturer of a certain commodity is $6000, and the cost of material is $1.00 per unit. If not more than 4500 units are manufactured per month, labor cost is $0.40 per unit; but for each unit over 4500, the manufacturer must pay time-and-a-half for labor. The manufacturer can sell 4000 units per month at $7.00 per unit and estimates that monthly sales will rise by 100 for each $0.10 reduction in price. Find (a) the total cost function, (b) the price function, and (c) the number of units that should be produced each month for maximum profit.

C **16.** The ZEE Company makes zingos, which it markets at a price $p(x) = 10 - 0.001x$ dollars, where x is the number produced each month. Its total monthly cost is $C(x) = 200 + 4x - 0.01x^2$. At peak production it can make 300 units. What is its maximum monthly profit and what level of production gives this profit?

C **17.** If the company of Problem 16 expands its facilities so that it can produce up to 450 units each month, its monthly cost function takes the form $C(x) = 800 + 3x - 0.01x^2$ for $300 < x \le 450$. Find the production level that maximizes monthly profit and evaluate this profit. Sketch the graph of the monthly profit function $P(x)$ on $0 \le x \le 450$. (See Example 5.)

18. Let us suppose that a manufacturer has m employees, who produce a total of x units of product per week. These are sold at a price $p = p(x)$. Then the total weekly revenue $R(x) = x \cdot p$ can be thought of as depending on m. The derivative dR/dm is called the *marginal revenue product*. It is (approximately) the change in revenue when the manufacturer adds one employee. Show that

$$\frac{dR}{dm} = \frac{dx}{dm}\left(p + x\frac{dp}{dx}\right)$$

Hint: Use the Product Rule and then the Chain Rule.

C **19.** Refer to Problem 18. A manufacturer has determined that m employees can produce $x = 5m^2/\sqrt{m^2 + 13}$ units in a week, which it can then sell at a price $p = 10x - (0.1)x^2$ dollars. Determine the marginal revenue product when $m = 6$.

20. To be successful, a retail store must control its inventory. Overstocking results in excessive interest costs, extra warehouse rental, and the danger of obsolescence. Too small an inventory involves more paper work in reordering, extra delivery charges, and the greater likelihood of running out of stock. Carvers Appliance Outlet estimates that it costs $20 to hold a microwave oven in stock for a year. To reorder a lot of ovens costs $200 plus $3 for each oven. What lot size will result in the smallest inventory cost? Assume that Carvers sells 1000 ovens per year and that ordering lots of size x means that, on the average, $x/2$ ovens will be in stock.

21. Suppose that a store expects to sell N units of a particular item per year, that it costs A dollars to keep one unit in stock for a year, and that to reorder a lot of size x costs $(F + Bx)$ dollars. Show that $x = \sqrt{2FN/A}$ is the lot size that minimizes inventory cost. See Problem 20.

22. If expected sales of an item quadruple, what will happen to the ideal lot size? See Problem 21.

4.6
Limits at Infinity, Infinite Limits

The concept of the "infinite" has inspired and bedeviled mathematicians from time immemorial. The deepest problems and profoundest paradoxes of mathematics are often intertwined with use of this word. Yet mathematical progress can in part be measured in terms of understanding of the role of infinity. We have already used the symbols ∞ and $-\infty$ in our notation for certain intervals. Thus, $(3, \infty)$ is our way of denoting the set of all real numbers greater than 3. Please note that we have never referred to ∞ as a number.

CHAPTER 4/Applications of the Derivative

$$g(x) = \frac{x}{1 + x^2}$$

FIGURE 1

x	$\dfrac{x}{1 + x^2}$
10	.099
100	.010
1000	.001
10000	.0001
↓	↓
∞	?

FIGURE 2

For example, we have never added it to a number or divided it by a number. We will use the symbols ∞ and $-\infty$ in a new way in this section, but they will still not represent numbers.

AN EXAMPLE   Consider the function $g(x) = x/(1 + x^2)$. It was graphed rather carefully in Section 4.2; a smaller version of that graph is shown in Figure 1. We ask this question: What happens to $g(x)$ as x gets larger and larger? In symbols, we ask for the value of $\lim_{x \to \infty} g(x)$.

When we write $x \to \infty$, we are *not* implying that somewhere far, far to the right on the x-axis, there is a number—bigger than all other numbers—which x is approaching. Rather, we use $x \to \infty$ as a shorthand way of saying that x gets larger and larger without bound.

In the table in Figure 2, we have listed values of $g(x) = x/(1 + x^2)$ for several values of x. It appears that $g(x)$ gets smaller and smaller as x gets larger and larger. We write

$$\lim_{x \to \infty} \frac{x}{1 + x^2} = 0$$

Experimenting with large negative numbers would lead us to write

$$\lim_{x \to -\infty} \frac{x}{1 + x^2} = 0$$

RIGOROUS DEFINITIONS OF LIMITS AS $x \to \pm \infty$ In analogy with our ε, δ definition for ordinary limits, we make the following definition.

> *Definition*
>
> **(Limit as $x \to \infty$).** Let f be defined on $[c, \infty)$ for some number c. We say that $\lim_{x \to \infty} f(x) = L$ if for each $\varepsilon > 0$, there is a corresponding number M such that
>
> $$x > M \quad \Rightarrow \quad |f(x) - L| < \varepsilon$$

You will note that M can depend on ε. In general, the smaller ε is, the larger M will have to be. The graph in Figure 3 may help you understand what we are saying.

FIGURE 3

> *Definition*
>
> **(Limit as $x \to -\infty$).** Let f be defined on $(-\infty, c]$ for some number c. We say that $\lim_{x \to -\infty} f(x) = L$ if for each $\varepsilon > 0$, there is a corresponding number M such that
>
> $$x < M \quad \Rightarrow \quad |f(x) - L| < \varepsilon$$

EXAMPLE 1   Show that if k is a positive integer, then

$$\lim_{x \to \infty} \frac{1}{x^k} = 0 \quad \text{and} \quad \lim_{x \to -\infty} \frac{1}{x^k} = 0$$

Solution   Let $\varepsilon > 0$ be given. Choose $M = \sqrt[k]{1/\varepsilon}$. Then $x > M$ implies

$$\left| \frac{1}{x^k} - 0 \right| = \frac{1}{x^k} < \frac{1}{M^k} = \varepsilon$$

The proof of the second statement is similar.   ∎

Having given the definitions of these new limits, we must face the question of whether the Main Limit Theorem (Theorem 2.6A) holds for them. The answer is yes, and the proof is similar to the original one. Note how we use this theorem in the following examples.

EXAMPLE 2   Prove that $\displaystyle\lim_{x \to \infty} \frac{x}{1 + x^2} = 0$.

Solution   Here we use a standard trick: dividing numerator and denominator by the highest power of x that appears in the denominator, namely, x^2.

$$\lim_{x \to \infty} \frac{x}{1 + x^2} = \lim_{x \to \infty} \frac{\dfrac{x}{x^2}}{\dfrac{1 + x^2}{x^2}} = \lim_{x \to \infty} \frac{\dfrac{1}{x}}{\dfrac{1}{x^2} + 1}$$

$$= \frac{\displaystyle\lim_{x \to \infty} \frac{1}{x}}{\displaystyle\lim_{x \to \infty} \frac{1}{x^2} + \lim_{x \to \infty} 1} = \frac{0}{0 + 1} = 0$$ ∎

EXAMPLE 3   Find $\displaystyle\lim_{x \to \infty} \frac{2 - 3x + x^2}{7 + 4x - 5x^2}$.

Solution   Again, we divide numerator and denominator by x^2.

$$\lim_{x \to \infty} \frac{2 - 3x + x^2}{7 + 4x - 5x^2} = \lim_{x \to \infty} \frac{2/x^2 - 3/x + 1}{7/x^2 + 4/x - 5}$$

$$= \frac{\displaystyle\lim_{x \to \infty} (2/x^2 - 3/x + 1)}{\displaystyle\lim_{x \to \infty} (7/x^2 + 4/x - 5)} = \frac{0 - 0 + 1}{0 + 0 - 5} = -\frac{1}{5}$$ ∎

EXAMPLE 4   Find $\displaystyle\lim_{x \to -\infty} \frac{2x^3}{1 + x^3}$.

Solution Divide numerator and denominator by x^3.

$$\lim_{x \to -\infty} \frac{2x^3}{1 + x^3} = \lim_{x \to -\infty} \frac{2}{1/x^3 + 1} = \frac{2}{0 + 1} = 2 \qquad \blacksquare$$

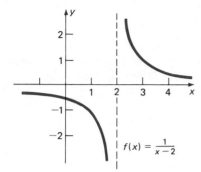

FIGURE 4

INFINITE LIMITS Consider the graph of $f(x) = 1/(x - 2)$, which is shown in Figure 4. It makes no sense to ask for $\lim_{x \to 2} 1/(x - 2)$ but we think it is reasonable to write

$$\lim_{x \to 2^-} \frac{1}{x - 2} = -\infty \qquad \lim_{x \to 2^+} \frac{1}{x - 2} = \infty$$

Here is the definition that relates to this situation.

Definition

(Infinite Limits). We say that $\lim_{x \to c^+} f(x) = \infty$ if for each positive number M, there corresponds a $\delta > 0$ such that

$$0 < x - c < \delta \quad \Rightarrow \quad f(x) > M$$

There are corresponding definitions of

$$\lim_{x \to c^+} f(x) = -\infty, \ \lim_{x \to c^-} f(x) = \infty, \text{ and } \lim_{x \to c^-} f(x) = -\infty.$$

EXAMPLE 5 Find $\lim_{x \to 1^-} \dfrac{1}{(x - 1)^2}$ and $\lim_{x \to 1^+} \dfrac{1}{(x - 1)^2}$.

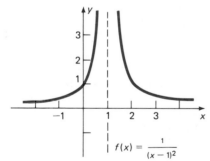

FIGURE 5

Solution The graph of $f(x) = 1/(x - 1)^2$ is shown in Figure 5. We think it is quite clear that

$$\lim_{x \to 1^-} \frac{1}{(x - 1)^2} = \infty \qquad \lim_{x \to 1^+} \frac{1}{(x - 1)^2} = \infty$$

Since both limits are ∞, we could also write

$$\lim_{x \to 1} \frac{1}{(x - 1)^2} = \infty \qquad \blacksquare$$

EXAMPLE 6 Find $\lim_{x \to 2^+} \dfrac{x + 1}{x^2 - 5x + 6}$.

Solution

$$\lim_{x \to 2^+} \frac{x + 1}{x^2 - 5x + 6} = \lim_{x \to 2^+} \frac{x + 1}{(x - 3)(x - 2)}$$

As $x \to 2^+$, we see that $x + 1 \to 3$, $x - 3 \to -1$, and $x - 2 \to 0^+$; thus, the numerator is approaching 3, but the denominator is negative and approaching 0. We conclude that

$$\lim_{x \to 2^+} \frac{x + 1}{(x - 3)(x - 2)} = -\infty$$

■

RELATION TO ASYMPTOTES Asymptotes were discussed briefly in Section 2.1, but now we can say more about them. The line $x = c$ is a **vertical asymptote** of the graph of $y = f(x)$ if any of the following four statements is true.

1. $\lim\limits_{x \to c^+} f(x) = \infty$
2. $\lim\limits_{x \to c^+} f(x) = -\infty$
3. $\lim\limits_{x \to c^-} f(x) = \infty$
4. $\lim\limits_{x \to c^-} f(x) = -\infty$

Thus, in Example 5, the line $x = 1$ is a vertical asymptote. Likewise, the lines $x = 2$ and $x = 3$ are vertical asymptotes in Example 6.

In a similar vein, the line $y = b$ is a **horizontal asymptote** of the graph of $y = f(x)$ if either

$$\lim_{x \to \infty} f(x) = b \quad \text{or} \quad \lim_{x \to -\infty} f(x) = b$$

The line $y = 0$ is a horizontal asymptote in Example 5.

EXAMPLE 7 Find the vertical and horizontal asymptotes of the graph of $y = f(x)$ if

$$f(x) = \frac{2x}{x - 1}$$

Solution We expect a vertical asymptote at a point where the denominator is zero, and we are right since

$$\lim_{x \to 1^+} \frac{2x}{x - 1} = \infty \quad \text{and} \quad \lim_{x \to 1^-} \frac{2x}{x - 1} = -\infty$$

On the other hand,

$$\lim_{x \to \infty} \frac{2x}{x - 1} = \lim_{x \to \infty} \frac{2}{1 - 1/x} = 2 \quad \text{and} \quad \lim_{x \to -\infty} \frac{2x}{x - 1} = 2$$

and so $y = 2$ is a horizontal asymptote. The graph of $y = 2x/(x - 1)$ is shown in Figure 6. ■

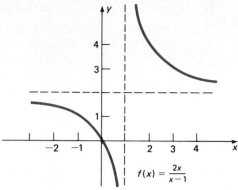

FIGURE 6

PROBLEM SET 4.6

In Problems 1–32, find the limits.

1. $\lim\limits_{x \to \infty} \dfrac{3 - 2x}{x + 5}$

2. $\lim\limits_{x \to -\infty} \dfrac{5x + 1}{x - 1}$

3. $\lim\limits_{x \to -\infty} \dfrac{2x^2 - x + 5}{5x^2 + 6x - 1}$

4. $\lim\limits_{x \to 0} \dfrac{2x + 7}{x^2 - x}$

5. $\lim\limits_{x \to \infty} \dfrac{2x^3 - 3x^2 + 1}{5x^3 - 4x + 7}$

6. $\lim\limits_{x \to \infty} \dfrac{(3x - 2)(2x + 4)}{(2x + 1)(x + 2)}$

7. $\lim\limits_{x \to -\infty} \dfrac{3x^3 - 4x + 1}{(x^2 + 1)(x^2 - 1)}$

8. $\lim\limits_{x \to \infty} \dfrac{3x\sqrt{x} + 3x + 1}{x^2 - x + 11}$

9. $\lim\limits_{x \to \infty} \sqrt[3]{\dfrac{1 + 8x^2}{x^2 + 4}}$

10. $\lim\limits_{x \to \infty} \sqrt{\dfrac{x^2 + x + 3}{(x - 1)(x + 1)}}$

11. $\lim\limits_{x \to \infty} \dfrac{2x + 1}{\sqrt{x^2 + 3}}$. *Hint*: Divide numerator and denominator by x. Note that for $x > 0$, $\sqrt{x^2 + 3}/x = \sqrt{(x^2 + 3)/x^2}$.

12. $\lim\limits_{x \to \infty} \dfrac{\sqrt{2x + 1}}{x + 4}$

13. $\lim\limits_{x \to \infty} (\sqrt{2x^2 + 3} - \sqrt{2x^2 - 5})$. *Hint*: Multiply and divide by $\sqrt{2x^2 + 3} + \sqrt{2x^2 - 5}$.

14. $\lim\limits_{x \to \infty} (\sqrt{x^2 + 2x} - x)$

15. $\lim\limits_{y \to -\infty} \dfrac{9y^3 + 1}{y^2 - 2y + 2}$. *Hint*: Divide numerator and denominator by y^2.

16. $\lim\limits_{x \to \infty} \dfrac{a_0 x^n + a_1 x^{n-1} + \cdots + a_{n-1} x + a_n}{b_0 x^n + b_1 x^{n-1} + \cdots + b_{n-1} x + b_n}$,

where $a_0 \neq 0$, $b_0 \neq 0$, and n is in $\mathbb{N}$.

17. $\lim\limits_{t \to 3^+} \dfrac{3 + t}{3 - t}$

18. $\lim\limits_{t \to 3^-} \dfrac{3 + t}{3 - t}$

19. $\lim\limits_{x \to 2^+} \dfrac{3}{x^2 - 4}$

20. $\lim\limits_{x \to 3^-} \dfrac{x^2}{x^2 - 9}$

21. $\lim\limits_{x \to (3/2)^-} \dfrac{4x + 1}{2x - 3}$

22. $\lim\limits_{x \to (3/2)^+} \dfrac{2x + 1}{2x^2 - x - 3}$

23. $\lim\limits_{x \to 3} \dfrac{2x}{x^2 - 6x + 9}$

24. $\lim\limits_{x \to 3} \dfrac{2x}{x - 3}$

25. $\lim\limits_{x \to 3^-} \dfrac{x^2 - x - 6}{x - 3}$

26. $\lim\limits_{x \to 2^+} \dfrac{x^2 + 2x - 8}{x^2 - 4}$

27. $\lim\limits_{x \to 0^+} \dfrac{[\![x]\!]}{x}$

28. $\lim\limits_{x \to 0^-} \dfrac{[\![x]\!]}{x}$

29. $\lim\limits_{x \to 0^-} \dfrac{|x|}{x}$

30. $\lim\limits_{x \to 0^+} \dfrac{|x|}{x}$

31. $\lim\limits_{x \to 0^-} \dfrac{1 + \cos x}{\sin x}$

32. $\lim\limits_{x \to \infty} \dfrac{\sin x}{x}$

In Problems 33–38, find the horizontal and vertical asymptotes for the graphs of the indicated functions. Then sketch their graphs.

33. $f(x) = \dfrac{3}{x + 1}$

34. $f(x) = \dfrac{3}{(x + 1)^2}$

35. $F(x) = \dfrac{2x}{x - 3}$

36. $F(x) = \dfrac{3}{9 - x^2}$

37. $g(x) = \dfrac{14}{2x^2 + 7}$ **38.** $g(x) = \dfrac{2x}{\sqrt{x^2 + 5}}$

39. The line $y = ax + b$ is called an **oblique asymptote** to the graph of $y = f(x)$ if either $\lim\limits_{x \to \infty} [f(x) - (ax + b)] = 0$ or $\lim\limits_{x \to -\infty} [f(x) - (ax + b)] = 0$. Find the oblique asymptote for

$$f(x) = \frac{2x^4 + 3x^3 - 2x - 4}{x^3 - 1}$$

Hint: Begin by dividing the denominator into the numerator.

40. Find the oblique asymptote for

$$f(x) = \frac{3x^3 + 4x^2 - x + 1}{x^2 + 1}.$$

41. Using the symbols M and δ, give precise definitions of each expression.

(a) $\lim\limits_{x \to c^+} f(x) = -\infty$ (b) $\lim\limits_{x \to c^-} f(x) = \infty$

42. Using the symbols M and N, give precise definitions of each expression.

(a) $\lim\limits_{x \to \infty} f(x) = \infty$ (b) $\lim\limits_{x \to -\infty} f(x) = \infty$

43. Give a rigorous proof that if $\lim\limits_{x \to \infty} f(x) = A$ and $\lim\limits_{x \to \infty} g(x) = B$, then

$$\lim_{x \to \infty} [f(x) + g(x)] = A + B$$

44. We have given meaning to $\lim\limits_{x \to A} f(x)$ for $A = a, a^-, a^+, -\infty, \infty$. Moreover, in each case, this limit may be L (finite), $-\infty$, ∞, or may fail to exist in any sense. Make a table illustrating each of the 20 possible cases.

45. Find each of the following limits or indicate that it does not exist even in the infinite sense.

(a) $\lim\limits_{x \to \infty} \sin x$ (b) $\lim\limits_{x \to \infty} \sin \dfrac{1}{x}$

(c) $\lim\limits_{x \to \infty} x \sin \dfrac{1}{x}$ (d) $\lim\limits_{x \to \infty} x^{3/2} \sin \dfrac{1}{x}$

(e) $\lim\limits_{x \to \infty} x^{-1/2} \sin x$ (f) $\lim\limits_{x \to \infty} \sin\left(\dfrac{\pi}{6} + \dfrac{1}{x}\right)$

(g) $\lim\limits_{x \to \infty} \sin\left(x + \dfrac{1}{x}\right)$

(h) $\lim\limits_{x \to \infty} \left[\sin\left(x + \dfrac{1}{x}\right) - \sin x\right]$

46. Let f be differentiable for $x > a$. Prove or disprove.

(a) $\lim\limits_{x \to \infty} f(x) = 0 \;\Rightarrow\; \lim\limits_{x \to \infty} f'(x) = 0$

(b) $\lim\limits_{x \to a^+} f(x) = \infty \;\Rightarrow\; \lim\limits_{x \to a^+} f'(x) = -\infty$

4.7 Sophisticated Graphing

Our treatment of graphing in Section 1.7 was elementary. We proposed plotting enough points so the essential features of the graph were clear. We mentioned that symmetries of the graph could reduce the effort involved. We suggested that one should be alert to possible asymptotes. But if the equation to be graphed is complicated or if we want a very accurate graph, the techniques of Chapter 1 are inadequate.

Calculus provides a powerful tool for analyzing the fine structure of a graph, especially in identifying those points where the character of the graph changes. We can locate local maximum points, local minimum points, and inflection points; we can determine precisely where the graph is increasing or where it is concave up. Inclusion of all these ideas in our graphing procedure is the program for this section.

POLYNOMIALS A polynomial of degree 1 or 2 is trivial to graph; one of degree 50 would be next to impossible. If the degree is of modest size, such as 3 to 6, we can use the tools of calculus to great advantage.

EXAMPLE 1 Sketch the graph of $f(x) = \dfrac{3x^5 - 20x^3}{32}$.

Solution Since $f(-x) = -f(x)$, f is an odd function, and therefore its graph is symmetric with respect to the origin. Setting $f(x) = 0$, we find the x-intercepts to be 0 and $\pm\sqrt{20/3} \approx \pm 2.6$. We can go this far without calculus.

When we differentiate f, we obtain

$$f'(x) = \frac{15x^4 - 60x^2}{32} = \frac{15x^2(x-2)(x+2)}{32}$$

FIGURE 1

Thus, the critical points are -2, 0, and 2; we quickly discover (Figure 1) that $f'(x) > 0$ on $(-\infty, -2)$ and $(2, \infty)$ and that $f'(x) < 0$ on $(-2, 0)$ and $(0, 2)$. These facts tell us where f is increasing and where it is decreasing; they also confirm that $f(-2) = 2$ is a local maximum value and that $f(2) = -2$ is a local minimum value.

Differentiating again, we get

$$f''(x) = \frac{60x^3 - 120x}{32} = \frac{15x(x - \sqrt{2})(x + \sqrt{2})}{8}$$

FIGURE 2

By studying the sign of $f''(x)$ (Figure 2), we deduce that f is concave upward on $(-\sqrt{2}, 0)$ and $(\sqrt{2}, \infty)$ and concave downward on $(-\infty, -\sqrt{2})$ and $(0, \sqrt{2})$. Thus, there are three points of inflection, namely, $(-\sqrt{2}, 7\sqrt{2}/8) \approx (-1.4, 1.2)$, $(0, 0)$, and $(\sqrt{2}, -7\sqrt{2}/8) \approx (1.4, -1.2)$.

Much of this information is collected in the chart of Figure 3, which we use to sketch the graph directly below it.

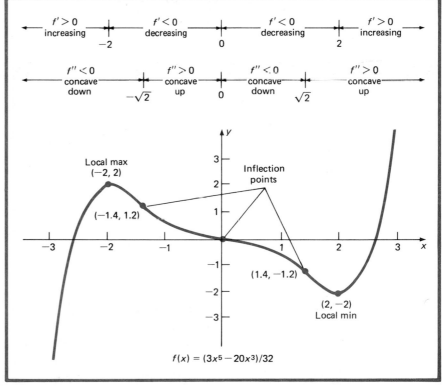

FIGURE 3

RATIONAL FUNCTIONS A rational function, being the quotient of two polynomial functions, is considerably more complicated to graph than a polynomial. In particular, we can expect dramatic behavior wherever the denominator is zero.

EXAMPLE 2 Sketch the graph of $f(x) = \dfrac{x^2 - 2x + 4}{x - 2}$.

Solution This function is neither even nor odd, so we do not expect any of the usual symmetries. There are no x-intercepts, since the solutions to $x^2 - 2x + 4 = 0$ are not real numbers. The y-intercept is -2. We anticipate a vertical asymptote at $x = 2$. In fact,

$$\lim_{x \to 2^-} \frac{x^2 - 2x + 4}{x - 2} = -\infty \qquad \lim_{x \to 2^+} \frac{x^2 - 2x + 4}{x - 2} = \infty$$

Differentiation twice gives

$$f'(x) = \frac{x(x - 4)}{(x - 2)^2} \qquad f''(x) = \frac{8}{(x - 2)^3}$$

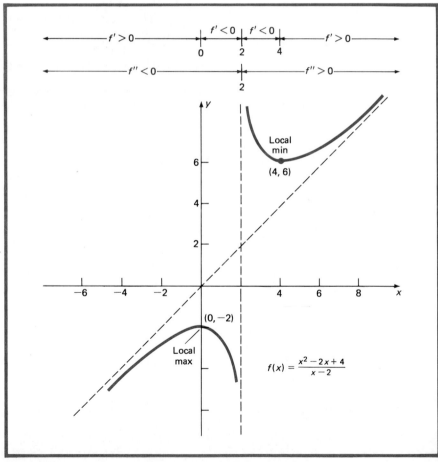

FIGURE 4

Thus, $f'(x) > 0$ on $(-\infty, 0) \cup (4, \infty)$ and $f'(x) < 0$ on $(0, 2) \cup (2, 4)$. Also, $f''(x) > 0$ on $(2, \infty)$ and $f''(x) < 0$ on $(-\infty, 2)$. Note that $f''(x)$ is never 0; there are no inflection points. On the other hand, $f(0) = -2$ and $f(4) = 6$ give local maximum and minimum values, respectively.

It is a good idea to check on the behavior of $f(x)$ for large $|x|$. Since

$$f(x) = \frac{x^2 - 2x + 4}{x - 2} = x + \frac{4}{x - 2}$$

the graph of $y = f(x)$ gets closer and closer to the line $y = x$ as $|x|$ gets larger and larger. We call this line an **oblique asymptote** for the graph (see Problem 39 of Section 4.6).

With all this information, we are able to sketch a rather accurate graph (Figure 4). ∎

ALGEBRAIC FUNCTIONS   There is an endless variety of algebraic functions. Here is one example.

EXAMPLE 3   Analyze the function

$$F(x) = \frac{\sqrt{x}(x - 5)^2}{4}$$

and sketch its graph.

Solution   The domain of F is $[0, \infty)$ and the range is $[0, \infty)$, so the graph of F is confined to the first quadrant. The x-intercepts are 0 and 5; the y-intercept is 0.

From

$$F'(x) = \frac{5(x - 1)(x - 5)}{8\sqrt{x}}, \qquad x > 0$$

we find the stationary points 1 and 5. Since $F'(x) > 0$ on $(0, 1)$ and $(5, \infty)$ while $F'(x) < 0$ on $(1, 5)$, we conclude that $F(1) = 4$ is a local maximum value and $F(5) = 0$ is a local minimum value.

So far, it has been clear sailing. But on calculating the second derivative, we obtain

$$F''(x) = \frac{5(3x^2 - 6x - 5)}{16x^{3/2}}, \qquad x > 0$$

which is quite complicated. However, $3x^2 - 6x - 5 = 0$ has one solution in $(0, \infty)$, $1 + 2\sqrt{6}/3 \approx 2.6$.

Using the test points 1 and 3, we conclude that $f''(x) < 0$ on $(0, 2.6)$ and $f''(x) > 0$ on $(2.6, \infty)$. It then follows that the point $(2.6, 2.3)$ is an inflection point.

As x grows large, $F(x)$ grows without bound and much faster than any linear function; there are no asymptotes. The graph is sketched in Figure 5. ∎

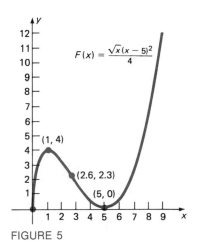

$$F(x) = \frac{\sqrt{x}(x - 5)^2}{4}$$

$(1, 4)$

$(2.6, 2.3)$

$(5, 0)$

FIGURE 5

SUMMARY OF THE METHOD   In graphing functions, there is no substitute for common sense. However, the following procedure will be helpful in most cases.

Step 1   Make a preliminary analysis as follows.
(a) Check the *domain* and *range* of the function to see if any regions of the plane are excluded.
(b) Test for *symmetry* with respect to the *y*-axis and the origin. (Is the function even or odd?)
(c) Find the *intercepts*.
(d) Use the first derivative to find the critical points and to find out where the graph is *increasing* and *decreasing*.
(e) Test the critical points for *local maxima* and *minima*.
(f) Use the second derivative to find out where the graph is *concave upward* and *concave downward* and to locate *inflection points*.
(g) Find the *asymptotes*.
Step 2   Plot a few points (including all critical points and inflection points).
Step 3   Sketch the graph.

PROBLEM SET   4.7

In Problems 1–16, make an analysis as suggested above and then sketch the graph.

1. $f(x) = x^3 - 4x$

2. $f(x) = x^3 - 3x^2 + 3x - 3$

3. $F(x) = \dfrac{x^4 - 18x^2 + 20}{20}$

4. $F(x) = \dfrac{x^4(x^2 - 6)}{8}$

5. $g(x) = \dfrac{4x}{x^2 + 2}$ 6. $g(x) = \dfrac{(x + 2)^2}{x}$

7. $h(x) = \dfrac{x}{x + 1}$ 8. $h(x) = \dfrac{x^2}{x^2 + 1}$

9. $f(x) = \dfrac{x^2}{x^2 - 4}$ 10. $f(x) = \dfrac{x}{x^2 - 4}$

11. $g(x) = 2x\sqrt{x + 3}$ 12. $g(x) = \sqrt{4 - x} + 1$

13. $H(x) = x|x|$ 14. $H(x) = x^2|x|$

15. $f(x) = |\sin x|$ 16. $f(x) = \sqrt{\sin x}$

17. Sketch a possible graph of a function f that has all the following properties:
(a) f is everywhere continuous;
(b) $f(2) = -3$, $f(6) = 1$;
(c) $f'(2) = 0$, $f'(x) > 0$ for $x \neq 2$, $f'(6) = 3$;
(d) $f''(6) = 0$, $f''(x) > 0$ for $2 < x < 6$, $f''(x) < 0$ for $x > 6$.

18. Sketch a possible graph of a function f that has all the following properties:
(a) f is everywhere continuous;
(b) $f(-3) = 1$;
(c) $f'(x) < 0$ for $x < -3$, $f'(x) > 0$ for $x > -3$, $f''(x) < 0$ for $x \neq -3$.

19. Sketch a possible graph of a function f that has all the following properties:
(a) f is everywhere continuous;
(b) $f(-4) = -3$, $f(0) = 0$, $f(3) = 2$;
(c) $f'(-4) = 0$, $f'(3) = 0$, $f'(x) > 0$ for $x < -4$, $f'(x) > 0$ for $-4 < x < 3$, $f'(x) < 0$ for $x > 3$;
(d) $f''(-4) = 0$, $f''(0) = 0$, $f''(x) < 0$ for $x < -4$, $f''(x) > 0$ for $-4 < x < 0$, $f''(x) < 0$ for $x > 0$.

20. Let f be a continuous function with $f(-3) = f(0) = 2$. If the graph of $y = f'(x)$ is as shown in Figure 6, sketch a possible graph for $y = f(x)$.

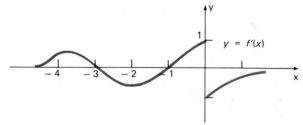

FIGURE 6

21. Let f be a continuous function with $f(0) = f(2) = 0$. If the graph of $y = f'(x)$ is as shown in Figure 7, sketch a possible graph for $y = f(x)$.

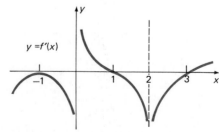

FIGURE 7

4.8
The Mean Value Theorem

The Mean Value Theorem is the midwife of calculus—not very important or glamorous by itself, but often helping to deliver other theorems that are of major significance. From now on, you will see the phrase "by the Mean Value Theorem" quite regularly, and later in this section we will use it to prove an important theorem left dangling back in Section 4.2.

In geometric language, the Mean Value Theorem is easy to state and understand. It says that if the graph of a continuous function has a non-vertical tangent line at every point between A and B, then there is at least one point C on the graph between A and B at which the tangent line is parallel to the secant line AB. In Figure 1, there is just one such point C; in Figure 2, there are several.

FIGURE 1

THE THEOREM PROVED First we state the theorem in the language of functions; then we prove it.

Theorem A

(Mean Value Theorem for Derivatives). If f is continuous on a closed interval $[a, b]$ and differentiable on its interior (a, b), then there is at least one number c in (a, b) where

$$\frac{f(b) - f(a)}{b - a} = f'(c)$$

or, equivalently, where

$$f(b) - f(a) = f'(c)(b - a)$$

FIGURE 2

Proof Our proof rests on a careful analysis of the function $s(x) = f(x) - g(x)$, introduced in Figure 3. Here $y = g(x)$ is the equation of the line through $(a, f(a))$ and $(b, f(b))$. Since this line has slope $[f(b) - f(a)]/(b - a)$ and goes through $(a, f(a))$, the point-slope form for its equation is

$$g(x) - f(a) = \frac{f(b) - f(a)}{b - a}(x - a)$$

This, in turn, yields a formula for $s(x)$, namely,

$$s(x) = f(x) - f(a) - \frac{f(b) - f(a)}{b - a}(x - a)$$

Note immediately that $s(b) = s(a) = 0$ and that for x in (a, b)

$$s'(x) = f'(x) - \frac{f(b) - f(a)}{b - a}$$

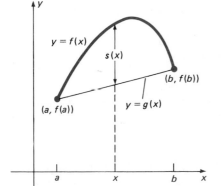

FIGURE 3

Now we make a crucial observation. If we knew that there was a number c in (a, b) satisfying $s'(c) = 0$, we would be all done. For then the last equation would say

$$0 = f'(c) - \frac{f(b) - f(a)}{b - a}$$

which is equivalent to the conclusion of the theorem.

To see that $s'(c) = 0$ for some c in (a, b), reason as follows. Clearly s is continuous on $[a, b]$, being the difference of two continuous functions. Thus, by the Max-Min Existence Theorem (Theorem 4.1A), s must attain both a maximum and a minimum value on $[a, b]$. If both of these values happen to be 0, then $s(x)$ is identically 0 on $[a, b]$, and consequently $s'(x) = 0$ for all x in (a, b), much more than we need.

If either the maximum value or the minimum value is different from 0, then that value is attained at an interior point c, since $s(a) = s(b) = 0$. Now s has a derivative at each point of (a, b), and so by the Critical Point Theorem (Theorem 4.1B), $s'(c) = 0$. That is all we needed to know. ∎

THE THEOREM ILLUSTRATED

EXAMPLE 1 Find the number c guaranteed by the Mean Value Theorem for $f(x) = 2\sqrt{x}$ on $[1, 4]$.

Solution

$$f'(x) = 2 \cdot \frac{1}{2} x^{-1/2} = \frac{1}{\sqrt{x}}$$

and

$$\frac{f(4) - f(1)}{4 - 1} = \frac{4 - 2}{3} = \frac{2}{3}$$

Thus, we must solve

$$\frac{1}{\sqrt{c}} = \frac{2}{3}$$

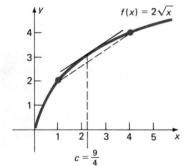

FIGURE 4

The single solution is $c = \frac{9}{4}$ (Figure 4). ∎

EXAMPLE 2 Let $f(x) = x^3 - x^2 - x + 1$ on $[-1, 2]$. Find all numbers c satisfying the conclusion to the Mean Value Theorem.

Solution

$$f'(x) = 3x^2 - 2x - 1$$

and

$$\frac{f(2) - f(-1)}{2 - (-1)} = \frac{3 - 0}{3} = 1$$

Therefore, we must solve

$$3c^2 - 2c - 1 = 1$$

or, equivalently,

$$3c^2 - 2c - 2 = 0$$

By the Quadratic Formula, there are two solutions, $(2 \pm \sqrt{4 + 24})/6$, which correspond to $c_1 = -0.55$ and $c_2 = 1.22$. Both numbers are in the interval $(-1, 2)$. The appropriate graph is shown in Figure 5. ∎

$f(x) = x^3 - x^2 - x + 1$

$c_1 = -.55 \qquad c_2 = 1.22$

FIGURE 5

EXAMPLE 3 Let $f(x) = x^{2/3}$ on $[-8, 27]$. Show that the conclusion to the Mean Value Theorem fails and figure out why.

Solution

$$f'(x) = \frac{2}{3} x^{-1/3}, \qquad x \neq 0$$

and

$$\frac{f(27) - f(-8)}{27 - (-8)} = \frac{9 - 4}{35} = \frac{1}{7}$$

We must solve

$$\frac{2}{3} c^{-1/3} = \frac{1}{7}$$

which gives

$$c = \left(\frac{14}{3}\right)^3 \approx 102$$

But $c = 102$ is not in the interval $(-8, 27)$ as required. The problem is, of course, that $f(x)$ is not differentiable everywhere on $(-8, 27)$; $f'(0)$ fails to exist (see Figure 6). ∎

$f(x) = x^{2/3}$

FIGURE 6

THE THEOREM USED Back in Section 4.2, we promised a rigorous proof of the Monotonicity Theorem (Theorem 4.2A). This is the theorem that relates whether a function is increasing or decreasing to the sign of its derivative.

Proof of the Monotonicity Theorem We suppose that f is continuous on I and that $f'(x) > 0$ at each point x in the interior of I. Consider any two points x_1 and x_2 of I with $x_1 < x_2$. By the Mean Value Theorem applied on

FIGURE 7

the interval $[x_1, x_2]$, there is a number c in (x_1, x_2) satisfying

$$f(x_2) - f(x_1) = f'(c)(x_2 - x_1)$$

Since $f'(c) > 0$, we see that $f(x_2) - f(x_1) > 0$—that is, $f(x_2) > f(x_1)$. This is what we mean when we say f is increasing on I.

The case where $f'(x) < 0$ on I is handled similarly. ∎

Our next theorem will be used repeatedly in the next chapter. In words, it says that *two functions with the same derivative must differ by a constant, possibly the zero constant* (see Figure 7).

Theorem B

If $F'(x) = G'(x)$ for all x in (a, b), then there is a constant C such that

$$F(x) = G(x) + C$$

for all x in (a, b).

Proof   Let $H(x) = F(x) - G(x)$. Then

$$H'(x) = F'(x) - G'(x) = 0$$

for all x in (a, b). Choose x_1 as some (fixed) point in (a, b) and let x be any other point there. The function H satisfies the hypotheses of the Mean Value Theorem on the closed interval with endpoints x_1 and x. Thus there is a number c between x_1 and x such that

$$H(x) - H(x_1) = H'(c)(x - x_1)$$

But $H'(c) = 0$ by hypothesis. Therefore, $H(x) - H(x_1) = 0$ or, equivalently, $H(x) = H(x_1)$ for all x in (a, b). Since $H(x) = F(x) - G(x)$, we conclude that $F(x) - G(x) = H(x_1)$. Now let $C = H(x_1)$, and we have the conclusion $F(x) = G(x) + C$. ∎

PROBLEM SET 4.8

In each of Problems 1–10, a function is defined and a closed interval is given. Decide whether the Mean Value Theorem applies to the given function on the given interval—if so, find all possible values of c; if not, state the reason. In each problem, sketch the graph of the given function on the given interval.

1. $f(x) = x^2 + 2x$; $[-2, 2]$

2. $f(x) = x^2 + 3x - 1$; $[-3, 1]$

3. $g(x) = \dfrac{x^3}{3}$; $[-2, 2]$

4. $g(x) = \frac{1}{3}(x^3 + x - 4)$; $[-1, 2]$

5. $F(t) = \dfrac{t + 3}{t - 3}$; $[-1, 4]$

6. $F(x) = \dfrac{x + 1}{x - 1}$; $[\frac{3}{2}, 5]$

7. $h(x) = x^{2/3}$; $[0, 2]$

8. $h(x) = x^{2/3}$; $[-2, 2]$

9. $\phi(x) = x + \dfrac{1}{x}$; $[-1, \frac{1}{2}]$

10. $\phi(x) = x + \dfrac{1}{x}$; $[\frac{1}{2}, \frac{3}{2}]$

11. Johnny traveled 112 miles in 2 hours and claimed that he never exceeded 55 miles per hour. Use the Mean Value Theorem to prove that he lied. *Hint*: Let $f(t)$ be the distance traveled in time t.

12. For the function graphed on $[0, 8]$ in Figure 8, find (approximately) all points c that satisfy the conclusion to the Mean Value Theorem.

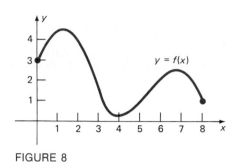

FIGURE 8

13. (Rolle's Theorem) *If f is continuous on $[a, b]$ and differentiable on (a, b) and if $f(a) = f(b)$, then there is at least one number c in (a, b) such that $f'(c) = 0$.* Show that Rolle's Theorem is just a special case of the Mean Value Theorem. (Michel Rolle (1652–1719) was a French mathematician.)

14. Show that if f is the quadratic function defined by $f(x) = \alpha x^2 + \beta x + \gamma$, $\alpha \neq 0$, then the number c of the Mean Value Theorem is always the midpoint of the given interval $[a, b]$.

15. Prove: If f is continuous on (a, b) and if $f'(x)$ exists and satisfies $f'(x) > 0$ except at one point x_0 in (a, b), then f is increasing on (a, b). *Hint*: Consider f on each of the intervals $(a, x_0]$ and $[x_0, b)$ separately.

16. Use Problem 15 to show that each of the following is increasing on $(-\infty, \infty)$.
(a) $f(x) = x^3$ (b) $f(x) = x^5$
(c) $f(x) = \begin{cases} x^3 & x \leq 0 \\ x & x > 0 \end{cases}$

17. Prove that if $F'(x) = 0$ for all x in (a, b), then there is a constant C such that $F(x) = C$ for all x in (a, b). *Hint*: Let $G(x) = 0$ and apply Theorem B.

18. Suppose that you know that $\cos(0) = 1$, $\sin(0) = 0$, $D_x \cos x = -\sin x$, and $D_x \sin x = \cos x$, but nothing else about the sine and cosine functions. Show that $\cos^2 x + \sin^2 x = 1$. *Hint*: Let $F(x) = \cos^2 x + \sin^2 x$ and use Problem 17.

19. Prove that if $F'(x) = D$ for all x in (a, b), then there is a constant C such that $F(x) = Dx + C$ for all x in (a, b). *Hint*: Let $G(x) = Dx$ and apply Theorem B.

20. Suppose $F'(x) = 5$ and $F(0) = 4$. Find a formula for $F(x)$. *Hint*: See Problem 19.

21. Prove: Let f have a derivative on $[a, b]$. If $f(a)$ and $f(b)$ have opposite signs and if $f'(x) \neq 0$ for all x in (a, b), then the equation $f(x) = 0$ has one and only one solution between a and b. *Hint*: Use the Intermediate Value Theorem and Rolle's Theorem (Problem 13).

22. Show that $f(x) = 2x^3 - 9x^2 + 1 = 0$ has exactly one solution on each of the intervals $(-1, 0)$, $(0, 1)$, and $(4, 5)$. *Hint*: Apply Problem 21.

23. Prove: Let f have a derivative on an interval I. Between successive distinct zeros of f', there can be at most one zero of f. *Hint*: Try a proof by contradiction and use Rolle's Theorem (Problem 13).

24. Prove: Let g be continuous on $[a, b]$ and suppose $g''(x)$ exists for all x in (a, b). If there are three values of x in $[a, b]$ for which $g(x) = 0$, then there is at least one value of x in (a, b) such that $g''(x) = 0$.

25. Prove that if $|f'(x)| \leq M$ for all x in (a, b) and if x_1 and x_2 are any two points in (a, b), then

$$|f(x_2) - f(x_1)| \leq M|x_2 - x_1|$$

Note: A function satisfying the above inequality is said to satisfy a *Lipschitz condition* with constant M. (Rudolph Lipschitz (1832–1903) was a German mathematician.)

26. Show that $f(x) = \sin 2x$ satisfies a Lipschitz condition with constant 2 on the interval $(-\infty, \infty)$. See Problem 25.

27. A function f is called **nondecreasing** on an interval I if $x_1 < x_2 \Rightarrow f(x_1) \leq f(x_2)$ for x_1 and x_2 in I. Similarly, f is **nonincreasing** on I if $x_1 < x_2 \Rightarrow f(x_1) \geq f(x_2)$ for x_1 and x_2 in I.
(a) Sketch the graph of a function which is nondecreasing but not increasing.
(b) Sketch the graph of a function which is nonincreasing but not decreasing.

28. Prove that if f is continuous on I and if $f'(x)$ exists and satisfies $f'(x) \geq 0$ on the interior of I, then f is nondecreasing on I. Similarly, if $f'(x) \leq 0$, then f is nonincreasing on I.

29. Prove that if $f(x) \geq 0$ and $f'(x) \geq 0$ on I, then f^2 is nondecreasing on I.

30. Prove that if $g'(x) \leq h'(x)$ for all x in (a, b), then

$$x_1 < x_2 \Rightarrow g(x_2) - g(x_1) \leq h(x_2) - h(x_1)$$

for all x_1 and x_2 in (a, b). *Hint*: Apply Problem 28 with $f(x) = h(x) - g(x)$.

31. Use the Mean Value Theorem to prove that

$$\lim_{x \to \infty}(\sqrt{x+2} - \sqrt{x}) = 0$$

32. Use the Mean Value Theorem to show that

$$|\sin x - \sin y| \le |x - y|$$

33. Suppose that in a race, horse A and horse B finished in a dead heat. Prove that their speeds were identical at some instant of the race.

34. In Problem 33, suppose that the two horses crossed the finish line together at the same speed. Show that they had the same acceleration at some instant.

35. Use the Mean Value Theorem to show that the graph of a concave up function f is always above its tangent line, that is, show that

$$f(x) > f(c) + f'(c)(x - c), \qquad x \ne c$$

36. Prove that if $|f(y) - f(x)| \le M(y - x)^2$ for all x and y, then f is a constant function.

4.9 Chapter Review

Respond with true or false to each of the following assertions. Be prepared to defend your answer.

1. A continuous function defined on a closed interval must attain a maximum value on that interval.
2. If a differentiable function f attains a maximum value at an interior point c of its domain, then $f'(c) = 0$.
3. It is possible for a function to have an infinite number of critical points.
4. A continuous function that increases for all x must be differentiable everywhere.
5. If $f(x) = 3x^6 + 4x^4 + 2x^2$, then the graph of f is concave up on the whole real line.
6. If f is an increasing differentiable function on an interval I, then $f'(x) > 0$ for all x in I.
7. If $f'(x) > 0$ for all x in I, then f is increasing on I.
8. If $f''(c) = 0$, then f has an inflection point at $(c, f(c))$.
9. A quadratic function has no inflection points.
10. If $f'(x) > 0$ for all x in $[a, b]$, then f attains its maximum value on $[a, b]$ at b.
11. We use the symbol ∞ to denote a number that is bigger than any other number.
12. The graph of $y = \dfrac{x^2 - x - 6}{x - 3} = \dfrac{(x + 2)(x - 3)}{x - 3}$ has a vertical asymptote at $x = 3$.
13. The graph of $y = \dfrac{x^2 + 1}{1 - x^2}$ has a horizontal asymptote of $y = -1$.
14. The graph of $y = \dfrac{3x^2 + 2x + \sin x}{x}$ has an oblique asymptote of $y = 3x + 2$.
15. The function $f(x) = \sqrt{x}$ satisfies the hypotheses of the Mean Value Theorem on $[0, 2]$.
16. On the interval $[-1, 1]$, there will be just one point where the tangent line to $y = x^3$ is parallel to the secant line.
17. If $f'(x) = 0$ for all x in (a, b), then f is constant on this interval.
18. If $f'(c) = f''(c) = 0$, then $f(c)$ is neither a maximum nor minimum value.
19. The graph of $y = \sin x$ has infinitely many points of inflection.
20. Among rectangles of fixed area K, the one with maximum perimeter is a square.
21. If the graph of a differentiable function has three x-intercepts, then it must have at least two points where the tangent line is horizontal.
22. The sum of two increasing functions is an increasing function.
23. The product of two increasing functions is an increasing function.
24. If $f'(0) = 0$ and $f''(x) > 0$ for $x \ge 0$, then f is increasing on $[0, \infty)$.
25. If f is a differentiable function, then f is nondecreasing on (a, b) if and only if $f'(x) \ge 0$ on (a, b).
26. Two differentiable functions have the same derivative on (a, b) if and only if they differ by a constant on (a, b).

27. If $f''(x) > 0$ for all x, then the graph of $y = f(x)$ cannot have a horizontal asymptote.

28. A global maximum value is always a local maximum value.

29. A cubic function $f(x) = ax^3 + bx^2 + cx + d$, $a \neq 0$ can have at most one local maximum value on any open interval.

30. The linear function $f(x) = ax + b$, $a \neq 0$ has no minimum value on any open interval.

SAMPLE TEST PROBLEMS

In Problems 1–6, a function f and its domain are given. Determine the critical points, evaluate f at these points, and find the (global) maximum and minimum values.

1. $f(x) = \dfrac{1}{x^2}$; $[-2, -\frac{1}{2}]$

2. $f(x) = \dfrac{1}{x^2}$; $[-2, 0)$

3. $f(x) = 3x^4 - 4x^3$; $[-2, 3]$

4. $f(x) = x^2(x - 2)^{1/3}$; $[-1, 3]$

5. $f(x) = 2x^5 - 5x^4 + 7$; $[-1, 3]$

6. $f(x) = (x - 1)^3(x + 2)^2$; $[-2, 2]$

In Problems 7–10, a function f is given with domain $\mathbb{R}$. Indicate where f is increasing and where it is concave down.

7. $f(x) = x^3 - 3x + 3$

8. $f(x) = -2x^3 + 3x^2 + 12x + 1$

9. $f(x) = x^4 - 4x^5$

10. $f(x) = x^3 - \frac{6}{5}x^5$

11. Find where the function f, defined by $f(x) = x^2(x - 4)$, is increasing and where it is decreasing. Find the local extreme values of f. Find the point of inflection. Sketch the graph.

12. Find the maximum and the minimum values, if they exist, of the function defined by

$$f(x) = \frac{4}{x^2 + 1} + 2$$

In Problems 13–18, sketch the graph of the given function f, labeling all extrema (local and global) and inflection points and showing any asymptotes. Be sure to make use of f' and f''.

13. $f(x) = x^4 - 32x$

14. $f(x) = (x^2 - 1)^2$

15. $f(x) = x\sqrt{x - 3}$

16. $f(x) = \dfrac{x - 2}{x + 1}$

17. $f(x) = 3x^4 - 4x^3$

18. $f(x) = \dfrac{x^2 - 1}{x}$

19. Sketch a possible graph of a function F that has all the following properties:

(a) F is everywhere continuous;

(b) $F(-2) = 3$, $F(2) = -1$;

(c) $F'(x) = 0$ for $x > 2$;

(d) $F''(x) < 0$ for $x < 2$.

20. Sketch a possible graph of a function F that has all the following properties:

(a) F is everywhere continuous;

(b) $F(-1) = 6$, $F(3) = -2$;

(c) $F'(x) < 0$ for $x < -1$, $F'(-1) = F'(3) = -2$, $F'(7) = 0$;

(d) $F''(x) < 0$ for $x < -1$, $F''(x) = 0$ for $-1 < x < 3$, $F''(x) > 0$ for $x > 3$.

21. A long sheet of metal, 16 inches wide, is to be turned up at both sides to make a horizontal gutter with vertical sides. How many inches should be turned up at each side for maximum carrying capacity?

22. A fence, 8 feet high, is parallel to the wall of a building and 1 foot from the building. What is the shortest plank that can go over the fence, from the level ground, to prop the wall?

23. A page of a book is to contain 27 square inches of print. If the margins at the top, bottom, and one side are 2 inches and the margin at the other side is 1 inch, what size page would use the least paper?

24. A metal water trough with equal semicircular ends and open top is to have a capacity of 128π cubic feet (Figure 1). Determine its radius r and length h if the trough is to require the least material for its construction.

FIGURE 1

25. Find the maximum and the minimum of the function defined on the closed interval $[-2, 2]$ by

$$f(x) = \begin{cases} \frac{1}{4}(x^2 + 6x + 8) & \text{if } -2 \le x \le 0 \\ -\frac{1}{6}(x^2 + 4x - 12) & \text{if } 0 \le x \le 2 \end{cases}$$

Find where the graph is concave upward and where it is concave downward. Sketch the graph.

26. For each of the following functions, decide whether the Mean Value Theorem applies on the indicated interval I. If so, find all possible values of c; if not, tell why. Make a sketch.

(a) $f(x) = \dfrac{x^3}{3}$; $I = [-3, 3]$

(b) $F(x) = x^{3/5} + 1$; $I = [-1, 1]$

(c) $g(x) = \dfrac{x + 1}{x - 1}$; $I = [2, 3]$

27. Find the equations of the tangent lines at the inflection points of the graph of

$$y = x^4 - 6x^3 + 12x^2 - 3x + 1$$

28. Determine each limit (possibly ∞ or $-\infty$) or state that it does not have a limit.

(a) $\displaystyle\lim_{x \to \infty} \dfrac{3x^2 - 2x + 7}{2x^2 + 5x + 9}$

(b) $\displaystyle\lim_{x \to \infty} \dfrac{3x + 9}{\sqrt{2x^2 + 1}}$

(c) $\displaystyle\lim_{x \to -\infty} \dfrac{\sin x}{x}$

(d) $\displaystyle\lim_{x \to -\infty} \cos x$

(e) $\displaystyle\lim_{x \to 3^+} \dfrac{x + 3}{x^2 - 9}$

(f) $\displaystyle\lim_{x \to 2} \dfrac{x + 2}{x^2 - 4}$

(g) $\displaystyle\lim_{x \to 1^-} \dfrac{|x - 1|}{x - 1}$

(h) $\displaystyle\lim_{x \to \infty} \cos\left(\dfrac{1}{x}\right)$

29. Sketch a possible graph of a function G with all the following properties:

(a) $G(x)$ is continuous and $G''(x) > 0$ for all x in $(-\infty, 0) \cup (0, \infty)$;

(b) $G(-2) = G(2) = 3$;

(c) $\displaystyle\lim_{x \to -\infty} G(x) = 2$, $\displaystyle\lim_{x \to \infty} [G(x) - x] = 0$;

(d) $\displaystyle\lim_{x \to 0^+} G(x) = \lim_{x \to 0^-} G(x) = \infty$.

5 The Integral

With Bernhard Riemann, Dirichlet's successor at Göttingen, we reach the man who more than any other has influenced the course of modern mathematics.
Dirk J. Struik

Georg Friedrich Bernhard Riemann
1826–1866

Bernhard Riemann received his early education from his father, a German Protestant minister. When he went to college in 1846, it was to study theology and philology. Fortunately for mathematics, he chose Göttingen University, which was then and was to remain for 100 years the mathematical center of the world. There he came under the influence of W. E. Weber, a first-rate physicist, and Karl F. Gauss, the greatest mathematician of the day. One could hardly wish for better teachers. In 1851, he received his Ph.D. under Gauss, after which he stayed at Göttingen to teach. He died of tuberculosis 15 years later.

Riemann's life was short, only 39 years. He did not have time to produce the quantity of mathematics of a Cauchy or an Euler. But his work is amazing for its quality and its insight. His mathematical papers set new directions in complex function theory, initiated the profound study of what is today called topology, and in geometry started a development that was to culminate 50 years later in Einstein's theory of relativity.

We associate Riemann with this chapter because, though both Newton and Leibniz had a version of the integral and knew the Fundamental Theorem of the integral calculus, it was Riemann who gave us the modern definition of the definite integral. In his honor, it is called the Riemann integral.

5.1
Antiderivatives (Indefinite Integrals)

If I put on my shoes, I may take them off again. The second operation undoes the first, restoring the shoes to their original position. We say the two operations are *inverse operations*. Mathematics has many pairs of inverse operations: addition and subtraction, multiplication and division, exponentiation and root taking, and taking logarithms and finding antilogarithms. We have been studying differentiation; its inverse is called *antidifferentiation*.

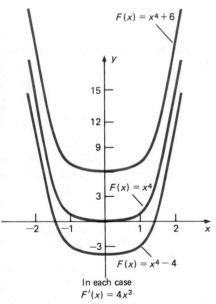

$F(x) = x^4 + 6$

$F(x) = x^4$

$F(x) = x^4 - 4$

In each case
$F'(x) = 4x^3$

FIGURE 1

Definition

We call F an **antiderivative** of f on the interval I if $DF = f$ on I—that is, if $F'(x) = f(x)$ for all x in I. (If x is an endpoint of I, $F'(x)$ need be only a one-sided derivative.)

We used *an* antiderivative rather than *the* antiderivative in our definition. You will soon see why.

EXAMPLE 1 Find an antiderivative of the function $f(x) = 4x^3$ on $(-\infty, \infty)$.

Solution We seek a function F satisfying $F'(x) = 4x^3$ for all real x. From our experience with differentiation, we know that $F(x) = x^4$ is one such function. ∎

A moment's thought will suggest other solutions to Example 1. The function $F(x) = x^4 + 6$ also satisfies $F'(x) = 4x^3$; it too is an antiderivative of $f(x) = 4x^3$. In fact, $F(x) = x^4 + C$, where C is any constant, is an antiderivative of $4x^3$ on $(-\infty, \infty)$ (see Figure 1).

Now we pose an important question. Is every antiderivative of $f(x) = 4x^3$ of the form $F(x) = x^4 + C$? The answer is yes. This follows from Theorem 4.8B, which says that two functions with the same derivative must differ by a constant.

Our conclusion is this. If a function f has an antiderivative, it will have a whole family of them and each member of this family can be obtained from one of them by the addition of an appropriate constant. We call this family of functions the **general antiderivative** of f. After we get used to this notion, we will often omit the adjective *general*.

EXAMPLE 2 Find the general antiderivative of $f(x) = x^2$ on $(-\infty, \infty)$.

Solution The function $F(x) = x^3$ will not do since its derivative is $3x^2$. But that suggests $F(x) = \frac{1}{3}x^3$, which satisfies $F'(x) = \frac{1}{3} \cdot 3x^2 = x^2$. However, the general antiderivative is $\frac{1}{3}x^3 + C$. ∎

NOTATION FOR ANTIDERIVATIVES Since we used the symbol D_x for the operation of taking a derivative, it would be natural to use A_x for the operation of finding the antiderivative. Thus,

$$A_x(x^2) = \tfrac{1}{3}x^3 + C$$

This is the notation used by several authors and it was, in fact, used in earlier editions of this book. However, Leibniz's original notation continues to enjoy overwhelming popularity, and we therefore choose to follow him. Rather than A_x, Leibniz used the symbol $\int \cdots dx$. He wrote

$$\int x^2 \, dx = \tfrac{1}{3}x^3 + C$$

and

$$\int 4x^3 \, dx = x^4 + C$$

Leibniz had reasons for using the elongated s, $\int$, and also for the dx, but these reasons will not make sense until later. For the moment, simply think of $\int \cdots dx$ as indicating the antiderivative with respect to x, just as D_x indicates the derivative with respect to x. Note that $D_x \int f(x)\,dx = f(x)$.

Theorem A

(Power Rule). If r is any rational number except -1, then

$$\int x^r \, dx = \frac{x^{r+1}}{r+1} + C$$

Proof To establish any result of the form

$$\int f(x) \, dx = F(x) + C$$

it is sufficient to show

$$D_x[F(x) + C] = f(x)$$

In our case,

$$D_x\left[\frac{x^{r+1}}{r+1} + C\right] = \frac{1}{r+1}(r+1)x^r = x^r$$

∎

We make two comments about Theorem A. First, it is meant to include the case $r = 0$, that is,

$$\int 1 \, dx = x + C$$

Second, since no interval I is specified, the conclusion is understood to be valid for any interval on which x^r is defined. In particular, we must exclude any interval containing the origin if $r < 0$. A similar understanding holds in what follows.

EXAMPLE 3 Find the general antiderivative of $f(x) = x^{4/3}$.

Solution

$$\int x^{4/3} \, dx = \frac{x^{7/3}}{\frac{7}{3}} + C = \tfrac{3}{7} x^{7/3} + C$$

Note that *to antidifferentiate a power of x we increase the exponent by 1 and divide by the new exponent.* ∎

Theorem B

$$\int \sin x \, dx = -\cos x + C \qquad \int \cos x \, dx = \sin x + C$$

Proof Simply note that $D_x(-\cos x) = \sin x$ and $D_x(\sin x) = \cos x$. ∎

There is more to be said about notation. Following Leibniz, we shall often use the term **indefinite integral** in place of antiderivative. To antidifferentiate is also to **integrate**. In the symbol $\int f(x) \, dx$, $\int$ is called the **integral sign** and $f(x)$ is called the **integrand**. Thus, we integrate the integrand and thereby obtain the indefinite integral. Perhaps Leibniz used the adjective *indefinite* to suggest that the indefinite integral always involves an arbitrary constant.

THE INDEFINITE INTEGRAL IS LINEAR Recall from Section 3.3 that D_x is a linear operator. This means two things.

1. $D_x[kf(x)] = kD_x f(x)$
2. $D_x[f(x) + g(x)] = D_x f(x) + D_x g(x)$

From these two properties, a third follows automatically.

3. $D_x[f(x) - g(x)] = D_x f(x) - D_x g(x)$

What is true for derivatives is true for indefinite integrals (antiderivatives).

> **Theorem C**
>
> **(Linearity of $\int \cdots dx$).** Let f and g have antiderivatives (indefinite integrals) and let k be a constant. Then:
>
> **(i)** $\int kf(x)\,dx = k\int f(x)\,dx$;
>
> **(ii)** $\int [f(x) + g(x)]\,dx = \int f(x)\,dx + \int g(x)\,dx$; and consequently
>
> **(iii)** $\int [f(x) - g(x)]\,dx = \int f(x)\,dx - \int g(x)\,dx$.

Proof   To show (i) and (ii), we simply differentiate the right side and observe that we get the integrand of the left side.

$$D_x\left[k\int f(x)\,dx \right] = kD_x \int f(x)\,dx = kf(x)$$

$$D_x\left[\int f(x)\,dx + \int g(x)\,dx \right] = D_x \int f(x)\,dx + D_x \int g(x)\,dx$$

$$= f(x) + g(x)$$

Property (iii) follows from (i) and (ii). ∎

EXAMPLE 4   Find   (a) $\int (3x^2 + 4x)\,dx$,   (b) $\int (u^{3/2} - 3u + 14)\,du$,   and (c) $\int (1/t^2 + \sqrt{t})\,dt$ by using the linearity of $\int$.

Solution

(a)
$$\int (3x^2 + 4x)\,dx = \int 3x^2\,dx + \int 4x\,dx$$

$$= 3\int x^2\,dx + 4\int x\,dx$$

$$= 3\left(\frac{x^3}{3} + C_1 \right) + 4\left(\frac{x^2}{2} + C_2 \right)$$

$$= x^3 + 2x^2 + (3C_1 + 4C_2)$$

$$= x^3 + 2x^2 + C$$

The appearance of two arbitrary constants is unnecessary. They are easily combined in one constant, C, a practice we consistently follow.

(b) Note the use of the variable u rather than x. That is fine as long as the corresponding differential symbol is du, since we then have a complete change of notation.

$$\int (u^{3/2} - 3u + 14)\,du = \int u^{3/2}\,du - 3\int u\,du + 14\int 1\,du$$

$$= \tfrac{2}{5}u^{5/2} - \tfrac{3}{2}u^2 + 14u + C$$

(c) $\displaystyle\int \left(\frac{1}{t^2} + \sqrt{t}\right) dt = \int (t^{-2} + t^{1/2}) \, dt = \int t^{-2} \, dt + \int t^{1/2} \, dt$

$$= \frac{t^{-1}}{-1} + \frac{t^{3/2}}{\frac{3}{2}} + C = -\frac{1}{t} + \frac{2}{3} t^{3/2} + C \qquad \blacksquare$$

GENERALIZED POWER RULE Recall the Chain Rule as applied to a power of a function. If $u = g(x)$ is a differentiable function and r is a rational number ($r \neq -1$), then

$$D_x\left[\frac{u^{r+1}}{r+1}\right] = u^r \cdot D_x u$$

or, in functional notation,

$$D_x\left(\frac{[g(x)]^{r+1}}{r+1}\right) = [g(x)]^r \cdot g'(x)$$

From this we obtain an important rule for indefinite integrals.

Theorem D

(Generalized Power Rule). Let g be a differentiable function and r a rational number different from -1. Then

$$\int [g(x)]^r g'(x) \, dx = \frac{[g(x)]^{r+1}}{r+1} + C$$

EXAMPLE 5 Find (a) $\int (x^4 + 3x)^{30}(4x^3 + 3) \, dx$ and (b) $\int \sin^{10}x \cos x \, dx$.

Solution

(a) Let $g(x) = x^4 + 3x$; then $g'(x) = 4x^3 + 3$. Thus, by Theorem D,

$$\int (x^4 + 3x)^{30}(4x^3 + 3) \, dx = \int [g(x)]^{30} g'(x) \, dx = \frac{[g(x)]^{31}}{31} + C$$

$$= \frac{(x^4 + 3x)^{31}}{31} + C$$

(b) Let $g(x) = \sin x$; then $g'(x) = \cos x$. Thus,

$$\int \sin^{10}x \cos x \, dx = \int [g(x)]^{10} g'(x) \, dx = \frac{[g(x)]^{11}}{11} + C$$

$$= \frac{\sin^{11}x}{11} + C \qquad \blacksquare$$

Now we can see why Leibniz used the differential dx in his notation $\int \cdots dx$. If we let $u = g(x)$, then $du = g'(x) \, dx$. The conclusion of Theorem D is, therefore,

$$\int u^r \, du = \frac{u^{r+1}}{r+1} + C, \qquad r \neq -1$$

which is the ordinary power rule with u as the variable. Thus, the generalized power rule is just the ordinary power rule applied to functions. But in applying it, we must always make sure that we have du to go with u^r. The following examples illustrate what we mean.

EXAMPLE 6   Find (a) $\int (x^3 + 6x)^5(6x^2 + 12) \, dx$, (b) $\int (x^2 + 4)^{10}x \, dx$, and (c) $\int (x^2/2 + 3)^2 x^2 \, dx$.

Solution

(a) Let $u = x^3 + 6x$;   then $du = (3x^2 + 6) \, dx$.   Thus, $(6x^2 + 12) \, dx = 2(3x^2 + 6) \, dx = 2 \, du$, and so

$$\int (x^3 + 6x)^5(6x^2 + 12) \, dx = \int u^5 2 \, du$$

$$= 2 \int u^5 \, du$$

$$= 2 \left[\frac{u^6}{6} + C \right]$$

$$= \frac{u^6}{3} + 2C$$

$$= \frac{(x^3 + 6x)^6}{3} + K$$

Two things should be noted about our solution. First, the fact that $(6x^2 + 12) \, dx$ is $2 \, du$ instead of du caused no trouble; the factor 2 could be moved in front of the integral sign by linearity. Second, we wound up with an arbitrary constant of $2C$. This is still an arbitrary constant; we called it K.

(b) Let $u = x^2 + 4$; then $du = 2x \, dx$. Thus,

$$\int (x^2 + 4)^{10}x \, dx = \int (x^2 + 4)^{10} \cdot \frac{1}{2} \cdot 2x \, dx$$

$$= \frac{1}{2} \int u^{10} \, du$$

$$= \frac{1}{2} \left(\frac{u^{11}}{11} + C \right)$$

$$= \frac{(x^2 + 4)^{11}}{22} + K$$

(c) Let $u = x^2/2 + 3$; then $du = x\,dx$. The method illustrated in (a) and (b) fails because $x^2\,dx = x(x\,dx) = x\,du$, and the x cannot be passed in front of the integral sign (that can be done only with a constant factor). However, by ordinary algebra,

$$\int \left(\frac{x^2}{2} + 3\right)^2 x^2\,dx = \int \left(\frac{x^4}{4} + 3x^2 + 9\right) x^2\,dx$$

$$= \int \left(\frac{x^6}{4} + 3x^4 + 9x^2\right) dx$$

$$= \frac{x^7}{28} + \frac{3x^5}{5} + 3x^3 + C \qquad \blacksquare$$

PROBLEM SET 5.1

Find the general antiderivative $F(x) + C$ for each of the following.

1. $f(x) = 4$
2. $f(x) = 2x - 4$
3. $f(x) = 3x^2 + \sqrt{2}$
4. $f(x) = 5x^4 + \pi$
5. $f(x) = x^{2/3}$
6. $f(x) = x^{-3/4}$
7. $f(x) = 6x^2 - 6x + 1$
8. $f(x) = 3x^2 + 10x - 7$
9. $f(x) = 18x^8 - 25x^4 + 3x^2$
10. $f(x) = x^2(20x^7 - 7x^4 + 6)$
11. $f(x) = \dfrac{4}{x^5} - \dfrac{3}{x^4}$
12. $f(x) = \dfrac{1}{x^3} + \dfrac{6}{x^7}$
13. $f(x) = \dfrac{4x^6 + 3x^5 - 8}{x^5}$
14. $f(x) = \dfrac{2x^3 - 3x^2 + 1}{x^2}$

In Problems 15–22, find the indefinite integrals.

15. $\displaystyle\int (x^3 + \sqrt{x})\,dx$
16. $\displaystyle\int (x^2 + 1)^2\,dx$
17. $\displaystyle\int (y^2 + 4y)^2\,dy$
18. $\displaystyle\int y^2(y^2 - 3)\,dy$
19. $\displaystyle\int \frac{x^4 - 2x^3 + 1}{x^2}\,dx$
20. $\displaystyle\int \frac{x^3 - 3x^2 + 1}{\sqrt{x}}\,dx$
21. $\displaystyle\int (3 \sin t - 2 \cos t)\,dt$
22. $\displaystyle\int (3t^2 - 2 \sin t)\,dt$

In Problems 23–38, use the methods of Examples 5 and 6 to find the indefinite integrals.

23. $\displaystyle\int (3x + 1)^4 3\,dx$
24. $\displaystyle\int (x^2 - 4)^3 2x\,dx$
25. $\displaystyle\int (5x^3 - 18)^7 15x^2\,dx$
26. $\displaystyle\int (x^2 - 3x + 2)^2(2x - 3)\,dx$
27. $\displaystyle\int 3x^4(2x^5 + 9)^3\,dx$
28. $\displaystyle\int 3x\sqrt{3x^2 + 7}\,dx$
29. $\displaystyle\int (5x^2 + 1)(5x^3 + 3x - 8)^6\,dx$
30. $\displaystyle\int (5x^2 + 1)\sqrt{5x^3 + 3x - 2}\,dx$
31. $\displaystyle\int 3t\sqrt[3]{2t^2 - 11}\,dt$
32. $\displaystyle\int \frac{3y}{\sqrt{2y^2 + 5}}\,dy$
33. $\displaystyle\int \sin^4 x \cos x\,dx$
34. $\displaystyle\int (\cos^4 2x)(-2 \sin 2x)\,dx$
35. $\displaystyle\int (\sin^5 x^2)(x \cos x^2)\,dx$
36. $\displaystyle\int \cos(3x + 1)\sin(3x + 1)\,dx$

37. $\displaystyle\int (x^2 + 1)^3 x^2\, dx$

38. $\displaystyle\int (x^4 - 1)x^2\, dx$

In Problems 39–44, $f''(x)$ is given. Find $f(x)$ by antidifferentiating twice. Note that in this case your answer should involve two arbitrary constants, one from each antidifferentiation. For example, if $f''(x) = x$, then $f'(x) = x^2/2 + C_1$ and $f(x) = x^3/6 + C_1 x + C_2$. The constants C_1 and C_2 cannot be combined.

39. $f''(x) = 3x + 1$　　　**40.** $f''(x) = -2x + 3$

41. $f''(x) = \sqrt{x}$　　　**42.** $f''(x) = x^{4/3}$

43. $f''(x) = \dfrac{x^4 + 1}{x^3}$　　**44.** $f''(x) = 2\sqrt[3]{x + 1}$

45. Prove the formula

$$\int [f(x)g'(x) + g(x)f'(x)]\, dx = f(x)g(x) + C$$

46. Prove the formula

$$\int \frac{g(x)f'(x) - f(x)g'(x)}{g^2(x)}\, dx = \frac{f(x)}{g(x)} + C$$

47. Use the formula from Problem 45 to find

$$\int \left[\frac{x^2}{2\sqrt{x - 1}} + 2x\sqrt{x - 1} \right] dx$$

48. Use the formula from Problem 45 to find

$$\int \left[\frac{-x^3}{(2x + 5)^{3/2}} + \frac{3x^2}{\sqrt{2x + 5}} \right] dx$$

49. Prove the formula

$$\int \frac{x^4 + 1}{x^2\sqrt{x^4 - 1}}\, dx = \frac{\sqrt{x^4 - 1}}{x} + C$$

50. Prove the formula

$$\int \frac{1}{\cos^2 3x}\, dx = \frac{1}{3}\frac{\sin 3x}{\cos 3x} + C$$

51. Find $\int f''(x)\, dx$ if $f(x) = x\sqrt{x^3 + 1}$.

52. Prove the formula

$$\int \frac{2g(x)f'(x) - f(x)g'(x)}{2[g(x)]^{3/2}}\, dx = \frac{f(x)}{\sqrt{g(x)}} + C$$

53. Prove the formula

$$\int f^{m-1}(x)g^{n-1}(x)[nf(x)g'(x) + mg(x)f'(x)]\, dx$$
$$= f^m(x)g^n(x) + C$$

54. Find the indefinite integral

$$\int \sin^3[(x^2 + 1)^4]\cos[(x^2 + 1)^4](x^2 + 1)^3 x\, dx$$

Hint: Let $u = \sin[(x^2 + 1)^4]$.

55. Find $\displaystyle\int |x|\, dx$

56. Find $\displaystyle\int \sin^2 x\, dx$

5.2
Introduction to Differential Equations

In the previous section, our task was to integrate (antidifferentiate) a function f to obtain a new function F. We wrote

$$\boxed{\int f(x)\, dx = F(x) + C}$$

and this was correct, provided $F'(x) = f(x)$. Now $F'(x) = f(x)$ is equivalent in differential language (Section 3.10) to $dF(x) = f(x)\, dx$. Thus, we may look on the boxed formula as

$$\boxed{\int dF(x) = F(x) + C}$$

From this perspective, we integrate the differential of a function to obtain the function (plus a constant). This was Leibniz's viewpoint; adopting it will help us in this section.

WHAT IS A DIFFERENTIAL EQUATION?   Consider first an example.

EXAMPLE 1   Find the xy-equation of the curve that passes through $(-1, 2)$ and whose slope at any point on the curve is equal to twice the abscissa (x-coordinate) of that point.

Solution   The condition that must hold at each point (x, y) on the curve is

$$\frac{dy}{dx} = 2x$$

We are looking for a function $y = f(x)$ that satisfies this equation and the additional condition that $y = 2$ when $x = -1$. We suggest two ways of looking at this problem.

METHOD 1   When the equation has the form $dy/dx = g(x)$, we observe that y must be an antiderivative of $g(x)$, that is,

$$y = \int g(x)\, dx$$

In our case,

$$y = \int 2x\, dx = x^2 + C$$

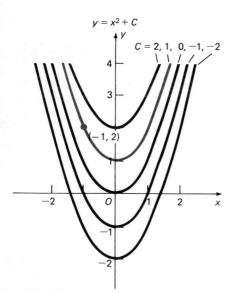

$y = x^2 + C$

$C = 2, 1, 0, -1, -2$

$(-1, 2)$

FIGURE 1

METHOD 2   Think of dy/dx as a quotient of two differentials. When we multiply both sides of $dy/dx = 2x$ by dx, we get

$$dy = 2x\, dx$$

Next we integrate both sides and simplify.

$$\int dy = \int 2x\, dx$$
$$y + C_1 = x^2 + C_2$$
$$y = x^2 + C_2 - C_1$$
$$y = x^2 + C$$

The second method works in a wide variety of problems, as we shall see.

The result $y = x^2 + C$ represents the family of curves illustrated in Figure 1. From this family, we must choose the one for which $y = 2$ when

212

$x = -1$; thus, we want

$$2 = (-1)^2 + C$$

We conclude that $C = 1$ and therefore that $y = x^2 + 1$. ∎

The equation $dy/dx = 2x$ is called a *differential equation*. Other examples are

$$\frac{dy}{dx} = 2xy + \sin x$$

$$y\,dy = (x^3 + 1)\,dx$$

$$\frac{d^2y}{dx^2} + 3\frac{dy}{dx} - 2xy = 0$$

Any equation in which the unknown is a function and which involves derivatives (or differentials) of this unknown function is called a **differential equation**. To solve a differential equation is to find the unknown function. In general, this is a difficult job and one about which many thick books have been written. Here we consider only the simplest case, namely, first-order separable differential equations. These are equations involving just the first derivative of the unknown function and are such that the variables can be separated.

SEPARATION OF VARIABLES   Consider the differential equation

$$\frac{dy}{dx} = \frac{x + 3x^2}{y^2}$$

If we multiply both sides by $y^2\,dx$, we obtain

$$y^2\,dy = (x + 3x^2)\,dx$$

In this form, the differential equation has its variables separated—that is, the y terms are on one side of the equation and the x terms are on the other. In separated form, we can solve it.

EXAMPLE 2   Solve the differential equation

$$\frac{dy}{dx} = \frac{x + 3x^2}{y^2}$$

Then find that solution for which $y = 6$ when $x = 0$.

Solution   As noted earlier, the given equation is equivalent to

$$y^2\,dy = (x + 3x^2)\,dx$$

Thus,

$$\int y^2 \, dy = \int (x + 3x^2) \, dx$$

$$\frac{y^3}{3} + C_1 = \frac{x^2}{2} + x^3 + C_2$$

$$y^3 = \frac{3x^2}{2} + 3x^3 + (3C_2 - 3C_1)$$

$$= \frac{3x^2}{2} + 3x^3 + C$$

$$y = \sqrt[3]{\frac{3x^2}{2} + 3x^3 + C}$$

To evaluate the constant C, we use the condition $y = 6$ when $x = 0$. This gives

$$6 = \sqrt[3]{C}$$

$$216 = C$$

Thus,

$$y = \sqrt[3]{\frac{3x^2}{2} + 3x^3 + 216}$$

The ultimate check on our work is to substitute this result in both sides of the original differential equation to see that this gives an equality. Substituting in the left side, we get

$$\frac{dy}{dx} = \frac{1}{3}\left(\frac{3x^2}{2} + 3x^3 + 216\right)^{-2/3}(3x + 9x^2)$$

$$= \frac{x + 3x^2}{(\frac{3}{2}x^2 + 3x^3 + 216)^{2/3}}$$

On the right side, we get

$$\frac{x + 3x^2}{y^2} = \frac{x + 3x^2}{(\frac{3}{2}x^2 + 3x^3 + 216)^{2/3}}$$

As expected, the two expressions are equal. ∎

MOTION PROBLEMS Recall that if $s(t)$, $v(t)$, and $a(t)$ represent the position, velocity, and acceleration, respectively, at time t of an object moving along a coordinate line, then

$$v(t) = s'(t) = \frac{ds}{dt}$$

$$a(t) = v'(t) = \frac{dv}{dt} = \frac{d^2s}{dt^2}$$

In earlier work (Section 3.7), we assumed that $s(t)$ was known, and from this we calculated $v(t)$ and $a(t)$. Now we want to consider the reverse process.

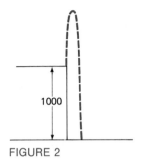

FIGURE 2

EXAMPLE 3 Near the surface of the earth, the acceleration of a falling body due to gravity is 32 feet per second per second, provided we assume that air resistance can be neglected. If an object is thrown upward from an initial height of 1000 feet (Figure 2) with a velocity of 50 feet per second, find its velocity and height 4 seconds later.

Solution Let us assume that the height s is measured positively in the upward direction. Then $v = ds/dt$ is initially positive (s is increasing) but $a = dv/dt$ is negative (the pull of gravity tends to decrease v). Thus, our starting point is the differential equation $dv/dt = -32$, with the additional conditions that $v = 50$ and $s = 1000$ when $t = 0$.

$$\frac{dv}{dt} = -32$$

$$v = \int -32 \, dt = -32t + C$$

Since $v = 50$ at $t = 0$, $C = 50$ and so

$$\boxed{v = -32t + 50}$$

Now $v = ds/dt$, and so we have the second differential equation

$$\frac{ds}{dt} = -32t + 50$$

When we integrate, we obtain

$$s = \int (-32t + 50) \, dt$$

$$s = -16t^2 + 50t + K$$

Since $s = 1000$ at $t = 0$, $K = 1000$ and

$$\boxed{s = -16t^2 + 50t + 1000}$$

Finally at $t = 4$,

$$v = -32(4) + 50 = -78 \text{ feet per second}$$

$$s = -16(4)^2 + 50(4) + 1000 = 944 \text{ feet} \qquad \blacksquare$$

We remark that if $v = v_0$ and $s = s_0$ at $t = 0$, the procedure of Example 3 leads to the well-known falling-body formulas:

$$a = -32$$

$$v = -32t + v_0$$

$$s = -16t^2 + v_0 t + s_0$$

EXAMPLE 4   The acceleration of an object moving along a coordinate line is given by $a(t) = (2t + 3)^{-3}$ in meters per second per second. If the velocity at $t = 0$ is 4 meters per second, find the velocity 2 seconds later.

Solution

$$\frac{dv}{dt} = (2t + 3)^{-3}$$

$$v = \int (2t + 3)^{-3} \, dt = \frac{1}{2} \int \underbrace{(2t + 3)}_{u} {}^{-3} \underbrace{2 \, dt}_{du}$$

$$= \frac{1}{2} \frac{(2t + 3)^{-2}}{-2} + C = -\frac{1}{4(2t + 3)^2} + C$$

Since $v = 4$ at $t = 0$,

$$4 = -\frac{1}{4(3)^2} + C$$

which gives $C = \frac{145}{36}$. Thus,

$$v = -\frac{1}{4(2t + 3)^2} + \frac{145}{36}$$

At $t = 2$,

$$v = -\frac{1}{4(49)} + \frac{145}{36} \approx 4.023 \text{ meters per second} \qquad \blacksquare$$

FIGURE 3

EXAMPLE 5 (optional)   **(Escape velocity).**   The gravitational attraction F exerted by the earth on an object of mass m at a distance s from the center of the earth is given by $F = -mgR^2/s^2$, where $-g$ ($g \approx 32$ feet per second per second) is the acceleration of gravity at the surface of the earth and R ($R \approx 3960$ miles) is the radius of the earth (Figure 3). Show that an object projected outward from the earth with an initial velocity $v_0 \geq \sqrt{2gR} \approx 6.93$ miles per second will not fall back to the earth. Neglect air resistance in making this calculation.

Solution According to Newton's Second Law, $F = ma$; that is,

$$F = m\frac{dv}{dt} = m\frac{dv}{ds}\frac{ds}{dt} = m\frac{dv}{ds}v$$

Thus,

$$mv\frac{dv}{ds} = -mg\frac{R^2}{s^2}$$

$$v\,dv = -gR^2 s^{-2}\,ds$$

$$\int v\,dv = -gR^2 \int s^{-2}\,ds$$

$$\frac{v^2}{2} = \frac{gR^2}{s} + C$$

Now $v = v_0$ when $s = R$, and so $C = \frac{1}{2}v_0^2 - gR$. Consequently

$$v^2 = \frac{2gR^2}{s} + v_0^2 - 2gR$$

Finally, since $2gR^2/s$ gets small with increasing s, we see that v remains positive if and only if $v_0 \ge \sqrt{2gR}$. ∎

PROBLEM SET 5.2

In Problems 1–4, show that the indicated function is a solution of the given differential equation; that is, substitute the indicated function for y to see that it works.

1. $\dfrac{dy}{dx} + \dfrac{x}{y} = 0$; $y = \sqrt{4 - x^2}$

2. $3y^2\dfrac{dy}{dx} + x = 0$; $y = \left(1 - \dfrac{x^2}{2}\right)^{1/3}$

3. $\dfrac{d^2y}{dx^2} + y = 0$; $y = C_1 \sin x + C_2 \cos x$

4. $\left(\dfrac{dy}{dx}\right)^2 = 1 - y^2$; $y = \sin(x + C)$

In Problems 5–14, first find the general solution (involving a constant C) for the given differential equation. Then find the particular solution that satisfies the indicated condition. (See Example 2.)

5. $\dfrac{dy}{dx} = 3x^2 + 1$; $y = 4$ at $x = 1$

6. $\dfrac{dy}{dx} = x^{-2} + 2x$; $y = 5$ at $x = 1$

7. $\dfrac{dy}{dx} = \dfrac{x}{2y}$; $y = 3$ at $x = 2$

8. $\dfrac{dy}{dx} = \sqrt[3]{\dfrac{x}{y}}$; $y = 8$ at $x = 1$

9. $\dfrac{dy}{dt} = t^3 y^2$; $y = 1$ at $t = 2$

10. $\dfrac{dy}{dt} = y^3$; $y = 1$ at $t = 1$

11. $\dfrac{ds}{dt} = 3t^2 + 4t - 1$; $s = 5$ at $t = 2$

12. $\dfrac{du}{dt} = u^2(t^2 - 3t)$; $u = 4$ at $t = 0$

13. $\dfrac{dy}{dx} = (2x + 1)^4$; $y = 6$ at $x = 0$

14. $\dfrac{dy}{dx} = -y^2 x(x^2 + 2)^4$; $y = 1$ at $x = 0$

15. Find the xy-equation of the curve through $(1,2)$ whose slope at any point is four times its x-coordinate (see Example 1).

16. Find the xy-equation of the curve through $(1,2)$ whose slope at any point is one-half the square of its y-coordinate.

In Problems 17–20, an object is moving along a coordinate line subject to the indicated acceleration a (in centimeters per second per second) with the initial velocity v_0 (in centimeters per second) and directed distance s_0 (in centimeters). Find both the velocity v and directed distance s after 2 seconds (see Example 4).

17. $a = t$; $v_0 = 2$, $s_0 = 0$

18. $a = (1 + t)^{-3}$; $v_0 = 4$, $s_0 = 6$

⊂ **19.** $a = \sqrt[3]{2t + 1}$; $v_0 = 0$, $s_0 = 10$

⊂ **20.** $a = (3t + 1)^{-3}$; $v_0 = 4$, $s_0 = 0$

21. A ball is thrown upward from the surface of the earth with an initial velocity of 96 feet per second. What is the maximum height it reaches? (See Example 3.)

22. A ball is thrown upward from the surface of a planet where the acceleration of gravity is k (a negative constant) feet per second per second. If the initial velocity is v_0, show that the maximum height is $-v_0^2/2k$.

⊂ **23.** On the surface of the moon, the acceleration of gravity is -5.28 feet per second per second. If an object is thrown upward from an initial height of 1000 feet with a velocity of 56 feet per second, find its velocity and height 4.5 seconds later.

⊂ **24.** What is the maximum height that the object of Problem 23 reaches?

25. The rate of change of volume V of a melting snowball is proportional to the surface area S of the ball—that is, $dV/dt = -kS$, where k is a positive constant. If the radius of the ball at $t = 0$ is $r = 2$ and at $t = 10$ is $r = 0.5$, show that $r = -\frac{3}{20}t + 2$.

26. From what height above the earth must a ball be dropped in order to strike the ground with a velocity of -136 feet per second?

⊂ **27.** Determine the escape velocity for each of the following celestial bodies (see Example 5). Here $g \approx 32$ feet per second per second.

	Acceleration of Gravity	Radius (miles)
Moon	$-0.165g$	1,080
Venus	$-0.85g$	3,800
Jupiter	$-2.6g$	43,000
Sun	$-28g$	432,000

28. If the brakes of a car, when fully applied, produce a constant deceleration of 11 feet per second per second, what is the shortest distance in which the car can be braked to a halt from a speed of 60 miles per hour?

29. What constant acceleration will cause a car to increase its velocity from 45 to 60 miles per hour in 10 seconds?

30. A block slides down an inclined plane with a constant acceleration of 8 feet per second per second. If the inclined plane is 75 feet long and the block reaches the bottom in 3.75 seconds, what was the initial velocity of the block?

31. A certain rocket shot straight up has an acceleration of $6t$ meters per second per second during the first 10 seconds after blast-off, after which the engine cuts out and the rocket is subject to gravitational acceleration of -10 meters per second per second. How high will the rocket go?

32. Starting at station A, a commuter train accelerates at 3 meters per second per second for 8 seconds, then travels at constant speed v_m for 100 seconds, and finally brakes (decelerates) to a stop at station B at 4 meters per second per second. Find (a) v_m and (b) the distance between A and B.

33. Starting from rest, a bus increases speed at constant acceleration a_1, then travels at constant speed v_m, and finally brakes to a stop at constant deceleration a_2. It took 4 minutes to travel the 2 miles between stop C and stop D and then 3 minutes to go the 1.4 miles between stop D and stop E.
(a) Sketch the graph of the speed v as a function of time t, $0 \le t \le 7$.
(b) Find the maximum speed v_m.
(c) If $a_1 = a_2 = a$, evaluate a.

34. A balloon left the ground rising at 4 feet per second. Sixteen seconds later, Helena threw a ball straight up to her friend Janet in the balloon. At what speed did she throw the ball if it just made it to Janet?

35. According to Torricelli's Law, the rate of change of the volume V of water in a draining tank is proportional to the square root of the water's depth. A cylindrical tank of radius $10/\sqrt{\pi}$ centimeters and height 16 centimeters, which was full initially, took 40 seconds to drain.
(a) Write the differential equation for V at time t and the two corresponding conditions.
(b) Solve the differential equation.
(c) Find the volume of water after 10 seconds.

⊂ **36.** The wolf population P in a certain state has been growing at a rate proportional to the cube root of the population size. The population was estimated at 1000 in 1970 and at 1700 in 1980.
(a) Write the differential equation for P at time t with the two corresponding conditions.
(b) Solve the differential equation.
(c) When will the population reach 4000?

5.3
Sums and Sigma Notation

FIGURE 1

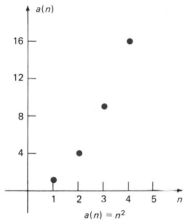

FIGURE 2

So far we have considered functions with domains that are intervals of real numbers. A typical example is $f(x) = x^2$, with—for example—domain the interval $[0, \infty)$. Its graph is shown in Figure 1. We used x as the domain variable, but we could as well have used s, t, u, or v. By custom, mathematicians use letters near the end of the alphabet to name variables that take values on an interval of the real line.

When we want to name a variable that takes only integer values, we typically use letters near the middle of the alphabet, such as i, j, k, m, and (especially) n. Thus, in this section we wish to consider the function determined by $a(n) = n^2$, where n takes positive integer values; its graph is shown in Figure 2. A function whose domain consists of just the positive integers (or some other subset of the integers) is called a **sequence**. In place of the standard functional notation $a(n)$, it is conventional to use a_n. Thus, we may say: Consider the sequence $\{a_n\}$ determined by $a_n = n^2$ and the sequence $\{b_n\}$ determined by $b_n = 1/n$. Sometimes we indicate a sequence by writing its first few values followed by dots as, for example,

$$a_1, a_2, a_3, a_4, \ldots$$

or even

$$1, 4, 9, 16, \ldots$$

Sequences will be studied in detail in Chapter 11. Here our main interest is in introducing notation for certain of their sums.

SIGMA NOTATION   Consider the sums:

$$1^2 + 2^2 + 3^2 + 4^2 + \cdots + 100^2$$

and

$$a_1 + a_2 + a_3 + a_4 + \cdots + a_n$$

To indicate these sums in a compact way, we write the first as

$$\sum_{i=1}^{100} i^2$$

and the second as

$$\sum_{i=1}^{n} a_i$$

Here $\sum$ (capital Greek sigma), which corresponds to our S, suggests that we are to sum (add) all numbers of the form indicated as the *index i* runs through the positive integers, starting with the integer shown below the $\sum$

and ending with the integer at the top. Thus,

$$\sum_{i=2}^{5} b_i = b_2 + b_3 + b_4 + b_5$$

$$\sum_{j=1}^{n} \frac{1}{j} = \frac{1}{1} + \frac{1}{2} + \frac{1}{3} + \cdots + \frac{1}{n}$$

$$\sum_{k=1}^{4} \frac{k}{k^2 + 1} = \frac{1}{1^2 + 1} + \frac{2}{2^2 + 1} + \frac{3}{3^2 + 1} + \frac{4}{4^2 + 1}$$

and, for $n \geq m$,

$$\sum_{i=m}^{n} F(i) = F(m) + F(m + 1) + F(m + 2) + \cdots + F(n)$$

If all the c_i in $\sum_{i=1}^{n} c_i$ have the same value, say c, then

$$\sum_{i=1}^{n} c_i = \underbrace{c + c + c + \cdots + c}_{n \text{ terms}} = nc$$

As a result, we adopt the convention

$$\boxed{\sum_{i=1}^{n} c = nc}$$

In particular,

$$\sum_{i=1}^{5} 2 = 5(2) = 10, \qquad \sum_{i=1}^{100} (-4) = 100(-4) = -400$$

PROPERTIES OF $\sum$ Thought of as an operator, $\sum$ operates on sequences, and it does so in a linear way.

Theorem A

(Linearity of $\sum$). Let $\{a_i\}$ and $\{b_i\}$ denote two sequences and let c be a constant. Then:

(i) $\displaystyle\sum_{i=1}^{n} ca_i = c \sum_{i=1}^{n} a_i;$

(ii) $\displaystyle\sum_{i=1}^{n} (a_i + b_i) = \sum_{i=1}^{n} a_i + \sum_{i=1}^{n} b_i;$ and consequently

(iii) $\displaystyle\sum_{i=1}^{n} (a_i - b_i) = \sum_{i=1}^{n} a_i - \sum_{i=1}^{n} b_i.$

Proof   The proofs are easy; we consider only (i).

$$\sum_{i=1}^{n} ca_i = ca_1 + ca_2 + \cdots + ca_n = c(a_1 + a_2 + \cdots + a_n)$$

$$= c \sum_{i=1}^{n} a_i \qquad \blacksquare$$

EXAMPLE 1   Suppose that $\displaystyle\sum_{i=1}^{100} a_i = 60$ and $\displaystyle\sum_{i=1}^{100} b_i = 11$. Calculate

$$\sum_{i=1}^{100} (2a_i - 3b_i + 4).$$

Solution

$$\sum_{i=1}^{100} (2a_i - 3b_i + 4) = \sum_{i=1}^{100} 2a_i - \sum_{i=1}^{100} 3b_i + \sum_{i=1}^{100} 4$$

$$= 2\sum_{i=1}^{100} a_i - 3\sum_{i=1}^{100} b_i + \sum_{i=1}^{100} 4$$

$$= 2(60) - 3(11) + 100(4) = 487 \qquad \blacksquare$$

EXAMPLE 2   **(Collapsing sums).**   Show that:

(a) $\displaystyle\sum_{i=1}^{n} (a_{i+1} - a_i) = a_{n+1} - a_1$;

(b) $\displaystyle\sum_{i=1}^{n} \left[(i+1)^2 - i^2\right] = (n+1)^2 - 1.$

Solution

(a) Here we should resist our inclination to apply linearity and instead write out the sum, hoping for some nice cancellations.

$$\sum_{i=1}^{n} (a_{i+1} - a_i) = (a_2 - a_1) + (a_3 - a_2) + (a_4 - a_3) + \cdots + (a_{n+1} - a_n)$$

$$= -a_1 + a_2 - a_2 + a_3 - a_3 + a_4 + \cdots - a_n + a_{n+1}$$

$$= -a_1 + a_{n+1} = a_{n+1} - a_1$$

(b) This follows immediately from part (a). $\qquad \blacksquare$

The symbol used for the index does not matter. Thus,

$$\sum_{i=1}^{n} a_i = \sum_{j=1}^{n} a_j = \sum_{k=1}^{n} a_k$$

and all of these are equal to $a_1 + a_2 + \cdots + a_n$. For this reason, the index is sometimes called a **dummy index**.

SOME SPECIAL SUMS In the next section, we will need to consider the sum of the first n positive integers, as well as the sums of their squares, cubes, and so on. There are nice formulas for these; proofs are indicated at the end of the section.

1. $\displaystyle\sum_{i=1}^{n} i = 1 + 2 + 3 + \cdots + n = \frac{n(n+1)}{2}$

2. $\displaystyle\sum_{i=1}^{n} i^2 = 1^2 + 2^2 + 3^2 + \cdots + n^2 = \frac{n(n+1)(2n+1)}{6}$

3. $\displaystyle\sum_{i=1}^{n} i^3 = 1^3 + 2^3 + 3^3 + \cdots + n^3 = \left[\frac{n(n+1)}{2}\right]^2$

4. $\displaystyle\sum_{i=1}^{n} i^4 = 1^4 + 2^4 + 3^4 + \cdots + n^4 = \frac{n(n+1)(6n^3 + 9n^2 + n - 1)}{30}$

EXAMPLE 3 Calculate (a) $\displaystyle\sum_{i=1}^{10} i$, (b) $\displaystyle\sum_{i=1}^{10} i^2$, and (c) $\displaystyle\sum_{i=2}^{10} i^4$.

Solution

(a) $\displaystyle\sum_{i=1}^{10} i = \frac{10(10+1)}{2} = 55$

(b) $\displaystyle\sum_{i=1}^{10} i^2 = \frac{10(10+1)(20+1)}{6} = 385$

(c) $\displaystyle\sum_{i=2}^{10} i^4 = \left(\sum_{i=1}^{10} i^4\right) - 1^4 = \frac{10(11)(6000 + 900 + 10 - 1)}{30} - 1$

$$= 25{,}332 \qquad\blacksquare$$

EXAMPLE 4 Calculate $\displaystyle\sum_{i=1}^{10} 2i(i - 5)$.

Solution We make use of Example 3 and linearity.

$$\sum_{i=1}^{10} 2i(i - 5) = \sum_{i=1}^{10} (2i^2 - 10i) = 2\sum_{i=1}^{10} i^2 - 10\sum_{i=1}^{10} i$$

$$= 2(385) - 10(55) = 220 \qquad\blacksquare$$

EXAMPLE 5 Find a formula for $\displaystyle\sum_{j=1}^{n} (j + 2)(j - 5)$.

Solution

$$\sum_{j=1}^{n} (j + 2)(j - 5) = \sum_{j=1}^{n} (j^2 - 3j - 10) = \sum_{j=1}^{n} j^2 - 3\sum_{j=1}^{n} j - \sum_{j=1}^{n} 10$$

$$= \frac{n(n+1)(2n+1)}{6} - 3\frac{n(n+1)}{2} - 10n$$

$$= \frac{n}{6}\left[2n^2 + 3n + 1 - 9n - 9 - 60\right]$$

$$= \frac{n(n^2 - 3n - 34)}{3} \qquad\blacksquare$$

PROOFS OF SPECIAL SUM FORMULAS Undoubtedly, many of our readers are expecting us to use *mathematical induction* to prove Formulas 1–4. Mathematical induction will certainly work (see Appendix A.1). However, we choose to offer alternative proofs that have the advantage of showing where the formulas come from.

To prove Formula 1, we start with the identity $(i + 1)^2 - i^2 = 2i + 1$, sum both sides, use linearity, and apply Example 2.

$$(i + 1)^2 - i^2 = 2i + 1$$

$$\sum_{i=1}^{n} [(i + 1)^2 - i^2] = \sum_{i=1}^{n} (2i + 1)$$

$$(n + 1)^2 - 1^2 = 2 \sum_{i=1}^{n} i + \sum_{i=1}^{n} 1$$

$$n^2 + 2n = 2 \sum_{i=1}^{n} i + n$$

$$\frac{n^2 + n}{2} = \sum_{i=1}^{n} i$$

We use Formula 1 and the same technique to obtain Formula 2.

$$(i + 1)^3 - i^3 = 3i^2 + 3i + 1$$

$$\sum_{i=1}^{n} [(i + 1)^3 - i^3] = \sum_{i=1}^{n} (3i^2 + 3i + 1)$$

$$(n + 1)^3 - 1^3 = 3 \sum_{i=1}^{n} i^2 + 3 \sum_{i=1}^{n} i + \sum_{i=1}^{n} 1$$

$$n^3 + 3n^2 + 3n = 3 \sum_{i=1}^{n} i^2 + 3 \frac{n(n + 1)}{2} + n$$

$$2n^3 + 6n^2 + 6n = 6 \sum_{i=1}^{n} i^2 + 3n^2 + 3n + 2n$$

$$\frac{2n^3 + 3n^2 + n}{6} = \sum_{i=1}^{n} i^2$$

$$\frac{n(n + 1)(2n + 1)}{6} = \sum_{i=1}^{n} i^2$$

The same technique works to establish Formulas 3 and 4.

PROBLEM SET 5.3

In Problems 1–8, find the value of the indicated sum.

1. $\displaystyle\sum_{k=1}^{5} (3k - 1)$

2. $\displaystyle\sum_{i=1}^{6} 2i^2$

3. $\displaystyle\sum_{i=3}^{5} \frac{2}{i + 1}$

4. $\displaystyle\sum_{j=2}^{6} (j + 1)^2$

5. $\displaystyle\sum_{i=1}^{5} (-1)^i 2^{i-1}$

6. $\displaystyle\sum_{i=2}^{4} \frac{(-1)^i}{i(2i + 1)}$

7. $\displaystyle\sum_{k=1}^{6} k \sin\left(\frac{k\pi}{2}\right)$

8. $\displaystyle\sum_{k=1}^{7} \cos k\pi$

In Problems 9–16, write the indicated sum in sigma notation.

9. $1 + 2 + 3 + \cdots + 98$

10. $2 + 4 + 6 + \cdots + 100$

11. $1 + \frac{1}{2} + \frac{1}{3} + \cdots + \frac{1}{69}$

12. $1 - \frac{1}{2} + \frac{1}{3} - \frac{1}{4} + \cdots - \frac{1}{50}$

13. $a_1 + a_2 + a_3 + \cdots + a_n$

14. $b_3 + b_4 + b_5 + \cdots + b_{22}$

15. $f(c_1) + f(c_2) + \cdots + f(c_n)$

16. $f(w_1)\Delta x + f(w_2)\Delta x + \cdots + f(w_n)\Delta x$

In Problems 17–20, suppose that $\sum_{i=1}^{10} a_i = 40$ and $\sum_{i=1}^{10} b_i = 50$. Calculate each of the following (see Example 1).

17. $\sum_{i=1}^{10} (2a_i + b_i)$

18. $\sum_{i=1}^{10} (3a_i - 2b_i)$

19. $\sum_{i=1}^{10} (4a_i - b_i + 2)$

20. $\sum_{k=1}^{10} (2b_k + 4)$

In Problems 21–24, find the value of each of the (collapsing) sums (see Example 2).

21. $\sum_{k=1}^{40} \left(\dfrac{1}{k} - \dfrac{1}{k+1} \right)$

22. $\sum_{k=1}^{10} (2^k - 2^{k-1})$

23. $\sum_{k=3}^{20} \left(\dfrac{1}{(k+1)^2} - \dfrac{1}{k^2} \right)$

24. $\sum_{k=3}^{m+1} (a_k - a_{k-1})$

In Problems 25–30, use Formulas 1–4, page 222, to find each sum (see Examples 3–5).

25. $\sum_{i=1}^{100} (3i - 2)$

26. $\sum_{i=1}^{10} [(i-1)(4i+3)]$

27. $\sum_{k=1}^{10} (k^3 - k^2)$

28. $\sum_{k=1}^{10} 5k^2(k+4)$

29. $\sum_{i=1}^{n} (2i^2 - 3i + 1)$

30. $\sum_{i=1}^{n} (2i - 3)^2$

Sometimes it is desirable to make a change of variable in the index for a sum. For example, the change of variable $k = i - 3$ gives

$$\sum_{i=4}^{13} (i-3)^3 = \sum_{k=1}^{10} k^3$$

For Problems 31–34, make the indicated change of variable in the index.

31. $\sum_{i=3}^{19} i(i-2); \quad k = i - 2$

32. $\sum_{k=5}^{14} k2^{k-4}; \quad i = k - 4$

33. $\sum_{k=0}^{10} \dfrac{k}{k+1}; \quad i = k + 1$

34. $\sum_{k=4}^{13} (k-3)\sin\left(\dfrac{\pi}{k-3} \right); \quad i = k - 3$

35. Evaluate $\sum_{i=1}^{10} f(w_i)\Delta x$ if $f(x) = 3x$, $w_i = i/5$, and $\Delta x = \frac{1}{5}$.

36. Prove the following formula for a **geometric sum**:

$$\sum_{k=0}^{n} ar^k = a + ar + ar^2 + \cdots + ar^n = \dfrac{a - ar^{n+1}}{1 - r} \quad (r \neq 1)$$

Hint: Let $S = a + ar + \cdots + ar^n$. Simplify $S - rS$ and solve for S.

37. Use Problem 36 to calculate each sum.

(a) $\sum_{k=1}^{10} (\frac{1}{2})^k$

(b) $\sum_{k=1}^{10} 2^k$

38. Add the two equalities below, solve for S, and thereby give another proof of Formula 1.

$$S = 1 + 2 + 3 + \cdots + (n-2) + (n-1) + n$$
$$S = n + (n-1) + (n-2) + \cdots + 3 + 2 + 1$$

39. Use a derivation like that in Problem 38 to obtain a formula for the **arithmetic sum**:

$$\sum_{k=0}^{n} (a + kd) = a + (a+d) + (a+2d) + \cdots + (a+nd)$$

40. Show that $\sum_{k=1}^{n} \dfrac{1}{\sqrt{k}} \geq \sqrt{n}$.

C 41. In statistics we define the mean $\bar{x}$ and the variance s^2 of a sequence of numbers $x_1, x_2, \ldots, x_n$ by

$$\bar{x} = \frac{1}{n}\sum_{i=1}^{n} x_i, \qquad s^2 = \frac{1}{n}\sum_{i=1}^{n}(x_i - \bar{x})^2$$

Find $\bar{x}$ and s^2 for the sequence of numbers 2, 5, 7, 8, 9, 10, 14.

42. Use the definitions in Problem 41 to show that each is true.

(a) $\sum_{i=1}^{n}(x_i - \bar{x}) = 0$

(b) $s^2 = \left(\dfrac{1}{n}\sum_{i=1}^{n} x_i^2 \right) - \bar{x}^2$

43. Let

$$A = \sum_{i=1}^{n} a_i^2, \qquad B = \sum_{i=1}^{n} a_i b_i, \qquad C = \sum_{i=1}^{n} b_i^2$$

Prove that $At^2 + 2Bt + C \geq 0$ for all real numbers t. *Hint*: Show

$$At^2 + 2Bt + C = \sum_{i=1}^{n}(a_i t + b_i)^2.$$

44. Prove the **Cauchy Inequality**:

$$\left(\sum_{i=1}^{n} a_i b_i \right)^2 \leq \left(\sum_{i=1}^{n} a_i^2 \right)\left(\sum_{i=1}^{n} b_i^2 \right)$$

Hint: In the notation of Problem 43, we must show that $B^2 - AC \leq 0$. Note that this is one-fourth the discriminant of the quadratic equation $At^2 + 2Bt + C = 0$.

45. Establish the following identity, which is known as *summation by parts*:

$$\sum_{i=1}^{n}(a_i - a_{i-1})b_{i-1} = a_n b_n - a_0 b_0 - \sum_{i=1}^{n} a_i(b_i - b_{i-1})$$

46. Use the identity $(i + 1)^4 - i^4 = 4i^3 + 6i^2 + 4i + 1$ to prove Formula 3, namely,

$$1^3 + 2^3 + \cdots + n^3 = \left[\frac{n(n + 1)}{2}\right]^2$$

47. Use the diagrams in Figure 3 to establish Formulas 1 and 3.

$1 + 2 + \ldots + n =$ $1^3 + 2^3 + \ldots + n^3 =$

FIGURE 3

48. Find a nice formula for

$$\frac{1}{1 \cdot 2} + \frac{1}{2 \cdot 3} + \frac{1}{3 \cdot 4} + \cdots + \frac{1}{n(n + 1)}$$

Hint: $\dfrac{1}{i(i + 1)} = \dfrac{1}{i} - \dfrac{1}{i + 1}$.

49. A grocer stacks oranges in a pyramid-like pile. If the bottom layer is rectangular with 10 rows of 16 oranges and the top layer has a single row of oranges, how many oranges are in the stack?

50. In the song *The Twelve Days of Christmas*, my true love gave me 1 gift on the first day, $1 + 2$ gifts on the second day, $1 + 2 + 3$ gifts on the third day, and so on for 12 days.

(a) Find the total number of gifts given in 12 days.
(b) Find a simple formula for T_n, the total number of gifts given during a Christmas of n days.

5.4
Introduction to Area

Two problems, both from geometry, motivate the two biggest ideas in calculus. The problem of the tangent line led us to the *derivative*. The problem of area will lead us to the *definite integral*.

For polygons (closed plane figures bounded by straight line segments), the area problem is hardly a problem at all. We start by defining the area of a rectangle to be its length times its width, and from this we successively derive the formulas for the area of a parallelogram, a triangle, and any polygon. The sequence of figures in Figure 1 suggests how this is done.

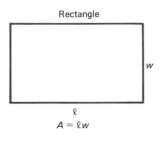

Rectangle
$A = \ell w$

Parallelogram
$A = bh$

Triangle
$A = \frac{1}{2} bh$

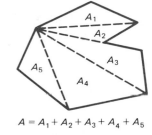

Polygon
$A = A_1 + A_2 + A_3 + A_4 + A_5$

FIGURE 1

Even in this simple setting, it is clear that we want area to satisfy five properties.

1. The area of a plane region is a nonnegative (real) number.
2. The area of a rectangle is the product of its length and width (both measured in the same units).

3. Congruent regions have equal areas.
4. The area of the union of two regions that overlap only in a line segment is the sum of the areas of the two regions.
5. If one region is contained in a second region, then the area of the first region is less than or equal to that of the second.

When we consider a region with a curved boundary, the problem of assigning area is significantly more difficult. However, over 2000 years ago, Archimedes provided the key to a solution. Consider, said he, a sequence of inscribed polygons that approximate the curved region with greater and greater accuracy. For example, for the circle of radius 1, consider regular inscribed polygons $P_1, P_2, P_3, \ldots$ with 4 sides, 8 sides, 16 sides, . . . , as shown in Figure 2. The area of the circle is the limit as $n \to \infty$ of the areas of P_n. Thus, if $A(F)$ denotes the area of a region F, then

$$A(\text{circle}) = \lim_{n \to \infty} A(P_n)$$

FIGURE 2

Archimedes went further, considering also circumscribed polygons $T_1, T_2, T_3, \ldots$ (Figure 3). He showed that you get the same value for the area of the circle of radius 1 (namely, $\pi \approx 3.14159$) whether you use inscribed or circumscribed polygons. It is just a small step from what he did to our modern treatment of area.

FIGURE 3

AREA BY INSCRIBED POLYGONS   Consider the region R bounded by the parabola $y = f(x) = x^2$, the x-axis, and the vertical line $x = 2$ (Figure 4). We refer to R as the region under the curve $y = x^2$ between $x = 0$ and $x = 2$. Our aim is to calculate its area $A(R)$.

Partition (as in Figure 5) the interval $[0, 2]$ into n subintervals, each of length $\Delta x = 2/n$, by means of points.

$$0 = x_0 < x_1 < x_2 < \cdots < x_{n-1} < x_n = 2$$

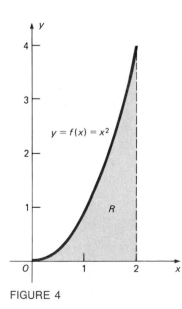

FIGURE 4

CHAPTER 5/The Integral

FIGURE 5

Thus,

$$x_0 = 0$$

$$x_1 = \Delta x = \frac{2}{n}$$

$$x_2 = 2 \cdot \Delta x = \frac{4}{n}$$

$$x_3 = 3 \cdot \Delta x = \frac{6}{n}$$

$$\cdots$$

$$x_i = i \cdot \Delta x = \frac{2i}{n}$$

$$\cdots$$

$$x_{n-1} = (n-1) \cdot \Delta x = \frac{(n-1)2}{n}$$

$$x_n = n \cdot \Delta x = n\left(\frac{2}{n}\right) = 2$$

Consider the rectangle with base $[x_{i-1}, x_i]$ and height $f(x_{i-1}) = x_{i-1}^2$. Its area is $f(x_{i-1})\Delta x$ (see the upper left part of Figure 6). The union R_n of all such rectangles forms the inscribed polygon shown in the lower right part of Figure 6.

The area $A(R_n)$ can be calculated.

$$A(R_n) = f(x_0)\Delta x + f(x_1)\Delta x + f(x_2)\Delta x + \cdots + f(x_{n-1})\Delta x$$

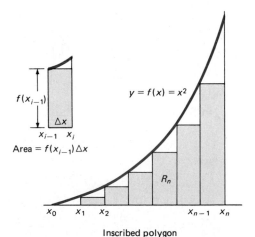

Inscribed polygon

FIGURE 6

Now

$$f(x_i)\,\Delta x = x_i^2\,\Delta x = \left(\frac{2i}{n}\right)^2 \cdot \frac{2}{n} = \left(\frac{8}{n^3}\right)i^2$$

Thus,

$$A(R_n) = \left[\frac{8}{n^3}(0^2) + \frac{8}{n^3}(1^2) + \frac{8}{n^3}(2^2) + \cdots + \frac{8}{n^3}(n-1)^2\right]$$

$$= \frac{8}{n^3}\left[1^2 + 2^2 + \cdots + (n-1)^2\right]$$

$$= \frac{8}{n^3}\left[\frac{(n-1)n(2n-1)}{6}\right] \quad \begin{array}{l}\text{(Formula 2, p. 222,}\\ \text{with } n-1 \text{ replacing } n\text{)}\end{array}$$

$$= \frac{8}{6}\left(\frac{2n^3 - 3n^2 + n}{n^3}\right)$$

$$= \frac{4}{3}\left(2 - \frac{3}{n} + \frac{1}{n^2}\right)$$

$$= \frac{8}{3} - \frac{4}{n} + \frac{4}{3n^2}$$

We conclude that

$$A(R) = \lim_{n \to \infty} A(R_n) = \lim_{n \to \infty}\left(\frac{8}{3} - \frac{4}{n} + \frac{4}{3n^2}\right) = \frac{8}{3}$$

The diagrams in Figure 7 should help you visualize what is happening as n gets larger and larger.

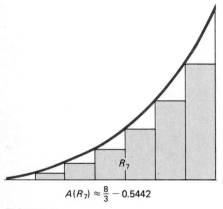

$A(R_7) \approx \frac{8}{3} - 0.5442$

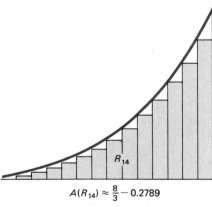

$A(R_{14}) \approx \frac{8}{3} - 0.2789$

$A(R_{28}) \approx \frac{8}{3} - 0.1412$

FIGURE 7

AREA BY CIRCUMSCRIBED POLYGONS   Perhaps you are still not convinced that $A(R) = \frac{8}{3}$. We can give more evidence. Consider the rectangle with base $[x_{i-1}, x_i]$ and height $f(x_i) = x_i^2$ (shown at the upper left in Figure 8). Its area is $f(x_i)\,\Delta x$. The union S_n of all such rectangles forms a circumscribed polygon for the region R, as shown at the lower right in Figure 8.

y = f(x) = x²

f(x_i)

Δx

x_{i-1} x_i

Area = f(x_i) Δx

x_0 x_1 x_2 x_{n-1} x_n

S_n

Circumscribed polygon

FIGURE 8

The area $A(S_n)$ is calculated in analogy with the calculation of $A(R_n)$.

$$A(S_n) = f(x_1)\Delta x + f(x_2)\Delta x + \cdots + f(x_n)\Delta x$$

As before, $f(x_i)\Delta x = x_i^2\,\Delta x = (8/n^3)i^2$, and so

$$A(S_n) = \left[\frac{8}{n^3}(1^2) + \frac{8}{n^3}(2^2) + \cdots + \frac{8}{n^3}(n^2)\right]$$

$$= \frac{8}{n^3}\left[1^2 + 2^2 + \cdots + n^2\right]$$

$$= \frac{8}{n^3}\left[\frac{n(n+1)(2n+1)}{6}\right] \quad \text{(Formula 2, p. 222)}$$

$$= \frac{8}{6}\left[\frac{2n^3 + 3n^2 + n}{n^3}\right]$$

$$= \frac{4}{3}\left[2 + \frac{3}{n} + \frac{1}{n^2}\right]$$

Thus,

$$\lim_{n\to\infty} A(S_n) - \lim_{n\to\infty}\frac{4}{3}\left[2 + \frac{3}{n} + \frac{1}{n^2}\right] = \frac{8}{3}$$

ANOTHER PROBLEM—SAME THEME   Suppose that an object is traveling along the t-axis in such a way that its velocity at time t is given by $v = f(t) = \frac{1}{4}t^3 + 1$ feet per second. How far did it travel between $t = 0$ and $t = 3$? This problem can be solved by the method of differential equations (Section 5.2), but we have something else in mind.

Our starting point is the familiar fact that if an object travels at constant velocity k over a time interval of length Δt, then the distance traveled is $k\,\Delta t$. But this is just the area of a rectangle—the one shown in Figure 9.

v = k

Δt

Distance = k Δt

FIGURE 9

Next consider the given problem, where $v = f(t) = \frac{1}{4}t^3 + 1$. The graph is shown in the left half of Figure 10. Partition the interval $[0, 3]$ into n subintervals of length $\Delta t = 3/n$ by means of points $0 = t_0 < t_1 < t_2 < \cdots < t_n = 3$. Then consider the corresponding circumscribed polygon S_n, displayed in the right half of Figure 10 (we could as well have considered the inscribed polygon). Its area, $A(S_n)$, should be a good approximation of the distance traveled, especially if Δt is small, since on each subinterval the actual velocity is almost equal to a constant (the value of v at the end of the subinterval). Moreover, this approximation should get better and better as n gets larger. We are led to the conclusion that the exact distance traveled is $\lim_{n \to \infty} A(S_n)$; that is, it is the area of the region under the velocity curve between $t = 0$ and $t = 3$.

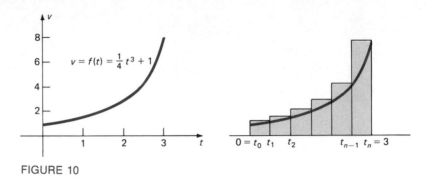

FIGURE 10

To calculate $A(S_n)$, note that $t_i = 3i/n$, and so the area of the ith rectangle is

$$f(t_i)\,\Delta t = \left[\frac{1}{4}\left(\frac{3i}{n}\right)^3 + 1\right]\frac{3}{n} = \frac{81}{4n^4}\,i^3 + \frac{3}{n}$$

Thus,

$$A(S_n) = f(t_1)\,\Delta t + f(t_2)\,\Delta t + \cdots + f(t_n)\,\Delta t$$

$$= \sum_{i=1}^{n} f(t_i)\,\Delta t$$

$$= \sum_{i=1}^{n} \left(\frac{81}{4n^4}\,i^3 + \frac{3}{n}\right)$$

$$= \frac{81}{4n^4} \sum_{i=1}^{n} i^3 + \sum_{i=1}^{n} \frac{3}{n}$$

$$= \frac{81}{4n^4}\left[\frac{n(n+1)}{2}\right]^2 + \frac{3}{n} \cdot n \qquad \text{(Formula 3, p. 222)}$$

$$= \frac{81}{16}\left[n^2\,\frac{(n^2 + 2n + 1)}{n^4}\right] + 3$$

$$= \frac{81}{16}\left(1 + \frac{2}{n} + \frac{1}{n^2}\right) + 3$$

CHAPTER 5/The Integral

We conclude that

$$\lim_{n \to \infty} A(S_n) = \frac{81}{16} + 3 = \frac{129}{16} \approx 8.06$$

The object traveled about 8.06 feet between $t = 0$ and $t = 3$.

What was true in this example is true for any object moving with positive velocity. *The distance traveled is the area under the velocity curve.*

PROBLEM SET 5.4

In Problems 1–6, find the area of the indicated inscribed or circumscribed polygon.

1.

2.

3.

4.

5.

6.

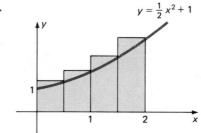

In Problems 7–10, sketch the graph of the given function over the interval $[a, b]$, then divide $[a, b]$ into n equal subintervals, and finally calculate the area of the corresponding circumscribed polygon.

7. $f(x) = 2x + 3$; $a = -1$, $b = 2$, $n = 3$

8. $f(x) = 3x - 2$; $a = 1$, $b = 3$, $n = 4$

[C] **9.** $f(x) = x^2 + 2$; $a = 0$, $b = 2$, $n = 6$

[C] **10.** $f(x) = 2x^2 + 1$; $a = -1$, $b = 1$, $n = 8$

In Problems 11–16, find the area of the region under the curve $y = f(x)$ over the interval $[a, b]$. To do this, divide the interval $[a, b]$ into n equal subintervals, calculate the area of the corresponding circumscribed polygon, and then let $n \to \infty$. (See the example for $y = x^2$ in the text.)

11. $y = x + 1$; $a = 0$, $b = 2$

12. $y = \frac{1}{2}x^2 + 1$; $a = 0$, $b = 2$

13. $y = 3x + 1$; $a = 1$, $b = 4$. *Hint:* $x_i = 1 + \dfrac{3i}{n}$.

14. $y = x^2$; $a = 1$, $b = 4$

15. $y = x^3$; $a = 0$, $b = 1$

16. $y = x^3 + x$; $a = 0$, $b = 1$

17. Suppose that an object is traveling along the t-axis in such a way that its velocity at time t seconds is $v = t + 1$ feet per second. How far did it travel between $t = 0$ and $t = 2$? *Hint*: See the discussion of this problem at the end of this section and use the result of Problem 11.

18. Follow the directions of Problem 17 given that $v = \frac{1}{2}t^2 + 1$. You will need the result of Problem 12.

19. Let A_a^b denote the area under the curve $y = x^2$ over the interval $[a, b]$.
(a) Prove that $A_0^b = b^3/3$. *Hint*: $\Delta x = b/n$ so $x_i = ib/n$; use circumscribed polygons.
(b) Show that $A_a^b = b^3/3 - a^3/3$. Assume $a \geq 0$.

20. Suppose that an object, moving along the t-axis, has velocity $v = t^2$ meters per second at time t seconds. How far did it travel between $t = 3$ and $t = 5$? See Problem 19.

21. Use the results of Problem 19 to calculate the area under the curve $y = x^2$ over each of the following intervals.
(a) $[0, 5]$ (b) $[1, 4]$ (c) $[2, 5]$

22. From Formulas 1–4 of Section 5.3, you might guess that

$$1^m + 2^m + 3^m + \cdots + n^m = \frac{n^{m+1}}{m+1} + C_n$$

where C_n is a polynomial in n of degree m. Assume this is true (which it is) and, for $a \geq 0$, let $A_a^b(x^m)$ be the area under the curve $y = x^m$ over the interval $[a, b]$.

(a) Prove that $A_0^b(x^m) = \dfrac{b^{m+1}}{(m+1)}$.

(b) Show that $A_a^b(x^m) = \dfrac{b^{m+1}}{m+1} - \dfrac{a^{m+1}}{m+1}$

23. Use the results of Problem 22 to calculate each of the following areas.
(a) $A_0^2(x^3)$ (b) $A_1^2(x^3)$ (c) $A_1^2(x^5)$ (d) $A_0^2(x^9)$

24. Derive the formulas $A_n = \frac{1}{2}nr^2 \sin(2\pi/n)$ and $B_n = nr^2 \tan(\pi/n)$ for the areas of the inscribed and circumscribed regular n-sided polygons for a circle of radius r. Then show that $\lim_{n \to \infty} A_n$ and $\lim_{n \to \infty} B_n$ are both πr^2.

5.5
The Definite
Integral

All the preparations have been made; we are ready to define the definite integral. Both Newton and Leibniz introduced early versions of this concept. However, it was Riemann who gave us the modern definition. In formulating this definition, we are guided by the ideas we discussed in the previous section. The first notion is that of a Riemann sum.

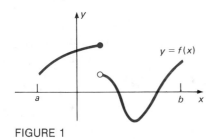

FIGURE 1

RIEMANN SUMS   Consider a function f defined on a closed interval $[a, b]$. It may have both positive and negative values on the interval and it does not even need to be continuous. Its graph might look something like the one in Figure 1.

Consider a partition P of the interval $[a, b]$ into n subintervals (not necessarily of equal length) by means of points $a = x_0 < x_1 < x_2 < \cdots < x_{n-1} < x_n = b$ and let $\Delta x_i = x_i - x_{i-1}$. On each subinterval $[x_{i-1}, x_i]$, pick an arbitrary point $\bar{x}_i$ (which may be an endpoint); we call it a *sample point* for the ith subinterval. An example of these constructions is shown in Figure 2 for $n = 6$.

A Partition of $[a, b]$ with Sample Points $\bar{x}_i$

FIGURE 2

Form the sum

$$R_P = \sum_{i=1}^{n} f(\bar{x}_i)\,\Delta x_i$$

We call R_P a **Riemann sum** for f corresponding to the partition P. Its geometric interpretation is shown in Figure 3. Note that the contributions from rectangles below the x-axis are the negatives of their areas.

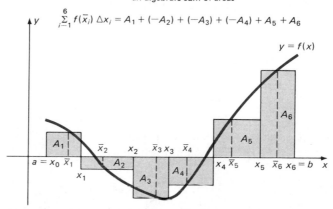

A Riemann sum interpreted as an algebraic sum of areas

$$\sum_{i=1}^{6} f(\bar{x}_i)\,\Delta x_i = A_1 + (-A_2) + (-A_3) + (-A_4) + A_5 + A_6$$

FIGURE 3

EXAMPLE 1 Evaluate the Riemann sum R_P for

$$f(x) = (x+1)(x-2)(x-4) = x^3 - 5x^2 + 2x + 8$$

on the interval $[0, 5]$ using the partition P with partition points $0 < 1.1 < 2 < 3.2 < 4 < 5$ and the corresponding sample points $\bar{x}_1 = 0.5$, $\bar{x}_2 = 1.5$, $\bar{x}_3 = 2.5$, $\bar{x}_4 = 3.6$, and $\bar{x}_5 = 5$.

Solution

FIGURE 4

$$
\begin{aligned}
R_P &= \sum_{i=1}^{5} f(\bar{x}_i)\,\Delta x_i \\
&= f(\bar{x}_1)\,\Delta x_1 + f(\bar{x}_2)\,\Delta x_2 + f(\bar{x}_3)\,\Delta x_3 + f(\bar{x}_4)\,\Delta x_4 + f(\bar{x}_5)\,\Delta x_5 \\
&= f(0.5)(1.1 - 0) + f(1.5)(2 - 1.1) + f(2.5)(3.2 - 2) \\
&\quad + f(3.6)(4 - 3.2) + f(5)(5 - 4) \\
&= (7.875)(1.1) + (3.125)(0.9) + (-2.625)(1.2) \\
&\quad + (-2.944)(0.8) + 18(1) \\
&= 23.9698
\end{aligned}
$$

The corresponding geometric picture appears in Figure 4. ∎

FIGURE 5

EXAMPLE 2 Evaluate the Riemann sum for $f(x) = x^2 + 1$ on the interval $[-1, 2]$ using the equally spaced partition points $-1 < -0.5 < 0 < 0.5 < 1 < 1.5 < 2$, with the sample point $\bar{x}_i$ being the midpoint of the ith subinterval.

Solution Note the picture in Figure 5.

$$R_P = \sum_{i=1}^{6} f(\bar{x}_i)\Delta x_i$$

$$= [f(-0.75) + f(-0.25) + f(0.25) + f(0.75) + f(1.25) + f(1.75)](0.5)$$

$$= [1.5625 + 1.0625 + 1.0625 + 1.5625 + 2.5625 + 4.0625](0.5)$$

$$= 5.9375 \qquad \blacksquare$$

DEFINITION OF THE DEFINITE INTEGRAL Suppose now that P, Δx_i, and $\bar{x}_i$ have the meanings discussed above. Also let $|P|$, called the **norm** of P, denote the length of the longest of the subintervals of the partition P. For instance, in Example 1, $|P| = 3.2 - 2 = 1.2$; in Example 2, $|P| = 0.5$.

Definition

(The definite integral). Let f be a function that is defined on the closed interval $[a, b]$. If

$$\lim_{|P| \to 0} \sum_{i=1}^{n} f(\bar{x}_i)\Delta x_i$$

exists, we say f is **integrable** on $[a, b]$. Moreover, $\displaystyle\int_a^b f(x)\,dx$, called the **definite integral** (or Riemann integral) of f from a to b, is then given by

$$\int_a^b f(x)\,dx = \lim_{|P| \to 0} \sum_{i=1}^{n} f(\bar{x}_i)\Delta x_i$$

The heart of the definition is the final line. The concept captured in that equation grows out of our discussion of area in the previous section. However, we have considerably modified the notion presented there. For example, we now allow f to be negative on part or all of $[a, b]$, we use partitions with subintervals that may be of unequal length, and we allow $\bar{x}_i$ to be any point on the ith subinterval. Since we have made these changes, it is important to state precisely how the definite integral relates to area. In general, $\displaystyle\int_a^b f(x)\,dx$ gives the *signed area* of the region trapped between the curve $y = f(x)$ and the x-axis on the interval $[a, b]$, meaning that a plus sign is attached to areas of parts above the x-axis and a minus sign is attached to

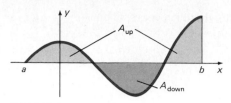

FIGURE 6

areas of parts below the x-axis. In symbols,

$$\int_a^b f(x)\,dx = A_{\text{up}} - A_{\text{down}}$$

where A_{up} and A_{down} are as shown in Figure 6.

The meaning of the word *limit* in the definition of the definite integral is more general than in earlier usage and should be explained. The equality

$$\lim_{|P| \to 0} \sum_{i=1}^{n} f(\bar{x}_i)\,\Delta x_i = L$$

means that corresponding to each $\varepsilon > 0$, there is a $\delta > 0$ such that

$$\left| \sum_{i=1}^{n} f(\bar{x}_i)\,\Delta x_i - L \right| < \varepsilon$$

for all Riemann sums $\sum_{i=1}^{n} f(\bar{x}_i)\,\Delta x_i$ for f on $[a,b]$ for which the norm $|P|$ of the associated partition is less than δ. In this case, we say that the indicated limit exists and has the value L.

That was a mouthful, and we are not going to digest it very well now. We simply assert that the usual limit theorems also hold for this kind of limit.

Returning to the symbol $\int_a^b f(x)\,dx$, we might call a the lower endpoint and b the upper endpoint for the integral. However, most authors use the terminology **lower limit** of integration and **upper limit** of integration, which is fine provided we realize that this usage of the word *limit* has nothing to do with its more technical meaning.

In our definition of $\int_a^b f(x)\,dx$, we implicitly assumed that $a < b$. We remove that restriction with the following definitions.

$$\int_a^a f(x)\,dx = 0$$

$$\int_a^b f(x)\,dx = -\int_b^a f(x)\,dx, \qquad a > b$$

Thus,

$$\int_2^2 x^3\,dx = 0, \qquad \int_6^2 x^3\,dx = -\int_2^6 x^3\,dx$$

Finally, we point out that x is a **dummy variable** in the symbol $\int_a^b f(x)\,dx$. By this we mean that x can be replaced by any other letter (provided, of course, that it is replaced in each place where it occurs). Thus,

$$\int_a^b f(x)\,dx = \int_a^b f(t)\,dt = \int_a^b f(u)\,du$$

WHAT FUNCTIONS ARE INTEGRABLE?   Not every function is integrable. For example, the unbounded function

$$f(x) = \begin{cases} \dfrac{1}{x^2} & \text{if } x \neq 0 \\ 1 & \text{if } x = 0 \end{cases}$$

which is graphed in Figure 7, is not integrable on $[-2, 2]$. This is because the contribution to any Riemann sum from the subinterval containing $x = 0$ can be made arbitrarily large by choosing the corresponding sample point $\bar{x}_i$ sufficiently close to zero. In fact, this reasoning shows that *any function that is integrable on $[a, b]$ must be bounded there*; that is, there must exist a constant M such that $|f(x)| \leq M$ for all x in $[a, b]$.

Even some bounded functions can fail to be integrable, but they have to be pretty complicated (see Problem 22 for one example). By all odds, Theorem A (below) is the most important theorem about integrability. Unfortunately, it is too difficult to prove here; we leave that for advanced calculus books.

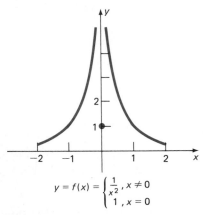

$$y = f(x) = \begin{cases} \dfrac{1}{x^2}, & x \neq 0 \\ 1, & x = 0 \end{cases}$$

FIGURE 7

> **Theorem A**
>
> **(Integrability Theorem).** If f is bounded on $[a, b]$ and if it is continuous there except at a finite number of points, then f is integrable on $[a, b]$. In particular, if f is continuous on the whole interval $[a, b]$, it is integrable on $[a, b]$.

As a consequence of this theorem, the following functions are integrable on every closed interval $[a, b]$.

1. Polynomial functions.
2. Sine and cosine functions.
3. Rational functions, provided the interval $[a, b]$ contains no points where a denominator is 0.

CALCULATING DEFINITE INTEGRALS   Knowing that a function is integrable allows us to calculate its integral by using a **regular partition** (equallength subintervals) and by picking the sample points $\bar{x}_i$ in any way convenient for us. Examples 3 and 4 involve polynomials, which we just learned are integrable.

EXAMPLE 3   Evaluate $\displaystyle\int_{-2}^{3} (x + 3)\,dx$.

Solution Partition the interval $[-2, 3]$ into n equal subintervals, each of length $\Delta x = 5/n$. In each subinterval $[x_{i-1}, x_i]$ use $\bar{x}_i = x_i$ as the sample point. Then

$$x_0 = -2$$

$$x_1 = -2 + \Delta x = -2 + \frac{5}{n}$$

$$x_2 = -2 + 2\Delta x = -2 + 2\left(\frac{5}{n}\right)$$

$$\vdots$$

$$x_i = -2 + i\Delta x = -2 + i\left(\frac{5}{n}\right)$$

$$\vdots$$

$$x_n = -2 + n\Delta x = -2 + n\left(\frac{5}{n}\right) = 3$$

Thus, $f(x_i) = x_i + 3 = 1 + i(5/n)$, and so

$$\sum_{i=1}^{n} f(\bar{x}_i)\Delta x_i = \sum_{i=1}^{n} f(x_i)\Delta x$$

$$= \sum_{i=1}^{n} \left[1 + i\left(\frac{5}{n}\right)\right]\frac{5}{n}$$

$$= \frac{5}{n}\sum_{i=1}^{n} 1 + \frac{25}{n^2}\sum_{i=1}^{n} i$$

$$= \frac{5}{n}(n) + \frac{25}{n^2}\left[\frac{n(n+1)}{2}\right] \quad \text{(Formula 1, p. 222)}$$

$$= 5 + \frac{25}{2}\left(1 + \frac{1}{n}\right)$$

Since P is a regular partition, $|P| \to 0$ is equivalent to $n \to \infty$. We conclude that

$$\int_{-2}^{3} (x + 3)\,dx = \lim_{|P|\to 0}\sum_{i=1}^{n} f(\bar{x}_i)\Delta x_i$$

$$= \lim_{n\to\infty}\left[5 + \frac{25}{2}\left(1 + \frac{1}{n}\right)\right]$$

$$= \frac{35}{2}$$

The result is not surprising since it corresponds to the area of the trapezoid shown in Figure 8. We could have calculated it from a grade-school formula, but for college students it is more fun doing it the hard way. ∎

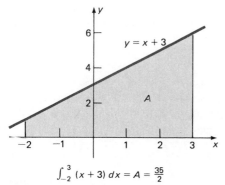

$$\int_{-2}^{3} (x + 3)\,dx = A = \frac{35}{2}$$

FIGURE 8

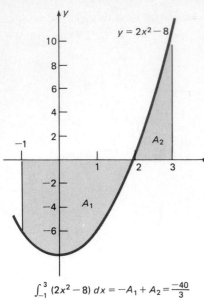

$y = 2x^2 - 8$

A_2

A_1

$\int_{-1}^{3} (2x^2 - 8)\, dx = -A_1 + A_2 = \frac{-40}{3}$

FIGURE 9

EXAMPLE 4 Evaluate $\displaystyle\int_{-1}^{3} (2x^2 - 8)\, dx$.

Solution No grade-school formulas will help here, though the integral does correspond to $-A_1 + A_2$, where A_1 and A_2 are the areas of the regions below and above the x-axis in Figure 9.

Let P be a regular partition of $[-1, 3]$ into n equal subintervals, each of length $\Delta x = 4/n$. In each subinterval $[x_{i-1}, x_i]$, choose $\bar{x}_i$ to be the right endpoint, so $\bar{x}_i = x_i$. Then

$$x_i = -1 + i\,\Delta x = -1 + i\left(\frac{4}{n}\right)$$

and

$$f(x_i) = 2x_i^2 - 8 = 2\left[-1 + i\left(\frac{4}{n}\right)\right]^2 - 8$$

$$= -6 - \frac{16i}{n} + \frac{32i^2}{n^2}$$

Consequently,

$$\sum_{i=1}^{n} f(\bar{x}_i)\,\Delta x_i = \sum_{i=1}^{n} f(x_i)\,\Delta x$$

$$= \sum_{i=1}^{n} \left[-6 - \frac{16}{n}i + \frac{32}{n^2}i^2\right]\frac{4}{n}$$

$$= -\frac{24}{n}\sum_{i=1}^{n} 1 - \frac{64}{n^2}\sum_{i=1}^{n} i + \frac{128}{n^3}\sum_{i=1}^{n} i^2$$

$$= \frac{-24}{n}(n) - \frac{64}{n^2}\frac{n(n+1)}{2} + \frac{128}{n^3}\frac{n(n+1)(2n+1)}{6}$$

$$= -24 - 32\left(1 + \frac{1}{n}\right) + \frac{128}{6}\left(2 + \frac{3}{n} + \frac{1}{n^2}\right)$$

We conclude that

$$\int_{-1}^{3} (2x^2 - 8)\, dx = \lim_{|P| \to 0} \sum_{i=1}^{n} f(\bar{x}_i)\,\Delta x_i$$

$$= \lim_{n \to \infty}\left[-24 - 32\left(1 + \frac{1}{n}\right) + \frac{128}{6}\left(2 + \frac{3}{n} + \frac{1}{n^2}\right)\right]$$

$$= -24 - 32 + \frac{128}{3} = \frac{-40}{3}$$

That the answer is negative is not surprising, since the region below the x-axis is larger than the one above the x-axis (Figure 9). ∎

In Problems 1 and 2, calculate the indicated Riemann sum (see Example 1).

1.

2.

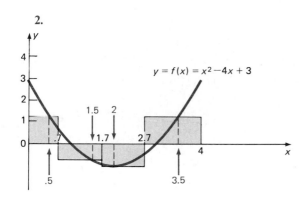

In Problems 3–6, calculate the Riemann sum $\sum_{i=1}^{n} f(\bar{x}_i)\,\Delta x_i$ for the given data.

3. $f(x) = x - 1$; $P: 3 < 3.75 < 4.25 < 5.5 < 6 < 7$; $\bar{x}_1 = 3, \bar{x}_2 = 4, \bar{x}_3 = 4.75, \bar{x}_4 = 6, \bar{x}_5 = 6.5$

4. $f(x) = -x/2 + 3$; $P: -3 < -1.3 < 0 < 0.9 < 2$; $\bar{x}_1 = -2, x_2 = -0.5, \bar{x}_3 - 0, \bar{x}_4 = 2$.

C **5.** $f(x) = x^2/2 + 1$; $[-1, 2]$ is divided into six equal subintervals, $\bar{x}_i$ is midpoint (scc Example 2).

C **6.** $f(x) = x^3/3 + 1$; $[0, 2]$ is divided into eight equal subintervals, $\bar{x}_i$ is right endpoint.

In Problems 7–10, use the given values of a and b and express the given limit as a definite integral.

7. $\lim\limits_{|P|\to 0} \sum\limits_{i=1}^{n} (\bar{x}_i)^2 \,\Delta x_i$; $a = 1, b = 3$

8. $\lim\limits_{|P|\to 0} \sum\limits_{i=1}^{n} (\bar{x}_i + 1)^3 \,\Delta x_i$; $a = 0, b = 2$

9. $\lim\limits_{|P|\to 0} \sum\limits_{i=1}^{n} \dfrac{\bar{x}_i}{1 + \bar{x}_i} \,\Delta x_i$; $a = 0, b = 3$

10. $\lim\limits_{|P|\to 0} \sum\limits_{i=1}^{n} (\cos^2 \bar{x}_i)\Delta x_i$; $a = 0, b = \pi$

In Problems 11–16, evaluate the definite integrals using the definition, as in Examples 3 and 4.

11. $\displaystyle\int_{0}^{4} (2x + 3)\,dx$ **12.** $\displaystyle\int_{0}^{4} (x^2 + 2)\,dx$

Hint: Use $\bar{x}_i = \dfrac{4i}{n}$.

13. $\displaystyle\int_{-1}^{2} (x^2 - 1)\,dx$ **14.** $\displaystyle\int_{-1}^{2} (2x^2 + 1)\,dx$

Hint: Use $\bar{x}_i = -1 + \dfrac{3i}{n}$.

15. $\displaystyle\int_{0}^{4} (x^2 - 2x)\,dx$ **16.** $\displaystyle\int_{-1}^{2} (x^2 + x + 1)\,dx$

In Problems 17–20, calculate $\displaystyle\int_{0}^{5} f(x)\,dx$ by using appropriate area formulas from plane geometry. Begin by graphing the given function.

17. $f(x) = \begin{cases} x & \text{if } 0 \le x < 1 \\ 1 & \text{if } 1 \le x \le 3 \\ x - 4 & \text{if } 3 < x \le 5 \end{cases}$

18. $f(x) = \begin{cases} x + 2 & \text{if } 0 \le x < 2 \\ 6 - x & \text{if } 2 \le x \le 5 \end{cases}$

19. $f(x) = \begin{cases} \sqrt{4 - x^2} & \text{if } 0 \le x \le 2 \\ 2 & \text{if } 2 < x \le 5 \end{cases}$

20. $f(x) = \begin{cases} -\sqrt{9 - x^2} & \text{if } 0 \le x \le 3 \\ 2 & \text{if } 3 < x \le 5 \end{cases}$

21. Which of the following functions are integrable on $[-2, 2]$ and which are not? Give reasons.

(a) $f(x) = x^3 + \sin x$

(b) $f(x) = \dfrac{1}{x + 3}$

(c) $f(x) = \dfrac{1}{x - 1}, f(1) = 3$

(d) $f(x) = \tan x$

(e) $f(x) = \begin{cases} \sin \dfrac{1}{x} & \text{if } x \ne 0 \\ 0 & \text{if } x = 0 \end{cases}$

(f) $f(x) = \begin{cases} x^2 & \text{if } -2 \le x \le 0 \\ 1 & \text{if } 0 < x \le 2 \end{cases}$

22. Let $f(x) = \begin{cases} 1 & \text{if } x \text{ is rational} \\ 0 & \text{if } x \text{ is irrational} \end{cases}$

Prove that f is not integrable on $[0, 1]$. *Hint:* Show that no matter how small the norm of the partition, $|P|$, the Riemann sum $\sum_{i=1}^{n} f(\bar{x}_i) \Delta x_i$ can be made to have value either 0 or 1.

23. Recall that $[\![x]\!]$ denotes the greatest integer less than or equal to x. Calculate each of the following integrals. You may use geometric reasoning and the fact that $\int_0^b x^2 \, dx = b^3/3$. (The latter is shown in Problem 26.)

(a) $\int_{-3}^{3} [\![x]\!] \, dx$

(b) $\int_{-3}^{3} [\![x]\!]^2 \, dx$

(c) $\int_{-3}^{3} (x - [\![x]\!]) \, dx$

(d) $\int_{-3}^{3} (x - [\![x]\!])^2 \, dx$

(e) $\int_{-3}^{3} |x| \, dx$

(f) $\int_{-3}^{3} x|x| \, dx$

(g) $\int_{-1}^{2} |x| [\![x]\!] \, dx$

(h) $\int_{-1}^{2} x^2 [\![x]\!] \, dx$

24. Let f be an odd function and g be an even function, and suppose $\int_0^1 |f(x)| \, dx = \int_0^1 g(x) \, dx = 3$. Use geometric reasoning to calculate each of the following.

(a) $\int_{-1}^{1} f(x) \, dx$

(b) $\int_{-1}^{1} g(x) \, dx$

(c) $\int_{-1}^{1} |f(x)| \, dx$

(d) $\int_{-1}^{1} [-g(x)] \, dx$

(e) $\int_{-1}^{1} x g(x) \, dx$

(f) $\int_{-1}^{1} f^3(x) g(x) \, dx$

25. Show that $\int_a^b x \, dx = \frac{1}{2}(b^2 - a^2)$ by completing the following argument. For the partition $a = x_0 < x_1 < \cdots < x_n = b$, choose $\bar{x}_i = \frac{1}{2}(x_{i-1} + x_i)$. Then, $R_p = \sum_{i=1}^{n} \bar{x}_i \Delta x_i = \frac{1}{2} \sum_{i=1}^{n} (x_i + x_{i-1})(x_i - x_{i-1})$. Now simplify R_p (collapsing sum) and take a limit.

26. Show that $\int_a^b x^2 \, dx = \frac{1}{3}(b^3 - a^3)$ by an argument like that in Problem 25 but using $\bar{x}_i = [\frac{1}{3}(x_{i-1}^2 + x_{i-1} x_i + x_i^2)]^{1/2}$

5.6
The Fundamental
Theorem of Calculus

We have been able to evaluate a few definite integrals directly from the definition, but only because we have nice formulas for $1 + 2 + \cdots + n$, $1^2 + 2^2 + \cdots + n^2$, and so on. Calculating definite integrals this way is always tedious, usually difficult, and sometimes practically impossible. Fortunately, there is a better way; that is the subject of this section.

We have credited Isaac Newton and Gottfried Leibniz with the simultaneous, but independent, discovery of calculus. Yet the concepts of the slope of the tangent line (derivative) and the area of a curved region (definite integral) were known earlier. Why then do Newton and Leibniz figure so prominently in the history of calculus? They do so because they understood and exploited the intimate relationship that exists between antiderivatives and definite integrals, a relationship that enables us to compute easily the exact values of many definite integrals without ever using Riemann sums. This connection is so important that it is called the *Fundamental Theorem of Calculus.*

THE FUNDAMENTAL THEOREM You have met several fundamental theorems before in your mathematical career. The Fundamental Theorem of Arithmetic says that a composite whole number factors uniquely into a product of primes. The Fundamental Theorem of Algebra says that an nth-degree polynomial equation has exactly n solutions, counting multiplicities.

Any theorem with this title should be studied carefully and then permanently committed to memory.

Theorem A

(Fundamental Theorem of Calculus). Let f be continuous (hence integrable) on $[a, b]$ and let F be any antiderivative of f there. Then,

$$\int_a^b f(x)\, dx = F(b) - F(a)$$

Proof Let $P: a = x_0 < x_1 < x_2 < \cdots < x_{n-1} < x_n = b$ be an arbitrary partition of $[a, b]$. Then the standard "subtract-and-add" trick gives

$$F(b) - F(a) = F(x_n) - F(x_{n-1}) + F(x_{n-1}) - F(x_{n-2}) + \cdots + F(x_1) - F(x_0)$$

$$= \sum_{i=1}^{n} \left[F(x_i) - F(x_{i-1}) \right]$$

By the Mean Value Theorem for Derivatives (Theorem 4.8A) applied to F on the interval $[x_{i-1}, x_i]$,

$$F(x_i) - F(x_{i-1}) = F'(\bar{x}_i)(x_i - x_{i-1}) = f(\bar{x}_i)\, \Delta x_i$$

for some choice of $\bar{x}_i$ in the open interval (x_{i-1}, x_i). Thus,

$$F(b) - F(a) = \sum_{i=1}^{n} f(\bar{x}_i)\, \Delta x_i$$

On the left we have a constant; on the right we have a Riemann sum for f on $[a, b]$. When we take limits of both sides as $|P| \to 0$, we obtain

$$F(b) - F(a) = \lim_{|P| \to 0} \sum_{i=1}^{n} f(\bar{x}_i)\, \Delta x_i = \int_a^b f(x)\, dx \qquad ■$$

EXAMPLE 1 Show that $\displaystyle\int_a^b k\, dx = k(b - a)$, k a constant.

Solution $F(x) = kx$ is an antiderivative of $f(x) = k$. Thus, by the Fundamental Theorem,

$$\int_a^b k\, dx = F(b) - F(a) = kb - ka = k(b - a) \qquad ■$$

EXAMPLE 2 Show that $\displaystyle\int_a^b x\, dx = \frac{b^2}{2} - \frac{a^2}{2}$.

Solution $F(x) = x^2/2$ is an antiderivative of $f(x) = x$. Therefore,

$$\int_a^b x\, dx = F(b) - F(a) = \frac{b^2}{2} - \frac{a^2}{2} \qquad ■$$

EXAMPLE 3 Show that if r is a rational number different from -1, then

$$\int_a^b x^r \, dx = \frac{b^{r+1}}{r+1} - \frac{a^{r+1}}{r+1}$$

Solution $F(x) = x^{r+1}/(r+1)$ is an antiderivative of $f(x) = x^r$. Thus, by the Fundamental Theorem,

$$\int_a^b x^r \, dx = F(b) - F(a) = \frac{b^{r+1}}{r+1} - \frac{a^{r+1}}{r+1}$$

Technical point: If $r < 0$, we require that 0 is not in $[a, b]$. Why? ■

It is convenient to introduce a special symbol for $F(b) - F(a)$. We write

$$F(b) - F(a) = \left[F(x) \right]_a^b$$

Thus, for example,

$$\int_2^5 x^2 \, dx = \left[\frac{x^3}{3} \right]_2^5 = \frac{125}{3} - \frac{8}{3} = \frac{117}{3} = 39$$

EXAMPLE 4 Evaluate $\displaystyle\int_{-1}^{2} (4x - 6x^2) \, dx$.

Solution

$$\int_{-1}^{2} (4x - 6x^2) \, dx = \left[2x^2 - 2x^3 \right]_{-1}^{2}$$

$$= (8 - 16) - (2 + 2) = -12 \qquad ■$$

EXAMPLE 5 Evaluate $\displaystyle\int_{1}^{8} (x^{1/3} + x^{4/3}) \, dx$.

Solution

$$\int_{1}^{8} (x^{1/3} + x^{4/3}) \, dx = \left[\tfrac{3}{4}x^{4/3} + \tfrac{3}{7}x^{7/3} \right]_1^8$$

$$= (\tfrac{3}{4} \cdot 16 + \tfrac{3}{7} \cdot 128) - (\tfrac{3}{4} \cdot 1 + \tfrac{3}{7} \cdot 1)$$

$$= \tfrac{45}{4} + \tfrac{381}{7} \approx 65.68 \qquad ■$$

EXAMPLE 6 Evaluate $\displaystyle\int_{0}^{\pi} 3 \sin x \, dx$.

Solution

$$\int_{0}^{\pi} 3 \sin x \, dx = \left[-3 \cos x \right]_0^\pi = 3 + 3 = 6 \qquad ■$$

In terms of the symbol for indefinite integrals, we may write the conclusion of the Fundamental Theorem as

$$\int_a^b f(x)\,dx = \left[\int f(x)\,dx\right]_a^b$$

The nontrivial part of applying the theorem is always to find the indefinite integral $\int f(x)\,dx$. To do that we may want to use the substitution technique (Generalized Power Rule) we learned in Section 5.1.

EXAMPLE 7 Evaluate $\displaystyle\int_0^4 \sqrt{x^2 + x}\,(2x + 1)\,dx.$

Solution Let $u = x^2 + x$; then $du = (2x + 1)\,dx.$

Thus,

$$\int \sqrt{x^2 + x}\,(2x + 1)\,dx = \int u^{1/2}\,du = \tfrac{2}{3}u^{3/2} + C$$
$$= \tfrac{2}{3}(x^2 + x)^{3/2} + C$$

Therefore, by the Fundamental Theorem,

$$\int_0^4 \sqrt{x^2 + x}\,(2x + 1)\,dx = \left[\tfrac{2}{3}(x^2 + x)^{3/2} + C\right]_0^4$$
$$= \left[\tfrac{2}{3}(20)^{3/2} + C\right] - [0 + C]$$
$$= \tfrac{2}{3}(20)^{3/2} \approx 59.63 \qquad \blacksquare$$

Note that the C of the indefinite integration cancels out, as it always will, in the definite integration. That is why in the statement of the Fundamental Theorem we could use the phrase *any antiderivative*. In particular, we may always choose $C = 0$ in applying the Fundamental Theorem.

EXAMPLE 8 Evaluate $\displaystyle\int_0^{\pi/4} \sin^3 2x \cos 2x\,dx.$

Solution Let $u = \sin 2x$; then $du = 2 \cos 2x\,dx.$ Thus,

$$\int \sin^3 2x \cos 2x\,dx = \frac{1}{2}\int (\sin^3 2x)(2 \cos 2x)\,dx = \frac{1}{2}\int u^3\,du$$
$$= \frac{1}{2}\frac{u^4}{4} + C = \frac{\sin^4 2x}{8} + C$$

Therefore, by the Fundamental Theorem,

$$\int_0^{\pi/4} \sin^3 2x \cos 2x\,dx = \left[\frac{\sin^4 2x}{8}\right]_0^{\pi/4} = \frac{1}{8} - 0 = \frac{1}{8} \qquad \blacksquare$$

THE DEFINITE INTEGRAL IS A LINEAR OPERATOR Earlier we learned that $D_x, \int \cdots dx$, and $\sum$ are linear operators. You can add $\int_a^b \cdots dx$ to the list.

Theorem B

(Linearity of the Definite Integral). Suppose that f and g are integrable on $[a, b]$ and that k is a constant. Then kf and $f + g$ are integrable and:

(i) $\displaystyle\int_a^b kf(x)\,dx = k\int_a^b f(x)\,dx;$

(ii) $\displaystyle\int_a^b [f(x) + g(x)]\,dx = \int_a^b f(x)\,dx + \int_a^b g(x)\,dx;$ and consequently

(iii) $\displaystyle\int_a^b [f(x) - g(x)]\,dx = \int_a^b f(x)\,dx - \int_a^b g(x)\,dx.$

Proof The proofs depend on the linearity of $\sum$ and the properties of limits. We show (ii).

$$\int_a^b [f(x) + g(x)]\,dx = \lim_{|P| \to 0} \sum_{i=1}^n [f(\bar{x}_i) + g(\bar{x}_i)]\,\Delta x_i$$

$$= \lim_{|P| \to 0} \left[\sum_{i=1}^n f(\bar{x}_i)\,\Delta x_i + \sum_{i=1}^n g(\bar{x}_i)\,\Delta x_i \right]$$

$$= \lim_{|P| \to 0} \sum_{i=1}^n f(\bar{x}_i)\,\Delta x_i + \lim_{|P| \to 0} \sum_{i=1}^n g(\bar{x}_i)\,\Delta x_i$$

$$= \int_a^b f(x)\,dx + \int_a^b g(x)\,dx \qquad \blacksquare$$

EXAMPLE 9 Evaluate $\displaystyle\int_{-1}^2 (4x - 6x^2)\,dx.$

Solution This is the same as Example 4, but now we will do it another way making use of the linearity of the definite integral.

$$\int_{-1}^2 (4x - 6x^2)\,dx = 4\int_{-1}^2 x\,dx - 6\int_{-1}^2 x^2\,dx$$

$$= 4\left[\frac{x^2}{2}\right]_{-1}^2 - 6\left[\frac{x^3}{3}\right]_{-1}^2$$

$$= 4\left(\frac{4}{2} - \frac{1}{2}\right) - 6\left(\frac{8}{3} + \frac{1}{3}\right) = -12 \qquad \blacksquare$$

EXAMPLE 10 Evaluate $\displaystyle\int_0^1 [x^2 + (x^2 + 1)^4 x]\,dx.$

Solution

$$\int_0^1 \left[x^2 + (x^2 + 1)^4 x \right] dx = \int_0^1 x^2\, dx + \int_0^1 (x^2 + 1)^4 x\, dx$$

The first integral is easy to do directly. To handle the second, we let $u = x^2 + 1$, so $du = 2x\, dx$ and write

$$\int (x^2 + 1)^4 x\, dx = \frac{1}{2} \int (x^2 + 1)^4 2x\, dx = \frac{1}{2} \int u^4\, du$$

$$= \frac{1}{2} \frac{u^5}{5} + C = \frac{(x^2 + 1)^5}{10} + C$$

Therefore,

$$\int_0^1 x^2\, dx + \int_0^1 (x^2 + 1)^4 x\, dx = \left[\frac{x^3}{3} \right]_0^1 + \left[\frac{(x^2 + 1)^5}{10} \right]_0^1$$

$$= \left(\frac{1}{3} - 0 \right) + \left(\frac{32}{10} - \frac{1}{10} \right)$$

$$= \frac{1}{3} + \frac{31}{10} = \frac{103}{30} \qquad \blacksquare$$

PROBLEM SET 5.6

In Problems 1–14, use the Fundamental Theorem of Calculus to evaluate each definite integral (see Examples 1–6).

1. $\displaystyle\int_0^2 x^3\, dx$

2. $\displaystyle\int_{-1}^2 x^4\, dx$

3. $\displaystyle\int_{-1}^2 (3x^2 - 2x + 3)\, dx$

4. $\displaystyle\int_1^2 (4x^3 + 7)\, dx$

5. $\displaystyle\int_1^4 \frac{1}{w^2}\, dw$

6. $\displaystyle\int_1^3 \frac{2}{t^3}\, dt$

7. $\displaystyle\int_0^4 \sqrt{t}\, dt$

8. $\displaystyle\int_1^8 \sqrt[3]{w}\, dw$

9. $\displaystyle\int_{-4}^{-2} \left(y^2 + \frac{1}{y^3} \right) dy$

10. $\displaystyle\int_1^4 \frac{s^4 - 8}{s^2}\, ds$

11. $\displaystyle\int_0^{\pi/2} \cos x\, dx$

12. $\displaystyle\int_{\pi/6}^{\pi/2} 2 \sin t\, dt$

13. $\displaystyle\int_0^1 (2x^4 - 3x^2 + 5)\, dx$

14. $\displaystyle\int_0^1 (x^{4/3} - 2x^{1/3})\, dx$

In Problems 15–30, use the Fundamental Theorem of Calculus combined with the Generalized Power Rule to evaluate the given definite integral (see Examples 7–10).

15. $\displaystyle\int_0^1 (x^2 + 1)^{10}(2x)\, dx$

16. $\displaystyle\int_{-1}^0 \sqrt{x^3 + 1}(3x^2)\, dx$

17. $\displaystyle\int_{-1}^3 \frac{1}{(t + 2)^2}\, dt$

18. $\displaystyle\int_2^{10} \sqrt{y - 1}\, dy$

19. $\displaystyle\int_5^8 \sqrt{3x + 1}\, dx$

20. $\displaystyle\int_1^7 \frac{1}{\sqrt{2x + 2}}\, dx$

21. $\displaystyle\int_{-3}^3 \sqrt{7 + 2t^2}(8t)\, dt$

22. $\displaystyle\int_1^3 \frac{x^2 + 1}{\sqrt{x^3 + 3x}}\, dx$

23. $\displaystyle\int_0^{\pi/2} \cos^2 x \sin x\, dx$

24. $\displaystyle\int_0^{\pi/2} \sin^2 3x \cos 3x\, dx$

25. $\displaystyle\int_0^{\pi/2} (2x + \sin x)\, dx$

26. $\displaystyle\int_0^{\pi/2} [4x + 3 + \cos x]\, dx$

27. $\displaystyle\int_0^4 [\sqrt{x} + \sqrt{2x + 1}]\, dx$

28. $\displaystyle\int_{-4}^{-1} \frac{1 - s^4}{2s^2}\, ds$

29. $\displaystyle\int_0^1 (x^2 + 2x)^2\, dx$

30. $\displaystyle\int_a^{8a} (a^{1/3} - x^{1/3})^3\, dx$

In Problems 31–34, first recognize the given limit as a definite integral and then evaluate that integral by the Fundamental Theorem of Calculus.

31. $\lim\limits_{n \to \infty} \sum\limits_{i=1}^{n} \left(\dfrac{3i}{n}\right)^2 \dfrac{3}{n}$

32. $\lim\limits_{n \to \infty} \sum\limits_{i=1}^{n} \left(\dfrac{2i}{n}\right)^3 \dfrac{2}{n}$

33. $\lim\limits_{n \to \infty} \sum\limits_{i=1}^{n} \left[\sin\left(\dfrac{\pi i}{n}\right)\right] \dfrac{\pi}{n}$

34. $\lim\limits_{n \to \infty} \sum\limits_{i=1}^{n} \left[1 + \dfrac{2i}{n} + \left(\dfrac{2i}{n}\right)^2\right] \dfrac{2}{n}$

35. Explain why $(1/n^3) \sum\limits_{i=1}^{n} i^2$ should be a good approximation to $\int_0^1 x^2 \, dx$ for large n. Now calculate the summation expression for $n = 10$ and the integral by the Fundamental Theorem of Calculus and compare their values.

Suppose that $\int_0^1 f(x)\,dx = 4$ and $\int_0^1 g(x)\,dx = -2$. Use the linearity of the integral to calculate each of the integrals in Problems 36–40.

36. $\int_0^1 3f(x)\,dx$

37. $\int_0^1 [f(x) - g(x)]\,dx$

38. $\int_0^1 [3f(x) + 2g(x)]\,dx$

39. $\int_0^1 [2g(x) - 3f(x)]\,dx$

40. $\int_1^0 [2f(x) - 3x^2]\,dx$

41. Evaluate $\int_{-2}^{4} (2[\![x]\!] - 3|x|)\,dx$

42. Show that $\frac{1}{2}x|x|$ is an antiderivative of $|x|$ and use this fact to get a nice formula for $\int_a^b |x|\,dx$.

43. Find a nice formula for $\int_0^b [\![x]\!]\,dx,\ b > 0$.

44. We claim that

$$\int_a^b x^n \, dx + \int_{a^n}^{b^n} \sqrt[n]{y}\, dy = b^{n+1} - a^{n+1}$$

(a) Use Figure 1 to justify this by a geometric argument.
(b) Prove the result using the Fundamental Theorem of Calculus.
(c) Show that $A_n = nB_n$.

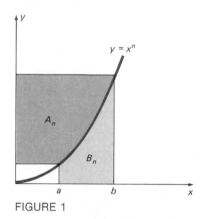

FIGURE 1

5.7
More Properties of the Definite Integral

Our definition of the definite integral was motivated by the problem of area for curved regions. Consider the two curved regions R_1 and R_2 in Figure 1 and let $R = R_1 \cup R_2$. It is clear that

$$A(R) = A(R_1 \cup R_2) = A(R_1) + A(R_2)$$

which suggests that

$$\int_a^c f(x)\,dx = \int_a^b f(x)\,dx + \int_b^c f(x)\,dx$$

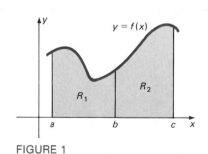

FIGURE 1

We quickly point out that this does not constitute a proof of the fact about integrals, since—first of all—our discussion of area in Section 5.4 was rather informal, and—second—our diagram supposes that f is positive, which it need not be. Nevertheless, definite integrals do satisfy this interval additive property and they do it no matter how the three points a, b, and c are arranged. We leave the rigorous proof to more advanced works (however, see Problem 41).

Theorem A

(Interval Additive Property). If f is integrable on an interval containing the three points a, b, and c, then

$$\int_a^c f(x)\,dx = \int_a^b f(x)\,dx + \int_b^c f(x)\,dx$$

no matter what the order of a, b, and c.

For example,

$$\int_0^2 x^2\,dx = \int_0^1 x^2\,dx + \int_1^2 x^2\,dx$$

which most people readily believe. But it is also true that

$$\int_0^2 x^2\,dx = \int_0^3 x^2\,dx + \int_3^2 x^2\,dx$$

which may seem surprising. If you mistrust the theorem, you might actually evaluate each of the above integrals to see that the equality holds.

COMPARISON PROPERTIES Consideration of the areas of the regions R_1 and R_2 in Figure 2 suggests another property of definite integrals.

FIGURE 2

Theorem B

(Comparison Property). If f and g are integrable on $[a,b]$ and if $f(x) \le g(x)$ for all x in $[a,b]$, then

$$\int_a^b f(x)\,dx \le \int_a^b g(x)\,dx$$

Proof Let $P\colon a = x_0 < x_1 < x_2 < \cdots < x_n = b$ be an arbitrary partition of $[a,b]$, and for each i let $\bar{x}_i$ be any sample point on the ith subinterval $[x_{i-1}, x_i]$. We may conclude successively that

$$f(\bar{x}_i) \le g(\bar{x}_i)$$

$$f(\bar{x}_i)\,\Delta x_i \le g(\bar{x}_i)\,\Delta x_i$$

$$\sum_{i=1}^n f(\bar{x}_i)\,\Delta x_i \le \sum_{i=1}^n g(\bar{x}_i)\,\Delta x_i$$

$$\lim_{|P| \to 0} \sum_{i=1}^n f(\bar{x}_i)\,\Delta x_i \le \lim_{|P| \to 0} \sum_{i=1}^n g(\bar{x}_i)\,\Delta x_i$$

$$\int_a^b f(x)\,dx \le \int_a^b g(x)\,dx \qquad \blacksquare$$

(Boundedness Property). If f is integrable on $[a, b]$ and if $m \leq f(x) \leq M$ for all x in $[a, b]$, then

$$m(b - a) \leq \int_a^b f(x)\, dx \leq M(b - a)$$

FIGURE 3

Proof The picture in Figure 3 helps us understand the theorem. Note that $m(b - a)$ is the area of the lower, small rectangle, $M(b - a)$ is the area of the large rectangle, and $\int_a^b f(x)\, dx$ is the area under the curve.

To prove the right-hand inequality, let $g(x) = M$ for all x in $[a, b]$. Then, by Theorem B,

$$\int_a^b f(x)\, dx \leq \int_a^b g(x)\, dx$$

However,

$$\int_a^b g(x)\, dx = \int_a^b M\, dx = [Mx]_a^b = M(b - a)$$

The left-hand inequality is handled similarly. ∎

FIGURE 4

DIFFERENTIATING A DEFINITE INTEGRAL WITH RESPECT TO A VARIABLE UPPER LIMIT The next theorem is important—so important that some authors call it the Second Fundamental Theorem of Calculus.

Let f be integrable on $[a, b]$ and let x be any point in $[a, b]$ (Figure 4). Then for each x, $\int_a^x f(t)\, dt$ is a unique number, and so

$$G(x) = \int_a^x f(t)\, dt$$

determines a function G with domain $[a, b]$. We call $\int_a^x f(t)\, dt$ an integral with variable upper limit. Note our use of t rather than x as the dummy variable to avoid confusion with the upper limit.

(Differentiating a Definite Integral). Let f be continuous on the closed interval $[a, b]$ and let x be a (variable) point in (a, b). Then

$$D_x\left[\int_a^x f(t)\, dt\right] = f(x)$$

CHAPTER 5/The Integral

Proof If $G(x) = \displaystyle\int_a^x f(t)\,dt$, we must show that

$$G'(x) = \lim_{h \to 0} \frac{G(x + h) - G(x)}{h} = f(x)$$

Now

$$G(x + h) - G(x) = \int_a^{x+h} f(t)\,dt - \int_a^x f(t)\,dt = \int_x^{x+h} f(t)\,dt$$

by Theorem A.

Assume for the moment that $h > 0$ and let m and M be the minimum value and maximum value, respectively, of f on the interval $[x, x + h]$ (Figure 5). By Theorem C,

$$mh \le \int_x^{x+h} f(t)\,dt \le Mh$$

FIGURE 5

or

$$mh \le G(x + h) - G(x) \le Mh$$

Dividing by h, we obtain

$$m \le \frac{G(x + h) - G(x)}{h} \le M$$

Now m and M really depend on h. Moreover, since f is continuous, both m and M must approach $f(x)$ as $h \to 0$. Thus, by the Squeeze Theorem,

$$\lim_{h \to 0} \frac{G(x + h) - G(x)}{h} = f(x)$$

The case where $h < 0$ is handled similarly. ∎

One theoretical consequence of this theorem is that every continuous function f has an antiderivative F given by

$$F(x) = \int_a^x f(t)\,dt$$

However, this fact is not helpful in getting a nice formula for any particular antiderivative.

EXAMPLE 1   Find $D_x\!\left[\displaystyle\int_1^x t^2\,dt\right]$ two ways.

Solution First, we will do it the hard way—that is, by evaluating the integral and then taking the derivative.

$$\int_1^x t^2\, dt = \left[\frac{t^3}{3}\right]_1^x = \frac{x^3}{3} - \frac{1}{3}$$

Therefore,

$$D_x\left[\int_1^x t^2\, dt\right] = D_x\left[\frac{x^3}{3} - \frac{1}{3}\right] = x^2$$

Now we do it the easy way. By Theorem D,

$$D_x\left[\int_1^x t^2\, dt\right] = x^2 \qquad\blacksquare$$

EXAMPLE 2 Find $D_x\left[\displaystyle\int_2^x \frac{t^{3/2}}{\sqrt{t^2+17}}\, dt\right]$.

Solution We challenge anyone to do this example by first evaluating the integral. However, by Theorem D, it is a trivial problem:

$$D_x\left[\int_2^x \frac{t^{3/2}}{\sqrt{t^2+17}}\, dt\right] = \frac{x^{3/2}}{\sqrt{x^2+17}} \qquad\blacksquare$$

EXAMPLE 3 Find $D_x\left[\displaystyle\int_x^4 \tan^2 u \cos u\, du\right], \dfrac{\pi}{2} < x < \dfrac{3\pi}{2}$.

Solution Use of the dummy variable u rather than t should not bother anyone. However, the fact that x is the lower limit, rather than the upper limit, is troublesome. Here is how we handle this difficulty.

$$D_x\left[\int_x^4 \tan^2 u \cos u\, du\right] = D_x\left[-\int_4^x \tan^2 u \cos u\, du\right]$$

$$= -D_x\left[\int_4^x \tan^2 u \cos u\, du\right] = -\tan^2 x \cos x$$

The interchange of the upper and lower limits is allowed if we prefix a minus sign (see the definition on p. 235). $\qquad\blacksquare$

EXAMPLE 4 Find $D_x\left[\displaystyle\int_0^{x^2} (3t-1)\right] dt$.

Solution Now we have a new complication; the upper limit is x^2 rather than x (we need x there in order to apply Theorem D). This problem is handled by the Chain Rule. We may think of the expression in brackets as

$$\int_0^u (3t-1)\, dt \qquad \text{where } u = x^2$$

By the Chain Rule, the derivative with respect to x of this composite function is

$$D_u\left[\int_0^u (3t - 1)\, dt\right] \cdot D_x u = (3u - 1)(2x) = (3x^2 - 1)(2x) = 6x^3 - 2x$$

If you are skeptical of what we have just done, try doing the problem by first evaluating the integral. ∎

EXAMPLE 5 Find $D_x\left[\displaystyle\int_{2x}^5 \sqrt{t^2 + 2}\, dt\right]$.

Solution Here we first interchange limits and then use Theorem D in conjunction with the Chain Rule.

$$D_x\left[\int_{2x}^5 \sqrt{t^2 + 2}\, dt\right] = D_x\left[-\int_5^{2x} \sqrt{t^2 + 2}\, dt\right]$$
$$= -\sqrt{(2x)^2 + 2}\,(2) = -2\sqrt{4x^2 + 2} \qquad ∎$$

THE MEAN VALUE THEOREM FOR INTEGRALS By this time you realize that the Mean Value Theorem for Derivatives plays an important role in calculus. There is a theorem by the same name for integrals, which is of less importance but still worth knowing. Geometrically, it says that in Figure 6, the area under the curve is equal to the area of the shaded rectangle.

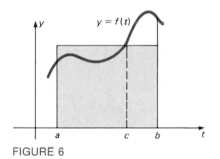

FIGURE 6

Theorem E

(Mean Value Theorem for Integrals). If f is continuous on $[a, b]$, there is a number c between a and b such that

$$\int_a^b f(t)\, dt = f(c)(b - a)$$

Proof Let

$$G(x) = \int_a^x f(t)\, dt \qquad a \le x \le b$$

By the Mean Value Theorem for Derivatives applied to G, there is a point c in (a, b) such that

$$G(b) - G(a) = G'(c)(b - a)$$

that is,

$$\int_a^b f(t)\, dt - 0 = G'(c)(b - a)$$

But by Theorem D, $G'(c) = f(c)$. The conclusion follows. ∎

Note that if we solve for $f(c)$ in the conclusion of Theorem E, we get

$$f(c) = \frac{\int_a^b f(t)\,dt}{b-a}$$

The number $\int_a^b f(x)\,dx/(b-a)$ is called the mean value, or **average value**, of f on $[a, b]$. To see why it has this name, consider a regular partition $P: a = x_0 < x_1 < x_2 < \cdots < x_n = b$ with $\Delta x = (b-a)/n$. The average of the n values $f(x_1), f(x_2), \ldots, f(x_n)$ is

$$\frac{f(x_1) + f(x_2) + \cdots + f(x_n)}{n} = \sum_{i=1}^n f(x_i)\frac{1}{n}$$

$$= \frac{1}{b-a}\sum_{i=1}^n f(x_i)\frac{b-a}{n}$$

$$= \frac{1}{b-a}\sum_{i=1}^n f(x_i)\,\Delta x$$

The sum in the last expression is a Riemann sum for f on $[a, b]$ and therefore approaches $\int_a^b f(x)\,dx$ as $n \to \infty$. Thus $\left(\int_a^b f(x)\,dx\right)\Big/(b-a)$ appears as the natural extension of the familiar notion of average value.

PROBLEM SET 5.7

Suppose $\int_0^1 f(x)\,dx = 2$, $\int_1^2 f(x)\,dx = 3$, $\int_0^1 g(x)\,dx = -1$, and $\int_0^2 g(x)\,dx = 4$. Use properties of definite integrals (linearity, interval additivity, and so on) to calculate each of the integrals in Problems 1–10.

1. $\int_0^2 f(x)\,dx$

2. $\int_0^2 [f(x) + 2g(x)]\,dx$

3. $\int_1^2 g(x)\,dx$

4. $\int_2^0 g(x)\,dx$

5. $\int_2^1 3f(x)\,dx$

6. $\int_1^1 g(x)\,dx$

7. $\int_0^2 [2f(x) - 3g(x)]\,dx$

8. $\int_0^1 [2f(x) + 3g(x) - 4]\,dx$

9. $\int_1^2 f(x)\,dx + \int_2^0 f(x)\,dx$

10. $\int_1^2 g(x)\,dx + \int_2^0 g(x)\,dx$

In Problems 11–20, find $G'(x)$ (see Examples 1–5).

11. $G(x) = \int_{-6}^x (2t + 1)\,dt$

12. $G(x) = \int_0^x (t^2 + t)\,dt$

13. $G(x) = \int_1^x \sqrt{1 + t^4}\,dt$

14. $G(x) = \int_0^x \sin^4 u \tan u\,du$, $-\pi/2 < x < \pi/2$

15. $G(x) = \int_x^{\pi/4} u \tan u\,du$, $-\pi/2 < x < \pi/2$

16. $G(x) = \int_x^1 x^2\sqrt{u^2 + 1}\,du$. (Be careful.)

17. $G(x) = \int_0^{\sin x} (u^2 + \cos u)\,du$

18. $G(x) = \int_1^{x^2+1} \sqrt{2 + \sin v}\,dv$

19. $G(x) = \int_x^{x^3} \sqrt{1 + s^4} \, ds$. *Hint:* $\int_x^{x^3} = \int_x^0 + \int_0^{x^3}$

20. $G(x) = \int_{\sin x}^{\cos x} u^2 \, du$

21. Show that the graph of $y = f(x)$ is concave up everywhere if

$$f(x) = \int_0^x \frac{s}{\sqrt{a^2 + s^2}} \, ds, \quad a \neq 0$$

Hint: Show $f''(x) > 0$ for all real x.

22. Find the interval on which the graph of $y = f(x)$ is concave up if

$$f(x) = \int_0^x \frac{1 + t}{1 + t^2} \, dt$$

In Problems 23–26, use the interval additive property (Theorem A) to evaluate $\int_0^4 f(x) \, dx$. Begin by drawing a graph of f.

23. $f(x) = \begin{cases} x^2 & \text{if } 0 \le x < 2 \\ x & \text{if } 2 \le x \le 4 \end{cases}$

24. $f(x) = \begin{cases} 1 & \text{if } 0 \le x < 1 \\ x & \text{if } 1 \le x < 2 \\ 4 - x & \text{if } 2 \le x \le 4 \end{cases}$

25. $f(x) = |x - 2|$

26. $f(x) = |\cos x|$

27. Show that $1 \le \int_0^1 \sqrt{1 + x^4} \, dx \le \frac{6}{5}$. *Hint:* $1 \le \sqrt{1 + x^4} \le 1 + x^4$; use the comparison property (Theorem B).

28. Let f be continuous on $[a, b]$ and thus integrable there. Show that

$$\left| \int_a^b f(x) \, dx \right| \le \int_a^b |f(x)| \, dx$$

Hint: $-|f(x)| \le f(x) \le |f(x)|$; use Theorem B.

In Problems 29–32, find the average value of the given function on the given interval.

29. $f(x) = 4x^3$; $[1, 3]$

30. $f(x) = \dfrac{x}{\sqrt{x^2 + 16}}$; $[0, 3]$

31. $f(x) = 2 + |x|$; $[-1, 3]$

32. $f(x) = \sin^2 x \cos x$; $[0, \pi/2]$

In Problems 33–38, decide whether the given statement is true or false. Then justify your answer.

33. If f is continuous and $f(x) \ge 0$ for all x in $[a, b]$, then $\int_a^b f(x) \, dx \ge 0$.

34. If $\int_a^b f(x) \, dx \ge 0$, then $f(x) \ge 0$ for all x in $[a, b]$.

35. If $\int_a^b f(x) \, dx = 0$, then $f(x) = 0$ for all x in $[a, b]$.

36. If $f(x) \ge 0$ and $\int_a^b f(x) \, dx = 0$, then $f(x) = 0$ for all x in $[a, b]$.

37. If $\int_a^b f(x) \, dx > \int_a^b g(x) \, dx$, then

$$\int_a^b [f(x) - g(x)] \, dx > 0.$$

38. If f and g are continuous and $f(x) > g(x)$ for all x in $[a, b]$, then $\left| \int_a^b f(x) \, dx \right| > \left| \int_a^b g(x) \, dx \right|$.

39. Let $s(t)$ and $v(t)$ denote the position and velocity at time t of an object moving along a coordinate line. In Section 3.1, we defined the average velocity on the interval $[a, b]$ as $[s(b) - s(a)]/(b - a)$; in this section, we defined it as $\int_a^b v(t) \, dt/(b - a)$. Show the two definitions are equivalent.

40. Suppose that f' is integrable and $|f'(x)| \le M$ for all x. Prove that $|f(x)| \le |f(a)| + M|x - a|$.

41. Prove Theorem A in the case where f is continuous on the interval $[a, c]$ and $a < b < c$ as follows. Let $P: a = x_0 < x_1 < \cdots < x_n = c$ be any partition of $[a, c]$ with b as one of the partition points and let $x_m = b$. Then

$$\sum_{i=1}^{n} f(\bar{x}_i) \Delta x_i = \sum_{i=1}^{m} f(\bar{x}_i) \Delta x_i + \sum_{i=m+1}^{n} f(\bar{x}_i) \Delta x_i$$

Finish.

42. Give an alternate proof of the Fundamental Theorem of Calculus (Theorem 5.6A) as follows: Let $G(x) = \int_a^x f(t) \, dt$ and let F be any antiderivative of f. Then $F'(x) = G'(x)$. What can we conclude about $G(x) - F(x)$? Now evaluate $G(a) - F(a)$ and then $G(b) - F(b)$.

43. Give a proof of Theorem E that does not use Theorem D. *Hint:* Use the Intermediate Value Theorem.

44. Use Theorem E to give an alternate proof of Theorem D.

Evaluating definite integrals is generally a two-step process. First, we find an indefinite integral; then we apply the Fundamental Theorem of Calculus. If the indefinite integration is easy, we can combine the two steps, as in

$$\int_1^2 x^2\,dx = \left[\frac{x^3}{3}\right]_1^2 = \frac{8}{3} - \frac{1}{3} = \frac{7}{3}$$

However, if the indefinite integration is complicated enough to require a substitution, we typically separate the two steps. Thus, to calculate

$$\int_0^4 x\sqrt{x^2 + 9}\,dx$$

we first write (using $u = x^2 + 9$ and $du = 2x\,dx$)

$$\int x\sqrt{x^2 + 9}\,dx = \tfrac{1}{2}\int \sqrt{x^2 + 9}(2x\,dx)$$

$$= \tfrac{1}{2}\int u^{1/2}\,du$$

$$= \tfrac{1}{2}\cdot\tfrac{2}{3}u^{3/2} + C$$

$$= \tfrac{1}{3}(x^2 + 9)^{3/2} + C$$

Then, by the Fundamental Theorem,

$$\int_0^4 x\sqrt{x^2 + 9}\,dx = [\tfrac{1}{3}(x^2 + 9)^{3/2}]_0^4 = \tfrac{125}{3} - \tfrac{27}{3} = \tfrac{98}{3}$$

The substitution method just illustrated generalizes in two ways. First, though it was introduced in Section 5.1 only for power functions, its application extends far beyond that use. Second, there is a way of using substitution directly in a definite integral. We take up these matters now.

THE METHOD OF SUBSTITUTION Consider the problem of finding

$$\int (2x + 3)\cos(x^2 + 3x)\,dx$$

If we let $u = x^2 + 3x$ so that $du = (2x + 3)\,dx$, the integral above transforms to $\int \cos u\,du$, which you will note is not the integral of a power function. However, formally,

$$\int (2x + 3)\cos(x^2 + 3x)\,dx = \int \cos u\,du$$

$$= \sin u + C$$

$$= \sin(x^2 + 3x) + C$$

In this example, we can easily check our answer by differentiating the result. But will the method of substitution always work? Yes, provided we can prove the following theorem.

Theorem A

(Substitution in Indefinite Integrals). Let g be a differentiable function and suppose that F is an antiderivative of f. Then if $u = g(x)$,

$$\int f(g(x))g'(x)\,dx = \int f(u)\,du = F(u) + C = F(g(x)) + C$$

Proof It is enough to show that the derivative of the right member is the integrand of the left member. But that is a simple application of the Chain Rule combined with the fact that $F' = f$.

$$D_x[F(g(x)) + C] = F'(g(x))g'(x) = f(g(x))g'(x) \qquad \blacksquare$$

EXAMPLE 1 Find $\displaystyle\int \frac{\sin\sqrt{x}}{\sqrt{x}}\,dx.$

Solution Let $u = \sqrt{x} = x^{1/2}$, so $du = \frac{1}{2}x^{-1/2}\,dx$. Then,

$$\int \frac{\sin\sqrt{x}}{\sqrt{x}}\,dx = 2\int \sin\sqrt{x}(\tfrac{1}{2}x^{-1/2}\,dx)$$

$$= 2\int \sin u\,du$$

$$= -2\cos u + C$$

$$= -2\cos\sqrt{x} + C \qquad \blacksquare$$

EXAMPLE 2 Evaluate $\displaystyle\int_0^{\sqrt{\pi/2}} x\sin^3(x^2)\cos(x^2)\,dx.$

Solution Let $u = \sin(x^2)$, so $du = 2x\cos(x^2)\,dx$. Then

$$\int x\sin^3(x^2)\cos(x^2)\,dx = \frac{1}{2}\int \sin^3(x^2) \cdot 2x\cos(x^2)\,dx$$

$$= \frac{1}{2}\int u^3\,du$$

$$= \frac{1}{2}\frac{u^4}{4} + C$$

$$= \frac{1}{8}\sin^4(x^2) + C$$

Thus, by the Fundamental Theorem of Calculus,

$$\int_0^{\sqrt{\pi/2}} x \sin^3(x^2)\cos(x^2)\,dx = \left[\frac{1}{8}\sin^4(x^2)\right]_0^{\sqrt{\pi/2}}$$

$$= \frac{1}{8}\sin^4\left(\frac{\pi}{4}\right) - \frac{1}{8}\cdot 0 = \frac{1}{32} \qquad \blacksquare$$

Note that in the two-step procedure illustrated in Example 2, we must be sure to express the indefinite integral in terms of x before we apply the Fundamental Theorem. This is because the limits 0 and $\sqrt{\pi}/2$ apply to x, not u. But what if, in making the substitution $u = \sin(x^2)$, we also made the corresponding changes in the limits of integration to $u = \sin(0^2) = 0$ and $u = \sin[(\sqrt{\pi}/2)^2] = \sqrt{2}/2$? Could we then finish the integration with u as variable? The answer is yes.

$$\int_0^{\sqrt{\pi/2}} x \sin^3(x^2)\cos(x^2)\,dx = \frac{1}{2}\int_0^{\sqrt{2}/2} u^3\,du = \left[\frac{1}{2}\frac{u^4}{4}\right]_0^{\sqrt{2}/2} = \frac{1}{32}$$

Here is the general result.

Theorem B

(Substitution in Definite Integrals). Let g have a continuous derivative on $[a,b]$ and let f be continuous on the range of g. Then

$$\int_a^b f(g(x))g'(x)\,dx = \int_{g(a)}^{g(b)} f(u)\,du$$

Proof Let F be an antiderivative of f (the existence of F is guaranteed by Theorem 5.7D). Then by the Fundamental Theorem,

$$\int_{g(a)}^{g(b)} f(u)\,du = \left[F(u)\right]_{g(a)}^{g(b)} = F(g(b)) - F(g(a))$$

On the other hand, by the Substitution Theorem for Indefinite Integrals (Theorem A),

$$\int f(g(x))g'(x)\,dx = F(g(x)) + C$$

and so, again by the Fundamental Theorem.

$$\int_a^b f(g(x))g'(x)\,dx = [F(g(x))]_a^b = F(g(b)) - F(g(a)) \qquad \blacksquare$$

EXAMPLE 3 Evaluate $\displaystyle\int_0^1 \frac{x+1}{(x^2+2x+6)^2}\,dx.$

Solution   Let $u = x^2 + 2x + 6$, so $du = (2x + 2)\,dx = 2(x + 1)\,dx$, and note that $u = 6$ when $x = 0$ and $u = 9$ when $x = 1$. Thus,

$$\int_0^1 \frac{x + 1}{(x^2 + 2x + 6)^2}\,dx = \frac{1}{2}\int_0^1 \frac{2x + 2}{(x^2 + 2x + 6)^2}\,dx$$

$$= \frac{1}{2}\int_6^9 u^{-2}\,du = \left[-\frac{1}{2}\frac{1}{u}\right]_6^9$$

$$= -\frac{1}{18} - \left(-\frac{1}{12}\right) = \frac{1}{36} \qquad \blacksquare$$

EXAMPLE 4   Evaluate $\displaystyle\int_{\pi^2/9}^{\pi^2/4} \frac{\cos\sqrt{x}}{\sqrt{x}}\,dx$.

Solution   Let $u = \sqrt{x}$, so $du = dx/(2\sqrt{x})$. Thus,

$$\int_{\pi^2/9}^{\pi^2/4} \frac{\cos\sqrt{x}}{\sqrt{x}}\,dx = 2\int_{\pi^2/9}^{\pi^2/4} \cos\sqrt{x}\cdot\frac{1}{2\sqrt{x}}\,dx$$

$$= 2\int_{\pi/3}^{\pi/2} \cos u\,du$$

$$= [2\sin u]_{\pi/3}^{\pi/2} = 2 - \sqrt{3}$$

Note the change in the limits of integration at the second equality. When $x = \pi^2/9$, $u = \sqrt{\pi^2/9} = \pi/3$; when $x = \pi^2/4$, $u = \pi/2$. $\qquad \blacksquare$

USE OF SYMMETRY   Recall that an even function is one satisfying $f(-x) = f(x)$, whereas an odd function satisfies $f(-x) = -f(x)$. The graph of the former is symmetric with respect to the y-axis; the graph of the latter is symmetric with respect to the origin. Here is a nice integration theorem for such functions.

Theorem C

(Symmetry Theorem).   If f is an even function, then

$$\int_{-a}^a f(x)\,dx = 2\int_0^a f(x)\,dx$$

If f is an odd function, then

$$\int_{-a}^a f(x)\,dx = 0$$

Even function
Left area = right area

FIGURE 1

Odd function
Left area neutralizes right area

FIGURE 2

Proof The geometric interpretation of this theorem is shown in Figures 1 and 2. To justify the results analytically, we first write

$$\int_{-a}^{a} f(x)\,dx = \int_{-a}^{0} f(x)\,dx + \int_{0}^{a} f(x)\,dx$$

In the first of the integrals on the right, we make the substitution $u = -x$, $du = -dx$. If f is even, $f(-x) = f(x)$ and

$$\int_{-a}^{0} f(x)\,dx = -\int_{-a}^{0} f(-x)(-dx) = -\int_{a}^{0} f(u)\,du$$

$$= \int_{0}^{a} f(u)\,du = \int_{0}^{a} f(x)\,dx$$

On the other hand, if f is odd, so $f(-x) = -f(x)$,

$$\int_{-a}^{0} f(x)\,dx = \int_{-a}^{0} -f(-x)\,dx = \int_{a}^{0} f(u)\,du$$

$$= -\int_{0}^{a} f(u)\,du = -\int_{0}^{a} f(x)\,dx \qquad \blacksquare$$

EXAMPLE 5   Evaluate $\displaystyle\int_{-\pi}^{\pi} \cos\!\left(\frac{x}{4}\right) dx$.

Solution   Since $\cos(-x/4) = \cos(x/4)$, $f(x) = \cos(x/4)$ is an even function. Thus,

$$\int_{-\pi}^{\pi} \cos\!\left(\frac{x}{4}\right) dx = 2\int_{0}^{\pi} \cos\!\left(\frac{x}{4}\right) dx = 8\int_{0}^{\pi} \cos\!\left(\frac{x}{4}\right)\cdot\frac{1}{4}\,dx$$

$$= 8\int_{0}^{\pi/4} \cos u\,du = [8\sin u]_{0}^{\pi/4} = 4\sqrt{2} \qquad \blacksquare$$

EXAMPLE 6   Evaluate $\displaystyle\int_{-5}^{5} \frac{x^5}{x^2 + 4}\,dx$.

Solution $f(x) = x^5/(x^2 + 4)$ is an odd function. Thus the above integral has the value 0. $\qquad \blacksquare$

EXAMPLE 7   Evaluate $\displaystyle\int_{-2}^{2} (x\sin^4 x + x^3 - x^4)\,dx$.

Solution   The first two terms in the integrand are odd, the last is even. Thus, we may write the integral as

$$\int_{-2}^{2} (x\sin^4 x + x^3)\,dx - \int_{-2}^{2} x^4\,dx = 0 - 2\int_{0}^{2} x^4\,dx$$

$$= \left[-2\frac{x^5}{5}\right]_{0}^{2} = -\frac{64}{5} \qquad \blacksquare$$

CHAPTER 5/The Integral

USE OF PERIODICITY   A function f is **periodic** if there is a number p such that

$$f(x + p) = f(x)$$

for all real numbers x in the domain of f. The smallest such positive number p is the **period** of a periodic function. The trigonometric functions are the prime examples of periodic functions.

Area (A) = Area (B)

FIGURE 3

Theorem D

If f is periodic with period p, then

$$\int_{a+p}^{b+p} f(x)\,dx = \int_{a}^{b} f(x)\,dx$$

Proof   The geometric interpretation can be seen in Figure 3. To prove the result, let $u = x - p$ so $x = u + p$ and $du = dx$. Then

$$\int_{a+p}^{b+p} f(x)\,dx = \int_{a}^{b} f(u + p)\,du = \int_{a}^{b} f(u)\,du = \int_{a}^{b} f(x)\,dx \qquad \blacksquare$$

EXAMPLE 8   Evaluate $\displaystyle\int_{0}^{2\pi} |\sin x|\,dx$.

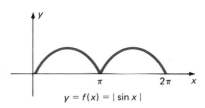

$y = f(x) = |\sin x|$

FIGURE 4

Solution   Note that $f(x) = |\sin x|$ is periodic with period π (Figure 4).

$$\int_{0}^{2\pi} |\sin x|\,dx = \int_{0}^{\pi} |\sin x|\,dx + \int_{\pi}^{2\pi} |\sin x|\,dx$$

$$= \int_{0}^{\pi} |\sin x|\,dx + \int_{0}^{\pi} |\sin x|\,dx$$

$$= 2\int_{0}^{\pi} \sin x\,dx = [-2\cos x]_{0}^{\pi} = 2 - (-2) = 4 \qquad \blacksquare$$

PROBLEM SET   5.8

Use the method of substitution to find each of the following indefinite integrals.

1. $\displaystyle\int \sqrt{3x + 2}\,dx$

2. $\displaystyle\int \sqrt[3]{2x - 4}\,dx$

3. $\displaystyle\int \cos(3x + 2)\,dx$

4. $\displaystyle\int \sin(2x - 4)\,dx$

5. $\displaystyle\int x\sqrt{x^2 + 4}\,dx$

6. $\displaystyle\int x^2(x^3 + 5)^9\,dx$

7. $\displaystyle\int x \sin(x^2 + 4)\,dx$

8. $\displaystyle\int x^2 \cos(x^3 + 5)\,dx$

9. $\displaystyle\int \frac{x \sin\sqrt{x^2 + 4}}{\sqrt{x^2 + 4}}\,dx$

10. $\displaystyle\int x^2(x^3 + 5)^8\cos[(x^3 + 5)^9]\,dx$

11. $\displaystyle\int x \cos(x^2 + 4)\sqrt{\sin(x^2 + 4)}\,dx$

12. $\displaystyle\int x^2 \sin(x^3 + 5)\cos^9(x^3 + 5)\,dx$

$\int x\,(x^2+4)^{\frac{1}{2}} \qquad u = x^2 + 4 \quad du = 2x$

$\int u^{\frac{1}{2}} \frac{1}{2} du$

$\frac{\frac{1}{2}u^{\frac{3}{2}}}{2 \cdot 3} \qquad \frac{u^{\frac{3}{2}}}{3}$

13. $\displaystyle\int \frac{(\sqrt{t}+4)^3}{\sqrt{t}}\,dt$

14. $\displaystyle\int \left(1+\frac{1}{t}\right)^{-2}\left(\frac{1}{t^2}\right)dt$

Use the method of substitution in definite integrals to evaluate each of the following (see Examples 3 and 4).

15. $\displaystyle\int_0^1 (3x+1)^3\,dx$

16. $\displaystyle\int_0^4 \sqrt{2t+1}\,dt$

17. $\displaystyle\int_0^2 \frac{t}{(t^2+9)^2}\,dt$

18. $\displaystyle\int_0^{\sqrt{5}} \sqrt{9-x^2}\,x\,dx$

19. $\displaystyle\int_0^1 \frac{x+2}{(x^2+4x+1)^2}\,dx$

20. $\displaystyle\int_0^2 \frac{x^2}{(9-x^3)^{3/2}}\,dx$

21. $\displaystyle\int_0^{\pi/6} \sin^3\theta\cos\theta\,d\theta$

22. $\displaystyle\int_0^{\pi/6} \frac{\sin\theta}{\cos^3\theta}\,d\theta$

23. $\displaystyle\int_0^1 \cos(3x-3)\,dx$

24. $\displaystyle\int_0^{1/2} \sin(2\pi x)\,dx$

25. $\displaystyle\int_0^1 x\sin(\pi x^2)\,dx$

26. $\displaystyle\int_0^{\pi/4} (\cos 2x+\sin 2x)\,dx$

27. $\displaystyle\int_0^{\pi/2} \sin x\,\sin(\cos x)\,dx$

28. $\displaystyle\int_0^1 x\cos^3(x^2)\sin(x^2)\,dx$

29. $\displaystyle\int_1^4 \frac{1}{\sqrt{t}(\sqrt{t}+1)^3}\,dt$

30. $\displaystyle\int_1^2 \left(1+\frac{1}{t}\right)^2\left(\frac{1}{t^2}\right)dt$

In Problems 31–38, use symmetry to help you evaluate the given integral.

31. $\displaystyle\int_{-\pi}^{\pi} (\sin x+\cos x)\,dx$

32. $\displaystyle\int_{-1}^{1} \frac{x^3}{(1+x^2)^4}\,dx$

33. $\displaystyle\int_{-\pi/2}^{\pi/2} \frac{\sin x}{1+\cos x}\,dx$

34. $\displaystyle\int_{-\sqrt[3]{\pi}}^{\sqrt[3]{\pi}} x^2\cos(x^3)\,dx$

35. $\displaystyle\int_{-\pi}^{\pi} (\sin x+\cos x)^2\,dx$

36. $\displaystyle\int_{-1}^{1} (1+x+x^2+x^3)\,dx$

37. $\displaystyle\int_{-1}^{1} (|x^3|+x^3)\,dx$

38. $\displaystyle\int_{-\pi}^{\pi} (x^5+|\sin x|)\,dx$

39. How does $\displaystyle\int_{-b}^{-a} f(x)\,dx$ compare with $\displaystyle\int_a^b f(x)\,dx$ in case f is an even function? An odd function?

40. Prove (by a substitution) that
$$\int_a^b f(-x)\,dx = \int_{-b}^{-a} f(x)\,dx$$

41. Use periodicity (see Example 8) to calculate $\displaystyle\int_0^{4\pi} |\cos x|\,dx$.

42. Calculate $\displaystyle\int_0^{4\pi} |\sin 2x|\,dx$.

43. If f is periodic with period p, then

$$\int_a^{a+p} f(x)\,dx = \int_0^p f(x)\,dx$$

Convince yourself that this is true by drawing a picture and then use the result to calculate $\displaystyle\int_1^{1+\pi} |\sin x|\,dx$.

44. Use the result in Problem 43 to calculate $\displaystyle\int_2^{2+\pi/2} |\sin 2x|\,dx$.

45. The temperature T on a certain day satisfied
$$T(t) = 70 + 8\sin\left[\frac{\pi}{12}(t-9)\right]$$
where t was the number of hours after midnight. Find the average temperature from 6 A.M. to 6 P.M.

46. Complete the generalization of the Pythagorean Theorem begun in Problem 37 of Section 1.5 by showing that $A+B=C$ in Figure 5, these being the areas of similar regions built on the two legs and the hypotenuse of a right triangle.

(a) Convince yourself that similarity means
$$g(x) = \frac{a}{c}f\left(\frac{c}{a}x\right) \quad\text{and}\quad h(x) = \frac{b}{c}f\left(\frac{c}{b}x\right)$$

(b) Show that $\displaystyle\int_0^a g(x)\,dx + \int_0^b h(x)\,dx = \int_0^c f(x)\,dx$

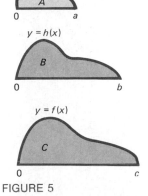

$y = q(x)$

A

0   a

$y = h(x)$

B

0   b

$y = f(x)$

C

0   c

FIGURE 5

47. Find the area of the shaded region shown in Figure 6.

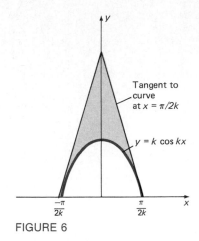

Tangent to curve at $x = \pi/2k$

$y = k \cos kx$

$\dfrac{-\pi}{2k}$ $\dfrac{\pi}{2k}$

FIGURE 6

48. Use the substitution $u = a + b - x$ to show that

$$\int_a^b f(x)\, dx = \int_a^b f(a + b - x)\, dx$$

Deduce that

$$\int_0^{\pi/2} \sin^n x\, dx = \int_0^{\pi/2} \cos^n x\, dx.$$

5.9   Chapter Review

TRUE-FALSE QUIZ

Respond with true or false to each of the following assertions. Be prepared to defend your answer.

1. The indefinite integral is a linear operator.

2. $\int \left[f(x)g'(x) + g(x)f'(x) \right] dx = f(x)g(x) + C$.

3. $y = \cos x$ is a solution to the differential equation $(dy/dx)^2 = 1 - y^2$.

4. $\int f'(x)\, dx = f(x)$ for every differentiable function f.

5. If $s = -16t^2 + v_0 t$ gives the height at time t of a ball thrown straight up from the surface of the earth, then the ball will hit the ground with velocity $-v_0$.

6. $\displaystyle\sum_{i=1}^{n} (a_i + a_{i-1}) = a_0 + a_n + 2\sum_{i=1}^{n-1} a_i$.

7. $\displaystyle\sum_{i=1}^{100} (2i - 1) = 10{,}000$.

8. If $\displaystyle\sum_{i=1}^{10} a_i^2 = 100$ and $\displaystyle\sum_{i=1}^{10} a_i = 20$, then $\displaystyle\sum_{i=1}^{10} (a_i + 1)^2 = 150$.

9. If f is bounded on $[a, b]$, then f is integrable there.

10. If f is integrable on $[a, b]$, then f must be bounded on $[a, b]$.

11. If $\displaystyle\int_a^b f(x)\, dx = 0$, then $f(x) = 0$ for all x in $[a, b]$.

12. If $\displaystyle\int_a^b [f(x)]^2\,dx = 0$ then $f(x) = 0$ for all x in $[a, b]$.

13. The operator lim is linear.

14. $\displaystyle\int_{-\pi}^{\pi} \sin^{13}x\,dx = 0.$

15. $\displaystyle\int_1^5 \sin^2 x\,dx = \int_1^7 \sin^2 x\,dx + \int_7^5 \sin^2 x\,dx.$

16. If f is continuous and positive everywhere, then $\displaystyle\int_c^d f(x)\,dx$ is positive.

17. $D_x\left[\displaystyle\int_0^{x^2} \frac{1}{1+t^2}\,dt\right] = \dfrac{1}{1+x^4}$

18. $\displaystyle\int_0^{2\pi} |\sin x|\,dx = \int_0^{2\pi} |\cos x|\,dx$

19. $\displaystyle\int_0^{2\pi} |\sin x|\,dx = 4\int_0^{\pi/2} \sin x\,dx$

20. If $f(x) = 4$ on $[0, 3]$, then every Riemann sum for f on the given interval has the value 12.

21. If $F'(x) = G'(x)$ for all x in $[a, b]$, then $F(b) - F(a) = G(b) - G(a)$.

22. If $f(x) = f(-x)$ for all x in $[-a, a]$, then $\displaystyle\int_{-a}^a f(x)\,dx = 0$.

23. If $F'(x) = f(x)$ for all x in $[0, b]$, then $\displaystyle\int_0^b f(x)\,dx = F(b)$.

24. $\displaystyle\int_{-99}^{99} (ax^3 + bx^2 + cx)\,dx = 2\int_0^{99} bx^2\,dx.$

25. If $f(x) \le g(x)$ on $[a, b]$, then $\displaystyle\int_a^b |f(x)|\,dx \le \int_a^b |g(x)|\,dx$.

26. If $f(x) \le g(x)$ on $[a, b]$, then $\left|\displaystyle\int_a^b f(x)\,dx\right| \le \left|\int_a^b g(x)\,dx\right|$.

27. $\left|\displaystyle\sum_{i=1}^n a_i\right| \le \sum_{i=1}^n |a_i|$

28. If f is continuous on $[a, b]$, then $\left|\displaystyle\int_a^b f(x)\,dx\right| \le \int_a^b |f(x)|\,dx$.

29. $\displaystyle\lim_{n\to\infty} \sum_{i=1}^n \sin\left(\frac{2i}{n}\right)\cdot\frac{2}{n} = \int_0^2 \sin x\,dx.$

30. To say that $|P| \to 0$ is the same as to say the number of subintervals in the partition P tends to ∞.

SAMPLE TEST PROBLEMS

In Problems 1–6, find the indicated integrals.

1. $\displaystyle\int_0^1 (x^3 - 3x^2 + 3\sqrt{x})\,dx$

2. $\displaystyle\int \frac{2x^4 - 3x^2 + 1}{x^2}\,dx$

3. $\displaystyle\int y\sqrt{y^2 - 4}\,dy$

4. $\displaystyle\int_0^{\pi/2} \cos^4 x \sin x\,dx$

5. $\displaystyle\int_0^2 \frac{t^3}{\sqrt{t^4 + 9}}\,dt$

6. $\displaystyle\int \frac{y^2 - 1}{(y^3 - 3y)^2}\,dy$

In Problems 7–10, solve the differential equation subject to the indicated condition.

7. $\dfrac{dy}{dx} = \dfrac{1}{\sqrt{x+1}}$; $y = 18$ at $x = 3$

8. $\dfrac{dy}{dt} = \sqrt{2t - 1}$; $y = -1$ at $t = \frac{1}{2}$

9. $\dfrac{dy}{dt} = t^2 y^4$; $y = 1$ at $t = 1$

10. $\dfrac{dy}{dx} = \dfrac{6x - x^3}{2y}$; $y = 3$ at $x = 0$

11. Find the equation of the curve through $(-2, -\frac{1}{3})$ if its slope at each x is the negative reciprocal of the slope of the curve with equation $xy = 2$.

12. If a particle moving on the x-axis has acceleration $a = 15\sqrt{t} + 8$ at time t and if $v_0 = -6$, $x_0 = -44$, find its position x at $t = 4$. Assume x is measured in feet and t in seconds.

13. A ball is thrown directly upward from a tower 448 feet high with an initial velocity of 48 feet per second. In how many seconds will it strike the ground and with what velocity? Assume $g = 32$ feet per second per second and neglect air resistance.

14. What constant acceleration will cause a car to increase its velocity from 45 to 60 miles per hour in 10 seconds?

15. Let P be a regular partition of the interval $[0, 2]$ into four equal subintervals, and let $f(x) = x^2 - 1$. Write out the Riemann sum for f on P, in which $\bar{x}_1$ is the right endpoint of each subinterval of P, $i = 1, 2, 3, 4$. Find the value of this Riemann sum and make a sketch.

16. If $f(x) = \int_{-2}^{x} \frac{1}{t+3}\, dt$, $-2 \le x$, find $f'(7)$.

17. Find $\int_0^3 (2 - \sqrt{x+1})^2\, dx$

18. If $f(x) = 3x^2\sqrt{x^3 - 4}$, find the average value of f on $[2, 5]$.

19. Find $\int_2^4 \frac{5x^2 - 1}{x^2}\, dx$.

20. Evaluate $\sum_{i=1}^{n} (3^i - 3^{i-1})$.

21. Evaluate $\sum_{i=1}^{10} (6i^2 - 8i)$.

22. Evaluate each sum.

(a) $\sum_{m=2}^{4} \left(\frac{1}{m}\right)$
(b) $\sum_{i=1}^{6} (2 - i)$

(c) $\sum_{k=0}^{4} \cos\left(\frac{k\pi}{4}\right)$

23. Write in sigma notation.

(a) $\frac{1}{2} + \frac{1}{3} + \frac{1}{4} + \cdots + \frac{1}{78}$
(b) $x^2 + 2x^4 + 3x^6 + 4x^8 + \cdots + 50x^{100}$

24. Sketch the region under the curve $y = 16 - x^2$ between $x = 0$ and $x = 3$, showing the inscribed polygon corresponding to a regular partition of $[0, 3]$ into n subintervals. Find a formula for the area of this polygon and then find the area under the curve by taking a limit.

25. If $\int_0^1 f(x)\, dx = 4$, $\int_0^2 f(x)\, dx = 2$, and $\int_0^2 g(x)\, dx = -3$, evaluate each integral.

(a) $\int_1^2 f(x)\, dx$
(b) $\int_1^0 f(x)\, dx$

(c) $\int_0^2 3f(u)\, du$
(d) $\int_0^2 [2g(x) - 3f(x)]\, dx$

(e) $\int_0^{-2} f(-x)\, dx$

26. Evaluate each integral.

(a) $\int_0^4 |x - 1|\, dx$
(b) $\int_0^4 [\![x]\!]\, dx$

(c) $\int_0^4 (x - [\![x]\!])\, dx$

Hint: In (a) and (b), first sketch a graph.

27. Suppose that $f(x) = f(-x)$, $f(x) \le 0$, $g(-x) = -g(x)$, $\int_0^2 f(x)\, dx = -4$, and $\int_0^2 g(x)\, dx = 5$. Evaluate each integral.

(a) $\int_{-2}^2 f(x)\, dx$
(b) $\int_{-2}^2 |f(x)|\, dx$

(c) $\int_{-2}^2 g(x)\, dx$
(d) $\int_{-2}^2 [f(x) + f(-x)]\, dx$

(e) $\int_0^2 [2g(x) + 3f(x)]\, dx$
(f) $\int_{-2}^0 g(x)\, dx$

28. Evaluate $\int_{-100}^{100} (x^3 + \sin^5 x)\, dx$

29. Find c of the Mean Value Theorem for Integrals for $f(x) = 3x^2$ on $[-4, -1]$.

30. Find $G'(x)$ for each function G.

(a) $G(x) = \int_1^x \frac{1}{t^2 + 1}\, dt$
(b) $G(x) = \int_1^{x^2} \frac{1}{t^2 + 1}\, dt$

(c) $G(x) = \int_x^{x^3} \frac{1}{t^2 + 1}\, dt$

31. Evaluate each of the following limits by recognizing it as a definite integral.

(a) $\lim_{n \to \infty} \sum_{i=1}^{n} \sqrt{\frac{4i}{n}} \cdot \frac{4}{n}$
(b) $\lim_{n \to \infty} \sum_{i=1}^{n} \left(1 + \frac{2i}{n}\right)^2 \frac{2}{n}$

32. Show that if $f(x) = \int_{2x}^{5x} \frac{1}{t}\, dt$, then f is a constant function on $(0, \infty)$.

6 Applications of the Integral

[The works of Archimedes] are without exception, monuments of mathematical exposition; the gradual revelation of the plan of attack, the masterly ordering of the propositions, the stern elimination of everything not immediately relevant to the purpose, the finish of the whole, are so impressive in their perfection as to create a feeling akin to awe in the mind of the reader.
Sir Thomas Heath

Archimedes
287–212B.C.

Archimedes of Syracuse was without doubt the greatest mathematician of antiquity. Of Greek ancestry, he received his education in Alexandria, the center of Greek learning and culture. In his own time he was famous as an inventor and practical scientist. He invented the Archimedean screw for pumping water, he stated the properties of pulleys and levers ("give me a place to stand, and I shall move the earth"), he built a mechanical model that duplicated the motion of the moon and the planets, and—to please the king of Syracuse—he found a way of deciding whether the king's crown was made of pure gold without melting it down (Archimedes's buoyancy principle).

Inventions and practical devices were for Archimedes mere diversions; his best writings and most penetrating thought were devoted to that part of mathematics now known as integral calculus. Using a method (the method of exhaustion) in which he summed a large number of very small quantities, he anticipated some of the results in this chapter. Among his contributions are formulas for the area of a circle, the area of a parabolic segment, the area of an ellipse, the volume and surface area of a sphere, and the volume of a cone and other solids of revolution. He is said to have requested of his friends that over his tombstone be placed a cylinder containing an inscribed sphere, labeled with the ratio of the volume of the cylinder to the volume of the sphere.

6.1
The Area of a
Plane Region

The brief discussion of area in Section 5.4 served to motivate the definition of the definite integral. With the latter notion now firmly established, we reverse directions and use the definite integral to calculate areas of regions of more and more complicated shapes. As is our practice, we begin with simple cases.

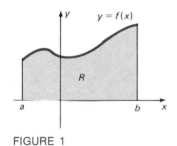

FIGURE 1

A REGION ABOVE THE X-AXIS Let $y = f(x)$ determine a curve in the xy-plane and suppose f is continuous and nonnegative on the interval $a \le x \le b$ (as in Figure 1). Consider the region R bounded by the graphs of $y = f(x)$, $x = a$, $x = b$, and $y = 0$. We refer to R as the region under $y = f(x)$ between $x = a$ and $x = b$. Its area $A(R)$ is given by

$$A(R) = \int_a^b f(x)\,dx$$

EXAMPLE 1 Find the area of the region R under $y = x^4 - 2x^3 + 2$ between $x = -1$ and $x = 2$.

Solution The graph of R is shown in Figure 2.

$$A(R) = \int_{-1}^{2} (x^4 - 2x^3 + 2)\,dx = \left[\frac{x^5}{5} - \frac{x^4}{2} + 2x\right]_{-1}^{2}$$

$$= \left(\frac{32}{5} - \frac{16}{2} + 4\right) - \left(-\frac{1}{5} - \frac{1}{2} - 2\right) = \frac{51}{10} \qquad \blacksquare$$

FIGURE 2

A REGION BELOW THE X-AXIS Area is a nonnegative number. If the graph of $y = f(x)$ is below the x-axis, then $\int_a^b f(x)\,dx$ is a negative number, and therefore cannot be an area. However, it is just the negative of the area of the region bounded by $y = f(x)$, $x = a$, $x = b$, and $y = 0$.

EXAMPLE 2 Find the area of the region R bounded by $y = x^2/3 - 4$, the x-axis, $x = -2$, and $x = 3$.

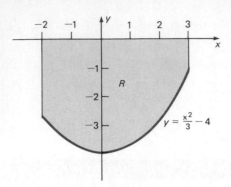

FIGURE 3

Solution The region R is shown in Figure 3.

$$A(R) = -\int_{-2}^{3}\left(\frac{x^2}{3} - 4\right)dx = \int_{-2}^{3}\left(-\frac{x^2}{3} + 4\right)dx$$

$$= \left[-\frac{x^3}{9} + 4x\right]_{-2}^{3} = \left(-\frac{27}{9} + 12\right) - \left(\frac{8}{9} - 8\right) = \frac{145}{9} \qquad \blacksquare$$

EXAMPLE 3 Find the area of the region R bounded by $y = x^3 - 3x^2 - x + 3$, the segment of the x-axis between $x = -1$ and $x = 2$, and the line $x = 2$.

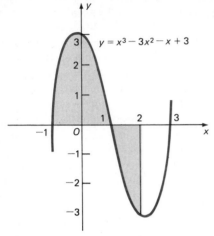

FIGURE 4

Solution The region R is shaded in Figure 4. Note that part of it is above the x-axis and part is below. The areas of these two parts must be calculated separately. You can check that the curve crosses the x-axis at -1, 1, and 3. Thus,

$$A(R) = \int_{-1}^{1}(x^3 - 3x^2 - x + 3)\,dx - \int_{1}^{2}(x^3 - 3x^2 - x + 3)\,dx$$

$$= \left[\frac{x^4}{4} - x^3 - \frac{x^2}{2} + 3x\right]_{-1}^{1} - \left[\frac{x^4}{4} - x^3 - \frac{x^2}{2} + 3x\right]_{1}^{2}$$

$$= 4 - \left(-\frac{7}{4}\right) = \frac{23}{4}$$

Notice that we could have written this area as one integral using the absolute value symbol.

$$A(R) = \int_{-1}^{2}\left|x^3 - 3x^2 - x + 3\right|dx$$

But this is no real simplification since, in order to evaluate this integral, we have to split it into two parts just as we did above. $\qquad \blacksquare$

A HELPFUL WAY OF THINKING So far so good. For simple regions of the type considered above, it is quite easy to write down the correct integral. When we consider more complicated regions (for example, regions between two curves), the task of selecting the right integral is more difficult. However, there is a way of thinking that can be very helpful. It goes back to the definition of area and of the definite integral. Here it is in five steps.

Step 1 Sketch the region.
Step 2 Slice it into thin pieces (strips); label a typical piece.
Step 3 Approximate the area of this typical piece, pretending it is a rectangle.
Step 4 Add up the approximations to the areas of the pieces.
Step 5 Take the limit as the width of the pieces approaches zero, thus getting a definite integral.

To illustrate we consider yet another simple example.

EXAMPLE 4 Set up the integral for the area of the region under $y = 1 + \sqrt{x}$ between $x = 0$ and $x = 4$ (Figure 5).

Solution

1. Sketch

2. Slice

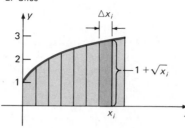

3. Approximate area of piece:
$$\Delta A_i \approx (1 + \sqrt{x_i}) \, \Delta x_i$$

4. Add up: $A \approx \Sigma \, (1 + \sqrt{x_i}) \, \Delta x_i$

5. Take limit: $A = \int_0^4 (1 + \sqrt{x}) \, dx$

FIGURE 5

Once we understand this five-step procedure, we can abbreviate it to three: *slice, approximate, integrate*. Think of the word *integrate* as meaning add up and take the limit as the piece width tends to zero. In this process $\sum \dots \Delta x$ transforms into $\int \dots dx$. Figure 6 gives the abbreviated form for the same problem.

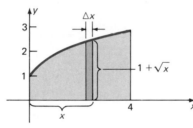

$$\Delta A \approx (1 + \sqrt{x}) \, \Delta x$$

$$A = \int_0^4 (1 + \sqrt{x}) \, dx$$

FIGURE 6

A REGION BETWEEN TWO CURVES   Consider curves $y = f(x)$ and $y = g(x)$ with $g(x) \le f(x)$ on $a \le x \le b$. They determine the region shown in Figure 7. We use the *slice, approximate, integrate* method to find its area. Be sure to note that $f(x) - g(x)$ gives the correct height for the thin slice, even when the graph of g goes below the x-axis. For in this case, $g(x)$ is negative; so subtracting $g(x)$ is the same as adding a positive number. You can check that $f(x) - g(x)$ also gives the correct height, even when both $f(x)$ and $g(x)$ are negative.

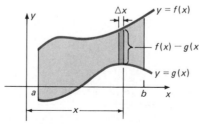

$$\Delta A \approx [f(x) - g(x)] \, \Delta x$$

$$A = \int_a^b [f(x) - g(x)] \, dx$$

FIGURE 7

EXAMPLE 5 Find the area of the region between the curves $y = x^4$ and $y = 2x - x^2$.

Solution We start by finding where the two curves intersect and sketching the required region. This means that we need to solve $2x - x^2 = x^4$, a fourth-degree equation, which would usually be difficult to solve. However, in this case, $x = 0$ and $x = 1$ are rather obvious solutions. Our sketch of the region, together with the appropriate approximation and the corresponding integral, is shown in Figure 8.

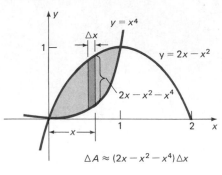

$$\Delta A \approx (2x - x^2 - x^4)\Delta x$$

$$A = \int_0^1 (2x - x^2 - x^4)\,dx$$

FIGURE 8

One job remains—to evaluate the integral.

$$\int_0^1 (2x - x^2 - x^4)\,dx = \left[x^2 - \frac{x^3}{3} - \frac{x^5}{5} \right]_0^1 = 1 - \frac{1}{3} - \frac{1}{5} = \frac{7}{15}$$ ∎

EXAMPLE 6 Find the area of the region between the parabola $y^2 = 4x$ and the line $4x - 3y = 4$.

Solution We will need the points of intersection of these two curves. The y-coordinates of these points can be found by writing the second equation as $4x = 3y + 4$ and then equating the two expressions for $4x$.

$$y^2 = 3y + 4$$
$$y^2 - 3y - 4 = 0$$
$$(y - 4)(y + 1) = 0$$
$$y = 4, -1$$

From this, we conclude that the points of intersection are $(4, 4)$ and $(\frac{1}{4}, -1)$. The required region is sketched in Figure 9.

Now imagine slicing this region vertically. We face a problem, because the lower boundary consists of two different curves. Slices at the extreme left extend from the lower branch of the parabola to its upper branch. For the rest of the region, slices extend from the line to the parabola. To do the problem with vertical slices requires that we first split our region into two parts, set up an integral for each part, and then evaluate both integrals.

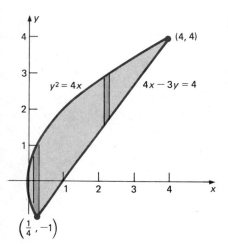

FIGURE 9

A far better approach is to slice the region horizontally as shown in Figure 10, thus using y rather than x as the integration variable. Note that horizontal slices always go from the parabola (at the left) to the line (at the right).

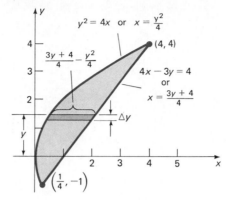

$$\Delta A \approx \left[\frac{3y+4}{4} - \frac{y^2}{4}\right]\Delta y \qquad A = \int_{-1}^{4}\left[\frac{3y+4}{4} - \frac{y^2}{4}\right]dy$$

FIGURE 10

$$A = \int_{-1}^{4}\left[\frac{3y+4-y^2}{4}\right]dy = \frac{1}{4}\int_{-1}^{4}(3y+4-y^2)\,dy$$

$$= \frac{1}{4}\left[\frac{3y^2}{2} + 4y - \frac{y^3}{3}\right]_{-1}^{4}$$

$$= \frac{1}{4}\left[\left(24 + 16 - \frac{64}{3}\right) - \left(\frac{3}{2} - 4 + \frac{1}{3}\right)\right]$$

$$= \frac{125}{24} \approx 5.21$$

There are two items to note: (1) The integrand resulting from a horizontal slicing involves y, not x; and (2) to get the integrand, solve both equations for x and subtract the smaller x-value from the larger. ∎

DISTANCE AND DISPLACEMENT Consider an object moving along a straight line with velocity $v(t)$ at time t. If $v(t) \geq 0$, then $\int_{a}^{b} v(t)\,dt$ gives the distance traveled during the time interval $a \leq t \leq b$ (see Section 5.4). However, if $v(t)$ is sometimes negative (which corresponds to the object moving in reverse), then

$$\int_{a}^{b} v(t)\,dt = s(b) - s(a)$$

measures the **displacement** of the object, that is, the directed distance from its starting position $s(a)$ to its ending position $s(b)$. To get the **total distance** that the object traveled during $a \leq t \leq b$, we must calculate $\int_{a}^{b} |v(t)|\,dt$, the area between the velocity curve and the t-axis. Problems 31–33 illustrate these ideas.

In Problems 1–10, use the three-step procedure (*slice*, *approximate*, *integrate*) to set up an integral (or integrals) for the area of the indicated region.

1.

$y = x^2 + 1$

2.

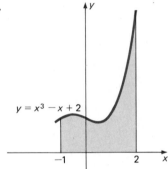

$y = x^3 - x + 2$

3.

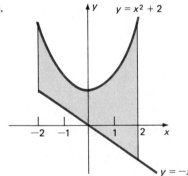

$y = x^2 + 2$

$y = -x$

4.

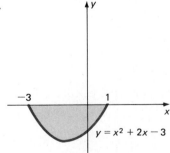

$y = x^2 + 2x - 3$

5.

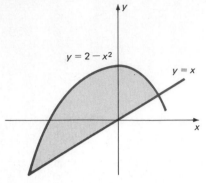

$y = 2 - x^2$

$y = x$

Hint: To find intersection points, solve $x = 2 - x^2$.

6.

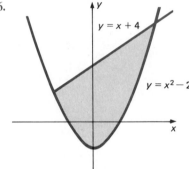

$y = x + 4$

$y = x^2 - 2$

7.

$y = x^3 - x^2 - 6x$

8.

$y = -x + 2$

$y = x^2$

9.

$y = x - 1$

$x = 3 - y^2$

10.

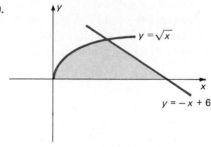

$y = \sqrt{x}$

$y = -x + 6$

In Problems 11–28, sketch the region bounded by the graphs of the given equations, show a typical rectangle, approximate its area, set up an integral, and calculate the area of the region.

11. $y = 4 - \frac{1}{3}x^2$, $y = 0$, between $x = 0$ and $x = 3$

12. $y = 4x - x^2$, $y = 0$, between $x = 1$ and $x = 3$

13. $y = x^2 - 2x - 3$, $y = 0$, between $x = 0$ and $x = 2$

14. $y = \frac{1}{2}(x^2 - 10)$, $y = 0$, between $x = -2$ and $x = 3$

15. $y = x^3$, $y = 0$, $x = -1$, $x = 2$

16. $y - \sqrt[3]{x}$, $y = 0$, $x = -1$, $x = 8$

17. $y = \sqrt{x - 4}$, $y = 0$, $x = 8$

18. $y = x^2 - 4x + 3$, $x - y - 1 = 0$

19. $y = x^2$, $y = x + 2$

20. $y = 2\sqrt{x}$, $y = 2x - 4$, $x = 0$

21. $y = x^2 - 4x$, $y = -x^2$

22. $y = x^2 - 2$, $y = 2x^2 + x - 4$

23. $x = 6y - y^2$, $x = 0$

24. $x = -y^2 + y + 2$, $x = 0$

25. $x = 4 - y^2$, $x + y - 2 = 0$

26. $x = y^2 - 3y$, $x - y + 3 = 0$

27. $y^2 - 2x = 0$, $y^2 + 4x - 12 = 0$

28. $x = y^4$, $x = 2 - y^4$

29. Sketch the region R bounded by $y = x + 6$, $y = x^3$, and $2y + x = 0$. Then find its area. *Hint:* You will have to divide R into two pieces.

30. Find the area of the triangle with vertices at $(-1, 4)$, $(2, -2)$, and $(5, 1)$ by integration.

31. An object moves along a line so that its velocity at time t is $v(t) = 3t^2 - 24t + 36$ feet per second (see Example 3 of Section 3.7). Find the displacement and total distance traveled by the object for $-1 \le t \le 9$.

32. Follow the directions of Problem 31 if $v(t) = \frac{1}{2} + \sin 2t$ and the interval is $0 \le t \le 3\pi/2$.

33. Starting at $s = 0$ when $t = 0$, an object moves along a line so that its velocity at time t is $v(t) = 2t - 4$ centimeters per second. How long will it take to get to $s = 12$? To travel a total distance of 12 centimeters?

34. Consider the curve $y = 1/x^2$ for $1 \le x \le 6$.

(a) Calculate the area under this curve.
(b) Determine c so that the line $x = c$ bisects the area of part (a).
(c) Determine d so that the line $y = d$ bisects the area of part (a).

35. Calculate areas A, B, C, and D in Figure 11. Check by calculating $A + B + C + D$ in one integration.

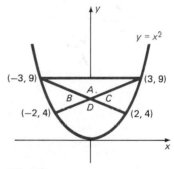

$y = x^2$

$(-3, 9)$ $(3, 9)$

A

B C

$(-2, 4)$ D $(2, 4)$

FIGURE 11

36. Prove Cavalieri's Principle. If two regions have the same width at every x in $[a, b]$, then they have the same area (see Figure 12).

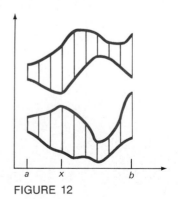

a x b

FIGURE 12

6.2
Volumes of Solids: Slabs, Disks, Washers

That the definite integral can be used to calculate areas is not surprising; it was invented for that purpose. But uses of the integral go far beyond that application. Almost any quantity that can be thought of as a result of chopping something into small pieces, approximating each piece, adding up, and taking the limit as the pieces shrink in size can be interpreted as a definite integral. In particular, this is true for the volumes of certain solids that we shall describe shortly.

What is volume? We start with simple solids called *right cylinders*, four of which are shown in Figure 1. In each case the solid is generated by moving a plane region (the base) through a distance h in a direction perpendicular to that region. And in each case, the volume of the solid is defined to be the area A of the base times the height h—that is,

$$V = A \cdot h$$

FIGURE 1

Next consider a solid which has the property that cross sections perpendicular to a given line have known area. In particular, suppose that line is the x-axis and that the area of the cross section at x is $A(x)$, $a \le x \le b$ (Figure 2). Partition the interval $[a, b]$ by inserting points $a = x_0 < x_1 < x_2 < \cdots < x_n = b$ and pass planes through these points perpendicular to the x-axis, thus slicing the solid into thin **slabs** (Figure 3). The "volume" ΔV_i of a slab should be approximately that of a cylinder, namely,

$$\Delta V_i \approx A(\bar{x}_i) \Delta x_i, \qquad x_{i-1} \le \bar{x}_i \le x_i$$

and the "volume" V of the solid should be given approximately by the Riemann sum

$$V \approx \sum_{i=1}^{n} A(\bar{x}_i) \Delta x_i$$

FIGURE 2

FIGURE 3

When we let the norm of the partition approach zero, we obtain a definite integral; this integral is defined to be the **volume** of the solid.

$$V = \int_a^b A(x)\, dx$$

Rather than routinely applying the boxed formula to obtain volumes, we suggest that in each problem you go through the process that led to it, at least in summary form. Just as for areas, we call this process *slice, approximate, integrate*. It is illustrated in the examples that follow.

FIGURE 4

SOLIDS OF REVOLUTION: METHOD OF DISKS   When a plane region, lying entirely on one side of a fixed line in its plane, is revolved about that line, it generates a **solid of revolution**. The fixed line is called the **axis** of the solid of revolution.

As an illustration, if the region bounded by a semicircle and its diameter is revolved about that diameter, it sweeps out a spherical solid. If the region inside a right triangle is revolved about one of its legs, it generates a conical solid (Figure 4). When a circular region is revolved about a line in its plane that does not intersect the circle (Figure 5), it sweeps out a torus (doughnut). In each case, it is possible to represent the volume as a definite integral.

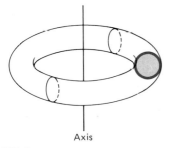

FIGURE 5

EXAMPLE 1   Find the volume of the solid of revolution obtained by revolving the plane region R bounded by $y = \sqrt{x}$, the x-axis, and the line $x = 4$ about the x-axis.

Solution   The region R, with a typical slice, is shown on the next page as the left part of Figure 6. When revolved about the x-axis, this region generates a solid of revolution and the slice generates a disk, a thin coin-shaped object. Recalling that the volume of a circular cylinder is $\pi r^2 h$, we approximate the volume ΔV of this disk, $\Delta V \approx \pi(\sqrt{x})^2\, \Delta x$, and then integrate.

$$V = \pi \int_0^4 x\, dx = \pi \left[\frac{x^2}{2} \right]_0^4 = \pi \frac{16}{2} = 8\pi \approx 25.13$$ ∎

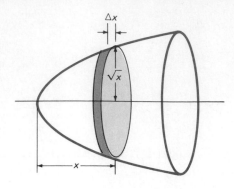

$$\Delta V \approx \pi (\sqrt{x})^2 \, \Delta x$$

$$V = \int_0^4 \pi x \, dx$$

FIGURE 6

EXAMPLE 2 Find the volume of the solid generated by revolving the region bounded by the curve $y = x^3$, the y-axis, and the line $y = 3$ about the y-axis (Figure 7).

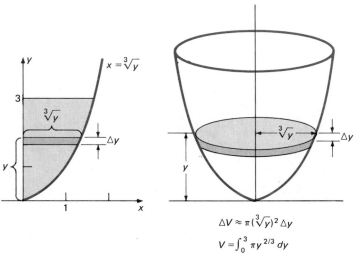

$$\Delta V \approx \pi (\sqrt[3]{y})^2 \, \Delta y$$

$$V = \int_0^3 \pi y^{2/3} \, dy$$

FIGURE 7

Solution Here y is the appropriate choice for the integration variable. Note that $y = x^3$ is equivalent to $x = \sqrt[3]{y}$ and $\Delta V \approx \pi (\sqrt[3]{y})^2 \Delta y$.

$$V = \pi \int_0^3 y^{2/3} \, dy = \pi \left[\frac{3}{5} y^{5/3} \right]_0^3 = \pi \frac{9\sqrt[3]{9}}{5} \approx 11.76 \qquad \blacksquare$$

METHOD OF WASHERS Sometimes, slicing a solid of revolution results in disks with holes in the middle. We call them **washers**. See the diagram and accompanying volume formula shown in Figure 8.

$$V = \pi (r_2^2 - r_1^2) h$$

FIGURE 8

EXAMPLE 3 Find the volume of the solid generated by revolving the region bounded by the parabolas $y = x^2$ and $y^2 = 8x$ about the x-axis.

Solution The key words are still *slice, approximate, integrate* (see Figure 9).

$$V = \pi \int_0^2 (8x - x^4)\, dx = \pi \left[\frac{8x^2}{2} - \frac{x^5}{5} \right]_0^2 = \frac{48\pi}{5} \approx 30.16$$

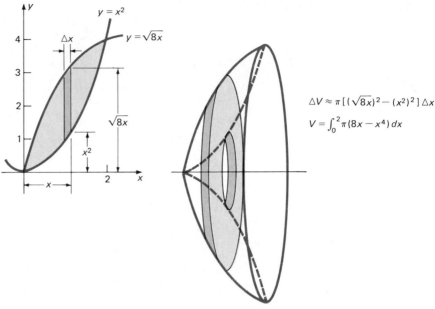

$$\Delta V \approx \pi [(\sqrt{8x})^2 - (x^2)^2]\, \Delta x$$
$$V = \int_0^2 \pi (8x - x^4)\, dx$$

FIGURE 9

EXAMPLE 4 The semicircular region bounded by $x = \sqrt{4 - y^2}$ and the y-axis is revolved about the line $x = -1$. Set up the integral that represents its volume.

Solution Here the outer radius of the washer is $\sqrt{4 - y^2} + 1$ and the inner radius is 1. Figure 10 exhibits the solution. The integral can be simplified.

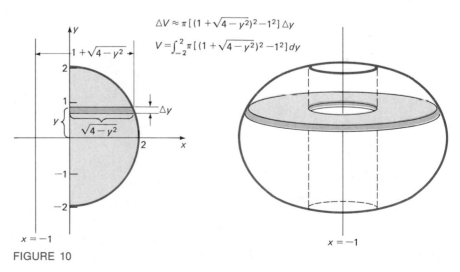

$$\Delta V \approx \pi [(1 + \sqrt{4 - y^2})^2 - 1^2]\, \Delta y$$
$$V = \int_{-2}^2 \pi [(1 + \sqrt{4 - y^2})^2 - 1^2]\, dy$$

FIGURE 10

The part above the x-axis has the same volume as the part below it (which manifests itself in an even integrand). Thus, we may integrate from 0 to 2 and double the result. Also, the integrand simplifies.

$$V = \pi \int_{-2}^{2} \left[(1 + \sqrt{4 - y^2})^2 - 1^2 \right] dy$$

$$= 2\pi \int_{0}^{2} \left[2\sqrt{4 - y^2} + 4 - y^2 \right] dy$$

Now see Problem 31 for a way of evaluating this integral. ∎

OTHER SOLIDS WITH KNOWN CROSS SECTIONS   So far, our solids have had circular cross sections. However, our method works just as well for solids whose cross sections are squares or triangles. In fact, all that is really needed is that the areas of the cross sections can be calculated.

EXAMPLE 5   Let the base of a solid be the first quadrant plane region bounded by $y = 1 - x^2/4$, the x-axis, and the y-axis. Suppose that cross sections perpendicular to the x-axis are squares. Find the volume of the solid.

Solution   When we slice this solid perpendicularly to the x-axis, we get thin square boxes (Figure 11).

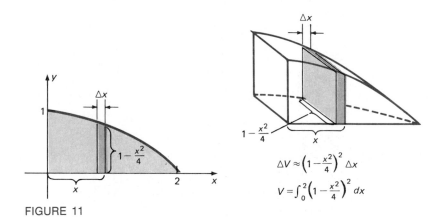

FIGURE 11

$$V = \int_{0}^{2} \left(1 - \frac{x^2}{2} + \frac{x^4}{16} \right) dx = \left[x - \frac{x^3}{6} + \frac{x^5}{80} \right]_{0}^{2}$$

$$= 2 - \frac{8}{6} + \frac{32}{80} = \frac{16}{15} \approx 1.07 \quad ∎$$

$$A = \frac{1}{2} u \frac{\sqrt{3}}{2} u = \frac{\sqrt{3}}{4} u^2$$

FIGURE 12

EXAMPLE 6   The base of a solid is the region between one arch of $y = \sin x$ and the x-axis. Each cross section perpendicular to the x-axis is an equilateral triangle sitting on this base. Find the volume of the solid.

Solution   We need the fact that the area of an equilateral triangle of side u is $\sqrt{3}u^2/4$ (see Figure 12). We proceed as shown in Figure 13. To perform the indicated integration, we use the half-angle formula $\sin^2 x = (1 - \cos 2x)/2$.

$$\Delta V \approx \left(\frac{\sqrt{3}}{4} \sin^2 x \right) \Delta x$$

$$V = \int_0^\pi \left(\frac{\sqrt{3}}{4} \sin^2 x \right) dx$$

FIGURE 13

$$V = \frac{\sqrt{3}}{4} \int_0^\pi \frac{1 - \cos 2x}{2} \, dx = \frac{\sqrt{3}}{8} \int_0^\pi (1 - \cos 2x) \, dx$$

$$= \frac{\sqrt{3}}{8} \left[\int_0^\pi 1 \, dx - \frac{1}{2} \int_0^\pi \cos 2x \cdot 2 \, dx \right]$$

$$= \frac{\sqrt{3}}{8} \left[x - \frac{1}{2} \sin 2x \right]_0^\pi = \frac{\sqrt{3}}{8} \pi \approx 0.68 \quad\blacksquare$$

PROBLEM SET   6.2

In Problems 1–4, find the volume of the solid generated when the indicated region is revolved about the specified axis; *slice, approximate, integrate.*

1. x-axis

2. x-axis

SECTION 6.2/Volumes of Solids: Slabs, Disks, Washers   **277**

3. (a) x-axis
   (b) y-axis

4. (a) x-axis
   (b) y-axis

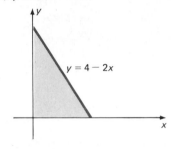

In Problems 5–10, sketch the region R bounded by the graphs of the given equations, showing a typical vertical rectangle. Then find the volume of the solid generated by revolving R about the x-axis.

5. $y = \dfrac{x^2}{4}$, $x = 4$, $y = 0$

6. $y = x^3$, $x = 2$, $y = 0$

7. $y = \dfrac{1}{x}$, $x = 1$, $x = 4$, $y = 0$

8. $y = x^{3/2}$, $y = 0$, between $x = 1$ and $x = 3$.

9. $y = \sqrt{4 - x^2}$, $y = 0$, between $x = -1$ and $x = 2$.

10. $y = x^{2/3}$, $y = 0$, between $x = 1$ and $x = 8$.

In Problems 11–16, sketch the region R bounded by the graphs of the given equations and show a typical horizontal rectangle. Find the volume of the solid generated by revolving R about the y-axis.

11. $x = y^2$, $x = 0$, $y = 2$

12. $x = \dfrac{2}{y}$, $y = 1$, $y = 6$, $x = 0$

13. $x = \sqrt{y}$, $y = 4$, $x = 0$

14. $x = y^{2/3}$, $y = 8$, $x = 0$

15. $x = y^{3/2}$, $y = 4$, $x = 0$

16. $x = \sqrt{9 - y^2}$, $x = 0$

17. Find the volume of the solid generated by revolving about the x-axis the region bounded by the upper half of the ellipse

$$\frac{x^2}{a^2} + \frac{y^2}{b^2} = 1$$

and the x-axis and thus find the volume of a *prolate spheroid*. Here a and b are positive constants, with $a > b$.

18. Find the volume of the solid generated by revolving about the x-axis the region bounded by the line $y = 4x$ and the parabola $y = 4x^2$.

19. Find the volume of the solid generated by revolving about the x-axis the region bounded by the line $x - 2y = 0$ and the parabola $y^2 - 2x = 0$.

20. Find the volume of the solid generated by revolving about the x-axis the region in the first quadrant bounded by the circle $x^2 + y^2 = r^2$, the x-axis, and the line $x = r - h$, $0 < h < r$, and thus find the volume of a spherical segment of height h, radius of sphere r.

21. Find the volume of the solid generated by revolving about the y-axis the region bounded by the line $y = 4x$ and the parabola $y = 4x^2$.

22. Find the volume of the solid generated by revolving about the line $y = 2$ the region in the first quadrant bounded by the parabolas $3x^2 - 16y + 48 = 0$ and $x^2 - 16y + 80 = 0$, and the y-axis.

23. The base of a solid is the region inside the circle $x^2 + y^2 = 4$. Find the volume of the solid if every cross section by a plane perpendicular to the x-axis is a square. *Hint*: See Examples 5 and 6.

24. Do Problem 23 assuming every cross section by a plane perpendicular to the x-axis is an isosceles triangle with base on the xy-plane and altitude 4. *Hint*: To complete the evaluation, interpret $\displaystyle\int_{-2}^{2} \sqrt{4 - x^2}\, dx$ as the area of a semicircle.

25. The base of a solid is bounded by one arch of $y = \sqrt{\cos x}$, $-\pi/2 \le x \le \pi/2$, and the x-axis. Each cross section perpendicular to the x-axis is a square sitting on this base. Find the volume of the solid.

26. The base of a solid is the region bounded by $y = 1 - x^2$ and $y = 1 - x^4$. Cross sections of the solid that are perpendicular to the x-axis are squares. Find the volume of the solid.

27. Find the volume of one octant (one-eighth) of the solid region common to two right circular cylinders of radius 1 whose axes intersect at right angles. *Hint*: Horizontal cross sections are squares. See Figure 14.

FIGURE 14

28. The base of a solid is the region R bounded by $y = \sqrt{x}$ and $y = x^2$. Each cross section perpendicular to the x-axis is a semicircle with diameter extending across R. Find the volume of the solid.

29. Find the volume of the solid generated by revolving the region in the first quadrant bounded by the curve $y^2 = x^3$, the line $x = 4$, and the x-axis

(a) about the line $x = 4$;   (b) about the line $y = 8$.

30. Find the volume of the solid generated by revolving the region bounded by the curve $y^2 = x^3$, the line $y = 8$, and the y-axis

(a) about the line $x = 4$;   (b) about the line $y = 8$.

31. Complete the evaluation of the integral in Example 4 by noting that

$$\int_0^2 \left[2\sqrt{4 - y^2} + 4 - y^2 \right] dy$$

$$= 2 \int_0^2 \sqrt{4 - y^2}\, dy + \int_0^2 (4 - y^2)\, dy$$

Now interpret the first integral as the area of a quarter circle.

32. An open barrel of radius r and height h is initially full of water. It is tilted until the water level coincides with a diameter of the base and just touches the rim of the top. Find the volume of water left in the barrel.

33. A wedge is cut from a right circular cylinder of radius r. The upper surface of the wedge is in a plane through a diameter of the circular base and makes an angle θ with the base. Find the volume of the wedge.

FIGURE 16

34. (The Water Clock)   A water tank is obtained by revolving the curve $y = kx^4$, $k > 0$, about the y-axis.

(a) Find $V(y)$, the volume of water in the tank as a function of its depth y.

(b) Water drains through a small hole according to Torricelli's Law $(dV/dt = -m\sqrt{y})$. Show that the water level falls at a constant rate.

35. Show that the volume of a general cone is $\frac{1}{3}Ah$, where A is the area of the base and h is the height. Use this result to give the formula for the volume of:

(a) a right circular cone of radius r and height h;

(b) a regular tetrahedron with edge length r.

FIGURE 15

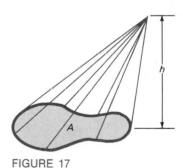

FIGURE 17

36. State the version of Cavalieri's Principle for volume (see Problem 36 of Section 6.1).

6.3
Volumes of Solids of Revolution: Shells

FIGURE 1

There is another method for finding the volume of a solid of revolution—the method of cylindrical shells. For many problems, it is easier to apply than the methods of disks or washers.

A cylindrical shell is a solid bounded by two concentric right circular cylinders (Figure 1). If the inner radius is r_1, the outer radius is r_2, and the height is h, then its volume is given by

$$V = (\text{area of base}) \cdot (\text{height})$$
$$= (\pi r_2^2 - \pi r_1^2)h$$
$$= \pi(r_2 + r_1)(r_2 - r_1)h$$
$$= 2\pi\left(\frac{r_2 + r_1}{2}\right)h(r_2 - r_1)$$

Thus,

$$V = 2\pi \cdot (\text{average radius}) \cdot (\text{height}) \cdot (\text{thickness})$$
$$= 2\pi r h \, \Delta r$$

Here is a good way to remember this formula: If the shell were very thin and flexible (like paper), we could slit it down the side, open it up to form a rectangular sheet, and then calculate its volume by the usual box formula (Figure 2).

FIGURE 2

THE METHOD OF SHELLS   Consider now a region of the type shown in Figure 3. Slice it vertically and revolve it about the y-axis. It will generate a solid of revolution and each slice will generate a piece that is approximately a cylindrical shell. To get the volume of this solid, we calculate the volume ΔV of a typical shell, add, and take the limit as the thickness of the shells tends to zero. The latter is, of course, an integral.

EXAMPLE 1   The region bounded by $y = 1/\sqrt{x}$, the x-axis, $x = 1$, and $x = 4$ is revolved about the y-axis. Find the volume of the resulting solid.

$$\Delta V \approx 2\pi x\, f(x)\, \Delta x$$

$$V = 2\pi \int_a^b x\, f(x)\, dx$$

FIGURE 3

Solution   The diagrams in Figure 3 are approximately correct. Thus,

$$V = 2\pi \int_1^4 x\, \frac{1}{\sqrt{x}}\, dx = 2\pi \int_1^4 x^{1/2}\, dx$$

$$= 2\pi \left[\frac{2}{3}\, x^{3/2}\right]_1^4 = 2\pi \left(\frac{2}{3}\cdot 8 - \frac{2}{3}\cdot 1\right) = \frac{28\pi}{3} \approx 29.32 \qquad \blacksquare$$

EXAMPLE 2   The region bounded by the line $y = (r/h)x$, the x-axis, and $x = h$ is revolved about the x-axis, thereby generating a cone (assume $r > 0, h > 0$). Find its volume by the disk method and by the shell method.

Solution   **(Disk method).**   Follow the steps in Figure 4.

$$V = \pi\, \frac{r^2}{h^2} \int_0^h x^2\, dx = \pi\, \frac{r^2}{h^2} \left[\frac{x^3}{3}\right]_0^h = \frac{\pi r^2 h^3}{3h^2} = \frac{1}{3}\, \pi r^2 h$$

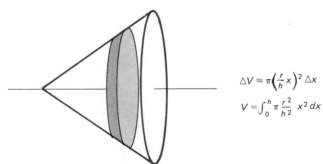

$$\Delta V \approx \pi \left(\frac{r}{h}x\right)^2 \Delta x$$

$$V = \int_0^h \pi\, \frac{r^2}{h^2}\, x^2\, dx$$

FIGURE 4

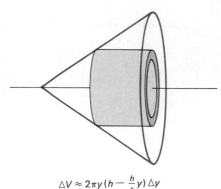

$$\Delta V \approx 2\pi y\left(h - \frac{h}{r}y\right)\Delta y$$

$$V = \int_0^r 2\pi y\left(h - \frac{h}{r}y\right)dy$$

FIGURE 5

(Shell method). See Figure 5.

$$V = \int_0^r 2\pi y\left(h - \frac{h}{r}\,y\right)dy = 2\pi h\int_0^r\left(y - \frac{1}{r}\,y^2\right)dy$$

$$= 2\pi h\left[\frac{y^2}{2} - \frac{y^3}{3r}\right]_0^r = 2\pi h\left[\frac{r^2}{2} - \frac{r^2}{3}\right] = \frac{1}{3}\,\pi r^2 h$$

Naturally, both methods yield the well-known formula for the volume of a cone. ■

EXAMPLE 3 Find the volume of the solid generated by revolving the region in the first quadrant that is above the parabola $y = x^2$ and below the parabola $y = 2 - x^2$ about the y-axis.

Solution One look at the region (left part of Figure 6) should convince you that horizontal slices leading to the disk method is not the best choice (because the right boundary consists of parts of two curves, making it necessary to use two integrals). However, vertical slices, resulting in cylindrical shells, will work fine.

$$V = \int_0^1 2\pi x(2 - 2x^2)\,dx = 4\pi\int_0^1 (x - x^3)\,dx$$

$$= 4\pi\left[\frac{x^2}{2} - \frac{x^4}{4}\right]_0^1 = 4\pi\left[\frac{1}{2} - \frac{1}{4}\right] = \pi \approx 3.14$$

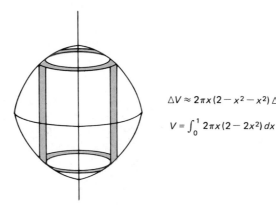

$$\Delta V \approx 2\pi x(2 - x^2 - x^2)\,\Delta x$$

$$V = \int_0^1 2\pi x(2 - 2x^2)\,dx$$

FIGURE 6

■

PUTTING IT ALL TOGETHER While most of us can draw a reasonably good plane figure, some of us do less well at drawing three-dimensional solids. But no law says we have to draw a solid in order to calculate its volume. Usually a plane figure will do, provided we can visualize the corresponding solid in our minds. In the example below, we are going to imagine revolving the region R of Figure 7 about various axes. Our job is to set up an integral for the volume of the resulting solid, and we are going to do it by looking at a plane figure. Be sure to study the example carefully.

$y = 3 + 2x - x^2$

R

FIGURE 7

EXAMPLE 4 Set up an integral for the volume of the solid that results when the region R shown in Figure 7 is revolved about (a) the x-axis, (b) the y-axis, (c) the line $y = -1$, and (d) the line $x = 4$.

Solution

(a)

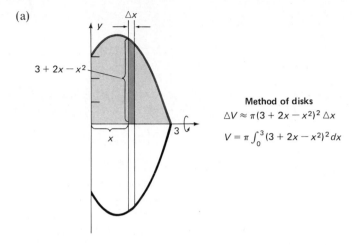

$3 + 2x - x^2$

Method of disks

$\triangle V \approx \pi (3 + 2x - x^2)^2 \triangle x$

$V = \pi \int_0^3 (3 + 2x - x^2)^2 \, dx$

(b)

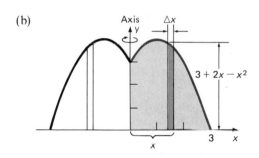

Axis

$3 + 2x - x^2$

Method of shells

$\triangle V \approx 2\pi x (3 + 2x - x^2) \triangle x$

$V = 2\pi \int_0^3 x(3 + 2x - x^2) \, dx$

(c)

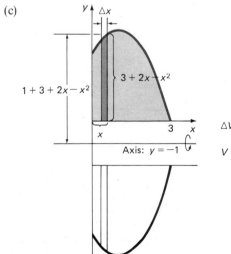

$1 + 3 + 2x - x^2$

$3 + 2x - x^2$

Axis: $y = -1$

Method of washers

$\triangle V \approx \pi \ [(4 + 2x - x^2)^2 - 1^2] \triangle x$

$V = \pi \int_0^3 [(4 + 2x - x^2)^2 - 1] \, dx$

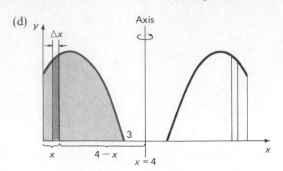

(d)

Axis

Method of shells

$$\Delta V \approx 2\pi(4-x)(3+2x-x^2)\,\Delta x$$

$$V = 2\pi \int_0^3 (4-x)(3+2x-x^2)\,dx$$

Note that in all four cases the limits of integration are the same; it is the original plane region that determines these limits. ∎

PROBLEM SET 6.3

In Problems 1–12, you are to find the volume of the solid generated when the region R bounded by the given curves is revolved about the indicated axis. Do this by performing the following steps.

(a) Sketch the region R.
(b) Show a typical rectangular slice properly labeled.
(c) Write a formula for the approximate volume of the shell generated by this slice.
(d) Set up the corresponding integral.
(e) Evaluate this integral.

1. $y = \dfrac{4}{x}$, $x = 1$, $x = 4$, $y = 0$; about the y-axis.

2. $y = x^2$, $x = 2$, $y = 0$; about the y-axis.

3. $y = \sqrt{x}$, $x = 4$, $y = 0$; about the y-axis.

4. $y = 4 - x^2\,(x \geq 0)$, $x = 0$, $y = 0$; about the y-axis.

5. $y = \sqrt{x}$, $x = 4$, $y = 0$; about the line $x = 4$.

6. $y = 4 - x^2\,(x \geq 0)$, $x = 0$, $y = 0$; about the line $x = 2$.

7. $y = \frac{1}{4}x^3 + 2$, $y = 2 - x$, $x = 2$; about the y-axis.

8. $y = x^2$, $y = 2x$; about the y-axis.

9. $x = y^2$, $y = 2$, $x = 0$; about the x-axis.

10. $x = \sqrt{2y} + 1$, $y = 2$, $x = 0$, $y = 0$; about the x-axis.

11. $x = y^2$, $y = 2$, $x = 0$; about the line $y = 2$.

12. $x = \sqrt{2y} + 1$, $y = 2$, $x = 0$, $y = 0$; about the line $y = 3$.

13. Consider the region R (Figure 8). Set up an integral for the volume of the solid obtained when R is revolved about each line.

(a) The x-axis (washers).
(b) The y-axis (shells).
(c) The line $x = a$ (shells).
(d) The line $x = b$ (shells).

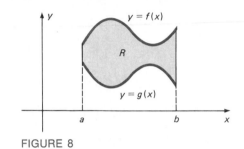

FIGURE 8

14. A region R is shown in Figure 9. Set up an integral for the volume of the solid obtained when R is revolved about each line.

(a) The y-axis (washers).
(b) The x-axis (shells).
(c) The line $y = 3$ (shells).

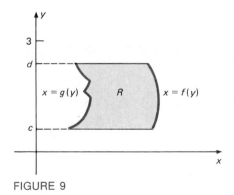

FIGURE 9

15. Sketch the region R bounded by $y = 1/x^3$, $x = 1$, $x = 3$, and $y = 0$. Set up (but do not evaluate) integrals for each of the following.

(a) Area of R.
(b) Volume of the solid obtained when R is revolved about the y-axis.

(c) Volume of the solid obtained when R is revolved about $y = -1$.

(d) Volume of the solid obtained when R is revolved about $x = 4$.

16. Follow the directions of Problem 15 for the region R bounded by $y = x^3 + 1$ and $y = 0$ and between $x = 0$ and $x = 2$.

17. Find the volume of the solid generated by revolving the region R bounded by the curves $x = \sqrt{y}$ and $x = y^3/32$ about the x-axis.

18. Follow the directions of Problem 17, but revolve R about the line $y = 4$.

19. A round hole of radius a is drilled through the center of a solid sphere of radius b (assume $b > a$). Find the volume of the solid that remains.

20. Set up the integral (using shells) for the volume of the torus obtained by revolving the region inside the circle $x^2 + y^2 = a^2$ about the line $x = b$, where $b > a$. Then evaluate this integral. *Hint:* It may help to think of part of this integral as an area.

21. The region in the first quadrant bounded by $x = 0$, $y = \sin(x^2)$, and $y = \cos(x^2)$ is revolved about the y-axis. Find the volume of the resulting solid.

22. The region bounded by $y = 2 + \sin x$, $y = 0$, $x = 0$, and $x = 2\pi$ is revolved about the y-axis. Find the volume that results. *Hint:* $\int x \sin x \, dx = \sin x - x \cos x + C$.

23. Let R be the region bounded by $y = x^2$ and $y = x$. Find the volume of the solid that results when R is revolved around: (a) the x-axis; (b) the y-axis; (c) the line $y = x$.

24. Suppose we know the formula $S = 4\pi r^2$ for the surface area of a sphere but do not know the corresponding formula for its volume V. Obtain this formula by slicing the solid sphere into thin concentric *spherical shells*. *Hint:* The volume ΔV of a thin spherical shell of outer radius x is $\Delta V \approx 4\pi x^2 \Delta x$.

FIGURE 10

25. Consider an irregular region of area S on the surface of a sphere of radius r. Find the volume of the solid that results when each point of this region is connected to the center of the sphere by a line segment. *Hint:* Use the method of spherical shells mentioned in Problem 24.

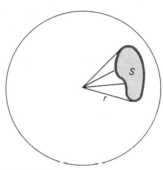

FIGURE 11

6.4
Length of a Plane Curve

FIGURE 1

How long is the spiral curve shown in Figure 1? If it were a piece of string, most of us would stretch it taut and measure it with a ruler. But if it is the graph of an equation, that is a little hard to do.

A little reflection suggests a prior question. What is a plane curve? We have used the term *curve* informally until now; it is time to be more precise. We begin with several examples.

The graph of $y = \sin x$, $0 \le x \le \pi$, is a plane curve (Figure 2). So is the graph of $x = y^2$, $-2 \le y \le 2$ (Figure 3). In both cases, the curve is the graph of a function, the first of the form $y = f(x)$, the second of the form $x = g(y)$. However, the spiral curve does not fit this pattern. Neither does the circle $x^2 + y^2 = a^2$, though in this case we could think of it as the graph of the two functions $y = f(x) = \sqrt{a^2 - x^2}$ and $y = g(x) = -\sqrt{a^2 - x^2}$.

The circle suggests another way of thinking about curves. Recall from trigonometry that $x = a \cos t$, $y = a \sin t$, $0 \le t \le 2\pi$, describe the circle

FIGURE 2

FIGURE 3

FIGURE 4

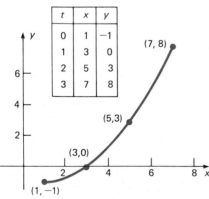

FIGURE 5

$x^2 + y^2 = a^2$ (Figure 4). Here t is an auxiliary variable, from now on called a **parameter**. Both x and y are expressed in terms of this parameter.

If we were to graph the parametric equations $x = t \cos t$, $y = t \sin t$, $0 \le t \le 5\pi$, we would get a curve something like the spiral with which we started. And we can even think of the sine curve and the parabola in parametric form. We write

$$y = \sin x, \qquad x = x, \qquad 0 \le x \le \pi$$
$$y = y, \qquad x = y^2, \qquad -2 \le y \le 2$$

In the first case, x is the parameter; in the second, y is the parameter.

Thus, for us, a **plane curve** is determined by a pair of parametric equations $x = f(t)$, $y = g(t)$, $a \le t \le b$, where we assume that f and g are continuous on the given interval. Think of t as a measuring time. As t increases from a to b, the point (x, y) traces out a curve in the plane. Here is another example.

EXAMPLE 1 Sketch the curve determined by the parametric equations $x = 2t + 1$, $y = t^2 - 1$, $0 \le t \le 3$.

Solution We make a three-column table of values, then plot the ordered pairs (x, y), and—finally—connect these points in the order of increasing t, as shown in Figure 5. ■

Actually, the definition we have given is too broad for the purposes we have in mind, so we immediately restrict it to what is called a *smooth curve*.

Definition

A plane curve is **smooth** if it is determined by a pair of parametric equations $x = f(t)$, $y = g(t)$, $a \le t \le b$, where f' and g' exist and are continuous on $[a, b]$, and $f'(t)$ and $g'(t)$ are not simultaneously zero on (a, b).

The adjective *smooth* is chosen to indicate that an object moving along the curve so that its position at time t is (x, y) would suffer no sudden changes of direction (continuity of f' and g' insures this) and would not stop or double back ($f'(t)$ and $g'(t)$ not simultaneously zero insures this).

LENGTH Finally, we are ready for the main question. What is meant by the length of the smooth curve given parametrically by $x = f(t)$, $y = g(t)$, $a \le t \le b$?

Partition the interval $[a, b]$ into n subintervals by means of points

$$a = t_0 < t_1 < t_2 < \cdots < t_n = b$$

This cuts the curve into n pieces with corresponding endpoints Q_0, Q_1, $Q_2, \ldots, Q_{n-1}, Q_n$, as shown in Figure 6.

FIGURE 6

Our idea is to approximate the curve by the indicated polygonal line, calculate its length, and then take the limit as the norm of the partition approaches zero. In particular, we approximate the "length" Δs_i of the typical piece (see Figure 6) by

$$\Delta w_i = \sqrt{(\Delta x_i)^2 + (\Delta y_i)^2}$$
$$= \sqrt{[f(t_i) - f(t_{i-1})]^2 + [g(t_i) - g(t_{i-1})]^2}$$

From the Mean Value Theorem for Derivatives, we know that there are points $\bar{t}_i$ and $\hat{t}_i$ in (t_{i-1}, t_i) such that

$$f(t_i) - f(t_{i-1}) = f'(\bar{t}_i)\Delta t_i$$
$$g(t_i) - g(t_{i-1}) = g'(\hat{t}_i)\Delta t_i$$

where $\Delta t_i = t_i - t_{i-1}$. Thus,

$$\Delta w_i = \sqrt{[f'(\bar{t}_i)\Delta t_i]^2 + [g'(\hat{t}_i)\Delta t_i]^2}$$
$$= \sqrt{[f'(\bar{t}_i)]^2 + [g'(\hat{t}_i)]^2}\,\Delta t_i$$

and the length of the polygonal line is

$$\sum_{i=1}^{n} \Delta w_i = \sum_{i=1}^{n} \sqrt{[f'(\bar{t}_i)]^2 + [g'(\hat{t}_i)]^2}\,\Delta t_i$$

The latter expression is almost a Riemann sum, the only difficulty being that $\bar{t}_i$ and $\hat{t}_i$ are not likely to be the same point. However, it is shown in advanced calculus books that in the limit, this makes no difference. Thus, we may define the length L of the curve to be the limit of the expression above as the norm of the partition approaches zero; that is,

$$L = \int_a^b \sqrt{[f'(t)]^2 + [g'(t)]^2}\,dt = \int_a^b \sqrt{\left(\frac{dx}{dt}\right)^2 + \left(\frac{dy}{dt}\right)^2}\,dt$$

Two special cases are of great interest. If this curve is given by $y = f(x)$, $a \leq x \leq b$, the boxed result takes the form

$$L = \int_a^b \sqrt{1 + \left(\frac{dy}{dx}\right)^2} \, dx$$

If it is given by $x = g(y)$, $c \leq y \leq d$, it becomes

$$L = \int_c^d \sqrt{1 + \left(\frac{dx}{dy}\right)^2} \, dy$$

Something would be amiss if these formulas failed to yield the familiar results for circles and line segments. They do, as the following two examples illustrate.

EXAMPLE 2 Find the circumference of the circle $x^2 + y^2 = a^2$.

Solution We write the equation of the circle in parametric form: $x = a \cos t$, $y = a \sin t$, $0 \leq t \leq 2\pi$. Then $dx/dt = -a \sin t$, $dy/dt = a \cos t$, and by the first of our formulas,

$$L = \int_0^{2\pi} \sqrt{a^2 \sin^2 t + a^2 \cos^2 t} \, dt = \int_0^{2\pi} a \, dt = [at]_0^{2\pi} = 2\pi a \qquad \blacksquare$$

EXAMPLE 3 Use the second length formula to find the length of the line segment from $A(0, 1)$ to $B(5, 13)$.

Solution The given line segment is shown in Figure 7. Note that the equation of the corresponding line is $y = \frac{12}{5}x + 1$, so $dy/dx = \frac{12}{5}$ and

$$L = \int_0^5 \sqrt{1 + \left(\frac{12}{5}\right)^2} \, dx = \int_0^5 \sqrt{\frac{5^2 + 12^2}{5^2}} \, dx = \frac{13}{5} \int_0^5 1 \, dx$$

$$= \left[\frac{13}{5} x\right]_0^5 = 13$$

This does agree with the result obtained by use of the distance formula. $\blacksquare$

EXAMPLE 4 Sketch the graph of the curve given parametrically by $x = 2 \cos t$, $y = 4 \sin t$, $0 \leq t \leq \pi$, and find its length.

Solution The graph (Figure 8) is drawn, as in Example 1, by first making a three-column table of values.

FIGURE 7

t	x	y
0	2	0
$\pi/6$	$\sqrt{3}$	2
$\pi/3$	1	$2\sqrt{3}$
$\pi/2$	0	4
$2\pi/3$	-1	$2\sqrt{3}$
$5\pi/6$	$-\sqrt{3}$	2
π	-2	0

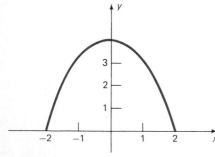

FIGURE 8

The length L of this curve is given by

$$L = \int_0^\pi \sqrt{(-2 \sin t)^2 + (4 \cos t)^2} \, dt = 2 \int_0^\pi \sqrt{\sin^2 t + 4 \cos^2 t} \, dt$$

$$= 2 \int_0^\pi \sqrt{1 + 3 \cos^2 t} \, dt$$

We are unable to carry the evaluation further because we cannot find an antiderivative for $\sqrt{1 + 3 \cos^2 t}$. In fact, it has been shown that this function does not have an antiderivative that can be written in terms of the elementary functions of calculus. If we want a value for the integral, we will have to find it by an approximation method (see Chapter 10). ■

<table>
</table>

ARC-LENGTH PROBLEMS

How do authors choose arc-length problems? Very, very carefully. Pick a curve $y = f(x)$ at random and it is almost certain that you will be unable to integrate $\sqrt{1 + [f'(x)]^2}$. For example, we can find the length of the curve $y = x^r$ by the Fundamental Theorem of Calculus only if $r = 1$ or $r = 1 + 1/n$ for some integer n. We have chosen the curve of Example 5 and those in the problem set very, very carefully.

The situation in Example 4 is quite typical. To get an integration that we can complete, we must pick our curves very carefully.

EXAMPLE 5 Find the length of the arc of the curve $y = x^{3/2}$ from the point $(1, 1)$ to the point $(4, 8)$ (see Figure 9).

Solution Since $dy/dx = \frac{3}{2} x^{1/2}$,

$$L = \int_1^4 \sqrt{1 + \left(\frac{3}{2} x^{1/2}\right)^2} \, dx = \int_1^4 \sqrt{1 + \frac{9}{4} x} \, dx$$

Let $u = 1 + \frac{9}{4} x$; then $du = \frac{9}{4} \, dx$. Hence,

$$\int \sqrt{1 + \frac{9}{4} x} \, dx = \frac{4}{9} \int \sqrt{u} \, du = \frac{4}{9} \frac{2}{3} u^{3/2} + C$$

$$= \frac{8}{27} \left(1 + \frac{9}{4} x\right)^{3/2} + C$$

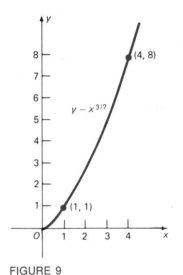

FIGURE 9

Therefore,

$$\int_1^4 \sqrt{1 + \frac{9}{4} x} \, dx = \left[\frac{8}{27} \left(1 + \frac{9}{4} x\right)^{3/2}\right]_1^4 = \frac{8}{27} \left(10^{3/2} - \frac{13^{3/2}}{8}\right) \approx 7.63 \quad ■$$

DIFFERENTIAL OF ARC LENGTH   Let f be continuously differentiable on $[a, b]$. For each x in (a, b), define $s(x)$ by

$$s(x) = \int_a^x \sqrt{1 + [f'(u)]^2} \, du$$

Then $s(x)$ gives the length of the arc of the curve $y = f(u)$ from the point $(a, f(a))$ to $(x, f(x))$ (see Figure 10). By the theorem on differentiating an integral with respect to its upper limit (Theorem 5.7D),

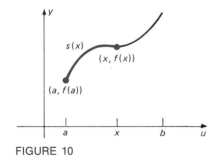

FIGURE 10

$$s'(x) = \frac{ds}{dx} = \sqrt{1 + [f'(x)]^2} = \sqrt{1 + \left(\frac{dy}{dx}\right)^2}$$

Thus, ds—the differential of arc length—can be written as

$$ds = \sqrt{1 + \left(\frac{dy}{dx}\right)^2}\, dx$$

In fact, depending on how a graph is parametrized, we are led to three formulas for ds, namely,

$$ds = \sqrt{1 + \left(\frac{dy}{dx}\right)^2}\, dx = \sqrt{1 + \left(\frac{dx}{dy}\right)^2}\, dy = \sqrt{\left(\frac{dx}{dt}\right)^2 + \left(\frac{dy}{dt}\right)^2}\, dt$$

Some people prefer to remember these formulas by writing (see Figure 11)

$$(ds)^2 = (dx)^2 + (dy)^2$$

FIGURE 11

The three forms arise by dividing and multiplying the right-hand side by $(dx)^2$, $(dy)^2$, and $(dt)^2$, respectively. For example,

$$(ds)^2 = \left[\frac{(dx)^2}{(dx)^2} + \frac{(dy)^2}{(dx)^2}\right](dx)^2 = \left[1 + \left(\frac{dy}{dx}\right)^2\right](dx)^2$$

which gives the first of the three formulas.

PROBLEM SET   6.4

1. Use an x-integration to find the length of the segment of the line $y = 3x + 5$ between $x = 1$ and $x = 4$. Check by the distance formula.

2. Use a y-integration to find the length of the segment of the line $2x - 4y + 6 = 0$ between $y = 0$ and $y = 2$. Check by the distance formula.

In Problems 3–8, find the length of the indicated curve.

3. $y = 2x^{3/2}$ between $x = 1/3$ and $x = 7$.

4. $y = \frac{2}{3}(x^2 + 1)^{3/2}$ between $x = 1$ and $x = 4$.

5. $y = (4 - x^{2/3})^{3/2}$ between $x = 1$ and $x = 8$.

6. $y = (x^4 + 3)/6x$ between $x = 1$ and $x = 4$.

7. $x = y^4/16 + 1/2y^2$ between $y = -2$ and $y = -1$.
Hint: Watch signs; $\sqrt{u^2} = -u$ when $u < 0$.

8. $30xy^3 - y^8 = 15$ between $y = 1$ and $y = 2$.

In Problems 9–12, sketch the graph of the given parametric equations and find its length.

9. $x = t^3$, $y = t^2$; $0 \le t \le 4$

10. $x = 3t^2 + 2$, $y = 2t^3 - 1$; $1 \le t \le 3$

11. $x = 3 \sin t$, $y = 3 \cos t - 3$; $0 \le t \le 2\pi$

12. $x = 4 \cos t + 5$, $y = 4 \sin t - 1$; $0 \le t \le 2\pi$

13. Sketch the graph of the four-cusped *hypocycloid* $x = a \sin^3 t$, $y = a \cos^3 t$, $0 \le t \le 2\pi$, and find its length. *Hint*: By symmetry, you can quadruple the length of the first quadrant portion.

14. A point P on the rim of a wheel of radius a is initially at the origin. As the wheel rolls to the right along the x-axis, P traces out a curve called a *cycloid* (see Figure 12). Derive parametric equations for the cycloid as follows.

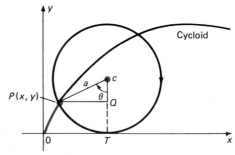

FIGURE 12

(a) Show that $\overline{OT} = a\theta$.

(b) Convince yourself that $\overline{PQ} = a\sin\theta$, $\overline{QC} = a\cos\theta$, $0 \le \theta \le \pi/2$.

(c) Show that $x = a(\theta - \sin\theta)$, $y = a(1 - \cos\theta)$.

15. Find the length of one arch of the cycloid of Problem 14. *Hint*: First show that

$$\left(\frac{dx}{d\theta}\right)^2 + \left(\frac{dy}{d\theta}\right)^2 = 4a^2 \sin^2\left(\frac{\theta}{2}\right)$$

16. Suppose the wheel of Problem 14 turns at a constant rate $\omega = d\theta/dt$, where t is time. Then $\theta = \omega t$.

(a) Show that the speed ds/dt of P along the cycloid is

$$\frac{ds}{dt} = 2a\omega\left|\sin\frac{\omega t}{2}\right|$$

(b) When is the speed a maximum and when is it a minimum?

(c) Explain why a bug on a wheel of a car going 60 miles per hour sometimes is traveling at 120 miles per hour.

17. Find the length of each curve.

(a) $y = \displaystyle\int_1^x \sqrt{u^3 - 1}\, du$, $1 \le x \le 2$

(b) $x = t - \sin t$, $y = 1 - \cos t$, $0 \le t \le 4\pi$

18. Find the length of each curve.

(a) $y = \displaystyle\int_{\pi/6}^x \sqrt{64\sin^2 u \cos^4 u - 1}\, du$, $\dfrac{\pi}{6} \le x \le \dfrac{\pi}{3}$

(b) $x = a\cos t + at\sin t$, $y = a\sin t - at\cos t$, $-1 \le t \le 1$

19. Show that $x = a\cos t$, $y = b\sin t$, $a > 0$, $b > 0$, $0 \le t \le 2\pi$, provide parametric equations for the ellipse $x^2/a^2 + y^2/b^2 = 1$. Then determine a so that $x^2/a^2 + y^2 = 1$ and one period of the cosine curve have the same length.

20. Consider $y = \dfrac{x^{r+1}}{r+1} + \dfrac{x^{1-r}}{4(r-1)}$, $r \ne \pm 1$.

(a) Show that $\sqrt{1 + (dy/dx)^2}$ simplifies nicely.

(b) Find the length of this curve on $1 \le x \le 2$ when $r = 2$.

6.5
Area of a Surface of Revolution

If a smooth plane curve is revolved about an axis in its plane, it generates a surface of revolution as illustrated in Figure 1. Our aim is to determine the area of such a surface.

To get started, we introduce the formula for the area of the frustum of a cone. A **frustum** of a cone is the surface between two planes perpendicular to the axis of the cone (shaded in Figure 2). If a frustum has base radii r_1 and r_2 and slant height l, then its area A is given by

$$A = 2\pi\left(\frac{r_1 + r_2}{2}\right)l = 2\pi(\text{average radius}) \cdot (\text{slant height})$$

The derivation of this result depends only on the formula for the area of a circle (see Problem 19).

Surface of revolution

FIGURE 1

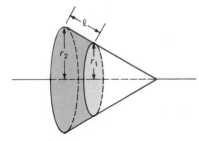

FIGURE 2

REVOLVING ABOUT THE x-AXIS Let $x = f(t)$, $y = g(t)$, $a \le t \le b$, determine a smooth curve in the upper half of the xy-plane, as shown in Figure 3. Partition the interval $[a,b]$ into n pieces by means of points $a = t_0 < t_1 < \cdots < t_n = b$, thereby also dividing the curve into n pieces. Let Δs_i denote the length of the typical piece and let y_i be the y-coordinate of a point on this piece. When the curve is revolved about the x-axis, it generates a surface, and the typical piece generates a narrow band. The "area" of this band ought to be approximately that of a frustum—that is, approximately $2\pi y_i \Delta s_i$. When we add the contributions of all the pieces and take the limit as the norm of the partition approaches zero, we get what we define to be

FIGURE 3

FIGURE 4

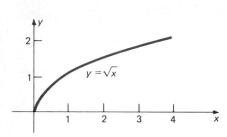

$y = \sqrt{x}$

FIGURE 5

the area of the surface of revolution. All this is indicated in Figure 4 and the boxed formula below.

$$A = \lim_{|P| \to 0} \sum_{i=1}^{n} 2\pi y_i \Delta s_i = \int_*^{**} 2\pi y \, ds$$

In using the formula for A, we must make an appropriate interpretation of y, ds, and the limits * and **. The discussion at the end of the previous section is helpful here. Thus, if the surface is obtained by revolving the curve $y = f(x)$, $a \le x \le b$, about the x-axis, the formula takes the form

$$A = 2\pi \int_*^{**} y \, ds = 2\pi \int_a^b f(x)\sqrt{1 + [f'(x)]^2} \, dx$$

EXAMPLE 1   Find the area of the surface of revolution generated by revolving the curve $y = \sqrt{x}$, $0 \le x \le 4$, about the x-axis (Figure 5).

Solution   Here, $f(x) = \sqrt{x}$ and $f'(x) = 1/(2\sqrt{x})$. Thus,

$$A = 2\pi \int_0^4 \sqrt{x}\sqrt{1 + \frac{1}{4x}} \, dx = 2\pi \int_0^4 \sqrt{x}\sqrt{\frac{4x + 1}{4x}} \, dx$$

$$= \pi \int_0^4 \sqrt{4x + 1} \, dx = \left[\pi \cdot \frac{1}{4} \cdot \frac{2}{3}(4x + 1)^{3/2} \right]_0^4$$

$$= \frac{\pi}{6}(17^{3/2} - 1^{3/2}) \approx 36.18 \qquad \blacksquare$$

If the curve is given parametrically by $x = f(t)$, $y = g(t)$, $a \le t \le b$, then the surface area formula becomes

$$A = 2\pi \int_*^{**} y \, ds = 2\pi \int_a^b g(t)\sqrt{[f'(t)]^2 + [g'(t)]^2} \, dt$$

EXAMPLE 2   Find the area of the surface generated when one arch of the cycloid

$$x = a(t - \sin t), \qquad y = a(1 - \cos t), \qquad 0 \le t \le 2\pi$$

is revolved about the x-axis.

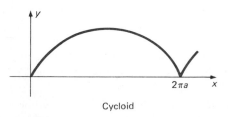

Cycloid

FIGURE 6

Solution   The cycloid was introduced in Problem 14 of Section 6.4. One arch of its graph is shown in Figure 6.

$$A = 2\pi \int_0^{2\pi} a(1 - \cos t)\sqrt{a^2(1 - \cos t)^2 + a^2 \sin^2 t} \, dt$$

$$= 2\pi a^2 \int_0^{2\pi} (1 - \cos t)\sqrt{2 - 2\cos t} \, dt = 2\sqrt{2}\pi a^2 \int_0^{2\pi} (1 - \cos t)^{3/2} \, dt$$

This integral looks complicated but, with the help of the half-angle formula $1 - \cos t = 2\sin^2(t/2)$, we press on.

$$A = 2\sqrt{2}\pi a^2 \int_0^{2\pi} \left[2\sin^2\left(\frac{t}{2}\right) \right]^{3/2} dt$$

$$= 8\pi a^2 \int_0^{2\pi} \sin^3\left(\frac{t}{2}\right) dt$$

$$= 8\pi a^2 \int_0^{2\pi} \left[1 - \cos^2\left(\frac{t}{2}\right) \right] \sin\left(\frac{t}{2}\right) dt$$

Now let $u = \cos(t/2)$, so that $du = -\frac{1}{2}\sin(t/2)\,dt$; note that $u = 1$ when $t = 0$ and $u = -1$ when $t = 2\pi$. Then

$$A = -16\pi a^2 \int_1^{-1} (1 - u^2)\,du = -16\pi a^2 \left[u - \frac{u^3}{3} \right]_1^{-1} = \frac{64\pi a^2}{3} \quad \blacksquare$$

FIGURE 7

REVOLVING ABOUT THE Y-AXIS   By analogy with the x-axis case and referring to Figure 7, we are led to the formula

$$\boxed{A = \int_*^{**} 2\pi x\,ds}$$

for the area of a surface obtained by revolving a curve about the y-axis.

EXAMPLE 3   Find the area of the surface obtained by revolving $x = \sqrt{a^2 - y^2}$, $-a \le y \le a$, about the y-axis (Figure 8).

Solution $x = g(y) = \sqrt{a^2 - y^2}$, so $g'(y) = -y/\sqrt{a^2 - y^2}$. Thus,

$$A = 2\pi \int_*^{**} x\,ds = 2\pi \int_{-a}^a \sqrt{a^2 - y^2} \sqrt{1 + \frac{y^2}{a^2 - y^2}}\,dy$$

$$= 2\pi \int_{-a}^a \sqrt{a^2}\,dy = \left[2\pi a y \right]_{-a}^a = 4\pi a^2$$

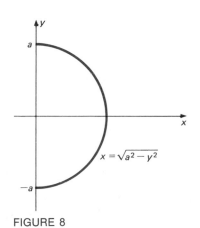

FIGURE 8

It is comforting to learn that the surface area of a sphere of radius a is $4\pi a^2$ in college calculus, just as it was in high school geometry. $\quad \blacksquare$

Since the curve in Example 3 was described by an equation of the form $x = f(y)$, it seemed most natural to express the complete integrand in terms of y, this resulting in a so-called y-integration. Other choices are possible. We could have expressed the complete integrand in terms of x, leading to an x-integration, or we could have used the parametric form $x = a \cos t$, $y = a \sin t$, for the semicircle, leading to a t-integration (see Problem 21).

EXAMPLE 4   Find the area of the surface obtained when $x = \sqrt{a^2 - y^2}$, $0 \le y \le a$, is revolved about the line $x = a$ (Figure 9).

FIGURE 9

Solution None of our formulas handles this problem directly; we return to basic principles. The advice *slice, approximate, integrate* is still useful.

$$\Delta A = 2\pi(a - x)\,\Delta s$$

$$A = 2\pi \int_{*}^{**} (a - x)\,ds$$

$$= 2\pi \int_{0}^{a} (a - \sqrt{a^2 - y^2})\sqrt{1 + \frac{y^2}{a^2 - y^2}}\,dy$$

$$= 2\pi \int_{0}^{a} \frac{a^2}{\sqrt{a^2 - y^2}}\,dy - 2\pi \int_{0}^{a} a\,dy$$

At this point in the course, we are unable to do the first integration. If instead of trying a y-integration, we use the parametric form $x = a\cos t$, $y = a\sin t$, $0 \le t \le \pi/2$, for the quarter-circle and then do a t-integration, we can complete the solution.

$$A = 2\pi \int_{*}^{**} (a - x)\,ds$$

$$= 2\pi \int_{0}^{\pi/2} (a - a\cos t)\sqrt{(-a\sin t)^2 + (a\cos t)^2}\,dt$$

$$= 2\pi a^2 \int_{0}^{\pi/2} (1 - \cos t)\,dt$$

$$= 2\pi a^2[t - \sin t]_{0}^{\pi/2} = 2\pi a^2[\pi/2 - 1] \qquad \blacksquare$$

PROBLEM SET 6.5

In Problems 1–10, find the area of the surface generated by revolving the given curve about the indicated axis.

1. $y = 6x$, $0 \le x \le 1$; about x-axis.

2. $x = 8y + 1$, $0 \le y \le 2$; about y-axis.

3. $y = \sqrt{25 - x^2}$, $-2 \le x \le 3$; about x-axis.

4. $y = x^3/3$, $1 \le x \le \sqrt{7}$; about x-axis.

5. $x = y^3$, $0 \le y \le 1$; about y-axis.

6. $y = x^2$, $0 \le x \le 2\sqrt{3}$; about y-axis.

7. $y = \frac{1}{2}x^2 - 1$, $0 \le x \le 2\sqrt{2}$; about y-axis.

8. $y = (x^6 + 2)/(8x^2)$, $1 \le x \le 3$; about x-axis.

9. $x = t$, $y = t^3$, $0 \le t \le 1$; about x-axis.

10. $x = 1 - t^2$, $y = 2t$, $0 \le t \le 1$; about x-axis.

In Problems 11–18, set up (but do not evaluate) an integral for the area of the surface generated by revolving the given curve about the indicated axis (see Example 4).

11. $y = \sin x$, $0 \le x \le \pi$; about $y = -1$.

12. $y = x^2 + 2x$, $1 \le x \le 3$; about $y = -2$.

13. $y = \sin x$, $0 \le x \le \pi$; about $x = 0$.

14. $y = \sqrt{x - 4}$, $4 \le x \le 8$; about $x = 2$.

15. $y = \sqrt{x^2 + x}$, $2 \le x \le 4$; about $x = 2$.

16. $y = x^4 + 2$, $0 \le x \le 2$; about $y = 2$.

17. $x = t^2$, $y = \sin t$, $0 \le t \le \pi$; about $y = -1$.

18. $x = a\cos t$, $y = a\sin t$, $0 \le t \le \pi/2$; about $x = -a$.

19. If the surface of a cone of slant height l and base radius r is cut along a lateral edge and laid flat, it becomes the sector of a circle of radius l and central angle θ (see Figure 10).

(a) Show that $\theta = 2\pi r/l$ radians.

(b) Use the formula $\frac{1}{2}l^2\theta$ for the area of a sector of radius l and central angle θ to show that the lateral surface area of a cone is πrl.

FIGURE 10

points of S are connected by line segments to the center of the earth (see Problem 25 of Section 6.3).

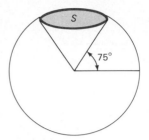

FIGURE 11

(c) Use the result of (b) to obtain the formula $A = 2\pi[(r_1 + r_2)/2]l$ for the lateral area of a frustum of a cone with base radii r_1 and r_2 and slant height l (see the discussion at the beginning of this section).

20. Show that the area of the part of the surface of a sphere of radius a between two parallel planes h units apart ($h < 2a$) is $2\pi ah$. Thus, show that if a right circular cylinder is circumscribed about a sphere, then two planes parallel to the base of the cylinder bound regions of the same area on the sphere and the cylinder.

21. Complete the solution to Example 3 in two new ways by setting up: (a) a t-integration with $x = a \cos t, y = a \sin t$ and (b) an x-integration. Be careful about limits.

C **22.** Assume the earth is a sphere of radius 4000 miles. Use the result of Problem 20 to find:

(a) the area of the polar cap S above the 75° parallel (Figure 11);
(b) the volume of the spherical sector obtained when the

23. The circle $x = a \cos t, \ y = a \sin t, \ 0 \le t \le 2\pi$, is revolved about the line $x = b, 0 < a < b$, thus generating a torus (doughnut). Find its surface area.

24. The ellipse $x^2/a^2 + y^2/b^2 = 1, \ a > b > 0$, is revolved about the x-axis. Find the surface area of the resulting football-shaped object called a *prolate spheroid*. (You will need Formula 54 from the back of the book to complete this derivation.) Evaluate when $b = \sqrt{2}a$.

25. Consider the curve $y = f(x) \ge 0, \ a \le x \le b$. Suppose this curve has length L_0 and, when revolved about the x-axis, generates a surface with area A_0.

(a) Express the corresponding area A_k for the curve $y = f(x) + k, k \ge 0$, in terms of L_0 and A_0.
(b) Use the result to find the surface area when the semicircle $y = 4 + \sqrt{9 - x^2}$ is revolved about the x-axis.

6.6 Work

We turn now to an application of the definite integral that does not concern length, area, or volume. In physics, we learn that if an object moves a distance d along a line while subjected to a *constant* force F in the direction of the motion, then the work W done by that force is given by

$$W = F \cdot d$$

that is,

$$\text{Work} = (\text{force}) \cdot (\text{distance})$$

If force is measured in pounds and distance in feet, then work is in foot-pounds. If force is in dynes (the force required to give a mass of 1 gram an acceleration of 1 centimeter per second per second) and distance is in centimeters, then work is in dyne-centimeters, also called ergs. If force is in newtons and distance is in meters, then work is in newton-meters, also called joules. For example, a man pushing a cart with a constant force of 150 pounds through a distance of 20 feet does $(150)(20) = 3000$ foot-pounds of

Force = 150 lb

—20 ft—

Work = (150)(20) = 3000 ft-lb

FIGURE 1

Δx

a b

x

$\Delta W \approx F(x)\,\Delta x$

$W = \int_a^b F(x)\,dx$

FIGURE 2

Natural length

0 1 2 3 4

Stretched x units

0 1 x

FIGURE 3

FIGURE 4

work (Figure 1). A woman lifting a weight (force) of 2 newtons a distance of 3 meters does $(2)(3) = 6$ joules of work.

In most practical situations, force is not constant, but rather varies as the object moves along the line. Suppose, in fact, that the object is being moved along the x-axis from a to b subject to a variable force of magnitude $F(x)$ at the point x, where F is a continuous function. Then how much work is done? Once again, the words *slice, approximate, integrate* lead us to an answer. Here *slice* means to partition the interval $[a, b]$ into small pieces; *approximate* means to suppose that on a typical piece from x to $x + \Delta x$, the force is constant with value $F(x)$, so that the bit of work done is $\Delta W \approx F(x)\,\Delta x$; *integrate* means to add up all the bits of work corresponding to the pieces Δx and then take the limit as the length of the pieces approaches zero (Figure 2). We conclude that the work done in moving the object from a to b is given by

$$W = \int_a^b F(x)\,dx$$

APPLICATION TO SPRINGS According to Hooke's Law in physics, the force $F(x)$ necessary to keep a spring stretched (or compressed) x units beyond (or short of) its natural length (Figure 3) is given by

$$F(x) = kx$$

Here, the constant k, the so-called spring constant, is positive and depends on the particular spring under consideration. The stiffer the spring, the greater the value of k.

EXAMPLE 1 If the natural length of a spring is 10 inches and if it takes a force of 3 pounds to keep it extended 2 inches, find the work done in stretching the spring from its natural length to a length of 15 inches.

Solution By Hooke's Law, the force $F(x)$ required to keep the spring stretched x inches is given by $F(x) = kx$. To evaluate k for this particular spring, we note that $F(2) = 3$. Thus, $k \cdot 2 = 3$, or $k = \frac{3}{2}$, and so

$$F(x) = \tfrac{3}{2}x$$

When the spring is at its natural length of 10 inches, $x = 0$; when it is 15 inches long, $x = 5$. Therefore, the work done in stretching the spring is

$$W = \int_0^5 \frac{3}{2}\,x\,dx = \left[\frac{3}{2} \cdot \frac{x^2}{2}\right]_0^5 = \frac{75}{4} = 18.75 \text{ inch-pounds} \quad\blacksquare$$

APPLICATION TO PUMPING A LIQUID To pump water out of a tank requires work, as anyone who has ever tried a hand pump will know (Figure 4). But how much work? That is the question.

While the problem does not quite fit within the previous discussion, its solution rests on the same basic principles. We illustrate.

EXAMPLE 2 A tank in the shape of a right circular cone (Figure 5) is full of water. If the height of the tank is 10 feet and the radius of its top is 4 feet, find the work done in pumping the water (a) over the top edge of the tank, and (b) to a height 10 feet above the top of the tank.

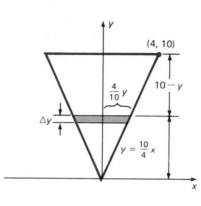

$$\Delta W \approx \delta\pi \left(\tfrac{4y}{10}\right)^2 (10 - y)\, \Delta y$$

$$W = \delta\pi \int_0^{10} \left(\tfrac{4y}{10}\right)^2 (10 - y)\, dy$$

FIGURE 5

Solution

(a) Place the tank in a coordinate system, as shown in Figure 5, both in a three-dimensional view and in a two-dimensional cross section. Imagine slicing the water into thin horizontal disks, each of which must be lifted over the edge of the tank. A disk of thickness Δy at height y has radius $4y/10$. Thus, its volume is approximately $\pi(4y/10)^2\,\Delta y$ and its weight is about $\delta\pi(4y/10)^2\,\Delta y$, where $\delta = 62.4$ is the (weight) density of water in pounds per cubic foot. The force necessary to lift this disk of water is its weight, and it must be lifted a distance $10 - y$. Thus, the work ΔW done on this disk is approximately

$$\Delta W = (\text{force}) \cdot (\text{distance}) \approx \delta\pi \left(\frac{4y}{10}\right)^2 \Delta y \cdot (10 - y)$$

$$W = \int_0^{10} \delta\pi \left(\frac{4y}{10}\right)^2 (10 - y)\, dy = \delta\pi \frac{4}{25} \int_0^{10} (10y^2 - y^3)\, dy$$

$$= \frac{(4\pi)(62.4)}{25}\left[\frac{10y^3}{3} - \frac{y^4}{4}\right]_0^{10} \approx 26{,}138 \text{ foot-pounds}$$

(b) Part b is just like part a, except that each disk of water must now be lifted a distance $20 - y$, rather than $10 - y$. Thus,

$$W = \delta\pi \int_0^{10} \left(\frac{4y}{10}\right)^2 (20 - y)\, dy = \frac{4\delta\pi}{25} \int_0^{10} (20y^2 - y^3)\, dy$$

$$= \frac{4(62.4)(\pi)}{25}\left[\frac{20y^3}{3} - \frac{y^4}{4}\right]_0^{10} \approx 130{,}690 \text{ foot-pounds}$$

Note that the limits are still 0 and 10 (not 0 and 20). Why? ∎

FIGURE 6

EXAMPLE 3 Find the work done in pumping the water over the rim of a tank, which is 50 feet long and has a semicircular end of radius 10 feet, if the tank is filled to a depth of 7 feet (Figure 6).

Solution We place the end of the tank in a coordinate system, as shown in Figure 7. A typical horizontal slice is shown both on this two-dimensional picture and on the three-dimensional one in Figure 6. This slice is approximately a thin box, so we calculate its volume by multiplying length, width, and thickness. Its weight is its density $\delta = 62.4$ times its volume. Finally we note that this slice must be lifted through a distance $-y$ (the minus sign results from the fact that y is negative in our diagram).

$$\Delta W \approx \delta (50) \left(2\sqrt{100 - y^2}\right) (\Delta y) (-y)$$

$$W = \delta \int_{-10}^{-3} 100\sqrt{100 - y^2}(-y)\, dy$$

FIGURE 7

$$W = 50\delta \int_{-10}^{-3} (100 - y^2)^{1/2}(-2y)\, dy = \left[(50\delta)(\tfrac{2}{3})(100 - y^2)^{3/2}\right]_{-10}^{-3}$$

$$= \tfrac{100}{3}(91)^{3/2}\delta \approx 1{,}805{,}606 \text{ foot-pounds.} \qquad \blacksquare$$

PROBLEM SET 6.6

1. A force of 8 pounds is required to keep a spring stretched $\frac{1}{2}$ foot beyond its normal length. Find the value of the spring constant and the work done in stretching the spring $\frac{1}{2}$ foot beyond its natural length.

2. For the spring of Problem 1, how much work is done in stretching the spring 1 foot?

3. A force of 200 dynes is required to keep a spring of natural length 10 centimeters compressed to a length of 8 centimeters. Find the work done in compressing the spring from its natural length to a length of 6 centimeters. (Hooke's Law applies to compressing as well as stretching.)

4. It requires 60 ergs (dyne-centimeters) of work to stretch a spring from a length of 8 centimeters to 9 centimeters and another 120 ergs to stretch it from 9 centimeters to 10 centimeters. Evaluate the spring constant and find the natural length of the spring.

5. For any spring obeying Hooke's Law, show that the work done in stretching a spring a distance d is given by $W = \frac{1}{2}kd^2$.

6. For a certain type of nonlinear spring, the force required to keep the spring stretched a distance s is given by the formula $F = ks^{4/3}$. If the force required to keep it stretched 8 inches is 1 pound, how much work is done in stretching this spring 27 inches?

7. A spring is such that the force required to keep it stretched s feet is given by $F = 11s$ pounds. How much work is done in stretching it 2 feet?

8. Two similar springs S_1 and S_2, each 3 feet long, are such that the force required to keep either of them stretched a distance of s feet is $F = 8s$ pounds. One end of one spring is fastened to an end of the other, and the combination is stretched between the walls of a room 12 feet wide (Figure 8). What work is done in moving the midpoint, P, 1 foot to the right?

FIGURE 8

In each of Problems 9–12, the vertical end of a tank is shown. Assume the tank is 10 feet long and that it is full of water and that the water is to be pumped to a height 5 feet above the top of the tank. Find the work done in emptying the tank.

9.

5 ft

4 ft

10.

3 ft

4 ft

11.

6 ft

4 ft

3 ft

12.

6 ft

13. Find the work done in pumping all the oil (density $\delta = 50$ pounds per cubic foot) over the edge of a cylindrical tank which stands on end. Assume that the radius of the base is 5 feet, that the height is 10 feet, and that the tank is full of oil.

14. Do Problem 13 assuming that the tank has circular cross sections of radius $5 + x$ feet at height x feet above the base.

15. A volume v of gas is confined in a cylinder, one end of which is closed by a movable piston. If A is the area in square inches of the face of the piston and x is the distance in inches from the cylinder head to the piston, then $v = Ax$.

The pressure of the confined gas is a continuous function p of the volume, and $p(v) = p(Ax)$ will be denoted by $f(x)$. Show that the work done by the piston in compressing the gas from a volume $v_1 = Ax_1$ to a volume $v_2 = Ax_2$ is

$$W = A \int_{x_2}^{x_1} f(x)\,dx$$

Hint: The total force on the face of the piston is $p(v) \cdot A = p(Ax) \cdot A = A \cdot f(x)$.

[C] **16.** A cylinder and piston, whose cross-sectional area is 1 square inch, contain 16 cubic inches of gas under a pressure of 40 pounds per square inch. If the pressure and the volume of the gas are related adiabatically (that is, without loss of heat) by the law $pv^{1.4} = c$ (a constant), how much work is done by the piston in compressing the gas to 2 cubic inches?

[C] **17.** If the area of the face of the piston in Problem 16 is 2 square inches, find the work done by the piston.

18. One cubic foot of air under a pressure of 80 pounds per square inch expands adiabatically to 4 cubic feet according to the law $pv^{1.4} = c$. Find the work done by the gas.

19. A cable weighing 2 pounds per foot is used to haul a 200-pound load to the top of a shaft that is 500 feet deep. How much work is done?

20. A 10-pound monkey hangs at the end of a 20-foot chain that weighs $\frac{1}{2}$ pound per foot. How much work does it do in climbing the chain to the top? Assume that the end of the chain is attached to the monkey.

21. A space capsule weighing 5000 pounds is propelled to an altitude of 200 miles above the surface of the earth. How much work is done against the force of gravity? Assume the earth is a sphere of radius 4000 miles and that the force of gravity is $f(x) = -k/x^2$, where x is the distance from the center of the earth to the capsule (the inverse-square law). Thus the lifting force required is k/x^2, and this equals 5000 when $x = 4000$.

22. According to Coulomb's Law, two like electrical charges repel each other with a force that is inversely proportional to the square of the distance between them. If the force of repulsion is 10 dynes when they are 2 centimeters apart, find the work done in bringing the charges from 5 centimeters apart to 1 centimeter apart.

23. A bucket weighing 100 pounds is filled with sand weighing 500 pounds. A crane lifts the bucket from the ground to a point 80 feet in the air at a rate of 2 feet per second, but sand simultaneously leaks out through a hole at 3 pounds per second. Neglecting friction and the weight of the cable, determine how much work is done. *Hint*: Begin by estimating ΔW, the work required to lift the bucket from y to $y + \Delta y$.

24. Center City has just built a new water tower (Figure 9). Its main elements consist of a spherical tank having an inner radius of 10 feet and weighing 10,000 pounds, four 40-foot pillars of linear density $100\cos(x^2/10,000)$ pounds per foot, with x measuring the number of feet from the base, and a 30-foot filler pipe weighting 40 pounds per foot. Initially, all elements were on the ground with the pillars and pipe lying flat. The tank was lifted to position and then the pillars and pipe were pivoted about their bases to the vertical. Find the work done.

FIGURE 9

25. The filler pipe for the water tower of Problem 24 has inner diameter 1 foot. How much work was done in filling the pipe and the tank with water assuming it was pumped from ground level up through the pipe into the tank?

26. A conical buoy weighs m pounds and floats with its vertex V down and h feet below the surface of the water (Figure 10). A boat crane lifts the buoy to the deck so that V is 15 feet above the water surface. How much work is done? *Hint:* You will need Archimedes's Principle, which says that the force required to hold the buoy y feet above its original position ($0 \le y \le h$) is equal to its weight minus the weight of the water displaced by the buoy.

FIGURE 10

27. Rather than lifting the buoy of Problem 26 and Figure 10 out of the water, suppose we attempt to push it down till its top is even with the water level. Assume $h = 8$, that the top is originally 2 feet above water level, and that the buoy weighs 300 pounds. How much work is required? *Hint:* You do not need to know a (the radius at water level), but it is helpful to know that $\delta(\frac{1}{3}\pi a^2)(8) = 300$. Archimedes's Principle says that the force needed to hold the buoy z feet ($0 \le z \le 2$) below floating position is equal to the weight of the additional water displaced.

6.7
Fluid Force

Imagine the tank shown in Figure 2 to be filled to a depth h with a fluid of density δ. Then the force on a horizontal rectangle of area A on the bottom is equal to the weight of the column of fluid that stands directly over that rectangle (Figure 1), that is,

$$F = \delta h A$$

$F = \delta h A$

FIGURE 1

It is a fact, first stated by Blaise Pascal (1623–1662), that the pressure (force per unit area) exerted by a fluid is the same in all directions. Thus, the pressure at all points on a surface, whether that surface is horizontal, vertical, or at some other angle, is the same—provided the points are at the same depth. In particular, the force against each of the three small rectangles in Figure 2 is approximately the same (assuming they have the same area). We say "approximately the same," because not all points of the two side rectangles are at the same depth—though the narrower these rectangles are, the closer this is to being true. It is this approximation that allows us to calculate the total force against one end of the tank.

FIGURE 2

EXAMPLE 1 Suppose that the vertical end of the tank has the shape shown in Figure 3 and that the tank is filled with water (δ = 62.4 pounds per cubic foot) to a depth of 5 feet. Find the total force against the end.

Solution Place the end of the tank in the coordinate system as shown in Figure 4. Note that the right edge has slope 3 and hence has equation $y - 0 = 3(x - 8)$ or, equivalently, $x = \frac{1}{3}y + 8$. The force against a narrow rectangle at depth $5 - y$ is approximately $\delta h A = \delta(5 - y)(\frac{1}{3}y + 8)\Delta y = \delta(40 - \frac{19}{3}y - \frac{1}{3}y^2)\Delta y$. Thus,

$$F = \delta \int_0^5 (40 - \tfrac{19}{3}y - \tfrac{1}{3}y^2)\, dy = \delta[40y - \tfrac{19}{6}y^2 - \tfrac{1}{9}y^3]_0^5$$

$$= 62.4(200 - \tfrac{475}{6} - \tfrac{125}{9}) \approx 6673 \text{ pounds}$$

FIGURE 3

FIGURE 4

$$\Delta F \approx \delta\,(5 - y)\left(\tfrac{1}{3}y + 8\right)\Delta y$$

$$F = \int_0^5 \delta\,(5 - y)\left(\tfrac{1}{3}y + 8\right)dy$$

■

EXAMPLE 2 A cylindrical barrel half full of oil is lying on its side (Figure 5). If each end is circular, 8 feet in diameter, find the total force against one end. Assume the density of oil is δ = 50 pounds per cubic foot.

Solution Place the circular end in the coordinate system, as shown in Figure 6. Then proceed as in Example 1.

$$F = \delta \int_{-4}^0 (16 - y^2)^{1/2}(-2y\, dy) = \delta[\tfrac{2}{3}(16 - y^2)^{3/2}]_{-4}^0$$

$$= (50)(\tfrac{2}{3})(16)^{3/2} \approx 2133 \text{ pounds}$$

FIGURE 5

FIGURE 6

$$\Delta F \approx \delta\,(-y)\,(2\sqrt{16 - y^2})\,\Delta y$$

$$F = \int_{-4}^0 \delta\,(-y)\,(2\sqrt{16 - y^2})\,dy$$

■

FIGURE 7

FIGURE 8

EXAMPLE 3 Suppose that the barrel of Example 2 is full of oil, including a filler pipe that extends 2 feet above the top of the barrel (Figure 7 shows an end view). Find the total force against one end of the barrel.

Solution It is a remarkable fact (Pascal's Principle) that the force against the shaded strip depends only on the height of the oil level above the strip (not on the size or shape of the filler pipe, assuming friction between the oil and the pipe can be ignored). Thus,

$$\Delta F \approx \delta(6 - y)(2\sqrt{16 - y^2})\Delta y$$

and

$$F = \int_{-4}^{4} \delta(6 - y)(2\sqrt{16 - y^2})\,dy$$

$$= 12\delta \int_{-4}^{4} \sqrt{16 - y^2}\,dy - 2\delta \int_{-4}^{4} y\sqrt{16 - y^2}\,dy$$

The first integrand is even and the second is odd. Therefore (Section 5.8),

$$F = 24\delta \int_{0}^{4} \sqrt{16 - y^2}\,dy$$

$$= 24\delta(\text{area of a quarter circle})$$

$$= (24)(50)(4\pi) \approx 15{,}080 \text{ pounds} \quad \blacksquare$$

EXAMPLE 4 The water side of a dam is a 200 foot by 100 foot rectangle inclined at 60° from the horizontal, as shown in Figure 8. Find the total force against the dam when the water level is at the top of the dam.

Solution Place the end of the dam in the coordinate system, as shown in Figure 9. Note that the vertical height of the dam is 100 sin 60° ≈ 86.6 feet.

$$F = (62.4)(200)(1.155)\int_{0}^{86.6} (86.6 - y)\,dy$$

$$= (62.4)(200)(1.155)\left[86.6y - \frac{y^2}{2}\right]_{0}^{86.6} \approx 54{,}100{,}000 \text{ pounds}$$

FIGURE 9

$\blacksquare$

In Problems 1–8, assume that the shaded region is part of a vertical side of a tank with water ($\delta = 62.4$ pounds per cubic foot) at the level shown. Find the total force against this region.

1.

Water level

3 ft

6 ft

2.

2 ft

3 ft

6 ft

3.

6 ft

6 ft   6 ft

4.

3 ft

3 ft

4 ft   5 ft

5.

y

(1, 1)

$y = x^2$

x

6.

y

(9, 3)

$y = \sqrt{|x|}$

x

7.

y

x

$\dfrac{x^2}{4} + y^2 = 1$

8.

2 ft

4 ft

9. Show that if a vertical dam in the shape of a rectangle is divided in half by means of a diagonal, the total force on one half of the dam is twice that on the other half. Assume the top edge of the dam is even with the surface of the water.

10. Find the total force on all sides of a cube of edge length 2 feet if its top is horizontal and 100 feet below the surface of a lake.

11. Find the total force against the bottom of the swimming pool shown in Figure 10, assuming it is full of water.

8 ft

4 ft

20 ft

10 ft

FIGURE 10

12. Find the total force against the lateral surface of a right circular cylinder of height 6 feet, which stands on its circular base of radius 5 feet, if it is filled with oil ($\delta = 50$ pounds per cubic foot).

13. Find the total force against the lateral surface of a conical tank of height 6 feet if the radius of the top is 3 feet and it is full of water.

14. Find the total force against the surface of an open hemispherical tank of radius a feet that is full of water.

15. The vertical end of a tank is bounded by the curves $x = \pm a \cos(\pi y^2 / 2b)$ and the x-axis, $-\sqrt{b} \le y \le 0$, $a > 0$, $b > 0$. If the tank is full of a liquid of weight density δ, find the total force against the end.

16. A circular plate of radius a feet makes an angle θ ($0 \le \theta \le \pi/2$) with the vertical and is submerged in water with its center h feet below the surface. Show that the total force against the upper side of the plate is $\delta h \pi a^2$, which equals pressure at the center times the area of the plate.

17. A vertical annular plate with inner radius a and outer radius b (in feet) is submerged in water with its center h feet below the surface as shown in Figure 11. Show that the total force against one side of this plate is $\delta h \pi (b^2 - a^2)$, which equals pressure at the center times the area of the plate.

18. On the basis of Problems 16 and 17, make a conjecture and check it on Problems 1 and 2.

FIGURE 11

6.8
Moments, Center of Mass

Suppose that two masses of sizes m_1 and m_2 are placed on a seesaw at distances d_1 and d_2 from the fulcrum and on opposite sides of it (Figure 1). The seesaw will balance if and only if $d_1 m_1 = d_2 m_2$.

A good mathematical model for this situation is obtained by replacing the seesaw with a horizontal coordinate line having its origin at the fulcrum (Figure 2). Then the coordinate x_1 of m_1 is $x_1 = -d_1$, that of m_2 is $x_2 = d_2$, and the condition for balance is

$$x_1 m_1 + x_2 m_2 = 0$$

FIGURE 1

FIGURE 2

The product of the mass m of a particle by its *directed* distance from a point (its lever arm) is called the **moment** of the particle with respect to that point (Figure 3). It measures the tendency of the mass to produce a rotation about that point. The condition for two masses along a line to balance at a point on that line is that the sum of their moments with respect to the point be zero.

The situation just described generalizes. The total moment M (with respect to the origin) of a system of n masses of sizes $m_1, m_2, \ldots, m_n$ located at points $x_1, x_2, \ldots, x_n$ along the x-axis is the sum of individual moments—that is,

Moment = (lever arm) · (mass)

$M = x \cdot m$

FIGURE 3

$$M = x_1 m_1 + x_2 m_2 + \cdots + x_n m_n = \sum_{i=1}^{n} x_i m_i$$

The condition for balance at the origin is that $M = 0$. Of course, we should not expect balance at the origin except in special circumstances. But surely any system of masses will balance somewhere. The question is where. What is the x-coordinate of the point where the fulcrum should be placed to make the system in Figure 4 balance?

FIGURE 4

Call the desired coordinate $\bar{x}$. The total moment with respect to it should be zero—that is,

$$(x_1 - \bar{x})m_1 + (x_2 - \bar{x})m_2 + \cdots + (x_n - \bar{x})m_n = 0$$

or

$$x_1 m_1 + x_2 m_2 + \cdots + x_n m_n = \bar{x}m_1 + \bar{x}m_2 + \cdots + \bar{x}m_n$$

When we solve for $\bar{x}$, we obtain

$$\bar{x} = \frac{M}{m} = \frac{\displaystyle\sum_{i=1}^{n} x_i m_i}{\displaystyle\sum_{i=1}^{n} m_i}$$

The point $\bar{x}$, called the **center of mass**, is the balance point. Notice that it is just the total moment with respect to the origin divided by the total mass.

FIGURE 5

EXAMPLE 1   Masses of 4, 2, 6, and 7 pounds are located at points 0, 1, 2, and 4, respectively, along the x-axis (Figure 5). Find the center of mass.

Solution

$$\bar{x} = \frac{(0)(4) + (1)(2) + (2)(6) + (4)(7)}{4 + 2 + 6 + 7} = \frac{42}{19} \approx 2.21$$

Your intuition should confirm that $x = 2.21$ is about right for the balance point.   ■

CONTINUOUS MASS DISTRIBUTION ALONG A LINE   Consider now a straight segment of thin wire of varying density (mass per unit length) for which we desire to find the balance point. We impose a coordinate line along the wire and suppose the density at x is $\delta(x)$. (For the meaning of density at a point, see Problem 19 of Section 3.1.) Following our standard procedure (*slice, approximate, integrate*), we first obtain the total mass m and then the total moment M with respect to the origin (Figure 6).

$$\triangle m \approx \delta(x)\,\triangle x \qquad \triangle M \approx x\,\delta(x)\,\triangle x$$

$$m = \int_a^b \delta(x)\,dx \qquad M = \int_a^b x\,\delta(x)\,dx$$

FIGURE 6

This leads to the formula

$$\bar{x} = \frac{M}{m} = \frac{\displaystyle\int_a^b x\delta(x)\,dx}{\displaystyle\int_a^b \delta(x)\,dx}$$

Two comments are in order. First, remember this formula by analogy with the point-mass formula

$$\frac{\sum x_i m_i}{\sum m_i} \sim \frac{\sum x\,\Delta m}{\sum \Delta m} \sim \frac{\int x\delta(x)\,dx}{\int \delta(x)\,dx}$$

Second, note that we have assumed that moments of small pieces of wire add together to give the total moment, just as was the case for point masses. This should seem reasonable to you if you imagine the mass of the typical piece of length Δx to be concentrated at the point x.

EXAMPLE 2   The density $\delta(x)$ of a wire at the point x centimeters from one end is given by $\delta(x) = 3x^2$ grams per centimeter. Find the center of mass of the piece between $x = 0$ and $x = 10$.

Solution   We expect $\bar{x}$ to be nearer 10 than 0, since the wire is much heavier (denser) toward the right end (Figure 7).

0   10

FIGURE 7

$$\bar{x} = \frac{\displaystyle\int_0^{10} x \cdot 3x^2\,dx}{\displaystyle\int_0^{10} 3x^2\,dx} = \frac{[3x^4/4]_0^{10}}{[x^3]_0^{10}} = \frac{7500}{1000} = 7.5 \text{ cm}$$ ∎

MASS DISTRIBUTIONS IN THE PLANE   Consider n point masses of sizes $m_1, m_2, \ldots, m_n$ situated at points $(x_1, y_1), (x_2, y_2), \ldots, (x_n, y_n)$ in the coordinate plane (Figure 8). Then the total moments M_y and M_x with respect to the y-axis and x-axis, respectively, are given by

$$M_y = \sum_{i=1}^n x_i m_i \qquad M_x = \sum_{i=1}^n y_i m_i$$

The coordinates $(\bar{x}, \bar{y})$ of the center of mass (balance point) are

$$\bar{x} = \frac{M_y}{m} = \frac{\displaystyle\sum_{i=1}^n x_i m_i}{\displaystyle\sum_{i=1}^n m_i} \qquad \bar{y} = \frac{M_x}{m} = \frac{\displaystyle\sum_{i=1}^n y_i m_i}{\displaystyle\sum_{i=1}^n m_i}$$

m_1
$\bullet$
(x_1, y_1)

m_n
$\bullet$
(x_n, y_n)

m_2
$\bullet$
(x_2, y_2)

m_4
$\bullet$
(x_4, y_4)

m_3
$\bullet$
(x_3, y_3)

FIGURE 8

CHAPTER 6/Applications of the Integral

(a)

(b)

FIGURE 9

EXAMPLE 3 Five particles, having masses 1, 4, 2, 3, and 2 units are located at points $(6, -1)$, $(2, 3)$, $(-4, 2)$, $(-7, 4)$, and $(2, -2)$, respectively. Find the center of mass.

Solution

$$\bar{x} = \frac{(6)(1) + (2)(4) + (-4)(2) + (-7)(3) + (2)(2)}{1 + 4 + 2 + 3 + 2} = \frac{-11}{12}$$

$$\bar{y} = \frac{(-1)(1) + (3)(4) + (2)(2) + (4)(3) + (-2)(2)}{1 + 4 + 2 + 3 + 2} = \frac{23}{12}$$ ∎

We next consider the problem of finding the center of mass of a lamina (thin planar sheet). For simplicity, we suppose it is homogeneous—that is, that it has constant mass density δ. For a homogeneous rectangular sheet, the center of mass is at the geometric center, as diagrams (a) and (b) in Figure 9 suggest.

Consider the homogeneous lamina bounded by $x = a$, $x = b$, $y = f(x)$, and $y = g(x)$, with $g(x) \le f(x)$. *Slice* this lamina into narrow strips parallel to the y-axis, which are therefore nearly rectangular in shape, and imagine the mass of each strip to be concentrated at its geometric center. Then *approximate* and *integrate* using what has come to be a standard procedure (Figure 10). From this we can calculate the coordinates $(\bar{x}, \bar{y})$ of the center of mass using the formulas

$$\bar{x} = \frac{M_y}{m} \qquad \bar{y} = \frac{M_x}{m}$$

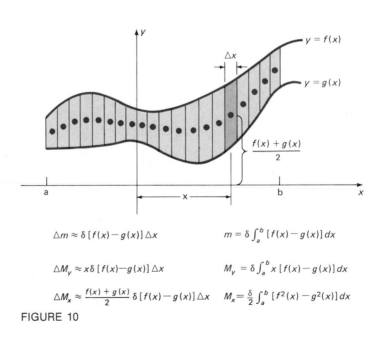

$$\Delta m \approx \delta\,[f(x) - g(x)]\,\Delta x \qquad m = \delta \int_a^b [f(x) - g(x)]\,dx$$

$$\Delta M_y \approx x\delta\,[f(x) - g(x)]\,\Delta x \qquad M_y = \delta \int_a^b x\,[f(x) - g(x)]\,dx$$

$$\Delta M_x \approx \frac{f(x) + g(x)}{2}\,\delta\,[f(x) - g(x)]\,\Delta x \qquad M_x = \frac{\delta}{2} \int_a^b [f^2(x) - g^2(x)]\,dx$$

FIGURE 10

When we do, the factor δ cancels between numerator and denominator, and we obtain

$$\bar{x} = \frac{\displaystyle\int_a^b x[f(x) - g(x)]\,dx}{\displaystyle\int_a^b [f(x) - g(x)]\,dx} \qquad \bar{y} = \frac{\dfrac{1}{2}\displaystyle\int_a^b [f^2(x) - g^2(x)]\,dx}{\displaystyle\int_a^b [f(x) - g(x)]\,dx}$$

Sometimes, slicing parallel to the x-axis works better than slicing parallel to the y-axis. This leads to formulas for $\bar{x}$ and $\bar{y}$ in which y is the variable of integration. Do not try to memorize all these formulas. It is much better to remember how they were derived.

The center of mass of a homogeneous lamina does not depend on its density or its mass, but only on the shape of the corresponding region in the plane. Thus, our problem becomes a geometric problem rather than a physical one. Accordingly, we often speak of the **centroid** of a planar region rather than the center of mass of a homogeneous lamina.

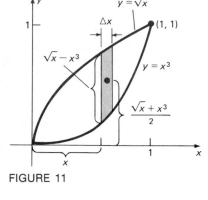

FIGURE 11

EXAMPLE 4 Find the centroid of the region bounded by the curves $y = x^3$ and $y = \sqrt{x}$.

Solution Note the diagram in Figure 11.

$$\bar{x} = \frac{\displaystyle\int_0^1 x[\sqrt{x} - x^3]\,dx}{\displaystyle\int_0^1 [\sqrt{x} - x^3]\,dx} = \frac{\left[\dfrac{2}{5}x^{5/2} - \dfrac{x^5}{5}\right]_0^1}{\left[\dfrac{2}{3}x^{3/2} - \dfrac{x^4}{4}\right]_0^1} = \frac{\dfrac{1}{5}}{\dfrac{5}{12}} = \frac{12}{25}$$

$$\bar{y} = \frac{\displaystyle\int_0^1 \dfrac{1}{2}(\sqrt{x} + x^3)(\sqrt{x} - x^3)\,dx}{\displaystyle\int_0^1 (\sqrt{x} - x^3)\,dx} = \frac{\dfrac{1}{2}\displaystyle\int_0^1 [(\sqrt{x})^2 - (x^3)^2]\,dx}{\displaystyle\int_0^1 (\sqrt{x} - x^3)\,dx}$$

$$= \frac{\dfrac{1}{2}\left[\dfrac{x^2}{2} - \dfrac{x^7}{7}\right]_0^1}{\dfrac{5}{12}} = \frac{\dfrac{5}{28}}{\dfrac{5}{12}} = \frac{3}{7}$$

These answers should seem reasonable to you. ■

FIGURE 12

EXAMPLE 5 Find the centroid of the region under the curve $y = \sin x$, $0 \le x \le \pi$ (Figure 12).

Solution This region is symmetric about the line $x = \pi/2$, from which we conclude (without an integration) that $\bar{x} = \pi/2$. In fact, it is both intuitively obvious and true that if a region has a vertical or horizontal line of symmetry, then the centroid will lie on that line.

Your intuition should also tell you that $\bar{y}$ will be less than $\frac{1}{2}$, since more of the area is near the x-axis. But to find this number exactly, we must calculate

$$\bar{y} = \frac{\displaystyle\int_0^\pi \frac{1}{2}\sin x \cdot \sin x\,dx}{\displaystyle\int_0^\pi \sin x\,dx} = \frac{\displaystyle\frac{1}{2}\int_0^\pi \sin^2 x\,dx}{\displaystyle\int_0^\pi \sin x\,dx}$$

The denominator is easy to calculate; it has value 2. To calculate the numerator, we use the half-angle formula $\sin^2 x = (1 - \cos 2x)/2$.

$$\int_0^\pi \sin^2 x\,dx = \frac{1}{2}\left(\int_0^\pi 1\,dx - \int_0^\pi \cos 2x\,dx \right)$$

$$= \frac{1}{2}\left[x - \frac{1}{2}\sin 2x \right]_0^\pi = \frac{\pi}{2}$$

Thus,

$$\bar{y} = \frac{\dfrac{1}{2}\cdot\dfrac{\pi}{2}}{2} \approx 0.39$$ ∎

THE THEOREM OF PAPPUS About A.D 300, the Greek geometer Pappus stated a novel result, which connects centroids with volumes of solids of revolution (Figure 13).

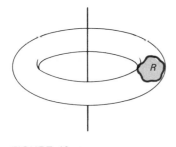

FIGURE 13

> *Theorem A*
>
> **(Pappus's Theorem).** If a region R, lying on one side of a line in its plane, is revolved about that line, then the volume of the resulting solid is equal to the area of R multiplied by the distance traveled by its centroid.

Rather than prove this theorem, which is really quite easy (see Problem 22), we illustrate it.

EXAMPLE 6 Illustrate the correctness of Pappus's Theorem for the region under $y = \sin x$, $0 \le x \le \pi$, when it is revolved about the x-axis (Figure 14).

Solution This is the region of Example 5, for which $\bar{y} = \pi/8$. The area A of this region is

$$A = \int_0^\pi \sin x\,dx = [-\cos x]_0^\pi = 2$$

FIGURE 14

The volume V of the corresponding solid of revolution is

$$V = \pi \int_0^\pi \sin^2 x\, dx = \frac{\pi}{2} \int_0^\pi [1 - \cos 2x]\, dx = \frac{\pi}{2} \left[x - \frac{1}{2}\sin 2x \right]_0^\pi = \frac{\pi^2}{2}$$

We must show that

$$A(2\pi\bar{y}) = V$$

But this amounts to showing

$$2\left(2\pi\frac{\pi}{8}\right) = \frac{\pi^2}{2}$$

which is clearly true.   ■

PROBLEM SET 6.8

1. Particles of mass $m_1 = 4$, $m_2 = 6$, and $m_3 = 9$ are located at $x_1 = 2$, $x_2 = -2$, and $x_3 = 1$ along a line. Where is the center of mass?

2. John and Mary—weighing 150 and 120 pounds, respectively—sit at opposite ends of a 12-foot teeter board with the fulcrum in the middle. Where should their 80-pound son Tom sit in order for the board to balance?

3. A straight wire 9 units long has density $\delta(x) = \sqrt{x}$ at a point x units from one end. Find the distance from this end to the center of mass.

4. Do Problem 3 if $\delta(x) = 1 + x^2$.

5. The masses and coordinates of a system of particles in the coordinate plane are given by the following: 3, (1, 1); 2, (7, 1); 4, (−2, −5); 6, (−1, 0); 2, (4, 6). Find the moments of this system with respect to the coordinate axes, and find the coordinates of the center of mass.

6. The masses and coordinates of a system of particles are given by the following: 3, (−3, 2); 6, (−2, −2); 2, (3, 5); 5, (4, 3); 1, (7, −1). Find the moments of this system with respect to the coordinate axes, and find the coordinates of the center of mass.

In Problems 7–10, divide the indicated region into rectangular pieces and assume that the moments M_x and M_y of the whole region can be found by adding the corresponding moments of the pieces. Use this to find the centroid of each region.

7.

8.

9.

10.

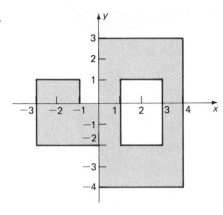

In Problems 11–18, find the centroid of the region bounded by the given curves. Make a sketch and use symmetry where possible.

11. $y = 4 - x^2$, $y = 0$

12. $y = \frac{1}{2}x^2$, $y = 0$, $x = 4$

13. $y = x^3$, $y = 0$, $x = 2$

14. $y = \frac{1}{2}(x^2 - 10)$, $y = 0$, and between $x = -2$ and $x = 3$

15. $y = 2x - 4$, $y = 2\sqrt{x}$, $x = 0$

16. $y = x^2$, $y = x + 2$

17. $x = y^2$, $x = 4$

18. $x = y^2 - 3y - 4$, $x = -y - 1$

19. Use Pappus's Theorem to find the volume of the solid obtained when the region bounded by $y = x^3$, $y = 0$, and $x = 2$ is revolved about the y-axis (see Problem 13 for the centroid). Do the same problem by cylindrical shells to check on your answer.

20. Use Pappus's Theorem to find the volume of the torus obtained when the region inside the circle $x^2 + y^2 = a^2$ is revolved about the line $x = 2a$.

21. Use Pappus's Theorem together with the known volume of a sphere to find the centroid of a semicircular region of radius a.

22. Prove Pappus's Theorem by assuming that the region of area A in Figure 15 is to be revolved about the y-axis. *Hint*: $V = 2\pi \displaystyle\int_a^b xh(x)\, dx$ and $\bar{x} = \displaystyle\int_a^b xh(x)\, dx / A$.

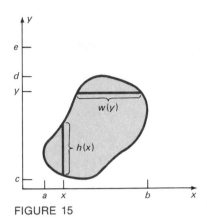

FIGURE 15

23. Prove: If a vertical plate of area A is submerged in a fluid of weight density δ with its centroid h units below the surface, then the total force against one side is $\delta h A$; that is, it is the pressure at the centroid times the area (compare with Problem 18 of Section 6.7). *Hint*: Use Figure 15 and assume the fluid level is at e.

24. Consider the triangle T of Figure 16.

(a) Show that $\bar{y} = h/3$ (and thus that the centroid of a triangle is at the intersection of the medians).
(b) Find the volume of the solid obtained when T is revolved around $y = k$ (Pappus's Theorem).
(c) Find the total force against T if it is submerged in a fluid of weight density δ with the surface at $y = k$ (Problem 23).

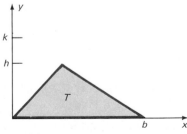

FIGURE 16

25. A regular polygon P of $2n$ sides is inscribed in a circle of radius r.

(a) Find the volume of the solid obtained when P is revolved about one of its sides.

(b) Find the total force against P if P is the vertical end of a tank full of a fluid of weight density δ (assume one of the corners is directly above the centroid).

(c) Check your answers by letting $n \to \infty$.

26. Let f be a nonnegative continuous function on $[0, 1]$.

(a) Show that $\int_0^\pi xf(\sin x)\,dx = (\pi/2)\int_0^\pi f(\sin x)\,dx$.

(b) Use part (a) to evaluate $\int_0^\pi x \sin x \cos^4 x\,dx$.

27. Let $0 \le f(x) \le g(x)$ for all x in $[0, 1]$ and let R and S be the regions under the graphs of f and g, respectively. Prove or disprove that $\bar{y}_R \le \bar{y}_S$.

6.9 Chapter Review

TRUE-FALSE QUIZ

Respond with true or false to each of the following assertions. Be prepared to defend your answer.

1. The area of the region bounded by $y = \cos x$, $y = 0$, $x = 0$, and $x = \pi$ is $\int_0^\pi \cos x\,dx$.

2. The area of a circle of radius a is $4\int_0^a \sqrt{a^2 - x^2}\,dx$.

3. The area of the region bounded by $y = f(x)$, $y = g(x)$, $x = a$, and $x = b$ is either $\int_a^b [f(x) - g(x)]\,dx$ or its negative.

4. All right cylinders whose bases have the same area and whose heights are the same have identical volumes.

5. If two solids with bases in the same plane have cross sections of the same area in all planes parallel to their bases, then they have the same volume.

6. If the radius of the base of a cone is doubled while the height is halved, the volume will remain the same.

7. To calculate the volume of the solid obtained by revolving the region bounded by $y = -x^2 + x$ and $y = 0$ about the y-axis, one should use the method of washers in preference to the method of shells.

8. The solids obtained by revolving the region of Problem 7 about $x = 0$ and $x = 1$ have the same volume.

9. Any smooth curve in the plane that lies entirely within the unit circle will have finite length.

10. The work required to stretch a spring 2 inches beyond its natural length is twice that required to stretch it 1 inch (assume Hooke's Law).

11. The force on a 2-inch-diameter round drain plug at the bottom of a 20-foot deep swimming pool is the same as the force on the bottom end of a 20-foot vertical 2-inch-diameter pipe that is filled with water.

12. It will require the same amount of work to empty a cone-shaped tank and a cylindrical tank of water by pumping it over the rim if both tanks have the same height and volume.

13. Two weights of 100 pounds at distances 10 and 15 feet from the fulcrum will just balance a 200-pound weight on the other side of the fulcrum and 12.5 feet from it.

14. If $\bar{x}$ is the center of mass of a system of masses $m_1, m_2, \ldots, m_n$ distributed along a line at points with coordinates $x_1, x_2, \ldots, x_n$, respectively, then $\sum_{i=1}^n (x_i - \bar{x})m_i = 0$.

15. The centroid of the region bounded by $y = \cos x$, $y = 0$, $x = 0$, and $x = 2\pi$ is at $(\pi, 0)$.

16. According to the theorem of Pappus, the volume of the solid obtained by revolving the region (of area 2) bounded by $y = \sin x$, $y = 0$, $x = 0$, and $x = \pi$ about the y-axis is $2(2\pi)\left(\dfrac{\pi}{2}\right) = 2\pi^2$.

17. The area of the region bounded by $y = \sqrt{x}$, $y = 0$, and $x = 9$ is $\int_0^3 (9 - y^2)\,dy$.

18. If the density of a wire is proportional to the square of the distance from its midpoint, then its center of mass is at the midpoint.

19. The centroid of a triangle with base on the x-axis has y-coordinate equal to one-third the altitude of the triangle.

20. The total force against a square gate in the face of a dam is divided equally between the two triangles above and below a diagonal.

SAMPLE TEST PROBLEMS

Problems 1–7 refer to the plane region R bounded by the curve $y = x - x^2$ and the x-axis (Figure 1).

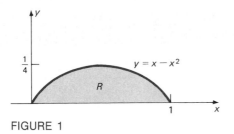

FIGURE 1

1. Find the area of R.

2. Find the volume of the solid S_1 generated by revolving the region R about the x-axis.

3. Use the cylindrical shell method to find the volume of the solid S_2 generated by revolving R about the y-axis.

4. Find the volume of the solid S_3 generated by revolving R about the line $y = -1$.

5. Find the volume of the solid S_4 generated by revolving R about the line $x = 2$.

6. Find the coordinates of the centroid of R.

7. Use Pappus's Theorem and Problems 1 and 6 to find the volume of the solids S_1, S_2, S_3, and S_4 above.

8. The natural length of a certain spring is 16 inches, and a force of 8 pounds is required to keep it stretched 4 inches. Find the work done in each case.

(a) Stretching it from a length of 18 inches to a length of 24 inches.
(b) Compressing it from its natural length to a length of 14 inches.

9. An upright cylindrical tank is 10 feet in diameter and 10 feet high. If water in the tank is 8 feet deep, how much work is done in pumping all the water over the edge of the top of the tank?

10. An object weighing 300 pounds is suspended from the top of a building by a uniform cable. If the cable is 100 feet long and weighs 120 pounds, how much work is done in pulling the object to the top?

11. A region R is bounded by the line $y = 3x$ and the parabola $y = x^2$. Find the area of R by (a) taking x as the integration variable, and (b) taking y as the integration variable.

12. Find the centroid of R in Problem 11.

13. Find the volume of the solid of revolution generated by revolving the region R of Problem 11 about the x-axis. Check by using Pappus's Theorem.

14. Find the volume of the solid generated by revolving the region R of Problem 11 about the y-axis. Check by using Pappus's Theorem.

15. Each end of a tank is an inverted isosceles triangle 10 feet wide and 8 feet deep (Figure 2(a)). What is the total force on one end when the tank is full of water?

FIGURE 2

16. Each end of a tank is an inverted right triangle with its 10-foot horizontal base at the top and the adjacent vertical side 8 feet long (Figure 2(b)). Find the force on one end when the tank is full of water.

17. Find the length of the arc of the curve $y = x^3/3 + 1/(4x)$ from $x = 1$ to $x = 3$.

18. Find the area of the surface generated by revolving the segment of the curve $y = x^3/6 + 1/(2x)$ from $x = 1$ to $x = 3$ about the x-axis.

19. A rectangular swimming pool is 30 feet long and 15 feet wide. If the depth of the water in the pool varies linearly from 3 feet at one end to 8 feet at the other end, find the total force of the water on a side of the pool.

20. Sketch the graph of the parametric equations

$$x = t^2, \qquad y = \tfrac{1}{3}(t^3 - 3t)$$

Then find the length of the loop of the resulting curve.

21. A solid with the semicircular base bounded by $y = \sqrt{4 - x^2}$ and $y = 0$ has cross sections perpendicular to the x-axis that are squares. Find the volume of this solid.

Each of Problems 22–27 refers to Figure 3.

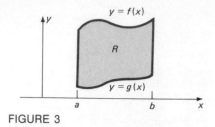

FIGURE 3

In each case, write an expression involving integrals that represents the required concept.

22. The area of R.

23. The volume of the solid obtained when R is revolved about the x-axis.

24. The volume of the solid obtained when R is revolved about $x = a$.

25. The moments M_x and M_y of a homogeneous lamina with shape R, assuming its density is δ.

26. The *total* length of the boundary of R.

27. The *total* surface area of the solid of Problem 23.

7 Transcendental Functions

Read Euler, read Euler, he is the master of us all.
P. S. Laplace

Leonhard Euler
1707–1783

Can you imagine anyone writing the equivalent of 75 books on mathematics? That is the contribution of Leonhard Euler, the dominant figure of eighteenth-century mathematics and the most prolific mathematical author of all time. Born near Basel, Switzerland, he studied under his fellow countryman Johann Bernoulli and was already publishing papers at age 18. He held positions at the University of Basel, St. Petersburg Academy of Sciences, and Berlin Academy of Sciences. When he died, it was said that all the mathematicians of Europe were his students.

Euler's interests spanned all of mathematics and physics. We have chosen him as representative of this chapter because of his contributions to the calculus of the transcendental (that is, nonalgebraic) functions. In particular, he introduced e as the base for natural logarithms, showed that e and e^2 are irrational, and discovered the remarkable relation $e^{i\pi} = -1$.

Blindness during the last 17 years of Euler's life seems not to have hampered his work. In part, this was due to his prodigious memory; he knew by heart the formulas of trigonometry and analysis, plus numerous poems and the entire Aeneid. He is said to have done a calculation to 50 decimal places in his head.

Euler was a devoted family man, who often spent his evenings constructing scientific games and reading the Bible to his 13 children. He was a truly remarkable human being.

7.1
The Natural Logarithm Function

The power of calculus, both that of derivatives and integrals, has been amply demonstrated. Yet we have only scratched the surface of potential applications. To dig deeper, we need to expand the class of functions with which we can work. That is the object of this chapter.

We begin by asking you to notice a peculiar gap in our knowledge of derivatives.

$$D_x(x^3/3) = x^2$$
$$D_x(x^2/2) = x^1$$
$$D_x(x) = 1 = x^0$$
$$D_x(????) = x^{-1}$$
$$D_x(-x^{-1}) = x^{-2}$$
$$D_x(-x^{-2}/2) = x^{-3}$$

Is there no function whose derivative is $1/x$? Alternatively, is there no antiderivative $\int (1/x)\,dx$? We will most certainly reach this conclusion if we restrict attention to the functions studied so far. However, we are about to launch the business of creating new functions.

The first function to be introduced is chosen to fill the gap noticed above. We call it the natural logarithm function and it does have something to do with the logarithm studied in elementary algebra, but that will only appear later. For the time being, just accept the fact that we are going to define a new function and study its properties.

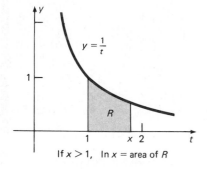

If $x > 1$, $\ln x$ = area of R

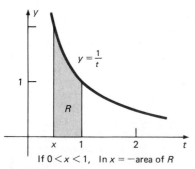

If $0 < x < 1$, $\ln x = -$area of R

FIGURE 1

Definition

The **natural logarithm function**, denoted by ln, is defined by

$$\ln x = \int_1^x \frac{1}{t}\,dt, \qquad x > 0$$

Its domain is the set of positive real numbers.

The diagrams in Figure 1 indicate the geometric meaning of $\ln x$. It measures the area under the curve $y = 1/t$ between 1 and x if $x > 1$ and the negative of this area if $0 < x < 1$. Also $\ln 1 = 0$. No one can doubt that $\ln x$ is well defined for $x > 0$. And what is the derivative of this new function? Just exactly what we want.

THE DERIVATIVE OF THE NATURAL LOGARITHM Recall that the derivative of an integral with respect to its upper limit is the integrand evaluated at the upper limit (Theorem 5.7D). Thus

$$D_x \ln x = \frac{1}{x}, \qquad x > 0$$

This can be combined with the Chain Rule. If $u = f(x) > 0$ and if f is differentiable, then

$$D_x \ln u = \frac{1}{u} D_x u$$

EXAMPLE 1 Find $D_x \ln \sqrt{x}$.

Solution Let $u = \sqrt{x} = x^{1/2}$. Then

$$D_x \ln \sqrt{x} = \frac{1}{x^{1/2}} \cdot \frac{1}{2} x^{-1/2} = \frac{1}{2x}$$ ∎

EXAMPLE 2 Find $D_x \ln(x^2 - x - 2)$.

Solution This problem makes sense, provided $x^2 - x - 2 > 0$. Now $x^2 - x - 2 = (x - 2)(x + 1)$, which is positive provided $x < -1$ or $x > 2$. Thus, the domain of $\ln(x^2 - x - 2)$ is $(-\infty, -1) \cup (2, \infty)$. On this domain,

$$D_x \ln(x^2 - x - 2) = \frac{1}{x^2 - x - 2} D_x(x^2 - x - 2) = \frac{2x - 1}{x^2 - x - 2}$$ ∎

EXAMPLE 3 Show that

$$D_x \ln|x| = \frac{1}{x}, \qquad x \neq 0$$

Solution Two cases are to be considered. If $x > 0$, $|x| = x$ and

$$D_x \ln|x| = D_x \ln x = \frac{1}{x}$$

If $x < 0$, $|x| = -x$, and so

$$D_x \ln|x| = D_x \ln(-x) = \frac{1}{-x} D_x(-x) = \left(\frac{1}{-x}\right)(-1) = \frac{1}{x}$$ ∎

Corresponding to each differentiation formula, there is an integration formula. Thus, Example 3 gives

$$\int \frac{1}{x} \, dx = \ln|x| + C, \qquad x \neq 0$$

or, with u replacing x,

$$\int \frac{1}{u} \, du = \ln|u| + C, \qquad u \neq 0$$

This fills the long-standing gap in the Power Rule: $\int u^r \, du = u^{r+1}/(r + 1)$, $r \neq -1$.

EXAMPLE 4 Find $\int \dfrac{5}{2x + 7} \, dx$.

Solution Let $u = 2x + 7$, so $du = 2 \, dx$. Then

$$\int \frac{5}{2x + 7} \, dx = \frac{5}{2} \int \frac{1}{2x + 7} \, 2 \, dx = \frac{5}{2} \int \frac{1}{u} \, du$$

$$= \frac{5}{2} \ln|u| + C = \frac{5}{2} \ln|2x + 7| + C \qquad \blacksquare$$

EXAMPLE 5 Evaluate $\int_{-1}^{3} \dfrac{x}{10 - x^2} \, dx$.

Solution Let $u = 10 - x^2$, so $du = -2x \, dx$. Then

$$\int \frac{x}{10 - x^2} \, dx = -\frac{1}{2} \int \frac{-2x}{10 - x^2} \, dx = -\frac{1}{2} \int \frac{1}{u} \, du$$

$$= -\frac{1}{2} \ln|u| + C = -\frac{1}{2} \ln|10 - x^2| + C$$

Thus, by the Fundamental Theorem of Calculus,

$$\int_{-1}^{3} \frac{x}{10 - x^2} \, dx = \left[-\frac{1}{2} \ln|10 - x^2| \right]_{-1}^{3}$$

$$= -\frac{1}{2} \ln 1 + \frac{1}{2} \ln 9 = \frac{1}{2} \ln 9$$

For the above calculation to be valid, $10 - x^2$ must never be 0 on the interval $[-1, 3]$. It is easy to see that this is true. $\qquad \blacksquare$

PROPERTIES OF THE NATURAL LOGARITHM   You should be reminded of common logarithms by the results of our next theorem.

Theorem A

If a and b are positive numbers and r is any rational number, then

(i) $\ln 1 = 0$;

(ii) $\ln ab = \ln a + \ln b$;

(iii) $\ln \dfrac{a}{b} = \ln a - \ln b$;

(iv) $\ln a^r = r \ln a$.

Proof

(i) $\ln 1 = \displaystyle\int_1^1 \frac{1}{t}\, dt = 0$

(ii)   Since, for $x > 0$,

$$D_x \ln ax = \frac{1}{ax} \cdot a = \frac{1}{x}$$

and

$$D_x \ln x = \frac{1}{x}$$

it follows (Theorem 4.8B) that

$$\ln ax = \ln x + C$$

To evaluate C, let $x = 1$, obtaining $\ln a = C$. Thus,

$$\ln ax = \ln x + \ln a$$

Finally, let $x = b$.

(iii) Replace a by $1/b$ in (ii) to obtain

$$\ln \frac{1}{b} + \ln b = \ln\!\left(\frac{1}{b} \cdot b\right) = \ln 1 = 0$$

Thus,

$$\ln \frac{1}{b} = -\ln b$$

Applying (ii) again, we get

$$\ln \frac{a}{b} = \ln\!\left(a \cdot \frac{1}{b}\right) = \ln a + \ln \frac{1}{b} = \ln a - \ln b$$

(iv) Since, for $x > 0$,

$$D_x(\ln x^r) = \frac{1}{x^r} \cdot rx^{r-1} = \frac{r}{x}$$

and

$$D_x(r \ln x) = r \cdot \frac{1}{x} = \frac{r}{x}$$

it follows by the theorem we used above in (ii) that

$$\ln x^r = r \ln x + C$$

Let $x = 1$, which gives $C = 0$. Thus,

$$\ln x^r = r \ln x$$

a result equivalent to (iv). ∎

EXAMPLE 6   Find dy/dx if $y = \ln \sqrt[3]{(x-1)/x^2}$, $x > 1$.

Solution   Our problem is easier if we first use the properties of natural logarithms to simplify y.

$$y = \ln\left(\frac{x-1}{x^2}\right)^{1/3} = \frac{1}{3}\ln\left(\frac{x-1}{x^2}\right)$$

$$= \frac{1}{3}\left[\ln(x-1) - \ln x^2\right] = \frac{1}{3}\left[\ln(x-1) - 2 \ln x\right]$$

Thus,

$$\frac{dy}{dx} = \frac{1}{3}\left[\frac{1}{x-1} - \frac{2}{x}\right] = \frac{2-x}{3x^2 - 3x}$$ ∎

LOGARITHMIC DIFFERENTIATION   The labor of differentiating expressions involving quotients, products, or powers can often be substantially reduced by first applying the natural logarithm function and using its properties. This method, called **logarithmic differentiation**, is illustrated in Example 7.

EXAMPLE 7   Differentiate $y = \dfrac{\sqrt{1-x^2}}{(x+1)^{2/3}}$.

Solution   First we take natural logarithms; then we differentiate implicitly with respect to x (recall Section 3.8).

$$\ln y = \frac{1}{2}\ln(1 - x^2) - \frac{2}{3}\ln(x + 1)$$

$$\frac{1}{y}\frac{dy}{dx} = \frac{-2x}{2(1 - x^2)} - \frac{2}{3(x + 1)} = \frac{-(x + 2)}{3(1 - x^2)}$$

Thus,

$$\frac{dy}{dx} = \frac{-y(x + 2)}{3(1 - x^2)} = \frac{-\sqrt{1 - x^2}(x + 2)}{3(x + 1)^{2/3}(1 - x^2)}$$

$$= \frac{-(x + 2)}{3(x + 1)^{2/3}(1 - x^2)^{1/2}} \qquad \blacksquare$$

Example 7 could have been done directly without first taking logarithms, and we suggest you try it.

THE GRAPH OF THE NATURAL LOGARITHM   The domain of $\ln x$ consists of the positive real numbers, so the graph of $y = \ln x$ is in the right half-plane. Also, for $x > 0$,

$$D_x \ln x = \frac{1}{x} > 0$$

and

$$D_x^2 \ln x = -\frac{1}{x^2} < 0$$

The first tells us that the graph is continuous (why?) and rises as x increases; the second tells us that the graph is everywhere concave downward. In Problems 39 and 40, it will be shown that

$$\lim_{x \to \infty} \ln x = \infty$$

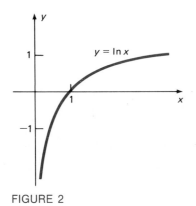

and

$$\lim_{x \to 0^+} \ln x = -\infty$$

Finally, $\ln 1 = 0$. These facts imply that the graph of $y = \ln x$ is something like what is shown in Figure 2.

Values of $\ln x$ have been tabulated by methods that will be explained later. For example,

FIGURE 2

$$\ln 2 \approx 0.6931$$

$$\ln 3 \approx 1.0986$$

$$\ln 10 \approx 2.3026$$

Table III of the appendix gives a fairly complete listing. Also, your calculator probably has a ln button.

1. Use the approximations $\ln 2 = 0.693$ and $\ln 3 = 1.099$ together with the properties stated in Theorem A to calculate approximations to each of the following. For example, $\ln 6 = \ln(2 \cdot 3) = \ln 2 + \ln 3$.

(a) $\ln 6$ (b) $\ln 1.5$ (c) $\ln 81$

(d) $\ln \sqrt{2}$ (e) $\ln \left(\frac{1}{36}\right)$ (f) $\ln 48$

$\boxed{\text{C}}$ **2.** Use your calculator to make the computations in Problem 1 directly.

In Problems 3–14, find the indicated derivative (see Examples 1 and 2). Assume in each case that x is restricted so that $\ln$ is defined.

3. $D_x \ln(x^2 - 5x + 6)$ **4.** $D_x \ln(2x^3 + 1)$

5. $D_x \ln(x - 5)^4$ **6.** $D_x \ln \sqrt{3x - 25}$

7. $\dfrac{dy}{dx}$ if $y = x \ln x$ **8.** $\dfrac{dy}{dx}$ if $y = \dfrac{\ln x}{x^2}$

9. $\dfrac{dy}{dx}$ if $y = \ln x^3 + (\ln x)^3$

10. $\dfrac{dy}{dx}$ if $y = \dfrac{1}{\ln x} + \ln\left(\dfrac{1}{x}\right)$

11. $f'(x)$ if $f(x) = \ln(x + \sqrt{x^2 - 1})$

12. $f'(x)$ if $f(x) = \ln(x + \sqrt{x^2 + 1})$

13. $f'(100)$ if $f(x) = \ln \sqrt[3]{x}$

14. $f'\left(\dfrac{\pi}{2}\right)$ if $f(x) = \ln(\sin x)$

In Problems 15–22, find the integrals (see Examples 4 and 5).

15. $\displaystyle\int \dfrac{4}{2x + 1}\, dx$ **16.** $\displaystyle\int \dfrac{2}{4x - 3}\, dx$

17. $\displaystyle\int \dfrac{4x + 2}{x^2 + x + 5}\, dx$ **18.** $\displaystyle\int \dfrac{x}{x^2 + 4}\, dx$

19. $\displaystyle\int \dfrac{\ln x}{x}\, dx$ **20.** $\displaystyle\int \dfrac{2}{x(\ln x)^2}\, dx$

21. $\displaystyle\int_0^3 \dfrac{x^3}{x^4 + 1}\, dx$ **22.** $\displaystyle\int_0^1 \dfrac{x + 1}{x^2 + 2x + 2}\, dx$

In Problems 23–26, use Theorem A to write the expressions as single logarithms.

23. $2 \ln(x + 1) - \ln x$

24. $\frac{1}{2} \ln(x - 9) + \frac{1}{2} \ln x$

25. $\ln(x - 2) - \ln(x + 2) + 2 \ln x$

26. $\ln(x^2 - 9) - 2 \ln(x - 3) - \ln(x + 3)$

In Problems 27–30, find dy/dx by logarithmic differentiation (see Example 7).

27. $y = \dfrac{x + 11}{\sqrt{x^3 - 4}}$

28. $y = (x^2 + 3x)(x - 2)(x^2 + 1)$

29. $y = \dfrac{\sqrt{x + 13}}{(x - 4)\sqrt[3]{2x + 1}}$

30. $y = \dfrac{(x^2 + 3)^{2/3}(3x + 2)^2}{\sqrt{x + 1}}$

In Problems 31–34, make use of the known graph of $y = \ln x$ to sketch the graphs of the equations.

31. $y = \ln|x|$ **32.** $y = \ln \sqrt{x}$

33. $y = \ln\left(\dfrac{1}{x}\right)$ **34.** $y = \ln(x - 2)$

35. Sketch the graph of $y = \ln \cos x + \ln \sec x$ on $(-\pi/2, \pi/2)$, but think before you do it.

36. Explain why $\displaystyle\lim_{x \to 0} \ln \dfrac{\sin x}{x} = 0$.

37. Find all local extreme values of $f(x) = 2x^2 \ln x - x^2$ on its domain.

38. The rate of transmission in a telegraph cable is observed to be proportional to $x^2 \ln(1/x)$, where x is the ratio of the radius of the core to the thickness of the insulation $(0 < x < 1)$. What value of x gives the maximum rate of transmission?

39. Use the fact that $\ln 4 > 1$ to show that $\ln 4^m > m$ for $m > 0$. Conclude that $\ln x$ can be made as large as desired by choosing x sufficiently large. What does this imply about $\displaystyle\lim_{x \to \infty} \ln x$?

40. Use the fact that $\ln x = -\ln(1/x)$ and Problem 39 to show that $\displaystyle\lim_{x \to 0^+} \ln x = -\infty$.

41. Solve for x: $\displaystyle\int_{1/3}^x \dfrac{1}{t}\, dt = 2 \int_1^x \dfrac{1}{t}\, dt$

42. Prove.

(a) Since $1/t < 1/\sqrt{t}$ for $t > 1$, $\ln x < 2(\sqrt{x} - 1)$ for $x > 1$.

(b) $\displaystyle\lim_{x \to \infty} (\ln x)/x = 0$.

43. Calculate

$$\lim_{n \to \infty} \left[\frac{1}{n+1} + \frac{1}{n+2} + \cdots + \frac{1}{2n} \right]$$

by writing the expression in brackets as

$$\left[\frac{1}{1+1/n} + \frac{1}{1+2/n} + \cdots + \frac{1}{1+n/n} \right] \frac{1}{n}$$

and recognizing the latter as a Riemann sum.

C **44.** A famous theorem (the Prime Number Theorem) says that the number of primes less than n for large n is approximately $n/(\ln n)$. About how many primes are there less than 1,000,000?

45. Find and simplify $f'(1)$.

(a) $f(x) = \ln \left(\dfrac{ax - b}{ax + b} \right)^c$, where $c = \dfrac{a^2 - b^2}{2ab}$.

(b) $f(x) = \displaystyle\int_1^u \cos^2 t \, dt$, where $u = \ln(x^2 + x - 1)$.

46. Evaluate

(a) $\displaystyle\int_0^{\pi/3} \tan x \, dx$

(b) $\displaystyle\int_{\pi/4}^{\pi/3} \sec x \csc x \, dx$

47. The region bounded by $y = (x^2 + 4)^{-1}$, $y = 0$, $x = 1$, and $x = 4$ is revolved about the y-axis, generating a solid. Find its volume.

48. Find the length of the curve $y = x^2/4 - \ln \sqrt{x}$, $1 \le x \le 2$.

49. By appealing to the graph of $y = 1/x$, show that

$$\frac{1}{2} + \frac{1}{3} + \cdots + \frac{1}{n} < \ln n < 1 + \frac{1}{2} + \frac{1}{3} + \cdots + \frac{1}{n}$$

50. Prove Napier's Inequality, which says that for $0 < x < y$,

$$\frac{1}{y} < \frac{\ln y - \ln x}{y - x} < \frac{1}{x}$$

7.2
Inverse Functions and Their Derivatives

Our stated aim for this chapter is to expand the number of functions in our arsenal. One way to manufacture new functions is to take old ones and "reverse" them. In doing this for the natural logarithm function, we will be led to the natural exponential function, the subject of Section 7.3. In this section, we study the general problem of reversing (or inverting) a function. Here is the idea.

A function f takes a number x from its domain D and assigns to it a single value y from its range R. If we are lucky, as in the case of the two functions graphed in Figures 1 and 2, we can reverse f; that is, for any given y

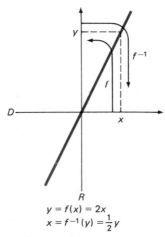

$y = f(x) = 2x$
$x = f^{-1}(y) = \frac{1}{2}y$

FIGURE 1

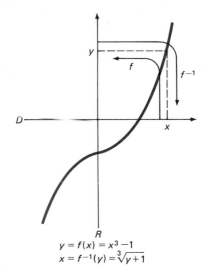

$y = f(x) = x^3 - 1$
$x = f^{-1}(y) = \sqrt[3]{y + 1}$

FIGURE 2

$y = f(x) = x^2$
No inverse

FIGURE 3

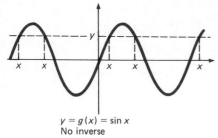

$y = g(x) = \sin x$
No inverse

FIGURE 4

in R, we can unambiguously go back and find the x from which it came. This new function that takes y and assigns x to it is denoted by f^{-1}. Note that its domain is R and its range is D. It is called the **inverse** of f, or simply f-inverse. Here we are using the superscript -1 in a new way. The symbol f^{-1} does not denote $1/f$, as you might have expected. We, and all mathematicians, use it to name the inverse function.

Sometimes, we can give a formula for f^{-1}. If $y = f(x) = 2x$, then $x = f^{-1}(y) = \frac{1}{2}y$ (see Figure 1). Similarly, if $y = f(x) = x^3 - 1$, then $x = f^{-1}(y) = \sqrt[3]{y+1}$ (Figure 2). In each case, we simply solve the equation that determines f for x in terms of y. The result is $x = f^{-1}(y)$.

But life is more complicated than the two examples above indicate. Not every function can be reversed in an unambiguous way. Consider $y = f(x) = x^2$, for example. For a given y there are *two* x's that correspond to it (Figure 3). The function $y = g(x) = \sin x$ is even worse. For each y, there are infinitely many x's that correspond to it (Figure 4). Such functions do not have inverses; at least, they do not unless we somehow restrict the set of x-values, a subject we will take up later.

EXISTENCE OF INVERSE FUNCTIONS   It would be nice to have a simple criterion for deciding whether a function f has an inverse. One such criterion is that the function be **one-to-one**; that is, $x_1 \neq x_2$ implies $f(x_1) \neq f(x_2)$. This is equivalent to the geometric condition that every horizontal line meet the graph of $y = f(x)$ in at most one point. But in a given situation, this criterion may be very hard to apply, since it demands that we have complete knowledge of the graph. A more practical criterion that covers most examples that arise in this book is that a function be **strictly monotonic**. By this we mean that it is either increasing or decreasing on its domain.

Theorem A

If f is strictly monotonic on its domain, then f has an inverse.

This is a practical result, because we have an easy way of deciding whether a function f is strictly monotonic. We simply examine the sign of f'.

EXAMPLE 1   Show that $f(x) = x^5 + 2x + 1$ has an inverse.

Solution $f'(x) = 5x^4 + 2 > 0$ for all x. Thus f is increasing on the whole real line and so has an inverse there.   ∎

We do not claim that we can always give a formula for f^{-1}. In the example just considered, this would require that we be able to solve $y = x^5 + 2x + 1$ for x, a task beyond our capabilities.

There is a way of salvaging the notion of inverse for functions that do not have inverses on their natural domain. We simply *restrict the domain* to a set on which the graph is either increasing or decreasing. Thus for $y = f(x) = x^2$, we may restrict the domain to $x \geq 0$ ($x \leq 0$ would also work). For $y = g(x) = \sin x$, we restrict the domain to the interval $[-\pi/2, \pi/2]$.

CHAPTER 7/Transcendental Functions

Then both functions have inverses (see Figure 5), and we can even give a formula for the first one, namely, $f^{-1}(y) = \sqrt{y}$.

Domain restricted to $x \geqslant 0$

Domain restricted to $\left[-\frac{\pi}{2}, \frac{\pi}{2}\right]$

FIGURE 5

If f has an inverse f^{-1}, then f^{-1} also has an inverse, namely, f. Thus, we may call f and f^{-1} a pair of inverse functions. One function undoes (or reverses) what the other did—that is,

$$f^{-1}(f(x)) = x \quad \text{and} \quad f(f^{-1}(y)) = y$$

EXAMPLE 2 Show that $f(x) = 2x + 6$ has an inverse, find a formula for $f^{-1}(y)$, and verify the results in the box above.

Solution Since f is an increasing function, it has an inverse. To find $f^{-1}(y)$, we solve $y = 2x + 6$ for x, which gives $x = (y - 6)/2 = f^{-1}(y)$. Finally, note that

$$f^{-1}(f(x)) = f^{-1}(2x + 6) = \frac{(2x + 6) - 6}{2} = x$$

and

$$f(f^{-1}(y)) = f\left(\frac{y - 6}{2}\right) = 2\frac{y - 6}{2} + 6 = y$$

∎

THE GRAPH OF $y = f^{-1}(x)$ Suppose f has an inverse. Then

$$x = f^{-1}(y) \quad \Leftrightarrow \quad y = f(x)$$

Consequently, $y = f(x)$ and $x = f^{-1}(y)$ determine the same (x, y) pairs, and so have identical graphs. However, it is conventional to use x as the domain variable for functions, so we now inquire about the graph of $y = f^{-1}(x)$ (note that we have interchanged the roles of x and y). A little thought convinces us that to interchange the roles of x and y on a graph is to reflect the graph across the line $y = x$. *Thus the graph of $y = f^{-1}(x)$ is just the graph of $y = f(x)$ reflected across the line $y = x$* (Figure 6).

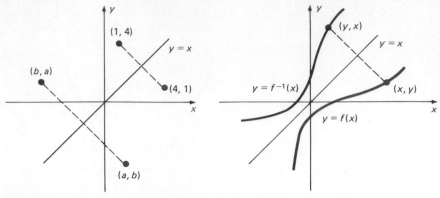

FIGURE 6

A related matter is that of finding a formula for $f^{-1}(x)$. To do it, we first find $f^{-1}(y)$ and then replace y by x in the resulting formula. Thus, we propose the following three-step process for finding $f^{-1}(x)$.

Step 1 Solve the equation $y = f(x)$ for x in terms of y.
Step 2 Use $f^{-1}(y)$ to name the resulting expression in y.
Step 3 Replace y by x to get the formula for $f^{-1}(x)$.

EXAMPLE 3 Find a formula for $f^{-1}(x)$ if $y = f(x) = x/(1 - x)$.

Solution Perhaps we should first show that f has an inverse. However, if we can actually carry out the first step in the three-step process and get a *single* x for each y, then f^{-1} does exist. (Note that when we try this for $y = g(x) = x^2$, we get $x = \pm\sqrt{y}$, which immediately shows that g^{-1} does not exist unless, of course, we have restricted the domain to eliminate one of the two signs $\pm$.) Here are the three steps for our example.

Step 1
$$y = \frac{x}{1 - x}$$
$$(1 - x)y = x$$
$$y - xy = x$$
$$x + xy = y$$
$$x(1 + y) = y$$
$$x = \frac{y}{1 + y}$$

Step 2 $\quad f^{-1}(y) = \dfrac{y}{1 + y}$

Step 3 $\quad f^{-1}(x) = \dfrac{x}{1 + x}$ ∎

DERIVATIVES OF INVERSE FUNCTIONS We conclude this section by investigating the relationship between the derivative of a function and the derivative of its inverse. Consider first what happens to a line l_1 when it is reflected across the line $y = x$. As the left half of Figure 7 makes clear, l_1 is reflected into a line l_2; moreover, their respective slopes m_1 and m_2 are related

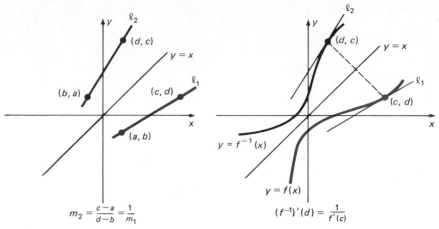

$$m_2 = \frac{c-a}{d-b} = \frac{1}{m_1}$$

$$(f^{-1})'(d) = \frac{1}{f'(c)}$$

FIGURE 7

by $m_2 = 1/m_1$, provided $m_1 \neq 0$. If l_1 happens to be the tangent line to the graph of f at the point (c, d), then l_2 is the tangent line to the graph of f^{-1} at the point (d, c) (see the right half of Figure 7). We are led to the conclusion that

$$(f^{-1})'(d) = m_2 = \frac{1}{m_1} = \frac{1}{f'(c)}$$

Pictures are sometimes deceptive, so we claim only to have made the following result plausible. For a formal proof, see any advanced calculus book.

Theorem B

(Inverse Function Theorem). Let f be differentiable and strictly monotonic on an interval I. If $f'(x) \neq 0$ at a certain x in I, then f^{-1} is differentiable at the corresponding point $y = f(x)$ in the range of f and

$$(f^{-1})'(y) = \frac{1}{f'(x)}$$

The conclusion to Theorem B is often written symbolically as

$$\frac{dx}{dy} = \frac{1}{dy/dx}$$

EXAMPLE 4 Let $y = f(x) = x^5 + 2x + 1$, as in Example 1. Find $(f^{-1})'(4)$.

Solution Even though we cannot find a formula for f^{-1} in this case, we note that $y = 4$ corresponds to $x = 1$ and, since $f'(x) = 5x^4 + 2$,

$$(f^{-1})'(4) = \frac{1}{f'(1)} = \frac{1}{5+2} = \frac{1}{7} \qquad \blacksquare$$

In Problems 1–6, the graph of $y = f(x)$ is shown. In each case, decide whether f has an inverse, and if so, estimate $f^{-1}(2)$.

1.

2.

3.

4.

5.

6.

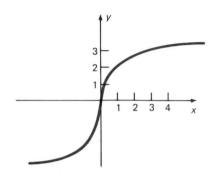

In Problems 7–14, show that f has an inverse by showing that it is strictly monotonic (see Example 1).

7. $f(x) = -3x^5 - x$

8. $f(x) = x^7 + 5x^3$

9. $f(x) = \tan x = \dfrac{\sin x}{\cos x}, \ -\dfrac{\pi}{2} < x < \dfrac{\pi}{2}$

10. $f(x) = \cos x, \ 0 \le x \le \pi$

11. $f(x) = (x + 1)^2, \ x \le -1$

12. $f(x) = x^2 + x + 5, \ x \ge -\dfrac{1}{2}$

13. $f(x) = \displaystyle\int_0^x \sqrt{t^2 + 2}\,dt$ *Hint:* Theorem 5.7D.

14. $f(x) = \displaystyle\int_1^x \sin^4 t\,dt$

In Problems 15–28, find a formula for $f^{-1}(x)$ and then verify that $f^{-1}(f(x)) = x$ and $f(f^{-1}(x)) = x$ (see Examples 2 and 3).

15. $f(x) = 3x - 1$

16. $f(x) = -\dfrac{x}{4} + 5$

17. $f(x) = \sqrt{2x + 5}$

18. $f(x) = -\sqrt{2 - x}$

19. $f(x) = \dfrac{1}{x - 5}$

20. $f(x) = \dfrac{1}{\sqrt{x - 3}}$

21. $f(x) = x^2,\ x \le 0$

22. $f(x) = (x + 1)^2,\ x \ge -1$

23. $f(x) = (x - 4)^3$ **24.** $f(x) = x^{3/2},\ x \ge 0$

25. $f(x) = \dfrac{2x - 2}{x + 3}$ **26.** $f(x) = \dfrac{3x + 2}{2x + 1}$

27. $f(x) = \dfrac{x^3 + 1}{x^3 + 2}$ **28.** $f(x) = \left(\dfrac{2x + 1}{3x - 1}\right)^3$

In Problems 29 and 30, restrict the domain of f so that f has an inverse, yet keeping its range as large as possible. Then find $f^{-1}(x)$. *Suggestion*: First graph f.

29. $f(x) = 2x^2 + x - 4$ **30.** $f(x) = x^2 - 3x + 1$

In each of Problems 31–34, the graph of $y = f(x)$ is shown. Sketch the graph of $y = f^{-1}(x)$ and estimate $(f^{-1})'(3)$.

31.

32.

33.

34.

In Problems 35–38, find $(f^{-1})'(2)$ by using Theorem B (see Example 4). Note that you can find the x corresponding to $y = 2$ by inspection.

35. $f(x) = 3x^5 + x - 2$ **36.** $f(x) = x^5 + 5x - 4$

37. $f(x) = 2 \tan x,\ -\dfrac{\pi}{2} < x < \dfrac{\pi}{2}$

38. $f(x) = \sqrt{x + 1}$

39. Suppose that both f and g have inverses and that $h(x) = (f \circ g)(x) = f(g(x))$. Show that h has an inverse given by $h^{-1} = g^{-1} \circ f^{-1}$.

40. Verify the result of Problem 39 for $f(x) = 1/x$, $g(x) = 3x + 2$.

41. If $f(x) = \displaystyle\int_0^x \sqrt{1 + \cos^2 t}\, dt$, then f has an inverse. (Why?) Let $A = f(\pi/2)$ and $B = f(5\pi/6)$. Find (a) $(f^{-1})'(A)$, (b) $(f^{-1})'(B)$, and (c) $(f^{-1})'(0)$.

42. Let $f(x) = \dfrac{ax + b}{cx + d}$ and assume $bc - ad \ne 0$.

(a) Find the formula for $f^{-1}(x)$.
(b) Why is the condition $bc - ad \ne 0$ needed?
(c) What condition on a, b, c, and d will make $f = f^{-1}$?

43. Suppose f is continuous and strictly increasing on $[0, 1]$ with $f(0) = 0$ and $f(1) = 1$. If $\displaystyle\int_0^1 f(x)\, dx = \tfrac{2}{5}$, calculate $\displaystyle\int_0^1 f^{-1}(y)\, dy$. *Hint*: Draw a picture.

44. Let f be continuous and strictly increasing on $[0, \infty)$ with $f(0) = 0$ and $f(x) \to \infty$ as $x \to \infty$. Use geometric reasoning to establish **Young's Inequality**. For $a > 0$, $b > 0$,

$$ab \le \int_0^a f(x)\, dx + \int_0^b f^{-1}(y)\, dy$$

What is the condition for equality?

45. Let $p > 1$, $q > 1$, and $1/p + 1/q = 1$. Show that the inverse of $f(x) = x^{p-1}$ is $f^{-1}(y) = y^{q-1}$ and use this together with Problem 44 to prove **Minkowski's Inequality**:

$$ab \le \dfrac{a^p}{p} + \dfrac{b^q}{q}, \qquad a > 0, b > 0$$

7.3
The Natural Exponential Function

The graph of $y = f(x) = \ln x$ was obtained at the end of Section 7.1 and is reproduced in Figure 1. The natural logarithm function is differentiable (hence continuous) and increasing on its domain $D = (0, \infty)$; its range is $R = (-\infty, \infty)$. It is, in fact, precisely the kind of function studied in Section 7.2, and therefore has an inverse $\ln^{-1}$ with domain $(-\infty, \infty)$ and range $(0, \infty)$. This function is so important that it is given a special name and a special symbol.

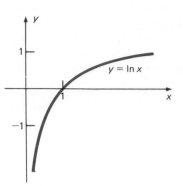

FIGURE 1

Definition

The inverse of ln is called the **natural exponential function** and is denoted by exp. Thus

$$x = \exp y \quad \Leftrightarrow \quad y = \ln x$$

It follows immediately from this definition that:

(i) $\exp(\ln x) = x, \quad x > 0$

(ii) $\ln(\exp y) = y, \quad$ all y

Since exp and ln are inverse functions, the graph of $y = \exp x$ is just the graph of $y = \ln x$ reflected across the line $y = x$ (Figure 2).

But why the name *exponential function*? You will see.

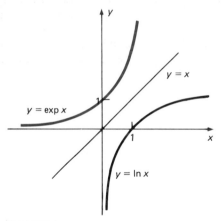

FIGURE 2

PROPERTIES OF THE EXPONENTIAL FUNCTION   We begin by introducing a new number, which—like π—is so important in mathematics that it gets a special symbol, namely, e. The letter e is appropriate since Leonhard Euler first recognized the significance of this number.

Definition

The letter e denotes the unique positive real number such that $\ln e = 1$.

Since $\ln e = 1$, it is also true that $\exp 1 = e$. The number e, like π, is irrational. Its decimal expansion is known to thousands of places; the first few digits are

$$e \approx 2.718281828459045$$

Now we make a crucial observation, one that depends only on facts already demonstrated: (i) above and Theorem 7.1A. If r is any rational number

$$e^r = \exp(\ln e^r) = \exp(r \ln e) = \exp r$$

Let us emphasize the result. For rational r, *exp r is identical with e^r.* What was introduced in the most abstract way as the inverse of the natural logarithm, which itself was defined by an integral, has turned out to be a simple power.

But what if r is irrational? Here we remind you of a gap in all elementary algebra books. Never are irrational powers defined in anything approaching a rigorous manner. What is meant by $e^{\sqrt{2}}$? You will have a hard time pinning that number down, based on elementary algebra. But we must pin it down if we are to talk of such things as $D_x e^x$. Guided by what we learned above, we simply define e^x for all x (rational or irrational) by

$$e^x = \exp x$$

Note that (i) and (ii) at the beginning of this section now take the form:

(i)'	$e^{\ln x} = x,$	$x > 0$
(ii)'	$\ln(e^y) = y,$	all y

Also, we can easily prove two of the familiar laws of exponents.

Theorem A

Let a and b be any real numbers. Then $e^a e^b = e^{a+b}$ and $e^a/e^b = e^{a-b}$.

Proof To prove the first, we write

$$\begin{aligned}
e^a e^b &= \exp(\ln e^a e^b) && \text{(by (ii))} \\
&= \exp(\ln e^a + \ln e^b) && \text{(Theorem 7.1A)} \\
&= \exp(a + b) && \text{(by (ii)')} \\
&= e^{a+b}
\end{aligned}$$

The second fact is proved similarly. ∎

THE DERIVATIVE OF e^x Since exp and ln are inverses, we know from Theorem 7.2B that $\exp x = e^x$ is differentiable. To find a formula for $D_x e^x$, we could use that theorem. Alternatively, let $y = e^x$, so that

$$x = \ln y$$

Now differentiate both sides with respect to x, obtaining

$$1 = \frac{1}{y} D_x y \quad \text{(Chain Rule)}$$

Thus,

$$D_x y = y = e^x$$

We have proved that e^x is its own derivative; that is,

$$\boxed{D_x e^x = e^x}$$

If $u = f(x)$ is differentiable, then the Chain Rule yields

$$D_x e^u = e^u D_x u$$

EXAMPLE 1   Find $D_x e^{\sqrt{x}}$.

Solution   Using $u = \sqrt{x}$, we obtain

$$D_x e^{\sqrt{x}} = e^{\sqrt{x}} D_x \sqrt{x} = e^{\sqrt{x}} \cdot \frac{1}{2} x^{-1/2} = \frac{e^{\sqrt{x}}}{2\sqrt{x}} \qquad \blacksquare$$

EXAMPLE 2   Find $D_x e^{x^2 \ln x}$.

Solution

$$D_x e^{x^2 \ln x} = e^{x^2 \ln x} D_x (x^2 \ln x)$$

$$= e^{x^2 \ln x} \left(x^2 \cdot \frac{1}{x} + 2x \ln x \right)$$

$$= x e^{x^2 \ln x} (1 + \ln x^2) \qquad \blacksquare$$

EXAMPLE 3   Let $f(x) = xe^{x/2}$. Find where f is increasing and decreasing and where it is concave upward and downward. Also, identify all extreme values and points of inflection. Then, sketch the graph of f.

Solution

$$f'(x) = \frac{xe^{x/2}}{2} + e^{x/2} = e^{x/2} \left(\frac{x + 2}{2} \right)$$

and

$$f''(x) = \frac{e^{x/2}}{2} + \left(\frac{x + 2}{2} \right) \frac{e^{x/2}}{2} = e^{x/2} \left(\frac{x + 4}{4} \right)$$

Keeping in mind that $e^{x/2} > 0$ for all x, we see that $f'(x) < 0$ for $x < -2$, $f'(-2) = 0$, and $f'(x) > 0$ for all $x > -2$. Thus, f is decreasing on $(-\infty, -2]$, increasing on $[-2, \infty)$, and has its minimum value at $x = -2$ of $f(-2) = -2/e \approx -0.7$.

Also, $f''(x) < 0$ for $x < -4$, $f''(-4) = 0$, and $f''(x) > 0$ for $x > -4$; so the graph of f is concave downward on $(-\infty, -4)$, concave upward on $(-4, \infty)$, and has a point of inflection at $(-4, -4e^{-2}) \approx (-4, -0.54)$. This information, plus a few plotted points, yield the graph in Figure 3. $\blacksquare$

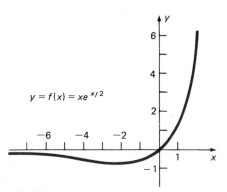

FIGURE 3

The derivative formula $D_x e^x = e^x$ automatically yields the integral formula $\int e^x\,dx = e^x + C$, or, with u replacing x,

$$\int e^u\,du = e^u + C$$

EXAMPLE 4 Find $\int e^{-4x}\,dx$

Solution Let $u = -4x$, so $du = -4\,dx$. Then,

$$\int e^{-4x}\,dx = -\tfrac{1}{4}\int e^{-4x}(-4\,dx) = -\tfrac{1}{4}\int e^u\,du = -\tfrac{1}{4}e^u + C$$
$$= -\tfrac{1}{4}e^{-4x} + C \qquad \blacksquare$$

EXAMPLE 5 Find $\int x^2 e^{-x^3}\,dx.$

Solution Let $u = -x^3$, so $du = -3x^2\,dx$. Then,

$$\int x^2 e^{-x^3}\,dx = -\tfrac{1}{3}\int e^{-x^3}(-3x^2\,dx)$$
$$= -\tfrac{1}{3}\int e^u\,du = -\tfrac{1}{3}e^u + C$$
$$= -\tfrac{1}{3}e^{-x^3} + C \qquad \blacksquare$$

EXAMPLE 6 Evaluate $\int_1^3 xe^{-3x^2}\,dx.$

Solution Let $u = -3x^2$, so $du = -6x\,dx$. Then,

$$\int xe^{-3x^2}\,dx = -\tfrac{1}{6}\int e^{-3x^2}(-6x\,dx) = -\tfrac{1}{6}\int e^u\,du$$
$$= -\tfrac{1}{6}e^u + C = -\tfrac{1}{6}e^{-3x^2} + C$$

Thus, by the Fundamental Theorem of Calculus,

$$\int_1^3 xe^{-3}\,dx = \left[-\frac{1}{6}e^{-3x^2}\right]_1^3 = -\frac{1}{6}(e^{-27} - e^{-3})$$
$$= \frac{e^{-3} - e^{-27}}{6} \approx 0.0082978$$

The last result can be obtained by using Table IV or a calculator. $\blacksquare$

While the symbol e^y will largely supplant exp y throughout the rest of this book, exp occurs frequently in scientific writing, especially when the exponent y is complicated. For example, in the study of statistics, one often encounters the normal curve, which is the graph of

$$f(x) = \frac{1}{\sigma \sqrt{2\pi}} \exp\left[-\frac{(x - \mu)^2}{2\sigma^2} \right]$$

PROBLEM SET 7.3

[C] **1.** Use your calculator or Tables II–IV of the appendix to calculate each of the following *Note:* On some calculators there is an $\boxed{e^x}$ button. On others you must press the $\boxed{\text{INV}}$ and $\boxed{\ln x}$ buttons.

(a) e^3 (b) e^{-4}

(c) $e^{2.1}$ (d) $e^{6.3}$

(e) $e^{\sqrt{2}}$ (f) $e^{\sin 4}$

(g) $e^{\cos(\ln 4)}$

[C] **2.** Calculate

(a) $e^{3 \ln 2}$,   (b) $e^{(\ln 64)/2}$.

Explain why your answers are not surprising.

In Problems 3–10, simplify the given expression.

3. $e^{2 \ln x}$ **4.** $e^{-\ln x}$

5. $\ln e^{\sin x}$ **6.** $\ln e^{-x+2}$

7. $\ln(x^2 e^{-2x})$ **8.** $e^{x + \ln x}$

9. $e^{\ln 2 + \ln x}$ **10.** $e^{\ln x - 2 \ln y}$

In Problems 11–22, find $D_x y$ (see Examples 1 and 2).

11. $y = e^{2x+1}$ **12.** $y = e^{3x^2 - x}$

13. $y = e^{\sqrt{x+1}}$ **14.** $y = e^{1/x^3}$

15. $y = e^{\ln x}$ **16.** $y = e^{(\ln x)/x}$

17. $y = x^2 e^x$ **18.** $y = e^{x^2 \ln x}$

19. $y = e^{\sqrt{x}} + \sqrt{e^x}$ **20.** $y = e^{1/x} + \dfrac{1}{e^x}$

21. $e^{xy} + y = 2$ *Hint:* Use implicit differentiation.

22. $xe^y + 2x - \ln y = 4$

23. Use your knowledge of the graph of $y = e^x$ to sketch the graphs of (a) $y = -e^x$ and (b) $y = e^{-x}$.

24. Explain why $a < b \Rightarrow e^{-a} > e^{-b}$.

In Problems 25–28, discuss and sketch the graph of the given function, as in Example 3.

25. $f(x) = x^2 e^{-x}$ **26.** $f(x) = e^x - x$

27. $f(x) = e^{-x^2}$ **28.** $f(x) = e^x + e^{-x}$

In Problems 29–36, find the integrals (see Examples 4–6).

29. $\displaystyle\int e^{3x+1}\, dx$ **30.** $\displaystyle\int xe^{x^2 - 3}\, dx$

31. $\displaystyle\int (x + 3)e^{x^2 + 6x}\, dx$ **32.** $\displaystyle\int \frac{e^x}{e^x - 1}\, dx$

33. $\displaystyle\int \frac{e^{-1/x}}{x^2}\, dx$ **34.** $\displaystyle\int e^{x + e^x}\, dx$

35. $\displaystyle\int_0^1 e^{2x+3}\, dx$ **36.** $\displaystyle\int_1^2 \frac{e^{3/x}}{x^2}\, dx$

37. Find the volume of the solid generated by revolving the region bounded by $y = e^x$, $y = 0$, $x = 0$, and $x = \ln 3$ about the x-axis.

38. The region bounded by $y = e^{-x^2}$, $y = 0$, $x = 0$, and $x = 1$ is revolved about the y-axis. Find the volume of the resulting solid.

39. Find the area of the region bounded by the curve $y = e^{-x}$ and the line through the points $(0, 1)$ and $(1, 1/e)$.

40. Show that $f(x) = \dfrac{x}{e^x - 1} - \ln(1 - e^{-x})$ is decreasing for $x > 0$.

[C] **41.** Stirling's Formula says that for large n, we can approximate $n! = 1 \cdot 2 \cdot 3 \cdots n$ by

$$n! \approx \sqrt{2\pi n}\left(\frac{n}{e}\right)^n$$

(a) Calculate 10! exactly and then approximately using the above formula.

(b) Approximate 60!.

[C] **42.** It will be shown later (Section 10.1) that for small x,

$$e^x \approx 1 + x + \frac{x^2}{2!} + \frac{x^3}{3!} + \frac{x^4}{4!}$$

$$= \left\{ \left[\left(\frac{x}{4} + 1 \right) \frac{x}{3} + 1 \right] \frac{x}{2} + 1 \right\} x + 1$$

Use this result to calculate $e^{0.3}$ and compare your result with what you get by calculating it directly.

43. Find the length of the curve given parametrically by $x = e^t \sin t$, $y = e^t \cos t$, $0 \le t \le \pi$.

C **44.** If customers arrive at a check-out counter at the average rate of k per minute, then (see books on probability theory) the probability that exactly n customers will arrive in a period of x minutes is given by the formula

$$P_n(x) = \frac{(kx)^n e^{-kx}}{n!}$$

Find the probability that exactly 8 customers will arrive during a 30-minute period if the average rate for this check-out counter is 1 customer every 4 minutes.

45. Let $f(x) = \dfrac{\ln x}{1 + (\ln x)^2}$ for x in $(0, \infty)$. Find:

(a) $\lim\limits_{x \to 0^+} f(x)$ and $\lim\limits_{x \to \infty} f(x)$;

(b) the maximum and minimum values of $f(x)$;

(c) $F'(\sqrt{e})$ if $F(x) = \displaystyle\int_1^{x^2} f(t)\,dt$.

46. Let R be the region bounded by $x = 0$, $y = e^x$, and the tangent line to $y = e^x$ that goes through the origin. Find:

(a) the area of R;

(b) the volume of the solid obtained when R is revolved around the x-axis.

47. Evaluate $\lim\limits_{n \to \infty} \dfrac{e^{1/n} + e^{2/n} + \cdots + e^{n/n}}{n}$.

48. The **normal curve** with mean μ and standard deviation σ is defined by

$$y = f(x) = \frac{1}{\sigma \sqrt{2\pi}} \exp\left[-\frac{1}{2}\left(\frac{x - \mu}{\sigma} \right)^2 \right]$$

Show that:

(a) its graph is symmetric about the line $x = \mu$;

(b) it has a maximum at $x = \mu$ and inflection points at $x = \mu \pm \sigma$.

7.4 General Exponential and Logarithmic Functions

We defined $e^{\sqrt{2}}$, e^π, and all other irrational powers of e in the previous section. But what about $2^{\sqrt{2}}$, π^π, π^e, $\sqrt{2}^\pi$, and similar irrational powers of other numbers? In fact, we want to give meaning to a^x for $a > 0$ and x any real number. Now if $r = p/q$ is a rational number, $a^r = (\sqrt[q]{a})^p$. But we also know that

$$a^r = \exp(\ln a^r) = \exp(r \ln a) = e^{r \ln a}$$

This suggests a definition.

> *Definition*
>
> For $a > 0$ and any real number x,
>
> $$a^x = e^{x \ln a}$$

Of course, this definition will be appropriate only if the usual properties of exponents are valid for it, a matter we take up shortly. To shore up our confidence in the definition, we use it to calculate 3^2 (with a little help from our calculator).

$$3^2 = e^{2 \ln 3} \approx e^{2(1.0986123)} \approx 9.0000002$$

The slight discrepancy is due to the round-off characteristics of our calculator.

Now we can fill a small gap in the properties of the natural logarithm left over from Section 7.1.

$$\ln(a^x) = \ln(e^{x \ln a}) = x \ln a$$

Thus Property (iv) of Theorem 7.1A holds for all real x, not just rational x as claimed there. We will need this fact in the proof of Theorem A below.

PROPERTIES OF a^x Our first theorem summarizes the familiar properties of exponents. They can all be proved now and in a completely rigorous manner.

Theorem A

If $a > 0$, $b > 0$, and x and y are real numbers,

(i) $a^x a^y = a^{x+y}$;

(ii) $\dfrac{a^x}{a^y} = a^{x-y}$;

(iii) $(a^x)^y = a^{xy}$;

(iv) $(ab)^x = a^x b^x$;

(v) $\left(\dfrac{a}{b}\right)^x = \dfrac{a^x}{b^x}$.

Proof We content ourselves with proving (ii) and (iii), leaving the others for you.

(ii) $\dfrac{a^x}{a^y} = e^{\ln(a^x/a^y)} = e^{\ln a^x - \ln a^y}$

$\qquad = e^{x \ln a - y \ln a} = e^{(x-y)\ln a} = a^{x-y}$

(iii) $(a^x)^y = e^{y \ln a^x} = e^{yx \ln a} = a^{yx} = a^{xy}$ ∎

Theorem B

$$D_x a^x = a^x \ln a$$

$$\int a^x \, dx = \left(\frac{1}{\ln a}\right) a^x + C, \qquad a \neq 1$$

Proof

$$D_x a^x = D_x(e^{x \ln a}) = e^{x \ln a} D_x(x \ln a)$$

$$= a^x \ln a$$

The integral formula follows immediately from the derivative formula. ∎

EXAMPLE 1 Find $D_x(3^{\sqrt{x}})$.

Solution We use the Chain Rule with $u = \sqrt{x}$.

$$D_x(3^{\sqrt{x}}) = 3^{\sqrt{x}} \ln 3 \cdot D_x \sqrt{x} = \frac{3^{\sqrt{x}} \ln 3}{2\sqrt{x}}$$ ∎

EXAMPLE 2 Find dy/dx if $y = (x^4 + 2)^5 + 5^{x^4 + 2}$.

Solution

$$\frac{dy}{dx} = 5(x^4 + 2)^4 \cdot 4x^3 + 5^{x^4 + 2} \ln 5 \cdot 4x^3$$

$$= 4x^3[5(x^4 + 2)^4 + 5^{x^4 + 2} \ln 5]$$

$$= 20x^3[(x^4 + 2)^4 + 5^{x^4 + 1} \ln 5]$$ ∎

EXAMPLE 3 Find $\displaystyle\int 2^{x^3} x^2 \, dx$.

Solution Let $u = x^3$, so $du = 3x^2 \, dx$. Then

$$\int 2^{x^3} x^2 \, dx = \frac{1}{3} \int 2^{x^3}(3x^2 \, dx) = \frac{1}{3} \int 2^u \, du$$

$$= \frac{1}{3} \frac{2^u}{\ln 2} + C = \frac{2^{x^3}}{3 \ln 2} + C$$ ∎

THE FUNCTION log$_a$ Finally, we are ready to make a connection with the logarithms you studied in algebra. We note that if $0 < a < 1$, $f(x) = a^x$ is a decreasing function; if $a > 1$, it is an increasing function, as you may check by considering the derivative. In either case, f has an inverse. We call this inverse the **logarithmic function to the base** a. This is equivalent to the following definition.

Definition

Let a be a positive number different from 1. Then

$$y = \log_a x \quad \Leftrightarrow \quad x = a^y$$

Historically, the most commonly used base was base 10 and the resulting logarithms were called **common logarithms**. But in calculus and all of advanced mathematics, the significant base is e. Notice that $\log_e$, being the inverse of $f(x) = e^x$, is just another symbol for ln; that is,

$$\log_e x = \ln x$$

FIGURE 1

We have come full circle (Figure 1). The function ln, which we introduced in Section 7.1, has turned out to be an ordinary logarithm but to a rather surprising base, e.

Now observe that if $y = \log_a x$ so $x = a^y$, then

$$\ln x = y \ln a$$

from which we conclude that

$$\boxed{\log_a x = \frac{\ln x}{\ln a}}$$

From this, it follows that $\log_a$ satisfies the usual properties associated with logarithms (see Theorem 7.1A). Also,

$$\boxed{D_x \log_a x = \frac{1}{x \ln a}}$$

EXAMPLE 4   If $y = \log_{10}(x^4 + 13)$, find $\dfrac{dy}{dx}$.

Solution   Let $u = x^4 + 13$ and apply the Chain Rule.

$$\frac{dy}{dx} = \frac{1}{(x^4 + 13)\ln 10} \cdot 4x^3 = \frac{4x^3}{(x^4 + 13)\ln 10} \qquad \blacksquare$$

THE FUNCTIONS a^x, x^a, AND x^x   Let a be a *constant* and x a *variable*. Do not confuse $f(x) = a^x$, an **exponential function**, with $g(x) = x^a$, a **power function**. And do not confuse their derivatives. We have just learned that

$$\boxed{D_x(a^x) = a^x \ln a}$$

What about $D_x(x^a)$? For a rational, we proved in Chapter 3 that

$$\boxed{D_x(x^a) = ax^{a-1}}$$

Now we assert that this is true even if a is irrational. To see this, write

$$D_x(x^a) = D_x(e^{a \ln x}) = e^{a \ln x} \cdot \frac{a}{x}$$

$$= x^a \cdot \frac{a}{x} = ax^{a-1}$$

Finally, we consider $f(x) = x^x$, a variable to a variable power. There is a formula for $D_x(x^x)$, but we do not recommend that you memorize it. Rather, we suggest that you learn two methods for finding it, as illustrated below.

EXAMPLE 5   If $y = x^x$, $x > 0$, find $D_x y$ by two different methods.

Solution

METHOD 1   We may write

$$y = x^x = e^{x \ln x}$$

Thus, by the Chain Rule,

$$D_x y = e^{x \ln x} D_x(x \ln x) = x^x \left(x \cdot \frac{1}{x} + \ln x \right) = x^x(1 + \ln x)$$

METHOD 2 **(Logarithmic differentiation)**

$$y = x^x$$

$$\ln y = x \ln x$$

$$\frac{1}{y} D_x y = x \cdot \frac{1}{x} + \ln x$$

$$D_x y = y(1 + \ln x) = x^x(1 + \ln x) \qquad \blacksquare$$

EXAMPLE 6   If $y = (x^2 + 1)^{\sin x}$, find $\dfrac{dy}{dx}$.

Solution   We use logarithmic differentiation.

$$\ln y = \sin x \ln(x^2 + 1)$$

$$\frac{1}{y} \frac{dy}{dx} = (\sin x) \frac{2x}{x^2 + 1} + (\cos x)\ln(x^2 + 1)$$

$$\frac{dy}{dx} = (x^2 + 1)^{\sin x} \left[\frac{2x \sin x}{x^2 + 1} + (\cos x)\ln(x^2 + 1) \right] \qquad \blacksquare$$

EXAMPLE 7   If $y = (x^2 + 1)^\pi + \pi^{\sin x}$, find dy/dx.

Solution

$$\frac{dy}{dx} = \pi(x^2 + 1)^{\pi - 1}(2x) + \pi^{\sin x} \ln \pi \cdot \cos x \qquad \blacksquare$$

EXAMPLE 8   Evaluate $\displaystyle\int_{1/2}^{1} \frac{5^{1/x}}{x^2} \, dx$.

Solution Let $u = 1/x$, so $du = (-1/x^2)\,dx$. Then

$$\int \frac{5^{1/x}}{x^2}\,dx = -\int 5^{1/x}\left(-\frac{1}{x^2}\,dx\right) = -\int 5^u\,du$$

$$= -\frac{5^u}{\ln 5} + C = -\frac{5^{1/x}}{\ln 5} + C$$

Thus, by the Fundamental Theorem of Calculus,

$$\int_{1/2}^{1} \frac{5^{1/x}}{x^2}\,dx = \left[-\frac{5^{1/x}}{\ln 5}\right]_{1/2}^{1} = \frac{1}{\ln 5}(5^2 - 5)$$

$$= \frac{20}{\ln 5} \approx 12.426699 \qquad\blacksquare$$

PROBLEM SET 7.4

In Problems 1–8, solve for x. *Hint*: $\log_a b = c \iff a^c = b$.

1. $\log_3 9 = x$ **2.** $\log_4 x = 2$

3. $\log_9 x = \frac{3}{2}$ **4.** $\log_x 81 = 4$

5. $2\log_{10}\left(\dfrac{x}{3}\right) = 1$ **6.** $\log_4\left(\dfrac{1}{x}\right) = 3$

7. $\log_2(x + 1) - \log_2 x = 2$

8. $\log_6(x + 1) + \log_6 x = 1$

<u>C</u> Use your calculator (or Table III) and $\log_a x = (\ln x)/(\ln a)$ to calculate each of the logarithms in Problems 9–12.

9. $\log_5 13$ **10.** $\log_6(0.12)$

11. $\log_{11}(8.16)^{1/5}$ **12.** $\log_{10}(91.2)^3$

<u>C</u> In Problems 13–16, use natural logarithms to solve each of the exponential equations. *Hint*: To solve $3^x = 11$, take $\ln$ of both sides, obtaining $x \ln 3 = \ln 11$; then $x = (\ln 11)/(\ln 3) \approx 2.1827$.

13. $2^x = 19$ **14.** $5^x = 12$

15. $4^{3x-1} = 5$ **16.** $12^{1/(x-1)} = 3$

Find the indicated derivative or integral (see Examples 1–4).

17. $D_x(5^{x^2})$ **18.** $D_x(3^{2x^4 - 4x})$

19. $D_x \log_2 e^x$ **20.** $D_x \log_{10}(x^2 + 9)$

21. $D_x[2^x \ln(x + 5)]$ **22.** $D_x \sqrt{\log_{10} x}$

23. $\displaystyle\int x 2^{x^2}\,dx$ **24.** $\displaystyle\int 10^{5x-1}\,dx$

25. $\displaystyle\int_1^4 \frac{5^{\sqrt{x}}}{\sqrt{x}}\,dx$ **26.** $\displaystyle\int_0^1 (10^{3x} + 10^{-3x})\,dx$

In Problems 27–32, find dy/dx. Be sure to distinguish between problems of the type a^x, x^a, and x^x as in Examples 5–7.

27. $y = 10^{(x^2)} + (x^2)^{10}$ **28.** $y = \sin^2 x + 2^{\sin x}$

29. $y = x^{\pi + 1} + (\pi + 1)^x$ **30.** $y = 2^{(e^x)} + (2^e)^x$

31. $y = (x^2 + 1)^{\ln x}$ **32.** $y = (\ln x^2)^{2x+3}$

33. If $f(x) = x^{\sin x}$, find $f'(1)$

<u>C</u> **34.** Let $f(x) = \pi^x$ and $g(x) = x^\pi$. Which is larger, $f(e)$ or $g(e)$? $f'(e)$ or $g'(e)$?

35. How are $\log_{1/2} x$ and $\log_2 x$ related?

36. Sketch the graphs of $\log_{1/3} x$ and $\log_3 x$ using the same coordinate axes.

<u>C</u> **37.** The magnitude M of an earthquake on the Richter scale is

$$M = 0.67 \log_{10}(0.37E) + 1.46$$

where E is the energy of the earthquake in kilowatt-hours. Find the energy of an earthquake of magnitude 7.

<u>C</u> **38.** The loudness of sound is measured in decibels in honor of Alexander Graham Bell (1847–1922), inventor of the telephone. If the variation in pressure is P pounds per square inch, then the loudness L in decibels is

$$L = 20 \log_{10}(121.3P)$$

Find the variation in pressure caused by a rock band at 115 decibels.

<u>C</u> **39.** In the equally tempered scale to which keyed instruments have been tuned since the days of J. S. Bach (1685–

1750), the frequencies of successive notes C, C♯, D, D♯, E, F, F♯, G, G♯, A, A♯, B, $\bar{C}$ form a geometric sequence (progression), with $\bar{C}$ having twice the frequency of C. What is the ratio r between the frequencies of successive notes? If the frequency of A is 440, find the frequency of $\bar{C}$.

40. Prove that $\log_2 3$ is irrational. *Hint*: Use proof by contradiction.

41. You are suspicious that the xy-data you have collected lie on either an exponential curve, $y = A \cdot b^x$, or a power curve, $y = A \cdot x^b$. To check, you plot $\ln y$ against x (using semilog graph paper) and also $\ln y$ against $\ln x$ (using log log graph paper). Explain how this will help you come to a conclusion.

42. (An Amusement) Given the problem of finding y' if $y = x^x$, student A did the following:

$$y = x^x$$
WRONG 1 $y' = x \cdot x^{x-1} \cdot 1$ $\left(\begin{array}{l}\text{misapplying the}\\\text{Power Rule}\end{array}\right)$
$$= x^x$$

Student B did this:

$$y = x^x$$
WRONG 2 $y' = x^x \cdot \ln x \cdot 1$ $\left(\begin{array}{l}\text{misapplying the}\\\text{Exponential}\\\text{Function Rule}\end{array}\right)$
$$= x^x \ln x$$

The sum $x^x + x^x \ln x$ is correct (Example 5), so

WRONG 1 + WRONG 2 = RIGHT

Show that the same procedure yields a correct answer for $y = f(x)^{g(x)}$.

43. Convince yourself that $f(x) = (x^x)^x$ and $g(x) = x^{(x^x)}$ are not the same function. Then find $f'(x)$ and $g'(x)$. *Note*: x^{x^x} means $x^{(x^x)}$.

44. Consider $f(x) = \dfrac{a^x - 1}{a^x + 1}$ for fixed a, $a > 0$, $a \neq 1$. Show that f has an inverse and find a formula for $f^{-1}(x)$.

45. For fixed $a > 1$, let $f(x) = x^a/a^x$ on $[0, \infty)$. Show:

(a) $\lim\limits_{x \to \infty} f(x) = 0$ (study $\ln f(x)$);
(b) $f(x)$ is maximized at $x_0 = a/\ln a$;
(c) $x^a = a^x$ has two positive solutions if $a \neq e$, only one such solution if $a = e$;
(d) $\pi^e < e^\pi$.

46. Let $f_u(x) = x^u e^{-x}$ for $x \geq 0$. Show that for any fixed $u > 0$:

(a) $f_u(x)$ attains its maximum at $x_0 = u$;
(b) $f_u(u) > f_u(u+1)$ and $f_{u+1}(u+1) > f_{u+1}(u)$ imply

$$\left(\frac{u+1}{u}\right)^u < e < \left(\frac{u+1}{u}\right)^{u+1}$$

(c) $\dfrac{u}{u+1}\, e < \left(\dfrac{u+1}{u}\right)^u < e$.

Conclude from (c) that $\lim\limits_{u \to \infty}\left(1 + \dfrac{1}{u}\right)^u = e$.

7.5
Exponential Growth and Decay

At the beginning of 1975, the world's population was about 4 billion. It is said that by the year 2000, it will reach 6.6 billion. How are such predictions made?

To treat the problem mathematically, let $y = f(t)$ denote the size of the population at time t, where t is the number of years after 1975. Actually $f(t)$ is an integer, and its graph "jumps" when someone is born or dies. However, for a large population, these jumps are so small relative to the total population that we will not go far wrong if we pretend that f is a nice differentiable function.

It seems reasonable to suppose that the increase Δy in population (births minus deaths) during a short time period Δt is proportional to the size of the population at the beginning of the period and to the length of that period. Thus, $\Delta y = ky\,\Delta t$, or

$$\frac{\Delta y}{\Delta t} = ky$$

In its limiting form, this gives the differential equation

$$\frac{dy}{dt} = ky$$

If $k > 0$, the population is growing; if $k < 0$, it is shrinking. For world population, history indicates that k is about 0.0198 (assuming that t is measured in years), though some statisticians report a slightly lower figure.

SOLVING THE DIFFERENTIAL EQUATION We began our study of differential equations in Section 5.2, and you might refer to that section now. We want to solve $dy/dt = ky$ subject to the condition that $y = y_0$ when $t = 0$. Separating variables and integrating, we obtain

$$\frac{dy}{y} = k\,dt$$

$$\int \frac{dy}{y} = \int k\,dt$$

$$\ln y = kt + C$$

The condition $y = y_0$ at $t = 0$ gives $C = \ln y_0$. Thus,

$$\ln y - \ln y_0 = kt$$

or

$$\ln \frac{y}{y_0} = kt$$

Changing to exponential form yields

$$\frac{y}{y_0} = e^{kt}$$

or, finally,

$$y = y_0 e^{kt}$$

Returning to the problem of world population, we choose to measure time t in years after January 1, 1975, and y in billions of people. Thus, $y_0 = 4$ and, since $k = 0.0198$,

$$y = 4e^{0.0198t}$$

By the year 2000, when $t = 25$, we can predict that y will be about

$$y = 4e^{0.0198(25)} \approx 6.6 \text{ billion}$$

Naturally, we used our calculator to make the last calculation.

EXAMPLE 1 How long will it take world population to double under the assumptions above?

Solution The question is equivalent to asking how many years after 1975 it will take for the population to reach 8 billion. We need to solve

$$8 = 4e^{0.0198t}$$

for t. After dividing both sides by 4, we take logarithms.

$$\ln 2 = 0.0198t$$

$$t = \frac{\ln 2}{0.0198} \approx 35 \text{ years} \qquad \blacksquare$$

If world population will double in the first 35 years after 1975, it will double in any 35-year period; so, for example, it will quadruple in 70 years. More generally, if an exponentially growing quantity doubles in an initial interval of length T, it will double in *any* interval of length T, since

$$\frac{y(t + T)}{y(t)} = \frac{y_0 e^{k(t + T)}}{y_0 e^{kt}} = \frac{y_0 e^{kT}}{y_0} = 2$$

We call the number T the **doubling time**.

EXAMPLE 2 The number of bacteria in a rapidly growing culture was estimated to be 10,000 at noon and 40,000 after 2 hours. How many bacteria will there be at 5 P.M.?

Solution We assume that the differential equation $dy/dt = ky$ is applicable, so $y = y_0 e^{kt}$. Now we have two conditions ($y_0 = 10,000$ and $y = 40,000$ at $t = 2$), from which we conclude that

$$40,000 = 10,000 e^{k(2)}$$

or

$$4 = e^{2k}$$

Taking logarithms yields

$$\ln 4 = 2k$$

or

$$k = \tfrac{1}{2} \ln 4 = \ln \sqrt{4} = \ln 2 \approx 0.693$$

Thus,

$$y = 10,000 e^{0.693t}$$

and, at $t = 5$, this gives

$$y = 10{,}000e^{0.693(5)} \approx 320{,}000 \qquad \blacksquare$$

The exponential model for population growth is flawed since it projects faster and faster growth indefinitely far into the future (Figure 1). In most cases (including that of world population), the limited amount of space and resources will eventually force a slowing of the growth rate. This suggests another model for population growth, called the **logistic model**, in which we assume the rate of growth is proportional both to the population size y and to the difference $L - y$, where L is the maximum population that can be supported. This leads to the differential equation

$$\frac{dy}{dt} = ky(L - y)$$

Note that for small y, $dy/dt \approx kLy$ which gives exponential type growth. But as y nears L, dy/dt gets smaller and smaller, curtailing the growth and producing a growth curve like Figure 2. This model is explored in Problems 23 and 24 of this section and again in Problems 26 and 27 of Section 8.5.

RADIOACTIVE DECAY   Not everything grows. In particular, the radioactive elements decay, and they do it at a rate proportional to the amount present. Thus, their change rates also satisfy the differential equation

$$\frac{dy}{dt} = ky$$

but now with k negative. It is still true that $y = y_0 e^{kt}$ is the solution to this equation. A typical graph appears in Figure 3.

EXAMPLE 3   Carbon-14, one of the three isotopes of carbon, is radioactive and decays at a rate proportional to the amount present. Its **half-life** is 5730 years; that is, it takes 5730 years for a given amount of carbon-14 to decay to one-half its original size. If 10 grams were present originally, how much will be left after 2000 years?

Solution   The half-life of 5730 allows us to determine k, since it implies that

$$\frac{1}{2} = 1e^{k(5730)}$$

or after taking logarithms,

$$-\ln 2 = 5730k$$

$$k = \frac{-\ln 2}{5730} \approx -0.000121$$

FIGURE 1

FIGURE 2

FIGURE 3

Thus,

$$y = 10e^{-0.000121t}$$

At $t = 2000$, this gives

$$y = 10e^{-0.000121(2000)} \approx 7.85 \text{ grams} \qquad \blacksquare$$

In Problem 13, we show how Example 3 may be used to determine the age of fossils and other once-living things.

COMPOUND INTEREST If we put \$100 in the bank at 12% interest compounded monthly, it will be worth \$100(1.01) at the end of 1 month, \$100(1.01)^2 at the end of 2 months, and \$100(1.01)^{12} at the end of 12 months, or 1 year. More generally if we put A_0 dollars in the bank at $100r$ percent compounded n times per year, it will be worth $A(t)$ dollars at the end of t years, where

$$A(t) = A_0 \left(1 + \frac{r}{n} \right)^{nt}$$

EXAMPLE 4 Suppose that John put \$500 in the bank at 13% interest, compounded daily. How much will it be worth at the end of 2 years?

Solution Here $r = 0.13$ and $n = 365$, so

$$A = 500 \left(1 + \frac{0.13}{365} \right)^{365(2)} \approx \$648.43$$

Of course, we did this calculation on a calculator using the $\boxed{y^x}$ key. $\qquad \blacksquare$

Now let us consider what happens when interest is **compounded continuously**—that is, when n, the number of compounding periods in a year, tends to infinity. Then we claim

$$A(t) = \lim_{n \to \infty} A_0 \left(1 + \frac{r}{n} \right)^{nt} = A_0 \lim_{n \to \infty} \left[\left(1 + \frac{r}{n} \right)^{n/r} \right]^{rt}$$

$$= A_0 \left[\lim_{h \to 0} (1 + h)^{1/h} \right]^{rt} = A_0 e^{rt}$$

Here we replaced r/n by h and noted that $n \to \infty$ corresponds to $h \to 0$. But the big step is knowing that the expression in brackets is the number e. That is important enough to call it a theorem.

Theorem A

$$\lim_{h \to 0} (1 + h)^{1/h} = e$$

Proof First recall that if $f(x) = \ln x$, then $f'(x) = 1/x$ and, in particular, $f'(1) = 1$. But by the definition of derivative,

$$1 = f'(1) = \lim_{h \to 0} \frac{f(1+h) - f(1)}{h} = \lim_{h \to 0} \frac{\ln(1+h) - \ln 1}{h}$$

$$= \lim_{h \to 0} \frac{1}{h} \ln(1+h) = \lim_{h \to 0} \ln(1+h)^{1/h}$$

Thus, $\lim_{h \to 0} \ln(1+h)^{1/h} = 1$, a result we will use in a moment. Now, $g(x) = e^x$ is a continuous function, and it therefore follows that we can take the limit into the exponent in the argument below.

$$\lim_{h \to 0} (1+h)^{1/h} = \lim_{h \to 0} \exp[\ln(1+h)^{1/h}] = \exp\left[\lim_{h \to 0} \ln(1+h)^{1/h}\right]$$

$$= \exp 1 = e \qquad \blacksquare$$

For another proof of Theorem A, see Problem 46 of Section 7.4.

EXAMPLE 5 Suppose the bank of Example 4 compounded interest continuously. How much would John then have at the end of 2 years?

Solution

$$A(t) = A_0 e^{rt} = 500 e^{(0.13)(2)} \approx \$648.47$$

Note that, though some banks try to get advertising mileage out of offering continuous compounding of interest, the difference in yields between continuous and daily compounding (which most banks offer) is minuscule.

$\blacksquare$

Here is another approach to the problem of continuous compounding of interest. Let A be the value at time t of A_0 dollars invested at the interest rate r. To say that interest is compounded continuously is to say that the instantaneous rate of change of A with respect to time is rA—that is,

$$\frac{dA}{dt} = rA$$

This differential equation was solved at the beginning of the section; its solution is $A = A_0 e^{rt}$.

PROBLEM SET 7.5 _____

In most of these problems, you will need to use a calculator (or Tables III and IV) to complete the calculation. In Problems 1–4, solve the given differential equation subject to the given condition. Note that $y(a)$ denotes the value of y at $t = a$.

1. $\dfrac{dy}{dt} = -5y$, $y(0) = 4$

2. $\dfrac{dy}{dt} = 6y$, $y(0) = 5$

3. $\dfrac{dy}{dt} = 0.006y$, $y(10) = 2$

4. $\dfrac{dy}{dx} = -0.003y$, $y(-2) = 4$

5. A bacterial population grows at a rate proportional to its size. Initially, it is 10,000 and after 10 days it is 24,000. What is the population after 25 days? See Example 2.

6. How long will it take the population of Problem 5 to double? See Example 1.

7. How long will it take the population of Problem 5 to triple? See Example 1.

8. The population of the United States was 4 million in 1790 and 180 million in 1960. If the rate of growth is assumed proportional to the number present, what estimate would you give for the population in 2020?

9. The population of a certain country is growing at 3.2% per year; that is, if it is A at the beginning of a year, it is $1.032A$ at the end of that year. Assuming it is 4.5 million now, what will it be at the end of 1 year? 2 years? 10 years? 100 years?

10. Determine the proportionality constant k in $dy/dt = ky$ for Problem 9. Then use $y = 4.5e^{kt}$ to find the population after 100 years.

11. A radioactive substance has a half-life of 810 years. If there were 10 grams initially, how much would be left after 300 years?

12. If a radioactive substance loses 15% of its radioactivity in 3 days, what is its half-life?

13. (Carbon Dating) All living things contain carbon-12, which is stable, and carbon-14, which is radioactive. While a plant or animal is alive, the ratio of these two isotopes of carbon remains unchanged since the carbon-14 is constantly renewed; after death, no more carbon-14 is absorbed. The half-life of carbon-14 is 5730 years. If charred logs of an old fort showed only 70% of the carbon-14 expected in living matter, when did the fort burn down? Assume the fort burned soon after it was built of freshly sawn logs.

14. Human hair from a grave in Africa proved to have only 51% of the carbon-14 of living tissue. When was the body buried? See Problem 13.

15. Newton's law of cooling states that the rate at which an object cools is proportional to the difference in temperature between the object and the surrounding medium. Thus, if an object is taken from an oven at 300°F and left to cool in a room at 75°F, its temperature T after t hours will satisfy the differential equation

$$\frac{dT}{dt} = k(T - 75)$$

If the temperature fell to 200°F in $\frac{1}{2}$ hour, what will it be after 3 hours?

16. A thermometer registered -20°C outside and then was brought in the house where the temperature was 24°C.

After 5 minutes, it registered 0°C. When will it register 20°C? See Problem 15.

17. If $375 is put in the bank today, what will it be worth at the end of 2 years if interest is at 9.5% and is compounded as specified?

(a) Annually   (b) Monthly

(c) Daily   (d) Continuously

Hint: See Examples 4 and 5.

18. Do Problem 17 assuming the interest rate is 14.4%.

19. How long does it take money to double in value for the specified interest?

(a) 12% compounded monthly

(b) 12% compounded continuously

20. Inflation between 1977 and 1981 ran at about 11.5% per year. On this basis, what would you expect a car that cost $4000 in 1977 to cost in 1981?

21. Manhattan Island is said to have been bought from the Indians by Peter Minuit in 1626 for $24. Suppose Minuit had instead put the $24 in the bank at 6% interest compounded continuously. What was that $24 worth in 1984? It would be interesting to compare this result with the actual value of Manhattan Island in 1984.

22. If Methusaleh's parents had put $100 in the bank for him at birth and left it there, what would Methusaleh have had at his death (969 years later) if interest was 8% compounded annually?

23. It will be shown later that for small x, $\ln(1 + x) \approx x$. Use this fact to show that the doubling time for money invested at p percent compounded annually is about $70/p$ years.

24. The equation for logistic growth is

$$\frac{dy}{dt} = ky(L - y)$$

Show that this differential equation has the solution

$$y = \frac{Ly_0}{y_0 + (L - y_0)e^{-Lkt}}$$

Hint: $\dfrac{1}{y(L - y)} = \dfrac{1}{Ly} + \dfrac{1}{L(L - y)}$.

25. Sketch the graph of the solution in Problem 24 when $y_0 = 4$, $L = 16$, $k = 0.00186$ (a *logistic model* for world population—see the discussion at the beginning of this section). Note that $\lim\limits_{t \to \infty} y = 16$.

26. Find each of the following limits.

(a) $\lim\limits_{x \to 0}(1 + x)^{1000}$ (b) $\lim\limits_{x \to 0}(1)^{1/x}$

(c) $\lim\limits_{x \to 0^+} (1 + \varepsilon)^{1/x}$, $\varepsilon > 0$ (d) $\lim\limits_{x \to 0^-} (1 + \varepsilon)^{1/x}$, $\varepsilon > 0$

(e) $\lim\limits_{x \to 0} (1 + x)^{1/x}$

27. Use the fact that $e = \lim\limits_{h \to 0} (1 + h)^{1/h}$ to find each limit.

(a) $\lim\limits_{x \to 0} (1 - x)^{1/x}$ *Hint*: $(1 - x)^{1/x} = [(1 - x)^{1/-x}]^{-1}$

(b) $\lim\limits_{x \to 0} (1 + 3x)^{1/x}$ (c) $\lim\limits_{n \to \infty} \left(\dfrac{n + 2}{n}\right)^n$

(d) $\lim\limits_{n \to \infty} \left(\dfrac{n - 1}{n}\right)^{2n}$

28. Show that the differential equation

$$\frac{dy}{dt} = ay + b$$

has solution

$$y = \left(y_0 + \frac{b}{a}\right)e^{at} - \frac{b}{a}$$

Assume $a \neq 0$.

29. Consider a state which had a population of 10 million in 1985, a natural exponential growth rate of 1.2% per year, and immigration from other states and countries of 60,000 per year. Use the differential equation of Problem 28 to model this situation and predict the population in 2010. Take $a = 0.012$.

30. Important news is said to diffuse through an adult population of fixed size L at a time rate proportional to the number of people who have not heard the news. Five days after a scandal in city hall was reported, a poll showed that half the people had heard it. How long will it take for 99% of the people to hear it?

31. Assume (1) world population will continue to grow exponentially with growth constant $k = 0.0198$ indefinitely into the future, (2) it takes $\frac{1}{2}$ acre of land to supply food for a person, and (3) there are 13,500,000 square miles of arable land in the world. How long will it be before the world reaches the maximum population? *Note*: There were 4 billion people in 1975 and 1 square mile is 640 acres.

32. Let E be a differentiable function satisfying $E(u + v) = E(u)E(v)$ for all u and v. Find a formula for $E(x)$. *Hint*: First find $E'(x)$.

7.6 The Inverse Trigonometric Functions

The six basic trigonometric functions (sine, cosine, tangent, cotangent, secant, and cosecant) were defined in Section 2.3 and we have used them occasionally in examples and problems. With respect to the notion of inverse, they are miserable functions, since for each y in their range, there are infinitely many x's that correspond to it (Figure 1). Nonetheless, we are going to introduce a notion of inverse for them. That this is possible rests on a procedure called **restricting the domain**, which was discussed briefly in Section 7.2.

INVERSE SINE AND INVERSE COSINE   In the case of sine and cosine, we restrict the domain, keeping the range as large as possible while insisting that the resulting function have an inverse. This can be done in many ways, but the agreed procedure is shown in Figures 2 and 3. We show also the graph

$y = \sin x$

FIGURE 1

FIGURE 2

FIGURE 3

of the corresponding inverse function, obtained—as usual—by reflecting across the line $y = x$.

We formalize what we have shown in a definition.

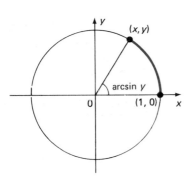

FIGURE 4

Definition

To obtain inverses for sine and cosine, we restrict their domains to $[-\pi/2, \pi/2]$ and $[0, \pi]$, respectively. Thus

$$x = \sin^{-1}y \quad \Leftrightarrow \quad y = \sin x \quad \text{and} \quad \frac{-\pi}{2} \le x \le \frac{\pi}{2}$$

$$x = \cos^{-1}y \quad \Leftrightarrow \quad y = \cos x \quad \text{and} \quad 0 \le x \le \pi$$

The symbol arcsin is often used for $\sin^{-1}$, and arccos is similarly used for $\cos^{-1}$. Think of arcsin as meaning "the arc whose sine is," or "the angle whose sine is" (Figure 4).

EXAMPLE 1 Calculate (a) $\sin^{-1}(\sqrt{2}/2)$, (b) $\sin^{-1}(-\frac{1}{2})$, (c) $\cos^{-1}(\sqrt{3}/2)$, (d) $\cos^{-1}(-\frac{1}{2})$, (e) $\cos(\cos^{-1}0.6)$, and (f) $\sin^{-1}(\sin 3\pi/2)$.

Solution

(a) $\sin^{-1}\left(\dfrac{\sqrt{2}}{2}\right) = \dfrac{\pi}{4}$ (b) $\sin^{-1}\left(-\dfrac{1}{2}\right) = -\dfrac{\pi}{6}$

(c) $\cos^{-1}\left(\dfrac{\sqrt{3}}{2}\right) = \dfrac{\pi}{6}$ (d) $\cos^{-1}\left(-\dfrac{1}{2}\right) = \dfrac{2\pi}{3}$

(e) $\cos(\cos^{-1}0.6) = 0.6$ (f) $\sin^{-1}\left(\sin \dfrac{3\pi}{2}\right) = -\dfrac{\pi}{2}$

The only one of these that is tricky is (f). Note that it would be wrong to give $3\pi/2$ as the answer, since $\sin^{-1}y$ is always in the interval $[-\pi/2, \pi/2]$. Work the problem in steps, as follows.

$$\sin^{-1}\left(\sin \frac{3\pi}{2}\right) = \sin^{-1}(-1) = -\pi/2$$

■

EXAMPLE 2 Calculate (a) $\cos^{-1}(-0.61)$, (b) $\sin^{-1}(-0.87)$, (c) $\sin^{-1}(1.21)$, and (d) $\sin^{-1}(\sin 4.13)$.

Solution Use a calculator in radian mode. It has been programmed to give answers that are consistent with the definitions we have given.

(a) $\cos^{-1}(-0.61) = 2.2268569$
(b) $\sin^{-1}(-0.87) = -1.0552023$
(c) Your machine should indicate an error, since $\sin^{-1}(1.21)$ does not exist.
(d) $\sin^{-1}(\sin 4.13) = -0.9884073$ ∎

INVERSE TANGENT In Figure 5, we have shown the graph of the tangent function, its restricted domain, and the graph of $y = \tan^{-1}x$.

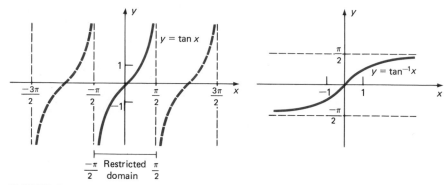

FIGURE 5

Definition

To obtain an inverse for tangent, we restrict the domain to $(-\pi/2, \pi/2)$. Thus,

$$x = \tan^{-1}y \quad \Leftrightarrow \quad y = \tan x \quad \text{and} \quad -\frac{\pi}{2} < x < \frac{\pi}{2}$$

There is a standard way to restrict the domain of the cotangent function, namely to $(0, \pi)$, so that it has an inverse. However, this function does not play a significant role in calculus, so we will say no more about it.

EXAMPLE 3 Calculate (a) $\tan^{-1}(1)$, (b) $\tan^{-1}(-\sqrt{3})$, (c) $\tan^{-1}(-0.145)$, and (d) $\tan^{-1}(\tan 5.236)$.

Solution

(a) $\tan^{-1}(1) = \frac{\pi}{4}$

(b) $\tan^{-1}(-\sqrt{3}) = -\dfrac{\pi}{3}$

(c) $\tan^{-1}(-0.145) = -0.1439964$

(d) $\tan^{-1}(\tan 5.236) = -1.0471853$

The results in (c) and (d) were obtained on a calculator. ■

INVERSE SECANT   Once again we begin by graphing the secant function, restricting its domain to obtain an inverse, and then graphing $y = \sec^{-1}x$ (Figure 6).

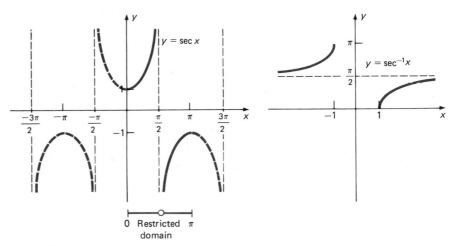

FIGURE 6

Definition

To obtain an inverse for secant, we restrict the domain to

$$[0, \pi/2) \cup (\pi/2, \pi].$$

Thus,

$$x = \sec^{-1}y \quad \Leftrightarrow \quad y = \sec x \quad \text{and} \quad 0 \le x \le \pi, \, x \ne \dfrac{\pi}{2}$$

Some authors restrict the domain of the secant in a different way. Thus, if you refer to another book, you must check that author's definition. We will have no need to define $\csc^{-1}$, though this can also be done.

EXAMPLE 4   Calculate (a) $\sec^{-1}(-1)$, (b) $\sec^{-1}(2)$, and (c) $\sec^{-1}(-1.32)$.

Solution   Most of us have trouble remembering our secants; moreover, most calculators fail to have a secant button. Therefore, we suggest that you remember that $\sec x = 1/\cos x$. From this, it follows that

$$\sec^{-1}y = \cos^{-1}\left(\frac{1}{y}\right)$$

and this allows us to use known facts about the cosine.

(a) $\sec^{-1}(-1) = \cos^{-1}(-1) = \pi$

(b) $\sec^{-1}(2) = \cos^{-1}\left(\frac{1}{2}\right) = \frac{\pi}{3}$

(c) $\sec^{-1}(-1.32) = \cos^{-1}\left(-\frac{1}{1.32}\right) = \cos^{-1}(-0.7575758)$

$$= 2.4303875 \qquad \blacksquare$$

FOUR USEFUL IDENTITIES   In the next section, we will need each of the following identities. You can recall them by reference to the triangles in the margin (Figure 7).

(i) $\sin(\cos^{-1}x) = \sqrt{1 - x^2}$

(ii) $\cos(\sin^{-1}x) = \sqrt{1 - x^2}$

(iii) $\sec(\tan^{-1}x) = \sqrt{1 + x^2}$

(iv) $\tan(\sec^{-1}x) = \pm\sqrt{x^2 - 1}$

To prove (i), recall that $\sin^2\theta + \cos^2\theta = 1$; in particular, if $0 \le \theta \le \pi$, then

$$\sin\theta = \sqrt{1 - \cos^2\theta}$$

Now apply this with $\theta = \cos^{-1}x$ and use the fact that $\cos(\cos^{-1}x) = x$ to get

$$\sin(\cos^{-1}x) = \sqrt{1 - \cos^2(\cos^{-1}x)} = \sqrt{1 - x^2}$$

Identity (ii) is proved in a completely similar manner. To prove (iii) and (iv), use the identity $\sec^2\theta = 1 + \tan^2\theta$ in place of $\sin^2\theta + \cos^2\theta = 1$. We note that we cannot eliminate the $\pm$ in (iv). If $x \ge 1$, we must use the plus sign; if $x \le -1$, we must use the minus sign.

EXAMPLE 5   Calculate $\sin[2\cos^{-1}(\tfrac{2}{3})]$.

Solution   Recall the double-angle identity $\sin 2\theta = 2\sin\theta\cos\theta$. Thus,

$$\sin\left[2\cos^{-1}\left(\frac{2}{3}\right)\right] = 2\sin\left[\cos^{-1}\left(\frac{2}{3}\right)\right]\cos\left[\cos^{-1}\left(\frac{2}{3}\right)\right]$$

$$= 2 \cdot \sqrt{1 - \left(\frac{2}{3}\right)^2} \cdot \frac{2}{3} = \frac{4\sqrt{5}}{9} \qquad \blacksquare$$

FIGURE 7

EXAMPLE 6 Show that $\cos(2 \tan^{-1}x) = \dfrac{1 - x^2}{1 + x^2}$.

Solution We apply the double-angle identity $\cos 2\theta = 2 \cos^2\theta - 1$, with $\theta = \tan^{-1}x$. Thus,

$$\cos(2 \tan^{-1}x) = \cos 2\theta$$
$$= 2 \cos^2\theta - 1$$
$$= \frac{2}{\sec^2\theta} - 1$$
$$= \frac{2}{1 + \tan^2\theta} - 1$$
$$= \frac{2}{1 + x^2} - 1$$
$$= \frac{1 - x^2}{1 + x^2} \qquad \blacksquare$$

PROBLEM SET 7.6

In Problems 1–12, find exact values without use of a calculator.

1. $\sin^{-1}\left(\dfrac{\sqrt{3}}{2}\right)$

2. $\cos^{-1}(\tfrac{1}{2})$

3. $\arcsin\left(-\dfrac{\sqrt{2}}{2}\right)$

4. $\arccos\left(-\dfrac{\sqrt{2}}{2}\right)$

5. $\tan^{-1}(-\sqrt{3})$

6. $\sec^{-1}(-\tfrac{1}{2})$

7. $\arccos(-\tfrac{1}{2})$

8. $\arctan\left(-\dfrac{\sqrt{3}}{3}\right)$

9. $\sec^{-1}(-2)$

10. $\sin^{-1}(-1)$

11. $\sin(\sin^{-1}0.541)$

12. $\sin(\cos^{-1}0.6)$

[C] In Problems 13–24, use a calculator (or tables) to approximate each value.

13. $\sin^{-1}(0.1259)$

14. $\cos^{-1}(0.6341)$

15. $\arctan(-5.123)$

16. $\arcsin(-0.5431)$

17. $\cos(\arctan 3.125)$

18. $\sec(\arcsin 0.5246)$

19. $\sec^{-1}(-2.314)$

20. $\tan^{-1}(-59.34)$

21. $\cos(\sin(\tan^{-1}2.312))$

22. $\sin^2(\ln(\cos 0.5695))$

23. $\dfrac{\sin^{-1}(-0.9812)}{1 + \cos^3(5.321)}$

24. $\dfrac{2 \arccos(-0.1567)}{\tan 5.193 + \tan^{-1}5.193}$

In Problems 25–30, express θ in terms of x using the inverse trigonometric functions $\sin^{-1}$, $\cos^{-1}$, $\tan^{-1}$, and $\sec^{-1}$.

25.

26.

27.

28.
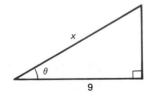

SECTION 7.6/The Inverse Trigonometric Functions

353

29.

30.

Calculate each of the following without using a calculator (see Example 5).

31. $\cos[2 \sin^{-1}(-\frac{2}{3})]$

32. $\tan[2 \tan^{-1}(\frac{1}{3})]$

33. $\sin[\cos^{-1}(\frac{3}{5}) + \cos^{-1}(\frac{5}{13})]$

34. $\cos[\cos^{-1}(\frac{4}{5}) + \sin^{-1}(\frac{12}{13})]$

In Problems 35–40, show that the equations are identities (see Example 6).

35. $\tan(\sin^{-1}x) = \dfrac{x}{\sqrt{1-x^2}}$

36. $\sin(\tan^{-1}x) = \dfrac{x}{\sqrt{1+x^2}}$

37. $\cos(2 \sin^{-1}x) = 1 - 2x^2$

38. $\tan(2 \tan^{-1}x) = \dfrac{2x}{1-x^2}$

39. $\cos(\tan^{-1}x) = \dfrac{1}{\sqrt{1+x^2}}$

40. $\sec(2 \cos^{-1}x) = \dfrac{1}{2x^2-1}$

41. Find each limit.

(a) $\displaystyle \lim_{x \to \infty} \tan^{-1}x$

(b) $\displaystyle \lim_{x \to -\infty} \tan^{-1}x$

42. Find each limit.

(a) $\displaystyle \lim_{x \to \infty} \sec^{-1}x$

(b) $\displaystyle \lim_{x \to -\infty} \sec^{-1}x$

43. Sketch the graph of $y = \cot^{-1}x$, assuming it has been obtained by restricting the domain of the cotangent to $(0, \pi)$.

$\boxed{C}$ **44.** A picture 5 feet high is hung on a wall so that its bottom is 8 feet from the floor, as shown in Figure 8. A viewer with eye level at 5.4 feet stands b feet from the wall. Express θ, the vertical angle subtended by the picture at her eye, in terms of b and then find θ if $b = 12.9$ feet.

FIGURE 8

45. Find formulas for $f^{-1}(x)$ for each of the following functions f, first indicating how you would restrict the domain so that f has an inverse. For example, if $f(x) = 3 \sin 2x$ and we restrict the domain to $-\pi/4 \le x \le \pi/4$, then $f^{-1}(x) = \frac{1}{2} \sin^{-1}(x/3)$.

(a) $f(x) = 3 \cos 2x$ (b) $f(x) = 2 \sin 3x$

(c) $f(x) = \frac{1}{2} \tan x$ (d) $f(x) = \sin \dfrac{1}{x}$

46. Show by repeated use of the addition formula $\tan(x + y) = (\tan x + \tan y)/(1 - \tan x \tan y)$ that

$$\frac{\pi}{4} = 3 \tan^{-1}\left(\frac{1}{4}\right) + \tan^{-1}\left(\frac{5}{99}\right)$$

47. Verify that

$$\frac{\pi}{4} = 4 \tan^{-1}\left(\frac{1}{5}\right) - \tan^{-1}\left(\frac{1}{239}\right)$$

a result discovered by John Machin in 1706 and used by him to calculate the first 100 decimal places of π.

48. Sketch the graph of $y = \sin^{-1}(\sin x)$ for $-\pi/2 \le x \le 6\pi$ and from it deduce that

$$D_x \sin^{-1}(\sin x) = \frac{\cos x}{|\cos x|}$$

49. Without using calculus, find a formula for the area of the shaded region in Figure 9 in terms of a and b. Note that the center of the larger circle is on the rim of the smaller.

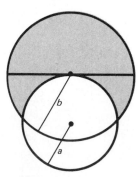

FIGURE 9

7.7
Derivatives of Trigonometric Functions

We learned in Section 3.4 that $D_x \sin x = \cos x$ and $D_x \cos x = -\sin x$. The other four trigonometric functions are defined in terms of sine and cosine by

$$\tan x = \frac{\sin x}{\cos x} \qquad \cot x = \frac{\cos x}{\sin x}$$

$$\sec x = \frac{1}{\cos x} \qquad \csc x = \frac{1}{\sin x}$$

These facts, together with the Quotient Rule for derivatives, allow us to find the derivatives of each of these functions. For example,

$$D_x \cot x = D_x\left(\frac{\cos x}{\sin x}\right) = \frac{\sin x(-\sin x) - \cos x \cos x}{\sin^2 x}$$

$$= \frac{-1}{\sin^2 x} = -\csc^2 x$$

Here is a summary of the six derivative formulas. They should be memorized.

$$\boxed{\begin{array}{ll} D_x \sin x = \cos x & D_x \cos x = -\sin x \\[2mm] D_x \tan x = \sec^2 x & D_x \cot x = -\csc^2 x \\[2mm] D_x \sec x = \sec x \tan x & D_x \csc x = -\csc x \cot x \end{array}}$$

COMPOSITE FUNCTIONS We can combine the rules above with the Chain Rule and with the other differentiation rules in more and more complicated ways. For example, if $u = f(x)$ is differentiable, then

$$D_x \sin u = \cos u \cdot D_x u$$

Here are several examples.

EXAMPLE 1   Find $D_x \sin(3x^2 + 4)$.

Solution   Let $u = 3x^2 + 4$.

$$D_x \sin(3x^2 + 4) = D_x \sin u = \cos u D_x u$$

$$= [\cos(3x^2 + 4)]6x = 6x \cos(3x^2 + 4) \qquad \blacksquare$$

EXAMPLE 2   Find $D_x \tan^2(9x)$.

Solution   This requires a double use of the Chain Rule.

$$D_x \tan^2(9x) = 2 \tan(9x) D_x \tan(9x)$$

$$= 2 \tan(9x)\sec^2(9x) D_x(9x)$$

$$= 2 \tan(9x)\sec^2(9x) \cdot 9$$

$$= 18 \tan(9x)\sec^2(9x) \qquad \blacksquare$$

EXAMPLE 3 If $y = (\sin^2 x)/(1 - \cot x)$, find dy/dx.

Solution

$$\frac{dy}{dx} = \frac{(1 - \cot x)D_x(\sin^2 x) - \sin^2 x D_x(1 - \cot x)}{(1 - \cot x)^2}$$

$$= \frac{(1 - \cot x)2 \sin x \cos x - \sin^2 x \csc^2 x}{(1 - \cot x)^2}$$

$$= \frac{2 \sin x \cos x - 2 \cos^2 x - 1}{(1 - \cot x)^2} \qquad ■$$

INVERSE TRIGONOMETRIC FUNCTIONS From the Inverse Function Theorem (Theorem 7.2B), we conclude that $\sin^{-1}$, $\cos^{-1}$, $\tan^{-1}$, and $\sec^{-1}$ are differentiable. Our aim is to find formulas for their derivatives. We state the results and then show from where they come.

(i) $\qquad D_x \sin^{-1} x = \dfrac{1}{\sqrt{1 - x^2}}, \qquad -1 < x < 1$

(ii) $\qquad D_x \cos^{-1} x = \dfrac{-1}{\sqrt{1 - x^2}}, \qquad -1 < x < 1$

(iii) $\qquad D_x \tan^{-1} x = \dfrac{1}{1 + x^2}$

(iv) $\qquad D_x \sec^{-1} x = \dfrac{1}{|x|\sqrt{x^2 - 1}}, \qquad |x| > 1$

Our derivations follow the same pattern in each case. To see (i), let $y = \sin^{-1} x$, so that

$$x = \sin y$$

Now differentiate both sides with respect to x, using the Chain Rule on the right-hand side. Then

$$1 = \cos y\, D_x y = \cos(\sin^{-1} x)D_x(\sin^{-1} x)$$
$$= \sqrt{1 - x^2}\, D_x(\sin^{-1} x)$$

At the last step, we used Identity (ii) from the end of Section 7.6. We conclude that $D_x(\sin^{-1} x) = 1/\sqrt{1 - x^2}$.

Results (ii), (iii), and (iv) are proved similarly, but (iv) has a little twist— so we will go through its demonstration. Let $y = \sec^{-1} x$, so

$$x = \sec y$$

Differentiating both sides with respect to x, we obtain

$$1 = \sec y \tan y D_x y$$

$$= \sec(\sec^{-1}x)\tan(\sec^{-1}x)D_x(\sec^{-1}x)$$

$$= x(\pm\sqrt{x^2 - 1})D_x(\sec^{-1}x)$$

At the last step, we used Identity (iv) from near the end of Section 7.6. But as explained in that section, we use the plus sign if $x > 1$ and the minus sign if $x < -1$. Thus,

$$1 = \pm x\sqrt{x^2 - 1}D_x(\sec^{-1}x)$$

$$= |x|\sqrt{x^2 - 1}D_x(\sec^{-1}x)$$

The desired result follows immediately.

EXAMPLE 4   Find $D_x \sin^{-1}(3x - 1)$.

Solution

$$D_x \sin^{-1}(3x - 1) = \frac{1}{\sqrt{1 - (3x - 1)^2}} D_x(3x - 1)$$

$$= \frac{3}{\sqrt{-9x^2 + 6x}} \quad \blacksquare$$

EXAMPLE 5   Find $D_x \tan^{-1}\sqrt{x + 1}$.

Solution

$$D_x \tan^{-1}\sqrt{x + 1} = \frac{1}{1 + (\sqrt{x + 1})^2} D_x\sqrt{x + 1}$$

$$= \frac{1}{x + 2} \cdot \frac{1}{2}(x + 1)^{-1/2}$$

$$= \frac{1}{2(x + 2)\sqrt{x + 1}} \quad \blacksquare$$

Of course, every differentiation formula leads to an integration formula, a matter we will say much more about in the next chapter. In particular:

(i) $\displaystyle\int \frac{1}{\sqrt{1 - x^2}} \, dx = \sin^{-1}x + C$;

(ii) $\displaystyle\int \frac{1}{1 + x^2} \, dx = \tan^{-1}x + C$;

(iii) $\displaystyle\int \frac{1}{x\sqrt{x^2 - 1}} \, dx = \sec^{-1}|x| + C$.

EXAMPLE 6 Evaluate $\int_0^{1/2} \dfrac{dx}{\sqrt{1-x^2}}$.

Solution

$$\int_0^{1/2} \frac{1}{\sqrt{1-x^2}}\, dx = [\sin^{-1}x]_0^{1/2} = \sin^{-1}\frac{1}{2} - \sin^{-1}0$$

$$= \frac{\pi}{6} - 0 = \frac{\pi}{6} \qquad \blacksquare$$

EXAMPLE 7 A man standing on top of a vertical cliff is 200 feet above a lake. As he watches, a motorboat moves directly away from the foot of the cliff at a rate of 25 feet per second. How fast is the angle of depression of his line of sight changing when the boat is 150 feet from the foot of the cliff?

Solution The essential details are shown in Figure 1. Note that θ, the angle of depression, is

$$\theta = \tan^{-1}\left(\frac{200}{x}\right)$$

Thus,

$$\frac{d\theta}{dt} = \frac{1}{1+(200/x)^2} \cdot \frac{-200}{x^2} \cdot \frac{dx}{dt} = \frac{-200}{x^2+40{,}000} \cdot \frac{dx}{dt}$$

When we substitute $x = 150$ and $dx/dt = 25$, we obtain $d\theta/dt = -0.08$ radians per second. $\qquad \blacksquare$

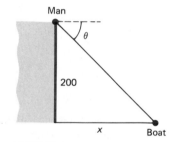

FIGURE 1

PROBLEM SET 7.7

In Problems 1–22, find dy/dx.

1. $y = \cos^2(x-2)$

2. $y = \sin\sqrt{2+7x}$

3. $y = \cot x \csc x$

4. $y = \dfrac{x}{1-\tan x}$

5. $y = e^{\cot x}$

6. $y = e^x \cos x$

7. $y = (\tan x)e^{\tan x}$

8. $y = \ln(\sec x + \tan x)$

9. $y = \sin^{-1}(x^2)$

10. $y = \frac{1}{2}\sin^{-1}(e^x)$

11. $y = 7\cos^{-1}\sqrt{2x}$

12. $y = (3x-1)\cos^{-1}(x^2)$

13. $y = \tan^{-1}(e^x)$

14. $y = e^x \sin^{-1}x$

15. $y = x\tan^{-1}x$

16. $y = (\tan^{-1}x)^2$

17. $y = \sin(\tan^{-1}x)$

18. $y = \tan(\sin^{-1}x)$

19. $y = \sec^{-1}(x^2)$

20. $y = (\sec^{-1}x)^2$

21. $y = \tan^{-1}\left(\dfrac{1-x}{1+x}\right)$

22. $y = (1+\sin^{-1}x)^6$

In Problems 23–36, find each of the integrals.

23. $\displaystyle\int x\sin(x^2)\,dx$

24. $\displaystyle\int \sin 2x \cos 2x\, dx$

25. $\displaystyle \int \tan x \, dx = \int \frac{\sin x}{\cos x} \, dx$

26. $\displaystyle \int \cot x \, dx$

27. $\displaystyle \int \frac{\sec^2 x}{\tan x} \, dx$

28. $\displaystyle \int \frac{\sin x}{\cos^2 x} \, dx$

29. $\displaystyle \int_0^1 e^{2x} \cos(e^{2x}) \, dx$

30. $\displaystyle \int_0^{\pi/2} \sin^2 x \cos x \, dx$

31. $\displaystyle \int_0^{\sqrt{2}/2} \frac{1}{\sqrt{1 - x^2}} \, dx$

32. $\displaystyle \int_{\sqrt{2}}^2 \frac{dx}{x\sqrt{x^2 - 1}}$

33. $\displaystyle \int_{-1}^1 \frac{1}{1 + x^2} \, dx$

34. $\displaystyle \int_0^{\pi/2} \frac{\sin \theta}{1 + \cos^2 \theta} \, d\theta$

35. $\displaystyle \int \frac{1}{1 + 4x^2} \, dx$

36. $\displaystyle \int \frac{e^x}{1 + e^{2x}} \, dx$

37. Show that

$$\int \frac{dx}{\sqrt{a^2 - x^2}} = \sin^{-1} \frac{x}{a} + C, \qquad a > 0$$

by writing $a^2 - x^2 = a^2[1 - (x/a)^2]$ and making the substitution $u = x/a$.

38. Show the result in Problem 37 by differentiating the right side to get the integrand.

39. Show

$$\int \frac{dx}{a^2 + x^2} = \frac{1}{a} \tan^{-1} \frac{x}{a} + C, \qquad a \neq 0$$

40. Show

$$\int \frac{dx}{x\sqrt{x^2 - a^2}} = \frac{1}{a} \sec^{-1} \frac{|x|}{a}, \qquad a > 0$$

41. Show, by differentiating the right side, that

$$\int \sqrt{a^2 - x^2} \, dx = \frac{x}{2}\sqrt{a^2 - x^2} + \frac{a^2}{2} \sin^{-1} \frac{x}{a} + C, \qquad a > 0$$

42. Use the result of Problem 41 to show

$$\int_{-a}^a \sqrt{a^2 - x^2} \, dx = \frac{\pi a^2}{2}$$

Why doesn't this surprise you?

43. Find the area of the region bounded by $x^2 y + y - 4 = 0$, the coordinate axes, and the line $x = 1$.

44. Find the volume of the solid generated by revolving about the x-axis the region bounded by $y = 5(x^2 + 1)^{-1/2}$, $y = 0$, $x = 0$, and $x = 4$.

45. The lower edge of a wall hanging, 10 feet in height, is 2 feet above the observer's eye level. Find the ideal distance b to stand from the wall for viewing the hanging; that is, find b that maximizes the angle subtended at the viewer's eye. (Compare with Problem 44 of Section 7.6.)

46. Express $d\theta/dt$ in terms of x, dx/dt, and the constants a and b.

(a)

(b)

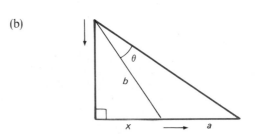

47. The structural steel work of a new office building is finished. Across the street, 60 feet from the ground floor of the freight elevator shaft in the building, a spectator is standing and watching the freight elevator ascend at a constant rate of 15 feet per second. How fast is the angle of elevation of the spectator's line of sight to the elevator increasing 6 seconds after his line of sight passes the horizontal?

48. An airplane is flying at a constant altitude of 2 miles and a constant speed of 600 miles per hour on a straight course that will take it directly over an observer on the ground. How fast is the angle of elevation of the observer's line of sight increasing when the distance from her to the plane is 3 miles? Give your result in radians per minute.

49. A revolving beacon light is located on an island and is 2 miles away from the nearest point P of the straight shoreline of the mainland. The beacon throws a spot of light that moves along the shoreline as the beacon revolves. If the speed of the spot of light on the shoreline is 5π miles per minute when the spot is 1 mile from P, how fast is the beacon revolving?

50. A man on a dock is pulling in a rope attached to a rowboat at a rate of 5 feet per second. If the man's hands are 8 feet higher than the point where the rope is attached to the boat, how fast is the angle of depression of the rope changing when there are still 17 feet of rope out?

51. A visitor from outer space is approaching the earth (radius = 6376 kilometers) at 2 kilometers per second. How fast is the angle θ subtended by the earth at her eye increasing when she is 3000 kilometers from the surface?

52. An object starts at the origin and moves along the y-axis so that its velocity in centimeters per second is $dy/dt = (0.2)\cos^2(0.1y)$. How long will it take to get to $y = 2.5\pi$ centimeters? To get to $y = 5\pi$ centimeters?

7.8
The Hyperbolic Functions and Their Inverses

In both mathematics and science, certain combinations of e^x and e^{-x} occur so often that they are given special names.

Definition

(The hyperbolic functions). The hyperbolic sine, hyperbolic cosine, and four related functions are defined by

$$\sinh x = \frac{1}{2}(e^x - e^{-x}) \qquad \cosh x = \frac{1}{2}(e^x + e^{-x})$$

$$\tanh x = \frac{\sinh x}{\cosh x} \qquad \coth x = \frac{\cosh x}{\sinh x}$$

$$\operatorname{sech} x = \frac{1}{\cosh x} \qquad \operatorname{csch} x = \frac{1}{\sinh x}$$

The terminology suggests that there must be some connection with the trigonometric functions; there is. First, the fundamental identity for the hyperbolic functions (reminiscent of $\cos^2 x + \sin^2 x = 1$ in trigonometry) is

$$\boxed{\cosh^2 x - \sinh^2 x = 1}$$

$x^2 + y^2 = 1$

FIGURE 1

To verify it, we write

$$\cosh^2 x - \sinh^2 x = \frac{e^{2x} + 2 + e^{-2x}}{4} - \frac{e^{2x} - 2 + e^{-2x}}{4} = 1$$

Second, recall that the trigonometric functions are intimately related to the unit circle (Figure 1), so much so that they are sometimes called the circular functions. In fact, the parametric equations $x = \cos t$, $y = \sin t$ describe the unit circle. In parallel fashion, the parametric equations $x = \cosh t$, $y = \sinh t$ describe the right branch of the unit hyperbola $x^2 - y^2 = 1$ (Figure 2). Moreover, in both cases the parameter t is related to the shaded area A by $t = 2A$, though this is not obvious in the second case (Problem 32).

Since $\sinh(-x) = -\sinh x$, sinh is an odd function; $\cosh(-x) = \cosh x$, so cosh is an even function. Correspondingly, the graph of $y = \sinh x$ is symmetric with respect to the origin and the graph of $y = \cosh x$ is symmetric with respect to the y-axis. The graphs are shown in Figure 3.

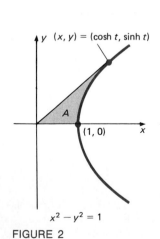

$x^2 - y^2 = 1$

FIGURE 2

360 CHAPTER 7/Transcendental Functions

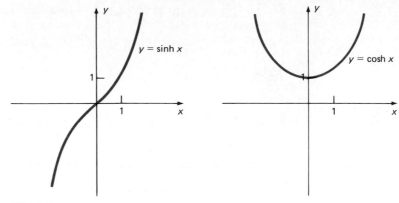

FIGURE 3

DERIVATIVES OF HYPERBOLIC FUNCTIONS   If we can find

$$D_x \sinh x \quad \text{and} \quad D_x \cosh x$$

the derivatives of the other four hyperbolic functions are easily found by the quotient rule.

$$D_x \sinh x = D_x \left(\frac{e^x - e^{-x}}{2} \right) = \frac{e^x + e^{-x}}{2} = \cosh x$$

and

$$D_x \cosh x = D_x \left(\frac{e^x + e^{-x}}{2} \right) = \frac{e^x - e^{-x}}{2} = \sinh x$$

Note that these facts confirm the character of the graphs we drew. For example, since $D_x(\sinh x) = \cosh x > 0$, the graph of hyperbolic sine is always increasing. Similarly $D_x^2(\cosh x) = \cosh x > 0$, which means that the graph of hyperbolic cosine is concave upward.

Here is a list of the six differentiation formulas.

$$
\begin{array}{ll}
D_x \sinh x = \cosh x & D_x \cosh x = \sinh x \\
D_x \tanh x = \text{sech}^2 x & D_x \coth x = -\text{csch}^2 x \\
D_x \text{sech } x = -\text{sech } x \tanh x \\
D_x \text{csch } x = -\text{csch } x \coth x
\end{array}
$$

EXAMPLE 1   Find $D_x \tanh(\sin x)$.

Solution

$$D_x \tanh(\sin x) = \text{sech}^2(\sin x) D_x(\sin x)$$
$$= \cos x \cdot \text{sech}^2(\sin x) \qquad \blacksquare$$

EXAMPLE 2 Find $D_x \cosh^2(3x - 1)$.

Solution We apply the Chain Rule twice.

$$D_x \cosh^2(3x - 1) = 2 \cosh(3x - 1)D_x \cosh(3x - 1)$$
$$= 2 \cosh(3x - 1)\sinh(3x - 1)D_x(3x - 1)$$
$$= 6 \cosh(3x - 1)\sinh(3x - 1) \qquad \blacksquare$$

EXAMPLE 3 Find $\int \tanh x \, dx$.

Solution Let $u = \cosh x$, so $du = \sinh x \, dx$.

$$\int \tanh x \, dx = \int \frac{\sinh x}{\cosh x} \, dx = \int \frac{1}{u} \, du$$
$$= \ln|u| + C = \ln|\cosh x| + C = \ln(\cosh x) + C$$

We could drop the absolute value signs because $\cosh x > 0$. $\qquad \blacksquare$

INVERSE HYPERBOLIC FUNCTIONS Since hyperbolic sine and hyperbolic tangent have positive derivatives, they are increasing functions and automatically have inverses. To obtain inverses for hyperbolic cosine and hyperbolic secant, we restrict their domains to $x \geq 0$. Thus,

$$x = \sinh^{-1} y \quad \Leftrightarrow \quad y = \sinh x$$
$$x = \cosh^{-1} y \quad \Leftrightarrow \quad y = \cosh x \quad \text{and} \quad x \geq 0$$
$$x = \tanh^{-1} y \quad \Leftrightarrow \quad y = \tanh x$$
$$x = \operatorname{sech}^{-1} y \quad \Leftrightarrow \quad y = \operatorname{sech} x \quad \text{and} \quad x \geq 0$$

Since the hyperbolic functions are defined in terms of e^x and e^{-x}, it is not too surprising that the inverse hyperbolic functions can be expressed in terms of the natural logarithm. For example, consider $y = \cosh x$ for $x \geq 0$; that is, consider

$$y = \frac{e^x + e^{-x}}{2}, \qquad x \geq 0$$

and let us try to solve for x. Multiplying both members by $2e^x$, we get $2ye^x = e^{2x} + 1$, or

$$(e^x)^2 - 2ye^x + 1 = 0, \qquad x \geq 0$$

If we solve this quadratic equation in e^x, we obtain

$$e^x = \frac{2y \pm \sqrt{(2y)^2 - 4}}{2} = y \pm \sqrt{y^2 - 1}$$

or, after taking natural logarithms,

$$x = \ln(y \pm \sqrt{y^2 - 1})$$

The condition $x \geq 0$ forces us to choose the plus sign, so

$$x = \cosh^{-1} y = \ln(y + \sqrt{y^2 - 1})$$

Similar arguments apply to each of the inverse hyperbolic functions. We obtain the following results (note that the roles of x and y have been interchanged).

$$\sinh^{-1} x = \ln(x + \sqrt{x^2 + 1})$$

$$\cosh^{-1} x = \ln(x + \sqrt{x^2 - 1}), \qquad x \geq 1$$

$$\tanh^{-1} x = \frac{1}{2} \ln \frac{1 + x}{1 - x}, \qquad -1 < x < 1$$

$$\text{sech}^{-1} x = \ln\left(\frac{1 + \sqrt{1 - x^2}}{x}\right), \qquad 0 < x \leq 1$$

Each of these functions is differentiable. In fact,

$$D_x \sinh^{-1} x = \frac{1}{\sqrt{x^2 + 1}}$$

$$D_x \cosh^{-1} x = \frac{1}{\sqrt{x^2 - 1}}, \qquad x > 1$$

$$D_x \tanh^{-1} x = \frac{1}{1 - x^2}, \qquad -1 < x < 1$$

$$D_x \text{sech}^{-1} x = \frac{-1}{x\sqrt{1 - x^2}}, \qquad 0 < x < 1$$

EXAMPLE 4 Show that $D_x \sinh^{-1} x = 1/\sqrt{x^2 + 1}$ by two different methods.

Solution

METHOD 1   Let $y = \sinh^{-1} x$, so

$$x = \sinh y$$

Now differentiate both sides with respect to x.

$$1 = (\cosh y) D_x y$$

Thus,

$$D_x y = D_x(\sinh^{-1} x) = \frac{1}{\cosh y} = \frac{1}{\sqrt{1 + \sinh^2 y}} = \frac{1}{\sqrt{1 + x^2}}$$

METHOD 2   Use the logarithmic expression for $\sinh^{-1}x$.

$$D_x(\sinh^{-1}x) = D_x \ln(x + \sqrt{x^2 + 1})$$

$$= \frac{1}{x + \sqrt{x^2 + 1}} D_x(x + \sqrt{x^2 + 1})$$

$$= \frac{1}{x + \sqrt{x^2 + 1}} \left(1 + \frac{x}{\sqrt{x^2 + 1}}\right)$$

$$= \frac{1}{\sqrt{x^2 + 1}} \qquad \blacksquare$$

$y = a \cosh \dfrac{x}{a}$

$(0, a)$

The catenary

FIGURE 4

APPLICATIONS: THE CATENARY   If a homogeneous flexible cable or chain is suspended between two fiixed points at the same height, it forms a curve called a **catenary** (Figure 4). Furthermore (see Problem 29), a catenary can be placed in a coordinate system so that its equation takes the form

$$y = a \cosh \frac{x}{a}$$

EXAMPLE 5   Find the length of the catenary $y = a \cosh(x/a)$ between $x = -a$ and $x = a$.

Solution   The desired length (see Section 6.4) is given by

$$\int_{-a}^{a} \sqrt{1 + \left(\frac{dy}{dx}\right)^2}\, dx = \int_{-a}^{a} \sqrt{1 + \sinh^2\left(\frac{x}{a}\right)}\, dx$$

$$= \int_{-a}^{a} \sqrt{\cosh^2\left(\frac{x}{a}\right)}\, dx$$

$$= 2 \int_{0}^{a} \cosh\left(\frac{x}{a}\right) dx$$

$$= 2a \int_{0}^{a} \cosh\left(\frac{x}{a}\right)\left(\frac{1}{a} dx\right)$$

$$= \left[2a \sinh \frac{x}{a}\right]_{0}^{a}$$

$$= 2a \sinh 1 \approx 2.35a \qquad \blacksquare$$

PROBLEM SET   7.8

In Problems 1–6, verify that the given equations are identities.

1. $e^x = \cosh x + \sinh x$

2. $e^{-x} = \cosh x - \sinh x$

3. $\sinh(x + y) = \sinh x \cosh y + \cosh x \sinh y$

4. $\cosh(x + y) = \cosh x \cosh y + \sinh x \sinh y$

5. $\tanh(x + y) = \dfrac{\tanh x + \tanh y}{1 + \tanh x \tanh y}$

6. $\sinh 2x = 2 \sinh x \cosh x$

In Problems 7–20, find $D_x y$.

7. $y = \sinh^2 x$ **8.** $y = 5 \sinh^3 x$

9. $y = \cosh(x^2 - 1)$ **10.** $y = \ln(\coth x)$

11. $y = x^2 \sinh x$

12. $y = e^x \cosh x$

13. $y = \sinh 4x \cosh 2x$

14. $y = \sinh 3x \cosh 5x$

15. $y = \cosh^{-1}(x^3)$

16. $y = \tanh^{-1}(2x^5 - 1)$

17. $y = x \sinh^{-1}(2x)$

18. $y = \ln(\sinh^{-1} x)$

19. $y = \sinh(\cos x)$

20. $y = \sinh^{-1}(\sin x)$

In Problems 21–24, find the integrals.

21. $\displaystyle\int x \cosh(x^2 + 3)\, dx$

22. $\displaystyle\int \frac{\sinh \sqrt{x}}{\sqrt{x}}\, dx$

23. $\displaystyle\int e^x \sinh e^x\, dx$

24. $\displaystyle\int \tanh x \ln(\cosh x)\, dx$

25. Find the area of the region bounded by $y = \cosh 2x$, $y = 0$, $x = 0$, and $x = \ln 3$.

26. Find the area of the region bounded by $y = \sinh x$, $y = 0$, and $x = \ln 2$.

27. The region bounded by $y = \cosh x$, $y = 0$, $x = 0$, and $x = 1$ is revolved about the x-axis. Find the volume of the resulting solid. *Hint*: $\cosh^2 x = (1 + \cosh 2x)/2$.

28. The curve $y = \cosh x$, $0 \le x \le 1$, is revolved about the x-axis. Find the area of the resulting surface.

29. To derive the equation of a hanging cable (catenary), we consider the section AP from the lowest point A to a general point $P(x, y)$ (see Figure 5) and imagine the rest of the cable to have been removed.

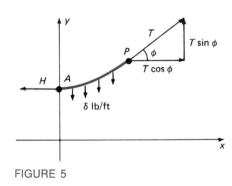

FIGURE 5

The forces acting on the cable are:

1. H = horizontal tension pulling at A;
2. T = tangential tension pulling at P;

3. $W = \delta s$ = weight of s feet of cable of density δ pounds per foot.

To be in equilibrium, the horizontal and vertical components of T must just balance H and W, respectively. Thus, $T \cos \phi = H$ and $T \sin \phi = W = \delta s$, and so

$$\frac{T \sin \phi}{T \cos \phi} = \tan \phi = \frac{\delta s}{H}$$

But since $\tan \phi = dy/dx$, we get

$$\frac{dy}{dx} = \frac{\delta s}{H}$$

and therefore

$$\frac{d^2 y}{dx^2} = \frac{\delta}{H}\frac{ds}{dx} = \frac{\delta}{H}\sqrt{1 + \left(\frac{dy}{dx}\right)^2}$$

Now show that $y = a \cosh(x/a) + C$ satisfies this differential equation with $a = H/\delta$.

C 30. Call the graph of $y = b - a \cosh(x/a)$ an inverted catenary and imagine it to be an arch sitting on the x-axis. Show that if the width of this arch along the x-axis is $2a$, then each of the following is true.

(a) $b = a \cosh 1 \approx 1.54308a$.

(b) The height of the arch is approximately $0.54308a$.

(c) The height of an arch of width 48 is approximately 13.

C 31. A farmer built a large hayshed of length 100 feet and width 48 feet. A cross section has the shape of an inverted catenary (see Problem 30) with equation $y = 37 - 24 \cosh(x/24)$.

(a) Draw a picture of this shed.

(b) Find the volume of the shed.

(c) Find the surface area of the roof of the shed.

32. Show that $A = t/2$, where A denotes the area in Figure 2 of this section. *Hint*: At some point you will need to use Formula 44 from the back of the book.

33. Demonstrate for r any real number:

(a) $(\sinh x + \cosh x)^r = \sinh rx + \cosh rx$

(b) $(\cosh x - \sinh x)^r = \cosh rx - \sinh rx$

34. The **gudermannian** of t is defined by

$$\text{gd}(t) = \tan^{-1}(\sinh t)$$

Show:

(a) gd is odd and increasing with an inflection point at the origin;

(b) $\text{gd}(t) = \sin^{-1}(\tanh t) = \displaystyle\int_0^t \text{sech } u\, du$

7.9 Chapter Review

Respond with true or false to each of the following assertions. Be prepared to defend your answer.

1. $\ln|x|$ is defined for all real x.
2. The graph of $y = \ln x$ has no inflection points.
3. $\displaystyle\int_1^{e^3} \frac{1}{t}\,dt = 3$
4. The graph of an invertible function $y = f(x)$ is intersected exactly once by every horizontal line.
5. The domain of $\ln^{-1}$ is the set of all real numbers.
6. $\ln x/\ln y = \ln x - \ln y$
7. $(\ln x)^4 = 4 \ln x$
8. $\ln(2e^{x+1}) - \ln(2e^x) = 1$ for all x in $\mathbb{R}$.
9. The functions $f(x) = 4 + e^x$ and $g(x) = \ln(x - 4)$ are a pair of inverse functions.
10. $\exp x + \exp y = \exp(x + y)$.
11. $\displaystyle\lim_{x \to 0^+} (\ln \sin x - \ln x) = 0$.
12. $\pi^{\sqrt{2}} = e^{\sqrt{2}\,\ln \pi}$
13. $\dfrac{d}{dx}(\ln \pi) = \dfrac{1}{\pi}$
14. $\displaystyle\int \frac{1}{x}\,dx = \ln 3|x| + C$
15. $D_x(x^e) = ex^{e-1}$
16. If $f(x) \cdot \exp[g(x)] = 0$ for $x = x_0$, then $f(x_0) = 0$.
17. $D_x(x^x) = x^x \ln x$.
18. The graphs of $y = \sin x$ and $y = \cos x$ intersect at right angles.
19. $\text{Arcsin}(\sin x) = x$ for all real numbers x.
20. $\tan^{-1}x = \sin^{-1}x/\cos^{-1}x$.
21. $\cosh(\ln 3) = \frac{5}{6}$
22. $\displaystyle\lim_{x \to -\infty} \tan^{-1}x = -\frac{\pi}{2}$
23. $\sin^{-1}(\cosh x)$ is defined for all real x.
24. $f(x) = \tanh x$ is an odd function.
25. Both $y = \sinh x$ and $y = \cosh x$ satisfy the differential equation $y'' + y = 0$.
26. $\ln(3^{100}) > 100$.
27. $\ln(2x^2 - 18) - \ln(x - 3) - \ln(x + 3) = \ln 2$ for all x in $\mathbb{R}$.
28. If y is growing exponentially and if y triples between $t = 0$ and $t = t_1$, then y will also triple between $t = 2t_1$ and $t = 3t_1$.
29. It is to a saver's advantage to have money invested at 11% compounded continuously rather than 12% compounded monthly.
30. If $D_x(a^x) = a^x$ with $a > 0$, then $a = e$.

In Problems 1–24, differentiate the indicated functions.

1. $\ln \dfrac{x^4}{2}$

2. $\sin^2(x^3)$

3. $e^{x^2 - 4x}$

4. $\log_{10}(x^5 - 1)$

5. $\tan(\ln e^x)$

6. $e^{\ln \cot x}$

7. $2 \tanh \sqrt{x}$

8. $\tanh^{-1}(\sin x)$

9. $\sinh^{-1}(\tan x)$

10. $2 \sin^{-1} \sqrt{3x}$

11. $\sec^{-1} e^x$

12. $\ln \sin^2\left(\dfrac{x}{2}\right)$

13. $3 \ln(e^{5x} + 1)$

14. $\ln(2x^3 - 4x + 5)$

15. $\cos e^{\sqrt{x}}$

16. $\ln(\tanh x)$

17. $2 \cos^{-1} \sqrt{x}$

18. $4^{3x} + (3x)^4$

19. $2 \csc e^{\ln \sqrt{x}}$

20. $(\log_{10} 2x)^{2/3}$

21. $4 \tan 5x \sec 5x$

22. $x \tan^{-1} \dfrac{x^2}{2}$

23. x^{1+x}

24. $(1 + x^2)^e$

In Problems 25–34, find the antiderivatives of the indicated functions and verify your results by differentiation.

25. e^{3x-1}

26. $6 \cot 3x$

27. $e^x \sin e^x$

28. $\dfrac{6x + 3}{x^2 + x - 5}$

29. $\dfrac{e^{x+2}}{e^{x+3} + 1}$

30. $4x \cos x^2$

31. $\dfrac{4}{\sqrt{1 - 4x^2}}$

32. $\dfrac{\cos x}{1 + \sin^2 x}$

33. $\dfrac{-1}{x + x(\ln x)^2}$

34. $\operatorname{sech}^2(x - 3)$

In Problems 35 and 36, find the intervals on which f is increasing and the intervals on which f is decreasing. Find where the graph of f is concave upward and where it is concave downward. Find any extreme values and points of inflection. Then sketch the graph of f.

35. $f(x) = \sin x + \cos x,\ \dfrac{-\pi}{2} \le x \le \dfrac{\pi}{2}$

36. $f(x) = \dfrac{x^2}{e^x},\ x \in \mathbb{R}$

37. Let $f(x) = x^5 + 2x^3 + 4x,\ -\infty < x < \infty$

(a) Prove that f has an inverse $g = f^{-1}$.
(b) Evaluate $g(7) = f^{-1}(7)$.
(c) Evaluate $g'(7)$.

38. A certain radioactive substance has a half-life of 10 years. How long will it take for 100 grams to decay to 1 gram?

39. If \$100 is put in the bank today at 12% interest, how much will it be worth at the end of 1 year if interest is compounded as indicated?

(a) Annually   (b) Monthly
(c) Daily   (d) Continuously

40. An airplane is flying horizontally at an altitude of 500 feet at the speed of 300 feet per second directly away from a searchlight on the ground. The searchlight is kept directed at the plane. At what rate is the angle between the light beam and the ground changing when this angle is 30°?

41. Find the equation of the tangent line to $y = (\cos x)^{\sin x}$ at $(0, 1)$.

42. A town grew exponentially from 10,000 in 1970 to 14,000 in 1980. Assuming the same type of growth continues, what will the population be in 2000?

8 Techniques of Integration

We have seen how to find the derivatives of quantities. Now we shall show inversely how the integrals of the derivatives are found, that is, those quantities from which the derivatives originate.
Johann Bernoulli

Johann Bernoulli
1667–1748

Johann Bernoulli is perhaps the most famous of a whole family of productive mathematicians. At least eight prominent eighteenth-century mathematicians bear the Swiss name of Bernoulli, and it is said that there are still active mathematicians with this name. However, so much mathematical talent in one family may not be an undivided blessing, as we shall see.

Johann and his brother Jacques were, after Newton and Leibniz, the most important of the founding fathers of calculus. The two brothers competed vigorously and often bitterly for recognition, though they continued to correspond with each other and with Leibniz about mathematics. Johann would claim results which were discovered by his brother (or by others), and Jacques would react in kind. Leibniz's attempts to mediate served only to entangle him in the arguments. Even Johann's son Daniel was drawn into the conflict and was forced to leave home when he won a prize for which his father had competed. It is an unhappy record for a truly remarkable family.

The Bernoullis tackled all kinds of basic problems in calculus, including inflection points, length of curves, infinite series, and techniques of integration. Johann wrote the first textbook on calculus in 1691 and 1692, but the part on integral calculus was not published until 1742 and that on differential calculus until 1924. Instead, in 1696, Guillaume F. A. de l'Hôpital, Johann's student, published the first calculus text. It was a slightly altered form of his teacher's work. Perhaps Johann's influence is best seen in that of another and more famous student, Leonhard Euler.

8.1
Integration by Substitution

Our arsenal of functions now includes all the so-called elementary functions. These are the constant functions, the power functions, the logarithmic and exponential functions, the trigonometric and inverse trigonometric functions, and all functions obtained from them by addition, subtraction, multiplication, division, and composition. Thus,

$$f(x) = \frac{e^x + e^{-x}}{2} = \cosh x$$

$$g(x) = (1 + \cos^4 x)^{1/2}$$

$$h(x) = \frac{3^{x^2 - 2x}}{\ln(x^2 + 1)} - \sin[\cos(\cosh x)]$$

are elementary functions.

Differentiation of an elementary function is straightforward, requiring only a systematic use of the rules we have learned. And the result is always an elementary function. Integration (antidifferentiation) is a far different matter. It involves a few techniques and a large bag of tricks; what is worse, it does not always yield an elementary function. For example, it is known that the antiderivatives of e^{-x^2} and $(\sin x)/x$ are not elementary functions.

The two principal techniques for integration are *substitution* (Sections 8.1–8.3) and *integration by parts* (Section 8.4). The method of substitution was introduced in Sections 5.1 and 5.8; we have used it occasionally in the intervening chapters. Here, we review the method and apply it in a wide variety of situations.

Finally in Section 8.5, we discuss the problem of integrating a rational function, that is, a quotient of two polynomial functions. We learn that, in theory, we can always perform such an integration and the result will be an elementary function.

STANDARD FORMS   Effective use of the method of substitution depends on the ready availability of a list of known integrals. One such list (but too long to memorize) appears inside the back cover of this book. The short list at the top of the next page is so useful that we think every calculus student should memorize it.

Constants, Powers	**1.** $\int k\,du = ku + C$ $\qquad$ **2.** $\int u^r\,du = \begin{cases} \dfrac{u^{r+1}}{r+1} + C & r \neq -1 \\ \ln\|u\| + C & r = -1 \end{cases}$
Exponentials	**3.** $\int e^u\,du = e^u + C$ $\qquad$ **4.** $\int a^u\,du = \dfrac{a^u}{\ln a} + C, \quad a \neq 1, a > 0$
Trigonometric Functions	**5.** $\int \sin u\,du = -\cos u + C$ $\qquad$ **6.** $\int \cos u\,du = \sin u + C$
	7. $\int \sec^2 u\,du = \tan u + C$ $\qquad$ **8.** $\int \csc^2 u\,du = -\cot u + C$
	9. $\int \sec u \tan u\,du = \sec u + C$ $\qquad$ **10.** $\int \csc u \cot u\,du = -\csc u + C$
	11. $\int \tan u\,du = -\ln\|\cos u\| + C$ $\qquad$ **12.** $\int \cot u\,du = \ln\|\sin u\| + C$
Algebraic Functions	**13.** $\int \dfrac{du}{\sqrt{a^2 - u^2}} = \sin^{-1}\left(\dfrac{u}{a}\right) + C$ $\qquad$ **14.** $\int \dfrac{du}{a^2 + u^2} = \dfrac{1}{a}\tan^{-1}\left(\dfrac{u}{a}\right) + C$
	15. $\int \dfrac{du}{u\sqrt{u^2 - a^2}} = \dfrac{1}{a}\sec^{-1}\left(\dfrac{\|u\|}{a}\right) + C = \dfrac{1}{a}\cos^{-1}\left(\dfrac{a}{\|u\|}\right) + C$

SUBSTITUTION IN INDEFINITE INTEGRALS Suppose that you face an indefinite integration. If it is a standard form, simply write the answer. If not, look for a substitution that will change it to a standard form. If the first substitution you try does not work, try another. Skill at this, like most worthwhile activities, depends on practice.

The method of substitution rests on Theorem 5.8A, here restated in words that suggest the process to go through.

Theorem A

(Substitution). To find $\int f(x)\,dx$, we may substitute $u = g(x)$, g being differentiable. If this changes $f(x)\,dx$ to $h(u)\,du$ and if H is an antiderivative of h, then

$$\int f(x)\,dx = \int h(u)\,du = H(u) + C = H(g(x)) + C$$

EXAMPLE 1 Find $\int \dfrac{x}{\cos^2(x^2)}\,dx$.

Solution Stare at this integral for a few moments. You should be reminded of the standard form $\int \sec^2 u\,du$. Let $u = x^2$, $du = 2x\,dx$. Then

$$\int \frac{x}{\cos^2(x^2)}\,dx = \frac{1}{2}\int \frac{1}{\cos^2(x^2)} \cdot 2x\,dx = \frac{1}{2}\int \sec^2 u\,du$$

$$= \frac{1}{2}\tan u + C = \frac{1}{2}\tan(x^2) + C \qquad \blacksquare$$

EXAMPLE 2 Find $\displaystyle\int \frac{3}{\sqrt{5 - 9x^2}}\,dx$.

Solution Think of $\displaystyle\int \frac{du}{\sqrt{a^2 - u^2}}$. Let $u = 3x$, so $du = 3\,dx$. Then,

$$\int \frac{3}{\sqrt{5 - 9x^2}}\,dx = \int \frac{1}{\sqrt{5 - u^2}}\,du = \sin^{-1}\left(\frac{u}{\sqrt{5}}\right) + C$$

$$= \sin^{-1}\left(\frac{3x}{\sqrt{5}}\right) + C \qquad\blacksquare$$

EXAMPLE 3 Find $\displaystyle\int \frac{6e^{1/x}}{x^2}\,dx$.

Solution Think of $\int e^u\,du$. Let $u = 1/x$, so $du = (-1/x^2)\,dx$. Then,

$$\int \frac{6e^{1/x}}{x^2}\,dx = -6\int e^{1/x}\left(\frac{-1}{x^2}\,dx\right) = -6\int e^u\,du$$

$$= -6e^u + C = -6e^{1/x} + C \qquad\blacksquare$$

EXAMPLE 4 Find $\displaystyle\int \frac{e^x}{4 + 9e^{2x}}\,dx$.

Solution Think of $\displaystyle\int \frac{1}{a^2 + u^2}\,du$. Let $u = 3e^x$, so $du = 3e^x\,dx$. Then,

$$\int \frac{e^x}{4 + 9e^{2x}}\,dx - \frac{1}{3}\int \frac{1}{4 + 9e^{2x}}\,(3e^x\,dx) = \frac{1}{3}\int \frac{1}{4 + u^2}\,du$$

$$= \frac{1}{3}\cdot\frac{1}{2}\tan^{-1}\left(\frac{u}{2}\right) + C = \frac{1}{6}\tan^{-1}\left(\frac{3e^x}{2}\right) + C \qquad\blacksquare$$

No law says you have to write out the *u*-substitution. If you can do it in your head, that is fine. Here are two illustrations of what we mean.

EXAMPLE 5 Find $\displaystyle\int x^3\sqrt{x^4 + 11}\,dx$.

Solution Mentally, substitute $u = x^4 + 11$.

$$\int x^3\sqrt{x^4 + 11}\,dx = \frac{1}{4}\int (x^4 + 11)^{1/2}(4x^3\,dx)$$

$$= \frac{1}{4}\int (x^4 + 11)^{1/2}\,d(x^4 + 11)$$

$$= \frac{1}{6}(x^4 + 11)^{3/2} + C \qquad\blacksquare$$

EXAMPLE 6 Find $\displaystyle\int \frac{a^{\tan t}}{\cos^2 t}\, dt$.

Solution Mentally, substitute $u = \tan t$.

$$\int \frac{a^{\tan t}}{\cos^2 t}\, dt = \int a^{\tan t}\sec^2 t\, dt$$

$$= \int a^{\tan t}\, d(\tan t) = \frac{a^{\tan t}}{\ln a} + C \qquad\blacksquare$$

MANIPULATING THE INTEGRAND   Before trying a substitution, it is often necessary to rewrite the integrand in a more convenient form.

EXAMPLE 7 Find $\displaystyle\int \frac{7}{x^2 - 6x + 25}\, dx$.

Solution Integrals with quadratic expressions in the denominator can often be reduced to standard forms by *completing the square*. Recall that $x^2 + bx$ becomes a perfect square by the addition of $(b/2)^2$.

$$\int \frac{7}{x^2 - 6x + 25}\, dx = \int \frac{7}{x^2 - 6x + 9 + 16}\, dx$$

$$= 7\int \frac{1}{(x-3)^2 + 4^2}\, d(x-3)$$

$$= \frac{7}{4}\tan^{-1}\!\left(\frac{x-3}{4}\right) + C$$

We made the mental substitution $u = x - 3$ at the final stage. $\blacksquare$

EXAMPLE 8 Find $\displaystyle\int \frac{x^2 - x}{x + 1}\, dx$.

$$
\begin{array}{r}
x - 2 \\
x + 1\ \overline{\smash{\big)}\ x^2 - x} \\
\underline{x^2 + x} \\
-2x \\
\underline{-2x - 2} \\
2
\end{array}
$$

FIGURE 1

Solution When the integrand is the quotient of two polynomials (that is, a rational function) and the numerator is of equal or greater degree than the denominator, always *divide the denominator into the numerator first*. By long division (Figure 1),

$$\frac{x^2 - x}{x + 1} = x - 2 + \frac{2}{x + 1}$$

Hence

$$\int \frac{x^2 - x}{x + 1}\, dx = \int (x - 2)\, dx + 2\int \frac{1}{x + 1}\, dx$$

$$= \frac{x^2}{2} - 2x + 2\int \frac{1}{x + 1}\, d(x + 1)$$

$$= \frac{x^2}{2} - 2x + 2\ln|x + 1| + C \qquad\blacksquare$$

EXAMPLE 9 Find $\int \sec x \, dx$.

Solution The manipulations in Examples 7 and 8 seemed quite natural and reasonable. But the one we now use is more like a magician pulling a rabbit out of a hat.

$$\int \sec x \, dx = \int \sec x \, \frac{\sec x + \tan x}{\sec x + \tan x} \, dx$$

$$= \int \frac{\sec^2 x + \sec x \tan x}{\sec x + \tan x} \, dx$$

$$= \int \frac{1}{\sec x + \tan x} \, d(\sec x + \tan x)$$

$$= \ln|\sec x + \tan x| + C \qquad \blacksquare$$

See Problem 72 for another way to do Example 9.

SUBSTITUTION IN DEFINITE INTEGRALS This topic was covered in Section 5.8. It is just like substitution in indefinite integrals, but we must remember to make the appropriate change in the limits of integration.

EXAMPLE 10 Evaluate $\int_2^5 t\sqrt{t^2 - 4} \, dt$.

Solution Let $u = t^2 - 4$, so $du = 2t \, dt$; note that $u = 0$ when $t = 2$ and $u = 21$ when $t = 5$. Thus,

$$\int_2^5 t\sqrt{t^2 - 4} \, dt = \frac{1}{2} \int_2^5 (t^2 - 4)^{1/2}(2t \, dt)$$

$$= \frac{1}{2} \int_0^{21} u^{1/2} \, du$$

$$= \left[\frac{1}{3} u^{3/2} \right]_0^{21} = \frac{1}{3} (21)^{3/2} \approx 32.08 \qquad \blacksquare$$

USE OF TABLES OF INTEGRALS Our list of standard forms is very short (15 formulas); the list inside the back cover of this book is longer (113 formulas) and potentially more useful. Notice that the integrals listed there are grouped according to type. We illustrate the use of this list.

EXAMPLE 11 Find $\int \sqrt{6x - x^2} \, dx$ and $\int_0^{\pi/2} (\cos x)\sqrt{6 \sin x - \sin^2 x} \, dx$.

Solution We use Formula 102 with $a = 3$.

$$\int \sqrt{6x - x^2} \, dx = \frac{x - 3}{2} \sqrt{6x - x^2} + \frac{9}{2} \sin^{-1}\left(\frac{x - 3}{3} \right) + C$$

In the second integral, let $u = \sin x$, so $du = \cos x \, dx$. Then

$$\int_0^{\pi/2} \cos x \sqrt{6 \sin x - \sin^2 x} \, dx = \int_0^1 \sqrt{6u - u^2} \, du$$

$$= \left[\frac{u - 3}{2} \sqrt{6u - u^2} + \frac{9}{2} \sin^{-1}\left(\frac{u - 3}{3}\right) \right]_0^1$$

$$= -\sqrt{5} + \frac{9}{2} \sin^{-1}\left(\frac{-2}{3}\right) - \frac{9}{2} \sin^{-1}(-1)$$

$$\approx 1.55 \qquad \blacksquare$$

Much more extensive tables of integrals may be found in most libraries. One of the better known is *Standard Mathematical Tables*, published by the Chemical Rubber Company.

PROBLEM SET 8.1 _____

In Problems 1–58, perform the indicated integrations.

1. $\displaystyle\int (x - 1)^4 \, dx$

2. $\displaystyle\int \sqrt{2x} \, dx$

3. $\displaystyle\int x(x^2 + 1)^4 \, dx$

4. $\displaystyle\int x\sqrt{x^2 + 2} \, dx$

5. $\displaystyle\int \frac{dx}{x^2 + 1}$

6. $\displaystyle\int \frac{e^x}{1 + 2e^x} \, dx$

7. $\displaystyle\int \frac{x}{x^2 + 1} \, dx$

8. $\displaystyle\int \frac{x^2}{x^2 + 1} \, dx$

9. $\displaystyle\int 3t\sqrt{2 + t^2} \, dt$

10. $\displaystyle\int \frac{dt}{\sqrt{1 + t}}$

11. $\displaystyle\int \frac{\tan z}{\cos z} \, dz$

12. $\displaystyle\int \frac{e^{\sin z}}{\sec z} \, dz$

13. $\displaystyle\int \frac{\cos \sqrt{x}}{\sqrt{x}} \, dx$

14. $\displaystyle\int \frac{x \, dx}{\sqrt{1 - x^4}}$

15. $\displaystyle\int_0^{\pi/2} \frac{\cos x}{1 + \sin^2 x} \, dx$

16. $\displaystyle\int_0^{3/4} \frac{\cos \sqrt{1 - x}}{\sqrt{1 - x}} \, dx$

17. $\displaystyle\int \frac{2x^2 + x}{x + 1} \, dx$

18. $\displaystyle\int \frac{x^3 + 2x^2}{x - 2} \, dx$

19. $\displaystyle\int \frac{\sqrt{\tan x}}{1 - \sin^2 x} \, dx$

20. $\displaystyle\int e^{\sin \theta \cos \theta} \cos 2\theta \, d\theta$

21. $\displaystyle\int \frac{\cos(\ln 4x^2) \, dx}{x}$

22. $\displaystyle\int \frac{\csc^2(\ln x)}{x} \, dx$

23. $\displaystyle\int \frac{5e^x}{\sqrt{1 - e^{2x}}} \, dx$

24. $\displaystyle\int \frac{x}{x^4 + 1} \, dx$

25. $\displaystyle\int \frac{5e^{2x}}{\sqrt{1 - e^{2x}}} \, dx$

26. $\displaystyle\int \frac{x^3}{x^4 + 1} \, dx$

27. $\displaystyle\int_0^1 x10^{x^2} \, dx$

28. $\displaystyle\int_0^{\pi/6} 2^{\sin x} \cos x \, dx$

29. $\displaystyle\int \frac{\sin x - \cos x}{\sin x} \, dx$

30. $\displaystyle\int \frac{\sin(4t - 1)}{1 - \sin^2(4t - 1)} \, dt$

31. $\displaystyle\int \frac{z + 2}{\cot(z^2 + 4z - 3)} \, dz$

32. $\displaystyle\int \csc 2t \, dt$

33. $\displaystyle\int e^x \sec e^x \, dx$

34. $\displaystyle\int e^x \sec^2(e^x) \, dx$

35. $\displaystyle\int \frac{\sec^3 x + e^{\sin x}}{\sec x} \, dx$

36. $\displaystyle\int \frac{(6t - 1)\sin \sqrt{3t^2 - t - 1}}{\sqrt{3t^2 - t - 1}} \, dt$

37. $\displaystyle\int \frac{t^2 \cos(t^3 - 2)}{\sin^2(t^3 - 2)} \, dt$

38. $\displaystyle\int \frac{1 + \cos 2x}{\sin^2 2x} \, dx$

39. $\displaystyle\int \frac{t^2 \cos^2(t^3 - 2)}{\sin^2(t^3 - 2)} \, dt$

40. $\displaystyle\int \frac{\csc^2 2t}{\sqrt{1 + \cot 2t}} \, dt$

41. $\displaystyle\int \frac{e^{\tan^{-1} 2t}}{1 + 4t^2} \, dt$

42. $\displaystyle\int (t + 1)e^{-t^2 - 2t - 5} \, dt$

43. $\displaystyle\int \frac{y}{\sqrt{16 - 9y^4}} \, dy$

44. $\displaystyle\int \frac{\sec^2 2y}{9 + \tan^2 2y} \, dy$

45. $\displaystyle\int \frac{\sec x \tan x}{1 + \sec^2 x} \, dx$

46. $\displaystyle\int \frac{5}{\sqrt{9 - 4x^2}} \, dx$

47. $\displaystyle\int \frac{e^{3t}}{\sqrt{4 - e^{6t}}} \, dt$

48. $\displaystyle\int \frac{dt}{2t\sqrt{4t^2 - 1}}$

49. $\displaystyle\int_0^{\pi/2} \frac{\sin x}{16 + \cos^2 x} \, dx$

50. $\displaystyle\int_0^1 \frac{e^{2x} - e^{-2x}}{e^{2x} + e^{-2x}} \, dx$

51. $\displaystyle\int \frac{1}{x^2 + 2x + 5} \, dx$

52. $\displaystyle\int \frac{1}{x^2 - 4x + 9} \, dx$

53. $\displaystyle\int \frac{dx}{9x^2 + 18x + 10}$

54. $\displaystyle\int \frac{dx}{\sqrt{16 + 6x - x^2}}$

55. $\displaystyle\int \frac{x + 1}{9x^2 + 18x + 10} \, dx$

56. $\displaystyle\int \frac{3 - x}{\sqrt{16 + 6x - x^2}} \, dx$

57. $\displaystyle\int \frac{dt}{t\sqrt{2t^2 - 9}}$

58. $\displaystyle\int \frac{\tan x}{\sqrt{\sec^2 x - 4}} \, dx$

In Problems 59–70, use the table of integrals on the back end pages, perhaps combined with a substitution, to perform the indicated integrations.

59. $\displaystyle\int x\sqrt{3x + 2} \, dx$

60. $\displaystyle\int 2t\sqrt{3 - 4t} \, dt$

61. $\displaystyle\int \frac{dx}{9 - 16x^2}$

62. $\displaystyle\int \frac{dx}{5x^2 - 11}$

63. $\displaystyle\int x^2\sqrt{9 - 2x^2} \, dx$

64. $\displaystyle\int \frac{\sqrt{16 - 3t^2}}{t} \, dt$

65. $\displaystyle\int \frac{dx}{\sqrt{5 + 3x^2}}$

66. $\displaystyle\int t^2\sqrt{3 + 5t^2} \, dt$

67. $\displaystyle\int \frac{dt}{\sqrt{t^2 + 2t - 3}}$

68. $\displaystyle\int \frac{\sqrt{x^2 + 2x - 3}}{x + 1} \, dx$

69. $\displaystyle\int \frac{\sin t \cos t}{\sqrt{3 \sin t + 5}} \, dt$

70. $\displaystyle\int \frac{\sin x}{\cos x\sqrt{5 - 4\cos x}} \, dx$

71. Find the length of the curve $y = \ln(\cos x)$ between $x = 0$ and $x = \pi/4$.

72. Establish the identity

$$\sec x = \frac{\sin x}{\cos x} + \frac{\cos x}{1 + \sin x}$$

and then use it to give a new derivation of the formula for $\int \sec x \, dx$.

73. Evaluate $\displaystyle\int_0^{2\pi} \frac{x|\sin x|}{1 + \cos^2 x} \, dx$. *Hint*: Make the substitution $u = x - \pi$ in the definite integral and then use symmetry properties.

74. Let R be the region bounded by $y = \sin x$ and $y = \cos x$ between $x = -\pi/4$ and $x = 3\pi/4$. Find the volume of the solid gotten when R is revolved about $x = -\pi/4$. *Hint*: Use cylindrical shells to write a single integral, make the substitution $u = x - \pi/4$, and apply symmetry properties.

8.2
Some Trigonometric Integrals

When we combine the method of substitution with a clever use of trigonometric identities, we can integrate a wide variety of trigonometric forms. We consider five commonly encountered types.

1. $\displaystyle\int \sin^n x \, dx$ and $\displaystyle\int \cos^n x \, dx$

2. $\displaystyle\int \sin^m x \cos^n x \, dx$

3. $\displaystyle\int \tan^n x \, dx$ and $\displaystyle\int \cot^n x \, dx$

4. $\displaystyle\int \tan^m x \sec^n x \, dx$ and $\displaystyle\int \cot^m x \csc^n x \, dx$

5. $\displaystyle\int \sin mx \cos nx \, dx, \int \sin mx \sin nx \, dx, \int \cos mx \cos nx \, dx$

TYPE 1 ($\int \sin^n x \, dx$, $\int \cos^n x \, dx$)   Consider first the case where n is an odd positive integer. After taking out either the factor $\sin x$ or $\cos x$, use the identity $\sin^2 x + \cos^2 x = 1$.

EXAMPLE 1 (*n* **odd**). Find $\int \sin^5 x\,dx$.

Solution

$$\int \sin^5 x\,dx = \int \sin^4 x \sin x\,dx$$

$$= \int (1 - \cos^2 x)^2 \sin x\,dx$$

$$= \int (1 - 2\cos^2 x + \cos^4 x)\sin x\,dx$$

$$= -\int (1 - 2\cos^2 x + \cos^4 x)\,d(\cos x)$$

$$= -\cos x + \tfrac{2}{3}\cos^3 x - \tfrac{1}{5}\cos^5 x + C \qquad \blacksquare$$

If *n* is an even positive integer, we make use of the half-angle formulas

$$\sin^2 x = \frac{1 - \cos 2x}{2}, \qquad \cos^2 x = \frac{1 + \cos 2x}{2}$$

EXAMPLE 2 (*n* **even**). Find $\int \sin^2 x\,dx$ and $\int \cos^4 x\,dx$.

Solution

$$\int \sin^2 x\,dx = \int \frac{1 - \cos 2x}{2}\,dx$$

$$= \frac{1}{2}\int dx - \frac{1}{4}\int (\cos 2x)(2)\,dx$$

$$= \frac{1}{2}\int dx - \frac{1}{4}\int \cos 2x\,d(2x)$$

$$= \frac{1}{2}x - \frac{1}{4}\sin 2x + C$$

$$\int \cos^4 x\,dx = \int \left(\frac{1 + \cos 2x}{2}\right)^2 dx$$

$$= \frac{1}{4}\int (1 + 2\cos 2x + \cos^2 2x)\,dx$$

$$= \frac{1}{4}\int dx + \frac{1}{4}\int (\cos 2x)(2)\,dx + \frac{1}{8}\int (1 + \cos 4x)\,dx$$

$$= \frac{3}{8}\int dx + \frac{1}{4}\int \cos 2x\,d(2x) + \frac{1}{32}\int \cos 4x\,d(4x)$$

$$= \frac{3}{8}x + \frac{1}{4}\sin 2x + \frac{1}{32}\sin 4x + C \qquad \blacksquare$$

TYPE 2 ($\int \sin^m x \cos^n x \, dx$) If either m or n is an odd positive integer and the other exponent is any number, we factor out $\sin x$ or $\cos x$ and use the identity $\sin^2 x + \cos^2 x = 1$.

EXAMPLE 3 (*m* or *n* odd). Find $\int \sin^3 x \cos^{-4} x \, dx$.

Solution

$$\int \sin^3 x \cos^{-4} x \, dx = \int (1 - \cos^2 x)(\cos^{-4} x)(\sin x) \, dx$$

$$= -\int (\cos^{-4} x - \cos^{-2} x) \, d(\cos x)$$

$$= -\left[\frac{(\cos x)^{-3}}{-3} - \frac{(\cos x)^{-1}}{-1} \right] + C$$

$$= \frac{1}{3} \sec^3 x - \sec x + C \qquad \blacksquare$$

If both m and n are even positive integers, we use half-angle formulas to reduce the degree of the integrand.

EXAMPLE 4 (**Both *m* and *n* even**). Find $\int \sin^2 x \cos^4 x \, dx$.

Solution

$$\int \sin^2 x \cos^4 x \, dx$$

$$= \int \left(\frac{1 - \cos 2x}{2} \right) \left(\frac{1 + \cos 2x}{2} \right)^2 dx$$

$$= \frac{1}{8} \int (1 + \cos 2x - \cos^2 2x - \cos^3 2x) \, dx$$

$$= \frac{1}{8} \int \left[1 + \cos 2x - \frac{1}{2}(1 + \cos 4x) - (1 - \sin^2 2x)\cos 2x \right] dx$$

$$= \frac{1}{8} \int \left[\frac{1}{2} - \frac{1}{2} \cos 4x + \sin^2 2x \cos 2x \right] dx$$

$$= \frac{1}{8} \left[\int \frac{1}{2} \, dx - \frac{1}{8} \int \cos 4x \, d(4x) + \frac{1}{2} \int \sin^2 2x \, d(\sin 2x) \right]$$

$$= \frac{1}{8} \left[\frac{1}{2} x - \frac{1}{8} \sin 4x + \frac{1}{6} \sin^3 2x \right] + C \qquad \blacksquare$$

TYPE 3 ($\int \tan^n x \, dx$, $\int \cot^n x \, dx$) In the tangent case, factor out $\tan^2 x = \sec^2 x - 1$; in the cotangent case, factor out $\cot^2 x = \csc^2 x - 1$.

EXAMPLE 5 Find $\displaystyle\int \cot^4 x\, dx$.

Solution

$$\int \cot^4 x\, dx = \int \cot^2 x\,(\csc^2 x - 1)\, dx$$

$$= \int \cot^2 x\, \csc^2 x\, dx - \int \cot^2 x\, dx$$

$$= -\int \cot^2 x\, d(\cot x) - \int (\csc^2 x - 1)\, dx$$

$$= -\tfrac{1}{3} \cot^3 x + \cot x + x + C \qquad \blacksquare$$

EXAMPLE 6 Find $\displaystyle\int \tan^5 x\, dx$.

Solution

$$\int \tan^5 x\, dx = \int \tan^3 x\,(\sec^2 x - 1)\, dx$$

$$= \int \tan^3 x\, \sec^2 x\, dx - \int \tan^3 x\, dx$$

$$= \int \tan^3 x\, d(\tan x) - \int \tan x\,(\sec^2 x - 1)\, dx$$

$$= \int \tan^3 x\, d(\tan x) - \int \tan x\, d(\tan x) + \int \tan x\, dx$$

$$= \tfrac{1}{4} \tan^4 x - \tfrac{1}{2} \tan^2 x - \ln|\cos x| + C \qquad \blacksquare$$

TYPE 4 ($\int \tan^m x\, \sec^n x\, dx$, $\int \cot^m x\, \csc^n x\, dx$)

EXAMPLE 7 (**n even, m any number**). Find $\displaystyle\int \tan^{-3/2} x\, \sec^4 x\, dx$.

Solution

$$\int \tan^{-3/2} x\, \sec^4 x\, dx = \int (\tan^{-3/2} x)(1 + \tan^2 x)\sec^2 x\, dx$$

$$= \int (\tan^{-3/2} x)\sec^2 x\, dx + \int (\tan^{1/2} x)\sec^2 x\, dx$$

$$= -2 \tan^{-1/2} x + \tfrac{2}{3} \tan^{3/2} x + C \qquad \blacksquare$$

EXAMPLE 8 (***m* odd, *n* any number**). Find $\int \tan^3 x \, \sec^{-1/2} x \, dx$.

Solution

$$\int \tan^3 x \, \sec^{-1/2} x \, dx = \int (\tan^2 x)(\sec^{-3/2} x)(\sec x \tan x) \, dx$$

$$= \int (\sec^2 x - 1)\sec^{-3/2} x \, d(\sec x)$$

$$= \int \sec^{1/2} x \, d(\sec x) - \int \sec^{-3/2} x \, d(\sec x)$$

$$= \tfrac{2}{3} \sec^{3/2} x + 2 \sec^{-1/2} x + C \qquad \blacksquare$$

TYPE 5 ($\int$ sin *mx* cos *nx* dx, $\int$ sin *mx* sin *nx* dx, $\int$ cos *mx* cos *nx* dx)
Integrals of this type occur in alternating current theory, heat-transfer problems, and any place where Fourier series are used. To handle these integrals, we use the product identities.

$$\sin mx \cos nx = \tfrac{1}{2}\big[\sin(m + n)x + \sin(m - n)x\big]$$

$$\sin mx \sin nx = -\tfrac{1}{2}\big[\cos(m + n)x - \cos(m - n)x\big]$$

$$\cos mx \cos nx = \tfrac{1}{2}\big[\cos(m + n)x + \cos(m - n)x\big]$$

EXAMPLE 9 Find $\int \sin 2x \cos 3x \, dx$.

Solution

$$\int \sin 2x \cos 3x \, dx = \frac{1}{2} \int \big[\sin 5x + \sin(-x)\big] \, dx$$

$$= \frac{1}{10} \int \sin 5x \, d(5x) - \frac{1}{2} \int \sin x \, dx$$

$$= -\frac{1}{10} \cos 5x + \frac{1}{2} \cos x + C \qquad \blacksquare$$

EXAMPLE 10 If *m* and *n* are positive integers, show that

$$\int_{-\pi}^{\pi} \sin mx \sin nx \, dx = \begin{cases} 0 & \text{if } n \neq m \\ \pi & \text{if } n = m \end{cases}$$

Solution If $m \neq n$,

$$\int_{-\pi}^{\pi} \sin mx \sin nx \, dx = -\frac{1}{2} \int_{-\pi}^{\pi} \big[\cos(m + n)x - \cos(m - n)x\big] \, dx$$

$$= -\frac{1}{2}\left[\frac{1}{m + n} \sin(m + n)x - \frac{1}{m - n} \sin(m - n)x\right]_{-\pi}^{\pi}$$

$$= 0$$

If $m = n$,

$$\int_{-\pi}^{\pi} \sin mx \sin nx \, dx = -\frac{1}{2} \int_{-\pi}^{\pi} [\cos 2mx - 1] \, dx$$

$$= -\frac{1}{2} \left[\frac{1}{2m} \sin 2mx - x \right]_{-\pi}^{\pi}$$

$$= -\frac{1}{2} [-2\pi] = \pi \qquad \blacksquare$$

PROBLEM SET 8.2

In Problems 1–30, perform the indicated integrations.

1. $\int \cos^2 x \, dx$

2. $\int \sin^4 5x \, dx$

3. $\int \cos^3 x \, dx$

4. $\int \sin^3 x \, dx$

5. $\int_0^{\pi/2} \sin^5 t \, dt$

6. $\int_0^{\pi/2} \cos^6 t \, dt$

7. $\int \tan^3 y \, dy$

8. $\int \tan^6 2x \, dx$

9. $\int \sin^7 3x \cos^2 3x \, dx$

10. $\int (\sin^3 t) \sqrt{\cos t} \, dt$

11. $\int \cos^3 \theta \sin^{-2} \theta \, d\theta$

12. $\int \sin^{1/2} \theta \cos^3 \theta \, d\theta$

13. $\int \sin^4 2t \cos^4 2t \, dt$

14. $\int \sin^6 t \cos^2 t \, dt$

15. $\int \tan^3 3y \sec^3 3y \, dy$

16. $\int \cos^4\left(\frac{w}{2}\right) \sin^2\left(\frac{w}{2}\right) dw$

17. $\int \cot x \csc^3 x \, dx$

18. $\int \cot x \sec^2 x \, dx$

19. $\int \tan^{-3} t \sec^2 t \, dt$

20. $\int \tan^5 t \sec^{-3/2} t \, dt$

21. $\int \sin 4y \cos 5y \, dy$

22. $\int \cos y \cos 4y \, dy$

23. $\int \cot^4 2x \, dx$

24. $\int \csc^4 3y \, dy$

25. $\int \sec^4 7x \, dx$

26. $\int \cot^6 4w \, dw$

27. $\int \cot^3 x \, dx$

28. $\int \sin 3t \sin t \, dt$

29. $\int (\tan x + \cot x)^2 \, dx$

30. $\int_{-\pi}^{\pi} \cos mx \cos nx \, dx, \quad m \neq n$

31. The graph of the region bounded by $y = x + \sin x$, $y = 0$, $x = \pi$, is revolved about the x-axis. Find the volume of the resulting solid.

32. The parametric equations $x = t - \sin t$, $y = 1 - \cos t$, $0 \leq t \leq 2\pi$, determine a curve. Find the area of the surface that results when this curve is revolved about the x-axis.

33. Let $f(x) = \sum_{n=1}^{N} a_n \sin(nx)$. Use Example 10 to show each of the following.

(a) $\frac{1}{\pi} \int_{-\pi}^{\pi} f(x) \sin(mx) \, dx = \begin{cases} a_m & \text{if } m \leq N \\ 0 & \text{if } m > N \end{cases}$

(b) $\frac{1}{\pi} \int_{-\pi}^{\pi} f^2(x) \, dx = \sum_{n=1}^{N} a_n^2$

Note: These results hint at a very large subject called *Fourier series*, which has applications to heat, vibrating strings, and other physical phenomena.

34. Show that

$$\lim_{n \to \infty} \cos \frac{x}{2} \cos \frac{x}{4} \cos \frac{x}{8} \cdots \cos \frac{x}{2^n} = \frac{\sin x}{x}$$

by completing the following steps.

(a) $\cos \frac{x}{2} \cos \frac{x}{4} \cdots \cos \frac{x}{2^n}$

$$= \left[\cos \frac{1}{2^n} x + \cos \frac{3}{2^n} x + \cdots + \cos \frac{2^n - 1}{2^n} x \right] \frac{1}{2^{n-1}}$$

(See Problem 36 of Section 2.3.)

(b) Recognize a Riemann sum leading to a definite integral.

(c) Evaluate the definite integral.

35. Use the result of Problem 34 to obtain a famous formula of François Vieta (1540–1603), namely

$$\frac{2}{\pi} = \frac{\sqrt{2}}{2} \cdot \frac{\sqrt{2 + \sqrt{2}}}{2} \cdot \frac{\sqrt{2 + \sqrt{2 + \sqrt{2}}}}{2} \cdots$$

8.3
Rationalizing Substitutions

Radicals in an integrand are always troublesome and we usually try to get rid of them. Often an appropriate substitution will do the job.

INTEGRANDS INVOLVING $\sqrt[n]{ax + b}$ If $\sqrt[n]{ax + b}$ appears in an integral, the substitution $u = \sqrt[n]{ax + b}$ will eliminate the radical.

EXAMPLE 1   Find $\displaystyle\int \frac{dx}{x - \sqrt{x}}$.

Solution   Let $u = \sqrt{x}$ so $u^2 = x$ and $2u\, du = dx$. Then

$$\int \frac{dx}{x - \sqrt{x}} = \int \frac{2u}{u^2 - u}\, du = 2 \int \frac{1}{u - 1}\, du = 2 \int \frac{1}{u - 1}\, d(u - 1)$$

$$= 2 \ln|u - 1| + C = 2 \ln|\sqrt{x} - 1| + C \qquad \blacksquare$$

EXAMPLE 2   Find $\displaystyle\int x \sqrt[3]{x - 4}\, dx$.

Solution   Let $u = \sqrt[3]{x - 4}$, so $u^3 = x - 4$ and $3u^2\, du = dx$. Then

$$\int x \sqrt[3]{x - 4}\, dx = \int (u^3 + 4)u \cdot 3u^2\, du = 3 \int (u^6 + 4u^3)\, du$$

$$= 3 \left[\frac{u^7}{7} + u^4 \right] + C = \frac{3}{7}(x - 4)^{7/3} + 3(x - 4)^{4/3} + C \qquad \blacksquare$$

EXAMPLE 3   Find $\displaystyle\int x \sqrt[5]{(x + 1)^2}\, dx$.

Solution   Let $u = (x + 1)^{1/5}$, so $u^5 = x + 1$ and $5u^4\, du = dx$. Then

$$\int x(x + 1)^{2/5}\, dx = \int (u^5 - 1)u^2 \cdot 5u^4\, du$$

$$= 5 \int (u^{11} - u^6)\, du = \tfrac{5}{12}u^{12} - \tfrac{5}{7}u^7 + C$$

$$= \tfrac{5}{12}(x + 1)^{12/5} - \tfrac{5}{7}(x + 1)^{7/5} + C \qquad \blacksquare$$

INTEGRANDS INVOLVING $\sqrt{a^2 - x^2}$, $\sqrt{a^2 + x^2}$, AND $\sqrt{x^2 - a^2}$ To eliminate these radicals, we make the trigonometric substitutions

1. $x = a \sin t,$
2. $x = a \tan t,$
3. $x = a \sec t,$

respectively. To see the effect of these substitutions, observe that:

1. $a^2 - x^2 = a^2 - a^2 \sin^2 t = a^2(1 - \sin^2 t) = a^2 \cos^2 t$;
2. $a^2 + x^2 = a^2 + a^2 \tan^2 t = a^2(1 + \tan^2 t) = a^2 \sec^2 t$;
3. $x^2 - a^2 = a^2 \sec^2 t - a^2 = a^2(\sec^2 t - 1) = a^2 \tan^2 t$.

If we restrict t so these substitutions are invertible (that is, so sine, tangent, and secant have inverses, as in Section 7.6), then:

1. $\sqrt{a^2 - x^2} = a|\cos t| = a \cos t$ (because $-\pi/2 \le t \le \pi/2$);
2. $\sqrt{a^2 + x^2} = a|\sec t| = a \sec t$ (because $-\pi/2 < t < \pi/2$);
3. $\sqrt{x^2 - a^2} = a|\tan t| = \pm a \tan t$ (because $0 \le t \le \pi, t \neq \pi/2$).

EXAMPLE 4 Find $\displaystyle\int \sqrt{a^2 - x^2}\, dx$.

Solution We make the substitution

$$x = a \sin t, \qquad -\frac{\pi}{2} \le t \le \frac{\pi}{2}$$

Then $dx = a \cos t\, dt$ and $\sqrt{a^2 - x^2} = a \cos t$. Thus,

$$\int \sqrt{a^2 - x^2}\, dx = \int a \cos t \cdot a \cos t\, dt = a^2 \int \cos^2 t\, dt$$

$$= \frac{a^2}{2} \int (1 + \cos 2t)\, dt$$

$$= \frac{a^2}{2}\left(t + \frac{1}{2}\sin 2t\right) + C$$

$$= \frac{a^2}{2}(t + \sin t \cos t) + C$$

Now $x = a \sin t$ is equivalent to $x/a = \sin t$ and, since t was restricted so sine is invertible,

$$t = \sin^{-1}\left(\frac{x}{a}\right)$$

Also, from an identity at the end of Section 7.6,

$$\cos t = \cos\left[\sin^{-1}\left(\frac{x}{a}\right)\right] = \sqrt{1 - \frac{x^2}{a^2}} = \frac{1}{a}\sqrt{a^2 - x^2}$$

a fact that you can also see from Figure 1. Thus,

$$\int \sqrt{a^2 - x^2}\, dx = \frac{a^2}{2}\sin^{-1}\left(\frac{x}{a}\right) + \frac{x}{2}\sqrt{a^2 - x^2} + C \qquad \blacksquare$$

$x = a \sin t$

FIGURE 1

$y = \sqrt{a^2 - x^2}$

A

$-a$ a x

$A = \int_{-a}^{a} \sqrt{a^2 - x^2}\, dx = \frac{\pi a^2}{2}$

FIGURE 2

The result in Example 4 is interesting. In particular, it allows us to calculate the following definite integral, which represents the area of a semicircle (Figure 2).

$$\int_{-a}^{a} \sqrt{a^2 - x^2}\, dx = \left[\frac{a^2}{2} \sin^{-1}\left(\frac{x}{a}\right) + \frac{x}{2}\sqrt{a^2 - x^2} \right]_{-a}^{a} = \frac{a^2}{2}\left[\frac{\pi}{2} + \frac{\pi}{2} \right] = \frac{\pi a^2}{2}$$

EXAMPLE 5 Find $\displaystyle\int \frac{\sqrt{4 - x^2}}{x^2}\, dx$.

Solution Let $x = 2 \sin t$, $-\pi/2 \le t \le \pi/2$. Then $dx = 2 \cos t\, dt$ and $\sqrt{4 - x^2} = 2 \cos t$. Thus,

$$\int \frac{\sqrt{4 - x^2}}{x^2}\, dx = \int \frac{2 \cos t}{4 \sin^2 t}(2 \cos t)\, dt$$

$$= \int \cot^2 t\, dt = \int (\csc^2 t - 1)\, dt$$

$$= -\cot t - t + C$$

Now $\sin t = x/2$, from which $t = \sin^{-1}(x/2)$. Also (see Figure 3), $\cot t = \sqrt{4 - x^2}/x$. Thus,

$$\int \frac{\sqrt{4 - x^2}}{x^2}\, dx = -\frac{\sqrt{4 - x^2}}{x} - \sin^{-1}\left(\frac{x}{2}\right) + C \qquad \blacksquare$$

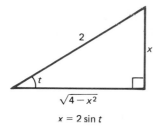

2

x

t

$\sqrt{4 - x^2}$

$x = 2 \sin t$

FIGURE 3

EXAMPLE 6 Find $\displaystyle\int \frac{dx}{\sqrt{9 + x^2}}$.

Solution Let $x = 3 \tan t$, $-\pi/2 < t < \pi/2$. Then $dx = 3 \sec^2 t\, dt$ and $\sqrt{9 + x^2} = 3 \sec t$.

$$\int \frac{dx}{\sqrt{9 + x^2}} = \int \frac{3 \sec^2 t}{3 \sec t}\, dt = \int \sec t\, dt$$

$$= \ln|\sec t + \tan t| + C$$

The last step—namely, the integration of $\sec t$—was handled in Example 9 of Section 8.1. Now $\tan t = x/3$, which suggests the sketch in Figure 4, from which we conclude that $\sec t = \sqrt{9 + x^2}/3$. Finally,

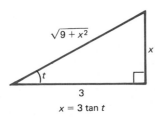

$\sqrt{9 + x^2}$

x

t

3

$x = 3 \tan t$

FIGURE 4

$$\int \frac{dx}{\sqrt{9 + x^2}} = \ln\left| \frac{\sqrt{9 + x^2} + x}{3} \right| + C$$

$$= \ln\left|\sqrt{9 + x^2} + x\right| - \ln 3 + C$$

$$= \ln\left|\sqrt{9 + x^2} + x\right| + K \qquad \blacksquare$$

EXAMPLE 7 Calculate $\displaystyle\int_2^4 \frac{\sqrt{x^2 - 4}}{x}\, dx.$

Solution Let $x = 2 \sec t$, where $0 \le t < \pi/2$. Note that the restriction of t to the interval above is acceptable, since x is in the interval $2 \le x \le 4$. This is important because it allows us to remove the absolute value sign that normally appears when we simplify $\sqrt{x^2 - a^2}$. In our case,

$$\sqrt{x^2 - 4} = \sqrt{4\sec^2 t - 4} = \sqrt{4\tan^2 t} = 2|\tan t| = 2\tan t$$

We now use the theorem on substituting in a definite integral (which requires changing the limits of integration) to write

$$\int_2^4 \frac{\sqrt{x^2 - 4}}{x}\, dx = \int_0^{\pi/3} \frac{2\tan t}{2\sec t}\, 2\sec t \tan t\, dt$$

$$= \int_0^{\pi/3} 2\tan^2 t\, dt = 2\int_0^{\pi/3} (\sec^2 t - 1)\, dt$$

$$= 2[\tan t - t]_0^{\pi/3} = 2\sqrt{3} - \frac{2\pi}{3} \approx 1.37 \qquad \blacksquare$$

COMPLETING THE SQUARE When a quadratic expression of the type $x^2 + Bx + C$ appears under a radical, completing the square will prepare it for a trigonometric substitution.

EXAMPLE 8 Find $\displaystyle\int \frac{dx}{\sqrt{x^2 + 2x + 26}}$ and $\displaystyle\int \frac{2x}{\sqrt{x^2 + 2x + 26}}\, dx.$

Solution $x^2 + 2x + 26 = x^2 + 2x + 1 + 25 = (x + 1)^2 + 25$. Let $u = x + 1$ and $du = dx$. Then,

$$\int \frac{dx}{\sqrt{x^2 + 2x + 26}} = \int \frac{du}{\sqrt{u^2 + 25}}$$

Next let $u = 5 \tan t$, $-\pi/2 < t < \pi/2$. Then $du = 5\sec^2 t\, dt$ and $\sqrt{u^2 + 25} = \sqrt{25(\tan^2 t + 1)} = 5\sec t$, so

$$\int \frac{du}{\sqrt{u^2 + 25}} = \int \frac{5\sec^2 t\, dt}{5\sec t} = \int \sec t\, dt$$

$$= \ln|\sec t + \tan t| + C$$

$$= \ln\left|\frac{\sqrt{u^2 + 25}}{5} + \frac{u}{5}\right| + C \qquad \text{(Figure 5)}$$

$$= \ln\left|\sqrt{u^2 + 25} + u\right| - \ln 5 + C$$

$$= \ln\left|\sqrt{x^2 + 2x + 26} + x + 1\right| + K$$

$u = 5 \tan t$

FIGURE 5

CHAPTER 8/Techniques of Integration

To handle the second integral, we write

$$\int \frac{2x}{\sqrt{x^2 + 2x + 26}} \, dx = \int \frac{2x + 2}{\sqrt{x^2 + 2x + 26}} \, dx - 2\int \frac{1}{\sqrt{x^2 + 2x + 26}} \, dx$$

The first of the integrals on the right is handled by the substitution $u = x^2 + 2x + 26$; the second was just done. We obtain

$$\int \frac{2x}{\sqrt{x^2 + 2x + 26}} \, dx$$
$$= 2\sqrt{x^2 + 2x + 26} - 2\ln\left|\sqrt{x^2 + 2x + 26} + x + 1\right| + K$$

∎

PROBLEM SET 8.3

In Problems 1–20, perform the indicated integrations.

1. $\int x\sqrt{x + 3} \, dx$

2. $\int x\sqrt[3]{x + 4} \, dx$

3. $\int \frac{t \, dt}{\sqrt{2t + 7}}$

4. $\int \frac{x^2 + 2x}{\sqrt{x + 1}} \, dx$

5. $\int_1^4 \frac{dx}{\sqrt{x + 2}}$

6. $\int_0^4 \frac{\sqrt{t}}{t + 1} \, dt$

7. $\int 7t(2t + 1)^{3/2} \, dt$

8. $\int x(x + 2)^{2/3} \, dx$

9. $\int \frac{\sqrt{1 - x^2}}{x} \, dx$

10. $\int \frac{x^2 \, dx}{\sqrt{9 - x^2}}$

11. $\int \frac{dx}{x\sqrt{x^2 + 9}}$

12. $\int \frac{dx}{(x^2 + 9)^{3/2}}$

13. $\int_5^8 \frac{dx}{x^2\sqrt{x^2 - 16}}$

14. $\int_2^5 \frac{\sqrt{t^2 - 4}}{t^3} \, dt$

15. $\int \frac{t}{\sqrt{4 - t^2}} \, dt$

16. $\int \frac{l}{\sqrt{l^2 + 9}} \, dt$

17. $\int \frac{2x - 3}{\sqrt{4 - x^2}} \, dx$

18. $\int \frac{2x + 1}{\sqrt{x^2 + 9}} \, dx$

19. $\int \frac{y^3 \, dy}{(y^2 + 4)^{3/2}}$

20. $\int \frac{y^2 \, dy}{(9 - y^2)^{5/2}}$

In Problems 21–30, use the method of completing the square, along with a trigonometric substitution if needed, to do the indicated integration.

21. $\int \frac{dx}{\sqrt{x^2 + 2x + 5}}$

22. $\int \frac{dx}{\sqrt{x^2 + 4x + 5}}$

23. $\int \frac{3x}{\sqrt{x^2 + 2x + 5}} \, dx$

24. $\int \frac{2x - 1}{\sqrt{x^2 + 4x + 5}} \, dx$

25. $\int \sqrt{5 - 4x - x^2} \, dx$

26. $\int \frac{dx}{\sqrt{16 + 6x - x^2}}$

27. $\int \frac{dx}{\sqrt{4x - x^2}}$

28. $\int \frac{x}{\sqrt{4x - x^2}} \, dx$

29. $\int \frac{2x + 1}{x^2 + 2x + 2} \, dx$

30. $\int \frac{2x - 1}{x^2 - 6x + 18} \, dx$

31. The region bounded by $y = 1/(x^2 + 2x + 5)$, $y = 0$, $x = 0$, and $x = 1$ is revolved about the x-axis. Find the volume of the resulting solid.

32. The region of Problem 31 is revolved about the y-axis. Find the volume of the resulting solid.

33. Find $\int \frac{x \, dx}{x^2 + 9}$ by (a) an algebraic substitution and (b) a trigonometric substitution. Then reconcile your answers.

34. Find $\int_0^3 \frac{x^3 \, dx}{\sqrt{9 + x^2}}$ by making the substitutions

$$u = \sqrt{9 + x^2}, \quad u^2 = 9 + x^2, \quad 2u \, du = 2x \, dx$$

35. Find $\int \frac{\sqrt{4 - x^2}}{x} \, dx$ by (a) the substitution $u = \sqrt{4 - x^2}$ and (b) a trigonometric substitution. Then reconcile your answers.

36. Two circles of radius b intersect as shown in Figure 6 with their centers $2a$ apart ($0 \leq a \leq b$). Find the area of the region in which they overlap.

FIGURE 6

37. Hippocrates of Chios (ca. 430 B.C.) showed that the two shaded regions in Figure 7 have the same area (he squared the lune). Note that C is the center of the lower arc. Show Hippocrates's result (a) using calculus and (b) without calculus.

FIGURE 7

38. Generalize Problem 37 by finding a formula for the area of the shaded lune shown in Figure 8.

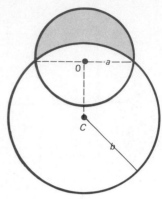

FIGURE 8

39. Starting at $(a, 0)$, an object is pulled along by a string of length a with the pulling end moving along the positive y-axis. The path of the object is a curve called a **tractrix** and has the property that the string is always tangent to the curve (see Figure 9). Set up a differential equation for the curve and solve it.

FIGURE 9

8.4
Integration by Parts

If integration by substitution fails, it may be possible to use a double substitution, better known as *integration by parts*. This method is based on the integration of the formula for the derivative of a product of two functions. Let $u = u(x)$ and $v = v(x)$. Then

$$D_x[u(x)v(x)] = u(x)v'(x) + v(x)u'(x)$$

and, by integrating both members of this equation, we obtain

$$u(x)v(x) = \int u(x)v'(x)\,dx + \int v(x)u'(x)\,dx$$

or

$$\int u(x)v'(x)\,dx = u(x)v(x) - \int v(x)u'(x)\,dx$$

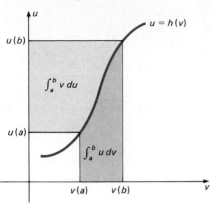

A Geometric Interpretation
of
Integration by Parts

$$\int_a^b u\,dv = u(b)v(b) - u(a)v(a) - \int_a^b v\,du$$

FIGURE 1

Since $dv = v'(x)\,dx$ and $du = u'(x)\,dx$, the preceding equation is usually written symbolically as follows.

Integration by Parts in Indefinite Integrals

$$\int u\,dv = uv - \int v\,du$$

The corresponding formula for definite integrals is

$$\int_a^b u(x)v'(x)\,dx = [u(x)v(x)]_a^b - \int_a^b v(x)u'(x)\,dx$$

We abbreviate this as follows (see also Figure 1).

Integration by Parts in Definite Integrals

$$\int_a^b u\,dv = [uv]_a^b - \int_a^b v\,du$$

These formulas allow us to shift the problem of integrating $u\,dv$ to that of integrating $v\,du$. Success depends on the proper choice of u and dv, which comes with practice.

SIMPLE EXAMPLES

EXAMPLE 1   Find $\int x \cos x\,dx$.

Solution   We wish to write $x \cos x\,dx$ as $u\,dv$. One possibility is to let $u = x$ and $dv = \cos x\,dx$. Then $du = dx$ and $v = \int \cos x\,dx = \sin x$ (we can omit the arbitrary constant at this stage). Here is a summary of this double substitution in a convenient format.

$$u = x \qquad dv = \cos x\,dx$$
$$du = dx \qquad v = \sin x$$

The integration-by-parts formula gives

$$\int \underbrace{x}_{u}\,\underbrace{\cos x\,dx}_{dv} = \underbrace{x}_{u}\,\underbrace{\sin x}_{v} - \int \underbrace{\sin x}_{v}\,\underbrace{dx}_{du}$$
$$= x \sin x + \cos x + C$$

We were successful on our first try. Another substitution would be

$$u = \cos x \qquad dv = x\,dx$$
$$du = -\sin x\,dx \qquad v = \frac{x^2}{2}$$

The integration-by-parts formula gives

$$\int \underbrace{\cos x}_{u} \underbrace{x\, dx}_{dv} = \underbrace{(\cos x)}_{u} \underbrace{\frac{x^2}{2}}_{v} - \int \underbrace{\frac{x^2}{2}}_{v} \underbrace{(-\sin x\, dx)}_{du}$$

which is correct but not helpful. The new integral on the right-hand side is more complicated than the original one. Thus, we see the importance of a wise choice for u and dv. ∎

EXAMPLE 2 Find $\displaystyle\int_1^2 \ln x\, dx$.

Solution We make the double substitution

$$u = \ln x \qquad dv = dx$$

$$du = \left(\frac{1}{x}\right) dx \qquad v = x$$

Then

$$\int_1^2 \ln x\, dx = \left[x \ln x\right]_1^2 - \int_1^2 x\, \frac{1}{x}\, dx$$

$$= 2 \ln 2 - \int_1^2 dx$$

$$= 2 \ln 2 - 1 \qquad\qquad ∎$$

EXAMPLE 3 Find $\displaystyle\int \arcsin x\, dx$.

Solution We make the double substitution

$$u = \arcsin x \qquad dv = dx$$

$$du = \frac{1}{\sqrt{1 - x^2}}\, dx \qquad v = x$$

Then

$$\int \arcsin x\, dx = x \arcsin x - \int \frac{x}{\sqrt{1 - x^2}}\, dx$$

$$= x \arcsin x + \frac{1}{2} \int (1 - x^2)^{-1/2}(-2x\, dx)$$

$$= x \arcsin x + \frac{1}{2} \cdot 2(1 - x^2)^{1/2} + C$$

$$= x \arcsin x + \sqrt{1 - x^2} + C \qquad\qquad ∎$$

REPEATED INTEGRATION BY PARTS Sometimes it is necessary to apply integration by parts several times.

EXAMPLE 4 Find $\int x^2 \sin x \, dx$.

Solution Let

$$u = x^2 \qquad dv = \sin x \, dx$$
$$du = 2x \, dx \qquad v = -\cos x$$

Then

$$\int x^2 \sin x \, dx = -x^2 \cos x + 2 \int x \cos x \, dx$$

We have improved our situation (the exponent on x has gone from 2 to 1), which suggests reapplying integration by parts to the integral on the right. Actually, we did this integration in Example 1, so we will make use of the result obtained there.

$$\int x^2 \sin x \, dx = -x^2 \cos x + 2(x \sin x + \cos x + C)$$

$$= -x^2 \cos x + 2x \sin x + 2 \cos x + K \qquad \blacksquare$$

EXAMPLE 5 Find $\int e^x \sin x \, dx$.

Solution Take $u = e^x$ and $dv = \sin x \, dx$. Then $du = e^x \, dx$ and $v = -\cos x$. Thus,

$$\int e^x \sin x \, dx = -e^x \cos x + \int e^x \cos x \, dx$$

which does not seem to have improved things. But do not give up; let us integrate by parts again. In the right integral, let $u = e^x$ and $dv = \cos x \, dx$, so $du = e^x \, dx$ and $v = \sin x$. Then

$$\int e^x \cos x \, dx = e^x \sin x - \int e^x \sin x \, dx$$

When we substitute this in our first result, we get

$$\int e^x \sin x \, dx = -e^x \cos x + e^x \sin x - \int e^x \sin x \, dx$$

By transposing the last term to the left side and combining terms, we obtain

$$2 \int e^x \sin x \, dx = e^x(\sin x - \cos x) + C$$

from which

$$\int e^x \sin x \, dx = \tfrac{1}{2} e^x (\sin x - \cos x) + K$$ ∎

The fact that the integral we wanted to find reappeared on the right side is what made Example 5 work. Here is another illustration of this technique.

EXAMPLE 6   Find $\int \sec^3 \theta \, d\theta$.

Solution   Let $u = \sec \theta$ and $dv = \sec^2 \theta \, d\theta$. Then $du = \sec \theta \tan \theta \, d\theta$, $v = \tan \theta$, and

$$\int \sec^3 \theta \, d\theta = \sec \theta \tan \theta - \int \sec \theta \tan^2 \theta \, d\theta$$

$$= \sec \theta \tan \theta - \int \sec \theta \, (\sec^2 \theta - 1) \, d\theta$$

$$= \sec \theta \tan \theta - \int \sec^3 \theta \, d\theta + \int \sec \theta \, d\theta$$

Consequently,

$$2 \int \sec^3 \theta \, d\theta = \sec \theta \tan \theta + \int \sec \theta \, d\theta$$

$$= \sec \theta \tan \theta + \ln|\sec \theta + \tan \theta| + C$$

the latter by Example 9 of Section 8.1. Finally,

$$\int \sec^3 \theta \, d\theta = \tfrac{1}{2} \sec \theta \tan \theta + \tfrac{1}{2} \ln|\sec \theta + \tan \theta| + K$$ ∎

See Problem 35 for another way to do Example 6.

REDUCTION FORMULAS   A formula of the form

$$\int f^n(x) \, dx = g(x) + \int f^k(x) \, dx,$$

where $k < n$ is called a **reduction formula** (the exponent on f is reduced). Such formulas can often be obtained via integration by parts.

EXAMPLE 7   Derive a reduction formula for $\int \sin^n x \, dx$.

Solution   Let $u = \sin^{n-1} x$ and $dv = \sin x \, dx$. Then

$$du = (n-1)\sin^{n-2} x \cos x \, dx \quad \text{and} \quad v = -\cos x$$

from which

$$\int \sin^n x \, dx = -\sin^{n-1} x \cos x + (n-1) \int \sin^{n-2} x \cos^2 x \, dx$$

If we replace $\cos^2 x$ by $1 - \sin^2 x$ in the last integral, we obtain

$$\int \sin^n x \, dx = -\sin^{n-1} x \cos x + (n-1) \int \sin^{n-2} x \, dx - (n-1) \int \sin^n x \, dx$$

After combining the first and last integrals above, and solving for $\int \sin^n x \, dx$, we get the reduction formula (valid for $n \geq 2$)

$$\int \sin^n x \, dx = \frac{-\sin^{n-1} x \cos x}{n} + \frac{n-1}{n} \int \sin^{n-2} x \, dx \qquad \blacksquare$$

EXAMPLE 8 Use the reduction formula above to evaluate

$$\int_0^{\pi/2} \sin^8 x \, dx$$

Solution Note first that

$$\int_0^{\pi/2} \sin^n x \, dx = \left[\frac{-\sin^{n-1} x \cos x}{n} \right]_0^{\pi/2} + \frac{n-1}{n} \int_0^{\pi/2} \sin^{n-2} x \, dx$$

$$= 0 + \frac{n-1}{n} \int_0^{\pi/2} \sin^{n-2} x \, dx$$

Thus,

$$\int_0^{\pi/2} \sin^8 x \, dx = \frac{7}{8} \int_0^{\pi/2} \sin^6 x \, dx$$

$$= \frac{7}{8} \cdot \frac{5}{6} \int_0^{\pi/2} \sin^4 x \, dx$$

$$= \frac{7}{8} \cdot \frac{5}{6} \cdot \frac{3}{4} \int_0^{\pi/2} \sin^2 x \, dx$$

$$= \frac{7}{8} \cdot \frac{5}{6} \cdot \frac{3}{4} \cdot \frac{1}{2} \int_0^{\pi/2} 1 \, dx$$

$$= \frac{7}{8} \cdot \frac{5}{6} \cdot \frac{3}{4} \cdot \frac{1}{2} \cdot \frac{\pi}{2} = \frac{35}{256} \pi \qquad \blacksquare$$

The general formula for $\int_0^{\pi/2} \sin^n x \, dx$ can be found in a similar way (Formula 113 at the back of the book).

In Problems 1–18, use integration by parts to perform the indicated integrations.

1. $\int xe^x \, dx$

2. $\int xe^{3x} \, dx$

3. $\int x \sin 3x \, dx$

4. $\int \ln 3x \, dx$

5. $\int \arctan x \, dx$

6. $\int x\sqrt{x+1} \, dx$

7. $\int t \sec^2 5t \, dt$

8. $\int \arctan(1/t) \, dt$

9. $\int \sqrt{x} \ln x \, dx$

10. $\int z^3 \ln z \, dz$

11. $\int x \arctan x \, dx$

12. $\int t \cos 4t \, dt$

13. $\int w \ln w \, dw$

14. $\int x \cos^2 x \sin x \, dx$

15. $\int_{\pi/6}^{\pi/2} x \csc^2 x \, dx$

16. $\int_{\pi/4}^{\pi/2} \csc^3 x \, dx$

17. $\int xa^x \, dx$

18. $\int x \sin^3 x \, dx$. *Hint*: $\sin^3 x = (1 - \cos^2 x)\sin x$. Use Problem 14.

In Problems 19–24, use the formula for integration by parts *twice* to perform the indicated integrations (see Examples 4 and 5).

19. $\int x^2 e^x \, dx$

20. $\int \ln^2 x \, dx$

21. $\int e^t \cos t \, dt$

22. $\int x^2 \cos x \, dx$

23. $\int \sin(\ln x) \, dx$

24. $\int (\ln x)^3 \, dx$. *Hint*: Use Problem 20.

25. Find the area of the region bounded by the curve $y = \ln x$, the x-axis, and the line $x = e$.

26. Find the volume of the solid generated by revolving the region of Problem 25 about the x-axis.

27. Find the area of the region bounded by the curves $y = 3xe^{-x/3}$, $y = 0$, and $x = 9$. Make a sketch.

28. Find the volume of the solid generated by revolving the region described in Problem 27 about the x-axis.

29. Derive the reduction formula (see Example 7).

$$\int \cos^n x \, dx = \frac{\cos^{n-1} x \sin x}{n} + \frac{n-1}{n} \int \cos^{n-2} x \, dx$$

30. Use Problem 29 to find a nice reduction formula for $\int_0^{\pi/2} \cos^n x \, dx$.

31. Evaluate $\int_0^{\pi/2} \sin^7 x \, dx$ (see Example 8).

32. Evaluate $\int_0^{\pi/2} \cos^6 x \, dx$ (see Problem 30).

33. Derive the reduction formula

$$\int x^n e^x \, dx = x^n e^x - n \int x^{n-1} e^x \, dx$$

and use it to find $\int x^3 e^x \, dx$.

34. If $p(x)$ is a polynomial of degree n and $G_1, G_2, \ldots, G_{n+1}$ are successive antiderivatives of g, then by repeated integration by parts

$$\int p(x)g(x) \, dx = p(x)G_1(x) - p'(x)G_2(x) + p''(x)G_3(x) - \cdots$$
$$+ (-1)^n p^{(n)}(x)G_{n+1}(x) + C$$

Use this result to find each of the following.

(a) $\int (x^3 - 2x)e^x \, dx$

(b) $\int (x^2 - 3x + 1)\sin x \, dx$

35. Establish the identity

$$2 \sec^3 x \, dx = \sec x \, dx + d(\sec x \tan x)$$

and use it to give a new derivation of the formula for $\int \sec^3 x \, dx$ (Example 6).

36. Evaluate $\int_0^{2\pi} \sin^n x \, dx$. *Hint*: First rewrite this expression in terms of $\int_0^{\pi/2} \sin^n x \, dx$ and then use Formula 113 from the back of the book.

37. The graph of $y = x \sin x$ for $x \geq 0$ is sketched in Figure 2.

(a) Find a formula for the area of the nth hump.

(b) The second hump is revolved about the y-axis. Find the volume of the resulting solid.

First hump

Second hump

FIGURE 2

38. The Fourier coefficient, $a_n = \dfrac{1}{\pi} \displaystyle\int_{-\pi}^{\pi} f(x)\sin nx \, dx$,

plays an important role in applied mathematics. Show that

if $f'(x)$ is continuous on $[-\pi, \pi]$, then $\displaystyle\lim_{n \to \infty} a_n = 0$. *Hint:* Integration by parts.

39. Let $G_n = \sqrt[n]{(n+1)(n+2)\cdots(n+n)}$. Show that $\displaystyle\lim_{n \to \infty}(G_n/n) = 4/e$. *Hint:* Consider $\ln(G_n/n)$, recognize it as a Riemann sum, and use Example 2.

40. Find the error in the following "proof" that $0 = 1$. In $\int (1/t)\,dt$, set $u = 1/t$ and $dv = dt$. Then $du = -t^{-2}\,dt$, and $uv = 1$. Integration by parts gives

$$\int (1/t)\,dt = 1 - \int (-1/t)\,dt$$

or $0 = 1$.

8.5
Integration of Rational Functions

A **rational function** is by definition the quotient of two polynomial functions. Examples are

$$f(x) = \frac{2}{(x+1)^3}, \qquad g(x) = \frac{2x+2}{x^2-4x+8}, \qquad h(x) = \frac{x^5+2x^3-x+1}{x^3+5x}$$

Of these, f and g are **proper rational functions**, meaning that the degree of the numerator is less than that of the denominator. An improper (not proper) rational function can always be written as a sum of a polynomial function and a proper rational function. Thus, for example,

$$h(x) = \frac{x^5+2x^3-x+1}{x^3+5x} = x^2 - 3 + \frac{14x+1}{x^3+5x}$$

$$
\begin{array}{r}
x^2 - 3 \\
x^3 + 5x \overline{\smash{\big)}\ x^5 + 2x^3 - x + 1} \\
\underline{x^5 + 5x^3} \\
-3x^3 - x \\
\underline{-3x^3 - 15x} \\
14x + 1
\end{array}
$$

FIGURE 1

a result obtained by long division (Figure 1). Since polynomials are easy to integrate, the problem of integrating rational functions is really that of integrating proper rational functions. But can we always integrate proper rational functions? In theory, the answer is yes, though the practical details may overwhelm us. Consider first the cases of f and g above.

EXAMPLE 1 Find $\displaystyle\int \frac{2}{(x+1)^3}\,dx$.

Solution Think of the substitution $u = x + 1$.

$$\int \frac{2}{(x+1)^3}\,dx = 2 \int (x+1)^{-3} d(x+1) = \frac{2(x+1)^{-2}}{-2} + C$$

$$= \frac{-1}{(x+1)^2} + C \qquad \blacksquare$$

EXAMPLE 2 Find $\displaystyle\int \frac{2x+2}{x^2-4x+8}\,dx.$

Solution Think first of the substitution $u = x^2 - 4x + 8$ for which $du = (2x-4)\,dx$. Then write

$$\int \frac{2x+2}{x^2-4x+8}\,dx = \int \frac{2x-4}{x^2-4x+8}\,dx + \int \frac{6}{x^2-4x+8}\,dx$$

$$= \ln|x^2-4x+8| + 6\int \frac{1}{x^2-4x+8}\,dx$$

In the second integral, complete the square.

$$\int \frac{1}{x^2-4x+8}\,dx = \int \frac{1}{x^2-4x+4+4}\,dx = \int \frac{1}{(x-2)^2+4}\,dx$$

$$= \int \frac{1}{(x-2)^2+4}\,d(x-2) = \frac{1}{2}\tan^{-1}\left(\frac{x-2}{2}\right) + C$$

We conclude that

$$\int \frac{2x+2}{x^2-4x+8}\,dx = \ln|x^2-4x+8| + 3\tan^{-1}\left(\frac{x-1}{2}\right) + K \quad\blacksquare$$

It is a remarkable fact that any proper rational function can be written as a sum of *simple* proper rational functions something like those illustrated in Examples 1 and 2. We must be more precise.

PARTIAL FRACTION DECOMPOSITION (LINEAR FACTORS)
To add fractions is a standard algebraic exercise. For example,

$$\frac{2}{x-1} + \frac{3}{x+1} = \frac{5x-1}{(x-1)(x+1)} = \frac{5x-1}{x^2-1}$$

It is the reverse process of decomposing a fraction into a sum of simpler fractions that interests us now. We focus on the denominator and consider cases.

EXAMPLE 3 **(Distinct linear factors).** Decompose $(3x-1)/(x^2-x-6)$ and then find its indefinite integral.

Solution Since the denominator factors as $(x+2)(x-3)$, it seems reasonable to hope for a decomposition of the following form.

(1) $$\frac{3x-1}{(x+2)(x-3)} = \frac{A}{x+2} + \frac{B}{x-3}$$

Our job is, of course, to determine A and B so (1) is an identity, a task we find easier after we have multiplied both sides by $(x + 2)(x - 3)$. We obtain

(2) $$3x - 1 = A(x - 3) + B(x + 2)$$

or, equivalently,

(3) $$3x - 1 = (A + B)x + (-3A + 2B)$$

However (3) is an identity if and only if coefficients of like powers of x on both sides are equal, that is,

$$A + B = 3$$
$$-3A + 2B = -1$$

By solving this pair of equations for A and B, we obtain $A = \frac{7}{5}$, $B = \frac{8}{5}$. Consequently,

$$\frac{3x - 1}{x^2 - x - 6} = \frac{3x - 1}{(x + 2)(x - 3)} = \frac{\frac{7}{5}}{x + 2} + \frac{\frac{8}{5}}{x - 3}$$

and

$$\int \frac{3x - 1}{x^2 - x - 6}\, dx = \frac{7}{5} \int \frac{1}{x + 2}\, dx + \frac{8}{5} \int \frac{1}{x - 3}\, dx$$

$$= \frac{7}{5} \ln|x + 2| + \frac{8}{5} \ln|x - 3| + C \qquad \blacksquare$$

If there was anything difficult about this process, it was the determination of A and B. We found their values by "brute force"; there is an easier way. In (2), which we wish to be an identity, substitute the convenient values $x = 3$ and $x = -2$, obtaining

$$8 = A \cdot 0 + B \cdot 5$$
$$-7 = A \cdot (-5) + B \cdot 0$$

This immediately gives $B = \frac{8}{5}$ and $A = \frac{7}{5}$.

You have just witnessed an odd, but correct, mathematical maneuver. Equation (1) turns out to be an identity (true for all x except -2 and 3) if and only if the essentially equivalent equation (2) is true precisely at -2 and 3. Ask yourself why this is so. Ultimately it depends on the fact that the two sides of equation (2), both linear polynomials, are identical if they have the same values at any two points.

SOLVE THIS D.E.

"...how, often, there is little resemblance between a differential equation and its solution. Who would suppose that an expression as simple as

$$\frac{dy}{dx} = \frac{1}{a^2 - x^2}$$

could be transformed into

$$y = \frac{1}{2a} \log_e \left(\frac{a + x}{a - x} \right) + C$$

This resembles the transformation of a chrysalis into a butterfly."

Silvanus P. Thompson

The method of partial fractions makes this an easy transformation. Do you see how it is done?

EXAMPLE 4 **(Distinct linear factors).** Find $\displaystyle\int \frac{5x + 3}{x^3 - 2x^2 - 3x}\, dx.$

Solution Since the denominator factors as $x(x + 1)(x - 3)$, we write

$$\frac{5x + 3}{x(x + 1)(x - 3)} = \frac{A}{x} + \frac{B}{x + 1} + \frac{C}{x - 3}$$

and seek to determine A, B, and C. Clearing of fractions gives

$$5x + 3 = A(x + 1)(x - 3) + Bx(x - 3) + Cx(x + 1)$$

Substitution of the values $x = 0$, $x = -1$, and $x = 3$ results in

$$3 = A(-3)$$
$$-2 = B(4)$$
$$18 = C(12)$$

or $A = -1$, $B = -\frac{1}{2}$, $C = \frac{3}{2}$. Thus,

$$\int \frac{5x + 3}{x^3 - 2x^2 - 3x}\, dx = -\int \frac{1}{x}\, dx - \frac{1}{2}\int \frac{1}{x + 1}\, dx + \frac{3}{2}\int \frac{1}{x - 3}\, dx$$

$$= -\ln|x| - \frac{1}{2}\ln|x + 1| + \frac{3}{2}\ln|x - 3| + C \quad\blacksquare$$

EXAMPLE 5 **(Repeated linear factors).** Find $\displaystyle\int \frac{x}{(x - 3)^2}\, dx.$

Solution Now the decomposition takes the form

$$\frac{x}{(x - 3)^2} = \frac{A}{x - 3} + \frac{B}{(x - 3)^2}$$

with A and B to be determined. After clearing of fractions, we get

$$x = A(x - 3) + B$$

If we now substitute the convenient value $x = 3$ and any other value, such as $x = 0$, we obtain $B = 3$ and $A = 1$. Thus,

$$\int \frac{x}{(x - 3)^2}\, dx = \int \frac{1}{x - 3}\, dx + 3\int \frac{1}{(x - 3)^2}\, dx$$

$$= \ln|x - 3| - \frac{3}{x - 3} + C \quad\blacksquare$$

EXAMPLE 6 (Some distinct, some repeated linear factors). Find

$$\int \frac{3x^2 - 8x + 13}{(x + 3)(x - 1)^2} \, dx.$$

Solution   We decompose the integrand in the following way.

$$\frac{3x^2 - 8x + 13}{(x + 3)(x - 1)^2} = \frac{A}{x + 3} + \frac{B}{x - 1} + \frac{C}{(x - 1)^2}$$

Clearing of fractions changes this to

$$3x^2 - 8x + 13 = A(x - 1)^2 + B(x + 3)(x - 1) + C(x + 3)$$

Substitution of $x = 1$, $x = -3$, and $x = 0$ yields $C = 2$, $A = 4$, and $B = -1$. Thus,

$$\int \frac{3x^2 - 8x + 13}{(x + 3)(x - 1)^2} \, dx = 4 \int \frac{dx}{x + 3} - \int \frac{dx}{x - 1} + 2 \int \frac{dx}{(x - 1)^2}$$

$$= 4 \ln|x + 3| - \ln|x - 1| - \frac{2}{x - 1} + C \qquad \blacksquare$$

Be sure to note the inclusion of the two fractions $B/(x - 1)$ and $C/(x - 1)^2$ in the decomposition above. The general rule for repeated linear factors is this: For each factor $(ax + b)^k$ of the denominator, there are k terms in the partial fraction decomposition, namely,

$$\frac{A_1}{ax + b} + \frac{A_2}{(ax + b)^2} + \frac{A_3}{(ax + b)^3} + \cdots + \frac{A_k}{(ax + b)^k}$$

PARTIAL FRACTION DECOMPOSITION (QUADRATIC FACTORS)

In factoring the denominator of a fraction, we may well get some quadratic factors (such as $x^2 + 1$), which cannot be factored into linear factors without introducing complex numbers.

EXAMPLE 7 (A single quadratic factor). Decompose $\dfrac{6x^2 - 3x + 1}{(4x + 1)(x^2 + 1)}$ and then find its indefinite integral.

Solution   The best we can hope for is a decomposition of the form

$$\frac{6x^2 - 3x + 1}{(4x + 1)(x^2 + 1)} = \frac{A}{4x + 1} + \frac{Bx + C}{x^2 + 1}$$

To determine the constants A, B, and C, we multiply both sides by $(4x + 1)(x^2 + 1)$ and obtain

$$6x^2 - 3x + 1 = A(x^2 + 1) + (Bx + C)(4x + 1)$$

Substitution of $x = -\frac{1}{4}$, $x = 0$, and $x = 1$ yields

$$\frac{6}{16} + \frac{3}{4} + 1 = A(\frac{17}{16}) \qquad \Rightarrow \quad A = 2$$

$$1 = 2 + C \qquad \Rightarrow \quad C = -1$$

$$4 = 4 + (B - 1)5 \qquad \Rightarrow \quad B = 1$$

Thus,

$$\int \frac{6x^2 - 3x + 1}{(4x + 1)(x^2 + 1)}\, dx = \int \frac{2}{4x + 1}\, dx + \int \frac{x - 1}{x^2 + 1}\, dx$$

$$= \frac{1}{2} \int \frac{4\, dx}{4x + 1} + \frac{1}{2} \int \frac{2x\, dx}{x^2 + 1} - \int \frac{dx}{x^2 + 1}$$

$$= \frac{1}{2} \ln|4x + 1| + \frac{1}{2} \ln(x^2 + 1) - \tan^{-1}x + C \quad \blacksquare$$

EXAMPLE 8 **(A repeated quadratic factor).** Find $\displaystyle \int \frac{6x^2 - 15x + 22}{(x + 3)(x^2 + 2)^2}\, dx.$

Solution   Here the appropriate decomposition is

$$\frac{6x^2 - 15x + 22}{(x + 3)(x^2 + 2)^2} = \frac{A}{x + 3} + \frac{Bx + C}{x^2 + 2} + \frac{Dx + E}{(x^2 + 2)^2}$$

After considerable work, we discover that $A = 1$, $B = -1$, $C = 3$, $D = -5$, and $E = 0$. Thus,

$$\int \frac{6x^2 - 15x + 22}{(x + 3)(x^2 + 2)^2}\, dx$$

$$= \int \frac{dx}{x + 3} - \int \frac{x - 3}{x^2 + 2}\, dx - 5 \int \frac{x}{(x^2 + 2)^2}\, dx$$

$$= \int \frac{dx}{x + 3} - \frac{1}{2} \int \frac{2x}{x^2 + 2}\, dx + 3 \int \frac{dx}{x^2 + 2} - \frac{5}{2} \int \frac{2x\, dx}{(x^2 + 2)^2}$$

$$= \ln|x + 3| - \frac{1}{2} \ln(x^2 + 2) + \frac{3}{\sqrt{2}} \tan^{-1}\left(\frac{x}{\sqrt{2}}\right) + \frac{5}{2(x^2 + 2)} + C$$

$$\blacksquare$$

SUMMARY   To decompose a rational function $f(x) = p(x)/q(x)$ into partial fractions, proceed as follows.

Step 1   If $f(x)$ is improper—that is, if $p(x)$ is of degree at least that of $q(x)$—divide $p(x)$ by $q(x)$, obtaining

$$f(x) = \text{a polynomial} + \frac{N(x)}{D(x)}$$

Step 2 Factor $D(x)$ into a product of linear and irreducible quadratic factors with real coefficients. By a theorem of algebra, this is always (theoretically) possible.

Step 3 For each factor of the form $(ax + b)^k$, expect the decomposition to have the terms

$$\frac{A_1}{(ax + b)} + \frac{A_2}{(ax + b)^2} + \cdots + \frac{A_k}{(ax + b)^k}$$

Step 4 For each factor of the form $(ax^2 + bx + c)^m$, expect the decomposition to have the terms

$$\frac{B_1 x + C_1}{ax^2 + bx + c} + \frac{B_2 x + C_2}{(ax^2 + bx + c)^2} + \cdots + \frac{B_m x + C_m}{(ax^2 + bx + c)^m}$$

Step 5 Set $N(x)/D(x)$ equal to the sum of all the terms found in Steps 3 and 4. The number of constants to be determined should equal the degree of the denominator, $D(x)$.

Step 6 Multiply both sides of the equation found in Step 5 by $D(x)$ and solve for the unknown constants. This can be done by either of two methods: (1) Equate coefficients of like-degree terms; or (2) assign convenient values to the variable x.

PROBLEM SET 8.5

In Problems 1–22, use the method of partial fractions to perform the required integration.

1. $\displaystyle\int \frac{2}{x^2 + 2x}\, dx$

2. $\displaystyle\int \frac{2}{x^2 - 1}\, dx$

3. $\displaystyle\int \frac{5x + 3}{x^2 - 9}\, dx$

4. $\displaystyle\int \frac{x - 6}{x^2 - 2x}\, dx$

5. $\displaystyle\int \frac{x - 11}{x^2 + 3x - 4}\, dx$

6. $\displaystyle\int \frac{3x - 13}{x^2 + 3x\,10}\, dx$

7. $\displaystyle\int \frac{2x + 21}{2x^2 + 9x - 5}\, dx$

8. $\displaystyle\int \frac{17x - 3}{3x^2 + x - 2}\, dx$

9. $\displaystyle\int \frac{2x^2 + x - 4}{x^3 - x^2 - 2x}\, dx$

10. $\displaystyle\int \frac{6x^2 + 22x - 23}{(2x - 1)(x^2 + x - 6)}\, dx$

11. $\displaystyle\int \frac{3x^3}{x^2 + x - 2}\, dx$

12. $\displaystyle\int \frac{x^4 + 8x^2 + 8}{x^3 - 4x}\, dx$

13. $\displaystyle\int \frac{x + 1}{(x - 3)^2}\, dx$

14. $\displaystyle\int \frac{5x + 7}{x^2 + 4x + 4}\, dx$

15. $\displaystyle\int \frac{3x^2 - 21x + 32}{x^3 - 8x^2 + 16x}\, dx$

16. $\displaystyle\int \frac{x^2 + 19x + 10}{2x^4 + 5x^3}\, dx$

17. $\displaystyle\int \frac{2x^2 + x - 8}{x^3 + 4x}\, dx$

18. $\displaystyle\int \frac{2x^2 - 3x - 36}{(2x - 1)(x^2 + 9)}\, dx$

19. $\displaystyle\int \frac{x^3 - 8x^2 - 1}{(x + 3)(x - 2)(x^2 + 1)}\, dx$

20. $\displaystyle\int \frac{20x - 11}{(3x + 2)(x^2 - 4x + 5)}\, dx$

21. $\displaystyle\int \frac{x^3 - 4x}{(x^2 + 1)^2}\, dx$

22. $\displaystyle\int \frac{2x^3 + 5x^2 + 16x}{x^5 + 8x^3 + 16x}\, dx$ *Hint:* To integrate

$$\int (x^2 + 4)^{-2}\, dx, \text{ let } x = 2 \tan \theta.$$

23. Evaluate $\displaystyle\int_4^6 \frac{x - 17}{x^2 + x - 12}\, dx$.

24. Find the value of $\displaystyle\int_1^5 \frac{3x + 13}{x^2 + 4x + 3}\, dx$.

25. The Law of Mass Action in chemistry results in the differential equation

$$\frac{dx}{dt} = k(a - x)(b - x) \qquad k > 0, \quad a > 0, \quad b > 0$$

where x is the amount of a substance at time t resulting from the reaction of two others. Assume that $x = 0$ when $t = 0$.

(a) Solve this differential equation in the case $b > a$.

(b) Show that $x \to a$ as $t \to \infty$ (if $b > a$).

(c) Suppose $a = 2$ and $b = 4$ and that 1 gram of the substance is formed in 20 minutes. How much will be present in 1 hour?

(d) Solve the differential equation if $a = b$.

[C] **26.** In many population growth problems, there is an upper limit beyond which the population cannot grow. Let us suppose that the earth will not support a population of more than 16 billion and that there were 2 billion people in 1925 and 4 billion people in 1975. Then if y is the population t years after 1925, an appropriate model is the differential equation

$$\frac{dy}{dt} = ky(16 - y)$$

(a) Solve this differential equation.

(b) Find the population in 2015.

(c) When will the population be 9 billion?

This model, called the logistic model, was discussed in Section 7.5.

[C] **27.** Do Problem 26 assuming the upper limit for the population is 10 billion.

28. The differential equation

$$\frac{dy}{dt} = k(y - m)(M - y)$$

with $k > 0$ and $0 \le m < y_0 < M$ has been used to model some growth problems. Solve the equation and find $\lim_{t \to \infty} y$.

29. As a model for the production of trypsin from trypsinogen in digestion, biochemists have proposed the model

$$\frac{dy}{dt} = k(A - y)(B + y)$$

where $k > 0$, A is the initial amount of trypsinogen, and B is the initial amount of trypsin. Solve this differential equation.

30. Evaluate $\displaystyle\int_{\pi/6}^{\pi/2} \frac{\cos x}{\sin x(\sin^2 x + 1)^2}\, dx$.

8.6 Chapter Review

TRUE-FALSE QUIZ

Respond with true or false to each of the following assertions. Be prepared to defend your answer.

1. To find $\displaystyle\int x \sin(x^2)\, dx$, try the substitution $u = x^2$.

2. To find $\displaystyle\int \frac{x}{9 + x^4}\, dx$, try the substitution $u = x^2$.

3. To find $\displaystyle\int \frac{x^3}{9 + x^4}\, dx$, try the substitution $u = x^2$.

4. To find $\displaystyle\int \frac{2x - 3}{x^2 - 3x + 5}\, dx$, begin by completing the square of the denominator.

5. To find $\displaystyle\int \frac{3}{x^2 - 3x + 5}\, dx$, begin by completing the square of the denominator.

6. To find $\displaystyle\int \frac{1}{\sqrt{4 - 5x^2}}\, dx$, try the substitution $u = \sqrt{5}\, x$.

7. To find $\displaystyle\int \frac{t + 2}{t^3 - 9t}\, dt$, try a partial fraction decomposition.

8. To find $\displaystyle\int \frac{t^4}{t^2 - 1}\, dt$, use integration by parts.

9. To find $\displaystyle\int \sin^6 x \cos^2 x\, dx$, use half-angle formulas.

10. To find $\displaystyle\int \frac{e^x}{1 + e^x}\, dx$, use integration by parts.

11. To find $\int \dfrac{x+2}{\sqrt{-x^2-4x}}\,dx$, use a trigonometric substitution.

12. To find $\int x^2 \sqrt[3]{3-2x}\,dx$, let $u = \sqrt[3]{3-2x}$.

13. To find $\int \sin^2 x \cos^5 x\,dx$, rewrite the integrand as $\sin^2 x(1-\sin^2 x)^2 \cos x$.

14. To find $\int \dfrac{1}{x^2\sqrt{9-x^2}}\,dx$, try a trigonometric substitution.

15. To find $\int x^2 \ln x\,dx$, try integration by parts.

16. To find $\int \sin 2x \cos 4x\,dx$, use half-angle formulas.

17. $\dfrac{x^2}{x^2-1}$ can be expressed in the form $\dfrac{A}{x-1}+\dfrac{B}{x+1}$.

18. $\dfrac{x^2+2}{x(x^2-1)}$ can be expressed in the form $\dfrac{A}{x}+\dfrac{B}{x-1}+\dfrac{C}{x+1}$.

19. $\dfrac{x^2+2}{x(x^2+1)}$ can be expressed in the form $\dfrac{A}{x}+\dfrac{Bx+C}{x^2+1}$.

20. $\dfrac{x+2}{x^2(x^2-1)}$ can be expressed in the form $\dfrac{A}{x^2}+\dfrac{B}{x-1}+\dfrac{C}{x+1}$.

21. The integration by parts formula is the integral version of the Product Rule for derivatives.

22. To complete the square of ax^2+bx, add $(b/2)^2$.

23. Any polynomial with real coefficients can be factored into a product of linear polynomials with real coefficients.

24. Two polynomials in x have the same values for all x only if the coefficients of like degree terms are identical.

25. If $a_n x^n + a_{n-1}x^{n-1} + \cdots + a_1 x + a_0 = 0$ has $n+1$ solutions, then all the coefficients are 0.

SAMPLE TEST PROBLEMS

In Problems 1–42, perform the indicated integrations by any correct method.

1. $\displaystyle\int_0^4 \dfrac{t}{\sqrt{9+t^2}}\,dt$

2. $\displaystyle\int \cot^2(2\theta)\,d\theta$

3. $\displaystyle\int_0^{\pi/2} e^{\cos x}\sin x\,dx$

4. $\displaystyle\int_0^{\pi/4} x\sin 2x\,dx$

5. $\displaystyle\int \dfrac{y^3+y}{y+1}\,dy$

6. $\displaystyle\int \sin^3(2t)\,dt$

7. $\displaystyle\int \dfrac{(y-2)}{y^2-4y+2}\,dy$

8. $\displaystyle\int_0^{3/2} \dfrac{dy}{\sqrt{2y+1}}$

9. $\displaystyle\int \dfrac{e^{2t}}{e^t-2}\,dt$

10. $\displaystyle\int \dfrac{\sin x+\cos x}{\tan x}\,dx$

11. $\displaystyle\int \dfrac{dx}{\sqrt{16+4x-2x^2}}$

12. $\displaystyle\int x^2 e^x\,dx$

13. $\displaystyle\int \dfrac{dy}{\sqrt{2+3y^2}}$

14. $\displaystyle\int \dfrac{w^3}{1-w^2}\,dw$

15. $\displaystyle\int \dfrac{\tan x}{\ln|\cos x|}\,dx$

16. $\displaystyle\int \dfrac{3\,dt}{t^3-1}$

17. $\displaystyle\int \sinh x\,dx$

18. $\displaystyle\int \dfrac{(\ln y)^5}{y}\,dy$

19. $\displaystyle\int x\cot^2 x\,dx$

20. $\displaystyle\int \dfrac{\sin\sqrt{x}}{\sqrt{x}}\,dx$

21. $\displaystyle\int \dfrac{\ln t^2}{t}\,dt$

22. $\displaystyle\int \ln(y^2+9)\,dy$

23. $\displaystyle\int e^{t/3}\sin 3t\,dt$

24. $\displaystyle\int \dfrac{t+9}{t^3+9t}\,dt$

25. $\displaystyle\int \sin\dfrac{3x}{2}\cos\dfrac{x}{2}\,dx$

26. $\displaystyle\int \cos^4\!\left(\dfrac{x}{2}\right)dx$

27. $\displaystyle\int \tan^3 2x \sec 2x\,dx$

28. $\displaystyle\int \dfrac{\sqrt{x}}{1+\sqrt{x}}\,dx$

29. $\displaystyle\int \tan^{3/2} x \sec^4 x\,dx$

30. $\displaystyle\int \dfrac{dt}{t(t^{1/6}+1)}$

31. $\displaystyle\int \frac{e^{2y}\,dy}{\sqrt{9 - e^{2y}}}$

32. $\displaystyle\int \cos^5 x \sqrt{\sin x}\,dx$

33. $\displaystyle\int e^{\ln(3\cos x)}\,dx$

34. $\displaystyle\int \frac{\sqrt{9 - y^2}}{y}\,dy$

35. $\displaystyle\int \frac{e^{4x}}{1 + e^{8x}}\,dx$

36. $\displaystyle\int \frac{\sqrt{x^2 + a^2}}{x^4}\,dx$

37. $\displaystyle\int \frac{w}{\sqrt{w + 5}}\,dw$

38. $\displaystyle\int \frac{\sin t\,dt}{\sqrt{1 + \cos t}}$

39. $\displaystyle\int \frac{\sin y \cos y}{9 + \cos^4 y}\,dy$

40. $\displaystyle\int \frac{dx}{\sqrt{1 - 6x - x^2}}$

41. $\displaystyle\int \frac{4x^2 + 3x + 6}{x^2(x^2 + 3)}\,dx$

42. $\displaystyle\int \frac{dx}{(16 + x^2)^{3/2}}$

43. Find the area of the region bounded by the x-axis, the curve $y = 18/(x^2\sqrt{x^2 + 9})$, and the lines $x = \sqrt{3}$ and $x = 3\sqrt{3}$.

44. Find the area of the region bounded by the curve $s = t/(t - 1)^2$, $s = 0$, $t = -6$, and $t = 0$.

45. Find the volume of the solid generated by revolving the region

$$\left\{ (x, y) \colon -3 \le x \le -1, \frac{6}{x\sqrt{x + 4}} \le y \le 0 \right\}$$

about the x-axis. Make a sketch.

46. Find the length of the segment of the curve $y = \ln(\sin x)$ from $x = \pi/6$ to $x = \pi/3$.

9 Indeterminate Forms and Improper Integrals

Ordinary analysis deals only with finite quantities; this one penetrates as far as infinity itself.

l'Hôpital

Guillaume F. A. de l'Hôpital
1661–1704

Guillaume François Antoine de l'Hôpital is one of the lesser lights in the mathematical firmament, and yet he is immortalized by a result that goes by his name (l'Hôpital's Rule). Moreover, that result is not his own; it is due to his teacher Johann Bernoulli. And thereby hangs a tale that is worth retelling.

L'Hôpital was born to parents of French high nobility. With some mathematical talent and more money, he agreed to support Johann Bernoulli if the latter would allow him to publish Bernoulli's discoveries. After l'Hôpital's death, Bernoulli tried to claim these discoveries as his own, but evidence was lacking. It was not until 1955 that letters between Bernoulli and l'Hôpital specifying these arrangements were found, finally justifying Bernoulli's claim.

Among l'Hôpital's contributions was the first textbook on differential calculus, published in 1696. This book uses a kind of language common to all early workers in calculus, the language of infinitesimals. What is an infinitesimal? According to l'Hôpital, if two quantities differ by an infinitesimal, they can be considered to be equal. This seems to say that two quantities can simultaneously be equal and unequal. In the hands of talented people like Newton, Leibniz, Bernoulli, and Euler, such ambiguous notions did not seem to hinder—and may even have helped—in the discovery of new mathematics. Fortunately for mere mortals, infinitesimals were banished from mathematics during the nineteenth century and were replaced with the rigorous notion of limit. However, mathematicians have recently resurrected an acceptable version of these long-despised quantities in a branch of mathematics called nonstandard analysis.

Here are three familiar limit problems:

$$\lim_{x \to 0} \frac{\sin x}{x}, \qquad \lim_{x \to 3} \frac{x^2 - 9}{x^2 - x - 6}, \qquad \lim_{x \to a} \frac{f(x) - f(a)}{x - a}$$

The first was treated at length in Section 3.4 and the third actually defines the derivative $f'(a)$. The three limits have a common feature. In each case, a quotient is involved and, in each case, both numerator and denominator have 0 as their limits. An attempt to apply the quotient rule for limits leads to the nonsensical answer 0/0. In fact, the quotient rule does not apply, since that rule requires that the limit of the denominator be different from 0. We are not saying that these limits do not exist, only that the quotient rule will not determine them.

You may recall that an intricate geometric argument led us to the conclusion $\lim_{x \to 0}(\sin x)/x = 1$. On the other hand, the algebraic technique of factoring yields

$$\lim_{x \to 3} \frac{x^2 - 9}{x^2 - x - 6} = \lim_{x \to 3} \frac{(x - 3)(x + 3)}{(x - 3)(x + 2)} = \lim_{x \to 3} \frac{x + 3}{x + 2} = \frac{6}{5}$$

Would it not be nice to have a standard procedure for handling all problems of this type? That is too much to hope for. However, there is a simple rule that works beautifully on a wide variety of such problems. It is known as *l'Hôpital's Rule* (pronounced low-pee-tăl).

L'HÔPITAL'S RULE   In 1696, Guillaume François Antoine de l'Hôpital published the first textbook on differential calculus; it included the following rule, which he had learned from his teacher Johann Bernoulli.

Theorem A

(L'Hôpital's Rule for forms of type 0/0).   Suppose that $\lim_{x \to u} f(x) = \lim_{x \to u} g(x) = 0$. If $\lim_{x \to u}[f'(x)/g'(x)]$ exists in either the finite or infinite sense (that is, if this limit is a finite number or $-\infty$ or $+\infty$), then

$$\lim_{x \to u} \frac{f(x)}{g(x)} = \lim_{x \to u} \frac{f'(x)}{g'(x)}$$

Here u may stand for any of the symbols a, a^-, a^+, $-\infty$, or $+\infty$.

Before attempting to prove this theorem, we illustrate it. Note that l'Hôpital's Rule allows us to replace one limit by another, which may be simpler and—in particular—may not have the 0/0 form.

EXAMPLE 1   Use l'Hôpital's Rule to show that

$$\lim_{x \to 0} \frac{\sin x}{x} = 1 \quad \text{and} \quad \lim_{x \to 0} \frac{1 - \cos x}{x} = 0$$

Solution   We worked pretty hard to demonstrate these two facts in Section 3.4. After noting that both limits have the 0/0 form, we can now establish the desired results in two lines (but see Problem 25). By l'Hôpital's Rule,

$$\lim_{x \to 0} \frac{\sin x}{x} = \lim_{x \to 0} \frac{D_x \sin x}{D_x x} = \lim_{x \to 0} \frac{\cos x}{1} = 1$$

$$\lim_{x \to 0} \frac{1 - \cos x}{x} = \lim_{x \to 0} \frac{D_x(1 - \cos x)}{D_x x} = \lim_{x \to 0} \frac{\sin x}{1} = 0 \qquad \blacksquare$$

EXAMPLE 2   Find $\lim\limits_{x \to 3} \dfrac{x^2 - 9}{x^2 - x - 6}$ and $\lim\limits_{x \to 2^+} \dfrac{x^2 + 3x - 10}{x^2 - 4x + 4}$.

Solution   Both limits have the 0/0 form, so by l'Hôpital's Rule,

$$\lim_{x \to 3} \frac{x^2 - 9}{x^2 - x - 6} = \lim_{x \to 3} \frac{2x}{2x - 1} = \frac{6}{5}$$

$$\lim_{x \to 2^+} \frac{x^2 + 3x - 10}{x^2 - 4x + 4} = \lim_{x \to 2^+} \frac{2x + 3}{2x - 4} = \infty$$

The first of these limits was handled in the introduction by the method of factoring. Of course, we get the same answer either way. $\qquad \blacksquare$

EXAMPLE 3   Find $\lim\limits_{x \to 0} \dfrac{\tan 2x}{\ln(1 + x)}$.

Solution   Both numerator and denominator have limit 0. Hence,

$$\lim_{x \to 0} \frac{\tan 2x}{\ln(1 + x)} = \lim_{x \to 0} \frac{2 \sec^2 2x}{1/(1 + x)} = \frac{2}{1} = 2 \qquad \blacksquare$$

Sometimes $\lim f'(x)/g'(x)$ also has the indeterminate form 0/0. Then we may apply l'Hôpital's Rule again, as we now illustrate.

EXAMPLE 4   Find $\lim\limits_{x \to 0} \dfrac{\sin x - x}{x^3}$.

Solution By l'Hôpital's Rule applied three times in succession,

$$\lim_{x\to 0} \frac{\sin x - x}{x^3} \overset{\text{\textcircled{L}}}{=} \lim_{x\to 0} \frac{\cos x - 1}{3x^2}$$

$$\overset{\text{\textcircled{L}}}{=} \lim_{x\to 0} \frac{-\sin x}{6x}$$

$$\overset{\text{\textcircled{L}}}{=} \lim_{x\to 0} \frac{-\cos x}{6}$$

$$= -\frac{1}{6} \qquad\blacksquare$$

Just because we have an elegant rule does not mean we should use it indiscriminately. In particular, we must always make sure it applies. Otherwise we will be led into all kinds of errors, as we now illustrate.

EXAMPLE 5 Find $\lim\limits_{x\to 0} \dfrac{1 - \cos x}{x^2 + 3x}$.

Solution

$$\lim_{x\to 0} \frac{1 - \cos x}{x^2 + 3x} \overset{\text{\textcircled{L}}}{=} \lim_{x\to 0} \frac{\sin x}{2x + 3} \overset{\text{\textcircled{L}}}{=} \lim_{x\to 0} \frac{\cos x}{2} = \frac{1}{2} \qquad \text{WRONG}$$

The first application of l'Hôpital's Rule was correct; the second was not since, at that stage, the limit did not have the 0/0 form. Here is what we should have done.

$$\lim_{x\to 0} \frac{1 - \cos x}{x^2 + 3x} \overset{\text{\textcircled{L}}}{=} \lim_{x\to 0} \frac{\sin x}{2x + 3} = 0 \qquad \text{RIGHT}$$

We stop differentiating as soon as either the numerator or denominator has a nonzero limit. $\qquad\blacksquare$

Even if l'Hôpital's Rule applies, it may not help us; witness the following example.

EXAMPLE 6 Find $\lim\limits_{x\to \infty} \dfrac{e^{-x}}{x^{-1}}$.

Solution We may apply l'Hôpital's Rule indefinitely.

$$\lim_{x \to \infty} \frac{e^{-x}}{x^{-1}} = \lim_{x \to \infty} \frac{e^{-x}}{x^{-2}} = \lim_{x \to \infty} \frac{e^{-x}}{2x^{-3}} = \cdots$$

Clearly, we are only complicating the problem. A better approach is to do a bit of algebra first.

$$\lim_{x \to \infty} \frac{e^{-x}}{x^{-1}} = \lim_{x \to \infty} \frac{x}{e^x}$$

Written this way, the limit is indeterminate of the form ∞/∞, the subject of the next section. However, you should be able to guess that the limit is 0 by considering how much faster e^x grows than x (see Figure 1). A rigorous demonstration will come later (Example 1 of Section 9.2). ∎

CAUCHY'S MEAN VALUE THEOREM The modern proof of l'Hôpital's Rule depends on an extension of the Mean Value Theorem due to Augustin-Louis Cauchy (1789–1857).

Theorem B

(Cauchy's Mean Value Theorem). Let the functions f and g be differentiable on (a, b) and continuous on $[a, b]$. If $g'(x) \neq 0$ for all x in (a, b), then there exists a number c in (a, b) such that

$$\frac{f(b) - f(a)}{g(b) - g(a)} = \frac{f'(c)}{g'(c)}$$

Note that this theorem reduces to the ordinary Mean Value Theorem (Theorem 4.8A) when $g(x) = x$.

Proof It is tempting to apply the ordinary Mean Value Theorem to both numerator and denominator of the left side of the conclusion. If we do this, we obtain

(1) $f(b) - f(a) = f'(c_1)(b - a)$

and

(2) $g(b) - g(a) = g'(c_2)(b - a)$

for appropriate choices of c_1 and c_2. If only c_1 and c_2 were equal, we could divide the first equality by the second and be done; but there is no reason to hope for such a coincidence. However, this attempt is not a complete failure since (2) yields the valuable information that $g(b) - g(a) \neq 0$, a fact we will need later (this follows from the hypothesis that $g'(x) \neq 0$ for all x in (a, b)).

FIGURE 1

Recall that the proof of the Mean Value Theorem (Theorem 4.8A) rested on the introduction of an auxiliary function s. If we try to mimic that proof, we are led to the following choice for $s(x)$. Let

$$s(x) = f(x) - f(a) - \frac{f(b) - f(a)}{g(b) - g(a)} [g(x) - g(a)]$$

No division by zero is involved since we earlier established that $g(b) - g(a) \neq 0$. Note further that $s(a) = 0 = s(b)$. Also, s is continuous on $[a, b]$ and differentiable on (a, b), this following from the corresponding facts for f and g. Thus, by the Mean Value Theorem, there is a number c in (a, b) such that

$$s'(c) = \frac{s(b) - s(a)}{b - a} = \frac{0 - 0}{b - a} = 0$$

but

$$s'(c) = f'(c) - \frac{f(b) - f(a)}{g(b) - g(a)} g'(c) = 0$$

or

$$\frac{f'(c)}{g'(c)} = \frac{f(b) - f(a)}{g(b) - g(a)}$$

which is what we wished to prove. ∎

PROOF OF L'HÔPITAL'S RULE

Proof   Refer back to Theorem A, which actually states several theorems at once. We will prove only the case where L is finite and lim is the one-sided limit $\lim_{x \to a^+}$.

The hypotheses for Theorem A imply more than they say explicitly. In particular, the existence of $\lim_{x \to a^+} [f'(x)/g'(x)]$ implies that both $f'(x)$ and $g'(x)$ exist in at least a small interval $(a, b]$ and that $g'(x) \neq 0$ there. At a, we do not even know that f and g are defined, but we do know that $\lim_{x \to a^+} f(x) = 0$ and $\lim_{x \to a^+} g(x) = 0$. Thus, we may define (or redefine if necessary) both $f(a)$ and $g(a)$ to be zero, thereby making both f and g (right) continuous at a. All of this is to say that f and g satisfy the hypotheses of Cauchy's Mean Value Theorem on $[a, b]$. Consequently, there is a number c in (a, b) such that

$$\frac{f(b) - f(a)}{g(b) - g(a)} = \frac{f'(c)}{g'(c)}$$

or, since $f(a) = 0 = g(a)$,

$$\frac{f(b)}{g(b)} = \frac{f'(c)}{g'(c)}$$

When we let $b \to a^+$, thereby forcing $c \to a^+$, we obtain

$$\lim_{b \to a^+} \frac{f(b)}{g(b)} = \lim_{c \to a^+} \frac{f'(c)}{g'(c)}$$

which is equivalent to what we wanted to prove.

A very similar proof works for the case of left-hand limits, and thus for two-sided limits. The proofs for the cases where a or L is infinite are harder, and we omit them. ∎

PROBLEM SET 9.1

In Problems 1–24, find the indicated limit. Make sure l'Hôpital's Rule applies before you use it.

1. $\displaystyle\lim_{x \to 0} \frac{\sin x - 2x}{x}$

2. $\displaystyle\lim_{x \to (1/2)\pi} \frac{\cos x}{x - \frac{1}{2}\pi}$

3. $\displaystyle\lim_{x \to 0} \frac{x - 2\sin x}{\tan x}$

4. $\displaystyle\lim_{x \to 0} \frac{\sin^{-1} x}{3\tan^{-1} x}$

5. $\displaystyle\lim_{x \to -1} \frac{x^2 + 5x + 4}{x^2 - 4x - 5}$

6. $\displaystyle\lim_{x \to 0} \frac{x^3 - 3x^2 + 5x}{x^3 - x}$

7. $\displaystyle\lim_{x \to 1^+} \frac{x^2 - 2x + 2}{x^2 - 1}$

8. $\displaystyle\lim_{x \to 1} \frac{\ln x}{x^2 - 1}$

9. $\displaystyle\lim_{x \to (1/2)\pi} \frac{\ln \sin x}{\frac{1}{2}\pi - x}$

10. $\displaystyle\lim_{x \to 0} \frac{e^x - e^{-x}}{4\sin x}$

11. $\displaystyle\lim_{t \to 1} \frac{\sqrt{t} - t}{\ln t}$

12. $\displaystyle\lim_{x \to 0^+} \frac{8^{\sqrt{x}} - 1}{3^{\sqrt{x}} - 1}$

13. $\displaystyle\lim_{x \to 0} \frac{\ln \cos 3x}{2x^2}$

14. $\displaystyle\lim_{x \to 0^+} \frac{2\sin x}{\sqrt{x}}$

15. $\displaystyle\lim_{x \to 0} \frac{\tan x - x}{\sin x - x}$

16. $\displaystyle\lim_{x \to 0} \frac{\tan x - \sin x}{x^2 \tan x}$

17. $\displaystyle\lim_{x \to 0^+} \frac{x^2}{\sin x - x}$

18. $\displaystyle\lim_{x \to 0} \frac{e^x - \ln(1 + x) - 1}{x^2}$

19. $\displaystyle\lim_{x \to 0} \frac{\tan^{-1} x - x}{8x^3}$

20. $\displaystyle\lim_{x \to 0} \frac{\cosh x - 1}{x^2}$

21. $\displaystyle\lim_{x \to 0^+} \frac{1 - \cos x - x\sin x}{2 - 2\cos x - \sin^2 x}$

22. $\displaystyle\lim_{x \to 0^-} \frac{\sin x + \tan x}{e^x + e^{-x} - 2}$

23. $\displaystyle\lim_{x \to 0} \frac{\displaystyle\int_0^x \sqrt{1 + \sin t}\, dt}{x}$

24. $\displaystyle\lim_{x \to 0^+} \frac{\displaystyle\int_0^x \sqrt{t}\cos t\, dt}{x^2}$

25. In Section 3.4, we worked very hard to prove that $\displaystyle\lim_{x \to 0}(\sin x)/x = 1$; l'Hôpital's Rule allows us to show this fact in one line. However, even if we had l'Hôpital's Rule available at that stage, it would not have helped us. Explain why.

26. Find $\displaystyle\lim_{x \to 0} \frac{x^2 \sin(1/x)}{\tan x}$

Hint: Begin by deciding why l'Hôpital's Rule is not applicable. Then find the limit by other means.

27. For Figure 2, compute the following limits.

(a) $\displaystyle\lim_{t \to 0^+} \frac{\text{area of triangle } ABC}{\text{area of curved region } ABC}$

(b) $\displaystyle\lim_{t \to 0^+} \frac{\text{area of curved region } BCD}{\text{area of curved region } ABC}$

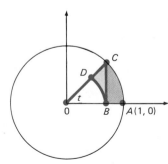

FIGURE 2

28. In Figure 3, $\overline{CD} = \overset{\frown}{DE} = \overline{DF} = t$. Find each limit.

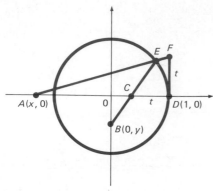

FIGURE 3

(a) $\lim\limits_{t \to 0^+} y$

(b) $\lim\limits_{t \to 0^+} x$

29. In Problem 24 of Section 6.5, you were asked to show that the surface area of the prolate spheroid gotten by rotating the ellipse $x^2/a^2 + y^2/b^2 = 1$ $(a > b)$ about the x-axis is

$$A = 2\pi b^2 + 2\pi ab \left[\frac{a}{\sqrt{a^2 - b^2}} \arcsin \frac{\sqrt{a^2 - b^2}}{a} \right]$$

What should A approach as $a \to b^+$? Use l'Hôpital's Rule to show that this does happen.

30. Determine constants a, b, and c so that

$$\lim\limits_{x \to 1} \frac{ax^4 + bx^3 + 1}{(x - 1)\sin \pi x} = c$$

9.2
Other Indeterminate Forms

In the solution to Example 6 of the previous section, we faced the following limit problem.

$$\lim\limits_{x \to \infty} \frac{x}{e^x}$$

This is typical of a class of problems of the form $\lim\limits_{x \to \infty} f(x)/g(x)$, where both numerator and denominator are growing indefinitely large; we call it an indeterminate form of type ∞/∞. It turns out that l'Hôpital's Rule also applies in this situation, that is,

$$\lim\limits_{x \to \infty} \frac{f(x)}{g(x)} = \lim\limits_{x \to \infty} \frac{f'(x)}{g'(x)}$$

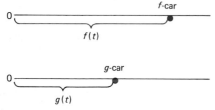

FIGURE 1

A rigorous proof is quite painful, but there is an intuitive way of seeing that the result has to be true. Imagine that $f(t)$ and $g(t)$ represent the positions of two cars on the t-axis at time t (Figure 1). These two cars, the f-car and the g-car, are on endless journeys with respective velocities $f'(t)$ and $g'(t)$. Now, if

$$\lim\limits_{x \to \infty} \frac{f'(t)}{g'(t)} = L$$

then ultimately the f-car travels about L times as fast as the g-car. It is therefore reasonable to say that in the long run, it will travel about L times as far; that is,

$$\lim\limits_{x \to \infty} \frac{f(t)}{g(t)} = L$$

We do not call this a proof, but it does lend plausibility to a result that we now state formally.

Theorem A

(L'Hôpital's Rule for forms of type ∞/∞). Suppose that $\lim\limits_{x \to u} |f(x)| = \lim\limits_{x \to u} |g(x)| = \infty$. If $\lim\limits_{x \to u} [f'(x)/g'(x)]$ exists in either the finite or infinite sense, then

$$\lim_{x \to u} \frac{f(x)}{g(x)} = \lim_{x \to u} \frac{f'(x)}{g'(x)}$$

Here u may stand for any of the symbols a, a^-, a^+, $-\infty$, or $+\infty$.

THE INDETERMINATE FORM ∞/∞   We use Theorem A to finish Example 6 of the previous section.

EXAMPLE 1   Find $\lim\limits_{x \to \infty} \dfrac{x}{e^x}$.

Solution   Both x and e^x tend to ∞ as $x \to \infty$. Hence, by l'Hôpital's Rule,

$$\lim_{x \to \infty} \frac{x}{e^x} = \lim_{x \to \infty} \frac{D_x x}{D_x e^x} = \lim_{x \to \infty} \frac{1}{e^x} = 0 \qquad \blacksquare$$

Here is a general result of the same type.

EXAMPLE 2   Show that if a is any positive real number, $\lim\limits_{x \to \infty} \dfrac{x^a}{e^x} = 0$.

Solution   Suppose first that $a = 2.5$. Then three applications of l'Hôpital's Rule give

$$\lim_{x \to \infty} \frac{x^{2.5}}{e^x} \overset{\text{\textcircled{L}}}{=} \lim_{x \to \infty} \frac{2.5x^{1.5}}{e^x} \overset{\text{\textcircled{L}}}{=} \lim_{x \to \infty} \frac{(2.5)(1.5)x^{0.5}}{e^x} \overset{\text{\textcircled{L}}}{=} \lim_{x \to \infty} \frac{(2.5)(1.5)(0.5)}{x^{0.5}e^x} = 0$$

A similar argument works for any $a > 0$. $\qquad \blacksquare$

EXAMPLE 3   Show that if a is any positive real number, $\lim\limits_{x \to \infty} \dfrac{\ln x}{x^a} = 0$.

Solution   Both $\ln x$ and x^a tend to ∞ as $x \to \infty$. Hence by one application of l'Hôpital's Rule,

$$\lim_{x \to \infty} \frac{\ln x}{x^a} \overset{\text{\textcircled{L}}}{=} \lim_{x \to \infty} \frac{1/x}{ax^{a-1}} = \lim_{x \to \infty} \frac{1}{ax^a} = 0 \qquad \blacksquare$$

Examples 2 and 3 say something that is worth remembering, namely, *for large x, e^x grows faster than any constant power of x, while $\ln x$ grows more slowly than any constant power of x.* For example, e^x grows faster than x^{100} and $\ln x$ grows more slowly than $\sqrt[100]{x}$. The chart in the margin offers additional illustration.

EXAMPLE 4 Find $\displaystyle\lim_{x \to 0^+} \frac{\ln x}{\cot x}$.

Solution As $x \to 0^+$, $\ln x \to -\infty$ and $\cot x \to \infty$, so l'Hôpital's Rule applies.

$$\lim_{x \to 0^+} \frac{\ln x}{\cot x} \overset{L}{=} \lim_{x \to 0^+} \left[\frac{1/x}{-\csc^2 x} \right]$$

This is still indeterminate as it stands, but rather than apply l'Hôpital's Rule again (which only makes things worse), we rewrite the bracketed expression as

$$\frac{1/x}{-\csc^2 x} = -\frac{\sin^2 x}{x} = -\sin x \frac{\sin x}{x}$$

Thus,

$$\lim_{x \to 0^+} \frac{\ln x}{\cot x} = \lim_{x \to 0^+} \left[-\sin x \frac{\sin x}{x} \right] = 0 \cdot 1 = 0 \qquad \blacksquare$$

THE INDETERMINATE FORMS $0 \cdot \infty$ AND $\infty - \infty$ Suppose that $A(x) \to 0$, but $B(x) \to \infty$. What is going to happen to the product $A(x)B(x)$? Two competing forces are at work, tending to pull the product in opposite directions. Which will win this battle, A or B or neither? It depends on whether one is stronger (that is, doing its job at a faster rate) or if they are evenly matched. L'Hôpital's Rule will help us decide, but only after we have transformed the problem to a 0/0 or ∞/∞ form.

EXAMPLE 5 Find $\displaystyle\lim_{x \to \pi/2} (\tan x \cdot \ln \sin x)$.

Solution Since $\displaystyle\lim_{x \to \pi/2} \ln \sin x = 0$ and $\displaystyle\lim_{x \to \pi/2} |\tan x| = \infty$, this is a $0 \cdot \infty$ indeterminate form. We can rewrite it as a 0/0 form by the simple device of changing $\tan x$ to $1/\cot x$. Thus,

$$\lim_{x \to \pi/2} (\tan x \cdot \ln \sin x) = \lim_{x \to \pi/2} \frac{\ln \sin x}{\cot x}$$

$$\overset{L}{=} \lim_{x \to \pi/2} \frac{\dfrac{1}{\sin x} \cdot \cos x}{-\csc^2 x} = \lim_{x \to \pi/2} (-\cos x \cdot \sin x) = 0 \qquad \blacksquare$$

EXAMPLE 6   Find $\lim\limits_{x \to 1^+} \left(\dfrac{x}{x-1} - \dfrac{1}{\ln x} \right)$.

Solution   The first term is growing without bound; so is the second. We say that the limit is an $\infty - \infty$ indeterminate form. Again a battle rages. L'Hôpital's Rule will determine the result, but only after we rewrite the problem in a form to which the rule applies. In this case, the two fractions must be combined, a procedure that changes the problem to a 0/0 form. A double application of l'Hôpital's Rule yields

$$\lim_{x \to 1^+} \left(\frac{x}{x-1} - \frac{1}{\ln x} \right) = \lim_{x \to 1^+} \frac{x \ln x - x + 1}{(x-1)\ln x} \overset{\text{\textcircled{L}}}{=} \lim_{x \to 1^+} \frac{x \cdot 1/x + \ln x - 1}{(x-1)(1/x) + \ln x}$$

$$= \lim_{x \to 1^+} \frac{x \ln x}{x - 1 + x \ln x} \overset{\text{\textcircled{L}}}{=} \lim_{x \to 1^+} \frac{1 + \ln x}{2 + \ln x} = \frac{1}{2} \quad \blacksquare$$

THE INDETERMINATE FORMS 0^0, ∞^0, 1^∞   We turn now to three indeterminate forms of exponential type. Here the trick is to consider not the original expression, but rather its logarithm. Usually l'Hôpital's Rule will apply to the logarithm.

EXAMPLE 7   Find $\lim\limits_{x \to 0^+} (x + 1)^{\cot x}$.

Solution   This takes the indeterminate form 1^∞. Let $y = (x+1)^{\cot x}$, so

$$\ln y = \cot x \ln(x + 1) = \frac{\ln(x + 1)}{\tan x}$$

By l'Hôpital's Rule for 0/0 forms,

$$\lim_{x \to 0^+} \ln y = \lim_{x \to 0^+} \frac{\ln(x + 1)}{\tan x} \overset{\text{\textcircled{L}}}{=} \lim_{x \to 0^+} \frac{\dfrac{1}{x + 1}}{\sec^2 x} = 1$$

Now $y = e^{\ln y}$, and since the exponential function $f(x) = e^x$ is continuous,

$$\lim_{x \to 0^+} y = \lim_{x \to 0^+} \exp(\ln y) = \exp\left(\lim_{x \to 0^+} \ln y \right) = \exp 1 = e \quad \blacksquare$$

EXAMPLE 8   Find $\lim\limits_{x \to \pi/2^-} (\tan x)^{\cos x}$.

Solution   This has the indeterminate form ∞^0. Put $y = (\tan x)^{\cos x}$, so

$$\ln y = \cos x \cdot \ln \tan x = \frac{\ln \tan x}{\sec x}$$

Then,

$$\lim_{x \to \pi/2^-} \ln y = \lim_{x \to \pi/2^-} \frac{\ln \tan x}{\sec x} \overset{L}{=} \lim_{x \to \pi/2^-} \frac{\dfrac{1}{\tan x} \cdot \sec^2 x}{\sec x \tan x}$$

$$= \lim_{x \to \pi/2^-} \frac{\sec x}{\tan^2 x} = \lim_{x \to \pi/2^-} \frac{\cos x}{\sin^2 x} = 0$$

Therefore,

$$\lim_{x \to \pi/2^-} y = e^0 = 1 \qquad \blacksquare$$

SUMMARY   We have classified certain limit problems as indeterminate forms, using the seven symbols $0/0$, ∞/∞, $0 \cdot \infty$, $\infty - \infty$, 0^0, ∞^0, and 1^∞. Each involves a competition of opposing forces, which means that the result is not obvious. However, with the help of l'Hôpital's Rule, which applies directly only to the $0/0$ and ∞/∞ forms, we can usually determine the correct limit.

There are many other possibilities symbolized by, for example, $0/\infty$, $\infty/0$, $\infty + \infty$, $\infty \cdot \infty$, 0^∞, and ∞^∞. Why don't we call these indeterminate forms? Because, in each of these cases, the forces are in collusion, not competition.

EXAMPLE 9   Find $\lim\limits_{x \to 0^+} \sin x^{\cot x}$.

Solution   We might call this a 0^∞ form, but it is not indeterminate. Note that $\sin x$ is approaching zero, and raising it to the exponent $\cot x$, an increasingly large number, serves only to make it approach zero faster. Thus,

$$\lim_{x \to 0^+} (\sin x)^{\cot x} = 0 \qquad \blacksquare$$

PROBLEM SET   9.2

Find the limits in Problems 1–40. Be sure l'Hôpital's Rule applies before you use it.

1. $\lim\limits_{x \to \infty} \dfrac{\ln(x^{100})}{x}$

2. $\lim\limits_{x \to \infty} \dfrac{\ln x}{2^x}$

3. $\lim\limits_{x \to \infty} \dfrac{x^{10}}{e^x}$

4. $\lim\limits_{x \to \infty} \dfrac{2x}{\ln(3x + e^x)}$

5. $\lim\limits_{x \to \pi/2} \dfrac{\sec x + 1}{\tan x}$

6. $\lim\limits_{x \to 0^+} \dfrac{\ln \sin x}{\ln \tan x}$

7. $\lim\limits_{x \to \infty} \dfrac{\ln(\ln x)}{\ln x}$

8. $\lim\limits_{x \to (\frac{1}{2})^-} \dfrac{\ln(1 - 2x)}{\tan \pi x}$

9. $\lim\limits_{x \to 0^+} \dfrac{\cot x}{\ln x}$

10. $\lim\limits_{x \to 0} \dfrac{\csc x}{\cot^2 x}$

11. $\lim\limits_{x \to 0}(2x \ln x^2)$

12. $\lim\limits_{x \to 0} x^2 \csc x$

13. $\lim\limits_{x \to 0}(\csc x - \cot x)$

14. $\lim\limits_{x \to \pi/2} (\sec x - \tan x)$

15. $\lim\limits_{x \to 0^+} (2x)^{x^2}$

16. $\lim\limits_{x \to 0}(\cos x)^{\cot x}$

17. $\lim\limits_{x \to (\frac{1}{2})\pi^-} (\cos x)^{\tan x}$

18. $\lim\limits_{x \to 0}\left(\csc^2 x - \dfrac{1}{x^2} \right)$

19. $\lim\limits_{x \to 0}(x + e^{x/2})^{2/x}$

20. $\lim\limits_{x \to (\frac{1}{2})\pi^-} (\cos x)^{x - \pi/2}$

21. $\lim\limits_{x \to \pi/2} (\sin x)^{\cos x}$

22. $\lim\limits_{x \to \infty} x^x$

23. $\lim\limits_{x \to \infty} x^{1/x}$

24. $\lim\limits_{x \to 0}(\cos x)^{1/x^2}$

25. $\lim_{x \to 0}(\cos x)^{2+x}$

26. $\lim_{x \to -\infty}(e^{-x} - x)$

27. $\lim_{x \to 0^+}(\sin x)^x$

28. $\lim_{x \to 0}(\cos x - \sin x)^{1/x}$

29. $\lim_{x \to 0}\left(\csc x - \dfrac{1}{x}\right)$

30. $\lim_{x \to \infty}\left(1 + \dfrac{1}{x}\right)^x$

31. $\lim_{x \to 0^+}(1 + 2e^x)^{1/x}$

32. $\lim_{x \to 1}\left(\dfrac{1}{x-1} - \dfrac{x}{\ln x}\right)$

33. $\lim_{x \to 0}(\cos x)^{1/x}$

34. $\lim_{x \to 0^+}(x^{1/2} \ln x)$

35. $\lim_{x \to \infty} e^{\cos x}$

36. $\lim_{x \to \infty}[\ln(x+1) - \ln(x-1)]$

37. $\lim_{x \to 0^+}\dfrac{x}{\ln x}$

38. $\lim_{x \to 0^+}(\ln x \cot x)$

39. $\lim_{x \to \infty}\dfrac{\displaystyle\int_1^x \sqrt{1 + e^{-t}}\, dt}{x}$

40. $\lim_{x \to 1^+}\dfrac{\displaystyle\int_1^x \sin t\, dt}{x - 1}$

41. Find each limit.

(a) $\lim_{n \to \infty} \sqrt[n]{a}$

(b) $\lim_{n \to \infty} \sqrt[n]{n}$

(c) $\lim_{n \to \infty} n(\sqrt[n]{a} - 1)$

(d) $\lim_{n \to \infty} n(\sqrt[n]{n} - 1)$

Hint: Transform to problems involving a continuous variable x.

42. Find each limit.

(a) $\lim_{x \to 0^+} x^x$

(b) $\lim_{x \to 0^+} (x^x)^x$

(c) $\lim_{x \to 0^+} x^{(x^x)}$

(d) $\lim_{x \to 0^+} ((x^x)^x)^x$

(e) $\lim_{x \to 0^+} x^{(x^{(x^x)})}$

43. Graph $y = x^{1/x}$ for $x > 0$. Show what happens for very small x and very large x. Indicate the maximum value.

44. Find each limit.

(a) $\lim_{x \to 0^+} (1^x + 2^x)^{1/x}$

(b) $\lim_{x \to 0^-} (1^x + 2^x)^{1/x}$

(c) $\lim_{x \to \infty} (1^x + 2^x)^{1/x}$

(d) $\lim_{x \to -\infty} (1^x + 2^x)^{1/x}$

45. For $k \geq 0$, find

$$\lim_{n \to \infty} \frac{1^k + 2^k + \cdots + n^k}{n^{k+1}}$$

Hint: Though this has the ∞/∞ form, l'Hôpital's Rule is not helpful. Think of another often-used trick.

9.3
Improper Integrals: Infinite Limits

In the definition of $\displaystyle\int_a^b f(x)\, dx$, it was assumed that the interval $[a, b]$ was finite. However, there are many applications in physics, economics, and probability in which we wish to allow a or b (or both) to be infinite. We must therefore find a way to give meaning to symbols like

$$\int_0^\infty \frac{1}{1 + x^2}\, dx, \qquad \int_{-\infty}^{-1} xe^{-x^2}\, dx \qquad \int_{-\infty}^\infty x^2 e^{-x^2}\, dx$$

These integrals are called **improper integrals** with infinite limits.

ONE INFINITE LIMIT The graph of $f(x) = e^{-x}$ on $[0, \infty)$ is shown in Figure 1. The integral $\displaystyle\int_0^b e^{-x}\, dx$ makes perfectly good sense no matter how big we make b; in fact, we can evaluate it explicitly.

$$\int_0^b e^{-x}\, dx = -\int_0^b e^{-x}(-dx) = \left[-e^{-x}\right]_0^b = 1 - e^{-b}$$

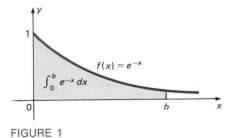

FIGURE 1

Now $\lim_{b \to \infty}(1 - e^{-b}) = 1$, so it seems natural to define

$$\int_0^\infty e^{-x}\,dx = 1$$

Here is the general definition.

Definition

$$\int_{-\infty}^b f(x)\,dx = \lim_{a \to -\infty} \int_a^b f(x)\,dx$$

$$\int_a^\infty f(x)\,dx = \lim_{b \to \infty} \int_a^b f(x)\,dx$$

If the limits on the right exist and have finite values, then we say the corresponding improper integrals **converge** and have those values. Otherwise, the integrals are said to **diverge**.

EXAMPLE 1   Find, if possible, $\displaystyle\int_{-\infty}^{-1} xe^{-x^2}\,dx$.

Solution

$$\int_a^{-1} xe^{-x^2}\,dx = -\frac{1}{2}\int_a^{-1} e^{-x^2}(-2x\,dx) = \left[-\frac{1}{2}e^{-x^2}\right]_a^{-1}$$

$$= -\frac{1}{2}e^{-1} + \frac{1}{2}e^{-a^2}$$

Thus,

$$\int_{-\infty}^{-1} xe^{-x^2}\,dx = \lim_{a \to -\infty}\left[-\frac{1}{2}e^{-1} + \frac{1}{2}e^{-a^2}\right] = -\frac{1}{2e}$$

We say the integral converges and has value $-1/2e$.   ∎

EXAMPLE 2   Find, if possible, $\displaystyle\int_0^\infty \sin x\,dx$.

Solution

$$\int_0^\infty \sin x\,dx = \lim_{b \to \infty}\int_0^b \sin x\,dx = \lim_{b \to \infty}\left[-\cos x\right]_0^b$$

$$= \lim_{b \to \infty}\left[1 - \cos b\right]$$

The latter limit does not exist; we conclude that the given integral diverges. Think about the geometric meaning of $\displaystyle\int_0^\infty \sin x\,dx$ to confirm this result.   ∎

EXAMPLE 3 According to Newton's Inverse-Square Law, the force exerted by the earth on a space capsule is $-k/x^2$, where x is the distance (in miles, for instance) from the capsule to the center of the earth. The force $F(x)$ required to lift the capsule is therefore $F(x) = k/x^2$. How much work is done in propelling a 1000-pound capsule out of the earth's gravitational field?

Solution We can evaluate k by noting that at $x = 3960$ miles (the radius of the earth), $F = 1000$ pounds. This yields $k = 1000(3960)^2 \approx 1.568 \times 10^{10}$. The work done in mile-pounds is therefore

$$1.568 \times 10^{10} \int_{3960}^{\infty} \frac{1}{x^2}\, dx = \lim_{b \to \infty} 1.568 \times 10^{10} \left[-\frac{1}{x} \right]_{3960}^{b}$$

$$= \lim_{b \to \infty} 1.568 \times 10^{10} \left[-\frac{1}{b} + \frac{1}{3960} \right]$$

$$= \frac{1.568 \times 10^{10}}{3960} \approx 3.96 \times 10^6 \qquad \blacksquare$$

BOTH LIMITS INFINITE First we need a definition.

Definition

If both $\displaystyle\int_{-\infty}^{0} f(x)\, dx$ and $\displaystyle\int_{0}^{\infty} f(x)\, dx$ converge, then $\displaystyle\int_{-\infty}^{\infty} f(x)\, dx$ is said to converge and have value

$$\int_{-\infty}^{\infty} f(x)\, dx = \int_{-\infty}^{0} f(x)\, dx + \int_{0}^{\infty} f(x)\, dx$$

Otherwise, $\displaystyle\int_{-\infty}^{\infty} f(x)\, dx$ diverges.

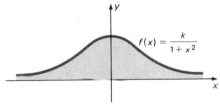

$$f(x) = \frac{k}{1 + x^2}$$

FIGURE 2

EXAMPLE 4 The function $f(x) = k/(1 + x^2)$ occurs in probability theory, where it is called the Cauchy density function. The constant k is to be chosen so that the total area under the curve on $(-\infty, \infty)$ is 1 (see Figure 2). Determine k.

Solution We want

$$\int_{-\infty}^{\infty} \frac{k}{1 + x^2}\, dx = 1$$

Now,

$$\int_{0}^{\infty} \frac{k}{1 + x^2}\, dx = \lim_{b \to \infty} \int_{0}^{b} \frac{k}{1 + x^2}\, dx$$

$$= \lim_{b \to \infty} \left[k \tan^{-1} x \right]_{0}^{b}$$

$$= \lim_{b \to \infty} k \tan^{-1} b = k \frac{\pi}{2}$$

Similarly, from symmetry,

$$\int_{-\infty}^{0} \frac{k}{1 + x^2}\, dx = k\frac{\pi}{2}$$

Thus,

$$\int_{-\infty}^{\infty} \frac{k}{1 + x^2}\, dx = \int_{-\infty}^{0} \frac{k}{1 + x^2}\, dx + \int_{0}^{\infty} \frac{k}{1 + x^2}\, dx = k\pi$$

We conclude that $k = 1/\pi$. ∎

EXAMPLE 5   The most important density function in probability theory is the standard **normal** density function, defined by

$$f(x) = \frac{1}{\sqrt{2\pi}}\, e^{-x^2/2}$$

It is surprisingly difficult to show that

$$\frac{1}{\sqrt{2\pi}} \int_{-\infty}^{\infty} e^{-x^2/2}\, dx = 1$$

though we will do so later (Section 16.4). Use this fact to show that this density function has mean 0 and variance 1; that is, show each of the following.

(a) $\dfrac{1}{\sqrt{2\pi}} \displaystyle\int_{-\infty}^{\infty} xe^{-x^2/2}\, dx = 0$
  (b) $\dfrac{1}{\sqrt{2\pi}} \displaystyle\int_{-\infty}^{\infty} x^2 e^{-x^2/2}\, dx = 1$

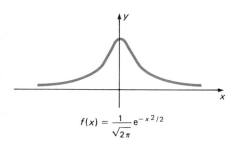

$f(x) = \dfrac{1}{\sqrt{2\pi}}\, e^{-x^2/2}$

FIGURE 3

Solution

(a) Since $e^{-x^2/2}$ is an even function, the graph of $y = f(x)$ is symmetric with respect to the y-axis (Figure 3). Thus,

$$\frac{1}{\sqrt{2\pi}} \int_{0}^{\infty} e^{-x^2/2}\, dx = \frac{1}{2}$$

a fact we will need shortly. Now

$$\frac{1}{\sqrt{2\pi}} \int_{0}^{\infty} xe^{-x^2/2}\, dx = \lim_{b \to \infty}\left[-\frac{1}{\sqrt{2\pi}} \int_{0}^{b} e^{-x^2/2}(-x)\, dx \right]$$

$$= \lim_{b \to \infty}\left[-\frac{1}{\sqrt{2\pi}}\, e^{-x^2/2} \right]_{0}^{b}$$

$$= \frac{1}{\sqrt{2\pi}}$$

Since $xe^{-x^2/2}$ is an odd function,

$$\frac{1}{\sqrt{2\pi}} \int_{-\infty}^{0} xe^{-x^2/2}\, dx = -\frac{1}{\sqrt{2\pi}} \int_{0}^{\infty} xe^{-x^2/2}\, dx = -\frac{1}{\sqrt{2\pi}}$$

which establishes (a).

(b) The demonstration of (b) uses integration by parts and l'Hôpital's Rule.

$$\frac{1}{\sqrt{2\pi}} \int_{0}^{\infty} x^2 e^{-x^2/2}\, dx = \lim_{b \to \infty} \frac{1}{\sqrt{2\pi}} \int_{0}^{b} (-x)(-e^{-x^2/2}x)\, dx$$

$$= \lim_{b \to \infty} \frac{1}{\sqrt{2\pi}} \left([-xe^{-x^2/2}]_{0}^{b} + \int_{0}^{b} e^{-x^2/2}\, dx \right)$$

$$= \frac{1}{\sqrt{2\pi}} \left(0 + \int_{0}^{\infty} e^{-x^2/2}\, dx \right) = \frac{1}{2}$$

Since $x^2 e^{-x^2/2}$ is an even function, we get a similar contribution to the left of zero, and so

$$\frac{1}{\sqrt{2\pi}} \int_{-\infty}^{\infty} x^2 e^{-x^2/2}\, dx = 1 \qquad \blacksquare$$

THE PARADOX OF GABRIEL'S HORN   Let the curve $y = 1/x$ on $[1, \infty)$ be revolved about the x-axis, thereby generating a surface called Gabriel's horn (Figure 4). We claim:

FIGURE 4

1. the volume V of this horn is finite;
2. the surface area A of the horn is infinite.

To put the results in practical terms, they seem to say that the horn can be filled with a finite amount of paint, and yet there is not enough to paint its inside surface. Before we try to unravel this paradox, let us establish (1) and (2).

$$V = \int_{1}^{\infty} \pi \left(\frac{1}{x}\right)^2 dx = \lim_{b \to \infty} \pi \int_{1}^{b} x^{-2}\, dx$$

$$= \lim_{b \to \infty} \left[-\frac{\pi}{x} \right]_{1}^{b} = \pi$$

$$A = \int_{1}^{\infty} 2\pi y\, ds = \int_{1}^{\infty} 2\pi y \sqrt{1 + y'^2}\, dx$$

$$= 2\pi \int_{1}^{\infty} \frac{1}{x} \sqrt{1 + \left(\frac{-1}{x^2}\right)^2}\, dx$$

$$= \lim_{b \to \infty} 2\pi \int_{1}^{b} \frac{\sqrt{x^4 + 1}}{x^3}\, dx$$

Now,

$$\frac{\sqrt{x^4 + 1}}{x^3} > \frac{\sqrt{x^4}}{x^3} = \frac{1}{x}$$

Thus,

$$\int_1^b \frac{\sqrt{x^4 + 1}}{x^3} \, dx > \int_1^b \frac{1}{x} \, dx = \ln b$$

and since $\ln b \to \infty$ as $b \to \infty$, we conclude that A is infinite.

Is something wrong with our mathematics? No. Imagine the horn to be slit down the side, opened up, and laid flat. Given a finite amount of paint, we could not possibly paint this surface with a paint coat of *uniform* thickness. However, we could do it if we allowed the paint coat to get thinner and thinner as we moved farther and farther from the horn's fat end. And, of course, that is exactly what happens when we fill the unslit horn with π cubic units of paint. (Imaginary paint can be spread to arbitrary thinness.)

This problem involved study of two integrals of the form $\int_1^\infty 1/x^p \, dx$.

For later reference, we now analyze this integral for all values of p.

EXAMPLE 6 Show that $\int_1^\infty 1/x^p$ diverges for $p \le 1$ and converges for $p > 1$.

Solution We showed in our solution to Gabriel's horn that the integral diverges for $p = 1$. If $p \ne 1$,

$$\int_1^\infty \frac{1}{x^p} \, dx = \lim_{b \to \infty} \int_1^b x^{-p} \, dx = \lim_{b \to \infty} \left[\frac{x^{-p+1}}{-p+1} \right]_1^b$$

$$= \lim_{b \to \infty} \left[\frac{1}{1-p} \right] \left[\frac{1}{b^{p-1}} - 1 \right] = \begin{cases} \infty & \text{if } p < 1 \\ \dfrac{1}{p-1} & \text{if } p > 1 \end{cases}$$

The conclusion follows. ∎

PROBLEM SET 9.3

In Problems 1–24, evaluate the given improper integral or show that it is divergent.

1. $\displaystyle\int_1^\infty e^x \, dx$

2. $\displaystyle\int_{-\infty}^{-2} \frac{dx}{x^5}$

3. $\displaystyle\int_4^\infty x e^{-x^2} \, dx$

4. $\displaystyle\int_{-\infty}^0 e^{3x} \, dx$

5. $\displaystyle\int_3^\infty \frac{x \, dx}{\sqrt{9 + x^2}}$

6. $\displaystyle\int_1^\infty \frac{dx}{\sqrt{3x}}$

7. $\displaystyle\int_1^\infty \frac{dx}{x^{1.01}}$

8. $\displaystyle\int_2^\infty \frac{x}{1 + x^2} \, dx$

9. $\displaystyle\int_1^\infty \frac{dx}{x^{0.99}}$

10. $\displaystyle\int_2^\infty \frac{x}{(1 + x^2)^2} \, dx$

11. $\displaystyle\int_2^\infty \frac{dx}{x \ln x}$

12. $\displaystyle\int_1^\infty \frac{\ln x}{x} \, dx$

13. $\displaystyle\int_2^\infty \frac{dx}{x(\ln x)^2}$

14. $\displaystyle\int_0^\infty x e^{-x} \, dx$

15. $\displaystyle\int_{-\infty}^0 \frac{dx}{(2x - 1)^3}$

16. $\displaystyle\int_2^\infty \frac{dx}{(1 - x)^{2/3}}$

17. $\displaystyle\int_{-\infty}^\infty \frac{x}{\sqrt{x^2 + 4}} \, dx$

18. $\displaystyle\int_{-\infty}^\infty \frac{x}{(x^2 + 4)^2} \, dx$

19. $\displaystyle\int_{-\infty}^{\infty} \frac{1}{x^2 + 2x + 5}\, dx$ **20.** $\displaystyle\int_{-\infty}^{\infty} \frac{x}{e^{|x|}}\, dx$

21. $\displaystyle\int_{-\infty}^{\infty} \operatorname{sech} x\, dx.$

Hint: $\displaystyle\int \operatorname{sech} x\, dx = \tan^{-1}(\sinh x) + C.$

22. $\displaystyle\int_{1}^{\infty} \operatorname{csch} x\, dx$

23. $\displaystyle\int_{0}^{\infty} e^{-x} \cos x\, dx.$ *Hint:* Use table of integrals.

24. $\displaystyle\int_{0}^{\infty} e^{-x} \sin x\, dx$

25. Find the area of the region under the curve $y = 2/(4x^2 - 1)$ to the right of $x = 1$. *Hint:* Use partial fractions.

26. Find the area of the region under the curve $y = 1/(x^2 + x)$ to the right of $x = 1$.

27. Suppose that Newton's law for the force of gravity had the form $-k/x$ rather than $-k/x^2$ (see Example 3). Show that it would then be impossible to send anything out of the earth's gravitational field.

28. If a 1000-pound capsule weighs only 165 pounds on the moon (radius 1080 miles), how much work is done in propelling this capsule out of the moon's gravitational field (see Example 3)?

29. Suppose a company expects its annual profits t years from now to be $f(t)$ dollars and that interest is considered to be compounded continuously at an annual rate r. Then the present value of all future profits can be shown to be

$$FP = \int_{0}^{\infty} e^{-rt} f(t)\, dt$$

Find FP if $r = 0.08$ and $f(t) = 100{,}000$.

30. Do Problem 29 assuming $f(t) = 100{,}000 + 1000t$.

31. In probability theory, *waiting times* tend to have the exponential density given for $\alpha > 0$ by $f(x) = \alpha e^{-\alpha x}$ on $[0, \infty)$. Show each of the following.

(a) $\displaystyle\int_{0}^{\infty} f(x)\, dx = 1$

(b) $\displaystyle\int_{0}^{\infty} xf(x)\, dx = 1/\alpha$ (the mean)

32. In electromagnetic theory, the magnetic potential u at a point on the axis of a circular coil is given by

$$u = Ar \int_{a}^{\infty} \frac{dx}{(r^2 + x^2)^{3/2}}$$

where A, r, and a are constants. Evaluate u.

33. There is a subtlety in the definition of $\displaystyle\int_{-\infty}^{\infty} f(x)\, dx$ illustrated by the following. Show that (a) $\displaystyle\int_{-\infty}^{\infty} \sin x\, dx$ diverges and (b) $\displaystyle\lim_{a \to \infty} \int_{-a}^{a} \sin x\, dx = 0.$

34. Consider an infinitely long wire coinciding with the positive x-axis and having mass density $\delta(x) = (1 + x^2)^{-1}$ at x, $0 \leq x < \infty$.

(a) Calculate the total mass of the wire (Example 4).
(b) Show that this wire does not have a center of mass. (In probability language, the Cauchy distribution does not have a mean.)

35. Give an example of a region in the first quadrant that gives a solid of finite volume when revolved about the x-axis but gives a solid of infinite volume when revolved about the y-axis.

36. Let f be a nonnegative continuous function defined on $0 \leq x < \infty$ with $\displaystyle\int_{0}^{\infty} f(x)\, dx < \infty$. Show:

(a) if $\displaystyle\lim_{x \to \infty} f(x)$ exists, it must be 0;
(b) it is possible that $\displaystyle\lim_{x \to \infty} f(x)$ does not exist.

9.4
Improper Integrals: Infinite Integrands

Considering the many complicated integrations we have done, here is one that looks simple enough.

$$\int_{-2}^{1} \frac{1}{x^2}\, dx = \left[-\frac{1}{x} \right]_{-2}^{1} = -1 - \frac{1}{2} = -\frac{3}{2} \qquad ???$$

$$y = \frac{1}{x^2}$$

FIGURE 1

One glance at the graph in Figure 1 tells us that something is terribly wrong. The answer (if there is one) has to be a positive number. (Why?)

Where is our mistake? To answer, we refer back to Section 5.5, especially page 236. Recall that for a function to be integrable in the standard (or proper) sense, it must be bounded. Our function, $f(x) = 1/x^2$, is not bounded, so it is not integrable in the proper sense. We say that $\int_{-2}^{1} x^{-2}\, dx$ is an improper integral with an infinite integrand (*unbounded integrand* is a more accurate but less colorful term).

Until now, we have carefully avoided infinite integrands in all our examples and problems. We could continue to do this, but that would be to avoid a kind of integral which has important applications. Our task for this section is to define and analyze this new kind of integral.

INTEGRANDS WHICH ARE INFINITE AT AN ENDPOINT We give the definition for the case where f tends to infinity at the right endpoint of the interval of integration. There is a completely analogous definition for the case where the left endpoint is the troublesome one.

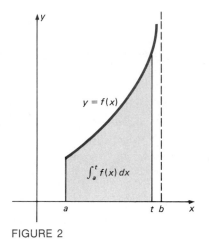

$$y = f(x)$$

$$\int_a^t f(x)\, dx$$

FIGURE 2

Definition

Let f be continuous on the half-open interval $[a, b)$ and suppose $\lim\limits_{x \to b^-} |f(x)| = \infty$. Then,

$$\int_a^b f(x)\, dx = \lim_{t \to b^-} \int_a^t f(x)\, dx$$

provided this limit exists and is finite, in which case we say the integral converges. Otherwise, we say it diverges.

Note the geometric interpretation shown in Figure 2.

EXAMPLE 1 Evaluate, if possible, the improper integral $\int_0^2 \dfrac{dx}{\sqrt{4 - x^2}}$.

Solution Note that the integrand tends to infinity at 2.

$$\int_0^2 \frac{dx}{\sqrt{4 - x^2}} = \lim_{t \to 2^-} \int_0^t \frac{dx}{\sqrt{4 - x^2}} = \lim_{t \to 2^-} \left[\sin^{-1}\left(\frac{x}{2}\right) \right]_0^t$$

$$= \lim_{t \to 2^-} \left[\sin^{-1}\left(\frac{t}{2}\right) - \sin^{-1}\left(\frac{0}{2}\right) \right] = \frac{\pi}{2} \qquad \blacksquare$$

EXAMPLE 2 Evaluate, if possible, $\int_0^{16} \dfrac{1}{\sqrt[4]{x}}\, dx$.

Solution

$$\int_0^{16} x^{-1/4}\,dx = \lim_{t\to 0^+}\int_t^{16} x^{-1/4}\,dx = \lim_{t\to 0^+}\left[\frac{4}{3}x^{3/4}\right]_t^{16}$$

$$= \lim_{t\to 0^+}\left[\frac{32}{3}-\frac{4}{3}t^{3/4}\right] = \frac{32}{3} \qquad \blacksquare$$

EXAMPLE 3   Evaluate, if possible, $\displaystyle\int_0^1 \frac{1}{x}\,dx.$

Solution

$$\int_0^1 \frac{1}{x}\,dx = \lim_{t\to 0^+}\int_t^1 \frac{1}{x}\,dx = \lim_{t\to 0^+}\left[\ln x\right]_t^1$$

$$= \lim_{t\to 0^+}\left[-\ln t\right] = \infty$$

We conclude that the integral diverges. $\qquad \blacksquare$

EXAMPLE 4   Show that $\displaystyle\int_0^1 \frac{1}{x^p}\,dx$ converges if $p < 1$, but diverges if $p \geq 1$.

Solution   Example 3 took care of the case $p = 1$. If $p \neq 1$,

$$\int_0^1 \frac{1}{x^p}\,dx = \lim_{t\to 0^+}\int_t^1 x^{-p}\,dx = \lim_{t\to 0^+}\left[\frac{x^{-p+1}}{-p+1}\right]_t^1$$

$$= \lim_{t\to 0^+}\left[\frac{1}{1-p}-\frac{1}{1-p}\cdot\frac{1}{t^{p-1}}\right] = \begin{cases} \dfrac{1}{1-p} & \text{if } p < 1 \\[2mm] \infty & \text{if } p > 1 \end{cases} \qquad \blacksquare$$

EXAMPLE 5   Sketch the graph of the hypocycloid of four cusps, $x^{2/3} + y^{2/3} = 1$, and find its perimeter.

Solution   The graph is shown in Figure 3. To find the perimeter, it is enough to find the length L of the first quadrant portion and quadruple it. We use the length formula from Section 6.4,

$$L = \int_0^1 \sqrt{1+(y')^2}\,dx$$

By implicit differentiation of $x^{2/3} + y^{2/3} = 1$, we obtain

$$\frac{2}{3}x^{-1/3}+\frac{2}{3}y^{-1/3}y' = 0$$

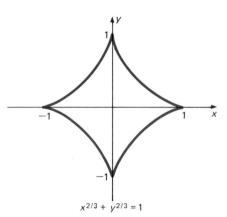

$x^{2/3} + y^{2/3} = 1$

FIGURE 3

or

$$y' = -\frac{y^{1/3}}{x^{1/3}}$$

Thus,

$$1 + (y')^2 = 1 + \frac{y^{2/3}}{x^{2/3}} = 1 + \frac{1 - x^{2/3}}{x^{2/3}} = \frac{1}{x^{2/3}}$$

and so

$$L = \int_0^1 \sqrt{1 + (y')^2}\, dx = \int_0^1 \frac{1}{x^{1/3}}\, dx$$

The value at this improper integral can be read from the solution to Example 4; it is $L = 1/(1 - \frac{1}{3}) = \frac{3}{2}$. We conclude that the hypocycloid has perimeter $4L = 6$. ∎

INTEGRANDS THAT ARE INFINITE AT AN INTERIOR POINT The integral $\int_{-2}^{1} 1/x^2\, dx$ of our introduction illustrates the behavior we have in mind. Note that its integrand tends to infinity at $x = 0$. Here is the appropriate definition.

Definition

Let f be continuous on $[a, b]$ except at a number c, where $a < c < b$, and suppose $\lim_{x \to c} |f(x)| = \infty$. Then we define

$$\int_a^b f(x)\, dx = \int_a^c f(x)\, dx + \int_c^b f(x)\, dx$$

provided both integrals on the right converge. Otherwise, we say $\int_a^b f(x)\, dx$ diverges.

EXAMPLE 6 Show that $\int_{-2}^{1} 1/x^2\, dx$ diverges.

Solution

$$\int_{-2}^{1} \frac{1}{x^2}\, dx = \int_{-2}^{0} \frac{1}{x^2}\, dx + \int_{0}^{1} \frac{1}{x^2}\, dx$$

The second of the integrals on the right diverges by Example 4. This is enough to give the conclusion. ∎

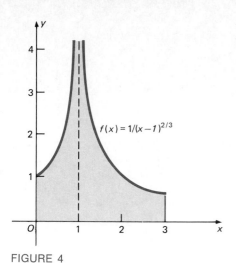

FIGURE 4

EXAMPLE 7 Evaluate, if possible, the improper integral $\int_0^3 \dfrac{dx}{(x-1)^{2/3}}$.

Solution The integrand tends to infinity at $x = 1$ (see Figure 4). Thus,

$$\int_0^3 \frac{dx}{(x-1)^{2/3}} = \int_0^1 \frac{dx}{(x-1)^{2/3}} + \int_1^3 \frac{dx}{(x-1)^{2/3}}$$

$$= \lim_{t \to 1^-} \int_0^t \frac{dx}{(x-1)^{2/3}} + \lim_{s \to 1^+} \int_s^3 \frac{dx}{(x-1)^{2/3}}$$

$$= \lim_{t \to 1^-} \left[3(x-1)^{1/3} \right]_0^t + \lim_{s \to 1^+} \left[3(x-1)^{1/3} \right]_s^3$$

$$= 3 \lim_{t \to 1^-} \left[(t-1)^{1/3} + 1 \right] + 3 \lim_{s \to 1^+} \left[2^{1/3} - (s-1)^{1/3} \right]$$

$$= 3 + 3(2^{1/3}) \approx 6.78 \qquad \blacksquare$$

PROBLEM SET 9.4

In Problems 1–16, evaluate the given improper integral or show that it diverges.

1. $\displaystyle\int_1^2 \frac{dx}{(x-1)^{1/3}}$

2. $\displaystyle\int_3^7 \frac{dx}{\sqrt{x-3}}$

3. $\displaystyle\int_0^1 \frac{dx}{\sqrt{1-x^2}}$

4. $\displaystyle\int_0^9 \frac{dx}{\sqrt{9-x}}$

5. $\displaystyle\int_{-3}^2 \frac{1}{x^4}\,dx$

6. $\displaystyle\int_{-1}^{27} x^{-2/3}\,dx$

7. $\displaystyle\int_2^4 \frac{dx}{(3-x)^{2/3}}$

8. $\displaystyle\int_0^3 \frac{x}{9-x^2}\,dx$

9. $\displaystyle\int_{-3}^0 \frac{x}{(x^2-4)^{2/3}}\,dx$

10. $\displaystyle\int_0^3 \frac{x}{\sqrt{9-x^2}}\,dx$

11. $\displaystyle\int_{-2}^{-1} \frac{dx}{(x+1)^{4/3}}$

12. $\displaystyle\int_0^2 \frac{3}{x^2+x-2}\,dx$

13. $\displaystyle\int_0^{\pi/4} \tan 2x\,dx$

14. $\displaystyle\int_0^{\pi/2} \csc x\,dx$

15. $\displaystyle\int_0^1 \frac{\ln x}{x}\,dx$

16. $\displaystyle\int_0^1 \ln x\,dx$

17. If $f(x)$ tends to infinity at both a and b, then we define

$$\int_a^b f(x)\,dx = \int_a^c f(x)\,dx + \int_c^b f(x)\,dx,$$

where c is any point between a and b, provided of course that both the latter integrals converge. Otherwise, we say the given integral diverges. Use this to evaluate

$$\int_{-3}^3 \frac{x}{\sqrt{9-x^2}}\,dx \text{ or show that it diverges.}$$

18. Evaluate $\displaystyle\int_{-2}^2 \frac{1}{4-x^2}\,dx$ or show that it diverges. See Problem 17.

19. Show that $\displaystyle\int_0^\infty \frac{1}{x^p}\,dx$ diverges for all p. *Hint*: Write

$$\int_0^\infty = \int_0^1 + \int_1^\infty.$$

20. Suppose f is continuous on $[0, \infty)$ except at $x = 1$, where $\lim_{x \to 1} |f(x)| = \infty$. How would you define $\displaystyle\int_0^\infty f(x)\,dx$?

21. Find the area of the region between the curves $y = (x-8)^{-2/3}$ and $y = 0$ for $0 \le x < 8$.

22. Find the area of the region between the curves $y = 1/x$ and $y = 1/(x^3 + x)$ for $0 < x \le 1$.

23. Let R be the region in the first quadrant below the curve $y = x^{-2/3}$ and to the left of $x = 1$.

(a) Show that the area of R is finite by finding its value.
(b) Show that the volume of the solid generated by revolving R about the x-axis is infinite.

24. Find b so that $\displaystyle\int_0^b \ln x\,dx = 0$.

25. Is $\int_0^1 \dfrac{\sin x}{x}\,dx$ an improper integral? Explain.

26. **(Comparison Test)** If $0 \le f(x) \le g(x)$ on $[a, \infty)$, it can be shown that the convergence of $\int_a^\infty g(x)\,dx$ implies the convergence of $\int_a^\infty f(x)\,dx$ and the divergence of $\int_a^\infty f(x)\,dx$ implies the divergence of $\int_a^\infty g(x)\,dx$. Use this to show that $\int_1^\infty \dfrac{1}{x^4(1+x^4)}\,dx$ converges. *Hint*: On $[1, \infty)$, $1/x^4(1+x^4) \le 1/x^4$.

27. Use the comparison test of Problem 26 to show that $\int_1^\infty e^{-x^2}\,dx$ converges. *Hint*: $e^{-x^2} \le e^{-x}$ on $[1, \infty)$.

28. Formulate a comparison test for improper integrals with infinite integrands.

29. (a) Use Example 2 of Section 9.2 to show that for any positive number n, there is a number M such that

$$0 < \frac{x^{n-1}}{e^x} \le \frac{1}{x^2} \qquad \text{for } x \ge M$$

(b) Use (a) and Problem 26 to show that $\int_1^\infty x^{n-1}e^{-x}\,dx$ converges.

30. Using Problem 28, prove that $\int_0^1 x^{n-1}e^{-x}\,dx$ converges for $n > 0$.

31. **(Gamma Function)** Let $\Gamma(n) = \int_0^\infty x^{n-1}e^{-x}\,dx$, $n > 0$. This integral converges by Problems 29 and 30. Show each of the following.

(a) $\Gamma(1) = 1$
(b) $\Gamma(n+1) = n\Gamma(n)$
(c) $\Gamma(n+1) = n!$, n a positive integer.

32. Suppose $0 < p < q$ and $\int_0^\infty \dfrac{1}{x^p + x^q}\,dx$ converges. What can you say about p and q?

33. By interpreting each of the following integrals as an area and then calculating this area by a y-integration, evaluate:

(a) $\displaystyle\int_0^1 \sqrt{\dfrac{1-x}{x}}\,dx;$ 　　(b) $\displaystyle\int_{-1}^1 \sqrt{\dfrac{1+x}{1-x}}\,dx.$

9.5 Chapter Review

TRUE-FALSE QUIZ

Respond with true or false to each of the following assertions. Be prepared to defend your answer.

1. $\lim\limits_{x \to \infty} \dfrac{x^{100}}{e^x} = 0.$

2. $\lim\limits_{x \to \infty} \dfrac{x^{1/10}}{\ln x} = \infty.$

3. $\lim\limits_{x \to \infty} \dfrac{1000x^4 + 1000}{0.001x^4 + 1} = \infty.$

4. $\lim\limits_{x \to \infty} xe^{-1/x} = 0.$

5. If $\lim\limits_{x \to a} f(x) = \lim\limits_{x \to a} g(x) = \infty$, then $\lim\limits_{x \to a} \dfrac{f(x)}{g(x)} = 1.$

6. If $\lim\limits_{x \to a} f(x) = 1$ and $\lim\limits_{x \to a} g(x) = \infty$, then $\lim\limits_{x \to a}[f(x)]^{g(x)} = 1.$

7. If $\lim\limits_{x \to a} f(x) = 1$, then $\lim\limits_{n \to \infty}\left\{\lim\limits_{x \to a}[f(x)]^n\right\} = 1.$

8. If $\lim\limits_{x \to a} f(x) = 0$ and $\lim\limits_{x \to a} g(x) = \infty$, then $\lim\limits_{x \to a}[f(x)]^{g(x)} = 0.$ (Assume $f(x) \ge 0$ for $x \ne a$.)

9. If $\lim\limits_{x \to a} f(x) = -1$ and $\lim\limits_{x \to a} g(x) = \infty$, then $\lim\limits_{x \to a}[f(x)g(x)] = -\infty.$

10. If $\lim_{x \to a} f(x) = 0$ and $\lim_{x \to a} g(x) = \infty$, then $\lim_{x \to a} [f(x)g(x)] = 0$.

11. If $\lim_{x \to \infty} \dfrac{f(x)}{g(x)} = 3$, then $\lim_{x \to \infty} [f(x) - 3g(x)] = 0$.

12. If $\lim_{x \to a} f(x) = 2$ and $\lim_{x \to a} g(x) = 0$, then $\lim_{x \to a} \dfrac{f(x)}{|g(x)|} = \infty$. (Assume $g(x) \neq 0$ for $x \neq a$.)

13. If $\lim_{x \to \infty} \ln f(x) = 2$, then $\lim_{x \to \infty} f(x) = e^2$.

14. If $f(x) \neq 0$ for $x \neq a$ and $\lim_{x \to a} f(x) = 0$, then $\lim_{x \to a} [1 + f(x)]^{1/f(x)} = e$.

15. If $p(x)$ is a polynomial, then $\lim_{x \to \infty} \dfrac{p(x)}{e^x} = 0$.

16. If $p(x)$ is a polynomial, then $\lim_{x \to 0} \dfrac{p(x)}{e^x} = p(0)$.

17. If $f(x)$ and $g(x)$ are both differentiable and $\lim_{x \to 0} \dfrac{f'(x)}{g'(x)} = L$, then $\lim_{x \to 0} \dfrac{f(x)}{g(x)} = L$.

18. $\displaystyle\int_0^1 \dfrac{1}{x^{1.001}}\, dx$ converges.

19. $\displaystyle\int_0^\infty \dfrac{1}{x^p}\, dx$ diverges for all $p > 0$.

20. If f is continuous on $[0, \infty]$ and $\lim_{x \to \infty} f(x) = 0$, then $\displaystyle\int_0^\infty f(x)\, dx$ converges.

21. If f is an even function and $\displaystyle\int_0^\infty f(x)\, dx$ converges, then $\displaystyle\int_{-\infty}^\infty f(x)\, dx$ converges.

22. If $\lim_{b \to \infty} \displaystyle\int_{-b}^b f(x)\, dx$ exists and is finite, then $\displaystyle\int_{-\infty}^\infty f(x)\, dx$ converges.

23. If f' is continuous on $[0, \infty)$ and $\lim_{x \to \infty} f(x) = 0$, then $\displaystyle\int_0^\infty f'(x)\, dx$ converges.

24. If $0 \leq f(x) \leq e^{-x}$ on $[0, \infty)$, then $\displaystyle\int_0^\infty f(x)\, dx$ converges.

25. $\displaystyle\int_0^{\pi/4} \dfrac{\tan x}{x}\, dx$ is an improper integral.

SAMPLE TEST PROBLEMS

Find the limits in Problems 1–18.

1. $\lim_{x \to 0} \dfrac{4x}{\tan x}$

2. $\lim_{x \to 0} \dfrac{\tan 2x}{\sin 3x}$

3. $\lim_{x \to 0} \dfrac{\sin x - \tan x}{\frac{1}{3}x^2}$

4. $\lim_{x \to 0} \dfrac{\cos x}{x^2}$

5. $\lim_{x \to 0} 2x \cot x$

6. $\lim_{x \to 1^-} \dfrac{\ln(1 - x)}{\cot \pi x}$

7. $\lim_{t \to \infty} \dfrac{\ln t}{t^2}$

8. $\lim_{x \to \infty} \dfrac{2x^3}{\ln x}$

9. $\lim_{x \to 0^+} (\sin x)^{1/x}$

10. $\lim_{s \to 0^+} s \ln s$

11. $\lim_{x \to 0^+} x^x$

12. $\lim_{x \to 0} (1 + \sin x)^{2/x}$

13. $\lim_{x \to 0^+} \sqrt{x} \ln x$

14. $\lim_{t \to \infty} t^{1/t}$

15. $\lim_{x \to 0^+} \left(\dfrac{1}{\sin x} - \dfrac{1}{x} \right)$

16. $\lim_{x \to \pi/2} \dfrac{\tan 3x}{\tan x}$

17. $\lim_{x \to \pi/2} (\sin x)^{\tan x}$

18. $\lim_{x \to \pi/2} \left(x \tan x - \dfrac{\pi}{2} \sec x \right)$

In Problems 19–34, evaluate the given improper integral or show that it diverges.

19. $\displaystyle\int_0^\infty \dfrac{dx}{(x + 1)^2}$

20. $\displaystyle\int_0^\infty \dfrac{dx}{1 + x^2}$

21. $\displaystyle\int_{-\infty}^1 e^{2x}\, dx$

22. $\displaystyle\int_{-1}^1 \dfrac{dx}{1 - x}$

23. $\displaystyle\int_0^\infty \frac{dx}{x+1}$

24. $\displaystyle\int_{1/2}^2 \frac{dx}{x(\ln x)^{1/5}}$

25. $\displaystyle\int_1^\infty \frac{dx}{x^2 + x^4}\,dx$

26. $\displaystyle\int_{-\infty}^1 \frac{dx}{(2-x)^2}$

27. $\displaystyle\int_{-2}^0 \frac{dx}{2x+3}$

28. $\displaystyle\int_1^4 \frac{dx}{\sqrt{x-1}}$

29. $\displaystyle\int_2^\infty \frac{dx}{x(\ln x)^2}$

30. $\displaystyle\int_0^\infty \frac{dx}{e^{x/2}}$

31. $\displaystyle\int_3^5 \frac{dx}{(4-x)^{2/3}}$

32. $\displaystyle\int_2^\infty xe^{-x^2}\,dx$

33. $\displaystyle\int_{-\infty}^\infty \frac{x}{(x^2+1)}\,dx$

34. $\displaystyle\int_{-\infty}^\infty \frac{x}{1+x^4}\,dx$

In Problems 35–38, use a comparison test (see Problem 26 of Section 9.4) to decide whether each of the following converges or diverges.

35. $\displaystyle\int_1^\infty \frac{dx}{\sqrt{x^6+x}}$

36. $\displaystyle\int_1^\infty \frac{\ln x}{e^{2x}}\,dx$

37. $\displaystyle\int_3^\infty \frac{\ln x}{x}\,dx$

38. $\displaystyle\int_1^\infty \frac{\ln x}{x^3}\,dx$

10 Numerical Methods, Approximations

Reason is the slow and tortuous method by which those who do not know the truth discover it.
The heart has its own reasons which reason does not know.
Blaise Pascal

Blaise Pascal
1623–1662

Blaise Pascal made contributions to calculus, but that is not the main reason for choosing him as representative of this chapter. It is rather because at age 19 he invented the first adding machine. This device, with its hand-turned wheels, is the primitive ancestor of today's electronic calculators and computers. And it is these modern computing devices that make the methods of this chapter especially useful and practical.

Pascal was born at Clermont, France, a sickly child who was plagued by poor health throughout his short life. At age 12, he began studying geometry; it is in this field that he made his greatest mathematical contributions. With his countryman, Pierre Fermat, he is credited with initiating the serious study of probability theory. The name

Pascal's triangle is given to the array of numbers consisting of the binomial coefficients. His ideas influenced Leibniz and, through him, the founding of the calculus.

Even more than his scientific achievements, Pascal is immortalized by his religious writings. His *Pensées* is a profound literary classic. While he thought that only reason was the basis for knowledge in science and mathematics, he said that the mysteries of faith are hidden from reason and should be accepted on the authority of the Bible. And in the final analysis, Pascal would insist that faith is superior to reason.

10.1
Taylor's Approximation to Functions

So far in this book we have emphasized what might be called exact methods. There have been some exceptions, however. The simplest example is illustrated by rounding decimals, as when we write $\frac{1}{3} \approx 0.333$ or $\pi \approx 3.1416$. More notable was our discussion of differentials in Section 3.10. There we used the differential dy to approximate the actual change Δy in $y = f(x)$ when x changed by an amount Δx. That example, in fact, illustrates the kind of approximation methods we want to highlight in this chapter.

Two factors contribute to the importance of approximation methods. First is the fact that many of the mathematical entities that occur in applications cannot be calculated by exact methods. We mention, for example, the integrals $\int_0^b \sin(x^2)\,dx$, which is used extensively in optics, and $\int_a^b e^{-x^2}\,dx$, which plays a key role in statistics. Second, the invention of high-speed electronic computers and calculators has made approximate numerical methods practical. In fact, it is often easier to calculate something approximately using a calculator (and get the answer to desired accuracy) than to use exact methods, even when the latter are available.

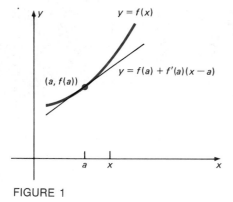

FIGURE 1

LINEAR APPROXIMATION   The idea behind the differential approximation introduced in Section 3.10 was to approximate a curve near a point by its tangent line at that point. Our idea here is exactly the same, only we do not use differential notation. Examine the diagram in Figure 1. The equation of the tangent line to the curve $y = f(x)$ at $(a, f(a))$ is

$$y = f(a) + f'(a)(x - a)$$

This leads directly to the linear approximation

$$\boxed{f(x) \approx f(a) + f'(a)(x - a)}$$

The linear polynomial $P_1(x) = f(a) + f'(a)(x - a)$ is called the **Taylor polynomial of order 1 based at a** for $f(x)$, after the English mathematician Brook Taylor (1685–1731). Clearly, we can expect $P_1(x)$ to be a good approximation to $f(x)$ only near $x = a$.

EXAMPLE 1   Find $P_1(x)$ based at $a = 1$ for $f(x) = \ln x$ and use it to approximate $\ln(0.9)$ and $\ln(1.5)$.

Solution Since $f(x) = \ln x$, $f'(x) = 1/x$; thus, $f(1) = 0$ and $f'(1) = 1$. Therefore,

$$P_1(x) = 0 + 1(x - 1) = x - 1$$

Consequently (see Figure 2),

$$\ln x \approx x - 1$$

and

$$\ln(0.9) \approx 0.9 - 1 = -0.1$$

$$\ln(1.5) \approx 1.5 - 1 = 0.5$$

These approximations should be compared with the correct four-place values of -0.1054 and 0.4055. As expected, the approximation is much better in the first case, since 0.9 is closer to 1 than is 1.5. ∎

FIGURE 2

TAYLOR POLYNOMIALS OF ORDER n Polynomials are the easiest of all functions to evaluate, since they involve only the three arithmetic operations of addition, subtraction, and multiplication. This is why we use polynomials to approximate other functions. If a linear polynomial gives a certain approximation to $f(x)$, we should expect a quadratic polynomial (with its curved graph) to do better, a cubic polynomial to do still better, and so on. Our aim is to find the polynomial of degree n that "best" approximates $f(x)$. We consider the quadratic case first.

A significant observation to make about the linear case is that f and its approximation P_1, as well as their derivatives f' and P_1', agree at $x = a$. In generalizing to the quadratic polynomial P_2, we impose three conditions, namely,

$$f(a) = P_2(a), \qquad f'(a) = P_2'(a), \qquad f''(a) = P_2''(a)$$

The unique quadratic polynomial satisfying these conditions (the **Taylor polynomial of order 2 based at a**) is

$$P_2(x) = f(a) + f'(a)(x - a) + \frac{f''(a)}{2}(x - a)^2$$

as you may check. The corresponding quadratic approximation is

$$\boxed{f(x) \approx f(a) + f'(a)(x - a) + \frac{f''(a)}{2}(x - a)^2}$$

EXAMPLE 2 Find $P_2(x)$ based at $a = 1$ for $f(x) = \ln x$ and use it to approximate $\ln(0.9)$ and $\ln(1.5)$.

Solution Here $f(x) = \ln x$, $f'(x) = 1/x$, and $f''(x) = -1/x^2$, and so $f(1) = 0$, $f'(1) = 1$, and $f''(1) = -1$. Thus,

$$P_2(x) = 0 + 1(x - 1) - \tfrac{1}{2}(x - 1)^2$$

Consequently,

$$\ln x \approx (x - 1) - \tfrac{1}{2}(x - 1)^2$$

and

$$\ln(0.9) \approx (0.9 - 1) - \tfrac{1}{2}(0.9 - 1)^2 = -0.1050$$
$$\ln(1.5) \approx (1.5 - 1) - \tfrac{1}{2}(1.5 - 1)^2 = 0.3750$$

As expected, these are better approximations than we got in the linear case (Example 1). Note the diagram in Figure 3. ∎

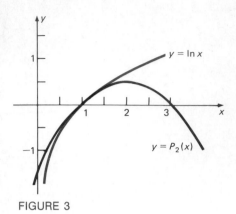

FIGURE 3

With a little effort, we could now derive $P_3(x)$. And from this, it is a simple matter to guess the form of the **Taylor polynomial of order n based at a**; that is, the nth order polynomial P_n, which—together with its first n derivatives—agrees with f and its derivatives at $x = a$. It is

$$P_n(x) = f(a) + f'(a)(x - a) + \frac{f''(a)}{2!}(x - a)^2 + \cdots + \frac{f^{(n)}(a)}{n!}(x - a)^n$$

Here $n!$ (read **n factorial**) is the standard symbol for the product of the first n positive integers—that is,

$$n! = n(n - 1)(n - 2) \cdots 3 \cdot 2 \cdot 1$$

The corresponding approximation is

$$\boxed{f(x) \approx f(a) + f'(a)(x - a) + \frac{f''(a)}{2!}(x - a)^2 + \cdots + \frac{f^{(n)}(a)}{n!}(x - a)^n}$$

MACLAURIN POLYNOMIALS In case $a = 0$, the Taylor polynomial of order n simplifies to the **Maclaurin polynomial of order n**, named in honor of the Scottish mathematician Colin Maclaurin (1698–1746). It gives a particularly useful approximation valid near $x = 0$, namely,

$$\boxed{f(x) \approx f(0) + f'(0)x + \frac{f''(0)}{2!}x^2 + \cdots + \frac{f^{(n)}(0)}{n!}x^n}$$

EXAMPLE 3 Find the Maclaurin polynomials of order n for e^x and $\cos x$. Then approximate $e^{0.2}$ and $\cos(0.2)$ using $n = 4$.

Solution The calculation of the required derivatives is shown in the table.

		At $x = 0$		At $x = 0$
$f(x)$	e^x	1	$\cos x$	1
$f'(x)$	e^x	1	$-\sin x$	0
$f''(x)$	e^x	1	$-\cos x$	-1
$f^{(3)}(x)$	e^x	1	$\sin x$	0
$f^{(4)}(x)$	e^x	1	$\cos x$	1
$f^{(5)}(x)$	e^x	1	$-\sin x$	0
$\vdots$	$\vdots$	$\vdots$	$\vdots$	$\vdots$

It follows that

$$e^x \approx 1 + x + \frac{1}{2!}x^2 + \frac{1}{3!}x^3 + \frac{1}{4!}x^4 + \cdots + \frac{1}{n!}x^n$$

$$\cos x \approx 1 - \frac{1}{2!}x^2 + \frac{1}{4!}x^4 - \cdots + (-1)^{n/2}\frac{1}{n!}x^n \qquad (n \text{ even})$$

Thus, using $n = 4$ and $x = 0.2$, we obtain

$$e^{0.2} \approx 1 + 0.2 + \frac{(0.2)^2}{2} + \frac{(0.2)^3}{6} + \frac{(0.2)^4}{24} = 1.2214$$

$$\cos(0.2) \approx 1 - \frac{(0.2)^2}{2} + \frac{(0.2)^4}{24} \approx 0.9800667$$

These results should be compared with the correct seven-place values of 1.2214028 and 0.9800666. ∎

To give you a visual idea of how the Maclaurin polynomials provide approximation to cos x, we have sketched the graphs of $P_1(x)$ through $P_5(x)$ and $P_8(x)$ in Figure 4.

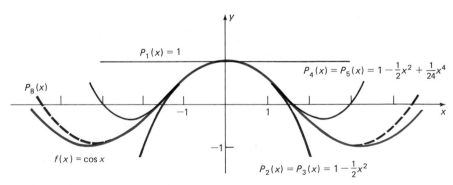

Maclaurin approximations to $f(x) = \cos x$

FIGURE 4

HORNER'S METHOD FOR EVALUATING POLYNOMIALS To evaluate the polynomial

$$p(x) = a_n x^n + a_{n-1} x^{n-1} + \cdots + a_1 x + a_0$$

at $x = c$ is, of course, to evaluate

$$p(c) = a_n c^n + a_{n-1} c^{n-1} + \cdots + a_1 c + a_0$$

One way to do this on a calculator is to proceed directly from left to right using the $\boxed{y^x}$ key. However, on most calculators this key is both slow and notably inaccurate.

A better procedure goes by the name **Horner's Method**. Begin by writing $p(x)$ in nested form.

$$p(x) = \{\ldots [(a_n x + a_{n-1})x + a_{n-2}]x + \cdots + a_1\}x + a_0$$

For example,

$$2x^4 + 3x^3 - 10x^2 + 2x - 5 = \{[(2x + 3)x - 10]x + 2\}x - 5$$

which avoids exponents completely. The whole calculation of $p(c)$ is easily done on a calculator without even using parentheses, as we now illustrate.

EXAMPLE 4 Evaluate $3x^3 + 2x^2 - 4x + 8$ at $x = 2.123$.

Solution Note that

$$3x^3 + 2x^2 - 4x + 8 = [(3x + 2)x - 4]x + 8$$

Let $\boxed{S}$ denote the key (or keys) required to store a number and $\boxed{R}$ the corresponding key (or keys) to recall it. Then, on an algebraic logic calculator, we may perform the required evaluation by pressing in succession

$$2.123 \ \boxed{S} \ \ 3$$
$$\boxed{X} \ \boxed{R} \ \boxed{+} \ 2 \ \boxed{=}$$
$$\boxed{X} \ \boxed{R} \ \boxed{-} \ 4 \ \boxed{=}$$
$$\boxed{X} \ \boxed{R} \ \boxed{+} \ 8 \ \boxed{=}$$

Our calculator responds with 37.228163. ∎

EXAMPLE 5 Evaluate $2x^5 - 4x^4 + 5x^2 - 12x - 9$ at $x = 1.95$.

Solution Press the following keys in succession.

$$1.95 \ \boxed{S} \ \ 2$$
$$\boxed{X} \ \boxed{R} \ \boxed{-} \ 4 \ \boxed{=}$$
$$\boxed{X} \ \boxed{R} \ \boxed{+} \ 0 \ \boxed{=}$$
$$\boxed{X} \ \boxed{R} \ \boxed{+} \ 5 \ \boxed{=}$$
$$\boxed{X} \ \boxed{R} \ \boxed{-} \ 12 \ \boxed{=}$$
$$\boxed{X} \ \boxed{R} \ \boxed{-} \ 9 \ \boxed{=}$$

SYNTHETIC DIVISION

Those who remember the synthetic division method of evaluating a polynomial will recognize it as Horner's Method in disguise. Check that the steps in the evaluation at $x = 2$ of

$$3x^3 + 2x^2 - 4x + 8$$

by synthetic division are exactly those in Horner's Method.

$$
\begin{array}{r|rrrr}
  & 3 & 2 & -4 & 8 \\
  &   & 6 & 16 & 24 \\
\hline
2 & 3 & 8 & 12 & \boxed{32} \\
\end{array}
$$

This yields the answer -14.833401. Note the insertion of 0 corresponding to the missing term in x^3. ∎

The calculations in Examples 4 and 5 are even more easily done on a reverse Polish logic calculator. If one has to do many evaluations for the same polynomial, then the advantage of having a programmable calculator becomes obvious. In any case, practice making calculations of the above type on your calculator.

PROBLEM SET 10.1

[C] In Problems 1–4, calculate $f(1.23)$ first by using the $\boxed{y^x}$ key and then by using Horner's Method.

1. $f(x) = 2x^3 - 3x^2 - 2x + 5$

2. $f(x) = 4x^3 + 1.2x^2 - 3x - 6$

3. $f(x) = 2x^4 - 3x^3 - 2x^2 + 5x - 2$

4. $f(x) = 3x^4 - 11x^2 + 2x - 3$

[C] 5. Calculate $p(3.456)$ if $p(x) = x^5 - 3x^4 + 2x^3 - 4x^2 + 5x + 1$ using Horner's Method.

[C] 6. Calculate $p(6.321)$ if $p(x) = x^5 - 3.12x^4 + 2.53x^2 - 6.32$.

[C] In Problems 7–14, find the Maclaurin polynomial of order 4 for $f(x)$ and use it to approximate $f(0.23)$.

7. $f(x) = e^{2x}$ 8. $f(x) = e^{-3x}$

9. $f(x) = \sin 2x$ 10. $f(x) = \tan x$

11. $f(x) = \ln(1 + x)$ 12. $f(x) = \sqrt{1 + x}$

13. $f(x) = \tan^{-1} x$ 14. $f(x) = \sinh x$

In Problems 15–20, find the Taylor polynomial of order 3 based at a for the given function.

15. $e^x; a = 2$ 16. $\sin x; a = \dfrac{\pi}{6}$

17. $\tan x; a = \dfrac{\pi}{4}$ 18. $\sec x; a = \dfrac{\pi}{6}$

19. $\tan^{-1} x; a = 1$ 20. $\sqrt{x}; a = 1$

21. Find the Taylor polynomial of order 3 based at 1 for $f(x) = x^3 - 2x^2 + 3x + 5$ and show that it is an exact representation of $f(x)$.

22. Find the Taylor polynomial of order 4 based at 2 for $f(x) = x^4$ and show that it represents $f(x)$ exactly.

23. Find the Maclaurin polynomial of order n for $f(x) = 1/(1 - x)$. Then use it with $n = 4$ to approximate each of the following.

(a) $f(0.1)$ (b) $f(0.5)$ (c) $f(0.9)$ (d) $f(2)$

This example should convince you that the Maclaurin approximation can be exceedingly poor if x is far from zero.

[C] 24. Find the Maclaurin polynomial of order n (n odd) for $\sin x$. Then use it with $n = 5$ to approximate each of the following.

(a) $\sin(0.1)$ (b) $\sin(0.5)$

(c) $\sin(1)$ (d) $\sin(10)$

Compare your answers with those given by your calculator. What conclusion do you draw?

25. Use a Maclaurin polynomial to obtain the estimate $A \approx r^2 t^3/12$ for the area of the shaded region in Figure 5. First express A exactly; then approximate.

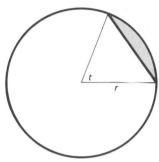

FIGURE 5

26. If an object of rest mass m_0 has velocity v, then (according to the theory of relativity) its mass m is given by $m = m_0/\sqrt{1 - v^2/c^2}$, where c is the velocity of light. Show how physicists get the approximation

$$m \approx m_0 + \frac{m_0}{2}\left(\frac{v}{c}\right)^2$$

27. If money is invested at interest rate r compounded monthly, it will double in n years, where n satisfies

$$\left(1 + \frac{r}{12}\right)^{12n} = 2$$

(a) Show that

$$n = \ln 2 \left[\frac{1}{12 \ln(1 + r/12)} \right]$$

(b) Use the Maclaurin polynomial of order 2 for $\ln(1 + x)$ and a partial fraction decomposition to obtain the approximation

$$n \approx \frac{0.693}{r} + 0.029$$

C (c) Some people use the *rule of 70*, $n \approx 70/100r$, to approximate n. Fill in the table to compare the values obtained from these three formulas.

r	n (exact)	n (approx.)	n (rule 70)
0.05			
0.10			
0.15			
0.20			

28. The author of a biology text claimed that the smallest positive solution to $x = 1 - e^{-(1 + k)x}$ is approximately $x = 2k$, provided k is very small. Show how she reached this conclusion and check on it for $k = 0.01$.

10.2
Estimating the Errors

In Example 3 of the previous section, we used the Maclaurin polynomial of order 4 to approximate $\cos(0.2)$ as follows:

$$\cos(0.2) \approx 1 - \frac{1}{2!}(0.2)^2 + \frac{1}{4!}(0.2)^4 \approx 0.9800667$$

with "first error" marked above the $\frac{1}{2!}(0.2)^2$ term and "second error" marked above the 0.9800667.

This example illustrates the two kinds of errors that occur in approximation processes. First, there is the **error of the method**. In this case, we approximated $\cos x$ by a fourth-degree polynomial. Second, there is the **error of calculation**. This includes errors due to round-off, as when we replaced the unending decimal $0.0000666\ldots$ by 0.0000667 in the last term above. It may also include errors due to the characteristics of the calculating device we are using. For example, the $\boxed{y^x}$ key tends to be inaccurate on many hand-held calculators.

Now notice a sad fact of the numerical analyst's life. We can reduce the error of the method by taking more terms in the Maclaurin expansion. But taking more terms means more calculations, which potentially increases the error of calculation. To be a good numerical analyst is to know how to compromise between these two types of error. That unfortunately is more of an art than a science. However, we can say something definite about the first type of error, the subject to which we now turn.

THE ERROR IN THE METHOD   For the problem of approximating a function by its Taylor polynomial, we can actually give a formula for the error. This formula is due to the French-Italian mathematician, Joseph-Louis Lagrange (1736–1813).

Theorem A

(Taylor's Formula with Remainder). Let f be a function whose $(n + 1)$th derivative, $f^{(n+1)}(x)$, exists for each x in an open interval I containing a. Then for each x in I,

$$f(x) = f(a) + f'(a)(x - a) + \frac{f''(a)}{2!}(x - a)^2 + \cdots$$

$$+ \frac{f^{(n)}(a)}{n!}(x - a)^n + R_n(x)$$

where the remainder (or error) $R_n(x)$ is given by the formula

$$R_n(x) = \frac{f^{(n+1)}(c)}{(n + 1)!}(x - a)^{n+1}$$

and c is some point between x and a.

We postpone the proof until later in the section, choosing to begin with several illustrations. The case $a = 0$ occurs most often in practice, and Taylor's Formula is then called Maclaurin's Formula.

EXAMPLE 1 Approximate $e^{0.8}$ with an error of less than 0.001.

Solution For $f(x) = e^x$, Maclaurin's Formula gives the remainder

$$R_n(x) = \frac{f^{(n+1)}(c)}{(n + 1)!}x^{n+1} = \frac{e^c}{(n + 1)!}x^{n+1}$$

and so

$$R_n(0.8) = \frac{e^c}{(n + 1)!}(0.8)^{n+1}$$

where $0 < c < 0.8$. Now, $e^c < e^{0.8} < 3$ and $(0.8)^{n+1} < (1)^{n+1}$, and so

$$|R_n(0.8)| < \frac{3(1)^{n+1}}{(n + 1)!} = \frac{3}{(n + 1)!}$$

It is easy to check that $3/(n + 1)! < 0.001$ when $n \geq 6$, and so we can obtain the desired accuracy by using the Maclaurin polynomial of order 6.

$$e^{0.8} \approx 1 + (0.8) + \frac{(0.8)^2}{2!} + \frac{(0.8)^3}{3!} + \frac{(0.8)^4}{4!} + \frac{(0.8)^5}{5!} + \frac{(0.8)^6}{6!}$$

Our calculator gives 2.2254948.

Can we be sure that this answer is within 0.001 of the true result? Certainly the error of the method is less than 0.001. But could the error of calculation have distorted our answer? Possibly so; however, so few calculations are involved that we feel confident in reporting an answer of 2.2255 accurate within 0.001. ∎

EXAMPLE 2 Use the Taylor polynomial of order 4 based at $a = 1$ to approximate $\ln(0.9)$ and give an estimate for the maximum error that is made.

Solution We will need the first five derivatives of $f(x) = \ln x$.

$$f(x) = \ln x \qquad\qquad f(1) = 0$$
$$f'(x) = x^{-1} \qquad\qquad f'(1) = 1$$
$$f''(x) = -x^{-2} \qquad\qquad f''(1) = -1$$
$$f'''(x) = 2x^{-3} \qquad\qquad f'''(1) = 2$$
$$f^{(4)}(x) = -6x^{-4} \qquad f^{(4)}(1) = -6$$
$$f^{(5)}(x) = 24x^{-5} \qquad f^{(5)}(c) = \frac{24}{c^5}$$

Thus, by Taylor's Formula,

$$\ln(x) = (x - 1) - \frac{1}{2}(x - 1)^2 + \frac{1}{3}(x - 1)^3 - \frac{1}{4}(x - 1)^4 + R_4(x)$$

and

$$\ln(0.9) = -0.1 - \frac{1}{2}(0.1)^2 - \frac{1}{3}(0.1)^3 - \frac{1}{4}(0.1)^4 + R_4(0.9)$$
$$\approx -0.1053583 + R_4(0.9)$$

Also,

$$R_4(0.9) = \frac{24}{c^5}\frac{(-0.1)^5}{5!} = \frac{(-0.1)^5}{5c^5}$$

where $0.9 < c < 1$, and so (using the fact that a fraction gets bigger when its denominator is made smaller)

$$|R_4(0.9)| < \frac{(0.1)^5}{5(0.9)^5} \approx 0.0000034$$

We conclude that $\ln(0.9) = -0.1053583$ with an error of less than 0.0000034. In saying this, we assumed that the error of calculation was insignificant. ∎

EXAMPLE 3 Use a Taylor polynomial of order 2 to approximate $\cos 62°$ and then estimate the error of the approximation.

Solution Since $62°$ is near $60°$ (whose cosine and sine are known), we use the Taylor polynomial based at $a = \pi/3$.

$$f(x) = \cos x \qquad f\left(\frac{\pi}{3}\right) = \frac{1}{2}$$

$$f'(x) = -\sin x \qquad f'\left(\frac{\pi}{3}\right) = -\frac{\sqrt{3}}{2}$$

$$f''(x) = -\cos x \qquad f''\left(\frac{\pi}{3}\right) = -\frac{1}{2}$$

$$f'''(x) = \sin x \qquad f'''(c) = \sin c$$

Now

$$62° = \frac{\pi}{3} + \frac{\pi}{90} \text{ radians}$$

Thus,

$$\cos x = \frac{1}{2} - \frac{\sqrt{3}}{2}\left(x - \frac{\pi}{3}\right) - \frac{1}{4}\left(x - \frac{\pi}{3}\right)^2 + R_2(x)$$

and

$$\cos\left(\frac{\pi}{3} + \frac{\pi}{90}\right) = \frac{1}{2} - \frac{\sqrt{3}}{2}\left(\frac{\pi}{90}\right) - \frac{1}{4}\left(\frac{\pi}{90}\right)^2 + R_2\left(\frac{\pi}{3} + \frac{\pi}{90}\right)$$

$$\approx 0.4694654 + R_2$$

Now

$$|R_2| = \left|\frac{\sin c}{3!}\left(\frac{\pi}{90}\right)^3\right| < \frac{1}{6}\left(\frac{\pi}{90}\right)^3 \approx 0.0000071$$

Again the number of calculations is small, so we feel safe in reporting $\cos 62° = 0.4694654$ with an error of less than 0.0000071. ∎

BOUNDING $|R_n|$ The precise value of R_n is almost never obtainable, since we do not know c, only that c lies on a certain interval. Our task is therefore to find the maximum possible value of $|R_n|$ for c in the given interval. To do this exactly is often difficult, so we usually content ourselves with getting a good upper bound for $|R_n|$. This involves a sensible use of inequalities. Our chief tools are the triangle inequality $|a \pm b| \le |a| + |b|$ and the fact that a fraction gets larger when we make its numerator larger or its denominator smaller.

EXAMPLE 4   If c is known to be in $[2, 4]$, give a good bound for the maximum value of

$$\left| \frac{c^2 - \sin c}{c} \right|$$

Solution

$$\left| \frac{c^2 - \sin c}{c} \right| = \frac{|c^2 - \sin c|}{|c|} \leq \frac{|c^2| + |\sin c|}{|c|} \leq \frac{4^2 + 1}{2} = 8.5$$

A different and better bound is obtained as follows.

$$\left| \frac{c^2 - \sin c}{c} \right| = \left| c - \frac{\sin c}{c} \right| \leq |c| + \left| \frac{\sin c}{c} \right| \leq 4 + \frac{1}{2} = 4.5 \qquad \blacksquare$$

PROOF OF TAYLOR'S FORMULA   Recall that $R_n(x)$ is defined on I by

$$R_n(x) = f(x) - f(a) - f'(a)(x - a) - \frac{f''(a)(x - a)^2}{2!} - \cdots - \frac{f^{(n)}(a)}{n!}(x - a)^n$$

Now think of x as a constant and define a new function g on I by

$$g(t) = f(x) - f(t) - f'(t)(x - t) - \frac{f''(t)(x - t)^2}{2!} - \cdots$$

$$- \frac{f^{(n)}(t)}{n!}(x - t)^n - R_n(x)\frac{(x - t)^{n+1}}{(x - a)^{n+1}}$$

Clearly $g(x) = 0$, and using the definition of $R_n(x)$, it is almost as easy to see that $g(a) = 0$ (Figure 1). By the Mean Value Theorem for Derivatives applied to g (on the interval with endpoints a and x), there is a point c between a and x such that $g'(c) = 0$. Now if we differentiate the expression for $g(t)$ with respect to t (keeping x fixed), we find that most of the terms cancel. The result is

$$g'(t) = -\frac{f^{(n+1)}(t)}{n!}(x - t)^n + R_n(x)(n + 1)\frac{(x - t)^n}{(x - a)^{n+1}}$$

When we set $g'(c) = 0$, we obtain Taylor's Formula for $R_n(x)$.

THE ERROR OF CALCULATION   In all of our examples so far, we have assumed that the error of calculation is small enough so that it can be ignored. We will ordinarily make that assumption in this book, since our problems will always involve a small number of calculations. We feel obligated, however, to warn you that when high-speed computers are used to do thousands or millions of operations, these errors of calculation may well accumulate and distort an answer.

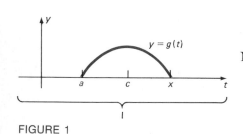

FIGURE 1

There are two sources of calculation errors that may be significant even in using a hand-held calculator. Consider calculating

$$a + b_1 + b_2 + b_3 + \cdots + b_m$$

where a is very much larger than any of the b's; for example $a = 10{,}000{,}000$ and $b_i = 0.4$, $i = 1, 2, \ldots, m$. If we use an eight-digit calculator and proceed from left to right, first adding b_1 to a, then adding b_2 to the result, and so on, we will simply get $10{,}000{,}000$ at each stage. Yet a sum of just 25 of the b's ought to affect the seventh digit of our answer. The moral here is that in adding a large number of small terms to one or more large ones, it is wise to find the sum of the small terms first.

A more likely source of calculation error is due to the loss of significant digits in a subtraction of nearly equal numbers. For example, subtracting 0.823421 from 0.823445, each with six significant digits, results in 0.000024, which has only two significant digits. That this can cause trouble is easily illustrated by calculating a numerical approximation to a derivative.

Consider calculating $f'(2)$ for $f(x) = x^4$ by using larger and larger values of n (hence smaller and smaller values of $h = 10^{-n}$) in the difference quotient

$$\frac{f(2 + h) - f(2)}{h} = \frac{(2 + 10^{-n})^4 - 2^4}{10^{-n}}$$

Theoretically, as n increases the result should get closer and closer to the correct value, namely, 32. But note what happens on one eight-digit calculator when n gets too large.

n	$(2 + 10^{-n})^4 - 2^4$	$[(2 + 10^{-n})^4 - 2^4]10^n$
2	0.32240801	32.240801
3	0.03202401	32.024010
4	0.00320024	32.002400
5	0.00032000	32.000000
6	0.00003200	32.000000
7	0.00000320	32.000000
8	0.00000032	32.000000
9	0.00000003	32.000000
10	0.00000000	0.000000
⋮	⋮	⋮

PROBLEM SET 10.2

As illustrated in Example 4, find a good estimate for the maximum value of the given expression, c being in the given interval. Answers may vary depending on the technique used.

1. $\left| e^c + e^{-c} \right|$; $[0, 2]$

2. $\left| \tan c + \sec c \right|$; $\left[\dfrac{-\pi}{4}, 0 \right]$

3. $\left| \dfrac{2}{\cos c} \right|$; $\left[\dfrac{2\pi}{3}, \pi \right]$

4. $\left| \dfrac{4}{c + 2} \right|$; $[-1.5, 2]$

5. $\left| \dfrac{e^{-c}}{c + 4} \right|$; $[-8, -6]$

6. $\left| \dfrac{\sin c}{c + 1} \right|$; $[0, 1]$

7. $\left| \dfrac{c^2 + \cos c}{16 \ln c} \right|$; $[2, 3]$

8. $\left| \dfrac{c^4 - c}{\sin c} \right|$; $[1, 3]$

In Problems 9–12, find a formula for $R_6(x)$, the remainder for the Taylor polynomial of order 6 based at a. Then estimate $|R_6(0.5)|$, that is, give a good upper bound for it. See Examples 1 and 2.

9. $\ln(1 + x)$; $a = 0$

10. e^{-x}; $a = 0$

11. $\sin x$; $a = 1$

12. $\dfrac{1}{x-2}$; $a = 1$

13. Determine the order n of the Maclaurin polynomial for e^x that is required to approximate e to five decimal places—that is, so $|R_n(1)| \le 0.000005$ (see Example 1).

14. Find the third-order Maclaurin polynomial for $(1+x)^{3/2}$ and estimate the error $R_3(x)$ if $-0.1 \le x \le 0$.

15. Find the third-order Maclaurin polynomial for $(1+x)^{-1/2}$ and estimate the error $R_3(x)$ if $-0.05 \le x \le 0.05$.

16. Find the fourth-order Maclaurin polynomial for $\ln[(1+x)/(1-x)]$ and estimate the error $R_3(x)$ for $-0.5 \le x \le 0.5$.

17. Note that the fourth-order Maclaurin polynomial for $\sin x$ is really of third degree since the coefficient of x^4 is 0. Thus,

$$\sin x = x - \frac{x^3}{6} + R_4(x)$$

Show that if $0 \le x \le 0.5$, $|R_4(x)| \le 0.0002605$. Use this result to approximate $\displaystyle\int_0^{0.5} \sin x\, dx$ and give an estimate of the error.

18. In analogy with Problem 17,

$$\cos x = 1 - \frac{x^2}{2} + \frac{x^4}{24} + R_5(x)$$

If $0 \le x \le 1$, give a good estimate for $|R_5(x)|$. Then use your result to calculate $\displaystyle\int_0^1 \cos x\, dx$ with an estimate for the error made.

19. Expand $x^4 - 3x^3 + 2x^2 + x - 2$ in a Taylor polynomial of order 4 based at 1 and show that $R_4(x) = 0$ for all x.

20. Show that an nth degree polynomial can be expressed exactly as an nth degree polynomial in $x - a$ (see Problem 19).

[C] **21.** Calculate $\sin 43° = \sin 43\pi/180$ by using the Taylor polynomial of order 3 based at $\pi/4$ for $\sin x$. Estimate the error made. See Example 3.

[C] **22.** Calculate $\cos 63°$ by the method illustrated in Example 3. Choose n large enough so $|R_n| \le 0.0005$.

[C] **23.** Show that if x is in $[0, \pi/2]$, the error in using

$$\sin x \approx x - \frac{x^3}{3!} + \frac{x^5}{5!} - \frac{x^7}{7!} + \frac{x^9}{9!}$$

is less than 5×10^{-6} and, therefore, that this formula is good enough to build a five-place sine table.

24. Use Maclaurin's Formula rather than l'Hôpital's Rule to find:

(a) $\displaystyle\lim_{x \to 0} \frac{\sin x - x + x^3/6}{x^5}$

(b) $\displaystyle\lim_{x \to 0} \frac{\cos x - 1 + x^2/2 - x^4/24}{x^6}$

25. Let $g(x) = p(x) + x^{n+1}f(x)$, where $p(x)$ is a polynomial of degree at most n and f has derivatives through order n. Show that $p(x)$ is the Maclaurin polynomial of order n for g.

26. Recall that the Second Derivative Test (Section 4.3) for maxima and minima fails if $f''(c) = 0$. Prove the following generalization, which may help in this case. Suppose that

$$f'(c) = f''(c) = f'''(c) = \cdots = f^{(n)}(c) = 0$$

where n is odd and $f^{(n+1)}(x)$ is continuous near c.
(i) If $f^{(n+1)}(c) < 0$, $f(c)$ is a local maximum value.
(ii) If $f^{(n+1)}(c) > 0$, $f(c)$ is a local minimum value.
Test this result on $f(x) = x^4$.

10.3
Numerical Integration

We know that if f is continuous on a closed interval $[a, b]$, then the definite integral $\displaystyle\int_a^b f(x)\, dx$ must exist. Existence is one thing; evaluation is a very different matter. There are many definite integrals that cannot be evaluated by the methods we have learned—that is, by use of the Fundamental Theorem of Calculus. This is because the indefinite integrals of such integrands as

$$e^{-x^2}, \qquad \sin(x^2), \qquad \sqrt{1-x^4}, \qquad \frac{\sin x}{x}$$

cannot be expressed in terms of elementary functions, that is, in terms of functions studied in a first calculus course. Even when elementary indefinite integrals can be found, it is often advantageous to use the approximation methods soon to be discussed, since they are well suited to use on a calculator or computer. We present two such methods: the Trapezoidal Rule and the Parabolic Rule.

THE TRAPEZOIDAL RULE   Consider the graph of $y = f(x)$ on $[a, b]$; it might look something like the curve in Figure 1. Partition the interval $[a, b]$ into n subintervals, each of length $h = (b - a)/n$, by means of points $a = x_0 < x_1 < x_2 < \cdots < x_n = b$. Join the pairs of points $(x_{i-1}, f(x_{i-1}))$ and $(x_i, f(x_i))$ by line segments, as shown in the figure, thus forming n trapezoids.

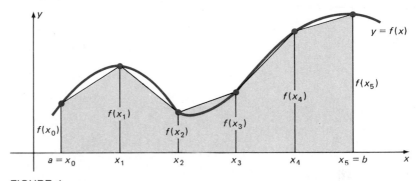

FIGURE 1

Recalling the area formula shown in Figure 2, we can write the area of the ith trapezoid as

$$A_i = \frac{h}{2}\left[f(x_{i-1}) + f(x_i)\right]$$

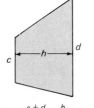

$A = h\frac{c+d}{2} = \frac{h}{2}(c + d)$

FIGURE 2

More accurately, we should say *signed* area, since A_i will be negative for a subinterval where f is negative. The definite integral $\displaystyle\int_a^b f(x)\,dx$ is approximately equal to $A_1 + A_2 + \cdots + A_n$, that is, to

$$\frac{h}{2}\left[f(x_0) + f(x_1)\right] + \frac{h}{2}\left[f(x_1) + f(x_2)\right] + \cdots + \frac{h}{2}\left[f(x_{n-1}) + f(x_n)\right]$$

This simplifies to the Trapezoidal Rule.

TRAPEZOIDAL RULE

$$\int_a^b f(x)\,dx \approx \frac{h}{2}\left[f(x_0) + 2f(x_1) + 2f(x_2) + \cdots + 2f(x_{n-1}) + f(x_n)\right]$$

We illustrate this rule first for a definite integral where we know the exact answer.

EXAMPLE 1   Use the Trapezoidal Rule to find an approximate value for

$$\int_1^3 x^4 \, dx,$$

taking $n = 8$.

Solution   Since $n = 8$, $h = (3 - 1)/8 = 0.25$, and

$$
\begin{aligned}
x_0 &= 1.00 & f(x_0) &= (1.00)^4 = & 1.0000 \\
x_1 &= 1.25 & f(x_1) &= (1.25)^4 \approx & 2.4414 \\
x_2 &= 1.50 & f(x_2) &= (1.50)^4 = & 5.0625 \\
x_3 &= 1.75 & f(x_3) &= (1.75)^4 \approx & 9.3789 \\
x_4 &= 2.00 & f(x_4) &= (2.00)^4 = & 16.0000 \\
x_5 &= 2.25 & f(x_5) &= (2.25)^4 \approx & 25.6289 \\
x_6 &= 2.50 & f(x_6) &= (2.50)^4 = & 39.0625 \\
x_7 &= 2.75 & f(x_7) &= (2.75)^4 \approx & 57.1914 \\
x_8 &= 3.00 & f(x_8) &= (3.00)^4 = & 81.0000
\end{aligned}
$$

Thus,

$$\int_1^3 x^4 \, dx \approx \frac{0.25}{2} \left[1.0000 + 2(2.4414) + \cdots + 2(57.1914) + 81.0000 \right]$$

$$= 48.9414$$

This may be compared with the exact value

$$\int_1^3 x^4 \, dx = \left[\frac{x^5}{5} \right]_1^3 = \frac{242}{5} = 48.4000$$

Presumably we could get a better approximation by taking n larger; this would be easy to do using a computer. However, while taking n larger reduces the error of the method, it at least potentially increases the error of calculation. It would be unwise, for example, to take $n = 1,000,000$, since the potential round-off errors would more than compensate for the fact that the error of the method would be minuscule. We will have more to say about errors shortly. ∎

EXAMPLE 2   Use the Trapezoidal Rule with $n = 6$ to approximate

$$\int_0^1 e^{-x^2} \, dx$$

Solution   Since $h = \frac{1}{6}$ and $f(x) = e^{-x^2}$,

$$x_0 = 0.0000 \qquad f(x_0) = 1.0000$$
$$x_1 \approx 0.1667 \qquad f(x_1) \approx 0.9726$$
$$x_2 \approx 0.3333 \qquad f(x_2) \approx 0.8948$$
$$x_3 = 0.5000 \qquad f(x_3) \approx 0.7788$$
$$x_4 \approx 0.6667 \qquad f(x_4) \approx 0.6412$$
$$x_5 \approx 0.8333 \qquad f(x_5) \approx 0.4994$$
$$x_6 = 1.0000 \qquad f(x_6) \approx 0.3679$$

$$\int_0^1 e^{-x^2}\, dx \approx \tfrac{1}{12}\left[1.000 + 2(0.9726) + \cdots + 2(0.4994) + 0.3679\right]$$

$$= 0.7451$$

As we mentioned earlier, this integral, important in statistical theory, cannot be evaluated by a direct application of the Fundamental Theorem of Calculus. ∎

THE ERROR IN THE TRAPEZOIDAL RULE   In any practical use of the Trapezoidal Rule, we need to have some idea of the size of the error involved. Fortunately, we can give a formula for the error of the method.

Theorem A

Suppose that f'' exists on $[a, b]$. Then,

$$\int_a^b f(x)\, dx = \frac{h}{2}\left[f(x_0) + 2f(x_1) + \cdots + 2f(x_{n-1}) + f(x_n)\right] + E_n$$

where the error E_n is given by

$$E_n = -\frac{(b-a)^3}{12n^2}\, f''(c)$$

and c is some point between a and b.

We omit the proof of this theorem, which may be found in more advanced books (such as J. M. H. Olmsted, *Advanced Calculus* (New York: Prentice-Hall, 1961) pp. 118–19), choosing rather to illustrate its use.

EXAMPLE 3   Estimate the maximum possible error in Example 2.

Solution   Since $f(x) = e^{-x^2}$, $f'(x) = -2xe^{-x^2}$, and $f''(x) = e^{-x^2}(4x^2 - 2)$. We could actually find the maximum value of $f''(x)$ on $[0, 1]$, but it is sufficient to bound it there using properties of absolute value.

$$|f''(x)| = |e^{-x^2}(4x^2 - 2)| = e^{-x^2}|4x^2 - 2|$$
$$\leq e^{-x^2}(4x^2 + 2) \leq 1(4 + 2) = 6$$

Thus,

$$|E_n| = \frac{(b-a)^3}{12n^2} |f''(c)| \leq \frac{1^3}{12(6^2)} (6) = \frac{1}{72} \approx 0.0139$$

This is, of course, a bound for the error of the method. However, n is so small that the error of calculation can safely be ignored. We feel confident in reporting

$$\int_0^1 e^{-x^2} dx = 0.7451 \pm 0.0139$$

A slightly more sophisticated analysis (use the fact that $4x^2 - 2$ is increasing on $[0,1]$) shows that $|4x^2 - 2| \leq 2$ on $[0,1]$, and hence $|E_n| \leq 0.0047$. ∎

EXAMPLE 4 How large must n be to insure that the error of the method in Example 2 is less than 0.0001?

Solution From Example 3,

$$|E_n| = \frac{(b-a)^3}{12n^2} |f''(c)| \leq \frac{1^3 \cdot 6}{12n^2} = \frac{1}{2n^2}$$

We want $1/2n^2 < 0.0001$, which holds if $n \geq 71$. ∎

THE PARABOLIC RULE (SIMPSON'S RULE) In the Trapezoidal Rule, we approximated the curve $y = f(x)$ by line segments. It seems likely that we could do better using parabolic segments. Just as before, partition the interval $[a, b]$ into n subintervals of length $h = (b - a)/n$, but this time with n an *even* number. Then fit parabolic segments to neighboring triples of points, as shown in Figure 3.

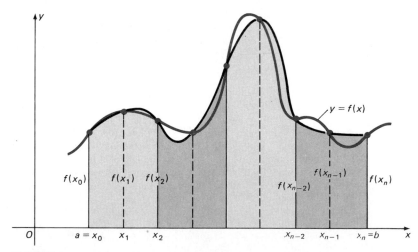

FIGURE 3

CHAPTER 10/Numerical Methods, Approximations

Using the area formula in Figure 4 (see Problem 15 for the derivation) leads to an approximation called the **Parabolic Rule**. It is also called **Simpson's Rule**, after the English mathematician Thomas Simpson (1710–1761).

PARABOLIC RULE

$$\int_a^b f(x)\,dx \approx \frac{h}{3}\left[f(x_0) + 4f(x_1) + 2f(x_2) + \cdots + 4f(x_{n-1}) + f(x_n)\right]$$

The pattern of coefficients is 1, 4, 2, 4, 2, 4, 2, ..., 2, 4, 1.

EXAMPLE 5 Use the Parabolic Rule with $n = 8$ to approximate

$$\int_1^3 x^4\,dx$$

Parabola

$A = \frac{h}{3}(c + 4d + e)$

FIGURE 4

Solution We can make use of the calculations in Example 1. We obtain

$$\int_1^3 x^4\,dx \approx \frac{0.25}{3}\left[1.0000 + 4(2.4414) + 2(5.0625) + 4(9.3789) + 2(16.0000)\right.$$
$$\left. + 4(25.6289) + 2(39.0625) + 4(57.1914) + 81.0000\right]$$
$$\approx 48.4010 \qquad \blacksquare$$

As expected, the Parabolic Rule gives an answer closer to the true value 48.4000 than did the Trapezoidal Rule, which gave 48.9414. Because it requires so little extra work, we prefer the Parabolic Rule over the Trapezoidal Rule in most problems. That its error term is generally smaller is borne out by the following theorem (note the factor of n^4 in the denominator). For a proof, see J. M. H. Olmsted, *Advanced Calculus* (New York: Prentice-Hall, 1961), pp. 118–19.

Theorem B

Suppose that the fourth derivative $f^{(4)}(x)$ exists on $[a, b]$. Then the error E_n in the Parabolic Rule is given by

$$E_n = -\frac{(b-a)^5}{180n^4}f^{(4)}(c)$$

for some c between a and b.

EXAMPLE 6 Use the Parabolic Rule with $n = 8$ to approximate

$$\int_1^2 (1 + x)^{-1}\,dx$$

and find a bound for the error made.

Solution Since $h = \frac{1}{8}$ and $f(x) = 1/(1 + x)$, we calculate the results shown in the table.

i	x_i	$f(x_i)$	c_i	$c_i f(x_i)$
0	1	0.50000	1	0.50000
1	1.125	0.47059	4	1.88236
2	1.25	0.44444	2	0.88888
3	1.375	0.42105	4	1.68420
4	1.5	0.40000	2	0.80000
5	1.625	0.38095	4	1.52380
6	1.75	0.36364	2	0.72728
7	1.875	0.34783	4	1.39132
8	2	0.33333	1	0.33333
				Sum = 9.73117

Thus,

$$\int_1^2 \frac{dx}{1 + x} \approx \frac{1}{24} (9.73117) \approx 0.4055$$

To find a bound for E_8, we first calculate $f^{(4)}(x)$ to be $24/(1 + x)^5$, and then

$$E_8 = -\frac{(2 - 1)^5}{180(8^4)} \cdot \frac{24}{(1 + c)^5}$$

Consequently

$$|E_8| \leq \frac{24}{180(8^4)(1 + 1)^5} \approx 0.00000102$$

Assuming the calculation errors to be negligible, we feel very confident in reporting that the answer 0.4055 is accurate to four decimal places. ∎

PROBLEM SET 10.3

[C] In Problems 1–4, use the Trapezoidal Rule and the Parabolic Rule, both with $n = 8$, to approximate each integral. Then calculate the integral using the Fundamental Theorem of Calculus.

1. $\int_1^2 \frac{1}{x^2} \, dx$

2. $\int_1^2 \frac{1}{x} \, dx$

3. $\int_0^4 \sqrt{x} \, dx$

4. $\int_0^2 x\sqrt{x^2 + 4} \, dx$

[C] **5.** Use the Trapezoidal Rule with $n = 2, 6, 12$ to approximate $\int_0^\pi \sin x \, dx$. Note how these approximations do get closer to the actual value, which is 2.

[C] **6.** Follow the instructions of Problem 5 using the Parabolic Rule.

[C] **7.** Approximate π by calculating

$$\int_0^1 \frac{4}{1 + x^2} \, dx$$

using the Parabolic Rule with $n = 10$.

[C] **8.** Use the Trapezoidal Rule with $n = 10$ to approximate $\int_0^1 \cos(\sin x) \, dx$.

[C] In Problems 9–12, determine an n so that the Trapezoidal Rule will approximate the integral with an error E_n

satisfying $|E_n| \le 0.01$ (see Example 4). Then, using that n, approximate the integral.

9. $\displaystyle\int_0^{1.2} e^{-x^2}\,dx$ **10.** $\displaystyle\int_0^{0.5} e^{x^2}\,dx$

11. $\displaystyle\int_1^2 \sqrt{\sin x}\,dx$ **12.** $\displaystyle\int_1^2 \sin\sqrt{x}\,dx$

[C] In Problems 13 and 14, determine an n so that the Parabolic Rule will approximate the integral with an error E_n satisfying $|E_n| \le 0.005$. Then, using that n, approximate the integral.

13. $\displaystyle\int_2^4 \frac{1 + x}{1 - x}\,dx$ **14.** $\displaystyle\int_2^4 \ln x\,dx$

15. Let $f(x) = ax^2 + bx + c$. Show that

$$\int_{m-h}^{m+h} f(x)\,dx \quad \text{and} \quad (h/3)\big[f(m-h) + 4f(m) + f(m+h)\big]$$

both have the value $(h/3)[a(6m^2 + 2h^2) + b(6m) + 6c]$. This establishes the area formula on which the Parabolic Rule is based.

16. Show that the Parabolic Rule is exact for any cubic polynomial in two different ways.

(a) By direct calculation.
(b) By showing that $E_n = 0$.

17. We know that $\ln 2 = \displaystyle\int_1^2 (1/x)\,dx$. If we wish to estimate $\ln 2$ using the Trapezoidal Rule with an error less than 10^{-10}, how large an n would we need?

18. Answer the question of Problem 17 for the Parabolic Rule.

19. Show that the Parabolic Rule gives the exact value of $\displaystyle\int_{-a}^{a} x^k\,dx$ provided k is odd.

20. It is interesting that a modified version of the Trapezoidal Rule turns out to be in general more accurate than the Parabolic Rule. This version says that

$$\int_a^b f(x)\,dx \approx T - \frac{[f'(b) - f'(a)]h^2}{12}$$

where T is the standard trapezoidal estimate.

(a) Use this formula with $n = 8$ to estimate $\displaystyle\int_1^3 x^4\,dx$ and note its remarkable accuracy (see Example 1 for T and also the exact value of this integral).

(b) Use this formula with $n = 12$ to estimate $\displaystyle\int_0^\pi \sin x\,dx$ (the true value is 2 and T was calculated in Problem 5).

That this rule is so accurate does not seem so surprising after one shows that it is exact for all cubic polynomials, a result we will not ask you to prove since it is quite tedious.

21. Figure 5 shows the depth in feet of the water in a river measured at 20-foot intervals across the width of the river. If the river flows at 4 miles per hour, how much water (in cubic feet) flows past the place where these measurements were taken in one day? Use the Parabolic Rule.

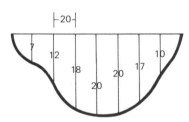

FIGURE 5

10.4
Solving Equations
Numerically

In mathematics and science, we often need to find the roots (solutions) of an equation $f(x) = 0$. To be sure, if $f(x)$ is a linear or quadratic polynomial, formulas for writing exact solutions exist and are well known. But for other algebraic equations, and certainly for transcendental equations, formulas for exact solutions are rarely available. What can be done in such cases?

There is a general method of solving problems known to all resourceful people. It might be called "muddling through," or "trial and error." Given a cup of tea, we add sugar a bit at a time until it tastes just right. Given a stopper too large for a hole, we whittle it down until it fits. We change the solution a bit at a time, improving the accuracy, until we are satisfied. Mathematicians call it the *method of successive approximations*, or the *method of iterations*.

In this section, we present two such methods for solving equations: the Bisection Method and Newton's Method. Both are designed to find the real roots of $f(x) = 0$. Both require many computations. You will want to keep your calculator handy.

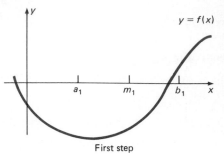

FIGURE 1

THE BISECTION METHOD   This method has two great virtues—simplicity and reliability. It also has a major vice—the large number of steps to achieve desired accuracy (otherwise known as slowness of convergence).

Begin the process by sketching the graph of f which is assumed to be a continuous function. A real root r of $f(x) = 0$ is a point (technically, the x-coordinate of a point) where the graph crosses the x-axis. As a first step in pinning down this point, locate two points, $a_1 < b_1$, at which you are sure that f has opposite signs. (Try choosing a_1 and b_1 on opposite sides of your best guess at r.) The Intermediate Value Theorem guarantees the existence of a root between a_1 and b_1. Now evaluate f at the midpoint $m_1 = (a_1 + b_1)/2$ of $[a_1, b_1]$. The number m_1 is our first approximation to r (see Figure 1).

Either $f(m_1) = 0$, in which case we are done, or $f(m_1)$ differs in sign from $f(a_1)$ or $f(b_1)$. Denote the one of the subintervals $[a_1, m_1]$ or $[m_1, b_1]$ on which the sign change occurs by the symbol $[a_2, b_2]$ and evaluate f at its midpoint $m_2 = (a_2 + b_2)/2$. The number m_2 is our second approximation to r (Figure 2).

Repeat the process, thus determining a sequence of approximations $m_1, m_2, m_3, \ldots$ and subintervals $[a_1, b_1], [a_2, b_2], [a_3, b_3], \ldots$, each subinterval containing the root r and each half the length of its predecessor. Stop when r is determined to desired accuracy—that is, when $(b_n - a_n)/2$ is less than the allowable error.

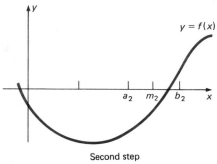

FIGURE 2

To streamline the process for a calculator, we first suppose that the graph of $y = f(x)$ is rising as it crosses the x-axis; if not, consider instead the graph of $y = -f(x)$, which crosses at the same point. Then let

$$m_n = \frac{a_n + b_n}{2}, \qquad h_n = \frac{b_n - a_n}{2}$$

and note that m_n is the midpoint of $[a_n, b_n]$ and h_n is its half-length. Assuming that m_n and h_n are known, using the following *algorithm* to determine m_{n+1} and h_{n+1}.

1. Calculate $f(m_n)$ and if $f(m_n) = 0$, stop.

2. Put $h_{n+1} = \dfrac{h_n}{2}$.

3. If $f(m_n) < 0$, put $m_{n+1} = m_n + h_{n+1}$.

4. If $f(m_n) > 0$, put $m_{n+1} = m_n - h_{n+1}$.

EXAMPLE 1   Determine the real root r of $f(x) = x^3 - 3x - 5 = 0$ to accuracy within 0.0000001.

FIGURE 3

Solution   We first sketch the graph of $y = x^3 - 3x - 5$ (Figure 3) and, noting that it crosses the x-axis between 2 and 3, we use $a_1 = 2$, $b_1 = 3$, $m_1 = 2.5$, and $h_1 = 0.5$. Then we construct the following table.

n	h_n	m_n	$f(m_n)$
1	0.5	2.5	3.125
2	0.25	2.25	−0.359
3	0.125	2.375	1.271
4	0.0625	2.3125	0.429
5	0.03125	2.28125	0.02811
6	0.015625	2.265625	−0.16729
7	0.0078125	2.2734375	−0.07001
8	0.0039063	2.2773438	−0.02106
9	0.0019532	2.2792969	0.00350
10	0.0009766	2.2783203	−0.00878
11	0.0004883	2.2788086	−0.00264
12	0.0002442	2.2790528	0.00043
13	0.0001221	2.2789307	−0.00111
14	0.0000611	2.2789918	−0.00034
15	0.0000306	2.2790224	0.00005
16	0.0000153	2.2790071	−0.00015
17	0.0000077	2.2790148	−0.00005
18	0.0000039	2.2790187	−0.000001
19	0.0000020	2.2790207	0.000024
20	0.0000010	2.2790197	0.000011
21	0.0000005	2.2790192	0.000005
22	0.0000003	2.2790189	0.0000014
23	0.0000002	2.2790187	−0.0000011
24	0.0000001	2.2790188	0.0000001

We conclude that $r = 2.2790188$ with an error of at most 0.0000001. ∎

Example 1 illustrates the shortcoming of the Bisection Method. The approximations $m_1, m_2, m_3, \ldots$ converge very slowly to the root r. But they do converge. The method works and we have at step n a good bound for the error $E_n = r - m_n$, namely, $|E_n| \leq h_n$.

NEWTON'S METHOD   We are still considering the problem of solving the equation $f(x) = 0$ for a root r. Suppose that f is differentiable, so that the graph of $y = f(x)$ has a tangent line at each point. If we can find a first approximation x_1 to r by graphing or any other means, then a better approximation x_2 ought to lie at the intersection of the tangent at $(x_1, f(x_1))$ with the x-axis (see Figure 4). Using x_2 as an approximation, we can then find a still better approximation x_3, and so on.

The process can be mechanized so that it is easy to do on a calculator. The equation of the tangent at $(x_1, f(x_1))$ is

$$y - f(x_1) = f'(x_1)(x - x_1)$$

and its x-intercept x_2 is found by setting $y = 0$ and solving for x. The result is

$$x_2 = x_1 - \frac{f(x_1)}{f'(x_1)}$$

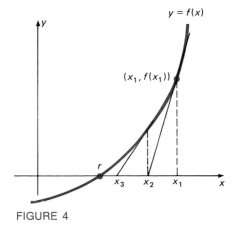

FIGURE 4

More generally we have the **algorithm**, also called a *recursion formula* or an *iteration scheme*,

$$x_{n+1} = x_n - \frac{f(x_n)}{f'(x_n)}$$

EXAMPLE 2 Use Newton's Method to find the real root r of $f(x) = x^3 - 3x - 5 = 0$ to seven decimal places.

Solution This is the same equation considered in Example 1. Let's use $x_1 = 2.5$ as our first approximation to r, as we did there. Since $f(x) = x^3 - 3x - 5$ and $f'(x) = 3x^2 - 3$, the algorithm is

$$x_{n+1} = x_n - \frac{x_n^3 - 3x_n - 5}{3x_n^2 - 3} = \frac{2x_n^3 + 5}{3x_n^2 - 3}$$

We obtain the data in the table below.

n	x_n
1	2.5
2	2.30
3	2.2793
4	2.2790188
5	2.2790188

After just four steps, we get a repetition of the first eight digits. We feel confident in reporting that $r \approx 2.2790188$, with perhaps some question about the last digit. ∎

EXAMPLE 3 Use Newton's Method to find the real root r of $f(x) = x - e^{-x} = 0$ to seven decimal places.

Solution The graph of $y = x - e^{-x}$ is sketched in Figure 5. We use $x_1 = 0.5$ and $x_{n+1} = x_n - (x_n - e^{-x_n})/(1 + e^{-x_n}) = (x_n + 1)/(e^{x_n} + 1)$ to obtain the table below.

n	x_n
1	0.5
2	0.566
3	0.56714
4	0.5671433
5	0.5671433

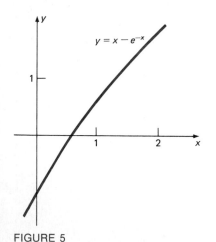

$y = x - e^{-x}$

FIGURE 5

After just four steps, we get a repetition of the seven digits after the decimal point. We conclude that $r \approx 0.5671433$. ∎

FIGURE 6

CONVERGENCE OF NEWTON'S METHOD   It is not always obvious that Newton's Method yields approximations that converge to the root r, though our two examples give evidence for that assertion. Could it be that we were just lucky in our choice of these two examples? As a matter of fact, the method does not always lead to convergence, as the diagram in Figure 6 shows (see also Problem 21). The difficulty in this case is that x_1 is not close enough to r to get a convergent process started. Another obvious difficulty arises if $f'(x)$ is zero at or near r since $f'(x_n)$ occurs in the denominator of the algorithm. However, we have the following theorem.

Theorem A

Let f be twice differentiable on an interval I, having as its midpoint a root r of $f(x) = 0$. Suppose there are positive numbers m and M such that $|f'(x)| \geq m$ and $|f''(x)| \leq M$ on I. If x_1 is in I and sufficiently close to r ($|x_1 - r| < 2m/M$ will do), then

(i) $|x_{n+1} - r| \leq \dfrac{M}{2m}(x_n - r)^2$;

(ii) x_n converges to r as $n \to \infty$.

Perhaps we should point out that we have used the word *converge* without giving it a precise definition. For a sequence $\{x_n\}$ to converge to r means that $\lim\limits_{n \to \infty} x_n = r$. We will study this notion in detail in the next chapter.

Proof   From Taylor's Formula with Remainder (Theorem 10.2A), there is a number c between x_n and r such that

$$f(r) = f(x_n) + f'(x_n)(r - x_n) + \frac{f''(c)}{2}(r - x_n)^2$$

After dividing both sides by $f'(x_n)$ and using the fact that $f(r) = 0$, we obtain

$$0 = \frac{f(x_n)}{f'(x_n)} + r - x_n + \frac{f''(c)}{2f'(x_n)}(r - x_n)^2$$

and then, successively,

$$x_n - \frac{f(x_n)}{f'(x_n)} - r = \frac{f''(c)}{2f'(x_n)}(r - x_n)^2$$

$$|x_{n+1} - r| = \left|\frac{f''(c)}{2f'(x_n)}\right|(x_n - r)^2$$

$$|x_{n+1} - r| \leq \frac{M}{2m}(x_n - r)^2$$

which is (i).

From (i), one may show by induction (Problem 18) that

$$|x_n - r| \leq \frac{2m}{M} \left(\frac{M}{2m} |x_1 - r| \right)^{2^{n-1}}$$

Since $(M/2m)|x_1 - r| < 1$, the right side of the last inequality approaches 0 as $n \to \infty$. This implies that $|x_n - r|$ also tends to 0 as $n \to \infty$, which is equivalent to (ii). ∎

The speed of convergence of Newton's Method is truly remarkable, tending in fact to double the number of decimal places of accuracy at each step. To see why this is so, suppose $M/2m \leq 2$. Then if the error $|x_n - r|$ at the nth step is less than 0.005, the error $|x_{n+1} - r|$ at the next step satisfies (by (i))

$$|x_{n+1} - r| \leq \frac{M}{2m} |x_n - r|^2 \leq 2(0.005)^2 = 0.00005$$

Thus the accuracy of x_n to two decimal places is doubled to accuracy of x_{n+1} to four decimal places. Of course, we should not expect quite such spectacular results if $M/2m$ is substantially greater than 2.

PROBLEM SET 10.4

C In Problems 1–4, use the Bisection Method to find the real root of the given equation on the given interval. Your answers should be accurate to two decimal places.

1. $x^3 + 3x - 6 = 0$; $[1, 2]$

2. $x^4 + 4x^3 + 1 = 0$; $[-1, 0]$

3. $\cos x - e^{-x} = 0$; $[1, 2]$

4. $x - 2 + \ln x = 0$; $[1, 2]$

C In Problems 5–14, use Newton's Method to approximate the indicated root of the given equation accurate to five decimal places. Begin by sketching a graph.

5. The largest root of $x^3 + 6x^2 + 9x + 2 = 0$.

6. The real root of $7x^3 + x - 6 = 0$.

7. The root of $x - 2 + \ln x = 0$ (see Problem 4).

8. The smallest positive root of $\cos x - e^{-x} = 0$ (see Problem 3).

9. The root of $\cos x = x$.

10. The root of $x \ln x = 1$.

11. All real roots of $x^4 - 8x^3 + 22x^2 - 24x + 6 = 0$.

12. All real roots of $x^4 + x^3 - 3x^2 + 4x - 28 = 0$.

13. The positive root of $2x^2 - \sin^{-1} x = 0$.

14. The positive root of $2 \tan^{-1} x = x$.

C **15.** Use Newton's Method to calculate $\sqrt[3]{6}$ to five decimal places. *Hint:* Solve $x^3 - 6 = 0$.

C **16.** Use Newton's Method to calculate $\sqrt[4]{47}$ to five decimal places.

C **17.** Where on $(\pi, 2\pi)$ does $(\sin x)/x$ attain a minimum and what is its minimum value?

18. Show by induction that if

$$|x_{n+1} - r| \leq \frac{M}{2m} (x_n - r)^2, \qquad n = 1, 2, \ldots$$

then

$$|x_n - r| \leq \frac{2m}{M} \left(\frac{M}{2m} |x_1 - r| \right)^{2^{n-1}}, \qquad n = 1, 2, \ldots$$

C **19.** Suppose we use Newton's Method to find the positive root of $x^2 - 2 = 0$—that is, to approximate $\sqrt{2}$. Suppose further that we know this root is on the interval $[1, 2]$. Calculate m and M of Theorem A. Use the second inequality of Problem 18 to estimate $|x_6 - \sqrt{2}|$, given that $x_1 = 1.5$.

C **20.** How large should we take n in Problem 19 to make sure $|x_n - \sqrt{2}| \leq 5 \times 10^{-41}$?

C **21.** Consider finding the real root of $(1 + \ln x)/x = 0$ by Newton's Method. Show that this leads to the algorithm

$$x_{n+1} = 2x_n + \frac{x_n}{\ln x_n}$$

Apply this algorithm with $x_1 = 1.2$. Next try it with $x_1 = 0.5$. Finally, graph $y = (1 + \ln x)/x$ to understand your results.

22. Sketch the graph of $y = x^{1/3}$. Obviously, its only x-intercept is zero. Convince yourself that Newton's Method fails to converge. Explain this failure.

23. In installment buying, one would like to figure out the real interest rate (effective rate), but unfortunately this involves solving a complicated equation. If one buys an item worth $\$P$ today and agrees to pay for it with payments of $\$R$ at the end of each month for k months, then

$$P = \frac{R}{i}\left[1 - \frac{1}{(1+i)^k}\right]$$

where i is the interest rate per month. Tom bought a used car for $2000 and agreed to pay for it with $100 payments at the end of each of the next 24 months.

(a) Show that i satisfies the equation

$$20i(1 + i)^{24} - (1 + i)^{24} + 1 = 0$$

(b) Derive Newton's Algorithm for this equation, namely,

$$i_{n+1} = i_n - \left[\frac{20i_n^2 + 19i_n - 1 + (1 + i_n)^{-23}}{500i_n - 4}\right]$$

C (c) Find i accurate to five decimal places starting with $i_1 = 0.012$ and then give the annual rate r as a percent ($r = 1200i$).

24. In applying Newton's Algorithm to solve $f(x) = 0$, one can usually tell by simply looking at the numbers $x_1, x_2, x_3, \ldots$ whether the sequence is converging. But even if it converges, say to $\bar{x}$, can we be sure that $\bar{x}$ is a solution? Show that the answer is yes provided f and f' are continuous at $\bar{x}$ and $f'(\bar{x}) \neq 0$.

10.5
Fixed-Point
Methods

We offer next a method of solving equations that is so simple it has no right to work. Yet it does work in a large number of cases. Moreover, this method has a host of applications in advanced mathematics.

Suppose an equation that interests us can be written in the form $x = g(x)$. To solve this equation is to find a number r that is unchanged by the function g. We call such a number a **fixed point** of g. To find this number, we propose the following algorithm. Make a first guess x_1. Then let $x_2 = g(x_1)$, $x_3 = g(x_2)$, and, in general,

$$x_{n+1} = g(x_n)$$

If we are lucky, x_n will converge to the root r as $n \to \infty$.

THE METHOD ILLUSTRATED   We begin with an example treated in the previous section (Example 3).

EXAMPLE 1   Solve $x - e^{-x} = 0$ using the algorithm above.

Solution   We write the equation as $x = e^{-x}$ and apply the algorithm $x_{n+1} = e^{-x_n}$ with $x_1 = 0.5$. The results (obtained on a hand-held calculator) are shown in the table at the top of the next page.

n	x_n	n	x_n	n	x_n
1	0.5	10	0.5675596	19	0.5671408
2	0.6065307	11	0.5669072	20	0.5671447
3	0.5452392	12	0.5672772	21	0.5671425
4	0.5797031	13	0.5670674	22	0.5671438
5	0.5600646	14	0.5671864	23	0.5671430
6	0.5711722	15	0.5671189	24	0.5671435
7	0.5648630	16	0.5671572	25	0.5671432
8	0.5684381	17	0.5671354	26	0.5671433
9	0.5664095	18	0.5671478	27	0.5671433

While it took 27 steps to get a repetition of the first seven digits, the process did converge and to the right value. Moreover, it was extremely easy to carry out on a calculator. ∎

EXAMPLE 2 Solve $x = 2 \cos x$.

Solution We use the algorithm $x_{n+1} = 2 \cos x_n$. The graph in Figure 1 suggests an initial guess of $x_1 = 1$. Here are our results, again obtained on a hand-held calculator.

FIGURE 1

n	x_n	n	x_n
1	1	6	1.4394614
2	1.0806046	7	0.2619155
3	0.9415902	8	1.9317916
4	1.1770062	9	−0.7064109
5	0.7673820	10	1.5213931

Quite clearly the process is unstable, even though our initial guess is very close to the actual root.

Let's take a different tack. Rewrite the equation $x = 2 \cos x$ as $x = (x + 2 \cos x)/2$ and use the algorithm

$$x_{n+1} = \frac{x_n + 2 \cos x_n}{2}$$

This yields the data in the table below.

n	x_n	n	x_n	n	x_n
1	1	7	1.0298054	13	1.0298665
2	1.0403023	8	1.0298883	14	1.0298666
3	1.0261107	9	1.0298588	15	1.0298665
4	1.0312046	10	1.0298693	16	1.0298666
5	1.0293881	11	1.0298655		
6	1.0300374	12	1.0298668		

The process has converged (the oscillation in the last digit is probably due to round-off errors). ∎

Now we raise an obvious question. Why did the second algorithm yield a convergent process, whereas the first one failed to do so? And can we be sure that we have obtained the correct answer in the second case? To this latter question we can offer an affirmative answer. The nature of the calculation suggests that when we get a repetition of the first seven digits, we have a solution to at least six-digit accuracy. We will answer the first question after we have considered another example.

EXAMPLE 3   Solve $x^3 + 6x - 3 = 0$ by the Fixed-Point Method.

Solution   The given equation is equivalent to $x = (-x^3 + 3)/6$, so we use the algorithm

$$x_{n+1} = \frac{-x_n^3 + 3}{6}$$

The graph in Figure 2 suggests an initial value of $x_1 = 0.5$, but we also consider what happens with $x_1 = 1.5$, $x_1 = 2.2$, and $x_1 = 2.7$. The results are shown in the table.

FIGURE 2

n	x_n	n	x_n	n	x_n	n	x_n
1	0.5	1	1.5	1	2.2	1	2.7
2	0.4791667	2	−0.0625	2	−1.2744667	2	−2.7805
3	0.4816638	3	0.5000407	3	0.8451745	3	4.0827578
4	0.4813757	4	0.4791616	4	0.3993792	4	−10.842521
5	0.4814091	5	0.4816644	5	0.4893829	5	212.9416
6	0.4814052	6	0.4813756	6	0.4804658	6	−16909274.5
7	0.4814057	7	0.4814091	7	0.4815143		
8	0.4814056	8	0.4814052	8	0.4813930		
9	0.4814056	9	0.4814057	9	0.4814071		
		10	0.4814056	10	0.4814054		
		11	0.4814056	11	0.4814056		
				12	0.4814056		

It appears that if our initial guess x_1 is close enough to the fixed point r, the process will converge, but if we begin too far away from r, it will diverge. ∎

CONVERGENCE OF THE METHOD   Sometimes the method works; sometimes it fails. We can get a pretty good idea about which is which by executing an appropriate number of steps of the algorithm. But wouldn't it be nice to have a way of telling in advance whether there will be convergence? And to be sure of our conclusion?

To get a feeling for the problem, let's look at it geometrically. Note that we can get x_2 from x_1 by locating $g(x_1)$, sending it horizontally to the line $y = x$, and projecting down to the x-axis (see Figure 3). When we use this process repeatedly, we are faced with one of the situations in Figure 4.

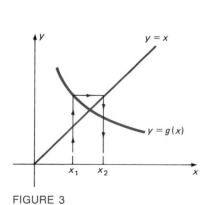

FIGURE 3

What determines success or failure, convergence or divergence? It appears to depend on the slope of the curve $y = g(x)$, namely, $g'(x)$, near the root r. If $|g'(x)|$ is too large, the method fails; if $|g'(x)|$ is small enough, the method works. Here is a general result.

Method works

Method fails

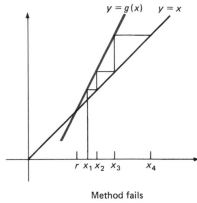

Method works

Method fails

FIGURE 4

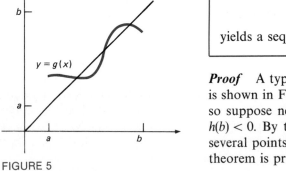

FIGURE 5

Theorem A

(Fixed-Point Theorem). Let g be a continuous function taking $[a, b]$ into itself—that is, which satisfies $a \leq g(x) \leq b$ whenever $a \leq x \leq b$. Then g has at least one fixed point r on $[a, b]$. If, in addition, g is differentiable and satisfies $|g'(x)| \leq M < 1$ for all x in $[a, b]$, M a constant, then the fixed point is unique and the algorithm

$$x_{n+1} = g(x_n), \qquad x_1 \text{ in } [a, b]$$

yields a sequence which converges to r as $n \to \infty$.

Proof A typical graph of a continuous function mapping $[a, b]$ into $[a, b]$ is shown in Figure 5. If either $g(a) = a$ or $g(b) = b$, we have our fixed point, so suppose neither is true. Let $h(x) = g(x) - x$ and note that $h(a) > 0$ and $h(b) < 0$. By the Intermediate Value Theorem, there is a point r (possibly several points) such that $h(r) = 0$, that is, $r = g(r)$. The first assertion of our theorem is proved.

CHAPTER 10/Numerical Methods, Approximations

Next suppose $|g'(x)| \le M < 1$ for all x in $[a, b]$ and let r be a fixed point of g. By the Mean Value Theorem for Derivatives, we may write

$$g(x) - g(r) = g'(c)(x - r)$$

with c some point between x and r. Thus,

$$|g(x) - g(r)| = |g'(c)| |x - r| \le M|x - r|$$

Applying this inequality successively to $x_1, x_2, \ldots$ yields

$$|x_2 - r| = |g(x_1) - g(r)| \le M|x_1 - r|$$
$$|x_3 - r| = |g(x_2) - g(r)| \le M|x_2 - r| \le M^2|x_1 - r|$$
$$|x_4 - r| = |g(x_3) - g(r)| \le M|x_3 - r| \le M^3|x_1 - r|$$
$$\vdots$$
$$|x_n - r| = |g(x_{n-1}) - g(r)| \le M|x_{n-1} - r| \le M^{n-1}|x_1 - r|$$

Since $M^{n-1} \to 0$ as $n \to \infty$, we conclude that $x_n \to r$ as $n \to \infty$.

Finally, if r and s are two fixed points of g, we have just shown that $x_n \to r$ and $x_n \to s$ as $n \to \infty$. This is impossible unless $r = s$. Thus there is only one fixed point. ∎

Now we can understand the behavior in Example 2. If $g(x) = 2 \cos x$, then $|g'(x)| = |-2 \sin x|$, which is greater than 1 in a neighborhood of the fixed point $x \approx 1.03$. On the other hand, if $g(x) = (x + 2 \cos x)/2$, then $|g'(x)| = |\frac{1}{2} - \sin x| < 1$ near $x = 1.03$. We should not expect convergence in the first case; we can guarantee it in the second.

In Example 3, $g(x) = (x^3 + 3)/6$ and $g'(x) = x^2/2$. Clearly $|g'(x)| \le 1$ near the fixed point $x \approx 0.48$. We can be confident of convergence so long as we choose x_1 on the interval where $|g'(x)| \le 1$. Actually, our experiments in Example 3 showed that we can start as far away as $x_1 = 2.2$ (which is more than we have a right to expect), but $x_1 = 2.7$ is too far away.

One final remark: The closer $|g'(x)|$ is to zero near the root, the faster will be the convergence of the fixed-point algorithm.

EXAMPLE 4 The equation $x^3 - 3x + 1 = 0$ has three real roots (see Figure 6). Use the fixed-point method to find the root between 1 and 2.

Solution Proceeding as in Example 3, we write

$$x = \frac{x^3 + 1}{3} = g(x)$$

but unfortunately $g'(x) = x^2 \ge 1$ on $[1, 2]$. Another way to write the equation is

$$x = -\frac{1}{x^2 - 3} = g(x)$$

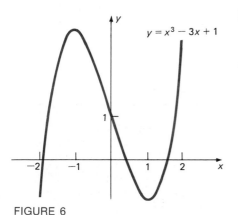

$y = x^3 - 3x + 1$

FIGURE 6

n	x_n
1	1.5
2	1.5555556
3	1.5153061
⋮	⋮
21	1.5320411
22	1.5321234
⋮	⋮
33	1.5320871
34	1.5320902
⋮	⋮
43	1.5320888
44	1.5320889
45	1.5320889

Now $g'(x) = 2x/(x^2 - 3)^2$. This too can be greater than 1 on the interval $[1, 2]$ (for example, $g'(1.5) \approx 5.33$). But there are still other possibilities. Consider

$$x = \frac{3}{x} - \frac{1}{x^2} = g(x)$$

for which $g'(x) = (-3x + 2)/x^3$. After a bit of work (we study g''), we see that g' is increasing on $[1, 2]$, varying from $g'(1) = -1$ to $g'(2) = -\frac{1}{2}$. Thus, $|g'(x)|$ is strictly less than 1 as long as we stay strictly away from the left endpoint $x = 1$.

With the algorithm

$$x_{n+1} = \frac{3}{x_n} - \frac{1}{x_n^2}$$

we obtain the data in the table in the margin. ∎

PROBLEM SET 10.5

C   In Problems 1–4, use the fixed-point algorithm with x_1 as indicated to solve the equations to five decimal places.

1. $x = \dfrac{1}{10} e^{-2x}; \; x_1 = 1$

2. $x = 3 \tan^{-1} x; \; x_1 = 2$

3. $x = \sqrt{2.5 + x}; \; x_1 = 1$

4. $x = \sqrt{3 + x}; \; x_1 = 50$

C   5. Consider the equation $x = 2(x - x^2) = g(x)$.
(a) Sketch the graphs of $y = x$ and $y = g(x)$ using the same coordinate system, and thereby approximately locate the positive root of $x = g(x)$.
(b) Try solving the equation by the fixed-point algorithm starting with $x_1 = 0.6$.
(c) Solve the equation algebraically.
(d) Find $g'(x)$ and evaluate it at the root.

C   6. Follow the direction of Problem 5 for $x = 4(x - x^2) = g(x)$. Explain your results.

C   7. Consider the equation $x = 2 \sin \pi x = g(x)$.
(a) Sketch the graphs of $y = x$ and $y = g(x)$.
(b) Try solving the equation by the fixed-point algorithm.
(c) Find $g'(x)$ and use it to explain your results.

C   8. Follow the directions of Problem 7 for $x = \frac{1}{2} \sin \pi x = g(x)$.

C   9. Consider $x = 2 \sin \pi x$ of Problem 7 again.
(a) Show that it can be rewritten as $x = \frac{5}{6}x + \frac{1}{3} \sin \pi x = g(x)$.
(b) Now solve the latter equation by the fixed-point algorithm.
(c) Why is the convergence so rapid? *Hint*: Evaluate $g'(x)$ at the root.

C   10. Consider $x = 4(x - x^2)$ of Problem 6.
(a) Rewrite this equation so that the fixed-point algorithm will converge (see Problem 9).
(b) Use this algorithm to solve the equation.

C   11. Find the positive root of $x^3 - x^2 - x - 1 = 0$. *Hint*: See Example 4.

12. Evaluate

$$\sqrt{5 + \sqrt{5 + \sqrt{5 + \cdots}}}$$

Hint: Consider solving $x = \sqrt{5 + x}$ by the fixed-point algorithm, starting with $x_1 = 0$. What are x_2, x_3, x_4, and x_5? Are you sure that this fixed-point algorithm converges? To what does it converge? (You can find this number algebraically if you wish.)

C   13. Note that $\sqrt{a}$ is a solution of $x = \frac{1}{2}(x + a/x) = g(x)$. Use the fixed-point algorithm to find $\sqrt{\pi}$. Calculate $g'(a)$ to see why the convergence is so rapid.

C   14. Kepler's equation $x = m + E \sin x$ is important in astronomy. Use the fixed-point algorithm to solve this equation when $m = 0.8$ and $E = 0.2$.

C   15. If an item selling today for P dollars is purchased on the installment plan with monthly payments of R dollars at the end of each of the next k months with interest at the rate of i per month, then

$$P = \frac{R}{i}\left[1 - (1 + i)^{-k}\right]$$

(a) A new car costing $10,000 is purchased on the installment plan with payments of R dollars at the end of each of the next 48 months and stated interest of 18% (which means $i = 0.18/12 = 0.015$). Use algebra to find R.

(b) Suppose the monthly payment in (a) is $300. Then what is i? *Hint*: Use the fixed-point algorithm with $i_1 = 0.015$.

C **16.** A television set costing $500 is purchased on the installment plan with payments of $30 at the end of each of the next 24 months. What is i, the monthly interest rate? See Problem 15.

C **17.** Consider the equation $x = x - f(x)/f'(x)$ and suppose $f'(x) \neq 0$ in an interval $[a, b]$.

(a) Show that if r is in $[a, b]$, then r is a root if and only if $f(r) = 0$.
(b) Show that Newton's Method is a special case of the fixed-point method, in which $g'(r) = 0$.

C **18.** Sketch graphs to convince yourself that each of the following equations has a unique solution. Decide whether the Fixed-Point Method will work and, if so, use it. Otherwise, solve by Newton's Method.

(a) $\sin^{-1} x = \dfrac{1}{\sin x}$ (b) $\cos^{-1} x = \dfrac{1}{\cos x}$

(c) $\tan^{-1} x = \dfrac{1}{\tan x}$

The iterative method that we have called the Fixed-Point Method leads to an area of contemporary research and serves as a possible model for turbulence, one of the least-understood phenomena in science. The remaining problems will introduce you to this exciting area, but you will need a programmable calculator or a computer to carry out the required calculations. Each problem deals with the equation

(*) $\qquad x = f(x) = \lambda(x - x^2)$

which was introduced in Problems 5 and 6 for $\lambda = 2$ and $\lambda = 4$. To find its positive root, consider the corresponding fixed-point iteration

$$x_{n+1} = \lambda(x_n - x_n^2)$$

as we gradually increase λ from 2 to 4.

19. ($\lambda = 2.5$) Sketch $y = x$ and $y = 2.5(x - x^2)$ using the same axes and solve equation (*) by iteration.

20. Solve equation (*) for $\lambda = 2.5$ by simple algebra, thus confirming your answer to Problem 19.

21. ($\lambda = 3.1$) Sketch $y = x$ and $y = f(x) = 3.1(x - x^2)$ using the same axes and attempt to solve (*) by iteration (note that $|f'(x)| > 1$ at the root). You will find that x_n bounces back and forth but gets closer and closer to two values r_1 and r_2, called *attractors*. Find r_1 and r_2 to five decimal places. Superimpose the graph of $y = g(x) = f(f(x)) = 9.61x - 39.40x^2 + 59.58x^3 - 29.79x^4$ on your earlier graph and observe that r_1 and r_2 appear to be the two roots of $x = g(x)$, where $|g'(x)| < 1$.

22. Note that $f(r_1) = r_2$ and $f(r_2) = r_1$. Use this to show that $g'(r_1) = g'(r_2)$.

23. ($\lambda = 3.5$) In this case, use the iteration to obtain four attractors, s_1, s_2, s_3, and s_4. Guess to what equation they are the solutions.

24. ($\lambda = 3.56$) Use the iteration to get eight attractors.

25. ($\lambda = 3.57$) As you keep increasing λ by smaller and smaller amounts, you will double the number of attractors at each stage until at approximately 3.57 you should get chaos. Beyond $\lambda = 3.57$, other strange things happen, which you may want to investigate.

26. Try similar experiments on $x = \lambda \sin \pi x$ (see Problems 7 and 8).

For very readable accounts of this strange phenomenon, see:

Robert M. May, *Simple mathematical models with very complicated dynamics*, *Nature*, vol. 261 (1976), 459–467.

Douglas R. Hofstadter, *Strange attractors: mathematical patterns delicately poised between order and chaos*, *Scientific American*, vol. 245 (November 1981), 22–43.

10.6 Chapter Review

TRUE-FALSE QUIZ

Respond with true or false to each of the following assertions. Be prepared to defend your answer.

1. If $P(x)$ is the Maclaurin polynomial of order 2 for $f(x)$, then $P(0) = f(0)$, $P'(0) = f'(0)$, and $P''(0) = f''(0)$.

2. The Taylor polynomial of order n based at a for $f(x)$ is unique; that is, $f(x)$ has only one such polynomial.

3. $f(x) = x^{5/2}$ has a second-order Maclaurin polynomial.

4. The Maclaurin polynomial of order 3 for $f(x) = 2x^3 - x^2 + 7x - 11$ is an exact representation of $f(x)$.

5. The Maclaurin polynomial of order 16 for $\cos x$ involves only even powers of x.

6. If $f'(0)$ exists for an even function, then $f'(0) = 0$.

7. Taylor's Formula with Remainder contains the Mean Value Theorem for Derivatives as a special case.

8. With a calculator and the formula $[f(a + h) - f(a)]/h$ one can approximate $f'(a)$ to any desired degree of accuracy by taking h small enough.

9. We can always express the indefinite integral of an elementary function in terms of elementary functions.

10. The Trapezoidal Rule with $n = 10$ will give a value for $\int_0^5 x^3 \, dx$ that is smaller than the true value.

11. The Parabolic Rule with $n = 10$ will give the exact value of $\int_0^5 x^3 \, dx$.

12. With a computer and the Parabolic Rule one can always approximate $\int_a^b f(x) \, dx$ to any desired degree of accuracy by taking h small enough.

13. The function $f(x) = e^{-x^2} + x^2 + \sin(x + 1)$ satisfies $|f(x)| \leq 6$ on $[-1, 2]$.

14. If $f(x) = ax^2 + bx + c$, then $\int_{-2}^2 f(x) \, dx = \frac{2}{3}[f(-2) + 4f(0) + f(2)]$.

15. If f is continuous on $[a, b]$ and $f(a)f(b) < 0$, then $f(x) = 0$ has a root between a and b.

16. One of the virtues of the Bisection Method is its rapid convergence.

17. Newton's Method would converge for the function $f(x) = x^{1/3}$.

18. If $f'(x) > 1$ on an open interval containing a root r of $x = f(x)$, then the Fixed-Point Method will fail to converge to r (unless the first guess happened to be r).

19. The Fixed-Point Method will work to find the largest root of $x = 5(x - x^2) + 0.01$.

20. The Fixed-Point Method will certainly converge for $x = \frac{1}{2}\left(x + \frac{a}{x}\right)$ if $a > 0$ and the first guess is greater than $\sqrt{a/3}$.

SAMPLE TEST PROBLEMS

[C] **1.** Use Horner's Method to calculate $p(2.31)$ if $p(x) = 3x^4 - 2x^3 - 5x^2 + 7x - 3$.

2. Find the Maclaurin polynomial of order 4 for $f(x)$ and use it to approximate $f(0.1)$.

(a) $f(x) = xe^x$ (b) $f(x) = \cosh x$

3. Find the Taylor polynomial of order 3 based at 2 for $g(x) = x^3 - 2x^2 + 5x - 7$ and show that it is an exact representation of $g(x)$.

4. Use the result of Problem 3 to calculate $g(2.1)$.

5. Find the Taylor polynomial of order 4 based at 1 for $f(x) = 1/(x + 1)$.

6. Obtain an expression for the error term $R_4(x)$ in Problem 5 and estimate its size if $x = 1.2$.

7. Find the Maclaurin polynomial of order 4 for $f(x) = \sin^2 x = \frac{1}{2}(1 - \cos 2x)$ and estimate the error $R_4(x)$ if $|x| \leq 0.2$. *Note*: A sharper estimate is obtained if you observe that $R_4(x) = R_5(x)$ and then bound $R_5(x)$.

[C] **8.** If $f(x) = \ln x$, then $f^{(n)}(x) = (-1)^{n-1}(n-1)!/x^n$. Thus, the Taylor polynomial of order n based at 1 for $\ln x$ is

$$\ln x = (x - 1) - \frac{1}{2}(x - 1)^2 + \frac{1}{3}(x - 1)^3 + \cdots$$

$$+ \frac{(-1)^{n-1}}{n}(x - 1)^n + R_n(x)$$

How large would n have to be for us to know $|R_n(x)| \leq 0.00005$ if $0.8 \leq x \leq 1.2$?

[C] **9.** Refer to Problem 8. Use the Taylor polynomial of order 5 based at 1 to find

$$\int_{0.8}^{1.2} \ln x \, dx$$

and give an estimate for the error that is made.

[C] **10.** Use the Trapezoidal Rule with $n = 8$ to calculate

$$\int_{0.8}^{1.2} \ln x \, dx$$

and give an error estimate.

C 11. Use the Parabolic Rule with $n = 8$ to calculate

$$\int_{0.8}^{1.2} \ln x \, dx$$

and give an error estimate.

C 12. Compute

$$\int_{0.8}^{1.2} \ln x \, dx$$

using the Fundamental Theorem of Calculus. *Hint*: $D_x[x \ln x - x] = \ln x$.

C 13. Use Newton's Method to solve $3x - \cos 2x = 0$ accurate to six decimal places. Use $x_1 = 0.5$.

C 14. Use the Fixed-Point Method to solve $3x - \cos 2x = 0$, starting with $x_1 = 0.5$.

C 15. Use Newton's Method to find the solution of $x - \tan x = 0$ in the interval $(\pi, 2\pi)$ accurate to four decimal places. *Hint*: Sketch graphs of $y = x$ and $y = \tan x$ using the same axes to get a good initial guess for x_1.

C 16. Try the Fixed-Point Method on the equation of Problem 15. Why doesn't it work?

C 17. Use Newton's Method to find the largest solution of $e^x - \sin x = 0$. *Hint*: Begin by sketching $y = e^x$ and $y = \sin x$ to get an initial guess, x_1.

11 Infinite Series

It is true that a mathematician, who is not somewhat of a poet, will never be a perfect mathematician.
K. Weierstrass

Karl Weierstrass
1815–1897

Many names are associated with infinite series. Newton, Leibniz, the Bernoullis, Taylor, Maclaurin, Euler, and Lagrange used and misused series in their work. Perhaps no other subject has led to more mathematical controversy, this because all the early mathematicians failed to distinguish carefully between convergent and divergent series. In fact, Cauchy (1789–1857) was the first person to give a precise definition of convergence and to prove a number of the convergence tests in this chapter. Still later, Karl Weierstrass developed a complete theory of series of functions and established the legitimacy of such operations as integrating and differentiating term by term.

A German by birth, Weierstrass studied law at the University of Bonn but failed to get a degree (partly because of his beer-drinking antics). He did pass the state examinations for teachers and for 15 years taught such subjects as writing and gymnastics, while studying mathematics at night. From an obscure position in a small town, he was soon doing work in mathematics comparable to the best in Europe. The few results he published brought him an invitation to teach first at the Technical Institute in Berlin and then at the University of Berlin. From there his influence spread throughout the mathematical world.

Weierstrass was a careful methodical thinker. He insisted on complete rigor in all of mathematics and set a standard which is admired and imitated to this day.

11.1
Infinite Sequences

The word *sequence* was introduced in Chapter 5 and reappeared in Chapter 10. In simple language, a sequence

$$a_1, a_2, a_3, a_4, \ldots$$

is an ordered arrangement of numbers, one for each positive integer. More formally, an **infinite sequence** is a function whose domain is the set of positive integers. We may denote a sequence by $a_1, a_2, a_3, \ldots$, by $\{a_n\}_{n=1}^{\infty}$, or simply by $\{a_n\}$. Occasionally, we will extend the notion slightly by allowing the domain to consist of all integers greater than or equal to a specified integer, as in $b_0, b_1, b_2, \ldots$ and $c_8, c_9, c_{10}, \ldots$, which are also denoted by $\{b_n\}_{n=0}^{\infty}$ and $\{c_n\}_{n=8}^{\infty}$.

A sequence may be specified by giving enough initial terms to establish a pattern, as in

$$1, 4, 7, 10, 13, \ldots$$

by an **explicit formula** for the nth term, as in

$$a_n = 3n - 2, \qquad n \geq 1$$

or by a **recursion formula**

$$a_n = a_{n-1} + 3, \qquad n \geq 2, a_1 = 1$$

Note that each of our three illustrations describes the same sequence. Here are four more explicit formulas and the first few terms of the sequences they generate.

(1) $a_n = 1 - \dfrac{1}{n},$ $n \geq 1$: $0, \dfrac{1}{2}, \dfrac{2}{3}, \dfrac{3}{4}, \dfrac{4}{5}, \ldots$

(2) $b_n = 1 + (-1)^n \dfrac{1}{n},$ $n \geq 1$: $0, \dfrac{3}{2}, \dfrac{2}{3}, \dfrac{5}{4}, \dfrac{4}{5}, \dfrac{7}{6}, \dfrac{6}{7}, \ldots$

(3) $c_n = (-1)^n + \dfrac{1}{n},$ $n \geq 1$: $0, \dfrac{3}{2}, \dfrac{-2}{3}, \dfrac{5}{4}, \dfrac{-4}{5}, \dfrac{7}{6}, \dfrac{-6}{7}, \ldots$

(4) $d_n = 0.999,$ $n \geq 1$: $0.999, 0.999, 0.999, 0.999, \ldots$

FIGURE 1

CONVERGENCE Consider the four sequences defined above. Each of them has values that pile up near 1 (see the diagrams in Figure 1). But do they all *converge* to 1? The correct response is that sequences $\{a_n\}$ and $\{b_n\}$ converge to 1, but $\{c_n\}$ and $\{d_n\}$ do not.

For a sequence to converge to 1 means first that values of the sequence should get close to 1. But they must do more than get close; they must remain close, which rules out sequence $\{c_n\}$. And close means arbitrarily close, that is, within *any* specified degree of accuracy, which rules out sequence $\{d_n\}$. While sequence $\{d_n\}$ does not converge to 1, it is correct to say that it converges to 0.999. Sequence $\{c_n\}$ does not converge at all; we say it diverges.

Here is a formal definition; it should sound vaguely familiar.

Definition

We say that the sequence $\{a_n\}$ **converges** to L or has limit L and write

$$\lim_{n \to \infty} a_n = L$$

if for each positive number ε, there is a corresponding positive number N such that

$$n \geq N \quad \Rightarrow \quad |a_n - L| < \varepsilon$$

A sequence that fails to converge to any finite number L is said to **diverge**, or to be divergent.

To see a relationship with something studied earlier (Section 4.6), consider graphing $a_n = 1 - 1/n$ and $a(x) = 1 - 1/x$. The only difference is that in the sequence case, the domain is restricted to the positive integers. In the first case, we write $\lim_{n \to \infty} a_n = 1$; in the second, $\lim_{x \to \infty} a(x) = 1$. Note the interpretation of ε and N in the diagrams in Figure 2.

EXAMPLE 1 Show that if p is a positive integer, then

$$\lim_{n \to \infty} \frac{1}{n^p} = 0$$

$$n \geqslant N \Rightarrow |a_n - 1| < \epsilon$$

$$x \geqslant N \Rightarrow |a(x) - 1| < \epsilon$$

FIGURE 2

Solution This is almost obvious from earlier work, but we can give a formal demonstration. Let an arbitrary $\varepsilon > 0$ be given. Choose N to be any number greater than $\sqrt[p]{1/\varepsilon}$. Then $n \geq N$ implies

$$|a_n - L| = \left| \frac{1}{n^p} - 0 \right| = \frac{1}{n^p} \leq \frac{1}{N^p} < \frac{1}{(\sqrt[p]{1/\varepsilon})^p} = \varepsilon \qquad \blacksquare$$

All the familiar limit theorems hold for sequences. We state them without proof.

Theorem A

Let $\{a_n\}$ and $\{b_n\}$ be convergent sequences and k a constant. Then:

1. $\lim\limits_{n \to \infty} k = k;$
2. $\lim\limits_{n \to \infty} ka_n - k \lim\limits_{n \to \infty} a_n;$
3. $\lim\limits_{n \to \infty} (a_n \pm b_n) = \lim\limits_{n \to \infty} a_n \pm \lim\limits_{n \to \infty} b_n;$
4. $\lim\limits_{n \to \infty} (a_n \cdot b_n) = \lim\limits_{n \to \infty} a_n \cdot \lim\limits_{n \to \infty} b_n;$
5. $\lim\limits_{n \to \infty} \dfrac{a_n}{b_n} = \dfrac{\lim\limits_{n \to \infty} a_n}{\lim\limits_{n \to \infty} b_n}$ provided $\lim\limits_{n \to \infty} b_n \neq 0.$

EXAMPLE 2 Find $\lim\limits_{n \to \infty} \dfrac{3n^2}{7n^2 + 1}$.

Solution To decide what is happening to a quotient of two polynomials in n as n gets large, it is wise to divide numerator and denominator by the largest power of n that occurs in the denominator. This justifies our first step below; the others are justified by appealing to statements from Theorem A as indicated by the circled numbers.

$$\lim_{n \to \infty} \frac{3n^2}{7n^2 + 1} = \lim_{n \to \infty} \frac{3}{7 + (1/n^2)}$$

$$\overset{\text{⑤}}{=} \frac{\lim\limits_{n \to \infty} 3}{\lim\limits_{n \to \infty} [7 + (1/n^2)]}$$

$$\overset{\text{③}}{=} \frac{\lim\limits_{n \to \infty} 3}{\lim\limits_{n \to \infty} 7 + \lim\limits_{n \to \infty} 1/n^2}$$

$$\overset{\text{①}}{=} \frac{3}{7 + \lim\limits_{n \to \infty} 1/n^2} = \frac{3}{7 + 0} = \frac{3}{7}$$

By this time, the limit theorems are so familiar that we will normally jump directly from the first step to the final result. $\qquad \blacksquare$

EXAMPLE 3 Does the sequence $\{(\ln n)/e^n\}$ converge, and if so, to what number?

Solution Here and in many sequence problems, it is convenient to use the following almost obvious fact.

$$\text{If } \lim_{x \to \infty} f(x) = L, \text{ then } \lim_{n \to \infty} f(n) = L.$$

This is convenient because we can apply l'Hôpital's Rule to the continuous variable problem. In particular, by l'Hôpital's Rule,

$$\lim_{x \to \infty} \frac{\ln x}{e^x} = \lim_{x \to \infty} \frac{1/x}{e^x} = 0$$

Thus,

$$\lim_{n \to \infty} \frac{\ln n}{e^n} = 0$$

that is, $\{(\ln n)/e^n\}$ converges to 0. ∎

Here's another theorem that we've seen before in a slightly different guise (Theorem 2.6C).

Theorem B

(Squeeze Theorem). Suppose $\{a_n\}$ and $\{c_n\}$ both converge to L and that $a_n \le b_n \le c_n$ for $n \ge K$ (K a fixed integer). Then $\{b_n\}$ also converges to L.

EXAMPLE 4 Show that $\lim\limits_{n \to \infty} \dfrac{\sin^3 n}{n} = 0$.

Solution For $n \ge 1$, $-1/n \le (\sin^3 n)/n \le 1/n$. Since $\lim\limits_{n \to \infty}(-1/n) = 0$ and $\lim\limits_{n \to \infty}(1/n) = 0$, the result follows by the Squeeze Theorem. ∎

Theorem C

If $\lim\limits_{n \to \infty} |a_n| = 0$, then $\lim\limits_{n \to \infty} a_n = 0$.

Proof Since $-|a_n| \le a_n \le |a_n|$, the result follows from the Squeeze Theorem. ∎

What happens to the numbers in the sequence $\{0.999^n\}$ as $n \to \infty$? We suggest that you calculate 0.999^n for $n = 10, 100, 1000$, and $10,000$ on your calculator to make a good guess. Then note the following example.

EXAMPLE 5 Show that if $-1 < r < 1$, then $\lim\limits_{n \to \infty} r^n = 0$.

Solution   If $r = 0$, the result is trivial, so suppose otherwise. Then $1/|r| > 1$, and so $1/|r| = 1 + p$ for some number $p > 0$. By the Binomial Theorem,

$$\frac{1}{|r|^n} = (1 + p)^n = 1 + pn + \text{(positive terms)} \geq pn$$

Thus,

$$0 \leq |r|^n \leq \frac{1}{pn}$$

Since $\lim\limits_{n \to \infty}(1/pn) = (1/p)\lim\limits_{n \to \infty}(1/n) = 0$, it follows from the Squeeze Theorem that $\lim\limits_{n \to \infty}|r|^n = 0$, or—equivalently—$\lim\limits_{n \to \infty}|r^n| = 0$. By Theorem C, $\lim\limits_{n \to \infty} r^n = 0$. ∎

What if $r > 1$; for example, $r = 1.5$? Then r^n will march off toward ∞. In this case, we write

$$\lim_{n \to \infty} r^n = \infty, \qquad r > 1$$

However, we say the sequence $\{r^n\}$ diverges. To converge, it would have to approach a *finite* limit. The sequence $\{r^n\}$ also diverges when $r \leq -1$.

MONOTONIC SEQUENCES   Consider now an arbitrary **nondecreasing sequence** $\{a_n\}$, by which we mean $a_n \leq a_{n+1}$, $n \geq 1$. One example is the sequence $a_n = n^2$; another is $a_n = 1 - 1/n$. If you think about it a little, you may convince yourself that such a sequence can only do one of two things. Either it marches off to infinity or, if it cannot do that because it is bounded above, then it must bump against a lid (see Figure 3). Here is the formal statement of this very important result.

$$\lim_{n \to \infty} a_n = A$$

FIGURE 3

Theorem D

(Monotonic Sequence Theorem). If U is an upper bound for a nondecreasing sequence $\{a_n\}$, then the sequence converges to a limit A that is less than or equal to U. Similarly, if L is a lower bound for a nonincreasing sequence $\{b_n\}$, then the sequence $\{b_n\}$ converges to a limit B that is greater than or equal to L.

The phrase **monotonic sequence** is used to describe either a nondecreasing or nonincreasing sequence; hence the name for this theorem.

Theorem D describes a very deep property of the real number system. It is equivalent to the *completeness property* of the real numbers, which in simple language says that the real line has no "holes" in it (see Problems 47–48). It is this property that distinguishes the real number line from the rational number line (which is full of holes). A great deal more could be said about this topic; we hope Theorem D appeals to your intuition and that you will accept it on faith until you take a more advanced course.

We make one more comment about Theorem D. It is not necessary that the sequences $\{a_n\}$ and $\{b_n\}$ be monotonic initially, only that they be monotonic from some point on—that is, for $n \geq K$. In fact, *the convergence or divergence of a sequence does not depend on the character of the initial terms but rather on what is true for large n.*

EXAMPLE 6   Show that the sequence $b_n = n^2/2^n$ converges by using Theorem D.

Solution   The first few terms of this sequence are

$$\frac{1}{2}, 1, \frac{9}{8}, 1, \frac{25}{32}, \frac{36}{64}, \frac{49}{128}, \cdots$$

For $n \geq 3$ the sequence appears to be decreasing $(b_n > b_{n+1})$, a fact that we now establish. Each of the following inequalities is equivalent to the others.

$$\frac{n^2}{2^n} > \frac{(n+1)^2}{2^{n+1}}$$

$$n^2 > \frac{(n+1)^2}{2}$$

$$2n^2 > n^2 + 2n + 1$$

$$n^2 - 2n > 1$$

$$n(n-2) > 1$$

The last inequality is clearly true for $n \geq 3$. Since the sequence is decreasing (a stronger condition than nonincreasing) and is bounded below by zero, the Monotonic Sequence Theorem guarantees that it has a limit.

It would be easy using l'Hôpital's Rule to show that the limit is zero. ∎

PROBLEM SET   11.1

In Problems 1–20, an explicit formula for a_n is given. Write out the first five terms, determine if the sequence converges or diverges, and if it converges, find $\lim\limits_{n \to \infty} a_n$.

1. $a_n = \dfrac{n}{2n-1}$

2. $a_n = \dfrac{3n+1}{n+2}$

3. $a_n = \dfrac{4n^2+1}{n^2-2n+3}$

4. $a_n = \dfrac{3n^2+2}{n+4}$

5. $a_n = \dfrac{n+4}{2n^2+1}$

6. $a_n = \dfrac{\sqrt{n}}{n+1}$

7. $a_n = (-1)^n \dfrac{n}{n+1}$

8. $a_n = \dfrac{n\sin(n\pi/2)}{2n+1}$

9. $a_n = \dfrac{\sin(n\pi/2)}{n}$

10. $a_n = e^{-n}\cos n$

11. $a_n = \dfrac{e^n}{n^2}$

12. $a_n = \dfrac{e^n}{2^n}$

13. $a_n = \dfrac{(-\pi)^n}{4^n}$

14. $a_n = (\tfrac{1}{2})^n + 2^n$

15. $a_n = 1 + (0.9)^n$

16. $a_n = \dfrac{n^3}{e^n}$

17. $a_n = \dfrac{\ln n}{n}$

18. $a_n = \dfrac{\ln(1/n)}{\sqrt{n}}$

19. $a_n = \left(1 + \dfrac{1}{n}\right)^n$ **20.** $a_n = n^{1/n}$

Hint: Theorem 7.5A

In Problems 21–30, find an explicit formula $a_n = \underline{\hspace{1cm}}$ for each sequence, determine if the sequence converges or diverges, and if it converges, find $\lim\limits_{n \to \infty} a_n$.

21. $\frac{1}{2}, \frac{2}{3}, \frac{3}{4}, \frac{4}{5}, \ldots$

22. $\dfrac{1}{2^2}, \dfrac{2}{2^3}, \dfrac{3}{2^4}, \dfrac{4}{2^5}, \ldots$

23. $-1, \frac{2}{3}, -\frac{3}{5}, \frac{4}{7}, -\frac{5}{9}, \ldots$

24. $1, \dfrac{1}{1 - \frac{1}{2}}, \dfrac{1}{1 - \frac{2}{3}}, \dfrac{1}{1 - \frac{3}{4}}, \ldots$

25. $1, \dfrac{2}{2^2 - 1^2}, \dfrac{3}{3^2 - 2^2}, \dfrac{4}{4^2 - 3^2}, \ldots$

26. $\dfrac{1}{2 - \frac{1}{2}}, \dfrac{2}{3 - \frac{1}{3}}, \dfrac{3}{4 - \frac{1}{4}}, \dfrac{4}{5 - \frac{1}{5}}, \ldots$

27. $\sin 1, 2 \sin \frac{1}{2}, 3 \sin \frac{1}{3}, 4 \sin \frac{1}{4}, \ldots$

28. $-\frac{1}{3}, \frac{4}{9}, -\frac{9}{27}, \frac{16}{81}, \ldots$

29. $2, 1, \dfrac{2^3}{3^2}, \dfrac{2^4}{4^2}, \dfrac{2^5}{5^2}, \ldots$

30. $1 - \frac{1}{2}, \frac{1}{2} - \frac{1}{3}, \frac{1}{3} - \frac{1}{4}, \frac{1}{4} - \frac{1}{5}, \ldots$

In Problems 31–36, write the first four terms of the sequence $\{a_n\}$. Then use Theorem D to show that the sequence converges.

31. $a_n = \dfrac{4n - 3}{2^n}$

32. $a_n = \dfrac{n}{n + 1}\left(2 - \dfrac{1}{n^2}\right)$

33. $a_n = \left(1 - \dfrac{1}{4}\right)\left(1 - \dfrac{1}{9}\right)\cdots\left(1 - \dfrac{1}{n^2}\right), \ n \geq 2$

34. $a_n = 1 + \dfrac{1}{2!} + \dfrac{1}{3!} + \cdots + \dfrac{1}{n!}$

35. $a_{n+1} = 1 + \frac{1}{2}a_n, \ a_1 = 1$

36. $a_{n+1} = \dfrac{1}{2}\left(a_n + \dfrac{2}{a_n}\right), \ a_1 = 2$

[C] **37.** Assuming $u_{n+1} = \sqrt{3 + u_n}, \ u_1 = \sqrt{3}$, determines a convergent sequence, find $\lim\limits_{n \to \infty} u_n$ to four decimal places, using a calculator.

38. Show that $\{u_n\}$ of Problem 37 is bounded above and increasing. Conclude from Theorem D that $\{u_n\}$ converges. *Hint*: Use mathematical induction.

39. Find $\lim\limits_{n \to \infty} u_n$ of Problem 37 algebraically. *Hint*: Let $u = \lim\limits_{n \to \infty} u_n$. Then since $u_{n+1} = \sqrt{3 + u_n}$, $u = \sqrt{3 + u}$. Now square both sides and solve for u.

40. Use the technique of Problem 39 to find $\lim\limits_{n \to \infty} a_n$ of Problem 36.

[C] **41.** Assuming $u_{n+1} = 1.1^{u_n}, \ u_1 = 0$, determines a convergent sequence, find $\lim\limits_{n \to \infty} u_n$ to four decimal places using a calculator.

42. Show that $\{u_n\}$ of Problem 41 is increasing and bounded above by 2.

43. Find

$$\lim_{n \to \infty} \sum_{k=1}^{n} \left(\sin \frac{k}{n}\right)\frac{1}{n}$$

Hint: Write an equivalent definite integral.

44. Show that

$$\lim_{n \to \infty} \sum_{k=1}^{n} \left[\frac{1}{1 + \left(\dfrac{k}{n}\right)^2}\right]\frac{1}{n} = \frac{\pi}{4}$$

45. Using the definition of limit, prove that $\lim\limits_{n \to \infty} n/(n + 1) = 1$; that is, for a given $\varepsilon > 0$ find N such that $n \geq N \Rightarrow |n/(n + 1) - 1| < \varepsilon$.

46. As in Problem 45, prove that $\lim\limits_{n \to \infty} n/(n^2 + 1) = 0$.

47. Let $S = \{x : x \text{ rational}, \ x^2 < 2\}$. Convince yourself that S does not have a *least* upper bound in the rational numbers, but does have such a bound in the real numbers.

48. The *completeness property* of the real numbers says that every set of real numbers which is bounded above has a least upper bound, which is a real number. This property is usually taken as an axiom for the real numbers. Prove Theorem D using this property.

49. Prove that if $\lim\limits_{n \to \infty} a_n = 0$ and $\{b_n\}$ is bounded, then $\lim\limits_{n \to \infty} a_n b_n = 0$.

50. Prove that if $\{a_n\}$ converges and $\{b_n\}$ diverges, then $\{a_n + b_n\}$ diverges.

51. If $\{a_n\}$ and $\{b_n\}$ both diverge, does it follow that $\{a_n + b_n\}$ diverges?

52. A famous sequence $\{f_n\}$, called the *Fibonacci Sequence* after Leonardo Fibonacci, who introduced it around A.D. 1200, is defined by the recursion formula

$$f_{n+2} = f_{n+1} + f_n, \qquad f_1 = f_2 = 1$$

(a) Find f_3 through f_{10}.

(b) Let $\phi = \frac{1}{2}(1 + \sqrt{5}) \approx 1.618034$. The Greeks called this number the *golden ratio*, claiming that a rectangle whose dimensions were in this ratio was perfect. It can be shown that

$$f_n = \frac{1}{\sqrt{5}}\left[\left(\frac{1+\sqrt{5}}{2}\right)^n - \left(\frac{1-\sqrt{5}}{2}\right)^n\right] = \frac{1}{\sqrt{5}}\left[\phi^n - (-1)^n \phi^{-n}\right]$$

and you should check that this gives the right result for $n = 1$ and $n = 2$. The general result can be proved by induction (it is a nice challenge). More in line with this section, use this explicit formula to prove that $\lim_{n \to \infty} f_{n+1}/f_n = \phi$.

(c) Show, using the limit just proved, that ϕ satisfies the equation $x^2 - x - 1 = 0$. Then in another interesting twist, use the Quadratic Formula to show that the two roots of this equation are ϕ and $-1/\phi$, the two numbers that occur in the explicit formula for f_n.

53. Consider an equilateral triangle containing $1 + 2 + 3 + \cdots + n = n(n+1)/2$ circles each of diameter 1 and stacked as indicated in Figure 4 for the case $n = 4$. Find $\lim_{n \to \infty} A_n/B_n$, where A_n is the total area of the circles and B_n is the area of the triangle.

FIGURE 4

11.2
Infinite Series

In a famous paradox announced some 2400 years ago, Zeno of Elea said that a runner cannot finish a race because he must first cover half the distance, then half the remaining distance, then half the still remaining distance, and so on, forever. Since the runner's time is finite, he cannot traverse the infinite number of segments of the course. Yet we all know that runners do finish races.

Imagine a race course to be 1 mile long. The segments of Zeno's argument would then have length $\frac{1}{2}$ mile, $\frac{1}{4}$ mile, $\frac{1}{8}$ mile, and so on (Figure 1). In mathematical language, finishing the race would amount to evaluating the sum

$$\frac{1}{2} + \frac{1}{4} + \frac{1}{8} + \frac{1}{16} + \frac{1}{32} + \cdots$$

which might seem impossible. But wait. Up to now, the word *sum* has been defined only for the addition of a finite number of terms. The indicated "infinite sum" has, as yet, no meaning for us.

Consider the partial sums

FIGURE 1

$$S_1 = \frac{1}{2}$$

$$S_2 = \frac{1}{2} + \frac{1}{4} = \frac{3}{4}$$

$$S_3 = \frac{1}{2} + \frac{1}{4} + \frac{1}{8} = \frac{7}{8}$$

$$S_n = \frac{1}{2} + \frac{1}{4} + \frac{1}{8} + \cdots + \frac{1}{2^n} = 1 - \frac{1}{2^n}$$

Clearly these partial sums get increasingly close to 1. In fact

$$\lim_{n \to \infty} S_n = \lim_{n \to \infty} \left(1 - \frac{1}{2^n}\right) = 1$$

This we define to be the value of the infinite sum.

More generally, consider

$$a_1 + a_2 + a_3 + a_4 + \cdots$$

which is also denoted by $\sum_{k=1}^{\infty} a_k$ or $\sum a_k$ and is called an **infinite series** (or series for short). Then S_n, the **nth partial sum**, is given by

$$S_n = a_1 + a_2 + a_3 + \cdots + a_n = \sum_{k=1}^{n} a_k$$

We make the following formal definition.

Definition

The infinite series $\sum_{k=1}^{\infty} a_k$ **converges** and has sum S if the sequence of partial sums $\{S_n\}$ converges to S. If $\{S_n\}$ diverges, then the series **diverges**. A divergent series has no sum.

GEOMETRIC SERIES   A series of the form

$$\sum_{k=1}^{\infty} ar^{k-1} = a + ar + ar^2 + ar^3 + \cdots$$

where $a \neq 0$ is called a **geometric series**.

EXAMPLE 1   Show that a geometric series converges with sum $S = a/(1 - r)$ if $|r| < 1$, but diverges if $|r| \geq 1$.

Solution   Let $S_n = a + ar + ar^2 + \cdots + ar^{n-1}$. If $r = 1$, $S_n = na$, which grows without bound, and so $\{S_n\}$ diverges. If $r \neq 1$, we may write

$$S_n - rS_n = (a + ar + \cdots + ar^{n-1}) - (ar + ar^2 + \cdots + ar^n) = a - ar^n$$

and so

$$S_n = \frac{a - ar^n}{1 - r} = \frac{a}{1 - r} - \frac{a}{1 - r} r^n$$

If $|r| < 1$, then $\lim_{n \to \infty} r^n = 0$ (Section 11.1, Example 5) and thus

$$S = \lim_{n \to \infty} S_n = \frac{a}{1 - r}$$

If $|r| > 1$ or $r = -1$, the sequence $\{r^n\}$ diverges, and consequently so does $\{S_n\}$. ∎

EXAMPLE 2 Use the result of Example 1 to sum the following two geometric series.

(a) $\dfrac{4}{3} + \dfrac{4}{9} + \dfrac{4}{27} + \dfrac{4}{81} + \cdots$

(b) $0.515151\ldots = \dfrac{51}{100} + \dfrac{51}{10,000} + \dfrac{51}{1,000,000} + \cdots$

Solution

(a)
$$S = \frac{a}{1-r} = \frac{\frac{4}{3}}{1 - \frac{1}{3}} = \frac{\frac{4}{3}}{\frac{2}{3}} = 2$$

(b)
$$S = \frac{\frac{51}{100}}{1 - \frac{1}{100}} = \frac{\frac{51}{100}}{\frac{99}{100}} = \frac{51}{99} = \frac{17}{33}$$

Incidently, the procedure in (b) shows that any repeating decimal represents a rational number. ∎

EXAMPLE 3 The diagram in Figure 2 represents an equilateral triangle containing infinitely many circles, tangent to the triangle and each other and reaching into the corners. What fraction of the area of the triangle is occupied by the circles?

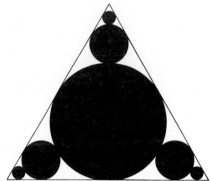

FIGURE 2

Solution Suppose for convenience that the triangle has sides of length $2\sqrt{3}$, which gives it an altitude of 3. Concentrate attention on the vertical stack of circles. With a bit of geometric reasoning (the center of the large circle is two-thirds of the way from the upper vertex to the base), we see that the radii of these circles are $1, \frac{1}{3}, \frac{1}{9}, \ldots$ and conclude that the vertical stack has area

$$\pi\left[1^2 + \left(\frac{1}{3}\right)^2 + \left(\frac{1}{9}\right)^2 + \left(\frac{1}{27}\right)^2 + \cdots \right]$$

$$= \pi\left[1 + \frac{1}{9} + \frac{1}{81} + \frac{1}{729} + \cdots \right] = \pi\left[\frac{1}{1 - \frac{1}{9}} \right] = \frac{9\pi}{8}$$

The total area of all the circles is three times this number minus twice the area of the big circle, that is, $27\pi/8 - 2\pi$, or $11\pi/8$. Since the triangle has area $3\sqrt{3}$, the fraction of this area occupied by the circles is

$$\frac{11\pi}{24\sqrt{3}} \approx 0.83$$
∎

Consider the geometric series $a + ar + ar^2 + \cdots + ar^{n-1} + \cdots$ once more. Its nth term a_n is given by $a_n = ar^{n-1}$. Example 1 shows that a geometric series converges if and only if $\lim\limits_{n \to \infty} a_n = 0$. Could this possibly be true of all series? The answer is no, though half of the statement is correct.

> **Theorem A**
>
> (***n*th-Term Test for Divergence**). If the series $\displaystyle\sum_{n=1}^{\infty} a_n$ converges, then $\displaystyle\lim_{n\to\infty} a_n = 0$. Equivalently, if $\displaystyle\lim_{n\to\infty} a_n \neq 0$ (or $\displaystyle\lim_{n\to\infty} a_n$ does not exist), the series diverges.

Proof Let S_n be the nth partial sum and $S = \displaystyle\lim_{n\to\infty} S_n$. Since $a_n = S_n - S_{n-1}$, it follows that

$$\lim_{n\to\infty} a_n = \lim_{n\to\infty} S_n - \lim_{n\to\infty} S_{n-1} = S - S = 0 \qquad \blacksquare$$

EXAMPLE 4 Show that $\displaystyle\sum_{n=1}^{\infty} \frac{n^3}{3n^3 + 2n^2}$ diverges.

Solution

$$\lim_{n\to\infty} a_n = \lim_{n\to\infty} \frac{n^3}{3n^3 + 2n^2} = \lim_{n\to\infty} \frac{1}{3 + 2/n} = \frac{1}{3}$$

Thus by the nth-Term Test, the series diverges. $\qquad \blacksquare$

THE HARMONIC SERIES Students invariably want to turn Theorem A around and make it say that $a_n \to 0$ implies convergence of $\sum a_n$. The **harmonic series**

$$\sum_{n=1}^{\infty} \frac{1}{n} = 1 + \frac{1}{2} + \frac{1}{3} + \cdots + \frac{1}{n} + \cdots$$

shows that this is false. Clearly $\displaystyle\lim_{n\to\infty} a_n = \lim_{n\to\infty} (1/n) = 0$. However, the series diverges, as we now show.

EXAMPLE 5 Show that the harmonic series diverges.

Solution We show that S_n grows without bound. Imagine n to be large and write

$$\begin{aligned}
S_n &= 1 + \frac{1}{2} + \frac{1}{3} + \frac{1}{4} + \frac{1}{5} + \cdots + \frac{1}{n} \\[2mm]
&= 1 + \frac{1}{2} + \left(\frac{1}{3} + \frac{1}{4}\right) + \left(\frac{1}{5} + \frac{1}{6} + \frac{1}{7} + \frac{1}{8}\right) + \left(\frac{1}{9} + \cdots + \frac{1}{16}\right) + \cdots + \frac{1}{n} \\[2mm]
&> 1 + \frac{1}{2} + \frac{2}{4} + \frac{4}{8} + \frac{8}{16} + \cdots + \frac{1}{n} \\[2mm]
&= 1 + \frac{1}{2} + \frac{1}{2} + \frac{1}{2} + \frac{1}{2} + \cdots + \frac{1}{n}
\end{aligned}$$

It is clear that by taking n sufficiently large, we can introduce as many $\frac{1}{2}$'s into the last expression as we wish. Thus, $\{S_n\}$ diverges; hence, so does the harmonic series. ∎

COLLAPSING SERIES   A geometric series is one of the few series where we can actually calculate S_n; a **collapsing series** is another (see Example 2 of Section 5.3).

EXAMPLE 6   Show that the following series converges and find its sum.

$$\sum_{k=1}^{\infty} \frac{1}{(k+2)(k+3)}$$

Solution   Use a partial fraction decomposition to write

$$\frac{1}{(k+2)(k+3)} = \frac{1}{k+2} - \frac{1}{k+3}$$

Then,

$$S_n = \sum_{k=1}^{n} \left(\frac{1}{k+2} - \frac{1}{k+3} \right) = \left(\frac{1}{3} - \frac{1}{4} \right) + \left(\frac{1}{4} - \frac{1}{5} \right) + \cdots + \left(\frac{1}{n+2} - \frac{1}{n+3} \right)$$

$$= \frac{1}{3} - \frac{1}{n+3}$$

Therefore,

$$\lim_{n \to \infty} S_n = \frac{1}{3}$$

The series converges and has sum $\frac{1}{3}$. ∎

PROPERTIES OF CONVERGENT SERIES   Convergent series behave much like finite sums; what you expect to be true usually is true.

Theorem B

(Linearity). If $\displaystyle\sum_{k=1}^{\infty} a_k$ and $\displaystyle\sum_{k=1}^{\infty} b_k$ both converge and c is a constant, then $\displaystyle\sum_{k=1}^{\infty} ca_k$ and $\displaystyle\sum_{k=1}^{\infty} (a_k + b_k)$ also converge and, moreover,

(i) $\displaystyle\sum_{k=1}^{\infty} ca_k = c \sum_{k=1}^{\infty} a_k$;

(ii) $\displaystyle\sum_{k=1}^{\infty} (a_k + b_k) = \sum_{k=1}^{\infty} a_k + \sum_{k=1}^{\infty} b_k$.

Proof   This theorem introduces a subtle shift in language. The symbol $\displaystyle\sum_{k=1}^{\infty} a_k$ is now being used both for the infinite series $a_1 + a_2 + \cdots$ and for the sum of this series, which is a number.

By hypothesis, $\lim\limits_{n \to \infty} \sum\limits_{k=1}^{n} a_k$ and $\lim\limits_{n \to \infty} \sum\limits_{k=1}^{n} b_k$ both exist. Thus, use the properties of sums with finitely many terms and the properties of limits.

(i) $\displaystyle\sum_{k=1}^{\infty} ca_k = \lim_{n \to \infty} \sum_{k=1}^{n} ca_k = \lim_{n \to \infty} c \sum_{k=1}^{n} a_k$

$\qquad\qquad = c \lim_{n \to \infty} \sum_{k=1}^{n} a_k = c \sum_{k=1}^{\infty} a_k$

(ii) $\displaystyle\sum_{k=1}^{\infty} (a_k + b_k) = \lim_{n \to \infty} \sum_{k=1}^{n} (a_k + b_k) = \lim_{n \to \infty} \left[\sum_{k=1}^{n} a_k + \sum_{k=1}^{n} b_k \right]$

$\qquad\qquad = \lim_{n \to \infty} \sum_{k=1}^{n} a_k + \lim_{n \to \infty} \sum_{k=1}^{n} b_k = \sum_{k=1}^{\infty} a_k + \sum_{k=1}^{\infty} b_k$ ■

EXAMPLE 7 Calculate $\displaystyle\sum_{k=1}^{\infty} \left[3(\tfrac{1}{8})^k - 5(\tfrac{1}{3})^k \right]$.

Solution By Theorem B and Example 1,

$$\sum_{k=1}^{\infty} \left[3 \left(\frac{1}{8} \right)^k - 5 \left(\frac{1}{3} \right)^k \right] = 3 \sum_{k=1}^{\infty} \left(\frac{1}{8} \right)^k - 5 \sum_{k=1}^{\infty} \left(\frac{1}{3} \right)^k$$

$$= 3 \frac{\frac{1}{8}}{1 - \frac{1}{8}} - 5 \frac{\frac{1}{3}}{1 - \frac{1}{3}} = \frac{3}{7} - \frac{5}{2} = -\frac{29}{14}$$ ■

Theorem C

If $\displaystyle\sum_{k=1}^{\infty} a_k$ diverges and $c \neq 0$, then $\displaystyle\sum_{k=1}^{\infty} ca_k$ diverges.

We leave the proof of this theorem to you (Problem 35). It implies, for example, that

$$\sum_{k=1}^{\infty} \frac{1}{3k} = \sum_{k=1}^{\infty} \frac{1}{3} \cdot \frac{1}{k}$$

diverges, since we know that the harmonic series diverges.

The Associative Law of Addition allows us to group terms in a finite sum in any way we please. For example,

$$2 + 7 + 3 + 4 + 5 = (2 + 7) + (3 + 4) + 5 = 2 + (7 + 3) + (4 + 5)$$

But notice what happens to the infinite series

$$1 - 1 + 1 - 1 + 1 - 1 + \cdots + (-1)^{n+1} + \cdots$$

when we group terms in two different ways.

$$(1 - 1) + (1 - 1) + (1 - 1) + \cdots = 0 + 0 + 0 + \cdots = 0$$

$$1 - (1 - 1) - (1 - 1) - (1 - 1) - \cdots = 1 - 0 - 0 - 0 - \cdots = 1$$

The original series was divergent (since $\lim_{n \to \infty} a_n \neq 0$), whereas the first series of grouped terms converges with sum 0 and the second series of grouped terms converges with sum 1. The contradictions that can arise from grouping are pretty obvious; fortunately, if the original series converges, there are no problems at all.

Theorem D

(Grouping). The terms of a convergent series can be grouped in any way (provided the order of the terms is maintained) and the new series will converge with the same sum as the original series.

Proof Let $\sum a_n$ be the original convergent series and let $\{S_n\}$ be its sequence of partial sums. If $\sum b_m$ is a series formed by grouping the terms of $\sum a_n$ and if $\{T_m\}$ is its sequence of partial sums, then each T_m is one of the S_n's. For example, T_4 might be

$$T_4 = a_1 + (a_2 + a_3) + (a_4 + a_5 + a_6) + (a_7 + a_8)$$

in which case $T_4 = S_8$. Thus, $\{T_m\}$ is a "subsequence" of $\{S_n\}$. A moment's thought should convince you that if $S_n \to S$, then $T_m \to S$. ∎

PROBLEM SET 11.2

In Problems 1–14, indicate whether the given series converges or diverges. If it converges, find its sum. *Hint*: It may help you to write out the first few terms of the series.

1. $\sum_{k=1}^{\infty} (\frac{1}{5})^k$

2. $\sum_{k=1}^{\infty} (-\frac{1}{3})^{k-1}$

3. $\sum_{k=0}^{\infty} [2(\frac{1}{3})^k + 3(\frac{1}{6})^k]$

4. $\sum_{k=1}^{\infty} [3(\frac{1}{4})^k - 2(\frac{1}{5})^k]$

5. $\sum_{k=1}^{\infty} \frac{k-3}{k}$

6. $\sum_{k=1}^{\infty} (\frac{4}{3})^k$

7. $\sum_{k=1}^{\infty} \left(\frac{1}{k} - \frac{1}{k+1}\right)$. *Hint*: Example 6.

8. $\sum_{k=1}^{\infty} \frac{2}{k}$

9. $\sum_{k=1}^{\infty} \frac{k!}{10^k}$

10. $\sum_{k=3}^{\infty} \frac{2}{(k-1)(k)}$

11. $\sum_{k=3}^{\infty} \left(\frac{e}{\pi}\right)^{k-1}$

12. $\sum_{k=1}^{\infty} \frac{3^{k+1}}{5^{k-1}}$

13. $\sum_{k=1}^{\infty} \left(\frac{3}{(k+1)^2} - \frac{3}{k^2}\right)$

14. $\sum_{k=4}^{\infty} \frac{4}{k-3}$

In Problems 15–20, write the given decimal as an infinite series, then find the sum of the series, and finally use the result to write the decimal as a ratio of two integers (see Example 2).

15. $0.22222\ldots$

16. $0.21212121\ldots$

17. $0.013013013\ldots$

18. $0.125125125\ldots$

19. $0.49999\ldots$

20. $0.36717171\ldots$

21. Evaluate $\sum_{k=0}^{\infty} r(1-r)^k$, $0 < r < 2$.

22. Evaluate $\sum_{k=0}^{\infty} (-1)^k x^k$, $-1 < x < 1$.

23. Show that $\sum_{k=1}^{\infty} \ln \frac{k}{k+1}$ diverges. *Hint*: Obtain a formula for S_n.

24. Show that $\sum_{k=2}^{\infty} \ln\left(1 - \frac{1}{k^2}\right) = -\ln 2$.

25. A ball is dropped from a height of 100 feet. Each time it hits the floor, it rebounds to $\frac{2}{3}$ its previous height. Find the total distance it travels.

26. Three people, A, B, and C, divide an apple as follows. First they divide it in fourths, each taking a quarter. Then they divide the left over quarter in fourths, each taking a quarter, and so on. Show that each gets a third of the apple.

27. Suppose the government pumps an extra $1 billion into the economy. Assume that each business and individual saves 25% of its income and spends the rest, so that of the initial $1 billion, 75% is respent by individuals and businesses. Of that amount, 75% is spent, and so forth. What is the total increase in spending due to the government action? (This is called the *multiplier effect* in economics.)

28. Do Problem 27 assuming that only 10% of the income is saved at each stage.

29. Assume square $ABCD$ (Figure 3) has sides of length 1 and that E, F, G, and H are midpoints of their respective sides. If the indicated pattern is continued indefinitely, what will be the area of the painted region?

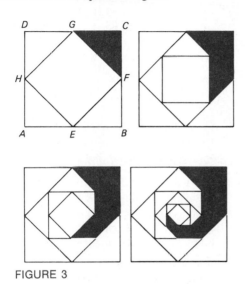

FIGURE 3

30. If the pattern shown in Figure 4 is continued indefinitely, what fraction of the original square will be painted?

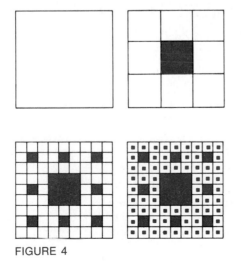

FIGURE 4

31. Each triangle in the descending chain (Figure 5) has its vertices at the midpoints of the sides of the next larger one. If the indicated pattern of painting is continued indefinitely, what fraction of the area of the original triangle will be painted? Does the original triangle need to be equilateral for this to be true?

FIGURE 5

32. Circles are inscribed in the triangles of Problem 31 as indicated in Figure 6. If the original triangle is equilateral, what fraction of the area is painted?

FIGURE 6

33. In another version of Zeno's paradox, Achilles can run ten times as fast as the tortoise, but the tortoise has a 100-yard headstart. Achilles cannot catch the tortoise, says Zeno, because when he runs 100 yards, the tortoise will have moved 10 yards ahead, when Achilles runs another 10 yards, the tortoise will have moved 1 yard ahead, and so on. Convince Zeno that Achilles will catch the tortoise and tell him exactly how many yards Achilles will have to run to do it.

34. Tom and Joel are good runners, both able to run at a constant speed of 10 miles per hour. Their amazing dog Trot can do even better; he runs at 20 miles per hour. Starting from towns 60 miles apart, Tom and Joel run toward each other while Trot runs back and forth between them. How far does Trot run by the time the boys meet? Assume Trot started with Tom running toward Joel and that he is able to make instant turnarounds. Solve the problem two ways.

(a) Use a geometric series.
(b) Find a trivial way to do the problem.

35. Prove: If $\sum_{k=1}^{\infty} a_k$ diverges, so does $\sum_{k=1}^{\infty} ca_k$ for $c \neq 0$.

36. Use Problem 35 to conclude that $\frac{1}{2} + \frac{1}{4} + \frac{1}{6} + \frac{1}{8} + \cdots$ diverges.

37. Suppose one has an unlimited supply of identical blocks each 1 unit long.

(a) Convince yourself that they may be stacked as in Figure 7 without toppling. *Hint*: Consider centers of mass.

(b) How far can one make the top block protrude to the right of the bottom block using this method of stacking?

FIGURE 7

[C] **38.** How large must N be in order for $S_N = \sum_{k=1}^{N} (1/k)$ just to exceed 4? *Note*: Computer calculations show that for S_N to exceed 20, $N = 272,400,600$, and for S_N to exceed 100, $N \approx 1.5 \times 10^{43}$.

39. Prove that if $\sum a_n$ diverges and $\sum b_n$ converges, then $\sum (a_n + b_n)$ diverges.

40. Show that it is possible for $\sum a_n$ and $\sum b_n$ both to diverge, and yet for $\sum (a_n + b_n)$ to converge.

41. By looking at the region in Figure 8 first vertically and then horizontally, conclude that

$$1 + \tfrac{1}{2} + \tfrac{1}{4} + \tfrac{1}{8} + \cdots = \tfrac{1}{2} + \tfrac{2}{4} + \tfrac{3}{8} + \tfrac{4}{16} + \cdots$$

and use this fact to calculate:

(a) $\sum_{k=1}^{\infty} \dfrac{k}{2^k}$

(b) $\bar{x}$, the horizontal coordinate of the centroid of region

FIGURE 8

42. Let r be a fixed number with $|r| < 1$. Then it can be shown that $\sum_{k=1}^{\infty} kr^k$ converges, say with sum S. Use the properties of $\sum$ to show that

$$(1 - r)S = \sum_{k=1}^{\infty} r^k$$

and then obtain a formula for S thus generalizing Problem 41(a).

43. Many drugs are eliminated from the body in an exponential manner. Thus, if a drug is given in dosages of size C at time intervals of length t, the amount A_n of the drug in the body just after the $(n + 1)$st dose is

$$A_n = C + Ce^{-kt} + Ce^{-2kt} + \cdots + Ce^{-nkt}$$

where k is a positive constant that depends on the type of drug.

(a) Derive a formula for A, the amount of drug in the body just after a dose if a person has been on the drug for a very long time (assume an infinitely long time).

(b) Evaluate A if it is known that one-half of a dose is eliminated from the body in 6 hours and doses of size 2 milligrams are given every 12 hours.

44. Find the sum of the series

$$\sum_{k=1}^{\infty} \frac{2^k}{(2^{k+1} - 1)(2^k - 1)}$$

45. Evaluate $\sum_{k=1}^{\infty} \dfrac{1}{f_k f_{k+2}}$, where $\{f_k\}$ is the Fibonacci sequence introduced in Problem 52 of Section 11.1. *Hint*: First show

$$\frac{1}{f_k f_{k+2}} = \frac{1}{f_k f_{k+1}} - \frac{1}{f_{k+1} f_{k+2}}$$

11.3
Positive Series: The Integral Test

IMPORTANT REMINDERS

$$a_1, a_2, a_3, \ldots$$

is a *sequence*.

$$a_1 + a_2 + a_3 + \cdots$$

is a *series*.

$$S_n = a_1 + a_2 + a_3 + \cdots + a_n$$

is the *n*th *partial sum* of the series.

$$S_1, S_2, S_3, \ldots$$

is the *sequence of partial sums* of the series. The series *converges* if and only if

$$S = \lim_{n \to \infty} S_n$$

exists and is finite, in which case S is called the *sum* of series.

We introduced some very big ideas in Section 11.2, but we illustrated them mainly for two very special types of series, namely, geometric series and collapsing series. These are series for which we can give exact formulas for the partial sums S_n, something that we can rarely do for most other types of series. Our task now is to begin a study of very general infinite series.

There are always two important questions to ask about a series.

1. Does the series converge?
2. If it converges, what is its sum?

How shall we answer these questions? Someone is sure to suggest that we use a computer. To answer the first question, simply add up more and more terms of the series watching the numbers you get as partial sums. If these numbers settle down on a fixed number S, the series converges. And in that case, S is the sum of the series, answering the second question. This response is plain wrong for question 1 and only partially adequate for question 2. Let us see why.

Consider the harmonic series

$$1 + \tfrac{1}{2} + \tfrac{1}{3} + \tfrac{1}{4} + \cdots$$

introduced in Section 11.2 and discussed in Example 5 and Problem 38 of that section. We know that this series diverges, but a computer would never help us to discover this fact. The partial sums S_n of this series grow without bound, but they grow so slowly that it takes over 272 million terms for S_n to reach 20 and over 10^{43} terms for S_n to reach 100. Because of the inherent limitation in the number of digits it can handle, any computer would rather soon give repeated values for S_n, suggesting wrongly that the S_n's were converging. What is true for the harmonic series is true for any slowly diverging series. We state it emphatically: A computer is no substitute for good mathematical tests of convergence and divergence, a subject to which we now turn.

In this and the next section, we restrict our attention to series with positive (or at least nonnegative) terms. With this restriction, we will be able to give some remarkably simple convergence tests. Tests for series with terms of arbitrary sign will follow in Section 11.5.

BOUNDED PARTIAL SUMS   Our first result flows directly from the Monotonic Sequence Theorem.

Theorem A

(Bounded-Sum Test). A series $\sum a_k$ of nonnegative terms converges if and only if its partial sums are bounded above.

Proof   As usual, let $S_n = a_1 + a_2 + \cdots + a_n$. Since $a_k \geq 0$, $S_{n+1} \geq S_n$; that is, $\{S_n\}$ is a nondecreasing sequence. Thus, by Theorem 11.1D, the sequence

$\{S_n\}$ will converge provided there is a number U such that $S_n \leq U$ for all n. Otherwise, the S_n's will grow without bound, in which case $\{S_n\}$ diverges.

∎

EXAMPLE 1 Show that the series $\dfrac{1}{1!} + \dfrac{1}{2!} + \dfrac{1}{3!} + \cdots$ converges.

Solution   We aim to show that the partial sums S_n are bounded above. Note first that

$$n! = 1 \cdot 2 \cdot 3 \cdots n \geq 1 \cdot 2 \cdot 2 \cdots 2 = 2^{n-1}$$

and so $1/n! \leq 1/2^{n-1}$. Thus,

$$S_n = \frac{1}{1!} + \frac{1}{2!} + \frac{1}{3!} + \cdots + \frac{1}{n!}$$

$$\leq 1 + \frac{1}{2} + \frac{1}{4} + \cdots + \frac{1}{2^{n-1}}$$

These latter terms come from a geometric series with $r = \frac{1}{2}$. They can be added by a formula in Example 1 of Section 11.2. We obtain

$$S_n \leq \frac{1 - (\frac{1}{2})^n}{1 - \frac{1}{2}} = 2\left[1 - \left(\frac{1}{2}\right)^n\right] < 2$$

Thus, by the Bounded-Sum Test, the given series converges. The argument also shows that its sum S is at most 2. Later we will show that $S = e - 1 \approx 1.71828$.

∎

SERIES AND IMPROPER INTEGRALS   The behavior of $\displaystyle\sum_{k=1}^{\infty} f(k)$ and $\displaystyle\int_{1}^{\infty} f(x)\,dx$ with respect to convergence is similar and gives a very powerful test.

Theorem B

(Integral Test). Let f be a continuous, positive, nonincreasing function on the interval $[1, \infty)$ and suppose $a_k = f(k)$ for all positive integers k. Then the infinite series

$$\sum_{k=1}^{\infty} a_k$$

converges if and only if the improper integral

$$\int_{1}^{\infty} f(x)\,dx$$

converges.

We remark that the integer 1 may be replaced by any positive integer M throughout this theorem (see Example 4).

Proof The diagrams in Figure 1 indicate how we may interpret the partial sums of the series $\sum a_k$ as areas and thereby relate the series to a corresponding integral. Note that the area of each rectangle is equal to its height, since the width is 1 in each case. From these diagrams, we easily see that

$$\sum_{k=2}^{n} a_k \leq \int_{1}^{n} f(x)\,dx \leq \sum_{k=1}^{n-1} a_k$$

 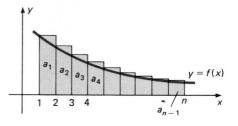

FIGURE 1

Now suppose $\displaystyle\int_{1}^{\infty} f(x)\,dx$ converges. Then, by the left inequality above,

$$S_n = a_1 + \sum_{k=2}^{n} a_k \leq a_1 + \int_{1}^{n} f(x)\,dx \leq a_1 + \int_{1}^{\infty} f(x)\,dx$$

Therefore, by the Bounded-Sum Test, $\displaystyle\sum_{k=1}^{\infty} a_k$ converges.

On the other hand, suppose $\displaystyle\sum_{k=1}^{\infty} a_k$ converges. Then, by the right inequality above, if $t \leq n$,

$$\int_{1}^{t} f(x)\,dx \leq \int_{1}^{n} f(x)\,dx \leq \sum_{k=1}^{n-1} a_k \leq \sum_{k=1}^{\infty} a_k$$

Since $\displaystyle\int_{1}^{t} f(x)\,dx$ increases with t and is bounded above, $\displaystyle\lim_{t \to \infty} \int_{1}^{t} f(x)\,dx$ must exist; that is, $\displaystyle\int_{1}^{\infty} f(x)\,dx$ converges. ∎

The conclusion to Theorem B is often stated this way. *The series $\displaystyle\sum_{k=1}^{\infty} a_k$ and the improper integral $\displaystyle\int_{1}^{\infty} f(x)\,dx$ converge or diverge together.* You should see that this is equivalent to our statement.

EXAMPLE 2 (*p*-Series Test). The series

$$\sum_{k=1}^{\infty} \frac{1}{k^p} = 1 + \frac{1}{2^p} + \frac{1}{3^p} + \frac{1}{4^p} + \cdots$$

where p is a constant, is called a **p-series**. Show each of the following.
(a) The p-series converges if $p > 1$.
(b) The p-series diverges if $p \leq 1$.

Solution If $p \geq 0$, the function $f(x) = 1/x^p$ is continuous, positive, and non-increasing on $[1, \infty)$ and $f(k) = 1/k^p$. Thus, by the Integral Test, $\sum (1/k^p)$ converges if and only if $\lim\limits_{t \to \infty} \int_1^t x^{-p} \, dx$ exists (as a finite number).

If $p \neq 1$,

$$\int_1^t x^{-p} \, dx = \left[\frac{x^{1-p}}{1-p} \right]_1^t = \frac{t^{1-p} - 1}{1 - p}$$

If $p = 1$,

$$\int_1^t x^{-1} \, dx = [\ln x]_1^t = \ln t$$

Since $\lim\limits_{t \to \infty} t^{1-p} = 0$ if $p > 1$ and $\lim\limits_{t \to \infty} t^{1-p} = \infty$ if $p < 1$ and since $\lim\limits_{t \to \infty} \ln t = \infty$, we conclude that the p-series converges if $p > 1$ and diverges if $0 \leq p \leq 1$.

We still have the case $p < 0$ to consider. In that case, the nth term of $\sum (1/k^p)$, namely, $1/n^p$ does not even tend toward 0. Thus by the nth-Term Test, the series diverges.

Note that the case $p = 1$ gives the harmonic series, which was treated in Section 11.2. Our results here and there are consistent. The harmonic series diverges. ∎

EXAMPLE 3 Does $\sum\limits_{k=4}^{\infty} \dfrac{1}{k^{1.001}}$ converge or diverge?

Solution By the p-Series Test, $\sum\limits_{k=1}^{\infty} (1/k^{1.001})$ converges. *The insertion or deletion of a finite number of terms in a series cannot affect its convergence or divergence (though it may affect the sum).* Thus, the given series converges. ∎

EXAMPLE 4 Determine whether $\sum\limits_{k=2}^{\infty} \dfrac{1}{k \ln k}$ converges or diverges.

Solution The hypotheses of the Integral Test are satisfied for $f(x) = 1/(x \ln x)$ on $[2, \infty)$. That the interval is $[2, \infty)$ rather than $[1, \infty)$ is inconsequential, as we noted right after Theorem B. Now,

$$\int_2^{\infty} \frac{1}{x \ln x} \, dx = \lim_{t \to \infty} \int_2^t \frac{1}{\ln x} \, d(\ln x) = \lim_{t \to \infty} [\ln \ln x]_2^t = \infty$$

Thus, $\sum 1/(k \ln k)$ diverges. ∎

EXAMPLE 5 By means of an improper integral, find a good upper bound for the error in using the sum of the first five terms of the convergent series

$$\sum_{n=1}^{\infty} \frac{n}{e^{n^2}}$$

to approximate the sum of the series.

FIGURE 2

Solution The error E is

$$E = \sum_{n=6}^{\infty} \frac{n}{e^{n^2}}$$

The function $f(x) = x/e^{x^2}$ is continuous and nonincreasing on $[5, \infty)$ (see Figure 2). Thus,

$$E = \sum_{n=6}^{\infty} \frac{n}{e^{n^2}} < \int_{5}^{\infty} xe^{-x^2}\,dx$$

$$= \lim_{t \to \infty} -\frac{1}{2} \int_{5}^{t} e^{-x^2}(-2x\,dx)$$

$$= \lim_{t \to \infty} -\frac{1}{2}\left[e^{-x^2} \right]_{5}^{t} = \frac{1}{2} e^{-25} \approx 6.94 \times 10^{-12} \qquad \blacksquare$$

PROBLEM SET 11.3

Use the Integral Test to decide the convergence or divergence of each of the following series.

1. $\sum_{k=1}^{\infty} \dfrac{1}{k+2}$

2. $\sum_{k=4}^{\infty} \dfrac{2}{k-3}$

3. $\sum_{k=1}^{\infty} \dfrac{k}{1+k^2}$

4. $\sum_{k=1}^{\infty} \dfrac{2}{1+k^2}$

5. $\sum_{k=1}^{\infty} \dfrac{1}{\sqrt{k+2}}$

6. $\sum_{k=1}^{\infty} \dfrac{1}{(k+2)^2}$

7. $\sum_{k=1}^{\infty} \dfrac{1}{10k+3}$

8. $\sum_{k=1}^{\infty} \dfrac{k}{e^k}$

9. $\sum_{k=1}^{\infty} \dfrac{1}{(4+3k)^{3/2}}$

10. $\sum_{k=1}^{\infty} \dfrac{k^2}{1+k^3}$

11. $\sum_{k=1}^{\infty} ke^{-k^2}$

12. $\sum_{k=2}^{\infty} \dfrac{1}{k(\ln k)^2}$

In Problems 13–22, use any test developed so far including those in Section 11.2 to decide about the convergence or divergence of the series. Give a reason for your conclusion.

13. $\sum_{k=1}^{\infty} \dfrac{k^2+1}{k^2+5}$

14. $\sum_{k=1}^{\infty} \left(\dfrac{3}{\pi} \right)^k$

15. $\sum_{k=1}^{\infty} \left[\left(\dfrac{1}{2} \right)^k + \dfrac{k-1}{2k+1} \right]$

16. $\sum_{k=1}^{\infty} \left(\dfrac{1}{k^2} + \dfrac{1}{2^k} \right)$

17. $\sum_{k=1}^{\infty} \sin\left(\dfrac{k\pi}{2} \right)$

18. $\sum_{k=1}^{\infty} k \sin \dfrac{1}{k}$

19. $\sum_{k=1}^{\infty} k^2 e^{-k^3}$

20. $\sum_{k=1}^{\infty} \left(\dfrac{1}{k} - \dfrac{1}{k+1} \right)$

21. $\sum_{k=1}^{\infty} \dfrac{\tan^{-1} k}{1+k^2}$

22. $\sum_{k=1}^{\infty} \dfrac{1}{1+4k^2}$

In Problems 23–26, estimate the error that is made by approximating the sum of the given series by the sum of the first five terms (see Example 5).

23. $\sum_{k=1}^{\infty} \dfrac{k}{e^k}$

24. $\sum_{k=1}^{\infty} \dfrac{1}{k\sqrt{k}}$

25. $\sum_{k=1}^{\infty} \dfrac{1}{1+k^2}$

26. $\displaystyle\sum_{k=1}^{\infty} \frac{1}{k(k+1)} = \sum_{k=1}^{\infty}\left(\frac{1}{k} - \frac{1}{k+1}\right)$

27. For what values of p does $\displaystyle\sum_{n=2}^{\infty} 1/[n(\ln n)^p]$ converge?

28. Does $\displaystyle\sum_{n=3}^{\infty} 1/[n \cdot \ln n \cdot \ln(\ln n)]$ converge or diverge?

29. Use pictures, as in the proof of the Integral Test, to show

$$\ln(n+1) < 1 + \frac{1}{2} + \frac{1}{3} + \cdots + \frac{1}{n} < 1 + \ln n$$

Hint: $\displaystyle\int_1^n (1/x)\,dx = \ln n.$

30. Show, using Problem 29, that the sequence

$$B_n = 1 + \frac{1}{2} + \frac{1}{3} + \cdots + \frac{1}{n} - \ln(n+1)$$

is increasing and bounded above by 1.

31. From Problem 30, conclude that $\lim_{n\to\infty} B_n$ exists; call it γ (γ is known as *Euler's constant*). It is easy to show that $\gamma \approx 0.5772$. Decide whether γ is rational or irrational (but before you work too hard on this, consult the answer key).

32. Use Problem 29 to get good upper and lower bounds for the sum of the first 10 million terms of the harmonic series.

33. From Problem 31, we infer that

$$1 + \frac{1}{2} + \frac{1}{3} + \cdots + \frac{1}{n} \approx \gamma + \ln(n+1)$$

Use this to estimate the number of terms of the harmonic series that are needed to get a sum greater than 20 and compare with the result reported in Problem 38 of Section 11.2.

34. Now that we have shown the existence of Euler's constant the hard way (Problems 29–31), we will solve a much more general problem the easy way and watch γ appear out of thin air, so to speak. Let f be continuous and decreasing on $[1, \infty)$ and let

$$B_n = f(1) + f(2) + \cdots + f(n) - \int_1^{n+1} f(x)\,dx$$

Note that B_n is the area of the shaded region in Figure 3.

(a) Why is it obvious that B_n increases with n?
(b) Show that $B_n \leq f(1)$. *Hint:* Simply shift all the little shaded pieces leftward into the outlined rectangle.
(c) Conclude that $\lim_{n\to\infty} B_n$ exists.
(d) How do we get γ out of this?

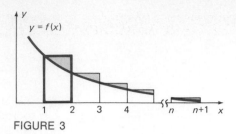

FIGURE 3

35. Let f be continuous, increasing, and concave down on $[1, \infty)$ as in Figure 4. Further, let A_n be the area of the shaded region. Show that A_n is increasing with n, that $A_n \leq T$ where T is the area of the heavily outlined triangle, and thus that $\lim_{n\to\infty} A_n$ exists.

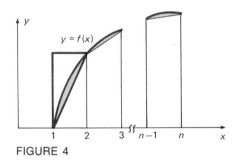

FIGURE 4

36. Specialize f of Problem 35 to $f(x) = \ln x$.

(a) Show that

$$A_n = \int_1^n \ln x\,dx - \left[\frac{\ln 1 + \ln 2}{2} + \cdots + \frac{\ln(n-1) + \ln n}{2}\right]$$

$$= n \ln n - n + 1 - \ln n! + \ln\sqrt{n}$$

$$= 1 + \ln\frac{(n/e)^n \sqrt{n}}{n!}$$

(b) Conclude from (a) and Problem 35 that

$$k = \lim_{n\to\infty} \frac{n!}{(n/e)^n \sqrt{n}}$$

exists. It can be shown that $k = \sqrt{2\pi}$.
(c) This means that $n! \approx \sqrt{2\pi n}(n/e)^n$, which is called **Stirling's Formula**. Use it to approximate 15! and compare it with the value your calculator gives for 15!

11.4
Positive Series: Other Tests

We have completely analyzed the convergence and divergence of two series—the geometric series and the p-series.

$$\sum_{n=1}^{\infty} r^n \quad \text{converges if } -1 < r < 1, \text{ diverges otherwise}$$

$$\sum_{n=1}^{\infty} \frac{1}{n^p} \quad \text{converges if } p > 1, \text{ diverges otherwise}$$

These series provide standards, or models, against which we can measure other series. Keep in mind that we are still considering series whose terms are positive (or at least nonnegative).

COMPARING ONE SERIES WITH ANOTHER A series with terms less than the corresponding terms of a convergent series ought to converge; a series with terms greater than the corresponding terms of a divergent series ought to diverge. What ought to be true is true.

Theorem A

(Ordinary Comparison Test). Suppose $0 \le a_n \le b_n$ for $n \ge N$.

(i) If $\sum b_n$ converges, so does $\sum a_n$.
(ii) If $\sum a_n$ diverges, so does $\sum b_n$.

Proof We suppose $N = 1$; the case $N > 1$ is only slightly harder. To prove (i), let $S_n = a_1 + a_2 + \cdots + a_n$ and note that $\{S_n\}$ is a nondecreasing sequence. If $\sum b_n$ converges—for instance, with sum B—then

$$S_n \le b_1 + b_2 + \cdots + b_n \le \sum_{n=1}^{\infty} b_n = B$$

By the Bounded-Sum Test (Theorem 11.3A), $\sum a_n$ converges.

Property (ii) follows from (i); for if $\sum b_n$ converged, then $\sum a_n$ would have to converge. ∎

EXAMPLE 1 Does $\sum \dfrac{n}{5n^2 - 4}$ converge or diverge?

Solution A good guess would be that it diverges, since the nth term behaves like $1/5n$ for large n. In fact,

$$\frac{n}{5n^2 - 4} > \frac{n}{5n^2} = \frac{1}{5} \cdot \frac{1}{n}$$

We know $\sum \dfrac{1}{5} \cdot \dfrac{1}{n}$ diverges since it is one-fifth of the harmonic series (Theorem 11.2C). Thus, by the Ordinary Comparison Test, the given series also diverges. ∎

EXAMPLE 2 Does $\sum \dfrac{n}{2^n(n+1)}$ converge or diverge?

Solution A good guess would be that it converges, since the nth term behaves like $(1/2)^n$ for large n. To substantiate our guess, we note that

$$\frac{n}{2^n(n+1)} = \left(\frac{1}{2}\right)^n \frac{n}{n+1} < \left(\frac{1}{2}\right)^n$$

Since $\sum \left(\frac{1}{2}\right)^n$ converges—it is a geometric series with $r = \frac{1}{2}$—we conclude that the given series converges. ∎

If there is a problem in applying the Ordinary Comparison Test, it is in finding exactly the right known series with which to compare the series to be tested. Suppose we wish to determine the convergence or divergence of

$$\sum_{n=3}^{\infty} \frac{1}{(n-2)^2} = \sum_{n=3}^{\infty} \frac{1}{n^2 - 4n + 4}$$

Our inclination is to compare $1/(n-2)^2$ with $1/n^2$, but unfortunately

$$\frac{1}{(n-2)^2} > \frac{1}{n^2}$$

which gives no test at all (the inequality goes the wrong way for what we want). After some experimenting, we discover that

$$\frac{1}{(n-2)^2} \leq \frac{9}{n^2}$$

for $n \geq 3$; since $\sum 9/n^2$ converges, so does $\sum 1/(n-2)^2$.

Can we avoid these contortions with inequalities? Our intuition tells us that $\sum a_n$ and $\sum b_n$ converge or diverge together, provided a_n and b_n are approximately the same size for large n (give or take a multiplicative constant). That is the essential content of our next theorem.

Theorem B

(Limit Comparison Test). Suppose $a_n \geq 0$, $b_n > 0$, and

$$\lim_{n \to \infty} \frac{a_n}{b_n} = L$$

If $0 < L < \infty$, then $\sum a_n$ and $\sum b_n$ converge or diverge together. If $L = 0$ and $\sum b_n$ converges, then $\sum a_n$ converges.

Proof Begin by taking $\varepsilon = L/2$ in the definition of limit of a sequence (Section 11.1). There is a number N such that $n \geq N \quad \Rightarrow \quad |(a_n/b_n) - L| < L/2$; that is,

$$-\frac{L}{2} < \frac{a_n}{b_n} - L < \frac{L}{2}$$

This inequality is equivalent (by adding L throughout) to

$$\frac{L}{2} < \frac{a_n}{b_n} < \frac{3L}{2}$$

Hence for $n \geq N$,

$$b_n < \frac{2}{L} a_n \quad \text{and} \quad a_n < \frac{3L}{2} b_n$$

These two inequalities, together with the Ordinary Comparison Test, show that $\sum a_n$ and $\sum b_n$ converge or diverge together. We leave the proof of the final statement of the theorem to the reader (Problem 37). ∎

EXAMPLE 3 Determine the convergence or divergence of each series.

(a) $\displaystyle\sum_{n=1}^{\infty} \frac{3n - 2}{n^3 - 2n^2 + 11}$ (b) $\displaystyle\sum_{n=1}^{\infty} \frac{1}{\sqrt{n^2 + 19n}}$

Solution We apply the Limit Comparison Test, but we still must decide to what we should compare the nth term. We ask what the nth term is like for large n, which we determine by looking at the largest-degree terms in numerator and denominator. In the first case, the nth term is like $3/n^2$; in the second, it is like $1/n$.

(a) $\displaystyle\lim_{n\to\infty} \frac{a_n}{b_n} = \lim_{n\to\infty} \frac{(3n - 2)/(n^3 - 2n^2 + 11)}{3/n^2} = \lim_{n\to\infty} \frac{3n^3 - 2n^2}{3n^3 - 6n^2 + 33} = 1$

(b) $\displaystyle\lim_{n\to\infty} \frac{a_n}{b_n} = \lim_{n\to\infty} \frac{1/\sqrt{n^2 + 19n}}{1/n} = \lim_{n\to\infty} \sqrt{\frac{n^2}{n^2 + 19n}} = 1$

We conclude that the series in (a) converges, and that the series in (b) diverges. ∎

EXAMPLE 4 Does $\displaystyle\sum_{n=1}^{\infty} \frac{\ln n}{n^2}$ converge or diverge?

Solution To what shall we compare $(\ln n)/n^2$? If we try $1/n^2$, we get

$$\lim_{n\to\infty} \frac{a_n}{b_n} = \lim_{n\to\infty} \frac{\ln n}{n^2} \div \frac{1}{n^2} = \lim_{n\to\infty} \ln n = \infty$$

The test fails. On the other hand, if we use $1/n$, we get

$$\lim_{n\to\infty} \frac{a_n}{b_n} = \lim_{n\to\infty} \frac{\ln n}{n^2} \div \frac{1}{n} = \lim_{n\to\infty} \frac{\ln n}{n} = 0$$

Again, the test fails. Possibly something between $1/n^2$ and $1/n$ will work, such as $1/n^{3/2}$.

$$\lim_{n\to\infty} \frac{a_n}{b_n} = \lim_{n\to\infty} \frac{\ln n}{n^2} \div \frac{1}{n^{3/2}} = \lim_{n\to\infty} \frac{\ln n}{\sqrt{n}} = 0$$

(The last equality follows from l'Hôpital's Rule.) We conclude from the second part of the Limit Comparison Test that $\sum (\ln n)/n^2$ converges (since $\sum 1/n^{3/2}$ converges). ∎

COMPARING A SERIES WITH ITSELF To apply the comparison tests requires real insight. We must choose wisely among known series to find one that is just right for comparison with the series we wish to test. Wouldn't it be nice if we could somehow compare a series with itself and thereby determine convergence or divergence? Roughly speaking, this is what we do in the Ratio Test.

Theorem C

(Ratio Test). Let $\sum a_n$ be a series of positive terms and suppose

$$\lim_{n\to\infty} \frac{a_{n+1}}{a_n} = \rho$$

(i) If $\rho < 1$, the series converges.
(ii) If $\rho > 1$, the series diverges.
(iii) If $\rho = 1$, the test is inconclusive.

Proof Here is what is behind the Ratio Test. Since $\lim_{n\to\infty} a_{n+1}/a_n = \rho$, $a_{n+1} \approx \rho a_n$; that is, the series behaves like a geometric series with ratio ρ. A geometric series converges when its ratio is less than 1 and diverges when its ratio is greater than 1. Tying down this argument is the task before us.

(i) Since $\rho < 1$, we may choose a number r such that $\rho < r < 1$—for example, $r = (\rho + 1)/2$. Next choose N so large that $n \geq N$ implies $a_{n+1}/a_n < r$. Then,

$$a_{N+1} < ra_N$$
$$a_{N+2} < ra_{N+1} < r^2 a_N$$
$$a_{N+3} < ra_{N+2} < r^3 a_N$$
$$\vdots$$

Since $ra_N + r^2 a_N + r^3 a_N + \cdots$ is a geometric series with $0 < r < 1$, it converges. By the Ordinary Comparison Test, $\sum\limits_{n=N+1}^{\infty} a_n$ converges, and hence so does $\sum\limits_{n=1}^{\infty} a_n$.

(ii) Since $\rho > 1$, there is a number N such that $a_{n+1}/a_n > 1$ for all $n \geq N$. Thus

$$a_{N+1} > a_N$$

$$a_{N+2} > a_{N+1} > a_N$$

$$\vdots$$

Hence, $a_n > a_N > 0$ for all $n > N$, which means that $\lim\limits_{n \to \infty} a_n$ cannot be zero. By the nth-Term Test for Divergence, $\sum a_n$ diverges.

(iii) We know that $\sum 1/n$ diverges, whereas $\sum 1/n^2$ converges. For the first series,

$$\lim_{n \to \infty} \frac{a_{n+1}}{a_n} = \lim_{n \to \infty} \frac{1}{n+1} \div \frac{1}{n} = \lim_{n \to \infty} \frac{n}{n+1} = 1$$

For the second series,

$$\lim_{n \to \infty} \frac{a_{n+1}}{a_n} = \lim_{n \to \infty} \frac{1}{(n+1)^2} \div \frac{1}{n^2} = \lim_{n \to \infty} \frac{n^2}{(n+1)^2} = 1$$

Thus, the Ratio Test does not distinguish between convergence and divergence when $\rho = 1$. ∎

The Ratio Test will always fail for a series whose nth term is a rational expression in n, since in this case $\rho = 1$ (the cases $a_n = 1/n$ and $a_n = 1/n^2$ were treated above). However for a series whose nth term involves $n!$ or r^n, the Ratio Test usually works beautifully.

EXAMPLE 5 Test for convergence or divergence: $\sum\limits_{n=1}^{\infty} \dfrac{2^n}{n!}$.

Solution

$$\rho = \lim_{n \to \infty} \frac{a_{n+1}}{a_n} = \lim_{n \to \infty} \frac{2^{n+1}}{(n+1)!} \frac{n!}{2^n} = \lim_{n \to \infty} \frac{2}{n+1} = 0$$

We conclude by the Ratio Test that the series converges. ∎

EXAMPLE 6 Test for convergence or divergence: $\sum\limits_{n=1}^{\infty} \dfrac{2^n}{n^{20}}$.

Solution

$$\rho = \lim_{n \to \infty} \frac{a_{n+1}}{a_n} = \lim_{n \to \infty} \frac{2^{n+1}}{(n+1)^{20}} \frac{n^{20}}{2^n}$$

$$= \lim_{n \to \infty} \left(\frac{n}{n+1}\right)^{20} \cdot 2 = 2$$

We conclude that the given series diverges. ∎

EXAMPLE 7 Test for convergence or divergence: $\sum_{n=1}^{\infty} \frac{n!}{n^n}$.

Solution We will need the fact that

$$\lim_{n \to \infty} \left(1 + \frac{1}{n}\right)^n = e$$

which follows from Theorem 7.5A. Taking this as known, we may write

$$\rho = \lim_{n \to \infty} \frac{a_{n+1}}{a_n} = \lim_{n \to \infty} \frac{(n+1)!}{(n+1)^{n+1}} \frac{n^n}{n!} = \lim_{n \to \infty} \left(\frac{n}{n+1}\right)^n$$

$$= \lim_{n \to \infty} \frac{1}{((n+1)/n)^n} = \lim_{n \to \infty} \frac{1}{(1 + 1/n)^n} = \frac{1}{e} < 1$$

Therefore, the given series converges. ∎

SUMMARY To test a series $\sum a_n$ of positive terms for convergence or divergence, look carefully at a_n.

1. If $\lim_{n \to \infty} a_n \neq 0$, conclude from the nth-Term Test that the series diverges.
2. If a_n involves $n!$, r^n, or n^n, try the Ratio Test.
3. If a_n involves only constant powers of n, try the Limit Comparison Test. In particular, if a_n is a rational expression in n, use this test with b_n as the quotient of the leading terms from numerator and denominator.
4. If the tests above do not work, try the Ordinary Comparison Test, the Integral Test, or the Bounded-Sum Test.
5. Some series require a clever manipulation or a neat trick to determine convergence or divergence.

PROBLEM SET 11.4 _____

In Problems 1–4, use the Limit Comparison Test to determine convergence or divergence.

1. $\sum_{n=1}^{\infty} \frac{n}{n^2 + 2n + 3}$

2. $\sum_{n=1}^{\infty} \frac{3n+1}{n^3 - 4}$

3. $\sum_{n=1}^{\infty} \frac{1}{n\sqrt{n+1}}$

4. $\sum_{n=1}^{\infty} \frac{\sqrt{2n+1}}{n^2}$

In Problems 5–10, use the Ratio Test to determine convergence or divergence.

5. $\sum_{n=1}^{\infty} \frac{8^n}{n!}$

6. $\sum_{n=1}^{\infty} \frac{5^n}{n^5}$

7. $\sum_{n=1}^{\infty} \frac{n!}{n^{100}}$

8. $\sum_{n=1}^{\infty} n(\tfrac{1}{3})^n$

9. $\sum_{n=1}^{\infty} \frac{n^3}{(2n)!}$

10. $\sum_{k=1}^{\infty} \frac{3^k + k}{k!}$

In Problems 11–34, determine convergence or divergence for each of the series. Indicate the test you used.

11. $\sum_{n=1}^{\infty} \frac{n}{n + 200}$

12. $\sum_{n=1}^{\infty} \frac{5+n}{n!}$

13. $\sum_{n=1}^{\infty} \frac{n+3}{n^2\sqrt{n}}$

14. $\sum_{n=1}^{\infty} \frac{\sqrt{n+1}}{n^2 + 1}$

15. $\sum_{n=1}^{\infty} \frac{n^2}{n!}$

16. $\sum_{n=1}^{\infty} \frac{\ln n}{2^n}$

17. $\sum_{n=1}^{\infty} \frac{4n^3 + 3n}{n^5 - 4n^2 + 1}$

18. $\sum_{n=1}^{\infty} \frac{n^2 + 1}{3^n}$

19. $\frac{1}{1 \cdot 2} + \frac{1}{2 \cdot 3} + \frac{1}{3 \cdot 4} + \frac{1}{4 \cdot 5} + \cdots$.

Hint: $a_n = \dfrac{1}{n(n+1)}$.

20. $\dfrac{1}{2^2} + \dfrac{2}{3^2} + \dfrac{3}{4^2} + \dfrac{4}{5^2} + \cdots$

21. $\dfrac{2}{1 \cdot 3 \cdot 4} + \dfrac{3}{2 \cdot 4 \cdot 5} + \dfrac{4}{3 \cdot 5 \cdot 6} + \dfrac{5}{4 \cdot 6 \cdot 7} + \cdots$

22. $\dfrac{1}{1^2 + 1} + \dfrac{2}{2^2 + 1} + \dfrac{3}{3^2 + 1} + \dfrac{4}{4^2 + 1} + \cdots$

23. $\dfrac{1}{3} + \dfrac{2}{3^2} + \dfrac{3}{3^3} + \dfrac{4}{3^4} + \cdots$

24. $3 + \dfrac{3^2}{2!} + \dfrac{3^3}{3!} + \dfrac{3^4}{4!} + \cdots$

25. $1 + \dfrac{1}{2\sqrt{2}} + \dfrac{1}{3\sqrt{3}} + \dfrac{1}{4\sqrt{4}} + \cdots$

26. $\dfrac{\ln 2}{2^2} + \dfrac{\ln 3}{3^2} + \dfrac{\ln 4}{4^2} + \dfrac{\ln 5}{5^2} + \cdots$

27. $\displaystyle\sum_{n=1}^{\infty} \dfrac{1}{2 + \sin^2 n}$ **28.** $\displaystyle\sum_{n=1}^{\infty} \dfrac{5}{3^n + 1}$

29. $\displaystyle\sum_{n=1}^{\infty} \dfrac{4 + \cos n}{n^3}$ **30.** $\displaystyle\sum_{n=1}^{\infty} \dfrac{5^{2n}}{n!}$

31. $\displaystyle\sum_{n=1}^{\infty} \dfrac{n^n}{(2n)!}$ **32.** $\displaystyle\sum_{n=2}^{\infty} \left(1 - \dfrac{1}{n}\right)^n$

33. $\displaystyle\sum_{n=1}^{\infty} \dfrac{4^n + n}{n!}$ **34.** $\displaystyle\sum_{n=1}^{\infty} \dfrac{n}{2 + n5^n}$

35. Let $a_n > 0$ and suppose $\sum a_n$ converges. Prove that $\sum a_n^2$ converges.

36. Prove that $\lim_{n \to \infty}(n!/n^n) = 0$ by considering the series $\sum n!/n^n$. *Hint*: Example 7, followed by nth-Term Test.

37. Prove that if $a_n \geq 0, b_n > 0$, $\lim_{n \to \infty} a_n/b_n = 0$, and $\sum b_n$ converges, then $\sum a_n$ converges.

38. Prove that if $a_n \geq 0$, $b_n > 0$, $\lim_{n \to \infty} a_n/b_n = \infty$, and $\sum b_n$ diverges, then $\sum a_n$ diverges.

39. Suppose $\lim_{n \to \infty} na_n = 1$. Prove $\sum a_n$ diverges.

40. Prove that if $\sum a_n$ is a convergent series of positive terms, then $\sum \ln(1 + a_n)$ converges.

41. (Root Test) Prove that if $a_n > 0$ and $\lim_{n \to \infty}(a_n)^{1/n} = R$, then $\sum a_n$ converges if $R < 1$ and diverges if $R > 1$.

42. Test for convergence or divergence using the Root Test.

(a) $\displaystyle\sum_{n=2}^{\infty} \left(\dfrac{1}{\ln n}\right)^n$ (b) $\displaystyle\sum_{n=1}^{\infty} \left(\dfrac{n}{3n + 2}\right)^n$

(c) $\displaystyle\sum_{n=1}^{\infty} \left(\dfrac{1}{2} + \dfrac{1}{n}\right)^n$

43. Test for convergence or divergence. In some cases, a clever manipulation using the properties of logarithms will simplify the problem.

(a) $\displaystyle\sum_{n=1}^{\infty} \ln\left(1 + \dfrac{1}{n}\right)$ (b) $\displaystyle\sum_{n=1}^{\infty} \ln\left[\dfrac{(n + 1)^2}{n(n + 2)}\right]$

(c) $\displaystyle\sum_{n=2}^{\infty} \dfrac{1}{(\ln n)^{\ln n}}$ (d) $\displaystyle\sum_{n=2}^{\infty} \dfrac{1}{[\ln(\ln n)]^{\ln n}}$

(e) $\displaystyle\sum_{n=2}^{\infty} \dfrac{1}{(\ln n)^4}$ (f) $\displaystyle\sum_{n=1}^{\infty} \left[\dfrac{\ln n}{n}\right]^2$

44. Let $p(n)$ and $q(n)$ be polynomials in n with nonnegative coefficients. Give simple conditions that determine the convergence or divergence of $\displaystyle\sum_{n=1}^{\infty} \dfrac{p(n)}{q(n)}$.

45. Give conditions on p which determine the convergence or divergence of $\displaystyle\sum_{n=1}^{\infty} \dfrac{1}{n^p}\left(1 + \dfrac{1}{2^p} + \dfrac{1}{3^p} + \cdots + \dfrac{1}{n^p}\right)$.

46. Test for convergence or divergence.

(a) $\displaystyle\sum_{n=1}^{\infty} \sin^2\left(\dfrac{1}{n}\right)$ (b) $\displaystyle\sum_{n=1}^{\infty} \tan\left(\dfrac{1}{n}\right)$

(c) $\displaystyle\sum_{n=1}^{\infty} \sqrt{n}\left[1 - \cos\left(\dfrac{1}{n}\right)\right]$

11.5
Alternating Series, Absolute Convergence

In the last two sections, we have considered series of nonnegative terms. Now we remove that restriction, allowing some terms to be negative. In particular, we study **alternating series**—that is, series of the form

$$a_1 - a_2 + a_3 - a_4 + \cdots$$

where $a_n > 0$ for all n. An important example is the **alternating harmonic series**

$$1 - \tfrac{1}{2} + \tfrac{1}{3} - \tfrac{1}{4} + \cdots$$

FIGURE 1

n even

n odd

FIGURE 2

We have seen that the harmonic series diverges; we shall soon see that the alternating harmonic series converges.

A CONVERGENCE TEST Let us suppose that the sequence $\{a_n\}$ is decreasing; that is, $a_n > a_{n+1}$ for all n. Also let S_n have its usual meaning,

$$S_1 = a_1$$
$$S_2 = a_1 - a_2 = S_1 - a_2$$
$$S_3 = a_1 - a_2 + a_3 = S_2 + a_3$$
$$S_4 = a_1 - a_2 + a_3 - a_4 = S_3 - a_4$$

and so on. A geometric interpretation of these partial sums is shown in Figure 1. Note that the even numbered terms $S_2, S_4, S_6, \ldots$ are increasing and bounded above and hence must converge to a limit, call it S'. Similarly, the odd numbered terms $S_1, S_3, S_5, \ldots$ are decreasing and bounded below. They also converge, say to S''.

Both S' and S'' are between S_n and S_{n+1} for all n (see Figure 2) and so

$$|S'' - S'| \le |S_{n+1} - S_n| = a_{n+1}$$

Thus, the condition $a_{n+1} \to 0$ as $n \to \infty$ will guarantee that $S' = S''$ and, consequently, the convergence of the series to their common value, which we call S. Finally, we note that since S is between S_n and S_{n+1},

$$|S - S_n| \le |S_{n+1} - S_n| = a_{n+1}$$

that is, the error made by using S_n as an approximation to the sum S of the whole series is not more than the magnitude of the first neglected term. We have proved the following theorem.

Theorem A

(Alternating-Series Test). Let

$$a_1 - a_2 + a_3 - a_4 + \cdots$$

be an alternating series with $a_n > a_{n+1} > 0$. If $\lim\limits_{n \to \infty} a_n = 0$, then the series converges. Moreover, the error made by using the sum S_n of the first n terms to approximate the sum S of the series is not more than a_{n+1}.

EXAMPLE 1 Show that the alternating harmonic series

$$1 - \tfrac{1}{2} + \tfrac{1}{3} - \tfrac{1}{4} + \cdots$$

converges. How many terms of this series would we need to take in order to get a partial sum S_n within 0.01 of the sum S of the whole series?

Solution The alternating harmonic series satisfies the hypotheses of Theorem A, and so converges. We want $|S - S_n| \leq 0.01$, and this will hold if $a_{n+1} \leq 0.01$. Since $a_{n+1} = 1/(n+1)$, we require $1/(n+1) \leq 0.01$, which is satisfied if $n \geq 99$. Thus we need to take 99 terms to make sure we have the desired accuracy. This gives you an idea of how slowly the alternating harmonic series converges. (See Problem 45 for a clever way to find the exact sum of this series.) ∎

EXAMPLE 2 Show that

$$\frac{1}{1!} - \frac{1}{2!} + \frac{1}{3!} - \frac{1}{4!} + \cdots$$

converges. Calculate S_5 and estimate the error made by using this as a value for the sum of the whole series.

Solution Theorem A applies and guarantees convergence.

$$S_5 = 1 - \frac{1}{2} + \frac{1}{6} - \frac{1}{24} + \frac{1}{120} \approx 0.6333$$

$$|S - S_5| \leq a_6 = \frac{1}{6!} \approx 0.0014$$

∎

EXAMPLE 3 Show that $\displaystyle\sum_{n=1}^{\infty} (-1)^{n+1} \frac{n^2}{2^n}$ converges.

Solution To get a feeling for this series, we write the first few terms:

$$\tfrac{1}{2} - 1 + \tfrac{9}{8} - 1 + \tfrac{25}{32} - \tfrac{36}{64} + \cdots$$

The series is alternating and $\lim_{n \to \infty} n^2/2^n = 0$ (l'Hôpital's Rule), but unfortunately the terms are not decreasing initially. However, they do appear to be decreasing after the first two terms; that is good enough, since what happens at the beginning of a series never affects convergence or divergence. To show that the sequence $\{n^2/2^n\}$ is decreasing from the third term on, consider the function

$$f(x) = \frac{x^2}{2^x}$$

Note that if $x \geq 3$, the derivative

$$f'(x) = \frac{2x \cdot 2^x - x^2 2^x \ln 2}{2^{2x}} = \frac{x 2^x (2 - x \ln 2)}{2^{2x}}$$

$$\approx \frac{x(2 - 0.69x)}{2^x} < 0$$

Thus, f is decreasing on $[3, \infty)$, and so $\{n^2/2^n\}$ is decreasing for $n \geq 3$. For a different demonstration of this last fact, see Example 6 of Section 11.1. ∎

ABSOLUTE CONVERGENCE   Does a series like

$$1 + \tfrac{1}{4} - \tfrac{1}{9} + \tfrac{1}{16} + \tfrac{1}{25} - \tfrac{1}{36} + \cdots$$

in which there is a pattern of two positive terms followed by one negative term, converge or diverge? The Alternating-Series Test does not apply. However, since the corresponding series of all positive terms

$$1 + \tfrac{1}{4} + \tfrac{1}{9} + \tfrac{1}{16} + \tfrac{1}{25} + \tfrac{1}{36} + \cdots$$

converges (*p*-series with $p = 2$), it seems plausible to think that the same series with some terms negative should converge (even better). That is the content of our next theorem.

Theorem B

(Absolute Convergence Test). If $\sum |u_n|$ converges, then $\sum u_n$ converges.

Proof   We use a trick. Let $v_n = u_n + |u_n|$, so

$$u_n = v_n - |u_n|$$

Now $0 \le v_n \le 2|u_n|$, and so $\sum v_n$ converges by the Ordinary Comparison Test. It follows from the Linearity Theorem (Theorem 11.2B) that $\sum u_n = \sum (v_n - |u_n|)$ converges. ∎

A series $\sum u_n$ is said to **converge absolutely** if $\sum |u_n|$ converges. Theorem B asserts that absolute convergence implies convergence. All our tests for convergence of series of positive terms are automatically tests for the absolute convergence of a series in which some terms are negative. In particular, this is true of the Ratio Test, which we now restate.

Theorem C

(Absolute Ratio Test). Let $\sum u_n$ be a series of nonzero terms and suppose

$$\lim_{n \to \infty} \frac{|u_{n+1}|}{|u_n|} = \rho$$

(i)   If $\rho < 1$, the series converges absolutely (hence converges).
(ii) If $\rho > 1$, the series diverges.
(iii) If $\rho = 1$, the test is inconclusive.

Proof   Only (ii) requires proof. From the original Ratio Test, we could conclude that $\sum |u_n|$ diverges, but here we are claiming more—namely, that $\sum u_n$ diverges. Since

$$\lim_{n \to \infty} \frac{|u_{n+1}|}{|u_n|} > 1$$

it follows that for n sufficiently large, say $n \geq N$, $|u_{n+1}| > |u_n|$. This, in turn, implies that $|u_n| > |u_N| > 0$ for all $n \geq N$ and so $\lim_{n \to \infty} u_n$ cannot be 0. We conclude by the nth-Term Test that $\sum u_n$ diverges. ∎

EXAMPLE 4 Show that $\displaystyle\sum_{n=1}^{\infty} (-1)^{n+1} \frac{3^n}{n!}$ converges absolutely.

Solution

$$\rho = \lim_{n \to \infty} \frac{|u_{n+1}|}{|u_n|} = \lim_{n \to \infty} \frac{3^{n+1}}{(n+1)!} \div \frac{3^n}{n!}$$

$$= \lim_{n \to \infty} \frac{3}{n+1} = 0$$

We conclude from the Absolute Ratio Test that the series converges absolutely (and therefore converges). ∎

EXAMPLE 5 Test for the convergence or divergence of $\displaystyle\sum_{n=1}^{\infty} \frac{\cos(n!)}{n^2}$.

Solution If you write out the first hundred terms of this series, you will discover that the signs of the terms vary in a rather random way. The series is in fact a difficult one to analyze directly. However,

$$\left| \frac{\cos(n!)}{n^2} \right| \leq \frac{1}{n^2}$$

and so the series converges absolutely by the Ordinary Comparison Test. We conclude from Theorem B that the series converges. ∎

CONDITIONAL CONVERGENCE A common error is to try to turn Theorem B around. It does *not* say that convergence implies absolute convergence. That is clearly false; witness the alternating harmonic series. We know that

$$1 - \tfrac{1}{2} + \tfrac{1}{3} - \tfrac{1}{4} + \cdots$$

converges but that

$$1 + \tfrac{1}{2} + \tfrac{1}{3} + \tfrac{1}{4} + \cdots$$

diverges. A series $\sum u_n$ is called **conditionally convergent** if $\sum u_n$ converges but $\sum |u_n|$ diverges. The alternating harmonic series is the premier example of a conditionally convergent series, but there are many others.

EXAMPLE 6 Show that $\displaystyle\sum_{n=1}^{\infty} (-1)^{n+1} \frac{1}{\sqrt{n}}$ is conditionally convergent.

Solution $\displaystyle\sum_{n=1}^{\infty} (-1)^{n+1}[1/\sqrt{n}]$ converges by the Alternating-Series Test. However, $\displaystyle\sum_{n=1}^{\infty} 1/\sqrt{n}$ diverges, since it is a p-series with $p = \tfrac{1}{2}$. ∎

Absolutely convergent series behave much better than do conditionally convergent ones. Here is a nice (but hard-to-prove) theorem about absolutely convergent series. It is spectacularly false for conditionally convergent series (see Problems 35–38).

Theorem D

(Rearrangement Theorem). The terms of an absolutely convergent series can be rearranged without affecting either the convergence or the sum of the series.

For example, the series

$$1 + \tfrac{1}{4} - \tfrac{1}{9} + \tfrac{1}{16} + \tfrac{1}{25} - \tfrac{1}{36} + \tfrac{1}{49} + \tfrac{1}{64} - \tfrac{1}{81} + \cdots$$

converges absolutely. The rearrangement

$$1 + \tfrac{1}{4} + \tfrac{1}{16} - \tfrac{1}{9} + \tfrac{1}{25} + \tfrac{1}{49} + \tfrac{1}{64} - \tfrac{1}{36} + \cdots$$

converges and has the same sum as the original series.

PROBLEM SET 11.5

In Problems 1–6, show that each of the alternating series converges and then estimate the error made by using the partial sum S_9 as an approximation to the sum S of the series (see Examples 1–3).

1. $\displaystyle\sum_{n=1}^{\infty} (-1)^{n+1} \frac{2}{3n+1}$ **2.** $\displaystyle\sum_{n=1}^{\infty} (-1)^{n+1} \frac{1}{\sqrt{n}}$

3. $\displaystyle\sum_{n=1}^{\infty} (-1)^{n+1} \frac{1}{\ln(n+1)}$ **4.** $\displaystyle\sum_{n=1}^{\infty} (-1)^{n+1} \frac{n}{n^2+1}$

5. $\displaystyle\sum_{n=1}^{\infty} (-1)^{n+1} \frac{\ln n}{n}$ **6.** $\displaystyle\sum_{n=1}^{\infty} (-1)^{n+1} \frac{\ln n}{\sqrt{n}}$

In Problems 7–12, show that each of the series converges absolutely.

7. $\displaystyle\sum_{n=1}^{\infty} \left(-\tfrac{3}{4}\right)^n$ **8.** $\displaystyle\sum_{n=1}^{\infty} (-1)^n \frac{1}{n\sqrt{n}}$

9. $\displaystyle\sum_{n=1}^{\infty} (-1)^{n+1} \frac{n}{2^n}$ **10.** $\displaystyle\sum_{n=1}^{\infty} (-1)^{n+1} \frac{n^2}{e^n}$

11. $\displaystyle\sum_{n=1}^{\infty} (-1)^{n+1} \frac{1}{n(n+1)}$ **12.** $\displaystyle\sum_{n=1}^{\infty} (-1)^{n+1} \frac{2^n}{n!}$

In Problems 13–30, classify the series as absolutely convergent, conditionally convergent, or divergent.

13. $\displaystyle\sum_{n=1}^{\infty} (-1)^{n+1} \frac{1}{5n}$ **14.** $\displaystyle\sum_{n=1}^{\infty} (-1)^{n+1} \frac{1}{5n^{1.1}}$

15. $\displaystyle\sum_{n=1}^{\infty} (-1)^{n+1} \frac{n}{10n+1}$

16. $\displaystyle\sum_{n=1}^{\infty} (-1)^{n+1} \frac{n}{10n^{1.1}+1}$

17. $\displaystyle\sum_{n=2}^{\infty} (-1)^n \frac{1}{n \ln n}$ **18.** $\displaystyle\sum_{n=1}^{\infty} (-1)^{n+1} \frac{1}{n\sqrt{n}}$

19. $\displaystyle\sum_{n=1}^{\infty} (-1)^{n+1} \frac{n^4}{2^n}$ **20.** $\displaystyle\sum_{n=2}^{\infty} (-1)^n \frac{1}{\sqrt{n^2-1}}$

21. $\displaystyle\sum_{n=1}^{\infty} (-1)^{n+1} \frac{n}{n^2+1}$ **22.** $\displaystyle\sum_{n=1}^{\infty} (-1)^{n+1} \frac{n-1}{n}$

23. $\displaystyle\sum_{n=1}^{\infty} \frac{\cos n\pi}{n}$ **24.** $\displaystyle\sum_{n=1}^{\infty} \frac{\sin(n\pi/2)}{n^2}$

25. $\displaystyle\sum_{n=1}^{\infty} \frac{\sin n}{n\sqrt{n}}$ **26.** $\displaystyle\sum_{n=1}^{\infty} n \sin\left(\frac{1}{n}\right)$

27. $\displaystyle\sum_{n=1}^{\infty} (-1)^{n+1} \frac{1}{\sqrt{n(n+1)}}$

28. $\displaystyle\sum_{n=1}^{\infty} \frac{(-1)^{n+1}}{\sqrt{n+1}+\sqrt{n}}$

29. $\displaystyle\sum_{n=1}^{\infty} \frac{(-3)^{n+1}}{n^2}$

30. $\displaystyle\sum_{n=1}^{\infty} (-1)^{n+1} \sin \frac{\pi}{n}$

31. Prove that if $\sum a_n$ diverges, so does $\sum |a_n|$.

32. Give an example of two series $\sum a_n$ and $\sum b_n$, both convergent, such that $\sum a_n b_n$ diverges.

33. Show that the positive terms of the alternating harmonic series form a divergent series. Show the same for the negative terms.

34. Show that the results in Problem 33 hold for any conditionally convergent series.

35. Show that the alternating harmonic series

$$1 - \tfrac{1}{2} + \tfrac{1}{3} - \tfrac{1}{4} + \tfrac{1}{5} - \tfrac{1}{6} + \cdots$$

(whose sum is actually $\ln 2 \approx 0.69$) can be rearranged to converge to 1.3 by using the following steps.
(a) Take enough of the positive terms $1 + \tfrac{1}{3} + \tfrac{1}{5} + \cdots$ to just exceed 1.3.
(b) Now add enough of the negative terms $-\tfrac{1}{2} - \tfrac{1}{4} - \tfrac{1}{6} - \cdots$ so that the partial sum S_n falls just below 1.3.
(c) Add just enough more positive terms to again exceed 1.3, and so on.

$\boxed{C}$ **36.** Use your calculator to help you find the first 20 terms of the series described in Problem 35. Calculate S_{20}.

37. Convince yourself that a conditionally convergent series can be rearranged to converge to any given number.

38. Show that a conditionally convergent series can be rearranged so as to diverge.

39. Show that $\lim_{n \to \infty} a_n = 0$ is not sufficient to guarantee the convergence of the alternating series $\sum (-1)^{n+1} a_n$. *Hint:* Alternate the terms of $\sum 1/n$ and $\sum -1/n^2$.

40. Discuss the convergence or divergence of

$$\frac{1}{\sqrt{2}-1} - \frac{1}{\sqrt{2}+1} + \frac{1}{\sqrt{3}-1} - \frac{1}{\sqrt{3}+1} + \frac{1}{\sqrt{4}+1}$$
$$- \frac{1}{\sqrt{4}-1} + \cdots$$

41. Prove that if $\displaystyle\sum_{k=1}^{\infty} a_k^2$ and $\displaystyle\sum_{k=1}^{\infty} b_k^2$ both converge, then $\displaystyle\sum_{k=1}^{\infty} a_k b_k$ converges absolutely. *Hint:* First show that $2|a_k b_k| \le a_k^2 + b_k^2$.

42. Sketch the graph of $y = (\sin x)/x$ and then show that $\displaystyle\int_0^{\infty} (\sin x)/x \, dx$ converges.

43. Show that $\displaystyle\int_0^{\infty} |\sin x|/x \, dx$ diverges.

44. Show that the graph of $y = x \sin \dfrac{\pi}{x}$ on $(0, 1]$ has infinite length.

45. Note that

$$1 - \frac{1}{2} + \frac{1}{3} - \frac{1}{4} + \cdots - \frac{1}{2n} = 1 + \frac{1}{2} + \frac{1}{3} + \cdots + \frac{1}{2n}$$
$$- \left(1 + \frac{1}{2} + \frac{1}{3} + \cdots + \frac{1}{n} \right)$$
$$= \frac{1}{n+1} + \frac{1}{n+2} + \cdots + \frac{1}{2n}$$

Recognize the latter expression as a Riemann sum and use it to find the sum of the alternating harmonic series.

11.6
Power Series

So far, we have been studying what might be called *series of constants*—that is, series of the form $\sum u_n$, where each u_n is a number. Now we consider *series of functions*, series of the form $\sum u_n(x)$. A typical example of such a series is

$$\sum_{n=1}^{\infty} \frac{\sin nx}{n^2} = \frac{\sin x}{1} + \frac{\sin 2x}{4} + \frac{\sin 3x}{9} + \cdots$$

Of course, as soon as we substitute a value for x (such as $x = 2.1$), we are back to familiar territory; we have a series of constants.

There are two important questions to ask about a series of functions.

1. For what x's does the series converge?
2. To what function does it converge; that is, what is the sum $S(x)$ of the series?

The general situation is a proper subject for an advanced calculus course. However, even in elementary calculus, we can learn a good deal about the special case of a power series. A **power series in x** has the form

$$\sum_{n=0}^{\infty} a_n x^n = a_0 + a_1 x + a_2 x^2 + \cdots$$

(Here we interpret $a_0 x^0$ to be a_0 even if $x = 0$.) We can immediately answer our two questions for one such power series.

EXAMPLE 1   For what x's does the power series

$$\sum_{n=0}^{\infty} a x^n = a + a x + a x^2 + a x^3 + \cdots$$

converge and what is its sum? Assume $a \neq 0$.

Solution   We actually studied this series in Section 11.2 (with r in place of x) and called it a geometric series. It converges for $-1 < x < 1$ and has sum $S(x)$ given by

$$S(x) = \frac{a}{1 - x}, \qquad -1 < x < 1 \qquad \blacksquare$$

Convergence set

FIGURE 1

THE CONVERGENCE SET   We call the set on which a power series converges its **convergence set**. What kind of set can be a convergence set? Example 1 shows that it can be an open interval (see Figure 1). Are there other possibilities?

EXAMPLE 2   What is the convergence set for

$$\sum_{n=0}^{\infty} \frac{x^n}{(n+1)2^n} = 1 + \frac{1}{2}\frac{x}{2} + \frac{1}{3}\frac{x^2}{2^2} + \frac{1}{4}\frac{x^3}{2^3} + \cdots$$

Solution   Note that some of the terms may be negative (if x is negative). Let's test for absolute convergence using the Absolute Ratio Test (Theorem 11.5C).

$$\rho = \lim_{n \to \infty} \left| \frac{x^{n+1}}{(n+2)2^{n+1}} \div \frac{x^n}{(n+1)2^n} \right| = \lim_{n \to \infty} \frac{|x|}{2} \cdot \frac{n+1}{n+2} = \frac{|x|}{2}$$

The series converges absolutely (hence converges) when $\rho = |x|/2 < 1$ and diverges when $|x|/2 > 1$. Consequently, it converges when $|x| < 2$ and diverges when $|x| > 2$.

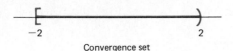

−2

2

Convergence set

FIGURE 2

If $x = 2$ or $x = -2$, the Ratio Test fails. However, when $x = 2$, the series is the harmonic series, which diverges; and when $x = -2$, it is the alternating harmonic series, which converges. We conclude that the convergence set for the given series is the interval $-2 \leq x < 2$ (Figure 2). ∎

EXAMPLE 3 Find the convergence set for $\sum\limits_{n=0}^{\infty} \dfrac{x^n}{n!}$.

Solution

$$\rho = \lim_{n \to \infty} \left| \frac{x^{n+1}}{(n+1)!} \div \frac{x^n}{n!} \right| = \lim_{n \to \infty} \frac{|x|}{n+1} = 0$$

Convergence set

FIGURE 3

We conclude from the Absolute Ratio Test that the series converges for all x (Figure 3). ∎

EXAMPLE 4 Find the convergence set for $\sum\limits_{n=0}^{\infty} n! x^n$.

Solution

$$\rho = \lim_{n \to \infty} \left| \frac{(n+1)! x^{n+1}}{n! x^n} \right| = \lim_{n \to \infty} (n+1)|x| = \begin{cases} 0 & \text{if } x = 0 \\ \infty & \text{if } x \neq 0 \end{cases}$$

0

Convergence set

FIGURE 4

We conclude that the series converges only at $x = 0$ (Figure 4). ∎

In each of our examples, the convergence set was an interval (a degenerate interval in the last example). This will always be the case. For example, it is impossible for a power series to have a convergence set consisting of two disconnected parts (like $[0, 1] \cup [2, 3]$). Our next theorem tells the whole story.

Theorem A

The convergence set for a power series $\sum a_n x^n$ is always an interval of one of the following three types.

(i) The single point $x = 0$.
(ii) An interval $(-R, R)$, plus possibly one or both endpoints.
(iii) The whole real line.

In (i), (ii), and (iii), the series is said to have **radius of convergence** 0, R, and ∞, respectively.

Proof Suppose that the series converges at $x = x_1 \neq 0$. Then $\lim\limits_{n \to \infty} a_n x_1^n = 0$, and so there is certainly a number N such that $|a_n x_1^n| < 1$ for $n \geq N$. Then,

for any x for which $|x| < |x_1|$,

$$|a_n x^n| = |a_n x_1^n| \left|\frac{x}{x_1}\right|^n < \left|\frac{x}{x_1}\right|^n$$

this holding for $n \geq N$. Now $\sum |x/x_1|^n$ converges, since it is a geometric series with ratio less than 1. Thus, by the Ordinary Comparison Test, $\sum |a_n x^n|$ converges. We have shown that if a power series converges at x_1, it converges (absolutely) for all x such that $|x| < |x_1|$.

On the other hand, suppose a power series diverges at x_2. Then it must diverge for all x for which $|x| > |x_2|$. For if it converged at x_1 such that $|x_1| > |x_2|$, then, by what we have already shown, it would converge at x_2, contrary to hypothesis.

These two paragraphs together eliminate all convergence sets except the three types mentioned in the theorem. ∎

Actually we have proved slightly more than we have claimed in Theorem A, and it is worth stating as another theorem.

Theorem B

A power series $\sum a_n x^n$ converges absolutely on the interior of its interval of convergence.

Of course, it might even converge absolutely at the endpoints of the interval of convergence, but of that we cannot be sure; witness Example 2.

POWER SERIES IN $x - a$ A series of the form

$$\sum a_n(x - a)^n = a_0 + a_1(x - a) + a_2(x - a)^2 + \cdots$$

is called a **power series in $x - a$**. All that we have said about power series in x applies equally well for series in $x - a$. In particular, its convergence set is always one of the following kinds of intervals.

1. The single point $x = a$.
2. An interval $(a - R, a + R)$, plus possibly one or both endpoints (Figure 5).
3. The whole real line.

$a - R \qquad a \qquad a + R$

FIGURE 5

EXAMPLE 5   Find the convergence set for $\displaystyle\sum_{n=0}^{\infty} \frac{(x - 1)^n}{(n + 1)^2}$.

Solution   We apply the Absolute Ratio Test.

$$\rho = \lim_{n \to \infty} \left|\frac{(x - 1)^{n+1}}{(n + 2)^2} \div \frac{(x - 1)^n}{(n + 1)^2}\right| = \lim_{n \to \infty} |x - 1| \frac{(n + 1)^2}{(n + 2)^2}$$

$$= |x - 1|$$

FIGURE 6

Convergence set

Thus, the series converges if $|x - 1| < 1$—that is, if $0 < x < 2$; it diverges if $|x - 1| > 1$. It also converges (even absolutely) at both of the endpoints 0 and 2, as we see by substitution of these values. The convergence set is the closed interval $[0, 2]$ (Figure 6). ∎

EXAMPLE 6 Determine the convergence set for

$$\frac{x + 2}{3} + \frac{(x + 2)^2 \ln 2}{2 \cdot 9} + \frac{(x + 2)^3 \ln 3}{3 \cdot 27} + \frac{(x + 2^4 \ln 4)}{4 \cdot 81} + \cdots$$

Solution The nth term is $u_n = \dfrac{(x + 2)^n \ln n}{n \cdot 3^n}$. Thus,

$$\rho = \lim_{n \to \infty} \left| \frac{(x + 2)^{n+1} \ln(n + 1)}{(n + 1)3^{n+1}} \cdot \frac{n3^n}{(x + 2)^n \ln n} \right|$$

$$= \frac{|x + 2|}{3} \lim_{n \to \infty} \frac{n}{n + 1} \frac{\ln(n + 1)}{\ln n} = \frac{|x + 2|}{3}$$

We know the series converges when $\rho < 1$, that is, when $|x + 2| < 3$ or, equivalently, $-5 < x < 1$, but we must check the endpoints -5 and 1.
At $x = -5$,

$$u_n = \frac{(-3)^n \ln n}{n3^n} = (-1)^n \frac{\ln n}{n}$$

and $\sum (-1)^n (\ln n)/n$ converges by the Alternating Series Test.
At $x = 1$, $u_n = (\ln n)/n$ and $\sum (\ln n)/n$ diverges by comparison with the harmonic series.
We conclude that the given series converges on the interval $-5 \le x < 1$. ∎

PROBLEM SET 11.6

In Problems 1–20, find the convergence set of the given power series. *Hint:* First find a formula for the nth term; then use the Absolute Ratio Test.

1. $\dfrac{x}{1 \cdot 2} - \dfrac{x^2}{2 \cdot 3} + \dfrac{x^3}{3 \cdot 4} - \dfrac{x^4}{4 \cdot 5} + \dfrac{x^5}{5 \cdot 6} - \cdots$

2. $1 + x + \dfrac{x^2}{2!} + \dfrac{x^3}{3!} + \dfrac{x^4}{4!} + \cdots$

3. $x - \dfrac{x^3}{3!} + \dfrac{x^5}{5!} - \dfrac{x^7}{7!} + \dfrac{x^9}{9!} - \cdots$

4. $1 - \dfrac{x^2}{2!} + \dfrac{x^4}{4!} - \dfrac{x^6}{6!} + \dfrac{x^8}{8!} - \dfrac{x^{10}}{10!} + \cdots$

5. $x + 2x^2 + 3x^3 + 4x^4 + \cdots$

6. $x + 2^2 x^2 + 3^2 x^3 + 4^2 x^4 + \cdots$

7. $1 - x + \dfrac{x^2}{2} - \dfrac{x^3}{3} + \dfrac{x^4}{4} - \cdots$

8. $1 + x + \dfrac{x^2}{\sqrt{2}} + \dfrac{x^3}{\sqrt{3}} + \dfrac{x^4}{\sqrt{4}} + \dfrac{x^5}{\sqrt{5}} + \cdots$

9. $1 - \dfrac{x}{1 \cdot 3} + \dfrac{x^2}{2 \cdot 4} - \dfrac{x^3}{3 \cdot 5} + \dfrac{x^4}{4 \cdot 6} - \cdots$

10. $\dfrac{x}{2^2 - 1} + \dfrac{x^2}{3^2 - 1} + \dfrac{x^3}{4^2 - 1} + \dfrac{x^4}{5^2 - 1} + \cdots$

11. $1 - \dfrac{x}{2} + \dfrac{x^2}{2^2} - \dfrac{x^3}{2^3} + \dfrac{x^4}{2^4} - \cdots$

12. $1 + 2x + 2^2 x^2 + 2^3 x^3 + 2^4 x^4 + \cdots$

13. $1 + 2x + \dfrac{2^2 x^2}{2!} + \dfrac{2^3 x^3}{3!} + \dfrac{2^4 x^4}{4!} + \cdots$

14. $\dfrac{x}{2} + \dfrac{2x^2}{3} + \dfrac{3x^2}{4} + \dfrac{4x^4}{5} + \dfrac{5x^5}{6} + \cdots$

15. $\dfrac{(x-1)}{1} + \dfrac{(x-1)^2}{2} + \dfrac{(x-1)^3}{3} + \dfrac{(x-1)^4}{4} + \cdots$

16. $1 + (x+2) + \dfrac{(x+2)^2}{2!} + \dfrac{(x+2)^3}{3!} + \cdots$

17. $1 + \dfrac{(x+1)}{2} + \dfrac{(x+1)^2}{2^2} + \dfrac{(x+1)^3}{2^3} + \cdots$

18. $\dfrac{(x-2)}{1^2} + \dfrac{(x-2)^2}{2^2} + \dfrac{(x-2)^3}{3^2} + \dfrac{(x-2)^4}{4^2} + \cdots$

19. $\dfrac{(x+5)}{1 \cdot 2} + \dfrac{(x+5)^2}{2 \cdot 3} + \dfrac{(x+5)^3}{3 \cdot 4} + \dfrac{(x+5)^4}{4 \cdot 5} + \cdots$

20. $(x+3) - 2(x+3)^2 + 3(x+3)^3 - 4(x+3)^4 + \cdots$

21. From Example 3, we know that $\sum x^n / n!$ converges for all x. Why can we conclude that $\lim\limits_{n \to \infty} x^n / n! = 0$ for all x?

22. Let k be an arbitrary number and $-1 < x < 1$. Prove that

$$\lim_{n \to \infty} \frac{k(k-1)(k-2) \cdots (k-n)}{n!} x^n = 0$$

Hint: See Problem 21.

23. Find the radius of convergence of

$$\sum_{n=1}^{\infty} \frac{1 \cdot 2 \cdot 3 \cdots n}{1 \cdot 3 \cdot 5 \cdots (2n-1)} x^{2n+1}$$

24. Find the radius of convergence of

$$\sum_{n=0}^{\infty} \frac{(pn)!}{(n!)^p} x^n$$

where p is a positive integer.

25. Find the sum $S(x)$ of $\sum\limits_{n=0}^{\infty} (x-3)^n$. Where is it valid?

26. Suppose $\sum\limits_{n=0}^{\infty} a_n (x-3)^n$ converges at $x = -1$. Why can you conclude that it converges at $x = 6$? Can you be sure it converges at $x = 7$? Explain.

27. Find the convergence set for each series.

(a) $\displaystyle\sum_{n=1}^{\infty} \frac{(3x+1)^n}{n \cdot 2^n}$ (b) $\displaystyle\sum_{n=1}^{\infty} (-1)^n \frac{(2x-3)^n}{4^n \sqrt{n}}$

28. Refer to Problem 52 of Section 11.1, where the Fibonacci sequence $f_1, f_2, f_3, \ldots$ was defined. Find the radius of convergence of $\sum\limits_{n=1}^{\infty} f_n x^n$.

29. Suppose that $a_{n+3} = a_n$ and let $S(x) = \sum\limits_{n=0}^{\infty} a_n x^n$. Show that the series converges for $|x| < 1$ and give a formula for $S(x)$.

30. Follow the directions of Problem 29 for the case where $a_{n+p} = a_n$ for some fixed positive integer p.

11.7
Operations on
Power Series

We know from the previous section that the convergence set of a power series $\sum a_n x^n$ is an interval I. This interval is the domain for a new function $S(x)$, the sum of the series. The most obvious question to ask about $S(x)$ is whether we can give a simple formula for it. We have done this for one series, a geometric series.

$$\sum_{n=0}^{\infty} ax^n = \frac{a}{1-x}, \qquad -1 < x < 1$$

Actually, there is little reason to hope that the sum of an arbitrarily given power series will be one of the elementary functions studied earlier in this book, though we will make a little progress in that direction in this section and more in Section 11.8.

A better question to ask now is whether we can say anything about the properties of $S(x)$. For example, is it differentiable? Is it integrable? The answer to both questions is yes.

TERM-BY-TERM DIFFERENTIATION AND INTEGRATION Think of a power series as a polynomial with infinitely many terms. It behaves just like a polynomial under both integration and differentiation; these operations can be performed term by term, as follows.

Theorem A

Suppose that $S(x)$ is the sum of a power series on an interval I; that is,

$$S(x) = \sum_{n=0}^{\infty} a_n x^n = a_0 + a_1 x + a_2 x^2 + a_3 x^3 + \cdots$$

Then, if x is interior to I,

(i) $S'(x) = \sum_{n=0}^{\infty} D_x(a_n x^n) = \sum_{n=1}^{\infty} n a_n x^{n-1}$
$$= a_1 + 2a_2 x + 3a_3 x^2 + \cdots$$

(ii) $\int_0^x S(t)\,dt = \sum_{n=0}^{\infty} \int_0^x a_n t^n\,dt = \sum_{n=0}^{\infty} \frac{a_n}{n+1} x^{n+1}$
$$= a_0 x + \frac{1}{2} a_1 x^2 + \frac{1}{3} a_2 x^3 + \frac{1}{4} a_3 x^4 + \cdots$$

The theorem entails several things. It asserts that S is both differentiable and integrable, it shows how the derivative and integral may be calculated, and it implies that the radius of convergence of both the differentiated and integrated series is the same as for the original series (though it says nothing about the endpoints of the interval of convergence). The theorem is hard to prove. We leave the proof to more advanced books.

EXAMPLE 1 Apply Theorem A to the geometric series

$$\frac{1}{1-x} = 1 + x + x^2 + x^3 + \cdots, \qquad -1 < x < 1$$

to obtain formulas for two new series.

Solution Differentiating term by term yields

$$\frac{1}{(1-x)^2} = 1 + 2x + 3x^2 + 4x^3 + \cdots, \qquad -1 < x < 1$$

Integrating term by term gives

$$\int_0^x \frac{1}{1-t}\,dt = \int_0^x 1\,dt + \int_0^x t\,dt + \int_0^x t^2\,dt + \cdots$$

that is,

$$-\ln(1-x) = x + \frac{x^2}{2} + \frac{x^3}{3} + \cdots, \qquad -1 < x < 1$$

If we replace x by $-x$ in the latter and multiply both sides by -1, we obtain

$$\ln(1 + x) = x - \frac{x^2}{2} + \frac{x^3}{3} - \frac{x^4}{4} + \cdots, \qquad -1 < x < 1$$

From Problem 45 of Section 11.5, we learn that this result is valid at the endpoint $x = 1$ (also see the note in the margin). ∎

EXAMPLE 2   Find the power series representation for $\tan^{-1} x$.

Solution   Recall that

$$\tan^{-1} x = \int_0^x \frac{1}{1 + t^2}\, dt$$

From the geometric series for $1/(1 - x)$ with x replaced by $-t^2$, we get

$$\frac{1}{1 + t^2} = 1 - t^2 + t^4 - t^6 + \cdots, \qquad -1 < t < 1$$

Thus,

$$\tan^{-1} x = \int_0^x (1 - t^2 + t^4 - t^6 + \cdots)\, dt$$

that is,

$$\tan^{-1} x = x - \frac{x^3}{3} + \frac{x^5}{5} - \frac{x^7}{7} + \cdots, \qquad -1 < x < 1$$

(By the note in the margin, this also holds at $x = \pm 1$.) ∎

EXAMPLE 3   Find a formula for the sum of the series

$$S(x) = 1 + x + \frac{x^2}{2!} + \frac{x^3}{3!} + \cdots$$

Solution   We saw earlier (Section 11.6, Example 3) that this series converges for all x. Differentiating term by term, we obtain

$$S'(x) = 1 + x + \frac{x^2}{2!} + \frac{x^3}{3!} + \cdots$$

that is, $S'(x) = S(x)$ for all x. Furthermore, $S(0) = 1$. This differential equation has the unique solution $S(x) = e^x$ (see Section 7.5). Thus,

$$e^x = 1 + x + \frac{x^2}{2!} + \frac{x^3}{3!} + \cdots$$

∎

EXAMPLE 4 Obtain the power series representation for e^{-x^2}.

Solution Simply substitute $-x^2$ for x in the series for e^x.

$$e^{-x^2} = 1 - x^2 + \frac{x^4}{2!} - \frac{x^6}{3!} + \cdots \qquad \blacksquare$$

ALGEBRAIC OPERATIONS Power series behave like polynomials under the operations of addition and subtraction, as we know from Theorem 11.2B. The same is true for multiplication and division, as we now illustrate.

EXAMPLE 5 Multiply and divide the power series for $\ln(1 + x)$ by that for e^x.

Solution We refer to Examples 1 and 3 for the required series. The key to multiplication is to first find the constant term, then the x-term, then the x^2-term, and so on. We arrange our work as follows.

$$0 + x - \frac{x^2}{2} + \frac{x^3}{3} - \frac{x^4}{4} + \cdots$$

$$1 + x + \frac{x^2}{2!} + \frac{x^3}{3!} + \frac{x^4}{4!} + \cdots$$

$$0 + (0 + 1)x + \left(0 + 1 - \frac{1}{2}\right)x^2 + \left(0 + \frac{1}{2!} - \frac{1}{2} + \frac{1}{3}\right)x^3$$

$$+ \left(0 + \frac{1}{3!} - \frac{1}{2!2} + \frac{1}{3} - \frac{1}{4}\right)x^4 + \cdots$$

$$= 0 + x + \frac{1}{2}x^2 + \frac{1}{3}x^3 + 0 \cdot x^4 + \cdots$$

Here is how division is done.

$$
1 + x + \frac{1}{2}x^2 + \frac{1}{6}x^3 + \cdots \overline{\smash{\big)}\,
\begin{array}{l}
x - \frac{3}{2}x^2 + \frac{4}{3}x^3 - \quad x^4 + \cdots \\
\hline
x - \frac{1}{2}x^2 + \frac{1}{3}x^3 - \frac{1}{4}x^4 + \cdots \\
x + \quad x^2 + \frac{1}{2}x^3 + \frac{1}{6}x^4 + \cdots \\
\hline
\quad -\frac{3}{2}x^2 - \frac{1}{6}x^3 - \frac{5}{12}x^4 + \cdots \\
\quad -\frac{3}{2}x^2 - \frac{3}{2}x^3 - \frac{3}{4}x^4 + \cdots \\
\hline
\quad\quad\quad \frac{4}{3}x^3 + \frac{1}{3}x^4 + \cdots \\
\quad\quad\quad \frac{4}{3}x^3 + \frac{4}{3}x^4 + \cdots \\
\hline
\quad\quad\quad\quad\quad -x^4 + \cdots
\end{array}}
$$

$\blacksquare$

The real question relative to Example 5 is whether the two series we have obtained converge to $[\ln(1 + x)]e^x$ and $[\ln(1 + x)]/e^x$, respectively. Our next theorem, stated without proof, answers this question.

Let $f(x) = \sum a_n x^n$ and $g(x) = \sum b_n x^n$, with both of these series converging at least for $|x| < r$. If the operations of addition, subtraction, and multiplication are performed on these series as if they were polynomials, the resulting series will converge for $|x| < r$ and represent $f(x) + g(x)$, $f(x) - g(x)$, and $f(x) \cdot g(x)$, respectively. If $b_0 \neq 0$, the corresponding result holds for division, but we can guarantee its validity only for $|x|$ sufficiently small.

We mention that the operation of substituting one power series in another is also legitimate for $|x|$ sufficiently small, provided the constant term of the substituted series is zero. Here is an illustration.

EXAMPLE 6 Find the power series for $e^{\tan^{-1}x}$ through terms of degree 4.

Solution Since

$$e^u = 1 + u + \frac{u^2}{2!} + \frac{u^3}{3!} + \frac{u^4}{4!} + \cdots$$

$$e^{\tan^{-1}x} = 1 + \tan^{-1}x + \frac{(\tan^{-1}x)^2}{2!} + \frac{(\tan^{-1}x)^3}{3!} + \frac{(\tan^{-1}x)^4}{4!} + \cdots$$

Now substitute the series for $\tan^{-1}x$ from Example 2 and combine like terms.

$$e^{\tan^{-1}x} = 1 + \left(x - \frac{x^3}{3} + \cdots\right) + \frac{\left(x - \frac{x^3}{3} + \cdots\right)^2}{2!} + \frac{\left(x - \frac{x^3}{3} + \cdots\right)^3}{3!}$$

$$+ \frac{\left(x - \frac{x^3}{3} + \cdots\right)^4}{4!} + \cdots$$

$$= 1 + \left(x - \frac{x^3}{3} + \cdots\right) + \frac{(x^2 - \frac{2}{3}x^4 + \cdots)}{2} + \frac{(x^3 + \cdots)}{6}$$

$$+ \frac{(x^4 + \cdots)}{24} + \cdots$$

$$= 1 + x + \frac{x^2}{2} - \frac{x^3}{6} - \frac{7x^4}{24} + \cdots$$ ∎

POWER SERIES IN $x - a$ We have stated the theorems of this section for power series in x, but with obvious modifications they are equally valid for power series in $x - a$.

CHAPTER 11/Infinite Series

In Problems 1–10, find the power series representation for $f(x)$ and specify the radius of convergence. Each is somehow related to a geometric series (see Examples 1 and 2).

1. $f(x) = \dfrac{1}{1 + x} = \dfrac{1}{1 - (-x)}$

2. $f(x) = \dfrac{1}{(1 + x)^2}$ *Hint*: Differentiate Problem 1.

3. $f(x) = \dfrac{1}{(1 - x)^3}$

4. $f(x) = \dfrac{x}{(1 + x)^2}$

5. $f(x) = \dfrac{1}{2 - 3x} = \dfrac{\frac{1}{2}}{1 - \frac{3}{2}x}$

6. $f(x) = \dfrac{1}{3 + 2x}$

7. $f(x) = \dfrac{x^2}{1 - x^4}$ **8.** $f(x) = \dfrac{x^3}{2 - x^3}$

9. $f(x) = \displaystyle\int_0^x \ln(1 + t)\, dt$ **10.** $f(x) = \displaystyle\int_0^x \tan^{-1} t\, dt$

11. Obtain the power series in x for $\ln[(1 + x)/(1 - x)]$ and specify its radius of convergence. *Hint*: $\ln[(1 + x)/(1 - x)] = \ln(1 + x) - \ln(1 - x)$.

12. Show that any positive number M can be represented by $(1 + x)/(1 - x)$, where x lies within the interval of convergence of the series of Problem 11. Hence conclude that the natural logarithm of any positive number can be found by means of this series. Find $\ln 8$ this way to three decimal places.

In Problems 13–16, use the result of Example 3 to find the power series in x for the given functions.

13. $f(x) = e^{-x}$ **14.** $f(x) = xe^{x^2}$

15. $f(x) = e^x + e^{-x}$ **16.** $f(x) = e^{2x} - 1 - 2x$

In Problems 17–24, use the methods of Example 5 to find power series in x for each function f.

17. $f(x) = e^{-x} \cdot \dfrac{1}{1 - x}$ **18.** $f(x) = e^x \tan^{-1} x$

19. $f(x) = \dfrac{\tan^{-1} x}{e^x}$ **20.** $f(x) = \dfrac{e^x}{1 + \ln(1 + x)}$

21. $f(x) = (\tan^{-1} x)(1 + x^2 + x^4)$

22. $f(x) = \dfrac{\tan^{-1} x}{1 + x^2 + x^4}$

23. $f(x) = \displaystyle\int_0^x \dfrac{e^t}{1 + t}\, dt$ **24.** $f(x) = \displaystyle\int_0^x \dfrac{\tan^{-1} t}{t}\, dt$

25. Find the sum of each of the following series by recognizing how it is related to something familiar.

(a) $x - x^2 + x^3 - x^4 + x^5 - \cdots$

(b) $\dfrac{1}{2!} + \dfrac{x}{3!} + \dfrac{x^2}{4!} + \dfrac{x^3}{5!} + \cdots$

(c) $2x + \dfrac{4x^2}{2} + \dfrac{8x^3}{3} + \dfrac{16x^4}{4} + \cdots$

26. Follow the directions of Problem 25.

(a) $1 + x^2 + x^4 + x^6 + x^8 + \cdots$

(b) $\cos x + \cos^2 x + \cos^3 x + \cos^4 x + \cdots$

(c) $\dfrac{x^2}{2} + \dfrac{x^4}{4} + \dfrac{x^6}{6} + \dfrac{x^8}{8} + \cdots$

27. Find the sum of $\displaystyle\sum_{n=1}^{\infty} nx^n$.

28. Find the sum of $\displaystyle\sum_{n=1}^{\infty} n(n + 1)x^n$.

29. Use the method of substitution (Example 6) to find power series through terms of degree 3.

(a) $\tan^{-1}(e^x - 1)$ (b) $e^{e^x - 1}$

30. Suppose that $f(x) = \displaystyle\sum_{n=0}^{\infty} a_n x^n = \sum_{n=0}^{\infty} b_n x^n$ for $|x| < R$. Show that $a_n = b_n$ for all n. *Hint*: Let $x = 0$; then differentiate and let $x = 0$ again. Continue.

31. Find the power series representation of $x/(x^2 - 3x + 2)$. *Hint*: Use partial fractions.

32. Let $y = y(x) = x - \dfrac{x^3}{3!} + \dfrac{x^5}{5!} - \dfrac{x^7}{7!} + \cdots$. Show that y satisfies the differential equation $y'' + y = 0$ with the conditions $y(0) = 0$ and $y'(0) = 1$. From this, guess at a simple formula for y.

33. Let $\{f_n\}$ be the Fibonacci sequence defined by

$$f_{n+2} = f_{n+1} + f_n, \qquad f_0 = 0, f_1 = 1$$

(See Problem 52 of Section 11.1 and Problem 28 of Section 11.6.) If $F(x) = \displaystyle\sum_{n=0}^{\infty} f_n x^n$, show that

$$F(x) - xF(x) - x^2 F(x) = x$$

and then use this fact to obtain a simple formula for $F(x)$.

34. Let $y = y(x) = \displaystyle\sum_{n=0}^{\infty} \dfrac{f_n}{n!} x^n$, where f_n is as in Problem 33. Show that y satisfies the differential equation $y'' - y' - y = 0$.

$\boxed{\text{C}}$ **35.** Did you ever wonder how people find the decimal expansion of π to a large number of places? One method depends on the following identity (see Problem 47 of Section 7.6).

$$\pi = 16 \tan^{-1}(\tfrac{1}{5}) - 4 \tan^{-1}(\tfrac{1}{239})$$

Find the first 6 digits of π using this identity and the series for $\tan^{-1} x$. (You will need terms through $x^9/9$ for $\tan^{-1}(\tfrac{1}{5})$ but only the first term for $\tan^{-1}(1/239)$.) In 1706, John Machin used this method to calculate the first 100 digits of π, while in 1973, Jean Guilloud and Martine Bouyer found the first million digits using the related identity

$$\pi = 48 \tan^{-1}(\tfrac{1}{18}) + 32 \tan^{-1}(\tfrac{1}{57}) - 20 \tan^{-1}(\tfrac{1}{239})$$

In 1983, π was calculated to over 16 million digits by a somewhat different method. Of course, computers were used in these recent calculations.

36. The number e is readily calculated to as many digits as desired using the rapidly converging series

$$e = 1 + 1 + \frac{1}{2!} + \frac{1}{3!} + \frac{1}{4!} + \cdots$$

This series can also be used to show that e is irrational. Do so by completing the following argument. Suppose $e = p/q$, where p and q are positive integers. Choose $n > q$ and let

$$M = n!\left(e - 1 - 1 - \frac{1}{2!} - \frac{1}{3!} - \cdots - \frac{1}{n!}\right)$$

Now M is a positive integer. (Why?) Also

$$M = n!\left[\frac{1}{(n+1)!} + \frac{1}{(n+2)!} + \frac{1}{(n+3)!} + \cdots\right]$$

$$= \frac{1}{n+1} + \frac{1}{(n+1)(n+2)} + \frac{1}{(n+1)(n+2)(n+3)} + \cdots$$

$$< \frac{1}{n+1} + \frac{1}{(n+1)^2} + \frac{1}{(n+1)^3} + \cdots$$

$$= \frac{1}{n}$$

which gives a contradiction (to what?).

11.8
Taylor and
Maclaurin Series

The big question still dangling is this: Given a function f (for example, $\sin x$ or $\ln(\cos^2 x)$), can we represent it as a power series in x or, more generally, in $x - a$? More precisely, can we find numbers $c_0, c_1, c_2, c_3, \ldots$ such that

$$f(x) = c_0 + c_1(x - a) + c_2(x - a)^2 + c_3(x - a)^3 + \cdots$$

on some interval around a?

Suppose such a representation exists. Then, by the theorem on differentiating series (Theorem 11.7A),

$$f'(x) = c_1 + 2c_2(x - a) + 3c_3(x - a)^2 + 4c_4(x - a)^3 + \cdots$$

$$f''(x) = 2!c_2 + 3!c_3(x - a) + 4 \cdot 3c_4(x - a)^2 + \cdots$$

$$f'''(x) = 3!c_3 + 4!c_4(x - a) + 5 \cdot 4 \cdot 3c_5(x - a)^2 + \cdots$$

$$\vdots$$

When we substitute $x = a$ and solve for c_n, we get

$$c_0 = f(a)$$

$$c_1 = f'(a)$$

$$c_2 = \frac{f''(a)}{2!}$$

$$c_3 = \frac{f'''(a)}{3!}$$

and, more generally,

$$c_n = \frac{f^{(n)}(a)}{n!}$$

(To make this valid for $n = 0$, we assume $f^{(0)}(a)$ means $f(a)$ and $0! = 1$.) Thus, the coefficients c_n are determined by the function f. This also shows that a function f cannot be represented by two different power series in $x - a$, an important point that we have glossed over until now. We summarize in the following theorem.

Theorem A

(Uniqueness Theorem). Suppose f satisfies

$$f(x) = c_0 + c_1(x - a) + c_2(x - a)^2 + c_3(x - a)^3 + \cdots$$

for all x in some interval around a. Then,

$$c_n = \frac{f^{(n)}(a)}{n!}$$

Thus, a function cannot have more than one power series in $x - a$ that represents it.

The form of the coefficients c_n should remind us of Taylor's Formula. Because of this, we call the power series representation of a function in $x - a$ its **Taylor series**. If $a = 0$, the corresponding series is called the **Maclaurin series**.

CONVERGENCE OF THE TAYLOR SERIES   But still the big question remains. Given a function f, can we represent it in a power series in $x - a$ (which must necessarily be the Taylor series)? Here is the answer.

Theorem B

(Taylor's Theorem). Let f be a function with derivatives of all orders in some interval $(a - r, a + r)$. A necessary and sufficient condition that the Taylor series

$$f(a) + f'(a)(x - a) + \frac{f''(a)}{2!}(x - a)^2 + \frac{f'''(a)}{3!}(x - a)^3 + \cdots$$

represents the function f on that interval is that

$$\lim_{n \to \infty} R_n(x) = 0$$

where $R_n(x)$ is the remainder in Taylor's Formula; that is,

$$R_n(x) = \frac{f^{(n+1)}(c)}{(n + 1)!}(x - a)^{n+1}$$

c being some point in $(a - r, a + r)$.

Proof We need only recall Taylor's Formula (Theorem 10.2A),

$$f(x) = f(a) + f'(a)(x - a) + \cdots + \frac{f^{(n)}(a)}{n!}(x - a)^n + R_n(x)$$

and the result is obvious. ∎

Note that if $a = 0$, we get the Maclaurin series

$$f(0) + f'(0)x + \frac{f''(0)}{2!}x^2 + \frac{f'''(0)}{3!}x^3 + \cdots$$

EXAMPLE 1 Find the Maclaurin series for $\sin x$ and prove that it represents $\sin x$ for all x.

Solution

$$
\begin{array}{ll}
f(x) = \sin x & f(0) = 0 \\
f'(x) = \cos x & f'(0) = 1 \\
f''(x) = -\sin x & f''(0) = 0 \\
f'''(x) = -\cos x & f'''(0) = -1 \\
f^{(4)}(x) = \sin x & f^{(4)}(0) = 0 \\
\quad\vdots & \quad\vdots
\end{array}
$$

Thus,

$$\boxed{\sin x = x - \frac{x^3}{3!} + \frac{x^5}{5!} - \frac{x^7}{7!} + \cdots}$$

and this is valid for all x, provided we can show

$$\lim_{n \to \infty} R_n(x) = \lim_{n \to \infty} \frac{f^{(n+1)}(c)}{(n+1)!} x^{n+1} = 0$$

Now, $\left|f^{(n+1)}(x)\right| = |\cos x|$ or $\left|f^{(n+1)}(x)\right| = |\sin x|$ and so

$$|R_n(x)| \leq \frac{|x|^{n+1}}{(n+1)!}$$

But $\lim_{n \to \infty} x^n/n! = 0$, since $x^n/n!$ is the nth term of a convergent series (see Example 3 and Problem 21 of Section 11.6). As a consequence, we see that $\lim_{n \to \infty} R_n(x) = 0$. ∎

EXAMPLE 2 Find the Maclaurin series for $\cos x$ and show that it represents $\cos x$ for all x.

Solution We could proceed as in Example 1. However, it is easier to get the result by differentiating the series of that example (a valid procedure according to Theorem 11.7A). We obtain

$$\cos x = 1 - \frac{x^2}{2!} + \frac{x^4}{4!} - \frac{x^6}{6!} + \cdots$$

∎

EXAMPLE 3 Find the Maclaurin series for $f(x) = \cosh x$ in two different ways, and show that it represents $\cosh x$ for all x.

Solution

Method 1 is the direct method.

$$f(x) = \cosh x \qquad f(0) = 1$$
$$f'(x) = \sinh x \qquad f'(0) = 0$$
$$f''(x) = \cosh x \qquad f''(0) = 1$$
$$f'''(x) = \sinh x \qquad f'''(0) = 0$$
$$\vdots \qquad\qquad \vdots$$

Thus,

$$\cosh x = 1 + \frac{x^2}{2!} + \frac{x^4}{4!} + \frac{x^6}{6!} + \cdots$$

provided we can show $\lim_{n \to \infty} R_n(x) = 0$ for all x.

Now let B be an arbitrary number and suppose $|x| < B$. Then,

$$|\cosh x| = \left| \frac{e^x + e^{-x}}{2} \right| \leq \frac{e^x}{2} + \frac{e^{-x}}{2} \leq \frac{e^B}{2} + \frac{e^B}{2} = e^B$$

By similar reasoning, $|\sinh x| \leq e^B$. Since $f^{(n+1)}(x)$ is either $\cosh x$ or $\sinh x$, we conclude that

$$|R_n(x)| = \left| \frac{f^{(n+1)}(c)x^{n+1}}{(n+1)!} \right| \leq \frac{e^B |x|^{n+1}}{(n+1)!}$$

The latter expression tends to zero as $n \to \infty$, just as in Example 1.

Method 2. We use the fact that $\cosh x = (e^x + e^{-x})/2$. From Example 3 of Section 11.7,

$$e^x = 1 + x + \frac{x^2}{2!} + \frac{x^3}{3!} + \frac{x^4}{4!} + \cdots$$

$$e^{-x} = 1 - x + \frac{x^2}{2!} - \frac{x^3}{3!} + \frac{x^4}{4!} + \cdots$$

The previously obtained result follows by adding these two series and dividing by 2. ∎

EXAMPLE 4 Find the Maclaurin series for $\sinh x$ and show that it represents $\sinh x$ for all x.

Solution We do both jobs at once when we differentiate the series for $\cosh x$ (Example 3) term by term and use Theorem 11.7A.

$$\sinh x = x + \frac{x^3}{3!} + \frac{x^5}{5!} + \frac{x^7}{7!} + \cdots$$

∎

THE BINOMIAL SERIES We are all familiar with the Binomial Formula. For a positive integer p,

$$(1 + x)^p = 1 + \binom{p}{1}x + \binom{p}{2}x^2 + \cdots + \binom{p}{p}x^p$$

where

$$\binom{p}{k} = \frac{p(p-1)(p-2)\cdots(p-k+1)}{k!}$$

Note that the symbol $\binom{p}{k}$ makes sense for any real number p, provided k is a positive integer. This suggests a theorem.

Theorem C

(Binomial Series). For any real number p and for $|x| < 1$,

$$(1 + x)^p = 1 + \binom{p}{1}x + \binom{p}{2}x^2 + \binom{p}{3}x^3 + \cdots$$

with $\binom{p}{k}$ defined as above.

Partial proof Let $f(x) = (1 + x)^p$. Then,

$$f(x) = (1 + x)^p \qquad\qquad f(0) = 1$$
$$f'(x) = p(1 + x)^{p-1} \qquad\qquad f'(0) = p$$
$$f''(x) = p(p-1)(1 + x)^{p-2} \qquad\qquad f''(0) = p(p-1)$$
$$f'''(x) = p(p-1)(p-2)(1 + x)^{p-3} \qquad f'''(0) = p(p-1)(p-2)$$
$$\vdots \qquad\qquad\qquad\qquad \vdots$$

Thus, the Maclaurin series for $(1 + x)^p$ is as indicated in the theorem. To show that it represents $(1 + x)^p$, we need to show that $\lim_{n \to \infty} R_n(x) = 0$. That, unfortunately, is difficult, and we leave it for more advanced courses. (See Problem 36 for a completely different way to prove Theorem C.) ∎

If p is a positive integer, $\binom{p}{k} = 0$ for $k > p$, and so the infinite series collapses to a series with finitely many terms; the series reduces to the usual Binomial Formula in this case.

EXAMPLE 5 Represent $(1 - x)^{-2}$ in a Maclaurin series for $-1 < x < 1$.

Solution By Theorem C,

$$(1 + x)^{-2} = 1 + (-2)x + \frac{(-2)(-3)}{2!} x^2 + \frac{(-2)(-3)(-4)}{3!} x^3 + \cdots$$

$$= 1 - 2x + 3x^2 - 4x^3 + \cdots$$

Thus,

$$(1 - x)^{-2} = 1 + 2x + 3x^2 + 4x^3 + \cdots$$

Naturally, this agrees with a result we obtained by a different method in Example 1 of Section 11.7. ∎

EXAMPLE 6 Represent $\sqrt{1 + x}$ in a Maclaurin series and use it to approximate $\sqrt{1.1}$ to five decimal places.

Solution By Theorem C,

$$(1 + x)^{1/2} = 1 + \frac{1}{2} x + \frac{(\frac{1}{2})(-\frac{1}{2})}{2!} x^2 + \frac{(\frac{1}{2})(-\frac{1}{2})(-\frac{3}{2})}{3!} x^3$$

$$+ \frac{(\frac{1}{2})(-\frac{1}{2})(-\frac{3}{2})(-\frac{5}{2})}{4!} x^4 + \cdots$$

$$= 1 + \frac{1}{2} x - \frac{1}{8} x^2 + \frac{1}{16} x^3 - \frac{5}{128} x^4 + \cdots$$

Thus,

$$\sqrt{1.1} = 1 + \frac{0.1}{2} - \frac{0.01}{8} + \frac{0.001}{16} - \frac{5(0.0001)}{128} + \cdots$$

$$\approx 1.04881$$ ∎

EXAMPLE 7 Compute $\displaystyle\int_0^{0.4} \sqrt{1 + x^4}\, dx$ to five decimal places.

Solution From Example 6,

$$\sqrt{1 + x^4} = 1 + \frac{1}{2} x^4 - \frac{1}{8} x^8 + \frac{1}{16} x^{12} - \frac{5}{128} x^{16} + \cdots$$

Thus,

$$\int_0^{0.4} \sqrt{1 + x^4}\, dx = \left[x + \frac{x^5}{10} - \frac{x^9}{72} + \frac{x^{13}}{208} + \cdots \right]_0^{0.4} \approx 0.40102 \qquad \blacksquare$$

SUMMARY We conclude our discussion of series with a list of the important Maclaurin series we have found. These series will be useful in doing the problem set but, what is more significant, they find application throughout mathematics and science.

Important Maclaurin Series

1. $\dfrac{1}{1 - x} = 1 + x + x^2 + x^3 + x^4 + \cdots$ $\qquad\qquad -1 < x < 1$

2. $\ln(1 + x) = x - \dfrac{x^2}{2} + \dfrac{x^3}{3} - \dfrac{x^4}{4} + \dfrac{x^5}{5} - \cdots$ $\qquad -1 < x \le 1$

3. $\tan^{-1} x = x - \dfrac{x^3}{3} + \dfrac{x^5}{5} - \dfrac{x^7}{7} + \dfrac{x^9}{9} + \cdots$ $\qquad -1 \le x \le 1$

4. $e^x = 1 + x + \dfrac{x^2}{2!} + \dfrac{x^3}{3!} + \dfrac{x^4}{4!} + \cdots$

5. $\sin x = x - \dfrac{x^3}{3!} + \dfrac{x^5}{5!} - \dfrac{x^7}{7!} + \dfrac{x^9}{9!} - \cdots$

6. $\cos x = 1 - \dfrac{x^2}{2!} + \dfrac{x^4}{4!} - \dfrac{x^6}{6!} + \dfrac{x^8}{8!} - \cdots$

7. $\sinh x = x + \dfrac{x^3}{3!} + \dfrac{x^5}{5!} + \dfrac{x^7}{7!} + \dfrac{x^9}{9!} + \cdots$

8. $\cosh x = 1 + \dfrac{x^2}{2!} + \dfrac{x^4}{4!} + \dfrac{x^6}{6!} + \dfrac{x^8}{8!} + \cdots$

9. $(1 + x)^p = 1 + \binom{p}{1} x + \binom{p}{2} x^2 + \binom{p}{3} x^3 + \binom{p}{4} x^4 + \cdots$ $\quad -1 < x < 1$

PROBLEM SET 11.8

In Problems 1–16, find the terms through x^5 in the Maclaurin series for $f(x)$. *Hint*: It may be easiest to use known Maclaurin series and then perform multiplications, divisions, and so on. For example, $\tan x = (\sin x)/(\cos x)$.

1. $f(x) = \tan x$

2. $f(x) = \tanh x$

3. $f(x) = e^x \sin x$

4. $f(x) = e^{-x} \cos x$

5. $f(x) = \cos x \ln(1 + x)$

6. $f(x) = (\sin x)\sqrt{1 + x}$

7. $f(x) = e^x + x + \sin x$

8. $f(x) = \dfrac{\cos x - 1 + x^2/2}{x^4}$

9. $f(x) = \dfrac{1}{1 - x}\cosh x$

10. $f(x) = \dfrac{1}{1 + x}\ln\left(\dfrac{1}{1 + x}\right) = \dfrac{-\ln(1 + x)}{1 + x}$

11. $f(x) = \dfrac{1}{1 + x + x^2}$

12. $f(x) = \dfrac{1}{1 - \sin x}$

13. $f(x) = \sin^3 x$

14. $f(x) = x(\sin 2x + \sin 3x)$

15. $f(x) = x \sec(x^2) + \sin x$

16. $f(x) = \dfrac{\cos x}{\sqrt{1 + x}}$

17. $f(x) = (1 + x)^{3/2}$

18. $f(x) = (1 - x^2)^{2/3}$

In Problems 19–24, find the Taylor series in $x - a$ through the term in $(x - a)^3$.

19. e^x, $a = 1$

20. $\sin x$, $a = \dfrac{\pi}{6}$

21. $\cos x$, $a = \dfrac{\pi}{3}$

22. $\tan x$, $a = \dfrac{\pi}{4}$

23. $1 + x^2 + x^3$, $a = 1$

24. $2 - x + 3x^2 - x^3$, $a = -1$

25. Let $f(x) = \sum a_n x^n$ be an even function ($f(-x) = f(x)$) for x in $(-R, R)$. Prove that $a_n = 0$ if n is odd. *Hint*: Use the Uniqueness Theorem.

26. State and prove a theorem analogous to that in Problem 25 for odd functions.

27. Recall that

$$\sin^{-1} x = \int_0^x \frac{1}{\sqrt{1 - t^2}}\, dt$$

Find the first four nonzero terms in the Maclaurin series for $\sin^{-1} x$.

28. Given that

$$\sinh^{-1} x = \int_0^x \frac{1}{\sqrt{1 + t^2}}\, dt$$

find the first four nonzero terms in the Maclaurin series for $\sinh^{-1} x$.

C **29.** Calculate, accurate to four decimal places,

$$\int_0^1 \cos(x^2)\, dx.$$

C **30.** Calculate, accurate to five decimal places,

$$\int_0^{0.5} \sin\sqrt{x}\, dx.$$

31. By writing $1/x = 1/[1 - (1 - x)]$ and using the known expansion of $1/(1 - x)$, find the Taylor series for $1/x$ in powers of $x - 1$.

32. Let $f(x) = (1 + x)^{1/2} + (1 - x)^{1/2}$. Find the Maclaurin series for f and use it to find $f^{(4)}(0)$ and $f^{(51)}(0)$.

33. In each case, find the Maclaurin series for $f(x)$ by use of known series and then use it to calculate $f^{(4)}(0)$.

(a) $f(x) = e^{x + x^2}$

(b) $f(x) = e^{\sin x}$

(c) $f(x) = \displaystyle\int_0^x \frac{e^{t^2} - 1}{t^2}\, dt$

(d) $f(x) = e^{\cos x} = e \cdot e^{\cos x - 1}$

(e) $f(x) = \ln(\cos^2 x)$

34. One can sometimes find a Maclaurin series by the *method of equating coefficients*. For example, let

$$\tan x = \frac{\sin x}{\cos x} = a_0 + a_1 + a_2 x^2 + \cdots$$

Then multiply by $\cos x$ and replace $\sin x$ and $\cos x$ by their series to obtain

$$x - \frac{x^3}{6} + \cdots = (a_0 + a_1 x + a_2 x^2 + \cdots)\left(1 - \frac{x^2}{2} + \cdots\right)$$

$$= a_0 + a_1 + \left(a_2 - \frac{a_0}{2}\right)x^2 + \left(a_3 - \frac{a_1}{2}\right)x^3 + \cdots$$

Thus,

$$a_0 = 0, \quad a_1 = 1, \quad a_2 - \frac{a_0}{2} = 0, \quad a_3 - \frac{a_1}{2} = -\frac{1}{6}, \cdots$$

so

$$a_0 = 0, \quad a_1 = 1, \quad a_2 = 0, \quad a_3 = \tfrac{1}{3}, \cdots$$

and therefore

$$\tan x = 0 + x + 0 + \tfrac{1}{3}x^3 + \cdots$$

which agrees with Problem 1. Use this method to find the terms through x^4 in the series for $\sec x$.

35. Use the method of Problem 34 to find the terms through x^5 in the Maclaurin series for $\tanh x$.

36. Prove Theorem C as follows. Let

$$f(x) = 1 + \sum_{n=1}^{\infty} \binom{p}{n} x^n.$$

(a) Show the series converges for $|x| < 1$.
(b) Show $(1 + x)f'(x) = pf(x)$ and $f(0) = 1$.
(c) Solve this differential equation to get $f(x) = (1 + x)^p$.

37. Let

$$f(t) = \begin{cases} 0 & t < 0 \\ t^4 & t \geq 0 \end{cases}$$

Explain why $f(t)$ cannot be represented by a Maclaurin series. Also show that if $g(t)$ gives the distance traveled by a car that is stationary for $t < 0$ and moving ahead for $t \geq 0$, then $g(t)$ cannot be represented by a Maclaurin series.

38. Let

$$f(x) = \begin{cases} e^{-1/x^2} & x \neq 0 \\ 0 & x = 0 \end{cases}$$

(a) Show that $f'(0) = 0$ by using the definition of the derivative.

(b) Show that $f''(0) = 0$.
(c) Assuming the known fact that $f^{(n)}(0) = 0$ for all n, find the Maclaurin series for $f(x)$.
(d) Does the Maclaurin series represent $f(x)$?
(e) What is the remainder in Maclaurin's Formula in this case?

This shows that a Maclaurin series may exist and yet not represent the given function (the remainder does not tend to 0 as $n \to \infty$).

11.9 Chapter Review

TRUE-FALSE QUIZ

Respond with true or false to each of the following assertions. Be prepared to defend your answer.

1. If $0 \leq a_n \leq b_n$ for all n in $\mathbb{N}$ and $\lim_{n \to \infty} b_n$ exists, then $\lim_{n \to \infty} a_n$ exists.

2. For every positive integer n, it is true that $n! \leq n^n \leq (2n - 1)!$.

3. If $\lim_{n \to \infty} a_n = L$, then $\lim_{n \to \infty} a_{3n+4} = L$.

4. If $\lim_{n \to \infty} a_{2n} = L$ and $\lim_{n \to \infty} a_{3n} = L$, then $\lim_{n \to \infty} a_n = L$.

5. If $\lim_{n \to \infty} a_{mn} = L$ for every positive integer $m \geq 2$, then $\lim_{n \to \infty} a_n = L$.

6. If $\lim_{n \to \infty} a_{2n} = L$ and $\lim_{n \to \infty} a_{2n+1} = L$, then $\lim_{n \to \infty} a_n = L$.

7. If $\lim_{n \to \infty}(a_n - a_{n+1}) = 0$, then $\lim_{n \to \infty} a_n$ exists and is finite.

8. If $\{a_n\}$ and $\{b_n\}$ both diverge, then $\{a_n + b_n\}$ diverges.

9. If $\{a_n\}$ converges, then $\{a_n/n\}$ converges to 0.

10. If $\sum_{n=1}^{\infty} a_n$ converges, so does $\sum_{n=1}^{\infty} a_n^2$.

11. If $0 < a_{n+1} < a_n$ for all n in $\mathbb{N}$ and $\lim_{n \to \infty} a_n = 0$, then $\sum_{n=1}^{\infty} (-1)^{n+1} a_n$ converges and has sum S satisfying $0 < S < a_1$.

12. $\sum_{n=1}^{\infty} \left(\frac{1}{n}\right)^n$ converges and has sum S satisfying $1 < S < 2$.

13. If a series $\sum a_n$ diverges, then its sequence of partial sums is unbounded.

14. If $0 \leq a_n \leq b_n$ for all n in $\mathbb{N}$ and if $\sum_{n=1}^{\infty} b_n$ diverges, then $\sum_{n=1}^{\infty} a_n$ diverges.

15. The Ratio Test will not help in determining the convergence or divergence of $\sum_{n=1}^{\infty} \dfrac{2n + 3}{3n^4 + 2n^3 + 3n + 1}$.

16. If $a_n > 0$ for all n in $\mathbb{N}$ and $\sum_{n=1}^{\infty} a_n$ converges, then $\lim_{n \to \infty}(a_{n+1}/a_n) < 1$.

17. $\sum_{n=1}^{\infty} \left(1 - \frac{1}{n}\right)^n$ converges.

18. $\sum_{n=1}^{\infty} \dfrac{1}{\ln(n^4 + 1)}$ converges.

19. $\sum_{n=2}^{\infty} \dfrac{n + 1}{(n \ln n)^2}$ converges.

20. $\displaystyle\sum_{n=1}^{\infty} \frac{\sin^2(n\pi/2)}{n}$ converges.

21. If $0 \le a_{n+100} \le b_n$ for all n in $\mathbb{N}$ and $\displaystyle\sum_{n=1}^{\infty} b_n$ converges, then $\displaystyle\sum_{n=1}^{\infty} a_n$ converges.

22. If for some $c > 0$, $ca_n \ge 1/n$ for all n in $\mathbb{N}$, then $\displaystyle\sum_{n=1}^{\infty} a_n$ diverges.

23. $\frac{1}{3} + \left(\frac{1}{3}\right)^2 + \left(\frac{1}{3}\right)^3 + \cdots + \left(\frac{1}{3}\right)^{1000} < \frac{1}{2}$.

24. If $\displaystyle\sum_{n=1}^{\infty} a_n$ converges, then $\displaystyle\sum_{n=1}^{\infty} (-1)^n a_n$ converges.

25. If $b_n \le a_n \le 0$ for all n in $\mathbb{N}$ and $\displaystyle\sum_{n=1}^{\infty} b_n$ converges, then $\displaystyle\sum_{n=1}^{\infty} a_n$ converges.

26. If $0 \le a_n$ for all n in $\mathbb{N}$ and $\displaystyle\sum_{n=1}^{\infty} a_n$ converges, then $\displaystyle\sum_{n=1}^{\infty} (-1)^n a_n$ converges.

27. $\left| \displaystyle\sum_{n=1}^{\infty} (-1)^{n+1} \frac{1}{n} - \sum_{n=1}^{99} (-1)^{n+1} \frac{1}{n} \right| < 0.01.$

28. If $\displaystyle\sum_{n=1}^{\infty} a_n$ diverges, then $\displaystyle\sum_{n=1}^{\infty} |a_n|$ diverges.

29. If the power series $\displaystyle\sum_{n=0}^{\infty} a_n(x - 3)^n$ converges at $x = -1.1$, it also converges at $x = 7$.

30. If $\displaystyle\sum_{n=0}^{\infty} a_n x^n$ converges at $x = -2$, it also converges at $x = 2$.

31. If $f(x) = \displaystyle\sum_{n=0}^{\infty} a_n x^n$ and the series converges at $x = 1.5$, then $\displaystyle\int_0^1 f(x)\,dx = \sum_{n=0}^{\infty} a_n/(n + 1)$.

32. Every power series converges for at least two values of the variable.

33. If $f(0), f'(0), f''(0), \ldots$ all exist, then the Maclaurin series for $f(x)$ converges to $f(x)$ in a neighborhood of $x = 0$.

34. The function $f(x) = 1 + x + x^2 + x^3 + \cdots$ satisfies the differential equation $y' = y^2$ on the interval $(-1, 1)$.

35. The function $f(x) = \displaystyle\sum_{n=0}^{\infty} (-1)^n x^n/n!$ satisfies the differential equation $y' + y = 0$ on the whole real line.

SAMPLE TEST PROBLEMS

In Problems 1–8, determine whether the given sequence converges or diverges and, if it converges, find $\lim_{n \to \infty} a_n$.

1. $a_n = \dfrac{9n}{\sqrt{9n^2 + 1}}$

2. $a_n = \dfrac{\ln n}{\sqrt{n}}$

3. $a_n = \left(1 + \dfrac{4}{n}\right)^n$

4. $a_n = \dfrac{n!}{3^n}$

5. $a_n = \sqrt[n]{n}$

6. $a_n = \dfrac{1}{\sqrt[3]{n}} + \dfrac{1}{\sqrt[n]{3}}$

7. $a_n = \dfrac{\sin^2 n}{\sqrt{n}}$

8. $a_n = \cos\left(\dfrac{n\pi}{6}\right)$

In Problems 9–18, determine whether the given series converges or diverges and, if it converges, find its sum.

9. $\displaystyle\sum_{k=1}^{\infty} \left(\frac{1}{\sqrt{k}} - \frac{1}{\sqrt{k+1}}\right)$

10. $\displaystyle\sum_{k=1}^{\infty} \left(\frac{1}{k} - \frac{1}{k+2}\right)$

11. $\ln \frac{1}{2} + \ln \frac{2}{3} + \ln \frac{3}{4} + \cdots$

12. $\displaystyle\sum_{k=1}^{\infty} \cos k\pi$

13. $\displaystyle\sum_{k=0}^{\infty} e^{-2k}$

14. $\displaystyle\sum_{k=0}^{\infty} \left(\frac{3}{2^k} + \frac{4}{3^k}\right)$

15. $0.91919191\ldots = \displaystyle\sum_{k=1}^{\infty} 91\left(\frac{1}{100}\right)^k$

16. $\displaystyle\sum_{k=1}^{\infty} \left(\frac{1}{\ln 2}\right)^k$

17. $1 - \dfrac{2^2}{2!} + \dfrac{2^4}{4!} - \dfrac{2^6}{6!} + \cdots$

18. $1 - \dfrac{1}{1!} + \dfrac{1}{2!} - \dfrac{1}{3!} + \dfrac{1}{4!} - \cdots$

In Problems 19–32, indicate whether the given series converges or diverges and give a reason for your conclusion.

19. $\displaystyle\sum_{n=1}^{\infty} \frac{n}{1 + n^2}$

20. $\displaystyle\sum_{n=1}^{\infty} \frac{n + 5}{1 + n^3}$

21. $\displaystyle\sum_{n=1}^{\infty} (-1)^{n+1} \frac{1}{\sqrt[3]{n}}$

22. $\displaystyle\sum_{n=1}^{\infty} (-1)^{n+1} \frac{1}{\sqrt[n]{3}}$

23. $\displaystyle\sum_{n=1}^{\infty} \frac{2^n + 3^n}{4^n}$

24. $\displaystyle\sum_{n=1}^{\infty} \frac{n}{e^{n^2}}$

25. $\displaystyle\sum_{n=1}^{\infty} (-1)^{n+1} \frac{n + 1}{10n + 12}$

26. $\displaystyle\sum_{n=1}^{\infty} \frac{\sqrt{n}}{n^2 + 7}$

27. $\displaystyle\sum_{n=1}^{\infty} \frac{n^2}{n!}$

28. $\displaystyle\sum_{n=1}^{\infty} \frac{n^3 3^n}{(n + 1)!}$

29. $\displaystyle\sum_{n=1}^{\infty} \frac{2^n n!}{(n + 2)!}$

30. $\displaystyle\sum_{n=2}^{\infty} \left(1 - \frac{1}{n}\right)^n$

31. $\displaystyle\sum_{n=1}^{\infty} n^2 (\tfrac{2}{3})^n$

32. $\displaystyle\sum_{n=1}^{\infty} \frac{(-1)^n}{1 + \ln n}$

In Problems 33–36, state whether the given series is absolutely convergent, conditionally convergent, or divergent.

33. $\displaystyle\sum_{n=1}^{\infty} (-1)^n \frac{1}{3n - 1}$

34. $\displaystyle\sum_{n=1}^{\infty} \frac{(-1)^n n^3}{2^n}$

35. $\displaystyle\sum_{n=1}^{\infty} (-1)^n \frac{3^n}{2^{n+8}}$

36. $\displaystyle\sum_{n=2}^{\infty} \frac{(-1)^n \sqrt[n]{n}}{\ln n}$

In Problems 37–42, find the convergence set for the power series.

37. $\displaystyle\sum_{n=0}^{\infty} \frac{x^n}{n^3 + 1}$

38. $\displaystyle\sum_{n=0}^{\infty} \frac{(-2)^{n+1} x^n}{2n + 3}$

39. $\displaystyle\sum_{n=0}^{\infty} \frac{(-1)^n (x - 4)^n}{n + 1}$

40. $\displaystyle\sum_{n=0}^{\infty} \frac{3^n x^{3n}}{(3n)!}$

41. $\displaystyle\sum_{n=0}^{\infty} \frac{(x - 3)^n}{2^n + 1}$

42. $\displaystyle\sum_{n=0}^{\infty} \frac{n!(x + 1)^n}{3^n}$

43. By differentiating the geometric series

$$\frac{1}{1 + x} = 1 - x + x^2 - x^3 + x^4 - \cdots, \qquad |x| < 1,$$

find a power series that represents $1/(1 + x)^2$. What is its interval of convergence?

44. Find a power series that represents $1/(1 + x)^3$ on the interval $(-1, 1)$.

45. Find the Maclaurin series for $\sin^2 x$. For what values of x does the series represent the function?

46. Find the first five terms of the Taylor series for e^x based at the point $x = 2$.

47. Write the Maclaurin series for $f(x) = \sin x + \cos x$. For what values of x does it represent f?

C **48.** Write the Maclaurin series for $f(x) = \cos x^2$ and use it to approximate

$$\int_0^1 \cos x^2 \, dx$$

How many terms of the series are needed to compute the value of this integral correct to four decimal places?

C **49.** Calculate the following integral correct to five decimal places.

$$\int_0^{0.2} \frac{e^x - 1}{x} \, dx$$

50. How many terms do we have to take in the convergent series

$$1 - \frac{1}{\sqrt{2}} + \frac{1}{\sqrt{3}} - \frac{1}{\sqrt{4}} + \frac{1}{\sqrt{5}} - \frac{1}{\sqrt{6}} + \cdots$$

to be sure that we have approximated its sum to within 0.001?

51. Estimate the maximum error made in approximating $\cos x$ by $1 - x^2/2$ for $-0.1 \le x \le 0.1$.

52. Use the simplest method you can think of to find the first three nonzero terms of the Maclaurin series for each of the following.

(a) $\dfrac{1}{1 - x^3}$

(b) $\sqrt{1 + x^2}$

(c) $e^{-x} - 1 + x$

(d) $x \sec x$

(e) $e^{-x} \sin x$

(f) $\dfrac{1}{1 + \sin x}$

12 Conics and Polar Coordinates

The chief aim of all investigations of the external world should be to discover the rational order and harmony which has been imposed on it by God and which He revealed to us in the language of mathematics.

J. Kepler

Johannes Kepler
1571–1630

Johannes Kepler was born in Weil, Germany, to a drunken father. As a boy he contracted smallpox, which left him with crippled hands and poor eyesight. Difficulties followed Kepler throughout life. His first wife and several of his children died. His mother was accused of witchcraft and Kepler endured persecution during the religious upheavals of the time. In addition to all this, he was often plagued by inadequate income. Yet he persevered in his scientific work with devotion and imagination.

At the University of Tübingen, he studied mathematics and astronomy. Appointed an assistant to Tycho Brahe at the observatory at Prague, he acquired accurate data on the orbits of the planets. He was convinced that God had designed the world in an aesthetically pleasing and mathematically simple way. This view attracted him to the beauty and harmony of Copernicus's heliocentric system, which put the sun rather than the earth at the center of the universe.

Kepler's greatest contributions were his three laws of planetary motion: (1) Planets move in ellipses with the sun at one focus, (2) the line from the sun to the planet sweeps out equal areas in equal times, and (3) the square of the period of revolution is proportional to the cube of the major diameter. These laws, which Kepler hypothesized on the basis of observation and aesthetic necessity, were later shown by Isaac Newton to be consequences of the Inverse Square Law of attraction between two masses.

12.1
The Parabola

Take a right circular cone with two nappes and pass planes through it at various angles, as shown in Figure 1. As sections, you will obtain curves called, respectively, an ellipse, a parabola, and a hyperbola. (You may also obtain various limiting forms: a circle, a point, intersecting lines, and one line.) These curves are called *conic sections*, or simply *conics*. This definition, which is due to the Greeks, is cumbersome and we shall immediately adopt a different one. It can be shown that the two notions are consistent.

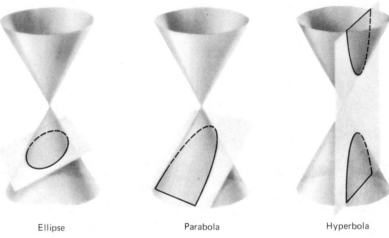

Ellipse   Parabola   Hyperbola

FIGURE 1

FIGURE 2

In the plane let *l* be a fixed line (the **directrix**) and *F* be a fixed point (the **focus**) not on the line, as in Figure 2. The set of points *P* for which the ratio of the distance $|PF|$ from the focus to the distance $|PL|$ from the line is a positive constant *e* (the **eccentricity**)—that is, which satisfies

$$\boxed{|PF| = e|PL|}$$

is called a **conic**. If $0 < e < 1$, it is an **ellipse**; if $e = 1$, it is a **parabola**; if $e > 1$, it is a **hyperbola**.

When we draw the curves corresponding to $e = \frac{1}{2}$, $e = 1$, and $e = 2$, we get the three curves shown in Figure 3.

Ellipse $\left(e = \frac{1}{2}\right)$　　　Parabola (e = 1)　　　Hyperbola (e = 2)

FIGURE 3

In each case, the curves are symmetric with respect to the line through the focus perpendicular to the directrix. We call this line the **major axis** (or simply the *axis*) of the conic. A point where the conic crosses the axis is called a **vertex**. The parabola has one vertex, while the ellipse and hyperbola have two vertices.

THE PARABOLA (e = 1)　　A **parabola** is the set of points P that are equidistant from the directix l and the focus F—that is, which satisfy

$$|PF| = |PL|$$

FIGURE 4

From this definition, we wish to derive the xy-equation, and we want it to be as simple as possible. The position of the coordinate axes has no effect on the curve, but it does affect the simplicity of the curve's equation. Since a parabola is symmetric with respect to its axis, it is natural to place one of the coordinate axes—for instance, the x-axis—along the axis. Let the focus F be to the right of the origin, say at $(p, 0)$, and the directrix to the left with equation $x = -p$. Then the vertex is at the origin. All this is shown in Figure 4.

From the condition $|PF| = |PL|$ and the distance formula, we get

$$\sqrt{(x - p)^2 + (y - 0)^2} = \sqrt{(x + p)^2 + (y - y)^2}$$

After squaring both sides and simplifying, we obtain

$$\boxed{y^2 = 4px}$$

This is called the **standard equation** of a horizontal parabola (horizontal axis) opening to the right. Note that $p > 0$ and that p is the distance from the focus to the vertex.

EXAMPLE 1　Find the focus and directrix of the parabola with equation $y^2 = 12x$.

Solution　Since $y^2 = 4(3)x$, we see that $p = 3$. The focus is at $(3, 0)$; the directrix is the line $x = -3$. ∎

There are three variants of the standard equation. If we interchange the roles of x and y, we obtain the equation $x^2 = 4py$. It is the equation of a vertical parabola with focus at $(0, p)$ and directrix $y = -p$. Finally, introducing a minus sign on one side of the equation causes the parabola to open in the opposite direction. All four cases are shown in Figure 5.

FIGURE 5

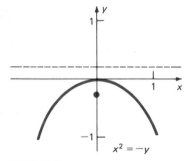

FIGURE 6

EXAMPLE 2 Determine the focus and directrix of the parabola $x^2 = -y$ and sketch the graph.

Solution We write $x^2 = -4(\frac{1}{4})y$, from which we conclude that $p = \frac{1}{4}$. The form of the equation tells us that the parabola is vertical and opens down. The focus is at $(0, -\frac{1}{4})$; the directrix is the line $y = \frac{1}{4}$. The graph is shown in Figure 6. ∎

EXAMPLE 3 Find the equation of the parabola with vertex at the origin and focus at $(0, 5)$.

Solution The parabola opens up and $p = 5$. The equation is $x^2 = 4(5)y$—that is, $x^2 = 20y$. ∎

EXAMPLE 4 Find the equation of the parabola with vertex at the origin that goes through $(-2, 4)$ and opens left. Sketch the graph.

Solution The equation has the form $y^2 = -4px$. Because $(-2, 4)$ is on the graph, $(4)^2 = -4p(-2)$, from which $p = 2$. The desired equation is $y^2 = -8x$ and its graph is sketched in Figure 7. ∎

FIGURE 7

THE OPTICAL PROPERTY A simple geometric property of a parabola is the basis of many important applications. If F is the focus and P is any point on the parabola, the tangent line at P makes equal angles with FP and the line GP, which is parallel to the axis of the parabola (see Figure 8). A principle from physics says that when a light ray strikes a reflecting surface, the angle of incidence is equal to the angle of reflection. It follows that if a parabola is revolved about its axis to form a hollow reflecting shell, all light rays from

524 CHAPTER 12/Conics and Polar Coordinates

FIGURE 8

the focus after hitting the shell are reflected outward parallel to the axis. This property of the parabola is used in designing search lights, with the light source placed at the focus. Conversely, it is used in certain telescopes where incoming parallel rays from a distant star are focused at a single point.

EXAMPLE 5   Prove the optical property of the parabola.

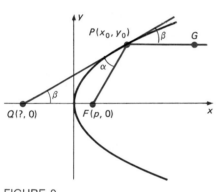

FIGURE 9

Solution   In Figure 9, let QP be the tangent line at P and let GP be the line parallel to the x-axis. We must show that $\alpha = \beta$. After noting that $\angle FQP = \beta$, we reduce the problem to showing that triangle FQP is isosceles.

First, we obtain the x-coordinate of Q. Differentiating $y^2 = 4px$ implicitly gives $2y'y = 4p$, from which we conclude that the slope of the tangent line at $P(x_0, y_0)$ is $2p/y_0$. The equation of this line is

$$y - y_0 = \frac{2p}{y_0}(x - x_0)$$

Setting $y = 0$ and solving for x gives $-y_0 = (2p/y_0)(x - x_0)$, or $x - x_0 = -y_0^2/2p$. Now $y_0^2 = 4px_0$, which gives $x - x_0 = -2x_0$—that is, $x = -x_0$; Q has coordinates $(-x_0, 0)$.

To show that the segments FP and FQ have equal length, we use the distance formula

$$|FP| = \sqrt{(x_0 - p)^2 + y_0^2} = \sqrt{x_0^2 - 2x_0p + p^2 + 4px_0}$$
$$= \sqrt{x_0^2 + 2x_0p + p^2} = x_0 + p = |FQ| \qquad \blacksquare$$

Sound obeys the same laws of reflection as light, and parabolic microphones are used to pick up and concentrate sounds from, for example, a distant part of a football stadium. Radar and radio telescopes are also based on these same principles.

There are many other applications of parabolas. For example, the path of a projectile is a parabola if air resistance and other minor factors are neglected. The cable of an evenly loaded suspension bridge takes the form of a parabola. Arches are often parabolic. The paths of a few comets are parabolic.

In Problems 1–8, find the coordinates of the focus and the equation of the directrix for each parabola. Make a sketch, showing the parabola, its focus, and its directrix.

1. $y^2 = 16x$

2. $y^2 = -28x$

3. $x^2 = -16y$

4. $x^2 = 12y$

5. $y^2 = 2x$

6. $y^2 - 7x = 0$

7. $x^2 - 6y = 0$

8. $2x^2 - 7y = 0$

In Problems 9–14, find the standard equation of each parabola from the given information. Assume the vertex is at the origin.

9. Focus is $(3, 0)$.

10. Directrix is $x = 2$.

11. Directrix is $y + 4 = 0$.

12. Focus is $(0, -\frac{1}{3})$.

13. Focus is $(-5, 0)$.

14. Directrix is $y = \frac{5}{2}$.

15. Find the equation of the parabola with vertex at the origin and axis along the x-axis if the parabola passes through the point $(3, -1)$. Make a sketch.

16. Find the equation of the parabola through the point $(-2, 4)$ if its vertex is at the origin and its axis is along the x-axis. Make a sketch.

17. Find the equation of the parabola through the point $(6, -5)$ if its vertex is at the origin and its axis is along the y-axis. Make a sketch.

18. Find the equation of the parabola whose vertex is the origin and whose axis is the y-axis if the parabola passes through the point $(-3, 5)$. Make a sketch.

In Problems 19–26, find the equations of the tangent and the normal to the given parabola at the given point. Sketch the parabola, the tangent, and the normal in 19–22.

19. $y^2 = 16x$, $(1, -4)$

20. $x^2 = -10y$, $(2\sqrt{5}, -2)$

21. $x^2 = 2y$, $(4, 8)$

22. $y^2 = -9x$, $(-1, -3)$

23. $y^2 = -15x$, $(-3, -3\sqrt{5})$

24. $x^2 = 4y$, $(4, 4)$

25. $x^2 = -6y$, $(3\sqrt{2}, -3)$

26. $y^2 = 20x$, $(2, -2\sqrt{10})$

27. The slope of the tangent to the parabola $y^2 = 5x$ at a certain point on the parabola is $\sqrt{5}/4$. Find the coordinates of that point. Make a sketch.

28. The slope of the tangent to the parabola $x^2 = -14y$ at a certain point on the parabola is $-2\sqrt{7}/7$, Find the coordinates of that point.

29. Find the equation of the tangent to the parabola $y^2 = -18x$ that is parallel to the line $3x - 2y + 4 = 0$.

30. Any line segment through the focus of a parabola, with endpoints on the parabola, is a **focal chord**. Prove that the tangents to a parabola at the extremities of any focal chord intersect on the directrix.

31. Prove that the tangents to a parabola at the extremities of any focal chord are perpendicular to each other (see Problem 30).

32. A chord of a parabola that is perpendicular to the axis and 1 unit from the vertex has length 1 unit. How far is it from the vertex to the focus?

33. Prove that the vertex is the point on a parabola closest to the focus.

34. A space ship from outer space is sighted from earth moving on a parabolic path with the earth at the focus. When the line from the earth to the space ship first makes an angle of 90° with the axis of the parabola, it is measured to be 40 million miles away. How close will the space ship come to the earth (see Problem 33)? Treat the earth as a point.

C **35.** Work Problem 34 assuming the angle is 75° rather than 90°.

36. The cables for the central span of a suspension bridge take the shape of a parabola (see Problem 41). If the towers are 800 meters apart and the cables are attached to them at points 400 meters above the floor of the bridge, how long must the vertical strut be that is 100 meters from the tower? Assume that the cable touches the floor at the midpoint of the bridge (Figure 10).

FIGURE 10

37. The focal chord that is perpendicular to the axis of a parabola is called the **latus rectum**. For the parabola $y^2 = 4px$ in Figure 11, let F be the focus, R be any point on the parabola to the left of the latus rectum, and G be the intersection of the latus rectum with the line through R parallel to the axis. Find $|FR| + |RG|$ and note that it is a constant.

FIGURE 11

38. Show that the set of points equidistant from a circle and a line outside the circle is a parabola.

39. Show that the focal chord of the parabola $y^2 = 4px$ with endpoints (x_1, y_1) and (x_2, y_2) has length $x_1 + x_2 + 2p$. Specialize to find the length L of the latus rectum.

40. For the parabola $y^2 = 4px$ in Figure 12, P is any of its points except the vertex, PB is the normal line at P, PA is perpendicular to the axis of the parabola, A and B are on the axis. Find $|AB|$ and note that it is a constant.

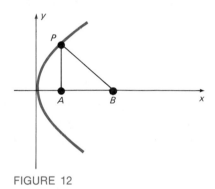

FIGURE 12

41. Consider a bridge deck weighing δ pounds per lineal foot and supported by a cable, which is assumed to be of negligible weight compared to the bridge deck. The section

OP from the lowest point (the origin) to a general point $P(x, y)$ is shown in Figure 13.

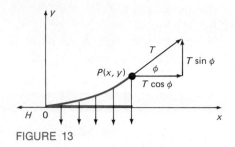

FIGURE 13

The forces acting on this section of cable are:

$$H = \text{horizontal tension pulling at } O$$
$$T = \text{tangential tension pulling at } P$$
$$W = \delta x = \text{weight of } x \text{ feet of bridge deck}$$

For equilibrium, the horizontal and vertical components of T must balance H and W, respectively. Thus,

$$\frac{T \sin \phi}{T \cos \phi} = \tan \phi = \frac{\delta x}{H}$$

that is,

$$\frac{dy}{dx} = \frac{\delta x}{H}, \qquad y(0) = 0$$

Solve this differential equation to show that the cable hangs in the shape of a parabola. (Compare this result with that for the unloaded hanging cable of Problem 29 of Section 7.8.)

42. Consider the parabola $y = x^2$ over the interval $[a, b]$ and let $c = (a + b)/2$ be the midpoint of $[a, b]$, d be the midpoint of $[a, c]$, and e be the midpoint of $[c, b]$. Let T_1 be the triangle with vertices on the parabola at a, c, and b, and let T_2 be the union of the two triangles with vertices on the parabola at a, d, c and c, e, b, respectively (Figure 14).

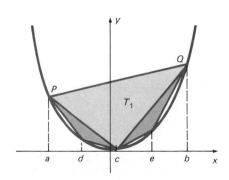

FIGURE 14

Continue to build triangles on triangles in this manner, thus obtaining sets $T_3, T_4, \ldots$.

(a) Show that $A(T_1) = (b - a)^3/8$.
(b) Show that $A(T_2) = A(T_1)/4$.
(c) Let S be the parabolic segment cut off by the chord PQ.

Show that the area of P satisfies

$$A(S) = A(T_1) + A(T_2) + A(T_3) + \cdots = (\tfrac{4}{3})A(T_1)$$

This is a famous result of Archimedes, which he obtained without coordinates.
(d) Use these results to show that the area under $y = x^2$ between a and b is $b^3/3 - a^3/3$.

12.2
Ellipses and Hyperbolas

Recall that the conic determined by the condition $|PF| = e|PL|$ is an **ellipse** if $0 < e < 1$ and a **hyperbola** if $e > 1$ (see the introduction to Section 12.1). In either case, the conic has two vertices, which we label A' and A. Call the point on the major axis midway between A' and A the **center** of the conic. Ellipses and hyperbolas are symmetric with respect to their centers (as we shall demonstrate soon) and are, therefore, called *central conics*.

To derive the equation of a central conic, place the x-axis along the major axis with the origin at the center. We may suppose the focus to be $F(c, 0)$, the directrix $x = k$, and the vertices $A'(-a, 0)$ and $A(a, 0)$, with c, k, and a all positive. The two possible arrangements are shown in Figures 1 and 2.

The defining condition $|PF| = e|PL|$ applied first with $P = A$ and then with $P = A'$ yields

$$a - c = e(k - a) = ek - ea$$
$$a + c = e(k + a) = ek + ea$$

When these two equations are solved for c and k, we get

$$c = ea \quad \text{and} \quad k = \frac{a}{e}$$

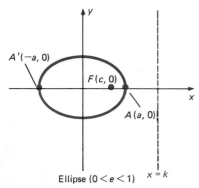

Ellipse $(0 < e < 1)$

FIGURE 1

Now let $P(x, y)$ be any point on the ellipse (or hyperbola). Then $L(a/e, y)$ is its projection on the directrix (see Figure 3 for the case of the ellipse). The condition $|PF| = e|PL|$ becomes

$$\sqrt{(x - ae)^2 + y^2} = e\sqrt{\left(x - \frac{a}{e}\right)^2}$$

Squaring both members and collecting terms, we obtain the equivalent equation (why is it equivalent?)

$$x^2 - 2aex + a^2e^2 + y^2 = e^2\left(x^2 - \frac{2a}{e}x + \frac{a^2}{e^2}\right)$$

or

$$(1 - e^2)x^2 + y^2 = a^2(1 - e^2)$$

or

$$\frac{x^2}{a^2} + \frac{y^2}{a^2(1 - e^2)} = 1$$

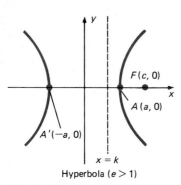

Hyperbola $(e > 1)$

FIGURE 2

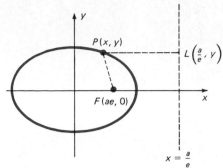

FIGURE 3

Because this last equation contains x and y only to even powers, it corresponds to a curve that is symmetric with respect to both the x- and y-axes and the origin. Also, because of this symmetry, there must be a second focus at $(-ae, 0)$ and a second directrix at $x = -a/e$. The axis containing the two vertices (and the two foci) is the **major axis** and the axis perpendicular to it (through the center) is the **minor axis**.

THE STANDARD EQUATION OF THE ELLIPSE   For the ellipse, $0 < e < 1$, and so $(1 - e^2)$ is positive. To simplify notation, let $b = a\sqrt{1 - e^2}$. Then the equation derived above takes the form

$$\boxed{\frac{x^2}{a^2} + \frac{y^2}{b^2} = 1}$$

which is called the **standard equation of the ellipse**. The number $2a$ is the **major diameter**, whereas $2b$ is the **minor diameter**. Moreover since $c = ae$, the numbers a, b, and c satisfy the Pythagorean relationship $a^2 = b^2 + c^2$, as is easily verified. All this is summarized in Figure 4. Note the role of the shaded triangle, which captures the condition $a^2 = b^2 + c^2$.

FIGURE 4

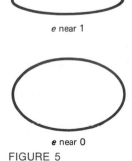

FIGURE 5

Consider the effect of changing the value of e. If e is near 1, then $b = a\sqrt{1 - e^2}$ is small relative to a; the ellipse is thin and very eccentric. On the other hand, if e is near 0 (near zero eccentricity), b is almost as large as a; the ellipse is fat and well rounded (Figure 5). In the limiting case where $b = a$, the equation takes the form

$$\frac{x^2}{a^2} + \frac{y^2}{a^2} = 1$$

which is equivalent to $x^2 + y^2 = a^2$. This is the equation of a circle of radius a centered at the origin.

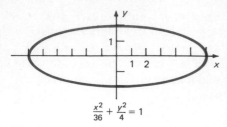

$$\frac{x^2}{36} + \frac{y^2}{4} = 1$$

FIGURE 6

EXAMPLE 1 Sketch the graph of

$$\frac{x^2}{36} + \frac{y^2}{4} = 1$$

and determine its foci and eccentricity.

Solution Since $a = 6$ and $b = 2$, we calculate

$$c = \sqrt{a^2 - b^2} = \sqrt{36 - 4} = 4\sqrt{2} \approx 5.66$$

The foci are at $(\pm c, 0) = (\pm 4\sqrt{2}, 0)$, and $e = c/a \approx 0.94$. The graph is sketched in Figure 6. ∎

We call the ellipses sketched so far *horizontal ellipses* because the major axis is the x-axis. If we interchange the roles of x and y, we obtain a vertical ellipse with equation

$$\frac{x^2}{b^2} + \frac{y^2}{a^2} = 1$$

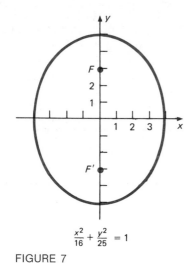

$$\frac{x^2}{16} + \frac{y^2}{25} = 1$$

FIGURE 7

EXAMPLE 2 Sketch the graph of

$$\frac{x^2}{16} + \frac{y^2}{25} = 1$$

and determine its foci and eccentricity.

Solution The larger square is now under y^2, which tells us that the major axis is vertical. Noting that $a = 5$ and $b = 4$, we conclude that $c = \sqrt{25 - 16} = 3$. Thus, the foci are $(0, \pm 3)$ and $e = c/a = \frac{3}{5} = 0.6$ (Figure 7). ∎

THE STANDARD EQUATION OF THE HYPERBOLA   For the hyperbola, $e > 1$ and so $e^2 - 1$ is positive. If we let $b = a\sqrt{e^2 - 1}$, then the equation $x^2/a^2 + y^2/(1 - e^2)a^2 = 1$, which was derived earlier, takes the form

$$\boxed{\frac{x^2}{a^2} - \frac{y^2}{b^2} = 1}$$

This is called the **standard equation of the hyperbola**. Since $c = ae$, we now obtain $c^2 = a^2 + b^2$. (Note how this differs from the corresponding relationship for an ellipse.)

To interpret b, observe that if we solve for y in terms of x, we get

$$y = \pm\frac{b}{a}\sqrt{x^2 - a^2}$$

530

For large x, $\sqrt{x^2 - a^2}$ behaves like x (that is, $(\sqrt{x^2 - a^2} - x) \to 0$ as $x \to \infty$; see Problem 38) and hence y behaves like

$$y = \pm \frac{b}{a} x$$

More precisely, the graph of the given hyperbola has these two lines as asymptotes.

The important facts for the hyperbola are summarized in Figure 8. Once again, there is an important triangle (shaded in our diagram); it determines the asymptotes mentioned above.

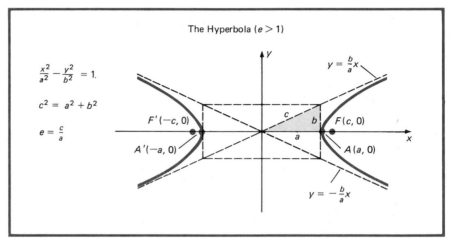

FIGURE 8

EXAMPLE 3 Sketch the graph of

$$\frac{x^2}{9} - \frac{y^2}{16} = 1$$

showing the asymptotes. What are the equations of the asymptotes? What are the foci?

Solution We begin by determining the fundamental triangle; it has horizontal leg 3 and vertical leg 4. After drawing it, we can indicate the asymptotes and sketch the graph (Figure 9). The asymptotes are $y = \pm \frac{4}{3} x$. Since $c = \sqrt{a^2 + b^2} = \sqrt{9 + 16} = 5$, the foci are at $(\pm 5, 0)$. ∎

Again, we should consider the effect of interchanging the roles of x and y. The equation takes the form

$$\frac{y^2}{a^2} - \frac{x^2}{b^2} = 1$$

This is the equation of a vertical hyperbola (vertical major axis). Its vertices are at $(0, \pm a)$; its foci are at $(0, \pm c)$.

For both the ellipse and the hyperbola, a is always the distance from the center to a vertex. For the ellipse, $a > b$; for the hyperbola, there is no such requirement.

FIGURE 9

FIGURE 10

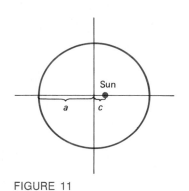

FIGURE 11

EXAMPLE 4 Determine the foci of

$$-\frac{x^2}{4} + \frac{y^2}{9} = 1$$

and sketch its graph.

Solution We note immediately that this is a vertical hyperbola, which is determined by the fact that the plus sign is associated with the y^2 term. Thus, $a = 3$, $b = 2$, and $c = \sqrt{9+4} = \sqrt{13} \approx 3.61$. The foci are at $(0, \pm\sqrt{13})$ (Figure 10). ∎

APPLICATIONS According to Johannes Kepler, the planets revolve around the sun in elliptical orbits, with the sun at one focus. Other examples of elliptical orbits are satellites orbiting the earth and electrons orbiting the nucleus of an atom.

EXAMPLE 5 The earth's maximum distance from the sun is 94.56 million miles, and its minimum distance is 91.45 million miles. What is the eccentricity of the orbit and what are the major and minor diameters?

Solution Using the notation in Figure 11, we see that

$$a + c = 94.56 \qquad a - c = 91.45$$

When we solve these equations for a and c, we obtain $a = 93.01$ and $c = 1.56$. Thus,

$$e = \frac{c}{a} = \frac{1.56}{93.01} \approx 0.017$$

and the major diameter and minor diameter (in millions of miles) are, respectively,

$$2a \approx 186.02 \qquad 2b = 2\sqrt{a^2 - c^2} \approx 185.99$$ ∎

Other applications of ellipses and hyperbolas result from the optical properties of these curves. They are discussed in the next section.

PROBLEM SET 12.2 _____

In Problems 1–8, name the conic (horizontal ellipse, vertical hyperbola, and so on) corresponding to the given equation.

1. $\dfrac{x^2}{4} + \dfrac{y^2}{9} = 1$

2. $\dfrac{x^2}{4} - \dfrac{y^2}{9} = 1$

3. $\dfrac{-x^2}{4} + \dfrac{y^2}{9} = 1$

4. $\dfrac{-x^2}{4} + \dfrac{y^2}{9} = -1$

5. $\dfrac{-x^2}{4} + \dfrac{y}{9} = 0$

6. $\dfrac{-x^2}{4} = \dfrac{y}{9}$

7. $4x^2 + 9y^2 = 4$

8. $4x^2 - y^2 = 4$

In Problems 9–16, sketch the graph of the given equation indicating vertices, foci, and asymptotes (if it is a hyperbola).

9. $\dfrac{x^2}{9} + \dfrac{y^2}{16} = 1$

10. $\dfrac{x^2}{10} - \dfrac{y^2}{4} = 1$

11. $\dfrac{-x^2}{16} + \dfrac{y^2}{9} = 1$

12. $\dfrac{x^2}{10} + \dfrac{y^2}{4} = 1$

13. $x^2 + 4y^2 = 36$ **14.** $25x^2 + 4y^2 = 100$

15. $4x^2 - 25y^2 = 100$ **16.** $x^2 - 4y^2 = 4$

In Problems 17–30, find the equation of the given conic. Assume the center is at the origin.

17. The ellipse with a focus at $(-3, 0)$ and a vertex at $(6, 0)$.

18. The ellipse with a focus at $(6, 0)$ and eccentricity $\frac{2}{3}$.

19. The ellipse with a focus at $(0, -5)$ and eccentricity $\frac{1}{3}$.

20. The ellipse with a focus at $(0, 3)$ and minor diameter 8.

21. The ellipse with a vertex at $(5, 0)$ and passing through $(2, 3)$.

22. The hyperbola with a focus at $(5, 0)$ and a vertex at $(4, 0)$.

23. The hyperbola with a vertex at $(0, -4)$ and a focus at $(0, -5)$.

24. The hyperbola with a vertex at $(0, -3)$ and eccentricity $\frac{3}{2}$.

25. The hyperbola with asymptotes $2x \pm 4y = 0$ and a vertex at $(8, 0)$.

26. The vertical hyperbola with eccentricity $\sqrt{6}/2$ which passes through $(2, 4)$.

27. The ellipse with foci $(\pm 2, 0)$ and directrices $x = \pm 8$.

28. The hyperbola with foci $(\pm 4, 0)$ and directrices $x = \pm 1$.

29. The hyperbola whose asymptotes are $x \pm 2y = 0$ and that goes through the point $(4, 3)$.

30. The horizontal ellipse which goes through $(-5, 1)$ and $(-4, -2)$.

31. A door has the shape of an elliptical arch (a half-ellipse) that is 10 feet wide and 4 feet high at the center. A box 2 feet high is to be pushed through the door. How wide can the box be?

32. How high is the arch of Problem 31 at a distance 2 feet from the center?

33. How long is the *latus rectum* (chord through the focus perpendicular to the major axis) for the ellipse $x^2/a^2 + y^2/b^2 = 1$?

34. Determine the length of the latus rectum (see Problem 33) of the hyperbola $x^2/a^2 - y^2/b^2 = 1$.

$\boxed{C}$ **35.** Halley's comet has an elliptical orbit with major and minor diameters of 36.18 AU and 9.12 AU, respectively

(1 AU is 1 astronomical unit, the earth's mean distance from sun). What is its closest approach to the sun (assuming the sun is at a focus)?

$\boxed{C}$ **36.** The orbit of the comet Kahoutek is an ellipse with eccentricity $e = 0.999925$ with the sun at a focus. If its minimum distance to the sun is 0.13 AU, what is its maximum distance from the sun?

$\boxed{C}$ **37.** In October, 1957, Russia launched Sputnik I. Its elliptical orbit around the earth reached maximum and minimum distances from the earth of 583 miles and 132 miles, respectively. Assuming the earth is a sphere of radius 4000 miles, find the eccentricity of the orbit.

38. Show that $(\sqrt{x^2 - a^2} - x) \to 0$ as $x \to \infty$. *Hint*: Rationalize the numerator.

39. For an ellipse, let p and q be the distances from a focus to the two vertices. Show that $b = \sqrt{pq}$, $2b$ being the minor diameter.

40. The wheel in Figure 12 is turning at t radians per second so that Q has coordinates $(a \cos t, a \sin t)$. Find the coordinates (x, y) of R at time t and show that it is traveling an elliptical path. *Note*: PQR is a right triangle.

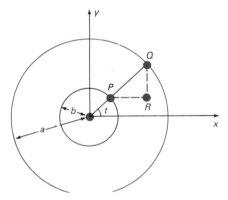

FIGURE 12

41. Let P be a point on a ladder of length $a + b$, P being a units from the top end. As the ladder slides with its top end on the y-axis and its bottom end on the x-axis, P traces out a curve. Find the equation of this curve.

42. Show that a line through a focus of a hyperbola and perpendicular to an asymptote intersects that asymptote on the directrix nearest the focus.

43. If a horizontal hyperbola and a vertical hyperbola have the same asymptotes, show that their eccentricities e and E satisfy $e^{-2} + E^{-2} = 1$.

44. Let C be the curve of intersection of a right circular cylinder and a plane making an angle $\phi(0 < \phi < \pi/2)$ with the axis of the cylinder. Show that C is an ellipse.

12.3
More on Ellipses and Hyperbolas

Ellipse: $|PF'| + |PF| = 2a$

FIGURE 1

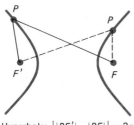

Hyperbola: $\left||PF'| - |PF|\right| = 2a$

FIGURE 2

In Section 12.1, we gave the so-called eccentricity definition of the ellipse and hyperbola. The condition $|PF| = e|PL|$ determines an ellipse if $0 < e < 1$ and a hyperbola if $e > 1$. This definition allowed us to treat these two curves in a unified way. Many authors prefer to introduce these curves via the following alternate definitions.

An **ellipse** is the set of all points P in the plane, the sum of whose distances from two fixed points (the foci) is a given positive constant $2a$.

A **hyperbola** is the set of all points P in the plane, the difference of whose distances from two fixed points (the foci) is a given positive constant $2a$. Here, the word *difference* is taken to mean the larger distance minus the smaller distance.

To interpret these definitions geometrically, study Figures 1 and 2. For the ellipse, imagine a string of length $2a$ tacked down at its two end-points. If a pencil is held tight against the string with its tip at P, it can be used to trace the ellipse (also see Problem 29). For this reason, we refer to the properties described in the new definitions as the *string properties* of the ellipse and hyperbola. For us, these properties should be consequences of the eccentricity definition. We derive them now.

DERIVATION OF STRING PROPERTIES We pointed out in Section 12.2 that the ellipse and hyperbola have two foci and two directrices. When these curves are placed in the coordinate system with major axis along the x-axis and center at the origin, the foci have coordinates $(\pm ae, 0)$ and the directrices have equations $x = \pm a/e$. These facts are indicated in Figure 3.

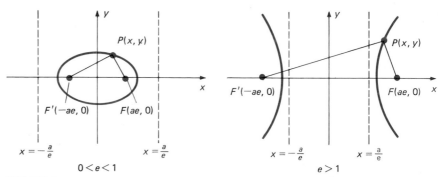

FIGURE 3

If we take an arbitrary point $P(x, y)$ on the ellipse, then—from the condition $|PF| = e|PL|$, applied first to the left focus and directrix and then to the right ones—we get

$$|PF'| = e\left(x + \frac{a}{e}\right) = ex + a \qquad |PF| = e\left(\frac{a}{e} - x\right) = a - ex$$

and so

$$\boxed{|PF'| + |PF| = 2a}$$

Next consider the hyperbola with $P(x, y)$ on its right branch, as shown in the right part of Figure 3. Then,

$$|PF'| = e\left(x + \frac{a}{e}\right) = ex + a \qquad |PF| = e\left(x - \frac{a}{e}\right) = ex - a$$

and so $|PF'| - |PF| = 2a$. If $P(x, y)$ had been on the left branch, we would have gotten $-2a$ in place of $2a$. In either case,

$$\boxed{\big||PF'| - |PF|\big| = 2a}$$

EXAMPLE 1 Find the equation of the set of points, the sum of whose distances from $(\pm 3, 0)$ is equal to 10.

Solution This is a horizontal ellipse with $a = 5$ and $c = 3$. Thus $b = \sqrt{a^2 - c^2} = 4$, and the equation is

$$\frac{x^2}{25} + \frac{y^2}{16} = 1 \qquad\blacksquare$$

EXAMPLE 2 Find the equation of the set of points, the difference of whose distances from $(0, \pm 6)$ is equal to 4.

Solution This is a vertical hyperbola with $a = 2$ and $c = 6$. Thus, $b = \sqrt{c^2 - a^2} = \sqrt{32} = 4\sqrt{2}$, and the equation is

$$-\frac{x^2}{32} + \frac{y^2}{4} = 1 \qquad\blacksquare$$

OPTICAL PROPERTIES Consider mirrors with the shapes of an ellipse and hyperbola, respectively. If a light ray emanating from one focus strikes the mirror, it will be reflected back to the other focus in the case of the ellipse and directly away from the other focus in the case of the hyperbola. These facts are shown in Figure 4.

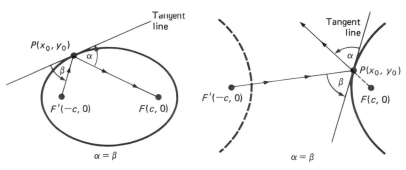

FIGURE 4

To demonstrate these optical properties (that is, to show $\alpha = \beta$ in Figure 4), we suppose the curves to be in standard position, so that their equations are $x^2/a^2 + y^2/b^2 = 1$ and $x^2/a^2 - y^2/b^2 = 1$, respectively. For the ellipse, we differentiate implicitly and then substitute (x_0, y_0), thereby obtaining the slope m of the tangent line.

$$\frac{2x}{a^2} + \frac{2yy'}{b^2} = 0$$

$$y' = -\frac{b^2}{a^2}\frac{x}{y}$$

$$m = -\frac{b^2}{a^2}\frac{x_0}{y_0}$$

The equation of the tangent line may be written successively as

$$y - y_0 = -\frac{b^2 x_0}{a^2 y_0}(x - x_0)$$

$$\frac{x_0}{a^2}(x - x_0) + \frac{y_0}{b^2}(y - y_0) = 0$$

$$\frac{x_0 x}{a^2} + \frac{y_0 y}{b^2} = \frac{x_0^2}{a^2} + \frac{y_0^2}{b^2} = 1$$

A similar derivation for the hyperbola leads to similar results. We summarize in the table below.

	Ellipse	**Hyperbola**
Equation	$\dfrac{x^2}{a^2} + \dfrac{y^2}{b^2} = 1$	$\dfrac{x^2}{a^2} - \dfrac{y^2}{b^2} = 1$
Slope of tangent at (x_0, y_0)	$m = \dfrac{-b^2 x_0}{a^2 y_0}$	$m = \dfrac{b^2 x_0}{a^2 y_0}$
Equation of tangent at (x_0, y_0)	$\dfrac{x_0 x}{a^2} + \dfrac{y_0 y}{b^2} = 1$	$\dfrac{x_0 x}{a^2} - \dfrac{y_0 y}{b^2} = 1$

To calculate $\tan \alpha$ for the ellipse, we recall a formula for the tangent of the counterclockwise angle from one line l_1 to another l in terms of their respective slopes m_1 and m, namely,

$$\tan \alpha = \frac{m - m_1}{1 + mm_1}$$

Thus,

$$\tan \alpha = \frac{\dfrac{-b^2 x_0}{a^2 y_0} - \dfrac{y_0 - 0}{x_0 - c}}{1 + \left(\dfrac{-b^2 x_0}{a^2 y_0}\right)\left(\dfrac{y_0 - 0}{x_0 - c}\right)} = \frac{-b^2 x_0 (x_0 - c) - a^2 y_0^2}{a^2 y_0 (x_0 - c) - b^2 x_0 y_0}$$

$$= \frac{b^2 c x_0 - (b^2 x_0^2 + a^2 y_0^2)}{(a^2 - b^2) x_0 y_0 - a^2 c y_0} = \frac{b^2 c x_0 - a^2 b^2}{c^2 x_0 y_0 - a^2 c y_0}$$

$$= \frac{b^2 (c x_0 - a^2)}{c y_0 (c x_0 - a^2)} = \frac{b^2}{c y_0}$$

The same calculation with c replaced by $-c$ gives

$$\tan(-\beta) = \frac{b^2}{-c y_0}$$

and so $\tan \beta = b^2 / c y_0$. We conclude that $\tan \alpha = \tan \beta$, and consequently $\alpha = \beta$.

A similar derivation establishes the corresponding result for the hyperbola.

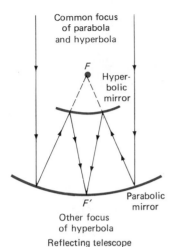

Common focus
of parabola
and hyperbola

F

Hyper-
bolic
mirror

Parabolic
mirror

F'

Other focus
of hyperbola

Reflecting telescope

FIGURE 5

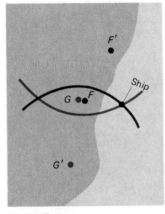

FIGURE 6

APPLICATIONS   The reflecting property of the ellipse is the basis of the "whispering gallery" effect that can be observed, for example, in the U.S. Capitol and the Mormon Tabernacle. A speaker standing at one focus can be heard whispering by a listener at the other focus, even though his or her voice is inaudible in other parts of the room.

The optical properties of the parabola and hyperbola are combined in one design for a reflecting telescope (Figure 5). The parallel rays from a star are finally focused at the eyepiece at F'.

The string property of the hyperbola is used in navigation. A ship at sea can determine the difference $2a$ in its distance from two fixed transmitters by measuring the difference in reception times of synchronized radio signals. This puts its path on a hyperbola, with the two transmitters as foci. If two pairs of transmitters are used, the ship must lie at the intersection of the two corresponding hyperbolas (see Figure 6). LORAN, a system of long-range navigation, is based on this principle.

PROBLEM SET   12.3

In Problems 1–4, find the equation of the set of points P satisfying the given conditions.

1. The sum of the distances of P from $(0, \pm 4)$ is 10.

2. The sum of the distances of P from $(\pm 12, 0)$ is 26.

3. The difference of the distances of P from $(\pm 5, 0)$ is 8.

4. The difference of the distances of P from $(0, \pm 13)$ is 10.

In Problems 5–12, find the equation of the tangent line to the given curve at the given point (see table on page 536).

5. $\dfrac{x^2}{9} + \dfrac{y^2}{27} = 1$ at $(\sqrt{6}, 3)$

6. $\dfrac{x^2}{16} + \dfrac{y^2}{24} = 1$ at $(-2, 3\sqrt{2})$

7. $\dfrac{x^2}{8} - \dfrac{y^2}{4} = 1$ at $(-4, 2)$

8. $-\dfrac{x^2}{36} + \dfrac{y^2}{8} = 1$ at $(6, 4)$

9. $x^2 + y^2 = 25$ at $(3, 4)$

10. $x^2 - y^2 = 9$ at $(5, 4)$

11. The curve of Problem 1 at $(0, 5)$.

12. The curve of Problem 2 at $(13, 0)$.

13. If two tangent lines to the ellipse $9x^2 + 4y^2 = 36$ intersect the y-axis at $(0, 6)$, find the points of tangency.

14. If the tangent lines to the hyperbola $9x^2 - y^2 = 36$ intersect the y-axis at $(0, 6)$, find the points of tangency.

15. The slope of the tangent to the hyperbola $2x^2 - 7y^2 - 35 = 0$ at a certain point on the hyperbola is $-\frac{2}{3}$. What are the coordinates of the point of tangency (two solutions)?

16. Find the equations of the tangents to the ellipse $x^2 + 2y^2 - 2 = 0$ that are parallel to the line $3x - 3\sqrt{2}y - 7 = 0$.

17. Find the area of the ellipse $b^2x^2 + a^2y^2 = a^2b^2$.

18. Find the volume of the solid obtained by revolving the ellipse $b^2x^2 + a^2y^2 = a^2b^2$ about the y-axis.

19. The region bounded by the hyperbola $b^2x^2 - a^2y^2 = a^2b^2$ and a vertical line through a focus is revolved about the x-axis. Find the volume of the resulting solid.

20. If the ellipse of Problem 18 is revolved about the x-axis, find the volume of the resulting solid.

21. Find the dimensions of the rectangle having the greatest possible area that can be inscribed in the ellipse $b^2x^2 + a^2y^2 = a^2b^2$. Assume that the sides of the rectangle are parallel to the axes of the ellipse.

22. Show that the point of contact of any tangent to a hyperbola is midway between the points in which the tangent intersects the asymptotes.

23. Find the point in the first quadrant where the two hyperbolas $25x^2 - 9y^2 = 225$ and $-25x^2 + 18y^2 = 450$ intersect.

24. Find the points of intersection of $x^2 + 4y^2 = 20$ and $x + 2y = 6$.

25. Sketch the design for a reflecting telescope that uses a parabola and an ellipse rather than the parabola and hyperbola described in the text.

26. A ball placed at a focus of an elliptical billiard table is shot with tremendous force so that it continues to bounce off the cushions indefinitely. What is its ultimate path? *Hint*: Draw a picture.

27. If the ball of Problem 26 is initially on the major axis between a focus and the neighboring vertex, what can you say about its path?

28. Show that an ellipse and a hyperbola with the same foci intersect at right angles. *Hint*: Draw a picture and use the optical properties.

29. Describe a string apparatus for constructing a hyperbola. (There are several possibilities).

30. Sound travels at u feet per second and a rifle bullet at $v > u$ feet per second. The sound of the firing of a rifle and the impact of the bullet hitting the target were heard simultaneously. If the rifle was at $A(-c, 0)$, the target was at $B(c, 0)$, and the listener was at $P(x, y)$, find the equation of the curve on which P lies (in terms of u, v, and c).

31. Listeners $A(-8, 0)$, $B(8, 0)$, and $C(8, 10)$ recorded the exact times at which they heard an explosion. If B and C heard the explosion at the same time and A heard it 12 seconds later, where was the explosion? Assume that distances are in kilometers and that sound travels $\frac{1}{3}$ kilometer per second.

12.4
Translation of Axes

So far we have placed the conics in the coordinate system in very special ways—always with the major axis along one of the coordinate axes and either the vertex (in the case of a parabola) or the center (in the case of an ellipse or hyperbola) at the origin. Now we place our conics in a more general position, though we still require that the major axis be parallel to one of the coordinate axes. Even this restriction will be removed in Section 12.5.

$(x - 2)^2 + (y - 3)^2 = 25$
or
$u^2 + v^2 = 25$

FIGURE 1

The case of a circle is instructive. The circle of radius 5 centered at $(2, 3)$ has equation

$$(x - 2)^2 + (y - 3)^2 = 25$$

or, in equivalent expanded form,

$$x^2 + y^2 - 4x - 6y = 12$$

The same circle with its center at the origin of the uv-coordinate system (Figure 1) has the simple equation

$$u^2 + v^2 = 25$$

The introduction of new axes does not change the shape or size of a curve, but it may greatly simplify its equation. It is this so-called translation of axes and the corresponding change of variables in an equation that we wish to investigate.

TRANSLATIONS If new axes are chosen in the plane, every point will have two sets of coordinates, the old ones, (x, y), relative to the old axes and the new ones, (u, v), relative to the new axes. The original coordinates are said to undergo a **transformation**. If the new axes are parallel, respectively, to the original axes and have the same directions, the transformation is called a **translation of axes**.

From Figure 2, it is easy to see how the new coordinates (u, v) relate to the old ones (x, y). Let (h, k) be the old coordinates of the new origin. Then,

FIGURE 2

$$\boxed{u = x - h, \qquad v = y - k}$$

or, equivalently,

$$\boxed{x = u + h, \qquad y = v + k}$$

EXAMPLE 1 Find the new coordinates of $P(-6, 5)$ after a translation of axes to a new origin at $(2, -4)$.

Solution Since $h = 2$ and $k = -4$, it follows that

$$u = x - h = -6 - 2 = -8 \qquad v = y - k = 5 - (-4) = 9$$

The new coordinates are $(-8, 9)$. ∎

EXAMPLE 2 Given the equation $4x^2 + y^2 + 40x - 2y + 97 = 0$, find the equation of its graph after a translation with new origin $(-5, 1)$.

Solution In the equation, we replace x by $u + h = u - 5$ and y by $v + k = v + 1$. We obtain

$$4(u - 5)^2 + (v + 1)^2 + 40(u - 5) - 2(v + 1) + 97 = 0$$

or

$$4u^2 - 40u + 100 + v^2 + 2v + 1 + 40u - 200 - 2v - 2 + 97 = 0$$

This simplifies to

$$4u^2 + v^2 = 4$$

or

$$u^2 + \frac{v^2}{4} = 1$$

which we recognize as the equation of an ellipse. ∎

COMPLETING THE SQUARE Given a complicated second-degree equation, how do we know what translation will simplify the equation and bring it to a recognizable form? Here a familiar algebraic process called **completing the square** provides the answer. In particular, we can use this process to eliminate the first-degree terms of any expression of the form

$$Ax^2 + Cy^2 + Dx + Ey + F = 0, \qquad A \neq 0, C \neq 0$$

EXAMPLE 3 Make a translation which will eliminate the first-degree terms of

$$4x^2 + 9y^2 + 8x - 90y + 193 = 0$$

and use this information to sketch the graph of the given equation.

Solution Recall that to complete the square of $x^2 + ax$, we must add $a^2/4$ (the square of half the coefficient of x). Using this, we rewrite the given equation as follows.

$$4(x^2 + 2x \quad\quad) + 9(y^2 - 10y \quad\quad) = -193$$
$$4(x^2 + 2x + 1) + 9(y^2 - 10y + 25) = -193 + 4 + 225$$
$$4(x + 1)^2 + 9(y - 5)^2 = 36$$
$$\frac{(x + 1)^2}{9} + \frac{(y - 5)^2}{4} = 1$$

The translation $u = x + 1$ and $v = y - 5$ transforms this to

$$\frac{u^2}{9} + \frac{v^2}{4} = 1$$

FIGURE 3

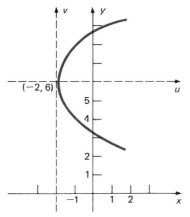

FIGURE 4

which is the standard form of a horizontal ellipse. The graph is shown in Figure 3. ∎

EXAMPLE 4   Use a translation to simplify

$$y^2 - 4x - 12y + 28 = 0$$

Then determine which conic it represents, list the important characteristics of this conic, and sketch its graph.

Solution   We complete the square.

$$y^2 - 12y \qquad = 4x - 28$$
$$y^2 - 12y + 36 = 4x - 28 + 36$$
$$(y - 6)^2 = 4(x + 2)$$

The translation $u = x + 2$, $v = y - 6$ transforms this to $v^2 = 4u$, which we recognize as a horizontal parabola opening right with $p = 1$ (Figure 4). ∎

GENERAL SECOND-DEGREE EQUATIONS   Now we ask an important question. Is the graph of an equation of the form

$$Ax^2 + Cy^2 + Dx + Ey + F = 0$$

always a conic? The answer is no, unless we admit certain limiting forms. The table below indicates the possibilities with a sample equation for each.

Conics	Limiting Forms
1. $(AC = 0)$ Parabola: $y^2 = 4x$ ⊂	Parallel lines: $y^2 = 4$ ═══ Single line: $y^2 = 0$ ──── Empty set: $y^2 = -1$
2. $(AC > 0)$ Ellipse: $\dfrac{x^2}{9} + \dfrac{y^2}{4} = 1$ ◯	Circle: $x^2 + y^2 = 4$ ◯ Point: $2x^2 + y^2 = 0$ • Empty set: $2x^2 + y^2 = -1$
3. $(AC < 0)$ Hyperbola: $\dfrac{x^2}{9} - \dfrac{y^2}{4} = 1$)(	Intersecting lines: $x^2 - y^2 = 0$ ✕

Thus, the graphs of the general quadratic equation above fall into three general categories but yield nine possibilities, including limiting forms.

EXAMPLE 5 Use a translation to simplify

$$4x^2 - y^2 - 8x - 6y - 5 = 0$$

and sketch its graph.

Solution We rewrite the equation as follows.

$$4(x^2 - 2x \quad) - (y^2 + 6y \quad) = 5$$
$$4(x^2 - 2x + 1) - (y^2 + 6y + 9) = 5 + 4 - 9$$
$$4(x - 1)^2 - (y + 3)^2 = 0$$

Let $u = x - 1$ and $v = y + 3$, which results in

$$4u^2 - v^2 = 0$$

or

$$(2u - v)(2u + v) = 0$$

This is the equation of two intersecting lines (Figure 5). ∎

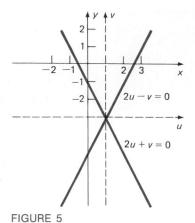

FIGURE 5

EXAMPLE 6 Write the equation of a hyperbola with foci at $(1, 1)$ and $(1, 11)$ and vertices at $(1, 3)$ and $(1, 9)$.

Solution The center is $(1, 6)$, midway between the vertices. Thus, $a = 3$ and $c = 5$, and so $b = \sqrt{c^2 - a^2} = 4$. The equation is

$$\frac{(y - 6)^2}{9} - \frac{(x - 1)^2}{16} = 1$$ ∎

SUMMARY Consider the general equation

$$Ax^2 + Cy^2 + Dx + Ey + F = 0$$

If both A and C are zero, we have the equation of a line (provided, of course, that D and E are not both zero). If at least one of A and C is different from zero, we may apply the process of completing the square. We obtain one of several forms, the most typical being:

(1) $$(y - k)^2 = \pm 4p(x - h)$$

(2) $$\frac{(x - h)^2}{a^2} + \frac{(y - k)^2}{b^2} = 1$$

(3) $$\frac{(x - h)^2}{a^2} - \frac{(y - k)^2}{b^2} = 1$$

These can be recognized even in this form as the equations of a horizontal parabola with vertex at (h, k), a horizontal ellipse (if $a^2 > b^2$) with center at (h, k), and a horizontal hyperbola with center at (h, k). But to remove any doubt, we may translate the axes by the substitutions $u = x - h$, $v = y - k$, thereby obtaining:

(1) $$v^2 = \pm 4pu$$

(2) $$\frac{u^2}{a^2} + \frac{v^2}{b^2} = 1$$

(3) $$\frac{u^2}{a^2} - \frac{v^2}{b^2} = 1$$

Our work may also yield these equations with u and v interchanged, or we may get one of the six limiting forms illustrated in the table on page 541. There are no other possibilities.

PROBLEM SET 12.4

In Problems 1–16, name the conic or limiting form represented by the given equation. Usually you will need to use the process of completing the square (see Examples 3–5).

1. $x^2 + y^2 - 2x + 4y + 4 = 0$

2. $x^2 + y^2 - 6x + 2y + 6 = 0$

3. $4x^2 + 9y^2 - 16x + 72y + 124 = 0$

4. $9x^2 - 16y^2 + 90x + 192y - 495 = 0$

5. $4x^2 + 9y^2 - 16x + 72y + 160 = 0$

6. $9x^2 + 16y^2 + 90x + 192y + 1000 = 0$

7. $y^2 - 10x - 8y - 14 = 0$

8. $4x^2 + 4y^2 + 16x - 20y - 10 = 0$

9. $x^2 + y^2 - 2x + 4y + 20 = 0$

10. $4x^2 - 4y^2 + 16x - 20y - 10 = 0$

11. $4x^2 - 4y^2 + 16x - 20y - 9 = 0$

12. $4x^2 - 16x + 16 = 0$

13. $4x^2 - 16x + 15 = 0$

14. $25x^2 + 4y^2 + 150x - 8y + 129 = 0$

15. $25x^2 - 4y^2 + 150x - 8y + 129 = 0$

16. $3x + 4y - 16 = 0$

In Problems 17–30, sketch the graph of the given equation.

17. $\dfrac{(x + 3)^2}{4} + \dfrac{(y + 2)^2}{16} = 1$

18. $(x + 3)^2 + (y - 4)^2 = 25$

19. $\dfrac{(x + 3)^2}{4} - \dfrac{(y + 2)^2}{16} = 1$

20. $4(x + 3) = (y + 2)^2$

21. $(x + 2)^2 = 8(y - 1)$

22. $(x + 2)^2 = 4$

23. $(y - 1)^2 = 16$

24. $\dfrac{(x + 3)^2}{4} + \dfrac{(y - 2)^2}{8} = 0$

25. $x^2 + 4y^2 - 2x + 16y + 1 = 0$

26. $25x^2 + 9y^2 + 150x - 18y + 9 = 0$

27. $9x^2 - 16y^2 + 54x + 64y - 127 = 0$

28. $x^2 - 4y^2 - 14x - 32y - 11 = 0$

29. $4x^2 + 16x - 16y + 32 = 0$

30. $x^2 - 4x + 8y = 0$

31. Find the focus and directrix of the parabola
$$2y^2 - 4y - 10x = 0.$$

32. Determine the distance between the vertices of
$$-9x^2 + 18x + 4y^2 + 24y = 9.$$

33. Find the foci of the ellipse
$$16(x - 1)^2 + 25(y + 2)^2 = 400.$$

34. Find the focus and directrix of the parabola
$$x^2 - 6x + 4y + 3 = 0.$$

In Problems 35–44, find the equation of the given conic.

35. The horizontal ellipse with center $(5,1)$, major diameter 10, minor diameter 8.

36. The hyperbola with center $(2, -1)$, vertex at $(4, -1)$, and focus at $(5, -1)$.

37. The parabola with vertex $(2,3)$ and focus $(2,5)$.

38. The ellipse with center $(2,3)$ passing through $(6,3)$ and $(2,5)$.

39. The hyperbola with vertices at $(0,0)$ and $(0,6)$ and a focus at $(0,8)$.

40. The ellipse with foci at $(2,0)$ and $(2,12)$ and a vertex at $(2,14)$.

41. The parabola with focus $(2,5)$ and directrix $x = 10$.

42. The parabola with focus $(2,5)$ and vertex $(2,6)$.

43. The ellipse with foci $(\pm 2, 2)$ that passes through the origin.

44. The hyperbola with foci $(0,0)$ and $(0,4)$ that passes through $(12,9)$.

45. A curve C goes through the three points, $(-1,2)$, $(0,0)$, and $(3,6)$. Find an equation for C if C is:
(a) a vertical parabola; (b) a horizontal parabola;
(c) a circle.

46. The ends of an elastic string with a knot at $K(x,y)$ are attached to a fixed point $A(a,b)$ and a point P on the rim of a wheel of radius r centered at $(0,0)$. As the wheel turns, K traces a curve C. Find the equation of C. Assume the string stays taut and stretches uniformly (that is, $\alpha = |KP|/|AP|$ is constant).

47. Name the conic $y^2 = Lx + Kx^2$ according to the value of K and then show that in every case, $|L|$ is the length of the latus rectum of the conic. Assume $L \neq 0$.

48. Show that the equations of the parabola and hyperbola with vertex $(a,0)$ and focus $(c,0)$, $c > a > 0$, can be written as $y^2 = 4(c - a)(x - a)$ and $y^2 = (b^2/a^2)(x^2 - a^2)$, respectively. Then use these expressions for y^2 to show that the parabola is always "inside" the right branch of the hyperbola.

12.5
Rotation of Axes

Consider the most general second-degree equation in x and y

$$Ax^2 + Bxy + Cy^2 + Dx + Ey + F = 0$$

The new feature is the appearance of the cross-product term Bxy. Is it still true that the graph is a conic or one of the limiting forms? The answer is yes, but the axis (or axes) of the conic are rotated with respect to the coordinate axes.

FIGURE 1

ROTATIONS Introduce a new pair of coordinate axes, the u- and v-axes, with the same origin as the x- and y-axes but rotated through an angle θ, as shown in Figure 1. A point P then has two sets of coordinates: (x, y) and (u, v). How are they related?

Let r denote the length of OP and let ϕ denote the angle from the positive u-axis to OP. Then, x, y, u, and v have the geometric interpretations shown in the diagram.

Looking at the right triangle OPM, we see that

$$\cos(\phi + \theta) = \frac{x}{r}$$

so

$$x = r\cos(\phi + \theta) = r(\cos\phi\cos\theta - \sin\phi\sin\theta)$$

$$= (r\cos\phi)\cos\theta - (r\sin\phi)\sin\theta$$

CHAPTER 12/Conics and Polar Coordinates

Consideration of triangle OPN shows that $u = r \cos \phi$ and $v = r \sin \phi$. Thus,

$$x = u \cos \theta - v \sin \theta$$

Similar reasoning leads to

$$y = u \sin \theta + v \cos \theta$$

These formulas determine a transformation called a **rotation of axes**.

EXAMPLE 1 Find the new equation that results from $xy = 1$ after a rotation of axes through $\theta = \pi/4$. Sketch the graph.

Solution The required substitutions are

$$x = u \cos \frac{\pi}{4} - v \sin \frac{\pi}{4} = \frac{\sqrt{2}}{2}(u - v)$$

$$y = u \sin \frac{\pi}{4} + v \cos \frac{\pi}{4} = \frac{\sqrt{2}}{2}(u + v)$$

The equation $xy = 1$ takes the form

$$\frac{\sqrt{2}}{2}(u - v) \frac{\sqrt{2}}{2}(u + v) = 1$$

which simplifies to

$$\frac{u^2}{2} - \frac{v^2}{2} = 1$$

This we recognize as the equation of a hyperbola with $a = b = \sqrt{2}$. Note that the cross-product term has disappeared as a result of the rotation. The choice of the angle $\theta = \pi/4$ was just right to make this happen. The graph is shown in Figure 2. ∎

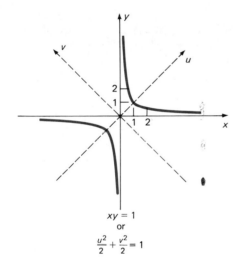

$xy = 1$
or
$\frac{u^2}{2} + \frac{v^2}{2} = 1$

FIGURE 2

DETERMINING THE ANGLE θ How do we know what rotation to make in order to eliminate the cross-product term? Consider the equation

$$Ax^2 + Bxy + Cy^2 + Dx + Ey + F = 0$$

If we make the substitutions

$$x = u \cos \theta - v \sin \theta$$
$$y = u \sin \theta + v \cos \theta$$

this equation takes the form

$$au^2 + buv + cv^2 + du + ev + f = 0$$

where a, b, c, d, e, and f are numbers that depend upon θ. We could find expressions for all of them, but we really care only about b. When we do the necessary algebra, we find

$$b = B(\cos^2\theta - \sin^2\theta) - 2(A - C)\sin\theta\cos\theta$$
$$= B\cos 2\theta - (A - C)\sin 2\theta$$

To make $b = 0$, we require

$$B\cos 2\theta = (A - C)\sin 2\theta$$

or

$$\boxed{\cot 2\theta = \frac{A - C}{B}}$$

   This formula answers our question. To eliminate the cross-product term, choose θ so it satisfies this formula. In the equation $xy = 1$ of Example 1, $A = 0$, $B = 1$, and $C = 0$, so we choose θ satisfying $\cot 2\theta = 0$. One angle that works is $\theta = \pi/4$. We could also use $\theta = 3\pi/4$ or $\theta = -5\pi/4$, but it is customary to choose a first-quadrant angle; that is, we choose 2θ satisfying $0 \leq 2\theta < \pi$, so that $0 \leq \theta < \pi/2$.

EXAMPLE 2   Make a rotation of axes to eliminate the cross-product term in

$$4x^2 + 2\sqrt{3}xy + 2y^2 + 10\sqrt{3}x + 10y = 5$$

Then sketch the graph

Solution

$$\cot 2\dot\theta = \frac{A - C}{B} = \frac{4 - 2}{2\sqrt{3}} = \frac{1}{\sqrt{3}}$$

which means that $2\theta = \pi/3$ and $\theta = \pi/6$. The appropriate substitutions are

$$x = u\frac{\sqrt{3}}{2} - v\frac{1}{2} = \frac{\sqrt{3}u - v}{2}$$

$$y = u\frac{1}{2} + v\frac{\sqrt{3}}{2} = \frac{u + \sqrt{3}v}{2}$$

Our equation transforms first to

$$4\frac{(\sqrt{3}u - v)^2}{4} + 2\sqrt{3}\frac{(\sqrt{3}u - v)(u + \sqrt{3}v)}{4} + 2\frac{(u + \sqrt{3}v)^2}{4}$$

$$+ 10\sqrt{3}\frac{\sqrt{3}u - v}{2} + 10\frac{u + \sqrt{3}v}{2} = 5$$

and, after simplifying, to

$$5u^2 + v^2 + 20u = 5$$

To put this equation in recognizable form, we complete the square.

$$5(u^2 + 4u + 4) + v^2 = 5 + 20$$

$$\frac{(u + 2)^2}{5} + \frac{v^2}{25} = 1$$

Finally, we identify this last equation as that of a vertical ellipse with center at $u = -2$ and $v = 0$ and with $a = 5$ and $b = \sqrt{5}$. Its graph is shown in Figure 3. ∎

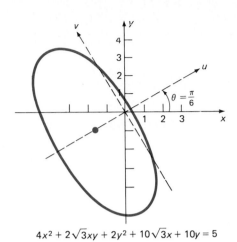

$$4x^2 + 2\sqrt{3}xy + 2y^2 + 10\sqrt{3}x + 10y = 5$$

or $\frac{(u + 2)^2}{5} + \frac{v^2}{25} = 1$

FIGURE 3

ROTATION THROUGH A NONSPECIAL ANGLE   Our two examples involved rotations through the special angles of $\pi/4$ and $\pi/6$, respectively. Nonspecial angles are handled through use of the half-angle formulas.

$$\sin\theta = \pm\sqrt{\frac{1 - \cos 2\theta}{2}} \qquad \cos\theta = \pm\sqrt{\frac{1 + \cos 2\theta}{2}}$$

EXAMPLE 3   Use a rotation to eliminate the cross-product term in

$$x^2 + 24xy + 8y^2 = 136$$

Then sketch the graph.

Solution   We choose θ satisfying

$$\cot 2\theta = \frac{A - C}{B} = \frac{1 - 8}{24} = -\frac{7}{24}$$

We take 2θ to be a second-quadrant angle with $(-7, 24)$ on its terminal side, a point with distance $r = \sqrt{(-7)^2 + (24)^2} = 25$ from the origin (Figure 4). It follows that $\cos 2\theta = -\frac{7}{25}$. Applying the half-angle formulas gives

FIGURE 4

$$\sin\theta = \sqrt{\frac{1 + \frac{7}{25}}{2}} = \frac{4}{5} \qquad \cos\theta = \sqrt{\frac{1 - \frac{7}{25}}{2}} = \frac{3}{5}$$

Next we use the rotation formulas

$$x = \frac{3}{5}u - \frac{4}{5}v \qquad y = \frac{4}{5}u + \frac{3}{5}v$$

When these expressions are substituted in our equation, we obtain

$$\left(\frac{3u - 4v}{5}\right)^2 + 24\left(\frac{3u - 4v}{5}\right)\left(\frac{4u + 3v}{5}\right) + 8\left(\frac{4u + 3v}{5}\right)^2 = 136$$

This simplifies to

$$425u^2 - 200v^2 = (136)(25)$$

or

$$\frac{u^2}{8} - \frac{v^2}{17} = 1$$

The graph of the last equation is a horizontal hyperbola in the uv-plane with center at the origin (Figure 5).

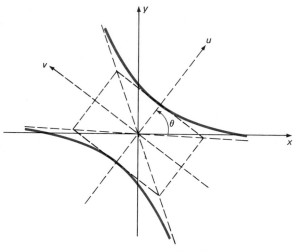

$$x^2 + 24xy + 8y^2 = 136 \quad \text{or} \quad \frac{u^2}{8} - \frac{v^2}{17} = 1$$

FIGURE 5

PROBLEM SET 12.5

In Problems 1–12, eliminate the cross-product term by a suitable rotation of axes and then, if necessary, translate axes (complete the squares) to put the equation in standard form. Finally graph the equation showing the rotated axes.

1. $3x^2 + 10xy + 3y^2 + 8 = 0$

2. $2x^2 + xy + 2y^2 = 90$

3. $4x^2 - 3xy = 18$

4. $4xy - 3y^2 = 64$

5. $x^2 - 2\sqrt{3}xy + 3y^2 - 12\sqrt{3}x - 12y = 0$

6. $x^2 + 2\sqrt{3}xy + 3y^2 + 8\sqrt{3}x - 8y = 0$

7. $13x^2 + 6\sqrt{3}xy + 7y^2 - 32 = 0$

8. $17x^2 + 12xy + 8y^2 + 17 = 0$

9. $9x^2 - 24xy + 16y^2 - 60x + 80y + 75 = 0$

10. $16x^2 + 24xy + 9y^2 - 20x - 15y - 150 = 0$

11. $5x^2 - 3xy + y^2 + 65x - 25y + 203 = 0$

12. $6x^2 - 5xy - 6y^2 + 78x + 52y + 26 = 0$

13. The graph of $x \cos \alpha + y \sin \alpha = d$ is a line. Show that the perpendicular distance from the origin to this line is $|d|$ by making a rotation of axes through the angle α.

14. Transform the equation $x^{1/2} + y^{1/2} = a^{1/2}$ by a rotation of axes through $45°$ and then square to eliminate radicals on variables. Identify the corresponding curve.

15. Solve the rotation formulas for u and v in terms of x and y.

16. Use the results of Problem 15 to find the uv-coordinates corresponding to $(x, y) = (5, -3)$ after a rotation of axes through $60°$.

17. Find the point of $x^2 + 14xy + 49y^2 = 100$ that is closest to the origin.

18. Recall that $Ax^2 + Bxy + Cy^2 + Dx + Ey + F = 0$ transforms to $au^2 + buv + cv^2 + du + ev + f = 0$ under a rotation of axes. Find formulas for a and c and show that $a + c = A + C$.

19. Show that $b^2 - 4ac = B^2 - 4AC$ (see Problem 18).

20. Use the result of Problem 19 to convince yourself that the graph of the general second-degree equation will be:

(a) a parabola if $B^2 - 4AC = 0$,
(b) an ellipse if $B^2 - 4AC < 0$,
(c) a hyperbola if $B^2 - 4AC > 0$,

or limiting forms of the above conics.

21. Let $Ax^2 + Bxy + Cy^2 = 1$ be transformed to $au^2 + cv^2 = 1$ by a rotation of axes and suppose $\Delta = 4AC - B^2 \neq 0$. Use Problems 18 and 19 to show that:

(a) $1/ac = 4/\Delta$,
(b) $1/a + 1/c = (4/\Delta)(A + C)$,
(c) $1/a$ and $1/c$ are the two values of
$$(2/\Delta)(A + C \pm \sqrt{(A - C)^2 + B^2})$$

22. Show that if $A + C$ and $\Delta = 4AC - B^2$ are both positive, then the graph of $Ax^2 + Bxy + Cy^2 = 1$ is an ellipse (or circle) with area $2\pi/\sqrt{\Delta}$. (Recall from Problem 17 of Section 12.3 that the area of the ellipse $x^2/p^2 + y^2/q^2 = 1$ is πpq.)

23. For what values of B is the graph of $x^2 + Bxy + y^2 = 1$:

(a) an ellipse,   (b) a circle,
(c) a hyperbola,   (d) two parallel lines?

24. Use the results of Problems 21 and 22 to find the distance between the foci and the area of the ellipse
$$25x^2 + 8xy + y^2 = 1$$

12.6
The Polar Coordinate System

Two Frenchmen, Pierre Fermat and René Descartes, introduced what we now call the *Cartesian*, or *rectangular*, coordinate system. Their idea was to specify each point P in the plane by giving two numbers (x, y), the directed distances from a pair of perpendicular axes (Figure 1). This notion is by now so familiar that we use it almost without thinking. Yet it is the fundamental idea in analytic geometry and makes possible the development of calculus as we have given it so far.

Giving the directed distances from a pair of perpendicular axes is not the only way to specify a point. Another way to do this is by giving the so-called polar coordinates.

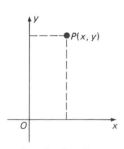

Cartesian Coordinates

FIGURE 1

POLAR COORDINATES   We start with a fixed half-line, called the **polar axis**, emanating from a fixed point 0, called the **pole** or **origin**. By custom, the polar axis is chosen to be horizontal and pointing to the right and may therefore be identified with the positive x-axis in the rectangular coordinate system. Any point P (other than the pole) is the intersection of a unique circle with center at 0 and a unique ray emanating from 0. If r is the radius

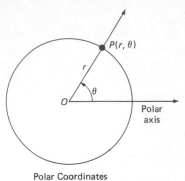

Polar Coordinates
FIGURE 2

of the circle and θ is one of the angles the ray makes with the polar axis, then (r, θ) is a pair of **polar coordinates** for P (Figure 2).

Points specified by polar coordinates are easiest to plot if we use polar graph paper. The grid on such paper consists of concentric circles and rays emanating from their common center. A simple version of a polar grid is shown in Figure 3, where we have also plotted several points.

FIGURE 3

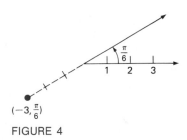

FIGURE 4

Notice a phenomenon that did not occur with Cartesian coordinates. Each point has many sets of polar coordinates, due to the fact that the angles $\theta + 2\pi n$, $n = 0, \pm1, \pm2, \ldots$, have the same terminal sides. For example, the point with polar coordinates $(4, \pi/2)$ also has coordinates $(4, 5\pi/2)$, $(4, 9\pi/2)$, $(4, -3\pi/2)$, and so on. This is even more true if we also allow r to be negative. In this case, (r, θ) is on the ray oppositely directed from the terminal side of θ and $|r|$ units from the origin. Thus, the point with polar coordinates $(-3, \pi/6)$ is as shown in Figure 4, and $(-4, 3\pi/2)$ is another set of coordinates for $(4, \pi/2)$. The origin has coordinates $(0, \theta)$, where θ is any angle.

POLAR EQUATIONS   Examples of polar equations are

$$r = 8 \sin \theta \quad \text{and} \quad r = \frac{2}{1 - \cos \theta}$$

Polar equations, like rectangular ones, are best visualized from their graphs. The **graph of a polar equation** is the set of points, each of which has at least one pair of polar coordinates that satisfy the equation. The most basic way to sketch a graph is to construct a table of values, plot the corresponding points, and then connect these points with a smooth curve.

EXAMPLE 1   Sketch the graph of the polar equation $r = 8 \sin \theta$.

Solution   We substitute multiples of $\pi/6$ for θ and calculate the corresponding r-values. Note that as θ increases from 0 to 2π, the graph is traced twice (Figure 5).   ∎

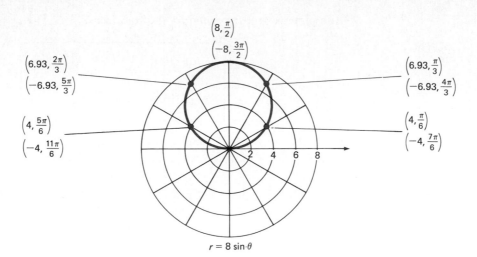

θ	r
0	0
$\pi/6$	4
$\pi/3$	6.93
$\pi/2$	8
$2\pi/3$	6.93
$5\pi/6$	4
π	0
$7\pi/6$	-4
$4\pi/3$	-6.93
$3\pi/2$	-8
$5\pi/3$	-6.93
$11\pi/6$	-4

$r = 8 \sin \theta$

FIGURE 5

EXAMPLE 2 Sketch the graph of $r = \dfrac{2}{1 - \cos\theta}$.

Solution See Figure 6.

θ	r
0	—
$\pi/4$	6.8
$\pi/2$	2
$3\pi/4$	1.2
π	1
$5\pi/4$	1.2
$3\pi/2$	2
$7\pi/4$	6.8
2π	—

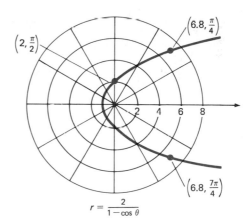

$r = \dfrac{2}{1 - \cos\theta}$

FIGURE 6

Note a phenomenon that does not occur with rectangular coordinates. The coordinates $(-2, 3\pi/2)$ do not satisfy the equation. Yet the point $P(-2, 3\pi/2)$ is on the graph, due to the fact that $(2, \pi/2)$ specifies the same point and does satisfy the equation. We conclude that *in polar coordinates, failure of a particular set of coordinates to satisfy a given equation is no guarantee that the corresponding point is not on the graph of that equation.* This fact causes many difficulties; we must learn to live with them. ∎

RELATION TO CARTESIAN COORDINATES   We suppose that the polar axis coincides with the positive x-axis of the Cartesian system. Then the polar coordinates (r, θ) of a point P and the Cartesian coordinates (x, y) of

FIGURE 7

the same point are related by the equations

$$x = r \cos \theta \qquad r^2 = x^2 + y^2$$

$$y = r \sin \theta \qquad \tan \theta = \frac{y}{x}$$

That this is true for a point P in the first quadrant is clear from Figure 7 and is easy to show for points in the other quadrants.

EXAMPLE 3   Find the Cartesian coordinates corresponding to $(4, \pi/6)$ and polar coordinates corresponding to $(-3, \sqrt{3})$.

$(-3, \sqrt{3})$

FIGURE 8

Solution   If $(r, \theta) = (4, \pi/6)$, then

$$x = 4 \cos \frac{\pi}{6} = 4 \cdot \frac{\sqrt{3}}{2} = 2\sqrt{3}$$

$$y = 4 \sin \frac{\pi}{6} = 4 \cdot \frac{1}{2} = 2$$

If $(x, y) = (-3, \sqrt{3})$, then (see Figure 8)

$$r^2 = (-3)^2 + (\sqrt{3})^2 = 12$$

$$\tan \theta = \frac{\sqrt{3}}{-3}$$

One value of (r, θ) is $(2\sqrt{3}, 5\pi/6)$. Another is $(-2\sqrt{3}, -\pi/6)$. ∎

Sometimes we can identify the graph of a polar equation by finding its equivalent Cartesian form. Here is an illustration.

EXAMPLE 4   Show that the graph of $r = 8 \sin \theta$ (Example 1) is a circle and that the graph of $r = 2/(1 - \cos \theta)$ (Example 2) is a parabola by changing to Cartesian coordinates.

Solution   If we multiply $r = 8 \sin \theta$ by r, we get

$$r^2 = 8r \sin \theta$$

which, in Cartesian coordinates, is

$$x^2 + y^2 = 8y$$

and may be written successively as

$$x^2 + y^2 - 8y = 0$$
$$x^2 + y^2 - 8y + 16 = 16$$
$$x^2 + (y - 4)^2 = 16$$

The latter is the equation of a circle of radius 4 centered at $(0, 4)$.

CAUTION

Since r can be 0, there is a potential danger in multiplying both sides of a polar equation by r or in dividing both sides by r. In the first case, we might add the pole to the graph; in the second, we might delete the pole from the graph. In Example 4, we multiplied both sides of $r = 8 \sin \theta$ by r but no harm was done since the pole was already on the graph as the point with θ-coordinate 0.

The second equation is handled by the following steps.

$$r = \frac{2}{1 - \cos \theta}$$

$$r - r \cos \theta = 2$$

$$r - x = 2$$

$$r = x + 2$$

$$r^2 = x^2 + 4x + 4$$

$$x^2 + y^2 = x^2 + 4x + 4$$

$$y^2 = 4(x + 1)$$

We recognize the last equation as that of a parabola with vertex at $(-1, 0)$ and focus at the origin. ∎

POLAR EQUATIONS FOR LINES, CIRCLES, AND CONICS If a line passes through the pole, it has the simple equation $\theta = \theta_0$. If the line does not go through the pole, it is some distance $d > 0$ from it. Let θ_0 be the angle from the polar axis to the perpendicular from the pole to the given line (Figure 9). Then, if $P(r, \theta)$ is any point on the line, $\cos(\theta - \theta_0) = d/r$, or

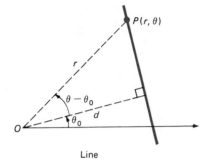

Line

FIGURE 9

$$\boxed{\text{Line:} \quad r = \frac{d}{\cos(\theta - \theta_0)}}$$

If a circle of radius a is centered at the pole, its equation is simply $r = a$. If it is centered at (r_0, θ_0), its equation is quite complicated unless we choose $r_0 = a$, as in Figure 10. Then, by the Law of Cosines, $a^2 = r^2 + a^2 - 2ra \cos(\theta - \theta_0)$, which simplifies to

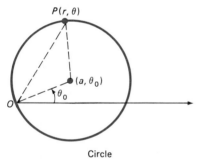

Circle

FIGURE 10

$$\boxed{\text{Circle:} \quad r = 2a \cos(\theta - \theta_0)}$$

The cases $\theta_0 = 0$ and $\theta_0 = \pi/2$ are particularly nice. The first gives $r = 2a \cos \theta$; the second gives $r = 2a \cos(\theta - \pi/2)$; that is, $r = 2a \sin \theta$. The latter should be compared with Example 1.

Finally, if a conic (ellipse, parabola, or hyperbola) is placed so that its focus is at the pole, and its directrix is d units away, as in Figure 11, then the familiar defining equation $|PF| = e|PL|$ takes the form

$$r = e[d - r \cos(\theta - \theta_0)]$$

or, equivalently,

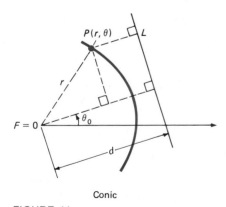

Conic

FIGURE 11

$$\boxed{\text{Conic:} \quad r = \frac{ed}{1 + e \cos(\theta - \theta_0)}}$$

Again, there is special interest in the cases $\theta_0 = 0$ and $\theta_0 = \pi/2$. Note in particular that if $e = 1$ and $\theta_0 = 0$, we have the equation of Example 2.

Our results are summarized in the chart as shown below.

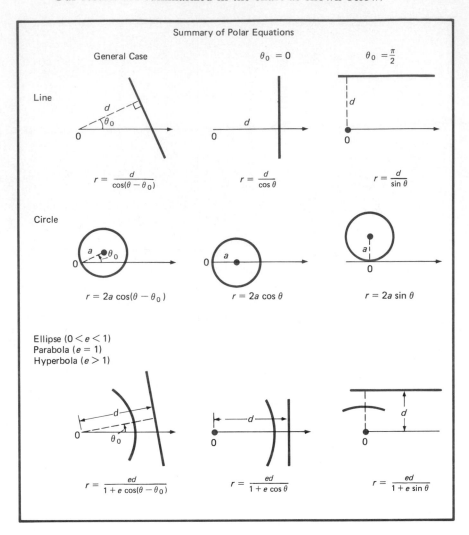

Summary of Polar Equations

General Case $\theta_0 = 0$ $\theta_0 = \frac{\pi}{2}$

Line

$$r = \frac{d}{\cos(\theta - \theta_0)}$$ $$r = \frac{d}{\cos \theta}$$ $$r = \frac{d}{\sin \theta}$$

Circle

$$r = 2a \cos(\theta - \theta_0)$$ $$r = 2a \cos \theta$$ $$r = 2a \sin \theta$$

Ellipse $(0 < e < 1)$
Parabola $(e = 1)$
Hyperbola $(e > 1)$

$$r = \frac{ed}{1 + e \cos(\theta - \theta_0)}$$ $$r = \frac{ed}{1 + e \cos \theta}$$ $$r = \frac{ed}{1 + e \sin \theta}$$

EXAMPLE 5 Find the equation of the horizontal ellipse with eccentricity $\frac{1}{2}$, focus at the pole, and vertical directrix 10 units to the right of the pole.

Solution

$$r = \frac{\frac{1}{2} \cdot 10}{1 + \frac{1}{2} \cos \theta} = \frac{10}{2 + \cos \theta} \qquad \blacksquare$$

EXAMPLE 6 Identify and sketch the graph of $r = \dfrac{7}{2 + 4 \sin \theta}$.

Solution We put this in standard form.

$$r = \frac{7}{2 + 4 \sin \theta} = \frac{\frac{7}{2}}{1 + 2 \sin \theta} = \frac{2(\frac{7}{4})}{1 + 2 \sin \theta}$$

which we recognize as the polar equation of a hyperbola with $e = 2$, focus at the pole, and horizontal directrix $\frac{7}{4}$ units above the polar axis (Figure 12).

$\blacksquare$

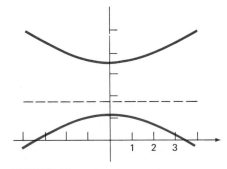

FIGURE 12

554   CHAPTER 12/Conics and Polar Coordinates

1. Plot the points whose polar coordinates are $(4, \frac{1}{3}\pi)$, $(2, \frac{1}{2}\pi)$, $(5, \frac{1}{6}\pi)$, $(0, \frac{11}{7}\pi)$, $(3, \frac{3}{2}\pi)$, $(\frac{7}{2}, \frac{2}{3}\pi)$, and $(4, 0)$.

2. Plot the points whose polar coordinates are $(3, \pi)$, $(7, 2\pi)$, $(2, \frac{11}{6}\pi)$, $(4, -\frac{1}{3}\pi)$, $(0, 0)$, $(0, \frac{2}{3}\pi)$, and $(3, -\frac{3}{2}\pi)$.

3. Plot the points whose polar coordinates are $(-5, \frac{1}{4}\pi)$, $(5, -\frac{3}{4}\pi)$, $(2, -\frac{1}{3}\pi)$, $(-2, \frac{2}{3}\pi)$, $(-6, 0)$, $(3, -\pi)$, $(-4, -\frac{2}{3}\pi)$, and $(-3, \pi)$.

4. Plot the points whose polar coordinates are $(2, \frac{5}{8}\pi)$, $(-4, \frac{17}{6}\pi)$, $(5, -\frac{7}{3}\pi)$, $(7, 1)$, $(-3, -2)$, $(3, 0)$, $(-3, -\pi)$, and $(0, -\frac{5}{4}\pi)$.

5. Plot the points whose polar coordinates follow. For each, give four other pairs of polar coordinates, two with positive r and two with negative r.

(a) $(4, \frac{1}{3}\pi)$ (b) $(-3, \frac{5}{4}\pi)$

(c) $(-5, \frac{1}{6}\pi)$ (d) $(7, -\frac{2}{3}\pi)$

6. Plot the points whose polar coordinates follow. For each, give four other pairs of polar coordinates, two with positive r and two with negative r.

(a) $(-2, \pi)$ (b) $(5, -\frac{1}{12}\pi)$

(c) $(4, \frac{19}{4}\pi)$ (d) $(-7, -\frac{3}{2}\pi)$

7. Find the Cartesian coordinates of the points in Problem 5.

8. Find the Cartesian coordinates of the points in Problem 6.

9. Find polar coordinates of the points whose Cartesian coordinates are as given.

(a) $(-2\sqrt{3}, -2)$ (b) $(1, \sqrt{3})$

(c) $(\sqrt{2}, -\sqrt{2})$ (d) $(0, 0)$

10. Find polar coordinates of the points whose Cartesian coordinates are as given.

(a) $(-\sqrt{2}/2, \sqrt{2}/2)$ (b) $(-7\sqrt{3}/2, 7/2)$

(c) $(0, -4)$ (d) $(5, -12)$

In each of Problems 11–16, sketch the graph of the given Cartesian equation and then find a polar equation for it.

11. $x - 4y + 2 = 0$ **12.** $x = 0$

13. $y = -5$ **14.** $x + y = 0$

15. $x^2 + y^2 = 16$ **16.** $y^2 = 4px$

In Problems 17–22, find the Cartesian equations of the graphs of the given polar equations.

17. $\theta = \frac{1}{3}\pi$ **18.** $r = 2$

19. $r \cos \theta + 6 = 0$ **20.** $r - 6 \cos \theta = 0$

21. $r \sin \theta - 4 = 0$

22. $r^2 - 8r \cos \theta - 4r \sin \theta + 11 = 0$

In Problems 23–36, name the curve with the given polar equation. If it is a conic, give its eccentricity. Sketch the graph.

23. $r = 6$ **24.** $\theta = \dfrac{2\pi}{3}$

25. $r = \dfrac{3}{\sin \theta}$ **26.** $r = \dfrac{-4}{\cos \theta}$

27. $r = 4 \sin \theta$ **28.** $r = -4 \cos \theta$

29. $r = \dfrac{4}{1 + \cos \theta}$ **30.** $r = \dfrac{4}{1 + 2 \sin \theta}$

31. $r = \dfrac{6}{2 + \sin \theta}$ **32.** $r = \dfrac{6}{4 - \cos \theta}$

33. $r = \dfrac{4}{2 + 2 \cos \theta}$ **34.** $r = \dfrac{4}{2 + 2 \cos(\theta - \pi/3)}$

35. $r = \dfrac{4}{\frac{1}{2} + \cos(\theta - \pi)}$ **36.** $r = \dfrac{4}{3 \cos(\theta - \pi/3)}$

37. Show that the polar equation of the circle with center (c, α) and radius a is $r^2 + c^2 - 2rc \cos(\theta - \alpha) = a^2$.

38. Prove that $r = a \sin \theta + b \cos \theta$ represents a circle and find its center and radius.

39. Find the length of the latus rectum for the general conic $r = ed/[1 + e \cos(\theta - \theta_0)]$ in terms of e and d.

40. Let r_1 and r_2 be the minimum and maximum distances (**perihelion** and **aphelion**) of the ellipse $r = ed/[1 + e \cos(\theta - \theta_0)]$ from a focus. Show that:

(a) $r_1 = ed/(1 + e)$, $r_2 = ed/(1 - e)$,

(b) major diameter $= 2ed/(1 - e^2)$ and minor diameter $= 2ed/\sqrt{1 - e^2}$.

41. The perihelion and aphelion for the astroid Icarus are 17 and 183 million miles, respectively. What is the eccentricity of its elliptical orbit?

42. The earth's orbit around the sun is an ellipse of eccentricity 0.0167 and major diameter 185.8 million miles. Find its perihelion.

43. The orbit of a certain comet is a parabola with the sun at a focus. The angle between the axis of the parabola (assumed pointing into the concave side) and a ray from the sun to the comet is $120°$ when the comet is 100 million miles from the sun. How close does the comet get to the sun?

44. The position of a comet with a highly eccentric elliptical orbit (e very near 1) is measured with respect to a fixed polar axis (sun is at focus but polar axis is not axis of ellipse) at two times, giving the two points $(4, \pi/2)$ and $(3, \pi/4)$

of the orbit. Here distances are measured in astronomical units (1 AU ≈ 93 million miles). For the part of the orbit near the sun, assume $e = 1$, so the orbit is given by $r = d/[1 + \cos(\theta - \theta_0)]$.

(a) The two points give two conditions for d and θ_0. Use them to show that $4.24 \cos \theta_0 - 3.76 \sin \theta_0 - 2 = 0$.
(b) Solve for θ_0 using Newton's method.
(c) How close does the comet get to the sun?

12.7
Graphs of Polar Equations

The polar equations considered in the previous section led to familiar graphs—mainly lines, circles, and conics. Now we turn our attention to more exotic graphs—cardioids, limaçons, roses, and spirals. The polar equations for these curves are still rather simple; the corresponding Cartesian equations are quite complicated. Thus, we see one of the advantages of having more than one coordinate system available. Some curves have simple equations in one system; other curves have simple equations in a second system. We will exploit this later in the book when we often begin the solution of a problem by choosing a convenient coordinate system.

Symmetry properties can help us draw a graph. Here are some *sufficient* tests for symmetry in polar coordinates. The diagrams in the margin will help you establish their validity.

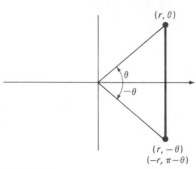

FIGURE 1

1. The graph of a polar equation is symmetric about the x-axis (the polar axis) if replacing (r, θ) by $(r, -\theta)$ (or by $(-r, \pi - \theta)$) produces an equivalent equation (Figure 1).
2. The graph of a polar equation is symmetric about the y-axis (the line $\theta = \pi/2$) if replacing (r, θ) by $(-r, -\theta)$ (or by $(r, \pi - \theta)$) produces an equivalent equation (Figure 2).
3. The graph of a polar equation is symmetric about the origin (pole) if replacing (r, θ) by $(-r, \theta)$ (or by $(r, \pi + \theta)$) produces an equivalent equation (Figure 3).

FIGURE 2

Because of the multiple representation of points in polar coordinates, symmetries may exist that are not identified by these three tests (see Problem 39).

CARDIOIDS AND LIMAÇONS   We consider equations of the form

$$r = a \pm b \cos \theta \qquad r = a \pm b \sin \theta$$

with a and b positive. Their graphs are called **limaçons**, with the special cases in which $a = b$ referred to as **cardioids**. Typical graphs are shown in Figure 4 on the next page.

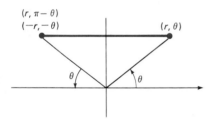

FIGURE 3

EXAMPLE 1   Analyze the equation $r = 2 + 4 \cos \theta$ for symmetry and sketch its graph.

Solution   Since cosine is an even function ($\cos(-\theta) = \cos \theta$), the graph is symmetric with respect to the x-axis. The other symmetry tests fail. A table of values and the graph appear in Figure 5.

a > b   a = b   a < b

FIGURE 4

θ	r
0	6
$\pi/6$	5.5
$\pi/3$	4
$\pi/2$	2
$7\pi/12$	1
$2\pi/3$	0
$3\pi/4$	-0.8
$5\pi/6$	-1.5
π	-2

FIGURE 5

$r = 2 + 4\cos\theta$

LEMNISCATES   The graphs of

$$r^2 = \pm a\cos 2\theta \qquad r^2 = \pm a\sin 2\theta$$

are figure-eight-shaped curves called **lemniscates**.

EXAMPLE 2   Analyze the equation $r^2 = 8\cos 2\theta$ for symmetry and sketch its graph.

Solution   Since $\cos(-2\theta) = \cos 2\theta$ and

$$\cos[2(\pi - \theta)] = \cos(2\pi - 2\theta) = \cos(-2\theta) = \cos 2\theta$$

the graph is symmetric with respect to both axes. Clearly, it is also symmetric with respect to the origin. A table of values and the graph are shown in Figure 6.

θ	r
0	± 2.8
$\pi/12$	± 2.6
$\pi/6$	$\pm|2$
$\pi/4$	0

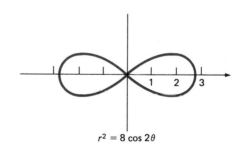

$r^2 = 8\cos 2\theta$

FIGURE 6

ROSES   Polar equations of the form

$$r = a\cos n\theta \qquad r = a\sin n\theta$$

represent flower-shaped curves called **roses**. The rose has n petals if n is odd and $2n$ petals if n is even.

EXAMPLE 3 Analyze $r = 4 \sin 2\theta$ for symmetry and sketch its graph.

Solution You can check that $r = 4 \sin 2\theta$ satisfies all three symmetry tests. For example, it meets Test 1 since

$$\sin 2(\pi - \theta) = \sin(2\pi - 2\theta) = -\sin 2\theta$$

and so replacing (r, θ) by $(-r, \pi - \theta)$ produces an equivalent equation.

To construct the correct graph, we begin by making a rather extensive table of values for $0 \le \theta \le \pi/2$ and a somewhat briefer one for $\pi/2 \le \theta \le 2\pi$. This table and the corresponding graph are shown in Figure 7. The arrows on the curve indicate the direction $P(r, \theta)$ moves as θ increases from 0 to 2π.

θ	r	θ	r
0	0	$2\pi/3$	-3.5
$\pi/12$	2	$5\pi/6$	-3.5
$\pi/8$	2.8	π	0
$\pi/6$	3.5	$7\pi/6$	3.5
$\pi/4$	4	$4\pi/3$	3.5
$\pi/3$	3.5	$3\pi/2$	0
$3\pi/8$	2.8	$5\pi/3$	-3.5
$5\pi/12$	2	$11\pi/6$	-3.5
$\pi/2$	0	2π	0

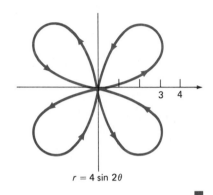

$r = 4 \sin 2\theta$

FIGURE 7

SPIRALS The graph of $r = a\theta$ is called a **spiral of Archimedes**; the graph of $r = ae^{b\theta}$ is called a **logarithmic spiral**.

EXAMPLE 4 Sketch the graph of $r = \theta$ for $\theta \ge 0$.

Solution We omit a table of values, but note that the graph crosses the polar axis at $(0, 0), (2\pi, 2\pi), (4\pi, 4\pi), \ldots$ and crosses its extension to the left at $(\pi, \pi), (3\pi, 3\pi), (5\pi, 5\pi), \ldots$, as in Figure 8.

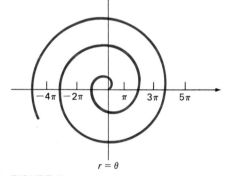

$r = \theta$

FIGURE 8

$r = \dfrac{\pi}{3}$

$(2, \pi/3)$

$r = 4 \cos \theta$

FIGURE 9

INTERSECTION OF CURVES IN POLAR COORDINATES In Cartesian coordinates, all points of intersection of two curves can be found by solving the equations of the curves simultaneously. But in polar coordinates, this is not always the case. This is because a point P has many pairs of polar coordinates, and one pair may satisfy the polar equation of one curve and a different pair may satisfy the polar equation of the other curve. For instance (see Figure 9), the circle $r = 4 \cos \theta$ intersects the line $\theta = \pi/3$ in two points, the pole and $(2, \pi/3)$, and yet only the latter is a common solution of the two equations. This is because the coordinates of the pole that satisfy the equation of the line are $(0, \pi/3)$ and those that satisfy the equation of the circle are $(0, \pi/2)$.

Our conclusion is this. In order to find all intersections of two curves whose polar equations are given, solve the equations simultaneously; then

CHAPTER 12/Conics and Polar Coordinates

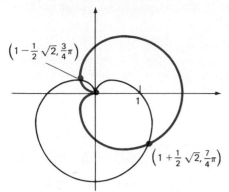

$\left(1 - \tfrac{1}{2}\sqrt{2}, \tfrac{3}{4}\pi\right)$

$\left(1 + \tfrac{1}{2}\sqrt{2}, \tfrac{7}{4}\pi\right)$

FIGURE 10

graph the two equations carefully to discover other possible points of intersection.

EXAMPLE 5 Find the points of intersection of the two cardioids $r = 1 + \cos \theta$ and $r = 1 - \sin \theta$.

Solution If we eliminate r between the two equations, we get $1 + \cos \theta = 1 - \sin \theta$. Thus $\cos \theta = -\sin \theta$, or $\tan \theta = -1$. We conclude that $\theta = \tfrac{3}{4}\pi$ and $\theta = \tfrac{7}{4}\pi$, which yields the two intersection points $(1 - \tfrac{1}{2}\sqrt{2}, \tfrac{3}{4}\pi)$ and $(1 + \tfrac{1}{2}\sqrt{2}, \tfrac{7}{4}\pi)$. The graphs in Figure 10 show, however, that we have missed a third intersection point, namely, the pole. The reason we missed it is that $r = 0$ in $r = 1 + \cos \theta$ when $\theta = \pi$, but $r = 0$ in $r = 1 - \sin \theta$ when $\theta = \pi/2$. ∎

PROBLEM SET 12.7

In Problems 1–32, sketch the graph of the given polar equation.

1. $\theta^2 - 1 = 0$

2. $(r - 2)\left(\theta + \dfrac{\pi}{4}\right) = 0$

3. $r \sin \theta + 6 = 0$

4. $r = -2 \sec \theta$

5. $r = 6 \sin \theta$

6. $r = 4 \cos \theta$

7. $r = \dfrac{4}{1 - \cos \theta}$

8. $r = \dfrac{2}{1 + \sin \theta}$

9. $r = 5 - 5 \sin \theta$ (cardioid)

10. $r = 4 - 4 \cos \theta$ (cardioid)

11. $r = 3 - 3 \cos \theta$ (cardioid)

12. $r = 4 + 4 \sin \theta$ (cardioid)

13. $r = 2 - 4 \cos \theta$ (limaçon)

14. $r = 4 - 2 \cos \theta$ (limaçon)

15. $r = 4 - 3 \sin \theta$ (limaçon)

16. $r = 3 - 4 \sin \theta$ (limaçon)

17. $r^2 = 9 \sin 2\theta$ (lemniscate)

18. $r^2 = 4 \cos 2\theta$ (lemniscate)

19. $r^2 = -16 \cos 2\theta$ (lemniscate)

20. $r^2 = -4 \sin 2\theta$ (lemniscate)

21. $r = 5 \cos 3\theta$ (three-leaved rose)

22. $r = 3 \sin 3\theta$ (three-leaved rose)

23. $r = 6 \sin 2\theta$ (four-leaved rose)

24. $r = 4 \cos 2\theta$ (four-leaved rose)

25. $r = 7 \cos 5\theta$ (five-leaved rose)

26. $r = 3 \sin 5\theta$ (five-leaved rose)

27. $r = \tfrac{1}{2}\theta, \theta \geq 0$ (spiral of Archimedes)

28. $r = 2\theta, \theta \geq 0$ (spiral of Archimedes)

29. $r = e^{\theta}, \theta \geq 0$ (logarithmic spiral)

30. $r = e^{\theta/2}, \theta \geq 0$ (logarithmic spiral)

31. $r = \dfrac{2}{\theta}, \theta > 0$ (reciprocal spiral)

32. $r = -\dfrac{1}{\theta}, \theta > 0$ (reciprocal spiral)

In Problems 33–38, sketch the given curves and find their points of intersection.

33. $r = 6, r = 4 + 4 \cos \theta$

34. $r = 1 - \cos \theta, r = 1 + \cos \theta$

35. $r = 3\sqrt{3} \cos \theta, r = 3 \sin \theta$

36. $r = 5, r = \dfrac{5}{1 - 2 \cos \theta}$

37. $r = 6 \sin \theta, r = \dfrac{6}{1 + 2 \sin \theta}$

38. $r^2 = 4 \cos 2\theta, r = 2\sqrt{2} \sin \theta$

39. The conditions for symmetry given in the text are sufficient conditions, not necessary conditions. Give an example of a polar equation $r = f(\theta)$ whose graph is symmetric with respect to the y-axis, even though replacing (r, θ) by either $(-r, -\theta)$ or $(r, \pi - \theta)$ fails to yield an equivalent equation.

40. Let a and b be fixed positive numbers and suppose OAP is a line segment with A on the line $x = a$ and $|AP| = b$. Find both the polar equation and the rectangular equation for the set of points P (called a *conchoid*) and sketch its graph.

41. Let F and F' be fixed points with polar coordinates $(a, 0)$ and $(-a, 0)$, respectively. Show that the set of points P

satisfying $|PF||PF'| = a^2$ is a lemniscate by finding its polar equation.

42. A line segment L of length $2a$ has its two endpoints on the x- and y-axes, respectively. The point P is on L and is such that OP is perpendicular to L. Show that the set of points P satisfying this condition is a four-leaved rose by finding its polar equation.

12.8
Calculus in Polar Coordinates

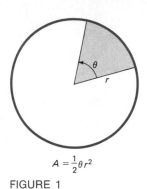

$A = \frac{1}{2}\theta r^2$

FIGURE 1

The two most basic problems in calculus are the determinations of the slope of a tangent line and the area of a curved region. Here we consider both problems, but in the context of polar coordinates. The area problem plays a larger role in the rest of the book, so we consider it first.

In Cartesian coordinates, the fundamental building block in area problems was the rectangle. In polar coordinates, it is the sector of a circle (a pie-shaped region like that in Figure 1). From the fact that the area of a circle is πr^2, we infer that the area of a sector with central angle θ radians is $(\theta/2\pi)\pi r^2$; that is,

$$\boxed{\text{Area of a sector:} \quad A = \frac{1}{2}\theta r^2}$$

AREA IN POLAR COORDINATES   To begin, let $r = f(\theta)$ determine a curve in the plane, where f is a continuous, nonnegative function for $\alpha \le \theta \le \beta$ and $\beta - \alpha \le 2\pi$. The curves $r = f(\theta)$, $\theta = \alpha$, and $\theta = \beta$ bound a region R (the one shown at the left in Figure 2), whose area $A(R)$ we wish to determine.

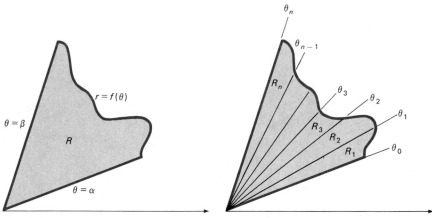

FIGURE 2

Partition the interval $[\alpha, \beta]$ into n subintervals by means of numbers $\alpha = \theta_0 < \theta_1 < \theta_2 < \cdots < \theta_n = \beta$, thereby slicing R into n smaller pie-shaped regions $R_1, R_2, \ldots, R_n$, as shown in the right half of Figure 2. Clearly $A(R) = A(R_1) + A(R_2) + \cdots + A(R_n)$.

FIGURE 3

We approximate the area $A(R_i)$ of the ith slice; in fact, we do it in two ways. On the ith interval $[\theta_{i-1}, \theta_i]$, f achieves its minimum value and maximum value—for instance, at u_i and v_i, respectively (Figure 3). Thus, if $\Delta\theta_i = \theta_i - \theta_{i-1}$,

$$\tfrac{1}{2}[f(u_i)]^2 \, \Delta\theta_i \le A(R_i) \le \tfrac{1}{2}[f(v_i)]^2 \, \Delta\theta_i$$

and so

$$\sum_{i=1}^{n} \tfrac{1}{2}[f(u_i)]^2 \, \Delta\theta_i \le \sum_{i=1}^{n} A(R_i) \le \sum_{i=1}^{n} \tfrac{1}{2}[f(v_i)]^2 \, \Delta\theta_i$$

The first and third members of this inequality are Riemann sums for the same integral, namely, $\displaystyle\int_{\alpha}^{\beta} \tfrac{1}{2}[f(\theta)]^2 \, d\theta$. When we let the norm of the partition tend toward zero, we obtain (Squeeze Principle) the area formula

$$A = \tfrac{1}{2} \int_{\alpha}^{\beta} [f(\theta)]^2 \, d\theta$$

This formula can, of course, be memorized. We prefer that you remember how it was derived. In fact, you will note that the three familiar words, *slice*, *approximate*, *integrate*, are the key to area problems also in polar coordinates. We illustrate what we mean.

EXAMPLE 1 Find the area of the region inside the limaçon $r = 2 + \cos\theta$.

Solution The graph is sketched in Figure 4; note that θ varies from 0 to 2π. We slice, approximate, integrate.

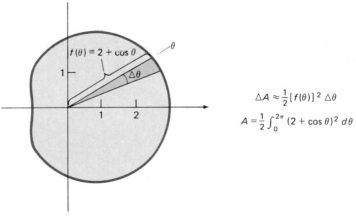

$$\Delta A \approx \frac{1}{2}[f(\theta)]^2 \, \Delta\theta$$

$$A = \frac{1}{2} \int_0^{2\pi} (2 + \cos\theta)^2 \, d\theta$$

FIGURE 4

SECTION 12.8/Calculus in Polar Coordinates

By symmetry, we can double the integral from 0 to π. Thus,

$$A = \int_0^\pi (2 + \cos \theta)^2 \, d\theta = \int_0^\pi (4 + 4\cos \theta + \cos^2\theta) \, d\theta$$

$$= \int_0^\pi 4 \, d\theta + 4 \int_0^\pi \cos \theta \, d\theta + \frac{1}{2} \int_0^\pi (1 + \cos 2\theta) \, d\theta$$

$$= \int_0^\pi \frac{9}{2} \, d\theta + 4 \int_0^\pi \cos \theta \, d\theta + \frac{1}{4} \int_0^\pi \cos 2\theta \cdot 2 \, d\theta$$

$$= \left[\frac{9}{2} \theta \right]_0^\pi + \left[4 \sin \theta \right]_0^\pi + \left[\frac{1}{4} \sin 2\theta \right]_0^\pi$$

$$= \frac{9\pi}{2} \qquad\blacksquare$$

EXAMPLE 2   Find the area of one leaf of the four-leaved rose $r = 4 \sin 2\theta$.

Solution   The complete rose was sketched in Example 3 of the previous section. Here we show only the first-quadrant leaf (Figure 5).

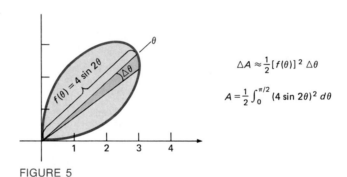

$$\Delta A \approx \frac{1}{2} [f(\theta)]^2 \, \Delta\theta$$

$$A = \frac{1}{2} \int_0^{\pi/2} (4 \sin 2\theta)^2 \, d\theta$$

FIGURE 5

$$A = \frac{1}{2} \int_0^{\pi/2} 16 \sin^2 2\theta \, d\theta = 8 \int_0^{\pi/2} \frac{1 - \cos 4\theta}{2} \, d\theta$$

$$= 4 \int_0^{\pi/2} d\theta - \int_0^{\pi/2} \cos 4\theta \cdot 4 \, d\theta$$

$$= \left[4\theta \right]_0^{\pi/2} - \left[\sin 4\theta \right]_0^{\pi/2} = 2\pi \qquad\blacksquare$$

EXAMPLE 3   Find the area of the region outside the cardioid $r = 1 + \cos \theta$ and inside the circle $r = \sqrt{3} \sin \theta$.

Solution   The graphs of the two curves are sketched in Figure 6. We will need the θ-coordinates of the points of intersection. Let's try solving the two equations simultaneously.

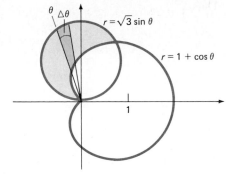

$$\triangle A \approx \frac{1}{2}[3\sin^2\theta - (1+\cos\theta)^2]\,\triangle\theta$$

$$A = \frac{1}{2}\int_{\pi/3}^{\pi}[3\sin^2\theta - (1+\cos\theta)^2]\,d\theta$$

FIGURE 6

$$1 + \cos\theta = \sqrt{3}\sin\theta$$

$$1 + 2\cos\theta + \cos^2\theta = 3\sin^2\theta$$

$$1 + 2\cos\theta + \cos^2\theta = 3(1 - \cos^2\theta)$$

$$4\cos^2\theta + 2\cos\theta - 2 = 0$$

$$2\cos^2\theta + \cos\theta - 1 = 0$$

$$(2\cos\theta - 1)(\cos\theta + 1) = 0$$

$$\cos\theta = \frac{1}{2}, -1 \qquad \theta = \frac{\pi}{3}, \pi$$

$$A = \frac{1}{2}\int_{\pi/3}^{\pi}\left[3\sin^2\theta - (1+\cos\theta)^2\right]d\theta$$

$$= \frac{1}{2}\int_{\pi/3}^{\pi}\left[3\sin^2\theta - 1 - 2\cos\theta - \cos^2\theta\right]d\theta$$

$$= \frac{1}{2}\int_{\pi/3}^{\pi}\left[\frac{3}{2}(1 - \cos 2\theta) - 1 - 2\cos\theta - \frac{1}{2}(1 + \cos 2\theta)\right]d\theta$$

$$= \frac{1}{2}\int_{\pi/3}^{\pi}\left[-2\cos\theta - 2\cos 2\theta\right]d\theta$$

$$= \frac{1}{2}\left[-2\sin\theta - \sin 2\theta\right]_{\pi/3}^{\pi}$$

$$= \frac{1}{2}\left[2\frac{\sqrt{3}}{2} + \frac{\sqrt{3}}{2}\right] = \frac{3\sqrt{3}}{4} \approx 1.299$$

TANGENTS IN POLAR COORDINATES In Cartesian coordinates, the slope m of the tangent line to a curve is given by $m = dy/dx$. We quickly reject $dr/d\theta$ as the corresponding slope formula in polar coordinates. Rather, if $r = f(\theta)$ determines the curve, we write

$$y = r\sin\theta = f(\theta)\sin\theta$$

$$x = r\cos\theta = f(\theta)\cos\theta$$

Thus,

$$\frac{dy}{dx} = \lim_{\Delta x \to 0} \frac{\Delta y}{\Delta x} = \lim_{\Delta \theta \to 0} \frac{\Delta y/\Delta \theta}{\Delta x/\Delta \theta} = \frac{dy/d\theta}{dx/d\theta}$$

that is,

$$m = \frac{f(\theta)\cos \theta + f'(\theta)\sin \theta}{-f(\theta)\sin \theta + f'(\theta)\cos \theta}$$

The formula just derived simplifies when the graph of $r = f(\theta)$ passes through the pole. Suppose, for example, that for some angle α, $r = f(\alpha) = 0$ and $f'(\alpha) \neq 0$. Then (at the pole) our formula for m is

$$m = \frac{f'(\alpha)\sin \alpha}{f'(\alpha)\cos \alpha} = \tan \alpha$$

Since the line $\theta = \alpha$ also has slope $\tan \alpha$, we conclude that this line is tangent to the curve at the pole. We infer the useful fact that *tangent lines at the pole can be found by solving the equation $f(\theta) = 0$.* We illustrate below.

EXAMPLE 4   Consider the polar equation $r = 4 \sin 3\theta$.
 (a) Find the slope of the tangent line at $\theta = \pi/6$ and $\theta = \pi/4$.
 (b) Find the tangent lines at the pole.
 (c) Sketch the graph.
 (d) Find the area of one leaf.

Solution

(a) $m = \dfrac{f(\theta)\cos \theta + f'(\theta)\sin \theta}{-f(\theta)\sin \theta + f'(\theta)\cos \theta} = \dfrac{4 \sin 3\theta \cos \theta + 12 \cos 3\theta \sin \theta}{-4 \sin 3\theta \sin \theta + 12 \cos 3\theta \cos \theta}$

At $\theta = \pi/6$,

$$m = \frac{4 \cdot 1 \cdot \dfrac{\sqrt{3}}{2} + 12 \cdot 0 \cdot \dfrac{1}{2}}{-4 \cdot 1 \cdot \dfrac{1}{2} + 12 \cdot 0 \cdot \dfrac{\sqrt{3}}{2}} = -\sqrt{3}$$

At $\theta = \pi/4$,

$$m = \frac{4 \cdot \dfrac{\sqrt{2}}{2} \cdot \dfrac{\sqrt{2}}{2} - 12 \cdot \dfrac{\sqrt{2}}{2} \cdot \dfrac{\sqrt{2}}{2}}{-4 \cdot \dfrac{\sqrt{2}}{2} \cdot \dfrac{\sqrt{2}}{2} - 12 \cdot \dfrac{\sqrt{2}}{2} \cdot \dfrac{\sqrt{2}}{2}} = \frac{2 - 6}{-2 - 6} = \frac{1}{2}$$

(b) We set $f(\theta) = 4 \sin 3\theta = 0$ and solve. This yields $\theta = 0$, $\theta = \pi/3$, $\theta = 2\pi/3$, $\theta = \pi$, $\theta = 4\pi/3$, and $\theta = 5\pi/3$.
(c) After noting that

$$\sin 3(\pi - \theta) = \sin(3\pi - 3\theta) = \sin 3\pi \cos 3\theta - \cos 3\pi \sin 3\theta$$

$$= \sin 3\theta$$

which implies symmetry with respect to the y-axis, we obtain a table of values and sketch the graph shown in Figure 7.

θ	r
0	0
$\pi/12$	2.8
$\pi/6$	4
$\pi/4$	2.8
$\pi/3$	0
$5\pi/12$	-2.8
$\pi/2$	-4

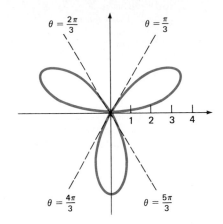

$r = 4 \sin 3\theta$

FIGURE 7

(d)
$$A = \frac{1}{2}\int_0^{\pi/3}(4\sin 3\theta)^2\,d\theta = 8\int_0^{\pi/3}\sin^2 3\theta\,d\theta$$

$$= 4\int_0^{\pi/3}(1 - \cos 6\theta)\,d\theta = 4\int_0^{\pi/3}d\theta - \frac{4}{6}\int_0^{\pi/3}\cos 6\theta \cdot 6\,d\theta$$

$$= \left[4\theta - \frac{2}{3}\sin 6\theta\right]_0^{\pi/3} = \frac{4\pi}{3} \qquad \blacksquare$$

PROBLEM SET 12.8

In each of Problems 1–10, sketch the graph of the given equation and find the area of the region bounded by it.

1. $r = a, a > 0$

2. $r = 2a \sin \theta, a > 0$

3. $r = 3 + \cos \theta$

4. $r = 4 + 3 \cos \theta$

5. $r = 4 - 4 \cos \theta$

6. $r = 7 - 7 \sin \theta$

7. $r = a(1 + \sin \theta), a > 0$

8. $r^2 = 5 \cos 2\theta$

9. $r^2 = 4 \sin 2\theta$

10. $r^2 = a \sin 2\theta, a > 0$

11. Sketch the limaçon $r = 2 - 4 \cos \theta$, and find the area of the region inside its small loop.

12. Sketch the limaçon $r = 3 - 6 \sin \theta$, and find the area of the region inside its small loop.

13. Sketch the limaçon $r = 2 - 4 \sin \theta$, and find the area of the region inside its large loop.

14. Sketch one leaf of the four-leaved rose $r = 3 \cos 2\theta$, and find the area of the region enclosed by it.

15. Sketch the three-leaved rose $r = 4 \cos 3\theta$, and find the area of the total region enclosed by it.

16. Sketch the three-leaved rose $r = 2 \sin 3\theta$, and find the area of the region bounded by it.

17. Find the area of the region between the two concentric circles $r = 7$ and $r = 10$.

18. Sketch the region that is inside the circle $r = 3 \sin \theta$ and outside the cardioid $r = 1 + \sin \theta$, and find its area.

19. Sketch the region that is outside the circle $r = 2$ and inside the lemniscate $r^2 = 8 \cos 2\theta$, and find its area.

20. Sketch the limaçon $r = 3 - 6 \sin \theta$ and find the area of the region that is inside its large loop and outside its small loop.

21. Sketch the region in the first quadrant that is inside the cardioid $r = 3 + 3 \cos \theta$ and outside the cardioid $r = 3 + 3 \sin \theta$, and find its area.

22. Sketch the region in the second quadrant that is inside the cardioid $r = 2 + 2 \sin \theta$ and outside the cardioid $r = 2 + 2 \cos \theta$, and find its area.

23. Find the slope of the tangent line to each of the following curves at $\theta = \pi/3$.

(a) $r = 2\cos\theta$
(c) $r = \sin 2\theta$
(b) $r = 1 + \sin\theta$
(d) $r = 4 - 3\cos\theta$

24. Find all points on the cardioid $r = a(1 + \cos\theta)$ where the tangent line is (a) horizontal, and (b) vertical.

25. Find all points on the limaçon $r = 1 - 2\sin\theta$ where the tangent line is horizontal.

26. Let $r = f(\theta)$, where f is continuous on the closed interval $[\alpha, \beta]$. Derive the following formula for the length L of the corresponding polar curve from $\theta = \alpha$ to $\theta = \beta$.

$$L = \int_{\alpha}^{\beta} \sqrt{[f(\theta)]^2 + [f'(\theta)]^2}\, d\theta$$

27. Use the formula of Problem 26 to find the perimeter of the cardioid $r = a(1 + \cos\theta)$.

28. Find the length of the logarithmic spiral $r = e^{\theta/2}$ from $\theta = 0$ to $\theta = 2\pi$.

29. Find the total area of the rose $r = a\cos n\theta$, n a positive integer.

30. Sketch the graph of the *strophoid* $r = \sec\theta - 2\cos\theta$ and find the area of its loop.

31. Consider the two circles $r = 2a\sin\theta$ and $r = 2b\cos\theta$ with a and b positive.

(a) Find the area of the region inside both circles.
(b) Show that the two circles intersect at right angles.

32. Assume that a planet of mass m is revolving around the sun (located at the pole) with constant angular momentum $mr^2\, d\theta/dt$. Deduce Kepler's Second Law: The line from the sun to the planet sweeps out equal areas in equal times.

33. (First Old Goat Problem) A goat is tethered to the edge of a circular pond of radius a by a rope of length ka ($0 < k \leq 2$). Use the method of this section to find its grazing area (the shaded area in Figure 8). *Note*: We solved this problem once before (Problem 49 of Section 7.6); you should be able to make your answers agree.

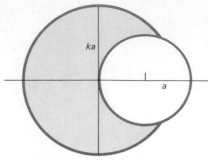

FIGURE 8

34. (Second Old Goat Problem) Do Problem 33 again but assume the pond has a fence around it so that in forming the wedge A, the rope wraps around the fence (Figure 9). *Hint*: If you are exceedingly ambitious, try the method of this section. Better, note that in the wedge A, $\Delta A \approx (\frac{1}{2})|PT|^2 \Delta\phi$, which leads to a Riemann sum for an integral. The final answer is $a^2(\pi k^2/2 + k^3/3)$, a result needed in Problem 35.

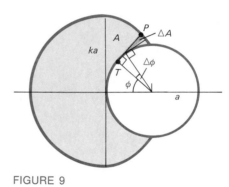

FIGURE 9

[C] **35.** (Third Old Goat Problem) An untethered goat grazes inside a yard enclosed by a circular fence of radius a; another grazes outside the fence tethered as in Problem 34. Find the length of the rope if the two goats have the same grazing area.

12.9 Chapter Review

TRUE-FALSE QUIZ

Respond with true or false to each of the following assertions. Be prepared to defend your answer.

1. The graph of $y = ax^2 + bx + c$ is a parabola for all choices of a, b, and c.

2. The vertex of a parabola is midway between the focus and the directrix.

3. A vertex of an ellipse is closer to a directrix than to a focus.

4. The point on a parabola closest to its focus is the vertex.

5. The hyperbolas $x^2/a^2 - y^2/b^2 = 1$ and $y^2/b^2 - x^2/a^2 = 1$ have the same asymptotes.

6. The circumference C of the ellipse $x^2/a^2 + y^2/b^2 = 1$ with $b < a$ satisfies $2\pi b < C < 2\pi a$.

7. The smaller the eccentricity e of an ellipse, the more nearly circular the ellipse is.

8. The ellipse $6x^2 + 4y^2 = 24$ has its foci on the x-axis.

9. The equation $x^2 - y^2 = 0$ represents a hyperbola.

10. The equation $(y^2 - 4x + 1)^2 = 0$ represents a parabola.

11. If $k \neq 0$, $x^2/a^2 - y^2/b^2 = k$ is the equation of a hyperbola.

12. If $k \neq 0$, $x^2/a^2 + y^2/b^2 = k$ is the equation of an ellipse.

13. The distance between the foci of the graph of $x^2/a^2 + y^2/b^2 = 1$ is $2\sqrt{a^2 - b^2}$.

14. The graph of $x^2/9 - y^2/8 = -2$ does not intersect the x-axis.

15. Light emanating from a point between a focus and the nearest vertex of an elliptical mirror will be reflected beyond the other focus.

16. The set of points equidistant from the circle $x^2 + y^2 = 1$ and the line $x = 3$ is a parabola.

17. From Kepler's law on sweeping out areas, we conclude that a planet in its elliptical orbit around the sun reaches its maximum speed at the vertex nearest the sun.

18. An ellipse that is drawn using a string of length 8 units attached to foci 2 units apart will have minor diameter of length $\sqrt{60}$ units.

19. The graph of $x^2 + y^2 + Cx + Dx + F = 0$ is either a circle, a point, or the empty set.

20. The graph of $2x^2 + y^2 + Cx + Dy + F = 0$ cannot be a single point.

21. The graph of $Ax^2 + Bxy + Cy^2 + Dx + Ey + F = 0$ is the intersection of a plane with a cone of two nappes for all choices of A, B, C, D, E, and F.

22. In an appropriate coordinate system, the intersection of a plane with a cone of two nappes will have an equation of the form $Ax^2 + Cy^2 + Dx + Ey + F = 0$.

23. The graph of a hyperbola must enter all four quadrants.

24. If one of the conic sections passes through the four points $(1,0)$, $(-1,0)$, $(0,1)$, and $(0,-1)$, it must be a circle.

25. The graph of the polar equation $r = 4\cos(\theta - \pi/3)$ is a circle.

26. Every point in the plane has infinitely many sets of polar coordinates.

27. All points of intersection of the graphs of the polar equations $r = f(\theta)$ and $r = g(\theta)$ can be found by solving these two equations simultaneously.

28. If f is an odd function, then the graph of $r = f(\theta)$ is symmetric with respect to the y-axis (the line $\theta = \pi/2$).

29. If f is an even function, then the graph of $r = f(\theta)$ is symmetric with respect to the x-axis (the line $\theta = 0$).

30. The graph of $r = 4\cos 3\theta$ is a rose of 3 leaves whose area is less than half that of the circle $r = 4$.

SAMPLE TEST PROBLEMS

1. From the numbered list, pick the correct response to put in the blanks that follow.

(1) No graph.
(2) A single point.
(3) A single line.
(4) Two parallel lines.
(5) Two intersecting lines.
(6) A circle.
(7) A parabola.
(8) An ellipse.
(9) A hyperbola.
(10) None of the above.

_____ (a) $x^2 - 4y^2 = 0$
_____ (b) $x^2 - 4y^2 = 0.01$
_____ (c) $x^2 - 4 = 0$
_____ (d) $x^2 - 4x + 4 = 0$
_____ (e) $x^2 + 4y^2 = 0$
_____ (f) $x^2 + 4y^2 = x$
_____ (g) $x^2 + 4y^2 = -x$
_____ (h) $x^2 + 4y^2 = -1$
_____ (i) $(x^2 + 4y - 1)^2 = 0$
_____ (j) $3x^2 + 4y^2 = -x^2 + 1$

In each of Problems 2–10, name the conic that has the given equation. Find its vertices and foci and sketch its graph.

2. $y^2 - 6x = 0$

3. $9x^2 + 4y^2 - 36 = 0$

4. $25x^2 - 36y^2 + 900 = 0$

5. $x^2 + 9y = 0$

6. $x^2 - 4y^2 - 16 = 0$

7. $9x^2 + 25y^2 - 225 = 0$

8. $9x^2 + 9y^2 - 225 = 0$

9. $r = \dfrac{5}{2 + 2\sin\theta}$

10. $r(2 + \cos\theta) = 3$

In each of Problems 11–18, find the Cartesian equation of the conic with the given properties.

11. Vertices $(\pm 4, 0)$ and eccentricity $\frac{1}{2}$.

12. Eccentricity 1, focus $(0, -3)$, and vertex $(0, 0)$.

13. Eccentricity 1, vertex $(0, 0)$, symmetric with respect to the x-axis, and passing through the point $(-1, 3)$.

14. Eccentricity $\frac{5}{3}$ and vertices $(0, \pm 3)$.

15. Vertices $(\pm 2, 0)$ and asymptotes $x \pm 2y = 0$.

16. Parabola with focus $(3, 2)$ and vertex $(3, 3)$.

17. Ellipse with center $(1, 2)$, and focus $(4, 2)$, and major diameter 10.

18. Hyperbola with vertices $(2, 0)$ and $(2, 6)$ and eccentricity $\frac{10}{3}$.

In Problems 19–22, use the process of completing the square to reduce the given equation to a standard form. Then name the corresponding curve and sketch its graph.

19. $4x^2 + 4y^2 - 24x + 36y + 81 = 0$

20. $4x^2 + 9y^2 - 24x - 36y + 36 = 0$

21. $x^2 + 8x + 6y + 28 = 0$

22. $3x^2 - 10y^2 + 36x - 20y + 68 = 0$

23. A rotation of axes through $\theta = 45°$ transforms $x^2 + 3xy + y^2 = 10$ into $ru^2 + sv^2 = 10$. Determine r and s, name the corresponding conic, and find the distance between its foci.

24. Determine the rotation angle θ needed to eliminate the cross-product term in $7x^2 + 8xy + y^2 = 9$. Then obtain the corresponding uv-equation and identify the conic it represents.

In Problems 25–36, analyze the given polar equation and sketch its graph.

25. $r = 6\cos\theta$

26. $r = \dfrac{5}{\sin\theta}$

27. $r = \cos 2\theta$

28. $r = \dfrac{3}{\cos\theta}$

29. $r = 4$

30. $r = 5 - 5\cos\theta$

31. $r = 4 - 3\cos\theta$

32. $r = 2 - 3\cos\theta$

33. $\theta = \frac{2}{3}\pi$

34. $r = 4\sin 3\theta$

35. $r^2 = 16\sin 2\theta$

36. $r = -\theta, \theta \geq 0$

37. Find a Cartesian equation of the graph of

$$r^2 - 6r(\cos\theta + \sin\theta) + 9 = 0$$

and then sketch the graph.

38. Find a Cartesian equation of the graph of

$$r^2\cos 2\theta = 9$$

and then sketch the graph.

39. Find the slope of the tangent to the graph of

$$r = 3 + 3\cos\theta$$

at the point on the graph where $\theta = \frac{1}{6}\pi$.

40. Sketch the graphs of

$$r = 5\sin\theta \quad \text{and} \quad r = 2 + \sin\theta$$

and find their points of intersection.

41. Find the area of the region bounded by the graph of

$$r = 5 - 5\cos\theta$$

(see Problem 30).

42. Find the area of the region that is outside the limaçon $r = 2 + \sin\theta$ and inside the circle $r = 5\sin\theta$ (see Problem 40).

43. A racing car on the elliptical race track $x^2/400 + y^2/100 = 1$ went out of control at the point $(16, 6)$ and thereafter continued on the tangent line till it hit a tree at $(14, k)$. Determine k.

13 Geometry in the Plane, Vectors

A versatile woman distinguished as a mathematician, linguist, philosopher, and somnambulist.
Howard Eves

Maria Gaetana Agnesi
1718–1799

Only two women appear in our calculus honor roll. The under-representation of women reflects a bias that long existed in Western Europe and continued into this century. Rarely were women encouraged to pursue academic excellence and those who did usually found academic careers barred to them. Fortunately some persisted in spite of the obstacles.

One who did was Maria Gaetana Agnesi. The oldest of 21 children, she was born into a wealthy and literate Italian family and to a father who was a mathematician. A child prodigy, she mastered Latin, Greek, Hebrew, and several modern languages by age 9. At 20, she began her most important work, a textbook on calculus. For its day, it was a marvel of clarity and the first comprehensive calculus textbook since l'Hôpital's early work. The book brought her many honors, including recognition by Empress Maria Theresa and Pope Benedict XIV.

The name Agnesi retains a place in the mathematical literature through one of Maria's minor contributions—her study of a curve then known as *versiera*, from the Latin *vertere*, to turn. Today it is known as the *witch of Agnesi*, because *versiera* is also the Italian word for *she devil*. This curve is discussed in Problem 35 of Section 13.1.

On the 100th anniversary of her death, Milan recognized Agnesi by naming a street in her honor. A cornerstone in the façade of Luogo Pio bears the inscription "erudite in Mathematics, glory of Italy and of her century."

13.1
Plane Curves: Parametric Representation

We gave the general definition of a plane curve in Section 6.4 in connection with our derivation of the arc length formula. A **plane curve** is determined by a pair of parametric equations

$$x = f(t), \qquad y = g(t), \qquad t \text{ in } I$$

Not simple, not closed

Simple, not closed

with f and g continuous on the interval I. Usually I is a closed interval $[a, b]$. Think of t, called the **parameter**, as measuring time. As t advances from a to b, the point (x, y) traces out the curve in the xy-plane. The points $P = (x(a), y(a))$ and $Q = (x(b), y(b))$ are the initial and final **endpoints**. If the endpoints coincide, the curve is **closed**. If distinct values of t yield distinct points in the plane (except possibly for $t = a$ and $t = b$), we say the curve is a **simple** curve (Figure 1).

ELIMINATING THE PARAMETER   To recognize a curve given by parametric equations, it may be desirable to eliminate the parameter. Sometimes this can be accomplished by solving one equation for t and substituting in the other (Example 1). Often we can make use of a familiar identity, as in Example 2.

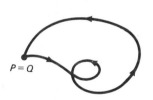

$P = Q$

Not simple, closed

EXAMPLE 1   Eliminate the parameter in

$$x = t^2 + 2t, \qquad y = t - 3, \qquad -2 \le t \le 3$$

Then identify the corresponding curve and sketch its graph.

Solution   From the second equation, $t = y + 3$. Substituting this expression for t in the first equation gives

$P = Q$

Simple and closed

FIGURE 1

$$x = (y + 3)^2 + 2(y + 3) = y^2 + 8y + 15$$

or

$$x + 1 = (y + 4)^2$$

This we recognize as a parabola with vertex at $(-1, -4)$ and opening to the right.

In graphing the given equation, we must be careful to display only that part of the parabola corresponding to $-2 \le t \le 3$. A table of values and the graph are shown in Figure 2. The arrow indicates the direction of increasing t.

t	x	y
-2	0	-5
-1	-1	-4
0	0	-3
1	3	-2
2	8	-1
3	15	0

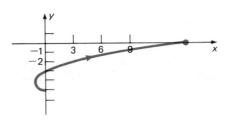

FIGURE 2

EXAMPLE 2 Show that

$$x = a \cos t, \qquad y = b \sin t, \qquad 0 \le t \le 2\pi$$

represents the ellipse shown in Figure 3.

Solution We solve the equations for $\cos t$ and $\sin t$, then square, and add.

$$\left(\frac{x}{a}\right)^2 + \left(\frac{y}{b}\right)^2 = \cos^2 t + \sin^2 t = 1$$

$$\frac{x^2}{a^2} + \frac{y^2}{b^2} = 1$$

A quick check of a few values for t convinces us that we do get the complete ellipse. In particular, $t = 0$ and $t = 2\pi$ give the same point, namely, $(a, 0)$. If $a = b$, we get the circle $x^2 + y^2 = a^2$.

Different pairs of parametric equations may have the same graph, as we now illustrate.

EXAMPLE 3 Show that each of the following pairs of parametric equations have the same graph, namely, the semicircle shown in Figure 4.

(a) $x = \sqrt{1 - t^2}$, $y = t$, $-1 \le t \le 1$

(b) $x = \cos t$, $y = \sin t$, $-\dfrac{\pi}{2} \le t \le \dfrac{\pi}{2}$

(c) $x = \dfrac{1 - t^2}{1 + t^2}$, $y = \dfrac{2t}{1 + t^2}$, $-1 \le t \le 1$

Solution In each case, we discover that

$$x^2 + y^2 = 1$$

It is then just a matter of checking a few values of t to make sure that the given intervals for t yield the same section of the circle.

Ellipse

FIGURE 3

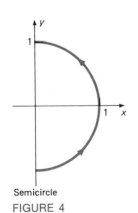

Semicircle

FIGURE 4

EXAMPLE 4 Show that each of the following pairs of parametric equations yields one branch of a hyperbola.

(a) $x = a \sec t$, $y = b \tan t$, $-\dfrac{\pi}{2} < t < \dfrac{\pi}{2}$

(b) $x = a \cosh t$, $y = b \sinh t$, $-\infty < t < \infty$

Assume in both cases that $a > 0$ and $b > 0$.

Solution
(a) In the first case,

$$\left(\frac{x}{a}\right)^2 - \left(\frac{y}{b}\right)^2 = \sec^2 t - \tan^2 t = 1$$

(b) In the second case,

$$\left(\frac{x}{a}\right)^2 - \left(\frac{y}{b}\right)^2 = \cosh^2 t - \sinh^2 t$$

$$= \left(\frac{e^t + e^{-t}}{2}\right)^2 - \left(\frac{e^t - e^{-t}}{2}\right)^2 = 1$$

Checking a few t-values shows that, in both cases, we obtain the branch of the hyperbola $x^2/a^2 - y^2/b^2 = 1$ shown in Figure 5. ∎

Hyperbola
(one branch)

FIGURE 5

THE CYCLOID   A cycloid is the curve traced by a point P on the rim of a wheel as the wheel rolls along a straight line without slipping (Figure 6). The Cartesian equation of a cycloid is so complicated that it is seldom used. But simple parametric equations are readily found, as shown in the next example.

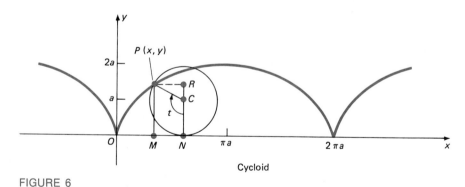

Cycloid

FIGURE 6

EXAMPLE 5   Find parametric equations for the cycloid.

Solution   Let the wheel roll along the x-axis with P initially at the origin. Denote the center of the wheel by C and let a be its radius. Choose for a parameter the radian measure t of the clockwise angle through which the line segment CP has turned from its vertical position when P was at the origin. All of this is shown in Figure 6.

Since $|ON| = $ arc $PN = at$,

$$x = |OM| = |ON| - |MN| = at - a \sin t = a(t - \sin t)$$

and

$$y = |MP| = |NR| = |NC| + |CR| = a - a \cos t = a(1 - \cos t)$$

Thus the parametric equations for the cycloid are

$$x = a(t - \sin t), \qquad y = a(1 - \cos t)$$ ■

FIGURE 7

The cycloid has a number of interesting applications, especially in mechanics. It is the "curve of fastest descent." If a particle, acted on only by gravity, is allowed to slide down some curve from a point A to a lower point B not on the same vertical line, it completes its journey in the shortest time when the curve is an inverted cycloid (Figure 7). Of course, the shortest distance is along the straight line segment AB, but the least time is used when the path is along a cycloid; this is because the acceleration when it is released depends on the steepness of descent, and along a cycloid it builds up velocity much more quickly than it does along a straight line.

FIGURE 8

Another interesting property is this: If L is the lowest point on an arch of an inverted cycloid, the time it takes a particle P to slide down the cycloid to L is the same no matter where P starts from on the inverted arch; thus, if several particles, P_1, P_2, and P_3, in different positions on the cycloid (Figure 8) start to slide at the same instant, all will reach the low point L at the same time.

In 1673, the Dutch astronomer Christian Huygens published a description of an ideal pendulum clock. Because the bob swings between cycloidal "cheeks," the path of the bob is a cycloid (Figure 9). This means that the period of the swing is independent of the amplitude, and so the period does not change as the clock's spring unwinds.

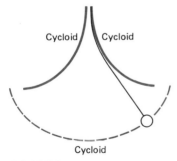

FIGURE 9

CALCULUS FOR FUNCTIONS DEFINED PARAMETRICALLY Can we find the slope of the tangent line to a curve given parametrically without first eliminating the parameter? The answer is yes, according to the following theorem.

> **Theorem A**
>
> Let f and g be continuously differentiable with $f'(t) \neq 0$ on $\alpha \leq t \leq \beta$. Then the parametric equations
>
> $$x = f(t), \qquad y = g(t)$$
>
> define y as a differentiable function of x and
>
> $$\frac{dy}{dx} = \frac{dy/dt}{dx/dt}$$

Proof Since $f'(t) \neq 0$ for $\alpha \leq t \leq \beta$, f is strictly monotonic and so has a differentiable inverse f^{-1} (see the Inverse Function Theorem (Theorem 7.2B)). Define F by $F = g \circ f^{-1}$, so that

$$y = g(t) = g(f^{-1}(x)) = F(x) = F(f(t))$$

Then by the Chain Rule,

$$\frac{dy}{dt} = F'(f(t)) \cdot f'(t) = \frac{dy}{dx} \cdot \frac{dx}{dt}$$

Since $dx/dt \neq 0$, we have

$$\frac{dy}{dx} = \frac{dy}{dt} \div \frac{dx}{dt} \qquad \blacksquare$$

EXAMPLE 6 Find the first two derivatives dy/dx and d^2y/dx^2 for the function determined by

$$x = 5 \cos t, \qquad y = 4 \sin t, \qquad 0 < t < 3$$

and evaluate them at $t = \pi/6$ (see Example 2).

Solution Let y' denote dy/dx. Then

$$\frac{dy}{dx} = \frac{dy}{dt} \div \frac{dx}{dt} = \frac{4 \cos t}{-5 \sin t} = -\frac{4}{5} \cot t$$

$$\frac{d^2y}{dx^2} = \frac{dy'}{dx} = \frac{dy'}{dt} \div \frac{dx}{dt} = \frac{\frac{4}{5} \csc^2 t}{-5 \sin t} = -\frac{4}{25} \csc^3 t$$

At $t = \pi/6$,

$$\frac{dy}{dx} = \frac{-4\sqrt{3}}{5}, \qquad \frac{d^2y}{dx^2} = \frac{-4}{25}(8) = \frac{-32}{25}$$

The first value is the slope of the tangent line to the ellipse $x^2/25 + y^2/16 = 1$ at the point $(5\sqrt{3}/2, 2)$. You can check that this is so by implicit differentiation. $\blacksquare$

Sometimes a definite integral involves two variables, such as x and y, in the integrand and differential, and y may be defined as a function of x by equations that give x and y in terms of a parameter such as t. In such cases, it is often convenient to evaluate the definite integral by expressing the integrand and the differential in terms of t and dt and adjusting the limits of integration before integrating with respect to t.

EXAMPLE 7 Evaluate (a) $\int_1^3 y \, dx$ and (b) $\int_1^3 xy^2 \, dx$, using $x = 2t - 1$ and $y = t^2 + 2$.

Solution From $x = 2t - 1$ we have $dx = 2 \, dt$; when $x = 1$, $t = 1$; and when $x = 3$, $t = 2$.

(a) $\displaystyle \int_1^3 y\,dx = \int_1^2 (t^2 + 2)2\,dt = 2\left[\frac{t^3}{3} + 2t\right]_1^2 = \frac{26}{3}$

(b) $\displaystyle \int_1^3 xy^2\,dx = \int_1^2 (2t - 1)(t^2 + 2)^2 2\,dt$

$$= 2\int_1^2 (2t^5 - t^4 + 8t^3 - 4t^2 + 8t - 4)\,dt = 86\tfrac{14}{15} \qquad \blacksquare$$

EXAMPLE 8 Find the area A under one arch of a cycloid (Figure 10). Also, find the length L of this arch.

Solution From Example 5, we know that we may represent one arch of the cycloid by

$$x = a(t - \sin t), \qquad y = a(1 - \cos t), \qquad 0 \le t \le 2\pi$$

Thus,

Cycloid (one arch)

FIGURE 10

$$A = \int_0^{2\pi a} y\,dx = \int_0^{2\pi} a(1 - \cos t)\,d[a(t - \sin t)]$$

$$= a^2 \int_0^{2\pi} (1 - \cos t)(1 - \cos t)\,dt$$

$$= a^2 \int_0^{2\pi} (1 - 2\cos t + \cos^2 t)\,dt$$

$$= a^2 \int_0^{2\pi} (1 - 2\cos t + \tfrac{1}{2} + \tfrac{1}{2}\cos 2t)\,dt$$

$$= a^2[\tfrac{3}{2}t - 2\sin t + \tfrac{1}{4}\sin 2t]_0^{2\pi} = 3\pi a^2$$

To calculate L, we recall the arc-length formula from Section 6.4,

$$L = \int_\alpha^\beta \sqrt{\left(\frac{dx}{dt}\right)^2 + \left(\frac{dy}{dt}\right)^2}\,dt$$

which in our case reduces to

$$L = \int_0^{2\pi} \sqrt{a^2(1 - \cos t)^2 + a^2(\sin^2 t)}\,dt$$

$$= a\int_0^{2\pi} \sqrt{2(1 - \cos t)}\,dt$$

$$= a\int_0^{2\pi} \sqrt{4\sin^2\frac{t}{2}}\,dt$$

$$= 2a\int_0^{2\pi} \sin\frac{t}{2}\,dt$$

$$= \left[-4a\cos\frac{t}{2}\right]_0^{2\pi} = 8a \qquad \blacksquare$$

In each of Problems 1–12, a parametric representation of a curve is given.

(a) Sketch the curve by assigning values to the parameter (see Example 1).

(b) Which of the following apply to the curve: simple or closed?

(c) Obtain the Cartesian equation of the curve by eliminating the parameter (see Examples 1–4).

1. $x = 2t$, $y = 3t$; t in $\mathbb{R}$

2. $x = 4t - 1$, $y = 2t$; $0 \le t \le 3$

3. $x = t - 4$, $y = \sqrt{t}$; $0 \le t \le 4$

4. $x = t$, $y = \dfrac{1}{t}$; $0 < t$

5. $x = t^2$, $y = t^3$; $-1 \le t \le 2$

6. $x = t^2 - 1$, $y = t^3 - t$; $-3 \le t \le 3$

7. $x = t^3 - 4t$, $y = t^2 - 4$; $-3 \le t \le 3$

8. $x = 3\sqrt{t - 3}$, $y = 2\sqrt{4 - t}$; $3 \le t \le 4$

9. $x = 3 \sin t$, $y = 5 \cos t$; $0 \le t \le 2\pi$

10. $x = 3 \sin \theta - 1$, $y = 2 \cos \theta + 2$; $0 \le \theta \le \pi$

11. $x = 4 \sin^4 t$, $y = 4 \cos^4 t$; $0 \le t \le \frac{1}{2}\pi$

12. $x = 2 \cos \theta$, $y = 2 \cos \frac{1}{2}\theta$; θ in $\mathbb{R}$

In Problems 13–18, find dy/dx and d^2y/dx^2 without eliminating the parameter.

13. $x = 3t^2$, $y = 2t^3$; $t \ne 0$

14. $x = 6t^2$, $y = t^3$; $t \ne 0$

15. $x = 2t - \dfrac{3}{t}$, $y = 2t + \dfrac{3}{t}$; $t \ne 0$

16. $x = 1 - \cos t$, $y = 2 + 3 \sin t$; $t \ne n\pi$

17. $x = 3 \tan t - 1$, $y = 5 \sec t + 2$; $t \ne \dfrac{(2n + 1)\pi}{2}$

18. $x = \dfrac{2}{1 + t^2}$, $y = \dfrac{2}{t(1 + t^2)}$; $t \ne 0$

In Problems 19–22, find the equation of the tangent to the given curve at the given point without eliminating the parameter. Make a sketch.

19. $x = t^2$, $y = t^3$; $t = 2$

20. $x = 3t$, $y = 8t^3$; $t = -\frac{1}{2}$

21. $x = 2 \sec t$, $y = 2 \tan t$; $t = -\dfrac{\pi}{6}$

22. $x = 2e^t$, $y = \frac{1}{3}e^{-t}$; $t = 0$

In Problems 23–24, evaluate the integrals.

23. $\displaystyle\int_0^1 (x^2 - 4y)\,dx$, where $x = t + 1$, $y = t^3 + 4$.

24. $\displaystyle\int_1^{\sqrt{3}} xy\,dy$, where $x = \sec t$, $y = \tan t$.

25. Find the area of the region between the curve $x = e^{2t}$, $y = e^{-t}$, and the x-axis from $t = 0$ to $t = \ln 5$. Make a sketch.

26. Find the area of the region bounded by the curve

$$x = t + \frac{1}{t}, \qquad y = t - \frac{1}{t}$$

and the line $3x - 10 = 0$, without eliminating the parameter. Make a sketch.

27. Modify the text discussion of the cycloid (and its accompanying diagram) to handle the case where the point P is $b < a$ units from the center of the wheel. Show that the corresponding parametric equations are

$$x = at - b \sin t \qquad y = a - b \cos t$$

Sketch the graph of these equations (called a **curtate cycloid**) when $a = 8$ and $b = 4$.

28. Follow the instructions of Problem 27 for the case $b > a$ (a flanged wheel, as on a train) showing that you get the same parametric equations. Sketch the graph of these equations (called a **prolate cycloid**) when $a = 6$ and $b = 8$.

29. The path of a projectile fired from level ground with a speed of v_0 feet per second at an angle α with the ground, is given by the parametric equations

$$x = (v_0 \cos \alpha)t \qquad y = -16t^2 + (v_0 \sin \alpha)t$$

(a) Show that the path is a parabola.

(b) Find the time of flight.

(c) Show that the range (horizontal distance traveled) is $(v_0^2/32)\sin 2\alpha$.

(d) For a given v_0, what value of α gives the largest possible range?

30. Each value of the parameter t in the equations of a cycloid,

$$x = a(t - \sin t), \qquad y = a(1 - \cos t)$$

determines a unique point $P(x, y)$ on the cycloid and also determines a unique position of the rolling circle that generates the cycloid. Let $P_1(x_1, y_1)$ be the point on the cycloid determined by the value t_1 of the parameter t. Prove that the tangent to the cycloid at P_1 passes through the highest point on the rolling circle when the circle is in the position determined by t_1. Through what point on the circle does the normal pass?

31. Let a circle of radius b roll, without slipping, inside a fixed circle of radius a, $a > b$. A point P on the circumference of the rolling circle traces out a curve called a **hypocycloid**. Find parametric equations of the hypocycloid. *Hint*: Place the origin O of Cartesian coordinates at the center of the fixed, larger circle and let the point $A(a, 0)$ be one position of the tracing point P. Denote by B the moving point of tangency of the two circles and let t, the radian measure of the angle AOB, be the parameter (see Figure 11).

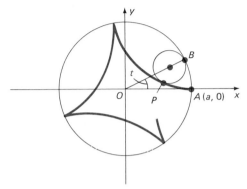

FIGURE 11

32. Show that if $b = a/4$ in Problem 31, the parametric equations of the hypocycloid may be simplified to

$$x = a \cos^3 t, \qquad y = a \sin^3 t$$

This is called a **hypocycloid of four cusps**. Sketch it carefully and show that its Cartesian equation is $x^{2/3} + y^{2/3} = a^{2/3}$.

33. The curve traced by a point on the circumference of a circle of radius b as it rolls without slipping on the outside of a fixed circle of radius a is called an **epicycloid**. Show that it has parametric equations

$$x = (a + b)\cos t - b \cos \frac{a + b}{b} t$$

$$y = (a + b)\sin t - b \sin \frac{a + b}{b} t$$

(See the hint in Problem 31.)

34. If $b = a$, the equations in Problem 33 are

$$x = 2a \cos t - a \cos 2t$$

$$y = 2a \sin t - a \sin 2t$$

Show that this special epicycloid is the cardioid $r = 2a(1 - \cos \theta)$, where the pole of the polar coordinate system is the point $(a, 0)$ in the Cartesian system and the polar axis

has the direction of the positive x-axis. *Hint*: Find a Cartesian equation of the epicycloid by eliminating the parameter t between the equations. Then show that the equations connecting the Cartesian and polar systems are

$$x = r \cos \theta + a \qquad y = r \sin \theta$$

and use these equations to transform the Cartesian equation into $r = 2a(1 - \cos \theta)$.

35. Consider a circle of radius a centered at $(0, a)$, as in Figure 12. Let a line OA intersect the line $y = 2a$ at A and the circle C at B. Finally, let P be the point of intersection of a horizontal line through B and a vertical line through A. As θ, the angle OA makes with the positive x-axis varies, P traces out a curve, the **witch of Agnesi**.

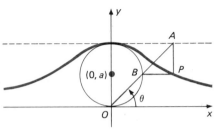

FIGURE 12

(a) Find parametric equations for the witch using θ as parameter.
(b) Find the Cartesian equation of the witch.

36. If $b = a/3$ in Problem 31, we obtain a hypocycloid of three cusps, called a **deltoid**, with parametric equations

$$x = \left(\frac{a}{3}\right)(2 \cos t + \cos 2t), \qquad y = \left(\frac{a}{3}\right)(2 \sin t - \sin 2t)$$

Find the length of the deltoid.

37. Find the length of the epicycloid of Problem 33 assuming a/b is an integer and $0 \le t \le 2\pi$.

38. Consider the ellipse $x^2/a^2 + y^2/b^2 = 1$.

(a) Show that its perimeter is

$$P = 4a \int_0^{\pi/2} \sqrt{1 - e^2 \cos^2 t}\, dt,$$

where e is the eccentricity.

C (b) The integral in part (a) is called an *elliptic integral*. It has been studied at great length, and it is known that the integrand does not have an elementary antiderivative, so we must turn to approximate methods to evaluate P. Do so when $a = 1$ and $e = \frac{1}{4}$ using the Parabolic Rule with $n = 4$. (Your answer should be near 2π. Why?)

13.2
Vectors in the Plane: Geometric Approach

FIGURE 1

Tail   Head

FIGURE 2

Equivalent
vectors

FIGURE 3

Many quantities that occur in science (for example, length, mass, volume, and electric charge) can be specified by giving a single number. These quantities (and the numbers that measure them) are called **scalars**. Other quantities, such as velocity, force, torque, and displacement, require both a magnitude and a direction for complete specification. We call such quantities **vectors** and represent them by arrows (directed line segments). The length of the arrow is the magnitude of the vector; its direction is the direction of the vector. The vector in Figure 1 has length 2.7 units and direction 30° north of east (or 30° from the positive x-axis).

Arrows that we draw, like those shot from a bow, have two ends. There is the feather end (the initial point), called the **tail**, and the pointed end (the terminal point), called the **head**, or tip (Figure 2). Two vectors are considered to be **equivalent** if they have the same magnitude and direction (Figure 3). We shall symbolize vectors by boldface letters, such as **u** and **v**. Since this is hard to accomplish in normal writing, you might use $\vec{u}$ and $\vec{v}$. The magnitude, or length, of a vector **u** is symbolized by $|\mathbf{u}|$.

OPERATIONS ON VECTORS   To find the **sum**, or resultant, of **u** and **v**, move **v** without changing its magnitude or direction until its tail coincides with the head of **u**. Then **u** + **v** is the vector connecting the tail of **u** to the head of **v**. This method (called the *Triangle Law*) is illustrated in the left half of Figure 4.

As an alternative way to find **u** + **v**, move **v** so that its tail coincides with that of **u**. Then **u** + **v** is the vector with this common tail and coinciding with the diagonal of the parallelogram that has **u** and **v** as sides. This method (called the *Parallelogram Law*) is illustrated on the right in Figure 4.

You should convince yourself that addition is commutative and associative, that is,

$$\mathbf{u} + \mathbf{v} = \mathbf{v} + \mathbf{u}$$

$$(\mathbf{u} + \mathbf{v}) + \mathbf{w} = \mathbf{u} + (\mathbf{v} + \mathbf{w})$$

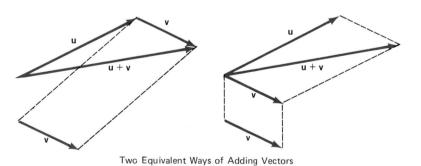

Two Equivalent Ways of Adding Vectors

FIGURE 4

FIGURE 5

If **u** is a vector, then 3**u** is the vector with the same direction as **u** but three times as long; $-2\mathbf{u}$ is twice as long but oppositely directed (Figure 5). In general, $c\mathbf{u}$ called a **scalar multiple** of **u**, has magnitude $|c|$ times that of **u** and is similarly or oppositely directed, depending on whether c is positive or negative. In particular, $(-1)\mathbf{u}$ (usually written $-\mathbf{u}$) has the same length as **u** but opposite direction. It is called the **negative** of **u** because when we add it to **u**, the result is a vector that has shriveled to a point. This latter vector (the only vector without a well-defined direction) is called the **zero vector** and is denoted by **0**. It is the identity element for addition; that is, $\mathbf{u} + \mathbf{0} = \mathbf{0} + \mathbf{u} = \mathbf{u}$. Finally, subtraction is defined by

$$\mathbf{u} - \mathbf{v} = \mathbf{u} + (-\mathbf{v})$$

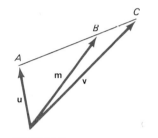

FIGURE 6

EXAMPLE 1 In Figure 6, express **w** in terms of **u** and **v**.

Solution Since $\mathbf{u} + \mathbf{w} = \mathbf{v}$, it follows that

$$\mathbf{w} = \mathbf{v} - \mathbf{u}$$

EXAMPLE 2 In Figure 7, $\overrightarrow{AB} = \frac{2}{3}\overrightarrow{AC}$. Express **m** in terms of **u** and **v**.

Solution

$$\mathbf{m} = \mathbf{u} + \overrightarrow{AB} = \mathbf{u} + \tfrac{2}{3}\overrightarrow{AC}$$
$$= \mathbf{u} + \tfrac{2}{3}(\mathbf{v} - \mathbf{u})$$
$$= \tfrac{1}{3}\mathbf{u} + \tfrac{2}{3}\mathbf{v}$$

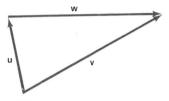

FIGURE 7

More generally, if $\overrightarrow{AB} = t\overrightarrow{AC}$ where $0 < t < 1$, then

$$\mathbf{m} = (1 - t)\mathbf{u} + t\mathbf{v}$$

The expression just obtained for **m** can also be written as

$$\mathbf{u} + t(\mathbf{v} - \mathbf{u})$$

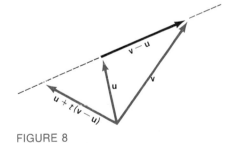

FIGURE 8

If we allow t to range over all scalars, we obtain the set of all vectors with heads on the line, shown in Figure 8. This fact will be important to us later in describing lines using vector language.

APPLICATIONS A force has both a magnitude and a direction. If two forces **u** and **v** act at a point, the resultant force at the point is the vector sum of the two forces.

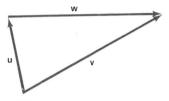

FIGURE 9

EXAMPLE 3 A weight of 200 newtons is supported by two wires, as shown in Figure 9. Find the magnitude of the tension in each wire.

FIGURE 10

Solution The weight **w** and the two tensions **u** and **v** are forces that behave as vectors (Figure 10). Each of these vectors can be expressed as a sum of a horizontal and a vertical component. For equilibrium, (1) the magnitude of the leftward force must equal the magnitude of the rightward force, and (2) the magnitude of the upward force must equal the magnitude of the downward force. Thus,

(1) $$|\mathbf{u}|\cos 33° = |\mathbf{v}|\cos 50°$$

(2) $$|\mathbf{u}|\sin 33° + |\mathbf{v}|\sin 50° = |\mathbf{w}| = 200$$

When we solve (1) for $|\mathbf{v}|$ and substitute in (2), we get

$$|\mathbf{u}|\sin 33° + \frac{|\mathbf{u}|\cos 33°}{\cos 50°}\sin 50° = 200$$

or

$$|\mathbf{u}| = \frac{200}{\sin 33° + \cos 33° \tan 50°} \approx 129.52 \text{ newtons}$$

Then,

$$|\mathbf{v}| = \frac{|\mathbf{u}|\cos 33°}{\cos 50°} \approx \frac{129.52 \cos 33°}{\cos 50°} \approx 168.99 \text{ newtons} \quad ■$$

Velocities have both direction and magnitude and add together as vectors. The magnitude of a velocity vector is called **speed**.

EXAMPLE 4   A river 0.62 miles wide is flowing at 6 miles per hour. Karen's boat travels at 20 miles per hour in still water. In what direction should she point her boat if she wants to go straight across the river? How long will it take her to get across by traveling in this direction?

Solution   Our first job is to determine α in Figure 11.

FIGURE 11

$$\sin \alpha = \frac{6}{20}$$

$$\alpha \approx 17.46°$$

Next, we determine $|\mathbf{w}|$, the speed of the boat in the direction of **w**.

$$|\mathbf{w}| \approx 20 \cos 17.46° \approx 19.08 \text{ miles per hour}$$

Finally, the time required to cross the river is

$$\frac{0.62}{19.08} \approx 0.0325 \text{ hours} = 1.95 \text{ minutes} \quad ■$$

In Problems 1–4, draw the vector **w**.

1. $\mathbf{w} = \mathbf{u} + \frac{3}{2}\mathbf{v}$

2. $\mathbf{w} = 2\mathbf{u} - 3\mathbf{v}$

3. $\mathbf{w} = \mathbf{u}_1 + \mathbf{u}_2 + \mathbf{u}_3$

4. $\mathbf{w} = \mathbf{u}_1 + \mathbf{u}_2 + \mathbf{u}_3$

5. Figure 12 is a parallelogram. Express **w** in terms of **u** and **v**.

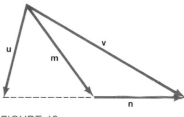

FIGURE 12

6. In the large triangle of Figure 13, **m** is a median (it bisects the side to which it is drawn). Express **m** and **n** in terms of **u** and **v**.

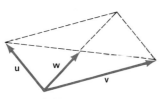

FIGURE 13

7. In Figure 14, $\mathbf{w} = -(\mathbf{u} + \mathbf{v})$ and $|\mathbf{u}| = |\mathbf{v}| = 1$. Find $|\mathbf{w}|$.

FIGURE 14

8. Do Problem 7 if the top angle is 90° and the two side angles are 135°.

[C] 9. In Figure 15, forces **u** and **v** each have magnitude 10 pounds. Find the magnitude and direction of the force **w** needed to counterbalance **u** and **v**.

FIGURE 15

10. John pushes on a post in the direction S 30° E (30° east of south) with a force of 50 pounds. Wayne pushes on the same post in the direction S 60° W with a force of 40 pounds. What is the magnitude and direction of the resultant force?

11. A 250-newton weight rests on a smooth (friction negligible) inclined plane that makes an angle of 30° with the horizontal. What force parallel to the plane will just keep the weight from sliding down the plane? *Hint*: Consider the downward force of 250 newtons to be the sum of two forces, one parallel to the plane and one perpendicular to it.

[C] 12. A body weighing 237.5 pounds is held in equilibrium by two ropes that make angles of 27.34° and 39.22°, respectively, with the vertical. Find the magnitude of the force exerted on the body by each rope.

[C] 13. A wind with velocity 58 miles per hour is blowing in the direction N 20° W. An airplane that flies at 425 miles per hour in still air is supposed to fly straight north. How should the airplane be headed and how fast will it then be flying with respect to the ground?

14. A ship is sailing due south at 20 miles per hour. A man walks west (that is, at right angles to the side of the ship) across the deck at 3 miles per hour. What is the magnitude and direction of his velocity relative to the surface of the water?

15. A pilot, flying in a wind blowing 80 miles per hour due south, discovers that she is heading due east when she points her plane in the direction N 60° E. Find the air speed (speed in still air) of the plane.

C 16. What heading and air speed are required for a plane to fly 837 miles per hour due north if a wind of 31.5 miles per hour is blowing in the direction S 11.5° E?

17. Prove, using vector methods, that the line segment joining the midpoints of two sides of a triangle is parallel to the third side.

18. Prove that the midpoints of the four sides of an arbitrary quadrilateral are the vertices of a parallelogram.

19. Let $\mathbf{v}_1, \mathbf{v}_2, \ldots, \mathbf{v}_n$ be the edges of a polygon arranged in cyclic order as shown in Figure 16. Show that $\mathbf{v}_1 + \mathbf{v}_2 + \cdots + \mathbf{v}_n = \mathbf{0}$.

FIGURE 16

20. Let n points be equally spaced on the rim of a circle and let $\mathbf{v}_1, \mathbf{v}_2, \ldots, \mathbf{v}_n$ be the vectors from the center of the circle to these n points. Show that $\mathbf{v}_1 + \mathbf{v}_2 + \cdots + \mathbf{v}_n = \mathbf{0}$.

21. Consider a horizontal triangular table with each vertex angle less than 120°. At each vertex are frictionless pulleys over which pass strings knotted at P and each with a weight W attached as shown in Figure 17. Show that at equilibrium the three angles at P are equal, that is, show that $\alpha + \beta = \alpha + \gamma = \beta + \gamma = 120°$.

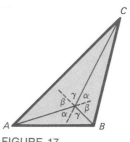

FIGURE 17

22. Show that the point P of the triangle of Problem 21 that minimizes $|AP| + |BP| + |CP|$ is the point where the three angles at P are equal. *Hint*: The system of Problem 21 is in equilibrium when the center of gravity of the three weights is lowest.

23. Let the weights at A, B, and C of Problem 21 be $3w$, $4w$, and $5w$, respectively. Determine the three angles at P at equilibrium. What geometric quantity (as in Problem 22) is now minimized?

24. A company will build a plant to manufacture refrigerators to be sold in cities A, B, and C in quantities a, b, and c, respectively, each year. Describe a physical experiment the company might use to determine the best location for the plant, that is, the location that will minimize delivery costs (see Problem 23).

13.3
Vectors in the Plane: Algebraic Approach

From the geometric perspective of the previous section, a vector can be described as a family of arrows all having the same length and direction (Figure 1). Our aim now is to place vectors in an algebraic context. Here is how we do it.

We begin by imposing a Cartesian coordinate system on the plane. Then, for a given vector $\mathbf{u}$, we pick as its representative the arrow that has its tail at the origin (Figure 2). This arrow is uniquely determined by the coordinates u_1 and u_2 of its head; that is, the vector $\mathbf{u}$ is completely described by

The vector **u**

FIGURE 1

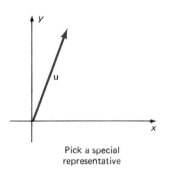

Pick a special representative

FIGURE 2

Identify **u** with the ordered pair $\langle u_1, u_2 \rangle$

FIGURE 3

the ordered pair $\langle u_1, u_2 \rangle$ (Figure 3). That being so, we consider $\langle u_1, u_2 \rangle$ to be the vector; it is the vector **u** in its algebraic clothes. Incidently, we use $\langle u_1, u_2 \rangle$ rather than (u_1, u_2) because the latter symbol already has two meanings—open intervals and points in the plane.

But why do we offer this new interpretation? There are two good answers. First the interplay between the geometric and algebraic aspects of vectors greatly enriches and clarifies the subject. Second, it is the algebraic viewpoint that most easily generalizes to higher dimensions. This is because it is almost as easy to talk about an ordered n-tuple $\langle u_1, u_2, \ldots, u_n \rangle$ as it is an ordered pair $\langle u_1, u_2 \rangle$.

OPERATIONS ON VECTORS   The numbers u_1 and u_2 are called **components** of $\mathbf{u} = \langle u_1, u_2 \rangle$. Two vectors $\mathbf{u} = \langle u_1, u_2 \rangle$ and $\mathbf{v} = \langle v_1, v_2 \rangle$ are equal if and only if $u_1 = v_1$ and $u_2 = v_2$. To add **u** and **v**, we add corresponding components—that is,

$$\mathbf{u} + \mathbf{v} = \langle u_1 + v_1, u_2 + v_2 \rangle$$

To multiply **u** by a scalar c, we multiply each component by c. Thus,

$$\mathbf{u}c = c\mathbf{u} = \langle cu_1, cu_2 \rangle$$

In particular,

$$-\mathbf{u} = \langle -u_1, -u_2 \rangle$$

and

$$\mathbf{0} = 0\mathbf{u} = \langle 0, 0 \rangle$$

Figure 4 shows that these definitions are equivalent to the earlier geometric ones.

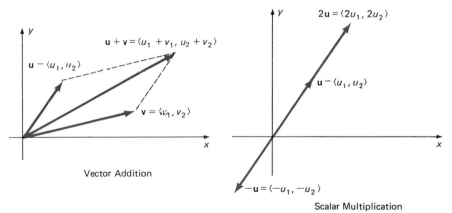

Vector Addition

Scalar Multiplication

FIGURE 4

Using the algebraic interpretation of vectors, it is almost trivial to establish the following rules for operating with vectors.

Theorem A

For any vectors **u**, **v**, and **w** and any scalars a and b, the following relationships hold.

1. $\mathbf{u} + \mathbf{v} = \mathbf{v} + \mathbf{u}$
2. $(\mathbf{u} + \mathbf{v}) + \mathbf{w} = \mathbf{u} + (\mathbf{v} + \mathbf{w})$
3. $\mathbf{u} + \mathbf{0} = \mathbf{0} + \mathbf{u} = \mathbf{u}$
4. $\mathbf{u} + (-\mathbf{u}) = \mathbf{0}$
5. $a(b\mathbf{u}) = (ab)\mathbf{u} = \mathbf{u}(ab)$
6. $a(\mathbf{u} + \mathbf{v}) = a\mathbf{u} + a\mathbf{v}$
7. $(a + b)\mathbf{u} = a\mathbf{u} + b\mathbf{u}$
8. $1\mathbf{u} = \mathbf{u}$

Proof We illustrate the proof by demonstrating Rule 6.

$$a(\mathbf{u} + \mathbf{v}) = a(\langle u_1, u_2 \rangle + \langle v_1, v_2 \rangle)$$
$$= a\langle u_1 + v_1, u_2 + v_2 \rangle$$
$$= \langle a(u_1 + v_1), a(u_2 + v_2) \rangle$$
$$= \langle au_1 + av_1, au_2 + av_2 \rangle$$
$$= \langle au_1, au_2 \rangle + \langle av_1, av_2 \rangle$$
$$= a\langle u_1, u_2 \rangle + a\langle v_1, v_2 \rangle$$
$$= a\mathbf{u} + a\mathbf{v} \qquad \blacksquare$$

LENGTH AND THE DOT PRODUCT   The **length** (or magnitude) $|\mathbf{u}|$ of the vector $\mathbf{u} = \langle u_1, u_2 \rangle$ is given by

$$\boxed{|\mathbf{u}| = \sqrt{u_1^2 + u_2^2}}$$

For example, if $\mathbf{u} = \langle 4, -2 \rangle$, then $|\mathbf{u}| = \sqrt{4^2 + (-2)^2} = 2\sqrt{5}$. If we multiply **u** by the scalar c, we multiply its length by $|c|$; that is,

$$|c\mathbf{u}| = |c|\,|\mathbf{u}|$$

Do not be confused by the apparent double usage of the symbol $|\ \ |$. The symbol $|c|$, called the *absolute value of c*, is the distance from the origin to c on the real line (Figure 5). The symbol $|\mathbf{u}|$, called the *length* of **u**, is the distance from the origin to the head of **u** in the plane (Figure 6).

EXAMPLE 1   Let $\mathbf{u} = \langle 4, -3 \rangle$. Find $|\mathbf{u}|$ and $|-2\mathbf{u}|$. Also find a vector **v** with the same direction as **u**, but with length 1.

FIGURE 5

FIGURE 6

Solution $|\mathbf{u}| = \sqrt{4^2 + (-3)^2} = 5$ and $|-2\mathbf{u}| = |-2|\,|\mathbf{u}| = 2 \cdot 5 = 10$. To find **v**, simply divide **u** by its length $|\mathbf{u}|$; that is,

$$\mathbf{v} = \frac{\mathbf{u}}{|\mathbf{u}|} = \frac{\langle 4, -3 \rangle}{5} = \frac{1}{5}\langle 4, -3 \rangle = \left\langle \frac{4}{5}, \frac{-3}{5} \right\rangle \qquad \blacksquare$$

We have discussed scalar multiplication—that is, the multiplication of a vector **u** by a scalar c. The result $c\mathbf{u}$ is always a vector. Now we introduce a multiplication for two vectors **u** and **v**. It is called the **dot product** and is symbolized by $\mathbf{u} \cdot \mathbf{v}$. We define it by the formula

$$\boxed{\mathbf{u} \cdot \mathbf{v} = u_1 v_1 + u_2 v_2}$$

Note that the dot product of two vectors is a scalar.

The properties of the dot product are easy to establish; we state them without proof.

Theorem B

If **u**, **v**, and **w** are vectors and c is a scalar, then these properties hold.

1. $\mathbf{u} \cdot \mathbf{v} = \mathbf{v} \cdot \mathbf{u}$
2. $\mathbf{u} \cdot (\mathbf{v} + \mathbf{w}) = \mathbf{u} \cdot \mathbf{v} + \mathbf{u} \cdot \mathbf{w}$
3. $c(\mathbf{u} \cdot \mathbf{v}) = (c\mathbf{u}) \cdot \mathbf{v} = \mathbf{u} \cdot (c\mathbf{v})$
4. $\mathbf{0} \cdot \mathbf{u} = 0$
5. $\mathbf{u} \cdot \mathbf{u} = |\mathbf{u}|^2$

To understand the significance of the dot product, we offer an alternative formula for it. If **u** and **v** are nonzero vectors, then

$$\boxed{\mathbf{u} \cdot \mathbf{v} = |\mathbf{u}|\,|\mathbf{v}|\cos \theta}$$

where θ is the angle between **u** and **v**. By the **angle between u and v**, we mean the smallest nonnegative angle between **u** and **v**, so $0 \le \theta \le \pi$.

To derive this formula, apply the Law of Cosines to the triangle in Figure 7.

$$|\mathbf{u} - \mathbf{v}|^2 = |\mathbf{u}|^2 + |\mathbf{v}|^2 - 2|\mathbf{u}|\,|\mathbf{v}|\cos \theta$$

On the other hand, from the properties of the dot product stated in Theorem B,

$$|\mathbf{u} - \mathbf{v}|^2 = (\mathbf{u} - \mathbf{v}) \cdot (\mathbf{u} - \mathbf{v}) = \mathbf{u} \cdot (\mathbf{u} - \mathbf{v}) - \mathbf{v} \cdot (\mathbf{u} - \mathbf{v})$$

$$= \mathbf{u} \cdot \mathbf{u} - \mathbf{u} \cdot \mathbf{v} - \mathbf{v} \cdot \mathbf{u} + \mathbf{v} \cdot \mathbf{v}$$

$$= |\mathbf{u}|^2 + |\mathbf{v}|^2 - 2\mathbf{u} \cdot \mathbf{v}$$

The result follows by equating these two expressions for $|\mathbf{u} - \mathbf{v}|^2$.

FIGURE 7

An extremely important consequence of the formula just obtained is the following theorem.

Theorem C

(Perpendicularity criterion). Two vectors **u** and **v** are perpendicular (orthogonal) if and only if their dot product, **u** · **v**, is 0.

Proof Two nonzero vectors are perpendicular if and only if the angle θ between them is $\pi/2$; that is, if and only if $\cos \theta = 0$. But $\cos \theta = 0$ if and only if **u** · **v** = 0. The result is valid for zero vectors, provided we agree that a zero vector is perpendicular to every vector. ∎

EXAMPLE 2 Find b so that **u** $= \langle 8, 6 \rangle$ and **v** $= \langle 3, b \rangle$ are perpendicular.

Solution

$$\mathbf{u} \cdot \mathbf{v} = (8)(3) + (6)(b) = 24 + 6b = 0$$

Thus, $b = -4$. ∎

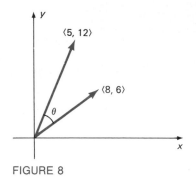

FIGURE 8

EXAMPLE 3 Find the angle between **u** $= \langle 8, 6 \rangle$ and **v** $= \langle 5, 12 \rangle$ (Figure 8).

Solution

$$\cos \theta = \frac{\mathbf{u} \cdot \mathbf{v}}{|\mathbf{u}||\mathbf{v}|} = \frac{(8)(5) + (6)(12)}{(10)(13)} = \frac{112}{130} \approx 0.862$$

Then,

$$\theta = \cos^{-1}(0.862) \approx 0.532 \text{ (or } 30.5°)$$ ∎

BASIS VECTORS Let **i** $= \langle 1, 0 \rangle$ and **j** $= \langle 0, 1 \rangle$ and note that these two vectors are perpendicular and of unit length. They are called **basis vectors**, because any vector **u** $= \langle u_1, u_2 \rangle$ can be represented in a unique way in terms of **i** and **j**. In fact,

FIGURE 9

$$\mathbf{u} = \langle u_1, u_2 \rangle = u_1 \langle 1, 0 \rangle + u_2 \langle 0, 1 \rangle = u_1 \mathbf{i} + u_2 \mathbf{j}$$

The geometric interpretation of this fact is shown in Figure 9.

EXAMPLE 4 If **u** is the arrow from the point $P(2, -1)$ to $Q(-3, 7)$, write **u** in the form $u_1 \mathbf{i} + u_2 \mathbf{j}$.

Solution We first translate this arrow so it emanates from the origin (Figure 10). This can always be accomplished by subtracting the components of the initial point from those of the terminal point. Thus, the corresponding algebraic vector is $\langle -3 - 2, 7 - (-1) \rangle = \langle -5, 8 \rangle$. We conclude that

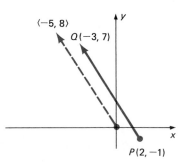

FIGURE 10

$$\mathbf{u} = -5\mathbf{i} + 8\mathbf{j}$$ ∎

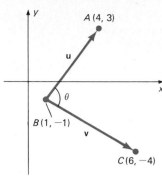

FIGURE 11

EXAMPLE 5 Find the measure of the angle ABC, where $A = (4, 3)$, $B = (1, -1)$, and $C = (6, -4)$, as in Figure 11.

Solution

$$\mathbf{u} = \overrightarrow{BA} = (4 - 1)\mathbf{i} + (3 + 1)\mathbf{j} = 3\mathbf{i} + 4\mathbf{j} = \langle 3, 4 \rangle$$

$$\mathbf{v} = \overrightarrow{BC} = (6 - 1)\mathbf{i} + (-4 + 1)\mathbf{j} = 5\mathbf{i} - 3\mathbf{j} = \langle 5, -3 \rangle$$

$$|\mathbf{u}| = \sqrt{3^2 + 4^2} = 5$$

$$|\mathbf{v}| = \sqrt{5^2 + (-3)^2} = \sqrt{34}$$

$$\mathbf{u} \cdot \mathbf{v} = (3)(5) + (4)(-3) = 3$$

$$\cos \theta = \frac{\mathbf{u} \cdot \mathbf{v}}{|\mathbf{u}|\,|\mathbf{v}|} = \frac{3}{5\sqrt{34}} \approx 0.1029$$

$$\theta \approx 1.468 \ (\text{or } 84.09°)$$

FIGURE 12

Let θ be the angle between $\mathbf{u}$ and $\mathbf{v}$. The scalar $|\mathbf{u}|\cos \theta$ is called the **scalar projection of u on v** for reasons that should be clear from Figure 12. It is positive, zero, or negative, depending on whether θ is acute, right, or obtuse.

The work done by a constant force $\mathbf{F}$ in moving an object along the line from P to Q is the magnitude of the force in the direction of the motion times the distance moved. Thus, if $\mathbf{D}$ is the vector from P to Q, the work done is

$$(\text{Scalar projection } \mathbf{F} \text{ on } \mathbf{D})|\mathbf{D}| = |\mathbf{F}|\cos \theta |\mathbf{D}|$$

that is,

$$\boxed{\text{Work} = \mathbf{F} \cdot \mathbf{D}}$$

EXAMPLE 6 A force $\mathbf{F} = 80\mathbf{i} + 50\mathbf{j}$ in pounds moves an object from $(1, 0)$ to $(7, 1)$, distances measured in feet (Figure 13). How much work is done?

Solution Let $\mathbf{D}$ be the vector from $(1, 0)$ to $(7, 1)$; that is, let $\mathbf{D} = 6\mathbf{i} + \mathbf{j}$. Then,

$$\text{Work} = \mathbf{F} \cdot \mathbf{D} = (80)(6) + (50)(1) = 530 \text{ foot-pounds}$$

FIGURE 13

PROBLEM SET 13.3

1. Let $\mathbf{a} = -3\mathbf{i} + 4\mathbf{j}$, $\mathbf{b} = 2\mathbf{i} - 3\mathbf{j}$, and $\mathbf{c} = -5\mathbf{j}$. Find each of the following.

(a) $2\mathbf{a} - 4\mathbf{b}$
(b) $\mathbf{a} \cdot \mathbf{b}$
(c) $\mathbf{a} \cdot (\mathbf{b} + \mathbf{c})$
(d) $(-2\mathbf{a} + 3\mathbf{b}) \cdot 5\mathbf{c}$
(e) $|\mathbf{a}|\mathbf{c} \cdot \mathbf{a}$
(f) $\mathbf{b} \cdot \mathbf{b} - |\mathbf{b}|$

2. Let $\mathbf{a} = \langle 4, -1 \rangle$, $\mathbf{b} = \langle 1, -1 \rangle$, and $\mathbf{c} = \langle 0, 6 \rangle$. Find each of the following.

(a) $-4\mathbf{a} + 3\mathbf{b}$
(b) $\mathbf{b} \cdot \mathbf{c}$
(c) $(\mathbf{a} + \mathbf{b}) \cdot \mathbf{c}$
(d) $2\mathbf{c} \cdot (3\mathbf{a} + 4\mathbf{b})$
(e) $|\mathbf{b}|\mathbf{b} \cdot \mathbf{a}$
(f) $|\mathbf{c}|^2 - \mathbf{c} \cdot \mathbf{c}$

3. Find the cosine of the angle between **a** and **b** and make a sketch.

(a) $\mathbf{a} = \langle 2, -3 \rangle$, $\mathbf{b} = \langle -1, 4 \rangle$

(b) $\mathbf{a} = \langle -5, -2 \rangle$, $\mathbf{b} = \langle 6, 0 \rangle$

(c) $\mathbf{a} = \langle -3, -1 \rangle$, $\mathbf{b} = \langle -2, -4 \rangle$

(d) $\mathbf{a} = \langle 4, -5 \rangle$, $\mathbf{b} = \langle -8, 10 \rangle$

4. Find the angle between **a** and **b** and make a sketch.

(a) $\mathbf{a} = 12\mathbf{i}$, $\mathbf{b} = -7\mathbf{i}$

(b) $\mathbf{a} = 4\mathbf{i} + 3\mathbf{j}$, $\mathbf{b} = 12\mathbf{i} + 9\mathbf{j}$

(c) $\mathbf{a} = -\mathbf{i} + 3\mathbf{j}$, $\mathbf{b} = 3\mathbf{i} - 9\mathbf{j}$

(d) $\mathbf{a} = \sqrt{3}\mathbf{i} + \mathbf{j}$, $\mathbf{b} = -3\mathbf{i} + \sqrt{3}\mathbf{j}$

5. Write the vector represented by $\overrightarrow{AB}$ in the form $\mathbf{a} = a_1\mathbf{i} + a_2\mathbf{j}$.

(a) $A(2, -1)$, $B(-3, 4)$

(b) $A(0, 5)$, $B(-6, 0)$

(c) $A(\sqrt{2}, -e)$, $B(0, 0)$

(d) $A(-7, 2)$, $B(-4, \frac{1}{3})$

6. Find a unit vector **u** in the direction of **a** and express it in the form $\mathbf{u} = u_1\mathbf{i} + u_2\mathbf{j}$.

(a) $\mathbf{a} = \langle -3, 4 \rangle$

(b) $\mathbf{a} = \langle 1, -7 \rangle$

(c) $\mathbf{a} = \langle 0, -2 \rangle$

(d) $\mathbf{a} = \langle -5, -12 \rangle$

7. If $\mathbf{u} + \mathbf{v}$ is perpendicular to $\mathbf{u} - \mathbf{v}$, what can you say about the relative magnitudes of **u** and **v**?

8. Show that $(\mathbf{u} + \mathbf{v}) \cdot (3\mathbf{u} - \mathbf{v}) = 3|\mathbf{u}|^2 - |\mathbf{v}|^2 + 2\mathbf{u} \cdot \mathbf{v}$.

In each of Problems 9–12, give a proof of the indicated property. Use $\mathbf{u} = \langle u_1, u_2 \rangle$, $\mathbf{v} = \langle u_1, v_2 \rangle$, and $\mathbf{w} = \langle w_1, w_2 \rangle$.

9. $(a + b)\mathbf{u} = a\mathbf{u} + b\mathbf{u}$

10. $\mathbf{u} \cdot (\mathbf{v} + \mathbf{w}) = \mathbf{u} \cdot \mathbf{v} + \mathbf{u} \cdot \mathbf{w}$

11. $c(\mathbf{u} \cdot \mathbf{v}) = (c\mathbf{u}) \cdot \mathbf{v}$

12. If $\mathbf{u} + \mathbf{v} = \mathbf{u}$, then $\mathbf{v} = \mathbf{0}$.

13. Find a vector that has the same direction as $3\mathbf{i} - 4\mathbf{j}$ and three times its length.

14. Find a vector that has the opposite direction to $5\mathbf{i} + 12\mathbf{j}$ and unit length.

15. Show that the vectors $\langle 6, 3 \rangle$ and $\langle -1, 2 \rangle$ are orthogonal (perpendicular).

16. Show that the vectors $\langle -5, \sqrt{3} \rangle$ and $\langle \sqrt{27}, 15 \rangle$ are orthogonal.

17. For what numbers c are $\langle c, 3 \rangle$ and $\langle c, -4 \rangle$ orthogonal?

18. For what numbers c are $2c\mathbf{i} - 4\mathbf{j}$ and $3\mathbf{i} + c\mathbf{j}$ orthogonal?

19. Given $\mathbf{a} = 3\mathbf{i} - 2\mathbf{j}$ and $\mathbf{b} = -\mathbf{i} + 4\mathbf{j}$, two noncollinear vectors (that is, vectors such that the angle θ between them satisfies $0 < \theta < \pi$), and another vector $\mathbf{r} = 7\mathbf{i} - 8\mathbf{j}$. Find scalars k and m such that $\mathbf{r} = k\mathbf{a} + m\mathbf{b}$.

20. Given $\mathbf{a} = -4\mathbf{i} + 3\mathbf{j}$ and $\mathbf{b} = 3\mathbf{i} - \mathbf{j}$ (two noncollinear vectors) and another vector $\mathbf{r} = 6\mathbf{i} - 7\mathbf{j}$. Find scalars k and m such that $\mathbf{r} = k\mathbf{a} + m\mathbf{b}$.

21. Let $\mathbf{a} = a_1\mathbf{i} + a_2\mathbf{j}$ and $\mathbf{b} = b_1\mathbf{i} + b_2\mathbf{j}$ be noncollinear vectors. If $\mathbf{r} = r_1\mathbf{i} + r_2\mathbf{j}$ is an arbitrarily chosen vector in the plane of **a** and **b**, find scalars k and m such that $\mathbf{r} = k\mathbf{a} + m\mathbf{b}$.

22. Show that the vector $\mathbf{n} = a\mathbf{i} + b\mathbf{j}$ is perpendicular to the line with equation $ax + by = c$. *Hint*: Let $P_1(x_1, y_1)$ and $P_2(x_2, y_2)$ be two points on the line. Show that $\mathbf{n} \cdot \overrightarrow{P_1P_2} = 0$.

23. Find the work done by the force $\mathbf{F} = 3\mathbf{i} + 10\mathbf{j}$ pounds in moving an object 10 feet north (that is, in the **j** direction).

24. Find the work done by a force of 100 dynes acting in the direction S 70° E in moving an object 50 centimeters straight east.

25. Find the work done by the force $\mathbf{F} = 3\mathbf{i} + 4\mathbf{j}$ pounds in moving an object from $(1, 0)$ to $(6, 8)$, distance in feet.

26. Find the work done by a force $\mathbf{F} = -5\mathbf{i} + 8\mathbf{j}$ newtons in moving an object 12 meters north.

27. Prove the **Cauchy-Schwarz Inequality**:

$$|\mathbf{u} \cdot \mathbf{v}| \le |\mathbf{u}| \, |\mathbf{v}|$$

28. Prove the **Triangle Inequality** (see Figure 14):

$$|\mathbf{u} + \mathbf{v}| \le |\mathbf{u}| + |\mathbf{v}|$$

Hint:

$$|\mathbf{u} + \mathbf{v}|^2 = (\mathbf{u} + \mathbf{v}) \cdot (\mathbf{u} + \mathbf{v}) = \mathbf{u} \cdot \mathbf{u} + 2\mathbf{u} \cdot \mathbf{v} + \mathbf{v} \cdot \mathbf{v}$$
$$= |\mathbf{u}|^2 + 2\mathbf{u} \cdot \mathbf{v} + |\mathbf{v}|^2$$

Now apply Problem 27.

FIGURE 14

29. What can you say about the nonzero vectors **u** and **v** in each case?

(a) $|\mathbf{u}|^2 + |\mathbf{v}|^2 = 2\mathbf{u} \cdot \mathbf{v}$

(b) $|\mathbf{u}|^2 = |\mathbf{v}|^2 = 2\mathbf{u} \cdot \mathbf{v}$

30. Show that the diagonals of a rhombus (parallelogram with equal edges) are perpendicular. *Hint*: Let **u** and **v** denote adjacent edges. Calculate $(\mathbf{u} + \mathbf{v}) \cdot (\mathbf{u} - \mathbf{v})$.

31. Let A and B be the ends of the diameter of a circle and C be any other point on the circle. Use vector methods to show that AC is perpendicular to BC.

32. Prove that $\mathbf{w} = |\mathbf{v}|\mathbf{u} + |\mathbf{u}|\mathbf{v}$ bisects the angle between $\mathbf{u}$ and $\mathbf{v}$.

33. Let $\mathbf{pr}_\mathbf{v}\mathbf{u}$ denote the **vector projection of u on v**; that is, $\mathbf{pr}_\mathbf{v}\mathbf{u}$ is the vector with the same direction as $\mathbf{v}$ having length

FIGURE 15

$|\mathbf{u}|\cos\theta$ (see Figure 15). Show that

$$\mathbf{pr}_\mathbf{v}\mathbf{u} = \frac{\mathbf{u}\cdot\mathbf{v}}{|\mathbf{v}|^2}\,\mathbf{v}$$

and use this to calculate $\mathbf{pr}_\mathbf{v}\mathbf{u}$ in each case.

(a) $\mathbf{u} = \langle 0,5\rangle$ and $\mathbf{v} = \langle 3,4\rangle$
(b) $\mathbf{u} = \langle -3,2\rangle$ and $\mathbf{v} = \langle 3,4\rangle$

34. Consider the line $ax + by + c = 0$.

(a) Show that $\mathbf{n} = a\mathbf{i} + b\mathbf{j}$ is perpendicular to this line.
(b) Derive the formula $d = |ax_0 + by_0 + c|/\sqrt{a^2 + b^2}$ for the distance from $P(x_0, y_0)$ to the line. *Hint*: Let $Q(x_1, y_1)$ be a point on the line. Find the magnitude of the scalar projection of $\overrightarrow{QP}$ on $\mathbf{n}$.

13.4
Vector-Valued Functions and Curvilinear Motion

Recall that a **function** f is a rule that associates with each member t of one set (the **domain**) a unique value $f(t)$ from a second set (Figure 1). The set of values so obtained is the **range** of the function. So far in this book, our functions have been real-valued functions (scalar-valued functions) of a real variable; that is, both the domain and range have been sets of real numbers. A typical example is $f(t) = t^2$, which associates with each real number t the real number t^2.

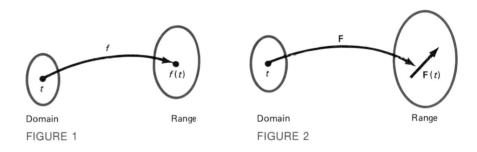

FIGURE 1   FIGURE 2

Now we offer the first of many generalizations (Figure 2). **A vector-valued function F** of a real variable t associates with each real number t a vector $\mathbf{F}(t)$. Thus,

$$\mathbf{F}(t) = f(t)\mathbf{i} + g(t)\mathbf{j} = \langle f(t), g(t)\rangle$$

where f and g are ordinary real-valued functions. A typical example is

$$\mathbf{F}(t) = t^2\mathbf{i} + e^t\mathbf{j} = \langle t^2, e^t\rangle$$

Note our use of a boldface letter; this helps us keep our vector functions and our scalar functions clearly distinguished.

FIGURE 3

CALCULUS FOR VECTOR FUNCTIONS   The most fundamental notion in calculus is that of limit. Intuitively, $\lim_{t \to c} \mathbf{F}(t) = \mathbf{L}$ means that the vector $\mathbf{F}(t)$ tends toward the vector $\mathbf{L}$ as t tends toward c. Alternatively it means that the vector $\mathbf{F}(t) - \mathbf{L}$ approaches $\mathbf{0}$ as $t \to c$ (Figure 3). The precise ε, δ definition is identical with that given for real-valued functions in Section 2.5.

Definition

To say that $\lim_{t \to c} \mathbf{F}(t) = \mathbf{L}$ means that for each given $\varepsilon > 0$ (no matter how small), there is a corresponding $\delta > 0$ such that $|\mathbf{F}(t) - \mathbf{L}| < \varepsilon$, provided that $0 < |t - c| < \delta$; that is,

$$0 < |t - c| < \delta \quad \Rightarrow \quad |\mathbf{F}(t) - \mathbf{L}| < \varepsilon$$

Only one word of explanation is necessary. The symbol $|\mathbf{F}(t) - \mathbf{L}|$ now stands for the length of the vector $\mathbf{F}(t) - \mathbf{L}$.

If we rephrase the definition of limit for $\mathbf{F}(t)$ in terms of its components $f(t)$ and $g(t)$, we obtain an important theorem, whose proof is given in the Appendix Section A.2, Theorem D.

Theorem A

Let $\mathbf{F}(t) = f(t)\mathbf{i} + g(t)\mathbf{j}$. Then $\mathbf{F}$ has a limit at c if and only if f and g have limits at c. In that case,

$$\lim_{t \to c} \mathbf{F}(t) = \left[\lim_{t \to c} f(t) \right] \mathbf{i} + \left[\lim_{t \to c} g(t) \right] \mathbf{j}$$

As you would expect, all the standard limit theorems hold. Also, continuity has its usual meaning; that is, F is **continuous** at c if $\lim_{t \to c} \mathbf{F}(t) = \mathbf{F}(c)$.

From Theorem A, it is clear that F is continuous at c if and only if both f and g are continuous there. Finally, the **derivative** $\mathbf{F}'(t)$ is defined just as for real-valued functions by

$$\mathbf{F}'(t) = \lim_{h \to 0} \frac{\mathbf{F}(t + h) - \mathbf{F}(t)}{h}$$

This also can be written in terms of components.

$$\mathbf{F}'(t) = \lim_{h \to 0} \frac{\left[f(t + h)\mathbf{i} + g(t + h)\mathbf{j} \right] - \left[f(t)\mathbf{i} + g(t)\mathbf{j} \right]}{h}$$

$$= \lim_{h \to 0} \frac{f(t + h) - f(t)}{h} \mathbf{i} + \lim_{h \to 0} \frac{g(t + h) - g(t)}{h} \mathbf{j}$$

$$= f'(t)\mathbf{i} + g'(t)\mathbf{j}$$

In summary, if $\mathbf{F}(t) = f(t)\mathbf{i} + g(t)\mathbf{j}$, then

$$\mathbf{F}'(t) = f'(t)\mathbf{i} + g'(t)\mathbf{j} = \langle f'(t), g'(t) \rangle$$

EXAMPLE 1 If $\mathbf{F}(t) = (t^2 + t)\mathbf{i} + e^t\mathbf{j}$, find $\mathbf{F}'(t)$, $\mathbf{F}''(t)$, and the angle θ between $\mathbf{F}'(0)$ and $\mathbf{F}''(0)$.

Solution $\mathbf{F}'(t) = (2t + 1)\mathbf{i} + e^t\mathbf{j}$ and $\mathbf{F}''(t) = 2\mathbf{i} + e^t\mathbf{j}$. Thus, $\mathbf{F}'(0) = \mathbf{i} + \mathbf{j}$, $\mathbf{F}''(0) = 2\mathbf{i} + \mathbf{j}$, and

$$\cos \theta = \frac{\mathbf{F}'(0) \cdot \mathbf{F}''(0)}{|\mathbf{F}'(0)| \, |\mathbf{F}''(0)|} = \frac{(1)(2) + (1)(1)}{\sqrt{1^2 + 1^2}\sqrt{2^2 + 1^2}} = \frac{3}{\sqrt{2}\sqrt{5}}$$

$$\theta \approx 0.3218 \quad \text{(about } 18.43°\text{)} \qquad \blacksquare$$

Here are the rules for differentiation.

Theorem B

(Differentiation formulas). Let $\mathbf{F}$ and $\mathbf{G}$ be differentiable, vector-valued functions, h a differentiable, real-valued function, and c a scalar. Then:

1. $D_t[\mathbf{F}(t) + \mathbf{G}(t)] = \mathbf{F}'(t) + \mathbf{G}'(t)$;
2. $D_t[c\mathbf{F}(t)] = c\mathbf{F}'(t)$;
3. $D_t[h(t)\mathbf{F}(t)] = h(t)\mathbf{F}'(t) + h'(t)\mathbf{F}(t)$;
4. $D_t[\mathbf{F}(t) \cdot \mathbf{G}(t)] = \mathbf{F}(t) \cdot \mathbf{G}'(t) + \mathbf{G}(t) \cdot \mathbf{F}'(t)$;
5. $D_t[\mathbf{F}(h(t))] = \mathbf{F}'(h(t))h'(t)$ (Chain Rule).

Proof We prove 4 and leave the other parts to the reader. Let

$$\mathbf{F}(t) = f_1(t)\mathbf{i} + f_2(t)\mathbf{j}, \qquad \mathbf{G}(t) = g_1(t)\mathbf{i} + g_2(t)\mathbf{j}$$

Then

$$\begin{aligned}
D_t[\mathbf{F}(t) \cdot \mathbf{G}(t)] &= D_t[f_1(t)g_1(t) + f_2(t)g_2(t)] \\
&= f_1(t)g_1'(t) + g_1(t)f_1'(t) + f_2(t)g_2'(t) + g_2(t)f_2'(t) \\
&= [f_1(t)g_1'(t) + f_2(t)g_2'(t)] + [g_1(t)f_1'(t) + g_2(t)f_2'(t)] \\
&= \mathbf{F}(t) \cdot \mathbf{G}'(t) + \mathbf{G}(t) \cdot \mathbf{F}'(t) \qquad \blacksquare
\end{aligned}$$

Since derivatives of vector-valued functions are found by differentiating components, it is natural to define integration in terms of components; that is, if $\mathbf{F}(t) = f(t)\mathbf{i} + g(t)\mathbf{j}$,

$$\int \mathbf{F}(t)\,dt = \left[\int f(t)\,dt\right]\mathbf{i} + \left[\int g(t)\,dt\right]\mathbf{j}$$

$$\int_a^b \mathbf{F}(t)\,dt = \left[\int_a^b f(t)\,dt\right]\mathbf{i} + \left[\int_a^b g(t)\,dt\right]\mathbf{j}$$

EXAMPLE 2 If $\mathbf{F}(t) = t^2\mathbf{i} + e^{-t}\mathbf{j}$, find (a) $D_t[t^3\mathbf{F}(t)]$, and (b) $\displaystyle\int_0^1 \mathbf{F}(t)\,dt$.

Solution

(a) $D_t[t^3\mathbf{F}(t)] = t^3(2t\mathbf{i} - e^{-t}\mathbf{j}) + 3t^2(t^2\mathbf{i} + e^{-t}\mathbf{j})$

$\qquad\qquad\quad = 5t^4\mathbf{i} + (3t^2 - t^3)e^{-t}\mathbf{j}$

(b) $\displaystyle\int_0^1 \mathbf{F}(t)\,dt = \left(\int_0^1 t^2\,dt\right)\mathbf{i} + \left(\int_0^1 e^{-t}\,dt\right)\mathbf{j}$

$\qquad\qquad\quad = \tfrac{1}{3}\mathbf{i} + (1 - e^{-1})\mathbf{j}$ ∎

CURVILINEAR MOTION   We are going to use the theory developed above for vector-valued functions to study the motion of a point in the plane. Let t measure time and suppose that the coordinates of a moving point P are given by the parametric equations $x = f(t)$, $y = g(t)$. Then the vector

$$\mathbf{r}(t) = f(t)\mathbf{i} + g(t)\mathbf{j}$$

assumed to emanate from the origin, is called the **position vector** of the point. As t varies, the head of $\mathbf{r}(t)$ traces the path of the moving point P (Figure 4). This is a curve and we call the corresponding motion **curvilinear motion**.

In analogy with linear (straight line) motion, we define the **velocity** $\mathbf{v}(t)$ and the **acceleration** $\mathbf{a}(t)$ of the moving point P by

$$\mathbf{v}(t) = \mathbf{r}'(t) = f'(t)\mathbf{i} + g'(t)\mathbf{j}$$

$$\mathbf{a}(t) = \mathbf{r}''(t) = f''(t)\mathbf{i} + g''(t)\mathbf{j}$$

Since

$$\mathbf{v}(t) = \lim_{h\to 0} \frac{\mathbf{r}(t + h) - \mathbf{r}(t)}{h}$$

it is clear (from Figure 5) that $\mathbf{v}(t)$ has the direction of the tangent line. The magnitude $|\mathbf{v}(t)|$ of the velocity is called **speed** of the moving point P. The acceleration vector $\mathbf{a}(t)$ points to the concave side of the curve (the side toward which the curve is bending).

FIGURE 4

  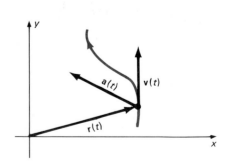

FIGURE 5

CHAPTER 13/Geometry in the Plane, Vectors

EXAMPLE 3 (Uniform circular motion). Suppose that a point P moves around a circle with center $(0,0)$ and radius r at a constant angular speed of ω radians per second. If its initial position is $(r,0)$, find its acceleration.

Solution The position vector at time t is

$$\mathbf{r}(t) = r\cos\omega t\mathbf{i} + r\sin\omega t\mathbf{j}$$

Thus,

$$\mathbf{v}(t) = -r\omega\sin\omega t\mathbf{i} + r\omega\cos\omega t\mathbf{j}$$
$$\mathbf{a}(t) = -r\omega^2\cos\omega t\mathbf{i} - r\omega^2\sin\omega t\mathbf{j}$$

Note that

$$\mathbf{a}(t) = -\omega^2\mathbf{r}(t)$$

Thus if we think of **a** as being based at P, it points to the origin and is perpendicular to **v** (Figure 6). ∎

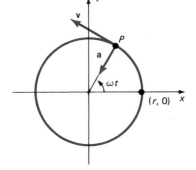

FIGURE 6

EXAMPLE 4 Parametric equations for a point P moving in the plane are $x = 3\cos t$ and $y = 2\sin t$, where t represents time.
(a) Graph the path of P.
(b) Find expressions for the velocity $\mathbf{v}(t)$, speed $|\mathbf{v}(t)|$, and acceleration $\mathbf{a}(t)$.
(c) Find the maximum and minimum values of the speed and where they occur.
(d) Show that the acceleration vector based at P always points to the origin.

Solution
(a) Since $x^2/9 + y^2/4 = 1$, the path is the ellipse shown in Figure 7.
(b) The position vector is

$$\mathbf{r}(t) = 3\cos t\mathbf{i} + 2\sin t\mathbf{j}$$

and so

$$\mathbf{v}(t) = -3\sin t\mathbf{i} + 2\cos t\mathbf{j}$$
$$|\mathbf{v}(t)| = \sqrt{9\sin^2 t + 4\cos^2 t} = \sqrt{5\sin^2 t + 4}$$
$$\mathbf{a}(t) = -3\cos t\mathbf{i} - 2\sin t\mathbf{j}$$

$\mathbf{r}(t) = 3\cos t\,\mathbf{i} + 2\sin t\,\mathbf{j}$

FIGURE 7

(c) Since the speed is given by $\sqrt{5\sin^2 t + 4}$, the maximum speed of 3 occurs when $\sin t = \pm 1$,—that is, when $t = \pi/2$ or $3\pi/2$. This corresponds to the points $(0, \pm 2)$ on the ellipse. Similarly, the minimum speed of 2 occurs when $\sin t = 0$, which corresponds to the points $(\pm 3, 0)$.
(d) Note that $\mathbf{a}(t) = -\mathbf{r}(t)$. Thus, if we base $\mathbf{a}(t)$ at P, this vector will point to and exactly reach the origin. We conclude that $|\mathbf{a}(t)|$ is largest at $(\pm 3, 0)$ and smallest at $(0, \pm 2)$. ∎

FIGURE 8

EXAMPLE 5 A projectile is shot from the origin at an angle θ from the positive x-axis with an initial speed of v_0 feet per second (Figure 8). Neglecting friction, find expressions for the velocity $\mathbf{v}(t)$ and position $\mathbf{r}(t)$, and show that the path is a parabola.

Solution The acceleration due to gravity is $\mathbf{a}(t) = -32\mathbf{j}$ feet per second per second. The initial conditions are $\mathbf{r}(0) = \mathbf{0}$ and $\mathbf{v}(0) = v_0 \cos \theta \mathbf{i} + v_0 \sin \theta \mathbf{j}$. Starting with $\mathbf{a}(t) = -32\mathbf{j}$, we integrate twice.

$$\mathbf{v}(t) = \int \mathbf{a}(t)\, dt = \int -32\, dt\, \mathbf{j} = -32t\mathbf{j} + \mathbf{C}$$

The condition $\mathbf{v}(0) = v_0 \cos \theta \mathbf{i} + v_0 \sin \theta \mathbf{j}$ allows us to evaluate $\mathbf{C}$ and gives $\mathbf{C} = v_0 \cos \theta \mathbf{i} + v_0 \sin \theta \mathbf{j}$. Thus,

$$\mathbf{v}(t) = (v_0 \cos \theta)\mathbf{i} + (v_0 \sin \theta - 32t)\mathbf{j}$$

and

$$\mathbf{r}(t) = \int \mathbf{v}(t)\, dt = (tv_0 \cos \theta)\mathbf{i} + (tv_0 \sin \theta - 16t^2)\mathbf{j} + \mathbf{K}$$

The condition $\mathbf{r}(0) = \mathbf{0}$ implies $\mathbf{K} = \mathbf{0}$, and so

$$\mathbf{r}(t) = (tv_0 \cos \theta)\mathbf{i} + (tv_0 \sin \theta - 16t^2)\mathbf{j}$$

To find the equation of the path, we eliminate the parameter t in the equations

$$x = (v_0 \cos \theta)t, \qquad y = (v_0 \sin \theta)t - 16t^2$$

Specifically, we solve the first equation for t and substitute in the second. The result is

$$y = (\tan \theta)x - \left(\frac{4}{v_0 \cos \theta}\right)^2 x^2$$

the equation of a parabola. ∎

PROBLEM SET 13.4

In Problems 1–8, find the required limit or indicate that it does not exist.

1. $\lim\limits_{t \to 1}[3t\mathbf{i} - t^2\mathbf{j}]$

2. $\lim\limits_{t \to 2}[(t-2)^2\mathbf{i} + 4t\mathbf{j}]$

3. $\lim\limits_{t \to 2}\left[\dfrac{t-2}{t^2-4}\mathbf{i} + \dfrac{t^2+t-6}{t-2}\mathbf{j}\right]$

4. $\lim\limits_{t \to 3}\left[\dfrac{t^2-9}{t+3}\mathbf{i} + \dfrac{t^2+t-6}{t^2-9}\mathbf{j}\right]$

5. $\lim\limits_{t \to 0}\left[\dfrac{\sin t}{t}\mathbf{i} + \dfrac{t}{e^t}\mathbf{j}\right]$

6. $\lim\limits_{t \to \infty}\left[\dfrac{\sin t}{t}\mathbf{i} + \dfrac{t^2+1}{2t^2-3t}\mathbf{j}\right]$

7. $\lim\limits_{t \to 0^+}\langle \ln(t^2), t \ln t\rangle$

8. $\lim_{t \to 0^-} \left\langle e^{1/t}, \dfrac{t}{|t|} \right\rangle$

9. When no domain is given in the definition of a vector-valued function, it is to be understood that the domain is the set of all (real) scalars to which correspond real vectors (that is, vectors with real components). Find the domain of each of the following vector-valued functions.

(a) $\mathbf{r}(t) = \dfrac{1}{t-2}\mathbf{i} + \sqrt{4+t}\,\mathbf{j}$

(b) $\mathbf{r}(t) = [\![t]\!]\mathbf{i} - \sqrt{t^2+1}\,\mathbf{j}$. ($[\![\ ]\!]$ denotes "greatest integer in.")

10. State the domain of each of the following vector-valued functions.

(a) $\mathbf{r}(t) = \ln(t^2+1)\mathbf{i} + (\tan^{-1}t)\mathbf{j}$

(b) $\mathbf{r}(t) = \ln(2t^{-1})\mathbf{i} - (6-t)^{1/2}\mathbf{j}$

11. For what values of t is each of the functions in Problem 9 continuous?

12. For what values of t is each of the functions in Problem 10 continuous?

13. Find $D_t\mathbf{r}(t)$ and $D_t^2\mathbf{r}(t)$ for each of the following.

(a) $\mathbf{r}(t) = (2t+3)^2\mathbf{i} - e^{2t}\mathbf{j}$ (b) $\mathbf{r}(t) = \cos 2t\mathbf{i} - \sin^3 t\mathbf{j}$

14. Find $\mathbf{r}'(t)$ and $\mathbf{r}''(t)$ for each of the following.

(a) $\mathbf{r}(t) = (e^t + e^{-t})\mathbf{i} - e^{t^2}\mathbf{j}$ (b) $\mathbf{r}(t) = \tan t\mathbf{i} - 2t^{5/3}\mathbf{j}$

15. If $\mathbf{r}(t) = e^{2t}\mathbf{i} + \ln(t^3)\mathbf{j}$, find $D_t[\mathbf{r}(t) \cdot \mathbf{r}'(t)]$.

16. If $\mathbf{r}(t) = \sin 3t\mathbf{i} - \cos 2t\mathbf{j}$, find $D_t[\mathbf{r}'(t) \cdot \mathbf{r}''(t)]$.

17. If $\mathbf{r}(t) = \sqrt{t-1}\,\mathbf{i} + \ln(2t^2)\mathbf{j}$ and $h(t) = e^{-3t}$, find $D_t[h(t)\mathbf{r}(t)]$.

18. If $\mathbf{r}(t) = \sin 2t\mathbf{i} + \cosh t\mathbf{j}$ and $h(t) = \ln(3t-2)$, find $D_t[h(t)\mathbf{r}(t)]$.

In Problems 19 and 20, $\mathbf{F}(t) = \mathbf{f}(g(t))$. Find $\mathbf{F}'(t)$ in terms of t.

19. $\mathbf{f}(u) = \cos u\mathbf{i} + e^{3u}\mathbf{j}$ and $g(t) = 3t^2 - 4$.

20. $\mathbf{f}(u) = u^2\mathbf{i} + \sin^2 u\mathbf{j}$ and $g(t) = \tan t$.

Evaluate the integrals in Problems 21 and 22.

21. $\displaystyle\int_0^1 (e^t\mathbf{i} + e^{-t}\mathbf{j})\,dt$

22. $\displaystyle\int_{-1}^1 [(1+t)^{3/2}\mathbf{i} + (1-t)^{3/2}\mathbf{j}]\,dt$

In each of Problems 23–30, the position of a moving particle at time t is given by $\mathbf{r}(t)$. Find the velocity and acceleration vectors, $\mathbf{v}(t)$ and $\mathbf{a}(t)$, and the speed at the given time $t = t_1$. Sketch a portion of the graph of $\mathbf{r}(t)$ containing the position P of the particle when $t = t_1$ and draw $\mathbf{v}(t_1)$ and $\mathbf{a}(t_1)$ with their initial points at P.

23. $\mathbf{r}(t) = e^{-t}\mathbf{i} + e^t\mathbf{j}$; $t_1 = 1$

24. $\mathbf{r}(t) = (3t^2 - 1)\mathbf{i} + t\mathbf{j}$; $t_1 = \frac{1}{2}$

25. $\mathbf{r}(t) = 2\cos t\mathbf{i} - 3\sin^2 t\mathbf{j}$; $t_1 = \dfrac{\pi}{3}$

26. $\mathbf{r}(t) = \tan t\mathbf{i} + \sin t\mathbf{j}$; $t_1 = \dfrac{\pi}{6}$

27. $\mathbf{r}(t) = 3t^2\mathbf{i} + t^3\mathbf{j}$; $t_1 = 2$

28. $\mathbf{r}(t) = a\sin t\mathbf{i} + b\cos t\mathbf{j}$; $t_1 = \dfrac{\pi}{4}$

29. $\mathbf{r}(t) = \cos t\mathbf{i} - 2\tan t\mathbf{j}$; $t_1 = -\dfrac{\pi}{4}$

30. $\mathbf{r}(t) = e^{t/2}\mathbf{i} + e^{-t}\mathbf{j}$; $t_1 = 2$

In Problems 31–34, use the given information to find the velocity vector $\mathbf{v}(t)$ and the position vector $\mathbf{r}(t)$ (see Example 5).

31. $\mathbf{a}(t) = -32\mathbf{j}$, $\mathbf{v}(0) = \mathbf{0}$, $\mathbf{r}(0) = \mathbf{0}$

32. $\mathbf{a}(t) = t\mathbf{j}$, $\mathbf{v}(0) = \mathbf{i} + 2\mathbf{j}$, $\mathbf{r}(0) = \mathbf{0}$

33. $\mathbf{a}(t) = \mathbf{i} + e^{-t}\mathbf{j}$, $\mathbf{v}(0) = 2\mathbf{i} + \mathbf{j}$, $\mathbf{r}(0) = \mathbf{i} + \mathbf{j}$

34. $\mathbf{a}(t) = -\cos t\mathbf{i} + \sin t\mathbf{j}$, $\mathbf{v}(0) = \mathbf{i}$, $\mathbf{r}(0) = \mathbf{i} + 3\mathbf{j}$

35. A point moves around the circle $x^2 + y^2 = 25$ at constant angular speed of 6 radians per second starting at $(5, 0)$. Find expressions for $\mathbf{r}(t)$, $\mathbf{v}(t)$, $|\mathbf{v}(t)|$, and $\mathbf{a}(t)$ (see Example 3).

36. A point moves so that its speed is constant, that is, $\mathbf{v}(t) \cdot \mathbf{v}(t) = c$ (a constant). Show that the velocity and acceleration vectors are always perpendicular to each other.

37. In Example 5, suppose $\theta = 30°$ and $v_0 = 96$ feet per second. When does the projectile hit the ground, with what speed, and how far from the origin?

38. A point moves in the plane with constant acceleration vector $\mathbf{a} = a\mathbf{j}$. Show that its path is a parabola or a straight line.

39. A point moves on the hyperbola $x^2 - y^2 = 1$ with position vector

$$\mathbf{r}(t) = \cosh \omega t\mathbf{i} + \sinh \omega t\mathbf{j}$$

where ω is a constant. Show that $\mathbf{a}(t) = c\mathbf{r}(t)$, where c is a positive constant.

40. A point moves on the ellipse $x^2/a^2 + y^2/b^2 = 1$ with position vector

$$\mathbf{r}(t) = a\cos \omega t\mathbf{i} + b\sin \omega t\mathbf{j}$$

where ω is a constant. Show that $\mathbf{a}(t) = c\mathbf{r}(t)$, where c is a negative constant (see Example 4).

41. If a batted baseball leaves the bat at an angle of 45° with the horizontal and is caught by an outfielder 300 feet from home plate, what was the initial velocity of the ball? (See Example 5.)

42. A small ball is rolled off a horizontal table 4 feet high with a speed of 20 feet per second. At what angle and with what speed will it hit the floor?

43. An open can is attached to the rim of a vertical spinning wheel of radius 60 centimeters with its open end pointing to the center of the wheel. In the can is a marble. At least what angular speed must be maintained if the marble is not to fall out? (See Problem 44 for the value of g.)

44. A body of mass m is revolving in a circular orbit of radius r with constant speed v around a body of mass M. From Example 3,

$$a = |\mathbf{a}(t)| = \omega^2 |\mathbf{r}(t)| = \frac{v^2}{r}$$

whereas from Newton's Inverse Square Law, $F = ma = GmM/r^2$ or

$$a = \frac{k}{r^2} \quad (k \text{ a constant})$$

Let T be the time for one revolution, $R \approx 3960$ miles (the radius of the earth), and $g \approx 32.17$ feet per second per second ≈ 980 centimeters per second per second (the acceleration of gravity at the earth's surface). Use these results to show each of the following.

(a) $v^2 = \dfrac{k}{r}$

(b) $T^2 = \left(\dfrac{4\pi^2}{k}\right) r^3$ (Kepler's Third Law for circular orbits)

(c) $k = gR^2$

[C] **45.** Apply the results of Problem 44 to the problem of the earth and a satellite by determining each of the following.

(a) The speed of a satellite that orbits the earth 200 miles above its surface.

(b) The distance from the center of the earth to a communications satellite in *synchronous* orbit (revolving so as to stay directly over a point of the earth).

[C] **46.** The moon orbits the earth every 27.32 days. Determine the distance from the earth to the moon in miles (see Problem 44).

13.5
Curvature and Acceleration

We want to introduce a number called *curvature*, which measures how sharply a curve bends. A line should have curvature 0, a sharply turning curve should have a large curvature (Figure 1). To lead up to an appropriate definition, we need to review some old ideas and introduce some new ones.

For $a \leq t \leq b$, let $\mathbf{r}(t) = f(t)\mathbf{i} + g(t)\mathbf{j} = \langle f(t), g(t) \rangle$ be the position vector for a point $P = P(t)$ in the plane. We suppose that $\mathbf{r}'(t)$ exists and is continuous, and that $\mathbf{r}'(t) \neq \mathbf{0}$ on the interval $[a, b]$. Then (see Section 6.4) as t increases, P traces out a *smooth* curve (Figure 2) and the length $s = h(t)$ of the path from $P(a)$ to $P(t)$ is given by

$$s = h(t) = \int_a^t \sqrt{[f'(u)]^2 + [g'(u)]^2}\, du = \int_a^t |\mathbf{r}'(u)|\, du$$

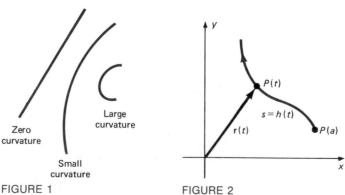

FIGURE 1

FIGURE 2

The speed of the moving point is

$$\frac{ds}{dt} = |\mathbf{r}'(t)| = |\mathbf{v}(t)|$$

Since $\mathbf{r}'(t) \neq \mathbf{0}$, it follows that $|\mathbf{v}(t)| > 0$, and so s increases as t increases. By the Inverse Function Theorem (Theorem 7.2B), $s = h(t)$ has an inverse $t = h^{-1}(s)$ and

$$\frac{dt}{ds} = \frac{1}{ds/dt} = \frac{1}{|\mathbf{v}(t)|}$$

Let $\mathbf{T}(t)$, called the **unit tangent vector** at $P(t)$, be defined by

$$\mathbf{T}(t) = \frac{\mathbf{r}'(t)}{|\mathbf{r}'(t)|} = \frac{\mathbf{v}(t)}{|\mathbf{v}(t)|}$$

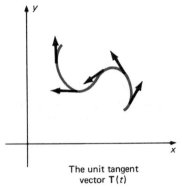

The unit tangent
vector $\mathbf{T}(t)$

FIGURE 3

As $P(t)$ moves along the curve, the unit vector $\mathbf{T}(t)$ changes its direction (Figure 3). The rate of change of $\mathbf{T}$ with respect to arc length s, namely $d\mathbf{T}/ds$, is called the **curvature vector** at P. Finally, we define the **curvature** κ (kappa) at P to be the magnitude of $d\mathbf{T}/ds$; that is, $\kappa = |d\mathbf{T}/ds|$. Now, by the Chain Rule,

$$\frac{d\mathbf{T}}{ds} = \frac{d\mathbf{T}}{dt}\frac{dt}{ds} = \frac{\mathbf{T}'(t)}{|\mathbf{v}(t)|}$$

Thus,

$$\kappa = \left|\frac{d\mathbf{T}}{ds}\right| = \frac{|\mathbf{T}'(t)|}{|\mathbf{v}(t)|} = \frac{|\mathbf{T}'(t)|}{|\mathbf{r}'(t)|}$$

SOME IMPORTANT EXAMPLES   To convince you that our definition of curvature is sensible, we illustrate.

EXAMPLE 1   Show that the curvature of a line is identically zero.

Solution   This follows immediately from the fact that $\mathbf{T}$ is a constant vector. But to illustrate vector methods, we give an algebraic demonstration. Let P and Q be two fixed points on the line and let $\mathbf{a} = \overrightarrow{OP}$ and $\mathbf{b} = \overrightarrow{PQ}$. Then the vector form for the equation of the line may be written as (Figure 4)

$$\mathbf{r} = \mathbf{r}(t) = \mathbf{a} + t\mathbf{b}$$

Thus,

$$\mathbf{v}(t) = \mathbf{r}'(t) = \mathbf{b}$$

$$\mathbf{T}(t) = \frac{\mathbf{b}}{|\mathbf{b}|}$$

$$\kappa = \frac{|\mathbf{T}'(t)|}{|\mathbf{v}(t)|} = \frac{0}{|\mathbf{b}|} = 0 \qquad \blacksquare$$

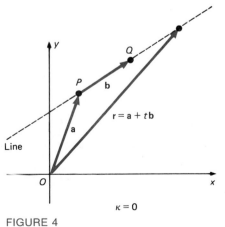

Line

$r = a + tb$

$\kappa = 0$

FIGURE 4

Circle

$$\kappa = \frac{1}{a}$$

FIGURE 5

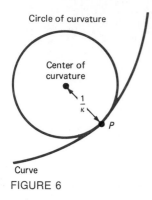

Circle of curvature

Center of curvature

$\frac{1}{\kappa}$

P

Curve

FIGURE 6

EXAMPLE 2 Show that at each point the curvature of a circle of radius a is $\kappa = 1/a$ (Figure 5).

Solution We may suppose the circle is centered at the origin. Then its vector equation may be written

$$\mathbf{r}(t) = a \cos t \mathbf{i} + a \sin t \mathbf{j}$$

Thus,

$$\mathbf{v}(t) = \mathbf{r}'(t) = -a \sin t \mathbf{i} + a \cos t \mathbf{j}$$

$$\mathbf{T}(t) = \frac{\mathbf{v}(t)}{|\mathbf{v}(t)|} = \frac{\mathbf{v}(t)}{a} = -\sin t \mathbf{i} + \cos t \mathbf{j}$$

$$\kappa = \frac{|\mathbf{T}'(t)|}{|\mathbf{v}(t)|} = \frac{|-\cos t \mathbf{i} - \sin t \mathbf{j}|}{a} = \frac{1}{a}$$

Since κ is the reciprocal of the radius, the larger the circle, the smaller is its curvature. ∎

The example of the circle leads to several new ideas. Let P be a point on a curve where $\kappa \neq 0$. Consider the circle that is tangent to the curve at P and has the same curvature there. Its center will lie on the concave side of the curve. This circle is called the **circle of curvature** (or osculating circle), its radius $R = 1/\kappa$ is the **radius of curvature**, and its center is the **center of curvature**. These notions are illustrated in Figure 6.

EXAMPLE 3 Find the curvature and radius of curvature for the hypocycloid

$$\mathbf{r} = 8 \cos^3 t \mathbf{i} + 8 \sin^3 t \mathbf{j}$$

at the point P where $t = \pi/12$. Then sketch the graph of this hypocycloid showing the circle of curvature at P.

Solution For $0 < t < \pi/2$,

$$\mathbf{v}(t) = \mathbf{r}'(t) = -24 \cos^2 t \sin t \mathbf{i} + 24 \sin^2 t \cos t \mathbf{j}$$

$$|\mathbf{v}(t)| = 24 \sin t \cos t$$

$$\mathbf{T}(t) = -\cos t \mathbf{i} + \sin t \mathbf{j}$$

$$\mathbf{T}'(t) = \sin t \mathbf{i} + \cos t \mathbf{j}$$

$$\kappa(t) = \frac{|\mathbf{T}'(t)|}{|\mathbf{v}(t)|} = \frac{|\sin t \mathbf{i} + \cos t \mathbf{j}|}{24 \sin t \cos t} = \frac{1}{24 \sin t \cos t}$$

$$= \frac{1}{12 \sin 2t}$$

$$\kappa\left(\frac{\pi}{12}\right) = \frac{1}{12 \cdot \frac{1}{2}} = \frac{1}{6}$$

$$R\left(\frac{\pi}{12}\right) = 6$$

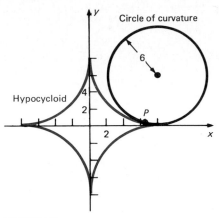

Circle of curvature

Hypocycloid

FIGURE 7

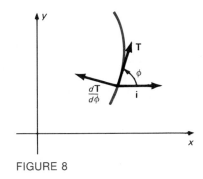

FIGURE 8

The required graph is shown in Figure 7. Note that P has the approximate coordinates $(7.21, 0.14)$. ∎

OTHER FORMULAS FOR CURVATURE   Let ϕ denote the angle measured counterclockwise from $\mathbf{i}$ to $\mathbf{T}$ (Figure 8). Then, $\mathbf{T} = \cos \phi \mathbf{i} + \sin \phi \mathbf{j}$, and so

$$\frac{d\mathbf{T}}{d\phi} = -\sin \phi \mathbf{i} + \cos \phi \mathbf{j}$$

Now $d\mathbf{T}/d\phi$ is a unit vector (length 1) and $\mathbf{T} \cdot d\mathbf{T}/d\phi = 0$. Moreover,

$$\kappa = \left| \frac{d\mathbf{T}}{ds} \right| = \left| \frac{d\mathbf{T}}{d\phi} \frac{d\phi}{ds} \right| = \left| \frac{d\mathbf{T}}{d\phi} \right| \left| \frac{d\phi}{ds} \right|$$

and consequently

$$\kappa = \left| \frac{d\phi}{ds} \right|$$

This formula for κ helps our intuitive understanding of curvature (it measures the rate of change of ϕ with respect to s), and it also allows us to give a fairly simple proof of the following important theorem.

Theorem A

Consider a curve with vector equation $\mathbf{r}(t) = f(t)\mathbf{i} + g(t)\mathbf{j}$, that is, with parametric equations $x = f(t)$ and $y = g(t)$. Then

$$\kappa = \frac{|x'y'' - y'x''|}{[x'^2 + y'^2]^{3/2}}$$

In particular, if the curve is the graph of $y = g(x)$, then

$$\kappa = \frac{|y''|}{[1 + y'^2]^{3/2}}$$

Proof   We might calculate κ directly from the formula $\kappa = |\mathbf{T}'(t)|/|\mathbf{r}'(t)|$, a task we propose in Problem 36. It is a good—but painful—exercise in differentiation and algebraic manipulation. Rather, we choose to use the formula $\kappa = |d\phi/ds|$ derived above. Refer to Figure 8, from which we see that

$$\tan \phi = \frac{dy}{dx} = \frac{dy/dt}{dx/dt} = \frac{y'}{x'}$$

Differentiate both sides of this equation with respect to t to obtain

$$\sec^2 \phi \frac{d\phi}{dt} = \frac{x'y'' - y'x''}{x'^2}$$

Then

$$\frac{d\phi}{dt} = \frac{x'y'' - y'x''}{x'^2 \sec^2\phi} = \frac{x'y'' - y'x''}{x'^2(1 + \tan^2\phi)}$$

$$= \frac{x'y'' - y'x''}{x'^2(1 + y'^2/x'^2)} = \frac{x'y'' - y'x''}{x'^2 + y'^2}$$

But

$$\kappa = \left|\frac{d\phi}{ds}\right| = \left|\frac{d\phi}{dt}\frac{dt}{ds}\right| = \left|\frac{d\phi}{dt} \div \frac{ds}{dt}\right| = \frac{|d\phi/dt|}{[x'^2 + y'^2]^{1/2}}$$

When we put these two results together, we obtain

$$\kappa = \frac{|x'y'' - y'x''|}{[x'^2 + y'^2]^{3/2}}$$

which is the first assertion of the theorem.

To obtain the second assertion, simply regard $y = g(x)$ as corresponding to the parametric equations $x = t$, $y = g(t)$, so that $x' = 1$ and $x'' = 0$. The conclusion follows. ∎

EXAMPLE 4 Find the curvature of the ellipse

$$x = 3\cos t, \qquad y = 2\sin t$$

at the points corresponding to $t = 0$ and $t = \pi/2$—that is, at $(3, 0)$ and $(0, 2)$. Sketch the ellipse showing the corresponding circles of curvature.

Solution From the given equations,

$$x' = -3\sin t \qquad y' = 2\cos t$$
$$x'' = -3\cos t \qquad y'' = -2\sin t$$

Thus,

$$\kappa = \kappa(t) = \frac{|x'y'' - y'x''|}{[x'^2 + y'^2]^{3/2}} = \frac{6\sin^2 t + 6\cos^2 t}{[9\sin^2 t + 4\cos^2 t]^{3/2}}$$

$$= \frac{6}{[5\sin^2 t + 4]^{3/2}}$$

Consequently

$$\kappa(0) = \frac{6}{4^{3/2}} = \frac{3}{4}$$

$$\kappa\left(\frac{\pi}{2}\right) = \frac{6}{9^{3/2}} = \frac{2}{9}$$

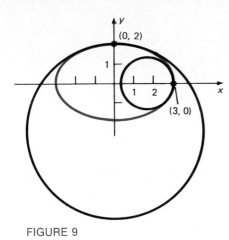

FIGURE 9

Note that $\kappa(0)$ is larger than $\kappa(\pi/2)$, as it should be. Figure 9 shows the circle of curvature at $(3,0)$, which has radius $\frac{4}{3}$, and the one at $(0,2)$, which has radius $\frac{9}{2}$.

EXAMPLE 5 Find the curvature of $y = \ln|\cos x|$ at $x = \pi/3$.

Solution We employ the second formula of Theorem A, noting that the primes now indicate differentiation with respect to x. Since $y' = \tan x$ and $y'' = \sec^2 x$,

$$\kappa = \frac{\sec^2 x}{(1 + \tan^2 x)^{3/2}} = \frac{\sec^2 x}{(\sec^2 x)^{3/2}} = |\cos x|$$

At $x = \pi/3$, $\kappa = \frac{1}{2}$.

NORMAL AND TANGENTIAL COMPONENTS OF ACCELERATION Let $P = P(t)$ be a point on a smooth curve. Define a new vector $\mathbf{N} = \mathbf{N}(t)$, called the **unit normal vector** at P, by

$$\mathbf{N} = \frac{d\mathbf{T}/ds}{|d\mathbf{T}/ds|} = \frac{1}{\kappa}\frac{d\mathbf{T}}{ds}$$

so that

$$\boxed{\frac{d\mathbf{T}}{ds} = \kappa\mathbf{N}}$$

Now $\mathbf{T} = \cos \phi\mathbf{i} + \sin \phi\mathbf{j}$, and so

$$\frac{d\mathbf{T}}{ds} = \frac{d\mathbf{T}}{d\phi}\frac{d\phi}{ds} = (-\sin \phi\mathbf{i} + \cos \phi\mathbf{j})\frac{d\phi}{ds}$$

from which it follows that $\mathbf{T} \cdot \mathbf{N} = 0$. Thus, $\mathbf{N}$ is a unit vector perpendicular to $\mathbf{T}$ and pointing (Problem 28) to the concave side of the curve.

We wish to express the acceleration vector $\mathbf{a}$ in terms of $\mathbf{T}$ and $\mathbf{N}$; that is, we want to decompose $\mathbf{a}$ into components tangent and normal to the curve. Since the velocity vector $\mathbf{v}$ satisfies

$$\mathbf{v} = |\mathbf{v}|\mathbf{T} = \frac{ds}{dt}\mathbf{T}$$

if follows that

$$\mathbf{a} = \frac{d\mathbf{v}}{dt} = \frac{d^2s}{dt^2}\mathbf{T} + \frac{ds}{dt}\frac{d\mathbf{T}}{dt}$$

$$= \frac{d^2s}{dt^2}\mathbf{T} + \frac{ds}{dt}\frac{d\mathbf{T}}{ds}\frac{ds}{dt}$$

Thus,

$$\mathbf{a} = \frac{d^2s}{dt^2}\mathbf{T} + \left(\frac{ds}{dt}\right)^2 \kappa\mathbf{N}$$

a result illustrated in Figure 10.

FIGURE 10

Let's interpret this result for the driver of a car who is pushing down on the accelerator as she travels along a straight road and then enters a curve. According to Newton's Law, force is mass times acceleration. Thus, the force on the car of mass m is

$$\mathbf{F} = m\mathbf{a} = m\frac{d^2s}{dt^2}\mathbf{T} + m\kappa\left(\frac{ds}{dt}\right)^2\mathbf{N}$$

The driver of mass m_0 also experiences a force. On the straight road where $\kappa = 0$, she feels only a push from the seat against her body of magnitude $m_0\,d^2s/dt^2$, this due to pushing down the accelerator. On entering the curve, she feels an additional inward push of magnitude $m_0\kappa\,(ds/dt)^2$. Note that this push is proportional both to the curvature and the square of the speed. The sharper the curve and the faster the driver is going, the stronger the push.

EXAMPLE 6   A particle moves so that its position vector is $\mathbf{r}(t) = t^2\mathbf{i} + \frac{1}{3}t^3\mathbf{j}$, $t \geq 0$. Express $\mathbf{a}(t)$ in terms of $\mathbf{T}$ and $\mathbf{N}$, and evaluate when $t = 2$.

Solution

$$\mathbf{v}(t) = 2t\mathbf{i} + t^2\mathbf{j}$$

$$\frac{ds}{dt} = |\mathbf{v}(t)| = \sqrt{4t^2 + t^4} = t\sqrt{4 + t^2}$$

$$\frac{d^2s}{dt^2} = \frac{4 + 2t^2}{\sqrt{4 + t^2}}$$

$$\kappa\left(\frac{ds}{dt}\right)^2 = \frac{|x'y'' - y'x''|}{[x'^2 + y'^2]^{3/2}}\left(\frac{ds}{dt}\right)^2 = \frac{|x'y'' - y'x''|}{ds/dt}$$

$$= \frac{|(2t)(2t) - (t^2)(2)|}{t\sqrt{4 + t^2}} = \frac{2t^2}{t\sqrt{4 + t^2}} = \frac{2t}{\sqrt{4 + t^2}}$$

Thus,

$$\mathbf{a}(t) = \frac{d^2s}{dt^2}\mathbf{T} + \kappa\left(\frac{ds}{dt}\right)^2\mathbf{N}$$

$$= \frac{4 + 2t^2}{\sqrt{4 + t^2}}\mathbf{T} + \frac{2t}{\sqrt{4 + t^2}}\mathbf{N}$$

and

$$\mathbf{a}(2) = \frac{12}{\sqrt{8}}\mathbf{T} + \frac{4}{\sqrt{8}}\mathbf{N} = 3\sqrt{2}\mathbf{T} + \sqrt{2}\mathbf{N} \qquad \blacksquare$$

CHAPTER 13/Geometry in the Plane, Vectors

We have expressed **a** in the form

$$\mathbf{a} = a_T\mathbf{T} + a_N\mathbf{N}$$

where the tangent and normal components are

$$a_T = \frac{d^2s}{dt^2} \quad \text{and} \quad a_N = \kappa\left(\frac{ds}{dt}\right)^2$$

To calculate a_N, it appears that we must calculate κ. However this can be avoided by noting that **T** and **N** are perpendicular and, therefore,

$$\boxed{|\mathbf{a}|^2 = a_T^2 + a_N^2}$$

We illustrate.

EXAMPLE 7 Without finding κ, find a_T and a_N for the motion given by $\mathbf{r} = t^2\mathbf{i} + \frac{1}{3}t^3\mathbf{j}$, $t \geq 0$.

Solution The motion is the same as that of Example 6, and so our calculation of a_T follows the lines of that example.

$$\mathbf{v} = 2t\mathbf{i} + t^2\mathbf{j}$$

$$\mathbf{a} = 2\mathbf{i} + 2t\mathbf{j}$$

$$\frac{ds}{dt} = |\mathbf{v}| = t\sqrt{4 + t^2}$$

$$a_T = \frac{d^2s}{dt^2} = \frac{4 + 2t^2}{\sqrt{4 + t^2}}$$

By the boxed formula above,

$$a_N^2 = |\mathbf{a}|^2 - a_T^2 = 4 + t^2 - \frac{(4 + 2t^2)^2}{4 + t^2} = \frac{4t^2}{4 + t^2}$$

$$a_N = \frac{2t}{\sqrt{4 + t^2}}$$ ∎

PROBLEM SET 13.5

In Problems 1–4, find the unit tangent vector $\mathbf{T}(t)$ and the curvature $\kappa(t)$ at the point where $t = t_1$. For calculating κ, we suggest using Theorem A, as in Example 4.

1. $\mathbf{r}(t) = 4t^2\mathbf{i} + 4t\mathbf{j}$; $t_1 = \frac{1}{2}$

2. $\mathbf{r}(t) = \frac{1}{2}t^2\mathbf{i} + \frac{1}{3}t^3\mathbf{j}$; $t_1 = 1$

3. $\mathbf{r}(t) = 4\cos t\mathbf{i} + 3\sin t\mathbf{j}$; $t_1 = \frac{\pi}{4}$

4. $\mathbf{r}(t) = e^t\sin t\mathbf{i} + e^t\cos t\mathbf{j}$; $t_1 = \frac{\pi}{2}$

In Problems 5–14, sketch the curve. Then for the given point, find the curvature and the radius of curvature. Finally, draw the circle of curvature at this point. *Hint*: For the curvature, you will use the second formula in Theorem A, as in Example 5.

5. $y = x^2$, $(1, 1)$ **6.** $y = x(x - 2)^2$, $(2, 0)$

7. $y^2 = x + 4$, $(-3, -1)$

8. $y = \ln x$, $(1, 0)$

9. $y = e^x - x$, $(0, 1)$

10. $y = \cos 2x$, $(\frac{1}{6}\pi, \frac{1}{2})$

11. $y = \cosh \frac{1}{2}x$, $(0, 1)$

12. $y = \ln \sin x$, $(\frac{1}{4}\pi, -\ln\sqrt{2})$

13. $y = \sin x$, $\left(\dfrac{\pi}{4}, \dfrac{\sqrt{2}}{2}\right)$

14. $y = \tan x$, $\left(\dfrac{\pi}{4}, 1\right)$

In Problems 15–18, find the point of the curve at which the curvature is a maximum.

15. $y = \ln x$

16. $y = e^x$

17. $y = \sin x$; $-\pi \le x \le \pi$

18. $y = \cosh x$

In Problems 19–22, find the tangential and normal components (a_T and a_N) of the acceleration vector at t. Then evaluate at $t = t_1$. See Examples 6 and 7.

19. $\mathbf{r}(t) = t^2\mathbf{i} + t\mathbf{j}$; $t_1 = 1$

20. $\mathbf{r}(t) = (2t + 1)\mathbf{i} + (t^2 - 2)\mathbf{j}$; $t_1 = -1$

21. $\mathbf{r}(t) = a \cos t\mathbf{i} + a \sin t\mathbf{j}$; $t_1 = \dfrac{\pi}{6}$

22. $\mathbf{r}(t) = 4 \cos^3 t\mathbf{i} + 4 \sin^3 t\mathbf{j}$; $t_1 = \dfrac{\pi}{3}$

23. Deduce from the definition

$$\mathbf{N} = \frac{d\mathbf{T}/ds}{|d\mathbf{T}/ds|}$$

the formula for $\mathbf{N}$ in terms of an arbitrary parameter t:

$$\mathbf{N} = \frac{d\mathbf{T}/dt}{|d\mathbf{T}/dt|}$$

24. Use the formula of Problem 23 to find $\mathbf{N}$ for the ellipse:

$$\mathbf{r}(t) = a \cos \omega t\mathbf{i} + b \sin \omega t\mathbf{j}$$

25. Sketch the path for a particle if its position vector is $\mathbf{r} = \sin t\mathbf{i} + \sin 2t\mathbf{j}$, $0 \le t \le 2\pi$ (you should get a figure eight). Where is the acceleration zero? Where does the acceleration vector point to the origin?

26. The position vector of a particle at time $t \ge 0$ is

$$\mathbf{r}(t) = (\cos t + t \sin t)\mathbf{i} + (\sin t - t \cos t)\mathbf{j}$$

(a) Show that the speed $ds/dt = t$.
(b) Show that $a_T = 1$ and $a_N = t$.

27. If for a particle, $a_T = 0$ for all t, what can you conclude about its speed? If $a_N = 0$ for all t, what can you conclude about its curvature?

28. Show that $\mathbf{N}$ points to the concave side of the curve. *Hint*: One method is to show

$$\mathbf{N} = (-\sin \phi\mathbf{i} + \cos \phi\mathbf{j}) \frac{d\phi/ds}{|d\phi/ds|}$$

Then consider the cases $d\phi/ds > 0$ (curve bends to the left) and $d\phi/ds < 0$ (curve bends to the right).

29. Show that $\mathbf{N}$ is perpendicular to $\mathbf{T}$ by differentiating $\mathbf{T} \cdot \mathbf{T} = 1$ with respect to s.

30. An object moves along the curve $y = \sin 2x$. Without doing any calculating, decide where $a_N = 0$.

31. A dog is running counterclockwise around the circle $x^2 + y^2 = 400$ (distances in feet). At the point $(-12, 16)$, it is running at 10 feet per second and is speeding up at 5 feet per second per second. Express its acceleration $\mathbf{a}$ at the point first in terms of $\mathbf{T}$ and $\mathbf{N}$ and then in terms of $\mathbf{i}$ and $\mathbf{j}$.

32. An object moves along the parabola $y = x^2$ with constant speed of 4. Express $\mathbf{a}$ at the point (x, x^2) in terms of $\mathbf{T}$ and $\mathbf{N}$.

33. A car traveling at constant speed v rounds a level curve, which we take to be a circle of radius R. If the car is to avoid sliding outward, the horizontal frictional force F exerted by the road on the tires must at least balance the centrifugal force pulling outward. The force F satisfies $F = \mu mg$, where μ is the *coefficient of friction*, m is the mass of the car, and g is the acceleration of gravity. Thus, $\mu mg \ge mv^2/R$. Show that v_R, the speed beyond which skidding will occur, satisfies

$$v_R = \sqrt{\mu g R}$$

and use this to determine v_R for a curve with $R = 400$ feet and $\mu = 0.4$. Use $g = 32$ feet per second per second.

34. Consider again the car of Problem 33. Suppose the curve is icy at its worst spot ($\mu = 0$) but is banked at angle θ from the horizontal (Figure 11). Let $\mathbf{F}$ be the force exerted

F: Force exerted by road

mv_R^2/R: Centifugal force

mg: Weight of car

FIGURE 11

by the road on the car. Then at the critical speed v_R, $mg = |\mathbf{F}|\cos\theta$ and $mv_R^2/R = |\mathbf{F}|\sin\theta$.

(a) Show that $v_R = \sqrt{Rg\tan\theta}$.

(b) Find v_R for a curve with $R = 400$ feet and $\theta = 10°$.

35. Derive the polar coordinate curvature formula

$$\kappa = \frac{|r^2 + 2r'^2 - rr''|}{(r^2 + r'^2)^{3/2}}$$

(derivatives are with respect to θ) and use it to find κ for:

(a) the circle $r = 4\cos\theta$;

(b) the cardioid $r = 1 + \cos\theta$ at $\theta = 0$.

36. Derive the first curvature formula in Theorem A by working directly with $\kappa = |\mathbf{T}'(t)|/|\mathbf{r}'(t)|$.

13.6 Chapter Review

TRUE-FALSE QUIZ

Respond with true or false to each of the following assertions. Be prepared to defend your answer.

1. The parametric representation of a curve is unique.
2. The graph of $x = 2t^3$, $y = t^3$ is a line.
3. If $x = f(t)$ and $y = g(t)$, then we can find a function h such that $y = h(x)$.
4. The curve with parametric representation $x = \ln t$ and $y = t^2 - 1$ passes through the origin.
5. If $x = f(t)$ and $y = g(t)$ and if both f'' and g'' exist, then $d^2y/dx^2 = g''(t)/f''(t)$ wherever $f''(t) \neq 0$.
6. A curve may have more than one tangent line at a point on the curve.
7. The vectors $2\mathbf{i} - 3\mathbf{j}$ and $6\mathbf{i} + 4\mathbf{j}$ are perpendicular.
8. If $\mathbf{u}$ and $\mathbf{v}$ are unit vectors, then the angle θ between them satisfies $\cos\theta = \mathbf{u}\cdot\mathbf{v}$.
9. The dot product for vectors satisfies the associative law.
10. If $\mathbf{u}$ and $\mathbf{v}$ are any two vectors, then $|\mathbf{u}\cdot\mathbf{v}| \leq |\mathbf{u}|\,|\mathbf{v}|$.
11. $|\mathbf{u}\cdot\mathbf{v}| = |\mathbf{u}|\,|\mathbf{v}|$ for nonzero vectors $\mathbf{u}$ and $\mathbf{v}$ if and only if $\mathbf{u}$ is a scalar multiple of $\mathbf{v}$.
12. If $|\mathbf{u}| = |\mathbf{v}| = |\mathbf{u} + \mathbf{v}|$, then $\mathbf{u} = \mathbf{v} = \mathbf{0}$.
13. If $\mathbf{u} + \mathbf{v}$ and $\mathbf{u} - \mathbf{v}$ are perpendicular, then $|\mathbf{u}| = |\mathbf{v}|$.
14. For any two vectors $\mathbf{u}$ and $\mathbf{v}$, $|\mathbf{u} + \mathbf{v}|^2 = |\mathbf{u}|^2 + |\mathbf{v}|^2 + 2\mathbf{u}\cdot\mathbf{v}$.
15. The vector-valued function $\langle f(t), g(t)\rangle$ is continuous at $t = a$ if and only if both f and g are continuous at $t = a$.
16. $D_t[\mathbf{F}(t)\cdot\mathbf{F}(t)] = 2\mathbf{F}(t)\cdot\mathbf{F}'(t)$.
17. The curvature of the curve determined by $x = 3t + 4$ and $y = 2t - 1$ is zero for all t.
18. The curvature of the curve determined by $x = 2\cos t$ and $y = 2\sin t$ is 2 for all t.
19. If $\mathbf{T} = \mathbf{T}(t)$ is a unit tangent vector to a smooth curve, then $\mathbf{T}(t)$ and $\mathbf{T}'(t)$ are perpendicular.
20. If $v = |\mathbf{v}|$ is the speed of a particle moving along a smooth curve, then $|dv/dt|$ is the magnitude of the acceleration.

SAMPLE TEST PROBLEMS

In Problems 1–4, a parametric representation of a curve is given. Eliminate the parameter to obtain the corresponding Cartesian equation. Sketch the given curve.

1. $x = 6t + 2$, $y = 2t$; $-\infty < t < \infty$

2. $x = 4t^2$, $y = 4t$; $-1 \leq t \leq 2$

3. $x = 4\sin t - 2$, $y = 3\cos t + 1$; $0 \leq t \leq 2\pi$

4. $x = 2\sec t$, $y = \tan t$; $-\dfrac{\pi}{2} < t < \dfrac{\pi}{2}$

In Problems 5 and 6, find the equation of the tangent line and the normal line at $t = 0$.

5. $x = 2t^3 - 4t + 7$, $y = t + \ln(t + 1)$

6. $x = 3e^{-t}$, $y = \frac{1}{2}e^t$

7. Find the length of the curve

$$x = \cos t + t\sin t$$
$$y = \sin t - t\cos t$$

from 0 to 2π. Make a sketch.

8. Find $\mathbf{u}$ and $|\mathbf{u}|$ if $\mathbf{u}$ is the vector from P_1 to P_2.

(a) $P_1 = (2,4)$, $P_2 = (-1,5)$

(b) $P_1 = (-3,0)$, $P_2 = (-4,5)$

9. Let $\mathbf{a} = \langle 2, -5 \rangle$, $\mathbf{b} = \langle 1,1 \rangle$, and $\mathbf{c} = \langle -6, 0 \rangle$. Find each of the following.

(a) $3\mathbf{a} - 2\mathbf{b}$

(b) $\mathbf{a} \cdot \mathbf{b}$

(c) $\mathbf{a} \cdot (\mathbf{b} + \mathbf{c})$

(d) $(4\mathbf{a} + 5\mathbf{b}) \cdot 3\mathbf{c}$

(e) $|\mathbf{c}|\mathbf{c} \cdot \mathbf{b}$

(f) $\mathbf{c} \cdot \mathbf{c} - |\mathbf{c}|$

10. Find the cosine of the angle between $\mathbf{a}$ and $\mathbf{b}$ and make a sketch.

(a) $\mathbf{a} = 3\mathbf{i} + 2\mathbf{j}$, $\mathbf{b} = -\mathbf{i} + 4\mathbf{j}$

(b) $\mathbf{a} = -5\mathbf{i} - 3\mathbf{j}$, $\mathbf{b} = 2\mathbf{i} - \mathbf{j}$

(c) $\mathbf{a} = \langle 7, 0 \rangle$, $\mathbf{b} = \langle 5, 1 \rangle$

11. Given $\mathbf{a} = -2\mathbf{i}$ and $\mathbf{b} = 3\mathbf{i} - 2\mathbf{j}$ and another vector $\mathbf{r} = 5\mathbf{i} - 4\mathbf{j}$, find scalars k and m such that $\mathbf{r} = k\mathbf{a} + m\mathbf{b}$.

12. Find a vector of length 3 which is parallel to the tangent line to $y = x^2$ at $(-1, 1)$.

13. Find a vector of length 10 that makes an angle of $150°$ with the positive x-axis.

14. Two forces $\mathbf{F}_1 = 2\mathbf{i} - 3\mathbf{j}$ and $\mathbf{F}_2 = 3\mathbf{i} + 12\mathbf{j}$ are applied at a point. What force $\mathbf{F}$ must be applied at the point to counteract the resultant of these two forces?

15. What heading and air speed are required for a plane to fly 450 miles per hour due north if a wind of 100 miles per hour is blowing in the direction N $60°$ E?

16. If $\mathbf{r}(t) = \langle e^{2t}, e^{-t} \rangle$ find each of the following.

(a) $\lim\limits_{t \to 0} \mathbf{r}(t)$

(b) $\lim\limits_{h \to 0} \dfrac{\mathbf{r}(0 + h) - \mathbf{r}(0)}{h}$

(c) $\displaystyle\int_0^{\ln 2} \mathbf{r}(t)\, dt$

(d) $D_t[\mathrm{tr}(t)]$

(e) $D_t[\mathbf{r}(3t + 10)]$

(f) $D_t[\mathbf{r}(t) \cdot \mathbf{r}'(t)]$

17. Find $\mathbf{r}'(t)$ and $\mathbf{r}''(t)$ for each of the following.

(a) $\mathbf{r}(t) = (\ln t)\mathbf{i} - 3t^2\mathbf{j}$

(b) $\mathbf{r}(t) = \sin t\,\mathbf{i} + \cos 2t\,\mathbf{j}$

(c) $\mathbf{r}(t) = \tan t\,\mathbf{i} - t^4\mathbf{j}$

18. Find the length of the arc $\mathbf{r}(t) = 4t^{3/2}\mathbf{i} + 3t\mathbf{j}$ from $t = 0$ to $t = 2$.

In Problems 19 and 20, the position of a moving particle at time t is given by $\mathbf{r}(t)$. Find the velocity and acceleration vectors, $\mathbf{v}(t)$ and $\mathbf{a}(t)$, and their values at the given time $t = t_1$. Also, find the speed of the particle for $t = t_1$.

19. $\mathbf{r}(t) = 2t^2\mathbf{i} + (4t + 2)\mathbf{j}$; $t_1 = -1$

20. $\mathbf{r}(t) = 4(1 - \sin t)\mathbf{i} + 4(t + \cos t)\mathbf{j}$; $t_1 = \frac{2}{3}\pi$.

21. Find the curvature κ of the given curve at P.

(a) $y = x^2 - x$ at $P(1, 0)$

(b) $\mathbf{r}(t) = (t + t^3)\mathbf{i} + (t + t^2)\mathbf{j}$ at $P(2, 2)$

(c) $y = a \cosh(x/a)$ at $P(a, a \cosh 1)$

22. Find the unit tangent vector $\mathbf{T}(t)$ for the curve $\mathbf{r}(t) = t\mathbf{i} + \frac{1}{3}t^3\mathbf{j}$. At the point $P(1)$ on the curve where $t = 1$, find $\mathbf{T}(1)$. Find the curvature $\kappa(1)$ of the curve at $P(1)$. Sketch the curve and draw the unit tangent vector $\mathbf{T}(1)$ with its initial point at $P(1)$.

23. If $\mathbf{r}(t) = (1 - t^2)\mathbf{i} + 2t\mathbf{j}$, find the tangential and the normal components, a_T and a_N, of the acceleration $\mathbf{a}$ at $P(0, 2)$.

14 Geometry in Space, Vectors

Such is the advantage of a well-constructed language that its simplified notation often becomes the source of profound theories.
P. S. Laplace

Josiah Willard Gibbs
1839–1903

With Josiah Willard Gibbs, we introduce the only American in our galaxy of stars who contributed to the development of calculus. American students are often surprised to learn that there was no significant mathematical activity in the western hemisphere until the late 1800s. It is true that the *American Journal of Mathematics* began publication in 1878, but its founding was due to the efforts of J. J. Sylvester, a British citizen who taught for a few years at Johns Hopkins University. In fact until after World War I, most American mathematicians either had emigrated from Europe or had received some of their training there.

Gibbs was born in New Haven, Connecticut, graduated from Yale, and studied in Paris, Berlin, and Heidelberg. He was offered a professorship of mathematical physics at Yale, the first such appointment in America, and a position which carried no salary for 10 years. Gibbs is best known for his work in applying thermodynamics to chemistry, but his place on our list rests on his use of vector methods. Beginning about 1880, he developed the symbolism and algebra of vectors. In 1901, the full treatment of his ideas was presented by one of his students, E. B. Wilson, in an influential book titled *Vector Analysis*. We explore Gibbs's ideas here and more fully in Chapter 17.

14.1
Cartesian Coordinates in Three-Space

We have reached an important transition point in our study of calculus. Until now, we have been traveling across that broad flat expanse known as the Euclidean plane, or two-space. The concepts of calculus have been applied to functions of a single variable, functions whose graphs can be drawn in the plane.

The mountains lie ahead. Our charted course winds through three-space and occasionally into *n*-space. We are going to study *multiple variable calculus*, the calculus that applies to functions of two or more variables. All the familiar ideas (such as limit, derivative, integral) are to be explored again from a loftier perspective.

To begin, consider three mutually perpendicular coordinate lines (the *x*-, *y*-, and *z*-axes) with their zero points at a common point *O*, called the *origin*. Although these lines can be oriented in any way one pleases, we follow a standard custom in thinking of the *y*- and *z*-axes as lying in the plane of the paper with their positive directions to the right and upward, respectively. The *x*-axis is then perpendicular to the paper, and we suppose its positive end to point toward us, thus forming a **right-handed system**. We call it right-handed because if the fingers of the right hand are cupped so they curve from the positive *x*-axis toward the positive *y*-axis, the thumb points in the direction of the positive *z*-axis (Figure 1).

The three axes determine three planes, the *yz*-, *xz*-, and *xy*-planes, which divide space into eight octants (Figure 2). To each point *P* in space corresponds an ordered triple of numbers (x, y, z), its **Cartesian coordinates**, which measure its directed distances from the three planes (Figure 3).

Left-handed system

Right-handed system

FIGURE 1

FIGURE 2

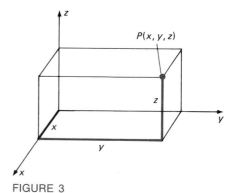

FIGURE 3

Plotting points in the first octant (the octant where all three coordinates are positive) is relatively easy. In Figures 4 and 5, we illustrate something more difficult by plotting two points from other octants, the points $P(2, -3, 4)$ and $Q(-3, 2, -5)$.

FIGURE 4   FIGURE 5

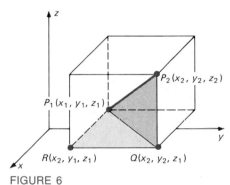

FIGURE 6

THE DISTANCE FORMULA Consider two points $P_1(x_1, y_1, z_1)$ and $P_2(x_2, y_2, z_2)$ in three-space ($x_1 \neq x_2$, $y_1 \neq y_2$, $z_1 \neq z_2$). They determine a rectangular box (parallelepiped), with P_1 and P_2 as opposite vertices and with edges parallel to the coordinate axes (Figure 6). The triangles P_1RQ and P_1QP_2 are right triangles, and—by the Pythagorean Theorem—

$$|P_1P_2|^2 = |P_1Q|^2 + |QP_2|^2$$

and

$$|P_1Q|^2 = |P_1R|^2 + |RQ|^2$$

Thus,

$$|P_1P_2|^2 = |P_1R|^2 + |RQ|^2 + |QP_2|^2$$
$$= (x_2 - x_1)^2 + (y_2 - y_1)^2 + (z_1 - z_2)^2$$

This gives us the **Distance Formula** in three-space.

$$\boxed{|P_1P_2| = \sqrt{(x_2 - x_1)^2 + (y_2 - y_1)^2 + (z_2 - z_1)^2}}$$

The formula is correct even if some of the coordinates are identical.

EXAMPLE 1 Find the distance between the points $P(2, -3, 4)$ and $Q(-3, 2, -5)$, which were plotted in Figures 4 and 5.

FIGURE 7

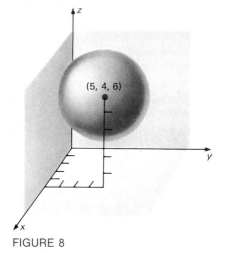

FIGURE 8

$$|PQ| = \sqrt{(-3-2)^2 + (2+3)^2 + (-5-4)^2} = \sqrt{131} \approx 11.45 \quad \blacksquare$$

SPHERES AND THEIR EQUATIONS It is a small step from the distance formula to the equation of a sphere. By a **sphere**, we mean the set of all points at constant distance (the radius) from a fixed point (the center). In fact, if (x, y, z) is a point on the sphere of radius r centered at (h, k, l), then (see Figure 7)

$$\boxed{(x-h)^2 + (y-k)^2 + (z-l)^2 = r^2}$$

We call this the **standard equation of a sphere**.

In expanded form, the boxed equation may be written as

$$x^2 + y^2 + z^2 + Gx + Hy + Iz + J = 0$$

Conversely, the graph of any equation of this form is either a sphere, a point (a degenerate sphere), or the empty set. To see why, consider the following example.

EXAMPLE 2 Find the center and radius of the sphere with equation

$$x^2 + y^2 + z^2 - 10x - 8y - 12z + 68 = 0$$

and sketch its graph.

Solution We use the process of completing the square.

$$(x^2 - 10x + \quad) + (y^2 - 8y + \quad) + (z^2 - 12z + \quad) = -68$$
$$(x^2 - 10x + 25) + (y^2 - 8y + 16) + (z^2 - 12z + 36) =$$
$$-68 + 25 + 16 + 36$$
$$(x-5)^2 + (y-4)^2 + (z-6)^2 = 9$$

Thus, the equation represents a sphere with center at $(5, 4, 6)$ and radius 3. Its graph is shown in Figure 8. $\quad \blacksquare$

If, after completing the square in Example 2, the equation had been

$$(x-5)^2 + (y-4)^2 + (z-6)^2 = 0$$

then the graph would be the single point $(5, 4, 6)$; if the right side were negative, the graph would be the empty set.

Another simple result that follows from the Distance Formula is the **Midpoint Formula**. If $P_1(x_1, y_1, z_1)$ and $P_2(x_2, y_2, z_2)$ are endpoints of a line segment, then the midpoint $M(m_1, m_2, m_3)$ has coordinates

$$m_1 = \frac{x_1 + x_2}{2} \qquad m_2 = \frac{y_1 + y_2}{2} \qquad m_3 = \frac{z_1 + z_2}{2}$$

In other words, to find the coordinates of the midpoint of a segment, simply take the average of corresponding coordinates of the endpoints.

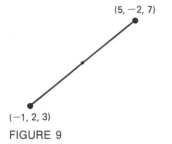

(5, −2, 7)

(−1, 2, 3)

FIGURE 9

EXAMPLE 3 Find the equation of the sphere that has the line segment joining $(-1, 2, 3)$ and $(5, -2, 7)$ as a diameter (Figure 9).

Solution The center of this sphere is at the midpoint of the segment, namely, at $(2, 0, 5)$; the radius r satisfies

$$r^2 = (5 - 2)^2 + (-2 - 0)^2 + (7 - 5)^2 = 17$$

We conclude that the equation of the sphere is

$$(x - 2)^2 + y^2 + (z - 5)^2 = 17 \qquad \blacksquare$$

GRAPHS IN THREE-SPACE It was natural to consider a quadratic equation first because of its relation to the distance formula. But presumably a **linear equation**—that is, an equation of the form

$$Ax + By + Cz = D, \qquad A^2 + B^2 + C^2 \neq 0$$

should be even easier to analyze. As a matter of fact we will show in the next section that the graph of a linear equation is a plane. Taking this for granted for now, let's consider how we might graph such an equation.

If, as will often be the case, the plane intersects the three axes, we begin by finding these intersection points; that is, we find the x-, y-, and z-intercepts. These three points determine the plane and allow us to draw the (coordinate plane) **traces**, which are the lines of intersection with the coordinate planes. Then with just a bit of artistry, we can shade in the plane.

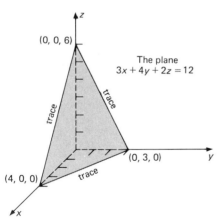

(0, 0, 6)

The plane
$3x + 4y + 2z = 12$

trace

trace

(0, 3, 0)

(4, 0, 0)

trace

x

y

z

FIGURE 10

EXAMPLE 4 Sketch the graph of $3x + 4y + 2z = 12$.

Solution To find the x-intercept, set y and z equal to zero and solve for x, obtaining $x = 4$. The corresponding point is $(4, 0, 0)$. Similarly, the y- and z-intercepts are $(0, 3, 0)$ and $(0, 0, 6)$. Next connect these points by line segments to get the traces. Then shade in (the first octant part of) the plane, thereby obtaining the result shown in Figure 10. $\blacksquare$

What if the plane does not intersect all three axes? This will happen, for example, if one of the variables in the equation of the plane is missing (that is, has a zero coefficient).

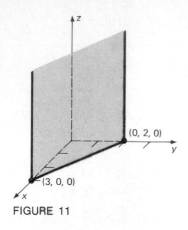

FIGURE 11

EXAMPLE 5 Sketch the graph of the linear equation

$$2x + 3y = 6$$

in three-space.

Solution The x- and y-intercepts are $(3, 0, 0)$ and $(0, 2, 0)$, respectively, and these points determine the trace in the xy-plane. The plane never crosses the z-axis (x and y cannot both be 0), and so the plane is parallel to the z-axis. We have sketched the graph in Figure 11. ■

Notice that in each of our examples, the graph of an equation in three-space was a surface. This contrasts with the two-space case, where the graph of an equation was usually a curve. We will have a good deal more to say about graphing equations and the corresponding surfaces in Section 14.6.

PROBLEM SET 14.1

1. Plot the points whose coordinates are $(3, 2, 4)$, $(2, 0, 3)$, $(-3, 4, 5)$, $(0, 4, 0)$, and $(-2, -6, -1)$. If appropriate, show the "box," as in Figures 4 and 5.

2. Follow the directions of Problem 1 for $(\sqrt{2}, -2, 2)$, $(0, \pi, -7)$, $(-3, \frac{1}{2}, 2)$, and $(0, 0, 3)$.

3. What is peculiar to the coordinates of all points in the xz-plane? On the y-axis?

4. What is peculiar to the coordinates of all points in the yz-plane? On the z-axis?

5. Find the distance between the following pairs of points.

(a) $(6, -2, 0)$ and $(2, 3, 7)$

(b) $(-1, 3, 2)$ and $(4, 0, -5)$

(c) $(\sqrt{2}, 0, \sqrt{3})$ and $(0, 2, 0)$

6. Show that $(4, 5, 2)$, $(1, 7, 3)$, and $(2, 4, 5)$ are vertices of an equilateral triangle.

7. Show that $(1, 0, 5)$, $(3, 6, 8)$, and $(7, 4, -7)$ are vertices of a right triangle. *Hint:* Only right triangles satisfy the Pythagorean Theorem.

8. Find the distance from $(3, 4, -2)$ to (a) the xy-plane, (b) the y-axis, and (c) the origin.

9. A rectangular box has its faces parallel to the coordinate planes and has $(2, 3, 4)$ and $(5, -2, 0)$ as endpoints of a main diagonal. Sketch the box and find the coordinates of all eight vertices.

10. $P(x, 4, z)$ is on a line through $Q(2, -5, 3)$ that is parallel to one of the coordinate axes. Which axis must it be, and what are x and z?

11. Write the equation of the sphere with the given center and radius.

(a) $(3, 1, 4)$; 5

(b) $(-6, 2, -3)$; 2

(c) $(-1, 0, 4)$; $\sqrt{6}$

12. Find the equation of the sphere whose center is $(2, 4, 5)$ and which is tangent to the xy-plane.

In Problems 13–16, complete the squares to find the center and radius of the sphere whose equation is given (see Example 2).

13. $x^2 + y^2 + z^2 - 12x + 14y - 8z + 1 = 0$

14. $x^2 + y^2 + z^2 + 2x - 6y - 10z + 34 = 0$

15. $4x^2 + 4y^2 + 4z^2 - 4x + 8y + 16z - 13 = 0$

16. $x^2 + y^2 + z^2 + 8x - 4y - 22z + 77 = 0$

In Problems 17–24, sketch the graphs of the given equations. Begin by sketching the traces in the coordinate planes (see Examples 4 and 5).

17. $2x + 6y + 3z = 12$

18. $3x - 4y + 2z = 24$

19. $x + 3y - z = 6$

20. $-3x + 2y + z = 6$

21. $x + 3y = 8$

22. $3x + 4z = 12$

23. $x^2 + y^2 + z^2 = 9$

24. $(x - 2)^2 + y^2 + z^2 = 4$

25. Find the equation of the sphere that has the line segment joining $(-2, 3, 6)$ and $(4, -1, 5)$ as a diameter (see Example 3).

26. Find the equations of the tangent spheres of equal radii whose centers are $(-3, 1, 2)$ and $(5, -3, 6)$.

27. Find the equation of the sphere that is tangent to the three coordinate planes, if its radius is 6 and its center is in the first octant.

28. Find the equation of the sphere with center $(1, 1, 4)$ that is tangent to the plane $x + y = 12$.

29. Describe the graph in 3-space of each of the following equations.

(a) $z = 2$

(b) $x = y$

(c) $xy = 0$

(d) $xyz = 0$

(e) $x^2 + y^2 = 4$

(f) $z = \sqrt{9 - x^2 - y^2}$

30. The sphere $(x - 1)^2 + (y + 2)^2 + (z + 1)^2 = 10$ intersects the plane $z = 2$ in a circle. Find its center and radius.

31. A point P moves so that its distance from $(1, 2, -3)$ is twice its distance from $(1, 2, 3)$. Show that P is on a sphere and find its center and radius.

32. A point P moves so that its distance from $(1, 2, -3)$ equals its distance from $(2, 3, 2)$. Find the equation of the plane on which P lies.

33. The balls $(x - 1)^2 + (y - 2)^2 + (z - 1)^2 \le 4$ and $(x - 2)^2 + (y - 4)^2 + (z - 3)^2 \le 4$ intersect in a solid. Find its surface area. *Hint*: See Problem 20 of Section 6.5.

34. Do Problem 33 assuming the second ball is $(x - 2)^2 + (y - 4)^2 + (z - 3)^2 \le 9$.

14.2
Vectors in
Three-Space

The material from Sections 13.2 and 13.3 on vectors in the plane can be repeated almost word for word for vectors in space. About the only difference is that a vector $\mathbf{u}$ now has three components; that is,

$$\mathbf{u} = \langle u_1, u_2, u_3 \rangle = u_1\mathbf{i} + u_2\mathbf{j} + u_3\mathbf{k}$$

Here $\mathbf{i}$, $\mathbf{j}$, and $\mathbf{k}$ are the standard unit vectors, called **basis vectors**, in the directions of the three positive coordinate axes (Figure 1). The **length** of $\mathbf{u}$, denoted by $|\mathbf{u}|$, is from the distance formula given by

$$\boxed{|\mathbf{u}| = \sqrt{u_1^2 + u_2^2 + u_3^2}}$$

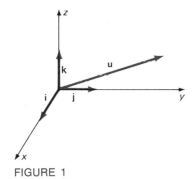

FIGURE 1

Vectors are added, multiplied by scalars, and subtracted just as in the plane, and the algebraic laws that are satisfied agree with those studied earlier. The **dot product** of $\mathbf{u} = \langle u_1, u_2, u_3 \rangle$ and $\mathbf{v} = \langle v_1, v_2, v_3 \rangle$ is defined by

$$\boxed{\mathbf{u} \cdot \mathbf{v} = u_1v_1 + u_2v_2 + u_3v_3}$$

and it has the geometric interpretation noted earlier, namely,

$$\boxed{\mathbf{u} \cdot \mathbf{v} = |\mathbf{u}|\,|\mathbf{v}|\cos\theta}$$

where θ is the angle between $\mathbf{u}$ and $\mathbf{v}$. Consequently, it continues to be true that two vectors are perpendicular if and only if their dot product is zero.

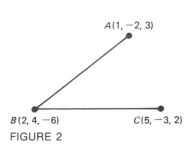

FIGURE 2

EXAMPLE 1 Find the angle ABC if $A = (1, -2, 3)$, $B = (2, 4, -6)$, and $C = (5, -3, 2)$ (Figure 2).

Solution First we determine vectors **u** and **v** (emanating from the origin) equivalent to $\overrightarrow{BA}$ and $\overrightarrow{BC}$. This is done by subtracting the coordinates of the initial points from those of the terminal points—that is,

$$\mathbf{u} = \langle 1 - 2, -2 - 4, 3 + 6 \rangle = \langle -1, -6, 9 \rangle$$
$$\mathbf{v} = \langle 5 - 2, -3 - 4, 2 + 6 \rangle = \langle 3, -7, 8 \rangle$$

Thus,

$$\cos \theta = \frac{\mathbf{u} \cdot \mathbf{v}}{|\mathbf{u}|\,|\mathbf{v}|} = \frac{(-1)(3) + (-6)(-7) + (9)(8)}{\sqrt{1 + 36 + 81}\,\sqrt{9 + 49 + 64}} \approx 0.9251$$

$$\theta \approx 0.3894 \quad \text{(about } 22.31°\text{)} \qquad \blacksquare$$

EXAMPLE 2 Express $\mathbf{u} = \langle 2, 4, 5 \rangle$ as the sum of a vector **m** parallel to $\mathbf{v} = \langle 2, -1, -2 \rangle$ and a vector **n** perpendicular to **v**.

Solution Figure 3 tells the story. First, we find $\mathbf{m} = \mathbf{pr_v u}$, the projection of **u** on **v** (see Problem 33 of Section 13.3).

$$\mathbf{m} = |\mathbf{m}| \frac{\mathbf{v}}{|\mathbf{v}|} = |\mathbf{u}| \cos \theta \frac{\mathbf{v}}{|\mathbf{v}|}$$

$$= \frac{\mathbf{u} \cdot \mathbf{v}}{|\mathbf{v}|} \frac{\mathbf{v}}{|\mathbf{v}|} = \frac{\mathbf{u} \cdot \mathbf{v}}{|\mathbf{v}|^2} \mathbf{v}$$

$$= \frac{(2)(2) + (4)(-1) + (5)(-2)}{4 + 1 + 4} \langle 2, -1, -2 \rangle$$

$$= \left\langle \frac{-20}{9}, \frac{10}{9}, \frac{20}{9} \right\rangle$$

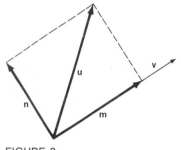

FIGURE 3

Then,

$$\mathbf{n} = \mathbf{u} - \mathbf{m} = \left\langle \frac{38}{9}, \frac{26}{9}, \frac{25}{9} \right\rangle$$

If you doubt that **m** and **n** are perpendicular, compute their dot product. You will get zero. $\qquad \blacksquare$

DIRECTION ANGLES AND COSINES The (smallest nonnegative) angles between a nonzero vector **a** and the basis vectors **i**, **j**, and **k** are the **direction angles** of **a**; they are designated by α, β, and γ, respectively (Figure 4). It is generally more convenient to work with the **direction cosines** $\cos \alpha$, $\cos \beta$, and $\cos \gamma$. If $\mathbf{a} = a_1\mathbf{i} + a_2\mathbf{j} + a_3\mathbf{k}$, then

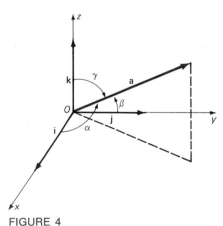

FIGURE 4

$$\cos \alpha = \frac{\mathbf{a} \cdot \mathbf{i}}{|\mathbf{a}|\,|\mathbf{i}|} = \frac{a_1}{|\mathbf{a}|}$$

and similarly

$$\cos \beta = \frac{a_2}{|\mathbf{a}|}, \qquad \cos \gamma = \frac{a_3}{|\mathbf{a}|}$$

Notice that

$$\cos^2\alpha + \cos^2\beta + \cos^2\gamma = 1$$

In fact, the vector $\langle \cos\alpha, \cos\beta, \cos\gamma \rangle$ is a unit vector with the same direction as the original vector $\mathbf{a}$.

EXAMPLE 3 Find the direction angles for the vector $\mathbf{a} = 4\mathbf{i} - 5\mathbf{j} + 3\mathbf{k}$.

Solution Since $|\mathbf{a}| = \sqrt{4^2 + (-5)^2 + 3^2} = 5\sqrt{2}$,

$$\cos\alpha = \frac{4}{5\sqrt{2}} = \frac{2\sqrt{2}}{5}, \qquad \cos\beta = \frac{-\sqrt{2}}{2}, \qquad \cos\gamma = \frac{3\sqrt{2}}{10}$$

and

$$\alpha \approx 55.55°, \qquad \beta = 135°, \qquad \gamma \approx 64.90° \qquad \blacksquare$$

EXAMPLE 4 Find a vector 5 units long that has $\alpha = 32°$ and $\beta = 100°$ as two of its direction angles.

Solution First we note that the third direction angle γ must satisfy

$$\cos^2\gamma = 1 - \cos^2 32° - \cos^2 100° \approx 0.25066$$

Thus,

$$\cos\gamma \approx \pm 0.50066$$

Two vectors meet the requirements of the problem. They are

$$5\langle \cos\alpha, \cos\beta, \cos\gamma \rangle \approx 5\langle 0.84805, -0.17365, 0.50066 \rangle$$
$$= \langle 4.2403, -0.8683, 2.5033 \rangle$$

and $\langle 4.2403, -0.8683, -2.5033 \rangle$. $\qquad \blacksquare$

PLANES One fruitful way to describe a plane is by using vector language. Let $\mathbf{n} = \langle A, B, C \rangle$ be a fixed nonzero vector and $P_1(x_1, y_1, z_1)$ be a fixed point. The set of points $P(x, y, z)$ satisfying $\overrightarrow{P_1P} \cdot \mathbf{n} = 0$ is the **plane** through P_1 perpendicular to $\mathbf{n}$. Since every plane contains a point and is perpendicular to some vector, a plane can be characterized in this way.

To get the Cartesian equation of the plane, write the vector $\overrightarrow{P_1P}$ in component form, that is,

$$\overrightarrow{P_1P} = \langle x - x_1, y - y_1, z - z_1 \rangle$$

Then, $\overrightarrow{P_1P} \cdot \mathbf{n} = 0$ is equivalent to

$$A(x - x_1) + B(y - y_1) + C(z - z_1) = 0$$

This equation (in which at least one of A, B, C is different from zero) is called the **standard form for the equation of a plane**.

If we remove the parentheses and simplify, the boxed equation takes the form of the general linear equation

$$Ax + By + Cz = D, \qquad A^2 + B^2 + C^2 \neq 0$$

Thus, every plane has a linear equation. Conversely, the graph of a linear equation in three-space is always a plane. To see the latter, let (x_1, y_1, z_1) satisfy the equation, that is,

$$Ax_1 + By_1 + Cz_1 = D$$

When we subtract this equation from the one above, we have the boxed equation, which we know represents a plane.

EXAMPLE 5 Find the equation of the plane through $(5, 1, -2)$ perpendicular to $\mathbf{n} = \langle 2, 4, 3 \rangle$. Then find the angle between this plane and the one with equation $3x - 4y + 7z = 5$.

Solution To perform the first task, simply apply the boxed result to the problem at hand. We obtain

$$2(x - 5) + 4(y - 1) + 3(z + 2) = 0$$

or, equivalently,

$$2x + 4y + 3z = 8$$

A vector $\mathbf{m}$ perpendicular to the second plane is $\mathbf{m} = \langle 3, -4, 7 \rangle$. The angle θ between two planes is the angle between their normals (Figure 5). Thus,

$$\cos \theta = \frac{\mathbf{m} \cdot \mathbf{n}}{|\mathbf{m}| \, |\mathbf{n}|} = \frac{(3)(2) + (-4)(4) + (7)(3)}{\sqrt{9 + 16 + 49}\sqrt{4 + 16 + 9}} \approx 0.2375$$

$$\theta = 76.26°$$

Actually there are two angles between two planes, but they are supplementary. The process just described will lead to one of them. The other, if desired, is obtained by subtracting the first value from 180°. In our case, it would be 103.74°. ∎

FIGURE 5

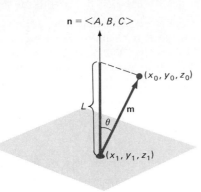

n = <A, B, C>

L

θ

m

(x_0, y_0, z_0)

(x_1, y_1, z_1)

FIGURE 6

EXAMPLE 6 Show that the distance L from the point (x_0, y_0, z_0) to the plane $Ax + By + Cz = D$ is given by the formula

$$L = \frac{|Ax_0 + By_0 + Cz_0 - D|}{\sqrt{A^2 + B^2 + C^2}}$$

Solution Let (x_1, y_1, z_1) be a point on the plane and let $\mathbf{m} = \langle x_0 - x_1, y_0 - y_1, z_0 - z_1 \rangle$ be the vector from (x_1, y_1, z_1) to (x_0, y_0, z_0), as in Figure 6. Now $\mathbf{n} = \langle A, B, C \rangle$ is a vector perpendicular to the given plane, though it might point in the opposite direction of that in our figure. The number L that we seek is the length of the projection of $\mathbf{m}$ on $\mathbf{n}$. Thus,

$$L = \big\|\mathbf{m}\big|\cos \theta\big| = \frac{|\mathbf{m} \cdot \mathbf{n}|}{|\mathbf{n}|}$$

$$= \frac{|A(x_0 - x_1) + B(y_0 - y_1) + C(z_0 - z_1)|}{\sqrt{A^2 + B^2 + C^2}}$$

$$= \frac{|Ax_0 + By_0 + Cz_0 - (Ax_1 + By_1 + Cz_1)|}{\sqrt{A^2 + B^2 + C^2}}$$

But (x_1, y_1, z_1) is on the plane and so

$$Ax_1 + By_1 + Cz_1 = D$$

Substitution of this result in the expression for L yields the desired formula. ∎

FIGURE 7

EXAMPLE 7 Find the distance between the parallel planes $3x - 4y + 5z = 9$ and $3x - 4y + 5z = 4$.

Solution The planes are parallel, since the vector $\langle 3, -4, 5 \rangle$ is perpendicular to both of them (Figure 7). The point $(1, 1, 2)$ is easily seen to be on the first plane. We find the distance L from $(1, 1, 2)$ to the second plane using the formula of Example 6.

$$L = \frac{|3(1) - 4(1) + 5(2) - 4|}{\sqrt{9 + 16 + 25}} = \frac{5}{5\sqrt{2}} \approx 0.7071 \qquad ∎$$

PROBLEM SET 14.2

1. For each pair of points P_1 and P_2 given below, sketch the directed line segment $\overrightarrow{P_1 P_2}$ and then write the corresponding vector in the form $a\mathbf{i} + b\mathbf{j} + c\mathbf{k}$.

(a) $P_1(1, 2, 3)$, $P_2(4, 5, 1)$

(b) $P_1(-2, 1, 1)$, $P_2(4, 3, 5)$

2. Follow the directions of Problem 1.

(a) $P_1(-2, 2, -1)$, $P_2(3, 0, 4)$

(b) $P_1(0, 2, 0)$, $P_2(-3, 1, 6)$

3. Find the length of and direction cosines for each of the following vectors.

(a) $4\mathbf{i} + \mathbf{j} + 2\mathbf{k}$

(b) $-3\mathbf{i} - 4\mathbf{j} + 5\mathbf{k}$

4. Follow the directions of Problem 3.

(a) $\langle 4, -2, -4 \rangle$

(b) $\langle -1, -2, 2 \rangle$

5. Find a unit vector with the same direction as $\langle 3, 1, -7 \rangle$. Also, find a vector of length 5 oriented in the opposite direction.

6. Find a vector of length 10 with direction opposite to $-2\mathbf{i} + 5\mathbf{j} - 3\mathbf{k}$.

7. Find the angle between $\langle 4, -3, -2 \rangle$ and $\langle -1, 2, 5 \rangle$.

8. Find the angle between $3\mathbf{i} - \mathbf{j} - 2\mathbf{k}$ and $-6\mathbf{i} + 2\mathbf{j} + 4\mathbf{k}$.

9. Find two vectors of length 10, each of which is perpendicular to both $4\mathbf{i} + 3\mathbf{j} + 6\mathbf{k}$ and $-2\mathbf{i} - 3\mathbf{j} - 2\mathbf{k}$.

10. Find all vectors perpendicular to both $\langle 1, -3, -2 \rangle$ and $\langle -3, 6, 5 \rangle$.

11. Find angle ABC if $A = (2, -3, 4)$, $B = (-2, 6, 1)$, and $C = (2, 0, 2)$ (see Example 1).

12. Show that triangle ABC is a right triangle if $A = (4, 3, 5)$, $B = (1, 1, 1)$, and $C = (-1, 10, -2)$. *Hint*: Check the angle at B.

13. Find the *scalar projection* of $\mathbf{u} = -4\mathbf{i} + \mathbf{j} - 2\mathbf{k}$ on $\mathbf{v} = \mathbf{i} + 3\mathbf{j} - 3\mathbf{k}$. The scalar projection is the signed magnitude of the vector projection (Example 2), that is, it is $|\mathbf{u}|\cos \theta = \mathbf{u} \cdot \mathbf{v}/|\mathbf{v}|$.

14. Find the scalar projection of $\mathbf{u} = 2\mathbf{i} - 5\mathbf{j} + \mathbf{k}$ on $\mathbf{v} = -3\mathbf{i} + \mathbf{j} + 7\mathbf{k}$.

15. If $\mathbf{u} = -3\mathbf{i} + \mathbf{j} - \sqrt{5}\mathbf{k}$ and $\mathbf{v} = 2\mathbf{i} + 4\mathbf{j} - \sqrt{5}\mathbf{k}$, express $\mathbf{u}$ as the sum of a vector $\mathbf{m}$ parallel to $\mathbf{v}$ and a vector $\mathbf{n}$ perpendicular to $\mathbf{v}$ (see Example 2).

16. Follow the directions of Problem 15 for $\mathbf{u} = \mathbf{i} + 3\mathbf{j} - 2\mathbf{k}$ and $\mathbf{v} = 2\mathbf{i} - 5\mathbf{j} - 2\mathbf{k}$.

17. Find the direction angles for each vector.

(a) $\mathbf{u} = 3\mathbf{i} - 2\mathbf{j} + \mathbf{k}$ (b) $\mathbf{u} = -2\mathbf{i} + 4\mathbf{j} - 6\mathbf{k}$

18. If $\alpha = 46°$ and $\beta = 108°$ are direction angles for a vector $\mathbf{u}$, find two possible values for the third angle (see Example 4).

19. A vector $\mathbf{u} = 2\mathbf{i} + 3\mathbf{j} + z\mathbf{k}$ emanating from the origin points into the first octant. If $|\mathbf{u}| = 5$, find z.

20. If $\mathbf{u} = 2\mathbf{i} + 3\mathbf{j} + z\mathbf{k}$ and $\mathbf{v} = 2\mathbf{i} + 6\mathbf{j} - 3\mathbf{k}$ are perpendicular, find z.

21. Find two perpendicular vectors $\mathbf{u}$ and $\mathbf{v}$ such that each is also perpendicular to $\mathbf{w} = \langle -4, 2, 5 \rangle$.

22. Find the vector emanating from the origin whose terminal point is the midpoint of the segment joining $(3, 2, -1)$ and $(5, -7, 2)$.

23. Which of the following do *not* make sense?

(a) $\mathbf{u} \cdot (\mathbf{v} \cdot \mathbf{w})$ (b) $(\mathbf{u} \cdot \mathbf{w}) + \mathbf{w}$
(c) $|\mathbf{u}|(\mathbf{v} \cdot \mathbf{w})$ (d) $(\mathbf{u} \cdot \mathbf{v})\mathbf{w}$
(e) $(|\mathbf{u}|\mathbf{v}) \cdot \mathbf{w}$ (f) $|\mathbf{u}| \cdot \mathbf{v}$

24. Let $\mathbf{a}$, $\mathbf{b}$, $\mathbf{c}$, and $\mathbf{d}$ be vectors emanating from the origin and terminating at A, B, C, and D, respectively. Use vector notation to express a necessary and sufficient condition that the figure $ABCD$ be a parallelogram.

25. Find the equation of the plane passing through P and perpendicular to $\mathbf{n}$ (see Example 5).

(a) $P(1, 2, -3)$, $\mathbf{n} = 2\mathbf{i} - 4\mathbf{j} + 3\mathbf{k}$

(b) $P(-2, -3, 4)$, $\mathbf{n} = 3\mathbf{i} - 2\mathbf{j} - \mathbf{k}$

26. Find the smaller of the angles between the planes $3x - 2y + 5z = 7$ and $4x - 2y - 3z = 2$ (see Example 5).

27. Find the smaller of the angles between the two planes of Problem 25.

28. Find the equation of a plane through $(-1, 2, -3)$ and parallel to the plane $2x + 4y - z = 6$.

29. Find the equation of the plane through $(-4, -1, 2)$ and parallel (a) to the xy-plane, (b) to the plane $2x - 3y - 4z = 0$.

30. Find the distance from $(1, -1, 2)$ to the plane $x + 3y + z = 7$ (see Example 6).

31. Find the distance from $(2, 6, 3)$ to the plane $-3x + 2y + z = 9$.

32. Find the distance between the parallel planes $-3x + 2y + z = 9$ and $6x - 4y - 2z = 19$ (see Example 7).

33. Find the distance between the parallel planes $5x - 3y - 2z = 5$ and $-5x + 3y + 2z = 7$.

34. Find the equation of the plane, each of whose points is equidistant from $(-2, 1, 4)$ and $(6, 1, -2)$.

35. Prove $|\mathbf{u} + \mathbf{v}|^2 + |\mathbf{u} - \mathbf{v}|^2 = 2|\mathbf{u}|^2 + 2|\mathbf{v}|^2$. *Hint*: $|\mathbf{w}|^2 = \mathbf{w} \cdot \mathbf{w}$.

36. Prove $\mathbf{u} \cdot \mathbf{v} = \frac{1}{4}|\mathbf{u} + \mathbf{v}|^2 - \frac{1}{4}|\mathbf{u} - \mathbf{v}|^2$.

37. Find the angle between a main diagonal of a cube and one of its faces.

38. Find a unit vector whose direction angles are equal.

39. Find the smallest angle between the main diagonals of a rectangular box 4 feet by 6 feet by 10 feet.

40. A constant force of $\mathbf{F} = 3\mathbf{i} - 6\mathbf{j} + 7\mathbf{k}$ pounds is applied to an object in moving it from $(2, 1, 3)$ to $(9, 4, 6)$, coordinates given in feet. Find the work done. Recall that $W = \mathbf{F} \cdot \mathbf{D}$ (see Section 13.3).

41. How much work is done by a force of 5 newtons acting in the direction $2\mathbf{i} + 2\mathbf{j} - \mathbf{k}$ in moving an object from $(0, 1, 2)$ to $(3, 5, 7)$, distances being measured in meters? (See Problem 40.)

42. A weight of 30 pounds is suspended by 3 wires with resulting tensions $3\mathbf{i} + 4\mathbf{j} + 15\mathbf{k}$, $-8\mathbf{i} - 2\mathbf{j} + 10\mathbf{k}$, and

$a\mathbf{i} + b\mathbf{j} + c\mathbf{k}$. Determine a, b, and c, assuming $\mathbf{k}$ points straight up.

43. Find the point one-fifth of the way from $(2, 3, -1)$ to $(7, -2, 9)$.

44. Suppose the three coordinate planes bounding the first octant are mirrors. A light ray with direction $a\mathbf{i} + b\mathbf{j} + c\mathbf{k}$ is reflected successively from the xy-plane, the xz-plane, and the yz-plane. Determine the direction of the ray after each reflection and state a nice conclusion concerning the final reflected ray.

45. Find the distance from the sphere $x^2 + y^2 + z^2 + 2x + 6y - 8z = 0$ to the plane $3x + 4y + z = 15$.

46. Refine the method of Example 7 by showing that the distance L between the parallel planes $Ax + By + Cz = D$ and $Ax + By + Cz = E$ is

$$L = \frac{|D - E|}{\sqrt{A^2 + B^2 + C^2}}$$

47. Let $\mathbf{a} = \langle a_1, a_2, a_3 \rangle$ and $\mathbf{b} = \langle b_1, b_2, b_3 \rangle$ be fixed vectors. Show that $(\mathbf{x} - \mathbf{a}) \cdot (\mathbf{x} - \mathbf{b}) = 0$ is the equation of a sphere and find its center and radius.

48. Show that the work done by a constant force $\mathbf{F}$ on an object that moves completely around a closed polygonal path is 0.

49. The medians of a triangle meet in a point P (the centroid by Problem 24 of Section 6.8) that is two-thirds of the way from a vertex to the midpoint of the opposite edge. Show that P is the head of the position vector $(\mathbf{a} + \mathbf{b} + \mathbf{c})/3$, where $\mathbf{a}$, $\mathbf{b}$, and $\mathbf{c}$ are the position vectors of the vertices and use this to find P if the vertices are $(2, 6, 5)$, $(4, -1, 2)$ and $(6, 1, 2)$.

50. Let $\mathbf{a}$, $\mathbf{b}$, $\mathbf{c}$, and $\mathbf{d}$ be the position vectors of the vertices of a tetrahedron. Show that the lines joining the vertices to the centroids of the opposite faces meet in a point P and give a nice vector formula for it, thus generalizing Problem 49.

14.3
The Cross Product

The dot product of two vectors is a scalar. We have explored some of its uses in earlier sections. Now we introduce the **cross product** (or vector product); it will also have many uses. The cross product $\mathbf{u} \times \mathbf{v}$ of $\mathbf{u} = \langle u_1, u_2, u_3 \rangle$ and $\mathbf{v} = \langle v_1, v_2, v_3 \rangle$ is defined by

$$\mathbf{u} \times \mathbf{v} = \langle u_2 v_3 - u_3 v_2, \; u_3 v_1 - u_1 v_3, \; u_1 v_2 - u_2 v_1 \rangle$$

In this form, the formula is hard to remember and its significance is not obvious. Note the one thing that is obvious. The cross product of two vectors is a vector.

To help us remember the formula for the cross product, we recall a subject from an earlier mathematics course, namely, *determinants*. First, the value of a 2×2 determinant is

$$\begin{vmatrix} a & b \\ c & d \end{vmatrix} = ad - bc$$

Then the value of a 3×3 determinant is (expanding according to the top row)

$$\begin{vmatrix} a_1 & a_2 & a_3 \\ b_1 & b_2 & b_3 \\ c_1 & c_2 & c_3 \end{vmatrix} = a_1 \begin{vmatrix} a_1 & a_2 & a_3 \\ b_1 & b_2 & b_3 \\ c_1 & c_2 & c_3 \end{vmatrix} - a_2 \begin{vmatrix} a_1 & a_2 & a_3 \\ b_1 & b_2 & b_3 \\ c_1 & c_2 & c_3 \end{vmatrix} + a_3 \begin{vmatrix} a_1 & a_2 & a_3 \\ b_1 & b_2 & b_3 \\ c_1 & c_2 & c_3 \end{vmatrix}$$

$$= a_1 \begin{vmatrix} b_2 & b_3 \\ c_2 & c_3 \end{vmatrix} - a_2 \begin{vmatrix} b_1 & b_3 \\ c_1 & c_3 \end{vmatrix} + a_3 \begin{vmatrix} b_1 & b_2 \\ c_1 & c_2 \end{vmatrix}$$

TORQUE

The cross product plays an important role in mechanics. Let O be a fixed point in a body and suppose a force $\mathbf{F}$ is applied at another point P of the body. Then, $\mathbf{F}$ tends to rotate the body about an axis through O and perpendicular to the plane of OP and $\mathbf{F}$. The vector

$$\tau = \overrightarrow{OP} \times \mathbf{F}$$

is called the **torque**. It points in the direction of the axis and has magnitude $|\overrightarrow{OP}||\mathbf{F}|\sin\theta$, which is just the moment of force about the axis due to $\mathbf{F}$.

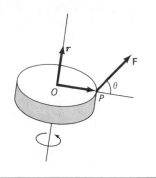

Using determinants, we may write the definition of $\mathbf{u} \times \mathbf{v}$ as

$$\mathbf{u} \times \mathbf{v} = \begin{vmatrix} \mathbf{i} & \mathbf{j} & \mathbf{k} \\ u_1 & u_2 & u_3 \\ v_1 & v_2 & v_3 \end{vmatrix} = \begin{vmatrix} u_2 & u_3 \\ v_2 & v_3 \end{vmatrix}\mathbf{i} - \begin{vmatrix} u_1 & u_3 \\ v_1 & v_3 \end{vmatrix}\mathbf{j} + \begin{vmatrix} u_1 & u_2 \\ v_1 & v_2 \end{vmatrix}\mathbf{k}$$

Note that the components of the left vector $\mathbf{u}$ go in the second row, and those of the right vector $\mathbf{v}$ go in the third row. This is important because if we interchange the positions of $\mathbf{u}$ and $\mathbf{v}$, we interchange the second and third rows of the determinant, and this changes the sign of the determinant's value, as you may check. Thus,

$$\mathbf{u} \times \mathbf{v} = -(\mathbf{v} \times \mathbf{u})$$

which is sometimes called the *anticommutative law*.

EXAMPLE 1 Let $\mathbf{u} = \langle 1, -2, -1 \rangle$ and $\mathbf{v} = \langle -2, 4, 1 \rangle$. Calculate $\mathbf{u} \times \mathbf{v}$ and $\mathbf{v} \times \mathbf{u}$ using the determinant definition.

Solution

$$\mathbf{u} \times \mathbf{v} = \begin{vmatrix} \mathbf{i} & \mathbf{j} & \mathbf{k} \\ 1 & -2 & -1 \\ -2 & 4 & 1 \end{vmatrix} = \mathbf{i}\begin{vmatrix} -2 & -1 \\ 4 & 1 \end{vmatrix} - \mathbf{j}\begin{vmatrix} 1 & -1 \\ -2 & 1 \end{vmatrix} + \mathbf{k}\begin{vmatrix} 1 & -2 \\ -2 & 4 \end{vmatrix}$$

$$= 2\mathbf{i} + \mathbf{j} + 0\mathbf{k}$$

$$\mathbf{v} \times \mathbf{u} = \begin{vmatrix} \mathbf{i} & \mathbf{j} & \mathbf{k} \\ -2 & 4 & 1 \\ 1 & -2 & -1 \end{vmatrix} = \mathbf{i}\begin{vmatrix} 4 & 1 \\ -2 & -1 \end{vmatrix} - \mathbf{j}\begin{vmatrix} -2 & 1 \\ 1 & -1 \end{vmatrix} + \mathbf{k}\begin{vmatrix} -2 & 4 \\ 1 & -2 \end{vmatrix}$$

$$= -2\mathbf{i} - \mathbf{j} + 0\mathbf{k} \quad \blacksquare$$

GEOMETRIC INTERPRETATION OF $\mathbf{u} \times \mathbf{v}$ Like the dot product, the cross product gains significance from its geometric interpretation.

Theorem A

Let $\mathbf{u}$ and $\mathbf{v}$ be vectors in three-space and θ be the angle between them. Then:

1. $\mathbf{u} \cdot (\mathbf{u} \times \mathbf{v}) = 0 = \mathbf{v} \cdot (\mathbf{u} \times \mathbf{v})$—that is, $\mathbf{u} \times \mathbf{v}$ is perpendicular to both $\mathbf{u}$ and $\mathbf{v}$;
2. $\mathbf{u}$, $\mathbf{v}$, and $\mathbf{u} \times \mathbf{v}$ form a right-handed triple;
3. $|\mathbf{u} \times \mathbf{v}| = |\mathbf{u}||\mathbf{v}|\sin\theta$.

Proof Let $\mathbf{u} = \langle u_1, u_2, u_3 \rangle$ and $\mathbf{v} = \langle v_1, v_2, v_3 \rangle$.

1. $\mathbf{u} \cdot (\mathbf{u} \times \mathbf{v}) = u_1(u_2 v_3 - u_3 v_2) + u_2(u_3 v_1 - u_1 v_3) + u_3(u_1 v_2 - u_2 v_1)$. When we remove parentheses, the six terms cancel in pairs. A similar event occurs when we expand $\mathbf{v} \cdot (\mathbf{u} \times \mathbf{v})$.

FIGURE 1

2. The meaning of right-handedness for the triple u, **v**, **u × v** is illustrated in Figure 1. There θ is the angle between **u** and **v**, and the hand is cupped in the direction of the rotation through θ that makes **u** coincide with **v**. It seems difficult to establish analytically that the indicated triple is right-handed, but you might check it with a few examples. Note in particular that since **i × j = k**, the triple **i**, **j**, **i × j** is right-handed.

3. We need Lagrange's identity

$$|\mathbf{u} \times \mathbf{v}|^2 = |\mathbf{u}|^2|\mathbf{v}|^2 - (\mathbf{u} \cdot \mathbf{v})^2$$

whose proof is a simple algebraic exercise (Problem 23). Taking it for granted, we may write

$$\begin{aligned}|\mathbf{u} \times \mathbf{v}|^2 &= |\mathbf{u}|^2|\mathbf{v}|^2 - (|\mathbf{u}|\,|\mathbf{v}|\cos\theta)^2 \\ &= |\mathbf{u}|^2|\mathbf{v}|^2(1 - \cos^2\theta) \\ &= |\mathbf{u}|^2|\mathbf{v}|^2\sin^2\theta\end{aligned}$$

Since $0 \le \theta \le \pi$, $\sin\theta \ge 0$. Thus, taking principal square roots yields

$$|\mathbf{u} \times \mathbf{v}| = |\mathbf{u}|\,|\mathbf{v}|\sin\theta \qquad \blacksquare$$

It is important that we have geometric interpretations of both **u · v** and **u × v**. It means that, while both products were originally defined in terms of components that depend on a choice of coordinate system, they are actually independent of coordinate systems. They are intrinsic geometric quantities, and you will get the same answers for **u · v** and **u × v** no matter how you introduce the coordinates used to compute them.

Here is a simple consequence of Theorem A; it depends on the fact that vectors are parallel if and only if the angle θ between them is either 0° or 180°.

Theorem B

Two vectors **u** and **v** in three-space are parallel if and only if **u × v = 0**.

APPLICATIONS   Our first application is to find the equation of the plane through three noncollinear points.

EXAMPLE 2   Find the equation of the plane (Figure 2) through the three points $P_1(1, -2, 3)$, $P_2(4, 1, -2)$, and $P_3(-2, -3, 0)$.

FIGURE 2

Solution   Let $\mathbf{u} = \overrightarrow{P_2P_1} = \langle -3, -3, 5 \rangle$ and $\mathbf{v} = \overrightarrow{P_2P_3} = \langle -6, -4, 2 \rangle$. Then,

$$\mathbf{u} \times \mathbf{v} = \begin{vmatrix} \mathbf{i} & \mathbf{j} & \mathbf{k} \\ -3 & -3 & 5 \\ -6 & -4 & 2 \end{vmatrix} = 14\mathbf{i} - 24\mathbf{j} - 6\mathbf{k}$$

The plane through $(4, 1, -2)$ with normal $14\mathbf{i} - 24\mathbf{j} - 6\mathbf{k}$ has equation (see Section 14.2)

$$14(x - 4) - 24(y - 1) - 6(z + 2) = 0$$

FIGURE 3

or

$$14x - 24y - 6z = 44 \qquad \blacksquare$$

EXAMPLE 3 Show that the area of a parallelogram with **a** and **b** as adjacent sides is $|\mathbf{a} \times \mathbf{b}|$.

Solution Recall that the area of a parallelogram is the product of the base times the height. Now look at Figure 3 and use the fact that $|\mathbf{a} \times \mathbf{b}| = |\mathbf{a}|\,|\mathbf{b}|\sin\theta$. $\blacksquare$

EXAMPLE 4 Let **a**, **b**, and **c** be vectors not all lying in the same plane. Show that the volume of the parallelepiped determined by **a**, **b**, and **c** is

$$V = |\mathbf{a} \cdot (\mathbf{b} \times \mathbf{c})| = \left| \begin{vmatrix} a_1 & a_2 & a_3 \\ b_1 & b_2 & b_3 \\ c_1 & c_2 & c_3 \end{vmatrix} \right|$$

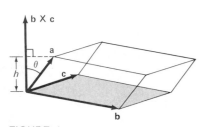

FIGURE 4

Solution Refer to Figure 4 and regard the parallelogram determined by **b** and **c** as the base of the parallelepiped. The area of this base is $|\mathbf{b} \times \mathbf{c}|$ by Example 3; the height h of the parallelepiped is the absolute value of the scalar projection of **a** on $\mathbf{b} \times \mathbf{c}$. Thus,

$$h = |\mathbf{a}|\,|\cos\theta| = \frac{|\mathbf{a}|\,|\mathbf{b} \times \mathbf{c}|\,|\cos\theta|}{|\mathbf{b} \times \mathbf{c}|} = \frac{|\mathbf{a} \cdot (\mathbf{b} \times \mathbf{c})|}{|\mathbf{b} \times \mathbf{c}|}$$

and

$$V = h|\mathbf{b} \times \mathbf{c}| = |\mathbf{a} \cdot (\mathbf{b} \times \mathbf{c})|$$

That V can also be expressed as a determinant is established by expanding $|\mathbf{a} \cdot (\mathbf{b} \times \mathbf{c})|$ in terms of components and then comparing it with the value of the indicated determinant. $\blacksquare$

ALGEBRAIC PROPERTIES The rules for calculating with cross products are summarized in the following theorem. Proving this theorem is a matter of writing everything out in terms of components and will be left as an exercise.

Theorem C

If **u**, **v**, and **w** are vectors in three-space and k is a scalar, then:

1. $\mathbf{u} \times \mathbf{v} = -(\mathbf{v} \times \mathbf{u})$ (anticommutative law);
2. $\mathbf{u} \times (\mathbf{v} + \mathbf{w}) = (\mathbf{u} \times \mathbf{v}) + (\mathbf{u} \times \mathbf{w})$ (left distributive law);
3. $k(\mathbf{u} \times \mathbf{v}) = (k\mathbf{u}) \times \mathbf{v} = \mathbf{u} \times (k\mathbf{v})$;
4. $\mathbf{u} \times \mathbf{0} = \mathbf{0} \times \mathbf{u} = \mathbf{0}, \quad \mathbf{u} \times \mathbf{u} = \mathbf{0}$;
5. $(\mathbf{u} \times \mathbf{v}) \cdot \mathbf{w} = \mathbf{u} \cdot (\mathbf{v} \times \mathbf{w})$;
6. $\mathbf{u} \times (\mathbf{v} \times \mathbf{w}) = (\mathbf{u} \cdot \mathbf{w})\mathbf{v} - (\mathbf{u} \cdot \mathbf{v})\mathbf{w}$.

FIGURE 5

Once the rules in Theorem C are mastered, complicated calculations with vectors can be done with ease. We illustrate by calculating a cross product in a new way. We will need the following simple but important products.

$$\mathbf{i} \times \mathbf{j} = \mathbf{k} \qquad \mathbf{j} \times \mathbf{k} = \mathbf{i} \qquad \mathbf{k} \times \mathbf{i} = \mathbf{j}$$

These results have a cyclic order, which can be remembered by appealing to Figure 5.

EXAMPLE 5 Calculate $\mathbf{u} \times \mathbf{v}$ if $\mathbf{u} = 3\mathbf{i} - 2\mathbf{j} + \mathbf{k}$ and $\mathbf{v} = 4\mathbf{i} + 2\mathbf{j} - 3\mathbf{k}$.

Solution We appeal to Theorem C, especially the distributive law and the anticommutative law.

$$\begin{aligned}
\mathbf{u} \times \mathbf{v} &= (3\mathbf{i} - 2\mathbf{j} + \mathbf{k}) \times (4\mathbf{i} + 2\mathbf{j} - 3\mathbf{k}) \\
&= 12(\mathbf{i} \times \mathbf{i}) + 6(\mathbf{i} \times \mathbf{j}) - 9(\mathbf{i} \times \mathbf{k}) - 8(\mathbf{j} \times \mathbf{i}) - 4(\mathbf{j} \times \mathbf{j}) \\
&\quad + 6(\mathbf{j} \times \mathbf{k}) + 4(\mathbf{k} \times \mathbf{i}) + 2(\mathbf{k} \times \mathbf{j}) - 3(\mathbf{k} \times \mathbf{k}) \\
&= 12(\mathbf{0}) + 6(\mathbf{k}) - 9(-\mathbf{j}) - 8(-\mathbf{k}) - 4(\mathbf{0}) \\
&\quad + 6(\mathbf{i}) + 4(\mathbf{j}) + 2(-\mathbf{i}) - 3(\mathbf{0}) \\
&= 4\mathbf{i} + 13\mathbf{j} + 14\mathbf{k}
\end{aligned}$$

Experts would do most of this in their heads; novices might find the determinant method easier. ∎

PROBLEM SET 14.3

1. Let $\mathbf{a} = -3\mathbf{i} + 2\mathbf{j} - 2\mathbf{k}$, $\mathbf{b} = \mathbf{i} - 4\mathbf{j} + 3\mathbf{k}$, and $\mathbf{c} = 2\mathbf{i} - \mathbf{j} + 4\mathbf{k}$. Find each of the following.

(a) $\mathbf{a} \times \mathbf{b}$ (b) $\mathbf{a} \times (\mathbf{b} + \mathbf{c})$

(c) $\mathbf{a} \cdot (\mathbf{b} \times \mathbf{c})$ (d) $\mathbf{a} \times (\mathbf{b} \times \mathbf{c})$

2. If $\mathbf{a} = \langle 4, -2, 1 \rangle$, $\mathbf{b} = \langle -2, 3, -2 \rangle$, and $\mathbf{c} = \langle -1, 4, 3 \rangle$, find each of the following.

(a) $\mathbf{a} \times \mathbf{b}$ (b) $\mathbf{a} \times (\mathbf{b} + \mathbf{c})$

(c) $\mathbf{a} \cdot (\mathbf{b} \times \mathbf{c})$ (d) $\mathbf{a} \times (\mathbf{b} \times \mathbf{c})$

3. Find all vectors perpendicular to both of the vectors $-2\mathbf{i} + \mathbf{j} - 4\mathbf{k}$ and $3\mathbf{i} - 4\mathbf{j} + 5\mathbf{k}$.

4. Find all vectors perpendicular to both of the vectors $\mathbf{i} - 4\mathbf{j} - 3\mathbf{k}$ and $4\mathbf{i} + \mathbf{j} - \mathbf{k}$.

5. Find the unit vectors perpendicular to the plane determined by the three points $(-1, 2, 0)$, $(5, -1, 3)$, and $(4, 0, -2)$.

6. Find the unit vectors perpendicular to the plane determined by the three points $(0, -1, 5)$, $(2, 2, 2)$, and $(-3, -4, 1)$.

7. Find the area of the parallelogram with $-2\mathbf{i} + \mathbf{j} + 4\mathbf{k}$ and $4\mathbf{i} - 2\mathbf{j} - 5\mathbf{k}$ as adjacent sides.

8. Find the area of the parallelogram with $2\mathbf{i} - 5\mathbf{j} + 2\mathbf{k}$ and $3\mathbf{i} - 3\mathbf{j} + 6\mathbf{k}$ as adjacent sides.

9. Find the area of the triangle with $(3, 2, -1)$, $(2, 4, 6)$, and $(-1, 2, 7)$ as vertices.

10. Find the area of the triangle with $(-1, 2, 1)$, $(0, 3, 0)$, and $(4, 5, 6)$ as vertices.

11. Find the equation of the plane through $(2, 5, 6)$, $(1, -1, 2)$, and $(4, 0, 6)$ (see Example 2).

12. Find the equation of the plane through $(5, 1, 6)$, $(-1, -2, -3)$, and $(4, -2, 1)$.

13. Find the equation of the plane through $(-1, -2, 3)$ and perpendicular to both the planes $x - 3y + 2z = 7$ and $2x - 2y - z = -3$.

14. Find the equation of the plane through $(2, -3, 2)$ and parallel to the plane of the vectors $4\mathbf{i} + 3\mathbf{j} - \mathbf{k}$ and $2\mathbf{i} - 5\mathbf{j} + 6\mathbf{k}$.

15. Find the equation of the plane through $(6, 2, -1)$ and perpendicular to the line of intersection of the planes $4x - 3y + 2z + 5 = 0$ and $3x + 2y - z + 11 = 0$.

16. Let $\mathbf{a}$ and $\mathbf{b}$ be nonzero vectors with the same initial point and let $\mathbf{c}$ be any nonzero vector. Show that $(\mathbf{a} \times \mathbf{b}) \times \mathbf{c}$ is a vector parallel to the plane of $\mathbf{a}$ and $\mathbf{b}$.

17. Find the volume of the parallelepiped with edges $\langle 2, 3, 4 \rangle$, $\langle 0, 4, -1 \rangle$, and $\langle 5, 1, 3 \rangle$ (see Example 4).

18. Find the volume of the parallelepiped with edges $3\mathbf{i} - 4\mathbf{j} + 2\mathbf{k}$, $-\mathbf{i} + 2\mathbf{j} + \mathbf{k}$, and $3\mathbf{i} - 2\mathbf{j} + 5\mathbf{k}$.

19. Let K be the parallelepiped determined by $\mathbf{u} = \langle 3, 2, 1 \rangle$, $\mathbf{v} = \langle 1, 1, 2 \rangle$, and $\mathbf{w} = \langle 1, 3, 3 \rangle$.

(a) Find the volume of K.
(b) Find the area of the face determined by $\mathbf{u}$ and $\mathbf{v}$.
(c) Find the angle between $\mathbf{u}$ and the plane containing the face determined by $\mathbf{v}$ and $\mathbf{w}$.

20. Show that if $\mathbf{a}$, $\mathbf{b}$, $\mathbf{c}$, and $\mathbf{d}$ all lie in the same plane, then

$$(\mathbf{a} \times \mathbf{b}) \times (\mathbf{c} \times \mathbf{d}) = \mathbf{0}$$

21. The volume of a tetrahedron is known to be $\frac{1}{3}$(area of base)(height). From this, show that the volume of the tetrahedron with edges $\mathbf{a}$, $\mathbf{b}$, and $\mathbf{c}$ is $\frac{1}{6}|\mathbf{a} \cdot (\mathbf{b} \times \mathbf{c})|$.

22. Find the volume of the tetrahedron with vertices $(-1, 2, 3)$, $(4, -1, 2)$, $(5, 6, 3)$, and $(1, 1, -2)$ (see Problem 21).

23. Prove **Lagrange's Identity**

$$|\mathbf{u} \times \mathbf{v}|^2 = |\mathbf{u}|^2|\mathbf{v}|^2 - (\mathbf{u} \cdot \mathbf{v})^2$$

Do this without using Theorem A.

24. Prove the left distributive law

$$\mathbf{u} \times (\mathbf{v} + \mathbf{w}) = (\mathbf{u} \times \mathbf{v}) + (\mathbf{u} \times \mathbf{w})$$

25. Use Problem 24 and the anticommutative law to prove the right distributive law.

26. If both $\mathbf{u} \times \mathbf{v} = \mathbf{0}$ and $\mathbf{u} \cdot \mathbf{v} = 0$, what can you conclude about $\mathbf{u}$ or $\mathbf{v}$?

27. Use Example 3 to develop a formula for the area of the triangle with vertices $P(a, 0, 0)$, $Q(0, b, 0)$, and $R(0, 0, c)$ shown in the top half of Figure 6.

28. Show that the triangle in the plane with vertices (x_1, y_1), (x_2, y_2), and (x_3, y_3) has area equal to one-half the absolute value of the determinant

$$\begin{vmatrix} x_1 & y_1 & 1 \\ x_2 & y_2 & 1 \\ x_3 & y_3 & 1 \end{vmatrix}$$

29. (A Pythagorean Theorem in Three-Space) As in Figure 6, let P, Q, R, and O be the vertices of a (right-angled) tetrahedron and let A, B, C, and D be the areas of the opposite faces, respectively. Show that $A^2 + B^2 + C^2 = D^2$.

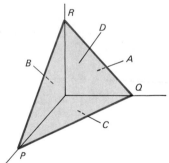

FIGURE 6

30. Let vectors $\mathbf{a}$, $\mathbf{b}$, and $\mathbf{c}$ with common initial point determine a tetrahedron and let $\mathbf{m}$, $\mathbf{n}$, $\mathbf{p}$, and $\mathbf{q}$ be vectors perpendicular to the four faces, pointing outward, and having length equal to the area of the corresponding face. Show that $\mathbf{m} + \mathbf{n} + \mathbf{p} + \mathbf{q} = \mathbf{0}$.

31. Let $\mathbf{a}$, $\mathbf{b}$, and $\mathbf{a} - \mathbf{b}$ denote the three edges of a triangle with lengths a, b, and c, respectively. Use Lagrange's Identity together with $2\mathbf{a} \cdot \mathbf{b} = |\mathbf{a}|^2 + |\mathbf{b}|^2 - |\mathbf{a} - \mathbf{b}|^2$ to prove **Heron's Formula** for the area A of a triangle, namely,

$$A = \sqrt{s(s - a)(s - b)(s - c)}$$

where s is the semiperimeter $(a + b + c)/2$.

32. Use the method of Example 5 to show directly that if $\mathbf{u} = u_1\mathbf{i} + u_2\mathbf{j} + u_3\mathbf{k}$ and $\mathbf{v} = v_1\mathbf{i} + v_2\mathbf{j} + v_3\mathbf{k}$, then

$$\mathbf{u} \times \mathbf{v} = (u_2v_3 - u_3v_2)\mathbf{i} + (u_3v_1 - u_1v_3)\mathbf{j} + (u_1v_2 - u_2v_1)\mathbf{k}$$

14.4
Lines and Curves in Three-Space

Our study of lines and curves in the plane extends easily to three-space. A **space curve** is determined by a triple of parametric equations

$$x = f(t), \qquad y = g(t), \qquad z = h(t), \qquad t \in I$$

with f, g, and h continuous on the interval I. In vector language a curve is specified by giving the position vector $\mathbf{r} = \mathbf{r}(t)$ of a point $P = P(t)$, namely,

$$\mathbf{r} = \mathbf{r}(t) = \langle f(t), g(t), h(t) \rangle = f(t)\mathbf{i} + g(t)\mathbf{j} + h(t)\mathbf{k}$$

The tip of $\mathbf{r}$ traces out the curve as t ranges over the interval I, as we see in Figure 1.

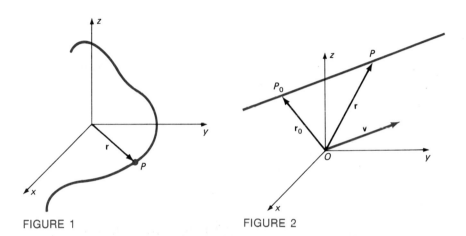

FIGURE 1   FIGURE 2

LINES The simplest of all curves is a line. A line is determined by a fixed point P_0 and a fixed vector $\mathbf{v} = a\mathbf{i} + b\mathbf{j} + c\mathbf{k}$. It is the set of all points P such that $\overrightarrow{P_0P}$ is parallel to $\mathbf{v}$—that is, that satisfy

$$\overrightarrow{P_0P} = t\mathbf{v}$$

for some real number t (Figure 2). If $\mathbf{r} = \overrightarrow{OP}$ and $\mathbf{r}_0 = \overrightarrow{OP_0}$ are the position vectors of P and P_0, respectively, then $\overrightarrow{P_0P} = \mathbf{r} - \mathbf{r}_0$ and the equation of the line can thus be written

$$\mathbf{r} = \mathbf{r}_0 + t\mathbf{v}$$

If we write $\mathbf{r} = \langle x, y, z \rangle$ and $\mathbf{r}_0 = \langle x_0, y_0, z_0 \rangle$ and equate components in the last equation above, we obtain

$$x = x_0 + at, \qquad y = y_0 + bt, \qquad z = z_0 + ct$$

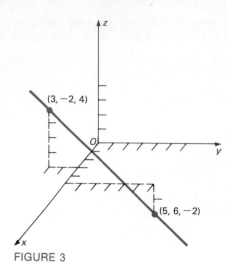

FIGURE 3

These are **parametric equations** of the line through (x_0, y_0, z_0) and parallel to $\mathbf{v} = \langle a, b, c \rangle$. The numbers a, b, and c are called **direction numbers** for the line. They are not unique; any nonzero constant multiples ka, kb, and kc are also direction numbers.

EXAMPLE 1 Find parametric equations for the line through $(3, -2, 4)$ and $(5, 6, -2)$ (see Figure 3).

Solution A vector parallel to the given line is

$$\mathbf{v} = \langle 5 - 3, 6 + 2, -2 - 4 \rangle = \langle 2, 8, -6 \rangle$$

If we choose (x_0, y_0, z_0) as $(3, -2, 4)$, we obtain the parametric equations

$$x = 3 + 2t, \qquad y = -2 + 8t, \qquad z = 4 - 6t$$

Note that $t = 0$ determines the point $(3, -2, 4)$, whereas $t = 1$ gives $(5, 6, -2)$. In fact, $0 \leq t \leq 1$ corresponds to the segment joining these two points. ∎

If we solve each of the parametric equations for t (assuming a, b, and c are all different from zero) and equate the results, we obtain the **symmetric equations** for the line through (x_0, y_0, z_0) with direction numbers a, b, c—namely

$$\boxed{\dfrac{x - x_0}{a} = \dfrac{y - y_0}{b} = \dfrac{z - z_0}{c}}$$

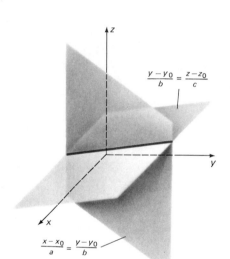

FIGURE 4

This is the conjunction of the two equations

$$\frac{x - x_0}{a} = \frac{y - y_0}{b} \quad \text{and} \quad \frac{y - y_0}{b} = \frac{z - z_0}{c}$$

both of which are the equations of planes (Figure 4); and of course, the intersection of two planes is a line.

EXAMPLE 2 Find the symmetric equations of the line that is parallel to the vector $\langle 4, -3, 2 \rangle$ and goes through $(2, 5, -1)$.

Solution

$$\frac{x - 2}{4} = \frac{y - 5}{-3} = \frac{z + 1}{2} \qquad\qquad\blacksquare$$

EXAMPLE 3 Find the symmetric equations of the line of intersection of the planes

$$2x - y - 5z = -14 \quad \text{and} \quad 4x + 5y + 4z = 28$$

FIGURE 5

Solution We begin by finding two points on the line. Any two points would do, but we choose to find the points where the line pierces the yz-plane and the xz-plane (Figure 5). The former is obtained by setting $x = 0$ and solving the resulting equations $-y - 5z = -14$ and $5y + 4z = 28$ simultaneously. This yields the point $(0, 4, 2)$. A similar procedure with $y = 0$ gives the point $(3, 0, 4)$. Consequently, a vector parallel to the required line is

$$\langle 3 - 0, 0 - 4, 4 - 2 \rangle = \langle 3, -4, 2 \rangle$$

Using $(3, 0, 4)$ for (x_0, y_0, z_0), we get

$$\frac{x - 3}{3} = \frac{y - 0}{-4} = \frac{z - 4}{2}$$

An alternative solution is based on the fact that the line of intersection of two planes is perpendicular to both of their normals. $\mathbf{u} = \langle 2, -1, -5 \rangle$ is a normal to the first plane; $\mathbf{v} = \langle 4, 5, 4 \rangle$ is a normal to the second. Since

$$\mathbf{u} \times \mathbf{v} = \begin{vmatrix} \mathbf{i} & \mathbf{j} & \mathbf{k} \\ 2 & -1 & -5 \\ 4 & 5 & 4 \end{vmatrix} = 21\mathbf{i} - 28\mathbf{j} + 14\mathbf{k}$$

the vector $\mathbf{w} = \langle 21, -28, 14 \rangle$ is parallel to the required line. This implies that $\frac{1}{7}\mathbf{w} = \langle 3, -4, 2 \rangle$ also has this property. Next, find any point on the line of intersection—for example, $(3, 0, 4)$—and proceed as in the earlier solution. ■

EXAMPLE 4 Find symmetric equations or parametric equations of the line through $(1, -2, 3)$ that is perpendicular to both the x-axis and the line

$$\frac{x - 4}{2} = \frac{y - 3}{-1} = \frac{z}{5}$$

Solution The x-axis and the given line have directions $\mathbf{u} = \langle 1, 0, 0 \rangle$ and $\mathbf{v} = \langle 2, -1, 5 \rangle$, respectively. A vector perpendicular to both $\mathbf{u}$ and $\mathbf{v}$ is

$$\mathbf{u} \times \mathbf{v} = \begin{vmatrix} \mathbf{i} & \mathbf{j} & \mathbf{k} \\ 1 & 0 & 0 \\ 2 & -1 & 5 \end{vmatrix} = 0\mathbf{i} - 5\mathbf{j} - \mathbf{k}$$

The required line is parallel to $\langle 0, -5, -1 \rangle$ and so also to $\langle 0, 5, 1 \rangle$. Since the first direction number is zero, the line does not have symmetric equations. Its parametric equations are

$$x = 1, \qquad y = -2 + 5t, \qquad z = 3 + t$$

■

$$\mathbf{r} = \mathbf{r}(t) = f(t)\mathbf{i} + g(t)\mathbf{j} + h(t)\mathbf{k}$$

be the position vector determining a curve in three-space (Figure 6). In complete analogy with what we did in the plane (Section 13.4), we define $\mathbf{r}'(t)$ by

$$\mathbf{r}'(t) = \lim_{h \to 0} \frac{\mathbf{r}(t + h) - \mathbf{r}(t)}{h}$$

It follows that $\mathbf{r}'(t)$, if it exists, has the direction of the tangent line to the curve at the point $P(t)$ corresponding to t. Moreover, $\mathbf{r}'(t)$ exists if and only if $f'(t)$, $g'(t)$, and $h'(t)$ exist and, in this case,

$$\mathbf{r}'(t) = f'(t)\mathbf{i} + g'(t)\mathbf{j} + h'(t)\mathbf{k}$$

Thus, $f'(t)$, $g'(t)$, and $h'(t)$ are direction numbers for the tangent line at P.

EXAMPLE 5 Find the symmetric equations for the tangent line to the curve determined by

$$\mathbf{r}(t) = t\mathbf{i} + \tfrac{1}{2}t^2\mathbf{j} + \tfrac{1}{3}t^3\mathbf{k}$$

at $P(2) = (2, 2, \tfrac{8}{3})$.

Solution

$$\mathbf{r}'(t) = \mathbf{i} + t\mathbf{j} + t^2\mathbf{k}$$

and

$$\mathbf{r}'(2) = \mathbf{i} + 2\mathbf{j} + 4\mathbf{k}$$

so the tangent line has direction $\langle 1, 2, 4 \rangle$. Its symmetric equations are

$$\frac{x - 2}{1} = \frac{y - 2}{2} = \frac{z - \tfrac{8}{3}}{4}$$
∎

FIGURE 6

PROBLEM SET 14.4

In Problems 1–4, find the parametric equations of the line through the given pair of points.

1. $(1, -2, 3)$, $(4, 5, 6)$

2. $(2, -1, 5)$, $(7, -2, 3)$

3. $(4, 2, 3)$, $(6, 2, -1)$

4. $(5, -3, -3)$, $(5, 4, 2)$

In Problems 5–8, write both the parametric equations and the symmetric equations for the line through the given point parallel to the given vector.

5. $(4, -6, 3)$, $\langle -2, 1, 5 \rangle$

6. $(-1, 3, 2)$, $\langle 4, 2, -1 \rangle$

7. $(2, 5, -4)$, $\langle -3, 4, 2 \rangle$

8. $(-3, 0, -3)$, $\langle 7, -6, 3 \rangle$

In Problems 9–12, find the symmetric equations of the line of intersection of the given pair of planes.

9. $5x + 2y - 5z = 5$, $10x + 6y - 5z = 25$

10. $x + y - z = 1$, $3x - 3y + 7z = 9$

11. $x + 4y + 2z = 13$, $2x - y - 2z = 10$

12. $x - 2y + z = 1$, $6x - 5y + 4z = 10$

13. Find the symmetric equations of the line through $(4, 0, 6)$ and perpendicular to the plane $x - 5y + 2z = 10$.

14. Find the symmetric equations of the line through $(-5, 7, -2)$ and perpendicular to both $\langle 2, 1, -3 \rangle$ and $\langle 5, 4, -1 \rangle$.

15. Find the parametric equations of the line through $(5, -3, 4)$ that intersects the z-axis at right angles.

16. Find the symmetric equations of the line through $(2, -4, 5)$ that is parallel to the plane $3x + y - 2z = 5$ and perpendicular to the line

$$\frac{x + 8}{2} = \frac{y - 5}{3} = \frac{z - 1}{-1}$$

17. Find the equation of the plane that contains the parallel lines

$$\begin{cases} x = -2 + 2t \\ y = 1 + 4t \\ z = 2 - t \end{cases} \quad \text{and} \quad \begin{cases} x = 2 - 2t \\ y = 3 - 4t \\ z = 1 + t \end{cases}$$

18. Show that the lines

$$\frac{x - 1}{-4} = \frac{y - 2}{3} = \frac{z - 4}{-2}$$

and

$$\frac{x - 2}{-1} = \frac{y - 1}{1} = \frac{z + 2}{6}$$

intersect and find the equation of the plane they determine.

19. Find the equation of the plane containing the line $x = 1 + 2t$, $y = -1 + 3t$, $z = 4 + t$ and the point $(1, -1, 5)$.

20. Find the equation of the plane containing the line $x = 3t$, $y = 1 + t$, $z = 2t$ and parallel to the intersection of the planes $2x - y + z = 0$ and $y + z + 1 = 0$.

21. Find the distance between the skew (nonintersecting) lines $x = 2 - t$, $y = 3 + 4t$, $z = 2t$ and $x = -1 + t$, $y = 2$, $z = -1 + 2t$ by using the following steps.

(a) Note by putting $t = 0$ that $(2, 3, 0)$ is on the first line.
(b) Find the equation of the plane π through $(2, 3, 0)$ parallel to both given lines (that is with normal perpendicular to both).
(c) Find a point Q on the second line.
(d) Find the distance from Q to the plane π. (See Example 6 of Section 14.2.)
See Problem 30 for another way to do this problem.

22. Find the distance between the skew lines $x = 1 + 2t$, $y = -3 + 4t$, $z = -1 - t$ and $x = 4 - 2t$, $y = 1 + 3t$, $z = 2t$ (see Problem 21).

23. Find the symmetric equations of the tangent line to the curve with equation

$$\mathbf{r}(t) = 2 \cos t\mathbf{i} + 6 \sin t\mathbf{j} + t\mathbf{k}$$

at $t = \pi/3$.

24. Find the parametric equations of the tangent line to the curve $x = 2t^2$, $y = 4t$, $z = t^3$ at $t = 1$.

25. Find the equation of the plane perpendicular to the curve $x = 3t$, $y = 2t^2$, $z = t^5$ at $t = -1$.

26. Find the equation of the plane perpendicular to the curve

$$\mathbf{r}(t) = t \sin t\mathbf{i} + 3t\mathbf{j} + 2t \cos t\mathbf{k}$$

at $t = \pi/2$.

27. Consider the curve $\mathbf{r} = \sin t \cos t\mathbf{i} + \sin^2 t\mathbf{j} + \cos t\mathbf{k}$. $0 \le t \le 2\pi$.

(a) Show that the curve lies on a sphere centered at the origin.
(b) Where does the tangent line at $t = \pi/6$ intersect the xy-plane?

28. (Point to Plane) Let P be a point on a plane with normal $\mathbf{n}$ and Q be a point off the plane (Figure 7). Show the distance d from Q to the plane is given by

$$d = \frac{|\overrightarrow{PQ} \cdot \mathbf{n}|}{|\mathbf{n}|}$$

and use this result to find the distance from $(4, -2, 3)$ to the plane $4x - 4y + 2z = 2$. Check with Example 6 of Section 14.2.

FIGURE 7

29. (Point to Line) Let P be a point on a line with direction $\mathbf{n}$ and Q a point off the line (Figure 8). Show that the distance d from Q to the line is given by

$$d = \frac{|\overrightarrow{PQ} \times \mathbf{n}|}{|\mathbf{n}|}$$

and use this result to find each distance in (a) and (b).

FIGURE 8

(a) From $Q(1, 0, -4)$ to the line $\dfrac{x - 3}{2} = \dfrac{y + 2}{-2} = \dfrac{z - 1}{1}$.

(b) From $Q(2, -1, 3)$ to the line $x = 1 + 2t$, $y = -1 + 3t$, $z = -6t$.

30. (Line to Line) Let P and Q be points on noninter-secting skew lines with directions $\mathbf{n}_1$ and $\mathbf{n}_2$ and let $\mathbf{n} = \mathbf{n}_1 \times \mathbf{n}_2$ (Figure 9). Show that the distance d between these lines is given by

$$d = \frac{|\overrightarrow{PQ} \cdot \mathbf{n}|}{|\mathbf{n}|}$$

and use this result to find the distance between each pair of lines in (a) and (b).

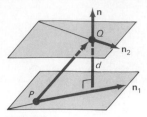

FIGURE 9

(a) $\dfrac{x - 3}{1} = \dfrac{y + 2}{1} = \dfrac{z - 1}{2}$ and $\dfrac{x + 4}{3} = \dfrac{y + 5}{4} = \dfrac{z}{5}$

(b) $x = 1 + 2t$, $y = -2 + 3t$, $z = -4t$ and $x = 3t$, $y = 1 + t$, $z = -5t$

14.5
Velocity, Acceleration, and Curvature

All that we did with curvilinear motion in the plane (Sections 13.4 and 13.5) generalizes in a natural way to three-space. Let

$$\mathbf{r}(t) = f(t)\mathbf{i} + g(t)\mathbf{j} + h(t)\mathbf{k}, \qquad a \le t \le b$$

be the position vector for a point $P = P(t)$ that is tracing out a curve as t increases (Figure 1). We suppose that $\mathbf{r}'(t)$ exists and is continuous and $\mathbf{r}'(t) \ne \mathbf{0}$, in which case the curve is said to be **smooth**. The length s of the arc from $P(a)$ to $P(t)$ is given by

$$s = \int_a^t |\mathbf{r}'(u)| \, du = \int_a^t \sqrt{[f'(u)]^2 + [g'(u)]^2 + [h'(u)]^2} \, du$$

If t measures time, we may define velocity, speed, and acceleration of the moving point P by

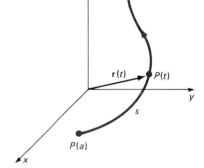

FIGURE 1

Velocity:	$\mathbf{v}(t) = \mathbf{r}'(t)$
Speed:	$\dfrac{ds}{dt} = |\mathbf{r}'(t)| = |\mathbf{v}(t)|$
Acceleration:	$\mathbf{a}(t) = \mathbf{r}''(t)$

AN EXAMPLE: THE CIRCULAR HELIX Suppose that a point P moves so that its position vector at time t is

$$\mathbf{r}(t) = a \cos t\,\mathbf{i} + a \sin t\,\mathbf{j} + ct\mathbf{k}$$

where a and c are positive constants. Then P traces a curve that winds around the right circular cylinder with parametric equations $x = a \cos t$, $y = a \sin t$ but spirals higher and higher because $z = ct$ increases with t. The curve is called a **circular helix**; part of it is shown in Figure 2.

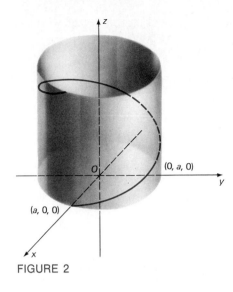

FIGURE 2

EXAMPLE 1 Find the arc length of the circular helix for $0 \le t \le 2\pi$.

Solution

$$s = \int_0^{2\pi} \sqrt{(-a \sin t)^2 + (a \cos t)^2 + c^2} \, dt$$

$$= \int_0^{2\pi} \sqrt{a^2 + c^2} \, dt = 2\pi \sqrt{a^2 + c^2}$$

Note that when $c = 0$, this reduces to $2\pi a$ (the circumference of a circle of radius a), as it should. ∎

EXAMPLE 2 For the motion $r(t) = a \cos t \mathbf{i} + a \sin t \mathbf{j} + ct\mathbf{k}$ described above, calculate the acceleration $\mathbf{a}$ at $t = 2\pi$.

Solution

$$\mathbf{v}(t) = \mathbf{r}'(t) = -a \sin t \mathbf{i} + a \cos t \mathbf{j} + c\mathbf{k}$$

$$\mathbf{a}(t) = \mathbf{r}''(t) = -a \cos t \mathbf{i} - a \sin t \mathbf{j}$$

$$\mathbf{a}(2\pi) = -a\mathbf{i}$$

∎

EXAMPLE 3 Beginning at $t = 0$, a bee flew so its position vector was $\mathbf{r} = t \cos t \mathbf{i} + t \sin t \mathbf{j} + t\mathbf{k}$ until $t = 4\pi$, at which time it went off on a tangent at the speed then attained. What was the total length of its flight path on the interval $0 \le t \le 4\pi + 3$?

Solution The path consists of a spiral part and a straight-line part with lengths L_1 and L_2, respectively. On the spiral part,

$$\mathbf{r}'(t) = (\cos t - t \sin t)\mathbf{i} + (\sin t + t \cos t)\mathbf{j} + \mathbf{k}$$

and

$$|\mathbf{r}'(t)| = [(\cos t - t \sin t)^2 + (\sin t + t \cos t)^2 + 1]^{1/2}$$
$$= \sqrt{2 + t^2}$$

Upon using Formula 44 at the end of the book, we find

$$L_1 = \int_0^{4\pi} \sqrt{2 + t^2} \, dt \approx 82.336$$

Also

$$L_2 = 3|\mathbf{r}'(4\pi)| = 3\sqrt{2 + 16\pi^2} \approx 37.937$$

We conclude that $L_1 + L_2 \approx 120.273$. ∎

CURVATURE   As we noted in Section 14.4, $\mathbf{v}(t) = \mathbf{r}'(t)$ is a vector with the same direction as the tangent to the curve at $P(t)$. Thus,

$$\mathbf{T} = \mathbf{T}(t) = \frac{\mathbf{v}(t)}{|\mathbf{v}(t)|}$$

is a **unit tangent vector** at $P(t)$. Since s denotes the arc length measured from some fixed point in the direction of increasing t, $d\mathbf{T}/ds$ measures the rate of change of direction of the tangent with respect to the distance along the curve. By the Chain Rule,

$$\frac{d\mathbf{T}}{ds} = \frac{d\mathbf{T}}{dt}\frac{dt}{ds} = \frac{\mathbf{T}'(t)}{|\mathbf{v}(t)|}$$

Thus, just as for plane curves, we may define the **curvature** κ of a space curve by

$$\kappa = \kappa(t) = \left|\frac{d\mathbf{T}}{ds}\right| = \frac{|\mathbf{T}'(t)|}{|\mathbf{v}(t)|}$$

EXAMPLE 4   Find the curvature of the circular helix

$$\mathbf{r}(t) = a\cos t\mathbf{i} + a\sin t\mathbf{j} + ct\mathbf{k}, \qquad a > 0$$

Solution

$$\mathbf{v}(t) = -a\sin t\mathbf{i} + a\cos t\mathbf{j} + c\mathbf{k}$$

$$\mathbf{T}(t) = \frac{\mathbf{v}(t)}{|\mathbf{v}(t)|} = \frac{1}{\sqrt{a^2 + c^2}}(-a\sin t\mathbf{i} + a\cos t\mathbf{j} + c\mathbf{k})$$

$$\kappa(t) = \frac{|\mathbf{T}'(t)|}{|\mathbf{v}(t)|} = \frac{1}{a^2 + c^2}|-a\cos t\mathbf{i} - a\sin t\mathbf{j}|$$

$$= \frac{a}{a^2 + c^2}$$

Thus, κ is a constant for the circular helix.   ∎

The radius of curvature R is the reciprocal of κ. In the example above, $R = (a^2 + c^2)/a$. This reduces to $R = a$ when $c = 0$, which corresponds to the fact that the motion is then along the circle $\mathbf{r} = a\cos t\mathbf{i} + a\sin t\mathbf{j}$ in the xy-plane. When c is large, R is large as we should expect.

COMPONENTS OF ACCELERATION   Just as in the plane case, we define the **principal unit normal vector** $\mathbf{N}$ at P by

$$\mathbf{N} = \frac{d\mathbf{T}/ds}{|d\mathbf{T}/ds|} = \frac{1}{\kappa}\frac{d\mathbf{T}}{ds}$$

It is obvious that $\mathbf{N}$ is a unit vector. That $\mathbf{N}$ is normal (perpendicular) to the curve follows by differentiating $\mathbf{T} \cdot \mathbf{T} = 1$ with respect to s. This gives

$$2\mathbf{T} \cdot \frac{d\mathbf{T}}{ds} = 0$$

which implies that $d\mathbf{T}/ds$ is perpendicular to $\mathbf{T}$. With these facts in hand, we can mimic a derivation given in the plane case (Section 13.5) to obtain

$$\boxed{\mathbf{a} = \frac{d^2s}{dt^2}\mathbf{T} + \left(\frac{ds}{dt}\right)^2 \kappa \mathbf{N} = a_T\mathbf{T} + a_N\mathbf{N}}$$

If we take the dot product of this last result with $\mathbf{T}$, we get

$$\mathbf{T} \cdot \mathbf{a} = a_T\mathbf{T} \cdot \mathbf{T} + a_N\mathbf{T} \cdot \mathbf{N} = a_T$$

or

$$\boxed{a_T = \mathbf{T} \cdot \mathbf{a} = \frac{\mathbf{r}' \cdot \mathbf{r}''}{|\mathbf{r}'|}}$$

If we take the cross product of $\mathbf{T}$ with $\mathbf{a}$, we get

$$\mathbf{T} \times \mathbf{a} = a_T(\mathbf{T} \times \mathbf{T}) + a_N(\mathbf{T} \times \mathbf{N}) = a_N(\mathbf{T} \times \mathbf{N})$$

and so

$$|\mathbf{T} \times \mathbf{a}| = a_N|\mathbf{T} \times \mathbf{N}| = a_N|\mathbf{T}|\,|\mathbf{N}|\sin 0 = a_N$$

or

$$\boxed{a_N = |\mathbf{T} \times \mathbf{a}| = \frac{|\mathbf{r}' \times \mathbf{r}''|}{|\mathbf{r}'|}}$$

Since $a_N = (ds/dt)^2\kappa = |\mathbf{r}'|^2\kappa$, we conclude that

$$\boxed{\kappa = \frac{|\mathbf{r}' \times \mathbf{r}''|}{|\mathbf{r}'|^3}}$$

EXAMPLE 5 At the point $(1, 1, \frac{1}{3})$, find $\mathbf{T}$, $\mathbf{N}$, a_T, a_N, and κ for the curvilinear motion

$$\mathbf{r}(t) = t\mathbf{i} + t^2\mathbf{j} + \tfrac{1}{3}t^3\mathbf{k}$$

Solution

$$\mathbf{r}'(t) = \mathbf{i} + 2t\mathbf{j} + t^2\mathbf{k}$$

$$\mathbf{r}''(t) = 2\mathbf{j} + 2t\mathbf{k}$$

At $t = 1$ (which gives the point $(1, 1, \frac{1}{3})$), we have

$$\mathbf{r}' = \mathbf{i} + 2\mathbf{j} + \mathbf{k}$$

$$\mathbf{r}'' = 2\mathbf{j} + 2\mathbf{k}$$

$$\mathbf{T} = \frac{\mathbf{r}'}{|\mathbf{r}'|} = \frac{\mathbf{i} + 2\mathbf{j} + \mathbf{k}}{\sqrt{6}}$$

$$a_T = \frac{\mathbf{r}' \cdot \mathbf{r}''}{|\mathbf{r}'|} = \frac{6}{\sqrt{6}}$$

$$a_N = \frac{|\mathbf{r}' \times \mathbf{r}''|}{|\mathbf{r}'|} = \frac{1}{\sqrt{6}} \left| \begin{matrix} \mathbf{i} & \mathbf{j} & \mathbf{k} \\ 1 & 2 & 1 \\ 0 & 2 & 2 \end{matrix} \right| = \frac{1}{\sqrt{6}} |2\mathbf{i} - 2\mathbf{j} + 2\mathbf{k}| = \sqrt{2}$$

$$\mathbf{N} = \frac{\mathbf{a} - a_T\mathbf{T}}{a_N} = \frac{(2\mathbf{j} + 2\mathbf{k}) - (\mathbf{i} + 2\mathbf{j} + \mathbf{k})}{\sqrt{2}} = \frac{-\mathbf{i} + \mathbf{k}}{\sqrt{2}}$$

$$\kappa = \frac{|\mathbf{r}' \times \mathbf{r}''|}{|\mathbf{r}'|^3} = \frac{a_N}{|\mathbf{r}'|^2} = \frac{\sqrt{2}}{6}$$

∎

FIGURE 3

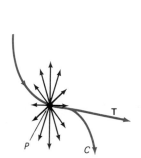

The moving trihedral

FIGURE 4

THE BINORMAL AT P Given a curve C and the unit tangent vector $\mathbf{T}$ at P, there are, of course, infinitely many unit vectors perpendicular to $\mathbf{T}$ at P (Figure 3). We picked one of them,

$$\mathbf{N} = \frac{d\mathbf{T}/ds}{|d\mathbf{T}/ds|}$$

and called it the principal normal. The vector

$$\mathbf{B} = \mathbf{T} \times \mathbf{N}$$

is called the **binormal**. It, too, is a unit vector and it is perpendicular to both $\mathbf{T}$ and $\mathbf{N}$. (Why?)

    If the unit tangent vector $\mathbf{T}$, the principal normal $\mathbf{N}$, and the binormal $\mathbf{B}$ have their initial points at P, they form a right-handed, mutually perpendicular triple of unit vectors known as the **trihedral** at P (Figure 4). This moving trihedral plays a crucial role in a subject called differential geometry. The plane of $\mathbf{T}$ and $\mathbf{N}$ is called the **osculating plane** at P.

PROBLEM SET 14.5

In Problems 1–6, find the velocity $\mathbf{v}$, acceleration $\mathbf{a}$, and speed s at the indicated time $t = t_1$.

1. $\mathbf{r}(t) = 3t\mathbf{i} + 4(t^2 - 1)\mathbf{j} + 2t^3\mathbf{k}$; $t_1 = 1$

2. $\mathbf{r}(t) = 4t^2\mathbf{i} + 9t\mathbf{j} + t^4\mathbf{k}$; $t_1 = 2$

3. $\mathbf{r}(t) = (2t - t^2)\mathbf{i} + 3t\mathbf{j} + (t^3 + 1)\mathbf{k}$; $t_1 = 1$

4. $\mathbf{r}(t) = 2 \sin t\mathbf{i} + 4 \cos t\mathbf{j} + e^{2t}\mathbf{k}$; $t_1 = \pi/6$

5. $\mathbf{r}(t) = \cos 2t\mathbf{i} + 2e^{-t}\mathbf{j} + 3 \sin t\mathbf{k}$; $t_1 = 0$

6. $\mathbf{r}(t) = t \sin t\mathbf{i} + e^t\mathbf{j} + \cos t\mathbf{k}; \ t_1 = 0$

7. Show that if the speed of a moving particle is constant, its acceleration vector is always perpendicular to its velocity vector.

8. Prove that $|\mathbf{r}(t)|$ is constant if and only if $\mathbf{r}(t) \cdot \mathbf{r}'(t) = 0$.

In Problems 9–14, find the length of the curve with the given vector equation.

9. $\mathbf{r}(t) = t \sin t\mathbf{i} + t \cos t\mathbf{j} + \sqrt{8}t\mathbf{k}; \ 0 \le t \le 4$

10. $\mathbf{r}(t) = \sqrt{6}t^2\mathbf{i} + \frac{2}{3}t^3\mathbf{j} + 6t\mathbf{k}; \ 3 \le t \le 6$

11. $\mathbf{r}(t) = 2t^{3/2}\mathbf{i} + (t - 3)\mathbf{j} + 3t\mathbf{k}; \ 6 \le t \le 10$

12. $\mathbf{r}(t) = e^{2t} \cos t\mathbf{i} + e^{2t} \sin t\mathbf{j} + e^{2t}\mathbf{k}; \ 0 \le t \le 5$

13. $\mathbf{r}(t) = \sinh 2t\mathbf{i} + \cosh 2t\mathbf{j} + 2t\mathbf{k}; \ -\frac{1}{2} \le t \le 1$

14. $\mathbf{r}(t) = \frac{1}{3}t^3\mathbf{i} + \frac{2}{9}(t^3 + 4)^{3/2}\mathbf{j} + \frac{2}{3}t^3\mathbf{k}; \ 0 \le t \le 3$

In Problems 15–20, find the curvature, the unit tangent vector, the principal normal, and the binormal at $t = t_1$.

15. $\mathbf{r}(t) = (t^2 - 1)\mathbf{i} + (2t + 3)\mathbf{j} + (t^2 - 4t)\mathbf{k}; \ t_1 = 2$

16. $\mathbf{r}(t) = \frac{1}{2}t^2\mathbf{i} + t\mathbf{j} + \frac{1}{3}t^3\mathbf{k}; \ t_1 = 2$

17. $x = \sin 3t, \ y = \cos 3t, \ z = t; \ t_1 = \pi/9$

18. $x = 8 \sin t, \ y = 8 \cos t, \ z = 4t; \ t_1 = \pi/6$

19. $\mathbf{r}(t) = e^t \sin t\mathbf{i} + e^t \cos t\mathbf{j} + e^t\mathbf{k}; \ t_1 = \pi/2$

20. $\mathbf{r}(t) = e^{-2t}\mathbf{i} + e^{2t}\mathbf{j} + 2\sqrt{2}t\mathbf{k}; \ t_1 = 0$

In Problems 21–24, find the tangential and normal vector components a_T and a_N of the acceleration vector at any time t.

21. $x = t, \ y = t^2, \ z = t^3$

22. $x = e^{-t}, \ y = 2t, \ z = e^t$

23. $\mathbf{r}(t) = t\mathbf{i} + \frac{1}{3}t^3\mathbf{j} + t^{-1}\mathbf{k}, \ t > 0$

24. $\mathbf{r}(t) = (\ln \sin t)\mathbf{i} + (\ln \cos t)\mathbf{j} + t\mathbf{k}, \ 0 < t < \pi/2$

[C] **25.** The DNA molecule in humans is a double helix, each with about 2.9×10^8 complete turns. Each helix has radius about 10 angstroms and rises about 34 angstroms on each complete turn (an angstrom is 10^{-8} centimeter). What is the total length of such a helix (see Example 1)?

26. A fly is crawling along a wire helix so that its position vector is $\mathbf{r} = 6 \cos \pi t\mathbf{i} + 6 \sin \pi t\mathbf{j} + 2t\mathbf{k}, \ t \ge 0$. At what point will the fly hit the sphere $x^2 + y^2 + z^2 = 100$ and how far did it travel in getting there (assuming it started when $t = 0$)?

27. A bee flew the spiral path

$$\mathbf{r} = 100e^{-t}(\cos t\mathbf{i} + \sin t\mathbf{j} + \mathbf{k})$$

Where was its final resting place and how far did it travel during its lifetime $0 \le t < \infty$?

28. Solve the following vector differential equations (c and $\mathbf{c}$ are constants), subject to $\mathbf{r} = \mathbf{r}_0$ at $t = 0$.

(a) $\dfrac{d\mathbf{r}}{dt} = \mathbf{0}$ (b) $\dfrac{d\mathbf{r}}{dt} = \mathbf{c}$

(c) $\dfrac{d^2\mathbf{r}}{dt^2} = \mathbf{c}$ subject also to $\mathbf{r}' = \mathbf{v}_0$ at $t = 0$

(d) $\dfrac{d\mathbf{r}}{dt} = c\mathbf{r}$

29. A bee was flying along a helical path so that its position vector at time t was $\mathbf{r} = \cos t\mathbf{i} + \sin t\mathbf{j} + 16t\mathbf{k}$. At $t = 12$, it had a heart attack and died instantly. Where did it land (that is, hit the xy-plane)? Assume distance is in feet, time is in seconds, and $g = 32$ feet per second per second. *Hint*: Measure time from the instant of the heart attack and use the result of Problem 28(c).

30. Where will the bee of Example 3 hit the plane $x + y = 30$?

31. Prove that for differentiable vector-valued functions $\mathbf{F}(t)$ and $\mathbf{G}(t)$,

$$\frac{d}{dt}[\mathbf{F}(t) \times \mathbf{G}(t)] = \mathbf{F}(t) \times \mathbf{G}'(t) + \mathbf{F}'(t) \times \mathbf{G}(t)$$

and use this to show that

$$\frac{d}{dt}[\mathbf{r}(t) \times \mathbf{r}'(t)] = \mathbf{r}(t) \times \mathbf{r}''(t)$$

32. Show that if an object moves subject only to a central force (that is, $\mathbf{r}''(t) = c\mathbf{r}(t)$), then the object moves in a plane. *Hint*: Show that $\mathbf{r}(t) \times \mathbf{r}'(t)$ is a constant.

33. The angular momentum $\mathbf{L}(t)$ and torque $\boldsymbol{\tau}(t)$ of a moving particle of mass m and position vector $\mathbf{r}(t)$ are

$$\mathbf{L}(t) = m\mathbf{r}(t) \times \mathbf{v}(t) \qquad \boldsymbol{\tau}(t) = m\mathbf{r}(t) \times \mathbf{a}(t)$$

Show that (a) $\mathbf{L}'(t) = \boldsymbol{\tau}(t)$, (b) if $\boldsymbol{\tau}(t) = \mathbf{0}$ for all t, then $\mathbf{L}(t)$ is constant (the law of conservation of angular momentum), and (c) angular momentum is conserved for a particle moving under a central force.

34. Let $\mathbf{r} = r \cos \theta\mathbf{i} + r \sin \theta\mathbf{j}$, where r and θ are polar coordinates of an object of mass m moving in the xy-plane. Show that the angular momentum $\mathbf{L}(t)$ is $(mr^2 \ d\theta/dt)\mathbf{k}$. *Note*: This result together with Problems 32 and 33 and Problem 32 of Section 12.8 establish Kepler's Second Law.

35. Find the equation of the osculating plane at $P(1, 1, \frac{1}{3})$ for the motion of Example 5.

36. Prove that $\mathbf{N} = \mathbf{B} \times \mathbf{T}$, where $\mathbf{T}$, $\mathbf{N}$, and $\mathbf{B}$ form a trihedral.

14.6
Surfaces in
Three-Space

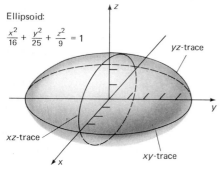

A cross section

FIGURE 1

The graph of an equation in three variables is normally a surface. We have met two examples already. The graph of $Ax + By + Cz = D$ is a plane; the graph of $(x - h)^2 + (y - k)^2 + (z - l)^2 = r^2$ is a sphere. Graphing surfaces can be very complicated. It is best accomplished by finding the intersections of the surface with well-chosen planes. These intersections are called **cross sections** (Figure 1); those with the coordinate planes are also called **traces**.

EXAMPLE 1   Sketch the graph of

$$\frac{x^2}{16} + \frac{y^2}{25} + \frac{z^2}{9} = 1$$

Solution   To find the trace in the xy-plane, we set $z = 0$ in the given equation. The graph of the resulting equation

$$\frac{x^2}{16} + \frac{y^2}{25} = 1$$

is an ellipse. The traces in the xz-plane and the yz-plane (obtained by setting $y = 0$ and $x = 0$, respectively) are also ellipses. These three traces are shown in Figure 2 and help provide a good visual image of the required surface (called an ellipsoid). ∎

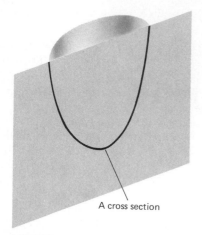

Ellipsoid:

$\frac{x^2}{16} + \frac{y^2}{25} + \frac{z^2}{9} = 1$

yz-trace

xz-trace

xy-trace

FIGURE 2

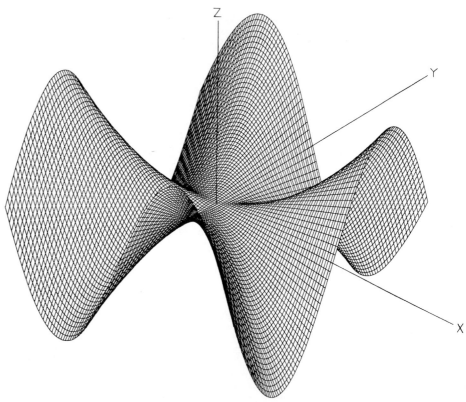

FIGURE 3

If the surface is very complicated, it may be useful to show the cross sections with many planes parallel to the coordinate planes. Here a computer with graphics capability can be very helpful. In Figure 3 on the previous page we have shown a typical computer-generated graph.

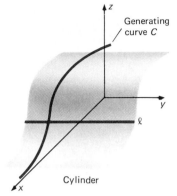

FIGURE 4

CYLINDERS You should be familiar with right circular cylinders from high school geometry. Here the word *cylinder* will denote a much more extensive class of surfaces.

Let C be a plane curve and let l be a line intersecting C that is not in the plane of C. The set of all points on lines that are parallel to l and that intersect C is called a **cylinder** (Figure 4).

Cylinders occur naturally when we graph an equation in three-space that involves just two variables. Consider as a first example,

$$\frac{y^2}{a^2} - \frac{x^2}{b^2} = 1$$

in which the variable z is missing. This equation determines a curve C in the xy-plane, namely, a hyperbola. Moreover, if $(x_1, y_1, 0)$ satisfies the equation, so does (x_1, y_1, z). As z runs through all real values, the point (x_1, y_1, z) traces out a line parallel to the z-axis. We conclude that the graph of the given equation is a cylinder, a hyperbolic cylinder (Figure 5).

A second example is the graph of $z = \sin y$ (Figure 6).

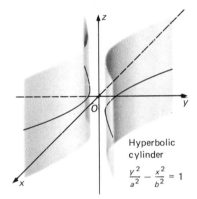

FIGURE 5

QUADRIC SURFACES If a surface is the graph in three-space of an equation of second degree, it is called a **quadric surface**. Plane sections of a quadric surface are conics.

The general second-degree equation has the form

$$Ax^2 + By^2 + Cz^2 + Dxy + Exz + Fyz + Gx + Hy + Iz + J = 0$$

It can be shown that any such equation can be reduced, by rotation and translation of coordinate axes, to one of the two forms

$$Ax^2 + By^2 + Cz^2 + J = 0$$

or

$$Ax^2 + By^2 + Iz = 0$$

FIGURE 6

The quadric surfaces represented by the first of these equations are symmetric with respect to the coordinate planes and the origin. They are called **central quadrics**.

In Figures 7 to 12 (pages 638–639), we show six general types of quadric surfaces. Study them carefully. The graphs were drawn by an artist; we do not expect that most of our readers will be able to duplicate them in doing the problems. A more reasonable drawing for most people to make is one like that shown with our next example on page 640.

QUADRIC SURFACES

ELLIPSOID $$\frac{x^2}{a^2} + \frac{y^2}{b^2} + \frac{z^2}{c^2} = 1$$	FIGURE 7
HYPERBOLOID OF ONE SHEET $$\frac{x^2}{a^2} + \frac{y^2}{b^2} - \frac{z^2}{c^2} = 1$$	FIGURE 8
HYPERBOLOID OF TWO SHEETS $$\frac{x^2}{a^2} - \frac{y^2}{b^2} - \frac{z^2}{c^2} = 1$$	FIGURE 9

QUADRIC SURFACES

ELLIPTIC PARABOLOID $z = \dfrac{x^2}{a^2} + \dfrac{y^2}{b^2}$	 FIGURE 10
HYPERBOLIC PARABOLOID $z = \dfrac{y^2}{b^2} - \dfrac{x^2}{a^2}$	 FIGURE 11
ELLIPTIC CONE $\dfrac{x^2}{a^2} + \dfrac{y^2}{b^2} - \dfrac{z^2}{c^2} = 0$	 FIGURE 12

EXAMPLE 2 Analyze the equation

$$\frac{x^2}{4} + \frac{y^2}{9} - \frac{z^2}{16} = 1$$

and sketch its graph.

Solution The traces in the three coordinate planes are obtained by setting $z = 0$, $y = 0$, and $x = 0$, respectively.

xy-plane: $\dfrac{x^2}{4} + \dfrac{y^2}{9} = 1$,   an ellipse

xz-plane: $\dfrac{x^2}{4} - \dfrac{z^2}{16} = 1$,   a hyperbola

yz-plane: $\dfrac{y^2}{9} - \dfrac{z^2}{16} = 1$,   a hyperbola

These traces are graphed in Figure 13. We have also shown the cross sections in the planes $z = 4$ and $z = -4$. Note that when we substitute $z = \pm 4$ in the original equation, we get

$$\frac{x^2}{4} + \frac{y^2}{9} - \frac{16}{16} = 1$$

which is equivalent to

$$\frac{x^2}{8} + \frac{y^2}{18} = 1$$

an ellipse. ∎

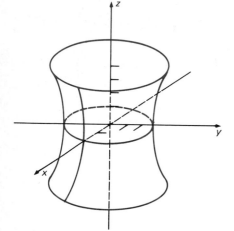

FIGURE 13

EXAMPLE 3 By inspection, name the graph of each of the following equations.

(a) $4x^2 + 4y^2 - 25z^2 + 100 = 0$
(b) $y^2 + z^2 - 12y = 0$
(c) $x^2 - z^2 = 0$
(d) $9x^2 + 4z^2 - 36y = 0$

Solution
(a) After dividing this equation by -100, it takes the form

$$-\frac{x^2}{25} - \frac{y^2}{25} + \frac{z^2}{4} = 1$$

Its graph is a hyperboloid of two sheets. It does not intersect the xy-plane but cross sections in planes parallel to this plane (and at least 2 units away) are circles.

(b) The variable x does not appear, so the graph is a cylinder parallel to the x-axis. Since the equation can be written in the form $(y - 6)^2 + z^2 = 36$, its graph is a circular cylinder.

(c) Since the variable y is missing, the graph is a cylinder. The given equation can be written $(x - z)(x + z) = 0$; so its graph consists of the two planes $x = z$ and $x = -z$.

(d) The equation can be rewritten as

$$\frac{x^2}{4} + \frac{z^2}{9} = y$$

which has an elliptic paraboloid as its graph. It is symmetric with respect to the y-axis. ∎

PROBLEM SET 14.6

Name and sketch the graph of each of the following equations in three-space.

1. $25x^2 + 16y^2 = 400$ **2.** $y^2 + z^2 = 9$

3. $2x + 5z - 12 = 0$ **4.** $z^2 = 6y$

5. $x^2 + y^2 - 8x + 2y + 13 = 0$

6. $4x^2 - 9z^2 = 0$

7. $9x^2 + 36y^2 + 16z^2 - 144 = 0$

8. $4x^2 - y^2 + 4z^2 - 4 = 0$

9. $4x^2 + 9y^2 - 12z = 0$ **10.** $x^2 - y^2 + z^2 = 0$

11. $y = e^z$ **12.** $5x + 8y = 0$

13. $x^2 - z^2 + y = 0$

14. $x^2 + y^2 - 4z^2 + 4 = 0$

15. $9x^2 + 4z^2 - 36y = 0$

16. $9x^2 + 25y^2 + 9z^2 = 225$

17. $5x + 8y - 2z = 10$ **18.** $y = \cos x$

19. $z = \sqrt{16 - x^2 - y^2}$ **20.** $z = \sqrt{x^2 + y^2 + 1}$

21. The graph of an equation in x, y, and z is symmetric with respect to the xy-plane if replacing z by $-z$ results in an equivalent equation. What condition leads to a graph that is symmetric with respect to each of the following?

(a) The yz-plane.   (b) The z-axis.

(c) The x-axis.   (d) The origin.

22. Which of the equations in Problems 1–20 has a graph which is symmetric with respect to each of the following?

(a) The xy-plane.   (b) The z-axis.

23. If the curve $z = x^2$ is revolved about the z-axis, the resulting surface has equation $z = x^2 + y^2$, obtained as a result of replacing x by $\sqrt{x^2 + y^2}$. If $y = 2x^2$ is revolved about the y-axis, what is the equation of the resulting surface?

24. Find the equation of the surface that results when the curve $z = 2y$ is revolved about the z-axis.

25. Find the equation of the surface that results when the curve $4x^2 + 3y^2 = 12$ is revolved about the y-axis.

26. Find the equation of the surface that results when the curve $4x^2 - 3y^2 = 12$ is revolved about the x-axis.

27. Find the coordinates of the foci of the ellipse which is the intersection of $z = x^2/4 + y^2/9$ with the plane $z = 4$.

28. Find the coordinates of the focus of the parabola which is the intersection of $z = x^2/4 + y^2/9$ with $x = 4$.

29. Find the area of the elliptical cross section cut from the surface $x^2/a^2 + y^2/b^2 + z^2/c^2 = 1$ by the plane $z = h$, $-c < h < c$. *Recall*: The area of the ellipse $x^2/A^2 + y^2/B^2 = 1$ is πAB.

30. Show that the volume of the solid bounded by the elliptic paraboloid $x^2/a^2 + y^2/b^2 = h - z$, $h > 0$, and the xy-plane is $\pi abh^2/2$, that is, the volume is one half the area of the base times the height. *Hint*: Use the method of slabs of Section 6.2.

31. Show that the projection in the xz-plane of the curve that is the intersection of the surfaces $y = 4 - x^2$ and $y = x^2 + z^2$ is an ellipse, and find its major and minor diameters.

32. Sketch the triangle in the plane $y = x$ that is above the plane $z = y/2$, below the plane $z = 2y$, and inside the cylinder $x^2 + y^2 = 8$. Then find the area of this triangle.

33. Show that the spiral $\mathbf{r} = t \cos t\mathbf{i} + t \sin t\mathbf{j} + t\mathbf{k}$ of Example 3 in Section 14.5 lies on the circular cone $x^2 + y^2 - z^2 = 0$. On what surface does the spiral $\mathbf{r} = 3t \cos t\mathbf{i} + t \sin t\mathbf{j} + t\mathbf{k}$ lie?

34. Show that the curve determined by $\mathbf{r} = t\mathbf{i} + t\mathbf{j} + t^2\mathbf{k}$ is a parabola and find the coordinates of its focus.

14.7
Cylindrical and
Spherical Coordinates

Giving the Cartesian (rectangular) coordinates (x, y, z) is just one of many ways of specifying the position of a point in three-space. Two other kinds of coordinates that play a significant role in calculus are cylindrical coordinates (r, θ, z) and spherical coordinates (ρ, θ, ϕ). The meaning of the three kinds of coordinates is illustrated for the same point P in Figure 1.

Cartesian Coordinates   Cylindrical Coordinates   Spherical Coordinates

FIGURE 1

The **cylindrical coordinate system** uses the polar coordinates r and θ (Section 12.6) in place of Cartesian coordinates x and y in the plane. The z-coordinate is the same as in Cartesian coordinates. We will usually require that $r \geq 0$ and we will restrict θ so that $0 \leq \theta < 2\pi$.

A point P has **spherical coordinates** (ρ, θ, ϕ) if ρ (the Greek letter rho) is the distance $|OP|$ from the origin to P, θ is the polar angle associated with the projection P' of P onto the xy-plane, and ϕ is the angle between the positive z-axis and the line segment OP. We require that

$$\rho \geq 0, \qquad 0 \leq \theta < 2\pi, \qquad 0 \leq \phi \leq \pi$$

Cylindrical and spherical coordinates are designed to help us with certain kinds of problems. We consider some of them now. Further uses will appear in Chapter 16.

CYLINDRICAL COORDINATES If a solid or a surface has an axis of symmetry, it is often wise to orient it so this axis is the z-axis and then use cylindrical coordinates. Note in particular the simplicity of the equation of a circular cylinder with z-axis symmetry (Figure 2) and also of a plane containing the z-axis (Figure 3). In Figure 3, we have allowed $r < 0$.

Cylindrical and Cartesian coordinates are related by the equations

$$
\boxed{
\begin{array}{lll}
x = r \cos \theta, & y = r \sin \theta, & z = z \\[2mm]
r^2 = x^2 + y^2, & \tan \theta = \dfrac{y}{x} &
\end{array}
}
$$

With these relationships, we can go back and forth between the two coordinate systems.

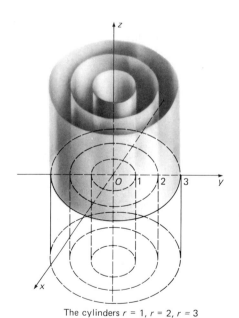

The cylinders $r = 1$, $r = 2$, $r = 3$

FIGURE 2

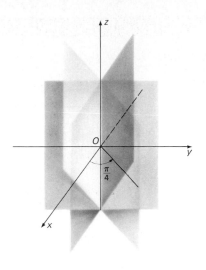

The planes $\theta = 0$

$\theta = \frac{\pi}{4}, \theta = \frac{\pi}{2}$

FIGURE 3

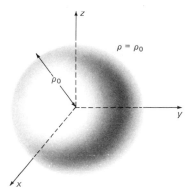

$\rho = \rho_0$

ρ_0

FIGURE 4

ϕ_0

$\phi = \phi_0$

FIGURE 5

EXAMPLE 1 Find (a) the Cartesian coordinates of the point with cylindrical coordinates $(4, 2\pi/3, 5)$ and (b) the cylindrical coordinates of the point with Cartesian coordinates $(-5, -5, 2)$.

Solution

(a) $x = 4 \cos \dfrac{2\pi}{3} = 4 \cdot \left(-\dfrac{1}{2}\right) = -2$

$y = 4 \sin \dfrac{2\pi}{3} = 4 \cdot \left(\dfrac{\sqrt{3}}{2}\right) = 2\sqrt{3}$

Thus, the Cartesian coordinates of $(4, 2\pi/3, 5)$ are $(-2, 2\sqrt{3}, 5)$.

(b) $r = \sqrt{(-5)^2 + (-5)^2} = 5\sqrt{2}$

$\tan \theta = \dfrac{-5}{-5} = 1, \quad$ so $\theta = \dfrac{5\pi}{4}$

The cylindrical coordinates of $(-5, -5, 2)$ are $(5\sqrt{2}, 5\pi/4, 2)$. ■

EXAMPLE 2 Find the equations in cylindrical coordinates of the paraboloid and cylinder whose Cartesian equations are $x^2 + y^2 = 4 - z$ and $x^2 + y^2 = 2x$.

Solution

Paraboloid: $\quad r^2 = 4 - z$

Cylinder $\quad\quad r^2 = 2r \cos \theta \quad$ or $\quad r = 2 \cos \theta$

The origin is not lost in the division by r because the origin satisfies $r = 2 \cos \theta$ with coordinates $(0, \pi/2)$. ■

EXAMPLE 3 Find the Cartesian equations of the surfaces whose equations in cylindrical coordinates are $r^2 + 4z^2 = 16$ and $r^2 \cos 2\theta = z$.

Solution Since $r^2 = x^2 + y^2$, the surface $r^2 + 4z^2 = 16$ has the Cartesian equation $x^2 + y^2 + 4z^2 = 16$, or $x^2/16 + y^2/16 + z^2/4 = 1$. Its graph is an ellipsoid.

Since $\cos 2\theta = \cos^2\theta - \sin^2\theta$, the second equation can be written $r^2 \cos^2\theta - r^2 \sin^2\theta = z$. In Cartesian coordinates it becomes $x^2 - y^2 = z$, the graph of which is a hyperbolic paraboloid. ■

SPHERICAL COORDINATES When a solid or a surface is symmetric with respect to a point, spherical coordinates are likely to play a simplifying role. In particular, a sphere centered at the origin (Figure 4) has the simple equation $\rho = \rho_0$. Also note the equation of a cone with axis along the z-axis and vertex at the origin (Figure 5), namely, $\phi = \phi_0$.

It is easy to determine the relationship between spherical and cylindrical coordinates and then between spherical and Cartesian coordinates.

$$r = \rho \sin \phi, \qquad\qquad \theta = \theta, \qquad\qquad\qquad z = \rho \cos \phi$$
$$x = \rho \sin \phi \cos \theta, \qquad y = \rho \sin \phi \sin \theta, \qquad z = \rho \cos \phi$$
$$\rho = \sqrt{x^2 + y^2 + z^2}$$

EXAMPLE 4   Find the Cartesian coordinates of the point P with spherical coordinates $(8, \pi/3, 2\pi/3)$.

Solution   We have plotted the point P in Figure 6.

$$x = 8 \sin \frac{2\pi}{3} \cos \frac{\pi}{3} = 8 \frac{\sqrt{3}}{2} \frac{1}{2} = 2\sqrt{3}$$

$$y = 8 \sin \frac{2\pi}{3} \sin \frac{\pi}{3} = 8 \frac{\sqrt{3}}{2} \frac{\sqrt{3}}{2} = 6$$

$$z = 8 \cos \frac{2\pi}{3} = 8\left(-\frac{1}{2}\right) = -4$$

Thus, P has Cartesian coordinates $(2\sqrt{3}, 6, -4)$.   ∎

FIGURE 6

EXAMPLE 5   Describe the graph of $\rho = 2 \cos \phi$.

Solution   We change to Cartesian coordinates. Multiply both sides by ρ to obtain

$$\rho^2 = 2\rho \cos \phi$$
$$x^2 + y^2 + z^2 = 2z$$
$$x^2 + y^2 + (z - 1)^2 = 1$$

The graph is a sphere of radius 1 centered at the point with Cartesian coordinates $(0, 0, 1)$.   ∎

EXAMPLE 6   Find the equation of the paraboloid $z = x^2 + y^2$ in spherical coordinates.

Solution   Substituting for x, y, and z yields

$$\rho \cos \phi = \rho^2 \sin^2\phi \cos^2\theta + \rho^2 \sin^2\phi \sin^2\theta$$
$$\rho \cos \phi = \rho^2 \sin^2\phi(\cos^2\theta + \sin^2\theta)$$
$$\rho \cos \phi = \rho^2 \sin^2\phi$$
$$\cos \phi = \rho \sin^2\phi$$
$$\rho = \cos \phi \csc^2\phi$$

Note that $\phi = \pi/2$ yields $\rho = 0$, which shows that we did not lose the origin when we canceled ρ at the fourth step.   ∎

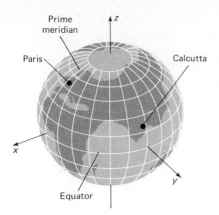

FIGURE 7

SPHERICAL COORDINATES IN GEOGRAPHY Geographers and navigators use a coordinate system very closely related to spherical coordinates, the longitude-latitude system. Suppose that the earth is a sphere with center at the origin, that the positive z-axis passes through the North Pole, and that the positive x-axis passes through the prime meridian (Figure 7). By convention, longitudes are specified in degrees east or west of the prime meridian and latitudes in degrees north or south of the equator. It is a simple matter to determine spherical coordinates from such data.

EXAMPLE 7 Assuming the earth to be a sphere of radius 3960 miles, find the great-circle distance from Paris (longitude 2.2° E, latitude 48.4° N) to Calcutta (longitude 88.2° E, latitude 22.3° N).

Solution We first calculate the spherical angles θ and ϕ for the two cities.

Paris: $\quad \theta = 2.2° \approx 0.0384$ radians

$\phi = 90° - 48.4° = 41.6° \approx 0.7261$ radians

Calcutta: $\quad \theta = 88.2° \approx 1.5394$ radians

$\phi = 90° - 22.3° = 67.7° \approx 1.1816$ radians

From these data and $\rho = 3960$ miles, we determine the Cartesian coordinates, as illustrated in Example 4.

Paris: $\quad P_1(2627.2, 100.9, 2961.3)$

Calcutta: $\quad P_2(115.1, 3662.0, 1502.6)$

FIGURE 8

Next, referring to Figure 8, we determine γ, the angle between $\overrightarrow{OP}_1$ and $\overrightarrow{OP}_2$.

$$\cos \gamma = \frac{\overrightarrow{OP}_1 \cdot \overrightarrow{OP}_2}{|\overrightarrow{OP}_1||\overrightarrow{OP}_2|} \approx \frac{(2627.2)(115.1) + (100.9)(3662) + (2961.3)(1502.6)}{(3960)(3960)}$$

$$\approx 0.3266$$

Thus, $\gamma \approx 1.2381$ radians and the great-circle distance d is

$$d = \rho\gamma \approx (3960)(1.2381) \approx 4903 \text{ miles} \qquad \blacksquare$$

PROBLEM SET 14.7

1. Change the following from cylindrical to Cartesian (rectangular) coordinates.

(a) $(6, \pi/6, -2)$ (b) $(4, 4\pi/3, -8)$

2. Change the following from spherical to Cartesian coordinates.

(a) $(8, \pi/4, \pi/6)$ (b) $(4, \pi/3, 3\pi/4)$

3. Change the following from Cartesian to spherical coordinates.

(a) $(2, -2\sqrt{3}, 4)$ (b) $(-\sqrt{2}, \sqrt{2}, 2\sqrt{3})$

4. Change the following from Cartesian to cylindrical coordinates.

(a) $(2, 2, 3)$ (b) $(4\sqrt{3}, -4, 6)$

In Problems 5–14, sketch the graph of the given cylindrical or spherical equation.

5. $r = 5$ **6.** $\rho = 5$

7. $\phi = \pi/6$ **8.** $\theta = \pi/6$

9. $r = 3 \cos \theta$ **10.** $r = 2 \sin 2\theta$

11. $\rho = 3 \cos \phi$ **12.** $\rho = \sec \phi$

13. $r^2 + z^2 = 9$ **14.** $r^2 \cos^2\theta + z^2 = 4$

In Problems 15–28, make the required change in the given equation.

15. $x^2 + y^2 = 9$ to cylindrical coordinates.

16. $x^2 - y^2 = 25$ to cylindrical coordinates.

17. $x^2 + y^2 + 4z^2 = 10$ to cylindrical coordinates.

18. $x^2 + y^2 + 4z^2 = 10$ to spherical coordinates.

19. $2x^2 + 2y^2 - 4z^2 = 0$ to spherical coordinates.

20. $x^2 - y^2 - z^2 = 1$ to spherical coordinates.

21. $r^2 + 2z^2 = 4$ to spherical coordinates.

22. $\rho = 2 \cos \phi$ to cylindrical coordinates.

23. $x + y = 4$ to cylindrical coordinates.

24. $x + y + z = 1$ to spherical coordinates.

25. $x^2 + y^2 = 9$ to spherical coordinates.

26. $r = 2 \sin \theta$ to Cartesian coordinates.

27. $r^2 \cos 2\theta = z$ to Cartesian coordinates.

28. $\rho \sin \phi = 1$ to Cartesian coordinates.

29. The parabola $z = 2x^2$ is revolved about the z-axis. Write the equation of the resulting surface in cylindrical coordinates.

30. The hyperbola $2x^2 - z^2 = 2$ is revolved about the z-axis. Write the equation of the resulting surface in cylindrical coordinates.

C 31. Find the great-circle distance from St. Paul (longitude 93.1° W, latitude 45° N) to Oslo (longitude 10.5° E, latitude 59.6° N). See Example 7.

C 32. Find the great-circle distance from New York (longitude 74° W, latitude 40.4° N) to Greenwich (longitude 0°, latitude 51.3° N).

C 33. Find the great-circle distance from St. Paul (longitude 93.1° W, latitude 45° N) to Turin, Italy (longitude 7.4° E, latitude 45° N).

C 34. What is the distance along the 45° parallel between St. Paul and Turin? See Problem 33.

C 35. How close does the great-circle route from St. Paul to Turin get to the North Pole? See Problem 33.

36. Let $(\rho_1, \theta_1, \phi_1)$ and $(\rho_2, \theta_2, \phi_2)$ be the spherical coordinates of two points and let d be the straight-line distance between them. Show that

$$d = \{(\rho_1 - \rho_2)^2 + 2\rho_1\rho_2[1 - \cos(\theta_1 - \theta_2)\sin\phi_1\sin\phi_2$$
$$- \cos\phi_1\cos\phi_2]\}^{1/2}$$

37. Let (a, θ_1, ϕ_1) and (a, θ_2, ϕ_2) be two points on the sphere $\rho = a$. Show (using Problem 36) that the great-circle distance between these points is $a\gamma$, where $0 \leq \gamma \leq \pi$ and

$$\cos\gamma = \cos(\theta_1 - \theta_2)\sin\phi_1\sin\phi_2 + \cos\phi_1\cos\phi_2$$

38. As you may have guessed, there is a simple formula for expressing great-circle distance directly in terms of longitude and latitude. Let (α_1, β_1) and (α_2, β_2) be the longitude-latitude coordinates of two points on the surface of the earth, where we interpret N and E as positive and S and W as negative. Show that the great-circle distance between these points is 3960γ miles, where $0 \leq \gamma \leq \pi$ and

$$\cos\gamma = \cos(\alpha_1 - \alpha_2)\cos\beta_1\cos\beta_2 + \sin\beta_1\sin\beta_2$$

C 39. Use Problem 38 to find the great-circle distance between each pair of places.

(a) New York and Greenwich (see Problem 32).
(b) St. Paul and Turin (see Problem 33).
(c) Turin and the South Pole (use $\alpha_1 = \alpha_2$).
(d) New York and Cape Town (18.4° E, 33.9° S).
(e) Two points on the equator with longitudes 100° E and 80° W, respectively.

40. It is easy to see that the graph of $\rho = 2a \cos \phi$ is a sphere of radius a sitting on the xy-plane at the origin. But what is the graph of $\rho = 2a \sin \phi$?

14.8 Chapter Review

TRUE-FALSE QUIZ

Respond with true or false to each of the following assertions. Be prepared to defend your answer.

1. Each point in three-space has a unique set of Cartesian coordinates.
2. The equation $x^2 + y^2 + z^2 - 4x + 9 = 0$ represents a sphere.
3. The linear equation $Ax + By + Cz = D$ represents a plane in three-space provided A, B, and C are not all zero.
4. In three-space, the equation $Ax + By = C$ represents a line.

5. The planes $3x - 2y + 4z = 12$ and $3x - 2y + 4z = -12$ are parallel and 24 units apart.

6. The vector $\langle 1, -2, 3 \rangle$ is parallel to the plane $2x - 4y + 6z = 5$.

7. The line $x = 2t - 1$, $y = 4t + 2$, $z = 6t - 5$ goes through the point $(0, 4, -2)$.

8. If $\mathbf{u} = a\mathbf{i} + b\mathbf{j} + c\mathbf{k}$ is a unit vector then a, b, and c are direction cosines for $\mathbf{u}$.

9. For any vector $\mathbf{u}$, $\|\mathbf{u}|\mathbf{u}|\| = |\mathbf{u}|^2$.

10. For any vector $\mathbf{u}$, $|\mathbf{u}| \cdot \mathbf{u} = \mathbf{u} \cdot |\mathbf{u}|$.

11. For any vectors $\mathbf{u}$ and $\mathbf{v}$, $|\mathbf{u} \times \mathbf{v}| = |\mathbf{v} \times \mathbf{u}|$.

12. If $\mathbf{u}$ is a scalar multiple of $\mathbf{v}$, then $\mathbf{u} \times \mathbf{v} = \mathbf{0}$.

13. The cross product of two unit vectors is a unit vector.

14. Multiplying each component of a vector $\mathbf{v}$ by the scalar a multiplies the length of $\mathbf{v}$ by a.

15. For any nonzero and nonperpendicular vectors $\mathbf{u}$ and $\mathbf{v}$ with angle θ between them, $|\mathbf{u} \times \mathbf{v}| \div (\mathbf{u} \cdot \mathbf{v}) = \tan \theta$.

16. If $\mathbf{u} \cdot \mathbf{v} = 0$ and $\mathbf{u} \times \mathbf{v} = \mathbf{0}$, then $\mathbf{u}$ or $\mathbf{v}$ is $\mathbf{0}$.

17. The volume of the parallelepiped determined by $2\mathbf{i}$, $2\mathbf{j}$, and $\mathbf{j} \times \mathbf{i}$ is 4.

18. For all vectors $\mathbf{u}$, $\mathbf{v}$, and $\mathbf{w}$, $\mathbf{u} \times (\mathbf{v} \times \mathbf{w}) = (\mathbf{u} \times \mathbf{v}) \times \mathbf{w}$.

19. If $a_1\mathbf{i} + a_2\mathbf{j} + a_3\mathbf{k}$ is a vector in the plane $b_1x + b_2y + b_3z = 0$, then $a_1b_1 + a_2b_2 + a_3b_3 = 0$

20. Any line can be represented by both parametric equations and by symmetric equations.

21. If $\mathbf{r}(t)$ is the position vector for a moving particle, then $D_t|\mathbf{r}(t)| = |\mathbf{r}'(t)|$.

22. The curve which is the intersection of the sphere $x^2 + y^2 + z^2 = 1$ and the plane $ax + by + cz = 0$ has constant curvature 1.

23. The graph of the equation $z = \rho$ is the z-axis (here ρ is a spherical coordinate).

24. The graph of $y = x^2$ in three-space is a paraboloid.

25. If we restrict ρ, θ, and ϕ by $\rho \geq 0$, $0 \leq \theta < 2\pi$, and $0 \leq \phi \leq \pi$, then each point in three-space has a unique set of spherical coordinates.

SAMPLE TEST PROBLEMS

1. Find the equation of the sphere with $(-2, 3, 3)$ and $(4, 1, 5)$ as endpoints of a diameter.

2. Find the center and radius of the sphere with equation $x^2 + y^2 + z^2 - 6x + 2y - 8z = 0$.

3. Sketch the two position vectors $\mathbf{a} = 2\mathbf{i} - \mathbf{j} + 2\mathbf{k}$ and $\mathbf{b} = 5\mathbf{i} + \mathbf{j} - 3\mathbf{k}$. Then find each of the following.

(a) Their lengths.
(b) Their direction cosines.
(c) The unit vector with the same direction as $\mathbf{a}$.
(d) The angle θ between $\mathbf{a}$ and $\mathbf{b}$.

4. Let $\mathbf{a} = 2\mathbf{i} - \mathbf{j} + \mathbf{k}$, $\mathbf{b} = -\mathbf{i} + 3\mathbf{j} + 2\mathbf{k}$, and $\mathbf{c} = \mathbf{i} + 2\mathbf{j} - \mathbf{k}$. Find each of the following.

(a) $\mathbf{a} \times \mathbf{b}$
(b) $\mathbf{a} \times (\mathbf{b} + \mathbf{c})$
(c) $\mathbf{a} \cdot (\mathbf{b} \times \mathbf{c})$
(d) $\mathbf{a} \times (\mathbf{b} \times \mathbf{c})$

5. Find all vectors that are perpendicular to both of the vectors $3\mathbf{i} + 3\mathbf{j} - \mathbf{k}$ and $-\mathbf{i} - 2\mathbf{j} + 4\mathbf{k}$.

6. Find the unit vectors that are perpendicular to the plane determined by the three points $(3, -6, 4)$, $(2, 1, 1)$, and $(5, 0, -2)$.

7. Write the equation of the plane through the point $(-5, 7, -2)$ that satisfies each condition.

(a) Parallel to the xz-plane.
(b) Perpendicular to the x-axis.
(c) Parallel to both the x- and y-axes.
(d) Parallel to the plane $3x - 4y + z = 7$.

8. A plane through the point $(2, -4, -5)$ is perpendicular to the line joining the points $(-1, 5, -7)$ and $(4, 1, 1)$.

(a) Write a vector equation of the plane.
(b) Find a Cartesian equation of the plane.
(c) Sketch the plane by drawing its traces.

9. Find the value of C if the plane $x + 5y + Cz + 6 = 0$ is perpendicular to the plane $4x - y + z - 17 = 0$.

10. Find a Cartesian equation of the plane through the three points $(2, 3, -1)$, $(-1, 5, 2)$, and $(-4, -2, 2)$.

11. Find parametric equations for the line through $(-2, 1, 5)$ and $(6, 2, -3)$.

12. Find the points where the line of intersection of the planes $x - 2y + 4z - 14 = 0$ and $-x + 2y - 5z + 30 = 0$ pierces the yz- and xz-planes.

13. Write the equation of the line in Problem 12 in parametric form.

14. Find symmetric equations of the line through $(4, 5, 8)$ and perpendicular to the plane $3x + 5y + 2z = 30$. Sketch the plane and the line.

15. Write a vector equation of the line through $(2, -2, 1)$ and $(-3, 2, 4)$.

16. Sketch the curve whose vector equation is $\mathbf{r}(t) = t\mathbf{i} + \frac{1}{2}t^2\mathbf{j} + \frac{1}{3}t^3\mathbf{k}$, $-2 \leq t \leq 3$.

17. Find the symmetric equations for the tangent line to the curve of Problem 16 at the point where $t = 2$. Also find the equation of the normal plane at this point.

18. Find $\mathbf{r}'(\pi/2)$, $\mathbf{T}(\pi/2)$, and $\mathbf{r}''(\pi/2)$ if

$$\mathbf{r}(t) = \langle t \cos t, t \sin t, 2t \rangle.$$

19. Find the length of the curve

$$\mathbf{r}(t) = e^t \sin t\mathbf{i} + e^t \cos t\mathbf{j} + e^t\mathbf{k}, \qquad 1 \leq t \leq 5$$

20. Suppose an object is moving so that its position vector at time t is

$$\mathbf{r}(t) = e^t\mathbf{i} + e^{-t}\mathbf{j} + 2t\mathbf{k}$$

Find $\mathbf{v}(t)$, $\mathbf{a}(t)$, and $\kappa(t)$, at $t = \ln 2$.

21. If $\mathbf{r}(t) = t\mathbf{i} + t^2\mathbf{j} + t^3\mathbf{k}$ is the position vector for a moving particle at time t, find the tangential and normal components, a_T and a_N, of the acceleration vector at $t = 1$.

For each of the equations in Problems 22–30, name and sketch the graph in three-space.

22. $x^2 + y^2 = 81$

23. $x^2 + y^2 + z^2 = 81$

24. $z^2 = 4y$

25. $x^2 + z^2 = 4y$

26. $3y - 6z - 12 = 0$

27. $3x + 3y - 6z - 12 = 0$

28. $x^2 + y^2 - z^2 - 1 = 0$

29. $3x^2 + 4y^2 + 9z^2 - 36 = 0$

30. $3x^2 + 4y^2 + 9z^2 + 36 = 0$

31. Write the following Cartesian equations in cylindrical coordinate form.

(a) $x^2 + y^2 = 9$ (b) $x^2 + 4y^2 = 16$

(c) $x^2 + y^2 = 9z$ (d) $x^2 + y^2 + 4z^2 = 10$

32. Find the Cartesian equation corresponding to each of the following cylindrical coordinate equations.

(a) $r^2 + z^2 = 9$ (b) $r^2 \cos^2\theta + z^2 = 4$

(c) $r^2 \cos 2\theta + z^2 = 1$

33. Write the following equations in spherical coordinate form.

(a) $x^2 + y^2 + z^2 = 4$ (b) $2x^2 + 2y^2 - 2z^2 = 0$

(c) $x^2 - y^2 - z^2 = 1$ (d) $x^2 + y^2 = z$

34. Find the (straight-line) distance between the points whose spherical coordinates are $(8, \pi/4, \pi/6)$ and $(4, \pi/3, 3\pi/4)$.

35. Find the distance between the parallel planes $2x - 3y + \sqrt{3}z = 4$ and $2x - 3y + \sqrt{3}z = 9$.

36. Find the acute angle between the planes $2x - 4y + z = 7$ and $3x + 2y - 5z = 9$.

37. Show that if the speed of a moving particle is constant, then its velocity and acceleration vectors are orthogonal.

15 The Derivative in *n*-Space

Lagrange is the lofty pyramid of the mathematical sciences.
Napoleon Bonaparte
The great masters of modern analysis are Lagrange, Laplace,
and Gauss, who were contemporaries.
W. W. R. Ball (1901)

Joseph-Louis Lagrange
1736–1813

Born in Turin, Lagrange was of mixed Italian and French ancestry. His father was rich, but managed to squander his entire fortune. In later life, Lagrange was to call this a lucky disaster for "if I had inherited a fortune I should probably not have cast my lot with mathematics."

Turned on to mathematics by reading an essay on calculus, he quickly mastered the subject. At age 19, he began perhaps his greatest work, *Mécanique analytique*, though it was not published until he was 52. This book, notable for a complete lack of diagrams, unified mechanics, turning it into what Hamilton called a "scientific poem."

When Lagrange sent some of his results to Euler, the famous mathematician immediately recognized the brilliance of the young man and postponed publishing some of his own related work so Lagrange could publish first—a rare example of academic unselfishness.

Lagrange's career was illustrious, including 20 years as court mathematician to Frederick the Great in Berlin and later the first professorship at the École Normale in Paris. Spanning the revolution, he was a favorite of Marie Antoinette and then of Napoleon. In Paris, he helped perfect the metric system of weights and measures.

His mathematical contributions were many, and included the method of Lagrange multipliers discussed in this chapter.

15.1
Functions of Two or More Variables

Two kinds of functions have so far made their appearance. The first, typified by $f(x) = x^2$, associates with the real number x another real number $f(x)$. We call it a real-valued function of a real variable. The second type of function, illustrated by $\mathbf{f}(x) = \langle x^3, e^x \rangle$, associates with the real number x a vector $\mathbf{f}(x)$. We call it a vector-valued function of a real variable.

Our interest shifts now to a **real-valued function of two real variables**; that is, a function f that assigns to each ordered pair (x, y) in some set D of the plane a (unique) real number $f(x, y)$. Examples are

$$(1) \qquad\qquad\qquad f(x, y) = x^2 + 3y^2$$

$$(2) \qquad\qquad\qquad g(x, y) = 2x\sqrt{y}$$

Note that $f(-1, 4) = (-1)^2 + 3(4)^2 = 49$ and $g(-1, 4) = 2(-1)\sqrt{4} = -4$.

The set D is called the **domain** of the function. If it is not specified, we take D to be the natural domain, that is, the set of all points (x, y) in the plane for which the function rule makes sense and gives a real number value. For $f(x, y) = x^2 + 3y^2$, the natural domain is the whole plane; for $g(x, y) = 2x\sqrt{y}$, it is $\{(x, y) : x \in \mathbb{R}, y \geq 0\}$. The **range** of a function is its set of values. If $z = f(x, y)$, we call x and y the **independent variables** and z the **dependent variable**.

All that we have said extends in a natural way to real-valued functions of three real variables (or even n real variables). We will feel free to use such functions without further comment.

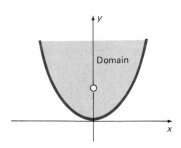

Domain

FIGURE 1

EXAMPLE 1 In the xy-plane, sketch the natural domain for

$$f(x, y) = \frac{\sqrt{y - x^2}}{x^2 + (y - 1)^2}$$

Solution For this rule to make sense, we must exclude $\{(x, y) : y < x^2\}$ and the point $(0, 1)$. The resulting domain is shown in Figure 1. ∎

GRAPHS By the **graph** of a function f of two variables, we mean the graph of the equation $z = f(x, y)$. This graph will normally be a surface (Figure 2) and, since to each (x, y) in the domain there corresponds just one value z, each vertical line intersects the surface in at most one point.

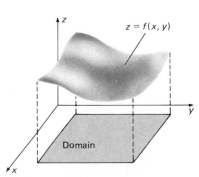

$z = f(x, y)$

Domain

FIGURE 2

CHAPTER 15/The Derivative in n-Space

$z = \frac{1}{3}\sqrt{36 - 9x^2 - 4y^2}$

FIGURE 3

EXAMPLE 2 Sketch the graph of $f(x, y) = \frac{1}{3}\sqrt{36 - 9x^2 - 4y^2}$.

Solution Let $z = \frac{1}{3}\sqrt{36 - 9x^2 - 4y^2}$ and note that $z \geq 0$. If we square both sides and simplify, we obtain the equation

$$9x^2 + 4y^2 + 9z^2 = 36$$

which we recognize as the equation of an ellipsoid (see Section 14.6). The graph of the given function is the upper half of this ellipsoid; it is shown in Figure 3. ∎

EXAMPLE 3 Sketch the graph of $z = f(x, y) = y^2 - x^2$.

Solution The graph is a hyperbolic paraboloid (see Section 14.6); it is sketched in Figure 4.

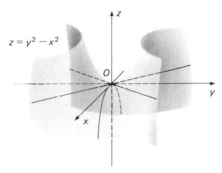

$z = y^2 - x^2$

FIGURE 4 ∎

LEVEL CURVES To sketch the surface corresponding to the graph of a function $z = f(x, y)$ of two variables is often very difficult. Map makers have given us another and usually simpler way to picture a surface, the so-called contour map. Each horizontal plane $z = c$ intersects the surface in a curve. The projection of this curve on the xy-plane is called a **level curve** (Figure 5),

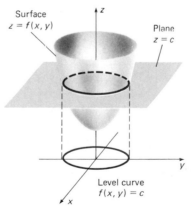

Surface
$z = f(x, y)$

Plane
$z = c$

Level curve
$f(x, y) = c$

FIGURE 5

Surface

8000 ft
7000 ft
6000 ft
5000 ft

Contour map
with level curves

FIGURE 6

and a collection of such curves is a **contour map**. We have shown a contour map for a hill-shaped surface in Figure 6.

EXAMPLE 4 Draw contour maps for the surfaces corresponding to $z = \frac{1}{3}\sqrt{36 - 9x^2 - 4y^2}$ and $z = y^2 - x^2$ (see Examples 2 and 3).

Solution The level curves of $z = \frac{1}{3}\sqrt{36 - 9x^2 - 4y^2}$ corresponding to $z = 0, 1, 1.5, 1.75, 2$ are shown in Figure 7. They are ellipses. Similarly, in Figure 8, we show the level curves of $z = y^2 - x^2$ for $z = -5, -4, -3, \ldots,$ 2, 3, 4. These curves are hyperbolas. ∎

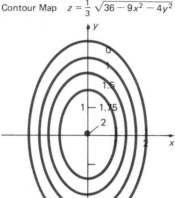

Contour Map $z = \frac{1}{3}\sqrt{36 - 9x^2 - 4y^2}$

FIGURE 7

Contour Map $z = y^2 - x^2$

FIGURE 8

Contour Map $z = xy$

FIGURE 9

EXAMPLE 5 Sketch a contour map for $z = f(x, y) = xy$.

Solution The level curves corresponding to $z = -4, -1, 0, 1, 4$ are shown in Figure 9. It can be shown that they are hyperbolas. Comparing the contour map of Figure 9 with that of Figure 8 suggests that the graph of $z = xy$ might be a hyperbolic paraboloid but with axes rotated through 45°. The suggestion is correct. ∎

COMPUTER GRAPHICS The modern electronic computer provides a powerful tool for visualizing surfaces in three-space. Although it would take us too far afield to discuss the vast subject of computer graphics, the computer-generated figures on pages 654 and 655 give some idea of what can be done.

PROBLEM SET   15.1

1. Let $f(x, y) = x^2y + \sqrt{y}$. Find each value.
(a) $f(2, 1)$ (b) $f(3, 0)$
(c) $f(1, 4)$ (d) $f(a, a^4)$
(e) $f(1/x, x^4)$ (f) $f(2, -4)$

What is the natural domain for this function?

2. Let $f(x, y) = y/x + xy$. Find each value.
(a) $f(1, 2)$ (b) $f(\frac{1}{4}, 4)$
(c) $f(4, \frac{1}{4})$ (d) $f(a, a)$
(e) $f(1/x, x^2)$ (f) $f(0, 0)$

What is the natural domain for this function?

CHAPTER 15/The Derivative in n-Space

3. Let $g(x, y, z) = x^2 \sin yz$. Find each value.

(a) $g(1, \pi, 2)$ (b) $g(2, 1, \pi/6)$

(c) $g(4, 2, \pi/4)$ $\boxed{C}$ (d) $g(\pi, \pi, \pi)$

$\boxed{C}$ (e) $g(1.2, 3.1, 4.2)$

4. Let $g(x, y, z) = \sqrt{x} \cos y + z^2$. Find each value.

(a) $g(4, 0, 2)$ (b) $g(-9, \pi, 3)$

(c) $g(2, \pi/3, -1)$ $\boxed{C}$ (d) $g(3, 6, 1.2)$

$\boxed{C}$ (e) $g(-2, 2, 3)$

5. Find $F(f(t), g(t))$ if $F(x, y) = x^2 y$ and $f(t) = t \cos t$, $g(t) = \sec^2 t$.

6. Find $F(f(t), g(t))$ if $F(x, y) = e^x + y^2$ and $f(t) = \ln t^2$, $g(t) = e^{t/2}$.

In Problems 7–16, sketch the graph of f.

7. $f(x, y) = 6$ **8.** $f(x, y) = 6 - x$

9. $f(x, y) = 6 - x - 2y$ **10.** $f(x, y) = 6 - x^2$

11. $f(x, y) = \sqrt{16 - x^2 - y^2}$

12. $f(x, y) = \sqrt{16 - 4x^2 - y^2}$

13. $f(x, y) = 3 - x^2 - y^2$ **14.** $f(x, y) = 2 - x - y^2$

15. $f(x, y) = e^{-(x^2 + y^2)}$ **16.** $f(x, y) = x^2/y, \; y > 0$

In Problems 17–22, sketch the level curve $z = k$ for the indicated values of k.

17. $z = \frac{1}{2}(x^2 + y^2)$, $k = 0, 2, 4, 6, 8$

18. $z = \dfrac{x}{y}$, $k = -2, -1, 0, 1, 2$

19. $z = \dfrac{x^2}{y}$, $k = -4, -1, 0, 1, 4$

20. $z = x^2 + y$, $k = -4, -1, 0, 1, 4$

21. $z = \dfrac{x^2 + y}{x + y^2}$, $k = 0, 1, 2, 4$

22. $z = y - \sin x$, $k = -2, -1, 0, 1, 2$

23. If $T(x, y)$ is the temperature at a point (x, y) in the plane, the level curves of T are called **isothermal curves**. Draw the isothermal curves for

$$T(x, y) = \dfrac{x^2}{x^2 + y^2}$$

corresponding to $T = \frac{1}{10}, \frac{1}{5}, \frac{1}{2}, 0$.

24. If $V(x, y)$ is the voltage at a point (x, y) in the plane, the level curves of V are called **equipotential curves**. Draw the equipotential curves for

$$V(x, y) = \dfrac{4}{\sqrt{(x - 2)^2 + (y + 3)^2}}$$

corresponding to $V = \frac{1}{2}, 1, 2, 4$.

In Problems 25–28, describe geometrically the domain of each of the indicated functions of three variables.

25. $f(x, y, z) = \sqrt{x^2 + y^2 + z^2 - 16}$

26. $f(x, y, z) = \sqrt{x^2 + y^2 - z^2 - 9}$

27. $f(x, y, z) = \sqrt{144 - 16x^2 - 9y^2 - 144z^2}$

28. $f(x, y, z) = \dfrac{(144 - 16x^2 - 16y^2 + 9z^2)^{3/2}}{xyz}$

A **level surface** for a function f of three variables is the graph of the set of points in three-dimensional space whose coordinates satisfy an equation $f(x, y, z) = k$, where k is a constant. Describe geometrically the level surfaces for the functions defined in Problems 29–33.

29. $f(x, y, z) = x^2 + y^2 + z^2; \; k > 0$

30. $f(x, y, z) = 100x^2 + 16y^2 + 25z^2; \; k > 0$

31. $f(x, y, z) = 16x^2 + 16y^2 - 9z^2; \; k \in \mathbb{R}$

32. $f(x, y, z) = 9x^2 - 4y^2 - z^2; \; k \in \mathbb{R}$

33. $f(x, y, z) = 4x^2 - 9y^2; \; k \in \mathbb{R}$

34. Sketch (as best you can) the graph of the monkey saddle $z = x(x^2 - 3y^2)$. Begin by noting where $z = 0$.

35. The contour map in Figure 10 shows level curves for a mountain 3000 feet high.

(a) What is special about the path to the top labeled AC? What is special about BC?

(b) Make good estimates of the total lengths of path AC and path BC.

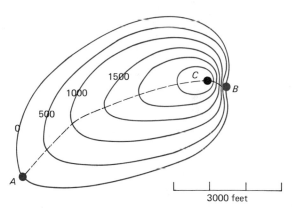

FIGURE 10

36. Put the missing axes on the computer graphs on pages 654 and 655. In each case the z-axis points straight up. Begin by noting the values of the function on the x- and y-axes.

37. Identify the graph of $f(x, y) = x^2 - x + 3y^2 + 12y - 13$, state where it attains its minimum value, and find this minimum value.

$$z = xy \cos(xy)$$

$$z = \left(\frac{3-x}{2}\right)\left(\frac{9}{4} - y^2\right)^4 \sin^2\left(\frac{\pi x}{2}\right)$$
$$= \left(1.5 - \frac{x}{2}\right)\left(2.25 - y^2\right)^4 \sin^2\left(\frac{\pi x}{2}\right)$$

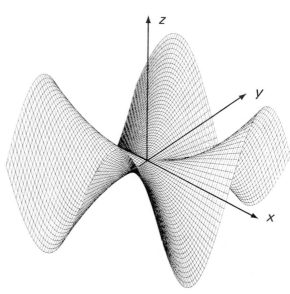

$$z = \begin{cases} xy \dfrac{x^2 - y^2}{x^2 + y^2} & \text{if } (x, y) \neq (0, 0) \\ 0 & \text{if } (x, y) = (0, 0) \end{cases}$$

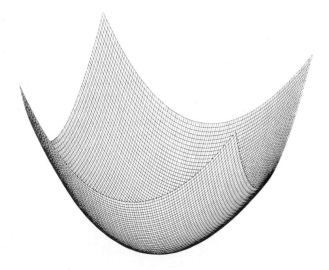

$$z = x^2 + y^2$$

The eight computer drawings on this page and the facing page were designed by Norton Starr, Amherst College, and are reprinted with his permission. © 1983 N*.

$$z = -10\sqrt{|xy|}$$

$$z = 2y^2 \sin(2x)$$

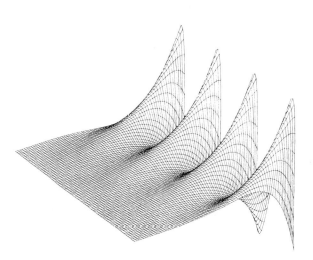

$$z = \cos\left[7.17307\left(x + \frac{3}{2}\right) + 2y - 3\right]e^{2y-3}$$

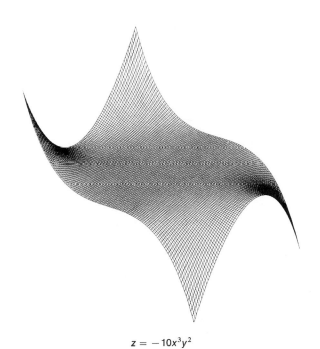

$$z = -10x^3y^2$$

15.2
Partial Derivatives

Suppose that f is a function of two variables x and y. If y is held constant, say $y = y_0$, then $f(x, y_0)$ is a function of the single variable x. Its derivative at $x = x_0$ is called the **partial derivative of f with respect to x** at (x_0, y_0) and is denoted by $f_x(x_0, y_0)$. Thus,

$$f_x(x_0, y_0) = \lim_{\Delta x \to 0} \frac{f(x_0 + \Delta x, y_0) - f(x_0, y_0)}{\Delta x}$$

Similarly, the partial derivative of f with respect to y at (x_0, y_0) is denoted by $f_y(x_0, y_0)$ and is given by

$$f_y(x_0, y_0) = \lim_{\Delta y \to 0} \frac{f(x_0, y_0 + \Delta y) - f(x_0, y_0)}{\Delta y}$$

Rather than calculate $f_x(x_0, y_0)$ and $f_y(x_0, y_0)$ directly from the boxed definitions, we typically find $f_x(x, y)$ and $f_y(x, y)$ using the standard rules for derivatives; then we substitute $x = x_0$ and $y = y_0$.

EXAMPLE 1   Find $f_x(1, 2)$ and $f_y(1, 2)$ if $f(x, y) = x^2 y + 3y^3$.

Solution   To find $f_x(x, y)$, we treat y as a constant and differentiate with respect to x, obtaining

$$f_x(x, y) = 2xy + 0$$

Thus,

$$f_x(1, 2) = 2 \cdot 1 \cdot 2 = 4$$

Similarly,

$$f_y(x, y) = x^2 + 9y^2$$

and so

$$f_y(1, 2) = 1^2 + 9 \cdot 2^2 = 37 \qquad \blacksquare$$

If $z = f(x, y)$, we use the following alternative notations.

$$f_x(x, y) = \frac{\partial z}{\partial x} = \frac{\partial f(x, y)}{\partial x} \qquad\qquad f_y(x, y) = \frac{\partial z}{\partial y} = \frac{\partial f(x, y)}{\partial y}$$

$$f_x(x_0, y_0) = \frac{\partial z}{\partial x}\bigg|_{(x_0, y_0)} \qquad\qquad f_y(x_0, y_0) = \frac{\partial z}{\partial y}\bigg|_{(x_0, y_0)}$$

The symbol ∂ is special to mathematics and is called the partial derivative sign.

EXAMPLE 2 If $z = x^2 \sin(xy^2)$, find $\partial z / \partial x$ and $\partial z / \partial y$.

Solution

$$\frac{\partial z}{\partial x} = x^2 \frac{\partial}{\partial x}\left[\sin(xy^2)\right] + \sin(xy^2)\frac{\partial}{\partial x}(x^2)$$

$$= x^2 \cos(xy^2)\frac{\partial}{\partial x}(xy^2) + \sin(xy^2)\cdot 2x$$

$$= x^2 \cos(xy^2)\cdot y^2 + 2x\sin(xy^2)$$

$$= x^2 y^2 \cos(xy^2) + 2x\sin(xy^2)$$

$$\frac{\partial z}{\partial y} = x^2 \cos(xy^2)\cdot 2xy = 2x^3 y \cos(xy^2)$$

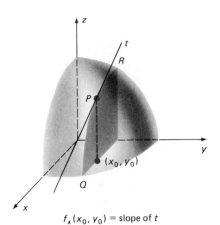

$f_x(x_0, y_0) = $ slope of t

FIGURE 1

$f_y(x_0, y_0) = $ slope of t

FIGURE 2

GEOMETRIC AND PHYSICAL INTERPRETATIONS Consider the surface whose equation is $z = f(x, y)$. The plane $y = y_0$ intersects this surface in the plane curve QPR (Figure 1) and the value of $f_x(x_0, y_0)$ is the slope of the tangent line to this curve at $P(x_0, y_0, f(x_0, y_0))$. Similarly, the plane $x = x_0$ cuts the surface in the plane curve LPM (Figure 2) and $f_y(x_0, y_0)$ is the slope of the tangent line to this curve at P.

Partial derivatives may also be interpreted as (instantaneous) rates of change. Suppose that a violin string is fixed at points A and B and vibrates in the xz-plane. Figure 3 shows the position of the string at a typical time t. If $z = f(x, t)$ denotes the height of the string at the point P with abscissa x at time t, then $\partial z / \partial x$ is the slope of the string at P and $\partial z / \partial t$ is the time rate of change of height of P along the indicated vertical line. In other words, $\partial z / \partial t$ is the vertical velocity of P.

EXAMPLE 3 The surface $z = f(x, y) = \sqrt{9 - 2x^2 - y^2}$ and the plane $y = 1$ intersect in a curve as in Figure 1. Find parametric equations for the tangent line at $(\sqrt{2}, 1, 2)$.

Solution

$$f_x(x, y) = \tfrac{1}{2}(9 - 2x^2 - y^2)^{-1/2}(-4x)$$

and so $f_x(\sqrt{2}, 1) = -\sqrt{2}$. This number is the slope of the tangent line to the curve at $(\sqrt{2}, 1, 2)$, that is, $-\sqrt{2}/1$ is the ratio of rise to run along the tangent line. It follows that this line has direction numbers $(1, 0, -\sqrt{2})$ and, since it goes through $(\sqrt{2}, 1, 2)$,

$$x = \sqrt{2} + t, \qquad y = 1, \qquad z = 2 - \sqrt{2}t$$

provide the required parametric equations.

Position of string at time t

$P = (x, f(x, t))$

FIGURE 3

EXAMPLE 4 The volume of a certain gas is related to its temperature T and its pressure P by the gas law $PV = 10T$, where V is measured in cubic inches, P in pounds per square inch, and T in degrees Celsius. If T is kept constant at 200, what is the rate of change of pressure with respect to volume at $V = 50$?

Solution Since $P = 10T/V$,

$$\frac{\partial P}{\partial V} = \frac{-10T}{V^2}$$

Thus,

$$\frac{\partial P}{\partial V}\bigg|_{T=200, V=50} = \frac{(-10)(200)}{(50)^2} = -\frac{4}{5} \qquad \blacksquare$$

HIGHER PARTIAL DERIVATIVES Since a partial derivative of a function of x and y is, in general, another function of these same two variables, it may be differentiated partially with respect to either x or y, resulting in four **second partial derivatives** of f:

$$f_{xx} = \frac{\partial}{\partial x}\left(\frac{\partial f}{\partial x}\right) = \frac{\partial^2 f}{\partial x^2} \qquad\qquad f_{yy} = \frac{\partial}{\partial y}\left(\frac{\partial f}{\partial y}\right) = \frac{\partial^2 f}{\partial y^2}$$

$$f_{xy} = (f_x)_y = \frac{\partial}{\partial y}\left(\frac{\partial f}{\partial x}\right) = \frac{\partial^2 f}{\partial y\,\partial x} \qquad f_{yx} = (f_y)_x = \frac{\partial}{\partial x}\left(\frac{\partial f}{\partial y}\right) = \frac{\partial^2 f}{\partial x\,\partial y}$$

EXAMPLE 5 Find the four second partial derivatives of

$$f(x, y) = xe^y - \sin(x/y) + x^3y^2.$$

Solution

$$f_x(x, y) = e^y - \frac{1}{y}\cos\left(\frac{x}{y}\right) + 3x^2y^2$$

$$f_y(x, y) = xe^y + \frac{x}{y^2}\cos\left(\frac{x}{y}\right) + 2x^3y$$

$$f_{xx}(x, y) = \frac{1}{y^2}\sin\left(\frac{x}{y}\right) + 6xy^2$$

$$f_{yy}(x, y) = xe^y + \frac{x^2}{y^4}\sin\left(\frac{x}{y}\right) - \frac{2x}{y^3}\cos\left(\frac{x}{y}\right) + 2x^3$$

$$f_{xy}(x, y) = e^y - \frac{x}{y^3}\sin\left(\frac{x}{y}\right) + \frac{1}{y^2}\cos\left(\frac{x}{y}\right) + 6x^2y$$

$$f_{yx}(x, y) = e^y - \frac{x}{y^3}\sin\left(\frac{x}{y}\right) + \frac{1}{y^2}\cos\left(\frac{x}{y}\right) + 6x^2y \qquad \blacksquare$$

Notice that in Example 5, $f_{xy} = f_{yx}$, which is usually the case for the functions of two variables encountered in a first course. A criterion for this will be given in Section 15.3 (Theorem B).

Partial derivatives of the third and higher orders are defined analogously, and the notation for them is similar. Thus, if f is a function of the two variables x and y, the third partial derivative of f obtained by differentiating f partially, first with respect to x and then twice with respect to y, will be indicated by

$$\frac{\partial}{\partial y}\left(\frac{\partial^2 f}{\partial y\,\partial x}\right) = \frac{\partial^3 f}{\partial y^2\,\partial x} = f_{xyy}$$

MORE THAN TWO VARIABLES   Let f be a function of three variables, x, y, and z. The **partial derivative of f with respect to x** at (x, y, z) is denoted by $f_x(x, y, z)$ or $\partial f(x, y, z)/\partial x$ and is defined by

$$f_x(x, y, z) = \lim_{\Delta x \to 0} \frac{f(x + \Delta x, y, z) - f(x, y, z)}{\Delta x}$$

Thus, $f_x(x, y, z)$ may be obtained by treating y and z as constants and differentiating with respect to x.

The partial derivatives with respect to y and z are defined in an analogous way.

EXAMPLE 6   If $f(x, y, z) = xy + 2yz + 3zx$, find f_x, f_y, and f_z.

Solution   To get f_x, we think of y and z as constants and differentiate with respect to the variable x. Thus,

$$f_x(x, y, z) = y + 3z$$

To find f_y, we treat x and z as constants and differentiate with respect to y:

$$f_y(x, y, z) = x + 2z$$

Similarly,

$$f_z(x, y, z) = 2y + 3x \qquad \blacksquare$$

EXAMPLE 7   If $f(x, y, z) = x \cos(y - z)$, find $\partial f/\partial x$, $\partial f/\partial y$, and $\partial f/\partial z$.

Solution

$$\frac{\partial}{\partial x}\left[x \cos(y - z)\right] = \cos(y - z)$$

$$\frac{\partial}{\partial y}\left[x \cos(y - z)\right] = -x \sin(y - z)$$

$$\frac{\partial}{\partial z}\left[x \cos(y - z)\right] = x \sin(y - z) \qquad \blacksquare$$

In Problems 1–16, find the first partial derivatives of the given functions with respect to each independent variable.

1. $f(x, y) = (2x - y)^4$

2. $f(x, y) = (4x - y^2)^{3/2}$

3. $f(x, y) = \dfrac{x^2 - y^2}{xy}$

4. $f(x, y) = e^x \cos y$

5. $f(x, y) = e^y \sin x$

6. $f(x, y) = (3x^2 + y^2)^{-1/3}$

7. $f(x, y) = \sqrt{x^2 - y^2}$

8. $f(u, v) = e^{uv}$

9. $g(x, y) = e^{-xy}$

10. $f(s, t) = \ln(s^2 - t^2)$

11. $f(x, y) = \tan^{-1}(4x - 7y)$

12. $F(w, z) = w \sin^{-1}\left(\dfrac{w}{z}\right)$

13. $f(x, y) = y \cos(x^2 + y^2)$

14. $f(s, t) = e^{t^2 - s^2}$

15. $F(x, y) = 2 \sin x \cos y$

16. $f(r, \theta) = 3r^3 \cos 2\theta$

In Problems 17–20, verify that

$$\frac{\partial^2 f}{\partial y\, \partial x} = \frac{\partial^2 f}{\partial x\, \partial y}$$

17. $f(x, y) = 2x^2 y^3 - x^3 y^5$

18. $f(x, y) = (x^3 + y^2)^5$

19. $f(x, y) = 3e^{2x} \cos y$

20. $f(x, y) = \tan^{-1} xy$

21. If $F(x, y) = \dfrac{2x - y}{xy}$, find $F_x(3, -2)$ and $F_y(3, -2)$.

22. If $F(x, y) = \ln(x^2 + xy + y^2)$, find $F_x(-1, 4)$ and $F_y(-1, 4)$.

23. If $f(x, y) = \tan^{-1}(y^2/x)$, find $f_x(\sqrt{5}, -2)$ and $f_y(\sqrt{5}, -2)$.

24. If $f(x, y) = e^y \cosh x$, find $f_x(-1, 1)$ and $f_y(-1, 1)$.

25. Find the slope of the tangent to the curve of intersection of the surface $36z = 4x^2 + 9y^2$ and the plane $x = 3$ at the point $(3, 2, 2)$.

26. Find the slope of the tangent to the curve of intersection of the surface $3z = \sqrt{36 - 9x^2 - 4y^2}$ and the plane $x = 1$ at the point $(1, -2, \sqrt{11/3})$.

27. Find the slope of the tangent to the curve of intersection of the surface $2z = \sqrt{9x^2 + 9y^2 - 36}$ and the plane $y = 1$ at the point $(2, 1, \tfrac{3}{2})$.

28. Find the slope of the tangent to the curve of intersection of the cylinder $4z = 5\sqrt{16 - x^2}$ and the plane $y = 3$ at the point $(2, 3, 5\sqrt{3}/2)$.

29. The volume V of a right circular cylinder is given by $V = \pi r^2 h$, where r is the radius and h is the height. If h is held fixed at $h = 10$ inches, find the rate of change of V with respect to r when $r = 6$ inches.

30. The temperature in degrees Celsius on a metal plate in the xy-plane is given by $T(x, y) = 4 + 2x^2 + y^3$. What is the rate of change of temperature with respect to distance (measured in feet) if we start moving from $(3, 2)$ in the direction of the positive y-axis?

31. According to the ideal gas law, the pressure, temperature, and volume of a gas are related by $PV = kT$, where k is a constant. Find the rate of change of pressure (pounds per square inch) with respect to temperature when the temperature is $40°C$ if the volume is kept fixed at 100 cubic inches.

32. Show that for the gas law of Problem 31,

$$V\frac{\partial P}{\partial V} + T\frac{\partial P}{\partial T} = 0 \quad \text{and} \quad \frac{\partial P}{\partial V}\frac{\partial V}{\partial T}\frac{\partial T}{\partial P} = -1$$

A function of two variables that satisfies Laplace's Equation,

$$\frac{\partial^2 f}{\partial x^2} + \frac{\partial^2 f}{\partial y^2} = 0$$

is said to be **harmonic**. (Pierre-Simon de Laplace (1749–1827) was a gifted French mathematician.) Show that the functions defined in Problems 33 and 34 are harmonic functions.

33. $f(x, y) = x^3 y - xy^3$

34. $f(x, y) = \ln(4x^2 + 4y^2)$

35. If $F(x, y) = 3x^4 y^5 - 2x^2 y^3$, find $\partial^3 F(x, y)/\partial y^3$.

36. If $f(x, y) = \cos(2x^2 - y^2)$, find $\partial^3 f(x, y)/\partial y\, \partial x^2$.

37. Express the following in ∂ notation.

(a) f_{yyy}

(b) f_{xxy}

(c) f_{xyyy}

38. Express the following in subscript notation.

(a) $\dfrac{\partial^3 f}{\partial x^2\, \partial y}$

(b) $\dfrac{\partial^4 f}{\partial x^2\, \partial y^2}$

(c) $\dfrac{\partial^5 f}{\partial x^3\, \partial y^2}$

39. If $f(x, y, z) = 3x^2 y - xyz + y^2 z^2$, find each of the following.

(a) $f_x(x, y, z)$

(b) $f_y(0, 1, 2)$

(c) $f_{xy}(x, y, z)$

40. If $f(x, y, z) = (x^3 + y^2 + z)^4$, find each of the following.

(a) $f_x(x, y, z)$

(b) $f_y(0, 1, 1)$

(c) $f_{zz}(x, y, z)$

41. If $f(x, y, z) = e^{-xyz} - \ln(xy - z^2)$, find $f_x(x, y, z)$.

42. If $f(x, y, z) = (xy/z)^{1/2}$, find $f_x(-2, -1, 8)$.

43. A bee was flying upward along the curve that is the intersection of $z = x^4 + xy^3 + 12$ with the plane $x = 1$. At

the point $(1, -2, 5)$, it went off on the tangent line. Where did the bee hit the xz-plane? (See Example 3.)

44. Let $A(x, y)$ be the area of a nondegenerate rectangle of dimensions x and y, the rectangle being inside a circle of radius 10. Determine the domain and range for this function.

45. The interval $[0, 1]$ is to be separated into three pieces by making cuts at x and y. Let $A(x, y)$ be the area of any nondegenerate triangle that can be formed from these three pieces. Determine the domain and range for this function.

46. The **wave equation** $c^2 \partial^2 u/\partial x^2 = \partial^2 u/\partial t^2$ and the **heat equation** $c \partial^2 u/\partial x^2 = \partial u/\partial t$ are two of the most important equations in physics (c is a constant). Show each of the following.

(a) $u = \cos x \cos ct$ and $u = e^x \cosh ct$ satisfy the wave equation.

(b) $u = e^{-ct} \sin x$ and $u = t^{-1/2} e^{-x^2/(4ct)}$ satisfy the heat equation.

47. For the contour map for $z = f(x, y)$ shown in Figure 4, estimate each value.

(a) $f_y(1, 1)$ (b) $f_x(-4, 2)$

(c) $f_x(-5, -2)$ (d) $f_y(0, -2)$

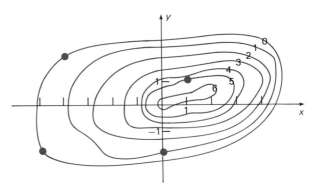

FIGURE 4

15.3
Limits and Continuity

FIGURE 1

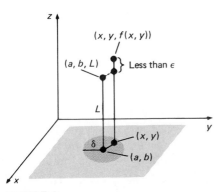

FIGURE 2

While partial derivative is a concept for functions of several variables, the theory needed so far is only that for functions of one variable (since all but one of the variables are held fixed). In particular, the notion of limit in Section 15.2 was the familiar notion treated in Section 2.5. To go further, we will need a more general notion of limit.

Our aim is to give meaning to the symbol

$$\lim_{(x, y) \to (a, b)} f(x, y) = L$$

It has the usual intuitive meaning: namely, the values of $f(x, y)$ get closer and closer to the number L as (x, y) approaches (a, b). The problem is that (x, y) can approach (a, b) in infinitely many different ways (Figure 1). We want a definition that gives the same L no matter what path (x, y) takes in approaching (a, b). Fortunately, the formal definition given first for real-valued functions of one variable (Section 2.5) and then for vector-valued functions of one variable (Section 13.4) is still appropriate, provided we interpret it correctly (see Figure 2).

Definition

To say that $\lim_{(x, y) \to (a, b)} f(x, y) = L$ means that for each $\varepsilon > 0$ (no matter how small) there is a corresponding $\delta > 0$ such that $|f(x, y) - L| < \varepsilon$ provided that $0 < |(x, y) - (a, b)| < \delta$.

Only the briefest explanation is necessary. To interpret $|(x, y) - (a, b)|$, think of (x, y) and (a, b) as vectors. Then

$$|(x, y) - (a, b)| = \sqrt{(x - a)^2 + (y - b)^2}$$

FIGURE 3

and the points satisfying $0 < |(x, y) - (a, b)| < \delta$ are those points inside a circle of radius δ excluding the center (a, b) (see Figure 3).

Note several aspects of this definition.

1. It ignores completely the notion of the path of approach to (a, b). This means that if different paths of approach lead to different L-values, then the limit does not exist.
2. The behavior of $f(x, y)$ at (a, b) is irrelevant; the function does not even have to be defined at (a, b). This follows from the restriction $0 < |(x, y) - (a, b)|$.
3. The definition is phrased so that it immediately extends to functions of three (or more) variables. Simply replace (x, y) and (a, b) by (x, y, z) and (a, b, c) wherever they occur.

What you expect to be true is generally true. For example,

$$\lim_{(x, y) \to (1, 2)} [x^2 y + 3y] = 1^2 \cdot 2 + 3 \cdot 2 = 8$$

$$\lim_{(x, y) \to (3, 4)} \sqrt{x^2 + y^2} = \sqrt{3^2 + 4^2} = 5$$

In fact, all the usual limit theorems are also valid for this new kind of limit. But the following example illustrates that we must be careful.

EXAMPLE 1 Show that the function f defined by

$$f(x, y) = \frac{x^2 - y^2}{x^2 + y^2}$$

has no limit at the origin.

Solution The function f is defined everywhere in the xy-plane except at the origin.

At all points on the x-axis, different from the origin, the value of f is

$$f(x, 0) = \frac{x^2 - 0}{x^2 + 0} = 1$$

Thus, the limit of $f(x, y)$ as (x, y) approaches $(0, 0)$ along the x-axis is

$$\lim_{(x, 0) \to (0, 0)} f(x, 0) = \lim_{(x, 0) \to (0, 0)} \frac{x^2 - 0}{x^2 + 0} = +1$$

Similarly, the limit of $f(x, y)$ as (x, y) approaches $(0, 0)$ along the y-axis is

$$\lim_{(0, y) \to (0, 0)} f(0, y) = \lim_{(0, y) \to (0, 0)} \frac{0 - y^2}{0 + y^2} = -1$$

Thus, we get different answers depending on how $(x, y) \to (0, 0)$. In fact, there are points arbitrarily close to $(0, 0)$ at which the value of f is 1 and other points equally close at which the value of f is -1. The limit cannot exist. ∎

See Problem 17 for an even more unusual example where a limit fails to exist and Problem 27 for another approach to Example 1.

CONTINUITY AT A POINT   To say that $f(x, y)$ is **continuous** at the point (a, b), we require that (1) f has a value at (a, b), (2) f has a limit at (a, b), and (3) the value of f at (a, b) is equal to the limit there. In summary, we require that

$$\lim_{(x, y) \to (a, b)} f(x, y) = f(a, b)$$

Intuitively this means that f has no jumps or wild fluctuations at (a, b).

As with functions of a single variable, sums, products, and quotients of continuous functions are continuous (provided, in the last case, that we avoid division by zero). It follows that *polynomial functions* of two variables are continuous everywhere, since they are sums and products of the continuous functions ax, by, and c, where a, b, and c are constants. For example, the function $f(x, y) = 5x^4y^2 - 2xy^3 + 4$ is continuous at all points in the xy-plane.

Rational functions of two variables are quotients of polynomial functions and thus are continuous wherever the denominator is not zero. To illustrate, $f(x, y) = (2x + 3y)/(y^2 - 4x)$ is continuous everywhere in the xy-plane except at points on the parabola $y^2 = 4x$.

As with functions of one variable, a continuous function of a continuous function is continuous.

Theorem A

(Composition of functions). If a function g of two variables is continuous at (a, b) and a function f of one variable is continuous at $g(a, b)$, then the composite function $f \circ g$, defined by $(f \circ g)(x, y) = f(g(x, y))$, is continuous at (a, b).

A neighborhood in 2-space

Proof of this theorem is similiar to the proof of Theorem 2.7D.

EXAMPLE 2   Show that $F(x, y) = \cos(x^3 - 4xy + y^2)$ is continuous at every point of the plane.

Solution   The function $g(x, y) = x^3 - 4xy + y^2$, being a polynomial, is continuous everywhere. Also $f(t) = \cos t$ is continuous at each number t in $\mathbb{R}$. We conclude from Theorem A, that $F(x, y) = f(g(x, y))$ is continuous at all (x, y) in the plane. ∎

CONTINUITY ON A SET   To say that $f(x, y)$ is continuous on a set S ought to mean that $f(x, y)$ is continuous at every point of the set. It does mean that, but there are some subtleties connected with this statement that need to be cleared up.

First we need to introduce some language relative to sets in the plane (and higher-dimensional spaces). By a **neighborhood** of radius δ of a point P, we mean the set of all points Q satisfying $|Q - P| < \delta$. In two-space, a neighborhood is the "inside" of a circle; in three-space, it is the inside of a sphere (Figure 4). A point P is an **interior point** of a set S if there is a neighborhood

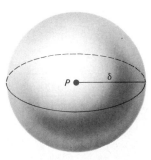

A neighborhood in 3-space

FIGURE 4

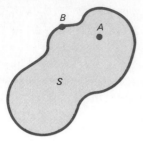

FIGURE 5

of P contained in S. The set of all interior points of S is the **interior** of S. On the other hand, P is a **boundary point** of S if every neighborhood of P contains points that are in S and points that are not in S. The set of all boundary points of S is called the **boundary** of S. In Figure 5, A is an interior point and B is a boundary point of S. Finally, a set is **open** if all its points are interior points and it is **closed** if it contains all its boundary points. The analogy with open and closed intervals in one-space should occur to you.

If S is an open set, to say that f is continuous on S means precisely that f is continuous at every point of S. On the other hand, if S contains some or all of its boundary points, we must be careful to give the right interpretation of continuity at such points (recall that in one-space, we had to talk about left and right continuity at the endpoints of an interval). To say that f is continuous at a boundary point P of S means that $f(Q)$ must approach $f(P)$ as Q approaches P through points of S.

Here is an example that will help clarify what we have said (see Figure 6). Let

$$f(x, y) = \begin{cases} 0 & \text{if } x^2 + y^2 \le 1 \\ 4 & \text{otherwise} \end{cases}$$

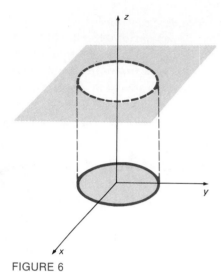

FIGURE 6

If S is the set $\{(x, y): x^2 + y^2 \le 1\}$, it is correct to say that $f(x, y)$ is continuous on S. On the other hand, it would be incorrect to say that $f(x, y)$ is continuous on the whole plane.

We said in Section 15.2 that for most functions of two variables studied in a first course, $f_{xy} = f_{yx}$; that is, the order of differentiation in mixed partial derivatives is immaterial. Now that continuity is understood, conditions for this to be true can be simply stated.

Theorem B

(Equality of mixed partials). Let f_{xy} and f_{yx} be continuous on an open set S. Then $f_{xy} = f_{yx}$ at each point of S.

A proof of this theorem is given in books on advanced calculus. A counterexample where continuity of f_{xy} is lacking is given in Problem 30.

Our discussion of continuity has dealt mainly with functions of two variables. We believe you can make the simple changes that are required to describe continuity for functions of three or more variables.

PROBLEM SET 15.3 _____

In Problems 1–8, find the indicated limit or state that it does not exist.

1. $\lim\limits_{(x, y) \to (1, 3)} (3x^2 y - xy^3)$

2. $\lim\limits_{(x, y) \to (-2, 1)} (xy^3 - xy + 3y^2)$

3. $\lim\limits_{(x, y) \to (2, \pi)} [x \cos^2 xy - \sin(xy/3)]$

4. $\lim\limits_{(x, y) \to (-1, 2)} \dfrac{xy - y^3}{(x + y + 1)^2}$

5. $\lim\limits_{(x,y)\to(0,0)} \dfrac{\sin(x^2 + y^2)}{3x^2 + 3y^2}$

6. $\lim\limits_{(x,y)\to(0,0)} \dfrac{\tan(x^2 + y^2)}{x^2 + y^2}$

7. $\lim\limits_{(x,y)\to(0,0)} \dfrac{x^2 + y^2}{x^4 - y^4}$

8. $\lim\limits_{(x,y)\to(0,0)} \dfrac{x^4 - y^4}{x^2 + y^2}$

In Problems 9–14, describe the largest set S on which it is correct to say that f is continuous.

9. $f(x, y) = \dfrac{x^3 + xy - 5}{x^2 + y^2 + 1}$

10. $f(x, y) = \ln(1 - x^2 - y^2)$

11. $f(x, y) = \dfrac{x^2 + 3xy + y^2}{y - x^2}$

12. $f(x, y) = \begin{cases} \dfrac{\sin(xy)}{xy} & xy \neq 0 \\ 1 & xy = 0 \end{cases}$

13. $f(x, y) = \sqrt{x - y + 1}$

14. $f(x, y) = (4 - x^2 - y^2)^{-1/2}$

15. Show that

$$\lim\limits_{(x,y)\to(0,0)} \dfrac{xy}{x^2 + y^2}$$

does not exist by considering the x-axis as one path to the origin and the line $y = x$ as another.

16. Show that

$$\lim\limits_{(x,y)\to(0,0)} \dfrac{xy + y^3}{x^2 + y^2}$$

does not exist.

17. Let $f(x, y) = x^2 y/(x^4 + y^2)$.

(a) Show that $f(x, y) \to 0$ as $(x, y) \to (0, 0)$ along any straight line $y = mx$.

(b) Show that $f(x, y) \to \frac{1}{2}$ as $(x, y) \to (0, 0)$ along the parabola $y = x^2$.

(c) What conclusion do you draw?

18. Show that $\lim\limits_{(x,y)\to(0,0)} [xy^2/(x^2 + y^2)] = 0$.

$Hint:$ $\left| \dfrac{xy^2}{x^2 + y^2} \right| \leq \dfrac{(\sqrt{x^2 + y^2})(x^2 + y^2)}{x^2 + y^2} = \sqrt{x^2 + y^2}.$

In Problems 19–24, sketch the indicated set. Describe the boundary of the set. Finally state whether the set is open, closed, or neither.

19. $\{(x,y) : 2 \leq x \leq 4, 1 \leq y \leq 5\}$

20. $\{(x,y) : x^2 + y^2 < 4\}$

21. $\{(x,y) : 0 < x^2 + y^2 \leq 4\}$

22. $\{(x,y) : 1 < x \leq 4\}$

23. $\{(x,y) : x > 0, y < \sin(1/x)\}$

24. $\{(x,y) : x = 0, y = 1/n, n \text{ a positive integer}\}$

25. Let

$$f(x, y) = \begin{cases} \dfrac{x^2 - 4y^2}{x - 2y} & x \neq 2y \\ g(x) & x = 2y \end{cases}$$

If f is continuous in the whole plane, find a formula for $g(x)$.

26. Prove that

$$\lim\limits_{(x,y)\to(a,b)} [f(x, y) + g(x, y)]$$

$$= \lim\limits_{(x,y)\to(a,b)} f(x, y) + \lim\limits_{(x,y)\to(a,b)} g(x, y)$$

provided the latter two limits exist.

27. Occasionally, it is easier to analyze the continuity of $f(x, y)$ by changing to polar coordinates. Let $f(x, y) = (x^2 - y^2)/(x^2 + y^2)$ for $(x, y) \neq (0,0)$ and $f(0,0) = 0$ (see Example 1). In polar coordinates,

$$f(x, y) = \dfrac{r^2 \cos^2\theta - r^2 \sin^2\theta}{r^2} = \cos 2\theta$$

which takes all values between -1 and 1 in every neighborhood of the origin. We conclude that $\lim\limits_{(x,y)\to(0,0)} f(x, y)$ does not exist and f is not continuous at the origin. Let each of the following functions have the value 0 at $(0,0)$. Which of them are continuous at $(0,0)$ and which are discontinuous there?

(a) $f(x, y) = \dfrac{xy}{\sqrt{x^2 + y^2}}$

(b) $f(x, y) = \dfrac{xy}{x^2 + y^2}$

(c) $f(x, y) = \dfrac{x^{7/3}}{x^2 + y^2}$

(d) $f(x, y) = xy \dfrac{x^2 - y^2}{x^2 + y^2}$

(e) $f(x, y) = \dfrac{x^2 y^2}{x^2 + y^4}$

(f) $f(x, y) = \dfrac{xy^2}{x^2 + y^4}$

28. Let $f(x, y)$ be the shortest distance that a raindrop landing at latitude x and longitude y in the state of Colorado must travel to reach the ocean. Where in Colorado is this function discontinuous?

29. Let H be the hemispherical shell $x^2 + y^2 + (z-1)^2 = 1$, $0 \le z < 1$, shown in Figure 7 and let $D = \{(x, y, z) : 1 \le z \le 2\}$. For each of the functions defined below, determine its set of discontinuities within D.

(a) $f(x, y, z)$ is the time required for a particle dropped from (x, y, z) to reach the level $z = 0$.

(b) $f(x, y, z)$ is the area of the inside of H (assumed opaque) that can be seen from (x, y, z).

(c) $f(x, y, z)$ is the area of the shadow of H on the xy-plane due to a point light source at (x, y, z).

(d) $f(x, y, z)$ is the distance along the shortest path from (x, y, z) to $(0, 0, 0)$ that does not penetrate H.

FIGURE 7

30. Let f, a function of n variables, be continuous on an open set D and suppose that P_0 is in D with $f(P_0) > 0$. Prove that there is a $\delta > 0$ such that $f(P) > 0$ in a neighborhood of P_0 with radius δ.

31. (The French Railroad) Paris is located at the origin of the xy-plane. Rail lines emanate from Paris along all rays, and these are the only rail lines. Determine the set of discontinuities of the following functions.

(a) $f(x, y)$ is the distance from (x, y) to $(1, 0)$ on the French railroad.

(b) $g(u, v, x, y)$ is the distance from (u, v) to (x, y) on the French railroad.

32. Let $f(x, y) = xy \dfrac{x^2 - y^2}{x^2 + y^2}$ if $(x, y) \ne 0$ and $f(0, 0) = 0$ (see page 654 for a computer graph of this function). Show that $f_{xy}(0, 0) \ne f_{yx}(0, 0)$ by completing the following steps.

(a) Show $f_x(0, y) = \lim\limits_{h \to 0} \dfrac{f(0 + h, y) - f(0, y)}{h} = -y$ for all y.

(b) Similarly, show that $f_y(x, 0) = x$ for all x.

(c) Show that $f_{yx}(0, 0) = \lim\limits_{h \to 0} \dfrac{f_y(0 + h, 0) - f_y(0, 0)}{h} = 1$.

(d) Similarly, show that $f_{xy}(0, 0) = -1$.

15.4
Differentiability

For a function of a single variable, differentiability of f at x meant the existence of the derivative $f'(x)$. This, in turn, was equivalent to the graph of f having a nonvertical tangent line at x.

Now we ask: What is the right concept of differentiability for a function of two variables? Surely it must correspond in a natural way to the existence of a tangent plane and clearly this requires more than the mere existence of the partial derivatives of f, for they reflect the behavior of f in only two directions. To emphasize this point, consider

$$f(x, y) = -10\sqrt{|xy|}$$

for which we have shown a computer-generated graph (Figure 1). Note that $f_x(0, 0)$ and $f_y(0, 0)$ both exist and equal 0; yet no one would claim that the graph has a tangent plane at the origin. The reason is, of course, that the graph of f is not well approximated there by any plane (in particular, the xy-plane) except in two directions. A tangent plane ought to approximate the graph very well in all directions.

Consider a second question. What plays the role of *the* derivative for a function of two variables? Again the partial derivatives fall short, if for no other reason than because there are two of them.

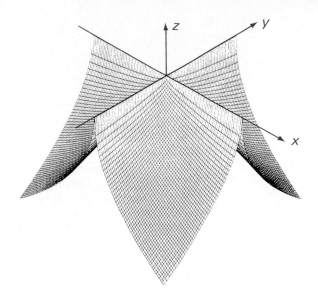

FIGURE 1

To answer these two questions, we start by downplaying the distinction between the point (x, y) and the vector $\langle x, y \rangle$. Thus, we write $\mathbf{p} = (x, y) = \langle x, y \rangle$ and $f(\mathbf{p}) = f(x, y)$. Recall that

$$(1) \qquad f'(x_0) = \lim_{x \to x_0} \frac{f(x) - f(x_0)}{x - x_0} = \lim_{h \to 0} \frac{f(x_0 + h) - f(x_0)}{h}$$

The analogue would seem to be

$$(2) \qquad f'(\mathbf{p}_0) = \lim_{\mathbf{p} \to \mathbf{p}_0} \frac{f(\mathbf{p}) - f(\mathbf{p}_0)}{\mathbf{p} - \mathbf{p}_0} = \lim_{\mathbf{h} \to \mathbf{0}} \frac{f(\mathbf{p}_0 + \mathbf{h}) - f(\mathbf{p}_0)}{\mathbf{h}}$$

but, unfortunately, the division by a vector makes no sense.

But let us not give up too quickly. The existence of the limit (1) means that there is a number $m = f'(x_0)$ such that

$$\left| \frac{f(x) - f(x_0)}{x - x_0} - m \right| = \varepsilon_1$$

where $\varepsilon_1 \to 0$ as $x \to x_0$. We may rewrite this as

$$\left| f(x) - f(x_0) - m(x - x_0) \right| = \varepsilon_1 \left| x - x_0 \right|$$

or (after replacing $\pm \varepsilon_1$ by ε) as

$$f(x) - f(x_0) - m(x - x_0) = \varepsilon \left| x - x_0 \right|$$

FIGURE 2

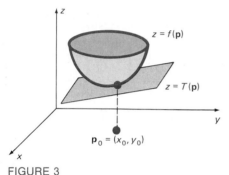

FIGURE 3

Thus, to say that f is differentiable at x_0 means that there is a number $m = f'(x_0)$ such that

$$(3) \qquad f(x) = f(x_0) + m(x - x_0) + |x - x_0|\varepsilon$$

where $\varepsilon \to 0$ as $x \to x_0$. Now, look at Figure 2 and note that $y = T(x) = f(x_0) + m(x - x_0)$ is the equation of the line through $(x_0, f(x_0))$ with slope m. Consequently, we can reformulate (3) to say that f is differentiable at x_0 provided there is a line $y = T(x)$ through $(x_0, f(x_0))$ that approximates $y = f(x)$ so well near x_0 that $|f(x) - T(x)|/|x - x_0| \to 0$ when $x \to x_0$. This line is what we have called the tangent line.

The way is now clear for an appropriate generalization to higher dimensions. To say that $f(\mathbf{p})$ is differentiable at $\mathbf{p}_0$ will mean that there is a vector $\mathbf{m}$ such that

$$(4) \qquad f(\mathbf{p}) = f(\mathbf{p}_0) + \mathbf{m} \cdot (\mathbf{p} - \mathbf{p}_0) + |\mathbf{p} - \mathbf{p}_0|\varepsilon$$

where $\varepsilon \to 0$ as $\mathbf{p} \to \mathbf{p}_0$. Here, $z = T(\mathbf{p}) = f(\mathbf{p}_0) + \mathbf{m} \cdot (\mathbf{p} - \mathbf{p}_0)$ is the equation of a plane through $(\mathbf{p}_0, f(\mathbf{p}_0))$, and so differentiability of $f(\mathbf{p})$ at $\mathbf{p}_0$ is equivalent to the existence of a plane $z = T(\mathbf{p})$ that approximates $z = f(\mathbf{p})$ in the sense indicated. Naturally, this plane is called the **tangent plane** (Figure 3).

We are finally ready to give a formal definition. We use the h-notation since this allows us to drop the subscript 0. Note also our use of the symbol $\varepsilon(\mathbf{h})$ to indicate that ε does depend on $\mathbf{h}$.

Definition

We say that f is **differentiable** at $\mathbf{p}$ if there exists a vector $\mathbf{m}$ such that

$$f(\mathbf{p} + \mathbf{h}) = f(\mathbf{p}) + \mathbf{m} \cdot \mathbf{h} + |\mathbf{h}|\varepsilon(\mathbf{h})$$

where $\varepsilon(\mathbf{h}) \to 0$ as $\mathbf{h} \to \mathbf{0}$.

It can be shown that if the vector $\mathbf{m}$ exists, it is unique (Problem 19). We call this vector the **gradient** of f at $\mathbf{p}$ and denote it by $\nabla f(\mathbf{p})$. Thus, if f is differentiable at $\mathbf{p}$, it has a gradient $\nabla f(\mathbf{p})$ and

$$\boxed{f(\mathbf{p} + \mathbf{h}) = f(\mathbf{p}) + \nabla f(\mathbf{p}) \cdot \mathbf{h} + |\mathbf{h}|\varepsilon(\mathbf{h})}$$

where $\varepsilon(\mathbf{h}) \to 0$ as $\mathbf{h} \to \mathbf{0}$.

We call your attention to several aspects of this definition.

1. While the derivative $f'(x)$ is a number, the gradient $\nabla f(\mathbf{p})$ is a vector.
2. The dot in $\nabla f(\mathbf{p}) \cdot \mathbf{h}$ indicates the dot product of two vectors.
3. The definition makes sense in any number of dimensions.

COMPUTING GRADIENTS To find a gradient from the definition is burdensome (as it was for derivatives). We need rules for calculation. Here is a very useful result, which we state without proof (but see Problem 20).

Theorem A

If f, a function of two variables, is differentiable at $\mathbf{p} = (x, y)$, then the first partial derivatives of f exist at $\mathbf{p}$ and

$$\nabla f(\mathbf{p}) = \frac{\partial f}{\partial x}(\mathbf{p})\mathbf{i} + \frac{\partial f}{\partial y}(\mathbf{p})\mathbf{j}$$

Similarly, if g, a function of three variables, is differentiable at $\mathbf{p} = (x, y, z)$, then

$$\nabla g(\mathbf{p}) = \frac{\partial g}{\partial x}(\mathbf{p})\mathbf{i} + \frac{\partial g}{\partial y}(\mathbf{p})\mathbf{j} + \frac{\partial g}{\partial z}(\mathbf{p})\mathbf{k}$$

To use Theorem A, we still need to know that f and g are differentiable, which is troublesome. Fortunately, there is a simple criterion, proved in most books on advanced calculus.

Theorem B

If f has first partial derivatives in a neighborhood of $\mathbf{p}$ and if these partial derivatives are continuous at $\mathbf{p}$, then f is differentiable at $\mathbf{p}$.

Theorems A and B together provide just what we need to handle most problems.

EXAMPLE 1 Show that $f(x, y) = xe^y + x^2 y$ is differentiable everywhere and calculate its gradient. Then find the equation $z = T(x, y)$ of the tangent plane at $(2, 0)$.

Solution We note first that

$$\frac{\partial f}{\partial x} = e^y + 2xy, \qquad \frac{\partial f}{\partial y} = xe^y + x^2$$

Both of these functions are continuous everywhere, and so by Theorem B, f is differentiable everywhere. Moreover, by Theorem A,

$$\nabla f(x, y) = (e^y + 2xy)\mathbf{i} + (xe^y + x^2)\mathbf{j}$$

Thus,

$$\nabla f(2, 0) = \mathbf{i} + 6\mathbf{j} = \langle 1, 6 \rangle$$

and the equation of the tangent plane is

$$z = f(2,0) + \nabla f(2,0) \cdot \langle x - 2, y \rangle$$
$$= 2 + \langle 1, 6 \rangle \cdot \langle x - 2, y \rangle$$
$$= 2 + x - 2 + 6y = x + 6y \qquad \blacksquare$$

EXAMPLE 2   For $f(x, y, z) = x \sin z + x^2 y$, find $\nabla f(1, 2, 0)$.

Solution   Since the partial derivatives

$$\frac{\partial f}{\partial x} = \sin z + 2xy, \qquad \frac{\partial f}{\partial y} = x^2, \qquad \frac{\partial f}{\partial z} = x \cos z$$

are all continuous, the gradient exists. Moreover at $(1, 2, 0)$, these partials have the values 4, 1, and 1, respectively. Thus,

$$\nabla f(1, 2, 0) = 4\mathbf{i} + \mathbf{j} + \mathbf{k} \qquad \blacksquare$$

RULES FOR GRADIENTS   In many respects, gradients behave like derivatives. Recall that D considered as an operator is linear. So is the operator ∇, often called the *del* operator.

Theorem C

∇ is a linear operator; that is:

(i) $\nabla[f(\mathbf{p}) + g(\mathbf{p})] = \nabla f(\mathbf{p}) + \nabla g(\mathbf{p})$;
(ii) $\nabla[\alpha f(\mathbf{p})] = \alpha \nabla f(\mathbf{p})$.

Also, we have the product rule.

(iii) $\nabla[f(\mathbf{p})g(\mathbf{p})] = f(\mathbf{p})\nabla g(\mathbf{p}) + g(\mathbf{p})\nabla f(\mathbf{p})$

Proof   All three results follow from the corresponding facts for partial derivatives. We prove (iii) in the two-variable case, suppressing the point $\mathbf{p}$ for brevity.

$$\nabla fg = \frac{\partial (fg)}{\partial x}\mathbf{i} + \frac{\partial (fg)}{\partial y}\mathbf{j}$$

$$= \left(f\frac{\partial g}{\partial x} + g\frac{\partial f}{\partial x} \right)\mathbf{i} + \left(f\frac{\partial g}{\partial y} + g\frac{\partial f}{\partial y} \right)\mathbf{j}$$

$$= f\left(\frac{\partial g}{\partial x}\mathbf{i} + \frac{\partial g}{\partial y}\mathbf{j} \right) + g\left(\frac{\partial f}{\partial x}\mathbf{i} + \frac{\partial f}{\partial y}\mathbf{j} \right)$$

$$= f\nabla g + g\nabla f \qquad \blacksquare$$

CONTINUITY VERSUS DIFFERENTIABILITY Recall for functions of one variable that differentiability implies continuity, but not vice versa. The same is true here.

Theorem D

If f is differentiable at $\mathbf{p}$, then f is continuous at $\mathbf{p}$.

Proof Since f is differentiable at $\mathbf{p}$,

$$f(\mathbf{p} + \mathbf{h}) - f(\mathbf{p}) = \nabla f(\mathbf{p}) \cdot \mathbf{h} + |\mathbf{h}| \varepsilon(\mathbf{h})$$

Thus,

$$|f(\mathbf{p} + \mathbf{h}) - f(\mathbf{p})| \leq |\nabla f(\mathbf{p}) \cdot \mathbf{h}| + |\mathbf{h}| \, |\varepsilon(\mathbf{h})|$$
$$= |\nabla f(\mathbf{p})| \, |\mathbf{h}| \, |\cos \theta| + |\mathbf{h}| \, |\varepsilon(\mathbf{h})|$$

Both of the latter terms approach 0 as $\mathbf{h} \to \mathbf{0}$, and so

$$\lim_{\mathbf{h} \to \mathbf{0}} f(\mathbf{p} + \mathbf{h}) = f(\mathbf{p})$$

This last equality is one way of formulating the continuity of f at $\mathbf{p}$. ∎

PROBLEM SET 15.4

In Problems 1–10, find the gradient ∇f.

1. $f(x, y) = x^2 y + 3xy$
2. $f(x, y) = x^3 y - y^3$
3. $f(x, y) = xe^{xy}$
4. $f(x, y) = x^2 y \cos y$
5. $f(x, y) = x^2 y / (x + y)$
6. $f(x, y) = \sin^3(x^2 y)$
7. $f(x, y, z) = \sqrt{x^2 + y^2 + z^2}$
8. $f(x, y, z) = x^2 y + y^2 z + z^2 x$
9. $f(x, y, z) = x^2 y e^{x - z}$
10. $f(x, y, z) = xz \ln(x + y + z)$

In Problems 11–14, find the gradient vector of the given function at the given point $\mathbf{p}$. Then find the equation of the tangent plane at $\mathbf{p}$ (see Example 1).

11. $f(x, y) = x^2 y - xy^2$, $\mathbf{p} = (-2, 3)$
12. $f(x, y) = x^3 y + 3xy^2$, $\mathbf{p} = (2, -2)$
13. $f(x, y) = \cos \pi x \sin \pi y + \sin 2\pi y$, $\mathbf{p} = (-1, \frac{1}{2})$
14. $f(x, y) = \dfrac{x^2}{y}$, $\mathbf{p} = (2, -1)$

In Problems 15 and 16, find the equation $w = T(x, y, z)$ of the tangent "hyperplane" at $\mathbf{p}$.

15. $f(x, y, z) = 3x^2 - 2y^2 + xz^2$, $\mathbf{p} = (1, 2, -1)$

16. $f(x, y, z) = xyz + x^2$, $\mathbf{p} = (2, 0, -3)$

17. Show that

$$\nabla \left(\frac{f}{g} \right) = \frac{g \nabla f - f \nabla g}{g^2}$$

18. Show that

$$\nabla(f^r) = rf^{r-1} \nabla f$$

19. Show that if

$$f(\mathbf{p} + \mathbf{h}) - f(\mathbf{p}) = \mathbf{q}_1 \cdot \mathbf{h} + |\mathbf{h}| \varepsilon_1(\mathbf{h})$$

and

$$f(\mathbf{p} + \mathbf{h}) - f(\mathbf{p}) = \mathbf{q}_2 \cdot \mathbf{h} + |\mathbf{h}| \varepsilon_2(\mathbf{h})$$

then $\mathbf{q}_1 = \mathbf{q}_2$. *Hint*: Calculate $|\mathbf{q}_1 - \mathbf{q}_2|$ by using

$$\mathbf{q}_1 \cdot \mathbf{h} - \mathbf{q}_2 \cdot \mathbf{h} = (\mathbf{q}_1 - \mathbf{q}_2) \cdot \mathbf{h} = |\mathbf{q}_1 - \mathbf{q}_2| \, |\mathbf{h}| \cos \theta$$

20. By letting $\mathbf{h} = (h, 0)$ and then $\mathbf{h} = (0, k)$ in the definition of the gradient, show that $\nabla f(x, y)$ must have the form asserted in Theorem A.

21. Refer to Figure 1. Find the equation of the tangent plane to $z = -10\sqrt{|xy|}$ at $(1, -1)$. *Recall*: $d|x|/dx = |x|/x$ for $x \neq 0$.

22. Mean Value Theorem (Several Variables) If f is differentiable at each point of the line segment from $\mathbf{a}$ to $\mathbf{b}$, then there exists on that line segment a point $\mathbf{c}$ between $\mathbf{a}$ and $\mathbf{b}$ such that

$$f(\mathbf{b}) - f(\mathbf{a}) = \nabla f(\mathbf{c}) \cdot (\mathbf{b} - \mathbf{a})$$

Taking this result for granted, show that if f is differentiable on a convex set S and if $\nabla f(\mathbf{p}) = \mathbf{0}$ on S, then f is constant on S. *Note*: A set S is convex if each pair of points in S can be connected by a line segment in S.

23. Use the result of Problem 22 to show that if $\nabla f(\mathbf{p}) = \nabla g(\mathbf{p})$ for all $\mathbf{p}$ in a convex set S, then f and g differ by a constant on S.

24. Find the most general function $f(\mathbf{p})$ satisfying $\nabla f(\mathbf{p}) = \mathbf{p}$.

25. Let

$$f(x, y) = \begin{cases} \dfrac{\sin(x^2 + y^2)}{x^2 + y^2} & \text{if } (x, y) \neq (0, 0) \\ 1 & \text{if } (x, y) = (0, 0) \end{cases}$$

Show that $f(x, y)$ is differentiable at $(0,0)$ with $\nabla f(0,0) = \langle 0, 0 \rangle$. *Hint*: Let $\mathbf{h} = (h, k)$ and note that we can write

$$f(\mathbf{0} + \mathbf{h}) = f(h, k)$$

$$= 1 + 0 + \frac{\sin(h^2 + k^2) - (h^2 + k^2)}{(h^2 + k^2)^{3/2}} \sqrt{h^2 + k^2}$$

What must be shown for the multiplier of $\sqrt{h^2 + k^2}$?

26. Is $f(x, y) = \sqrt{x^2 + y^2}$ differentiable at $(0, 0)$? Justify your answer.

15.5
Directional Derivatives and Gradients

Consider again a function $f(x, y)$ of two variables. The partial derivatives $f_x(x, y)$ and $f_y(x, y)$ measure the rate of change (and the slope of the tangent line) in directions parallel to the x- and y-axes. Our goal now is to study the rate of change of f in an arbitrary direction. This leads to the concept of the directional derivative, which in turn is related to the gradient.

It will be convenient to use vector notation. Let $\mathbf{p} = (x, y)$ and let $\mathbf{i}$ and $\mathbf{j}$ be the unit vectors in the positive x- and y-directions. Then the two partial derivatives at $\mathbf{p}$ may be written as follows.

$$f_x(\mathbf{p}) = \lim_{h \to 0} \frac{f(\mathbf{p} + h\mathbf{i}) - f(\mathbf{p})}{h}$$

$$f_y(\mathbf{p}) = \lim_{h \to 0} \frac{f(\mathbf{p} + h\mathbf{j}) - f(\mathbf{p})}{h}$$

To get the concept we are after, all we have to do is replace $\mathbf{i}$ or $\mathbf{j}$ by an arbitrary unit vector $\mathbf{u}$.

Definition

For each unit vector $\mathbf{u}$, let

$$D_{\mathbf{u}} f(\mathbf{p}) = \lim_{h \to 0} \frac{f(\mathbf{p} + h\mathbf{u}) - f(\mathbf{p})}{h}$$

This limit, if it exists, is called the **directional derivative** of f at $\mathbf{p}$ in the direction $\mathbf{u}$.

Slope $= D_{\mathbf{u}} f(x, y) = \tan \phi$

$z = f(x, y)$

FIGURE 1

Thus, $D_{\mathbf{i}} f(\mathbf{p}) = f_x(\mathbf{p})$ and $D_{\mathbf{j}} f(\mathbf{p}) = f_y(\mathbf{p})$. Since $\mathbf{p} = (x, y)$, we also use the notation $D_{\mathbf{u}} f(x, y)$. Figure 1 gives the geometric interpretation of $D_{\mathbf{u}} f(x_0, y_0)$. The vector $\mathbf{u}$ determines a line L in the xy-plane through (x_0, y_0). The plane through L perpendicular to the xy-plane intersects the surface $z = f(x, y)$ in a curve C. Its tangent at the point $(x_0, y_0, f(x_0, y_0))$ has slope $D_{\mathbf{u}}f(x_0, y_0)$. Another useful interpretation is that $D_{\mathbf{u}}f(x_0, y_0)$ measures the rate of change of f with respect to distance in the direction $\mathbf{u}$.

CONNECTION WITH THE GRADIENT   Recall from Section 15.4 that $\nabla f(\mathbf{p})$ is given by

$$\nabla f(\mathbf{p}) = f_x(\mathbf{p})\mathbf{i} + f_y(\mathbf{p})\mathbf{j}$$

Theorem A

Let f be differentiable at $\mathbf{p}$. Then f has a directional derivative at $\mathbf{p}$ in the direction of the unit vector $\mathbf{u} = u_1\mathbf{i} + u_2\mathbf{j}$ and

$$D_{\mathbf{u}}f(\mathbf{p}) = \mathbf{u} \cdot \nabla f(\mathbf{p})$$

that is,

$$D_{\mathbf{u}}f(x, y) = u_1 f_x(x, y) + u_2 f_y(x, y)$$

Proof   Since f is differentiable at $\mathbf{p}$,

$$f(\mathbf{p} + h\mathbf{u}) - f(\mathbf{p}) = h\mathbf{u} \cdot \nabla f(\mathbf{p}) + |h\mathbf{u}|\varepsilon(h\mathbf{u})$$

where $\varepsilon(h\mathbf{u}) \to 0$ as $h \to 0$. Thus,

$$\frac{f(\mathbf{p} + h\mathbf{u}) - f(\mathbf{p})}{h} = \mathbf{u} \cdot \nabla f(\mathbf{p}) \pm \varepsilon(h\mathbf{u})$$

The conclusion follows by taking limits as $h \to 0$.   ∎

EXAMPLE 1   If $f(x, y) = 4x^2 - xy + 3y^2$, find the directional derivative of f at $(2, -1)$ in the direction of the vector $\mathbf{a} = 4\mathbf{i} + 3\mathbf{j}$.

Solution   The unit vector $\mathbf{u}$ in the direction of $\mathbf{a}$ is $(\frac{4}{5})\mathbf{i} + (\frac{3}{5})\mathbf{j}$. Also, $f_x(x, y) = 8x - y$ and $f_y(x, y) = -x + 6y$; thus, $f_x(2, -1) = 17$ and $f_y(2, -1) = -8$. Consequently, by Theorem A,

$$D_{\mathbf{u}}f(2, -1) = \tfrac{4}{5}(17) + \tfrac{3}{5}(-8) = \tfrac{44}{5}$$ ∎

Although we will not go through the details, we assert that what we have done is valid for functions of three or more variables, with obvious modifications. We illustrate.

EXAMPLE 2   Find the directional derivative of the function $f(x, y, z) = xy \sin z$ at the point $(1, 2, \pi/2)$ in the direction of the vector $\mathbf{a} = \mathbf{i} + 2\mathbf{j} + 2\mathbf{k}$.

Solution The unit vector **u** in the direction of **a** is $\frac{1}{3}\mathbf{i} + \frac{2}{3}\mathbf{j} + \frac{2}{3}\mathbf{k}$. Also, $f_x(x, y, z) = y \sin z$, $f_y(x, y, z) = x \sin z$, and $f_z(x, y, z) = xy \cos z$, and so $f_x(1, 2, \pi/2) = 2$, $f_y(1, 2, \pi/2) = 1$, and $f_z(1, 2, \pi/2) = 0$. We conclude that

$$D_{\mathbf{u}}f\left(1, 2, \frac{\pi}{2}\right) = \frac{1}{3}(2) + \frac{2}{3}(1) + \frac{2}{3}(0) = \frac{4}{3} \qquad \blacksquare$$

MAXIMUM RATE OF CHANGE For a given function f at a given point **p**, it is natural to ask in what direction the function is changing most rapidly— that is, in what direction $D_{\mathbf{u}}f(\mathbf{p})$ is the largest. From the geometric formula for the dot product (Section 14.2), we may write

$$D_{\mathbf{u}}f(\mathbf{p}) = \mathbf{u} \cdot \nabla f(\mathbf{p}) = |\mathbf{u}|\,|\nabla f(\mathbf{p})|\cos\theta = |\nabla f(\mathbf{p})|\cos\theta$$

where θ is the angle between **u** and $\nabla f(\mathbf{p})$. Thus, $D_{\mathbf{u}}f(\mathbf{p})$ is maximized when $\theta = 0$ and minimized when $\theta = \pi$. We summarize as follows.

Theorem B

A function increases most rapidly at **p** in the direction of the gradient (with rate $|\nabla f(\mathbf{p})|$) and decreases most rapidly in the opposite direction (with rate $-|\nabla f(\mathbf{p})|$).

EXAMPLE 3 Suppose a bug is located on the hyperbolic paraboloid $z = y^2 - x^2$ at the point $(1, 1, 0)$, as in Figure 2. In what direction should it move for the steepest climb and what is the slope as it starts out?

FIGURE 2

Solution Let $f(x, y) = y^2 - x^2$. Since $f_x(x, y) = -2x$ and $f_y(x, y) = 2y$,

$$\nabla f(1, 1) = f_x(1, 1)\mathbf{i} + f_y(1, 1)\mathbf{j} = -2\mathbf{i} + 2\mathbf{j}$$

Thus the bug should move from $(1, 1, 0)$ in the direction $-2\mathbf{i} + 2\mathbf{j}$, where the slope will be $|-2\mathbf{i} + 2\mathbf{j}| = \sqrt{8}$. $\qquad \blacksquare$

LEVEL CURVES AND SURFACES Recall from Section 15.1 that the *level curves* of a surface $z = f(x, y)$ are the projections onto the xy-plane of the curves of intersection of the surface with planes $z = k$ that are parallel to

FIGURE 3

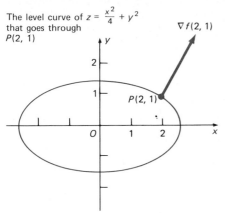

The level curve of $z = \dfrac{x^2}{4} + y^2$ that goes through $P(2, 1)$

$\nabla f(2, 1)$

$P(2, 1)$

FIGURE 4

DOES THIS HELP?	
$z = f(x, y)$	$w = f(x, y, z)$
The graph is a surface.	We cannot draw the graph since it would require four dimensional space.
$f(x, y) = k$ determines a *level curve* in the xy-plane.	$f(x, y, z) = k$ determines a *level surface* in xyz-space.
∇f is a vector perpendicular to the level curve.	∇f is a vector perpendicular to the level surface.

the xy-plane. The value of the function at all points on the same level curve is constant (Figure 3).

Denote by L the level curve of $f(x, y)$ that passes through an arbitrarily chosen point $P(x_0, y_0)$ in the domain of f and let the unit vector $\mathbf{u}$ be tangent to L at P. Since the value of f is the same at all points on the level curve L, its directional derivative $D_{\mathbf{u}}f(x_0, y_0)$, which is the rate of change of $f(x, y)$ in the direction $\mathbf{u}$, is zero when $\mathbf{u}$ is tangent to L. (This statement—which seems very clear intuitively—requires justification, a justification we omit since the result we want also follows from an argument to be given in Section 15.7.) Since

$$0 = D_{\mathbf{u}}f(x_0, y_0) = \nabla f(x_0, y_0) \cdot \mathbf{u}$$

we conclude that ∇f and $\mathbf{u}$ are perpendicular, a result worthy of theorem status.

> **Theorem C**
>
> The gradient of f at a point P is perpendicular to the level curve of f that goes through P.

EXAMPLE 4 For the paraboloid $z = x^2/4 + y^2$, find the equation of its level curve that passes through the point $P(2, 1)$ and sketch it. Find the gradient vector of the paraboloid at P and draw the gradient with its initial point at P.

Solution The level curve of the paraboloid that corresponds to the plane $z = k$ has the equation $x^2/4 + y^2 = k$. To find the value of k belonging to the level curve through P, we substitute $(2, 1)$ for (x, y) and obtain $k = 2$. Thus, the equation of the level curve that goes through P is the ellipse

$$\frac{x^2}{8} + \frac{y^2}{2} = 1$$

Next let $f(x, y) = x^2/4 + y^2$. Since $f_x(x, y) = x/2$ and $f_y(x, y) = 2y$, the gradient of the paraboloid at $P(2, 1)$ is

$$\nabla f(2, 1) = f_x(2, 1)\mathbf{i} + f_y(2, 1)\mathbf{j} = \mathbf{i} + 2\mathbf{j}$$

The level curve and the gradient at P are shown in Figure 4. ■

The concept of level curves for functions of two variables generalizes to level surfaces for functions of three variables. If f is a function of three variables, the surface $f(x, y, z) = k$, where k is a constant, is called a **level surface** for f. At all points on a level surface, the value of the function is the same, and the gradient vector of $f(x, y, z)$ at a point $P(x, y, z)$ in its domain is normal to the level surface of f that goes through P.

In problems of heat conduction in a homogeneous body where $w = f(x, y, z)$ gives the temperature at the point (x, y, z), the level surface $f(x, y, z) = k$ is called an *isothermal surface* because all points on it have the same temperature k. At any given point of the body, heat flows in the direction opposite to the gradient (that is, in the direction of the greatest decrease in temperature) and therefore perpendicular to the isothermal surface through the point. Again, if $w = f(x, y, z)$ gives the electrostatic potential (voltage) at any point in an electric potential field, the level surfaces of the function are called *equipotential surfaces*. All points on an equipotential surface have the same electrostatic potential, and the direction of flow of electricity is along the negative gradient—that is, in the direction of greatest drop in potential.

EXAMPLE 5 If the temperature at any point in a homogeneous body is given by $T = e^{xy} - xy^2 - x^2yz$, what is the direction of the greatest drop in temperature at the point $(1, -1, 2)$?

Solution The greatest decrease in temperature at $(1, -1, 2)$ is in the direction of the negative gradient at that point.

Since $\nabla T = (ye^{xy} - y^2 - 2xyz)\mathbf{i} + (xe^{xy} - 2xy - x^2z)\mathbf{j} + (-x^2y)\mathbf{k}$, $-\nabla T$ at $(1, -1, 2)$ is

$$(e^{-1} - 3)\mathbf{i} - e^{-1}\mathbf{j} - \mathbf{k} \qquad \blacksquare$$

PROBLEM SET 15.5

In Problems 1–8, find the directional derivative of f at the point **p** in the direction of **a**.

1. $f(x, y) = x^2y$; $\mathbf{p} = (1, 2)$; $\mathbf{a} = 3\mathbf{i} - 4\mathbf{j}$

2. $f(x, y) = y^2 \ln x$; $\mathbf{p} = (1, 4)$; $\mathbf{a} = \mathbf{i} - \mathbf{j}$

3. $f(x, y) = 2x^2 + xy - y^2$; $\mathbf{p} = (3, -2)$; $\mathbf{a} = \mathbf{i} - \mathbf{j}$

4. $f(x, y) = x^2 - 3xy + 2y^2$; $\mathbf{p} = (-1, 2)$; $\mathbf{a} = 2\mathbf{i} - \mathbf{j}$

5. $f(x, y) = e^x \sin y$; $\mathbf{p} = (0, \pi/4)$; $\mathbf{a} = \mathbf{i} + \sqrt{3}\mathbf{j}$

6. $f(x, y) = e^{-xy}$; $\mathbf{p} = (1, -1)$; $\mathbf{a} = -\mathbf{i} + \sqrt{3}\mathbf{j}$

7. $f(x, y, z) = x^3y - y^2z^2$;
   $\mathbf{p} = (-2, 1, 3)$; $\mathbf{a} = \mathbf{i} - 2\mathbf{j} + 2\mathbf{k}$

8. $f(x, y, z) = x^2 + y^2 + z^2$;
   $\mathbf{p} = (1, -1, 2)$; $\mathbf{a} = \sqrt{2}\mathbf{i} - \mathbf{j} - \mathbf{k}$

In Problems 9–12, find a unit vector in the direction in which f increases most rapidly at **p**. What is the rate of change in this direction?

9. $f(x, y) = x^3 - y^5$; $\mathbf{p} = (2, -1)$

10. $f(x, y) = e^y \sin x$; $\mathbf{p} = (5\pi/6, 0)$

11. $f(x, y, z) = x^2yz$; $\mathbf{p} = (1, -1, 2)$

12. $f(x, y, z) = xe^{yz}$; $\mathbf{p} = (2, 0, -4)$

13. In what direction **u** does $f(x, y) = 1 - x^2 - y^2$ decrease most rapidly at $\mathbf{p} = (-1, 2)$?

14. In what direction **u** does $f(x, y) = \sin(3x - y)$ decrease most rapidly at $\mathbf{p} = (\pi/6, \pi/4)$?

15. Sketch the level curve of $f(x, y) = y/x^2$ that goes through $\mathbf{p} = (1, 2)$. Calculate the gradient vector $\nabla f(\mathbf{p})$ and draw this vector, placing its initial point at **p**. What should be true about $\nabla f(\mathbf{p})$?

16. Follow the instructions of Problem 15 for $f(x, y) = x^2 + 4y^2$ and $\mathbf{p} = (2, 1)$.

17. Find the directional derivative of $f(x, y, z) = xy + z^2$ at $(1, 1, 1)$ in the direction toward $(5, -3, 3)$.

18. Find the directional derivative of $f(x, y) = e^{-x} \cos y$ at $(0, \pi/3)$ in the direction toward the origin.

19. The temperature at (x, y, z) of a ball centered at the origin is given by

$$T(x, y, z) = \frac{200}{5 + x^2 + y^2 + z^2}$$

(a) By inspection, decide where the ball is hottest.
(b) Find a vector pointing in the direction of greatest increase of temperature at $(1, -1, 1)$.
(c) Does the vector of (b) point toward the origin?

20. The temperature at (x, y, z) of a ball centered at the origin is $T(x, y, z) = 100e^{-(x^2+y^2+z^2)}$. Note that this ball is hottest at the origin. Show that the direction of greatest decrease in temperature is always a vector pointing away from the origin.

21. The elevation of a mountain above sea level at the point (x, y) is $f(x, y)$. A mountain climber at **p** notes that the slope in the easterly direction is $-\frac{1}{2}$ and the slope in the northerly direction is $-\frac{1}{4}$. In what direction should he move for fastest descent?

22. Given that $f_x(2, 4) = -3$ and $f_y(2, 4) = 8$, find the directional derivative of f at $(2, 4)$ in the direction toward $(5, 0)$.

23. The elevation of a mountain above sea level at (x, y) is $3000e^{-(x^2+2y^2)/100}$ meters. The positive x-axis points east and the positive y-axis points north. A climber is directly above $(10, 10)$. If the climber moves northwest, will she ascend or descend and at what slope?

24. If the temperature of a plate at the point (x, y) is $T(x, y) = 10 + x^2 - y^2$, find the path a heat-seeking particle (which always moves in the direction of greatest increase in temperature) would follow if it starts at $(-2, 1)$. *Hint*: The particle moves in the direction of the gradient

$$\nabla T = 2x\mathbf{i} - 2y\mathbf{j}$$

We may write the path in parametric form as

$$\mathbf{r}(t) = x(t)\mathbf{i} + y(t)\mathbf{j}$$

and we want $x(0) = -2$ and $y(0) = 1$. To move in the required direction means that $\mathbf{r}'(t)$ should be parallel to ∇T. This will be satisfied if

$$\frac{x'(t)}{2x(t)} = -\frac{y'(t)}{2y(t)}$$

together with the conditions $x(0) = -2$ and $y(0) = 1$. Now solve this differential equation and evaluate the arbitrary constant of integration.

25. Do Problem 24 assuming $T(x, y) = 20 - 2x^2 - y^2$.

26. The point $P(1, -1, -10)$ is on the surface $z = -10\sqrt{|xy|}$ (see Figure 1 of Section 15.4). Starting at P, in what direction $\mathbf{u} = u_1\mathbf{i} + u_2\mathbf{j}$ should one move in each case?

(a) To climb most rapidly.
(b) To stay at the same level.
(c) To climb at slope 1.

27. The temperature T in degrees Celsius at (x, y, z) is given by $T = 10/(x^2 + y^2 + z^2)$, where distances are in meters. A bee is flying away from the hot spot at the origin on a spiral path so that its position vector at time t seconds is $\mathbf{r} = t \cos \pi t \mathbf{i} + t \sin \pi t \mathbf{j} + t\mathbf{k}$. Determine the rate of change of T in each case.

(a) With respect to distance traveled at $t = 1$.
(b) With respect to time at $t = 1$. (Think of two ways to do this.)

28. Let $\mathbf{u} = (3\mathbf{i} - 4\mathbf{j})/5$ and $\mathbf{v} = (4\mathbf{i} + 3\mathbf{j})/5$ and suppose that at some point P, $D_{\mathbf{u}}f = -6$ and $D_{\mathbf{v}}f = 17$.

(a) Find ∇f at P.
(b) Note that $|\nabla f|^2 = (D_{\mathbf{u}}f)^2 + (D_{\mathbf{v}}f)^2$ in part (a). Show that this relation always holds if $\mathbf{u}$ and $\mathbf{v}$ are perpendicular.

29. Figure 5 shows the contour map for a hill 60 feet high, which we assume has equation $z = f(x, y)$.

(a) A raindrop landing on the hill above point A will reach the xy-plane at A', by following the path of steepest descent from A. Draw this path and use it to estimate A'.
(b) Do the same for point B.
(c) Estimate f_x at C, f_y at D, and $D_{\mathbf{u}}f$ at E, where $\mathbf{u} = (\mathbf{i} + \mathbf{j})/\sqrt{2}$.

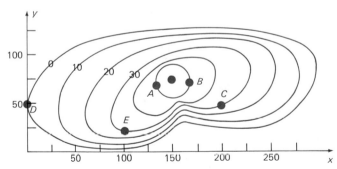

FIGURE 5

30. According to Theorem A, the differentiability of f at **p** implies the existence of $D_{\mathbf{u}}f(\mathbf{p})$ in all directions. Show that the converse is false by considering

$$f(x, y) = \begin{cases} 1 & \text{if } 0 < y < x^2 \\ 0 & \text{otherwise} \end{cases}$$

at the origin.

15.6
The Chain Rule

The Chain Rule for composite functions of one variable is by now familiar to all our readers. If $y = f(x(t))$, where both f and x are differentiable functions, then

$$\frac{dy}{dt} = \frac{dy}{dx}\frac{dx}{dt}$$

Our goal is to obtain generalizations for functions of several variables.

FIRST VERSION If $z = f(x, y)$, where x and y are functions of t, then it makes sense to ask for dz/dt, and there ought to be a formula for it.

Theorem A

(Chain Rule). Let $x = x(t)$ and $y = y(t)$ be differentiable at t and let $z = f(x, y)$ be differentiable at $(x(t), y(t))$. Then, $z = f(x(t), y(t))$ is differentiable at t and

$$\frac{dz}{dt} = \frac{\partial z}{\partial x}\frac{dx}{dt} + \frac{\partial z}{\partial y}\frac{dy}{dt}$$

BEAUTY AND GENERALITY

Does the general analogue of the one variable Chain Rule (Theorem A, Section 3.5) hold? Yes, and here is a particularly elegant statement of it. Let $\mathbb{R}^n$ denote Euclidean n-space, let g be a function from $\mathbb{R}$ to $\mathbb{R}^n$, and let f be a function from $\mathbb{R}^n$ to $\mathbb{R}$. If g is differentiable at t and if f is differentiable at $g(t)$, then the composite function $f \circ g$ is differentiable at $g(t)$ and

$$(f \circ g)'(t) = \nabla f(g(t)) \cdot g'(t)$$

All the machinery needed to demonstrate this is available; see if you can give the proof.

Proof   We mimic the one-variable proof of Appendix A.2, Theorem B. To simplify notation, let $\mathbf{p} = (x, y)$, $\Delta\mathbf{p} = (\Delta x, \Delta y)$, and $\Delta z = f(\mathbf{p} + \Delta\mathbf{p}) - f(\mathbf{p})$. Then, since f is differentiable,

$$\Delta z = f(\mathbf{p} + \Delta\mathbf{p}) - f(\mathbf{p}) = \nabla f(\mathbf{p}) \cdot \Delta\mathbf{p} + |\Delta\mathbf{p}|\varepsilon(\Delta\mathbf{p})$$
$$= f_x(\mathbf{p})\Delta x + f_y(\mathbf{p})\Delta y + |\Delta\mathbf{p}|\varepsilon(\Delta\mathbf{p})$$

with $\varepsilon(\Delta\mathbf{p}) \to 0$ as $\Delta\mathbf{p} \to \mathbf{0}$.

When we divide both sides by Δt, we obtain

(1)
$$\frac{\Delta z}{\Delta t} = f_x(\mathbf{p})\frac{\Delta x}{\Delta t} + f_y(\mathbf{p})\frac{\Delta y}{\Delta t} + \frac{|\Delta\mathbf{p}|}{\Delta t}\varepsilon(\Delta\mathbf{p})$$

Now

$$\left|\frac{|\Delta\mathbf{p}|}{\Delta t}\right| = \frac{\sqrt{(\Delta x)^2 + (\Delta y)^2}}{|\Delta t|} = \sqrt{\left(\frac{\Delta x}{\Delta t}\right)^2 + \left(\frac{\Delta y}{\Delta t}\right)^2}$$

and the latter approaches

$$\sqrt{\left(\frac{dx}{dt}\right)^2 + \left(\frac{dy}{dt}\right)^2}$$

as $\Delta t \to 0$. Also, when $\Delta t \to 0$, both Δx and Δy approach 0 (remember that $x(t)$ and $y(t)$ are continuous, being differentiable). It follows that $\Delta\mathbf{p} \to \mathbf{0}$.

Consequently, when we let $\Delta t \to 0$ in (1), we get

$$\frac{dz}{dt} = f_x(\mathbf{p})\frac{dx}{dt} + f_y(\mathbf{p})\frac{dy}{dt}$$

a result equivalent to the claimed assertion. ∎

EXAMPLE 1 Suppose $z = x^3 y$ where $x = 2t$ and $y = t^2$. Find dz/dt.

Solution

$$\frac{dz}{dt} = \frac{\partial z}{\partial x}\frac{dx}{dt} + \frac{\partial z}{\partial y}\frac{dy}{dt}$$

$$= (3x^2 y)(2) + (x^3)(2t)$$

$$= 6(2t)^2(t^2) + 2(2t)^3(t)$$

$$= 40t^4 \qquad ∎$$

We could have done Example 1 without use of the Chain Rule. By direct substitution,

$$z = x^3 y = (2t)^3 t^2 = 8t^5$$

and so $dz/dt = 40t^4$. However, the direct substitution method is often not available or not convenient—witness the next example.

EXAMPLE 2 As a solid right circular cylinder is heated, its radius r and height h increase; hence, so does its surface area S. Suppose that at the instant when $r = 10$ centimeters and $h = 100$ centimeters, r is increasing at 0.2 centimeters per hour and h is increasing at 0.5 centimeters per hour. How fast is S increasing at this instant?

Solution The formula for the total surface area of a cylinder (Figure 1) is

$$S = 2\pi rh + 2\pi r^2$$

FIGURE 1

Thus,

$$\frac{dS}{dt} = \frac{\partial S}{\partial r}\frac{dr}{dt} + \frac{\partial S}{\partial h}\frac{dh}{dt}$$

$$= (2\pi h + 4\pi r)(0.2) + (2\pi r)(0.5)$$

At $r = 10$ and $h = 100$,

$$\frac{dS}{dt} = (2\pi \cdot 100 + 4\pi \cdot 10)(0.2) + (2\pi \cdot 10)(0.5)$$

$$= 58\pi \text{ square centimeters per hour} \qquad ∎$$

The result in Theorem A extends readily to a function of three variables, as we now illustrate.

EXAMPLE 3 Suppose $w = x^2 y + y + xz$, where $x = \cos\theta$, $y = \sin\theta$, and $z = \theta^2$. Find $dw/d\theta$ and evaluate it at $\theta = \pi/3$.

Solution

$$\frac{dw}{d\theta} = \frac{\partial w}{\partial x}\frac{dx}{d\theta} + \frac{\partial w}{\partial y}\frac{dy}{d\theta} + \frac{\partial w}{\partial z}\frac{dz}{d\theta}$$

$$= (2xy + z)(-\sin\theta) + (x^2 + 1)(\cos\theta) + (x)(2\theta)$$

$$= -2\cos\theta\sin^2\theta - \theta^2\sin\theta + \cos^3\theta + \cos\theta + 2\theta\cos\theta$$

At $\theta = \pi/3$,

$$\frac{dw}{d\theta} = -2\cdot\frac{1}{2}\cdot\frac{3}{4} - \frac{\pi^2}{9}\cdot\frac{\sqrt{3}}{2} + \left(\frac{1}{4} + 1\right)\frac{1}{2} + \frac{2\pi}{3}\cdot\frac{1}{2}$$

$$= -\frac{1}{8} - \frac{\pi^2\sqrt{3}}{18} + \frac{\pi}{3} \qquad\blacksquare$$

SECOND VERSION   Suppose next that $z = f(x, y)$ where $x = x(s, t)$ and $y = y(s, t)$. Then it makes sense to ask for $\partial z/\partial s$ and $\partial z/\partial t$.

Theorem B

(Chain Rule). Let $x = x(s, t)$ and $y = y(s, t)$ have first partial derivatives at (s, t) and let $z = f(x, y)$ be differentiable at $(x(s, t),\ y(s, t))$. Then $z = f(x(s, t), y(s, t))$ has first partial derivatives given by:

(i) $\dfrac{\partial z}{\partial s} = \dfrac{\partial z}{\partial x}\dfrac{\partial x}{\partial s} + \dfrac{\partial z}{\partial y}\dfrac{\partial y}{\partial s};$

(ii) $\dfrac{\partial z}{\partial t} = \dfrac{\partial z}{\partial x}\dfrac{\partial x}{\partial t} + \dfrac{\partial z}{\partial y}\dfrac{\partial y}{\partial t}.$

Proof   If s is held fixed, then $x(s, t)$ and $y(s, t)$ become functions of t alone, which means that Theorem A applies. When we use this theorem with ∂ replacing d to indicate that s is fixed, we obtain the formula in (ii) for $\partial z/\partial t$. The formula for $\partial z/\partial s$ is obtained in a similar way by holding t fixed. $\blacksquare$

EXAMPLE 4   If $z = 3x^2 - y^2$ where $x = 2s + 7t$ and $y = 5st$, find $\partial z/\partial t$ and express it in terms of s and t.

Solution

$$\frac{\partial z}{\partial t} = \frac{\partial z}{\partial x}\frac{\partial x}{\partial t} + \frac{\partial z}{\partial y}\frac{\partial y}{\partial t}$$

$$= (6x)(7) + (-2y)(5s)$$

$$= 42(2s + 7t) - 10st(5s)$$

$$= 84s + 294t - 50s^2t \qquad\blacksquare$$

Here is the corresponding result for three intermediate variables illustrated in an example.

EXAMPLE 5 If $w = x^2 + y^2 + z^2 + xy$, where $x = st$, $y = s - t$, and $z = s + 2t$, find $\partial w/\partial t$.

Solution

$$\frac{\partial w}{\partial t} = \frac{\partial w}{\partial x}\frac{\partial x}{\partial t} + \frac{\partial w}{\partial y}\frac{\partial y}{\partial t} + \frac{\partial w}{\partial z}\frac{\partial z}{\partial t}$$

$$= (2x + y)(s) + (2y + x)(-1) + (2z)(2)$$

$$= (2st + s - t)(s) + (2s - 2t + st)(-1) + (2s + 4t)2$$

$$= 2s^2t + s^2 - 2st + 2s + 10t \qquad \blacksquare$$

IMPLICIT FUNCTIONS   Suppose that $F(x, y) = 0$ defines y implicitly as a function of x—for example, $y = g(x)$—but that the function g is difficult or impossible to determine. We can still find dy/dx. One method for doing this, implicit differentiation, was discussed in Section 3.8. Here is another method.

Let's differentiate both sides of $F(x, y) = 0$ with respect to x using the Chain Rule. We obtain

$$\frac{\partial F}{\partial x}\frac{dx}{dx} + \frac{\partial F}{\partial y}\frac{dy}{dx} = 0$$

Solving for dy/dx yields the formula

$$\boxed{\frac{dy}{dx} = -\frac{\partial F/\partial x}{\partial F/\partial y}}$$

EXAMPLE 6   Find dy/dx if $x^3 + x^2y - 10y^4 = 0$.

Solution   Let $F(x, y) = x^3 + x^2y - 10y^4$. Then

$$\frac{dy}{dx} = -\frac{\partial F/\partial x}{\partial F/\partial y} = -\frac{3x^2 + 2xy}{x^2 - 40y^3}$$

This may be compared with Example 3 of Section 3.8, where the same problem was done by implicit differentiation. $\qquad \blacksquare$

If z is an implicit function of x and y defined by the equation $F(x, y, z) = 0$, then differentiation of both sides with respect to x holding y fixed yields

$$\frac{\partial F}{\partial x}\frac{\partial x}{\partial x} + \frac{\partial F}{\partial y}\frac{\partial y}{\partial x} + \frac{\partial F}{\partial z}\frac{\partial z}{\partial x} = 0$$

If we solve for $\partial z/\partial x$ and note that $\partial y/\partial x = 0$, we get the first of the formulas below. A similar calculation holding x fixed and differentiating with respect to y produces the second formula.

$$\boxed{\frac{\partial z}{\partial x} = -\frac{\partial F/\partial x}{\partial F/\partial z}, \qquad \frac{\partial z}{\partial y} = -\frac{\partial F/\partial y}{\partial F/\partial z}}$$

EXAMPLE 7 If $F(x, y, z) = x^3 e^{y+z} - y \sin(x - z) = 0$ defines z implicitly as a function of x and y, find $\partial z/\partial x$.

Solution

$$\frac{\partial z}{\partial x} = -\frac{\partial F/\partial x}{\partial F/\partial z} = -\frac{3x^2 e^{y+z} - y \cos(x - z)}{x^3 e^{y+z} + y \cos(x - z)} \qquad \blacksquare$$

PROBLEM SET 15.6

In Problems 1–6, find dw/dt by using the Chain Rule. Express your final answer in terms of t.

1. $w = x^2 y^3$; $x = t^3$, $y = t^2$

2. $w = x^2 y - y^2 x$; $x = \cos t$, $y = \sin t$

3. $w = e^x \sin y + e^y \sin x$; $x = 3t$, $y = 2t$

4. $w = \ln(x/y)$; $x = \tan t$, $y = \sec^2 t$

5. $w = \sin(xyz^2)$; $x = t^3$, $y = t^2$, $z = t$

6. $w = xy + yz + xz$; $x = t^2$, $y = 1 - t^2$, $z = 1 - t$

In Problems 7–12, find $\partial w/\partial t$ by using the Chain Rule. Express your final answer in terms of s and t.

7. $w = x^2 y$; $x = st$, $y = s - t$

8. $w = x^2 - y \ln x$; $x = s/t$, $y = s^2 t$

9. $w = e^{x^2 + y^2}$; $x = s \sin t$, $y = t \sin s$

10. $w = \ln(x + y) - \ln(x - y)$; $x = te^s$, $y = e^{st}$

11. $w = \sqrt{x^2 + y^2 + z^2}$; $x = \cos st$, $y = \sin st$, $z = s^2 t$

12. $w = e^{xy+z}$; $x = s + t$, $y = s - t$, $z = t^2$

13. If $z = x^2 y$; $x = 2t + s$, $y = 1 - st^2$, find

$$\left. \frac{\partial z}{\partial t} \right|_{s=1, t=-2}$$

14. If $z = xy + x + y$, $x = r + s + t$, and $y = rst$, find

$$\left. \frac{\partial z}{\partial s} \right|_{r=1, s=-1, t=2}$$

15. If $w = u^2 - u \tan v$, $u = x$, and $v = \pi x$, find

$$\left. \frac{dw}{dx} \right|_{x=1/4}$$

16. If $w = x^2 y + z^2$, $x = \rho \cos \theta \sin \phi$, $y = \rho \sin \theta \sin \phi$, and $z = \rho \cos \phi$, find

$$\left. \frac{\partial w}{\partial \theta} \right|_{\rho=2, \theta=\pi, \phi=\pi/2}$$

17. The part of a tree normally sawed into lumber is the trunk, a solid shaped approximately like a right circular cylinder. If the radius of the trunk of a certain tree is growing $\frac{1}{2}$ inch per year and the height is increasing 8 inches per year, how fast is the volume increasing when the radius is 20 inches and the height is 400 inches? Express your answer in board feet per year (1 board foot = 1 inch by 12 inches by 12 inches).

18. The temperature of a metal plate at (x, y) is e^{-x-3y} degrees. A bug is walking northeast at a rate of $\sqrt{8}$ feet per minute (that is, $dx/dt = dy/dt = 2$). From the bug's point of view, how is the temperature changing with time as it crosses the origin?

19. A boy's toy boat slips from his grasp at the edge of a straight river. The stream carries it along at 5 feet per second. A crosswind blows it toward the opposite shore at 4 feet per second. If the boy runs along the shore at 3 feet per second following his boat, how fast is the boat moving away from him when $t = 3$ seconds?

20. Sand is pouring onto a conical pile in such a way that at a certain instant the height is 100 inches and increasing at 3 inches per minute and the radius is 40 inches and increasing at 2 inches per minute. How fast is the volume increasing at that instant?

In Problems 21–24, use the method of Example 6 to find dy/dx.

21. $x^3 + 2x^2 y - y^3 = 0$

22. $ye^{-x} + 5x - 17 = 0$

23. $x \sin y + y \cos x = 0$

24. $x^2 \cos y - y^2 \sin x = 0$

25. If $3x^2 z + y^3 - xyz^3 = 0$, find $\partial z/\partial x$ (Example 7).

26. If $ye^{-x} + z \sin x = 0$, find $\partial x/\partial z$ (Example 7).

27. If $T = f(x, y, z, w)$ and x, y, z, and w are each functions of s and t, write a chain rule for $\partial T/\partial s$.

28. Let $z = f(x, y)$, where $x = r \cos \theta$ and $y = r \sin \theta$. Show that

$$\left(\frac{\partial z}{\partial x} \right)^2 + \left(\frac{\partial z}{\partial y} \right)^2 = \left(\frac{\partial z}{\partial r} \right)^2 + \frac{1}{r^2} \left(\frac{\partial z}{\partial \theta} \right)^2$$

29. The wave equation of physics is the partial differential equation

$$\frac{\partial^2 y}{\partial t^2} = c^2 \frac{\partial^2 y}{\partial x^2}$$

where c is a constant. Show that if f is any twice differentiable function, then

$$y(x, t) = \tfrac{1}{2}[f(x - ct) + f(x + ct)]$$

satisfies this equation.

30. Show that if $w = f(r - s, s - t, t - r)$, then $\dfrac{\partial w}{\partial r} +$ $\dfrac{\partial w}{\partial s} + \dfrac{\partial w}{\partial t} = 0.$

31. Let $F(t) = \displaystyle\int_{g(t)}^{h(t)} f(u)\,du$, where f is continuous and g and h are differentiable. Show that

$$F'(t) = f(h(t))h'(t) - f(g(t))g'(t)$$

and use this result to find $F'(\sqrt{2})$, where

$$F(t) = \int_{\sin \sqrt{2}\pi t}^{t^2} \sqrt{9 + u^4}\,du$$

32. Call a function $f(x, y)$ *homogeneous of degree 1* if $f(tx, ty) = tf(x, y)$ for all $t > 0$. For example, $f(x, y) = x + ye^{y/x}$ satisfies this criterion. Prove Euler's Theorem that such a function satisfies

$$f(x, y) = x\frac{\partial f}{\partial x} + y\frac{\partial f}{\partial y}$$

Note: Let $f(x, y)$ denote the value of production from x units of capital and y units of labor, a homogeneous function (for example, doubling capital and labor doubles production). Euler's Theorem then asserts an important law of economics that may be phrased as follows: The value of production $f(x, y)$ equals the cost of capital plus the cost of labor provided they are paid for at their respective marginal rates $\partial f/\partial x$ and $\partial f/\partial y$.

C **33.** Leaving from the same point P, plane A flies due east while plane B flies N 50° E. At a certain instant, A is 200 miles from P flying at 450 miles per hour and B is 150 miles from P flying at 400 miles per hour. How fast are they separating at that instant?

34. Recall Newton's Law of Gravitation, which asserts that the magnitude F of the force of attraction between objects of mass M and m is $F = GMm/r^2$, where r is the distance between them and G is a universal constant. Let an object of mass M be located at the origin and suppose a second object of changing mass m (say from fuel consumption) is moving away from the origin so that its position vector is $\mathbf{r} = x\mathbf{i} + y\mathbf{j} + z\mathbf{k}$. Obtain a formula for dF/dt in terms of the time derivatives of m, x, y, and z.

15.7
Tangent Planes, Approximations

We introduced the notion of a tangent plane to a surface in Section 15.4 but dealt only with surfaces determined by equations of the form $z = f(x, y)$ (Figure 1). Now we want to consider the more general situation of a surface determined by $F(x, y, z) = k$. (Note that $z = f(x, y)$ can be written as $F(x, y, z) = f(x, y) - z = 0$.) Consider a curve on this surface passing through the point (x_0, y_0, z_0). If $x = x(t)$, $y = y(t)$, and $z = z(t)$ are parametric equations for this curve, then for all t,

$$F(x(t), y(t), z(t)) = k$$

By the Chain Rule,

$$\frac{dF}{dt} = \frac{\partial F}{\partial x}\frac{dx}{dt} + \frac{\partial F}{\partial y}\frac{dy}{dt} + \frac{\partial F}{\partial z}\frac{dz}{dt} = \frac{dk}{dt} = 0$$

We can express this in terms of the gradient of F and the derivative of the vector expression for the curve $\mathbf{r}(t) = x(t)\mathbf{i} + y(t)\mathbf{j} + z(t)\mathbf{k}$ as

$$\nabla F \cdot \frac{d\mathbf{r}}{dt} = 0$$

FIGURE 1

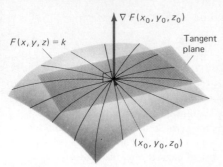

$F(x, y, z) = k$

$\nabla F(x_0, y_0, z_0)$

Tangent plane

(x_0, y_0, z_0)

FIGURE 2

As we learned earlier (Section 14.4), $d\mathbf{r}/dt$ is tangent to the curve. In summary, the gradient at (x_0, y_0, z_0) is perpendicular to the tangent line at this point.

The argument just given is valid for any curve through (x_0, y_0, z_0) that lies in the surface $F(x, y, z) = k$ (Figure 2). This suggests the following general definition.

Definition

Let $F(x, y, z) = k$ determine a surface and suppose that F is differentiable at a point $P(x_0, y_0, z_0)$ of this surface with $\nabla F(x_0, y_0, z_0) \neq \mathbf{0}$. Then the plane through P perpendicular to $\nabla F(x_0, y_0, z_0)$ is called the **tangent plane** to the surface at P.

As a consequence of this definition and Section 14.2, we can write the equation of the tangent plane.

Theorem A

(Tangent planes). For the surface $F(x, y, z) = k$, the equation of the tangent plane at (x_0, y_0, z_0) is $\nabla F(x_0, y_0, z_0) \cdot \langle x - x_0, y - y_0, z - z_0 \rangle = 0$, that is,

$$F_x(x_0, y_0, z_0)(x - x_0) + F_y(x_0, y_0, z_0)(y - y_0)$$
$$+ F_z(x_0, y_0, z_0)(z - z_0) = 0$$

In particular, for the surface $z = f(x, y)$, the equation of the tangent plane at $(x_0, y_0, f(x_0, y_0))$ is

$$z - z_0 = f_x(x_0, y_0)(x - x_0) + f_y(x_0, y_0)(y - y_0)$$

Proof The first statement is immediate and the second follows from it by considering $F(x, y, z) = f(x, y) - z$. ∎

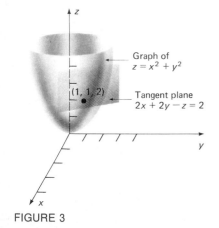

Graph of $z = x^2 + y^2$

$(1, 1, 2)$

Tangent plane $2x + 2y - z = 2$

FIGURE 3

EXAMPLE 1 Find the equation of the tangent plane (Figure 3) to $z = x^2 + y^2$ at the point $(1, 1, 2)$.

Solution Let $f(x, y) = x^2 + y^2$ and note that $\nabla f(x, y) = 2x\mathbf{i} + 2y\mathbf{j}$. Thus, $\nabla f(1, 1) = 2\mathbf{i} + 2\mathbf{j}$ and, from Theorem A, the required equation is

$$z - 2 = 2(x - 1) + 2(y - 1)$$

or

$$2x + 2y - z = 2 \qquad ∎$$

EXAMPLE 2 Find the equation of the tangent plane and the normal line to the surface $x^2 + y^2 + 2z^2 = 23$ at $(1, 2, 3)$.

Solution Let $F(x, y, z) = x^2 + y^2 + 2z^2$, so that $\nabla F(x, y, z) = 2x\mathbf{i} + 2y\mathbf{j} + 4z\mathbf{k}$ and $\nabla F(1, 2, 3) = 2\mathbf{i} + 4\mathbf{j} + 12\mathbf{k}$. According to Theorem A, the equation of the tangent plane at $(1, 2, 3)$ is

$$2(x - 1) + 4(y - 2) + 12(z - 3) = 0$$

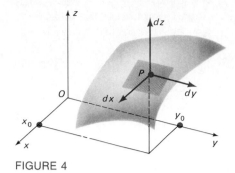

FIGURE 4

Similarly the symmetric equations of the normal line through (1, 2, 3) are

$$\frac{x-1}{2} = \frac{y-2}{4} = \frac{z-3}{12}$$ ∎

DIFFERENTIALS AND APPROXIMATIONS We suggest that the reader review Section 3.10, where the topic of differentials and approximations is treated for functions of one variable.

Let $z = f(x, y)$ and let $P(x_0, y_0, z_0)$ be a fixed point on the corresponding surface. Introduce new coordinate axes (the dx-, dy-, and dz-axes), parallel to the old axes, with P as origin (Figure 4). In the old system, the tangent plane at P has equation

$$z - z_0 = f_x(x_0, y_0)(x - x_0) + f_y(x_0, y_0)(y - y_0)$$

but in the new system this takes the simple form

$$dz = f_x(x_0, y_0)\, dx + f_y(x_0, y_0)\, dy$$

This suggests a definition.

Definition

Let $z = f(x, y)$, where f is a differentiable function, and let dx and dy (called the differentials of x and y) be variables. The **differential of the dependent variable**, dz, also called the total **differential of** f and written $df(x, y)$, is defined by

$$dz = df(x, y) = f_x(x, y)\, dx + f_y(x, y)\, dy$$

The significance of dz arises from the fact that if $dx = \Delta x$ and $dy = \Delta y$ represent small changes in x and y, respectively, then dz will be a good approximation to Δz, the corresponding change in z. This is illustrated in Figure 5 and, while dz does not appear to be a very good approximation to Δz, you can see that it will get better and better as Δx and Δy get smaller.

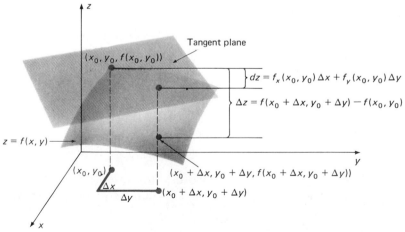

FIGURE 5

EXAMPLE 3 Let $z = f(x, y) = 2x^3 + xy - y^3$. Compute Δz and dz as (x, y) changes from $(2, 1)$ to $(2.03, 0.98)$.

Solution

$$\Delta z = f(2.03, 0.98) - f(2, 1)$$

$$= 2(2.03)^3 + (2.03)(0.98) - (0.98)^3 - [2(2)^3 + 2(1) - 1^3]$$

$$= 0.779062$$

$$dz = f_x(x, y)\,\Delta x + f_y(x, y)\,\Delta y$$

$$= (6x^2 + y)\,\Delta x + (x - 3y^2)\,\Delta y$$

At $(2, 1)$ with $\Delta x = 0.03$ and $\Delta y = -0.02$,

$$dz = (25)(0.03) + (-1)(-0.02) = 0.77 \qquad \blacksquare$$

EXAMPLE 4 The formula $P = k(T/V)$, where k is a constant, gives the pressure P of a confined gas of volume V and temperature T. Find, approximately, the maximum percentage error in P introduced by an error of $\pm 0.4\%$ in measuring the temperature and an error of $\pm 0.9\%$ in measuring the volume.

Solution   The error in P is ΔP, which we will approximate by dP. Thus,

$$|\Delta P| \approx |dP| = \left| \frac{\partial P}{\partial T}\,\Delta T + \frac{\partial P}{\partial V}\,\Delta V \right|$$

$$\leq \left| \frac{k}{V}\,(\pm 0.004T) \right| + \left| -\frac{kT}{V^2}\,(\pm 0.009V) \right|$$

$$= \frac{kT}{V}\,(0.004 + 0.009) = 0.013\,\frac{kT}{V} = 0.013P$$

The maximum relative error, $|\Delta P|/P$, is approximately 0.013, and the maximum percentage error is approximately 1.3%. $\qquad \blacksquare$

The subject of approximation goes far beyond what we have done here. Recall that, for functions of one variable, the first-order approximation

$$f(x) \approx f(x_0) + f'(x_0)(x - x_0)$$

was extended to the second-order Taylor approximation

$$f(x) \approx f(x_0) + f'(x_0)(x - x_0) + \tfrac{1}{2}f''(x_0)(x - x_0)^2$$

and then to nth-order Taylor approximations (see Section 10.1).

The analogues for functions of two variables are

$$f(x, y) \approx f(x_0, y_0) + f_x(x_0, y_0)(x - x_0) + f_y(x_0, y_0)(y - y_0)$$

and

$$f(x, y) \approx f(x_0, y_0) + f_x(x_0, y_0)(x - x_0) + f_y(x_0, y_0)(y - y_0)$$
$$+ \tfrac{1}{2}\left[f_{xx}(x_0, y_0)(x - x_0)^2 + 2f_{xy}(x_0, y_0)(x - x_0)(y - y_0)\right.$$
$$\left. + f_{yy}(x_0, y)(y - y_0)^2\right]$$

Moreover, these extend both to the nth-order case and to functions of more than two variables. The details are best left to higher-level books.

PROBLEM SET 15.7

In Problems 1–8, find the equation of the tangent plane to the given surface at the indicated point.

1. $x^2 + y^2 + z^2 = 16$; $(2, 3, \sqrt{3})$

2. $8x^2 + y^2 + 8z^2 = 16$; $(1, 2, \sqrt{2}/2)$

3. $x^2 - y^2 + z^2 + 1 = 0$; $(1, 3, \sqrt{7})$

4. $x^2 + y^2 - z^2 = 4$; $(2, 1, 1)$

5. $z = \dfrac{x^2}{4} + \dfrac{y^2}{4}$; $(2, 2, 2)$

6. $z = xe^{-2y}$; $(1, 0, 1)$

7. $z = 2e^{3y} \cos 2x$; $(\pi/3, 0, -1)$

8. $z = x^{1/2} + y^{1/2}$; $(1, 4, 3)$

In Problems 9–12, use the total differential dz to approximate the change in z as (x, y) moves from P to Q. Then use a calculator to find the corresponding exact change Δz (to the accuracy of your calculator). See Example 3.

C **9.** $z = 2x^2 y^3$; $P(1, 1)$, $Q(0.99, 1.02)$

C **10.** $z = x^2 - 5xy + y$; $P(2, 3)$, $Q(2.03, 2.98)$

C **11.** $z = \ln(x^2 y)$; $P(-2, 4)$, $Q(-1.98, 3.96)$

C **12.** $z = \tan^{-1} xy$; $P(-2, -0.5)$, $Q(-2.03, -0.51)$

13. Find all points on the surface $z = x^2 - 2xy - y^2 - 8x + 4y$, where the tangent plane is horizontal.

14. Find a point on the surface $z = 2x^2 + 3y^2$ where the tangent plane is parallel to the plane $8x - 3y - z = 0$.

15. Show that the surface $x^2 + 4y + z^2 = 0$ and $x^2 + y^2 + z^2 - 6z + 7 = 0$ are tangent to each other at $(0, -1, 2)$; that is, show that they have the same tangent plane at $(0, -1, 2)$.

16. Show that the surfaces $z = x^2 y$ and $y = \tfrac{1}{4}x^2 + \tfrac{3}{4}$ intersect at $(1, 1, 1)$ and have perpendicular tangent planes there.

17. Find a point on the surface $x^2 + 2y^2 + 3z^2 = 12$ where the tangent plane is perpendicular to the line with parametric equations: $x = 1 + 2t$, $y = 3 + 8t$, $z = 2 - 6t$.

18. Show that the equation of the tangent plane to the ellipsoid

$$\frac{x^2}{a^2} + \frac{y^2}{b^2} + \frac{z^2}{c^2} = 1$$

at (x_0, y_0, z_0) can be written in the form

$$\frac{x_0 x}{a^2} + \frac{y_0 y}{b^2} + \frac{z_0 z}{c^2} = 1$$

19. Find the parametric equations of the line that is tangent to the curve of intersection of the surfaces

$$f(x, y, z) = 9x^2 + 4y^2 + 4z^2 - 41 = 0$$

and

$$g(x, y, z) = 2x^2 - y^2 + 3z^2 - 10 = 0$$

at the point $(1, 2, 2)$. *Hint:* This line is perpendicular to $\nabla f(1, 2, 2)$ and $\nabla g(1, 2, 2)$.

20. Find the parametric equations of the line that is tangent to the curve of intersection of the surfaces $x = z^2$ and $y = z^3$ at $(1, 1, 1)$ (see Problem 19).

21. In determining the specific gravity of an object, its weight in air is found to be $A = 36$ pounds and its weight in water is $W = 20$ pounds, with a possible error in each measurement of 0.02 pound. Find, approximately, the maximum possible error in calculating its specific gravity S, where $S = A/(A - W)$.

22. Use differentials to find the approximate amount of copper in the four sides and bottom of a rectangular copper tank that is 6 feet long, 4 feet wide, and 3 feet deep *inside*, if the sheet copper is $\tfrac{1}{4}$ inch thick. *Hint:* Make a sketch.

23. The radius and height of a right circular cone are measured with errors of at most 2% and 3%, respectively. Use differentials to estimate the maximum percentage error in the calculated volume (see Example 4).

24. The period T of a pendulum of length L is given by $T = 2\pi\sqrt{L/g}$, where g is the acceleration of gravity. Show that $dT/T = \tfrac{1}{2}[dL/L - dg/g]$ and use this result to estimate the maximum percentage error in T due to an error of 0.5% in measuring L and 0.3% in measuring g.

25. The formula $1/R = 1/R_1 + 1/R_2$ determines the combined resistance R when resistors of resistance R_1 and R_2 are connected in parallel. Suppose that R_1 and R_2 were measured at 25 and 100 ohms, respectively, with possible errors in each measurement of 0.5 ohms. Calculate R and give an estimate for the maximum error in this value.

26. A bee sat at the point $(1, 2, 1)$ on the ellipsoid $x^2 + y^2 + 2z^2 = 6$ (distances in feet). At $t = 0$, it took off along the normal line at a speed of 4 feet per second. Where and when did it hit the plane $2x + 3y + z = 49$?

27. Show that the tangent plane to the surface $xyz = k$ at any point forms with the coordinate planes a tetrahedron of fixed volume and find this volume.

28. Find and simplify the equation of the tangent plane at (x_0, y_0, z_0) to the surface $\sqrt{x} + \sqrt{y} + \sqrt{z} = a$. Then show that the sum of the intercepts of this plane with the coordinate axes is a^2.

29. For the function $f(x, y) = \sqrt{x^2 + y^2}$, find the second-order Taylor approximation based at $(x_0, y_0) = (3, 4)$. Then estimate $f(3.1, 3.9)$ using (a) the first-order approximation, (b) the second-order approximation, and (c) your calculator directly.

15.8
Maxima and Minima

Our goal is to extend the notions of Chapter 4 to functions of several variables; a quick review of that chapter, especially Sections 4.1 and 4.3, will be helpful. The definitions given there extend almost without change, but for clarity we repeat them. In what follows, let $\mathbf{p} = (x, y)$ and $\mathbf{p}_0 = (x_0, y_0)$ be a variable point and a fixed point, respectively, in two-space (they could just as well be points in n-space).

Definition

Let $\mathbf{p}_0$ be a point in S, the domain of f.

(i) $f(\mathbf{p}_0)$ is a (global) **maximum value** of f on S if $f(\mathbf{p}_0) \geq f(\mathbf{p})$ for all $\mathbf{p}$ in S.

(ii) $f(\mathbf{p}_0)$ is a (global) **minimum value** of f on S if $f(\mathbf{p}_0) \leq f(\mathbf{p})$ for all $\mathbf{p}$ in S.

(iii) $f(\mathbf{p}_0)$ is a (global) **extreme value** of f on S if it is either a (global) maximum value or a (global) minimum value.

The same definitions hold with the word *global* replaced by *local* if, in (i) and (ii), we require only that the inequalities hold on $N \cap S$, where N is a neighborhood of $\mathbf{p}_0$.

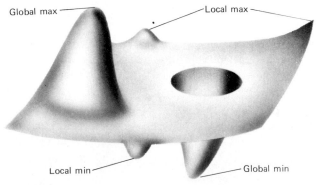

FIGURE 1

CHAPTER 15/The Derivative in n-Space

Figure 1 gives a geometric interpretation of the concepts we have defined. Note that a global maximum (or minimum) is automatically a local maximum (or minimum).

Our first theorem is a big one—difficult to prove, but intuitively clear.

Theorem A

(Max-Min Existence Theorem). If f is continuous on a closed bounded set S, then f attains both a (global) maximum value and a (global) minimum value there.

The proof may be found in most books on advanced calculus.

WHERE DO EXTREME VALUES OCCUR?   The situation is analogous to the one-variable case. The **critical points** of f on S are of three types.

1. **Boundary points**. See Section 15.3.
2. **Stationary points**. We call $\mathbf{p}_0$ a stationary point if $\mathbf{p}_0$ is an interior point of S where f is differentiable and $\nabla f(\mathbf{p}_0) = \mathbf{0}$. At such a point, the tangent plane is horizontal.
3. **Singular points**. We call $\mathbf{p}_0$ a singular point if $\mathbf{p}_0$ is an interior point of S where f is not differentiable—for example, a point where the graph of f has a sharp corner.

Now we can state another big theorem; we can actually prove this one.

Theorem B

(Critical Point Theorem). Let f be defined on a set S containing $\mathbf{p}_0$. If $f(\mathbf{p}_0)$ is an extreme value, then $\mathbf{p}_0$ must be a critical point; that is, either $\mathbf{p}_0$ is:

(i)   a boundary point of S; or
(ii)   a stationary point of f; or
(iii)   a singular point of f.

Proof   Suppose that $\mathbf{p}_0$ is neither a boundary point nor a singular point (so that $\mathbf{p}_0$ is an interior point where ∇f exists). We will be done if we can show that $\nabla f(\mathbf{p}_0) = \mathbf{0}$. For simplicity, set $\mathbf{p}_0 = (x_0, y_0)$; the higher-dimensional cases will follow in a similar fashion.

Since f has an extreme value at (x_0, y_0), the function $g(x) = f(x, y_0)$ has an extreme value at x_0. Moreover g is differentiable at x_0 since f is differentiable at (x_0, y_0) and therefore, by the Critical Point Theorem for functions of one variable (Theorem 4.1B),

$$g'(x_0) = f_x(x_0, y_0) = 0$$

Similarly, the function $h(y) = f(x_0, y)$ has an extreme value at y_0 and satisfies

$$h'(y_0) = f_y(x_0, y_0) = 0$$

The gradient is $\mathbf{0}$ since both partials are 0.　■

The theorem and its proof are valid whether the extreme values are global or local extreme values.

EXAMPLE 1　Find the local maximum or minimum values of $f(x, y) = x^2 - 2x + y^2/4$.

Solution　The given function is differentiable throughout its domain, the xy-plane. Thus, the only possible critical points are the stationary points obtained by setting $f_x(x, y)$ and $f_y(x, y)$ equal to zero. But $f_x(x, y) = 2x - 2$ and $f_y(x, y) = y/2$ are zero only when $x = 1$ and $y = 0$. It remains to decide whether $(1, 0)$ gives a maximum or a minimum or neither. We will develop a simple tool for this soon, but for now we must use a little ingenuity. Note that $f(1, 0) = -1$ and

$$f(x, y) = x^2 - 2x + \frac{y^2}{4} = x^2 - 2x + 1 + \frac{y^2}{4} - 1$$

$$= (x - 1)^2 + \frac{y^2}{4} - 1 \geq -1$$

Thus, $f(1, 0)$ is actually a global minimum for f. There are no local maximum values.　■

EXAMPLE 2　Find the local minimum or maximum values of $f(x, y) = -x^2/a^2 + y^2/b^2$.

Solution　The only critical points are obtained by setting $f_x(x, y) = -2x/a^2$ and $f_y(x, y) = 2y/b^2$ equal to zero. This yields the point $(0, 0)$, which gives neither a maximum nor minimum (see Figure 2). It is called a **saddle point**. The given function has no local extrema.　■

Example 2 illustrates the troublesome fact the $\nabla f(x_0, y_0) = \mathbf{0}$ does not guarantee that there is a local extremum at (x_0, y_0). Fortunately, there is a nice criterion for deciding what is happening at a stationary point—our next topic.

FIGURE 2

SUFFICIENT CONDITIONS FOR EXTREMA　You should think of the next theorem as being analogous to the Second Derivative Test for functions of one variable (Theorem 4.3B). The proof may be found in books on advanced calculus.

(Second-Partials Test.) Suppose that $f(x, y)$ has continuous second partial derivatives in a neighborhood of (x_0, y_0) and that $\nabla f(x_0, y_0) = \mathbf{0}$. Let

$$D = D(x_0, y_0) = f_{xx}(x_0, y_0)f_{yy}(x_0, y_0) - f_{xy}^2(x_0, y_0)$$

Then:

(i) if $D > 0$ and $f_{xx}(x_0, y_0) < 0$, $f(x_0, y_0)$ is a local maximum value;
(ii) if $D > 0$ and $f_{xx}(x_0, y_0) > 0$, $f(x_0, y_0)$ is a local minimum value;
(iii) if $D < 0$, $f(x_0, y_0)$ is not an extreme value ((x_0, y_0) is a saddle point);
(iv) if $D = 0$, the test is inconclusive.

EXAMPLE 3 Find the extrema, if any, of the function F defined by $F(x, y) = 3x^3 + y^2 - 9x + 4y$.

Solution Since $F_x(x, y) = 9x^2 - 9$ and $F_y(x, y) = 2y + 4$, the critical points, obtained by solving the simultaneous equations $F_x(x, y) = F_y(x, y) = 0$, are $(1, -2)$ and $(-1, -2)$.

Now $F_{xx}(x, y) = 18x$, $F_{yy}(x, y) = 2$, and $F_{xy} = 0$. Thus at the critical point $(1, -2)$, $D = F_{xx}(1, -2) \cdot F_{yy}(1, -2) - F_{xy}^2(1, -2) = 18(2) - 0 = 36 > 0$. Furthermore, $F_{xx}(1, -2) = 18 > 0$, and so by (ii), $F(1, -2) = -10$ is a local minimum value of F.

In testing the given function at the other critical point, $(-1, -2)$, we find $F_{xx}(-1, -2) = -18$, $F_{yy}(-1, -2) = 2$, and $F_{xy}(-1, -2) = 0$, which makes $D = -36 < 0$. Thus, by (iii), $(-1, 2)$ is a saddle point and $F(-1, -2)$ is not an extremum. ∎

EXAMPLE 4 Find the minimum distance between the origin and the surface $z^2 = x^2 y + 4$.

Solution Let $P(x, y, z)$ be any point on the surface. The square of the distance between the origin and P is $d^2 = x^2 + y^2 + z^2$. We seek the coordinates of P that makes d^2 (and hence d) a minimum.

Since P is on the surface, its coordinates satisfy the equation of the surface. Substituting $z^2 = x^2 y + 4$ in $d^2 = x^2 + y^2 + z^2$, we obtain d^2 as a function of two variables x and y:

$$d^2 = f(x, y) = x^2 + y^2 + x^2 y + 4$$

To find the critical points, we set $f_x(x, y) = 0$ and $f_y(x, y) = 0$, obtaining

$$2x + 2xy = 0 \quad \text{and} \quad 2y + x^2 = 0$$

By eliminating y between these equations, we get

$$2x - x^3 = 0$$

Thus, $x = 0$ or $x = \pm\sqrt{2}$. Substituting these values in the second of the equations, we obtain $y = 0$ and $y = -1$. Therefore, the critical points are $(0, 0)$, $(\sqrt{2}, -1)$, and $(-\sqrt{2}, -1)$.

To test each of these, we need $f_{xx}(x, y) = 2 + 2y$, $f_{yy}(x, y) = 2$, $f_{xy}(x, y) = 2x$, and $D(x, y) = f_{xx}f_{yy} - f_{xy}^2 = 4 + 4y - 4x^2$. Since $D(\pm\sqrt{2}, -1) = -8 < 0$, neither $(\sqrt{2}, -1)$ nor $(-\sqrt{2}, -1)$ yields an extremum. However, $D(0, 0) = 4 > 0$ and $f_{xx}(0, 0) = 2 > 0$; so $(0, 0)$ yields the minimum distance. Substituting $x = 0$ and $y = 0$ in the expression for d^2, we find $d^2 = 4$.

The minimum distance between the origin and the given surface is 2. ∎

None of your examples so far has involved boundary points. Here is one that does.

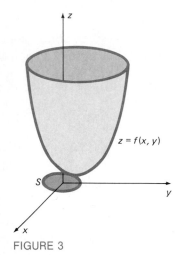

EXAMPLE 5 Find the maximum and minimum values of $f(x, y) = 2x^2 + y^2 - 4x - 2y + 5$ on the closed set $S = \{(x, y) : x^2 + \frac{1}{2}y^2 \leq 1\}$; see Figure 3.

Solution Since $f_x(x, y) = 4x - 4$ and $f_y(x, y) = 2y - 2$, the only possible interior critical point is $(1, 1)$. However, this point lies outside S, so it can be ignored.

The boundary of S is the ellipse $x^2 + \frac{1}{2}y^2 = 1$, which can be described parametrically by

$$x = \cos t, \qquad y = \sqrt{2} \sin t, \qquad 0 \leq t \leq 2\pi$$

We wish to maximize and minimize the function of one variable

$$g(t) = f(\cos t, \sqrt{2} \sin t), \qquad 0 \leq t \leq 2\pi$$

By the Chain Rule,

$$\begin{aligned}
g'(t) &= \frac{\partial f}{\partial x}\frac{dx}{dt} + \frac{\partial f}{\partial y}\frac{dy}{dt} \\
&= (4x - 4)(-\sin t) + (2y - 2)(\sqrt{2}\cos t) \\
&= (4\cos t - 4)(-\sin t) + (2\sqrt{2}\sin t - 2)(\sqrt{2}\cos t) \\
&= 4\sin t - 2\sqrt{2}\cos t
\end{aligned}$$

Setting $g'(t) = 0$ yields $\tan t = \sqrt{2}/2$ with two solutions $t_1 = \tan^{-1}(\sqrt{2}/2)$ and $t_2 = \pi + t_1$. Thus, g has the four critical points 0, t_1, t_2, and 2π on the interval $[0, 2\pi]$. These, in turn, determine the three points $(1, 0)$, $(2/\sqrt{6}, 2/\sqrt{6})$, and $(-2/\sqrt{6}, -2/\sqrt{6})$ on the boundary of S. The correspond-

FIGURE 3

ing values of f are

$$f(1,0) = 3$$

$$f\left(\frac{2}{\sqrt{6}}, \frac{2}{\sqrt{6}}\right) \approx 2.101$$

$$f\left(\frac{-2}{\sqrt{6}}, \frac{-2}{\sqrt{6}}\right) \approx 11.899$$

We conclude that the minimum value of f on S is 2.101 and the maximum value is 11.899. ∎

PROBLEM SET 15.8

In Problems 1–10, find all critical points. Indicate whether each such point gives a local maximum, a local minimum, or whether it is a saddle point. *Hint*: Use Theorem C.

1. $f(x, y) = x^2 + 4y^2 - 4x$

2. $f(x, y) = x^2 + 4y^2 - 2x + 8y - 1$

3. $f(x, y) = 2x^4 - x^2 + 3y^2$

4. $f(x, y) = xy^2 - 6x^2 - 3y^2$

5. $f(x, y) - xy$

6. $f(x, y) = x^3 + y^3 - 6xy$

7. $f(x, y) = xy + \frac{2}{x} + \frac{4}{y}$

8. $f(x, y) = e^{-(x^2 + y^2 - 4y)}$

9. $f(x, y) = \cos x + \cos y + \cos(x + y)$; $0 < x < \pi/2, 0 < y < \pi/2$

10. $f(x, y) = x^2 + a^2 - 2ax \cos y$; $-\pi < y < \pi$

In Problems 11–14, find the global maximum value and global minimum value of f on S and indicate where they occur.

11. $f(x, y) = 3x + 4y$;
$S = \{(x, y): 0 \le x \le 1, -1 \le y \le 1\}$

12. $f(x, y) = x^2 + y^2$;
$S = \{(x, y): -1 \le x \le 3, -1 \le y \le 4\}$

13. $f(x, y) = x^2 - y^2 + 1$; $S = \{(x, y): x^2 + y^2 \le 1\}$
(see Example 5).

14. $f(x, y) = x^2 - 6x + y^2 - 8y + 7$;
$S = \{(x, y): x^2 + y^2 \le 1\}$

15. Express a positive number N as a sum of three numbers such that the product of these three numbers is a maximum.

16. Use the methods of this section to find the shortest distance from the origin to the plane $x + 2y + 3z = 12$.

17. Find the shape of the closed rectangular box of volume V_0 with minimum surface area.

18. Find the shape of the rectangular box of volume V_0 for which the sum of the edge lengths is least.

19. A rectangular metal tank with open top is to hold 256 cubic feet of liquid. What are the dimensions of the tank that require the least material to build?

20. A rectangular box, whose edges are parallel to the coordinate axes, is inscribed in the ellipsoid $96x^2 + 4y^2 + 4z^2 = 36$. What is the greatest possible volume for such a box?

21. Find the three-dimensional vector with length 9, the sum of whose components is a maximum.

22. Find the minimum distance between the point $(1, 2, 0)$ and the quadric cone $z^2 = x^2 + y^2$.

23. An open gutter with cross section in the form of a trapezoid with equal base angles is to be made by bending up equal strips along both sides of a long piece of metal 12 inches wide. Find the base angles and the width of the sides for maximum carrying capacity.

24. Find the minimum distance between the lines having parametric equations $x = t - 1$, $y = 2t$, $z = t + 3$ and $x = 3s$, $y = s + 2$, $z = 2s - 1$.

25. Convince yourself that the maximum and minimum values of a linear function $f(x, y) = ax + by + c$ on a closed polygonal set will always occur at a vertex. Then use this fact to find each value.

(a) Maximum value of $2x + 3y + 4$ on the closed polygon with vertices $(-1, 2)$, $(0, 1)$, $(1, 0)$, $(-3, 0)$, and $(0, -4)$.

(b) Minimum value of $-3x + 2y + 1$ on the closed polygon with vertices $(-3, 0)$, $(0, 5)$, $(2, 3)$, $(4, 0)$, and $(1, -4)$.

26. Use the result of Problem 25 to maximize $2x + y$ subject to the constraints $4x + y \leq 8$, $2x + 3y \leq 14$, $x \geq 0$, and $y \geq 0$. *Hint*: Begin by graphing the set determined by the constraints.

27. Find the maximum and minimum values of $z = y^2 - x^2$ (Figure 2) on the closed triangle with vertices $(0, 0)$, $(1, 2)$, and $(2, -2)$.

28. (Least Squares) Given the n points $P_1(x_1, y_1)$, $P_2(x_2, y_2), \ldots, P_n(x_n, y_n)$ in the plane, we wish to find the line $y = mx + b$ such that the sum of the squares of the vertical distances from the points to the line is a minimum; that is, we wish to minimize

$$f(m, b) = \sum_{i=1}^{n} (y_i - mx_i - b)^2$$

Show that this minimum will occur where

$$m \sum x_i^2 + b \sum x_i = \sum x_i y_i$$
$$m \sum x_i + nb = \sum y_i$$

29. Find the least-squares line (Problem 28) for the data $(3, 2)$, $(4, 3)$, $(5, 4)$, $(6, 4)$, and $(7, 5)$.

30. Find the maximum and minimum values of $z = 2x^2 + y^2 - 4x - 2y + 5$ (Figure 3) on the set bounded by the closed triangle with vertices $(0, 0)$, $(4, 0)$, and $(0, 1)$.

31. Suppose the temperature T on the circular plate $\{(x, y) : x^2 + y^2 \leq 1\}$ is given by $T = 2x^2 + y^2 - y$. Find the hottest and coldest spots on the plate.

32. A wire of length k is to be cut into (at most) 3 pieces to form a circle and two squares, any of which may be degenerate. How should this be done to maximize and minimize the area thus enclosed? *Hint*: Reduce the problem to that of optimizing $x^2 + y^2 + z^2$ on the part of the plane $2\sqrt{\pi} x + 4y + 4z = k$ in the first octant. Then reason geometrically.

33. Find the shape of the triangle of largest area that can be inscribed in a circle of radius r. *Hint*: Let α, β, and γ be the central angles that subtend the three sides of the triangle. Show that the area of the triangle is $\frac{1}{2}r^2[\sin \alpha + \sin \beta - \sin(\alpha + \beta)]$. Maximize.

34. Let (a, b, c) be a fixed point in the first octant. Find the plane through this point that cuts off from the first octant the tetrahedron of minimum volume, and determine the resulting volume.

15.9
Lagrange's Method

We begin by distinguishing between two kinds of problems. To find the minimum value of $x^2 + 2y^2 + z^4 + 4$ is a *free extremum* problem. To find the minimum of $x^2 + 2y^2 + z^4 + 4$ subject to the condition that $x + 3y - z = 7$ is a *constrained extremum* problem. Many of the problems of the real world, especially those in economics, are of the latter type. For example, a manufacturer may wish to maximize profits, but is likely to be constrained by the amount of raw materials available, the size of its labor force, and so on.

Example 4 of the previous section was a constrained extremum problem. We were asked to find the minimum distance from the surface $z^2 = x^2 y + 4$ to the origin. We formulated the problem as that of minimizing $d^2 = x^2 + y^2 + z^2$ subject to the constraint $z^2 = x^2 y + 4$. We handled that problem by substituting the value for z^2 from the constraint in the expression for d^2 and then solving the resulting free extremum problem. However, it often happens that the constraint equation is not easily solved for one of the variables and, even if this can be done, another method may be more practical. This is the method of **Lagrange multipliers**, named after Joseph-Louis Lagrange.

GEOMETRIC INTERPRETATION OF THE METHOD Let us consider first the case where we wish to maximize or minimize $f(x, y)$ subject to the constraint $g(x, y) = 0$. Figure 1 suggests a geometric interpretation of the problem.

FIGURE 1

The level curves of f are the curves $f(x, y) = k$, where k is a constant. They are shown as black curves in Figure 1 for $k = 200, 300, \ldots, 700$. The graph of the constraint $g(x, y) = 0$ is also a curve; it is shown in color in Figure 1.

To maximize f subject to the constraint $g(x, y) = 0$ is to find the level curve with the greatest possible k that intersects the constraint curve. It is geometrically evident from Figure 1 that such a level curve is tangent to the constraint curve at a point $P_0(x_0, y_0)$ and therefore that the maximum value of f subject to the constraint $g(x, y) = 0$ is $f(x_0, y_0)$. The other point of tangency $P_1(x_1, y_1)$ gives the minimum value $f(x_1, y_1)$ of f subject to the constraint $g(x, y) = 0$.

Lagrange's method provides an algebraic procedure for finding the points P_0 and P_1. Since at such a point, the level curve and the constraint curve are tangent (that is, have a common tangent line), the two curves have a common perpendicular line. But at any point of a level curve, the gradient vector ∇f is perpendicular to the level curve (Section 15.5), and similarly ∇g is perpendicular to the constraint curve. Thus, ∇f and ∇g are parallel at P_0 and also at P_1; that is,

$$\nabla f(P_0) = \lambda_0 \nabla g(P_0) \quad \text{and} \quad \nabla f(P_1) = \lambda_1 \nabla g(P_1)$$

for some nonzero numbers λ_0 and λ_1.

The argument just given is admittedly an intuitive one but it can be made completely rigorous under appropriate hypotheses. Moreover, this argument works just as well for the problem of maximizing or minimizing $f(x, y, z)$ subject to the constraint $g(x, y, z) = 0$. We simply consider level surfaces rather than level curves. In fact, the result is valid in any number of variables.

All of this suggests the following formulation of the method of Lagrange multipliers.

(Lagrange's Method). To maximize or minimize $f(\mathbf{p})$ subject to the constraint $g(\mathbf{p}) = 0$, solve the system of equations

$$\nabla f(\mathbf{p}) = \lambda \nabla g(\mathbf{p}) \quad \text{and} \quad g(\mathbf{p}) = 0$$

for $\mathbf{p}$ and λ. Each such point $\mathbf{p}$ is a critical point for the constrained extremum problem and the corresponding λ is called a Lagrange multiplier.

APPLICATIONS We illustrate the method with several examples.

EXAMPLE 1 What is the greatest area that a rectangle can have if the length of its diagonal is 2?

Solution Place the rectangle in the first quadrant with two of its sides along the coordinate axes; then the vertex opposite the origin has coordinates (x, y), with x and y positive. The length of its diagonal is $\sqrt{x^2 + y^2} = 2$ and its area is xy.

Thus, we may formulate the problem to be that of maximizing $f(x, y) = xy$ subject to the constraint $g(x, y) = x^2 + y^2 - 4 = 0$. The corresponding gradients are

$$\nabla f(x, y) = f_x(x, y)\mathbf{i} + f_y(x, y)\mathbf{j} = y\mathbf{i} + x\mathbf{j}$$
$$\nabla g(x, y) = g_x(x, y)\mathbf{i} + g_y(x, y)\mathbf{j} = 2x\mathbf{i} + 2y\mathbf{j}$$

Lagrange's equations thus become

(1) $y = \lambda(2x)$

(2) $x = \lambda(2y)$

(3) $x^2 + y^2 = 4$

which we must solve simultaneously. If we multiply the first equation by y and the second by x, we get $y^2 = 2\lambda xy$ and $x^2 = 2\lambda xy$, from which

(4) $y^2 = x^2$

From (3) and (4), we find $x = \sqrt{2}$ and $y = \sqrt{2}$; and by substituting values in (1), we obtain $\lambda = \frac{1}{2}$. Thus the solution to equations (1)–(3), keeping x and y positive, is $x = \sqrt{2}$, $y = \sqrt{2}$, $\lambda = \frac{1}{2}$.

We conclude that the rectangle of greatest area with diagonal 2 is the square of side length $\sqrt{2}$. Its area is 2. A geometric interpretation of this problem is shown in Figure 2. ∎

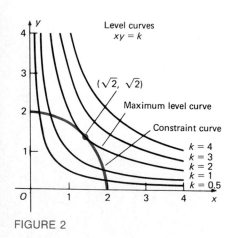

Level curves
$xy = k$

$(\sqrt{2}, \sqrt{2})$

Maximum level curve

Constraint curve

$k = 4$
$k = 3$
$k = 2$
$k = 1$
$k = 0.5$

FIGURE 2

EXAMPLE 2   Use Lagrange's method to find the maximum and minimum values of

$$f(x, y) = y^2 - x^2$$

on the ellipse $x^2/4 + y^2 = 1$.

Solution   Refer to Figure 2 of Section 15.8 for a graph of the hyperbolic paraboloid $z = f(x, y) = y^2 - x^2$. From this figure, we would certainly guess that the minimum value occurs at $(\pm 2, 0)$ and the maximum value at $(0, \pm 1)$. But let us justify this conjecture. We may write the constraint as $g(x, y) = x^2 + 4y^2 - 4 = 0$. Now

$$\nabla f = -2x\mathbf{i} + 2y\mathbf{j}$$

and

$$\nabla g = 2x\mathbf{i} + 8y\mathbf{j}$$

The Lagrange equations are

(1) $$-2x = \lambda 2x$$

(2) $$2y = \lambda 8y$$

(3) $$x^2 + 4y^2 = 4$$

Note from the third equation that x and y cannot both be 0. If $x \neq 0$, the first equation implies that $\lambda = -1$, and the second equation then requires that $y = 0$. We conclude from the third equation that $x = \pm 2$. We have thus obtained the critical points $(\pm 2, 0)$.

Exactly the same argument with $y \neq 0$ yields $\lambda = \frac{1}{4}$ from the second equation, then $x = 0$ from the first equation, and finally $y = \pm 1$ from the third equation. We conclude that $(0, \pm 1)$ are also critical points. Now for $f(x, y) = y^2 - x^2$,

$$f(2, 0) = -4$$
$$f(-2, 0) = -4$$
$$f(0, 1) = 1$$
$$f(0, -1) = 1$$

The minimum value of $f(x, y)$ on the given ellipse is -4; the maximum value is 1. ∎

EXAMPLE 3   Find the minimum of $f(x, y, z) = 3x + 2y + z + 5$, subject to the constraint $g(x, y, z) = 9x^2 + 4y^2 - z = 0$.

Solution   The gradients of f and g are $\nabla f(x, y, z) = 3\mathbf{i} + 2\mathbf{j} + \mathbf{k}$ and $\nabla g(x, y, z) = 18x\mathbf{i} + 8y\mathbf{j} - \mathbf{k}$. To find the critical points, we solve the equations

$$\nabla f(x, y, z) = \lambda \nabla g(x, y, z) \quad \text{and} \quad g(x, y, z) = 0$$

for (x, y, z, λ), in which λ is a Lagrange multiplier. This is equivalent, in the present problem, to solving the following system of four simultaneous equations in the four variables x, y, z, and λ.

(1) $$3 = 18x\lambda$$

(2) $$2 = 8y\lambda$$

(3) $$1 = -\lambda$$

(4) $$9x^2 + 4y^2 - z = 0$$

From (3), $\lambda = -1$. Substituting this result in equations (1) and (2), we get $x = -\frac{1}{6}$ and $y = -\frac{1}{4}$. By putting these values for x and y in equation 4, we obtain $z = \frac{1}{2}$. Thus the solution of the foregoing system of four simultaneous equations is $(-\frac{1}{6}, -\frac{1}{4}, \frac{1}{2}, -1)$, and the only critical point is $(-\frac{1}{6}, -\frac{1}{4}, \frac{1}{2})$. Therefore, the minimum of $f(x, y, z)$, subject to the constraint $g(x, y, z) = 0$, is $f(-\frac{1}{6}, -\frac{1}{4}, \frac{1}{2}) = 4\frac{1}{2}$. (How do we know that this value is a minimum rather than a maximum?) ∎

When more than one constraint is imposed on the variables of a function that is to be maximized or minimized, additional Lagrange multipliers are used (one for each constraint). For example, if we seek the extrema of a function f of three variables, subject to the two constraints $g(x, y, z) = 0$ and $h(x, y, z) = 0$, we solve the equations

$$\nabla f(x, y, z) = \lambda \nabla g(x, y, z) + \mu \nabla h(x, y, z), \qquad g(x, y, z) = 0, \qquad h(x, y, z) = 0$$

for x, y, z, λ, and μ, where λ and μ are Lagrange multipliers. This is equivalent to finding the solutions of the system of five simultaneous equations in the variables x, y, z, λ, and μ.

(1) $$f_x(x, y, z) = \lambda g_x(x, y, z) + \mu h_x(x, y, z)$$

(2) $$f_y(x, y, z) = \lambda g_y(x, y, z) + \mu h_y(x, y, z)$$

(3) $$f_z(x, y, z) = \lambda g_z(x, y, z) + \mu h_z(x, y, z)$$

(4) $$g(x, y, z) = 0$$

(5) $$h(x, y, z) = 0$$

From the solutions of this system we obtain the critical points.

EXAMPLE 4 Find the maximum and minimum values of $f(x, y, z) = x + 2y + 3z$ on the ellipse that is the intersection of the cylinder $x^2 + y^2 = 2$ and the plane $y + z = 1$ (see Figure 3).

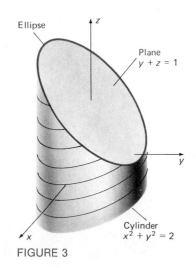

FIGURE 3

CHAPTER 15/The Derivative in n-Space

Solution We want to maximize and minimize $f(x, y, z)$ subject to $g(x, y, z) = x^2 + y^2 - 2 = 0$ and $h(x, y, z) = y + z - 1 = 0$. The corresponding Lagrange equations are

(1) $\qquad\qquad\qquad\qquad 1 = 2\lambda x$

(2) $\qquad\qquad\qquad\qquad 2 = 2\lambda y + \mu$

(3) $\qquad\qquad\qquad\qquad 3 = \mu$

(4) $\qquad\qquad x^2 + y^2 - 2 = 0$

(5) $\qquad\qquad\qquad y + z - 1 = 0$

From (1), $x = 1/2\lambda$; from (2) and (3), $y = -1/2\lambda$. Thus, from (4), $(1/2\lambda)^2 + (-1/2\lambda)^2 = 2$, which implies that $\lambda = \pm\frac{1}{2}$. The solution $\lambda = \frac{1}{2}$ yields the critical point $(x, y, z) = (1, -1, 2)$ and $\lambda = -\frac{1}{2}$ yields the critical point $(x, y, z) = (-1, 1, 0)$. We conclude that $f(1, -1, 2) = 5$ is the maximum value and $f(-1, 1, 0) = 1$ is the minimum value. ∎

PROBLEM SET 15.9

1. Find the minimum of $f(x, y) = x^2 + y^2$, subject to the constraint $g(x, y) = xy - 3 = 0$.

2. Find the maximum of $f(x, y) = xy$, subject to the constraint $g(x, y) = 4x^2 + 9y^2 - 36 = 0$.

3. Find the maximum of $f(x, y) = 4x^2 - 4xy + y^2$, subject to the constraint $x^2 + y^2 = 1$.

4. Find the minimum of $f(x, y) = x^2 + 4xy + y^2$, subject to the constraint $x - y - 6 = 0$.

5. Find the minimum of $f(x, y, z) = x^2 + y^2 + z^2$, subject to the constraint $x + 3y - 2z = 12$.

6. Find the minimum of $f(x, y, z) = 4x - 2y + 3z$, subject to the constraint $2x^2 + y^2 - 3z = 0$.

7. What are the dimensions of the rectangular box, open at the top, which has maximum volume when the surface area is 48?

8. Find the least distance between the origin and the plane $x + 3y - 2z = 4$.

9. The material for the bottom of a rectangular box costs three times as much per square foot as the material for the sides and top. Find the greatest capacity such a box can have if the total amount of money available for material is $12 and the material for the bottom costs $0.60 per square foot.

10. Find the least distance between the origin and the surface $x^2y - z^2 + 9 = 0$.

11. Find the maximum volume of a closed rectangular box with faces parallel to the coordinate planes inscribed

in the ellipsoid

$$\frac{x^2}{a^2} + \frac{y^2}{b^2} + \frac{z^2}{c^2} = 1$$

12. Find the maximum volume of the first octant rectangular box with faces parallel to the coordinate planes, one vertex at $(0, 0, 0)$, and diagonally opposite vertex on the plane

$$\frac{x}{a} + \frac{y}{b} + \frac{z}{c} = 1$$

13. Redo Problem 34 of Section 15.8 the easy way using Lagrange multipliers. *Hint*: Let the plane be $x/A + y/B + z/C = 1$.

14. Find the shape of the triangle of maximum perimeter that can be inscribed in a circle of radius r. *Hint*: Let α, β, and γ be as in Figure 4 and reduce the problem

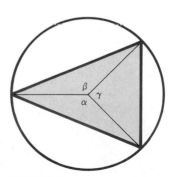

FIGURE 4

to maximizing $P = 2r(\sin \alpha/2 + \sin \beta/2 + \sin \gamma/2)$ subject to $\alpha + \beta + \gamma = 2\pi$.

15. Consider the Cobb-Douglas production model for a manufacturing process depending on three inputs x, y, and z with unit costs a, b, and c, respectively, given by

$$P = kx^\alpha y^\beta z^\gamma, \qquad \alpha > 0, \beta > 0, \gamma > 0, \alpha + \beta + \gamma = 1$$

subject to the cost constraint $ax + by + cz = d$. Determine x, y, and z to maximize the production P.

16. Find the minimum distance from the origin to the line of intersection of the two planes

$$x + y + z = 8 \quad \text{and} \quad 2x - y + 3z = 28$$

17. Find the maximum and minimum of $f(x, y, z) = -x + 2y + 2z$ on the ellipse $x^2 + y^2 = 2$, $y + 2z = 1$ (see Example 4).

18. Let $w = x_1 x_2 \cdots x_n$.

(a) Maximize w subject to $x_1 + x_2 + \cdots + x_n = 1$ and all $x_i > 0$.

(b) Use part (a) to deduce the famous **Geometric Mean–Arithmetic Mean Inequality** for positive numbers $a_1, a_2, \ldots, a_n$, namely,

$$\sqrt[n]{a_1 a_2 \cdots a_n} \leq \frac{(a_1 + a_2 + \cdots + a_n)}{n}$$

19. Maximize $w = a_1 x_1 + a_2 x_2 + \cdots + a_n x_n$, all $a_i > 0$, subject to $x_1^2 + x_2^2 + \cdots + x_n^2 = 1$.

15.10 Chapter Review

TRUE-FALSE QUIZ

Respond with true or false to each of the following assertions. Be prepared to defend your answer.

1. The level curves of $z = 2x^2 + 3y^2$ are ellipses.
2. If $f_x(0,0) = f_y(0,0)$, then $f(x, y)$ is continuous at the origin.
3. If $f_x(0,0)$ exists, then $g(x) = f(x, 0)$ is continuous at $x = 0$.
4. If $\lim\limits_{(x, y) \to (0, 0)} f(x, y) = L$, then $\lim\limits_{y \to 0} f(y, y) = L$.
5. If $f(x, y) = g(x)h(y)$ where g and h are continuous for all x and y, respectively, then f is continuous on the whole xy-plane.
6. If $f(x, y) = g(x)h(y)$ where both g and h are twice differentiable, then

$$\frac{\partial^2 f}{\partial x^2} + \frac{\partial^2 f}{\partial y^2} = g''(x)h(y) + g(x)h''(y)$$

7. If $f(x, y)$ and $g(x, y)$ have the same gradient, then they are identical functions.
8. The gradient of f is perpendicular to the graph of $z = f(x, y)$.
9. If f is differentiable and $\nabla f(a, b) = \mathbf{0}$, then the graph of $z = f(x, y)$ has a horizontal tangent plane at (a, b).
10. If $\nabla f(\mathbf{p}_0) = 0$, then f has an extreme value at $\mathbf{p}_0$.
11. If $T = e^y \sin x$ gives the temperature at a point (x, y) in the plane, then a heat-seeking object would move away from the origin in the direction $\mathbf{i}$.
12. The function $f(x, y) = \sqrt[3]{x^2 + y^4}$ has a global minimum value at the origin.
13. The function $f(x, y) = \sqrt[3]{x + y^4}$ has neither a global minimum nor a global maximum value.
14. If $f(x, y) = 4x + 4y$, then $|D_\mathbf{u} f(x, y)| \leq 4$.
15. If $D_\mathbf{u} f(x, y)$ exists, then $D_{-\mathbf{u}} f(x, y) = -D_\mathbf{u} f(x, y)$.
16. The set $\{(x, y): y = x, 0 \leq x \leq 1\}$ is a closed set in the plane.
17. If $f(x, y)$ is continuous on a closed bounded set S, then f attains a maximum value on S.
18. If $f(x, y)$ attains its maximum value at an interior point (x_0, y_0) of S, then $\nabla f(x_0, y_0) = \mathbf{0}$.
19. The function $f(x, y) = \sin(xy)$ does not attain a maximum value on the set $\{(x, y): x^2 + y^2 < 4\}$.
20. If $f_x(x_0, y_0)$ and $f_y(x_0, y_0)$ both exist, then f is differentiable at (x_0, y_0).

SAMPLE TEST PROBLEMS

1. Find and sketch the domain of each indicated function of two variables, showing clearly any points on the boundary of the domain that belong to the domain.

(a) $z = \sqrt{x^2 + 4y^2 - 100}$ (b) $z = -\sqrt{2x - y - 1}$

2. Sketch the level curves of $f(x, y) = (x + y^2)$ for $k = 0, 1, 2, 4$.

In Problems 3–6, find $\partial f/\partial x$, $\partial^2 f/\partial x^2$, and $\partial^2 f/\partial y \, \partial x$.

3. $f(x, y) = 3x^4 y^2 + 7x^2 y^7$

4. $f(x, y) = \cos^2 x - \sin^2 y$

5. $f(x, y) = e^{-y} \tan x$

6. $f(x, y) = e^{-x} \sin y$

7. If $F(x, y) = 5x^3 y^6 - xy^7$, find $\partial^3 F(x, y)/\partial x \, \partial y^2$.

8. If f is the function of three variables defined by $f(x, y, z) = xy^3 - 5x^2 yz^4$, find $f_x(2, -1, 1)$, $f_y(2, -1, 1)$, and $f_z(2, -1, 1)$.

9. Find the slope of the tangent to the curve of intersection of the surface $z = x^2 + y^2/4$ and the plane $x = 2$ at the point $(2, 2, 5)$.

10. For what points is the function defined by $f(x, y) = xy/(x^2 - y)$ continuous?

11. Does

$$\lim_{(x, y) \to (0, 0)} \frac{x - y}{x + y}$$

exist? Explain.

12. In each case, find the indicated limit or state that it does not exist.

(a) $\displaystyle\lim_{(x, y) \to (2, 2)} \frac{x^2 - 2y}{x^2 + 2y}$ (b) $\displaystyle\lim_{(x, y) \to (2, 2)} \frac{x^2 + 2y}{x^2 - 2y}$

(c) $\displaystyle\lim_{(x, y) \to (0, 0)} \frac{x^4 - 4y^4}{x^2 + 2y^2}$

13. Find $\nabla f(1, 2, -1)$.

(a) $f(x, y, z) = x^2 y z^3$ (b) $f(x, y, z) = y^2 \sin xz$

14. Find the directional derivative of $f(x, y) = \tan^{-1}(3xy)$. What is its value at the point $(4, 2)$ in the direction $\mathbf{u} = (\sqrt{3}/2)\mathbf{i} - (1/2)\mathbf{j}$?

15. Find the slope of the tangent line to the curve of intersection of the vertical plane $x - \sqrt{3}y + 2\sqrt{3} - 1 = 0$ and the surface $z = x^2 + y^2$ at the point $(1, 2, 5)$.

16. In what direction is $f(x, y) = 9x^4 + 4y^2$ increasing most rapidly at $(1, 2)$?

17. For $f(x, y) = x^2/2 + y^2$,

(a) find the equation of its level curve that goes through the point $(4, 1)$ in its domain;
(b) find the gradient vector ∇f at $(4, 1)$;
(c) draw the level curve and draw the gradient vector with its initial point at $(4, 1)$.

18. If $F(u, v) = \tan^{-1}(uv)$, $u = \sqrt{xy}$, $v = \sqrt{x} - \sqrt{y}$, find $\partial F/\partial x$ and $\partial F/\partial y$ in terms of u, v, x, and y.

19. If $f(u, v) = u/v$, $u = x^2 - 3y + 4z$, and $v = xyz$, find f_x, f_y, and f_z in terms of x, y, and z.

20. If $F(x, y) = x^3 - xy^2 - y^4$, $x = 2 \cos 3t$, and $y = 3 \sin t$, find dF/dt at $t = 0$.

21. If $F(x, y, z) = (5x^2 y/z^3)$, $x = t^{3/2} + 2$, $y = \ln 4t$, and $z = e^{3t}$, find dF/dt in terms of x, y, z, and t.

22. A triangle has vertices A, B, and C. The length of the side $c = AB$ is increasing at the rate of 3 inches per second, the side $b = AC$ is decreasing at 1 inch per second, and the included angle α is increasing at 0.1 radian per second. If $c = 10$ inches and $b = 8$ inches when $\alpha = \pi/6$, how fast is the area changing then?

23. Find the gradient vector of the ellipsoid $9x^2 + 4y^2 + 9z^2 = 34$ at the point $P(1, 2, -1)$. Write the equation of the tangent plane to this surface at the given point.

24. A right circular cylinder is measured to have a radius of 10 ± 0.02 inches and a height of $6 + 0.01$ inches. Calculate its volume and use differentials to give an estimate of the possible error.

25. If $f(x, y, z) = xy^2/(1 + z^2)$, use differentials to estimate $f(1.01, 1.98, 2.03)$.

26. Find the extrema of $f(x, y) = x^2 y - 6y^2 - 3x^2$.

27. A rectangular box whose edges are parallel to the coordinate axes is inscribed in the ellipsoid $36x^2 + 4y^2 + 9z^2 = 36$. What is the greatest possible volume for such a box?

28. Use Lagrange multipliers to find the maximum and the minimum of $f(x, y) = xy$, subject to the constraint $x^2 + y^2 = 1$.

29. Use Lagrange multipliers to find the dimensions of the right circular cylinder with maximum volume if its surface area is 24π.

16 The Integral in *n*-Space

This new integral of Lebesgue is proving itself a wonderful tool. I might compare it with a modern Krupp gun, so easily does it penetrate barriers which were impregnable.

E. B. Van Vleck

[Reprinted from *Bulletin of the American Mathematical Society*, vol. 23 (1916), pages 6 and 7, by permission of the American Mathematical Society.]

Henri Léon Lebesgue
1875–1941

Henri Lebesgue is the only twentieth-century mathematician in our list of contributors to calculus. We have included him mainly because we wanted to emphasize that calculus is not a dead subject; it is still being developed and expanded. In 1902, this brilliant young Frenchman completed his doctoral dissertation, titled *Integral, Length, and Area*. It opened the door to the modern theory of integration in one- and *n*-dimensions, a theory that all professional mathematicians meet in their graduate training. The Lebesgue integral gives a beautiful extension of the Riemann integral, agreeing with the latter whenever the Riemann integral exists, but making many more functions integrable.

It would take us too far afield to describe the Lebesgue integral, but we can explain a contribution Lebesgue made to the Riemann integral. We say that a set on the real line has measure zero if it can be enclosed in a finite or countable union of intervals whose total length is less than any given $\varepsilon > 0$. Any finite set has measure 0 but, surprisingly, so do the set of rational numbers and many other infinite sets. Lebesgue showed that a bounded function is Riemann integrable if and only if its set of discontinuities has measure zero.

Lebesgue's work also advanced the theory of multiple integrals. In his 1902 thesis, he was able to give simple conditions that allowed multiple integrals to be written as iterated integrals, results that were later perfected by his contemporary Guido Fubini (1879–1943).

16.1
Double Integrals over Rectangles

The Riemann integral for a function of one variable was introduced in Section 5.5, a section worth reviewing. Recall that we formed a partition P of the interval $[a, b]$ into subintervals of length Δx_k, $k = 1, 2, \ldots, n$, picked a sample point $\bar{x}_k$ from the kth subinterval, and then wrote

$$\int_a^b f(x)\, dx = \lim_{|P| \to 0} \sum_{k=1}^n f(\bar{x}_k)\, \Delta x_k$$

We proceed in a very similar fashion to define the integral for a function of two variables.

Let R be a rectangle with sides parallel to the coordinate axes; that is, let

$$R = \{(x, y) : a \le x \le b, c \le y \le d\}$$

Form a partition P of R by means of lines parallel to the x- and y-axes, as in Figure 1. This divides R into subrectangles, say n of them, which we denote by R_k, $k = 1, 2, \ldots, n$. Let Δx_k and Δy_k be the lengths of the sides of R_k and let $\Delta A_k = \Delta x_k \Delta y_k$ be its area. On R_k, pick a sample point $(\bar{x}_k, \bar{y}_k)$ and form the Riemann sum

$$\sum_{k=1}^n f(\bar{x}_k, \bar{y}_k)\, \Delta A_k$$

which corresponds (if $f(x, y) \ge 0$) to the sum of the volumes of n boxes (Figures 2 and 3). Letting the partition get finer and finer in such a way that all the R_k's get smaller will lead to the concept that we want.

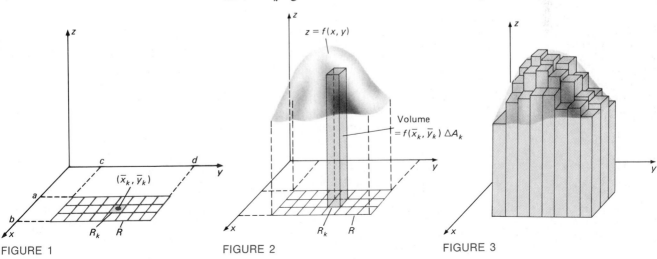

FIGURE 1

FIGURE 2

FIGURE 3

We are ready for a formal definition. We use the notation introduced above, with the additional proviso that the norm of the partition P—denoted by $|P|$—is the length of longest diagonal of any subrectangle in the partition.

Definition

(The double integral). Let f be a function of two variables that is defined on a closed rectangle R. If

$$\lim_{|P| \to 0} \sum_{k=1}^{n} f(\bar{x}_k, \bar{y}_k) \, \Delta A_k$$

exists, we say f is integrable on R. Moreover, $\iint_R f(x, y) \, dA$, called the **double integral** of f over R, is then given by

$$\iint_R f(x, y) \, dA = \lim_{|P| \to 0} \sum_{k=1}^{n} f(\bar{x}_k, \bar{y}_k) \, \Delta A_k$$

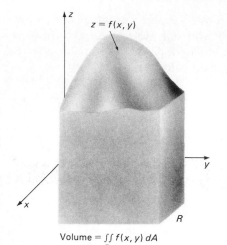

$z = f(x, y)$

Volume = $\iint_R f(x, y) \, dA$

FIGURE 4

Recall that if $f(x) \geq 0$, $\int_a^b f(x) \, dx$ represents the area of the region under the curve $y = f(x)$ between a and b. In a similar manner, if $f(x, y) \geq 0$, $\iint_R f(x, y) \, dA$ represents the **volume** of the solid under the surface $z = f(x, y)$ and above the rectangle R (Figure 4). In fact, we take this integral as the definition of the volume of such a solid.

THE EXISTENCE QUESTION   Not every function of two variables is integrable on a given rectangle R. The reasons are the same as in the one-variable case (Section 5.5). In particular, a function which is unbounded on R will always fail to be integrable. Fortunately, there is a natural generalization of Theorem 5.5A. Its proof is beyond the level of a first course.

Theorem A

(Integrability Theorem). If f is bounded on the closed rectangle R and if it is continuous there except on a finite number of smooth curves, then f is integrable on R. In particular, if f is continuous on all of R, then f is integrable there.

As a consequence, most of the common functions (provided they are bounded) are integrable on every rectangle. For example,

$$f(x, y) = e^{\sin(xy)} - y^3 \cos(x^2 y)$$

is integrable on every rectangle. On the other hand,

$$g(x, y) = \frac{x^2 y - 2x}{y - x^2}$$

Staircase Function

FIGURE 5

FIGURE 6

would fail to be integrable on any rectangle that intersected the parabola $y = x^2$. The "staircase" function of Figure 5 is integrable on R because its discontinuities occur along two line segments.

PROPERTIES OF THE DOUBLE INTEGRAL   The double integral inherits most of the properties of the single integral.

1. The double integral is linear; that is:

   (a) $\displaystyle\iint\limits_{R} kf(x, y)\, dA = k\iint\limits_{R} f(x, y)\, dA$;

   (b) $\displaystyle\iint\limits_{R} [f(x, y) + g(x, y)]\, dA = \iint\limits_{R} f(x, y)\, dA + \iint\limits_{R} g(x, y)\, dA$.

2. The double integral is additive on rectangles (Figure 6) that overlap only on a line segment.

$$\iint\limits_{R} f(x, y)\, dA = \iint\limits_{R_1} f(x, y)\, dA + \iint\limits_{R_2} f(x, y)\, dA$$

3. The comparison property holds. If $f(x, y) \le g(x, y)$ for all (x, y) in R, then

$$\iint\limits_{R} f(x, y)\, dA \le \iint\limits_{R} g(x, y)\, dA$$

All of these properties hold on more general sets than rectangles, but that is a matter we take up in Section 16.3.

EVALUATION OF DOUBLE INTEGRALS   This topic will receive major attention in the next section, where we will develop a powerful tool for evaluating double integrals. However, we can already evaluate a few integrals and we can approximate others.

Note first of all that if $f(x, y) = 1$ on R, then the double integral is the area of R, and from this it follows that

$$\iint\limits_{R} k\, dA = k\iint\limits_{R} 1\, dA = kA(R)$$

EXAMPLE 1   Let f be the staircase function of Figure 5—that is, let

$$f(x, y) = \begin{cases} 1 & 0 \le x \le 3,\, 0 \le y < 1 \\ 2 & 0 \le x \le 3,\, 1 \le y < 2 \\ 3 & 0 \le x \le 3,\, 2 \le y \le 3 \end{cases}$$

Calculate $\iint\limits_{R} f(x, y)\, dA$ where $R = \{(x, y) : 0 \le x \le 3,\, 0 \le y \le 3\}$.

Solution Introduce rectangles R_1, R_2, and R_3 as follows.

$$R_1 = \{(x, y) : 0 \le x \le 3, 0 \le y \le 1\}$$

$$R_2 = \{(x, y) : 0 \le x \le 3, 1 \le y \le 2\}$$

$$R_3 = \{(x, y) : 0 \le x \le 3, 2 \le y \le 3\}$$

Then, using the additivity property of the double integral, we obtain

$$\iint\limits_R f(x, y)\, dA = \iint\limits_{R_1} f(x, y)\, dA + \iint\limits_{R_2} f(x, y)\, dA + \iint\limits_{R_3} f(x, y)\, dA$$

$$= 1A(R_1) + 2A(R_2) + 3A(R_3)$$

$$= 1 \cdot 3 + 2 \cdot 3 + 3 \cdot 3 = 18$$

In this derivation, we also used the fact that the value of f on the boundary of a rectangle does not affect the value of the integral. ∎

Example 1 was a minor accomplishment, and to be honest we cannot do much more without more tools. However, we can always approximate a double integral by calculating a Riemann sum. In general, we can expect the approximation to be better the finer the partition we use.

EXAMPLE 2 Approximate $\iint\limits_R f(x, y)\, dA$, where

$$f(x, y) = \frac{64 - 8x + y^2}{16}$$

and

$$R = \{(x, y) : 0 \le x \le 4, 0 \le y \le 8\}$$

Do this by calculating the Riemann sum obtained by dividing R into eight equal squares and using the center of each square as the sample point (Figure 7).

Solution The required sample points and the corresponding values of the function are as follows.

$$(\bar{x}_1, \bar{y}_1) = (1, 1), \qquad f(\bar{x}_1, \bar{y}_1) = \tfrac{57}{16}$$

$$(\bar{x}_2, \bar{y}_2) = (1, 3), \qquad f(\bar{x}_2, \bar{y}_2) = \tfrac{65}{16}$$

$$(\bar{x}_3, \bar{y}_3) = (1, 5), \qquad f(\bar{x}_3, \bar{y}_3) = \tfrac{81}{16}$$

$$(\bar{x}_4, \bar{y}_4) = (1, 7), \qquad f(\bar{x}_4, \bar{y}_4) = \tfrac{105}{16}$$

$$(\bar{x}_5, \bar{y}_5) = (3, 1), \qquad f(\bar{x}_5, \bar{y}_5) = \tfrac{41}{16}$$

$$(\bar{x}_6, \bar{y}_6) = (3, 3), \qquad f(\bar{x}_6, \bar{y}_6) = \tfrac{49}{16}$$

$$(\bar{x}_7, \bar{y}_7) = (3, 5), \qquad f(\bar{x}_7, \bar{y}_7) = \tfrac{65}{16}$$

$$(\bar{x}_8, \bar{y}_8) = (3, 7), \qquad f(\bar{x}_8, \bar{y}_8) = \tfrac{89}{16}$$

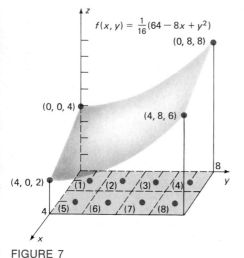

$f(x, y) = \frac{1}{16}(64 - 8x + y^2)$

FIGURE 7

Thus, since $\Delta A_k = 4$,

$$\iint\limits_{R} f(x, y)\, dA \approx \sum_{k=1}^{8} f(\bar{x}_k, \bar{y}_k)\, \Delta A_k$$

$$= 4 \sum_{k=1}^{8} f(\bar{x}_k, \bar{y}_k)$$

$$= \frac{4(57 + 65 + 81 + 105 + 41 + 49 + 65 + 89)}{16} = 138$$

In Section 16.2, we shall learn how to find the exact value of this integral. It is $138\frac{2}{3}$. ∎

PROBLEM SET 16.1

In Problems 1–4, let $R = \{(x, y): 1 \le x \le 4, 0 \le y \le 2\}$. Evaluate $\iint\limits_{R} f(x, y)\, dA$, where f is the given function (see Example 1).

1. $f(x, y) = \begin{cases} 2 & 1 \le x < 3, 0 \le y \le 2 \\ 3 & 3 \le x \le 4, 0 \le y \le 2 \end{cases}$

2. $f(x, y) = \begin{cases} -1 & 1 \le x \le 4, 0 \le y < 1 \\ 2 & 1 \le x \le 4, 1 \le y \le 2 \end{cases}$

3. $f(x, y) = \begin{cases} 2 & 1 \le x < 3, 0 \le y < 1 \\ 1 & 1 \le x < 3, 1 \le y \le 2 \\ 3 & 3 \le x \le 4, 0 \le y \le 2 \end{cases}$

4. $f(x, y) = \begin{cases} 2 & 1 \le x \le 4, 0 \le y < 1 \\ 3 & 1 \le x < 3, 1 \le y \le 2 \\ 1 & 3 \le x \le 4, 1 \le y \le 2 \end{cases}$

Suppose that $R = \{(x, y): 0 \le x \le 2, 0 \le y \le 2\}$, $R_1 = \{(x, y): 0 \le x \le 2, 0 \le y \le 1\}$, and $R_2 = \{(x, y): 0 \le x \le 2, 1 \le y \le 2\}$. Suppose, in addition, that $\iint\limits_{R} f(x, y)\, dA = 3$, $\iint\limits_{R} g(x, y)\, dA = 5$, and $\iint\limits_{R_1} g(x, y)\, dA = 2$. Use the properties of integrals to evaluate each of the following.

5. $\iint\limits_{R} [3f(x, y) - g(x, y)]\, dA$

6. $\iint\limits_{R} [2f(x, y) + 5g(x, y)]\, dA$

7. $\iint\limits_{R_2} g(x, y)\, dA$

8. $\iint\limits_{R_1} [2g(x, y) + 3]\, dA$

In Problems 9–14, $R = \{(x, y): 0 \le x \le 6, 0 \le y \le 4\}$ and P is the partition of R into six equal squares by the lines $x = 2$, $x = 4$, and $y = 2$. Approximate $\iint\limits_{R} f(x, y)\, dA$ by calculating the corresponding Riemann sum $\sum_{k=1}^{6} f(\bar{x}_k, \bar{y}_k)\, \Delta A_k$, assuming that $(\bar{x}_k, \bar{y}_k)$ are the centers of the six squares (see Example 2).

9. $f(x, y) = 12 - x - y$

10. $f(x, y) = 10 - y^2$

11. $f(x, y) = x^2 + 2y^2$

12. $f(x, y) = \frac{1}{6}(48 - 4x - 3y)$

$\boxed{C}$ **13.** $f(x, y) = \sqrt{x + y}$

$\boxed{C}$ **14.** $f(x, y) = e^{xy}$

15. Calculate $\iint\limits_{R} (6 - y)\, dA$, where $R = \{(x, y): 0 \le x \le 1, 0 \le y \le 1\}$. *Hint*: This integral represents the volume of a certain solid. Sketch this solid and calculate its volume from elementary principles.

16. Calculate $\iint\limits_{R} (1 + x)\, dA$, where $R = \{(x, y): 0 \le x \le 2, 0 \le y \le 1\}$. See the hint of Problem 15.

17. Use the comparison property of double integrals to show that if $f(x, y) \ge 0$ on R, then $\iint\limits_{R} f(x, y)\, dA \ge 0$.

18. Suppose $m \le f(x, y) \le M$ on R. Show that

$$mA(R) \le \iint\limits_{R} f(x, y)\, dA \le MA(R)$$

$\boxed{C}$ **19.** Let R be the rectangle shown in Figure 8. For the indicated partition into 12 equal squares, calculate the smallest and largest Riemann sums for $\iint\limits_{R} \sqrt{x^2 + y^2}\, dA$ and thereby obtain numbers c and C such that

$$c \le \iint\limits_{R} \sqrt{x^2 + y^2}\, dA \le C$$

FIGURE 8

20. Evaluate $\iint\limits_{R} x \cos^2(xy)\, dA$, where R is the rectangle of Figure 8. *Hint*: Does the graph of the integrand have any kind of symmetry?

21. Recall that $[\![x]\!]$ is the greatest integer in x. For R of Figure 8, evaluate:

(a) $\displaystyle\iint\limits_{R} [\![x]\!][\![y]\!]\, dA$ (b) $\displaystyle\iint\limits_{R} ([\![x]\!] + [\![y]\!])\, dA$

22. Suppose the rectangle of Figure 8 represents a thin plate (lamina) whose mass density at (x, y) is $\delta(x, y)$, say in grams per square centimeter. What does $\iint\limits_{R} \delta(x, y)\, dA$ represent?

23. Colorado is a rectangular state (assuming we ignore the curvature of the earth). Let $f(x, y)$ be the number of inches of rainfall during 1980 at the point (x, y) in that state. What does $\iint\limits_{\text{colorado}} f(x, y)\, dA$ represent? What does this number divided by the area of Colorado represent?

24. Let $f(x, y) = 1$ if both x and y are rational numbers and let $f(x, y) = 0$ otherwise. Show that $f(x, y)$ is not integrable over the rectangle R in Figure 8.

16.2
Iterated Integrals

Now we face in earnest the problem of evaluating $\iint\limits_{R} f(x, y)\, dA$, where R is the rectangle

$$R = \{(x, y) : a \le x \le b, c \le y \le d\}$$

Suppose for the time being that $f(x, y) \ge 0$ on R so that we may interpret the double integral as the volume V of the solid under the surface of Figure 1.

(1) $$V = \iint\limits_{R} f(x, y)\, dA$$

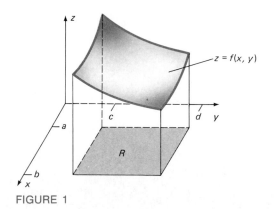

FIGURE 1

CHAPTER 16/The Integral in n-Space

There is another way to calculate the volume of this solid, which at least intuitively seems just as valid. Slice the solid into thin slabs by means of planes parallel to the xz-plane. A typical such slab is shown in Figure 2(a). The area of the face of this slab depends on how far it is from the xz-plane, that is, it depends on y; therefore, we denote this area by $A(y)$ (see Figure 2(b)).

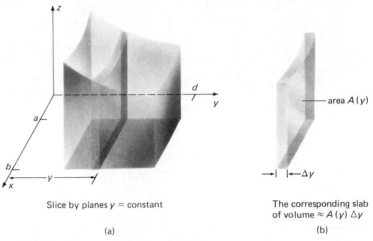

Slice by planes y = constant

The corresponding slab of volume $\approx A(y)\,\Delta y$

(a)

(b)

FIGURE 2

The volume ΔV of the slab is given approximately by

$$\Delta V \approx A(y)\,\Delta y$$

and, recalling our old motto (*slice, approximate, integrate*), we may write

$$V = \int_c^d A(y)\,dy$$

On the other hand, for fixed y we may calculate $A(y)$ by means of an ordinary single integral; in fact,

$$A(y) = \int_a^b f(x, y)\,dx$$

We conclude that

(2)
$$V = \int_c^d \left[\int_a^b f(x, y)\,dx \right] dy$$

an expression called an **iterated integral**.

When we equate the expressions for V from (1) and (2), we obtain the result we want.

$$\boxed{\iint\limits_R f(x, y)\,dA = \int_c^d \left[\int_a^b f(x, y)\,dx \right] dy}$$

If we had begun the process above by slicing the solid with planes parallel to the yz-plane, we would have obtained another iterated integral with the integrations occurring in the opposite order.

$$\iint\limits_{R} f(x, y)\, dA = \int_a^b \left[\int_c^d f(x, y)\, dy \right] dx$$

Two remarks are in order. First, while the two boxed results were derived under the assumption that f was nonnegative, they are valid in general. Second, the whole exercise was rather pointless unless iterated integrals can be evaluated. Fortunately, iterated integrals are often easy to evaluate as we demonstrate next.

EVALUATING ITERATED INTEGRALS   We begin with a simple example.

EXAMPLE 1   Evaluate $\int_0^3 \left[\int_1^2 (2x + 3y)\, dx \right] dy$.

Solution   In the inner integration y is a constant, so

$$\int_1^2 (2x + 3y)\, dx = [x^2 + 3yx]_1^2 = 4 + 6y - (1 + 3y) = 3 + 3y$$

Consequently,

$$\int_0^3 \left[\int_1^2 (2x + 3y)\, dx \right] dy = \int_0^3 [3 + 3y]\, dy = \left[3y + \tfrac{3}{2}y^2 \right]_0^3$$
$$= 9 + \tfrac{27}{2} = \tfrac{45}{2} \qquad \blacksquare$$

EXAMPLE 2   Evaluate $\int_1^2 \left[\int_0^3 (2x + 3y)\, dy \right] dx$.

Solution   Note that we have simply reversed the order of integration from Example 1; we expect the same answer as in that example.

$$\int_0^3 (2x + 3y)\, dy = [2xy + \tfrac{3}{2}y^2]_0^3$$
$$= 6x + \tfrac{27}{2}$$

Thus,

$$\int_1^2 \left[\int_0^3 (2x + 3y)\, dy \right] dx = \int_1^2 [6x + \tfrac{27}{2}]\, dx = \left[3x^2 + \tfrac{27}{2}x \right]_1^2$$
$$= 12 + 27 - (3 + \tfrac{27}{2}) = \tfrac{45}{2} \qquad \blacksquare$$

From now on, we shall usually omit the brackets in the iterated integral, allowing the order $dx\,dy$ or $dy\,dx$ to specify which integration is to be done first. Of course, the limits on the inner integral sign refer to the first variable to be integrated.

EXAMPLE 3 Evaluate $\displaystyle\int_0^8 \int_0^4 \tfrac{1}{16}[64 - 8x + y^2]\,dx\,dy$.

Solution Note that this iterated integral corresponds to the double integral of Example 2 of Section 16.1 for which we claimed the answer $138\tfrac{2}{3}$. Now

$$\tfrac{1}{16}\int_0^4 (64 - 8x + y^2)\,dx = \tfrac{1}{16}[64x - 4x^2 + y^2x]_0^4$$

$$= \tfrac{1}{16}[256 - 64 + 4y^2] = 12 + \tfrac{1}{4}y^2$$

The required iterated integral has the value

$$\int_0^8 \left(12 + \frac{1}{4}y^2\right)dy = \left[12y + \frac{y^3}{12}\right]_0^8 = 96 + \frac{512}{12}$$

$$= 96 + \frac{128}{3} = 138\tfrac{2}{3} \qquad\blacksquare$$

CALCULATING VOLUMES Now we can calculate volumes for a wide variety of solids.

EXAMPLE 4 Find the volume V of the solid under the surface $z = 4 - x^2 - y$ and over the rectangle $R = \{(x, y): 0 \le x \le 1, 0 \le y \le 2\}$ (see Figure 3).

Solution

$$V = \iint_R (4 - x^2 - y)\,dA = \int_0^2 \int_0^1 (4 - x^2 - y)\,dx\,dy$$

$$= \int_0^2 \left[4x - \frac{x^3}{3} - yx\right]_0^1 dy = \int_0^2 \left[4 - \frac{1}{3} - y\right]dy$$

$$= \left[4y - \frac{1}{3}y - \frac{1}{2}y^2\right]_0^2 = 8 \quad \frac{2}{3} - 2 - \frac{16}{3} \qquad\blacksquare$$

$z = 1 - x^2 - y$

$(1, 2)$

FIGURE 3

PROBLEM SET 16.2 _____

In Problems 1–12, evaluate each of the iterated integrals.

1. $\displaystyle\int_0^2 \int_1^3 x^2 y\,dy\,dx$

2. $\displaystyle\int_{-1}^4 \int_1^2 (x + y^2)\,dy\,dx$

3. $\displaystyle\int_1^2 \int_0^3 (xy + y^2)\,dx\,dy$

4. $\displaystyle\int_{-1}^1 \int_1^2 (x^2 + y^2)\,dx\,dy$

5. $\displaystyle\int_0^\pi \int_0^1 x \sin y \, dx \, dy$

6. $\displaystyle\int_0^{\ln 3} \int_0^{\ln 2} e^{x+y} \, dy \, dx$

7. $\displaystyle\int_0^{\pi/2} \int_0^1 x \sin xy \, dy \, dx$

8. $\displaystyle\int_0^1 \int_0^1 xe^{xy} \, dy \, dx$

9. $\displaystyle\int_0^3 \int_0^1 2x\sqrt{x^2 + y} \, dx \, dy$

10. $\displaystyle\int_0^1 \int_0^1 \frac{y}{(xy+1)^2} \, dx \, dy$

11. $\displaystyle\int_0^{\ln 3} \int_0^1 xye^{xy^2} \, dy \, dx$

12. $\displaystyle\int_0^1 \int_0^2 \frac{y}{1+x^2} \, dy \, dx$

Evaluate the indicated double integral over R.

13. $\displaystyle\iint_R xy^3 \, dA; \quad R = \{(x, y): 0 \le x \le 1, -1 \le y \le 1\}$

14. $\displaystyle\iint_R (x^2 + y^2) \, dA;$

$R = \{(x, y): -1 \le x \le 1, 0 \le y \le 2\}$

15. $\displaystyle\iint_R \sin(x + y) \, dA;$

$R = \{(x, y): 0 \le x \le \pi/2, 0 \le y \le \pi/2\}$

16. $\displaystyle\iint_R xy\sqrt{1 + x^2} \, dA;$

$R = \{(x, y): 0 \le x \le \sqrt{3}, 1 \le y \le 2\}$

In Problems 17–20, sketch the solid whose volume is the indicated iterated integral.

17. $\displaystyle\int_0^1 \int_0^2 \frac{x}{2} \, dx \, dy$

18. $\displaystyle\int_0^1 \int_0^1 (2 - x - y) \, dy \, dx$

19. $\displaystyle\int_0^2 \int_0^2 (x^2 + y^2) \, dy \, dx$

20. $\displaystyle\int_0^2 \int_0^2 (4 - y^2) \, dy \, dx$

In Problems 21–24, find the volume of the given solid.

21. The solid under the plane $z = x + y + 1$ over $R = \{(x, y): 0 \le x \le 1, 1 \le y \le 3\}$.

22. The solid under the plane $z = 2x + 3y$ and over $R = \{(x, y): 1 \le x \le 2, 0 \le y \le 4\}$.

23. The solid between $z = x^2 + y^2 + 2$ and $z = 1$ and lying above $R = \{(x, y): -1 \le x \le 1, 0 \le y \le 1\}$.

24. The solid in the first octant enclosed by $z = 4 - x^2$ and $y = 2$.

25. Show that if $f(x, y) = g(x)h(y)$, then

$$\int_a^b \int_c^d f(x, y) \, dy \, dx = \left[\int_a^b g(x) \, dx\right]\left[\int_c^d h(y) \, dy\right]$$

26. Use Problem 25 to evaluate

$$\int_0^{\sqrt{\ln 2}} \int_0^1 \frac{xye^{x^2}}{1 + y^2} \, dy \, dx$$

27. Evaluate

$$\int_0^1 \int_0^1 xye^{x^2 + y^2} \, dy \, dx$$

28. Find the volume of the solid trapped between the surface $z = \cos x \cos y$ and the xy-plane, where $-\pi \le x \le \pi$, $-\pi \le y \le \pi$.

29. Evaluate each of the following.

(a) $\displaystyle\int_{-2}^2 \int_{-1}^1 |x^2 y^3| \, dy \, dx$

(b) $\displaystyle\int_{-2}^2 \int_{-1}^1 [\![x^2]\!] y^3 \, dy \, dx$

(c) $\displaystyle\int_{-2}^2 \int_{-1}^1 [\![x^2]\!] |y^3| \, dy \, dx$

30. Evaluate $\displaystyle\int_0^{\sqrt{3}} \int_0^1 \frac{8x}{(x^2 + y^2 + 1)^2} \, dy \, dx$. *Hint*: Reverse the order of integration.

31. Prove the Cauchy-Schwarz Inequality for Integrals

$$\left[\int_a^b f(x)g(x) \, dx\right]^2 \le \int_a^b f^2(x) \, dx \int_a^b g^2(x) \, dx$$

Hint: Consider the double integral of

$$F(x, y) = [f(x)g(y) - f(y)g(x)]^2$$

over the rectangle $R = \{(x, y): a \le x \le b, a \le y \le b\}$.

32. Suppose f is increasing on $[a, b]$ and $\displaystyle\int_a^b f(x) \, dx > 0$. Prove that $\displaystyle\int_a^b xf(x) \, dx \Big/ \int_a^b f(x) \, dx > (a + b)/2$ and give a physical interpretation of this result. *Hint*: Let $F(x, y) = [y - x][f(y) - f(x)]$ and use the hint of Problem 31.

16.3
Double Integrals over Nonrectangular Regions

Consider now a closed bounded set S in the plane (Figure 1). Surround S by a rectangle R with sides parallel to the coordinate axes (Figure 2). Suppose $f(x, y)$ is defined on S and define (or redefine, if necessary) $f(x, y) = 0$ on the part of R outside of S (Figure 3). We say that f is integrable on S if it is integrable on R and write

$$\iint_S f(x, y) \, dA = \iint_R f(x, y) \, dA$$

We assert that the double integral on general sets S is (1) linear, (2) additive on sets that overlap only on smooth curves, and (3) satisfies the comparison property (see Section 16.1).

EVALUATION OF DOUBLE INTEGRALS OVER GENERAL SETS Sets with curved boundaries can be very complicated. For our purposes, it will be sufficient to consider so-called x-simple sets and y-simple sets (and finite unions of such sets). A set S is y-simple if it is simple in the y-direction, meaning that a line in this direction intersects S in a single interval (or point or not at all). Thus, a set S is **y-simple** (Figure 4) if there are functions ϕ_1 and ϕ_2 on $[a, b]$ such that

$$S = \{(x, y): \phi_1(x) \le y \le \phi_2(x), a \le x \le b\}$$

A set S is **x-simple** (Figure 5) if there are functions ψ_1 and ψ_2 on $[c, d]$ such that

$$S = \{(x, y): \psi_1(y) \le x \le \psi_2(y), c \le y \le d\}$$

Figure 6 exhibits a set that is neither x-simple nor y-simple.

Now suppose we wish to evaluate the double integral of a function $f(x, y)$ over a y-simple set S. We enclose S in a rectangle R (Figure 7) and

FIGURE 1

FIGURE 2

$z = f(x, y)$

$f(x, y) = 0$

FIGURE 3

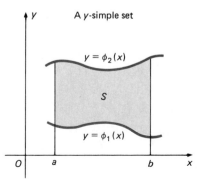

A y-simple set

$y = \phi_2(x)$

S

$y = \phi_1(x)$

O a b x

FIGURE 4

An x-simple set

$x = \psi_1(y)$ $x = \psi_2(y)$

d

S

c

O x

FIGURE 5

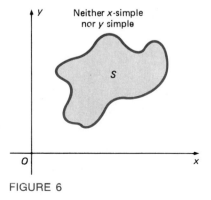

Neither x-simple nor y simple

S

O x

FIGURE 6

make $f(x, y) = 0$ outside of S. Then

$$\iint_S f(x, y)\, dA = \iint_R f(x, y)\, dA = \int_a^b \left[\int_c^d f(x, y)\, dy \right] dx$$

$$= \int_a^b \left[\int_{\phi_1(x)}^{\phi_2(x)} f(x, y)\, dy \right] dx$$

FIGURE 7

In summary,

$$\boxed{\iint_S f(x, y)\, dA = \int_a^b \int_{\phi_1(x)}^{\phi_2(x)} f(x, y)\, dy\, dx}$$

In the inner integration, x is held fixed; thus, that integration is along the heavy line of Figure 7. This integration yields the area $A(x)$ of the cross section shown in Figure 8. Finally, $A(x)$ is integrated from a to b.

If the set S is x-simple (Figure 5), similar reasoning leads to the formula

$$\boxed{\iint_S f(x, y)\, dA = \int_c^d \int_{\psi_1(y)}^{\psi_2(y)} f(x, y)\, dx\, dy}$$

FIGURE 8

If the set S is neither x-simple nor y-simple (Figure 6), it can usually be considered as a union of pieces which have one or the other of these properties. For example the annulus of Figure 9 is not simple in either direction, but it is the union of the two y-simple sets S_1 and S_2. The integrals on these pieces can be calculated and added together to obtain the integral over S.

EXAMPLES   For a little preliminary practice, we evaluate two iterated integrals, where the limits on the inner integral sign are variables.

EXAMPLE 1   Evaluate the iterated integral

$$\int_3^5 \int_{-x}^{x^2} (4x + 10y)\, dy\, dx$$

FIGURE 9

Solution   We first perform the inner integration with respect to y, temporarily thinking of x as constant, and obtain

$$\int_3^5 \int_{-x}^{x^2} (4x + 10y)\, dy\, dx = \int_3^5 \left[4xy + 5y^2 \right]_{-x}^{x^2} dx$$

$$= \int_3^5 \left[(4x^3 + 5x^4) - (-4x^2 + 5x^2) \right] dx$$

$$= \int_3^5 (5x^4 + 4x^3 - x^2)\, dx = \left[x^5 + x^4 - \frac{x^3}{3} \right]_3^5$$

$$= 3393\tfrac{1}{3} \qquad \blacksquare$$

Notice in iterated integrals that the outer integral always has constant limits.

EXAMPLE 2 Evaluate the iterated integral

$$\int_0^1 \int_0^{y^2} 2ye^x \, dx \, dy$$

Solution

$$\int_0^1 \int_0^{y^2} 2ye^x \, dx \, dy = \int_0^1 \left[\int_0^{y^2} 2ye^x \, dx \right] dy$$

$$= \int_0^1 \left[2ye^x \right]_0^{y^2} dy = \int_0^1 (2ye^{y^2} - 2ye^0) \, dy$$

$$= \int_0^1 e^{y^2} d(y^2) - 2 \int_0^1 y \, dy$$

$$= \left[e^{y^2} \right]_0^1 - 2 \left[\frac{y^2}{2} \right]_0^1 = e - 1 - 2\left(\frac{1}{2}\right) = e - 2 \quad ∎$$

We turn to the problem of calculating volumes by means of iterated integrals.

EXAMPLE 3 Use double integration to find the volume of the tetrahedron bounded by the coordinate planes and the plane $3x + 6y + 4z - 12 = 0$.

Solution Denote by S the triangular region in the xy-plane that forms the base of the tetrahedron (Figure 10). We seek the volume of the solid under the surface $z - \frac{3}{4}(4 - x - 2y)$ and above the region S.

The given plane intersects the xy-plane in the line $x + 2y - 4 = 0$, a segment of which belongs to the boundary of S. Since this equation can be written $y = 2 - x/2$ and $x = 4 - 2y$, S can be thought of as the y-simple set

$$S = \left\{ (x, y) : 0 \le x \le 4, 0 \le y \le 2 - \frac{x}{2} \right\}$$

or as the x-simple set

$$S = \{ (x, y) : 0 \le x \le 4 - 2y, 0 \le y \le 2 \}$$

We will treat S as a y-simple set; the final result would be the same either way, as you should verify.

The volume V of the solid is

$$V = \iint_S \tfrac{3}{4}(4 - x - 2y) \, dA$$

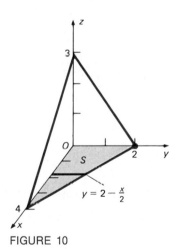

FIGURE 10

In writing this as an iterated integral, we fix x and integrate along a line (Figure 10) from $y = 0$ to $y = 2 - x/2$ and then integrate the result from $x = 0$ to $x = 4$. Thus,

$$
\begin{aligned}
V &= \int_0^4 \int_0^{2-x/2} \frac{3}{4}(4 - x - 2y)\, dy\, dx \\
&= \int_0^4 \left[\frac{3}{4} \int_0^{2-x/2} (4 - x - 2y)\, dy \right] dx \\
&= \int_0^4 \frac{3}{4} \left[4y - xy - y^2 \right]_0^{2-x/2} dx \\
&= \frac{3}{16} \int_0^4 (16 - 8x + x^2)\, dx \\
&= \frac{3}{16} \left[16x - 4x^2 + \frac{x^3}{3} \right]_0^4 = 4
\end{aligned}
$$

EXAMPLE 4 Find the volume of the solid in the first octant ($x \geq 0$, $y \geq 0$, $z \geq 0$) bounded by the circular paraboloid $z = x^2 + y^2$, the cylinder $x^2 + y^2 = 4$, and the coordinate planes (Figure 11).

Solution The region S in the first quadrant of the xy-plane is bounded by a quarter of the circle $x^2 + y^2 = 4$ and the lines $x = 0$ and $y = 0$. Although S can be thought of as either a y-simple or an x-simple region, we shall treat S as the latter and write its boundary curves as $x = \sqrt{4 - y^2}$, $x = 0$, and $y = 0$. Thus,

$$ S = \{(x, y) : 0 \leq x \leq \sqrt{4 - y^2},\ 0 \leq y \leq 2\} $$

Our goal is to calculate

$$ V = \iint_S (x^2 + y^2)\, dA $$

by means of an iterated integral. This time we first fix y and integrate along a line (Figure 11) from $x = 0$ to $x = \sqrt{4 - y^2}$, and then integrate the result from $y = 0$ to $y = 2$.

$$
V = \iint_S (x^2 + y^2)\, dA = \int_0^2 \int_0^{\sqrt{4-y^2}} (x^2 + y^2)\, dx\, dy
$$
$$
= \int_0^2 \left[\tfrac{1}{3}(4 - y^2)^{3/2} + y^2 \sqrt{4 - y^2} \right] dy
$$

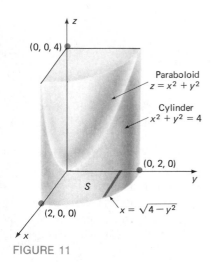

FIGURE 11

(0, 0, 4)

Paraboloid
$z = x^2 + y^2$

Cylinder
$x^2 + y^2 = 4$

(0, 2, 0)

S

(2, 0, 0)

$x = \sqrt{4 - y^2}$

By the trigonometric substitution $y = 2 \sin \theta$, the latter integral can be rewritten as

$$\int_0^{\pi/2} \left[\frac{8}{3} \cos^3\theta + 8 \sin^2\theta \cos \theta \right] 2 \cos \theta \, d\theta$$

$$= \int_0^{\pi/2} \left[\frac{16}{3} \cos^4\theta + 16 \sin^2\theta \cos^2\theta \right] d\theta$$

$$= \frac{16}{3} \int_0^{\pi/2} \cos^2\theta \, (1 - \sin^2\theta + 3 \sin^2\theta) \, d\theta$$

$$= \frac{16}{3} \int_0^{\pi/2} (\cos^2\theta + 2 \sin^2\theta \cos^2\theta) \, d\theta$$

$$= \frac{16}{3} \int_0^{\pi/2} (\cos^2\theta + \tfrac{1}{2} \sin^2 2\theta) \, d\theta$$

$$= \frac{16}{3} \int_0^{\pi/2} \left(\frac{1 + \cos 2\theta}{2} + \frac{1 + \cos 4\theta}{4} \right) d\theta = 2\pi \qquad \blacksquare$$

EXAMPLE 5 By changing the order of integration, evaluate

$$\int_0^4 \int_{x/2}^2 e^{y^2} \, dy \, dx$$

Solution The inner integral cannot be evaluated as it stands because e^{y^2} does not have an elementary antiderivative. However, we recognize that the given iterated integral is equal to

$$\iint_S e^{y^2} \, dA$$

where $S = \{(x, y) : x/2 \le y \le 2, 0 \le x \le 4\}$ (see Figure 12). If we write this double integral as an iterated integral with the x-integration performed first, we get

$$\int_0^2 \int_0^{2y} e^{y^2} \, dx \, dy = \int_0^2 [x e^{y^2}]_0^{2y} \, dy$$

$$= \int_0^2 2y e^{y^2} \, dy = [e^{y^2}]_0^2 = e^4 - 1 \qquad \blacksquare$$

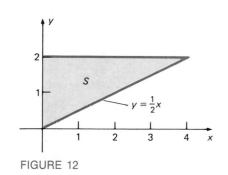

FIGURE 12

PROBLEM SET 16.3

Evaluate the iterated integrals in Problems 1–12.

1. $\displaystyle\int_0^1 \int_0^{3x} x^2 \, dy \, dx$

2. $\displaystyle\int_1^2 \int_0^{x-1} y \, dy \, dx$

3. $\displaystyle\int_{-1}^3 \int_0^{3y} (x^2 + y^2) \, dx \, dy$

4. $\displaystyle\int_{-3}^1 \int_0^x (x^2 - y^3) \, dy \, dx$

5. $\displaystyle\int_{1}^{3}\int_{-y}^{2y} xe^{y^3}\,dx\,dy$

6. $\displaystyle\int_{1}^{5}\int_{0}^{x} \frac{3}{x^2+y^2}\,dy\,dx$

7. $\displaystyle\int_{1/2}^{1}\int_{0}^{2x} \cos(\pi x^2)\,dy\,dx$

8. $\displaystyle\int_{0}^{\pi/4}\int_{\sqrt{2}}^{\sqrt{2}\cos\theta} r\,dr\,d\theta$

9. $\displaystyle\int_{0}^{\pi/9}\int_{\pi/4}^{3r} \sec^2\theta\,d\theta\,dr$

10. $\displaystyle\int_{0}^{\pi/2}\int_{0}^{\sin y} e^x\cos y\,dx\,dy$

11. $\displaystyle\int_{0}^{2}\int_{0}^{\sqrt{4-x^2}} (x+y)\,dy\,dx$

12. $\displaystyle\int_{\pi/6}^{\pi/2}\int_{0}^{\sin\theta} 6r\cos\theta\,dr\,d\theta$

In Problems 13–18, evaluate the given double integral by changing it to an iterated integral.

13. $\iint_{S} xy\,dA$; S is the region bounded by $y=x^2$ and $y=1$.

14. $\iint_{S} (x+y)\,dA$; S is the triangular region with vertices $(0,0)$, $(0,4)$, and $(1,4)$.

15. $\iint_{S} (x^2+2y)\,dA$; S is the region between $y=x^2$ and $y=\sqrt{x}$.

16. $\iint_{S} (x^2-xy)\,dA$; S is the region between $y=x$ and $y=3x-x^2$.

17. $\iint_{S} \dfrac{2}{1+x^2}\,dA$; S is the triangular region with vertices at $(0,0)$, $(2,2)$, and $(0,2)$.

18. $\iint_{S} x\,dA$; S is the region between $y=x$ and $y=x^3$. (Note that S has two parts.)

In Problems 19–30, sketch the indicated solid. Then find its volume by an iterated integration.

19. The tetrahedron bounded by the coordinate planes and the plane $z=6-2x-3y$.

20. The tetrahedron bounded by the coordinate planes and the plane $3x+4y+z-12=0$.

21. The wedge bounded by the coordinate planes and the planes $x=5$ and $y+2z-4=0$.

22. The solid in the first octant bounded by the coordinate planes and the planes $2x+y-4=0$ and $8x+y-4z=0$.

23. The solid in the first octant bounded by the surface $9x^2+4y^2=36$ and the plane $9x+4y-6z=0$.

24. The solid in the first octant bounded by the surface $z=9-x^2-y^2$ and the coordinate planes.

25. The solid in the first octant bounded by the cylinder $y=x^2$ and the planes $x=0$, $z=0$, and $y+z=1$.

26. The solid bounded by the parabolic cylinder $x^2=4y$ and the planes $z=0$ and $5y+9z-45=0$.

27. The solid in the first octant bounded by the cylinder $z=\tan x^2$ and the planes $x=y$, $x=1$, and $y=0$.

28. The solid in the first octant bounded by the surface $z=e^{x-y}$, the plane $x+y=1$, and the coordinate planes.

29. The solid in the first octant bounded by the surface $9z=36-9x^2-4y^2$ and the coordinate planes.

30. The solid in the first octant bounded by the circular cylinders $x^2+z^2=16$ and $y^2+z^2=16$, and the coordinate planes.

In Problems 31–36, write the given iterated integral as an iterated integral with the order of integration interchanged. *Hint*: Begin by sketching a region S, as in Example 5.

31. $\displaystyle\int_{0}^{1}\int_{0}^{x} f(x,y)\,dy\,dx$

32. $\displaystyle\int_{0}^{2}\int_{y^2}^{2y} f(x,y)\,dx\,dy$

33. $\displaystyle\int_{0}^{1}\int_{x^2}^{x^{1/4}} f(x,y)\,dy\,dx$

34. $\displaystyle\int_{1/2}^{1}\int_{x^3}^{x} f(x,y)\,dy\,dx$

35. $\displaystyle\int_{0}^{1}\int_{-y}^{y} f(x,y)\,dx\,dy$

36. $\displaystyle\int_{-1}^{0}\int_{-\sqrt{y+1}}^{\sqrt{y+1}} f(x,y)\,dx\,dy$

37. Evaluate $\iint_{S} xy^2\,dA$, where S is the region shown in Figure 13.

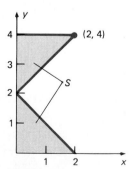

FIGURE 13

38. Evaluate $\iint\limits_{S} xy\,dA$, where S is the region in Figure 14.

FIGURE 14

39. Evaluate $\iint\limits_{S} \sin(y^3)\,dA$, where S is the region bounded by $y = \sqrt{x}, y = 2$, and $x = 0$. *Hint*: If one order of integration does not work, try the other.

40. Evaluate $\iint\limits_{S} \sin(xy^2)\,dA$, where S is the annulus $\{(x, y): 1 \le x^2 + y^2 \le 4\}$. *Hint*: Done without thinking, this problem is hard; using symmetry it is trivial.

41. Evaluate $\iint\limits_{S} (x^2 + x^4 y)\,dA$, where $S = \{(x, y): 1 \le x^2 + y^2 \le 4\}$. *Hint*: Use symmetry to reduce the problem to evaluating $4\left[\iint\limits_{S_1} x^2\,dA + \iint\limits_{S_2} x^2\,dA\right]$, where S_1 and S_2 are as in Figure 15.

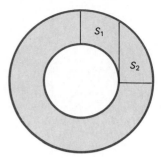

FIGURE 15

42. Evaluate $\iint\limits_{S} x^2\,dA$, where S is the region between the ellipse $x^2 + 2y^2 = 4$ and the circle $x^2 + y^2 = 4$.

16.4
Double Integrals in Polar Coordinates

Certain curves in the plane, such as circles, cardioids, and roses, are easier to describe in terms of polar coordinates than in Cartesian (rectangular) coordinates. Thus, we can expect that double integrals over regions enclosed by such curves are more easily evaluated using polar coordinates.

Let R have the shape shown in Figure 1, which we call a *polar rectangle* and will describe analytically in a moment. Let $z = f(x, y)$ determine a surface over R and suppose f is continuous and nonnegative. Then the volume V of the solid under this surface and above R (Figure 2) is given by

$$(1) \qquad V = \iint\limits_{R} f(x, y)\,dA$$

FIGURE 1

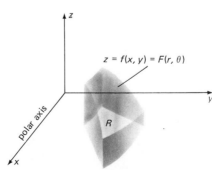

FIGURE 2

In polar coordinates, a polar rectangle R has the form

$$R = \{(r, \theta) : a \le r \le b, \alpha \le \theta \le \beta\}$$

where $a \ge 0$ and $\beta - \alpha \le 2\pi$. Also, the equation of the surface can be written as

$$z = f(x, y) = f(r \cos \theta, r \sin \theta) = F(r, \theta)$$

We are going to calculate the volume V in a new way using polar coordinates.

Partition R into smaller polar rectangles $R_1, R_2, \dots, R_n$ by means of a polar grid and let Δr_k and $\Delta \theta_k$ denote the dimensions of the typical piece R_k, as shown in Figure 3. The area $A(R_k)$ is given by (see Problem 30)

$$A(R_k) = \bar{r}_k \, \Delta r_k \, \Delta \theta_k$$

where $\bar{r}_k$ is the average radius of R_k. Thus,

$$V \approx \sum_{k=1}^{n} F(\bar{r}_k, \bar{\theta}_k) \bar{r}_k \, \Delta r_k \, \Delta \theta_k$$

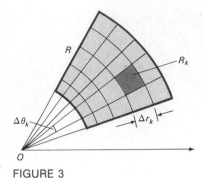

R

R_k

$\Delta \theta_k$ Δr_k

O

FIGURE 3

When we take the limit as the norm of the partition approaches zero, we ought to get the actual volume. This limit is a double integral.

$$(2) \qquad V = \iint\limits_{R} F(r, \theta) r \, dr \, d\theta = \iint\limits_{R} f(r \cos \theta, r \sin \theta) r \, dr \, d\theta$$

Now we have two expressions for V, namely, (1) and (2). Equating them yields

$$\boxed{\iint\limits_{R} f(x, y) \, dA = \iint\limits_{R} f(r \cos \theta, r \sin \theta) r \, dr \, d\theta}$$

The boxed result was derived under the assumption that f was nonnegative, but it is valid for very general functions, in particular for continuous functions of arbitrary sign.

ITERATED INTEGRALS The result announced above becomes useful when we write the polar double integral as an iterated integral, a statement we now illustrate.

EXAMPLE 1 Find the volume V of the solid above the polar rectangle (Figure 4).

$$R = \left\{(r, \theta) : 1 \le r \le 3, 0 \le \theta \le \frac{\pi}{4}\right\}$$

and under the surface $z = e^{x^2 + y^2}$.

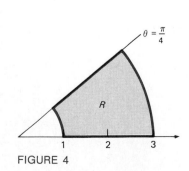

$\theta = \frac{\pi}{4}$

R

1   2   3

FIGURE 4

Solution Since $x^2 + y^2 = r^2$,

$$V = \iint\limits_{R} e^{x^2 + y^2} \, dA$$

$$= \int_0^{\pi/4} \left[\int_1^3 e^{r^2} r \, dr \right] d\theta$$

$$= \int_0^{\pi/4} \left[\frac{1}{2} e^{r^2} \right]_1^3 d\theta$$

$$= \int_0^{\pi/4} \frac{1}{2} (e^9 - e) \, d\theta = \frac{\pi}{8} (e^9 - e) \approx 3181$$

Without the help of polar coordinates, we could not have done this problem. Note how the extra factor of r was just what we needed in order to antidifferentiate e^{r^2}. ∎

GENERAL REGIONS Recall how we extended the double integral over an ordinary rectangle R to the integral over a general set S. We simply enclosed S in a rectangle and gave the function to be integrated the value zero outside of S. We can do the same thing for double polar integrals, except that we use polar rectangles rather than ordinary rectangles. Omitting the details, we simply assert that the boxed result stated earlier holds for general sets S.

Of special interest for polar integration are what we shall call r-simple and θ-simple sets. Call a set S an **r-simple** set if it has the form (Figure 5)

$$S = \{(r, \theta) : \phi_1(\theta) \le r \le \phi_2(\theta), \alpha \le \theta \le \beta\}$$

and call it **θ-simple** if it has the form (Figure 6)

$$S = \{(r, \theta) : a \le r \le b, \psi_1(r) \le \theta \le \psi_2(r)\}$$

An r-simple set

FIGURE 5

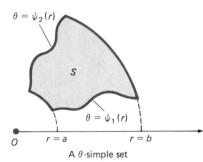

A θ-simple set

FIGURE 6

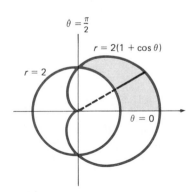

FIGURE 7

EXAMPLE 2 Evaluate

$$\iint\limits_{R} y \, dA$$

where S is the region in the first quadrant that is outside the circle $r = 2$ and inside the cardioid $r = 2(1 + \cos \theta)$ (see Figure 7).

Solution   Since S is an r-simple set, we write the given integral as an iterated polar integral, with r as the first variable of integration. In this inner integration, θ is held fixed; the integration is along the heavy line of Figure 7 from $r = 2$ to $r = 2(1 + \cos \theta)$.

$$\iint\limits_{S} y \, dA = \int_{0}^{\pi/2} \int_{2}^{2(1 + \cos \theta)} (r \sin \theta) r \, dr \, d\theta$$

$$= \int_{0}^{\pi/2} \left[\frac{r^3}{3} \sin \theta \right]_{2}^{2(1 + \cos \theta)} d\theta$$

$$= \frac{8}{3} \int_{0}^{\pi/2} \left[(1 + \cos \theta)^3 \sin \theta - \sin \theta \right] d\theta$$

$$= \frac{8}{3} \left[-\frac{1}{4} (1 + \cos \theta)^4 + \cos \theta \right]_{0}^{\pi/2}$$

$$= \frac{8}{3} \left[-\frac{1}{4} + 0 - (-4 + 1) \right] = \frac{22}{3} \qquad \blacksquare$$

EXAMPLE 3   Find the volume of the solid under the surface $z = x^2 + y^2$, above the xy-plane, and inside the cylinder $x^2 + y^2 = 2y$ (Figure 8).

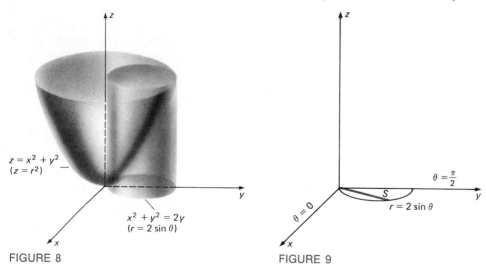

$z = x^2 + y^2$
$(z = r^2)$

$x^2 + y^2 = 2y$
$(r = 2 \sin \theta)$

FIGURE 8

$\theta = \dfrac{\pi}{2}$

S

$r = 2 \sin \theta$

$\theta = 0$

FIGURE 9

Solution   From symmetry, we can double the volume in the first octant. When we use $x = r \cos \theta$ and $y = r \sin \theta$, the equation of the surface becomes $z = r^2$ and that of the cylinder $r = 2 \sin \theta$. Let S denote the region shown in Figure 9. The required volume V is given by

$$V = 2 \iint\limits_{S} (x^2 + y^2) \, dA = 2 \int_{0}^{\pi/2} \int_{0}^{2 \sin \theta} r^2 r \, dr \, d\theta$$

$$= 2 \int_{0}^{\pi/2} \left[\frac{r^4}{4} \right]_{0}^{2 \sin \theta} d\theta = 8 \int_{0}^{\pi/2} \sin^4 \theta \, d\theta$$

$$= 8 \left(\frac{3}{8} \cdot \frac{\pi}{2} \right) = \frac{3\pi}{2}$$

The last integral was evaluated by means of Formula 113 in the table of integrals at the end of the book. ∎

A PROBABILITY INTEGRAL   In the theory of probability and statistics, it is essential to be able to calculate the improper integral

$$\int_{-\infty}^{\infty} e^{-x^2}\, dx = 2\int_0^{\infty} e^{-x^2}\, dx$$

EXAMPLE 4   Show that $I = \int_0^{\infty} e^{-x^2}\, dx = \sqrt{\pi}/2$.

Solution   We are going to sneak up on this problem in a roundabout, but decidedly ingenious, way. First recall that

$$I = \int_0^{\infty} e^{-x^2}\, dx = \lim_{b\to\infty} \int_0^b e^{-x^2}\, dx$$

Now let V_b be the volume of the solid that lies under the surface $z = e^{-x^2-y^2}$ and above the square with vertices $(\pm b, \pm b)$. Then,

$$V_b = \int_{-b}^b \int_{-b}^b e^{-x^2-y^2}\, dy\, dx = \int_{-b}^b e^{-x^2}\left[\int_{-b}^b e^{-y^2}\, dy\right] dx$$

$$= \int_{-b}^b e^{-x^2}\, dx \int_{-b}^b e^{-y^2}\, dy = \left[\int_{-b}^b e^{-x^2}\, dx\right]^2 = 4\left[\int_0^b e^{-x^2}\, dx\right]^2$$

It follows that the volume of the region under $z = e^{-x^2-y^2}$ above the whole xy-plane is

$$(1) \qquad V = \lim_{b\to\infty} V_b = \lim_{b\to\infty} 4\left[\int_0^b e^{-x^2}\, dx\right]^2$$

$$= 4\left[\int_0^{\infty} e^{-x^2}\, dx\right]^2 = 4I^2$$

On the other hand, we can also calculate V using polar coordinates. Here, V is the limit as $a \to \infty$ of V_a, the volume of the solid under the surface $z = e^{-x^2-y^2} = e^{-r^2}$ above the circular region of radius a centered at the origin.

$$(2) \qquad V = \lim_{a\to\infty} V_a = \lim_{a\to\infty} \int_0^{2\pi}\int_0^a e^{-r^2} r\, dr\, d\theta = \lim_{a\to\infty} \int_0^{2\pi}\left[-\tfrac{1}{2}e^{-r^2}\right]_0^a d\theta$$

$$= \lim_{a\to\infty} \tfrac{1}{2}\int_0^{2\pi}\left[1 - e^{-a^2}\right] d\theta = \lim_{a\to\infty} \pi\left[1 - e^{-a^2}\right] = \pi$$

Equating the two values obtained for V in (1) and (2) yields $4I^2 = \pi$, or $I = \tfrac{1}{2}\sqrt{\pi}$, as desired. ∎

Evaluate the following iterated integrals.

1. $\displaystyle\int_0^{\pi/2}\int_0^{\cos\theta} r^2\sin\theta\,dr\,d\theta$

2. $\displaystyle\int_0^{\pi/2}\int_0^{\sin\theta} r\,dr\,d\theta$

3. $\displaystyle\int_0^{\pi}\int_0^{\sin\theta} r^2\,dr\,d\theta$

4. $\displaystyle\int_0^{\pi}\int_0^{1-\cos\theta} r\sin\theta\,dr\,d\theta$

In Problems 5–10, find the area of the given region S by calculating $\iint_S r\,dr\,d\theta$. Be sure to make a sketch of the region first.

5. S is the region inside the circle $r = 4\cos\theta$ and outside the circle $r = 2$.

6. S is the smaller region bounded by $\theta = \pi/6$ and $r = 4\sin\theta$.

7. S is one leaf of the four-leaved rose $r = a\sin 2\theta$.

8. S is the region inside the cardioid $r = 6 - 6\sin\theta$.

9. S is the region inside the larger loop of the limaçon $r = 2 - 4\sin\theta$.

10. S is the region outside the circle $r = 2$ and inside the lemniscate $r^2 = 9\cos 2\theta$.

In Problems 11–18, evaluate by using polar coordinates. Sketch the region of integration first.

11. $\iint_S e^{x^2+y^2}\,dA$, where S is the region enclosed by $x^2 + y^2 = 4$.

12. $\iint_S \sqrt{4 - x^2 - y^2}\,dA$, where S is the first quadrant sector of the circle $x^2 + y^2 = 4$ between $y = 0$ and $y = x$.

13. $\displaystyle\iint_S \frac{1}{4 + x^2 + y^2}\,dA$, where S as in Problem 12.

14. $\iint_S y\,dA$, where S is the first quadrant polar rectangle inside $x^2 + y^2 = 4$ and outside $x^2 + y^2 = 1$.

15. $\displaystyle\int_0^1\int_0^{\sqrt{1-x^2}} (4 - x^2 - y^2)^{-1/2}\,dy\,dx$

16. $\displaystyle\int_0^1\int_0^{\sqrt{1-y^2}} \sin(x^2 + y^2)\,dx\,dy$

17. $\displaystyle\int_0^1\int_x^1 x^2\,dy\,dx$

18. $\displaystyle\int_1^2\int_0^{\sqrt{2x-x^2}} (x^2 + y^2)^{-1/2}\,dy\,dx$

19. Find the volume of the solid in the first octant under the paraboloid $z = x^2 + y^2$ and inside the cylinder $x^2 + y^2 = 9$ by using polar coordinates.

20. Find the volume of the solid bounded above by $2x^2 + 2y^2 + z^2 = 18$, below by $z = 0$, and on the inside of the lateral surface $x^2 + y^2 = 4$ using polar coordinates.

21. Switch to rectangular coordinates and then evaluate.

$$\int_{3\pi/4}^{4\pi/3}\int_0^{-5\sec\theta} r^3\sin^2\theta\,dr\,d\theta$$

22. Let $V = \iint_S \sin\sqrt{x^2 + y^2}\,dA$ and

$W = \iint_S \left|\sin\sqrt{x^2 + y^2}\right|\,dA$, where S is the region inside the circle $x^2 + y^2 = 4\pi^2$.

(a) Without calculation, determine the sign of V.
(b) Evaluate V.
(c) Evaluate W.

23. The centers of two balls of radius a are $2b$ units apart with $b \le a$. Find the volume of their intersection in terms of $d = b - a$.

24. The depth (in feet) of water distributed by a rotating lawn sprinkler in an hour is $ke^{-r/10}$, $0 \le r \le 10$, where r is the distance from the sprinkler and k is a constant. Determine k if 100 cubic feet of water are distributed in 1 hour.

25. Find the volume of the solid cut from the ball $r^2 + z^2 \le a^2$ by the cyclinder $r = a\sin\theta$.

26. Find the volume of the wedge cut from a tall right cylinder of radius a by a plane through a diameter of its base and making an angle $\alpha\,(0 < \alpha < \pi/2)$ with the base (compare Problem 33, Section 6.2).

27. Consider the ring A of height $2b$ obtained from a ball of radius a when a hole of radius $c\,(c < a)$ is bored through the center of the ball (left part of Figure 10). Show that the volume of A is $4\pi b^3/3$, which is remarkable for two reasons. It is independent of both a and c and it is the same as the volume of a sphere of radius b (right part of Figure 10).

28. There is a simple explanation for the remarkable result in Problem 27. Show that the cross-sectional areas of the two solids (ring A and ball B) in Figure 10 determined by a horizontal plane at height h are the same size, implying that the solids below this plane have the same volume. Find a nice formula for this volume in terms of h and b.

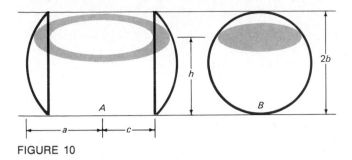

FIGURE 10

29. Show that

$$\int_0^\infty \int_0^\infty \frac{1}{(1 + x^2 + y^2)^2}\, dy\, dx = \frac{\pi}{4}$$

30. Recall the formula $A = \frac{1}{2}r^2\theta$ for the area of the sector of a circle of radius r and central angle θ radians (Section 12.8). Use this to obtain the formula

$$A = \frac{r_1 + r_2}{2}(r_2 - r_1)(\theta_2 - \theta_1)$$

for the area of the polar rectangle $\{(r, \theta): r_1 \leq r \leq r_2,\ \theta_1 \leq \theta \leq \theta_2\}$.

16.5
Applications of Double Integrals

The most obvious application of double integrals is in calculating volumes of solids. That use of double integrals has been amply illustrated, so now we turn to other applications (mass, center of mass, moment of inertia, and radius of gyration).

Consider a flat sheet that is so thin that we may consider it to be two-dimensional. In Section 6.8, we called such a sheet a lamina, but there we considered only laminas of constant density. Here we wish to study laminas of variable density—that is, laminas made of nonhomogeneous material (Figure 1).

Nonhomogeneous Material

FIGURE 1

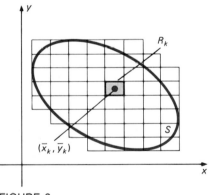

FIGURE 2

Suppose that a lamina covers a region S in the xy-plane and let the density (mass per unit area) at (x, y) be denoted by $\delta(x, y)$. Partition S into small rectangles $R_1, R_2, \ldots, R_k$ as shown in Figure 2. Pick a point $(\bar{x}_k, \bar{y}_k)$ on R_k. Then the mass of R_k is approximately $\delta(\bar{x}_k, \bar{y}_k)A(R_k)$ and the total mass of the lamina is approximately

$$m \approx \sum_{k=1}^{n} \delta(\bar{x}_k, \bar{y}_k)A(R_k)$$

The actual mass m is obtained by taking the limit of the above expression as the norm of the partition approaches zero which is, of course, a double integral.

$$m = \iint\limits_{S} \delta(x, y)\, dA$$

EXAMPLE 1   A lamina with density $\delta(x, y) = xy$ is bounded by the x-axis, the line $x = 8$, and the curve $y = x^{2/3}$ (Figure 3). Find its total mass.

Solution

$$m = \iint\limits_{S} xy\, dA = \int_{0}^{8} \int_{0}^{x^{2/3}} xy\, dy\, dx$$

$$= \int_{0}^{8} \left[\frac{xy^2}{2} \right]_{0}^{x^{2/3}} dx = \frac{1}{2} \int_{0}^{8} x^{7/3}\, dx$$

$$= \frac{1}{2} \left[\frac{3}{10} x^{10/3} \right]_{0}^{8} = \frac{768}{5} = 153.6$$

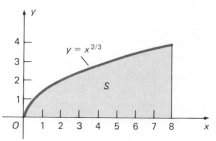

FIGURE 3

CENTER OF MASS   We suggest that you review the concept of center of mass from Section 6.8. There we learned that if $m_1, m_2, \ldots, m_n$ is a collection of point masses situated at $(x_1, y_1), (x_2, y_2), \ldots, (x_n, y_n)$, respectively, in the plane, then the total moments with respect to the y-axis and the x-axis are given by

$$M_y = \sum_{k=1}^{n} x_k m_k \qquad M_x = \sum_{k=1}^{n} y_k m_k$$

Moreover, the coordinates $(\bar{x}, \bar{y})$ of the center of mass (balance point) are

$$\bar{x} = \frac{M_y}{m} = \frac{\displaystyle\sum_{k=1}^{n} x_k m_k}{\displaystyle\sum_{k=1}^{n} m_k} \qquad \bar{y} = \frac{M_x}{m} = \frac{\displaystyle\sum_{k=1}^{n} y_k m_k}{\displaystyle\sum_{k=1}^{n} m_k}$$

Consider now a lamina of variable density $\delta(x, y)$ covering a region S in the xy-plane, as in Figure 1. Partition this lamina as in Figure 2 and assume as an approximation that the mass of each R_k is concentrated at $(\bar{x}_k, \bar{y}_k)$, $k = 1, 2, \ldots, n$. Finally, take the limit as the norm of the partition tends to zero. This leads to the formulas

$$\bar{x} = \frac{M_y}{m} = \frac{\displaystyle\iint\limits_{S} x\delta(x, y)\, dA}{\displaystyle\iint\limits_{S} \delta(x, y)\, dA} \qquad \bar{y} = \frac{M_x}{m} = \frac{\displaystyle\iint\limits_{S} y\delta(x, y)\, dA}{\displaystyle\iint\limits_{S} \delta(x, y)\, dA}$$

EXAMPLE 2 Find the center of mass of the lamina of Example 1.

Solution In Example 1, we showed that the mass m of this lamina is $\frac{768}{5}$. The moments M_y and M_x with respect to the y-axis and x-axis, respectively, are

$$M_y = \iint\limits_{S} x\delta(x, y)\, dA = \int_0^8 \int_0^{x^{2/3}} x^2 y\, dy\, dx$$

$$= \frac{1}{2} \int_0^8 x^{10/3}\, dx = \frac{12{,}288}{13} \approx 945.23$$

$$M_x = \iint\limits_{S} y\delta(x, y)\, dA = \int_0^8 \int_0^{x^{2/3}} xy^2\, dy\, dx$$

$$= \frac{1}{3} \int_0^8 x^3\, dx = \frac{1024}{3} \approx 341.33$$

We conclude that

$$\bar{x} = \frac{M_y}{m} = 6\tfrac{2}{13} \approx 6.15, \qquad \bar{y} = \frac{M_x}{m} = 2\tfrac{2}{9} \approx 2.22$$

Note the $(\bar{x}, \bar{y})$ is in the upper right portion of S; but that is to be expected since a lamina with density $\delta(x, y) = xy$ gets heavier as the distance from the x- and y-axes increases. ∎

EXAMPLE 3 Find the center of mass of a lamina in the shape of a quarter-circle of radius a whose density is proportional to the distance from the center of the circle (Figure 4).

Solution By hypothesis, $\delta(x, y) = k\sqrt{x^2 + y^2}$, where k is a constant. The shape of S suggests the use of polar coordinates.

$$m = \iint\limits_{S} k\sqrt{x^2 + y^2}\, dA = k \int_0^{\pi/2} \int_0^a r\, r\, dr\, d\theta$$

$$= k \int_0^{\pi/2} \frac{a^3}{3}\, d\theta = \frac{k\pi a^3}{6}$$

Also,

$$M_y = \iint\limits_{S} xk\sqrt{x^2 + y^2}\, dA = k \int_0^{\pi/2} \int_0^a (r\cos\theta) r^2\, dr\, d\theta$$

$$= k \int_0^{\pi/2} \frac{a^4}{4} \cos\theta\, d\theta = \left[\frac{ka^4}{4} \sin\theta \right]_0^{\pi/2} = \frac{ka^4}{4}$$

FIGURE 4

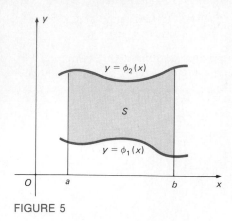

FIGURE 5

We conclude that

$$\bar{x} = \frac{M_y}{m} = \frac{ka^4/4}{k\pi a^3/6} = \frac{3a}{2\pi}$$

Because of the symmetry of the lamina, we recognize that $\bar{y} = \bar{x}$, so no further calculation is needed. ∎

A perceptive reader might well ask a question at this point. What if a lamina is homogeneous; that is, what if $\delta(x, y) = k$, a constant? Will the formulas derived in this section, which involve double integrals, agree with those of Section 6.8, which involved only single integrals? The answer is yes. To give a partial justification, consider calculating M_y for a y-simple region S (Figure 5)

$$M_y = \iint\limits_S xk\, dA = k \int_a^b \int_{\phi_1(x)}^{\phi_2(x)} x\, dy\, dx = k \int_a^b x[\phi_2(x) - \phi_1(x)]\, dx$$

The single integral on the right is the one given in Section 6.8.

MOMENT OF INERTIA   From physics, we learn that the kinetic energy, KE, of a particle of mass m and velocity v, moving in a straight line, is

(1) $$KE = \tfrac{1}{2}mv^2$$

If instead of moving in a straight line, the particle rotates about an axis with an angular velocity of ω radians per unit of time, its linear velocity is $v = r\omega$, where r is the radius of its circular path. When we substitute this in (1), we obtain

$$KE = \tfrac{1}{2}(r^2 m)\omega^2$$

The expression $r^2 m$ is called the **moment of inertia** of the particle and is denoted by I. Thus, for a rotating particle,

(2) $$KE = \tfrac{1}{2}I\omega^2$$

We conclude from (1) and (2) that the moment of inertia of a body in circular motion plays a role similar to the mass of a body in linear motion.

For a system of n particles in a plane with masses $m_1, m_2, \ldots, m_n$ and at distances $r_1, r_2, \ldots, r_n$ from line L, the moment of inertia of the system about L is defined to be

$$I = m_1 r_1^2 + m_2 r_2^2 + \cdots + m_n r_n^2 = \sum_{k=1}^{n} m_k r_k^2$$

In other words, we add the moments of inertia of the individual particles.

Now consider a lamina with density $\delta(x, y)$ covering a region S of the xy-plane (Figure 1). If we partition S as in Figure 2, approximate the moments of inertia of each piece R_k, add up, and take the limit, we are led to

CHAPTER 16/The Integral in n-Space

the following formulas. The **moments of inertia** (also called the second moments) of the lamina about the x-, y-, and z-axes are given by

$$I_x = \iint_S y^2 \delta(x, y) \, dA \qquad I_y = \iint_S x^2 \delta(x, y) \, dA$$

$$I_z = \iint_S (x^2 + y^2) \delta(x, y) \, dA = I_x + I_y$$

EXAMPLE 4 Find the moments of inertia about the x-, y-, and z-axes of the lamina of Example 1.

Solution

$$I_x = \iint_S xy^3 \, dA = \int_0^8 \int_0^{x^{2/3}} xy^3 \, dy \, dx = \frac{1}{4} \int_0^8 x^{11/3} \, dx = \frac{6144}{7} \approx 877.71$$

$$I_y = \iint_S x^3 y \, dA = \int_0^8 \int_0^{x^{2/3}} x^3 y \, dy \, dx = \frac{1}{2} \int_0^8 x^{13/3} \, dx = 6144$$

$$I_z = I_x + I_y = \frac{49{,}152}{7} \approx 7021.71 \qquad \blacksquare$$

Consider the problem of replacing a general mass system of total mass m by a single point mass m with the same moment of inertia I with respect to a line L (Figure 6). How far should this point be from L? The answer is $\bar{r}$ where $m\bar{r}^2 = I$. The number

$$\bar{r} = \sqrt{\frac{I}{m}}$$

is called the **radius of gyration** of the system. Thus the kinetic energy of the system rotating about L with angular velocity ω is

$$KE = \tfrac{1}{2}m\bar{r}^2\omega^2$$

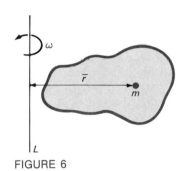

FIGURE 6

PROBLEM SET 16.5

In Problems 1–8, find the mass m and center of mass $(\bar{x}, \bar{y})$ of the lamina bounded by the given curves and with the indicated density.

1. $x = 0$, $x = 4$, $y = 0$, $y = 3$; $\delta(x, y) = y + 1$

2. $y = 0$, $y = \sqrt{4 - x^2}$; $\delta(x, y) = y$

3. $y = 0$, $y = \sin x$, $0 \le x \le \pi$; $\delta(x, y) = y$

4. $y = 1/x$, $y = x$, $y = 0$, $x = 2$; $\delta(x, y) = x$

5. $y = e^{-x}$, $y = 0$, $x = 0$, $x = 1$; $\delta(x, y) = y^2$

6. $y = e^x$, $y = 0$, $x = 0$, $x = 1$; $\delta(x, y) = 2 - x + y$

7. $r = 2 \sin \theta$; $\delta(r, \theta) = r$

8. $r = 1 + \cos \theta$; $\delta(r, \theta) = r$

In Problems 9–12, find the moments of inertia I_x, I_y, and I_z for the lamina bounded by the given curves and with the indicated density δ.

9. $y = \sqrt{x}$, $x = 9$, $y = 0$; $\delta(x, y) = x + y$

10. $y = x^2$, $y = 4$; $\delta(x, y) = y$

11. Square with vertices $(0,0)$, $(0,a)$, (a,a), $(a,0)$; $\delta(x, y) = x + y$

12. Triangle with vertices $(0,0)$, $(0,a)$, $(a,0)$; $\delta(x, y) = x^2 + y^2$

13. Find the radius of gyration of the lamina of Problem 11 with respect to the x-axis.

14. Find the radius of gyration of the lamina of Problem 12 with respect to the y-axis.

15. Find the moment of inertia and radius of gyration of a homogeneous (δ a constant) circular lamina of radius a with respect to a diameter.

16. Show that the moment of inertia of a homogeneous rectangular lamina with sides of length a and b about a perpendicular axis through its center of mass is

$$I = \tfrac{1}{12}(a^3 b + ab^3).$$

Here k is the constant density.

17. Find the moment of inertia of the lamina of Problem 15 about a line tangent to its boundary. *Hint*: Let the circle be $r = 2a \sin \theta$; then the tangent line is the x-axis. Formula 113 at the back of the book may help with the integration.

18. Consider the lamina S of constant density k bounded by the cardioid $r = a(1 + \sin \theta)$, as shown in Figure 7. Find its center of mass and moment of inertia with respect to the x-axis. *Hint*: Problem 7 of Section 12.8 gives the useful fact that S has area $3\pi a^2/2$; also, Formula 113 at the back of the book may prove helpful.

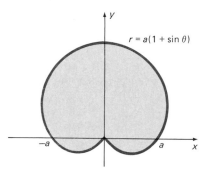

FIGURE 7

19. Find the center of mass of the part of the cardioid of Problem 18, that is, outside the circle $r = a$.

20. (Parallel Axis Theorem) Consider a lamina S of mass m together with parallel lines L and L' in the plane of S, the first line L passing through the center of mass of S. Show that if I and I' are the moments of inertia of S about L and L', then $I' = I + d^2 m$, where d is the distance between

L and L'. *Hint*: Assume S lies in the xy-plane, L is the y-axis, and L' is the line $x = -d$.

21. Refer to the lamina of Problem 11, for which we found $I_y = 5a^5/12$. Find (a) m, (b) $\bar{x}$, and (c) I_L, where L is a line through $(\bar{x}, \bar{y})$ parallel to the y-axis (see Problem 20).

22. Use the Parallel Axis Theorem together with Problem 15 to solve Problem 17 another way.

23. Find I_x, I_y, and I_z for the two-piece lamina of constant density k shown in Figure 8 (see Problems 15 and 20).

FIGURE 8

24. The Parallel Axis Theorem also holds for lines that are perpendicular to a lamina. Use this fact to find the moment of inertia of the rectangular lamina of Problem 16 about an axis perpendicular to the lamina and through a corner.

25. Let S_1 and S_2 be disjoint laminas in the xy-plane of mass m_1 and m_2 with centers of mass $(\bar{x}_1, \bar{y}_1)$ and $(\bar{x}_2, \bar{y}_2)$. Show that the center of mass $(\bar{x}, \bar{y})$ of the combined lamina $S_1 \cup S_2$ satisfies

$$\bar{x} = \bar{x}_1 \frac{m_1}{m_1 + m_2} + \bar{x}_2 \frac{m_2}{m_1 + m_2}$$

with a similar formula for $\bar{y}$. Conclude that in finding $(\bar{x}, \bar{y})$, the two laminas can be treated as if they were point masses at $(\bar{x}_1, \bar{y}_1)$ and $(\bar{x}_2, \bar{y}_2)$.

26. Let S_1 and S_2 be the homogeneous circular laminas of radius a and ta $(t > 0)$, centered at $(-a, a)$ and $(ta, 0)$, respectively. Use Problem 25 to find the center of mass of $S_1 \cup S_2$.

27. Let S be a lamina in the xy-plane with center of mass at the origin and let L be the line $ax + by = 0$, which goes through the origin. Show that the (signed) distance d of a point (x, y) from L is $d = (ax + by)/\sqrt{a^2 + b^2}$ and use this to conclude that the moment of S with respect to L is 0. *Note*: This shows that a lamina will balance on any line through its center of mass.

28. For the lamina of Example 3, find the equation of the balance line that makes an angle of $135°$ with the positive x-axis (see Problem 27). Write your answer in the form $Ax + By = C$.

16.6
Surface Area

In Section 6.5, we discussed surface area for surfaces of revolution. Now we consider the corresponding problem for more general surfaces.

We begin by considering a rectangle T, whose projection in the xy-plane is another rectangle R (see Figure 1). Clearly, their areas $A(T)$ and $A(R)$ are related by

$$(1) \qquad\qquad A(T) = A(R)\sec \gamma$$

where γ is the acute angle between the two rectangles.

Next consider any (nonvertical) plane making an acute angle γ with the xy-plane; that is, γ is the angle between the corresponding unit normal vectors $\mathbf{n}$ and $\mathbf{k}$. Let T be a parallelogram in this plane which projects onto a rectangle R in the xy-plane (Figure 2). If we partition R by means of lines parallel to and perpendicular to $\mathbf{n} \times \mathbf{k}$ into small rectangles R_k, the corresponding figures on T are rectangles T_k for which the formula (1) holds. It follows (by a limiting argument) that the formula holds for R and T as well.

$$A(R_k) = A(T_k) \cos \gamma$$

FIGURE 2

FIGURE 3

GENERAL SURFACES Let F be a function of three variables having continuous first partial derivatives F_x, F_y, and F_z, with $F_z \neq 0$. Consider the surface $F(x, y, z) = k$ and denote by G a part of this surface that projects onto a closed bounded region S in the xy-plane (Figure 3). We wish to give meaning to the concept *area of G* and find a formula for calculating it.

Form a partition P of S by means of lines parallel to the x- and y-axes. Let R_k, $k = 1, 2, \ldots, n$, denote the resulting rectangles that lie completely within S. For each k, let G_k be the part of the surface G that projects onto R_k and let P_k be the point of G_k that projects onto the corner of R_k with the

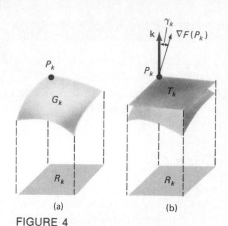

FIGURE 4

smallest x- and y-coordinates (Figure 4a). Finally, let T_k denote the parallelogram from the tangent plane at P_k that projects onto R_k (Figure 4b) and γ_k denote the acute angle that the upward normal at P_k makes with $\mathbf{k}$.

If the norm $|P|$ of the partition of S is small, the set of tangent parallelograms T_k will approximately conform to the surface G, and the smaller $|P|$ is taken, the better the conformation. This suggests the following definition of the area of G.

The area of the surface G is defined to be

$$(2) \qquad A(G) = \lim_{|P| \to 0} \sum_{k=1}^{n} A(T_k) = \lim_{|P| \to 0} \sum_{k=1}^{n} \sec \gamma_k A(R_k) = \iint_S \sec \gamma \, dA$$

where γ is the acute angle that the upward normal to the surface makes with the vertical vector $\mathbf{k}$.

To turn (2) into a usable formula, note first that the gradient vector ∇F is normal to the surface G, though it might point down rather than up. It follows (Section 14.2) that

$$\cos \gamma = \frac{|\nabla F \cdot \mathbf{k}|}{|\nabla F||\mathbf{k}|}$$

But $\nabla F = F_x \mathbf{i} + F_y \mathbf{j} + F_z \mathbf{k}$, and so

$$\sec \gamma = \frac{\sqrt{F_x^2 + F_y^2 + F_z^2}}{|F_z|}$$

In the special case where the surface is given by $z = f(x, y)$, we may write $F(x, y, z) = f(x, y) - z$ to obtain

$$\sec \gamma = \sqrt{f_x^2 + f_y^2 + 1}$$

This yields the important formula

$$\boxed{A(G) = \iint_S \sqrt{f_x^2 + f_y^2 + 1} \, dA}$$

A QUICKIE PROBLEM

John's house has a rectangular base and a gable roof. Peter's house has the same base with a pyramidal-type roof. The slopes of all parts of both roofs are the same. Peter and John are arguing about whose roof has the smaller area. Settle the argument.

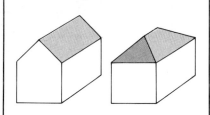

EXAMPLES We illustrate the boxed formula with two examples.

EXAMPLE 1 If S is the rectangular region in the xy-plane that is bounded by the lines $x = 0$, $x = 1$, $y = 0$, and $y = 2$, find the area of the part of the semicylindrical surface $z = \sqrt{4 - x^2}$ that projects onto S (Figure 5).

Solution Let $f(x, y) = \sqrt{4 - x^2}$. Then $f_x = -x/\sqrt{4 - x^2}$, $f_y = 0$, and

$$A(G) = \iint_S \sqrt{f_x^2 + f_y^2 + 1} \, dA = \iint_S \sqrt{\frac{x^2}{4 - x^2} + 1} \, dA = \iint_S \frac{2}{\sqrt{4 - x^2}} \, dA$$

$$= \int_0^1 \int_0^2 \frac{2}{\sqrt{4 - x^2}} \, dy \, dx = 4 \int_0^1 \frac{1}{\sqrt{4 - x^2}} \, dx = 4 \left[\sin^{-1} \frac{x}{2} \right]_0^1 = \frac{2\pi}{3} \quad \blacksquare$$

FIGURE 5

EXAMPLE 2 Find the area of the surface $z = x^2 + y^2$ below the plane $z = 9$.

Solution The designated part G of the surface projects onto the circular region S inside the circle $x^2 + y^2 = 9$ (Figure 6). Let $f(x, y) = x^2 + y^2$. Then $f_x = 2x$, $f_y = 2y$, and

$$A(G) = \iint_S \sqrt{4x^2 + 4y^2 + 1}\, dA$$

The shape of S suggests use of polar coordinates.

$$A(G) = \int_0^{2\pi} \int_0^3 \sqrt{4r^2 + 1}\, r\, dr\, d\theta$$

$$= \int_0^{2\pi} \frac{1}{8} \left[\frac{2}{3}(4r^2 + 1)^{3/2} \right]_0^3 d\theta$$

$$= \int_0^{2\pi} \tfrac{1}{12}(37^{3/2} - 1)\, d\theta = \frac{\pi}{6}(37^{3/2} - 1) \approx 117.32 \qquad \blacksquare$$

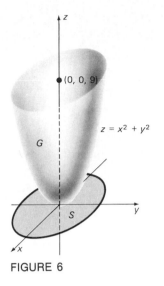

FIGURE 6

Our next example first appeared in Problem 20 of Section 6.5 It states the remarkable result that two planes perpendicular to the axis of a circular cylinder cut off equal areas on the cylinder and an inscribed sphere (Figure 7). Both areas are equal to $2\pi a h$, where a is the radius of the cylinder and h is the distance between the planes.

FIGURE 7

EXAMPLE 3 Show that the area of the surface G cut off from the sphere $x^2 + y^2 + z^2 = a^2$ by the planes $z = h_1$ and $z = h_2$ $(-a \le h_1 \le h_2 \le a)$ is

$$A(G) = 2\pi a(h_2 - h_1) = 2\pi a h$$

Solution We may assume $h_1 \geq 0$. Let $F(x, y, z) = x^2 + y^2 + z^2$. Then

$$\sec \gamma = \frac{\sqrt{F_x^2 + F_y^2 + F_z^2}}{|F_z|} = \frac{2\sqrt{x^2 + y^2 + z^2}}{2|z|} = \frac{a}{\sqrt{a^2 - x^2 - y^2}}$$

Referring to Figure 7, we obtain

$$A(G) = \iint_S \sec \gamma \, dA = \int_0^{2\pi} \int_b^c \frac{a}{\sqrt{a^2 - r^2}} r \, dr \, d\theta = \left[-2\pi a \sqrt{a^2 - r^2} \right]_b^c$$

$$= 2\pi a \left[\sqrt{a^2 - b^2} - \sqrt{a^2 - c^2} \right] = 2\pi a (h_2 - h_1) \qquad \blacksquare$$

PROBLEM SET 16.6

In Problems 1–14, find the area of the indicated surface. Make a sketch in each case.

1. The part of the plane $3x + 4y + 6z = 12$ that is above the rectangle in the xy-plane with vertices $(0, 0)$, $(2, 0)$, $(2, 1)$, and $(0, 1)$.

2. The part of the plane $3x - 2y + 6z = 12$ that is bounded by the planes $x = 0$, $y = 0$, and $3x + 2y = 12$.

3. The part of the surface $z = \sqrt{4 - y^2}$ that is directly above the square in the xy-plane with vertices $(1, 0)$, $(2, 0)$, $(2, 1)$, and $(1, 1)$.

4. The part of the surface $z = \sqrt{4 - y^2}$ in the first octant that is directly above the circle $x^2 + y^2 = 4$ in the xy-plane.

5. The part of the cylinder $x^2 + z^2 = 9$ that is directly over the rectangle in the xy-plane with vertices $(0, 0)$, $(2, 0)$, $(2, 3)$, and $(0, 3)$.

6. The part of the paraboloid $z = x^2 + y^2$ that is cut off by the plane $z = 4$.

7. The part of the conical surface $x^2 + y^2 = z^2$ that is directly over the triangle in the xy-plane with vertices $(0, 0)$, $(4, 0)$, and $(0, 4)$.

8. The part of the surface $z = x^2/4 + 4$ that is cut off by the planes $x = 0$, $x = 1$, $y = 0$, and $y = 2$.

9. The part of the sphere $x^2 + y^2 + z^2 = a^2$ inside the circular cylinder $x^2 + y^2 = b^2$, where $0 < b \leq a$.

10. The part of the sphere $x^2 + y^2 + z^2 = a^2$ inside the elliptic cylinder $b^2 x^2 + a^2 y^2 = a^2 b^2$, where $0 < b \leq a$.

11. The part of the sphere $x^2 + y^2 + z^2 = a^2$ inside the circular cylinder $x^2 + y^2 = ay$ ($r = a \sin \theta$ in polar coordinates), $a > 0$.

12. The part of the cylinder $x^2 + y^2 = ay$ inside the sphere $x^2 + y^2 + z^2 = a^2$, $a > 0$, *Hint:* Project to the yz-plane to get the region of integration.

13. The part of the saddle $az = x^2 - y^2$ inside the cylinder $x^2 + y^2 = a^2$, $a > 0$.

14. The surface of the solid that is the intersection of the two solid cylinders $x^2 + z^2 \leq a^2$ and $x^2 + y^2 \leq a^2$. *Hint:* You may need the integration formula $\int (1 + \sin \theta)^{-1} d\theta = -\tan[(\pi - 2\theta)/4] + C$.

Problems 15–17 are related to Example 3.

15. Find the centroid (center of mass) of the part of the homogeneous sphere $x^2 + y^2 + z^2 = a^2$ between the planes $z = h_1$ and $z = h_2$, $-a \leq h_1 < h_2 \leq a$.

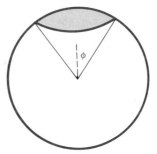

FIGURE 8

16. Show that the polar cap (Figure 8) on a sphere of radius a determined by the spherical angle ϕ has area $2\pi a^2 (1 - \cos \phi)$.

17. (Another Goat Problem) Four goats have grazing areas A, B, C, and D, respectively. The first three goats are each tethered by ropes of length b, the first on a flat plane, the second on the outside of a sphere of radius a, and the third on the inside of a sphere of radius a. The fourth goat must stay inside a ring of radius b that has been dropped over a sphere of radius a. Determine formulas for A, B, C, and D and arrange them in order of size. Assume $b < a$.

18. Let S be a planar region in three-space and let S_{xy}, S_{xz}, and S_{yz} be the projections on the three coordinate planes

(Figure 9). Show that

$$[A(S)]^2 = [A(S_{xy})]^2 + [A(S_{xz})]^2 + [A(S_{yz})]^2$$

FIGURE 9

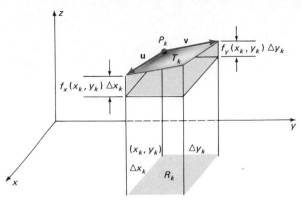

FIGURE 10

19. Assume the region S of Figure 9 lies in the plane $z = f(x, y) = ax + by + c$ and that S is above the xy-plane. Show that the volume of the solid cylinder under S is $A(S_{xy})f(\bar{x}, \bar{y})$, where $(\bar{x}, \bar{y})$ is the centroid of S_{xy}.

20. Let $z = f(x, y)$ and let T_k be the parallelogram in the tangent plane at P_k that projects onto a rectangle R_k in the xy-plane, as shown in Figure 10. Then the vectors $\mathbf{u}$ and

$\mathbf{v}$ from P_k, which form the sides of T_k, are

$$\mathbf{u} = \Delta x_k \mathbf{i} + f_x(x_k, y_k)\,\Delta x_k \mathbf{k}$$

$$\mathbf{v} = \Delta y_k \mathbf{j} + f_y(x_k, y_k)\,\Delta y_k \mathbf{k}$$

From Section 14.3, the area $A(T_k)$ is given by $|\mathbf{u} \times \mathbf{v}|$. Use this to show that

$$A(T_k) = \sqrt{[f_x(x_k, y_k)]^2 + [f_y(x_k, y_k)]^2 + 1}\ \Delta x_k \Delta y_k$$

and thereby give another derivation of the surface area formula in the text.

16.7
Triple Integrals (Cartesian Coordinates)

The concept embodied in single and double integrals extends in a natural way to triple, and even n-fold, integrals.

Consider a function f of three variables defined over a box-shaped region B with faces parallel to the coordinate planes. We can no longer graph f (four dimensions would be required), but we can picture B (Figure 1). Form

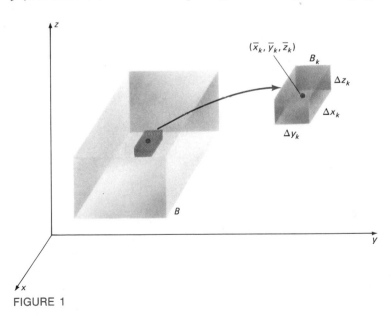

FIGURE 1

a partition P of B by passing planes through B parallel to the coordinate planes, thus cutting B into small sub-boxes $B_1, B_2, \ldots, B_n$; a typical one—B_k—is shown in Figure 1. On B_k, pick a sample point $(\bar{x}_k, \bar{y}_k, \bar{z}_k)$ and consider the Riemann sum

$$\sum_{k=1}^{n} f(\bar{x}_k, \bar{y}_k, \bar{z}_k)\,\Delta V_k$$

where $\Delta V_k = \Delta x_k\, \Delta y_k\, \Delta z_k$ is the volume of B_k. Let the norm of the partition $|P|$ be the length of the longest diagonal of all the sub-boxes. Then we define the **triple integral** by

$$\iiint\limits_{B} f(x, y, z)\,dV = \lim_{|P| \to 0} \sum_{k=1}^{n} f(\bar{x}_k, \bar{y}_k, \bar{z}_k)\,\Delta V_k$$

provided this limit exists.

The question of what kind of functions are integrable arises here, as it did for single and double integrals. It is certainly sufficient that f be continuous on B. Actually we can allow some discontinuities, for example, on a finite number of smooth surfaces. We do not prove this (a very difficult task), but we assert that it is true.

As you would expect, the triple integral has the standard properties: linearity, additivity on sets that overlap only on a boundary surface, and the comparison property. Finally, triple integrals can be written as triple iterated integrals, as we now illustrate.

EXAMPLE 1 Evaluate $\displaystyle\iiint\limits_{B} x^2 yz\,dV$, where B is the box

$$B = \{(x, y, z): 0 \le x \le 2,\, 0 \le y \le 1,\, 1 \le z \le 2\}$$

Solution

$$\iiint\limits_{B} x^2 yz\,dV = \int_0^2 \int_0^1 \int_1^2 x^2 yz\,dx\,dy\,dz$$

$$= \int_0^2 \int_0^1 \left[\frac{1}{3} x^3 yz\right]_1^2 dy\,dz = \int_0^2 \int_0^1 \frac{7}{3} yz\,dy\,dz$$

$$= \frac{7}{3} \int_0^2 \left[\frac{1}{2} y^2 z\right]_0^1 dz = \frac{7}{3} \int_0^2 \frac{1}{2} z\,dz$$

$$= \frac{7}{6} \left[\frac{z^2}{2}\right]_0^2 = \frac{7}{3}$$

There are six possible orders of integration. Any one of them will yield the answer $\frac{7}{3}$. ∎

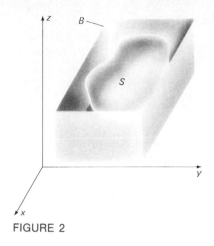

FIGURE 2

GENERAL REGIONS   Consider a closed bounded set S in three-space and enclose it in a box B, as shown in Figure 2. Let $f(x, y, z)$ be defined on S and give f the value zero outside of S. Then we define

$$\iiint\limits_{S} f(x, y, z)\, dV = \iiint\limits_{B} f(x, y, z)\, dV$$

The integral on the right was defined in our opening remarks, but that does not mean that it is easy to evaluate. In fact, if the set S is sufficiently complicated, we may not be able to make the evaluation.

Let S be a z-simple set (vertical lines intersect S in a single line segment) and let S_{xy} be its projection in the xy-plane (Figure 3). Then

$$\iiint\limits_{S} f(x, y, z)\, dV = \iint\limits_{S_{xy}} \left[\int_{\psi_1(x, y)}^{\psi_2(x, y)} f(x, y, z)\, dz \right] dA$$

If, in addition, S_{xy} is a y-simple set (as shown in Figure 3), we can rewrite the outer double integral as an iterated integral.

$$\iiint\limits_{S} f(x, y, z)\, dV = \int_{a_1}^{a_2} \int_{\phi_1(x)}^{\phi_2(x)} \int_{\psi_1(x, y)}^{\psi_2(x, y)} f(x, y, z)\, dz\, dy\, dx$$

FIGURE 3

Other orders of integration may be possible, depending on the shape of S, but in each case we should expect the limits on the inner integral to be functions of two variables, those on the middle integral to be functions of one variable, and those on the outer integral to be constants.

We give several examples. The first simply illustrates evaluation of a triple iterated integral.

EXAMPLE 1 Evaluate the iterated integral

$$\int_{-2}^{5} \int_{0}^{3x} \int_{y}^{x+2} 4 \, dz \, dy \, dx$$

Solution

$$\int_{-2}^{5} \int_{0}^{3x} \int_{y}^{x+2} 4 \, dz \, dy \, dx = \int_{-2}^{5} \int_{0}^{3x} \left(\int_{y}^{x+2} 4 \, dz \right) dy \, dx$$

$$= \int_{-2}^{5} \int_{0}^{3x} \left[4z \right]_{y}^{x+2} dy \, dx$$

$$= \int_{-2}^{5} \int_{0}^{3x} (4x - 4y + 8) \, dy \, dx$$

$$= \int_{-2}^{5} \left[4xy - 2y^2 + 8y \right]_{0}^{3x} dx$$

$$= \int_{-2}^{5} (-6x^2 + 24x) \, dx = -14 \qquad \blacksquare$$

EXAMPLE 2 Evaluate the triple integral of $f(x, y, z) = 2xyz$ over the solid region S that is bounded by the parabolic cylinder $z = 2 - \frac{1}{2}x^2$ and the planes $z = 0$, $y = x$, and $y = 0$.

Solution The solid region S is shown in Figure 4. The triple integral

$$\iiint_{S} 2xyz \, dV$$

can be evaluated by an iterated integral.

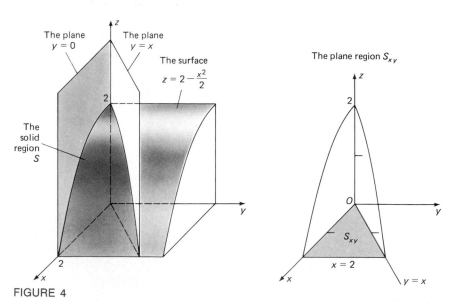

FIGURE 4

Note first that S is a z-simple set and that its projection S_{xy} in the xy-plane is y-simple (also x-simple). Thus,

$$\iiint\limits_S 2xyz\,dV = \int_0^2 \int_0^x \int_0^{2-x^2/2} 2xyz\,dz\,dy\,dx$$

$$= \int_0^2 \int_0^x \left[xyz^2\right]_0^{2-x^2/2} dy\,dx$$

$$= \int_0^2 \int_0^x (4xy - 2x^3y + \tfrac{1}{4}x^5y)\,dy\,dx$$

$$= \int_0^2 (2x^3 - x^5 + \tfrac{1}{8}x^7)\,dx = \tfrac{4}{3} \qquad ■$$

Many different orders of integration are possible in Example 2. We illustrate another way to do this problem.

EXAMPLE 3 Evaluate the integral of Example 2 by doing the integration in the order $dy\,dx\,dz$.

Solution Note that the solid S is y-simple and that it projects onto the plane set S_{xz} shown in Figure 5. Thus,

$$\iiint\limits_S 2xyz\,dV = \int_0^2 \int_0^{\sqrt{4-2z}} \int_0^x 2xyz\,dy\,dx\,dz$$

$$= \int_0^2 \int_0^{\sqrt{4-2z}} x^3z\,dx\,dz = \tfrac{1}{4}\int_0^2 (\sqrt{4-2z})^4 z\,dz$$

$$= \tfrac{1}{4}\int_0^2 (16z - 16z^2 + 4z^3)\,dz = \tfrac{4}{3} \qquad ■$$

MASS AND CENTER OF MASS The concepts of mass and center of mass generalize easily to solid regions. By now, the process that leads to the right formula is very familiar and can be summarized in our motto: *slice, approximate, integrate.* Figure 6 gives away the whole idea. The symbol $\delta(x, y, z)$ denotes the density (mass per unit volume) at (x, y, z).

The plane region S_{xz}

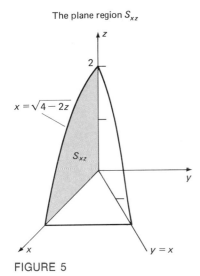

$x = \sqrt{4 - 2z}$

S_{xz}

$y = x$

FIGURE 5

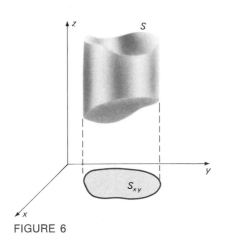

FIGURE 6

Mass of $B_k \approx \delta(\bar{x}_k, \bar{y}_k, \bar{z}_k)\,\triangle V_k$

Moment of B_k
about xy-plane $\approx \bar{z}_k\,\delta(\bar{x}_k, \bar{y}_k, \bar{z}_k)\,\triangle V_k$

The corresponding integral formulas for the mass m of the solid S, moment M_{xy} of S with respect to the xy-plane, and z-coordinate $\bar{z}$ of the center of mass are

$$m = \iiint\limits_{S} \delta(x, y, z)\, dV$$

$$M_{xy} = \iiint\limits_{S} z\delta(x, y, z)\, dV$$

$$\bar{z} = \frac{M_{xy}}{m}$$

There are similar formulas for M_{yz}, M_{xz}, $\bar{x}$, and $\bar{y}$.

EXAMPLE 4   Find the mass and center of mass of the solid S of Example 2, assuming that its density is proportional to the distance from its base in the xy-plane.

Solution   By hypothesis, $\delta(x, y, z) = kz$, where k is a constant. Thus,

$$m = \iiint\limits_{S} kz\, dV = \int_0^2 \int_0^x \int_0^{2-x^2/2} kz\, dz\, dy\, dx$$

$$= k \int_0^2 \int_0^x \frac{1}{2}\left(2 - \frac{x^2}{2}\right)^2 dy\, dx = k \int_0^2 \int_0^x \left(2 - x^2 + \frac{1}{8}x^4\right) dy\, dx$$

$$= k \int_0^2 \left(2x - x^3 + \frac{1}{8}x^5\right) dx = k\left[x^2 - \frac{x^4}{4} + \frac{x^6}{48}\right]_0^2 = \frac{4}{3}k$$

$$M_{xy} = \iiint\limits_{S} kz^2\, dV = \int_0^2 \int_0^x \int_0^{2-x^2/2} kz^2\, dz\, dy\, dx$$

$$= \frac{k}{3} \int_0^2 \int_0^x \left(2 - \frac{x^2}{2}\right)^3 dy\, dx$$

$$= \frac{k}{3} \int_0^2 \int_0^x \left(8 - 6x^2 + \frac{3}{2}x^4 - \frac{1}{8}x^6\right) dy\, dx$$

$$= \frac{k}{3} \int_0^2 \left(8x - 6x^3 + \frac{3}{2}x^5 - \frac{1}{8}x^7\right) dx$$

$$= \frac{k}{3}\left[4x^2 - \frac{3}{2}x^4 + \frac{1}{4}x^6 - \frac{1}{64}x^8\right]_0^2 = \frac{4}{3}k$$

$$M_{xz} = \iiint\limits_{S} kyz\, dV = \int_0^2 \int_0^x \int_0^{2-x^2/2} kyz\, dz\, dy\, dx$$

$$= k \int_0^2 \int_0^x \frac{1}{2}y\left(2 - \frac{x^2}{2}\right)^2 dy\, dx = k \int_0^2 \frac{1}{4}x^2\left(2 - \frac{x^2}{2}\right)^2 dx$$

$$= k \int_0^2 \left(x^2 - \frac{1}{2}x^4 + \frac{1}{16}x^6\right) dx = \frac{64}{105}k$$

$$M_{yz} = \iiint\limits_{S} kxz\, dV = \int_0^2 \int_0^x \int_0^{2-x^2/2} kxz\, dz\, dy\, dx = \frac{128}{105}k$$

$$\bar{z} = \frac{M_{xy}}{m} = \frac{4k/3}{4k/3} = 1$$

$$\bar{y} = \frac{M_{xz}}{m} = \frac{64k/105}{4k/3} = \frac{16}{35}$$

$$\bar{x} = \frac{M_{yz}}{m} = \frac{128k/105}{4k/3} = \frac{32}{35}$$

■

PROBLEM SET 16.7

In Problems 1–8, evaluate the iterated integrals.

1. $\displaystyle\int_{-3}^{7} \int_0^{2x} \int_y^{x-1} dz\, dy\, dx$

2. $\displaystyle\int_0^2 \int_{-1}^4 \int_0^{3y+x} dz\, dy\, dx$

3. $\displaystyle\int_1^4 \int_{z-1}^{2z} \int_0^{y+2z} dx\, dy\, dz$

4. $\displaystyle\int_0^5 \int_{-2}^4 \int_1^2 6xy^2z^3\, dx\, dy\, dz$

5. $\displaystyle\int_0^2 \int_1^z \int_0^{\sqrt{x/z}} 2xyz\, dy\, dx\, dz$

6. $\displaystyle\int_0^{\pi/2} \int_0^z \int_0^y \sin(x+y+z)\, dx\, dy\, dz$

7. $\displaystyle\int_{-2}^4 \int_{x-1}^{x+1} \int_0^{\sqrt{2y/x}} 3xyz\, dz\, dy\, dx$

8. $\displaystyle\int_0^{\pi/2} \int_{\sin 2z}^0 \int_0^{2yz} \sin\!\left(\frac{x}{y}\right) dx\, dy\, dz$

In Problems 9–18, sketch the solid S. Then write an iterated integral for

$$\iiint\limits_{S} f(x, y, z)\, dV$$

9. $S = \{(x, y, z) : 0 \le x \le 1, 0 \le y \le 3,$
$0 \le z \le \frac{1}{6}(12 - 3x - 2y)\}$

10. $S = \{(x, y, z) : 0 \le x \le \sqrt{4 - y^2},$
$0 \le y \le 2, 0 \le z \le 3\}$

11. $S = \{(x, y, z) : 0 \le x \le \frac{1}{2}y, 0 \le y \le 4, 0 \le z \le 2\}$

12. $S = \{(x, y, z) : 0 \le x \le \sqrt{y}, 0 \le y \le 4, 0 \le z \le \frac{3}{2}x\}$

13. $S = \{(x, y, z) : 0 \le x \le 3z,$
$0 \le y \le 4 - x - 2z, 0 \le z \le 2\}$

14. $S = \{(x, y, z) : 0 \le x \le y^2, 0 \le y \le \sqrt{z}, 0 \le z \le 1\}$

15. S is the tetrahedron with vertices $(0, 0, 0)$, $(3, 2, 0)$, $(0, 3, 0)$, and $(0, 0, 2)$.

16. S is the region in the first octant bounded by the surface $z = 9 - x^2 - y^2$ and the coordinate planes.

17. S is the region in the first octant bounded by the cylinder $y^2 + z^2 = 1$ and the planes $x = 1$ and $x = 4$.

18. S is the smaller region bounded by the cylinder $x^2 + y^2 - 2y = 0$ and the planes $x - y = 0, z = 0$, and $z = 3$.

In Problems 19–26, use triple iterated integrals to find the indicated quantities.

19. The volume of the solid in the first octant bounded by $y = 2x^2$ and $y + 4z = 8$.

20. The volume of the solid in the first octant bounded by the elliptic cylinder $y^2 + 64z^2 = 4$ and the plane $y = x$.

21. The volume of the solid bounded by the cylinders $x^2 = y$ and $z^2 = y$, and the plane $y = 1$.

22. The volume of the solid bounded by the cylinder $y = x^2 + 2$ and the planes $y = 4, z = 0$, and $3y - 4z = 0$.

23. The center of mass of the tetrahedron bounded by the planes $x + y + z = 1, x = 0, y = 0$, and $z = 0$ if the density is proportional to the sum of the coordinates of the point.

24. The center of mass of the solid bounded by the cylinder $x^2 + y^2 = 9$ and the planes $z = 0$ and $z = 4$ if the density is proportional to the square of the distance from the origin.

25. The center of mass of that part of the ball $\{(x, y, z) : x^2 + y^2 + z^2 \le a^2\}$ that lies in the first octant, assuming it has constant density.

26. The moment of inertia I_x about the x-axis of the solid bounded by the cylinder $y^2 + z^2 = 4$ and the planes

$x - y = 0$, $x = 0$, and $z = 0$ if the density $\delta(x, y, z) = z$. *Hint*: You will need to develop your own formula; *slice, approximate, integrate.*

In Problems 27–30, write the given iterated integral as an iterated integral with the indicated order of integration.

27. $\displaystyle\int_0^1 \int_0^{\sqrt{1-y^2}} \int_0^{\sqrt{1-y^2-z^2}} f(x, y, z)\, dx\, dz\, dy;\ dz\, dy\, dx$

28. $\displaystyle\int_0^2 \int_0^{4-2y} \int_0^{4-2y-z} f(x, y, z)\, dx\, dz\, dy;\ dz\, dy\, dx$

29. $\displaystyle\int_0^2 \int_0^{9-x^2} \int_0^{2-x} f(x, y, z)\, dz\, dy\, dx;\ dy\, dx\, dz$

30. $\displaystyle\int_0^2 \int_0^{9-x^2} \int_0^{2-x} f(x, y, z)\, dz\, dy\, dx;\ dz\, dx\, dy$

31. Consider the solid (Figure 7) in the first octant cut off from the square cylinder with sides $x = 0$, $x = 1$, $z = 0$, and $z = 1$ by the plane $2x + y + 2z = 6$. Find its volume in three ways.

(a) Hard way: by a $dz\, dy\, dx$ integration.
(b) Easier way: by a $dy\, dx\, dz$ integration.
(c) Easiest way: by Problem 19 of Section 16.6.

FIGURE 7

32. Assuming the density of the solid of Figure 7 is a constant k, find the moment of inertia of the solid with respect to the y-axis.

33. If the temperature at (x, y, z) is $T = 30 - z$ degrees, find the average temperature of the solid of Figure 7. Also find the z-coordinate of its centroid.

34. (Soda Can Problem) A full soda can of height h stands on the xy-plane. Punch a hole in the base and watch $\bar{z}$ (the z-coordinate of the center of mass) as the soda leaks away. Starting at $h/2$, $\bar{z}$ gradually drops to a minimum and then rises back to $h/2$ when the can is empty. Show that $\bar{z}$ is least when it coincides with the height of the soda. Would the same conclusion hold for a soda bottle? *Hint*: Don't calculate; think geometrically.

16.8
Triple Integrals (Cylindrical and Spherical Coordinates)

When a solid region S in three-space has an axis of symmetry, the evaluation of triple integrals over S is often facilitated by using cylindrical coordinates. Similarly if S is symmetric with respect to a point, spherical coordinates may be helpful. Cylindrical and spherical coordinates were introduced in Section 14.7, a section you may wish to review before going on.

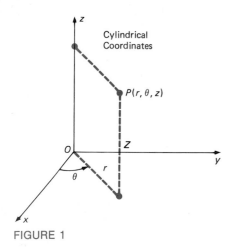

FIGURE 1

CYLINDRICAL COORDINATES   Figure 1 serves to remind us of the meaning of cylindrical coordinates and displays the symbols we will use. Cylindrical and Cartesian (rectangular) coordinates are related by the equations

$$x = r \cos \theta, \qquad y = r \sin \theta, \qquad x^2 + y^2 = r^2$$

As a result the function $f(x, y, z)$ transforms to

$$f(x, y, z) = f(r \cos \theta, r \sin \theta, z) = F(r, \theta, z)$$

when written in cylindrical coordinates.

Suppose now that we wish to evaluate $\iiint\limits_{S} f(x, y, z)\, dV$, where S is a solid region. Consider partitioning S by means of a cylindrical grid, where the typical volume element has the shape shown in Figure 2. Since this piece (called a *cylindrical wedge*) has volume $\Delta V_k = \bar{r}_k \Delta r_k \Delta \theta_k \Delta z_k$, the sum that

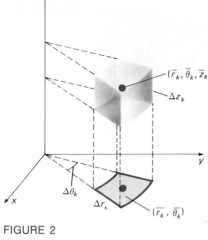

FIGURE 2

approximates the integral has the form

$$\sum_{k=1}^{n} F(\bar{r}_k, \bar{\theta}_k, \bar{z}_k)\bar{r}_k \Delta z_k \Delta r_k \Delta \theta_k$$

Taking the limit as the norm of the partition tends to zero leads to a new integral and suggests an important formula for changing from Cartesian to cylindrical coordinates in a triple integral.

Let S be a z-simple solid and suppose its projection S_{xy} in the xy-plane is r-simple, as shown in Figure 3. If f is continuous on S, then

$$\iiint\limits_{S} f(x, y, z)\, dV = \int_{\theta_1}^{\theta_2} \int_{r_1(\theta)}^{r_2(\theta)} \int_{g_1(r,\theta)}^{g_2(r,\theta)} f(r\cos\theta, r\sin\theta, z)r\, dz\, dr\, d\theta$$

The key fact to note is that the $dz\,dy\,dx$ of Cartesian coordinates becomes $r\,dz\,dr\,d\theta$ in cylindrical coordinates.

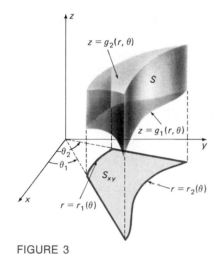

FIGURE 3

EXAMPLE 1 Find the mass and center of mass of a solid cylinder S, assuming the density is proportional to the distance from the base.

Solution With S oriented as shown in Figure 4, we can write the density function as $\delta(x, y, z) = kz$, where k is a constant. Then,

$$m = \iiint\limits_{S} \delta(x, y, z)\, dV = k \int_0^{2\pi} \int_0^a \int_0^h zr\, dz\, dr\, d\theta$$

$$= k \int_0^{2\pi} \int_0^a \frac{1}{2} h^2 r\, dr\, d\theta = \frac{1}{2} kh^2 \int_0^{2\pi} \int_0^a r\, dr\, d\theta$$

$$= \frac{1}{2} kh^2 \int_0^{2\pi} \frac{1}{2} a^2\, d\theta = \frac{1}{2} kh^2 \pi a^2$$

$$M_{xy} = \iiint\limits_{S} z\delta(x, y, z)\, dV = k \int_0^{2\pi} \int_0^a \int_0^h z^2 r\, dz\, dr\, d\theta$$

$$= k \int_0^{2\pi} \int_0^a \frac{1}{3} h^3 r\, dr\, d\theta = \frac{1}{3} kh^3 \int_0^{2\pi} \int_0^a r\, dr\, d\theta$$

$$= \frac{1}{3} kh^3 \pi a^2$$

$$\bar{z} = \frac{M_{xy}}{m} = \frac{\frac{1}{3}kh^3\pi a^2}{\frac{1}{2}kh^2\pi a^2} = \frac{2}{3} h$$

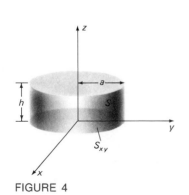

FIGURE 4

By symmetry, $\bar{x} = \bar{y} = 0$. ∎

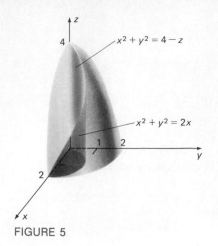

FIGURE 5

EXAMPLE 2 Find the volume of the solid region S bounded above by the paraboloid $z = 4 - x^2 - y^2$, below by $z = 0$, and laterally by $y = 0$ and the cylinder $x^2 + y^2 = 2x$, as shown in Figure 5.

Solution In cylindrical coordinates, the paraboloid is $z = 4 - r^2$ and the cylinder is $r = 2 \cos \theta$. Thus,

$$V = \iiint_S 1 \, dV = \int_0^{\pi/2} \int_0^{2 \cos \theta} \int_0^{4 - r^2} r \, dz \, dr \, d\theta$$

$$= \int_0^{\pi/2} \int_0^{2 \cos \theta} r(4 - r^2) \, dr \, d\theta = \int_0^{\pi/2} \left[2r^2 - \frac{1}{4} r^4 \right]_0^{2 \cos \theta} d\theta$$

$$= \int_0^{\pi/2} (8 \cos^2\theta - 4 \cos^4\theta) \, d\theta$$

$$= 8 \cdot \frac{1}{2} \cdot \frac{\pi}{2} - 4 \cdot \frac{3}{8} \cdot \frac{\pi}{2} = \frac{5\pi}{4}$$

We used Formula 113 from the table of integrals at the end of the book to make the last calculation. ∎

SPHERICAL COORDINATES Figure 6 serves to remind us of the meaning of spherical coordinates, which were introduced in Section 14.7. There we learned that the equations

$$x = \rho \sin \phi \cos \theta, \qquad y = \rho \sin \phi \sin \theta, \qquad z = \rho \cos \phi$$

relate spherical coordinates and Cartesian coordinates. Figure 7 exhibits the volume element in spherical coordinates (called a spherical wedge). Though

Spherical Coordinates

FIGURE 6

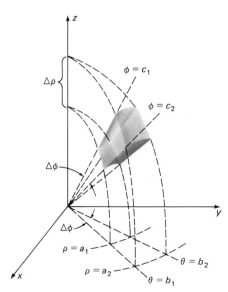

A Spherical Wedge

FIGURE 7

we omit the details, it can be shown that the volume of the indicated spherical wedge is

$$\Delta V = \bar{\rho}^2 \sin \bar{\phi} \, \Delta\rho \, \Delta\theta \, \Delta\phi$$

where $(\bar{\rho}, \bar{\theta}, \bar{\phi})$ is an appropriately chosen point in the wedge.

Partitioning a solid S by means of a spherical grid, forming the appropriate sum, and taking the limit leads to an integral in which $dz \, dy \, dx$ is replaced by $\rho^2 \sin \phi \, d\rho \, d\theta \, d\phi$.

$$\iiint\limits_S f(x, y, z) \, dV$$

$$= \iiint\limits_{\substack{\text{appropriate} \\ \text{limits}}} f(\rho \sin \phi \cos \theta, \rho \sin \phi \sin \theta, \rho \cos \phi)\rho^2 \sin \phi \, d\rho \, d\theta \, d\phi$$

EXAMPLE 3   Find the mass of a solid ball S if its density δ is proportional to the distance from the center.

Solution   Center the ball at the origin and let its radius be a. The density δ is given by $\delta = k\sqrt{x^2 + y^2 + z^2} = k\rho$. Thus, the mass m is given by

$$m = \iiint\limits_S \delta \, dV = k \int_0^\pi \int_0^{2\pi} \int_0^a \rho\rho^2 \sin \phi \, d\rho \, d\theta \, d\phi$$

$$= k\frac{a^4}{4} \int_0^\pi \int_0^{2\pi} \sin \phi \, d\theta \, d\phi = \frac{1}{2} k\pi a^4 \int_0^\pi \sin \phi \, d\phi$$

$$= k\pi a^4 \qquad\blacksquare$$

EXAMPLE 4   Find the volume and center of mass of the homogeneous solid S that is bounded above by the sphere $\rho = a$ and below by the cone $\phi = \alpha$, where a and α are constants (Figure 8).

Solution   The volume V is given by

$$V = \int_0^\alpha \int_0^{2\pi} \int_0^a \rho^2 \sin \phi \, d\rho \, d\theta \, d\phi$$

$$= \int_0^\alpha \int_0^{2\pi} \left(\frac{a^3}{3}\right)\sin \phi \, d\theta \, d\phi$$

$$= \frac{2\pi a^3}{3} \int_0^\alpha \sin \phi \, d\phi = \frac{2\pi a^3}{3}(1 - \cos \alpha)$$

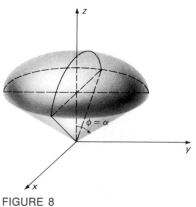

FIGURE 8

It follows that the mass m of the solid is

$$m = kV = \frac{2\pi a^3 k}{3}(1 - \cos \alpha)$$

where k is the constant density.

From symmetry, the center of mass is on the z-axis; that is, $\bar{x} = \bar{y} = 0$. To find $\bar{z}$, we first calculate M_{xy}.

$$M_{xy} = \iiint\limits_{S} kz\, dV = \int_0^\alpha \int_0^{2\pi} \int_0^a k(\rho \cos \phi)\rho^2 \sin \phi\, d\rho\, d\theta\, d\phi$$

$$= \int_0^\alpha \int_0^{2\pi} \int_0^a k\rho^3 \sin \phi \cos \phi\, d\rho\, d\theta\, d\phi$$

$$= \int_0^\alpha \int_0^{2\pi} \frac{1}{4} ka^4 \sin \phi \cos \phi\, d\theta\, d\phi$$

$$= \int_0^\alpha \frac{1}{2} \pi ka^4 \sin \phi \cos \phi\, d\phi = \frac{1}{4} \pi a^4 k \sin^2\alpha$$

Thus,

$$\bar{z} = \frac{\frac{1}{4}\pi a^4 k \sin^2\alpha}{\frac{2}{3}\pi a^3 k(1 - \cos \alpha)} = \frac{3a \sin^2\alpha}{8(1 - \cos \alpha)}$$

$$= \frac{3}{8} a(1 + \cos \alpha) \qquad \blacksquare$$

CHANGE OF VARIABLES IN MULTIPLE INTEGRALS The two boxed formulas of this section and the one in Section 16.4 are examples of change of variable formulas. They illustrate a very general result, which we wish to discuss briefly.

Suppose that

$$x = m(u, v) \qquad y = n(u, v)$$

relate the old variables x and y to new variables u and v. Define a function $J(u, v)$, called the **Jacobian**, by

$$J(u, v) = \begin{vmatrix} \dfrac{\partial x}{\partial u} & \dfrac{\partial x}{\partial v} \\[2ex] \dfrac{\partial y}{\partial u} & \dfrac{\partial y}{\partial v} \end{vmatrix}$$

Then, under suitable conditions on the functions m and n and with appropriate limits on the integral signs,

$$\boxed{\iint f(x, y)\, dx\, dy = \iint f[m(u, v), n(u, v)]|J(u, v)|\, du\, dv}$$

Note, for example, that if

$$x = r \cos \theta, \qquad y = r \sin \theta$$

then

$$J(r, \theta) = \begin{vmatrix} \dfrac{\partial x}{\partial r} & \dfrac{\partial x}{\partial \theta} \\[2ex] \dfrac{\partial y}{\partial r} & \dfrac{\partial y}{\partial \theta} \end{vmatrix} = \begin{vmatrix} \cos \theta & -r \sin \theta \\ \sin \theta & r \cos \theta \end{vmatrix} = r$$

This is exactly the extra factor that occurred when we changed from rectangular to polar coordinates in Section 16.4.

In the three-variable case, where

$$x = m(u, v, w), \qquad y = n(u, v, w), \qquad z = p(u, v, w)$$

the **Jacobian** $J(u, v, w)$ is defined by

$$J(u, v, w) = \begin{vmatrix} \dfrac{\partial x}{\partial u} & \dfrac{\partial x}{\partial v} & \dfrac{\partial x}{\partial w} \\[2ex] \dfrac{\partial y}{\partial u} & \dfrac{\partial y}{\partial v} & \dfrac{\partial y}{\partial w} \\[2ex] \dfrac{\partial z}{\partial u} & \dfrac{\partial z}{\partial v} & \dfrac{\partial z}{\partial w} \end{vmatrix}$$

Again, $|J(u, v, w)|$ is the extra factor that appears in the transformed integral. In Problems 22 and 23, we ask you to show that

$$J(r, \theta, z) = r, \qquad J(\rho, \theta, \phi) = \rho^2 \sin \phi$$

are the Jacobians corresponding to changing from rectangular to cylindrical and spherical coordinates, respectively. For more details about the change of variable problem, see any book on advanced calculus.

PROBLEM SET 16.8

In Problems 1–6, use cylindrical coordinates to find the indicated quantity.

1. The volume of the solid bounded by the paraboloid $z = x^2 + y^2$ and the plane $z = 4$.

2. The volume of the solid bounded above by the sphere $x^2 + y^2 + z^2 = 9$, below by the plane $z = 0$, and laterally by the cylinder $x^2 + y^2 = 4$.

3. The volume of the solid bounded above by the sphere $r^2 + z^2 = 5$ and below by the paraboloid $r^2 = 4z$.

4. The volume of the solid under the surface $z = xy$, above the xy-plane, and within the cylinder $x^2 + y^2 = 2x$.

5. The center of mass of the homogeneous solid bounded above by $z = 12 - 2x^2 - 2y^2$ and below by $z = x^2 + y^2$.

6. The center of mass of the homogeneous solid inside $x^2 + y^2 = 4$, outside $x^2 + y^2 = 1$, below $z = 12 - x^2 - y^2$, and above $z = 0$.

In Problems 7–14, use spherical coordinates to find the indicated quantity.

7. The mass of the solid inside the sphere $\rho = b$ and outside the sphere $\rho = a \, (a < b)$ if the density is proportional to the distance from the origin.

8. The mass of a solid inside a sphere of radius $2a$ and outside a circular cylinder of radius a whose axis is a diameter of the sphere, if the density is proportional to the square of the distance from the center of the sphere.

9. The center of mass of a solid hemisphere of radius a, if the density is proportional to the distance from the center of the sphere.

10. The center of mass of a solid hemisphere of radius a, if the density is proportional to the distance from the axis of symmetry.

11. The moment of inertia of the solid of Problem 10 with respect to its axis of symmetry.

12. The volume of the solid within the sphere $x^2 + y^2 + z^2 = 16$, outside the cone $z = \sqrt{x^2 + y^2}$, and above the xy-plane.

13. The volume of the smaller wedge cut from the unit sphere by two planes which meet at a diameter at an angle of $30°$.

14. $\displaystyle\int_{-3}^{3} \int_{-\sqrt{9-x^2}}^{\sqrt{9-x^2}} \int_{-\sqrt{9-x^2-z^2}}^{\sqrt{9-x^2-z^2}} (x^2 + y^2 + z^2)^{3/2}\, dy\, dz\, dx$

15. Find the volume of the solid bounded above by the plane $z = y$ and below by the paraboloid $z = x^2 + y^2$. *Hint*: In cylindrical coordinates the plane has equation $z = r\sin\theta$ and the paraboloid has equation $z = r^2$. Solve simultaneously to get the projection in the xy-plane.

16. Find the volume of the solid inside both of the spheres $\rho = 2\sqrt{2}\cos\phi$ and $\rho = 2$.

17. For a ball of radius a, find each average distance.

(a) From its center.
(b) From a diameter.
(c) From a point on its boundary (consider $\rho = 2a\cos\phi$).

18. For any homogeneous solid S, show that the average value of the linear function $f(x, y, z) = ax + by + cz + d$ on S is $f(\bar{x}, \bar{y}, \bar{z})$ where $(\bar{x}, \bar{y}, \bar{z})$ is the center of mass.

19. A homogeneous ball of radius a is centered at the origin. For the section S bounded by the half-planes $\theta = -\alpha$ and $\theta = \alpha$ (like a section of an orange), find each value.

(a) The x-coordinate of the center of mass.
(b) The average distance from the z-axis.

20. All balls in this problem have radius a, constant density k, and mass m. Find in terms of a and m the moment of inertia of each of the following.

(a) A ball about a diameter.
(b) A ball about a tangent line to its boundary (the Parallel Axis Theorem holds also for solids—see Problem 20 of Section 16.5).
(c) The two-ball solid of Figure 9 about the z-axis.

FIGURE 9

21. Suppose that the left ball in Figure 9 has density k and the right ball density ck. Find the y-coordinate of the center of mass of this two-ball solid (convince yourself that the analogue of Problem 25 of Section 16.5 is valid).

22. Show that the Jacobian for changing from Cartesian to cylindrical coordinates has value r.

23. Show that the Jacobian for changing from Cartesian to spherical coordinates has value $\rho^2 \sin\phi$.

24. Find the volume of the ellipsoid $x^2/a^2 + y^2/b^2 + z^2/c^2 \leq 1$ by making the change of variables $x = ua$, $y = vb$, and $z = cw$. Also find the moment of inertia of this solid about the z-axis assuming it has constant density k.

16.9 Chapter Review

TRUE-FALSE QUIZ

Respond with true or false to each of the following assertions. Be prepared to defend your answer.

1. $\displaystyle\int_{a}^{b} \int_{a}^{b} f(x)f(y)\, dy\, dx = \left[\int_{a}^{b} f(x)\, dx\right]^2$

2. $\displaystyle\int_0^1 \int_0^x f(x, y)\, dy\, dx = \int_0^1 \int_0^y f(x, y)\, dx\, dy$

3. $\displaystyle\int_0^2 \int_{-1}^1 \sin(x^3 y^2)\, dx\, dy = 0$

4. $\displaystyle\int_{-1}^1 \int_{-1}^1 e^{x^2 + 2y^2}\, dy\, dx = 4\int_0^1 \int_0^1 e^{x^2 + 2y^2}\, dy\, dx$

5. $\displaystyle\int_1^2 \int_0^2 \sin^2(x/y)\, dx\, dy \le 2.$

6. If f is continuous and nonnegative on R and $f(x_0, y_0) > 0$ where (x_0, y_0) is an interior point of R, then $\iint_R f(x, y)\, dA > 0$.

7. $\iint_R f(x, y)\, dA \le \iint_R g(x, y)\, dA$, then $f(x, y) \le g(x, y)$ on R.

8. If $f(x, y) \ge 0$ on R and $\iint_R f(x, y)\, dA = 0$, then $f(x, y) = 0$ for all (x, y) in R.

9. If $\delta(x, y) = k$ gives the density of a lamina at (x, y), the coordinates of the center of mass of the lamina do not involve k.

10. If $\delta(x, y) = y^2/(1 + x^2)$ gives the density of the lamina $\{(x, y): 0 \le x \le 1, 0 \le y \le 1\}$, we know without calculating that $\bar{x} < \frac{1}{2}$ and $\bar{y} > \frac{1}{2}$.

11. If $S = \{(x, y, z): 1 \le x^2 + y^2 + z^2 \le 16\}$, then $\iiint_S dV = 84\pi$.

12. If the top of a right circular cylinder of radius 1 is sliced off by a plane that makes an angle of $30°$ with the base of the cylinder, the area of the resulting slanted top is $2\sqrt{3}\pi/3$.

13. There are eight possible orders of integration for a triple iterated integral.

14. $\displaystyle\int_0^2 \int_0^{2\pi} \int_0^1 dr\, d\theta\, dz$ represents the volume of a right circular cylinder of radius 1 and height 2.

15. If $|f_x| \le 2$ and $|f_y| \le 2$, then the surface G determined by $z = f(x, y)$, $0 \le x \le 1$, $0 \le y \le 1$, has area at most 3.

SAMPLE TEST PROBLEMS

In Problems 1–4, evaluate the integrals.

1. $\displaystyle\int_0^1 \int_x^{\sqrt{x}} xy\, dy\, dx$

2. $\displaystyle\int_{-2}^2 \int_{-\sqrt{4-y^2}}^{\sqrt{4-y^2}} 2xy^2\, dx\, dy$

3. $\displaystyle\int_0^{\pi/2} \int_0^{2\sin\theta} r\cos\theta\, dr\, d\theta$

4. $\displaystyle\int_1^2 \int_3^x \int_0^{\sqrt{3y}} \frac{y}{y^2 + z^2}\, dz\, dy\, dx$

In Problems 5–8, rewrite the iterated integrals with the indicated order of integration. Make a sketch first.

5. $\displaystyle\int_0^1 \int_x^1 f(x, y)\, dy\, dx;\ dx\, dy$

6. $\displaystyle\int_0^1 \int_0^{\cos^{-1}y} f(x, y)\, dx\, dy;\ dy\, dx$

7. $\displaystyle\int_0^1 \int_0^{(1-x)/2} \int_0^{1-x-2y} f(x, y, z)\, dz\, dy\, dx;\ dx\, dz\, dy$

8. $\displaystyle\int_0^2 \int_{x^2}^4 \int_0^{4-y} f(x, y, z)\, dz\, dy\, dx;\ dx\, dy\, dz$

9. Write the triple iterated integrals for the volume of a sphere of radius a in each case.

(a) Cartesian coordinates.   (b) Cylindrical coordinates.
(c) Spherical coordinates.

10. Evaluate $\iint_S (x + y)\, dA$, where S is the region bounded by $y = \sin x$ and $y = 0$ between $x = 0$ and $x = \pi$.

11. Evaluate $\iiint_S z^2\, dV$, where S is the region bounded by $x^2 + z = 1$ and $y^2 + z = 1$ and the xy-plane.

12. Evaluate $\iint_S \dfrac{1}{x^2 + y^2}\, dA$ where S is the region between the circles $x^2 + y^2 = 4$ and $x^2 + y^2 = 9$.

13. Find the center of mass of the rectangular lamina bounded by $x = 1$, $x = 3$, $y = 0$, and $y = 2$ if the density is $\delta(x, y) = xy^2$.

14. Find the moment of inertia of the lamina of Problem 13 with respect to the x-axis.

15. Find the area of the surface of the cylinder $z^2 + y^2 = 9$ lying in the first octant between the planes $y = x$ and $y = 3x$.

16. Evaluate by changing to cylindrical or spherical coordinates

(a) $\displaystyle\int_0^3 \int_0^{\sqrt{9-x^2}} \int_0^2 \sqrt{x^2 + y^2}\, dz\, dy\, dx$

(b) $\displaystyle\int_0^2 \int_0^{\sqrt{4-x^2}} \int_0^{\sqrt{4-x^2-y^2}} z\sqrt{4 - x^2 - y^2}\, dz\, dy\, dx$

17. Find the mass of the solid between the spheres $x^2 + y^2 + z^2 = 1$ and $x^2 + y^2 + z^2 = 9$ if the density is proportional to the distance from the origin.

18. Find the center of mass of the homogeneous lamina bounded by the cardioid $r = 4(1 + \sin \theta)$.

19. Find the mass of the solid in the first octant under the plane $x/a + y/b + z/c = 1$ (a, b, c positive) if the density $\delta(x, y, z) = kx$.

20. Compute the volume of the solid bounded by $z = x^2 + y^2$, $z = 0$, and $x^2 + (y - 1)^2 = 1$.

17 Vector Calculus

Mathematics is the queen of the sciences and arithmetic the queen of mathematics. She often condescends to render service to astronomy and other natural sciences, but in all relations she is entitled to the first rank.
C. F. Gauss

Carl Friedrich Gauss
1777–1855

Gauss has been called the greatest mathematician since Newton and was known to his contemporaries as the Prince of Mathematicians. Fortunately, his genius was recognized in elementary school. One story has him at age 10 confounding his teacher by adding the integers 1 to 1000 almost instantaneously (he used a simple trick). At age 15, he began to entertain the idea of non-Euclidean geometry; at 18, he invented the method of least squares; and at 19, he settled a 2000-year-old question by constructing a 17-sided regular polygon with a ruler and compass.

Gauss spent most of his life at the astronomical observatory at Göttingen, a small city in the heart of Germany. There he established a mathematical tradition that soon made Göttingen and its university the mathematical center of the world, a tradition which lasted until Hitler managed to disperse its famous, but partly Jewish, faculty.

Gauss contributed to many branches of mathematics. His doctoral thesis of 1799 gave the first proof of the Fundamental Theorem of Algebra. His classic *Disquisitiones Arithmeticae* of 1801 is the most influential book on number theory of all time. In calculus, his landmark work on curved surfaces includes the Divergence Theorem, which appears in this chapter.

Like Newton, whom he greatly admired, Gauss applied mathematics to the problems of the physical world, especially in astronomy and physics. His work in magnetism is today acknowledged by assigning the unit *gauss* as a measure of field strength.

17.1
Vector Fields

All that we have done relates in some way to the concept of a function. This concept and the associated calculus have been steadily generalized as we have moved along.

1. Real-valued functions of one real variable (Chapters 1–12).
2. Vector-valued functions of one real variable (Chapters 13–14).
3. Real-valued functions of several real variables (Chapters 15–16).

You will correctly guess that the next step is to study vector-valued functions of several real variables, which is in fact the principal task of the present chapter. This is the final step in a first calculus course; more advanced courses will carry the process of generalization further—to complex-valued functions of complex variables, to real-valued functions of infinitely many variables, and so on.

Consider then a function $\mathbf{F}$ which associates with each point $\mathbf{p}$ in n-space a vector $\mathbf{F(p)}$. A typical example in 2-space is

$$\mathbf{F(p)} = \mathbf{F}(x, y) = -\tfrac{1}{2}y\mathbf{i} + \tfrac{1}{2}x\mathbf{j}$$

For historical reasons, we refer to such a function as a **vector field**, a name arising from a visual image that we now describe. Imagine that to each point $\mathbf{p}$ in a region of space is attached a vector $\mathbf{F(p)}$ emanating from $\mathbf{p}$. We cannot draw all of these vectors, but a representative sample can give us a good intuitive picture of a field. Figure 1 is just such a picture for the vector field $\mathbf{F}(x, y) = -\tfrac{1}{2}y\mathbf{i} + \tfrac{1}{2}x\mathbf{j}$ mentioned earlier. It is the velocity field of a wheel spin-

FIGURE 1

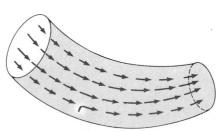

FIGURE 2

ning at a constant rate of $\frac{1}{2}$ radian per unit of time (see Example 2). Figure 2 might represent the velocity field for water flowing in a curved pipe.

Other vector fields that arise naturally in science are electric fields, magnetic fields, force fields, and gravitational fields. We consider only the case where these fields are independent of time, which we call **steady vector fields**. In contrast to a vector field, a function F that attaches a number to each point in space is called a **scalar field**. The function that gives the temperature at each point would be a good physical example of a scalar field.

EXAMPLE 1 Sketch a representative sample of vectors from the vector field

$$\mathbf{F}(x, y) = \frac{x\mathbf{i} + y\mathbf{j}}{\sqrt{x^2 + y^2}}$$

Solution $\mathbf{F}(x, y)$ is a unit vector pointing in the same direction as $x\mathbf{i} + y\mathbf{j}$—that is, away from the origin. Several such vectors are shown in Figure 3. ∎

EXAMPLE 2 Show that each vector from the vector field

$$\mathbf{F}(x, y) = -\tfrac{1}{2}y\mathbf{i} + \tfrac{1}{2}x\mathbf{j}$$

is tangent to a circle centered at the origin and has length equal to one-half the radius of that circle (see Figure 1).

Solution If $\mathbf{r} = x\mathbf{i} + y\mathbf{j}$ is the position vector of the point (x, y), then

$$\mathbf{r} \cdot \mathbf{F}(x, y) = -\tfrac{1}{2}xy + \tfrac{1}{2}xy = 0$$

Thus, $\mathbf{F}(x, y)$ is perpendicular to $\mathbf{r}$ and is therefore tangent to the circle of radius $|\mathbf{r}|$. Finally,

$$|\mathbf{F}(x, y)| = \sqrt{(-\tfrac{1}{2}y)^2 + (\tfrac{1}{2}x)^2} = \tfrac{1}{2}|\mathbf{r}|$$ ∎

According to Isaac Newton, the magnitude of the force of attraction between objects of mass M and m, respectively, is given by GMm/d^2, where d is the distance between the objects and G is a universal constant. This is the famous inverse square law of gravitational attraction. It supplies us with an important example of a force field.

EXAMPLE 3 Suppose that a spherical object of mass M (for example, the earth) is centered at the origin. Derive the formula for the gravitational field of force $\mathbf{F}(x, y, z)$ exerted by this mass on an object of mass m located at a point (x, y, z) in space. Then sketch this field.

Solution We assume that we may treat the object of mass M as a point mass located at the origin. Let $\mathbf{r} = x\mathbf{i} + y\mathbf{j} + z\mathbf{k}$. Then $\mathbf{F}$ has magnitude

$$|\mathbf{F}| = \frac{GMm}{|\mathbf{r}|^2}$$

FIGURE 3

FIGURE 4

The direction of **F** is toward the origin; that is, **F** has the direction of the unit vector $-\mathbf{r}/|\mathbf{r}|$. We conclude that

$$\mathbf{F}(x, y, z) = \frac{GMm}{|\mathbf{r}|^2}\left(\frac{-\mathbf{r}}{|\mathbf{r}|}\right) = -GMm\,\frac{\mathbf{r}}{|\mathbf{r}|^3}$$

This field is sketched in Figure 4. ∎

THE GRADIENT OF A SCALAR FIELD   Let $f(x, y, z)$ determine a scalar field and suppose that f is differentiable. Then the gradient of f, denoted by ∇f, is the vector field given by

$$\mathbf{F}(x, y, z) = \nabla f(x, y, z) = \frac{\partial f}{\partial x}\,\mathbf{i} + \frac{\partial f}{\partial y}\,\mathbf{j} + \frac{\partial f}{\partial z}\,\mathbf{k}$$

A field **F** that arises in this way is called a **conservative vector field**, and f is its **potential function** (the origin of these names will be clarified in Section 17.3). Such fields and their potential functions are important in physics. In particular, fields that obey the inverse square law (for example, electric fields and gravitational fields) are conservative, as we now show.

EXAMPLE 4   Let **F** result from an inverse square law; that is, let

$$\mathbf{F}(x, y, z) = -c\,\frac{\mathbf{r}}{|\mathbf{r}|^3} = -c\,\frac{x\mathbf{i} + y\mathbf{j} + z\mathbf{k}}{(x^2 + y^2 + z^2)^{3/2}}$$

where c is a constant (see Example 3). Show that

$$f(x, y, z) = \frac{c}{(x^2 + y^2 + z^2)^{1/2}} = c(x^2 + y^2 + z^2)^{-1/2}$$

is a potential function for **F**, and therefore that **F** is conservative (for $\mathbf{r} \neq \mathbf{0}$).

Solution

$$\begin{aligned}
\nabla f(x, y, z) &= \frac{\partial f}{\partial x}\,\mathbf{i} + \frac{\partial f}{\partial y}\,\mathbf{j} + \frac{\partial f}{\partial z}\,\mathbf{k} \\
&= \frac{-c}{2}(x^2 + y^2 + c^2)^{-3/2}(2x\mathbf{i} + 2y\mathbf{j} + 2z\mathbf{k}) \\
&= \mathbf{F}(x, y, z)
\end{aligned}$$
∎

Example 4 was really too easy since we gave the function f. A much harder and more significant problem is this. Given a vector field **F**, decide whether it is conservative and, if so, find its potential function. We discuss this problem in Section 17.3.

THE DIVERGENCE AND CURL OF A VECTOR FIELD With a given vector field

$$\mathbf{F}(x, y, z) = M(x, y, z)\mathbf{i} + N(x, y, z)\mathbf{j} + P(x, y, z)\mathbf{k}$$

are associated two other important fields. The first, called the **divergence** of **F**, is a scalar field; the second, called the **curl** of **F**, is a vector field.

Definition

(Div and curl). Let $\mathbf{F} = M\mathbf{i} + N\mathbf{j} + P\mathbf{k}$ be a vector field for which $\partial M/\partial x$, $\partial N/\partial y$, and $\partial P/\partial z$ exist. Then

$$\text{div } \mathbf{F} = \frac{\partial M}{\partial x} + \frac{\partial N}{\partial y} + \frac{\partial P}{\partial z}$$

$$\text{curl } \mathbf{F} = \left(\frac{\partial P}{\partial y} - \frac{\partial N}{\partial z}\right)\mathbf{i} + \left(\frac{\partial M}{\partial z} - \frac{\partial P}{\partial x}\right)\mathbf{j} + \left(\frac{\partial N}{\partial x} - \frac{\partial M}{\partial y}\right)\mathbf{k}$$

At this point it is hard to see the significance of these fields; that will be apparent later. Our interest now is in learning to calculate divergence and curl easily and in relating them to the gradient operator ∇. Recall that ∇ is the operator

$$\nabla = \frac{\partial}{\partial x}\mathbf{i} + \frac{\partial}{\partial y}\mathbf{j} + \frac{\partial}{\partial z}\mathbf{k}$$

When ∇ operates on a function f, it produces the gradient; that is,

$$\nabla f = \text{grad } f$$

Note that by a slight (but very helpful) abuse of notation, we can write

$$\nabla \cdot \mathbf{F} = \left(\frac{\partial}{\partial x}\mathbf{i} + \frac{\partial}{\partial y}\mathbf{j} + \frac{\partial}{\partial z}\mathbf{k}\right) \cdot (M\mathbf{i} + N\mathbf{j} + P\mathbf{k})$$

$$= \frac{\partial M}{\partial x} + \frac{\partial N}{\partial y} + \frac{\partial P}{\partial z} = \text{div } \mathbf{F}$$

$$\nabla \times \mathbf{F} = \begin{vmatrix} \mathbf{i} & \mathbf{j} & \mathbf{k} \\ \dfrac{\partial}{\partial x} & \dfrac{\partial}{\partial y} & \dfrac{\partial}{\partial z} \\ M & N & P \end{vmatrix}$$

$$= \mathbf{i}\left(\frac{\partial P}{\partial y} - \frac{\partial N}{\partial z}\right) - \mathbf{j}\left(\frac{\partial P}{\partial x} - \frac{\partial M}{\partial z}\right) + \mathbf{k}\left(\frac{\partial N}{\partial x} - \frac{\partial M}{\partial y}\right) = \text{curl } \mathbf{F}$$

Thus, grad f, div **F**, and curl **F** can all be written in terms of the operator ∇; this is, in fact, the way to remember how these fields are defined.

EXAMPLE 5 Let

$$\mathbf{F}(x, y, z) = x^2yz\mathbf{i} + 3xyz^3\mathbf{j} + (x^2 - z^2)\mathbf{k}$$

Find div **F** and curl **F**.

Solution

$$\operatorname{div} \mathbf{F} = \nabla \cdot \mathbf{F} = 2xyz + 3xz^3 - 2z$$

$$\operatorname{curl} \mathbf{F} = \nabla \times \mathbf{F} = \begin{vmatrix} \mathbf{i} & \mathbf{j} & \mathbf{k} \\ \dfrac{\partial}{\partial x} & \dfrac{\partial}{\partial y} & \dfrac{\partial}{\partial z} \\ x^2yz & 3xyz^3 & x^2 - z^2 \end{vmatrix}$$

$$= -\mathbf{i}(9xyz^2) - \mathbf{j}(2x - x^2y) + \mathbf{k}(3yz^3 - x^2z) \qquad \blacksquare$$

PROBLEM SET 17.1

In Problems 1–6, sketch a sample of vectors for the given vector field **F**.

1. $\mathbf{F}(x, y) = x\mathbf{i} + y\mathbf{j}$ 2. $\mathbf{F}(x, y) = x\mathbf{i} - y\mathbf{j}$

3. $\mathbf{F}(x, y) = -x\mathbf{i} + 2y\mathbf{j}$ 4. $\mathbf{F}(x, y) = 3x\mathbf{i} + y\mathbf{j}$

5. $\mathbf{F}(x, y, z) = x\mathbf{i} + 0\mathbf{j} + \mathbf{k}$

6. $\mathbf{F}(x, y, z) = -z\mathbf{k}$

In Problems 7–12, find ∇f.

7. $f(x, y, z) = x^2 - 3xy + 2z$

8. $f(x, y, z) = \sin(xyz)$

9. $f(x, y, z) = \ln|xyz|$

10. $f(x, y, z) = \frac{1}{2}(x^2 + y^2 + z^2)$

11. $f(x, y, z) = xe^y \cos z$

12. $f(x, y, z) = y^2 e^{-2z}$

In Problems 13–18, find div **F** and curl **F**.

13. $\mathbf{F}(x, y, z) = x^2\mathbf{i} - 2xy\mathbf{j} + yz^2\mathbf{k}$

14. $\mathbf{F}(x, y, z) = x^2\mathbf{i} + y^2\mathbf{j} + z^2\mathbf{k}$

15. $\mathbf{F}(x, y, z) = yz\mathbf{i} + xz\mathbf{j} + xy\mathbf{k}$

16. $\mathbf{F}(x, y, z) = \cos x\mathbf{i} + \sin y\mathbf{j} + 3\mathbf{k}$

17. $\mathbf{F}(x, y, z) = e^x \cos y\mathbf{i} + e^x \sin y\mathbf{j} + z\mathbf{k}$

18. $\mathbf{F}(x, y, z) = (y + z)\mathbf{i} + (x + z)\mathbf{j} + (x + y)\mathbf{k}$

19. Let f be a scalar field and **F** a vector field. Indicate which of the following are scalar fields, vector fields, or meaningless.

(a) div f (b) grad f
(c) grad **F**   (d) div(grad f)
(e) curl(grad f)   (f) grad(div **F**)
(g) curl(curl **F**)   (h) div(div **F**)
(i) grad(grad f)   (j) div(curl(grad f))
(k) curl(div(grad f))

20. Assuming that the required partial derivatives exist and are continuous, show that

(a) div(curl **F**) = 0;
(b) curl(grad f) = **0**;
(c) div(f**F**) = (f)(div **F**) + (grad f) · **F**;
(d) curl(f**F**) = (f)(curl **F**) + (grad f) × **F**.

21. Let $\mathbf{F}(x, y, z) = -c\mathbf{r}/|\mathbf{r}|^3$ be an inverse square law field (see Examples 3 and 4). Show that curl **F** = **0** and div **F** = 0. *Hint*: Use Problem 20 with $f = -c/|\mathbf{r}|^3$.

22. Let $\mathbf{F}(x, y, z) = -c\mathbf{r}/|\mathbf{r}|^m$, $c \neq 0, m \neq 3$. Show in contrast to Problem 21 that div **F** ≠ 0, though curl **F** = **0**.

23. Let $\mathbf{F}(x, y, z) = f(r)\mathbf{r}$, where $r = |\mathbf{r}| = \sqrt{x^2 + y^2 + z^2}$ and f is a differentiable scalar function (except possibly at $r = 0$). Show that curl **F** = **0** (except at $r = 0$). *Hint*: First show that grad $f = f'(r)\mathbf{r}/r$ and then apply Problem 20d.

24. Let $\mathbf{F}(x, y, z)$ be as in Problem 23. Show that if div **F** = 0, then $f(r) = cr^{-3}$, where c is a constant.

25. This problem relates to the interpretation of div and curl given in the margin on page 755. Consider the four velocity fields **F**, **G**, **H**, and **L**, which have for every z the configuration illustrated in Figure 5. Decide each of the following by geometric reasoning.

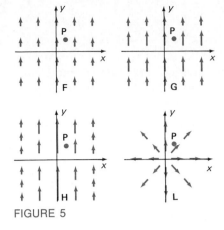
FIGURE 5

(a) Is the divergence at **p** positive, negative, or zero?
(b) Will a paddle wheel at **p** with vertical axis (Figure 4, Section 17.7) rotate clockwise, counterclockwise, or not at all?

Now let $\mathbf{F} = c\mathbf{j}$, $\mathbf{G} = e^{-y^2}\mathbf{j}$, $\mathbf{H} = e^{-x^2}\mathbf{j}$, and $\mathbf{L} = (x\mathbf{i} + y\mathbf{j})/\sqrt{x^2 + y^2}$ which might be modeled as in Figure 5.

(c) Calculate the divergence and curl for each of these fields and thereby confirm your answers in parts (a) and (b).

26. Consider the velocity field $\mathbf{v}(x, y, z) = -\omega y\mathbf{i} + \omega x\mathbf{j}$, $\omega > 0$, (see Example 2 and Figure 1). Note that $\mathbf{v}$ is perpendicular to $x\mathbf{i} + y\mathbf{j}$ and that $|\mathbf{v}| = \omega\sqrt{x^2 + y^2}$. Thus, $\mathbf{v}$ describes a fluid that is rotating (like a solid) about the z-axis with constant angular velocity ω. Show that div $\mathbf{v} = 0$ and curl $\mathbf{v} = 2\omega\mathbf{k}$.

27. An object of mass m, which is revolving in a circular orbit with constant angular velocity ω, is subject to the centrifugal force given by

$$\mathbf{F}(x, y, z) = m\omega^2\mathbf{r} = m\omega^2(x\mathbf{i} + y\mathbf{j} + z\mathbf{k})$$

Show that

$$f(x, y, z) = \tfrac{1}{2}m\omega^2(x^2 + y^2 + z^2)$$

is a potential function for $\mathbf{F}$.

28. The scalar function div(grad f) $= \nabla \cdot \nabla f$ (also written $\nabla^2 f$) is called the *Laplacian* and a function f satisfying $\nabla^2 f = 0$ is said to be *harmonic*, concepts important in physics. Show that $\nabla^2 f = f_{xx} + f_{yy} + f_{zz}$. Then find $\nabla^2 f$ for each of the following and decide which are harmonic.

(a) $f(x, y, z) = 2x^2 - y^2 - z^2$
(b) $f(x, y, z) = xyz$
(c) $f(x, y, z) = x^3 - 3xy^2 + 3z$
(d) $f(x, y, z) = (x^2 + y^2 + z^2)^{-1/2}$

29. Show that:

(a) div$(\mathbf{F} \times \mathbf{G}) = \mathbf{G} \cdot$ curl $\mathbf{F} - \mathbf{F} \cdot$ curl $\mathbf{G}$
(b) div$(\nabla f \times \nabla g) = 0$

30. By analogy with earlier definitions, define each of the following.

(a) $\displaystyle\lim_{(x, y, z) \to (a, b, c)} \mathbf{F}(x, y, z) = \mathbf{L}$

(b) $\mathbf{F}(x, y, z)$ is continuous at (a, b, c).

17.2
Line Integrals

One kind of generalization of the definite integral $\displaystyle\int_a^b f(x)\,dx$ is obtained by replacing the set $[a, b]$ over which we integrate by two- and three-dimensional sets. This led us to the double and triple integrals of Chapter 16. A very different generalization is obtained by replacing $[a, b]$ with a curve C. The resulting integral $\displaystyle\int_C f(x, y)\,ds$ is called a *line integral* but would more properly be called a *curve integral*.

Let C be a smooth plane curve; that is, let C be given parametrically by

$$x = x(t), \qquad y = y(t), \qquad a \le t \le b$$

where x' and y' are continuous and not simultaneously zero on $[a, b]$. We suppose that C is **positively oriented** (that is, that its positive direction corresponds to increasing values of t) and that C is traced only once as t varies from a to b. Thus, C has the initial point $A = (x(a), y(a))$ and the terminal point $B = (x(b)), y(b))$. Consider the partition P of the parameter interval $[a, b]$

obtained by inserting the points

$$a = t_0 < t_1 < t_2 < \cdots < t_n = b$$

This partition of $[a, b]$ results in a division of the curve C into n subarcs $P_{i-1}P_i$ in which the point P_i corresponds to t_i. Let Δs_i denote the length of the arc $P_{i-1}P_i$ and let $|P|$ be the norm of the partition P; that is, let $|P|$ be the largest $\Delta t_i = t_i - t_{i-1}$. Finally, choose a sample point $Q_i(\bar{x}_i, \bar{y}_i)$ on the subarc $P_{i-1}P_i$.

Now consider the Riemann sum

$$\sum_{i=1}^{n} f(\bar{x}_i, \bar{y}_i)\,\Delta s_i$$

If f is nonnegative, this sum approximates the area of the curved vertical curtain shown in Figure 1. If f is continuous on a region D containing the curve C, then this Riemann sum has a limit as $|P| \to 0$. This limit is called the **line integral of f along C from A to B with respect to arc length**; that is,

$$\int_C f(x, y)\,ds = \lim_{|P| \to 0} \sum_{i=1}^{n} f(\bar{x}_i, \bar{y}_i)\,\Delta s_i$$

It represents (for $f(x, y) \geq 0$) the exact area of the curved curtain of Figure 1.

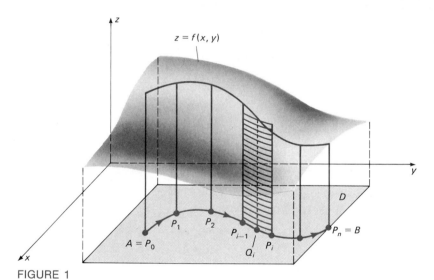

FIGURE 1

The definition does not provide a very good way of evaluating $\int_C f(x, y)\,ds$. That is best accomplished by expressing everything in terms of the parameter t and leads to an ordinary definite integral (recall Section 6.4).

$$\int_C f(x, y)\,ds = \int_a^b f(x(t), y(t))\sqrt{[x'(t)]^2 + [y'(t)]^2}\,dt$$

FIGURE 2

Of course, a curve can be parameterized in many different ways; fortunately, any parameterization results in the same value for $\int_C f(x, y)\, ds$.

The definition of a line integral can be extended to the case where C, though not smooth itself, is piecewise smooth—that is, consists of several smooth curves $C_1, C_2, \ldots, C_k$ joined together, as shown in Figure 2. We simply define the integral over C to be the sum of the integrals over the individual curves.

EXAMPLES AND APPLICATIONS   We begin with two examples where C is part of a circle.

EXAMPLE 1   Evaluate $\int_C x^2 y\, ds$, where C is determined by the parametric equations $x = 3 \cos t$, $y = 3 \sin t$, $0 \le t \le \pi/2$. Also show that the parameterization $x = \sqrt{9 - y^2}$, $y = y$, $0 \le y \le 3$, gives the same value.

Solution

$$\int_C x^2 y\, ds = \int_0^{\pi/2} (3 \cos t)^2 (3 \sin t) \sqrt{(-3 \sin t)^2 + (3 \cos t)^2}\, dt$$

$$= 81 \int_0^{\pi/2} \cos^2 t \sin t\, dt$$

$$= \left[-\frac{81}{3} \cos^3 t \right]_0^{\pi/2} = 27$$

For the second parameterization,

$$ds = \sqrt{1 + \left(\frac{dx}{dy}\right)^2}\, dy = \sqrt{1 + \frac{y^2}{9 - y^2}}\, dy = \frac{3}{\sqrt{9 - y^2}}\, dy$$

and

$$\int_C x^2 y\, ds = \int_0^3 (9 - y^2) y \frac{3}{\sqrt{9 - y^2}}\, dy = 3 \int_0^3 \sqrt{9 - y^2}\, y\, dy$$

$$= -\left[(9 - y^2)^{3/2} \right]_0^3 = 27 \qquad \blacksquare$$

EXAMPLE 2   A thin wire is bent in the shape of the semicircle

$$x = a \cos t, \qquad y = a \sin t, \qquad 0 \le t \le \pi, \qquad a > 0$$

If the density of the wire at a point is proportional to its distance from the x-axis, find the mass and center of mass of the wire.

Solution   Our old motto—*slice, approximate, integrate*—is still appropriate. The mass of a small piece of wire of length Δs (Figure 3) is approximately $\delta(x, y) \Delta s$ where $\delta(x, y) = ky$ is the density at (x, y) (k is a constant). Thus,

FIGURE 3

the mass m of the whole wire is

$$m = \int_C ky\,ds = \int_0^\pi ka\sin t\sqrt{a^2\sin^2 t + a^2\cos^2 t}\,dt$$

$$= ka^2 \int_0^\pi \sin t\,dt$$

$$= \left[-ka^2\cos t \right]_0^\pi = 2ka^2$$

The moment of the wire with respect to the x-axis is given by

$$M_x = \int_C y\,ky\,ds = \int_0^\pi ka^3\sin^2 t\,dt$$

$$= \frac{ka^3}{2}\int_0^\pi (1-\cos 2t)\,dt$$

$$= \frac{ka^3}{2}\left[t - \tfrac{1}{2}\sin 2t \right]_0^\pi = \frac{ka^3\pi}{2}$$

Thus,

$$\bar{y} = \frac{M_x}{m} = \frac{\frac{1}{2}ka^3\pi}{2ka^2} = \tfrac{1}{4}\pi a$$

From symmetry, $\bar{x} = 0$, so the center of mass is at $(0, \pi a/4)$. ∎

All that we have done extends easily to a smooth curve C in three-space. In particular, if C is given parametrically by

$$x = x(t), \qquad y = y(t), \qquad z = z(t), \qquad a \le t \le b$$

then

$$\boxed{\int_C f(x, y, z)\,ds = \int_a^b f(x(t), y(t), z(t))\sqrt{[x'(t)]^2 + [y'(t)]^2 + [z'(t)]^2}\,dt}$$

EXAMPLE 3   Find the mass of a wire of density $\delta(x, y, z) = kz$ if it has the shape of the helix C with parameterization

$$x = 3\cos t, \qquad y = 3\sin t, \qquad z = 4t, \qquad 0 \le t \le \pi$$

Solution

$$m = \int_C kz\,ds = k\int_0^\pi (4t)\sqrt{9\sin^2 t + 9\cos^2 t + 16}\,dt$$

$$= 20k\int_0^\pi t\,dt = \left[20k\,\frac{t^2}{2} \right]_0^\pi = 10k\pi^2$$

The units for m depend on those for length and density. ∎

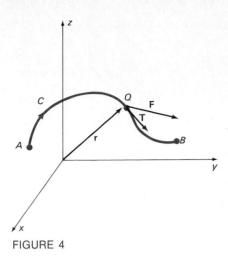

FIGURE 4

WORK Suppose that the force acting at a point (x, y, z) in space is given by the vector field

$$\mathbf{F}(x, y, z) = M(x, y, z)\mathbf{i} + N(x, y, z)\mathbf{j} + P(x, y, z)\mathbf{k}$$

where M, N, and P are continuous. We want to find the work W done by $\mathbf{F}$ in moving a particle along a smooth oriented curve C. Let $\mathbf{r} = x\mathbf{i} + y\mathbf{j} + z\mathbf{k}$ be the position vector for a point $Q(x, y, z)$ on the curve (Figure 4). If $\mathbf{T}$ is the unit tangent vector $d\mathbf{r}/ds$ at Q, then $\mathbf{F} \cdot \mathbf{T}$ is the tangential component of $\mathbf{F}$ at Q. The work done by $\mathbf{F}$ in moving the particle from Q a short distance Δs along the curve is approximately $\mathbf{F} \cdot \mathbf{T}\, \Delta s$, and consequently the work done in moving the particle from A to B along C is defined to be $\int_C \mathbf{F} \cdot \mathbf{T}\, ds$. From Section 13.5, we know that $\mathbf{T} = (d\mathbf{r}/dt)(dt/ds)$, and so we have the following alternative formulas for work.

$$W = \int_C \mathbf{F} \cdot \mathbf{T}\, ds = \int_C \mathbf{F} \cdot \frac{d\mathbf{r}}{dt}\, dt = \int_C \mathbf{F} \cdot d\mathbf{r}$$

To interpret the last expression, think of $\mathbf{F} \cdot d\mathbf{r}$ as representing the work done by $\mathbf{F}$ in moving a particle along the "infinitesimal" tangent vector $d\mathbf{r}$, a formulation preferred by physicists and many applied mathematicians.

There is still another expression for work that is often useful in calculations. If we agree to write $d\mathbf{r} = dx\mathbf{i} + dy\mathbf{j} + dz\mathbf{k}$, then

$$\mathbf{F} \cdot d\mathbf{r} = (M\mathbf{i} + N\mathbf{j} + P\mathbf{k}) \cdot (dx\mathbf{i} + dy\mathbf{j} + dz\mathbf{k}) = M\, dx + N\, dy + P\, dz$$

and

$$W = \int_C \mathbf{F} \cdot d\mathbf{r} = \int_C M\, dx + N\, dy + P\, dz$$

The integrals $\int_C M\, dx$, $\int_C N\, dy$, and $\int_C P\, dz$ are a special kind of line integral. They are defined just as $\int_C f\, ds$ was defined at the beginning of the section, except that Δs_i is replaced by Δx_i, Δy_i, and Δz_i, respectively. However, we point out that while Δs_i is always taken to be positive, Δx_i, Δy_i, and Δz_i may well be negative on a path C. The result of this is that a change in the orientation of C switches the sign of $\int_C M\, dx$, $\int_C N\, dy$, and $\int_C P\, dz$ while leaving that of $\int_C f\, ds$ unchanged (see Problem 31).

EXAMPLE 4 Find the work done by the inverse-square law force field

$$\mathbf{F}(x, y, z) = \frac{-c\mathbf{r}}{|\mathbf{r}|^3} = \frac{-c(x\mathbf{i} + y\mathbf{j} + z\mathbf{k})}{(x^2 + y^2 + z^2)^{3/2}} = M\mathbf{i} + N\mathbf{j} + P\mathbf{k}$$

FIGURE 5

in moving a particle along the straight-line curve C from $(0, 3, 0)$ to $(4, 3, 0)$ shown in Figure 5.

Solution Along C, $y = 3$ and $z = 0$, so $dy = dz = 0$. Using x as the parameter, we obtain

$$W = \int_C M \, dx + N \, dy + P \, dz = -c \int_C \frac{x \, dx + y \, dy + z \, dz}{(x^2 + y^2 + z^2)^{3/2}}$$

$$= -c \int_0^4 \frac{x}{(x^2 + 9)^{3/2}} \, dx = \left[\frac{c}{(x^2 + 9)^{1/2}} \right]_0^4 = \frac{-2c}{15}$$

Of course, appropriate units must be assigned, depending on those for length and force. ∎

Here is a planar version of the line integral introduced above.

EXAMPLE 5 Evaluate the line integral

$$\int_C (x^2 - y^2) \, dx + 2xy \, dy$$

along the curve C whose parametric equations are $x = t^2$, $y = t^3$, $0 \le t \le \frac{3}{2}$.

Solution Since $dx = 2t \, dt$ and $dy = 3t^2 \, dt$,

$$\int_C (x^2 - y^2) \, dx + 2xy \, dy = \int_0^{3/2} \left[(t^4 - t^6) 2t + 2t^5 (3t^2) \right] dt$$

$$= \int_0^{3/2} (2t^5 + 4t^7) \, dt = \frac{8505}{512} \approx 16.61$$ ∎

EXAMPLE 6 Evaluate $\int_C xy^2 \, dx + xy^2 \, dy$ along the path $C = C_1 \cup C_2$ shown in Figure 6. Also evaluate this integral along the straight path C_3 from $(0, 2)$ to $(3, 5)$.

Solution On C_1, $y = 2$, $dy = 0$, and

$$\int_{C_1} xy^2 \, dx + xy^2 \, dy = \int_0^3 4x \, dx = \left[2x^2 \right]_0^3 = 18$$

On C_2, $x = 3$, $dx = 0$, and

$$\int_{C_2} xy^2 \, dx + xy^2 \, dy = \int_2^5 3y^2 \, dy = \left[y^3 \right]_2^5 = 117$$

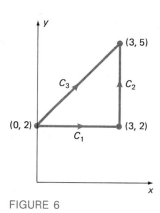

FIGURE 6

We conclude that

$$\int_C xy^2 \, dx + xy^2 \, dy = 18 + 117 = 135$$

On C_3, $y = x + 2$, $dy = dx$, and so

$$\int_{C_3} xy^2 \, dx + xy^2 \, dy = 2 \int_0^3 x(x+2)^2 \, dx$$

$$= 2 \int_0^3 (x^3 + 4x^2 + 4x) \, dx = 2 \left[\frac{x^4}{4} + \frac{4x^3}{3} + 2x^2 \right]_0^3$$

$$= \frac{297}{2}$$

Note that the two paths from $(0, 2)$ to $(3, 5)$ give different values for the integral. ∎

PROBLEM SET 17.2

In Problems 1–16, evaluate the line integrals.

1. $\int_C (x^3 + y) \, ds$; C is the curve $x = 3t$, $y = t^3$, $0 \le t \le 1$.

2. $\int_C xy^{2/5} \, ds$; C is the curve $x = \frac{1}{2}t$, $y = t^{5/2}$, $0 \le t \le 1$.

3. $\int_C (\sin x + \cos y) \, ds$; C is the line segment from $(0, 0)$ to $(\pi, 2\pi)$.

4. $\int_C xe^y \, ds$; C is the line segment from $(-1, 2)$ to $(1, 1)$.

5. $\int_C (2x + 9z) \, ds$; C is the curve $x = t$, $y = t^2$, $z = t^3$, $0 \le t \le 1$.

6. $\int_C (x^2 + y^2 + z^2) \, ds$; C is the curve $x = 4 \cos t$, $y = 4 \sin t$, $z = 3t$, $0 \le t \le 2\pi$.

7. $\int_C y \, dx + x^2 \, dy$; C is the curve $x = 2t$, $y = t^2 - 1$, $0 \le t \le 2$.

8. $\int_C y \, dx + x^2 \, dy$; C is the right-angle curve from $(0, -1)$ to $(4, -1)$ to $(4, 3)$.

9. $\int_C y^3 \, dx + x^3 \, dy$; C is the right-angle curve from $(-4, 1)$ to $(-4, -2)$ to $(2, -2)$.

10. $\int_C y^3 \, dx + x^3 \, dy$; C is the curve $x = 2t$, $y = t^2 - 3$, $-2 \le t \le 1$.

11. $\int_C (x + 2y) \, dx + (x - 2y) \, dy$; C is the line segment from $(1, 1)$ to $(3, -1)$.

12. $\int_C y \, dx + x \, dy$; C is the curve $y = x^2$, $0 \le x \le 1$.

13. $\int_C (x + y + z) \, dx + x \, dy - yz \, dz$; C is the line segment from $(1, 2, 1)$ or $(2, 1, 0)$.

14. $\int_C xz \, dx + (y + z) \, dy + x \, dz$; C is the curve $x = e^t$, $y = e^{-t}$, $z = e^{2t}$, $0 \le t \le 1$.

15. $\int_C (x+y+z) \, dx + (x - 2y + 3z) \, dy + (2x + y - z) \, dz$; C is the line-segment path from $(0, 0, 0)$ to $(2, 0, 0)$ to $(2, 3, 0)$ to $(2, 3, 4)$.

16. Same integral as in Problem 15; C is the line segment from $(0, 0, 0)$ to $(2, 3, 4)$.

17. Find the mass of a wire with the shape of the curve $y = x^2$ between $(-2, 4)$ and $(2, 4)$ if the density is given by $\delta(x, y) = k|x|$.

18. A wire of constant density has the shape of the helix $x = a \cos t$, $y = a \sin t$, $z = bt$, $0 \le t \le 3\pi$. Find its mass and center of mass.

In Problems 19–24, find the work done by the force field $\mathbf{F}$ in moving a particle along the curve C.

19. $\mathbf{F}(x, y) = (x^3 - y^3)\mathbf{i} + xy^2\mathbf{j}$; C is the curve $x = t^2$, $y = t^3$, $-1 \le t \le 0$.

20. $\mathbf{F}(x, y) = e^x\mathbf{i} - e^{-y}\mathbf{j}$; C is the curve $x = 3 \ln t$, $y = \ln 2t$, $1 \le t \le 5$.

21. $\mathbf{F}(x, y) = (x + y)\mathbf{i} + (x - y)\mathbf{j}$; C is the quarter-ellipse $x = a \cos t$, $y = b \sin t$, $0 \leq t \leq \pi/2$.

22. $\mathbf{F}(x, y, z) = (2x - y)\mathbf{i} + 2z\mathbf{j} + (y - z)\mathbf{k}$; C is the line segment from $(0, 0, 0)$ to $(1, 1, 1)$.

23. Same $\mathbf{F}$ as in Problem 22; C is the curve $x = \sin(\pi t/2)$, $y = \sin(\pi t/2)$, $z = t$, $0 \leq t \leq 1$.

24. $\mathbf{F}(x, y, z) = y\mathbf{i} + z\mathbf{j} + x\mathbf{k}$; C is the curve $x = t$, $y = t^2$, $z = t^3$, $0 \leq t \leq 2$.

25. Karen plans to paint both sides of a fence whose base is in the xy-plane with shape $x = 30 \cos^3 t$, $y = 30 \sin^3 t$, $0 \leq t \leq \pi/2$ and whose height at (x, y) is $1 + \frac{1}{3}y$, all measured in feet. Sketch a picture of the fence and decide how much paint she will need if a gallon covers 200 square feet.

26. A squirrel weighing 1.2 pounds climbed a cylindrical tree by following the helical path $x = \cos t$, $y = \sin t$, $z = 4t$, $0 \leq t \leq 8\pi$ (distance measured in feet). How much work did it do? Use a line integral, but then think of a trivial way to answer this question.

27. Use a line integral to find the area of the part cut out of the vertical square cylinder $|x| + |y| = a$ by the sphere $x^2 + y^2 + z^2 = a^2$. Check your answer by finding a trivial way to do this problem.

28. A wire of constant density k has the shape $|x| + |y| = a$. Find its moment of inertia with respect to the y-axis and with respect to the z-axis.

29. Use a line integral to find the area of that part of the cylinder $x^2 + y^2 = ay$ inside the sphere $x^2 + y^2 + z^2 = a^2$ (compare with Problem 12, Section 16.6). *Hint:* Use polar coordinates where $ds = [r^2 + (dr/d\theta)^2]^{1/2} \, d\theta$; see Problem 26, Section 12.8.

30. Two circular cylinders of radius a intersect so their axes meet at right angles. Use a line integral to find the area of the part from one cut off by the other (compare with Problem 14, Section 16.6).

31. Evaluate (a) $\int_C x^2 y \, ds$ using the parameterization $x = 3 \sin t$, $y = 3 \cos t$, $0 \leq t \leq \pi/2$, which reverses the orientation of C in Example 1 and (b) $\int_{C_4} xy^2 \, dx + xy^2 \, dy$ using the parameterization $x = 3 - t$, $y = 5 - t$, $0 \leq t \leq 3$, and note that C_4 has the reverse orientation of C_3 in Example 6. Orientation reversing parameterizations do not change the sign of $\int_C f \, ds$ but do change the sign of the other types of line integrals considered in this section.

17.3
Independence of Path

The basic tool in evaluating ordinary definite integrals is the Fundamental Theorem of Calculus. In symbols, it says that

$$\int_a^b f'(x) \, dx = f(b) - f(a)$$

Now we ask the question: Is there an analogous theorem for line integrals? The answer is yes.

In what follows, interpret $\mathbf{r}(t)$ as $x(t)\mathbf{i} + y(t)\mathbf{j}$ if the context is two-space and as $x(t)\mathbf{i} + y(t)\mathbf{j} + z(t)\mathbf{k}$ if it is three-space. Correspondingly, $f(\mathbf{r})$ will mean $f(x, y)$ in the first case and $f(x, y, z)$ in the second.

Theorem A

(Fundamental Theorem for Line Integrals). Let C be a piecewise smooth curve given parametrically by $\mathbf{r} = \mathbf{r}(t)$, $a \leq t \leq b$, which begins at $\mathbf{a} = \mathbf{r}(a)$ and ends at $\mathbf{b} = \mathbf{r}(b)$. If f is continuously differentiable on an open set containing C, then

$$\int_C \nabla f(\mathbf{r}) \cdot d\mathbf{r} = f(\mathbf{b}) - f(\mathbf{a})$$

Proof We suppose first that C is smooth. Then

$$\int_C \nabla f(\mathbf{r}) \cdot d\mathbf{r} = \int_a^b \left[\nabla f(\mathbf{r}(t)) \cdot \mathbf{r}'(t) \right] dt$$

$$= \int_a^b \frac{d}{dt} \left[f(\mathbf{r}(t)) \right] dt = f(\mathbf{r}(b)) - f(\mathbf{r}(a))$$

$$= f(\mathbf{b}) - f(\mathbf{a})$$

Note how we first wrote the line integral as an ordinary definite integral, then applied the Chain Rule, and finally used the Fundamental Theorem of Calculus.

If C is not smooth but only piecewise smooth, we simply apply the above result to the individual pieces. We leave the details to you. ∎

EXAMPLE 1 Recall from Example 4 of Section 17.1 that

$$f(x, y, z) = f(\mathbf{r}) = \frac{c}{|\mathbf{r}|} = \frac{c}{\sqrt{x^2 + y^2 + z^2}}.$$

is the potential function for the inverse square law field $F(\mathbf{r}) = -c\mathbf{r}/|\mathbf{r}|^3$. Calculate $\int_C \mathbf{F}(\mathbf{r}) \cdot d\mathbf{r}$ where C is any piecewise smooth curve from $(0, 3, 0)$ to $(4, 3, 0)$ that misses the origin.

Solution Since $\mathbf{F}(\mathbf{r}) = \nabla f(\mathbf{r})$,

$$\int_C \mathbf{F}(\mathbf{r}) \cdot d\mathbf{r} = \int_C \nabla f(\mathbf{r}) \cdot d\mathbf{r} = f(4, 3, 0) - f(0, 3, 0)$$

$$= \frac{c}{\sqrt{16 + 9}} - \frac{c}{\sqrt{9}} = \frac{-2c}{15}$$ ∎

Now compare Example 1 with Example 4 of the previous section. There we calculated the same integral, but for a specific curve C, namely, the line segment from $(0, 3, 0)$ to $(4, 3, 0)$. Surprisingly, we will get the same answer no matter what curve we take from $(0, 3, 0)$ to $(4, 3, 0)$. We say that the given line integral is **independent of path**.

A Connected Set

A Disconnected Set

FIGURE 1

CRITERIA FOR INDEPENDENCE OF PATH   Call a set D **connected** if any two points in D can be joined by a piecewise smooth curve lying entirely in D (Figure 1). Then call $\int_C \mathbf{F}(\mathbf{r}) \cdot d\mathbf{r}$ **independent of path in D** if for any two points A and B in D, the line integral has the same value for every path C in D which is positively oriented from A to B.

One consequence of Theorem A is that if **F** is the gradient of another function f, then $\int_C \mathbf{F}(\mathbf{r}) \cdot d\mathbf{r}$ is independent of path. The converse is also true.

(a)

(b)

FIGURE 2

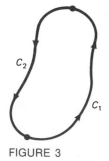

FIGURE 3

> **Theorem B**
>
> Let $\mathbf{F}(\mathbf{r})$ be continuous on an open connected set D. Then the line integral $\int_C \mathbf{F}(\mathbf{r}) \cdot d\mathbf{r}$ is independent of path if and only if $\mathbf{F}(\mathbf{r}) = \nabla f(\mathbf{r})$ for some scalar function f—that is, if and only if $\mathbf{F}$ is a conservative vector field on D.

Proof Theorem A takes care of the "if" statement. Suppose then that $\int_C \mathbf{F}(\mathbf{r}) \cdot d\mathbf{r}$ is independent of path in D. Our task is to construct a function f satisfying $\nabla f = \mathbf{F}$; that is, we must find a potential for the vector field $\mathbf{F}$. For simplicity, we restrict ourselves to the two-dimensional case where D is a plane set and $\mathbf{F}(\mathbf{r}) = M(x, y)\mathbf{i} + N(x, y)\mathbf{j}$.

Let (x_0, y_0) be a fixed point of D and let (x, y) be any other point of D. Choose a third point (x_1, y) in D and slightly to the left of (x, y) and join it to (x, y) by a horizontal segment in D. Then join (x_0, y_0) to (x_1, y) by a curve in D. (All this is possible because D is both open and connected; see Figure 2(a)). Finally, let C denote the path from (x_0, y_0) to (x, y) composed of these two pieces and define f by

$$f(x, y) = \int_C \mathbf{F}(\mathbf{r}) \cdot d\mathbf{r} = \int_{(x_0,\, y_0)}^{(x_1,\, y)} \mathbf{F}(\mathbf{r}) \cdot d\mathbf{r} + \int_{(x_1,\, y)}^{(x,\, y)} \mathbf{F}(\mathbf{r}) \cdot d\mathbf{r}$$

That we get a unique value is clear from the assumed independence of path.

The first integral on the right above does not depend on x; the second, which has y fixed, can be written as an ordinary definite integral using—for example—t as a parameter. It follows that

$$\frac{\partial f}{\partial x} = 0 + \frac{\partial}{\partial x} \int_{x_1}^{x} M(t, y)\, dt = M(x, y)$$

The last equality is a consequence of the theorem on differentiating an integral with respect to its upper limit (Theorem 5.7D).

A similar argument using Figure 2(b) shows that $\partial f / \partial y = N(x, y)$. We conclude that $\nabla f = M(x, y)\mathbf{i} + N(x, y)\mathbf{j} = \mathbf{F}$ as desired. ∎

The condition that $\int_C \mathbf{F}(\mathbf{r}) \cdot d\mathbf{r}$ is independent of path in D implies that if C is any closed, oriented curve in D, then $\int_C \mathbf{F}(\mathbf{r}) \cdot d\mathbf{r} = 0$. To see this, consider C to be composed of two oriented curves C_1 and C_2, as shown in Figure 3. Let $-C_2$ denote the curve C_2 with opposite orientation. Since C_1 and $-C_2$ have the same initial and terminal points, the independence of

CHAPTER 17/Vector Calculus

path guarantees that

$$\int_C \mathbf{F}(\mathbf{r}) \cdot d\mathbf{r} = \int_{C_1} \mathbf{F}(\mathbf{r}) \cdot d\mathbf{r} + \int_{C_2} \mathbf{F}(\mathbf{r}) \cdot d\mathbf{r}$$

$$= \int_{C_1} \mathbf{F}(\mathbf{r}) \cdot d\mathbf{r} - \int_{-C_2} \mathbf{F}(\mathbf{r}) \cdot d\mathbf{r}$$

$$= \int_{C_1} \mathbf{F}(\mathbf{r}) \cdot d\mathbf{r} - \int_{C_1} \mathbf{F}(\mathbf{r}) \cdot d\mathbf{r} = 0$$

The argument just given is reversible. Hence we have three equivalent conditions.

1. $\mathbf{F} = \nabla f$ for some function f ($\mathbf{F}$ is conservative).

2. $\displaystyle\int_C \mathbf{F}(\mathbf{r}) \cdot d\mathbf{r}$ is independent of path.

3. $\displaystyle\int_C \mathbf{F}(\mathbf{r}) \cdot d\mathbf{r} = 0$ for every closed path.

There is an interesting physical interpretation of Condition 3. The work done by a conservative force field as it moves a particle around a closed path is zero. In particular, this is true of both gravitational fields and electric fields, since they are conservative.

While Conditions 2 and 3 each imply that $\mathbf{F}$ is the gradient of a scalar function f, they are not particularly useful in this connection. A more useful criterion is given in the following theorem. We need, however, to impose an additional condition on D, namely, that it is **simply connected**. In two-space, this means that D has no "holes" and in three-space that it has no "tunnels" all the way through D. (For the technical definition, see any advanced calculus book.)

Theorem C

Let $\mathbf{F} = M\mathbf{i} + N\mathbf{j} + P\mathbf{k}$, where M, N, and P are continuous together with their first-order partial derivatives in an open connected set D which is also simply connected. Then $\mathbf{F}$ is conservative ($\mathbf{F} = \nabla f$) if and only if curl $\mathbf{F} = \mathbf{0}$, that is, if and only if

$$\frac{\partial M}{\partial y} = \frac{\partial N}{\partial x}, \qquad \frac{\partial M}{\partial z} = \frac{\partial P}{\partial x}, \qquad \frac{\partial N}{\partial z} = \frac{\partial P}{\partial y}$$

In particular, in the two-variable case where $\mathbf{F} = M\mathbf{i} + N\mathbf{j}$, $\mathbf{F}$ is conservative if and only if

$$\frac{\partial M}{\partial y} = \frac{\partial N}{\partial x}$$

The "only if" statement is easy to prove (Problem 17). We postpone the proof of the "if" statement until Example 4 of Section 17.7. The need for D to be simply connected is illustrated in Problem 25.

RECOVERING A FUNCTION FROM ITS GRADIENT   Suppose that we are given a vector field $\mathbf{F}$ satisfying the conditions of Theorem C. Then we know there is a function f satisfying $\nabla f = \mathbf{F}$. But how can we find f? We illustrate the answer first for a two-dimensional vector field.

EXAMPLE 2   Determine whether $\mathbf{F} = (4x^3 + 9x^2y^2)\mathbf{i} + (6x^3y + 6y^5)\mathbf{j}$ is conservative and, if so, find the function f of which it is the gradient.

Solution $M(x, y) = 4x^3 + 9x^2y^2$ and $N(x, y) = 6x^3y + 6y^5$. In this, the two-variable case, the conditions of Theorem C reduce to showing

$$\frac{\partial M}{\partial y} = \frac{\partial N}{\partial x}$$

Now

$$\frac{\partial M}{\partial y} = 18x^2y, \qquad \frac{\partial N}{\partial x} = 18x^2y$$

so the condition is satisfied and f must exist.
   To find f, we first note that

$$\nabla f = \frac{\partial f}{\partial x}\mathbf{i} + \frac{\partial f}{\partial y}\mathbf{j} = M\mathbf{i} + N\mathbf{j}$$

Thus,

(1) $$\frac{\partial f}{\partial x} = 4x^3 + 9x^2y^2, \qquad \frac{\partial f}{\partial y} = 6x^3y + 6y^5$$

If we antidifferentiate the left equation with respect to x, we obtain

(2) $$f(x, y) = x^4 + 3x^3y^2 + C_1(y)$$

in which the "constant" of integration C_1 may depend on y. But the partial with respect to y of the expression in (2) must equal $6x^3y + 6y^5$.

$$\frac{\partial f}{\partial y} = 6x^3y + C_1'(y) = 6x^3y + 6y^5$$

We conclude that $C_1'(y) = 6y^5$. Another antidifferentiation gives

$$C_1(y) = y^6 + C$$

where C is a constant (independent of both x and y). Substitution of this result in (1) yields

$$f(x, y) = x^4 + 3x^3y^2 + y^6 + C$$

Next we use the result of Example 2 to calculate a line integral.

EXAMPLE 3 Let $\mathbf{F(r)} = \mathbf{F}(x, y) = (4x^3 + 9x^2y^2)\mathbf{i} + (6x^3y + 6y^5)\mathbf{j}$. Calculate $\int_C \mathbf{F(r)} \cdot d\mathbf{r} = \int_C (4x^3 + 9x^2y^2) \, dx + (6x^3y + 6y^5) \, dy$, where C is any path from $(0, 0)$ to $(1, 2)$.

Solution Example 1 shows that $F = \nabla f$, where

$$f(x, y) = x^4 + 3x^3y^2 + y^6 + C$$

and thus the given line integral is independent of path. In fact, by Theorem A,

$$\int_C \mathbf{F(r)} \cdot d\mathbf{r} = \left[x^4 + 3x^3y^2 + y^6 + C \right]_{(0,0)}^{(1,2)}$$
$$= 1 + 12 + 64 = 77$$

EXAMPLE 4 Show that $\mathbf{F} = (e^x \cos y + yz)\mathbf{i} + (xz - e^x \sin y)\mathbf{j} + xy\mathbf{k}$ is conservative and find f such that $\mathbf{F} = \nabla f$.

Solution

$$M = e^x \cos y + yz, \qquad N = xz - e^x \sin y, \qquad P = xy$$

and so

$$\frac{\partial M}{\partial y} = -e^x \sin y + z = \frac{\partial N}{\partial x}, \qquad \frac{\partial M}{\partial z} = y = \frac{\partial P}{\partial x}, \qquad \frac{\partial N}{\partial z} = x = \frac{\partial P}{\partial y}$$

which are the conditions of Theorem C. Now

$$\frac{\partial f}{\partial x} = e^x \cos y + yz$$

(3)
$$\frac{\partial f}{\partial y} = xz - e^x \sin y$$

$$\frac{\partial f}{\partial z} = xy$$

When we antidifferentiate the first of these with respect to x, we get

(4)
$$f(x, y, z) = e^x \cos y + xyz + C_1(y, z)$$

Now differentiate (4) with respect to y and set the result equal to the second expression in (3).

$$-e^x \sin y + xz + \frac{\partial C_1}{\partial y} = xz - e^x \sin y$$

or

$$\frac{\partial C_1(y, z)}{\partial y} = 0$$

Antidifferentiating the latter with respect to y gives

$$C_1(y, z) = C_2(z)$$

which we in turn substitute in (4).

(5) $$f(x, y, z) = e^x \cos y + xyz + C_2(z)$$

When we differentiate (5) with respect to z and equate the result to the third expression in (3), we get

$$\frac{\partial f}{\partial z} = xy + C_2'(z) = xy$$

or $C_2'(z) = 0$ and $C_2(z) = C$. We conclude that

$$f(x, y, z) = e^x \cos y + xyz + C \qquad \blacksquare$$

CONSERVATION OF ENERGY   Let us make an application to physics and at the same time offer a reason for the name *conservative* force field. We will establish the Law of Conservation of Energy, which says that the sum of the kinetic energy and the potential energy of an object due to a conservative force is constant.

Suppose that an object of mass m is moving along a smooth curve C given by

$$\mathbf{r} = \mathbf{r}(t) = x(t)\mathbf{i} + y(t)\mathbf{j} + z(t)\mathbf{k}, \qquad a \le t \le b$$

under the influence of a conservative force $\mathbf{F}(\mathbf{r}) = \nabla f(\mathbf{r})$. From physics, we learn three facts about the object at time t.

1. $\mathbf{F}(\mathbf{r}(t)) = m\mathbf{a}(t) = m\mathbf{r}''(t)$ (Newton's Second Law).
2. $KE = \frac{1}{2}m|\mathbf{r}'(t)|^2$ (KE = kinetic energy).
3. $PE = -f(\mathbf{r})$ (PE = potential energy).

Thus,

$$\frac{d}{dt}(\text{KE} + \text{PE}) = \frac{d}{dt}\left[\frac{1}{2}m|\mathbf{r}'(t)|^2 - f(\mathbf{r})\right]$$

$$= \frac{m}{2}\frac{d}{dt}\left[\mathbf{r}'(t) \cdot \mathbf{r}'(t)\right] - \left[\frac{\partial f}{\partial x}\frac{dx}{dt} + \frac{\partial f}{\partial y}\frac{dy}{dt} + \frac{\partial f}{\partial z}\frac{dz}{dt}\right]$$

$$= m\mathbf{r}''(t) \cdot \mathbf{r}'(t) - \nabla f(\mathbf{r}) \cdot \mathbf{r}'(t)$$

$$= \left[m\mathbf{r}''(t) - \nabla f(\mathbf{r})\right] \cdot \mathbf{r}'(t)$$

$$= \left[\mathbf{F}(\mathbf{r}) - \mathbf{F}(\mathbf{r})\right] \cdot \mathbf{r}'(t) = 0$$

We conclude that KE + PE is constant.

PROBLEM SET   17.3

In Problems 1–10, determine whether the given field $\mathbf{F}$ is conservative. If so, find f so that $\mathbf{F} = \nabla f$; if not, state that $\mathbf{F}$ is not conservative. See Examples 2 and 4.

1. $\mathbf{F}(x, y) = (10x - 7y)\mathbf{i} - (7x - 2y)\mathbf{j}$

2. $\mathbf{F}(x, y) = (12x^2 + 3y^2 + 5y)\mathbf{i} + (6xy - 3y^2 + 5x)\mathbf{j}$

3. $\mathbf{F}(x, y) = (45x^4y^2 - 6y^6 + 3)\mathbf{i} + (18x^5y - 12xy^5 + 7)\mathbf{j}$

4. $\mathbf{F}(x, y) = (35x^4 - 3x^2y^4 + y^9)\mathbf{i} - (4x^3y^3 - 9xy^8)\mathbf{j}$

5. $\mathbf{F}(x, y) = \left(\dfrac{6x^2}{5y^2}\right)\mathbf{i} - \left(\dfrac{4x^3}{5y^3}\right)\mathbf{j}$

6. $\mathbf{F}(x, y) = 4y^2 \cos(xy^2)\mathbf{i} + 8x \cos(xy^2)\mathbf{j}$

7. $\mathbf{F}(x, y) = (2e^y - ye^x)\mathbf{i} + (2xe^y - e^x)\mathbf{j}$

8. $\mathbf{F}(x, y) = -e^{-x}\ln y\,\mathbf{i} + e^{-x}y^{-1}\mathbf{j}$

9. $\mathbf{F}(x, y, z) = 3x^2\mathbf{i} + 6y^2\mathbf{j} + 9z^2\mathbf{k}$

10. $\mathbf{F}(x, y, z) = (2xy + z^2)\mathbf{i} + x^2\mathbf{j} + (2xz + \pi \cos \pi z)\mathbf{k}$

In Problems 11–16, show that the given line integral is independent of path (use Theorem C) and then evaluate the integral (either by choosing a convenient path or, if you prefer, by finding a potential function f and applying Theorem A).

11. $\displaystyle\int_{(-1, 2)}^{(3, 1)} (y^2 + 2xy)\,dx + (x^2 + 2xy)\,dy$

12. $\displaystyle\int_{(0, 0)}^{(1, \pi/2)} e^x \sin y\,dx + e^x \cos y\,dy$

13. $\displaystyle\int_{(0, 0, 0)}^{(1, 1, 1)} (6xy^3 + 2z^2)\,dx + 9x^2y^2\,dy + (4xz + 1)\,dz$

Hint: Try the path consisting of line segments from $(0, 0, 0)$ to $(1, 0, 0)$ to $(1, 1, 0)$ to $(1, 1, 1)$.

14. $\displaystyle\int_{(0, 1, 0)}^{(1, 1, 1)} (yz + 1)\,dx + (xz + 1)\,dy + (xy + 1)\,dz$

15. $\displaystyle\int_{(0, 0, 0)}^{(-1, 0, \pi)} (y + z)\,dx + (x + z)\,dy + (x + y)\,dz$

16. $\displaystyle\int_{(0, 0, 0)}^{(\pi, \pi, 0)} (\cos x + 2yz)\,dx + (\sin y + 2xz)\,dy$
$$+ (z + 2xy)\,dz$$

17. Suppose $\nabla f(x, y, z) = M(x, y, z)\mathbf{i} + N(x, y, z)\mathbf{j} + P(x, y, z)\mathbf{k}$ where M, N, and P have continuous first-order partial derivatives in an open set D. Prove that

$$\frac{\partial M}{\partial y} = \frac{\partial N}{\partial x}, \quad \frac{\partial M}{\partial z} = \frac{\partial P}{\partial x}, \quad \frac{\partial N}{\partial z} = \frac{\partial P}{\partial y}$$

in D. *Hint*: Use Theorem 15.3B on f.

18. For each (x, y, z), let $\mathbf{F}(x, y, z)$ be a vector pointed toward the origin with magnitude inversely proportional to the distance from the origin; that is, let

$$\mathbf{F}(x, y, z) = \frac{-k(x\mathbf{i} + y\mathbf{i} + z\mathbf{k})}{(x^2 + y^2 + z^2)}$$

Show that $\mathbf{F}$ is conservative by finding a potential function for $\mathbf{F}$. *Hint*: If this looks like hard work, see Problem 20.

19. Follow the directions of Problem 18 for $\mathbf{F}(x, y, z)$ directed away from the origin with magnitude that is proportional to the distance from the origin.

20. Generalize Problems 18 and 19 by showing that if

$$\mathbf{F}(x, y, z) = [g(x^2 + y^2 + z^2)](x\mathbf{i} + y\mathbf{i} + z\mathbf{k})$$

where g is a continuous function of one variable, then $\mathbf{F}$ is conservative. *Hint*: Show that $\mathbf{F} = \nabla f$, where $f(x, y, z) = \frac{1}{2}h(x^2 + y^2 + z^2)$ and $h(u) = \int g(u)\,du$.

21. Suppose that an object of mass m is moved along a smooth curve C described by

$$\mathbf{r} = \mathbf{r}(t) = x(t)\mathbf{i} + y(t)\mathbf{j} + z(t)\mathbf{k}, \qquad a \leq t \leq b$$

while subject only to the continuous force $\mathbf{F}$. Show that the work done is equal to the change in the kinetic energy of the object, that is, show that

$$\int_C \mathbf{F} \cdot d\mathbf{r} = \frac{m}{2}\left[|\mathbf{r}'(b)|^2 - |\mathbf{r}'(a)|^2\right]$$

Hint: $\mathbf{F}(\mathbf{r}(t)) = m\mathbf{r}''(t)$.

22. Roberto moved a heavy object along the ground from A to B. The object was at rest at the beginning and at the end. Does Problem 21 imply that Roberto did no work? Explain.

23. We normally consider the gravitational force of the earth on an object of mass m to be given by the constant $\mathbf{F} = -gm\mathbf{k}$, but, of course, this is valid only in small regions near the earth's surface. Find the potential function f for $\mathbf{F}$ and use it to show that the work done by $\mathbf{F}$ when an object is moved from (x_1, y_1, z_1) to a nearby point (x_2, y_2, z_2) is $mg(z_1 - z_2)$.

$\boxed{C}$ **24.** The distance from the earth (mass m) to the sun (mass M) varies from a maximum (aphelion) of 152.1 million kilometers to a minimum (perihelion) of 147.1 million kilometers. Assume that Newton's Inverse Square Law $\mathbf{F} = -GMm\mathbf{r}/|\mathbf{r}|^3$ holds with $G = 6.67 \times 10^{-11}$ newton-meters²/kilogram², $M = 1.99 \times 10^{30}$ kilograms, and $m = 5.97 \times 10^{24}$ kilograms. How much work does $\mathbf{F}$ do in moving the earth in each case?

(a) From aphelion to perihelion.
(b) Around a complete orbit.

25. This problem shows the need for simple connectedness in the "if" statement of Theorem C. Let $\mathbf{F} = (y\mathbf{i} - x\mathbf{j})/(x^2 + y^2)$ on the set $D = \{(x, y): x^2 + y^2 \neq 0\}$. Show each of the following.

(a) The condition $\partial M/\partial y = \partial N/\partial x$ holds on D.

(b) $\mathbf{F}$ is not conservative on D.

Hint: To establish (b), show that $\displaystyle\int_C \mathbf{F} \cdot d\mathbf{r} = -2\pi$, where C is the circle with parametric equations $x = \cos t$, $y = \sin t$, $0 \leq t \leq 2\pi$.

26. Let $f(x, y) = -\tan^{-1}(y/x)$. Show that $\nabla f = (y\mathbf{i} - x\mathbf{j})/(x^2 + y^2)$, which is the vector function of Problem 25. Why doesn't the hint to that problem violate Theorem A?

17.4
Green's Theorem in the Plane

We begin with another look at the Fundamental Theorem of Calculus,

$$\int_a^b f'(x)\,dx = f(b) - f(a)$$

It says that the integral of a function over a set $S = [a, b]$ is equal to a related function (the antiderivative) evaluated in a certain way on the boundary of S, which in this case consists of just the two points a and b. In the remainder of this chapter, we are going to give three generalizations of this result: the theorems of Green, Gauss, and Stokes. These theorems are applied in physics, particularly in the study of heat, electricity, magnetism, and fluid flow. The first of these theorems is due to George Green (1793–1841), a self-taught English mathematical physicist.

We suppose that C is a simple closed curve (Section 13.1), which forms the boundary of a region S. Let C be oriented so that traversing C in its positive direction keeps S to the left (the counterclockwise orientation). The corresponding line integral of $\mathbf{F}(x, y) = M(x, y)\mathbf{i} + N(x, y)\mathbf{j}$ around C is denoted by

$$\oint_C M\,dx + N\,dy$$

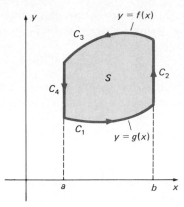

FIGURE 1

(Green's Theorem). Let C be a piecewise smooth, simple closed curve that forms the boundary of a region S in the xy-plane. If $M(x, y)$ and $N(x, y)$ are continuous and have continuous derivatives on S and its boundary C, then

$$\iint_S \left(\frac{\partial N}{\partial x} - \frac{\partial M}{\partial y} \right) dA = \oint_C M\, dx + N\, dy$$

Proof We prove the theorem for the case where S is both an x-simple and a y-simple set and then discuss extensions to the general case.

Since S is y-simple, it has the shape of Figure 1; that is,

$$S = \{(x, y) : g(x) \le y \le f(x), a \le x \le b\}$$

Its boundary C consists of four arcs C_1, C_2, C_3, and C_4 (C_2 or C_4 could be degenerate) and

$$\oint_C M\, dx = \int_{C_1} M\, dx + \int_{C_2} M\, dx + \int_{C_3} M\, dx + \int_{C_4} M\, dx$$

The integrals over C_2 and C_4 are zero, since on these curves x is constant, so that $dx = 0$. Thus,

$$\oint_C M\, dx = \int_a^b M(x, g(x))\, dx + \int_b^a M(x, f(x))\, dx$$

$$= -\int_a^b [M(x, f(x)) - M(x, g(x))]\, dx$$

$$= -\int_a^b \int_{g(x)}^{f(x)} \frac{\partial M(x, y)}{\partial y}\, dy\, dx$$

$$= -\iint_S \frac{\partial M}{\partial y}\, dA$$

Similarly, by treating S as an x-simple set, we obtain

$$\oint_C N\, dy = \iint_S \frac{\partial N}{\partial x}\, dA$$

We conclude that Green's Theorem holds on a set that is both x-simple and y-simple.

The result extends easily to a region S that decomposes into a union of regions $S_1, S_2, \ldots, S_k$, which are both x- and y-simple (Figure 2). We simply apply the theorem in the form we have proved to each of these sets

FIGURE 2

FIGURE 3

FIGURE 4

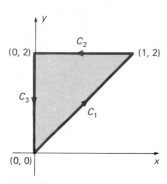

FIGURE 5

and then add the results. Note that the contributions of the line integrals cancel on boundaries shared by adjoining regions, since these boundaries are traversed twice but in opposite directions. ∎

Green's Theorem even holds for a region S with one or more holes (Figure 3), provided that each part of the boundary is oriented so S is always on the left as one traverses the curve in its positive direction. We simply decompose it into ordinary regions in the manner shown in Figure 4.

EXAMPLES AND APPLICATIONS Sometimes, Green's Theorem provides the simplest way of evaluating a line integral.

EXAMPLE 1 Let C be the boundary of the triangle with vertices $(0,0)$, $(1,2)$, and $(0,2)$ (Figure 5). Calculate

$$\oint_C 4x^2y\,dx + 2y\,dy$$

(a) by the direct method, (b) by Green's Theorem.

Solution (a) On C_1, $y = 2x$ and $dy = 2\,dx$, so

$$\int_{C_1} 4x^2y\,dx + 2y\,dy = \int_0^1 8x^3\,dx + 8x\,dx = \left[2x^4 + 4x^2\right]_0^1 = 6$$

Also,

$$\int_{C_2} 4x^2y\,dx + 2y\,dy = \int_1^0 8x^2\,dx = \left[\frac{8x^3}{3}\right]_1^0 = -\frac{8}{3}$$

$$\int_{C_3} 4x^2y\,dx + 2y\,dy = \int_2^0 2y\,dy = \left[y^2\right]_2^0 = -4$$

Thus,

$$\oint_C 4x^2y\,dx + 2y\,dy = 6 - \tfrac{8}{3} - 4 = -\tfrac{2}{3}$$

(b) By Green's Theorem,

$$\oint_C 4x^2y\,dx + 2y\,dy = \int_0^1 \int_{2x}^2 (0 - 4x^2)\,dy\,dx$$

$$= \int_0^1 \left[-4x^2y\right]_{2x}^2 dx = \int_0^1 (-8x^2 + 8x^3)\,dx$$

$$= \left[\frac{-8x^3}{3} + 2x^4\right]_0^1 = -\frac{2}{3}$$ ∎

CHAPTER 17/Vector Calculus

EXAMPLE 2 Show that if a region S in the plane has boundary C, where C is a piecewise smooth, simple closed curve, then the area of S is given by

$$A(S) = \tfrac{1}{2} \oint_C x \, dy - y \, dx$$

Solution Let $M(x, y) = -y/2$ and $N(x, y) = x/2$ and apply Green's Theorem.

$$\oint_C \left(-\frac{y}{2} dx + \frac{x}{2} dy \right) = \iint_S \left(\frac{1}{2} + \frac{1}{2} \right) dA = A(S)$$ ∎

EXAMPLE 3 Use the result of Example 2 to find the area enclosed by the ellipse $b^2x^2 + a^2y^2 = a^2b^2$.

Solution The given ellipse has parametric equations

$$x = a \cos t, \qquad y = b \sin t, \qquad 0 \le t \le 2t$$

Thus,

$$
\begin{aligned}
A(S) &= \tfrac{1}{2} \oint_C x \, dy - y \, dx \\[4pt]
&= \tfrac{1}{2} \int_0^{2\pi} (a \cos t)(b \cos t \, dt) - (b \sin t)(-a \sin t \, dt) \\[4pt]
&= \tfrac{1}{2} \int_0^{2\pi} ab(\cos^2 t + \sin^2 t) \, dt \\[4pt]
&= \tfrac{1}{2}ab \int_0^{2\pi} dt = \pi ab
\end{aligned}
$$ ∎

EXAMPLE 4 Use Green's Theorem to evaluate the line integral

$$\oint_C (x^3 + 2y) \, dx + (4x - 3y^2) \, dy$$

where C is the ellipse $b^2x^2 + a^2y^2 = a^2b^2$.

Solution Let $M(x, y) = x^3 + 2y$, $N(x, y) = 4x - 3y^2$ so that $\partial M/\partial y = 2$ and $\partial N/\partial x = 4$. By Green's Theorem and Example 3,

$$\oint (x^3 + 2y) \, dx + (4x - 3y^2) \, dy = \iint_S (4 - 2) \, dA$$

$$= 2A(S) = 2\pi ab$$ ∎

VECTOR FORMS OF GREEN'S THEOREM Our next goal is to restate Green's Theorem for the plane in its vector form and in two different ways. It is these forms that we will generalize later to two important theorems in 3-space.

We suppose that C is a smooth, simple closed curve in the xy-plane and that it has been given a counterclockwise orientation by means of its arc length parameterization $x = x(s)$ and $y = y(s)$. Then

$$\mathbf{T} = \frac{dx}{ds}\mathbf{i} + \frac{dy}{ds}\mathbf{j}$$

is a unit tangent vector and

$$\mathbf{n} = \frac{dy}{ds}\mathbf{i} - \frac{dx}{ds}\mathbf{j}$$

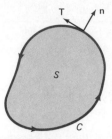

FIGURE 6

is a unit normal vector pointing out of the region S bounded by C (Figure 6). If $\mathbf{F}(x, y) = M(x, y)\mathbf{i} + N(x, y)\mathbf{j}$ is a vector field, then

$$\oint_C \mathbf{F} \cdot \mathbf{n}\, ds = \oint_C (M\mathbf{i} + N\mathbf{j}) \cdot \left(\frac{dy}{ds}\mathbf{i} - \frac{dx}{ds}\mathbf{j}\right) ds = \oint_C -N\, dx + M\, dy$$

$$= \iint_S \left(\frac{\partial M}{\partial x} + \frac{\partial N}{\partial y}\right) dA$$

The last equality comes from Green's Theorem. On the other hand,

$$\operatorname{div} \mathbf{F} = \nabla \cdot \mathbf{F} = \frac{\partial M}{\partial x} + \frac{\partial N}{\partial y}$$

We conclude that

$$\boxed{\oint_C \mathbf{F} \cdot \mathbf{n}\, ds = \iint_S \operatorname{div} \mathbf{F}\, dA = \iint_S \nabla \cdot \mathbf{F}\, dA}$$

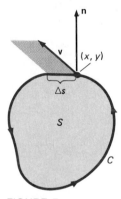

FIGURE 7

a result sometimes called Gauss's Divergence Theorem in the plane.

We give a physical interpretation to this last formula and thereby also understand the origin of the term *divergence*. Imagine a uniform layer of a fluid of constant density moving across the xy-plane, a layer so thin that we may consider it to be two-dimensional. We wish to compute the rate at which the fluid in a region S crosses its boundary curve C (Figure 7).

Let $\mathbf{F}(x, y) = \mathbf{v}(x, y)$ denote the velocity vector of the fluid at (x, y) and let Δs be the length of a short segment of the curve with initial point (x, y). The amount of fluid crossing this segment per unit of time is approximately the area of the parallelogram of Figure 7, namely, $\mathbf{v} \cdot \mathbf{n}\, \Delta s$. The (net) amount of fluid leaving S, called the **flux** of the vector field $\mathbf{F}$ across the curve C in the outward direction, is therefore

$$\boxed{\text{flux of } \mathbf{F} \text{ across } C = \oint_C \mathbf{F} \cdot \mathbf{n}\, ds}$$

Now consider a fixed point (x_0, y_0) in S and a small circle C_r of radius r around it. On S_r, the circular region with boundary C_r, div $\mathbf{F}$ will be approximately equal to its value div $\mathbf{F}(x_0, y_0)$ at the center (we are assuming that div $\mathbf{F}$ is continuous); so by Green's Theorem,

$$\text{flux of } \mathbf{F} \text{ across } C_r = \oint_{C_r} \mathbf{F} \cdot \mathbf{n}\, ds$$

$$= \iint_{S_r} \text{div } \mathbf{F}\, dA \approx \text{div } \mathbf{F}(x_0, y_0)(\pi r^2)$$

We conclude that div $\mathbf{F}(x_0, y_0)$ measures the rate at which the fluid is "diverging away" from (x_0, y). If div $\mathbf{F}(x_0, y) > 0$, there is a *source* of fluid at (x_0, y_0); if div $\mathbf{F}(x_0, y_0) < 0$, there is a *sink* for the fluid at (x_0, y_0). If the flux across the boundary of a region is zero, then the sources and sinks in the region must balance each other. On the other hand, if there are no sources or sinks in a region S, then div $\mathbf{F} = 0$ and, by Green's Theorem, there is a net flow of zero across the boundary of S.

There is another vector form for Green's Theorem. We redraw Figure 6 but now as a subset of three-space (Figure 8). If $\mathbf{F} = M\mathbf{i} + N\mathbf{j} + 0\mathbf{k}$, then Green's Theorem says

$$\oint_C \mathbf{F} \cdot \mathbf{T}\, ds = \oint_C M\, dx + N\, dy = \iint_S \left(\frac{\partial N}{\partial x} - \frac{\partial M}{\partial y} \right) dA$$

On the other hand,

$$\text{curl } \mathbf{F} = \nabla \times \mathbf{F} = \begin{vmatrix} \mathbf{i} & \mathbf{j} & \mathbf{k} \\ \dfrac{\partial}{\partial x} & \dfrac{\partial}{\partial y} & \dfrac{\partial}{\partial z} \\ M & N & 0 \end{vmatrix} = \left(\frac{\partial N}{\partial x} - \frac{\partial M}{\partial y} \right)\mathbf{k}$$

so that

$$(\text{curl } \mathbf{F}) \cdot \mathbf{k} = \left(\frac{\partial N}{\partial x} - \frac{\partial M}{\partial y} \right)$$

Green's Theorem thus takes the form

$$\boxed{\oint_C \mathbf{F} \cdot \mathbf{T}\, ds = \iint_S (\text{curl } \mathbf{F}) \cdot \mathbf{k}\, dA}$$

which is sometimes called Stoke's Theorem in the plane.

If we apply this result to a small circle C_r centered at (x_0, y_0), we obtain

$$\oint_{C_r} \mathbf{F} \cdot \mathbf{T}\, ds \approx (\text{curl } \mathbf{F}(x_0, y_0)) \cdot \mathbf{k}\, (\pi r^2)$$

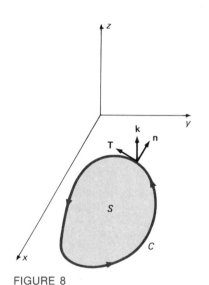

FIGURE 8

This says that the flow in the direction of the tangent to C_r (the "so-called" *circulation* of $\mathbf{F}$ around C_r) is measured by the curl of $\mathbf{F}$. In other words, curl $\mathbf{F}$ measures the tendency of the fluid to rotate about (x_0, y_0). If curl $\mathbf{F} = \mathbf{0}$ in a region S, the corresponding fluid flow is said to be *irrotational*.

EXAMPLE 5 The vector field $\mathbf{F}(x, y) = -\frac{1}{2}y\mathbf{i} + \frac{1}{2}x\mathbf{j} = M\mathbf{i} + N\mathbf{j}$ is the velocity field of a steady counterclockwise rotation of a wheel about the z-axis (see Example 2 of Section 17.1). Calculate $\oint_C \mathbf{F} \cdot \mathbf{n}\, ds$ and $\oint_C \mathbf{F} \cdot \mathbf{T}\, ds$ for any closed curve C in the xy-plane.

Solution If S is the region enclosed by C,

$$\oint_C \mathbf{F} \cdot \mathbf{n}\, ds = \iint_S \operatorname{div} \mathbf{F}\, dA = \iint_S \left(\frac{\partial M}{\partial x} + \frac{\partial N}{\partial y} \right) dA = 0$$

$$\oint_C \mathbf{F} \cdot \mathbf{T}\, ds = \iint_S (\operatorname{curl} \mathbf{F}) \cdot \mathbf{k}\, dA = \iint_S \left(\frac{\partial N}{\partial x} - \frac{\partial M}{\partial y} \right) dA$$

$$= \iint_S \left(\frac{1}{2} + \frac{1}{2} \right) dA = \text{area } S \qquad \blacksquare$$

PROBLEM SET 17.4

In Problems 1–6, use Green's Theorem to evaluate the given line integrals. Begin by sketching the region S.

1. $\oint_C 2xy\, dx + y^2\, dy$, where C is the closed curve formed by $y = x/2$ and $y = \sqrt{x}$ between $(0, 0)$ and $(4, 2)$.

2. $\oint_C \sqrt{y}\, dx + \sqrt{x}\, dy$, where C is the closed curve formed by $y = 0$, $x = 2$, and $y = x^2/2$.

3. $\oint_C (2x + y^2)\, dx + (x^2 + 2y)\, dy$, where C is the closed curve formed by $y = 0$, $x = 2$, and $y = x^3/4$.

4. $\oint_C xy\, dx + (x + y)\, dy$, where C is the triangle with vertices $(0, 0)$, $(2, 0)$, and $(0, 1)$.

5. $\oint_C (x^2 + 4xy)\, dx + (2x^2 + 3y)\, dy$, where C is the ellipse $9x^2 + 16y^2 = 144$.

6. $\oint_C (e^{3x} + 2y)\, dx + (x^2 + \sin y)\, dy$, where C is the rectangle with vertices $(2, 1)$, $(6, 1)$, $(6, 4)$, and $(2, 4)$.

In Problems 7–8, use the result of Example 2 to find the area of the indicated region S. Make a sketch.

7. S is bounded by the curves $y = 4x$ and $y = 2x^2$.

8. S is bounded by the curves $y = \frac{1}{2}x^3$ and $y = x^2$.

In Problems 9–12, use the vector forms of Green's Theorem to calculate (a) $\oint_C \mathbf{F} \cdot \mathbf{n}\, ds$, (b) $\oint_C \mathbf{F} \cdot \mathbf{T}\, ds$.

9. $\mathbf{F} = y^2\mathbf{i} + x^2\mathbf{j}$; C is the boundary of unit square with vertices $(0, 0)$, $(1, 0)$, $(1, 1)$, and $(0, 1)$.

10. $\mathbf{F} = ay\mathbf{i} + bx\mathbf{j}$; C as in Problem 9.

11. $\mathbf{F} = y^3\mathbf{i} + x^3\mathbf{j}$; C is the unit circle.

12. $\mathbf{F} = x\mathbf{i} + y\mathbf{j}$; C is the unit circle.

13. Suppose the integrals $\oint \mathbf{F} \cdot \mathbf{T}\, ds$ taken counterclockwise around the circles $x^2 + y^2 = 36$ and $x^2 + y^2 = 1$ are 30 and -20, respectively. Calculate $\iint_S (\operatorname{curl} \mathbf{F}) \cdot \mathbf{k}\, dA$, where S is the region between the circles.

14. If $\mathbf{F} = (x^2 + y^2)\mathbf{i} + 2xy\mathbf{j}$, find the flux of $\mathbf{F}$ across the boundary C of the unit square with vertices $(0, 0)$, $(1, 0)$,

(1, 1), and (0, 1); that is, calculate

$$\oint_C \mathbf{F} \cdot \mathbf{n}\,ds$$

15. Find the work done by $\mathbf{F} = (x^2 + y^2)\mathbf{i} - 2xy\mathbf{j}$ in moving a body counterclockwise around the curve C of Problem 14.

16. If $\mathbf{F} = (x^2 + y^2)\mathbf{i} + 2xy\mathbf{j}$, calculate the circulation of $\mathbf{F}$ around C of Problem 14; that is, calculate

$$\oint_C \mathbf{F} \cdot \mathbf{T}\,ds$$

17. Show that the work done by a constant force $\mathbf{F}$ in moving a body around a simple closed curve is 0.

18. Use Green's Theorem to prove the plane case of Theorem 17.3C; that is, show that $\partial N/\partial x = \partial M/\partial y$ implies $\oint_C M\,dx + N\,dy = 0$, which implies that $\mathbf{F} = M\mathbf{i} + N\mathbf{j}$ is conservative.

19. Let

$$\mathbf{F} = \frac{y}{x^2 + y^2}\mathbf{i} - \frac{x}{x^2 + y^2}\mathbf{j} = M\mathbf{i} + N\mathbf{j}$$

(a) Show that $\partial N/\partial x = \partial M/\partial y$.
(b) Show, by using the parameterization $x = \cos t$, $y = \sin t$, that $\oint_C M\,dx + N\,dy = -2\pi$, where C is the unit circle.
(c) Why doesn't this contradict Green's Theorem?

20. Let F be as in Problem 19. Calculate $\oint_C M\,dx + N\,dy$ where (a) C is the ellipse $x^2/9 + y^2/4 = 1$, (b) C is the square with vertices $(1, -1)$, $(1, 1)$, $(-1, 1)$, and $(-1, -1)$, and (c) C is the triangle with vertices $(1, 0)$, $(2, 0)$, and $(1, 1)$.

21. Let the piecewise smooth, simple closed curve C be the boundary of a region S in the xy-plane. Modify the argu-

ment in Example 2 to show that

$$A(S) = \oint_C (-y)\,dx = \oint_C x\,dy$$

22. Let S and C be as in Problem 21. Show that the moments M_x and M_y about the x- and y-axes are given by

$$M_x = -\frac{1}{2}\oint_C y^2\,dx, \qquad M_y = \frac{1}{2}\oint_C x^2\,dy$$

23. Calculate the area of the astroid $x^{2/3} + y^{2/3} = a^{2/3}$. *Hint*: Parameterize by $x = a\cos^3 t$, $y = a\sin^3 t$, $0 \le t \le 2\pi$.

24. Calculate the work done by $\mathbf{F} = 2y\mathbf{i} - 3x\mathbf{j}$ in moving an object around the astroid of Problem 23.

25. Let $\mathbf{F}(\mathbf{r}) = \mathbf{r}/|\mathbf{r}|^2 = (x\mathbf{i} + y\mathbf{j})/(x^2 + y^2)$.

(a) Show $\int_C \mathbf{F} \cdot \mathbf{n}\,ds = 2\pi$, where C is the circle centered at the origin of radius a and $\mathbf{n} = (x\mathbf{i} + y\mathbf{j})/\sqrt{x^2 + y^2}$ is the exterior unit normal to C.
(b) Show that div $\mathbf{F} = 0$.
(c) Explain why the results of (a) and (b) do not contradict the vector form of Green's Theorem on page 776.
(d) Show that if C is a smooth simple closed curve, then

$$\int_C \mathbf{F} \cdot \mathbf{n}\,ds$$

equals 2π or 0 according as the origin is inside or outside C.

26. (Area of a Polygon) Let $V_0(x_0, y_0)$, $V_1(x_1, y_1)$, ..., $V_n(x_n, y_n)$ be the vertices of a simple polygon P, labeled counterclockwise and with $V_0 = V_n$. Show each of the following.

(a) $\int_C x\,dy = \frac{1}{2}(x_1 + x_0)(y_1 - y_0)$, where C is the edge V_0V_1.

(b) Area $(P) = \sum_{i=1}^{n} \frac{x_i + x_{i-1}}{2}(y_i - y_{i-1})$.

(c) The area of a polygon with vertices having integral coordinates is always a multiple of $\frac{1}{2}$.
(d) The formula in (b) gives the correct answer for the polygon with vertices $(2, 0)$, $(2, -2)$, $(6, -2)$, $(6, 0)$, $(10, 4)$, and $(-2, 4)$.

17.5
Surface Integrals

A line integral generalizes the ordinary definite integral; in a similar way, a surface integral generalizes a double integral.

Let the surface G be the graph of $z = f(x, y)$ where (x, y) ranges over a rectangle R in the xy-plane. Let P be a partition of R into n subrectangles R_i; this results in a corresponding partition of the surface G into n pieces G_i (Figure 1). Choose a sample point $(\bar{x}_i, \bar{y}_i)$ in R_i and let $(\bar{x}_i, \bar{y}_i, \bar{z}_i) = (\bar{x}_i, \bar{y}_i, f(\bar{x}_i, \bar{y}_i))$ be the corresponding point on G_i. Then define the **surface**

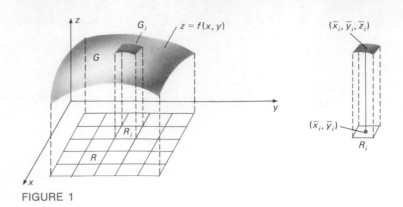

FIGURE 1

integral of g over G by

$$\iint\limits_{G} g(x, y, z)\, dS = \lim_{|P| \to 0} \sum_{i=1}^{n} g(\bar{x}_i, \bar{y}_i, \bar{z}_i)\, \Delta S_i$$

where ΔS_i is the area of G_i. Finally, extend the definition to the case where R is a general closed, bounded set in the xy-plane in the usual way (by giving g the value 0 outside of R).

EVALUATING SURFACE INTEGRALS   A definition is not enough; we need a practical way to evaluate a surface integral. The development in Section 16.6 suggests the correct result. There we showed that under appropriate hypotheses, the area of a small patch G_i of the surface satisfies

$$A(G_i) \approx \sec \gamma_i\, A(R_i) = \sec \gamma_i\, \Delta y_i\, \Delta x_i$$

where γ_i is the angle between $\mathbf{k}$ and the upward unit normal $\mathbf{n}$ (Figure 2). This leads to the following theorem.

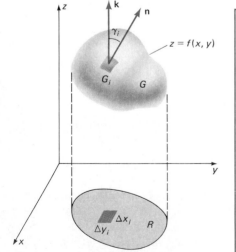

FIGURE 2

Theorem A

Let G be a surface given by $z = f(x, y)$, where (x, y) is in R. If f has continuous first-order partial derivatives and $g(x, y, z) = g(x, y, f(x, y))$ is continuous on R, then

$$\iint\limits_{G} g(x, y, z)\, dS = \iint\limits_{R} g(x, y, f(x, y))\sec \gamma\, dA$$

$$= \iint\limits_{R} g(x, y, f(x, y))\sqrt{f_x^2 + f_y^2 + 1}\, dy\, dx$$

where γ is the angle between the upward unit normal $\mathbf{n}$ at $(x, y, f(x, y))$ and the positive z-axis.

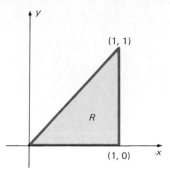

(1, 1)

R

(1, 0)   x

FIGURE 3

EXAMPLE 1 Evaluate $\iint_G (xy + z)\,dS$, where G is the part of the plane $2x - y + z = 3$ which is above the triangle R sketched in Figure 3.

Solution In this case, $z = 3 + y - 2x = f(x, y)$, $f_x = -2$, $f_y = 1$, and $g(x, y, z) = xy + 3 + y - 2x$. Thus,

$$\iint_G (xy + z)\,dS = \int_0^1 \int_0^x (xy + 3 + y - 2x)\sqrt{1^2 + (-2)^2 + 1}\,dy\,dx$$

$$= \sqrt{6} \int_0^1 \left[\frac{xy^2}{2} + 3y + \frac{y^2}{2} - 2xy\right]_0^x dx$$

$$= \sqrt{6} \int_0^1 \left[\frac{x^3}{2} + 3x - \frac{3x^2}{2}\right] dx = \frac{9\sqrt{6}}{8}$$ ■

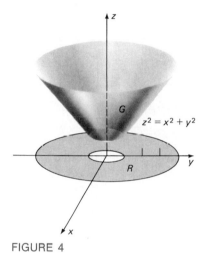

z

G

$z^2 = x^2 + y^2$

R

y

x

FIGURE 4

EXAMPLE 2 Evaluate $\iint_G xyz\,dS$, where G is the portion of the cone $z^2 = x^2 + y^2$ between the planes $z = 1$ and $z = 4$ (Figure 4).

Solution We may write

$$z = (x^2 + y^2)^{1/2} = f(x, y)$$

from which

$$f_x^2 + f_y^2 + 1 = \frac{x^2}{x^2 + y^2} + \frac{y^2}{x^2 + y^2} + 1 = 2$$

Thus,

$$\iint_G xyz\,dS = \iint_R xy\sqrt{x^2 + y^2}\,\sqrt{2}\,dy\,dx$$

After a change to polar coordinates, this becomes

$$\sqrt{2} \int_0^{2\pi} \int_1^4 (r\cos\theta)(r\sin\theta)r^2\,dr\,d\theta = \sqrt{2} \int_0^{2\pi} \left[\sin\theta\cos\theta\frac{r^5}{5}\right]_1^4 d\theta$$

$$= \frac{1023\sqrt{2}}{5} \left[\frac{\sin^2\theta}{2}\right]_0^{2\pi} = 0$$ ■

EXAMPLE 3 The partial spherical surface G with equation

$$z = f(x, y) = \sqrt{9 - x^2 - y^2}, \qquad 0 \le x^2 + y^2 \le 4$$

has a thin metal covering whose density at (x, y, z) is $\delta(x, y, z) = z$. Find the mass of this covering.

Solution Let R be the projection of G in the xy-plane; that is, let $R = \{(x, y): 0 \le x^2 + y^2 \le 4\}$. Then

$$m = \iint_G \delta(x, y, z)\, dS = \iint_R z\sqrt{f_x^2 + f_y^2 + 1}\, dA$$

$$= \iint_R z\sqrt{\frac{x^2}{9 - x^2 - y^2} + \frac{y^2}{9 - x^2 - y^2} + 1}\, dA$$

$$= \iint_R z\frac{3}{z}\, dA = 3(\pi 2^2) = 12\pi \qquad \blacksquare$$

Let the surface G be given by an equation of the form $y = h(x, z)$ and let R be its projection in the xz-plane (Figure 5). Then the appropriate formula for the surface integral is

$$\iint_G g(x, y, z)\, dS = \iint_R g(x, h(x, z), z)\sec \beta\, dA$$

$$= \iint_R g(x, h(x, z), z)\sqrt{h_x^2 + h_z^2 + 1}\, dx\, dz$$

There is a corresponding formula when the surface G is given by $x = k(y, z)$. It involves the projection of G on the yz-plane and the angle α that the outward normal $\mathbf{n}$ makes with $\mathbf{i}$.

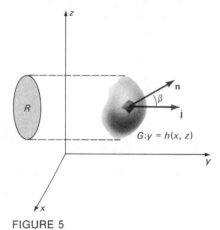

FIGURE 5

EXAMPLE 4 Evaluate $\iint_G (x^2 + z^2)\, dS$, where G is the part of the paraboloid $y = 1 - x^2 - z^2$ that projects onto $R = \{(x, z): 0 \le x^2 + z^2 \le 1\}$.

Solution

$$\iint_G (x^2 + z^2)\, dS = \iint_R (x^2 + z^2)\sqrt{4x^2 + 4z^2 + 1}\, dA$$

If we use polar coordinates, this becomes

$$\int_0^{2\pi} \int_0^1 r^2\sqrt{4r^2 + 1}\, r\, dr\, d\theta$$

In the inner integral, let $u = \sqrt{4r^2 + 1}$, so $u^2 = 4r^2 + 1$ and $u\, du = 4r\, dr$. We obtain

$$\frac{1}{16} \int_0^{2\pi} \int_1^{\sqrt{5}} (u^2 - 1)u^2\, du\, d\theta = \frac{(25\sqrt{5} + 1)\pi}{60} \approx 2.979 \qquad \blacksquare$$

FIGURE 6

FIGURE 7

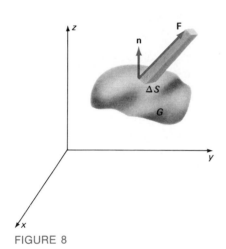

FIGURE 8

THE FLUX OF A VECTOR FIELD THROUGH A SURFACE For what we discuss now and for later applications, we need to limit further the kind of surfaces we consider. Most surfaces that arise in practice have two sides. However, it is surprisingly easy to construct a surface with just one side. Take a paper band (Figure 6), slit it at the dotted line, give one end a half twist, and paste it back together (Figure 7). You obtain a famous one-sided surface, called a Möbius band.

From now on, we consider only two-sided surfaces, so it will make sense to talk about a fluid flowing through the surface from one side to the other. We also suppose that the surface is *smooth*, meaning that it has a continuously-varying unit normal **n**. Let G be such a smooth, two-sided surface and assume that it is submerged in a fluid with a continuous velocity field $\mathbf{F}(x, y, z)$. If ΔS is the area of a small piece of G, then **F** is almost constant there, and the volume ΔV of fluid crossing this piece in the direction of the unit normal **n** (Figure 8) is

$$\Delta V \approx \mathbf{F} \cdot \mathbf{n}\,\Delta S$$

We conclude that

$$\text{Flux of } \mathbf{F} \text{ across } G = \iint_G \mathbf{F} \cdot \mathbf{n}\, dS$$

EXAMPLE 5 Find the upward flux of $\mathbf{F} = -y\mathbf{i} + x\mathbf{j} + 9\mathbf{k}$ across the part of the spherical surface G determined by

$$z = f(x, y) = \sqrt{9 - x^2 - y^2}, \qquad 0 \le x^2 + y^2 \le 4$$

Solution Note that the field **F** is a rotating stream flowing in the direction of the positive z-axis.

The equation of the surface may be written

$$H(x, y, z) = z - \sqrt{9 - x^2 - y^2} = z - f(x, y) = 0$$

and thus

$$\mathbf{n} = \frac{\nabla H}{|\nabla H|} = \frac{-f_x\mathbf{i} - f_y\mathbf{j} + \mathbf{k}}{\sqrt{f_x^2 + f_y^2 + 1}} = \frac{(x/z)\mathbf{i} + (y/z)\mathbf{j} + \mathbf{k}}{\sqrt{(x/z)^2 + (y/z)^2 + 1}}$$

is a unit vector normal to the surface. The vector $-\mathbf{n}$ is also normal to the surface, but, since we desire the normal unit vector that points upward, **n** is the right choice. A straightforward computation using the fact that $x^2 + y^2 + z^2 = 9$ gives

$$\mathbf{n} = \frac{(x/z)\mathbf{i} + (y/z)\mathbf{j} + \mathbf{k}}{3/z} = \frac{x}{3}\mathbf{i} + \frac{y}{3}\mathbf{j} + \frac{z}{3}\mathbf{k}$$

(A simple geometric argument will also give this result; the normal must point directly away from the origin.)

The flux of **F** across G is given by

$$
\text{flux} = \iint\limits_{G} \mathbf{F} \cdot \mathbf{n} \, dS
$$

$$
= \iint\limits_{G} (-y\mathbf{i} + x\mathbf{j} + 9\mathbf{k}) \cdot \left(\frac{x}{3}\mathbf{i} + \frac{y}{3}\mathbf{j} + \frac{z}{3}\mathbf{k} \right) dS
$$

$$
= \iint\limits_{G} 3z \, dS
$$

Finally we write this surface integral as a double integral, using the fact that R is a circle of radius 2 and that $\sec \gamma = \sqrt{f_x^2 + f_y^2 + 1} = 3/z$.

$$
\text{flux} = \iint\limits_{G} 3z \, dS = \iint\limits_{R} 3z \, \frac{3}{z} \, dA = 9(\pi \cdot 2^2) = 36\pi
$$

The total flux across G in one unit of time is 36π cubic units. ∎

An observant reader, after noting the cancellation that occurred in Example 5, will suspect that a theorem is lurking near.

Theorem B

Let G be a smooth, two-sided surface given by $z = f(x, y)$, where (x, y) is in R, and let **n** denote the upward unit normal on G. If f has continuous first-order partial derivatives and $\mathbf{F} = M\mathbf{i} + N\mathbf{j} + P\mathbf{k}$ is a continuous vector field, then the flux of **F** across G is given by

$$
\text{flux } \mathbf{F} = \iint\limits_{G} \mathbf{F} \cdot \mathbf{n} \, dS = \iint\limits_{R} [-Mf_x - Nf_y + P] \, dx \, dy
$$

Proof If we write $H(x, y, z) = z - f(x, y)$, we obtain

$$
\mathbf{n} = \frac{\nabla H}{|\nabla H|} = \frac{-f_x\mathbf{i} - f_y\mathbf{j} + \mathbf{k}}{\sqrt{f_x^2 + f_y^2 + 1}} = (-f_x\mathbf{i} - f_y\mathbf{j} + \mathbf{k})\cos \gamma
$$

where γ is the angle between the upward normal and the positive z-axis. It follows from Theorem A that

$$
\iint\limits_{G} \mathbf{F} \cdot \mathbf{n} \, dS = \iint\limits_{R} (M\mathbf{i} + N\mathbf{j} + P\mathbf{k}) \cdot (-f_x\mathbf{i} - f_y\mathbf{j} + \mathbf{k})\cos \gamma \sec \gamma \, dA
$$

$$
= \iint\limits_{R} (-Mf_x - Nf_y + P) \, dx \, dy \qquad ∎
$$

You might try reworking Example 5 using Theorem B. We offer a different example.

EXAMPLE 6 Evaluate the flux for the vector field $\mathbf{F} = x\mathbf{i} + y\mathbf{j} + z\mathbf{k}$ across the part G of the paraboloid $z = 1 - x^2 - y^2$ that lies above the xy-plane, taking $\mathbf{n}$ to be the upward normal.

Solution

$$f(x, y) = 1 - x^2 - y^2, \qquad f_x = -2x, \qquad f_y = -2y$$

$$-Mf_x - Nf_y + P = 2x^2 + 2y^2 + z$$

$$= 2x^2 + 2y^2 + 1 - x^2 - y^2 = 1 + x^2 + y^2$$

$$\iint_G \mathbf{F} \cdot \mathbf{n} \, dS = \iint_R (1 + x^2 + y^2) \, dx \, dy$$

$$= \int_0^{2\pi} \int_0^1 (1 + r^2) r \, dr \, d\theta = \frac{3}{2}\pi \qquad \blacksquare$$

PROBLEM SET 17.5

In Problems 1–8, evaluate $\iint_G g(x, y, z) \, dS$.

1. $g(x, y, z) = x^2 + y^2 + z$; $G:z = x + y + 1$, $0 \le x \le 1, 0 \le y \le 1$.

2. $g(x, y, z) = x$; $G:x + y + 2z = 4, 0 \le x \le 1$, $0 \le y \le 1$.

3. $g(x, y, z) = x + y$; $G:z = \sqrt{4 - x^2}, 0 \le x \le \sqrt{3}$, $0 \le y \le 1$.

4. $g(x, y, z) = 2y^2 + z$; $G:z = x^2 - y^2$, $0 \le x^2 + y^2 \le 1$.

5. $g(x, y, z) - \sqrt{4x^2 + 4y^2 + 1}$; G is the part of $z = x^2 + y^2$ below $y = z$.

6. $g(x, y, z) = y$; $G:z = 4 - y^2, 0 \le x \le 3, 0 \le y \le 2$.

7. $g(x, y, z) = x + y$; G is the surface of the cube $0 \le x \le 1, 0 \le y \le 1, 0 \le z \le 1$.

8. $g(x, y, z) = z$; G is the tetrahedron bounded by the coordinate planes and the plane $4x + 8y + 2z = 16$.

Use Theorem B to calculate the flux of $\mathbf{F}$ across G in Problems 9–12.

9. $\mathbf{F}(x, y, z) = -y\mathbf{i} + x\mathbf{j}$; G is the part of the plane $z = 8x - 4y - 5$ above the triangle with vertices $(0, 0, 0)$, $(0, 1, 0)$, and $(1, 0, 0)$.

10. $\mathbf{F}(x, y, z) = (9 - x^2)\mathbf{j}$; G is the part of the plane $2x + 3y + 6z = 6$ in the first octant.

11. $\mathbf{F}(x, y, z) = y\mathbf{i} - x\mathbf{j} + 2\mathbf{k}$; G is the surface determined by $z = \sqrt{1 - y^2}, 0 \le x \le 5$.

12. $\mathbf{F}(x, y, z) = 2\mathbf{i} + 5\mathbf{j} + 3\mathbf{k}$; G is the part of the cone $z = (x^2 + y^2)^{1/2}$ that is inside the cylinder $x^2 + y^2 = 1$.

13. Find the mass of the triangle with vertices $(a, 0, 0)$, $(0, a, 0)$, and $(0, 0, a)$ if its density δ satisfies $\delta(x, y, z) = kx^2$.

14. Find the mass of the surface $z = 1 - \frac{1}{2}(x^2 + y^2)$ over $0 \le x \le 1, 0 \le y \le 1, z = 0$, if $\delta(x, y, z) = kxy$.

15. Find the center of mass of the homogeneous triangle with vertices $(a, 0, 0)$, $(0, a, 0)$, and $(0, 0, a)$.

16. Refer to Example 3. The hemispherical surface $z = f(x, y) = \sqrt{9 - x^2 - y^2}$ has a thin metal covering with density $\delta(x, y, z) = z$. Find the mass of this covering. Note that Theorem A does not apply directly, since f_x and f_y are undefined on the boundary $x^2 + y^2 = 9$ of R. Therefore, proceed by letting R_ε be the region $0 \le x^2 + y^2 \le (3 - \varepsilon)^2$, make the calculation, and then let $\varepsilon \to 0$. Discover that you get the same answer as you would if you ignored this subtle point.

17. Let G be the sphere $x^2 + y^2 + z^2 = a^2$. Evaluate each of the following.

(a) $\displaystyle\iint_G z\,dS$

(b) $\displaystyle\iint_G \frac{x + y^3 + \sin z}{1 + z^4}\,dS$

(c) $\displaystyle\iint_G (x^2 + y^2 + z^2)\,dS$

(d) $\displaystyle\iint_G x^2\,dS$

(e) $\displaystyle\iint_G (x^2 + y^2)\,dS$

Hint: Use symmetry properties to make this a trivial problem.

18. The sphere $x^2 + y^2 + z^2 = a^2$ has constant area density k. Find each moment of inertia.

(a) About a diameter.
(b) About a tangent line (assume the Parallel Axis Theorem).

19. Find the total force against the surface of a tank full of a liquid of weight density k for each tank shape.

(a) A sphere of radius a.
(b) A hemisphere of radius a with a flat base.
(c) A vertical cylinder of radius a and height h.

Hint: The force against a small patch of area ΔG is approximately $kd\,\Delta G$, where d is the depth of the water at the patch.

20. Find the center of mass of the part of the sphere $x^2 + y^2 + z^2 = a^2$ between the planes $z = h_1$ and $z = h_2$, where $0 \leq h_1 \leq h_2 \leq a$. Do this by the methods of this section and then compare with Problem 15 of Section 16.6.

17.6
Gauss's Divergence Theorem

The theorems of Green, Gauss, and Stokes all relate an integral over a set S to another integral over the boundary of S. To emphasize the similarity in these theorems, we introduce the notation ∂S to stand for the boundary of S. Thus, one form of Green's Theorem (Section 17.4) can be written

$$\oint_{\partial S} \mathbf{F} \cdot \mathbf{n}\,ds = \iint_S \operatorname{div} \mathbf{F}\,dA$$

It says that the flux of $\mathbf{F}$ across the boundary ∂S of a closed bounded plane region S is equal to the double integral of div $\mathbf{F}$ over that region. Gauss's Theorem lifts this result up a dimension.

GAUSS'S THEOREM Let S be a closed bounded solid in three-space that is completely enclosed by a piecewise smooth surface ∂S (Figure 1).

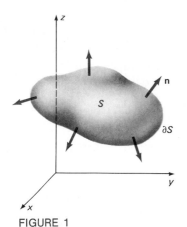

FIGURE 1

Theorem A

(Gauss's Theorem). Let $\mathbf{F} = M\mathbf{i} + N\mathbf{j} + P\mathbf{k}$ be a vector field such that M, N, and P have continuous first-order partial derivatives on a solid S with boundary ∂S. If $\mathbf{n}$ denotes the outer unit normal to ∂S, then

$$\iint_{\partial S} \mathbf{F} \cdot \mathbf{n}\,dS = \iiint_S \operatorname{div} \mathbf{F}\,dV$$

In other words, the flux of $\mathbf{F}$ across the boundary of a closed region in three-space is the triple integral of its divergence over the region.

It is useful both for some applications and for the proof to state the conclusion to Gauss's Theorem in its Cartesian (nonvector) form. We may write

$$\mathbf{n} = \cos \alpha \mathbf{i} + \cos \beta \mathbf{j} + \cos \gamma \mathbf{k}$$

where α, β, and γ are the direction angles for $\mathbf{n}$. Thus, $\mathbf{F} \cdot \mathbf{n} = M \cos \alpha + N \cos \beta + P \cos \gamma$, and so Gauss's formula becomes

$$\iint\limits_{\partial S} (M \cos \alpha + N \cos \beta + P \cos \gamma)\, dS = \iiint\limits_{S} \left(\frac{\partial M}{\partial x} + \frac{\partial N}{\partial y} + \frac{\partial P}{\partial z} \right) dV$$

Proof of Gauss's Theorem We first consider the case where the region S is x-simple, y-simple, and z-simple. It will be sufficient to show that

$$\iint\limits_{\partial S} M \cos \alpha\, dS = \iiint\limits_{S} \frac{\partial M}{\partial x}\, dV$$

$$\iint\limits_{\partial S} N \cos \beta\, dS = \iiint\limits_{S} \frac{\partial N}{\partial y}\, dV$$

$$\iint\limits_{\partial S} P \cos \gamma\, dS = \iiint\limits_{S} \frac{\partial P}{\partial z}\, dV$$

Since these demonstrations are similar, we show only the third.

Since S is z-simple, it can be described by the inequalities $f_1(x, y) \leq z \leq f_2(x, y)$. As in Figure 2, ∂S consists of three parts: S_1, corresponding to $z = f_1(x, y)$; S_2, corresponding to $z = f_2(x, y)$; and the lateral surface S_3, which may be empty. On S_3, $\cos \gamma = \cos 90° = 0$, so we can ignore its contribution. Also, from Theorem 17.5A,

$$\iint\limits_{S_2} P \cos \gamma\, dS = \iint\limits_{R} P(x, y, f_2(x, y))\, dx\, dy$$

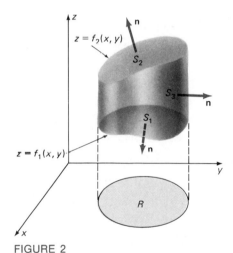

FIGURE 2

The result to which we just referred assumes that the normal $\mathbf{n}$ points upward. Hence, when we apply it to S_1, where $\mathbf{n}$ is a lower normal (Figure 2), we must reverse the sign.

$$\iint\limits_{S_1} P \cos \gamma\, dS = -\iint\limits_{R} P(x, y, f_1(x, y))\, dx\, dy$$

It follows that

$$\iint\limits_{\partial S} P \cos \gamma\, dS = \iint\limits_{R} \left[P(x, y, f_2(x, y)) - P(x, y, f_1(x, y)) \right] dx\, dy$$

$$= \iint\limits_{R} \left[\int_{f_1(x, y)}^{f_2(x, y)} \frac{\partial P}{\partial z}\, dz \right] dx\, dy$$

$$= \iiint\limits_{S} \frac{\partial P}{\partial z}\, dV$$

The result just proved extends easily to regions that are finite unions of the type considered. We omit the details. ∎

EXAMPLE 1 Verify Gauss's Theorem for $\mathbf{F} = x\mathbf{i} + y\mathbf{j} + z\mathbf{k}$ and $S = \{(x, y, z): x^2 + y^2 + z^2 \le a^2\}$ by independently calculating (a) $\iint_{\partial S} \mathbf{F} \cdot \mathbf{n}\, dS$ and (b) $\iiint_S \operatorname{div} \mathbf{F}\, dV$.

Solution (a) On S, $\mathbf{n} = (x\mathbf{i} + y\mathbf{j} + z\mathbf{k})/a$, and so $\mathbf{F} \cdot \mathbf{n} = (x^2 + y^2 + z^2)/a = a$. Thus,

$$\iint_{\partial S} \mathbf{F} \cdot \mathbf{n}\, dS = a \iint_{\partial S} dS = a(4\pi a^2) = 4\pi a^3$$

(b) Since $\operatorname{div} \mathbf{F} = 3$,

$$\iiint_S \operatorname{div} \mathbf{F}\, dV = 3 \iiint_S dV = 3\left(\frac{4\pi a^3}{3}\right) = 4\pi a^3 \qquad ∎$$

EXAMPLE 2 Compute the flux of the vector field $\mathbf{F} = x^2 y\mathbf{i} + 2xz\mathbf{j} + yz^3\mathbf{k}$ across the surface of the rectangular solid S determined by (Figure 3)

$$0 \le x \le 1, \qquad 0 \le y \le 2, \qquad 0 \le z \le 3$$

(a) by a direct method and (b) by Gauss's Theorem.

Solution (a) To calculate $\iint_{\partial S} \mathbf{F} \cdot \mathbf{n}\, dS$ directly, we calculate this integral over the six faces and add the results. On the face $x = 1$, $\mathbf{n} = \mathbf{i}$, and $\mathbf{F} \cdot \mathbf{n} = x^2 y = 1^2 y = y$, so $\iint_{x=1} \mathbf{F} \cdot \mathbf{n}\, dS = \int_0^3 \int_0^2 y\, dy\, dz = 6$. By similar calculations, we can construct the following table.

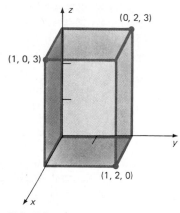

(0, 2, 3)

(1, 0, 3)

(1, 2, 0)

FIGURE 3

Face	$\mathbf{n}$	$\mathbf{F} \cdot \mathbf{n}$	$\iint_{face} \mathbf{F} \cdot \mathbf{n}\, dS$
$x = 1$	$\mathbf{i}$	y	6
$x = 0$	$-\mathbf{i}$	0	0
$y = 2$	$\mathbf{j}$	$2xz$	18
$y = 0$	$-\mathbf{j}$	$-2xz$	-18
$z = 3$	$\mathbf{k}$	$27y$	54
$z = 0$	$-\mathbf{k}$	0	0

Thus,

$$\iint_{\partial S} \mathbf{F} \cdot \mathbf{n}\, dS = 6 + 0 + 18 - 18 + 54 + 0 = 60$$

(b) By Gauss's Theorem,

$$\iint_{\partial S} \mathbf{F} \cdot \mathbf{n}\, dS = \iiint_{S} (2xy + 0 + 3yz^2)\, dV$$

$$= \int_0^1 \int_0^2 \int_0^3 (2xy + 3yz^2)\, dz\, dy\, dx = \int_0^1 \int_0^2 (6xy + 27y)\, dy\, dx$$

$$= \int_0^1 (12x + 54)\, dx = \left[6x^2 + 54x \right]_0^1 = 60 \quad \blacksquare$$

EXAMPLE 3 Let S be the solid cylinder bounded by $x^2 + y^2 = 4$, $z = 0$, and $z = 3$, and let $\mathbf{n}$ be the outer unit normal to the boundary ∂S (Figure 4). If $\mathbf{F} = (x^3 + \tan yz)\mathbf{i} + (y^3 - e^{xz})\mathbf{j} + (3z + x^3)\mathbf{k}$, find the flux of $\mathbf{F}$ across ∂S.

Solution Imagine the difficulty in trying to evaluate $\iint_{\partial S} \mathbf{F} \cdot \mathbf{n}\, dS$ directly. However,

$$\text{div } \mathbf{F} = 3x^2 + 3y^2 + 3 = 3(x^2 + y^2 + 1)$$

and so by Gauss's Theorem and a shift to cylindrical coordinates,

$$\iint_{\partial S} \mathbf{F} \cdot \mathbf{n}\, dS = 3 \iiint_{S} (x^2 + y^2 + 1)\, dV$$

$$= 3 \int_0^{2\pi} \int_0^2 \int_0^3 (r^2 + 1)r\, dz\, dr\, d\theta$$

$$= 9 \int_0^{2\pi} \int_0^2 (r^3 + r)\, dr\, d\theta$$

$$= 9 \int_0^{2\pi} 6\, d\theta = 108\pi \quad \blacksquare$$

FIGURE 4

EXTENSIONS AND APPLICATIONS So far we have implicitly assumed that the solid S has no holes in its interior and that its boundary ∂S consists of one connected surface. In fact, Gauss's Theorem holds for a solid with holes, like a chunk of Swiss cheese, provided we always require $\mathbf{n}$ to point away from the interior of the solid. For example, let S be the solid shell between two concentric spheres centered at the origin. Gauss's Theorem applies, provided we recognize that ∂S now consists of two surfaces (an outer surface where $\mathbf{n}$ points away from the origin and an inner surface where $\mathbf{n}$ points toward the origin).

EXAMPLE 4 Let S be the solid determined by

$$1 \le x^2 + y^2 + z^2 \le 4$$

and let $\mathbf{F} = x\mathbf{i} + (2y + z)\mathbf{j} + (z + x^2)\mathbf{k}$. Evaluate

$$\iint\limits_{\partial S} \mathbf{F} \cdot \mathbf{n} \, dS$$

Solution

$$\iint\limits_{\partial S} \mathbf{F} \cdot \mathbf{n} \, dS = \iiint\limits_{S} \text{div } \mathbf{F} \, dV$$

$$= \iiint\limits_{S} (1 + 2 + 1) \, dV$$

$$= 4 \left[\frac{4}{3}\pi(2^3) - \frac{4}{3}\pi(1^3) \right] = \frac{112\pi}{3} \qquad \blacksquare$$

Recall from Section 17.1 that the gravitational field $\mathbf{F}$ due to a point mass M at the origin has the form

$$\mathbf{F}(x, y, z) = -cM\frac{\mathbf{r}}{|\mathbf{r}|^3}$$

where $\mathbf{r} = x\mathbf{i} + y\mathbf{j} + z\mathbf{k}$ and c is a constant.

EXAMPLE 5 Let S be a solid region containing a point mass M at the origin in its interior and with the corresponding field $\mathbf{F} = -cM\mathbf{r}/|\mathbf{r}|^3$. Show that the flux of $\mathbf{F}$ across ∂S is $-4\pi cM$, regardless of the shape of S.

Solution Since $\mathbf{F}$ is discontinuous at the origin, Gauss's Theorem does not apply directly. However, let us imagine that a small solid sphere S_a centered at the origin and of radius a has been removed from S, leaving a solid W with outer boundary ∂S and inner boundary ∂S_a (Figure 5). When we apply Gauss's Theorem to W, we get

$$\iint\limits_{\partial S} \mathbf{F} \cdot \mathbf{n} \, dS + \iint\limits_{\partial S_a} \mathbf{F} \cdot \mathbf{n} \, dS = \iint\limits_{\partial W} \mathbf{F} \cdot \mathbf{n} \, dS$$

$$= \iiint\limits_{W} \text{div } \mathbf{F} \, dV$$

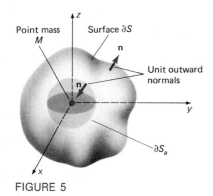

Point mass M

Surface ∂S

$\mathbf{n}$

Unit outward normals

$\mathbf{n}$

∂S_a

FIGURE 5

But div $\mathbf{F} = 0$, which is easy to check (Problem 21 of Section 17.1), and so

$$\iint\limits_{\partial S} \mathbf{F} \cdot \mathbf{n} \, dS = -\iint\limits_{\partial S_a} \mathbf{F} \cdot \mathbf{n} \, dS$$

CHAPTER 17/Vector Calculus

On the surface ∂S_a, $\mathbf{n} = -\mathbf{r}/|\mathbf{r}|$ and $|\mathbf{r}| = a$. Consequently,

$$-\iint_{\partial S_a} \mathbf{F} \cdot \mathbf{n}\, dS = -\iint_{\partial S_a} \left(-cM \frac{\mathbf{r}}{|\mathbf{r}|^3} \right) \cdot \left(-\frac{\mathbf{r}}{|\mathbf{r}|} \right) dS$$

$$= -cM \iint_{\partial S_a} \frac{\mathbf{r} \cdot \mathbf{r}}{a^4}\, dS$$

$$= -cM \iint_{\partial S_a} \frac{1}{a^2}\, dS$$

$$= \frac{-cM}{a^2} (4\pi a^2) = -4\pi cM \qquad \blacksquare$$

We can extend the result of Example 5 to the case where a solid S contains k point masses $M_1, M_2, \ldots, M_k$ in its interior. The result, known as *Gauss's Law*, gives the flux of $\mathbf{F}$ across ∂S as

$$\iint_{\partial S} \mathbf{F} \cdot \mathbf{n}\, dS = -4\pi c(M_1 + M_2 + \cdots + M_k)$$

Finally, Gauss's Law can be extended to a body B with a continuously distributed mass of size M by subdividing it into small pieces and approximating these pieces by point masses. The result is

$$\iint_{\partial S} \mathbf{F} \cdot \mathbf{n}\, dS = -4\pi cM$$

for any region S containing B.

PROBLEM SET 17.6

In Problems 1–10, use Gauss's Divergence Theorem to calculate $\iint_{\partial S} \mathbf{F} \cdot \mathbf{n}\, dS$.

1. $\mathbf{F}(x, y, z) = z\mathbf{i} + x\mathbf{j} + y\mathbf{k}$; S is the hemisphere $0 \leq z \leq \sqrt{9 - x^2 - y^2}$.

2. $\mathbf{F}(x, y, z) = x\mathbf{i} + 2y\mathbf{j} + 3z\mathbf{k}$; S is the cube $0 \leq x \leq 1$, $0 \leq y \leq 1$, $0 \leq z \leq 1$.

3. $\mathbf{F}(x, y, z) = x^2 yz\mathbf{i} + xy^2 z\mathbf{j} + xyz^2\mathbf{k}$; S is the box $0 \leq x \leq a$, $0 \leq y \leq b$, $0 \leq z \leq c$.

4. $\mathbf{F}(x, y, z) = 3x\mathbf{i} - 2y\mathbf{j} + 4z\mathbf{k}$; S is the sphere $x^2 + y^2 + z^2 \leq 9$.

5. $\mathbf{F}(x, y, z) = x^2\mathbf{i} + y^2\mathbf{j} + z^2\mathbf{k}$; S is the parabolic solid $0 \leq z \leq 4 - x^2 - y^2$.

6. $\mathbf{F}(x, y, z) = (x^2 + \cos yz)\mathbf{i} + (y - e^z)\mathbf{j} + (z^2 + x^2)\mathbf{k}$; S is the solid bounded by $x^2 + y^2 = 4$, $x + z = 2$, $z = 0$.

7. $\mathbf{F}(x, y, z) = (x + z^2)\mathbf{i} + (y - z^2)\mathbf{j} + x\mathbf{k}$; S is the solid $0 \leq y^2 + z^2 \leq 1$, $0 \leq x \leq 2$.

8. $\mathbf{F}(x, y, z) = x^2\mathbf{i} + y^2\mathbf{j} + z^2\mathbf{k}$; S is the solid enclosed by $x + y + z = 4$, $x = 0$, $y = 0$, $z = 0$.

9. $\mathbf{F}(x, y, z) = 2x\mathbf{i} + 3y\mathbf{j} + 4z\mathbf{k}$; S is the solid spherical shell $9 \leq x^2 + y^2 + z^2 \leq 25$.

10. $\mathbf{F}(x, y, z) = 2z\mathbf{i} + x\mathbf{j} + z^2\mathbf{k}$; S is the solid cylindrical shell $1 \le x^2 + y^2 \le 4$, $0 \le z \le 2$.

11. Let $\mathbf{F}(x, y, z) = x\mathbf{i} + y\mathbf{j} + z\mathbf{k}$ and let S be a solid to which Gauss's Divergence Theorem applies. Show that the volume of S is given by

$$V(S) = \tfrac{1}{3} \iint\limits_{\partial S} \mathbf{F} \cdot \mathbf{n}\, dS$$

12. Use the result of Problem 11 to verify the formula for the volume of a right circular cone of height h and radius a.

13. Consider the plane $ax + by + cz = d$, where a, b, c, and d are all positive. Use Problem 11 to show that the volume of the tetrahedron cut from the first octant by this plane is $dD/(3\sqrt{a^2 + b^2 + c^2})$, where D is the area of the part of the plane in the first octant.

14. Let $\mathbf{F}$ be a constant vector field. Show that

$$\iint\limits_{\partial S} \mathbf{F} \cdot \mathbf{n}\, dS = 0$$

for any "nice" solid S. What should we mean by "nice"?

15. Calculate $\iint\limits_{\partial S} \mathbf{F} \cdot \mathbf{n}\, dS$ for each of the following. Looked at the right way, all are quite easy and some are even trivial.

(a) $\mathbf{F} = (2x + yz)\mathbf{i} + 3y\mathbf{j} + z^2\mathbf{k}$; S is the ball $x^2 + y^2 + z^2 \le 1$.
(b) $\mathbf{F} = (x^2 + y^2 + z^2)^{5/3}(x\mathbf{i} + y\mathbf{j} + z\mathbf{k})$; S as in (a).
(c) $\mathbf{F} = x^2\mathbf{i} + y^2\mathbf{j} + z^2\mathbf{k}$; S is the ball $(x - 2)^2 + y^2 + z^2 \le 1$.
(d) $\mathbf{F} = x^2\mathbf{i}$; S is the cube $0 \le x \le 1$, $0 \le y \le 1$, $0 \le z \le 1$.
(e) $\mathbf{F} = (x + z)\mathbf{i} + (y + x)\mathbf{j} + (z + y)\mathbf{k}$; S is the tetrahedron cut from the first octant by the plane $3x + 4y + 2z = 12$.
(f) $\mathbf{F} = x^3\mathbf{i} + y^3\mathbf{j} + z^3\mathbf{k}$; S as in (a).
(g) $\mathbf{F} = (x\mathbf{i} + y\mathbf{j})\ln(x^2 + y^2)$; S is the solid cylinder $x^2 + y^2 \le 4$, $0 \le z \le 2$.

16. Calculate $\iint\limits_{\partial S} \mathbf{F} \cdot \mathbf{n}\, dS$. In each case, $\mathbf{r} = x\mathbf{i} + y\mathbf{j} + z\mathbf{k}$.

(a) $\mathbf{F} = \mathbf{r}/|\mathbf{r}|^3$; S is the ball $(x - 2)^2 + y^2 + z^2 \le 1$.
(b) $\mathbf{F} = \mathbf{r}/|\mathbf{r}|^3$; S is the ball $x^2 + y^2 + z^2 \le a^2$.
(c) $\mathbf{F} = \mathbf{r}/|\mathbf{r}|^2$; S as in (b).
(d) $\mathbf{F} = f(|\mathbf{r}|)\mathbf{r}$, f any scalar function; S as in (b).
(e) $\mathbf{F} = |\mathbf{r}|^n\mathbf{r}$, $n \ge 0$; S is the ball $x^2 + y^2 + z^2 \le az$ ($\rho \le a\cos\phi$ in spherical coordinates).

17. We have defined the laplacian of a scalar field by

$$\mathrm{lap}\, f = \nabla^2 f = \frac{\partial^2 f}{\partial x^2} + \frac{\partial^2 f}{\partial y^2} + \frac{\partial^2 f}{\partial z^2}$$

Show that

$$\iint\limits_{\partial S} D_\mathbf{n} f\, dS = \iiint\limits_{S} \nabla^2 f\, dV$$

Here $D_\mathbf{n} f$ is the directional derivative (Section 15.5).

18. Suppose $\nabla^2 f$ is identically zero in a region S. Show that

$$\iint\limits_{\partial S} f D_\mathbf{n} f\, dS = \iiint\limits_{S} |\nabla f|^2\, dV$$

19. Establish Green's First Identity

$$\iint\limits_{\partial S} f D_\mathbf{n} g\, dS = \iiint\limits_{S} (f\nabla^2 g + \nabla f \cdot \nabla g)\, dV$$

by applying Gauss's Divergence Theorem to $\mathbf{F} = f\nabla g$.

20. Establish Green's Second Identity

$$\iint\limits_{\partial S} (f D_\mathbf{n} g - g D_\mathbf{n} f)\, dS = \iiint\limits_{S} (f\nabla^2 g - g\nabla^2 f)\, dV$$

17.7
Stokes's Theorem

We showed in Section 17.4 that the conclusion to Green's Theorem could be written as

$$\oint_{\partial S} \mathbf{F} \cdot \mathbf{T}\, ds = \iint\limits_{S} (\mathrm{curl}\ \mathbf{F}) \cdot \mathbf{k}\, dA$$

As stated, it was a theorem for a plane set S bounded by a simple closed curve ∂S. We are going to generalize this result to the case where S is a

CHAPTER 17/Vector Calculus

FIGURE 1

curved surface in 3-space. In this form, the theorem is due to the Irish scientist George Gabriel Stokes (1819–1903).

We will need to put some restrictions on the surface S. First, we suppose that S is two-sided with a continuously varying unit normal $\mathbf{n}$ (the one-sided Möbius band of Section 17.5 is thereby eliminated from our discussion). Second, we require that the boundary ∂S be a piecewise smooth, simple closed curve, oriented consistently with $\mathbf{n}$. This means that if you stand near the edge of the surface with your head in the direction $\mathbf{n}$ and your eyes looking in the direction of the curve, the surface is to your left (Figure 1).

Theorem A

(Stokes's Theorem). Let S, ∂S, and $\mathbf{n}$ be as indicated above and suppose $\mathbf{F} = M\mathbf{i} + N\mathbf{j} + P\mathbf{k}$ is a vector field with M, N, and P having continuous first-order partial derivatives on S and its boundary ∂S. If $\mathbf{T}$ denotes the unit tangent vector to ∂S, then

$$\oint_{\partial S} \mathbf{F} \cdot \mathbf{T}\, ds = \iint_S (\operatorname{curl} \mathbf{F}) \cdot \mathbf{n}\, dS$$

EXAMPLES AND APPLICATIONS The proof of Stokes's Theorem is more appropriate for a course in advanced calculus. However, we can at least verify the theorem in an example.

EXAMPLE 1 Verify Stokes's Theorem for $\mathbf{F} = y\mathbf{i} - x\mathbf{j} + yz\mathbf{k}$ if S is the paraboloid $z = x^2 + y^2$ with the circle $x^2 + y^2 = 1$, $z = 1$ as its boundary (Figure 2).

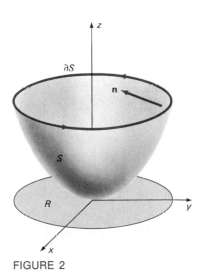

FIGURE 2

Solution We may describe ∂S by the parametric equations

$$x = \cos t, \qquad y = \sin t, \qquad z = 1$$

Then $dz = 0$ and (see Section 17.2)

$$\oint_{\partial S} \mathbf{F} \cdot \mathbf{T}\, ds = \oint_{\partial S} y\, dx - x\, dy = \int_0^{2\pi} \left[\sin t(-\sin t)\, dt - \cos t \cos t\, dt\right]$$

$$= -\int_0^{2\pi} \left[\sin^2 t + \cos^2 t\right] dt = -2\pi$$

On the other hand, to calculate $\iint_S (\operatorname{curl} \mathbf{F}) \cdot \mathbf{n}\, dS$, we first obtain

$$\operatorname{curl} \mathbf{F} = \nabla \times \mathbf{F} = \begin{vmatrix} \mathbf{i} & \mathbf{j} & \mathbf{k} \\ \dfrac{\partial}{\partial x} & \dfrac{\partial}{\partial y} & \dfrac{\partial}{\partial z} \\ y & -x & yz \end{vmatrix} = z\mathbf{i} + 0\mathbf{j} - 2\mathbf{k}$$

Then, by Theorem 17.5B,

$$\iint\limits_{S} (\text{curl } \mathbf{F}) \cdot \mathbf{n} \, dS = \iint\limits_{R} \left[-z(2x) - 0(2y) - 2 \right] dx \, dy$$

$$= -2 \iint\limits_{R} \left[xz + 1 \right] dx \, dy$$

$$= -2 \iint\limits_{R} \left[x(x^2 + y^2) + 1 \right] dx \, dy$$

$$= -2 \int_{0}^{2\pi} \int_{0}^{1} \left[r^3 \cos \theta + 1 \right] r \, dr \, d\theta$$

$$= -2 \int_{0}^{2\pi} \left[\tfrac{1}{5} \cos \theta + \tfrac{1}{2} \right] d\theta = -2\pi \qquad \blacksquare$$

EXAMPLE 2   Let S be the part of the spherical surface $x^2 + y^2 + (z - 4)^2 = 10$ below the plane $z = 1$, and let $\mathbf{F} = y\mathbf{i} - x\mathbf{j} + yz\mathbf{k}$. Use Stokes's Theorem to calculate

$$\iint\limits_{S} (\text{curl } \mathbf{F}) \cdot \mathbf{n} \, dS$$

where $\mathbf{n}$ is the upward unit normal.

Solution   Note that the field $\mathbf{F}$ is the same as that of Example 1 and also that S has the same circle as its boundary curve. We conclude that

$$\iint\limits_{S} (\text{curl } \mathbf{F}) \cdot \mathbf{n} \, dS = \oint_{\partial S} \mathbf{F} \cdot \mathbf{n} \, ds = -2\pi$$

In fact, we conclude that the flux of curl $\mathbf{F}$ is -2π for all surfaces S that have the circle ∂S of Figure 2 as their oriented boundary. $\qquad \blacksquare$

EXAMPLE 3   Use Stokes's Theorem to evaluate $\oint_{C} \mathbf{F} \cdot \mathbf{T} \, ds$, where $\mathbf{F} = 2z\mathbf{i} + (8x - 3y)\mathbf{j} + (3x + y)\mathbf{k}$ and C is the triangular curve of Figure 3.

Solution   We could let S be any surface with C as its oriented boundary, but it is to our advantage to choose the simplest such surface—namely, the flat planar triangle T. To determine $\mathbf{n}$ for this surface, we note that the vectors

$$\mathbf{A} = (0 - 1)\mathbf{i} + (0 - 0)\mathbf{j} + (2 - 0)\mathbf{k} = -\mathbf{i} + 2\mathbf{k}$$
$$\mathbf{B} = (0 - 1)\mathbf{i} + (1 - 0)\mathbf{j} + (0 - 0)\mathbf{k} = -\mathbf{i} + \mathbf{j}$$

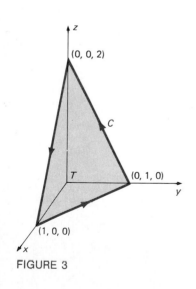

FIGURE 3

CHAPTER 17/Vector Calculus

lie on this surface and hence

$$N = A \times B = \begin{vmatrix} \mathbf{i} & \mathbf{j} & \mathbf{k} \\ -1 & 0 & 2 \\ -1 & 1 & 0 \end{vmatrix} = -2\mathbf{i} - 2\mathbf{j} - \mathbf{k}$$

is perpendicular to it. The upward unit normal $\mathbf{n}$ is therefore

$$\mathbf{n} = \frac{2\mathbf{i} + 2\mathbf{j} + \mathbf{k}}{\sqrt{4 + 4 + 1}} = \frac{2}{3}\mathbf{i} + \frac{2}{3}\mathbf{j} + \frac{1}{3}\mathbf{k}$$

Also

$$\text{curl } \mathbf{F} = \begin{vmatrix} \mathbf{i} & \mathbf{j} & \mathbf{k} \\ \dfrac{\partial}{\partial x} & \dfrac{\partial}{\partial y} & \dfrac{\partial}{\partial z} \\ 2z & 8x - 3y & 3x + y \end{vmatrix} = \mathbf{i} - \mathbf{j} + 8\mathbf{k}$$

and curl $\mathbf{F} \cdot \mathbf{n} = \frac{8}{3}$. We conclude that

$$\oint_C \mathbf{F} \cdot \mathbf{T}\, ds = \iint_T (\text{curl } \mathbf{F}) \cdot \mathbf{n}\, dS = \tfrac{8}{3}(\text{area of } T) = \tfrac{8}{3}(\tfrac{3}{2}) = 4 \qquad \blacksquare$$

EXAMPLE 4 Let the vector field $\mathbf{F}$ and the region D satisfy the hypotheses of Theorem 17.3C. Show that if curl $\mathbf{F} = \mathbf{0}$ in D, then $\mathbf{F}$ is conservative there.

Solution From the discussion in Section 17.3, we conclude that it is enough to show that $\oint_C \mathbf{F} \cdot d\mathbf{r} = 0$ for any simple closed path C in D. Let S be a surface having C as its boundary and oriented consistently with C (the simple connectedness of D can be shown to guarantee the existence of such a surface). Then from Stokes's Theorem,

$$\oint_C \mathbf{F} \cdot d\mathbf{r} = \oint_C \mathbf{F} \cdot \mathbf{T}\, ds = \iint_S (\text{curl } \mathbf{F}) \cdot \mathbf{n}\, dS = 0 \qquad \blacksquare$$

PHYSICAL INTERPRETATION OF THE CURL We offered an interpretation of the curl in Section 17.4. Now we can amplify that discussion. Let C be a circle of radius a centered at the point P. Then

$$\oint_C \mathbf{F} \cdot \mathbf{T}\, ds$$

FIGURE 4

is called the *circulation* of **F** around C and measures the tendency of a fluid with velocity field **F** to circulate around C. Now if **F** is continuous and C is very small, Stokes's Theorem gives

$$\oint_C \mathbf{F} \cdot \mathbf{T}\, ds = \iint_S (\operatorname{curl} \mathbf{F}) \cdot \mathbf{n}\, dS \approx [\operatorname{curl} \mathbf{F}(P)] \cdot \mathbf{n}\,(\pi a^2)$$

The expression on the right will have the largest magnitude if **n** has the same direction as curl $\mathbf{F}(P)$.

Suppose a small paddle wheel is placed in the fluid with center at P and axis having direction **n** (Figure 4). This wheel will rotate most rapidly if **n** has the direction of curl **F**. The direction of rotation will be that determined by the right-hand rule.

PROBLEM SET 17.7

In Problems 1–6, use Stokes's Theorem to calculate

$$\iint_S (\operatorname{curl} \mathbf{F}) \cdot \mathbf{n}\, dS.$$

1. $\mathbf{F} = x^2\mathbf{i} + y^2\mathbf{j} + z^2\mathbf{k}$; S is the hemisphere $z = \sqrt{1 - x^2 - y^2}$ and **n** is the upper normal.

2. $\mathbf{F} = xy\mathbf{i} + yz\mathbf{j} + xz\mathbf{k}$; S is the triangular surface with vertices $(0, 0, 0)$, $(1, 0, 0)$, and $(0, 2, 1)$ and **n** is the upper normal.

3. $\mathbf{F} = (y + z)\mathbf{i} + (x^2 + z^2)\mathbf{j} + y\mathbf{k}$; S is the half-cylinder $z = \sqrt{1 - x^2}$ between $y = 0$ and $y = 1$ and **n** is the upper normal.

4. $\mathbf{F} = xz^2\mathbf{i} + x^3\mathbf{j} + \cos xz\mathbf{k}$; S is the part of the ellipsoid $x^2 + y^2 + 3z^2 = 1$ below the xy-plane and **n** is the lower normal.

5. $\mathbf{F} = yz\mathbf{i} + 3xz\mathbf{j} + z^2\mathbf{k}$; S is the part of the sphere $x^2 + y^2 + z^2 = 16$ below the plane $z = 2$ and **n** is the outward normal.

6. $\mathbf{F} = (z - y)\mathbf{i} + (z + x)\mathbf{j} - (x + y)\mathbf{k}$; S is the part of the paraboloid $z = 1 - x^2 - y^2$ above the xy-plane and **n** is the upward normal.

In Problems 7–12, use Stokes's Theorem to calculate $\oint_C \mathbf{F} \cdot \mathbf{T}\, ds$.

7. $\mathbf{F} = 2z\mathbf{i} + x\mathbf{j} + 3y\mathbf{k}$; C is the ellipse that is the intersection of the plane $z = x$ and the cylinder $x^2 + y^2 = 4$, oriented clockwise as viewed from above.

8. $\mathbf{F} = y\mathbf{i} + z\mathbf{j} + x\mathbf{k}$; C is the triangular curve with vertices $(0, 0, 0)$, $(2, 0, 0)$, and $(0, 2, 2)$, oriented counterclockwise as viewed from above.

9. $\mathbf{F} = (y - x)\mathbf{i} + (x - z)\mathbf{j} + (x - y)\mathbf{k}$; C is the boundary of the plane $x + 2y + z = 2$ in the first octant, oriented clockwise as viewed from above.

10. $\mathbf{F} = y(x^2 + y^2)\mathbf{i} - x(x^2 + y^2)\mathbf{j}$; C is the rectangular path from $(0, 0, 0)$ to $(1, 0, 0)$ to $(1, 1, 1)$ to $(0, 1, 1)$ to $(0, 0, 0)$.

11. $\mathbf{F} = (z - y)\mathbf{i} + y\mathbf{j} + x\mathbf{k}$; C is the intersection of the cylinder $x^2 + y^2 = x$ with the sphere $x^2 + y^2 + z^2 = 1$, oriented counterclockwise as viewed from above.

12. $\mathbf{F} = (y - z)\mathbf{i} + (z - x)\mathbf{j} + (x - y)\mathbf{k}$; C is the ellipse which is the intersection of the plane $x + z = 1$ and the cylinder $x^2 + y^2 = 1$, oriented clockwise as viewed from above.

13. Suppose the surface S is determined by the formula $z = g(x, y)$. Show that the surface integral in Stokes's Theorem can be written as a double integral in the following way.

$$\iint_S (\operatorname{curl} \mathbf{F}) \cdot \mathbf{n}\, dS = \iint_{S_{xy}} (\operatorname{curl} \mathbf{F}) \cdot (-g_x\mathbf{i} - g_y\mathbf{j} + \mathbf{k})\, dA$$

where **n** is the upward normal to S and S_{xy} is the projection of S in the xy-plane.

14. Let $\mathbf{F} = x^2\mathbf{i} - 2xy\mathbf{j} + yz^2\mathbf{k}$ and ∂S be the boundary of the surface $z = xy$, $0 \le x \le 1$, $0 \le y \le 1$, oriented counterclockwise as viewed from above. Use Stokes's Theorem and Problem 13 to evaluate $\oint_{\partial S} \mathbf{F} \cdot \mathbf{T}\, ds$.

15. Let $\mathbf{F} = 2\mathbf{i} + xz\mathbf{j} + z^3\mathbf{k}$ and ∂S be the boundary of the surface $z = xy^2$, $0 \le x \le 1$, $0 \le y \le 1$, oriented counterclockwise as viewed from above. Evaluate $\oint_{\partial S} \mathbf{F} \cdot \mathbf{T}\, ds$.

16. Let $\mathbf{F} = 2\mathbf{i} + xz\mathbf{j} + z^3\mathbf{k}$ and ∂S be the boundary of the surface $z = x^2y^2$, $x^2 + y^2 \le a^2$, oriented counterclockwise as viewed from above. Evaluate $\oint_{\partial S} \mathbf{F} \cdot \mathbf{T}\, ds$.

17. Let $\mathbf{F} = 2z\mathbf{i} + 2y\mathbf{k}$ and let ∂S be the intersection of the cylinder $x^2 + y^2 = ay$ with the hemisphere $z = \sqrt{a^2 - x^2 - y^2}$, $a > 0$. Assuming distances in meters and force in newtons, find the work done by the force $\mathbf{F}$ in moving an object around ∂S in the counterclockwise direction as viewed from above.

18. A central force is one of the form $\mathbf{F} = f(|\mathbf{r}|)\mathbf{r}$, where f has a continuous derivative (except possibly at $|\mathbf{r}| = 0$). Show that the work done by such a force in moving an object around a closed path that misses the origin is 0.

19. Let S be a solid sphere (or any solid enclosed by a "nice" surface ∂S). Show that

$$\iint_{\partial S} (\text{curl } \mathbf{F}) \cdot \mathbf{n}\, dS = 0$$

(a) by using Stokes's Theorem;
(b) by using Gauss's Theorem. *Hint:* Show div(curl $\mathbf{F}$) = 0.

20. Show

$$\oint_{\partial S} (f\nabla g) \cdot \mathbf{T}\, ds = \iint_S (\nabla f \times \nabla g) \cdot \mathbf{n}\, dS.$$

17.8 Chapter Review

TRUE-FALSE QUIZ

Respond with true or false to each of the following assertions. Be prepared to defend your answer.

1. Inverse square law fields are conservative.
2. The divergence of a vector field is another vector field.
3. A physicist might be interested in both curl(grad f) and grad(curl $\mathbf{F}$).
4. If f has continuous second partial derivatives, then curl(grad f) = $\mathbf{0}$.
5. The work done by a conservative force field as it moves an object around a closed path is zero.
6. If $\int_C \mathbf{F}(\mathbf{r}) \cdot d\mathbf{r} = 0$ for every closed path in an open connected set D, then there is a function f such that $\nabla f = \mathbf{F}$ in D.
7. The field $\mathbf{F}(x, y, z) = (2x + 2y)\mathbf{i} + 2x\mathbf{j} + yz^2\mathbf{k}$ is conservative.
8. Green's theorem holds for a region S with a hole provided the complete boundary of S is oriented correctly.
9. The double integral is a special case of a surface integral.
10. A surface always has two sides.
11. If there are no sources or sinks in a region, then the net flow across the boundary of the region is zero.
12. If S is a sphere with outward normal $\mathbf{n}$ and $\mathbf{F}$ is a constant vector field, then

$$\iint_S (\mathbf{F} \cdot \mathbf{n})\, dS = 0$$

SAMPLE TEST PROBLEMS

1. Sketch a sample of vectors from the vector field $\mathbf{F}(x, y) = x\mathbf{i} + 2y\mathbf{j}$.

2. Find div $\mathbf{F}$, curl $\mathbf{F}$, grad(div $\mathbf{F}$), and div(curl $\mathbf{F}$) if $\mathbf{F}(x, y, z) = 2xyz\mathbf{i} - 3y^2\mathbf{j} + 2y^2z\mathbf{k}$.

3. We showed in Problem 20, Section 17.1, that

$$\text{curl}(f\mathbf{F}) = f(\text{curl } \mathbf{F}) + \nabla f \times \mathbf{F}$$

and in Problem 17, Section 17.3, that

$$\text{curl}(\nabla f) = \mathbf{0}$$

Use these facts to show that

$$\text{curl}(f\nabla f) = \mathbf{0}$$

4. Find a function f satisfying

(a) $\nabla f = (2xy + y)\mathbf{i} + (x^2 + x + \cos y)\mathbf{j}$;
(b) $\nabla f = (yz - e^{-x})\mathbf{i} + (xz + e^y)\mathbf{j} + xy\mathbf{k}$.

5. Calculate.

(a) $\int_C (1 - y^2)\,ds$; C is the quarter circle from $(0, -1)$ to $(1, 0)$, centered at the origin.

(b) $\int_C xy\,dx + z\cos x\,dy + z\,dz$; C is the curve $x = t$, $y = \cos t$, $z = \sin t$, $0 \le t \le \pi/2$.

6. Show that $\int_C y^2\,dx + 2xy\,dy$ is independent of path and use this to calculate the integral on any path from $(0, 0)$ to $(1, 2)$.

7. Find the work done by $\mathbf{F} = y^2\mathbf{i} + 2xy\mathbf{j}$ in moving an object from $(1, 1)$ to $(3, 4)$ (see Problem 6).

8. Evaluate (see Problem 4b).

$$\int_{(0,0,0)}^{(1,1,4)} (yz - e^{-x})\,dx + (xy + e^y)\,dy + xy\,dz$$

9. Evaluate $\oint_C xy\,dx + (x^2 + y^2)\,dy$ if

(a) C is the square path $(0,0)$ to $(1,0)$ to $(1,1)$ to $(0,1)$ to $(0,0)$;
(b) C is the triangular path $(0,0)$ to $(2,0)$ to $(2,1)$ to $(0,0)$;
(c) C is the circle $x^2 + y^2 = 1$ traversed in the clockwise direction.

10. Calculate the flux of $\mathbf{F} = x\mathbf{i} + y\mathbf{j}$ across the square curve C with vertices $(1, 1)$, $(-1, 1)$, $(-1, -1)$, and $(1, -1)$; that is, calculate $\oint_C \mathbf{F} \cdot \mathbf{n}\,ds$.

11. Calculate the flux of $\mathbf{F} = x\mathbf{i} + y\mathbf{j} + 3\mathbf{k}$ across the sphere $x^2 + y^2 + z^2 = 1$.

12. Evaluate $\iint_G xyz\,dS$, where G is the part of the plane $z = x + y$ above the triangular region with vertices $(0, 0, 0)$, $(1, 0, 0)$, and $(0, 2, 0)$.

13. Evaluate $\iint_G (\text{curl } \mathbf{F}) \cdot \mathbf{n}\,dS$, where

$$\mathbf{F} = x^3 y\mathbf{i} + e^y\mathbf{j} + z\tan\left(\frac{xyz}{4}\right)\mathbf{k}$$

and G is the part of the sphere $x^2 + y^2 + z^2 = 2$ above the plane $z = 1$ and $\mathbf{n}$ is the upward unit normal.

14. Evaluate $\iint_G \mathbf{F} \cdot \mathbf{n}\,dS$, where

$$\mathbf{F} = \sin x\mathbf{i} + (1 - \cos x)y\mathbf{j} + 4z\mathbf{k}$$

and G is the closed surface bounded by $z = \sqrt{9 - x^2 - y^2}$, and $z = 0$ with outward unit normal $\mathbf{n}$.

15. Let C be the circle that is the intersection of the plane $ax + by + z = 0$ ($a \ge 0$, $b \ge 0$) and the sphere $x^2 + y^2 + z^2 = 9$. For $\mathbf{F} = y\mathbf{i} - x\mathbf{j} + 3y\mathbf{k}$, evaluate

$$\oint_C \mathbf{F} \cdot \mathbf{T}\,ds$$

Hint: Use Stokes's Theorem.

18 Differential Equations

Perhaps the most dazzling mathematical genius to surface among women during the past two centuries was the highly gifted Russian, Sonya Corvin-Krukovsky Kovalevsky.
Lynn M. Osen, 1974

Sonya Kovalevsky
1850–1891

Three reasons dictated our choice of Sonya Kovalevsky to grace this final chapter. First, she epitomizes the bias against women that mars the history of mathematics. Second, she represents the great Russian mathematical tradition. Third, she contributed to the theory of differential equations.

Kovalevsky's first exposure to calculus came from the wallpaper of her room—sheets of lithographed lectures on differential and integral calculus. When she took her first formal calculus lessons at age 15, her teacher was astonished at her quickness, but she already knew all of the symbols and some of the concepts.

Prevented from attending Russian universities (which were closed to women), Sonya entered a platonic marriage to a paleontology student, with the sole purpose of gaining freedom to travel. Thus, in 1868 she and her husband moved to Heidelberg, Germany, where she studied mathematics. Learning of the great master Karl Weierstrass, she decided to go to Berlin, only to be advised that women were not accepted at the university there. Fortunately, Weierstrass agreed to give her private lessons and supervised her doctoral work on differential equations.

Unable to find a position in either Germany or Russia, she happily accepted Mittag-Leffler's offer of a lectureship at the newly founded University of Stockholm. There she distinguished herself as a teacher and a researcher. In 1888, she won the prestigious Prix Bordin of the French Academy of Sciences in a competition for papers submitted anonymously. At last her own country recognized her; she was the first woman to be elected a corresponding member of the Russian Academy of Sciences.

18.1
Linear First-Order Equations

We call an equation involving one or more derivatives of an unknown function a **differential equation**. In particular, an equation of the form

$$F(x, y, y^{(1)}, y^{(2)}, \ldots, y^{(n)}) = 0$$

in which $y^{(k)}$ denotes the kth derivative of y with respect to x, is called an **ordinary differential equation of order** n. Examples of order 1, 2, and 3 are

$$y' + 2 \sin x = 0$$

$$\frac{d^2 y}{dx^2} + 3x \frac{dy}{dx} - 2y = 0$$

$$\frac{d^3 y}{dx^3} + \left(\frac{dy}{dx}\right)^2 - e^x = 0$$

If, when $f(x)$ is substituted for y in the differential equation, the resulting equation is an identity for all x in some interval, then $f(x)$ is called a **solution** of the differential equation. Thus, $f(x) = 2 \cos x + 10$ is a solution to $y' + 2 \sin x = 0$ since

$$f'(x) + 2 \sin x = -2 \sin x + 2 \sin x = 0$$

for all x. We call $2 \cos x + C$ the **general solution** of the given equation, since it can be shown that every solution can be written in this form. In contrast, $2 \cos x + 10$ is called a **particular solution** of the equation.

Differential equations appeared earlier in this book, principally in two sections. In Section 5.2, we introduced the technique called *separation of variables* and used it to solve a wide variety of first-order equations. In Section 7.5, we solved the differential equation $y' = ky$ of exponential growth and decay.

In this chapter we consider only **linear** differential equations—that is, equations of the form

$$y^{(n)} + a_1(x)y^{(n-1)} + \cdots + a_{n-1}(x)y' + a_n(x)y = k(x)$$

(Note that y and all its derivatives occur to the first power.) This is called a linear equation because if it is written in operator notation,

$$[D_x^n + a_1(x)D_x^{n-1} + \cdots + a_{n-1}(x)D_x + a_n(x)]y = k(x)$$

the operator in brackets is a linear operator. Thus, if L denotes this operator and if f and g are functions and c is constant,

$$L(f + g) = L(f) + L(g)$$
$$L(cf) = cL(f)$$

That L has these properties follows readily from the corresponding properties for the derivative operators $D, D^2, \ldots, D^n$.

Of course, not all operators are linear; in fact, many important differential equations are nonlinear. A simple example is

$$\frac{dy}{dx} + y^2 = 0$$

The presence of the exponent 2 on y is enough to spoil the linearity, as you may check. The theory of nonlinear differential equations is both complicated and fascinating, but best left for more advanced courses.

THE GENERAL FIRST-ORDER LINEAR EQUATION   The equations we now consider can be put in the form

$$\frac{dy}{dx} + P(x)y = Q(x)$$

An equation of this type can always be solved in principle. We first multiply both sides by the **integrating factor**

$$e^{\int P(x)\,dx}$$

obtaining

$$e^{\int P(x)\,dx}\frac{dy}{dx} + e^{\int P(x)\,dx}P(x)y = Q(x)e^{\int P(x)\,dx}$$

Then we recognize the left side as the derivative of $ye^{\int P(x)\,dx}$, so that the equation takes the form

$$\frac{d}{dx}\left(ye^{\int P(x)\,dx}\right) = Q(x)e^{\int P(x)\,dx}$$

Integration of both sides yields

$$ye^{\int P(x)\,dx} = \int \left(Q(x)e^{\int P(x)\,dx}\right)dx$$

and thus

$$y = e^{-\int P(x)\,dx}\int \left(Q(x)e^{\int P(x)\,dx}\right)dx$$

It is not worth memorizing this final result; the process of getting there is easily recalled, and that is what we illustrate.

EXAMPLE 1 Solve

$$\frac{dy}{dx} + \frac{2}{x}\,y = \frac{\sin 3x}{x^2}$$

Solution Our integrating factor is

$$e^{\int (2/x)\,dx} = e^{2\ln|x|} = e^{\ln x^2} = x^2$$

(We take the arbitrary constant from the integration $\int P(x)\,dx$ to be 0, since it would cancel out in the end anyway.) Multiplying both sides of the original equation by x^2, we obtain

$$x^2 \frac{dy}{dx} + 2xy = \sin 3x$$

or, equivalently,

$$\frac{d}{dx}(x^2 y) = \sin 3x$$

Integration of both members yields

$$x^2 y = \int \sin 3x\,dx = -\tfrac{1}{3}\cos 3x + C$$

or

$$y = (-\tfrac{1}{3}\cos 3x + C)x^{-2} \qquad\blacksquare$$

EXAMPLE 2 Find the particular solution of

$$\frac{dy}{dx} - 3y = xe^{3x}$$

that satisfies $y = 4$ when $x = 0$.

Solution The appropriate integrating factor is

$$e^{\int -3\,dx} = e^{-3x}$$

Upon multiplication by this factor, our equation takes the form

$$\frac{d}{dx}(e^{-3x}y) = x$$

or

$$e^{-3x}y = \int x\,dx = \frac{1}{2}x^2 + C$$

Thus, the general solution is

$$y = \frac{1}{2} x^2 e^{3x} + C e^{3x}$$

Substitution of $y = 4$ when $x = 0$ makes $C = 4$. The desired particular solution is

$$y = \frac{1}{2} x^2 e^{3x} + 4 e^{3x}$$

∎

APPLICATIONS   We begin with a mixture problem, typical of many problems that arise in chemistry.

EXAMPLE 3   A tank initially contains 120 gallons of brine, holding 75 pounds of dissolved salt in solution. Salt water containing 1.2 pounds of salt per gallon is entering the tank at the rate of 2 gallons per minute and brine flows out at the same rate (Figure 1). If the mixture is kept uniform by constant stirring, find the amount of salt in the tank at the end of 1 hour.

FIGURE 1

Solution   Let y be the number of pounds of salt in the tank at the end of t minutes. From the brine flowing in, the tank gains 2.4 pounds of salt per minute; from that flowing out it loses $\frac{2}{120}y$ pounds per minute. Thus,

$$\frac{dy}{dt} = 2.4 - \frac{1}{60} y$$

subject to the condition $y = 75$ when $t = 0$. The equivalent equation

$$\frac{dy}{dt} + \frac{1}{60} y = 2.4$$

has the integrating factor $e^{t/60}$, and so

$$\frac{d}{dt} \left[y e^{t/60} \right] = 2.4 e^{t/60}$$

We conclude that

$$y e^{t/60} = \int 2.4 e^{t/60} \, dt = (60)(2.4) e^{t/60} + C$$

Substituting $y = 75$ when $t = 0$ yields $C = -69$, and so

$$y = e^{-t/60} \left[144 e^{t/60} - 69 \right] = 144 - 69 e^{-t/60}$$

At the end of 1 hour ($t = 60$),

$$y = 144 - 69 e^{-1} \approx 118.62 \text{ pounds}$$

Note that the limiting value for y as $t \to \infty$ is 144. This corresponds to the fact that the tank will ultimately take on the complexion of the brine entering the tank. One hundred twenty gallons of brine with a concentration of 1.2 pounds of salt per gallon will contain 144 pounds of salt. ∎

We turn next to an example from electricity. According to Kirchhoff's Law, a simple electrical circuit (Figure 2) containing a resistor with a resistance of R ohms and an inductor with an inductance of L henrys in series with a source of electromotive force (a battery or generator) that supplies a voltage of $E(t)$ volts at time t satisfies

$$L\frac{dI}{dt} + RI = E(t)$$

where I is the current measured in amperes. This is a linear equation, easily solved by the method of this section.

FIGURE 2

EXAMPLE 4   Consider a circuit (Figure 2) with $L = 2$ henrys, $R = 6$ ohms, and a battery supplying a constant voltage of 12 volts. If $I = 0$ at $t = 0$ (when the switch S is closed), find I at time t.

Solution   The differential equation is

$$2\frac{dI}{dt} + 6I = 12, \quad \text{or} \quad \frac{dI}{dt} + 3I = 6$$

Following our standard procedure (multiply by the integrating factor e^{3t}, integrate, and multiply by e^{-3t}), we obtain

$$I = e^{-3t}(2e^{3t} + C) = 2 + Ce^{-3t}$$

The initial condition, $I = 0$ at $t = 0$, gives $C = -2$; hence

$$I = 2 - 2e^{-3t}$$

As t increases, the current tends toward a current of 2 amps. ∎

EXAMPLE 5   If the battery of Example 4 is replaced by an alternating current generator that supplies a voltage of $E(t) = 12 \sin 9t$ volts, find I at time t and note its behavior for large t.

Solution   The differential equation may be written as

$$\frac{dI}{dt} + 3I = 6 \sin 9t$$

Multiplying by e^{3t} gives

$$\frac{d}{dt}(Ie^{3t}) = 6e^{3t} \sin 9t$$

By Formula 67 from the integral tables at the end of the book,

$$Ie^{3t} = 6 \int e^{3t} \sin 9t \, dt$$

$$= \frac{6e^{3t}}{9 + 81} (3 \sin 9t - 9 \cos 9t) + C$$

Thus,

$$I = \tfrac{1}{5} \sin 9t - \tfrac{3}{5} \cos 9t + Ce^{-3t}$$

Next, the initial condition $I = 0$ at $t = 0$ gives $C = \tfrac{3}{5}$. Finally, we obtain

$$I = \tfrac{1}{5} \sin 9t - \tfrac{3}{5} \cos 9t + \tfrac{3}{5}e^{-3t}$$

For large t, e^{-3t} is negligible, and

$$I \approx \tfrac{1}{5} \sin 9t - \tfrac{3}{5} \cos 9t$$

which is an alternating current with the same frequency as the imposed voltage. ∎

PROBLEM SET 18.1

In Problems 1–14, solve the differential equations.

1. $\dfrac{dy}{dx} + y = e^{-x}$

2. $(x + 1)\dfrac{dy}{dx} + y = x^2 - 1$

3. $(1 - x^2)\dfrac{dy}{dx} + xy = ax, \ |x| < 1$

4. $y' + y \tan x = \sec x$

5. $\dfrac{dy}{dx} - \dfrac{y}{x} = xe^x$

6. $y' - ay = f(x)$

7. $\dfrac{dy}{dx} + \dfrac{y}{x} = \dfrac{1}{x}$

8. $y' + \dfrac{2y}{x + 1} = (x + 1)^3$

9. $y' + yf(x) = f(x)$

10. $\dfrac{dy}{dx} + 2y = x$

11. $\dfrac{dy}{dx} - \dfrac{y}{x} = 3x^3; \ y = 3$ when $x = 1$.

12. $y' = e^{2x} - 3y; \ y = 1$ when $x = 0$.

13. $xy' + (1 + x)y = e^{-x}; \ y = 0$ when $x = 1$.

14. $\sin x \dfrac{dy}{dx} + 2y \cos x = \sin 2x; \ y = 2$ when $x = \dfrac{\pi}{6}$.

15. A tank contains 20 gallons of a solution, with 10 pounds of chemical A in the solution. At a certain instant, we begin pouring in a solution containing the same chemical in a concentration of 2 pounds per gallon. We pour at a rate of 3 gallons per minute while simultaneously draining off the resulting (well-stirred) solution at the same rate. Find the amount of chemical A in the tank after 20 minutes.

16. A tank initially contains 200 gallons of brine, with 50 pounds of salt in solution. Brine containing 2 pounds of salt per gallon is entering the tank at the rate of 4 gallons per minute and is flowing out at the same rate. If the mixture in the tank is kept uniform by constant stirring, find the amount of salt in the tank at the end of 40 minutes.

17. A tank initially contains 120 gallons of pure water. Brine with 1 pound of salt per gallon flows into the tank at 4 gallons per minute, and the well-stirred solution runs out at 6 gallons per minute. How much salt is in the tank after t minutes, $0 \le t \le 60$?

18. A tank initially contains 50 gallons of brine, with 30 pounds of salt in solution. Water runs into the tank at 3 gallons per minute and the well-stirred solution runs out at 2 gallons per minute. How long will it be until there are 25 pounds of salt in the tank?

19. Find the current I as a function of time in the circuit of Figure 3 if the switch S is closed and $I = 0$ at $t = 0$.

$R = 10^6 \ \Omega$

S

$L = 1 \ \text{H}$

$E = 1 \ \text{V}$

FIGURE 3

20. Find I as a function of time in the circuits of Figure 4, assuming the switch is closed and $I = 0$ at $t = 0$.

S

$L = 3.5 \ \text{H}$

$E = 120 \ \sin 377t$

(a)

S

$R = 1,000 \ \Omega$

$E = 120 \ \sin 377t$

(b)

FIGURE 4

21. Using Figure 5, find (a) I as a function of time t, assuming the switch is closed and $I = 0$ at $t = 0$, (b) the steady-state current as a function of time (let $t \to \infty$).

$R = 1000 \ \Omega$

$L = 3.5 \ \text{H}$

$E = 120 \ \sin 377t$

FIGURE 5

22. Suppose that tank 1 initially contains 100 gallons of solution, with 50 pounds of dissolved salt, and tank 2 contains 200 gallons, with 150 pounds of dissolved salt. Pure water flows into tank 1 at 2 gallons per minute, the well-mixed solution flows out and into tank 2 at the same rate, and—finally—the solution in tank 2 drains away also at the same rate. Let $x(t)$ and $y(t)$ denote the amounts of salt in tanks 1 and 2, respectively, at time t. Find $y(t)$. *Hint:* First find $x(t)$ and use it in setting up the differential equation for tank 2.

23. A tank of capacity 100 gallons is initially full of pure alcohol. The flow rate of the drain pipe is 5 gallons per minute; the flow rate of the filler pipe can be adjusted to c gallons per minute. An unlimited amount of 25% alcohol solution can be brought in through the filler pipe. Our goal is to reduce the amount of alcohol in the tank so that it will contain 100 gallons of 50% solution. Let T be the number of minutes required to accomplish the desired change.

(a) Evaluate T if $c = 5$ and both pipes are opened.
(b) Evaluate T if $c = 5$ and we first drain away a sufficient amount of the pure alcohol and then close the drain and open the filler pipe.
(c) For what values of c (if any) would strategy (b) give a faster time than (a)?
(d) Suppose $c = 4$. Determine the equation for T if we initially open both pipes and then close the drain.
(e) Solve the equation in (d), perhaps by Newton's Method.

24. The differential equation for a body falling under the influence of gravity near the earth's surface with air resistance proportional to the velocity v is $dv/dt = -g - av$, where $g = 32$ feet per second per second is the acceleration of gravity and $a > 0$ is the drag coefficient. Show each of the following.

(a) $v(t) = (v_0 - v_\infty)e^{-at} + v_\infty$, where $v_0 = v(0)$ and

$$v_\infty = -g/a = \lim_{t \to \infty} v(t)$$

the so-called terminal velocity.
(b) If $y(t)$ denotes the altitude, then

$$y(t) = y_0 + tv_\infty + (1/a)(v_0 - v_\infty)(1 - e^{-at})$$

25. A ball was thrown straight up from ground level with an initial velocity $v_0 = 120$ feet per second. Assuming a drag coefficient of $a = 0.05$, determine each of the following.

(a) The maximum altitude.
(b) The equation for T, the time when the ball hit the ground.
(c) The value of T (perhaps by Newton's Method).

26. Andrea bailed out of her plane at an altitude of 8000 feet, fell freely for 15 seconds, and then opened her parachute. Assume the drag coefficients are $a = 0.10$ for free fall and $a = 1.6$ with parachute. When did she hit the ground?

18.2
Second-Order
Homogeneous
Equations

A second-order linear differential equation has the form

$$y'' + a_1(x)y' + a_2(x)y = k(x)$$

In this section, we make two simplifying assumptions: (1) $a_1(x)$ and $a_2(x)$ are constants, and (2) $k(x)$ is identically zero (the homogeneous case). Thus, our initial task is to solve

$$y'' + a_1y' + a_2y = 0$$

To solve a first-order equation required one integration and led to a general solution with one arbitrary constant. By analogy, we might expect solving a second-order equation to involve two integrations and thus that the general solution would have two arbitrary constants. Our expectations are correct. In fact, a second-order homogeneous linear differential equation always has two fundamental solutions $u_1(x)$ and $u_2(x)$, which are **independent** of each other (that is, neither function is a constant multiple of the other). By the linearity of the operator $D^2 + a_1D + a_2$,

$$C_1u_1(x) + C_2u_2(x)$$

is also a solution. Moreover, it can be shown that every solution has this form.

THE AUXILIARY EQUATION Because $D_x(e^{rx}) = re^{rx}$, it seems likely that e^{rx} will be a solution to our differential equation for an appropriate choice of r. To test this possibility, we first write the equation in the operator form

(1) $$(D^2 + a_1D + a_2)y = 0$$

Now

$$\begin{aligned}(D^2 + a_1D + a_2)e^{rx} &= D^2(e^{rx}) + a_1D(e^{rx}) + a_2e^{rx}\\ &= r^2e^{rx} + a_1re^{rx} + a_2e^{rx}\\ &= e^{rx}(r^2 + a_1r + a_2)\end{aligned}$$

The latter expression is zero, provided

(2) $$r^2 + a_1r + a_2 = 0$$

Equation (2) is called the **auxiliary equation** for (1) (note the similarity in form). It is an ordinary quadratic equation and can be solved by factoring or, if necessary, by the Quadratic Formula. There are three cases to consider, corresponding to whether the auxiliary equation has two distinct, real roots, a single, repeated root, or two complex conjugate roots.

EXAMPLE 1 Find the general solution to $y'' + 7y' + 12y = 0$.

Solution The auxiliary equation

$$r^2 + 7r + 12 = (r + 3)(r + 4) = 0$$

has the two roots -3 and -4. Since e^{-3x} and e^{-4x} are independent solutions, the general solution to the differential equation is

$$y = C_1 e^{-3x} + C_2 e^{-4x}$$ ■

EXAMPLE 2 Find the solution of $y'' - 2y' - y = 0$ that satisfies $y(0) = 0$ and $y'(0) = \sqrt{2}$.

Solution The auxiliary equation $r^2 - 2r - 1 = 0$ is best solved by the Quadratic Formula.

$$r = \frac{-b \pm \sqrt{b^2 - 4ac}}{2a} = \frac{2 \pm \sqrt{4 + 4}}{2} = 1 \pm \sqrt{2}$$

The general solution to the differential equation is, therefore,

$$y = C_1 e^{(1 + \sqrt{2})x} + C_2 e^{(1 - \sqrt{2})x}$$

The condition $y(0) = 0$ implies that $C_2 = -C_1$. Then

$$y' = C_1(1 + \sqrt{2})e^{(1 + \sqrt{2})x} - C_1(1 - \sqrt{2})e^{(1 - \sqrt{2})x}$$

and so

$$\sqrt{2} = y'(0) = C_1(1 + \sqrt{2}) - C_1(1 - \sqrt{2}) = 2C_1\sqrt{2}$$

We conclude that $C_1 = \frac{1}{2}$ and

$$y = \tfrac{1}{2}e^{(1 + \sqrt{2})x} - \tfrac{1}{2}e^{(1 - \sqrt{2})x}$$ ■

This is all fine if the auxiliary equation has distinct real roots. But what if it has the form

$$r^2 - 2r_1 r + r_1^2 = (r - r_1)^2 = 0$$

CHAPTER 18/Differential Equations

Then our method produces the single fundamental solution $e^{r_1 x}$ and we must find another solution independent of this one. Such a solution is $xe^{r_1 x}$, as we now demonstrate.

$$(D^2 - 2r_1 D + r_1^2)xe^{r_1 x} = D^2(xe^{r_1 x}) - 2r_1 D(xe^{r_1 x}) + r_1^2 xe^{r_1 x}$$
$$= (xr_1^2 e^{r_1 x} + 2r_1 e^{r_1 x}) - 2r_1(xr_1 e^{r_1 x} + e^{r_1 x}) + r_1^2 xe^{r_1 x}$$
$$= 0$$

Theorem B

(A repeated root). If the auxiliary equation has the single repeated root r_1, then the general solution to $y'' + a_1 y' + a_2 y = 0$ is

$$y = C_1 e^{r_1 x} + C_2 xe^{r_1 x}$$

EXAMPLE 3   Solve $y'' - 6y' + 9y = 0$.

Solution   The auxiliary equation has 3 as a repeated root. Thus,

$$y = C_1 e^{3x} + C_2 xe^{3x} \qquad \blacksquare$$

Finally, we consider the case where the auxiliary equation has conjugate complex roots. The simple equation

$$(D^2 + \beta^2)y = 0$$

with auxiliary equation $r^2 + \beta^2 = 0$ and roots $\pm \beta i$ offers a hint. Its fundamental solutions are easily seen to be $\sin \beta x$ and $\cos \beta x$. You can check by direct differentiation that the general situation is as follows.

Theorem C

(Complex conjugate roots). If the auxiliary equation has complex conjugate roots $\alpha \pm \beta i$, then the general solution to $y'' + a_1 y' + a_2 y = 0$ is

$$y = C_1 e^{\alpha x} \cos \beta x + C_2 e^{\alpha x} \sin \beta x$$

EXAMPLE 4   Solve $y'' - 4y' + 13y = 0$.

Solution   The roots of the auxiliary equation $r^2 - 4r + 13 = 0$ are $2 \pm 3i$. Hence the general solution is

$$y = C_1 e^{2x} \cos 3x + C_2 e^{2x} \sin 3x \qquad \blacksquare$$

HIGHER-ORDER EQUATIONS All of what we have done extends to higher-order linear homogeneous equations with constant coefficients. To solve

$$y^{(n)} + a_1 y^{(n-1)} + \cdots + a_{n-1} y' + a_n y = 0$$

find the roots of the auxiliary equation

$$r^n + a_1 r^{n-1} + \cdots + a_{n-1} r + a_n = 0$$

and make the obvious generalizations of the second-order case. For example, if the auxiliary equation is

$$(r - r_1)(r - r_2)^3 [r - (\alpha + \beta i)][r - (\alpha - \beta i)] = 0$$

the general solution to the differential equation is

$$y = C_1 e^{r_1 x} + (C_2 + C_3 x + C_4 x^2) e^{r_2 x} + e^{\alpha x}[C_5 \cos \beta x + C_6 \sin \beta x]$$

EXAMPLE 5 Solve $\dfrac{d^4 y}{dx^4} - \dfrac{d^3 y}{dx^3} - 20 \dfrac{d^2 y}{dx^2} = 0.$

Solution The auxiliary equation is

$$r^4 - r^3 - 20 r^2 = r^2 (r - 5)(r + 4) = 0$$

with roots $-4, 5$, and a double root of 0. Hence, the general solution is

$$y = C_1 + C_2 x + C_3 e^{5x} + C_4 e^{-4x} \qquad \blacksquare$$

PROBLEM SET 18.2 _____

In Problems 1–16, solve the differential equations.

1. $y'' - 5y' + 6y = 0$

2. $y'' + 5y' - 6y = 0$

3. $y'' + 6y' - 7y = 0$; $y = 0$, $y' = 4$ at $x = 0$.

4. $y'' - 3y' - 10y = 0$; $y = 1$, $y' = 10$ at $x = 0$.

5. $y'' - 4y' + 4y = 0$

6. $y'' + 10y' + 25y = 0$

7. $y'' - 4y' + y = 0$

8. $y'' + 6y' - 2y = 0$

9. $y'' + 4y = 0$; $y = 2$ at $x = 0$, $y = 3$ at $x = \pi/4$.

10. $y'' + 9y = 0$; $y = 3$, $y' = 3$ at $x = \pi/3$.

11. $y'' + 2y' + 2y = 0$

12. $y'' + y' + y = 0$

13. $y^{(4)} + 3y''' - 4y'' = 0$

14. $y^{(4)} - y = 0$

15. $(D^4 + 3D^2 - 4)y = 0$

16. $[(D^2 + 1)(D^2 - D - 6)]y = 0$

17. Solve $y'' - 4y = 0$ and express your answer in terms of the hyperbolic functions cosh and sinh.

18. Show that the solution of

$$\frac{d^2 y}{dx^2} - 2b \frac{dy}{dx} - c^2 y = 0$$

can be written

$$y = e^{bx}(D_1 \cosh \sqrt{b^2 + c^2}\, x + D_2 \sinh \sqrt{b^2 + c^2}\, x)$$

19. Solve $y^{(4)} + 2y^{(3)} + 3y'' + 2y' + y = 0$. *Hint:* First show that the auxiliary equation is $(r^2 + r + 1)^2 = 0$.

20. Solve $y'' - 2y' + 2y = 0$ and express your answer in the form $ce^{\alpha x} \sin(\beta x + \gamma)$. *Hint:* Let $\sin \gamma = C_1/c$ and $\cos \gamma = C_2/c$, where $c = \sqrt{C_1^2 + C_2^2}$.

21. Solve $x^2 y'' + 5xy' + 4y = 0$ by first making the substitution $x = e^z$.

22. Show that the substitution $x = e^z$ transforms the Euler equation $ax^2 y'' + bxy' + cy = 0$ to a homogeneous linear equation with constant coefficients.

23. Recall that complex numbers have the form $a + bi$, where a and b are real. These numbers behave much like the real numbers with the proviso that $i^2 = -1$. Show each of the following.

(a) $e^{bi} = \cos b + i \sin b$
(b) $e^{a+bi} = e^a(\cos b + i \sin b)$
(c) $D_x e^{(\alpha + \beta i)x} = (\alpha + \beta i)e^{(\alpha + \beta i)x}$

Hint: For (a), use the Maclaurin series for e^u.

24. Let the roots of the auxiliary equation $r^2 + a_1 r + a_2 = 0$ be $\alpha \pm \beta i$. From Problem 23c, it follows just as in the real case that $y = c_1 e^{(\alpha + \beta i)x} + c_2 e^{(\alpha - \beta i)x}$ satisfies $(D^2 + a_1 D + a_2)y = 0$. Show that this solution can be rewritten in the form

$$y = C_1 e^{\alpha x} \cos \beta x + C_2 e^{\alpha x} \sin \beta x$$

giving another approach to Theorem C.

18.3
The Nonhomogeneous Equation

Consider the general nonhomogeneous linear equation with constant coefficients

$$y^{(n)} + a_1 y^{(n-1)} + \cdots + a_{n-1} y' + a_n y = k(x)$$

Solving this equation can be reduced to three steps.

1. Find the general solution

$$y_h = C_1 u_1(x) + C_2 u_2(x) + \cdots + C_n u_n(x)$$

to the homogeneous equation as described in the previous section.
2. Find a particular solution y_p to the nonhomogeneous equation.
3. Add the solutions from Steps 1 and 2. We state the result as a formal theorem.

Theorem A

If y_p is any particular solution to the nonhomogeneous equation

$$(*) \qquad L(y) = (D^n + a_1 D^{n-1} + \cdots + a_{n-1}D + a_n)y = k(x)$$

and if y_h is the general solution to the corresponding homogeneous equation, then

$$y = y_p + y_h$$

is the general solution of $(*)$.

Proof The linearity of the operator L is the key element in the proof. Let y_p and y_h be as described. Then

$$L(y_p + y_h) = L(y_p) + L(y_h) = k(x) + 0$$

and so $y = y_p + y_h$ is a solution to $(*)$.

Conversely, let y be any solution to (∗). Then

$$L(y - y_p) = L(y) - L(y_p) = k(x) - k(x) = 0$$

and so $y - y_p$ is a solution to the homogeneous equation. Consequently, $y = y_p + (y - y_p)$ can be written as y_p plus a solution to the homogeneous equation, as we wished to show. ∎

Now we apply this result to second-order equations.

THE METHOD OF UNDETERMINED COEFFICIENTS Consider the equation

$$y'' + a_1 y' + a_2 y = k(x)$$

It turns out that the functions $k(x)$ most apt to occur in applications are polynomials, exponentials, sines, and cosines. For these functions, we offer a procedure for finding y_p based on trial solutions.

If $k(x) =$	Try $y_p =$
$b_m x^m + \cdots + b_1 x + b_0$	$B_m x^m + \cdots + B_1 x + B_0$
$b e^{\alpha x}$	$B e^{\alpha x}$
$b \cos \beta x + c \sin \beta x$	$B \cos \beta x + C \sin \beta x$
Modification. If a term of $k(x)$ is a solution to the homogeneous equation, multiply the trial solution by x (or perhaps by a higher power of x).	

To illustrate the table, we suggest the appropriate trial solution y_p in six cases. The first three are straightforward; the last three are modified because a term on the right side of the differential equation is present in the solution to the homogeneous equation.

1. $y'' - 3y' - 4y = 3x^2 + 2$ $y_p = B_2 x^2 + B_1 x + B_0$
2. $y'' - 3y' - 4y = e^{2x}$ $y_p = B e^{2x}$
3. $y'' + 4y = 2 \sin x$ $y_p = B \cos x + C \sin x$
4. $y'' + 2y' = 3x^2 + 2$ $y_p = B_2 x^3 + B_1 x^2 + B_0 x$
   (2 is a solution to the homogeneous equation)
5. $y'' - 3y' - 4y = e^{4x}$ $y_p = B x e^{4x}$
   (e^{4x} is a solution to the homogeneous equation)
6. $y'' + 4y = \sin 2x$ $y_p = B x \cos 2x + C x \sin 2x$
   ($\sin 2x$ is a solution to the homogeneous equation)

Next we carry out the details in four specific examples.

EXAMPLE 1 Solve $y'' + y' - 2y = 2x^2 - 10x + 3$.

Solution The auxiliary equation $r^2 + r - 2 = 0$ has roots -2 and 1, and so

$$y_h = C_1 e^{-2x} + C_2 e^x$$

To find a particular solution to the nonhomogeneous equation, we try

$$y_p = Ax^2 + Bx + C$$

Substitution of this expression in the differential equation gives

$$2A + (2Ax + B) - 2(Ax^2 + Bx + C) = 2x^2 - 10x + 3$$

Equating coefficients of x^2, x, and 1, we find

$$-2A = 2, \qquad 2A - 2B = -10, \qquad 2A + B - 2C = 3$$

or $A = -1$, $B = 4$, and $C = -\frac{1}{2}$. Hence

$$y_p = -x^2 + 4x - \frac{1}{2}$$

and

$$y = -x^2 + 4x - \frac{1}{2} + C_1 e^{-2x} + C_2 e^x \qquad \blacksquare$$

EXAMPLE 2 Solve $y'' - 2y' - 3y = 8e^{3x}$.

Solution Since the auxiliary equation $r^2 - 2r - 3 = 0$ has roots -1 and 3, we have

$$y_h = C_1 e^{-x} + C_2 e^{3x}$$

Note that $k(x) = 8e^{3x}$ is a solution to the homogeneous equation. Thus, we use the *modified* trial solution

$$y_p = Bxe^{3x}$$

Substituting y_p in the differential equation gives

$$(Bxe^{3x})'' - 2(Bxe^{3x})' - 3Bxe^{3x} = 8e^{3x}$$

or

$$(9xBe^{3x} + 6Be^{3x}) - 2(3Bxe^{3x} + Be^{3x}) - 3Bxe^{3x} = 8e^{3x}$$

or finally

$$4Be^{3x} = 8e^{3x}$$

We conclude that $B = 2$ and

$$y = 2xe^{3x} + C_1 e^{-x} + C_2 e^{3x} \qquad \blacksquare$$

If 3 had been a double root of the auxiliary equation in Example 2, we would have used $Bx^2 e^{3x}$ as our trial solution.

EXAMPLE 3   Solve $y'' - 2y' - 3y = \cos 2x$.

Solution   The homogeneous equation agrees with that of Example 2, and so

$$y_h = C_1 e^{-x} + C_2 e^{3x}$$

For the trial solution y_p, we use

$$y_p = B \cos 2x + C \sin 2x$$

Now

$$Dy_p = -2B \sin 2x + 2C \cos 2x$$
$$D^2 y_p = -4B \cos 2x - 4C \sin 2x$$

Hence substitution of y_p in the differential equation gives (after collecting terms)

$$(-7B - 4C)\cos 2x + (4B - 7C)\sin 2x = \cos 2x$$

Thus, $-7B - 4C = 1$ and $4B - 7C = 0$, which imply that $B = -\frac{7}{65}$ and $C = -\frac{4}{65}$. We conclude that

$$y = -\tfrac{7}{65} \cos 2x - \tfrac{4}{65} \sin 2x + C_1 e^{-x} + C_2 e^{3x} \qquad \blacksquare$$

EXAMPLE 4   Solve $y'' - 2y' - 3y = 8e^{3x} + \cos 2x$.

Solution   Combining the results of Examples 2 and 3 and using the linearity of the operator $D^2 - 2D - 3$, we obtain

$$y = 2xe^{3x} - \tfrac{7}{65} \cos 2x - \tfrac{4}{65} \sin 2x + C_1 e^{-x} + C_2 e^{3x} \qquad \blacksquare$$

THE METHOD OF VARIATION OF PARAMETERS   A more general method than that of undetermined coefficients is the method of variation of parameters. If $u_1(x)$ and $u_2(x)$ are independent solutions to the homogeneous equation, then it can be shown (see Problem 23) that there is a particular solution to the nonhomogeneous equation of the form

$$y_p = v_1(x)u_1(x) + v_2(x)u_2(x)$$

in which

$$v_1' u_1 + v_2' u_2 = 0$$
$$v_1' u_1' + v_2' u_2' = k(x)$$

We show how this method works in an example.

EXAMPLE 5 Find the general solution of $y'' + y = \sec x$.

Solution The general solution to the homogeneous equation is

$$y_h = C_1 \cos x + C_2 \sin x$$

To find a particular solution to the nonhomogeneous equation, we set

$$y_p = v_1(x)\cos x + v_2(x)\sin x$$

and impose the conditions

$$v_1' \cos x + v_2' \sin x = 0$$

$$-v_1' \sin x + v_2' \cos x = \sec x$$

When we solve this system of equations for v_1' and v_2', we obtain $v_1' = -\tan x$ and $v_2' = 1$. Thus,

$$v_1(x) = \int -\tan x \, dx = \ln|\cos x|$$

$$v_2(x) = \int dx = x$$

(We can omit the arbitrary constants in the above integrations, since any solutions v_1 and v_2 will do.) A particular solution is therefore

$$y_p = (\ln|\cos x|)\cos x + x \sin x$$

a result which is easy to check by direct substitution in the original differential equation. We conclude that

$$y = (\ln|\cos x|)\cos x + x \sin x + C_1 \cos x + C_2 \sin x \qquad \blacksquare$$

PROBLEM SET 18.3

In Problems 1–16, use the method of undetermined coefficients to solve the differential equations.

1. $y'' - 9y = x$

2. $y'' + y' - 6y = 2x^2$

3. $y'' - 2y' + y = x^2 + x$

4. $y'' + y' = 4x$

5. $y'' - 5y' + 6y = e^x$

6. $y'' + 6y' + 9y = 2e^{-x}$

7. $y'' + 4y' + 3y = e^{-3x}$

8. $y'' + 2y' + 2y = 3e^{-2x}$

9. $y'' - y' - 2y = 2 \sin x$

10. $y'' + 4y' = \cos x$

11. $y'' + 4y = 2 \cos 2x$

12. $y'' + 9y = \sin 3x$

13. $y'' + 9y = \sin x + e^{2x}$

14. $y'' + y' = e^x + 3x$

15. $y'' - 5y' + 6y = 2e^x$; $y = 1$, $y' = 0$ when $x = 0$.

16. $y'' - 4y = 4 \sin x$; $y = 4$, $y' = 0$ when $x = 0$.

In Problems 17–22, solve the differential equations by variation of parameters.

17. $y'' - 3y' + 2y = 5x + 2$

18. $y'' - 4y = e^{2x}$

19. $y'' + y = \csc x \cot x$

20. $y'' + y = \cot x$

21. $y'' - 3y' + 2y = \dfrac{e^x}{e^x + 1}$

22. $y'' - 5y' + 6y = 2e^x$

23. Let $L(y) = y'' + by' + cy = 0$ have solutions u_1 and u_2 and let $y_p = v_1 u_1 + v_2 u_2$. Show that

$$L(y_p) = v_1(u_1'' + bu_1' + cu_1) + v_2(u_2'' + bu_2' + cu_2)$$
$$+ b(v_1' u_1 + v_2' u_2) + (v_1' u_1 + v_2' u_2)' + (v_1' u_1' + v_2' u_2')$$

Thus, if the conditions of the method of variation of parameters hold,

$$L(y_p) = (v_1)(0) + (v_2)(0) + (b)(0) + 0 + k(x) = k(x)$$

24. Solve $y'' + 4y = \sin^3 x$

18.4
Applications of Second-Order Equations

Many problems in physics lead to second-order linear differential equations. We first consider the problem of a vibrating spring, under various assumptions. Then we return to and generalize an earlier application to electric circuits.

A VIBRATING SPRING (SIMPLE HARMONIC MOTION) Consider a coiled spring weighted by an object A and hanging vertically from a support, as in Figure 1(a). We wish to consider the motion of the point P if the spring is pulled y_0 units below its equilibrium position (Figure 1(b)) and released. We assume friction to be negligible.

According to Hooke's Law, the force F tending to restore P to its equilibrium position at $y = 0$ satisfies $F = -ky$, where k is a constant depending on the characteristics of the spring and y is the y-coordinate of P. But by Newton's Second Law, $F = ma = (w/g)a$, where w is the weight of the object A, a is the acceleration of P, and g is the constant acceleration due to gravity ($g = 32$ feet per second per second). Thus,

$$\frac{w}{g}\frac{d^2y}{dt^2} = -ky, \qquad k > 0$$

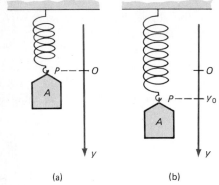

(a)   (b)

FIGURE 1

is the differential equation of the motion.

If we let $kg/w = B^2$, this equation takes the form

$$\frac{d^2y}{dt^2} + B^2 y = 0$$

and has the general solution

$$y = C_1 \cos Bt + C_2 \sin Bt$$

The conditions $y = y_0$ and $y' = 0$ at $t = 0$ determine the constants C_1 and C_2 to be y_0 and 0, respectively. Thus,

$$y = y_0 \cos Bt$$

Simple harmonic motion

y_0

Period $= \dfrac{2\pi}{B}$

FIGURE 2

We say that the spring is executing **simple harmonic motion** with amplitude y_0 and period $2\pi/B$ (Figure 2).

EXAMPLE 1   When an object weighing 5 pounds is attached to the lowest point P of a spring that hangs vertically, the spring is extended 6 inches. The 5-pound weight is replaced by a 20-pound weight, and the system is allowed to come to equilibrium. If the 20-pound weight is now pulled downward another 2 feet and then released, describe the motion of the lowest point P of the spring.

Solution   The first sentence of the example allows us to determine the spring constant. By Hooke's Law, $|F| = ks$, where s is the amount the spring is stretched, and so $5 = k(\frac{1}{2})$, or $k = 10$. Now put the origin at the equilibrium point after the 20-pound weight has been attached. From the derivation just before the example, we know that $y = y_0 \cos Bt$. In the present case, $y_0 = 2$ and $B^2 = kg/w = (10)(32)/20 = 16$. We conclude that

$$y = 2 \cos 4t$$

The motion of P is simple harmonic motion, with period $\frac{1}{2}\pi$ and amplitude 2 feet. That is, P oscillates up and down from 2 feet below 0 to 2 feet above 0 and then back to 2 feet below 0 every $\frac{1}{2}\pi \approx 1.57$ seconds. ■

DAMPED VIBRATIONS   So far, we have assumed a simplified situation, in which there is no friction either within the spring or resulting from the resistance of the air. We can take friction into account by assuming a retarding force that is proportional to the velocity dy/dt. The differential equation describing the motion then takes the form

$$\frac{w}{g}\frac{d^2y}{dt^2} = -ky - q\frac{dy}{dt}, \qquad k > 0, q > 0$$

By letting $E = qg/w$ and $B^2 = kg/w$, this equation can be written as

$$\frac{d^2y}{dt^2} + E\frac{dy}{dt} + B^2y = 0$$

an equation to which the methods of Section 18.2 apply. There are three cases to consider.

Case 1 ($E^2 - 4B^2 < 0$).   The roots of the auxiliary equation are complex conjugates; we will denote them by $-\alpha \pm \beta i$, where α and β are positive. The general solution is

$$y = e^{-\alpha t}(C_1 \cos \beta t + C_2 \sin \beta t)$$

which can be rewritten in the form (see Problem 15)

$$y = Ce^{-\alpha t} \sin(\beta t + \gamma)$$

The factor $e^{-\alpha t}$, called the **damping factor**, causes the amplitude of the motion to approach zero as $t \to \infty$ (Figure 3(a)).

(a)   (b)   (c)

FIGURE 3

Case 2 ($E^2 - 4B^2 = 0$). In this case, the auxiliary equation has a double root $-\alpha$ and the general solution of the differential equation is

$$y = C_1 e^{-\alpha t} + C_2 t e^{-\alpha t}$$

The motion described by this equation is said to be **critically damped**.

Case 3 ($E^2 - 4B^2 > 0$). The auxiliary equation has roots $-\alpha_1$ and $-\alpha_2$, and the general solution of the differential equation is

$$y = C_1 e^{-\alpha_1 t} + C_2 e^{-\alpha_2 t}$$

It describes a motion that is said to be **overdamped**.

The graphs in the critically damped and overdamped cases cross the t-axis at most once and may look something like Figure 3(b) or 3(c).

EXAMPLE 2 If a damping force with $q = 0.2$ is imposed on the system of Example 1, find the equation of motion.

Solution $E = qg/w = (0.2)(32)/20 = 0.32$ and $B^2 = (10)(32)/20 = 16$, and so we must solve

$$\frac{d^2y}{dt^2} + 0.32\frac{dy}{dt} + 16y = 0$$

The auxiliary equation $r^2 + 0.32r + 16 = 0$ has roots $r = -0.16 \pm \sqrt{15.9744}i \approx -0.16 \pm 4i$, and thus

$$y = e^{-0.16t}(C_1 \cos 4t + C_2 \sin 4t)$$

When we impose the conditions $y = 2$ and $y' = 0$ at $t = 0$, we find $C_1 = 2$ and $C_2 = 0.08$. Consequently,

$$y = e^{-0.16t}(2 \cos 4t + 0.08 \sin 4t) \qquad \blacksquare$$

FIGURE 4

ELECTRIC CIRCUITS Consider a circuit (Figure 4) with a resistor (R ohms), an inductor (L henrys), and a capacitor (C farads) in series with a source of

CHAPTER 18/Differential Equations

electromotive force supplying $E(t)$ volts. The new feature in comparison to the circuits of Section 18.1 is the presence of a capacitor (also called a condenser). Kirchhoff's Law in this situation says that the charge Q on the capacitor, measured in coulombs, satisfies

(1)
$$L \frac{d^2Q}{dt^2} + R \frac{dQ}{dt} + \frac{1}{C} Q = E(t)$$

The current $I = dQ/dt$, measured in amperes, satisfies the equation obtained by differentiating equation (1) with respect to t; that is,

(2)
$$L \frac{d^2I}{dt^2} + R \frac{dI}{dt} + \frac{1}{C} I = E'(t)$$

Either of these equations can be solved by the methods of Sections 18.2 and 18.3.

EXAMPLE 3   Find the charge Q and the current I as functions of time t in an RCL circuit (Figure 4) if $R = 16$, $L = 0.02$, $C = 2 \times 10^{-4}$, and $E = 12$. Assume that $Q = 0$ and $I = 0$ at $t = 0$ (when the switch is closed).

Solution   By Kirchhoff's Law as expressed in equation (1)

$$\frac{d^2Q}{dt^2} + 800 \frac{dQ}{dt} + 250{,}000Q = 600$$

The auxiliary equation has roots

$$\frac{-800 \pm \sqrt{640{,}000 - 1{,}000{,}000}}{2} = -400 \pm 300i$$

so

$$Q_h = e^{-400t}(C_1 \cos 300t + C_2 \sin 300t)$$

By inspection, a particular solution is $Q_p = 2.4 \times 10^{-3}$. Therefore, the general solution is

$$Q = 2.4 \times 10^{-3} + e^{-400t}(C_1 \cos 300t + C_2 \sin 300t)$$

When we impose the given initial conditions, we find $C_1 = -2.4 \times 10^{-3}$ and $C_2 = -3.2 \times 10^{-3}$. We conclude that

$$Q = 10^{-3}[2.4 - e^{-400t}(2.4 \cos 300t + 3.2 \sin 300t)]$$

and by differentiation that

$$I = \frac{dQ}{dt} = 2e^{-400t} \sin 300t$$

■

1. A spring with a spring constant k of 20 pounds per foot is loaded with a 10-pound weight and allowed to reach equilibrium. It is then raised 1 foot and released. Find the equation of motion and the period. Neglect friction.

2. A spring with a spring constant k of 100 pounds per foot is loaded with a 1-pound weight and brought to equilibrium. It is then stretched an additional 1 inch and released. Find the equation of motion, the amplitude, and the period. Neglect friction.

3. In Problem 1, what is the absolute value of the velocity of the moving weight as it passes through the equilibrium position?

4. A 10-pound weight stretches a spring 4 inches. This weight is removed and replaced with a 20-pound weight, which is then allowed to reach equilibrium. The weight is next raised 1 foot and released with an initial velocity of 2 feet per second downward. What is the equation of motion? Neglect friction.

5. A spring with a spring constant k of 20 pounds per foot is loaded with a 10-pound weight and allowed to reach equilibrium. It is then displaced 1 foot downward and released. If the weight experiences a retarding force in pounds equal to one-tenth the velocity, find the equation of motion.

6. Determine the motion in Problem 5 if the retarding force equals four times the velocity at every point.

7. In Problem 5, how long will it take the oscillations to diminish to one-tenth their original amplitude?

8. In Problem 5, what will be the equation of motion if the weight is given an upward velocity of 1 foot per second at the moment of release?

9. Using Figure 5, find the charge Q on the capacitor as a function of time if S is closed at $t = 0$. Assume that the capacitor is initially uncharged.

$R = 10^6 \ \Omega$

$C = 10^{-6}$ F

$E = 1$ V

FIGURE 5

10. Find the current I as a function of time in Problem 9 if the capacitor has an initial charge of 4 coulombs.

11. Use Figure 6.

S

$E = 120$ sin $377t$

$C = 2 \times 10^{-6}$ F

FIGURE 6

(a) Find Q as a function of time. Assume the capacitor initially uncharged.
(b) Find I as a function of time.

12. Using Figure 7, find the current as a function of time if the capacitor is initially uncharged and S is closed at $t = 0$. *Hint*: The current at $t = 0$ will equal 0, since the current through an inductance cannot change instantaneously.

S

$L = 10^{-2}$ H

$E = 20$ V

$C = 10^{-7}$ F

FIGURE 7

13. Use Figure 8, find the steady-state current as a function of time; that is, find a formula for I that is valid when t is very large ($t \to \infty$).

$R = 1{,}000 \ \Omega$

$L = 3.5$ H

$E = 120$ sin $377t$

$C = 2 \times 10^{-6}$ F

FIGURE 8

14. Suppose that an undamped spring is subjected to an external periodic force so that its differential equation has the form

$$\frac{d^2 y}{dt^2} + B^2 y = c \sin At, \qquad c > 0$$

(a) Show that the equation of motion for $A \neq B$ is

$$y = C_1 \cos Bt + C_2 \sin Bt + \frac{c}{B^2 - A^2} \sin At$$

CHAPTER 18/Differential Equations

(b) Solve the differential equation when $A = B$ (the resonance case).

(c) What happens to the amplitude of the motion in part b when $t \to \infty$?

15. Show that $C_1 \cos \beta t + C_2 \sin \beta t$ can be written in the form $C \sin(\beta t + \gamma)$. *Hint*: Let $C = \sqrt{C_1^2 + C_2^2}$, $\sin \gamma = C_1/C$, and $\cos \gamma = C_2/C$.

16. Show that the motion of part a of Problem 14 is periodic if B/A is rational.

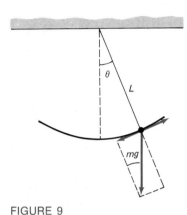

FIGURE 9

17. Refer to Figure 9, which shows a pendulum bob of mass m supported by a weightless wire of length L. Derive the equation of motion, that is, derive the differential equation satisfied by θ. *Suggestion*: Use the fact from Section 13.5 that the scalar tangential component of the acceleration is d^2s/dt^2, where s measures arc length in the counterclockwise direction.

18. The equation derived in Problem 17 is nonlinear, but—for small θ—it is customary to approximate it by the equation

$$\frac{d^2\theta}{dt^2} + \frac{g}{L}\theta = 0$$

Here $g = GM/R^2$, where G is a universal constant, M is the mass of the earth, and R is the distance from the pendulum to the center of the earth. Two clocks, with pendulums of length L_1 and L_2 and located at distances R_1 and R_2 from the center of the earth, have periods p_1 and p_2, respectively.

(a) Show that $\dfrac{p_1}{p_2} = \dfrac{R_1 \sqrt{L_1}}{R_2 \sqrt{L_2}}$.

(b) Find the height of a mountain if a clock which kept perfect time at sea level ($R = 3960$ miles) with $L = 81$ inches had to have its pendulum shortened to $L = 80.85$ inches to keep perfect time at the top of the mountain.

18.5 Chapter Review

TRUE-FALSE QUIZ

Respond with true or false to each of the following assertions. Be prepared to defend your answer.

1. $y'' + y^2 = 0$ is a linear differential equation.
2. $y'' + x^2 y = 0$ is a linear differential equation.
3. $y = \tan x + \sec x$ is a solution of $2y' - y^2 = 1$.
4. The general solution to $[D^2 + aD + b]^3 y = 0$ should involve 8 arbitrary constants.
5. An integrating factor for $y' + \dfrac{4}{x} y = e^x$ is x^4.
6. If $u_1(x)$ and $u_2(x)$ are two solutions to $y'' + a_1 y' + a_2 y = f(x)$, then $C_1 u_1(x) + C_2 u_2(x)$ is also a solution.
7. The general solution to $y''' + 3y'' + 3y' + y = 0$ is $y = C_1 e^{-x} + C_2 x e^{-x} + C_3 x^2 e^{-x}$.
8. If $u_1(x)$ and $u_2(x)$ are solutions to the linear differential equation $L(y) = f(x)$, then $u_1(x) - u_2(x)$ is a solution to $L(y) = 0$.
9. The equation $y'' + 9y = 2 \sin 3x$ has a particular solution of the form $y_p = B \sin 3x + C \cos 3x$.
10. An expression of the form $C_1 \cos \beta t + C_2 \sin \beta t$ can always be written in the form $C \sin(\beta t + \gamma)$.

SAMPLE TEST PROBLEMS

In Problems 1–19, solve the differential equations.

1. $\dfrac{dy}{dx} + \dfrac{y}{x} = 0$

2. $\dfrac{dy}{dx} - \dfrac{x^2 - 2y}{x} = 0$

3. $\dfrac{dy}{dx} + 2x(y - 1) = 0$; $y = 3$ when $x = 0$.

4. $\dfrac{dy}{dx} - ay = e^{ax}$

5. $\dfrac{dy}{dx} - 2y = e^x$

6. $\dfrac{dy}{dx} + y \tan x = 2 \sec x$

7. $\dfrac{d^2y}{dx^2} + 3\dfrac{dy}{dx} = e^x$

Suggestion: Let $u = dy/dx$.

8. $y'' - y = 0$

9. $y'' - 3y' + 2y = 0$; $y = 0$, $y' = 3$ when $x = 0$.

10. $4y'' + 12y' + 9y = 0$

11. $y'' - y = 1$

12. $y'' + 4y' + 4y = 3e^x$

13. $y'' + 4y' + 4y = e^{-2x}$

14. $y'' + 4y = 0$; $y = 0$, $y' = 2$ when $x = 0$.

15. $y'' + 6y' + 25y = 0$

16. $y'' + y = \sec x \tan x$

17. $y''' + 2y'' - 8y' = 0$

18. $y^{(4)} - 3y'' - 10y = 0$

19. $y^{(4)} - 4y'' + 4y = 0$

20. Suppose that glucose is infused into the bloodstream of a patient at the rate of 3 grams per minute, but that the patient's body converts and removes glucose from its blood at a rate proportional to the amount present (with constant of proportionality 0.02). If $Q(t)$ is the amount present at time t and $Q(0) = 120$,

(a) write the differential equation for Q;
(b) solve this differential equation;
(c) determine what happens to Q in the long run.

21. A spring with a spring constant k of 5 pounds per foot is loaded with a 10-pound weight and allowed to reach equilibrium. It is then raised 1 foot and released. What are the equation of motion, the amplitude, and the period? Neglect friction.

22. In Problem 21, what is the absolute value of the velocity of the moving weight as it passes through the equilibrium position?

23. Suppose the switch of the circuit in Figure 1 is closed at $t = 0$. Find I as a function of time if C is initially uncharged. (The current at $t = 0$ will equal zero, since current through an inductance cannot change instantaneously.)

FIGURE 1

Appendix

A.1 Mathematical Induction

Often in mathematics, we are faced with the task of wanting to establish that a certain proposition P_n is true for every integer $n \geq 1$ (or perhaps every integer $n \geq N$). Here are three examples.

1. P_n: $1^2 + 2^2 + 3^2 + \cdots + n^2 = \dfrac{n(n+1)(2n+1)}{6}.$

2. Q_n: $2^n > n + 20.$

3. R_n: $n^2 - n + 41$ is prime.

Proposition P_n is true for every positive integer and Q_n is true for every integer greater than or equal to 5 (as we will show soon). The third proposition, R_n, is interesting. Note that for $n = 1, 2, 3, \ldots$, the values of $n^2 - n + 41$ are 41, 43, 47, 53, 61, . . . (prime numbers so far). In fact, we will get a prime number for all n's through 40; but at $n = 41$, the formula yields the composite number $1681 = (41)(41)$. Showing the truth of a proposition for 40 (or 40 million) individual cases may make a proposition plausible, but it most certainly does not prove it is true for all n. The chasm between any finite number of cases and *all* cases is infinitely wide.

What is to be done? Is there a procedure for establishing that a proposition P_n is true for *all n*? An affirmative answer is provided by the **Principle of Mathematical Induction**, a favorite of all mathematics students.

> **(Principle of Mathematical Induction).** Let $\{P_n\}$ be a sequence of propositions (statements) satisfying these two conditions.
>
> **(i)** P_N is true (usually N will be 1).
> **(ii)** The truth of P_i implies the truth of P_{i+1}, $i \geq N$.
>
> Then, P_n is true for every integer $n \geq N$.

We do not prove this principle; it is often taken as an axiom, and we hope it seems obvious. After all, if the first domino falls and if each domino knocks over the next one, then the whole row of dominoes will fall. Our efforts will be directed toward illustrating how we use mathematical induction.

EXAMPLE 1 Prove that

$$P_n: \quad 1^2 + 2^2 + 3^2 + \cdots + n^2 = \frac{n(n+1)(2n+1)}{6}$$

is true for all $n \geq 1$.

Solution   First, we note that

$$P_1: \quad 1^2 = \frac{1(1+1)(2+1)}{6}$$

is a true statement.

Second, we demonstrate Implication (ii). We begin by writing the statements P_i and P_{i+1}.

$$P_i: \quad 1^2 + 2^2 + \cdots + i^2 = \frac{i(i+1)(2i+1)}{6}$$

$$P_{i+1}: \quad 1^2 + 2^2 + \cdots + i^2 + (i+1)^2 = \frac{(i+1)(i+2)(2i+3)}{6}$$

We must show that P_i implies P_{i+1}, so we assume P_i is true. Then the left side of P_{i+1} can be written as follows (* indicates where P_i is used).

$$[1^2 + 2^2 + \cdots + i^2] + (i+1)^2 \stackrel{*}{=} \frac{i(i+1)(2i+1)}{6} + (i+1)^2$$

$$= (i+1)\frac{2i^2 + i + 6i + 6}{6}$$

$$= (i+1)\frac{2i^2 + 7i + 6}{6}$$

$$= \frac{(i+1)(i+2)(2i+3)}{6}$$

Read from first expression to last expression, this chain of equalities is the statement P_{i+1}. Thus the truth of P_i does imply the truth of P_{i+1}. By the Principle of Mathematical Induction, P_n is true for each positive integer n. ∎

EXAMPLE 2   Prove P_n: $2^n > n + 20$ is true for each integer $n \geq 5$.

Solution   First, we note that P_5: $2^5 > 5 + 20$ is true. Second, we suppose that P_i: $2^i > i + 20$ is true and attempt to deduce from this that P_{i+1}: $2^{i+1} > i + 1 + 20$ is true. But

$$2^{i+1} = 2 \cdot 2^i \overset{*}{>} 2(i + 20) = 2i + 40 > i + 21$$

Read from left to right, this is proposition P_{i+1}. We conclude that P_n is true for $n \geq 5$. ■

EXAMPLE 3   Prove that

$$P_n: x - y \text{ is a factor of } x^n - y^n$$

is true for each integer $n \geq 1$.

Solution   Trivially, $x - y$ is a factor of $x - y$, so P_1 is true. Suppose that $x - y$ is a factor of $x^i - y^i$, that is, suppose that

$$x^i - y^i = Q(x, y)(x - y)$$

for some polynomial $Q(x, y)$. Then

$$
\begin{aligned}
x^{i+1} - y^{i+1} &= x^{i+1} - x^i y + x^i y - y^{i+1} \\
&= x^i(x - y) + y(x^i - y^i) \\
&\overset{*}{=} x^i(x - y) + yQ(x, y)(x - y) \\
&= [x^i + yQ(x, y)](x - y)
\end{aligned}
$$

which displays $x - y$ as a factor. Thus, the truth of P_i does imply the truth of P_{i+1}. We conclude by the Principle of Mathematical Induction that P_n is true for all $n \geq 1$. ■

PROBLEM SET   A.1

In Problems 1–8, use the Principle of Mathematical Induction to prove the given proposition is true for each integer $n \geq 1$.

1. $1 + 2 + 3 + \cdots + n = \dfrac{n(n + 1)}{2}$

2. $1 + 3 + 5 + \cdots + (2n - 1) = n^2$

3. $1 \cdot 2 + 2 \cdot 3 + 3 \cdot 4 + \cdots + n(n + 1) = \dfrac{n(n + 1)(n + 2)}{3}$

4. $1^2 + 3^2 + 5^2 + \cdots + (2n - 1)^2 = \dfrac{n(2n - 1)(2n + 1)}{3}$

5. $1^3 + 2^3 + 3^3 + \cdots + n^3 = \left[\dfrac{n(n + 1)}{2}\right]^2$

6. $1^4 + 2^4 + 3^4 + \cdots + n^4 = \dfrac{n(n + 1)(6n^3 + 9n^2 + n - 1)}{30}$

7. $n^3 - n$ is divisible by 6.

8. $n^3 + (n + 1)^3 + (n + 2)^3$ is divisible by 9.

In Problems 9–12, determine the first integer N for which the proposition is true for all $n \geq N$ and then prove the proposition for all $n \geq N$.

9. $3n + 25 < 3^n$ 10. $n - 100 > \log_{10} n$

11. $n^2 \leq 2^n$

12. $|\sin nx| \leq n|\sin x|$ for all x

In Problems 13–20, indicate what conclusion about P_n can be drawn from the given information.

13. P_5 is true and P_i true implies P_{i+2} true.

14. P_1 and P_2 are true and P_i true implies P_{i+2} true.

15. P_{30} is true and P_i true implies P_{i-1} true.

16. P_{30} is true and P_i true implies both P_{i+1} and P_{i-1} true.

17. P_1 is true and P_i true implies both P_{4i} and P_{i-1} true.

18. P_1 is true and P_{2i} true implies P_{2i+1} true.

19. P_1 and P_2 are true, P_i and P_{i+1} true imply P_{i+2} true.

20. P_1 is true and P_j true for $j \leq i$ imply P_{i+1} true.

In Problems 21–27, decide for what n's the given proposition is true and then use mathematical induction (perhaps in one of the alternative forms you may have discovered in Problems 13–20) to prove each of the following.

21. $x + y$ is a factor of $x^n + y^n$.

22. The sum of the measures of the interior angles of an n-sided convex (no holes or dents) polygon is $(n - 2)\pi$.

23. The number of diagonals of an n-sided convex polygon is $\dfrac{n(n - 3)}{2}$.

24. $\dfrac{1}{n+1} + \dfrac{1}{n+2} + \dfrac{1}{n+3} + \cdots + \dfrac{1}{2n} > \dfrac{3}{5}$

25. $\left(1 - \dfrac{1}{4}\right)\left(1 - \dfrac{1}{9}\right)\left(1 - \dfrac{1}{16}\right)\cdots\left(1 - \dfrac{1}{n^2}\right) = \dfrac{n+1}{2n}$

26. Let $f_0 = 0$, $f_1 = 1$, and $f_{n+2} = f_{n+1} + f_n$ for $n \geq 0$ (this is the Fibonacci sequence). Then,

$$f_n = \frac{1}{\sqrt{5}}\left[\left(\frac{1 + \sqrt{5}}{2}\right)^n - \left(\frac{1 - \sqrt{5}}{2}\right)^n\right]$$

27. Let $a_0 = 0$, $a_1 = a$, and $a_{n+2} = (a_{n+1} + a_n)/2$ for $n \geq 0$. Then,

$$a_n = \tfrac{2}{3}\left[1 - \left(-\tfrac{1}{2}\right)^n\right]$$

28. What is wrong with the following argument, which purports to show that all people in any set of n people are the same age? The statement is certainly true for a set consisting of one person. Suppose it is true for any set of i people and consider a set W of $i + 1$ people. We may think of W as the union of sets X and Y, each consisting of i people (draw a picture, for example, when W has 6 people). By supposition, each of these sets consists of identically aged people. But X and Y overlap (in $X \cap Y$) and so all members of $W = X \cup Y$ also are the same age.

A.2
Proofs of Several Theorems

Theorem A

(Main Limit Theorem). Let n be a positive integer, k be a constant, and f and g be functions which have limits at c. Then

1. $\lim\limits_{x \to c} k = k$;

2. $\lim\limits_{x \to c} x = c$;

3. $\lim\limits_{x \to c} kf(x) = k \lim\limits_{x \to c} f(x)$;

4. $\lim\limits_{x \to c}[f(x) + g(x)] = \lim\limits_{x \to c} f(x) + \lim\limits_{x \to c} g(x)$;

5. $\lim\limits_{x \to c}[f(x) - g(x)] = \lim\limits_{x \to c} f(x) - \lim\limits_{x \to c} g(x)$;

6. $\lim\limits_{x \to c}[f(x) \cdot g(x)] = \lim\limits_{x \to c} f(x) \cdot \lim\limits_{x \to c} g(x)$;

7. $\lim\limits_{x \to c} \dfrac{f(x)}{g(x)} = \dfrac{\lim\limits_{x \to c} f(x)}{\lim\limits_{x \to c} g(x)}$, provided $\lim\limits_{x \to c} g(x) \neq 0$;

8. $\lim\limits_{x \to c}[f(x)]^n = \left[\lim\limits_{x \to c} f(x)\right]^n$;

9. $\lim\limits_{x \to c} \sqrt[n]{f(x)} = \sqrt[n]{\lim\limits_{x \to c} f(x)}$, provided $\lim\limits_{x \to c} f(x) > 0$ when n is even.

Proof We proved parts 1–5 near the end of Section 2.6, so we should start with part 6. However, we choose first to prove a special case of part 8, namely,

$$\lim_{x \to c}[g(x)]^2 = \left[\lim_{x \to c} g(x)\right]^2$$

To see this, recall that we have proved that $\lim_{x \to c} x^2 = c^2$ (Example 6 of Section 2.5) and so $f(x) = x^2$ is continuous everywhere. Thus, by the Composite Limit Theorem (Theorem 2.7D),

$$\lim_{x \to c}[g(x)]^2 = \lim_{x \to c} f(g(x)) = f\left[\lim_{x \to c} g(x)\right] = \left[\lim_{x \to c} g(x)\right]^2$$

Next, write

$$f(x)g(x) = \tfrac{1}{4}\{[f(x) + g(x)]^2 - [f(x) - g(x)]^2\}$$

and apply parts 3, 4, and 5, plus what we have just proved. Part 6 falls out.

To prove part 7, apply the Composite Limit Theorem with $f(x) = 1/x$ and use Example 7 of Section 2.5. Then

$$\lim_{x \to c} \frac{1}{g(x)} = \lim_{x \to c} f(g(x)) = f\left(\lim_{x \to c} g(x)\right) = \frac{1}{\lim_{x \to c} g(x)}$$

Finally, by part 6,

$$\lim_{x \to c} \frac{f(x)}{g(x)} = \lim_{x \to c}\left[f(x) \cdot \frac{1}{g(x)}\right] = \lim_{x \to c} f(x) \cdot \lim_{x \to c} \frac{1}{g(x)}$$

from which the result follows.

Part 8 follows from repeated use of part 6 (technically, by mathematical induction).

We prove part 9 only for square roots. Let $f(x) = \sqrt{x}$, which is continuous for positive numbers by Example 4 of Section 2.5. By the Composite Limit Theorem,

$$\lim_{x \to c} \sqrt{g(x)} = \lim_{x \to c} f(g(x)) = f\left(\lim_{x \to c} g(x)\right) = \sqrt{\lim_{x \to c} g(x)}$$

which is equivalent to the desired result. ∎

Theorem B

(Chain Rule). If g is differentiable at a and f is differentiable at $g(a)$, then $f \circ g$ is differentiable at a and

$$(f \circ g)'(a) = f'(g(a))g'(a)$$

Proof We offer a proof that generalizes easily to higher dimensions (see Section 15.6).

By hypothesis, f is differentiable at $b = g(a)$; that is, there is a number $f'(b)$ such that

$$(1) \qquad \lim_{\Delta u \to 0} \frac{f(b + \Delta u) - f(b)}{\Delta u} = f'(b)$$

Define a function ε depending on Δu by

$$\varepsilon(\Delta u) = \frac{f(b + \Delta u) - f(b)}{\Delta u} - f'(b)$$

and multiply both sides by Δu to obtain

$$(2) \qquad f(b + \Delta u) - f(b) = f'(b)\,\Delta u + \Delta u\,\varepsilon(\Delta u)$$

The existence of the limit in (1) is equivalent to $\varepsilon(\Delta u) \to 0$ as $\Delta u \to 0$ in (2). If in (2), we replace Δu by $g(a + \Delta x) - g(a)$ and b by $g(a)$, we get

$$f(g(a + \Delta x)) - f(g(a)) = f'(g(a))[g(a + \Delta x) - g(a)]$$
$$+ [g(a + \Delta x) - g(a)]\varepsilon(\Delta u)$$

or upon dividing both sides by Δx,

$$(3) \qquad \frac{f(g(a + \Delta x)) - f(g(a))}{\Delta x} = f'(g(a))\frac{g(a + \Delta x) - g(a)}{\Delta x}$$
$$+ \frac{g(a + \Delta x) - g(a)}{\Delta x}\varepsilon(\Delta u)$$

In (3), let $\Delta x \to 0$. Since g is differentiable at a, it is continuous there, so $\Delta x \to 0$ forces $\Delta u \to 0$; this, in turn, makes $\varepsilon(\Delta u) \to 0$. We conclude that

$$\lim_{\Delta x \to 0} \frac{f(g(a + \Delta x)) - f(g(a))}{\Delta x} = f'(g(a)) \lim_{\Delta x \to 0} \frac{g(a + \Delta x) - g(a)}{\Delta x} + 0$$

that is, $f \circ g$ is differentiable at a and

$$(f \circ g)'(a) = f'(g(a))g'(a) \qquad \blacksquare$$

Theorem C

(Power Rule). If r is rational, then x^r is differentiable at any x that is in an open interval on which x^{r-1} is real and

$$D_x(x^r) = rx^{r-1}$$

Proof Consider first the case, where $r = 1/q$, q a positive integer. Recall that $a^q - b^q$ factors as

$$a^q - b^q = (a - b)(a^{q-1} + a^{q-2}b + \cdots + ab^{q-2} + b^{q-1})$$

so

$$\frac{a - b}{a^q - b^q} = \frac{1}{a^{q-1} + a^{q-2}b + \cdots + ab^{q-2} + b^{q-1}}$$

Thus, if $f(t) = t^{1/q}$,

$$f'(x) = \lim_{t \to x} \frac{t^{1/q} - x^{1/q}}{t - x} = \lim_{t \to x} \frac{t^{1/q} - x^{1/q}}{(t^{1/q})^q - (x^{1/q})^q}$$

$$= \lim_{t \to x} \frac{1}{t^{(q-1)/q} + t^{(q-2)/q}x^{1/q} + \cdots + x^{(q-1)/q}}$$

$$= \frac{1}{qx^{(q-1)/q}} = \frac{1}{q}x^{1/q - 1}$$

Now by the Chain Rule, with p an integer,

$$D_x(x^{p/q}) = D_x[(x^{1/q})^p] = p(x^{1/q})^{p-1}D_x(x^{1/q})$$

$$= px^{p/q - 1/q}\frac{1}{q}x^{1/q - 1} = \frac{p}{q}x^{p/q - 1} \qquad \blacksquare$$

Theorem D

(Vector limits). Let $\mathbf{F}(t) = f(t)\mathbf{i} + g(t)\mathbf{j}$. Then $\mathbf{F}$ has a limit at c if and only if f and g have limits at c. In that case,

$$\lim_{t \to c} \mathbf{F}(t) = \left[\lim_{t \to c} f(t)\right]\mathbf{i} + \left[\lim_{t \to c} g(t)\right]\mathbf{j}$$

Proof First, note that for any vector $\mathbf{u} = u_1\mathbf{i} + u_2\mathbf{j}$,

$$\boxed{|u_1| \leq |\mathbf{u}| \leq |u_1| + |u_2|}$$

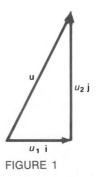

FIGURE 1

This fact is readily seen from Figure 1.

Now suppose $\lim_{t \to c} \mathbf{F}(t) = \mathbf{L} = a\mathbf{i} + b\mathbf{j}$. This means that for any $\varepsilon > 0$ there is a corresponding $\delta > 0$ such that

$$0 < |t - c| < \delta \quad \Rightarrow \quad |\mathbf{F}(t) - \mathbf{L}| < \varepsilon$$

But by the left part of the boxed inequality,

$$|f(t) - a| \leq |\mathbf{F}(t) - \mathbf{L}|$$

and so

$$0 < |t - c| < \delta \quad \Rightarrow \quad |f(t) - a| < \varepsilon$$

This shows that $\lim_{t \to c} f(t) = a$. A similar argument establishes that $\lim_{t \to c} g(t) = b$. The first half of our theorem is complete.

Conversely, suppose that

$$\lim_{t \to c} f(t) = a \qquad \lim_{t \to c} g(t) = b$$

and let $\mathbf{L} = a\mathbf{i} + b\mathbf{j}$. For any given $\varepsilon > 0$, there is a corresponding $\delta > 0$ such that $0 < |t - c| < \delta$ implies both

$$|f(t) - a| < \frac{\varepsilon}{2} \quad \text{and} \quad |g(t) - b| < \frac{\varepsilon}{2}$$

Hence, by the right part of the boxed inequality,

$$0 < |t - c| < \delta \quad \Rightarrow \quad |\mathbf{F}(t) - \mathbf{L}| \leq \frac{\varepsilon}{2} + \frac{\varepsilon}{2} = \varepsilon$$

Thus,

$$\lim_{t \to c} \mathbf{F}(t) = \mathbf{L} = a\mathbf{i} + b\mathbf{j} = \lim_{t \to c} f(t)\mathbf{i} + \lim_{t \to c} g(t)\mathbf{j} \qquad \blacksquare$$

A.3
A Backward Look

Those of you who have persisted to the end of this book are to be complemented. Now it is time to turn around, to survey the road we have traveled, to develop a bit of perspective. Just what have we been studying? What is calculus, this many-faceted subject that some have called the greatest invention of the human mind?

Suppose we divide school mathematics into three broad areas: geometry, algebra, and calculus. Then, one author has said that geometry is the study of *shape*, algebra is the study of *quantity*, and calculus is the study of *change*. It is true that the part of our subject known as differential calculus (including differential equations) is largely the study of rates of change. But what about integral calculus? It hardly fits under this rubric.

Another writer has suggested that we can distinguish between algebra and calculus by saying that algebra deals with *finite processes* (addition, multiplication, exponentiation, and so on), whereas calculus deals with *infinite processes* (differentiation, integration, summation of series, and so on).

We prefer to define calculus as the study of *limits*. The notion of limit pervades all of calculus; every major idea is determined by a certain kind of limit. The accompanying chart has the word limit at the center; the principal concepts (in ovals) emanate from it.

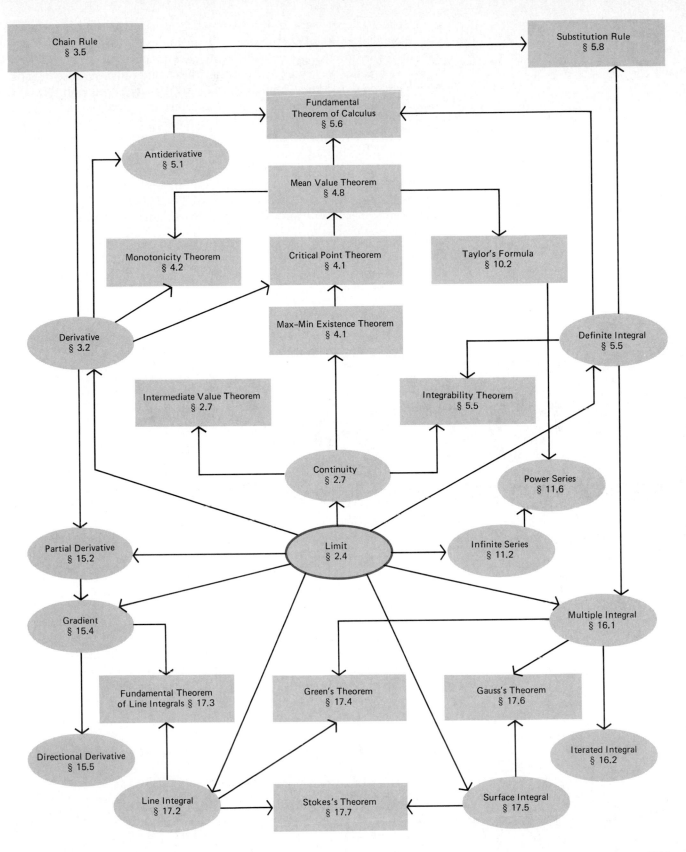

The machinery of calculus is embodied in its theorems. If we include the many that are stated in problems, the theorems in this book must number in the hundreds. Undoubtedly, you have managed to forget most of them—a practice we actually encourage. One should rather master the most important ones and know how they fit together. To help you with this task, our chart lists the principal theorems of calculus (in rectangles) and suggests how they relate to the concepts and to each other.

A student who understands what our chart represents is well prepared for those subjects that build on elementary calculus. And what are they? Suffice it to mention advanced calculus, ordinary and partial differential equations, integral equations, power series, Fourier series, real analysis, differential geometry, complex analysis, probability, Lebesgue integration, measure theory, calculus of variations, and abstract analysis. There is plenty to challenge the best of us.

A.4
Numerical Tables

TABLE I
Trigonometric Functions, Degree Measure

Angle	sin	tan	cot	cos	—
0.0°	0.0000	0.0000	—	1.0000	90.0°
0.5°	0.0087	0.0087	114.59	1.0000	89.5°
1.0°	0.0175	0.0175	57.290	0.9998	89.0°
1.5°	0.0262	0.0262	38.188	0.9997	88.5°
2.0°	0.0349	0.0349	28.636	0.9994	88.0°
2.5°	0.0436	0.0437	22.904	0.9990	87.5°
3.0°	0.0523	0.0524	19.081	0.9986	87.0°
3.5°	0.0610	0.0612	16.350	0.9981	86.5°
4.0°	0.0698	0.0699	14.301	0.9976	86.0°
4.5°	0.0785	0.0787	12.706	0.9969	85.5°
5.0°	0.0872	0.0875	11.430	0.9962	85.0°
5.5°	0.0958	0.0963	10.385	0.9954	84.5°
6.0°	0.1045	0.1051	9.5144	0.9945	84.0°
6.5°	0.1132	0.1139	8.7769	0.9936	83.5°
7.0°	0.1219	0.1228	8.1443	0.9925	83.0°
7.5°	0.1305	0.1317	7.5958	0.9914	82.5°
8.0°	0.1392	0.1405	7.1154	0.9903	82.0°
8.5°	0.1478	0.1495	6.6912	0.9890	81.5°
9.0°	0.1564	0.1584	6.3138	0.9877	81.0°
9.5°	0.1650	0.1673	5.9758	0.9863	80.5°
10.0°	0.1736	0.1763	5.6713	0.9848	80.0°
10.5°	0.1822	0.1853	5.3955	0.9833	79.5°
11.0°	0.1908	0.1944	5.1446	0.9816	79.0°
11.5°	0.1994	0.2035	4.9152	0.9799	78.5°
12.0°	0.2079	0.2126	4.7046	0.9781	78.0°
12.5°	0.2164	0.2217	4.5107	0.9763	77.5°
13.0°	0.2250	0.2309	4.3315	0.9744	77.0°
13.5°	0.2334	0.2401	4.1653	0.9724	76.5°
14.0°	0.2419	0.2493	4.0108	0.9703	76.0°
14.5°	0.2504	0.2586	3.8667	0.9681	75.5°
15.0°	0.2588	0.2679	3.7321	0.9659	75.0°
15.5°	0.2672	0.2773	3.6059	0.9636	74.5°
16.0°	0.2756	0.2867	3.4874	0.9613	74.0°
16.5°	0.2840	0.2962	3.3759	0.9588	73.5°
17.0°	0.2924	0.3057	3.2709	0.9563	73.0°
17.5°	0.3007	0.3153	3.1716	0.9537	72.5°
18.0°	0.3090	0.3249	3.0777	0.9511	72.0°
18.5°	0.3173	0.3346	2.9887	0.9483	71.5°
19.0°	0.3256	0.3443	2.9042	0.9455	71.0°
19.5°	0.3338	0.3541	2.8239	0.9426	70.5°
20.0°	0.3420	0.3640	2.7475	0.9397	70.0°
20.5°	0.3502	0.3739	2.6746	0.9367	69.5°
21.0°	0.3584	0.3839	2.6051	0.9336	69.0°
21.5°	0.3665	0.3939	2.5386	0.9304	68.5°
22.0°	0.3746	0.4040	2.4751	0.9272	68.0°
22.5°	0.3827	0.4142	2.4142	0.9239	67.5°
—	cos	cot	tan	sin	Angle

Angle	sin	tan	cot	cos	—
22.5°	0.3827	0.4142	2.4142	0.9239	67.5°
23.0°	0.3907	0.4245	2.3559	0.9205	67.0°
23.5°	0.3987	0.4348	2.2998	0.9171	66.5°
24.0°	0.4067	0.4452	2.2460	0.9135	66.0°
24.5°	0.4147	0.4557	2.1943	0.9100	65.5°
25.0°	0.4226	0.4663	2.1445	0.9063	65.0°
25.5°	0.4305	0.4770	2.0965	0.9026	64.5°
26.0°	0.4384	0.4877	2.0503	0.8988	64.0°
26.5°	0.4462	0.4986	2.0057	0.8949	63.5°
27.0°	0.4540	0.5095	1.9626	0.8910	63.0°
27.5°	0.4617	0.5206	1.9210	0.8870	62.5°
28.0°	0.4695	0.5317	1.8807	0.8829	62.0°
28.5°	0.4772	0.5430	1.8418	0.8788	61.5°
29.0°	0.4848	0.5543	1.8040	0.8746	61.0°
29.5°	0.4924	0.5658	1.7675	0.8704	60.5°
30.0°	0.5000	0.5774	1.7321	0.8660	60.0°
30.5°	0.5075	0.5890	1.6977	0.8616	59.5°
31.0°	0.5150	0.6009	1.6643	0.8572	59.0°
31.5°	0.5225	0.6128	1.6319	0.8526	58.5°
32.0°	0.5299	0.6249	1.6003	0.8480	58.0°
32.5°	0.5373	0.6371	1.5697	0.8434	57.5°
33.0°	0.5446	0.6494	1.5399	0.8387	57.0°
33.5°	0.5519	0.6619	1.5108	0.8339	56.5°
34.0°	0.5592	0.6745	1.4826	0.8290	56.0°
34.5°	0.5664	0.6873	1.4550	0.8241	55.5°
35.0°	0.5736	0.7002	1.4281	0.8192	55.0°
35.5°	0.5807	0.7133	1.4019	0.8141	54.5°
36.0°	0.5878	0.7265	1.3764	0.8090	54.0°
36.5°	0.5948	0.7400	1.3514	0.8039	53.5°
37.0°	0.6018	0.7536	1.3270	0.7986	53.0°
37.5°	0.6088	0.7673	1.3032	0.7934	52.5°
38.0°	0.6157	0.7813	1.2799	0.7880	52.0°
38.5°	0.6225	0.7954	1.2572	0.7826	51.5°
39.0°	0.6293	0.8098	1.2349	0.7771	51.0°
39.5°	0.6361	0.8243	1.2131	0.7716	50.5°
40.0°	0.6428	0.8391	1.1918	0.7660	50.0°
40.5°	0.6494	0.8541	1.1708	0.7604	49.5°
41.0°	0.6561	0.8693	1.1504	0.7547	49.0°
41.5°	0.6626	0.8847	1.1303	0.7490	48.5°
42.0°	0.6691	0.9004	1.1106	0.7431	48.0°
42.5°	0.6756	0.9163	1.0913	0.7373	47.5°
43.0°	0.6820	0.9325	1.0724	0.7314	47.0°
43.5°	0.6884	0.9490	1.0538	0.7254	46.5°
44.0°	0.6947	0.9657	1.0355	0.7193	46.0°
44.5°	0.7009	0.9827	1.0176	0.7133	45.5°
45.0°	0.7071	1.0000	1.0000	0.7071	45.0°
—	cos	cot	tan	sin	Angle

TABLE II
Trigonometric Functions, Radian Measure

Radians	sin	cos	tan	Radians	sin	cos	tan
0.00	0.0000	1.0000	0.0000	0.40	0.3894	0.9211	0.4228
0.01	0.0100	1.0000	0.0100	0.41	0.3986	0.9171	0.4346
0.02	0.0200	0.9998	0.0200	0.42	0.4078	0.9131	0.4466
0.03	0.0300	0.9996	0.0300	0.43	0.4169	0.9090	0.4586
0.04	0.0400	0.9992	0.0400	0.44	0.4259	0.9048	0.4708
0.05	0.0500	0.9988	0.0500	0.45	0.4350	0.9004	0.4831
0.06	0.0600	0.9982	0.0601	0.46	0.4439	0.8961	0.4954
0.07	0.0699	0.9976	0.0701	0.47	0.4529	0.8916	0.5080
0.08	0.0799	0.9968	0.0802	0.48	0.4618	0.8870	0.5206
0.09	0.0899	0.9960	0.0902	0.49	0.4706	0.8823	0.5334
0.10	0.0998	0.9950	0.1003	0.50	0.4794	0.8776	0.5463
0.11	0.1098	0.9940	0.1104	0.51	0.4882	0.8727	0.5594
0.12	0.1197	0.9928	0.1206	0.52	0.4969	0.8678	0.5726
0.13	0.1296	0.9916	0.1307	0.53	0.5055	0.8628	0.5859
0.14	0.1395	0.9902	0.1409	0.54	0.5141	0.8577	0.5994
0.15	0.1494	0.9888	0.1511	0.55	0.5227	0.8525	0.6131
0.16	0.1593	0.9872	0.1614	0.56	0.5312	0.8473	0.6269
0.17	0.1692	0.9856	0.1717	0.57	0.5396	0.8419	0.6410
0.18	0.1790	0.9838	0.1820	0.58	0.5480	0.8365	0.6552
0.19	0.1889	0.9820	0.1923	0.59	0.5564	0.8309	0.6696
0.20	0.1987	0.9801	0.2027	0.60	0.5646	0.8253	0.6841
0.21	0.2085	0.9780	0.2131	0.61	0.5729	0.8196	0.6989
0.22	0.2182	0.9759	0.2236	0.62	0.5810	0.8139	0.7139
0.23	0.2280	0.9737	0.2341	0.63	0.5891	0.8080	0.7291
0.24	0.2377	0.9713	0.2447	0.64	0.5972	0.8021	0.7445
0.25	0.2474	0.9689	0.2553	0.65	0.6052	0.7961	0.7602
0.26	0.2571	0.9664	0.2660	0.66	0.6131	0.7900	0.7761
0.27	0.2667	0.9638	0.2768	0.67	0.6210	0.7838	0.7923
0.28	0.2764	0.9611	0.2876	0.68	0.6288	0.7776	0.8087
0.29	0.2860	0.9582	0.2984	0.69	0.6365	0.7712	0.8253
0.30	0.2955	0.9553	0.3093	0.70	0.6442	0.7648	0.8423
0.31	0.3051	0.9523	0.3203	0.71	0.6518	0.7584	0.8595
0.32	0.3146	0.9492	0.3314	0.72	0.6594	0.7518	0.8771
0.33	0.3240	0.9460	0.3425	0.73	0.6669	0.7452	0.8949
0.34	0.3335	0.9428	0.3537	0.74	0.6743	0.7385	0.9131
0.35	0.3429	0.9394	0.3650	0.75	0.6816	0.7317	0.9316
0.36	0.3523	0.9359	0.3764	0.76	0.6889	0.7248	0.9505
0.37	0.3616	0.9323	0.3879	0.77	0.6961	0.7179	0.9697
0.38	0.3709	0.9287	0.3994	0.78	0.7033	0.7109	0.9893
0.39	0.3802	0.9249	0.4111	0.79	0.7104	0.7038	1.009

TABLE II
Trigonometric Functions, Radian Measure (cont.)

Radians	sin	cos	tan	Radians	sin	cos	tan
0.80	0.7174	0.6967	1.030	1.20	0.9320	0.3624	2.572
0.81	0.7243	0.6895	1.050	1.21	0.9356	0.3530	2.650
0.82	0.7311	0.6822	1.072	1.22	0.9391	0.3436	2.733
0.83	0.7379	0.6749	1.093	1.23	0.9425	0.3342	2.820
0.84	0.7446	0.6675	1.116	1.24	0.9458	0.3248	2.912
0.85	0.7513	0.6600	1.138	1.25	0.9490	0.3153	3.010
0.86	0.7578	0.6524	1.162	1.26	0.9521	0.3058	3.113
0.87	0.7643	0.6448	1.185	1.27	0.9551	0.2963	3.224
0.88	0.7707	0.6372	1.210	1.28	0.9580	0.2867	3.341
0.89	0.7771	0.6294	1.235	1.29	0.9608	0.2771	3.467
0.90	0.7833	0.6216	1.260	1.30	0.9636	0.2675	3.602
0.91	0.7895	0.6137	1.286	1.31	0.9662	0.2579	3.747
0.92	0.7956	0.6058	1.313	1.32	0.9687	0.2482	3.903
0.93	0.8016	0.5978	1.341	1.33	0.9711	0.2385	4.072
0.94	0.8076	0.5898	1.369	1.34	0.9735	0.2288	4.256
0.95	0.8134	0.5817	1.398	1.35	0.9757	0.2190	4.455
0.96	0.8192	0.5735	1.428	1.36	0.9779	0.2092	4.673
0.97	0.8249	0.5653	1.459	1.37	0.9799	0.1994	4.913
0.98	0.8305	0.5570	1.491	1.38	0.9819	0.1896	5.177
0.99	0.8360	0.5487	1.524	1.39	0.9837	0.1798	5.471
1.00	0.8415	0.5403	1.557	1.40	0.9854	0.1700	5.798
1.01	0.8468	0.5319	1.592	1.41	0.9871	0.1601	6.165
1.02	0.8521	0.5234	1.628	1.42	0.9887	0.1502	6.581
1.03	0.8573	0.5148	1.665	1.43	0.9901	0.1403	7.055
1.04	0.8624	0.5062	1.704	1.44	0.9915	0.1304	7.602
1.05	0.8674	0.4976	1.743	1.45	0.9927	0.1205	8.238
1.06	0.8724	0.4889	1.784	1.46	0.9939	0.1106	8.989
1.07	0.8772	0.4801	1.827	1.47	0.9949	0.1006	9.887
1.08	0.8820	0.4713	1.871	1.48	0.9959	0.0907	10.98
1.09	0.8866	0.4625	1.917	1.49	0.9967	0.0807	12.35
1.10	0.8912	0.4536	1.965	1.50	0.9975	0.0707	14.10
1.11	0.8957	0.4447	2.014	1.51	0.9982	0.0608	16.43
1.12	0.9001	0.4357	2.066	1.52	0.9987	0.0508	19.67
1.13	0.9044	0.4267	2.120	1.53	0.9992	0.0408	24.50
1.14	0.9086	0.4176	2.176	1.54	0.9995	0.0308	32.46
1.15	0.9128	0.4085	2.234	1.55	0.9998	0.0208	48.08
1.16	0.9168	0.3993	2.296	1.56	0.9999	0.0108	92.62
1.17	0.9208	0.3902	2.360	1.57	1.0000	0.0008	1256.
1.18	0.9246	0.3809	2.427				
1.19	0.9284	0.3717	2.498				

TABLE III
Natural Logarithms

	0.00	0.01	0.02	0.03	0.04	0.05	0.06	0.07	0.08	0.09
1.0	0.0000	0.0100	0.0198	0.0296	0.0392	0.0488	0.0583	0.0677	0.0770	0.0862
1.1	0.0953	0.1044	0.1133	0.1222	0.1310	0.1398	0.1484	0.1570	0.1655	0.1740
1.2	0.1823	0.1906	0.1989	0.2070	0.2151	0.2231	0.2311	0.2390	0.2469	0.2546
1.3	0.2624	0.2700	0.2776	0.2852	0.2927	0.3001	0.3075	0.3148	0.3221	0.3293
1.4	0.3365	0.3436	0.3507	0.3577	0.3646	0.3716	0.3784	0.3853	0.3920	0.3988
1.5	0.4055	0.4121	0.4187	0.4253	0.4318	0.4383	0.4447	0.4511	0.4574	0.4637
1.6	0.4700	0.4762	0.4824	0.4886	0.4947	0.5008	0.5068	0.5128	0.5188	0.5247
1.7	0.5306	0.5365	0.5423	0.5481	0.5539	0.5596	0.5653	0.5710	0.5766	0.5822
1.8	0.5878	0.5933	0.5988	0.6043	0.6098	0.6152	0.6206	0.6259	0.6313	0.6366
1.9	0.6419	0.6471	0.6523	0.6575	0.6627	0.6678	0.6729	0.6780	0.6831	0.6881
2.0	0.6931	0.6981	0.7031	0.7080	0.7130	0.7178	0.7227	0.7275	0.7324	0.7372
2.1	0.7419	0.7467	0.7514	0.7561	0.7608	0.7655	0.7701	0.7747	0.7793	0.7839
2.2	0.7885	0.7930	0.7975	0.8020	0.8065	0.8109	0.8154	0.8198	0.8242	0.8286
2.3	0.8329	0.8372	0.8416	0.8459	0.8502	0.8544	0.8587	0.8629	0.8671	0.8713
2.4	0.8755	0.8796	0.8838	0.8879	0.8920	0.8961	0.9002	0.9042	0.9083	0.9123
2.5	0.9163	0.9203	0.9243	0.9282	0.9322	0.9361	0.9400	0.9439	0.9478	0.9517
2.6	0.9555	0.9594	0.9632	0.9670	0.9708	0.9746	0.9783	0.9821	0.9858	0.9895
2.7	0.9933	0.9969	1.0006	1.0043	1.0080	1.0116	1.0152	1.0188	1.0225	1.0260
2.8	1.0296	1.0332	1.0367	1.0403	1.0438	1.0473	1.0508	1.0543	1.0578	1.0613
2.9	1.0647	1.0682	1.0716	1.0750	1.0784	1.0818	1.0852	1.0886	1.0919	1.0953
3.0	1.0986	1.1019	1.1053	1.1086	1.1119	1.1151	1.1184	1.1217	1.1249	1.1282
3.1	1.1314	1.1346	1.1378	1.1410	1.1442	1.1474	1.1506	1.1537	1.1569	1.1600
3.2	1.1632	1.1663	1.1694	1.1725	1.1756	1.1787	1.1817	1.1848	1.1878	1.1909
3.3	1.1939	1.1970	1.2000	1.2030	1.2060	1.2090	1.2119	1.2149	1.2179	1.2208
3.4	1.2238	1.2267	1.2296	1.2326	1.2355	1.2384	1.2413	1.2442	1.2470	1.2499
3.5	1.2528	1.2556	1.2585	1.2613	1.2641	1.2669	1.2698	1.2726	1.2754	1.2782
3.6	1.2809	1.2837	1.2865	1.2892	1.2920	1.2947	1.2975	1.3002	1.3029	1.3056
3.7	1.3083	1.3110	1.3137	1.3164	1.3191	1.3218	1.3244	1.3271	1.3297	1.3324
3.8	1.3350	1.3376	1.3403	1.3429	1.3455	1.3481	1.3507	1.3533	1.3558	1.3584
3.9	1.3610	1.3635	1.3661	1.3686	1.3712	1.3737	1.3762	1.3788	1.3813	1.3838
4.0	1.3863	1.3888	1.3913	1.3938	1.3962	1.3987	1.4012	1.4036	1.4061	1.4085
4.1	1.4110	1.4134	1.4159	1.4183	1.4207	1.4231	1.4255	1.4279	1.4303	1.4327
4.2	1.4351	1.4375	1.4398	1.4422	1.4446	1.4469	1.4493	1.4516	1.4540	1.4563
4.3	1.4586	1.4609	1.4633	1.4656	1.4679	1.4702	1.4725	1.4748	1.4770	1.4793
4.4	1.4816	1.4839	1.4861	1.4884	1.4907	1.4929	1.4952	1.4974	1.4996	1.5019
4.5	1.5041	1.5063	1.5085	1.5107	1.5129	1.5151	1.5173	1.5195	1.5217	1.5239
4.6	1.5261	1.5282	1.5304	1.5326	1.5347	1.5369	1.5390	1.5412	1.5433	1.5454
4.7	1.5476	1.5497	1.5518	1.5539	1.5560	1.5581	1.5602	1.5623	1.5644	1.5665
4.8	1.5686	1.5707	1.5728	1.5748	1.5769	1.5790	1.5810	1.5831	1.5851	1.5872
4.9	1.5892	1.5913	1.5933	1.5953	1.5974	1.5994	1.6014	1.6034	1.6054	1.6074
5.0	1.6094	1.6114	1.6134	1.6154	1.6174	1.6194	1.6214	1.6233	1.6253	1.6273
5.1	1.6292	1.6312	1.6332	1.6351	1.6371	1.6390	1.6409	1.6429	1.6448	1.6467
5.2	1.6487	1.6506	1.6525	1.6544	1.6563	1.6582	1.6601	1.6620	1.6639	1.6658
5.3	1.6677	1.6696	1.6715	1.6734	1.6752	1.6771	1.6790	1.6808	1.6827	1.6845
5.4	1.6864	1.6882	1.6901	1.6919	1.6938	1.6956	1.6974	1.6993	1.7011	1.7029

$$\ln (N \cdot 10^m) = \ln N + m \ln 10, \qquad \ln 10 = 2.3026$$

TABLE III
Natural Logarithms (cont.)

	0.00	0.01	0.02	0.03	0.04	0.05	0.06	0.07	0.08	0.09
5.5	1.7047	1.7066	1.7084	1.7102	1.7120	1.7138	1.7156	1.7174	1.7192	1.7210
5.6	1.7228	1.7246	1.7263	1.7281	1.7299	1.7317	1.7334	1.7352	1.7370	1.7387
5.7	1.7405	1.7422	1.7440	1.7457	1.7475	1.7492	1.7509	1.7527	1.7544	1.7561
5.8	1.7579	1.7596	1.7613	1.7630	1.7647	1.7664	1.7682	1.7699	1.7716	1.7733
5.9	1.7750	1.7766	1.7783	1.7800	1.7817	1.7834	1.7851	1.7867	1.7884	1.7901
6.0	1.7918	1.7934	1.7951	1.7967	1.7984	1.8001	1.8017	1.8034	1.8050	1.8066
6.1	1.8083	1.8099	1.8116	1.8132	1.8148	1.8165	1.8181	1.8197	1.8213	1.8229
6.2	1.8245	1.8262	1.8278	1.8294	1.8310	1.8326	1.8342	1.8358	1.8374	1.8390
6.3	1.8406	1.8421	1.8437	1.8453	1.8469	1.8485	1.8500	1.8516	1.8532	1.8547
6.4	1.8563	1.8579	1.8594	1.8610	1.8625	1.8641	1.8656	1.8672	1.8687	1.8703
6.5	1.8718	1.8733	1.8749	1.8764	1.8779	1.8795	1.8810	1.8825	1.8840	1.8856
6.6	1.8871	1.8886	1.8901	1.8916	1.8931	1.8946	1.8961	1.8976	1.8991	1.9006
6.7	1.9021	1.9036	1.9051	1.9066	1.9081	1.9095	1.9110	1.9125	1.9140	1.9155
6.8	1.9169	1.9184	1.9199	1.9213	1.9228	1.9242	1.9257	1.9272	1.9286	1.9301
6.9	1.9315	1.9330	1.9344	1.9359	1.9373	1.9387	1.9402	1.9416	1.9430	1.9445
7.0	1.9459	1.9473	1.9488	1.9502	1.9516	1.9530	1.9544	1.9559	1.9573	1.9587
7.1	1.9601	1.9615	1.9629	1.9643	1.9657	1.9671	1.9685	1.9699	1.9713	1.9727
7.2	1.9741	1.9755	1.9769	1.9782	1.9796	1.9810	1.9824	1.9838	1.9851	1.9865
7.3	1.9879	1.9892	1.9906	1.9920	1.9933	1.9947	1.9961	1.9974	1.9988	2.0001
7.4	2.0015	2.0028	2.0042	2.0055	2.0069	2.0082	2.0096	2.0109	2.0122	2.0136
7.5	2.0149	2.0162	2.0176	2.0189	2.0202	2.0215	2.0229	2.0242	2.0255	2.0268
7.6	2.0282	2.0295	2.0308	2.0321	2.0334	2.0347	2.0360	2.0373	2.0386	2.0399
7.7	2.0412	2.0425	2.0438	2.0451	2.0464	2.0477	2.0490	2.0503	2.0516	2.0528
7.8	2.0541	2.0554	2.0567	2.0580	2.0592	2.0605	2.0618	2.0631	2.0643	2.0656
7.9	2.0669	2.0681	2.0694	2.0707	2.0719	2.0732	2.0744	2.0757	2.0769	2.0782
8.0	2.0794	2.0807	2.0819	2.0832	2.0844	2.0857	2.0869	2.0882	2.0894	2.0906
8.1	2.0919	2.0931	2.0943	2.0956	2.0968	2.0980	2.0992	2.1005	2.1017	2.1029
8.2	2.1041	2.1054	2.1066	2.1078	2.1090	2.1102	2.1114	2.1126	2.1138	2.1150
8.3	2.1163	2.1175	2.1187	2.1190	2.1211	2.1223	2.1235	2.1247	2.1258	2.1270
8.4	2.1282	2.1294	2.1306	2.1318	2.1330	2.1342	2.1353	2.1365	2.1377	2.1389
8.5	2.1401	2.1412	2.1424	2.1436	2.1448	2.1459	2.1471	2.1483	2.1494	2.1506
8.6	2.1518	2.1529	2.1541	2.1552	2.1564	2.1576	2.1587	2.1599	2.1610	2.1622
8.7	2.1633	2.1645	2.1656	2.1668	2.1679	2.1691	2.1702	2.1713	2.1725	2.1736
8.8	2.1748	2.1759	2.1770	2.1782	2.1793	2.1804	2.1815	2.1827	2.1838	2.1849
8.9	2.1861	2.1872	2.1883	2.1894	2.1905	2.1917	2.1928	2.1939	2.1950	2.1961
9.0	2.1972	2.1983	2.1994	2.2006	2.2017	2.2028	2.2039	2.2050	2.2061	2.2072
9.1	2.2083	2.2094	2.2105	2.2116	2.2127	2.2138	2.2148	2.2159	2.2170	2.2181
9.2	2.2192	2.2203	2.2214	2.2225	2.2235	2.2246	2.2257	2.2268	2.2279	2.2289
9.3	2.2300	2.2311	2.2322	2.2332	2.2343	2.2354	2.2364	2.2375	2.2386	2.2396
9.4	2.2407	2.2418	2.2428	2.2439	2.2450	2.2460	2.2471	2.2481	2.2492	2.2502
9.5	2.2513	2.2523	2.2534	2.2544	2.2555	2.2565	2.2576	2.2586	2.2597	2.2607
9.6	2.2618	2.2628	2.2638	2.2649	2.2659	2.2670	2.2680	2.2690	2.2701	2.2711
9.7	2.2721	2.2732	2.2742	2.2752	2.2762	2.2773	2.2783	2.2793	2.2803	2.2814
9.8	2.2824	2.2834	2.2844	2.2854	2.2865	2.2875	2.2885	2.2895	2.2905	2.2915
9.9	2.2925	2.2935	2.2946	2.2956	2.2966	2.2976	2.2986	2.2996	2.3006	2.3016

TABLE IV
Exponential and Hyperbolic Functions

x	e^x	e^{-x}	$\sinh x$	$\cosh x$	$\tanh x$
0.00	1.0000	1.0000	0.0000	1.0000	0.0000
0.01	1.0101	0.9900	0.0100	1.0001	0.0100
0.02	1.0202	0.9802	0.0200	1.0002	0.0200
0.03	1.0305	0.9704	0.0300	1.0005	0.0300
0.04	1.0408	0.9608	0.0400	1.0008	0.0400
0.05	1.0513	0.9512	0.0500	1.0013	0.0500
0.06	1.0618	0.9418	0.0600	1.0018	0.0599
0.07	1.0725	0.9324	0.0701	1.0025	0.0699
0.08	1.0833	0.9231	0.0801	1.0032	0.0798
0.09	1.0942	0.9139	0.0901	1.0041	0.0898
0.10	1.1052	0.9048	0.1002	1.0050	0.0997
0.11	1.1163	0.8958	0.1102	1.0061	0.1096
0.12	1.1275	0.8869	0.1203	1.0072	0.1194
0.13	1.1388	0.8781	0.1304	1.0085	0.1293
0.14	1.1503	0.9694	0.1405	1.0098	0.1391
0.15	1.1618	0.8607	0.1506	1.0113	0.1489
0.16	1.1735	0.8521	0.1607	1.0128	0.1586
0.17	1.1853	0.8437	0.1708	1.0145	0.1684
0.18	1.1972	0.8353	0.1810	1.0162	0.1781
0.19	1.2092	0.8270	0.1911	1.0181	0.1877
0.20	1.2214	0.8187	0.2013	1.0201	0.1974
0.21	1.2337	0.8106	0.2115	1.0221	0.2070
0.22	1.2461	0.8025	0.2218	1.0243	0.2165
0.23	1.2586	0.7945	0.2320	1.0266	0.2260
0.24	1.2712	0.7866	0.2423	1.0289	0.2355
0.25	1.2840	0.7788	0.2526	1.0314	0.2449
0.26	1.2969	0.7711	0.2629	1.0340	0.2543
0.27	1.3100	0.7634	0.2733	1.0367	0.2636
0.28	1.3231	0.7558	0.2837	1.0395	0.2729
0.29	1.3364	0.7483	0.2941	1.0423	0.2821
0.30	1.3499	0.7408	0.3045	1.0453	0.2913
0.31	1.3634	0.7334	0.3150	1.0484	0.3004
0.32	1.3771	0.7261	0.3255	1.0516	0.3095
0.33	1.3910	0.7189	0.3360	1.0549	0.3185
0.34	1.4049	0.7118	0.3466	1.0584	0.3275
0.35	1.4191	0.7047	0.3572	1.0619	0.3364
0.36	1.4333	0.6977	0.3678	1.0655	0.3452
0.37	1.4477	0.6907	0.3785	1.0692	0.3540
0.38	1.4623	0.6839	0.3892	1.0731	0.3627
0.39	1.4770	0.6771	0.4000	1.0770	0.3714
0.40	1.4918	0.6703	0.4108	1.0811	0.3799
0.41	1.5068	0.6637	0.4216	1.0852	0.3885
0.42	1.5220	0.6570	0.4325	1.0895	0.3969
0.43	1.5373	0.6505	0.4434	1.0939	0.4053
0.44	1.5527	0.6440	0.4543	1.0984	0.4136

TABLE IV
Exponential and Hyperbolic Functions (cont.)

x	e^x	e^{-x}	$\sinh x$	$\cosh x$	$\tanh x$
0.45	1.5683	0.6376	0.4653	1.1030	0.4219
0.46	1.5841	0.6313	0.4764	1.1077	0.4301
0.47	1.6000	0.6250	0.4875	1.1125	0.4382
0.48	1.6161	0.6188	0.4986	1.1174	0.4462
0.49	1.6323	0.6126	0.5098	1.1225	0.4542
0.50	1.6487	0.6065	0.5211	1.1276	0.4621
0.51	1.6653	0.6005	0.5324	1.1329	0.4699
0.52	1.6820	0.5945	0.5438	1.1383	0.4777
0.53	1.6989	0.5886	0.5552	1.1438	0.4854
0.54	1.7160	0.5827	0.5666	1.1494	0.4930
0.55	1.7333	0.5769	0.5782	1.1551	0.5005
0.56	1.7507	0.5712	0.5897	1.1609	0.5080
0.57	1.7683	0.5655	0.6014	1.1669	0.5154
0.58	1.7860	0.5599	0.6131	1.1730	0.5227
0.59	1.8040	0.5543	0.6248	1.1792	0.5299
0.60	1.8221	0.5488	0.6367	1.1855	0.5370
0.61	1.8044	0.5434	0.6485	1.1919	0.5441
0.62	1.8589	0.5379	0.6605	1.1984	0.5511
0.63	1.8776	0.5326	0.6725	1.2051	0.5581
0.64	1.8965	0.5273	0.6846	1.2119	0.5649
0.65	1.9155	0.5220	0.6967	1.2188	0.5717
0.66	1.9348	0.5169	0.7090	1.2258	0.5784
0.67	1.9542	0.5117	0.7213	1.2330	0.5850
0.68	1.9739	0.5066	0.7336	1.2402	0.5915
0.69	1.9937	0.5016	0.7461	1.2476	0.5980
0.70	2.0138	0.4966	0.7586	1.2552	0.6044
0.71	2.0340	0.4916	0.7712	1.2628	0.6107
0.72	2.0544	0.4868	0.7838	1.2706	0.6169
0.73	2.0751	0.4819	0.7966	1.2785	0.6231
0.74	2.0959	0.4771	0.8094	1.2865	0.6291
0.75	2.1170	0.4724	0.8223	1.2947	0.6351
0.76	2.1383	0.4677	0.8353	1.3030	0.6411
0.77	2.1598	0.4630	0.8484	1.3114	0.6469
0.78	2.1815	0.4584	0.8615	1.3199	0.6527
0.79	2.2034	0.4538	0.8748	1.3286	0.6584
0.80	2.2255	0.4493	0.8881	1.3374	0.6640
0.81	2.2479	0.4449	0.9015	1.3464	0.6696
0.82	2.2705	0.4404	0.9150	1.3555	0.6751
0.83	2.2933	0.4360	0.9286	1.3647	0.6805
0.84	2.3164	0.4317	0.9423	1.3740	0.6858
0.85	2.3396	0.4274	0.9561	1.3835	0.6911
0.86	2.3632	0.4232	0.9700	1.3932	0.6963
0.87	2.3869	0.4190	0.9840	1.4029	0.7014
0.88	2.4109	0.4148	0.9981	1.4128	0.7064
0.89	2.4351	0.4107	1.0122	1.4229	0.7114

TABLE IV
Exponential and Hyperbolic Functions (cont.)

x	e^x	e^{-x}	$\sinh x$	$\cosh x$	$\tanh x$
0.90	2.4596	0.4066	1.0265	1.4331	0.7163
0.91	2.4843	0.4025	1.0409	1.4434	0.7211
0.92	2.5093	0.3985	1.0554	1.4539	0.7259
0.93	2.5345	0.3946	1.0700	1.4645	0.7306
0.94	2.5600	0.3906	1.0847	1.4753	0.7352
0.95	2.5857	0.3867	1.0995	1.4862	0.7398
0.96	2.6117	0.3829	1.1144	1.4973	0.7443
0.97	2.6379	0.3791	1.1294	1.5085	0.7487
0.98	2.6645	0.3753	1.1446	1.5199	0.7531
0.99	2.6912	0.3716	1.1598	1.5314	0.7574
1.00	2.7183	0.3679	1.1752	1.5431	0.7616
1.05	2.8577	0.3499	1.2539	1.6038	0.7818
1.10	3.0042	0.3329	1.3356	1.6685	0.8005
1.15	3.1582	0.3166	1.4208	1.7374	0.8178
1.20	3.3201	0.3012	1.5085	1.8107	0.8337
1.25	3.4903	0.2865	1.6019	1.8884	0.8483
1.30	3.6693	0.2725	1.6984	1.9709	0.8617
1.35	3.8574	0.2592	1.7991	2.0583	0.8741
1.40	4.0552	0.2466	1.9043	2.1509	0.8854
1.45	4.2631	0.2346	2.0143	2.2488	0.8957
1.50	4.4817	0.2231	2.1293	2.3524	0.9051
1.55	4.7115	0.2122	2.2496	2.4619	0.9138
1.60	4.9530	0.2019	2.3756	2.5775	0.9217
1.65	5.2070	0.1920	2.5075	2.6995	0.9289
1.70	5.4739	0.1827	2.6456	2.8283	0.9354
1.75	5.7546	0.1738	2.7904	2.9642	0.9414
1.80	6.0496	0.1653	2.9422	3.1075	0.9468
1.85	6.3598	0.1572	3.1013	3.2585	0.9517
1.90	6.6859	0.1496	3.2682	3.4177	0.9562
1.95	7.0287	0.1423	3.4432	3.5855	0.9603
2.00	7.3891	0.1353	3.6269	3.7622	0.9640
2.05	7.7679	0.1287	3.8196	3.9483	0.9674
2.10	8.1662	0.1225	4.0219	4.1443	0.9705
2.15	8.5849	0.1165	4.2342	4.3507	0.9732
2.20	9.0250	0.1108	4.4571	4.5679	0.9757
2.25	9.4877	0.1054	4.6912	4.7966	0.9780
2.30	9.9742	0.1003	4.9370	5.0372	0.9801
2.35	10.486	0.0954	5.1951	5.2905	0.9820
2.40	11.023	0.0907	5.4662	5.5569	0.9837
2.45	11.588	0.0863	5.7510	5.8373	0.9852
2.50	12.182	0.0821	6.0502	6.1323	0.9866
2.55	12.807	0.0781	6.3645	6.4426	0.9879
2.60	13.464	0.0743	6.6947	6.7690	0.9890
2.65	14.154	0.0707	7.0417	7.1123	0.9901
2.70	14.880	0.0672	7.4063	7.4735	0.9910

TABLE IV
Exponential and Hyperbolic Functions (cont.)

x	e^x	e^{-x}	$\sinh x$	$\cosh x$	$\tanh x$
2.75	15.643	0.0639	7.7894	7.8533	0.9919
2.80	16.445	0.0608	8.1919	8.2527	0.9926
2.85	17.288	0.0578	8.6150	8.6728	0.9933
2.90	18.174	0.0550	9.0596	9.1146	0.9940
2.95	19.106	0.0523	9.5268	9.5791	0.9945
3.00	20.086	0.0498	10.018	10.068	0.9951
3.05	21.115	0.0474	10.534	10.581	0.9955
3.10	22.198	0.0450	11.076	11.122	0.9959
3.15	23.336	0.0429	11.647	11.689	0.9963
3.20	24.533	0.0408	12.246	12.287	0.9967
3.25	25.790	0.0388	12.876	12.915	0.9970
3.30	27.113	0.0369	13.538	13.575	0.9973
3.35	28.503	0.0351	14.234	14.269	0.9975
3.40	29.964	0.0334	14.965	14.999	0.9978
3.45	31.500	0.0317	15.734	15.766	0.9980
3.50	33.115	0.0302	16.543	16.573	0.9982
3.55	34.813	0.0287	17.392	17.421	0.9983
3.60	36.598	0.0273	18.286	18.313	0.9985
3.65	38.475	0.0260	19.224	19.250	0.9986
3.70	40.447	0.0247	20.211	20.236	0.9988
3.75	42.521	0.0235	21.249	21.272	0.9989
3.80	44.701	0.0224	22.339	22.362	0.9990
3.85	46.993	0.0213	23.486	23.507	0.9991
3.90	49.402	0.0202	24.691	24.711	0.9992
3.95	51.935	0.0193	25.958	25.977	0.9993
4.00	54.598	0.0183	27.290	27.308	0.9993
4.10	60.340	0.0166	30.162	30.178	0.9995
4.20	66.686	0.0150	33.336	33.351	0.9996
4.30	73.700	0.0136	36.843	36.857	0.9996
4.40	81.451	0.0123	40.719	40.732	0.9997
4.50	90.017	0.0111	45.003	45.014	0.9998
4.60	99.484	0.0101	49.737	49.747	0.9998
4.70	109.95	0.0091	54.969	54.978	0.9998
4.80	121.51	0.0082	60.751	60.759	0.9999
4.90	134.29	0.0074	67.141	67.149	0.9999
5.00	148.41	0.0067	74.203	74.210	0.9999
5.20	181.27	0.0055	90.633	90.639	0.9999
5.40	221.41	0.0045	110.70	110.71	1.0000
5.60	270.43	0.0037	135.21	135.22	1.0000
5.80	330.30	0.0030	165.15	165.15	1.0000
6.00	403.43	0.0025	201.71	201.72	1.0000
7.00	1096.6	0.0009	548.32	548.32	1.0000
8.00	2981.0	0.0003	1490.5	1490.5	1.0000
9.00	8103.1	0.0001	4051.5	4051.5	1.0000
10.00	22026.	0.00005	11013.	11013.	1.0000

Answers to Odd-Numbered Problems

PROBLEM SET 1.1   *Page 5*

1. 10   **3.** -116 **5.** $-\frac{1}{12}$ **7.** $\frac{1}{24}$ **9.** $\frac{22}{189}$ **11.** $-\frac{5}{16}$ **13.** $\frac{3}{11}$ **15.** -1 **17.** -6 **19.** $\frac{36}{49}$ **21.** $4x^2 - 9$
23. $6x^2 - 15x - 9$ **25.** $9t^4 - 6t^3 + 7t^2 - 2t + 1$ **27.** $x + 2$ **29.** $(x^2 + 2x + 4)/2$ **31.** $2/x$ **33.** $(x - 2)/(x + 1)$
35. (a) 0;   (b) undefined;   (c) 0;   (d) undefined;   (e) 1;   (f) 0
37. (a) False;   (b) true;   (c) true;   (d) true;   (e) true;   (f) false.
39. $a < b \ \Rightarrow \ a + a < a + b < b + b \ \Rightarrow \ a < (a + b)/2 < b$ **41.** (a) $2^4 \cdot 3 \cdot 5$;   (b) $2 \cdot 5 \cdot 31$;   (c) $7 \cdot 17$;   (d) $2^3 \cdot 3^3 \cdot 5^2$
43. Follow the hint to get $2q^2 = p^2$. Use Problem 42 to write each side as a product of primes. On the left, 2 will appear an odd number of times; on the right, 2 will appear an even number of times. This contradicts the uniqueness of the prime factorization.
45. Let m/n and p/q be two rational numbers. Then $m/n + p/q = (mq + np)/nq$. The latter is a ratio of two integers and hence rational.
47. (a) Rational;   (b) rational;   (c) irrational;   (d) irrational;   (e) rational;   (f) irrational.
49. Suppose $\sqrt{m} = p/q$. Then $mq^2 = p^2$. Since m is not a perfect square, its prime factorization contains at least one prime an odd number of times. This prime occurs an odd number of times in mq^2 but an even number of times in p^2, contradicting unique factorization.
51. Suppose $\sqrt{2} - \sqrt{3} + \sqrt{6} = p/q$. Then $11 - 2\sqrt{6} + 4\sqrt{3} - 6\sqrt{3} = p^2/q^2$. This implies that $-2\sqrt{6} - 2\sqrt{3}$ is rational, contradicting Problem 50.

PROBLEM SET 1.2   *Page 9*

1. 0.875   **3.** 0.15   **5.** 3.6666…   **7.** $\frac{41}{333}$ **9.** $\frac{254}{99}$ **11.** $\frac{1}{5}$
13. Those rational numbers with terminating decimal expansions. Equivalently, those rational numbers with only factors of 2 and 5 in the denominators of their reduced forms.
15. Among many possible answers are: rational, 0.000001; irrational, 0.000001010010001 ….
17. One possible answer is 3.141591010010001 ….   **19.** No.   **21.** Irrational.   **23.** 14.071068   **25.** 0.04491545
27. 12.433228   **29.** 7.8761×10^4 **31.** 9.6221716   **33.** (a) 25.482842;   (b) 9.16925;   (c) 2046.9136
35. (a) -2;   (b) -2;   (c) $\frac{22}{9}$;   (d) 1;   (e) $\frac{3}{2}$;   (f) $\sqrt{2}$

1. (a)

(b)

(c)

(d)

(e)

(f)

3. $(-\infty, 12)$;

5. $\left[-\frac{5}{3}, \infty\right)$;

7. $(2, \infty)$;

9. $\left(-\frac{9}{2}, -2\right)$;

11. $\left[-\frac{2}{5}, \frac{3}{5}\right)$;

13. $\left(\frac{1}{2}, 3\right)$;

15. $(-4, 3)$;

17. $\left[-\frac{1}{3}, 4\right]$;

19. $(-\infty, -3) \cup \left(\frac{1}{2}, \infty\right)$;

21. $\left[-5, \frac{1}{2}\right)$;

23. $(-\infty, 0) \cup \left(\frac{1}{5}, \infty\right)$;

25. $\left(-\infty, \frac{2}{3}\right) \cup \left[\frac{3}{4}, \infty\right)$;

27. $(-\infty, -10) \cup (-4, \infty)$;

29. $\left[-\frac{7}{3}, -2\right] \cup \left[\frac{1}{2}, \infty\right)$;

31. $\left(-\frac{3}{2}, \frac{1}{3}\right) \cup \left(\frac{1}{3}, 5\right)$;

33. $(-\infty, -1) \cup (0, 6)$;

35. (a) $(-2, 1)$; (b) $(-2, \infty)$; (c) none.   37. (a) $[-3, -1] \cup [2, \infty)$; (b) $(-\infty, -2] \cup [2, \infty)$; (c) $(-2, -1) \cup (1, 2)$
39. $\left[\frac{60}{11}, \frac{120}{13}\right]$

1. $(-5, 3)$ 3. $\left(-4, \frac{4}{3}\right)$ 5. $[-12, 24]$ 7. $(-\infty, 2) \cup (5, \infty)$ 9. $(-\infty, -3] \cup [2, \infty)$ 11. $(-\infty, -5) \cup \left(-\frac{5}{3}, 0\right) \cup (0, \infty)$
13. $[(5 - \sqrt{57})/4, (5 + \sqrt{57})/4]$ 15. $(-\infty, (-1 - \sqrt{33})/8) \cup ((-1 + \sqrt{33})/8, \infty)$
17. $|x - 3| < 0.5 \quad \Rightarrow \quad 5|x - 3| < 2.5 \qquad$ (multiplied by 5)   19. $|x - 2| < \varepsilon/6 \quad \Rightarrow \quad 6|x - 2| < \varepsilon$
$\qquad\qquad\quad \Rightarrow \quad |5(x - 3)| < 2.5 \qquad (|a||b| = |ab|) \qquad\qquad\qquad\qquad\qquad\quad \Rightarrow \quad |6(x - 2)| < \varepsilon$
$\qquad\qquad\quad \Rightarrow \quad |5x - 15| < 2.5 \qquad\qquad\qquad\qquad\qquad\qquad\qquad\qquad\qquad\quad \Rightarrow \quad |6x - 12| < \varepsilon$
21. $\delta = \varepsilon/3$ (any smaller δ would work).   23. $\delta = \varepsilon/6$ (any smaller δ would work).   25. $(-\infty, -11.5) \cup (-4.75, \infty)$
27. $\left(-\frac{4}{5}, \frac{16}{3}\right)$
31. (a) $|a - b| = |a + (-b)| \leq |a| + |-b| = |a| + |b|$;
    (b) $|a| = |a - b + b| \leq |a - b| + |b| \quad \Rightarrow \quad |a| - |b| \leq |a - b|$;
    (c) $|a + b + c| \leq |a + b| + |c| \leq |a| + |b| + |c|$
33. $|(x - 2)/(x^2 + 9)| = |x - 2|/(x^2 + 9) \leq (|x| + 2)/(x^2 + 9) \leq (|x| + 2)/9$

35. $|x^4 + \frac{1}{2}x^3 + \frac{1}{4}x^2 + \frac{1}{8}x + \frac{1}{16}| \le |x^4| + \frac{1}{2}|x^3| + \frac{1}{4}|x^2| + \frac{1}{8}|x| + \frac{1}{16} \le 1 + \frac{1}{2} + \frac{1}{4} + \frac{1}{8} + \frac{1}{16} < 2$
37. $0 \le (a - 1/a)^2 = a^2 - 2 + 1/a^2 \;\Rightarrow\; a^2 + 1/a^2 \ge 2$
39. $0 < a < b \;\Rightarrow\; \sqrt{a} < \sqrt{b}$ (Problem 30) $\;\Rightarrow\; \sqrt{a}\sqrt{a} < \sqrt{a}\sqrt{b} < \sqrt{b}\sqrt{b}$ **41.** The hint does it.

PROBLEM SET 1.5   Page 24

1. 5 **3.** $2\sqrt{2}$ **5.** 3.339 **7.** Let $A = (5, 3)$, $B = (-2, 4)$, and $C = (10, 8)$. Then $d(A, B) = \sqrt{50}$ and $d(A, C) = \sqrt{50}$.
9. $(7, -1)$ and $(7, 3)$; $(-1, -1)$ and $(-1, 3)$; $(1, 1)$ and $(5, 1)$. **11.** $\sqrt{61}/2$ **13.** $(x - 1)^2 + (y + 2)^2 = 36$
15. $(x - 2)^2 + (y + 1)^2 = 25$ **17.** $(x - 1)^2 + (y - 5)^2 = 13$ **19.** $-2 \pm 4\sqrt{2}$ **21.** $(-1, 5)$; 1 **23.** $(6, 0)$; 1 **25.** $(-\frac{1}{2}, \frac{3}{2})$; $\frac{3}{2}$
27. $(x - 4)^2 + (y - 1)^2 = 4$; $(x - 4)^2 + (y - 1)^2 = 8$ **29.** Truck: \$1458.03; air: \$1344.75 **35.** 42.57.

PROBLEM SET 1.6   Page 30

1. $\frac{5}{2}$ **3.** $-\frac{2}{7}$ **5.** $-\frac{5}{3}$ **7.** 0.1920 **9.** $4x - y - 5 = 0$ **11.** $2x + y - 4 = 0$ **13.** $5x - 2y - 4 = 0$ **15.** $x + 0 \cdot y - 2 = 0$
17. $\frac{2}{3}$; $-\frac{4}{3}$ **19.** $-\frac{2}{3}$; 2
21. (a) $2x - y - 9 = 0$; (b) $x + 2y + 3 = 0$; (c) $2x + 3y + 3 = 0$; (d) $3x - 2y - 15 = 0$; (e) $3x + 4y + 3 = 0$;
(f) $x - 3 \doteq 0$; (g) $y + 3 = 0$
23. $2x - y - 4 = 0$ **25.** Above. **27.** $(-1, 2)$; $3x - 2y + 7 = 0$ **29.** $(3, 1)$; $4x + 3y - 15 = 0$ **31.** $\frac{7}{5}$ **33.** $\frac{18}{13}$ **35.** $\frac{6}{5}$
37. $V = 120,000 - 9600t$ **39.** $N = 700,000 + 12,000n$; 1,180,000
41. (a) There is a loss of \$2000 when no item is sold. (b) 450; The profit goes up by \$450 with each item sold.
43. (1) If $B = 0$, the equation can be rewritten as $x = -C/A$, the equation of a vertical line.
(2) If $B \ne 0$, we can write $y = (-A/B)x + (-C/B)$, which has the form $y = mx + b$.
45. The equation can be written as $(2 + k)x + (-1 + 3k)y + (4 - 6k) = 0$, which is the equation of a line. If (x_0, y_0) is on both lines, then $2x_0 - y_0 + 4 = 0$ and $x_0 + 3y_0 - 6 = 0$. Therefore $(2x_0 - y_0 + 4) + k(x_0 + 3y_0 - 6) = 0$.
47. $(3, 3)$ **49.** $x \pm \sqrt{3}y = 12$

PROBLEM SET 1.7   Page 35

1.

3.

5.

7.

9.

11.

13.

15.

17.

19.

21.

23.

25.

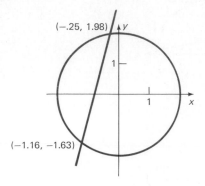

(−.25, 1.98)

(−1.16, −1.63)

27. $2\sqrt{5}$

29.

31.

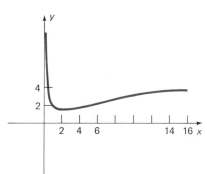

33. $d > 0 \Rightarrow$ graph crosses x-axis at two points; $d = 0 \Rightarrow$ graph touches x-axis at one point; $d < 0 \Rightarrow$ graph does not cross x-axis.

CHAPTER REVIEW 1.8   *Page 36*

True-False Quiz

1. False.   **2.** True.   **3.** False.   **4.** True.   **5.** False.   **6.** True.   **7.** False.   **8.** True.   **9.** True.   **10.** True.   **11.** True.
12. True.   **13.** True.   **14.** True.   **15.** False.   **16.** True.   **17.** True.   **18.** True.   **19.** True.   **20.** True.   **21.** True.
22. True.   **23.** True.   **24.** True.   **25.** False.   **26.** True.   **27.** False.   **28.** True.   **29.** False.   **30.** True.

Sample Test Problems

1. (a) $2, \frac{25}{4}, \frac{4}{25}$;   (b) $1, 9, 49$;   (c) $64, 8, \frac{1}{8}$ **2.** (a) $(mn + n + m)/(mn - n + m)$;   (b) $(x - 4)/(x - 8)$
3. $a/b + c/d = (ad + bc)/bd$ **4.** $4087/990$ **5.** $0.501001000100001\ldots$ **6.** 545.39 **7.** $(-2, \infty)$ **8.** $[\frac{1}{3}, 3]$ **9.** $(-3, \frac{1}{2})$
10. $(-\infty, \frac{1}{2}) \cup (2, \infty)$ **11.** $[-4, 3]$ **12.** $(-\frac{2}{3}, \frac{10}{3})$ **13.** $(-\infty, -\frac{1}{2}] \cup (1, \infty)$ **14.** $(-\infty, \frac{8}{5}] \cup [8, \infty)$
15. $\left|(2x^2 + 3x + 2)/(x^2 + 2)\right| = |2x^2 + 3x + 2|/(x^2 + 2) \le (|2x^2| + |3x| + 2)/2 = (2x^2 + 3|x| + 2)/2 \le 8$
16. Slope $AB = -\frac{4}{3}$ and slope $BC = \frac{3}{4}$; they are negative reciprocals.   **17.** $\sqrt{122}$ **18.** $(x - 6)^2 + (y - 2)^2 = 20$
19. $(4, -3)$; 5 **20.** 5
21. (a) $2x - 9y + 13 = 0$;   (b) $3x - 2y + 8 = 0$;   (c) $4x - 3y + 11 = 0$;   (d) $x + 2 = 0$;   (e) $x - y + 3 = 0$
22. If we label the three points A, B, and C, respectively, then slope $AB = \frac{4}{3}$ and slope $AC = \frac{4}{3}$.

23.

24.

25.

26.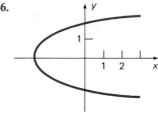

27. $(0, 4)$, $(3, 7)$ **28.** $x + 4y - 8 = 0$

PROBLEM SET 2.1 *Page 42*

1. (a) 0; (b) 3; (c) -1; (d) $k^2 - 1$; (e) 35; (f) $-\frac{3}{4}$; (g) $4t^2 - 1$; (h) $9x^2 - 1$; (i) $(1 - x^2)/x^2$
3. (a) -1; (b) -1000; (c) 100; (d) $1/(y^2 - 1)$; (e) $-1/(x + 1)$; (f) $x^2/(1 - x^2)$
5. (a) 1.1343898; (b) 138.49348; (c) 530.62951   **7.** (a) -3.2931062; (b) 1.1988766; (c) undefined.
9. (a) Does not define a function; (b) $f(x) = (4 - 3x)/(x + 1)$; (c) $f(x) = (x^2 - 1)/3$; (d) $f(x) = 3x/(1 - 3x)$
11. $2(2a + h)$ **13.** $-3/[(x + h - 2)(x - 2)]$ **15.** (a) $[-\frac{3}{2}, \infty)$; (b) $(-\infty, \frac{1}{4}) \cup (\frac{1}{4}, \infty)$; (c) $(-\infty, -3] \cup [3, \infty)$; (d) $[-5, 5]$
17. Even.

19. Neither.

21. Neither.

23. Odd.

25. Neither.

27. Even.

29. Neither.

31. Neither.

33. $T(x) = 2200 + 151x; \{0, 1, 2, \ldots, 100\}; u(x) = (2200/x) + 151; \{1, 2, \ldots, 100\}$ **35.** $E(x) = x - x^3$; estimate 0.6.
37. (a) $E(x) = 24 + 0.40x;$ (b) 240   **39.** $A(d) = (2d - \pi d^2)/4; 0 < d < 1/\pi$ **41.** $\{0, 2, 3, 6, 7, 9\}$

PROBLEM SET 2.2   *Page 49*

1. (a) $2 + \sqrt{5}$;   (b) 0;   (c) $2\sqrt{10}/3$;   (d) undefined;   (e) $\frac{3}{2}$;   (f) 1
3. (a) $(f + g)(x) = x^3 + 2 + 2/(x - 1); \{x : x \neq 1\}$;   (b) $(g/f)(x) = 2/[(x - 1)(x^3 + 2)]; \{x : x \neq 1, x \neq \sqrt[3]{-2}\}$;
   (c) $(f \circ g)(x) = 2 + 8/(x - 1)^3; \{x : x \neq 1\}$ (d) $(g \circ f)(x) = 2/(x^3 + 1); \{x : x \neq -1\}$
5. $(f \circ g)(x) = \sqrt{|x| - 4}; (g \circ f)(x) = \sqrt{x - 4}$ **7.** 0.0000195   **9.** 1.1133734
11. (a) $g(x) = \sqrt{x}; f(x) = x + 7$;   (b) $g(x) = x^{15}; f(x) = x^2 + x$
13. $p(x) = (f \circ g \circ h)(x)$, where $f(x) = \log x, g(x) = \sqrt{x}, h(x) = x^2 + 1$; or where $f(x) = \log x, g(x) = \sqrt{x + 1}, h(x) = x^2$.

15.

17.

19.

21.

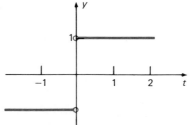

23. (a) Even; (b) odd; (c) even; (d) even; (e) odd. **25.** (a) *AF*; (b) *PF*; (c) *RF*; (d) *PF*; (e) *RF*; (f) *AF*

27. (a) $P = \sqrt{27 + \sqrt{t} + t}$; (b) 6.77 **29.** $D(t) = \begin{cases} 400t & \text{if } 0 \le t \le 1 \\ 100\sqrt{25t^2 - 18t + 9} & \text{if } t > 1 \end{cases}$ **33.** (a) $1/(1 - x)$; (b) x; (c) $1 - x$

PROBLEM SET 2.3 *Page 56*

1. (a) $4\pi/3$; (b) $-\pi/3$; (c) $-3\pi/4$; (d) 3π; (e) $10\pi/3$; (f) 4π; (g) $\pi/10$; (h) $\pi/8$; (i) $\pi/30$
3. (a) 0.58119464; (b) 8.2292274; (c) -6.8312187; (d) 0.26005406; (e) 0.07016224; (f) -0.026529
5. (a) 0.43676556; (b) 0.89957537; (c) 0.48552414; (d) -0.35320995; (e) 0.93554408; (f) -0.37754496
7. (a) 248.3004; (b) 1.2828166 **9.** (a) $\sqrt{3}/3$; (b) -1; (c) $-\sqrt{2}$; (d) 1; (e) 1; (f) -1
11. (a) $(1 + \sin z)(1 - \sin z) = 1 - \sin^2 z = \cos^2 z = 1/\sec^2 z$;
    (b) $(\sec t - 1)(\sec t + 1) = \sec^2 t - 1 = (1 + \tan^2 t) - 1 = \tan^2 t$;
    (c) $\sec t - \sin t \tan t = 1/\cos t - \sin t(\sin t/\cos t) = (1 - \sin^2 t)/\cos t = \cos^2 t/\cos t = \cos t$;
    (d) $(\sec^2 t - 1)/\sec^2 t = 1 - 1/\sec^2 t = 1 - \cos^2 t = \sin^2 t$;
    (e) $\cos t(\tan t + \cot t) = \cos t(\sin t/\cos t) + \cos t(\cos t/\sin t) = \sin t + \cos^2 t/\sin t = (\sin^2 t + \cos^2 t)/\sin t = 1/\sin t = \csc t$

13. (a)   (b)

(c)   (d)

15. (a) IV, $+$; (b) II, $-$; (c) II, $-$ **17.** (a) Even; (b) odd; (c) even; (d) odd; (e) even; (f) neither.
19. $\tan(t + \pi) = (\tan t + \tan \pi)/(1 - \tan t \tan \pi) = \tan t$ **21.** (a) 15 cm; (b) $12.5\pi/4 \approx 9.8$ cm
23. $5280/[(2.5)(2\pi)] \approx 336.1352$ rpm **25.** (a) $100\mu/(\sqrt{2} + \sqrt{2}\mu)$; (b) 50μ; (c) $50\mu/(0.540 + 0.841\mu)$; (d) 50
27. (a) $\pi/3$; (b) $5\pi/6$ **29.** (a) 0.1419; (b) 1.8925; (c) 1.7127 **31.** 25 cm² **33.** 5400π in.
35. $A = \frac{1}{2}r^2[\sin t + \pi \sin^2(t/2)]$

PROBLEM SET 2.4 *Page 62*

1. -2 **3.** 11 **5.** $\frac{1}{3}$ **7.** 5 **9.** -7 **11.** 6 **13.** 0 **15.** $-\frac{1}{3}$ **17.** $\frac{1}{2}$
19. $\frac{1}{6}$ (Your calculator may lead you astray on this one.) **21.** Does not exist. **23.** 0 **25.** Does not exist.
27. (a) 2; (b) 1; (c) does not exist; (d) 2.6; (e) 2; (f) does not exist; (g) 2; (h) 1
29. (a) 0; (b) 2; (c) does not exist; (d) 1 **31.** (a) 0; (b) does not exist; (c) 1; (d) $\frac{1}{2}$ **33.** Does not exist.
35. (a) Does not exist; (b) 0 **37.** $-1, 0, 1$ **39.** (a) Does not exist; (b) -1; (c) -3; (d) does not exist.

1. For each $\varepsilon > 0$, there is a $\delta > 0$ such that $0 < |t - a| < \delta \;\Rightarrow\; |f(t) - M| < \varepsilon$.

3. For each $\varepsilon > 0$, there is a $\delta > 0$ such that $0 < |z - d| < \delta \;\Rightarrow\; |h(z) - P| < \varepsilon$.

5. For each $\varepsilon > 0$, there is a $\delta > 0$ such that $0 < c - x < \delta \;\Rightarrow\; |f(x) - L| < \varepsilon$.

7. Let $\varepsilon > 0$ be given. Choose $\delta = \varepsilon/5$. Then $0 < |x - 3| < \delta \;\Rightarrow\; |(5x - 11) - 4| = |5x - 15| = 5|x - 3| < 5\delta = \varepsilon$.

9. Let $\varepsilon > 0$ be given. Choose $\delta = \varepsilon$. Then $0 < |x - 5| < \delta \;\Rightarrow\; |(x^2 - 25)/(x - 5) - 10| = |(x + 5) - 10| = |x - 5| < \delta = \varepsilon$.

11. Let $\varepsilon > 0$ be given. Choose $\delta = \varepsilon/2$. Then

$$0 < |x + 1| < \delta \;\Rightarrow\; |(2x^2 - x - 3)/(x + 1) + 5| = |(2x - 3) + 5| = 2|x + 1| < 2\delta = \varepsilon.$$

13. Let $\varepsilon > 0$ be given. Choose $\delta = \varepsilon$. Then

$$0 < |x - 2| < \delta \;\Rightarrow\; |\sqrt{2x} - 2| = |[(\sqrt{2x} - 2)(\sqrt{2x} + 2)]/(\sqrt{2x} + 2)| = |2x - 4|/(\sqrt{2x} + 2) < 2|x - 2|/2 = |x - 2| < \delta = \varepsilon.$$

15. Let $\varepsilon > 0$ be given. Choose $\delta = \min\{2, \varepsilon/2\}$. Then

$$0 < |x - 1| < \delta \;\Rightarrow\; |(2x + 3)(x^2 - 1)/(x^2 - 1) - 5| = |(2x + 3) - 5| = 2|x - 1| < 2\delta = \varepsilon.$$

17. Let $\varepsilon > 0$ be given. Choose $\delta = \min\{1, \varepsilon/5\}$. Then $0 < |x - 3| < \delta \;\Rightarrow\; |x^2 - 2x - 3| = |x - 3||x + 1| < \delta \cdot 5 \le \varepsilon$.

19. Suppose that $L \ne M$ and let $|L - M| = 2\varepsilon$. There are positive numbers δ_1 and δ_2 such that
(i) $0 < |x - c| < \delta_1 \;\Rightarrow\; |f(x) - L| < \varepsilon$ and
(ii) $0 < |x - c| < \delta_2 \;\Rightarrow\; |f(x) - M| < \varepsilon$. Let $\delta = \min\{\delta_1, \delta_2\}$.
Then $0 < |x - c| < \delta \;\Rightarrow\; 2\varepsilon = |L - M| = |L - f(x) + f(x) - M| \le |f(x) - L| + |f(x) - M| < \varepsilon + \varepsilon = 2\varepsilon$. Thus, $2\varepsilon < 2\varepsilon$, which is a contradiction.

21. Let $F(x) = x^4 \sin^2(1/x)$ and $G(x) = x^4$. Then $0 \le F(x) \le G(x)$ and, by Problem 18, $\lim_{x \to \infty} G(x) = 0$. The conclusion follows from Problem 20.

23. $\lim_{x \to 0^+} |x| = \lim_{x \to 0^+} x = 0$; $\lim_{x \to 0^-} |x| = \lim_{x \to 0^-} (-x) = 0$

25. Take $\varepsilon = 1$. There exists $\delta > 0$ such that $0 < |x - a| < \delta \;\Rightarrow\; |f(x) - L| < 1$. Thus,
$0 < |x - a| < \delta \;\Rightarrow\; L - 1 < f(x) < L + 1 \;\Rightarrow\; |f(x)| < |L| + 1$. Let $M = \max\{|f(a)|, |L| + 1\}$ and $(c, d) = (a - \delta, a + \delta)$.

27. (b) and (c)

1. $\lim_{x \to 3}(7x - 4) \overset{5}{=} \lim_{x \to 3} 7x - \lim_{x \to 3} 4 \overset{3,1}{=} 7 \lim_{x \to 3} x - 4 \overset{2}{=} 7 \cdot 3 - 4 = 17$

3. $\lim_{x - 2}[(x^2 + 1)(3x - 1)] \overset{6}{=} \lim_{x \to 2}(x^2 + 1) \cdot \lim_{x \to 2}(3x - 1) \overset{4,5}{=} \left(\lim_{x \to 2} x^2 + \lim_{x \to 2} 1\right)\left(\lim_{x \to 2} 3x - \lim_{x \to 2} 1\right)$

$\overset{8,1,3}{=} \left[\left(\lim_{x \to 2} x\right)^2 + 1\right] \cdot \left[3 \lim_{x \to 2} x - 1\right] \overset{2}{=} (2^2 + 1)(3 \cdot 2 - 1) = 25$

5. $\lim_{x \to 4} 2x/(3x^3 - 16) \overset{7}{=} \left(\lim_{x \to 4} 2x\right)\Big/\lim_{x \to 4}(3x^3 - 16) \overset{3,5}{=} \left(2 \lim_{x \to 4} x\right)\Big/\left(3 \lim_{x \to 4} x^3 - \lim_{x \to 4} 16\right) \overset{2,8,1}{=} 2 \cdot 4 \Big/ \left[3\left(\lim_{x \to 4} x\right)^3 - 16\right]$

$\overset{2}{=} 8/(3 \cdot 4^3 - 16) = 1/22$

7. $\lim_{x \to 3} \sqrt{3x - 5} \overset{9}{=} \sqrt{\lim_{x \to 3}(3x - 5)} \overset{5,3}{=} \sqrt{3 \lim_{x \to 3} x - \lim_{x \to 3} 5} \overset{2,1}{=} \sqrt{3 \cdot 3 - 5} = 2$

9. $\lim\limits_{t \to -2} (2t^3 + 15)^{13} \overset{\textcircled{8}}{=} \left[\lim\limits_{t \to -2} (2t^3 + 15)\right]^{13} \overset{\textcircled{4,3}}{=} \left[2 \lim\limits_{t \to -2} t^3 + \lim\limits_{t \to -2} 15\right]^{13} \overset{\textcircled{8,1}}{=} \left[2 \left(\lim\limits_{t \to -2} t\right)^3 + 15\right]^{13} \overset{\textcircled{2}}{=} [2(-2)^3 + 15]^{13} = -1$

11. $\lim\limits_{y \to 2}[(4y^3 + 8y)/(y + 4)]^{1/3} \overset{\textcircled{9}}{=} \left[\lim\limits_{y \to 2}(4y^3 + 8y)/(y + 4)\right]^{1/3} \overset{\textcircled{7}}{=} \left[\lim\limits_{y \to 2}(4y^3 + 8y)\Big/\lim\limits_{y \to 2}(y + 4)\right]^{1/3}$

$\overset{\textcircled{4,3,8}}{=} \left\{\left[4\left(\lim\limits_{y \to 2} y\right)^3 + 8 \lim\limits_{y \to 2} y\right]\Big/\left(\lim\limits_{y \to 2} y + \lim\limits_{y \to 2} 4\right)\right\}^{1/3} \overset{\textcircled{2,1}}{=} [(4 \cdot 2^3 + 8 \cdot 2)/(2 + 4)]^{1/3} = 2$

13. $\frac{37}{19}$ 15. 10   17. $-\frac{5}{6}$ 19. Does not exist.   21. 4   23. $\sqrt{10}$ 25. -6 27. 3   29. 20   31. $-\frac{1}{4}$

33. Let $\varepsilon > 0$ be given. There is a $\delta_1 > 0$ such that
$0 < |x - c| < \delta_1 \Rightarrow |g(x) - M| < 1 \Rightarrow |g(x)| \leq |g(x) - M| + |M| < 1 + |M|$.
Also, there are numbers δ_2 and δ_3 such that
$0 < |x - c| < \delta_2 \Rightarrow |g(x) - M| < \varepsilon/[2(|L| + 1)]$ and $0 < |x - c| < \delta_3 \Rightarrow |f(x) - L| < \varepsilon/[2(|M| + 1)]$.
Choose $\delta = \min(\delta_1, \delta_2, \delta_3)$. Then
$0 < |x - c| < \delta \Rightarrow |f(x)g(x) - LM| \leq |g(x)||f(x) - L| + |L||g(x) - M| < (|M| + 1)[\varepsilon/2(|M| + 1)] + |L|[\varepsilon/2(|L| + 1)] < \varepsilon$.

35. This follows from the fact that $|f(x) - L| = |f(x) - L - 0|$.

37. There is a $\delta > 0$ such that $0 < |x - c| < \delta \Rightarrow |f(x) - L| < L/2 \Rightarrow -L/2 < f(x) - L < L/2 \Rightarrow f(x) > L/2 > 0$.

39. (a) Let $f(x) = |x|/x$, $g(x) = -f(x)$, $c = 0$.   (b) Let $f(x) = g(x) = |x|/x$, $c = 0$.

41. 0   43. 0   45. $\frac{2}{5}$ 47. -1

49. Suppose, on the contrary, that $\lim\limits_{x \to a} f(x) = L$. Then, $\lim\limits_{x \to a} f(x)g(x) = L \cdot 0 = 0$. But $\lim\limits_{x \to a} f(x)g(x) = \lim\limits_{x \to a} 1 = 1$, a contradiction.

51. (a) 1;   (b) 0

PROBLEM SET 2.7   *Page 82*

1. Yes.   3. No, because $g(2)$ does not exist.   5. No, because $h(2)$ does not exist.   7. No, because $\lim\limits_{t \to 2} f(t)$ does not exist.

9. No, because $g(2)$ does not exist.   11. Yes.   13. Yes.   15. $f(3) = 6$ 17. $f(1) = 2$ 19. $\phi(-1) = -8$ 21. -2 and 3

23. None.   25. $(-\infty, -5]$ 27. None.   29. None.   31. All integers.

33.

35. The function $f(x) = x^3 + 3x - 2$ is continuous on $[0, 1]$ and $f(0) = -2$, $f(1) = 2$. Thus, there must be a number c in $(0, 1)$ such that $f(c) = 0$.

37. If we let $x = t + c$, then $\lim\limits_{t \to 0} f(t + c) = f(c)$ becomes $\lim\limits_{x \to c} f(x) = f(c)$.

39. If $f(0) = 0$ or $f(1) = 1$, we are done. Therefore, assume $f(0) > 0$ and $f(1) < 1$. Then $g(0) = -f(0) < 0$ and $g(1) = 1 - f(1) > 0$. Since g is continuous on $[0, 1]$, there is a number c in $(0, 1)$ such that $g(c) = 0$—that is, $c = f(c)$.

41. Let $f(x)$ be the x-coordinate of a point on the string after contraction.

43. Let $f(\theta)$ be the temperature at one end of the diameter minus the temperature at the other end. Note that if $f(0) = a$, then $f(\pi) = -a$.

45. Let $f(\theta)$ be the length in direction θ minus the length in direction $\theta + \pi/2$.

True-False Quiz

1. True.   **2.** False.   **3.** True.   **4.** True.   **5.** True.   **6.** False.   **7.** True.   **8.** False.   **9.** False.   **10.** True.   **11.** True.
12. False.   **13.** True.   **14.** True.   **15.** False.   **16.** False.   **17.** False.   **18.** False.   **19.** True.   **20.** True.   **21.** False.
22. True.   **23.** True.   **24.** False.   **25.** True.   **26.** True.   **27.** False.   **28.** False.   **29.** True.   **30.** True.

Sample Test Problems

1. (a) $-\frac{1}{2}$;   (b) 4;   (c) undefined;   (d) $-1/[t(t-1)]$;   (e) $-t^2/(1+t)$ **2.** (a) $\frac{3}{2}$;   (b) 3;   (c) 11;   (d) $-1/[2(2+h)]$
3. (a) $\{x : x \neq \pm 1\}$;   (b) $\{x : -2 \leq x \leq 2\}$;   (c) $\{x : x \neq -\frac{3}{2}\}$ **4.** (a) odd;   (b) even;   (c) odd;   (d) even.

5. (a)

(b)

(c)

6.

7. $V(x) = x(24 - 2x)(32 - 2x)$; $(0, 12)$ **8.** (a) $\frac{13}{2}$;   (b) $\frac{15}{2}$;   (c) $\frac{24}{5}$;   (d) $\frac{13}{4}$;   (e) 0;   (f) $\frac{109}{4}$

9.

10. (a) $(-\infty, 16]$;   (b) $[-2, 2]$;   (c) $(-\infty, 16]$ **11.** $k(x) = \sin x$; $h(x) = x^2$; $g(x) = 1 + x$; $f(x) = \sqrt{x}$
12. (a) $-\frac{1}{2}$;   (b) 0;   (c) 1;   (d) $\sqrt{3}/2$ **13.** (a) -0.8;   (b) -0.6;   (c) -0.96;   (d) $-\frac{4}{3} \approx -1.33$;   (e) 0.8;   (f) -0.8
14. $3 \sin t - 4 \sin^3 t$ **15.** 6π in.   **16.** 0   **17.** 2   **18.** Does not exist.   **19.** $\frac{1}{8}$ **20.** $\frac{4}{5}$ **21.** $\frac{1}{2}$ **22.** $\frac{3}{2}$ **23.** 4
24. Does not exist.   **25.** -1 **26.** 2   **27.** -1 **28.** (a) 0;   (b) 0;   (c) 1;   (d) -1 **29.** (a) $-1, 1$;   (b) $f(-1) = -1$
30. (a) For each $\varepsilon > 0$, there is a $\delta > 0$ such that $0 < |u - a| < \delta \implies |g(u) - M| < \varepsilon$.
   (b) For each $\varepsilon > 0$, there is a $\delta > 0$ such that $0 < a - x < \delta \implies |f(x) - L| < \varepsilon$.
31. (a) 14;   (b) -12;   (c) -2;   (d) -2;   (e) 5;   (f) 0

32.

33. $a = 2, b = -1$

34. Since $f(x) = x^5 - 4x^3 - 3x + 1$ is continuous on $[2, 3]$, $f(2) = -5 < 0$, and $f(3) = 127 > 0$, there is a number c in $(2, 3)$ such that $f(c) = 0$.

PROBLEM SET 3.1 *Page 93*

1. 4 **3.** -3 **5.** 2 **7.** (d) -6.01; (e) -6 **9.** $-7, 0, 1, 7$ **11.** $y - \frac{1}{2} = -\frac{1}{4}(x - 1)$
13. (a) 112 ft; (b) 112 ft/s; (c) 96.32 ft/s; (d) 96 ft/s **15.** (a) $1/(2\sqrt{c})$ ft/s; (b) 9 s
17. (a) 0.02005 g; (b) 2.005 g/hr; (c) 2 g/h **19.** (a) 49 g/cm; (b) 27 g/cm **21.** 3.91 g/wk
23. 29,170 gal/h; 116,700 gal/h **25.** 24π km^2/day

PROBLEM SET 3.2 *Page 100*

1. 5 **3.** -1 **5.** 5 **7.** $16x$ **9.** $2ax + b$ **11.** $3x^2 - 2$ **13.** $-3/(5x^2)$ **15.** $-12x/(x^2 + 1)^2$ **17.** $-7/(x - 4)^2$
19. $3/(2\sqrt{3x})$ **21.** $-\frac{3}{2}(x - 2)^{-3/2}$ **23.** $2x - 3$ **25.** $-5/(x - 5)^2$ **27.** $f(x) = 2x^3$; 5 **29.** $f(x) = x^2$; 2 **31.** $f(t) = t^2$; x
33. $f(x) = 2/x$; t **35.** $f(x) = \cos x$; x **37.** $f'(0) \approx -\frac{1}{2}$; $f'(2) \approx 1$; $f'(5) \approx 1$; $f'(7) \approx -2$

39.

41. (a) $2.4, 1, 1.8, -1$; (b) 0.45; (c) 5; (d) 3, 5; (e) 1, 3, 5; (f) 0; (g) $-0.8, 2, (5, 7)$
43. $m = 4$, $b = -4$ **45.** (a) m; (b) $-m$

PROBLEM SET 3.3 *Page 109*

1. $6x^2$ **3.** $2\pi x$ **5.** $9x^{-4}$ **7.** $8x^{-5}$ **9.** $-3x^{-6}$ **11.** $-3x^2 + 2$ **13.** $-4x^3 + 6x - 6$ **15.** $30x^5 - 15x^4 + 11$
17. $-15x^{-6} - 6x^{-4}$ **19.** $-2x^{-2} + 2x^{-3}$ **21.** $-\frac{1}{2}x^{-2} + 2$ **23.** $3x^2 + 1$ **25.** $8x + 4$ **27.** $5x^4 + 6x^2 + 2x$
29. $5x^4 + 42x^2 + 2x - 51$ **31.** $60x^3 - 30x^2 - 32x + 14$ **33.** $-6x/(3x^2 + 1)^2$ **35.** $-(8x - 3)/(4x^2 - 3x + 9)^2$
37. $2/(x + 1)^2$ **39.** $(6x^2 + 20x + 3)/(3x + 5)^2$ **41.** $(4x^2 + 4x - 5)/(2x + 1)^2$ **43.** $(x^2 - 1)/(x^2 + 1)^2$
45. (a) 23; (b) 4; (c) $-17/9$ **47.** $D[f(x)f(x)] = f(x)f'(x) + f(x)f'(x) = 2f(x)f'(x)$ **49.** $y = -2$ **51.** $(0, 0)$; $(\frac{2}{3}, -\frac{4}{27})$
53. (a) -24 ft/s; (b) 1.25 s **55.** $y - 3 = -2(x - 3)$; $y - 3 = 2(x - 1)$ **57.** $3\sqrt{5}$ **59.** 681 cm^3/week

PROBLEM SET 3.4 *Page 114*

1. $3\cos x + 5\sin x$ **3.** $-\csc^2 x$ **5.** $-\cot x \csc x$ **7.** $(\sin x + \cos x)^{-2}$ **9.** 0 **11.** $-(x \sin x + \cos x)/x^2$
13. $y - 0.84 = 0.54(x - 1)$ **15.** -60 ft/s **17.** $\frac{2}{3}$ **19.** 2 **21.** 0
23. The curves intersect at $(\pi/4, 1)$ with slopes 1 and -1, respectively. **27.** $\pi/3$; $71.57°$
29. (a) 1 is a common guess; (b) $\sin t(1 - \cos t)/(t - \sin t \cos t)$; (c) 0.75

PROBLEM SET 3.5 *Page 119*

1. $-135(2 - 9x)^{14}$ **3.** $(50x + 10)(5x^2 + 2x - 8)^4$ **5.** $(27x^2 - 54x + 99)(x^3 - 3x^2 + 11x)^8$
7. $-(36x^3 + 3)/(3x^4 + x - 8)^4$ **9.** $(-108x^3 - 9)(3x^4 + x - 8)^{-10}$ **11.** $(6x + 11)\cos(3x^2 + 11x)$ **13.** $3\sin^2 x \cos x$
15. $4(x^2 + 8x + 1)(x^2 - 1)^3/(x + 4)^5$ **17.** $17(2x + 5)^{-2} \cos[(3x - 1)/(2x + 5)]$ **19.** $(4x - 7)(24x + 10)$
21. $2(2x - 1)^2(x^2 - 3)(7x^2 - 2x - 9)$ **23.** $(3x^2 - 8x - 11)/(3x - 4)^2$ **25.** $(3x^2 + 2)(12x^3 - 68x)/(2x^2 - 5)^2$
27. $51(3t - 2)^2/(t + 5)^4$ **29.** $(3t - 2)^2(6t + 47)/(t + 5)^2$ **31.** $3\sin^2\theta \cos\theta$
33. $3\sin^2 x(\cos 2x \cos x + 2\sin x \sin 2x)/\cos^4 2x$ **35.** $\frac{48}{5}$ **37.** 1.4183109 **39.** $(8x + 12)\sin^3(x^2 + 3x)\cos(x^2 + 3x)$
41. $-3\sin t \sin^2(\cos t)\cos(\cos t)$ **43.** $-8\theta \cos\theta^2 \cos^3(\sin\theta^2)\sin(\sin\theta^2)$ **45.** $-2\cos 2x \cdot \sin(\sin 2x) \cdot \cos[\cos(\sin 2x)]$
47. $y - 32 = 224(x - 1)$ **49.** (a) $(10\cos 8\pi t, 10\sin 8\pi t)$; (b) 80π cm/s
51. (a) $(\cos 2\pi t, \sin 2\pi t)$; (b) $\sqrt{25 - \cos^2 2\pi t} + \sin 2\pi t$; (c) $(\pi \sin 4\pi t)(25 - \cos^2 2\pi t)^{-1/2} + 2\pi \cos 2\pi t$
53. (a) $2x|x^2 - 1|/(x^2 - 1)$; (b) $\cos x|\sin x|/\sin x$ **55.** (a) $2\sin^3 2x$; (b) $\sin^4 2x$ **57.** 16

1. 0.25   3. 0.00814082   5. $2x + \Delta x - 3$; $2x - 3$ 7. $1/[(x + \Delta x + 1)(x + 1)]$; $1/(x + 1)^2$ 9. $3(x^2 + 3x)^2(2x + 3)$
11. $2x \cos x^2$ 13. $4(x^2 + 1)^3[2x \cos x + (x^2 + 1) \sin x]/\cos^5 x$ 15. $2 \sin x \cos x \cos x^2 - 2x \sin^2 x \sin x^2$
17. $8x \sin^3(x^2 + 3)\cos(x^2 + 3)$ 19. $8x(x^2 - 2)^{-2} \sin[(2x^2 + 4)/(x^2 - 2)]$ 21. $(3 \sin t \cos t)(\sin t - \cos t)$
23. $-\pi(9r^2 + 22r + 6)$ 25. $\frac{375}{8}$ 27. (a) -5; (b) -11; (c) $\frac{5}{9}$; (d) 4   29. (a) 1; (b) -1
31. (a) 19,200 cm³/min; (b) 2880 cm²/min   33. $x = \sqrt{\pi}/2$

1. 6   3. $192(2x + 5)$ 5. $-27 \cos 3x$ 7. $-6/(x - 3)^4$ 9. 24   11. $\frac{1}{4}$ 13. 1860   15. $2\pi^2$ 19. (a) 0; (b) 0; (c) 0
21. $f''(-5) = -24$, $f''(3) = 24$ 23. (a) $v(t) = 12 - 4t$; $a(t) = -4$; (b) $t < 3$; (c) $t > 3$; (d) all t;

(e)

25. (a) $v(t) = 3t^2 - 18t + 24$; $a(t) = 6t - 18$; (b) $(-\infty, 2) \cup (4, \infty)$; (c) $(2, 4)$; (d) $(-\infty, 3)$;

(e)

27. (a) $v(t) = 2t - 16t^{-2}$; $a(t) = 2 + 32t^{-3}$; (b) $(2, \infty)$; (c) $(0, 2)$;

(d) Never for $t > 0$.

29. $v(1) = 11$, $v(4) = -16$ 31. (a) $t = \frac{3}{4}$; (b) $t = \frac{1}{2}$, $t = \frac{3}{4}$; (c) $t = 0$, $t = \frac{3}{2}$
33. (a) 48 ft/s; (b) $t = 1.5$ s; (c) 292 ft; (d) $t = 5.77$ s; (e) 136.7 ft/s   35. 581.3 ft/s   37. $(-\infty, -2) \cup (1, 4)$
39. (a) $ds/dt = ks$; (b) $d^2s/dt^2 > 0$; (c) not $d^2s/dt^2 < 0$ but $d^3s/dt^3 < 0$; (d) $d^2s/dt^2 = 10$ mi/h/min;
(e) $d^2s/dt^2 < 0$ and $|d^2s/dt^2|$ is small; (f) $ds/dt = k$
41. (a) If $C(t)$ is the cost at time t, $dC/dt > 0$, $d^2C/dt^2 > 0$.
(b) If $C(t)$ is the oil consumption at time t, then during the last two years, $dC/dt < 0$, $d^2C/dt^2 > 0$.
(c) If $P(t)$ is the world population at time t, then $dP/dt > 0$, but $d^2P/dt^2 < 0$.
(d) If $S(t)$ is the distance the car traveled in time t, then $S'(t) > 0$ and $S''(t) = k > 0$.
(e) $d\theta/dt > 0$, $d^2\theta/dt^2 > 0$;
(f) If $P(t)$ represents profit at time t, then $dP/dt > 0$, $d^2P/dt^2 < 0$.
(g) If $A(t)$ represents the company's assets, then $A'(t) < 0$ but $A''(t) > 0$.
43. $D_x^n(uv) = uD_x^n v + nD_x u D_x^{n-1} v + \frac{1}{2}n(n - 1)D_x^2 u D_x^{n-2} v + \cdots + nD_x^{n-1} u D_x v + vD_x^n u$; The coefficients are the binomial coefficients.

1. x/y 3. $-y/x$ 5. $(1 - y^2)/(2xy)$ 7. $(12x^2 + 11y^2)/(6y^2 - 22xy)$
9. $(y - 6\sqrt{2xy} - y^3\sqrt{2xy})/(3xy^2\sqrt{2xy} - 2y\sqrt{2xy} - x)$ 11. $(2x - y)/(x + \cos y)$ 13. $y - 2 = (-14/13)(x - 1)$
15. $y = 1$ 17. $y + 1 = \frac{1}{2}(x - 1)$ 19. $5x^{2/3} + \frac{1}{2}x^{-1/2}$ 21. $\frac{1}{3}x^{-2/3} - \frac{1}{3}x^{-4/3}$ 23. $\frac{1}{2}(3x - 2)(3x^2 - 4x)^{-3/4}$
25. $-\frac{2}{3}(3x^2 + 2)/(x^3 + 2x)^{5/3}$ 27. $(2x + \cos x)/(2\sqrt{x^2 + \sin x})$ 29. $-(2x \sin x + x^2 \cos x)/(3x^2 \sin x \sqrt[3]{x^2 \sin x})$
31. $-\frac{1}{2}(x + 1)\sin(x^2 + 2x)[1 + \cos(x^2 + 2x)]^{-3/4}$ 33. $ds/dt = -(s^2 + 3t^2)/(2st)$; $dt/ds = -2st/(s^2 + 3t^2)$
35. $y = \pm(\sqrt{3}/3)x$ 37. $y'' = 2xy/(x + 3y^2)^3$ 39. $y'' = -15$ 41. Normal line: $y = x$.   45. -1.107149 47. $y = 2x \pm 8$

PROBLEM SET 3.9 *Page 144*

1. 900 in.³/s   **3.** 214.66 mi/h   **5.** 627.38 mi/h   **7.** 0.408 ft/s   **9.** 0.0796 ft/s   **11.** 1 in./min   **13.** 1.018 in.²/s
15. 5π km/min   **17.** (a) $\frac{1}{2}$ ft/s in both cases;   (b) 5/2 ft/s;   (c) 0.0417 rad/s   **19.** 109.55 ft/s   **21.** -0.0163 ft/h
23. 200 in.³/min   **25.** (a) $-\frac{9}{8}$ ft/s;   (b) $-3\sqrt{3}/64$ ft/s²   **27.** (b) 3 h

PROBLEM SET 3.10 *Page 150*

1. $(4x - 3)\,dx$ **3.** $-24x^2(3 + 2x^3)^{-5}\,dx$ **5.** $(10x^4 + 4x^3)(4x^5 + 2x^4 - 5)^{-1/2}\,dx$ **7.** $\frac{4}{5}t(t^2 - 3)^{-3/5}\,dt$
9. (a) 0.75;   (b) 2.25

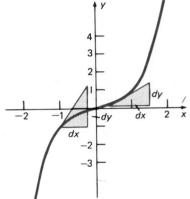

11. (a) 3.25;   (b) 0.9844   **13.** (a) $\Delta y = 2.25$; $dy = 2$;   (b) $\Delta y = -0.7056$; $dy = -0.72$ **15.** 20.05; calculator: 20.049938.
17. 2.996667; calculator: 2.996663.   **19.** $12.5\pi \approx 39.27$ cm³   **21.** 893.47 ft³   **23.** $4\pi \approx 12.57$ ft
25. $4000\pi/3 \pm 20\pi$, or 4188.8 ± 62.8 cm³   **27.** 79.097 ± 0.729 cm   **29.** $dy = 0.01$; $|\Delta y - dy| \le 0.000003$ **31.** 0.2 gal.

CHAPTER REVIEW 3.11 *Page 151*

True-False Quiz

1. False.   **2.** True.   **3.** True.   **4.** True.   **5.** False.   **6.** False.   **7.** True.   **8.** True.   **9.** False.   **10.** True.   **11.** True.
12. False.   **13.** False.   **14.** True.   **15.** Truc.   **16.** True.   **17.** True.   **18.** True.   **19.** True.   **20.** True.   **21.** True.
22. False.   **23.** False.   **24.** True.   **25.** True.   **26.** True.   **27.** True.   **28.** True.   **29.** True.   **30.** False.

Sample Test Problems

1. (a) $2x - 5$;   (b) $-1/(x - 3)^2$;   (c) $-1/2\sqrt{9 - x}$ **2.** (a) $3x^2$;   (b) $1/2\sqrt{x}$
3. (a) $f(x) = 3x^2$ at $x = 2$;   (b) $f(x) = \tan x$ at $x = \pi/4$;   (c) $f(p) = 3/p$ at $p = x$.
4. (a) -1;   (b) 1.7;   (c) 1.13;   (d) 2.4;   (e) -4 **5.** $3x^2 - 6x - 2x^{-3}$ **6.** $(-3x^2 + 10x + 3)/(x^2 + 1)^2$
7. $-2(3x + 2)^{-4/3}$ **8.** $(3t + 6)/\sqrt{2t + 6}$ **9.** $-x(x^2 + 4)^{-3/2}$ **10.** $2t\cos(t^2) - 2\sin t \cos t$ **11.** $-15\cos^2 5x \sin 5x$
12. $-8\sin 4t \sin(\cos 4t)\cos(\cos 4t)$ **13.** 672   **14.** 18   **15.** $(\frac{7}{4}, \frac{1}{16})$ **16.** 100π m²   **17.** 10π m³   **18.** $1/10\pi$ m/h
19. $\frac{1}{6}$ ft/min   **20.** (a) 4 s, 2.56 ft;   (b) 8 s, -128 ft/s   **21.** (a) $1 < t < 3$;   (b) $a(1) = -6$ ft/s², $a(3) = 6$ ft/s²;   (c) $t > 2$
22. (a) 0;   (b) $20!x^{-21}$ **23.** (a) $(x^2y^3 - x^2)/(y^2 - x^3y^2)$;   (b) $(2x - \sin xy - xy \cos xy)/(x^2 \cos xy)$
24. The respective slopes are 3 and $-\frac{1}{3}$.   **25.** $dy = (\pi + 4)(0.01) \approx 0.0714$ **26.** (a) 84;   (b) 23;   (c) 20;   (d) 26
27. $\frac{24}{5}$ ft/s   **28.** 103.53 mi/h   **29.** $D_x|\sin x| = |\sin x| \cot x, x \ne n\pi$

PROBLEM SET 4.1 *Page 160*

1. $\{0, 2, 3\}$; max $f(2) = 3$; min $f(0) = -1$ **3.** $\{-3, -2, 1, 3\}$; max $G(3) = 9$; min $G(1) = -\frac{7}{5}$
5. $\{-1, 1\}$; no max; min $f(1) = -1$ **7.** $\{-2, 0, 1\}$; max $g(0) = 1$; min $g(-2) = \frac{1}{5}$
9. $\{-1, \sqrt{2}, 4\}$; max $f(\sqrt{2}) = \sqrt{2}/4 \approx 0.354$; min $f(-1) = -\frac{1}{3}$ **11.** $\{1, 2, 5\}$; max $f(5) = 3$; min $f(2) = 0$
13. $\{-1, 0, 32\}$; max $g(32) = 4$; min $g(0) = 0$ **15.** $\{0\}$; no max; min $h(0) = 0$ **17.** Each number is 5.   **19.** 50 ft by 50 ft

21. 1024 in.³   **23.** 20 ft by 40 ft   **25.** $x = 40$ ft, $y = 100$ ft   **27.** 4 (horizontal) by 8 (vertical)   **29.** $\pi/3$ radians   **31.** 55 mi/h
33. $(2\sqrt{2}, 2)$, $(0, 0)$ **35.** $x = 1$, $y = 3$, $z = 3$

PROBLEM SET 4.2   *Page 167*

1. Inc. $[2, \infty)$; dec. $(-\infty, 2]$.   **3.** Inc. $(-\infty, \infty)$.   **5.** Inc. $[-1, \infty)$; dec. $(-\infty, -1]$.   **7.** Inc. $(-\infty, \infty)$.
9. Inc. $[0, \pi/4]$ and $[\pi/2, 3\pi/4]$; dec. $[\pi/4, \pi/2]$ and $[3\pi/4, \pi]$.   **11.** Up: $(-\infty, \infty)$; no inflection point.
13. Up: $(0, \infty)$; down: $(-\infty, 0)$; inflection point $(0, 0)$.
15. Up: $(-\infty, -1)$ and $(1, \infty)$; down: $(-1, 0)$ and $(0, 1)$; inflection points $(-1, 2)$ and $(1, 2)$.
17. Up $(-\infty, \infty)$; no inflection point.
19. Inc. $(-\infty, -1]$ and $[1, \infty)$; dec. $[-1, 1]$; up: $(0, \infty)$; down: $(-\infty, 0)$.

21. Inc. $[1, \infty)$; dec. $(-\infty, 1]$; up: $(-\infty, 0)$ and $(\frac{2}{3}, \infty)$; down: $(0, \frac{2}{3})$

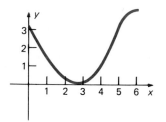

23. Inc. $(-\infty, -1]$ and $[1, \infty)$; dec. $[-1, 1]$; up: $(-\sqrt{2}/2, 0)$ and $(\sqrt{2}/2, \infty)$; down: $(-\infty, -\sqrt{2}/2)$ and $(0, \sqrt{2}/2)$.

25. Inc. $[0, \pi/2]$; dec. $[\pi/2, \pi]$; down: $(0, \pi)$.

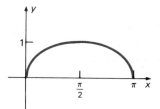

27. Inc. $[0, \frac{2}{5}]$; dec. $(-\infty, 0]$ and $[\frac{2}{5}, \infty)$; up: $(-\infty, -\frac{1}{5})$; down: $(-\frac{1}{5}, 0)$ and $(0, \infty)$.

29.

31.

33. If $f(x) = ax^2 + bx + c$, $a \neq 0$, then $f''(x) = 2a$, and thus f'' is of the same sign on $(-\infty, \infty)$.

35. $f'(x)$ must have the same sign throughout I, for otherwise $f'(x)$ would equal zero somewhere. If $f'(x) > 0$ everywhere, f is increasing; if $f'(x) < 0$ everywhere, f is decreasing throughout I.

37. (a) If $f(x) = x^2$, then $f'(x) = 2x > 0$ on $(0, \infty)$, and so f is increasing.
　　(b) If $f(x) = \sqrt{x}$, then $f'(x) = 1/2\sqrt{x} > 0$ on $(0, \infty)$, and so f is increasing.
　　(c) If $f(x) = 1/x$, then $f'(x) = -1/x^2 < 0$ on $(0, \infty)$, and so f is decreasing.

39. $a = \frac{39}{8}$, $b = \frac{13}{2}$　**41.** (a) none;　(b) $f(x) > 0$, $g(x) > 0$;　(c) none.

PROBLEM SET 4.3　*Page 171*

1. $\{0, 2\}$; local max $f(0) = 2$, local min $f(2) = -2$.　**3.** $\{\pi/3, 5\pi/3\}$; local max $f(5\pi/3) \approx 3.484$, local min $f(\pi/3) \approx -0.342$.

5. $\{0\}$; local min $g(0) = 1$.　**7.** $\{-2, 2\}$; local max $f(-2) = 3$, local min $f(2) = -5$.

9. $\{-\frac{3}{2}, 0\}$; local min $h(-\frac{3}{2}) = -\frac{27}{16}$.　**11.** $\{1\}$; local max $g(1) = 2$.　**13.** $\{-1, 1\}$; local max $f(-1) = -2$, local min $f(1) = 2$.

15. $\{2\pi/3, 4\pi/3\}$; local max $f(2\pi/3) = \sqrt{3}/3$, local min $f(4\pi/3) = -\sqrt{3}/3$.　**17.** Max $F(1) = 3$, min $F(9) = -9$.

19. No max, min $f(\tan^{-1}\frac{4}{3}) = 125$.　**21.** Min $g(8 + \sqrt[3]{128}) \approx 277.148$.　**23.** -2; 3

25.

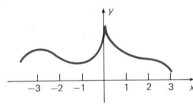

(a) $(-\infty, -3] \cup [-1, 0]$, $[-3, -1] \cup [0, \infty)$;
(b) $(-2, 0) \cup (0, 2)$, $(-\infty, -2) \cup (2, \infty)$;
(c) local max at $x = -3$ and at $x = 0$; local min at $x = -1$;
(d) at $x = \pm 2$.

PROBLEM SET 4.4　*Page 177*

1. $-2\sqrt{3}$ and $2\sqrt{3}$　**3.** $(9.75, \pm 2.208)$　**5.** $x = 15\sqrt{3}$, $y = 20\sqrt{3}$　**7.** $x = 10\sqrt{6}$, $y = 15\sqrt{6}$

9. Base: 20 ft by 20 ft; height: 30 ft.　**11.** Width: $4\sqrt{6}/3$; depth: $2\sqrt{6}$.　**13.** 2.268 mi from P.　**15.** At the town.

17. 8:09:03 A.M.　**19.** $4\pi\sqrt{3}r^3/9$　**21.** Height is twice radius.

23. For maximum, make only a square; for minimum, cut $400\sqrt{3}/(9 + 4\sqrt{3}) \approx 43.5$ cm from one end and make square of this piece.

25. Height of cylinder is twice its radius.　**27.** $r = \sqrt{A}$, $\theta = 2$ radians　**31.** $5\sqrt{5}$

33. (a) $x = 2a/3$;　(b) $x = 2a/3$;　(c) $x = 3a/4$

PROBLEM SET 4.5　*Page 183*

1. $C(x) = 8000 + 110x$　**3.** $P(x) = -8000 + 190x - x^2/2$　**5.** \$3.56, \$2.53　**7.** 200

9. (a) $R(x) = 20x + 4x^2 - \frac{1}{3}x^3$;　(b) $[0, 10]$;　(c) 4　**11.** $x_1 = 25$; $R(25) = \$639.29$; $R'(25) = \$0.06$

13. $p(x) = 8.4 - 0.0006x$; $p(7000) = \$4.20$

15. (a) $C(x) = \begin{cases} 6000 + 1.4x & \text{if } x \leq 4500 \\ 5100 + 1.6x & \text{if } x > 4500 \end{cases}$;　(b) $p(x) = 11 - 0.001x$　if $x \geq 4000$;　(c) 4700

17. $P(x) = \begin{cases} 0.009x^2 + 6x - 200 & \text{if } 0 \leq x \leq 300 \\ 0.009x^2 + 7x - 800 & \text{if } 300 < x \leq 450 \end{cases}$; max profit $P(450) = \$4172.50$.　**19.** \$1713 per employee.

1. -2 **3.** $\frac{2}{5}$ **5.** $\frac{2}{5}$ **7.** 0 **9.** 2 **11.** 2 **13.** 0 **15.** $-\infty$ **17.** $-\infty$ **19.** ∞ **21.** $-\infty$ **23.** ∞ **25.** 5 **27.** 0
29. -1 **31.** $-\infty$
33. $x = -1$, $y = 0$ **35.** $x = 3$, $y = 2$

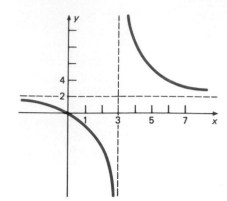

37. $y = 0$ **39.** $y = 2x + 3$

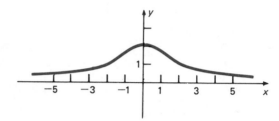

41. (a) For each negative number M, there is a corresponding $\delta > 0$ such that $0 < x - c < \delta \Rightarrow f(x) < M$.
  (b) For each positive number M, there is a corresponding $\delta > 0$ such that $0 < c - x < \delta \Rightarrow f(x) > M$.
43. Let $\varepsilon > 0$ be given. There are numbers M_1 and M_2 such that $x > M_1 \Rightarrow |f(x) - A| < \varepsilon/2$ and $x > M_2 \Rightarrow$
  $|g(x) - B| < \varepsilon/2$. Choose $M = \max\{M_1, M_2\}$. Then $x > M \Rightarrow |f(x) + g(x) - (A + B)| \le |f(x) - A| + |g(x) - B| <$
  $\varepsilon/2 + \varepsilon/2 = \varepsilon$.
45. (a) Does not exist;   **(b)** 0;   **(c)** 1;   **(d)** ∞;   **(e)** 0;   **(f)** $\frac{1}{2}$ **(g)** does not exist;   **(h)** 0

1. Symmetric with respect to origin.   **3.** Symmetric with respect to y-axis.

$(1.15, -3.08)$

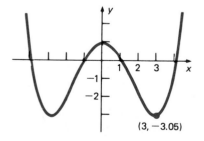

$(3, -3.05)$

5. Symmetric with respect to origin.

7.

9. Symmetric with respect to y-axis.

11.

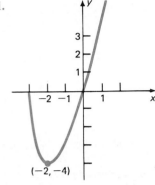

13. Symmetric with respect to origin.

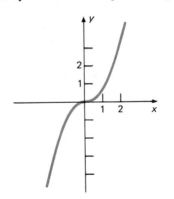

15. Symmetric with respect to y-axis.

17.

19.

21.

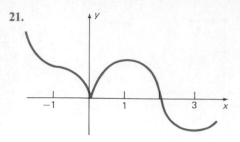

PROBLEM SET 4.8   *Page 198*

1. $c = 0$ **3.** $c = \pm 2\sqrt{3}/3$ **5.** Not continuous at $t = 3$ **7.** $c = \frac{16}{27}$ **9.** Not continuous at $x = 0$.
11. There is a number c in $(0, 2)$ such that $f'(c) = [f(2) - f(0)]/(2 - 0) = \frac{112}{2} = 56$; that is, the velocity must be $56\,\text{mi/hr}$ at some time between $t = 0$ and $t = 2$.
21. Since f has a derivative on $[a, b]$, it is continuous there. By the Intermediate Value Theorem, there is a solution to $f(x) = 0$ in (a, b). Suppose there were two such solutions, x_0 and x_1, with $x_0 < x_1$. Then by Rolle's Theorem applied to f on the interval $[x_0, x_1]$, there would be a number c in (x_0, x_1) such that $f'(c) = 0$, which contradicts one of the hypotheses.
23. Let a and b be successive distinct zeros of f' and suppose there are two zeros, x_0 and x_1, of f in (a, b). By Rolle's Theorem applied to f on the interval $[x_0, x_1]$, there is a zero c of f' in (x_0, x_1). This contradicts the hypothesis that a and b are successive zeros of f'.
25. We can assume $x_1 < x_2$. By the Mean Value Theorem applied to f on the interval $[x_1, x_2]$, there is a number c in (x_1, x_2) such that $f(x_2) - f(x_1) = f'(c)(x_2 - x_1)$. Thus, $|f(x_2) - f(x_1)| = |f'(c)||x_2 - x_1| \le M|x_2 - x_1|$.
27. A constant function will serve as an example for both (a) and (b).
29. Let x_1 and x_2 be two points of I with $x_1 < x_2$. Apply the Mean Value Theorem to f^2 on $[x_1, x_2]$. There is a number c in (x_1, x_2) such that $f^2(x_2) - f^2(x_1) = 2f(c)f'(c)(x_2 - x_1) \ge 0$. Thus, $f^2(x_2) > f^2(x_1)$ as required.
31. By the Mean Value Theorem applied to $f(t) = \sqrt{t}$ on the interval $[x, x + 2]$, there is a number c in $(x, x + 2)$ such that $\sqrt{x + 2} - \sqrt{x} = (1/2\sqrt{c})(x + 2 - x) = 1/\sqrt{c}$. Thus $\lim_{x \to \infty}(\sqrt{x + 2} - \sqrt{x}) = \lim_{c \to \infty}(1/\sqrt{c}) = 0$.
33. Let $h(t) = f(t) - g(t)$, where $f(t)$ and $g(t)$ are the distances of A and B from the starting line at time t. Apply the Mean Value Theorem.
35. From the Mean Value Theorem, $f(x) - f(c) = f'(d)(x - c)$ for some d between x and c. Thus, $f(x) = f(c) + f'(d)(x - c)$. Since f' is increasing, $f'(d) - f'(c)$ has the same sign as $x - c$, and so $[f'(d) - f'(c)][x - c] > 0$ for $x \ne c$. Finish the argument.

CHAPTER REVIEW 4.9   *Page 200*

True-False Quiz

1. True.   **2.** True.   **3.** True.   **4.** False.   **5.** True.   **6.** False.   **7.** True.   **8.** False.   **9.** True.   **10.** True.   **11.** False
12. False.   **13.** True.   **14.** True.   **15.** True.   **16.** False.   **17.** True.   **18.** False.   **19.** True.   **20.** False.   **21.** True.
22. True.   **23.** False.   **24.** True.   **25.** True.   **26.** True.   **27.** False.   **28.** True.   **29.** True   **30.** True.

Sample Test Problems

1. $\{-2, -\frac{1}{2}\}$; max $f(-\frac{1}{2}) = 4$, min $f(-2) = \frac{1}{4}$.   **2.** $\{-2\}$; no max, min $f(-2) = \frac{1}{4}$.
3. $\{-2, 0, 1, 3\}$; max $f(3) = 135$, min $f(1) = -1$ **4.** $\{-1, 0, \frac{12}{7}, 2, 3\}$; max $f(3) = 9$, min $f(\frac{12}{7}) = -1.936$.
5. $\{-1, 0, 2, 3\}$; max $f(3) = 88$, min $f(2) = -9$.   **6.** $\{-2, -\frac{4}{5}, 1, 2\}$; max $f(2) = 16$, min $f(-\frac{4}{5}) = -8.398$.
7. Inc. $(-\infty, -1]$ and $[1, \infty)$; concave down $(-\infty, 0)$.   **8.** Inc. $[-1, 2]$; concave down $(\frac{1}{2}, \infty)$.
9. Inc. $[0, \frac{1}{5}]$, concave down $(\frac{3}{20}, \infty)$.   **10.** Inc. $[-\sqrt{2}/2, \sqrt{2}/2]$, concave down $(-\frac{1}{2}, 0)$ and $(\frac{1}{2}, \infty)$.

11.

$(0, 0)$

$\left(\dfrac{4}{3}, \dfrac{-128}{27}\right)$

$\left(\dfrac{8}{3}, \dfrac{-256}{27}\right)$

12. Max $f(0) = 6$, no min.

13.

$(2, -48)$

14.

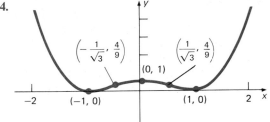

$\left(-\dfrac{1}{\sqrt{3}}, \dfrac{4}{9}\right)$ $\left(\dfrac{1}{\sqrt{3}}, \dfrac{4}{9}\right)$

$(0, 1)$

$(-1, 0)$ $(1, 0)$

15.

$(4, 4)$

16.

17.

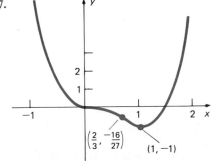

$\left(\dfrac{2}{3}, \dfrac{-16}{27}\right)$

$(1, -1)$

18.

19.

20.

21. 4 in.   **22.** $5\sqrt{5}$ ft   **23.** 7.5 in. by 10 in.   **24.** $r = 4\sqrt[3]{2}$ ft, $h = 8\sqrt[3]{2}$ ft
25. Max $f(0) = 2$; min $f(-2) = 0$; concave up $(-2, 0)$; concave down $(0, 2)$.

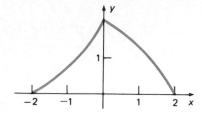

26. (a) $c = \pm\sqrt{3}$;   (b) $F'(0)$ does not exist;   (c) $c = 1 + \sqrt{2}$ **27.** $y - 5 = 7(x - 1)$, $y - 11 = 5(x - 2)$
28. (a) $\frac{3}{2}$;   (b) $3\sqrt{2}/2$;   (c) 0;   (d) no limit;   (e) ∞;   (f) no limit;   (g) -1;   (h) 1

29.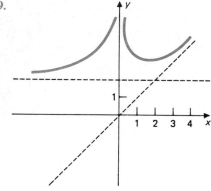

PROBLEM SET 5.1   *Page 210*

1. $4x + C$ **3.** $x^3 + \sqrt{2}x + C$ **5.** $\frac{3}{5}x^{5/3} + C$ **7.** $2x^3 - 3x^2 + x + C$ **9.** $2x^9 - 5x^5 + x^3 + C$ **11.** $-x^{-4} + x^{-3} + C$
13. $2x^2 + 3x + 2x^{-4} + C$ **15.** $\frac{1}{4}x^4 + \frac{2}{3}x^{3/2} + C$ **17.** $\frac{1}{5}y^5 + 2y^4 + \frac{16}{3}y^3 + C$ **19.** $\frac{1}{3}x^3 - x^2 - x^{-1} + C$
21. $-3\cos t - 2\sin t + C$ **23.** $(3x + 1)^5/5 + C$ **25.** $(5x^3 - 18)^8/8 + C$ **27.** $3(2x^5 + 9)^4/40 + C$
29. $(5x^3 + 3x - 8)^7/21 + C$ **31.** $\frac{9}{16}(2t^2 - 11)^{4/3} + C$ **33.** $\frac{1}{5}\sin^5 x + C$ **35.** $\frac{1}{12}\sin^6 x^2 + C$
37. $\frac{1}{9}x^9 + \frac{3}{7}x^7 + \frac{3}{5}x^5 + \frac{1}{3}x^3 + C$ **39.** $\frac{1}{2}x^3 + \frac{1}{2}x^2 + C_1 x + C_2$ **41.** $\frac{4}{15}x^{5/2} + C_1 x + C_2$ **43.** $\frac{1}{6}x^3 + \frac{1}{2}x^{-1} + C_1 x + C_2$
45. $D_x[f(x)g(x)] = f(x)g'(x) + g(x)f'(x)$ **47.** $x^2\sqrt{x - 1} + C$ **49.** Show $D_x[\sqrt{x^4 - 1}/x] = (x^4 + 1)/(x^2\sqrt{x^4 - 1})$.
51. $\int f''(x)\,dx = f'(x) + C = (5x^3 + 2)/(2\sqrt{x^3 + 1}) + C$ **55.** $\frac{1}{2}x|x| + C$

PROBLEM SET 5.2   *Page 217*

5. $y = x^3 + x + 2$ **7.** $y = \sqrt{\frac{1}{2}x^2 + 7}$ **9.** $y = -4/(t^4 - 20)$ **11.** $s = t^3 + 2t^2 - t - 9$ **13.** $y = [(2x + 1)^5 + 59]/10$
15. $y = 2x^2$ **17.** $v = 4$, $s = \frac{16}{3}$ **19.** $v \approx 2.831$, $s \approx 12.605$ **21.** 144 ft   **23.** 32.24 ft/s, 1198.54 ft

27. Moon: 1.47 mi/s; Venus: 6.26 mi/s; Jupiter: 36.81 mi/s; sun: 382.9 mi/s.   **29.** 5400 mi/h^2 **31.** 5500 m

33. (a)   (b) 0.6 mi/min;   (c) 0.9 mi/min^2

35. (a) $dV/dt = (k/10)\sqrt{V}$, $V = 1600$ at $t = 0$, $V = 0$ at $t = 40$;   (b) $V = (40 - t)^2$;   (c) 900 cm^3

PROBLEM SET 5.3   *Page 223*

1. 40   **3.** $\frac{37}{30}$ **5.** -11 **7.** 3   **9.** $\sum_{i=1}^{98} i$ **11.** $\sum_{i=1}^{69} 1/i$ **13.** $\sum_{i=1}^{n} a_i$ **15.** $\sum_{i=1}^{n} f(c_i)$ **17.** 130   **19.** 130   **21.** $\frac{40}{41}$ **23.** $-\frac{16}{147}$

25. 14,950   **27.** 2640   **29.** $n(n-1)(4n+1)/6$ **31.** $\sum_{k=1}^{17} (k+2)k$ **33.** $\sum_{i=1}^{11} (i-1)/i$ **35.** 6.6   **37.** (a) 1023/1024;   (b) 2046

39. $(n+1)(2a+nd)/2$ **41.** $\bar{x} = \frac{55}{7} \approx 7.8571$, $s^2 \approx 12.4082$ **49.** 715

PROBLEM SET 5.4   *Page 231*

1. 3.5   **3.** 3.75   **5.** 2.875   **7.** 15   **9.** 7.3704   **11.** $4 + (2/n)$; 4   **13.** $[12 + (\frac{27}{2})(1 + 1/n)]$; $\frac{51}{2}$ **15.** $(1 + 1/n)^2/4$; $\frac{1}{4}$

17. 4 ft   **21.** (a) $\frac{125}{3}$;   (b) 21;   (c) $\frac{117}{3}$ **23.** (a) 4;   (b) $\frac{15}{4}$;   (c) $\frac{31}{5}$;   (d) $\frac{512}{5}$

PROBLEM SET 5.5   *Page 239*

1. 5.625   **3.** 15.6875   **5.** 4.46875   **7.** $\int_1^3 x^2\,dx$ **9.** $\int_0^3 x/(1+x)\,dx$ **11.** 28   **13.** 0   **15.** $\frac{16}{3}$ **17.** $\frac{5}{2}$ **19.** $\pi + 6$

21. (a) Integrable, since f is continuous on $[-2, 2]$.   (b) Integrable, since f is continuous on $[-2, 2]$.   (c) Not integrable, since f is unbounded on $[-2, 2]$.   (d) Not integrable, since f is unbounded on $[-2, 2]$.   (e) Integrable, since f is bounded and has only one discontinuity on $[-2, 2]$.   (f) Integrable, since f is bounded and has only one discontinuity on $[-2, 2]$.

23. (a) -3;   (b) 19;   (c) 3;   (d) 2;   (e) 9;   (f) 0;   (g) 1;   (h) 2

PROBLEM SET 5.6   *Page 245*

1. 4   **3.** 15   **5.** $\frac{3}{4}$ **7.** $\frac{16}{3}$ **9.** $\frac{1783}{96} \approx 18.57$ **11.** 1   **13.** $\frac{22}{5}$ **15.** $\frac{2047}{11} \approx 186.09$ **17.** $\frac{4}{5}$ **19.** $\frac{122}{9} \approx 13.5556$ **21.** 0

23. $\frac{1}{3}$ **25.** $(\pi^2/4) + 1 \approx 3.47$ **27.** 14   **29.** $\frac{38}{15} \approx 2.533$ **31.** 9   **33.** 2   **35.** 0.385, 0.333   **37.** 6   **39.** -16 **41.** -24

43. $\frac{1}{2}[\![b]\!][\![b-1]\!] + [\![b]\!](b - [\![b]\!])$

PROBLEM SET 5.7   *Page 252*

1. 5   **3.** 5   **5.** -9 **7.** -2 **9.** -2 **11.** $2x + 1$ **13.** $\sqrt{1 + x^4}$ **15.** $-x \tan x$ **17.** $[\sin^2 x + \cos(\sin x)]\cos x$

19. $-\sqrt{1 + x^4} + 3x^2\sqrt{1 + x^{12}}$ **21.** $f''(x) = a^2/(a^2 + x^2)^{3/2}$ **23.** $\frac{26}{3}$ **25.** 4   **29.** 40   **31.** $\frac{13}{4}$

33. True; use Comparison Property.   **35.** False; $\int_{-1}^{1} x\,dx = 0$.   **37.** True; use Linearity Property.

PROBLEM SET 5.8   *Page 259*

1. $\frac{2}{9}(3x + 2)^{3/2} + C$ **3.** $\frac{1}{3}\sin(3x + 2) + C$ **5.** $\frac{1}{3}(x^2 + 4)^{3/2} + C$ **7.** $-\frac{1}{2}\cos(x^2 + 4) + C$ **9.** $-\cos\sqrt{x^2 + 4} + C$

11. $\frac{1}{3}\sin^{3/2}(x^2 + 4) + C$ **13.** $\frac{1}{2}(\sqrt{t} + 4)^4 + C$ **15.** $\frac{255}{12}$ **17.** $\frac{2}{117}$ **19.** $\frac{5}{12}$ **21.** $\frac{1}{64}$ **23.** $(\sin 3)/3$ **25.** $1/\pi$

27. $1 - \cos 1$ **29.** $\frac{5}{36}$ **31.** 0   **33.** 0   **35.** 2π **37.** $\frac{1}{2}$ **39.** They are equal. They are negatives of each other.

41. 8   **43.** 2   **45.** $70 + 8\sqrt{2}/\pi \approx 73.6°$ **47.** $\frac{1}{4}\pi^2 - 2 \approx 0.4674$

True-False Quiz

1. True. 2. True. 3. True. 4. False. 5. True. 6. True. 7. True. 8. True. 9. False 10. True. 11. False.
12. False. 13. True. 14. True 15. True. 16. False. 17. False. 18. True. 19. True. 20. True. 21. True.
22. False. 23. False. 24. True. 25. False. 26. False. 27. True. 28. True. 29. True. 30. False.

Sample Test Problems

1. $\frac{5}{4}$ 2. $\frac{2}{3}x^3 - 3x - x^{-1} + C$ 3. $\frac{1}{3}(y^2 - 4)^{3/2} + C$ 4. $\frac{1}{5}$ 5. 1 6. $-\frac{1}{3}(y^3 - 3y)^{-1} + C$ 7. $y = 2\sqrt{x+1} + 14$
8. $y = \frac{1}{3}(2t - 1)^{3/2} - 1$ 9. $y = (2 - t^3)^{-1/3}$ 10. $y = (3x^2 - \frac{1}{4}x^4 + 9)^{1/2}$ 11. $y = \frac{1}{6}x^3 + 1$ 12. 124
13. $t = 7$ s; $v = -176$ ft/s 14. 5400 mi/h^2 15. $\frac{7}{4}$ 16. $\frac{1}{10}$ 17. $\frac{5}{6}$ 18. 294 19. $\frac{39}{4}$ 20. $3^n - 1$ 21. 1870
22. (a) $\frac{13}{12}$; (b) -9; (c) 0 23. (a) $\sum_{i=2}^{78} 1/i$; (b) $\sum_{i=1}^{50} ix^{2i}$ 24. $48 - \frac{9}{2}(n + 1)(2n + 1)/n^2$; 39
25. (a) -2; (b) -4; (c) 6; (d) -12; (e) -2 26. (a) 5; (b) 6; (c) 2
27. (a) -8; (b) 8; (c) 0; (d) -16; (e) -2; (f) -5 28. 0 29. $c = -\sqrt{7}$
30. (a) $1/(x^2 + 1)$; (b) $2x/(x^4 + 1)$; (c) $3x^2/(x^6 + 1) - 1/(x^2 + 1)$ 31. $\int_0^4 \sqrt{x}\, dx = \frac{16}{3}$; (b) $\int_0^2 (1 + x)^2\, dx = \frac{26}{3}$
32. $f'(x) = (5/5x) - (2/2x) = 0 \Rightarrow f(x) = C$

PROBLEM SET 6.1 *Page 270*

1. $\int_{-1}^{2} (x^2 + 1)\, dx$ 3. $\int_{-2}^{2} (x^2 + 2 + x)\, dx$ 5. $\int_{-2}^{1} (2 - x^2 - x)\, dx$ 7. $\int_{-2}^{0} (x^3 - x^2 - 6x)\, dx + \int_0^3 -(x^3 - x^2 - 6x)\, dx$
9. $\int_{-2}^{1} (3 - y^2 - y - 1)\, dy$ 11. $\int_0^3 (4 - \frac{1}{3}x^2)\, dx = 9$ 13. $-\int_0^2 (x^2 - 2x - 3)\, dx = \frac{22}{3}$ 15. $-\int_{-1}^{0} x^3\, dx + \int_0^2 x^3\, dx = \frac{17}{4}$
17. $\int_4^8 \sqrt{x - 4}\, dx = \frac{16}{3}$ 19. $\int_{-1}^{2} (x + 2 - x^2)\, dx = \frac{9}{2}$ 21. $\int_0^2 (-x^2 - x^2 + 4x)\, dx = \frac{8}{3}$ 23. $\int_0^6 (6y - y^2)\, dy = 36$
25. $\int_{-1}^{2} (4 - y^2 - 2 + y)\, dy = \frac{9}{2}$ 27. $\int_{-2}^{2} ((12 - y^2)/4 - y^2/2)\, dy = 8$ 29. $\int_{-4}^{0} (x + 6 + \frac{1}{2}x)\, dx + \int_0^2 (x + 6 - x^3)\, dx = 22$
31. 130 ft, 194 ft,   33. $6, 2 + 2\sqrt{2} \approx 4.83$ s 35. $A = 9$, $B = C = \frac{37}{6}$, $D = \frac{44}{3}$

PROBLEM SET 6.2 *Page 277*

1. $\pi \int_0^2 (x^2 + 1)^2\, dx = 206\pi/15 \approx 43.14$ 3. (a) $\pi \int_0^2 (4 - x^2)^2\, dx \approx 53.62$; (b) $\pi \int_0^4 (\sqrt{4 - y})^2\, dy = 8\pi \approx 25.13$
5. $\pi \int_0^4 (\frac{1}{4}x^2)^2\, dx = 64\pi/5 \approx 40.21$ 7. $\pi \int_1^4 (1/x)^2\, dx = 3\pi/4 \approx 2.36$ 9. $\pi \int_{-1}^{2} (4 - x^2)\, dx = 9\pi \approx 28.27$
11. $\pi \int_0^2 y^4\, dy = 32\pi/5 \approx 20.11$ 13. $\pi \int_0^4 y\, dy = 8\pi \approx 25.13$ 15. $\pi \int_0^4 y^3\, dy = 64\pi \approx 201.06$ 17. $4\pi ab^2/3$
19. $64\pi/3 \approx 67.02$ 21. $2\pi/3 \approx 2.09$ 23. $\frac{128}{3}$ 25. 2 27. $\frac{2}{3}$ 29. (a) $1024\pi/35 \approx 91.91$; (b) $704\pi/5 \approx 442.34$
31. $2\pi + \frac{16}{3} \approx 11.62$ 33. $\frac{2}{3}r^3 \tan\theta$ 35. (a) $\frac{1}{3}\pi r^2 h$; (b) $\sqrt{2}r^3/12$

PROBLEM SET 6.3 *Page 284*

1. $2\pi \int_1^4 x(4/x)\, dx = 24\pi \approx 75.40$ 3. $2\pi \int_0^4 x\sqrt{x}\, dx = 128\pi/5 \approx 80.42$ 5. $2\pi \int_0^4 (4x^{1/2} - x^{3/2})\, dx = 256\pi/15 \approx 53.62$
7. $2\pi \int_0^2 x(\frac{1}{4}x^3 + x)\, dx = 128\pi/15 \approx 26.81$ 9. $2\pi \int_0^2 y(y^2)\, dy = 8\pi \approx 25.13$ 11. $2\pi \int_0^2 (2 - y)y^2\, dy = 8\pi/3 \approx 8.38$
13. (a) $\pi \int_a^b [f^2(x) - g^2(x)]\, dx$; (b) $2\pi \int_a^b x[f(x) - g(x)]\, dx$;
(c) $2\pi \int_a^b [x - a][f(x) - g(x)]\, dx$; (d) $2\pi \int_a^b [b - x][f(x) - g(x)]\, dx$
15. (a) $\int_1^3 x^{-3}\, dx$; (b) $2\pi \int_1^3 x \cdot x^{-3}\, dx$; (c) $\pi \int_1^3 [(x^{-3} + 1)^2 - 1]\, dx$; (d) $2\pi \int_1^3 (4 - x)x^{-3}\, dx$ 17. $64\pi/5 \approx 40.21$
19. $4\pi(b^2 - a^2)^{3/2}/3$ 21. $\pi(\sqrt{2} - 1)$ 23. (a) $2\pi/15$; (b) $\pi/6$; (c) $\sqrt{2}\pi/60$ 25. $\frac{1}{3}rS$

PROBLEM SET 6.4 *Page 290*

1. $3\sqrt{10} \approx 9.49$ 3. $\frac{112}{3}$ 5. 9 7. $\frac{21}{16}$ $8(37\sqrt{37} - 1)/27 \approx 66.39$ 11. 6π 13. $6a$ 15. $8a$
17. (a) $(2^{7/2} - 2)/5$; (b) 16 19. $a = \sqrt{2}$

PROBLEM SET 6.5 *Page 294*

1. $6\sqrt{37}\pi \approx 114.66$ 3. $50\pi \approx 157.08$ 5. $\pi(10\sqrt{10} - 1)/27 \approx 3.56$ 7. $52\pi/3 \approx 54.45$ 9. $\pi(10\sqrt{10} - 1)/27 \approx 3.56$
11. $2\pi \int_0^\pi (1 + \sin x)\sqrt{1 + \cos^2 x}\, dx$ 13. $2\pi \int_0^\pi x\sqrt{1 + \cos^2 x}\, dx$ 15. $2\pi \int_2^4 (x - 2)\sqrt{1 + (2x + 1)^2/4(x^2 + x)}\, dx$
17. $2\pi \int_0^\pi (1 + \sin t)\sqrt{4t^2 + \cos^2 t}\, dt$ 23. $4\pi^2 ab$ 25. (a) $A_k = A_0 + 2\pi k L_0$; (b) $36\pi + 24\pi^2$

PROBLEM SET 6.6 *Page 298*

1. $k = 16$, $W = \int_0^{1/2} 16x\,dx = 2$ ft-lb   3. 800 ergs   5. $\int_0^d kx\,dx = \frac{1}{2}kd^2$ 7. 22 ft-lb   9. 52,000 ft-lb   11. 76,128 ft-lb
13. $62{,}500\pi$ ft-lb   17. About 2076 in.-lb   19. 350,000 ft-lb   21. 9.5238×10^5 mi-lb   23. 43,200 ft-lb
25. 10,477,274 ft-lb   27. 264.8 ft-lb

PROBLEM SET 6.7 *Page 303*

1. 1684.8 lb   3. 1684.8 lb   5. 16.64 lb   7. 83.2 lb   11. 76,363 lb   13. 7890 lb   15. $2\delta ab/\pi$ 17. $\pi\delta h(b^2 - a^2)$

PROBLEM SET 6.8 *Page 310*

1. $\frac{5}{19}$ 3. 5.4 units   5. $M_x = -3$, $M_y = 11$, $\bar{x} = \frac{11}{17}$, $\bar{y} = -\frac{3}{17}$ 7. $(-\frac{3}{14}, \frac{1}{14})$ 9. $(\frac{9}{16}, \frac{31}{16})$
11. $(0, \frac{8}{5})$ 13. $(\frac{8}{5}, \frac{16}{7})$ 15. $(\frac{7}{5}, \frac{1}{2})$ 17. $(\frac{12}{5}, 0)$ 19. $64\pi/5$ 21. $(0, 4a/3\pi)$
25. (a) $4\pi nr^3 \sin(\pi/2n)\cos(\pi/2n)$;   (b) $\delta nr^3 \sin(\pi/n)$;   (c) $2\pi^2 r^3 = (2\pi r)(\pi r^2)$, $\delta\pi r^3 = (\delta r)(\pi r^2)$
27. It is false.

CHAPTER REVIEW 6.9 *Page 312*

True-False Quiz

1. False.   2. True.   3. False.   4. True.   5. True.   6. False.   7. False.   8. True.   9. False.   10. False.   11. True.
12. False.   13. True.   14. True.   15. True.   16. True.   17. True.   18. True.   19. True.   20. False

Sample Test Problems

1. $\int_0^1 (x - x^2)\,dx = \frac{1}{6}$ 2. $\pi\int_0^1 (x - x^2)^2\,dx = \pi/30$ 3. $2\pi\int_0^1 x(x - x^2)\,dx = \pi/6$ 4. $\pi\int_0^1 [(1 + x - x^2)^2 - 1]\,dx = 11\pi/30$
5. $2\pi\int_0^1 (2 - x)(x - x^2)\,dx = \pi/2$ 6. $(\frac{1}{2}, \frac{1}{10})$
7. $V(S_1) = \frac{1}{6} \cdot 2\pi \cdot \frac{1}{10} = \pi/30$; $V(S_2) = \frac{1}{6} \cdot 2\pi \cdot \frac{1}{2} = \pi/6$; $V(S_3) = \frac{1}{6} \cdot 2\pi \cdot \frac{11}{10} = 11\pi/30$; $V(S_4) = \frac{1}{6} \cdot 2\pi \cdot \frac{1}{2} = \pi/6$
8. (a) 60 in.-lb;   (b) 4 in.-lb   9. 235,242 ft-lb   10. 36,000 ft-lb   11. $\int_0^3 (3x - x^2)\,dx = \int_0^9 (\sqrt{y} - \frac{1}{3}y)\,dy = \frac{9}{2}$ 12. $(\frac{3}{2}, \frac{18}{5})$
13. $\pi\int_0^3 (9x^2 - x^4)\,dx = 162\pi/5$ 14. $2\pi\int_0^3 x(3x - x^2)\,dx = 27\pi/2$ 15. $62.4\int_0^8 \frac{5}{4}y(8 - y)\,dy = 6656$ lb   16. 6656 lb
17. $\frac{53}{6}$ 18. $208\pi/9$ 19. 30,264 lb   20. $4\sqrt{3}$ 21. $\frac{32}{3}$ 22. $\int_a^b [f(x) - g(x)]\,dx$ 23. $\pi\int_a^b [f^2(x) - g^2(x)]\,dx$
24. $2\pi\int_a^b (x - a)[f(x) - g(x)]\,dx$ 25. $M_x = \delta\int_a^b [(f(x) + g(x))/2][f(x) - g(x)]\,dx$; $M_y = \delta\int_a^b x[f(x) - g(x)]\,dx$
26. $\int_a^b \sqrt{1 + [f'(x)]^2}\,dx + \int_a^b \sqrt{1 + [g'(x)]^2}\,dx + f(a) - g(a) + f(b) - g(b)$
27. $2\pi\int_a^b f(x)\sqrt{1 + [f'(x)]^2}\,dx + 2\pi\int_a^b g(x)\sqrt{1 + [g'(x)]^2}\,dx + \pi[f^2(a) - g^2(a)] + \pi[f^2(b) - g^2(b)]$

PROBLEM SET 7.1 *Page 322*

1. (a) 1.792   (b) 0.406;   (c) 4.396   (d) 0.346;   (e) -3.584;   (f) 3.871   31. $(2x - 5)/(x^2 - 5x + 6)$ 5. $4/(x - 5)$
7. $1 + \ln x$ 9. $3[1 + (\ln x)^2]/x$ 11. $(x^2 - 1)^{-1/2}$ 13. $\frac{1}{300}$ 15. $2\ln|2x + 1| + C$ 17. $2\ln(x^2 + x + 5) + C$
19. $\frac{1}{2}(\ln x)^2 + C$ 21. $\frac{1}{4}\ln 82 \approx 1.102$ 23. $\ln[(x + 1)^2/x]$ 25. $\ln[x^2(x - 2)/(x + 2)]$ 27. $-\frac{1}{2}(x^3 + 33x^2 + 8)/(x^3 - 4)^{3/2}$
29. $-\frac{1}{6}(10x^2 + 219x - 118)(x + 13)^{-1/2}(x - 4)^{-2}(2x + 1)^{-4/3}$

31.

33.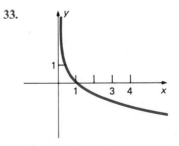

35. $y = 0$ 37. Local min $f(1) = -1$.   39. $\lim_{x \to \infty} \ln x = \infty$ 41. 3   43. $\int_1^2 1/x\,dx = \ln 2 \approx 0.693$

45. (a) 1;   (b) 3   47. $\pi\ln 4$

1. Yes; $f^{-1}(2) \approx 4.3$. 3. No. 5. Yes; $f^{-1}(2) \approx -1.8$ 7. $f'(x) = -15x^4 - 1 < 0$ for all x; f is decreasing.
9. $f'(x) = \sec^2 x > 0$ for $-\pi/2 < x < \pi/2$; f is increasing.
11. $f'(x) = 2(x + 1)$. Since f is continuous on $(-\infty, -1]$ and $f'(x) < 0$ on $(-\infty, -1)$, f is decreasing on $(-\infty, -1]$.
13. $f'(x) = (x^2 + 2)^{1/2} > 0$; f is increasing. 15. $f^{-1}(x) = \frac{1}{3}(x + 1)$ 17. $f^{-1}(x) = \frac{1}{2}(x^2 - 5)$, $x \geq 0$ 19. $f^{-1}(x) = x^{-1} + 5$
21. $f^{-1}(x) = -\sqrt{x}$ 23. $f^{-1}(x) = 4 + \sqrt[3]{x}$ 25. $f^{-1}(x) = (3x + 2)/(2 - x)$ 27. $f^{-1}(x) = [(1 - 2x)/(x - 1)]^{1/3}$
29. $f^{-1}(x) = \frac{1}{4}(\sqrt{8x + 33} - 1)$ 31. $(f^{-1})'(3) \approx 0.2$ 33. $(f^{-1})'(3) \approx -0.2$ 35. $\frac{1}{16}$ 37. $\frac{1}{4}$
41. (a) 1; (b) $2/\sqrt{5}$; (c) $1/\sqrt{2}$ 43. $\frac{3}{5}$

1. (a) 20.086; (b) 0.018316; (c) 8.1662; (d) 544.57; (e) 4.1133; (f) 0.46916; (g) 1.2014 3. x^2 5. $\sin x$
7. $2 \ln x - 2x$ 9. $2x$ 11. $2e^{2x+1}$ 13. $e^{\sqrt{x+1}}/2\sqrt{x+1}$ 15. 1 17. $(x^2 + 2x)e^x$ 19. $(e^{\sqrt{x}} + \sqrt{x}e^{x/2})/2\sqrt{x}$
21. $-ye^{xy}/(1 + xe^{xy})$

23.

25.

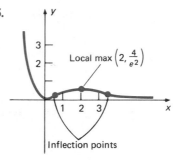

Local max $\left(2, \frac{4}{e^2}\right)$

Inflection points

27.

(0.71, 0.61)

29. $\frac{1}{3}e^{3x+1} + C$ 31. $\frac{1}{2}e^{x^2+6x} + C$ 33. $e^{-1/x} + C$ 35. $\frac{1}{2}(e^5 - e^3) \approx 64.16$ 37. $4\pi \approx 12.57$ 39. $(3 - e)/2e \approx 0.05182$
41. (a) 3,628,800; 3,598,696; (b) 8.3094×10^{81} 43. $\sqrt{2}(e^\pi - 1) \approx 31.31$
45. (a) Both 0; (b) max: $f(e) = \frac{1}{2}$, min: $f(1/e) = -\frac{1}{2}$; (c) $\sqrt{e}$ 47. $e - 1$

1. 2 3. 27 5. $3\sqrt{10}$ 7. $\frac{1}{3}$ 9. 1.594 11. 0.1751 13. 4.248 15. 0.7203 17. $(2x)5^{x^2} \ln 5$ 19. $1/\ln 2 \approx 1.443$
21. $2^x[(x + 5)^{-1} + (\ln 2)\ln(x + 5)]$ 23. $2^{x^2-1}/\ln 2 + C$ 25. $40/\ln 5 \approx 24.85$ 27. $2x(\ln 10)10^{x^2} + 20x^{19}$
29. $(\pi + 1)x^\pi + (\pi + 1)^x \ln(\pi + 1)$ 31. $(x^2 + 1)^{\ln x}[(2x \ln x)/(x^2 + 1) + \ln(x^2 + 1)/x]$ 33. $\sin 1 \approx 0.8415$
35. $\log_{1/2} x = -\log_2 x$ 37. 5.0171×10^8 kWhr 39. $\sqrt[12]{2} \approx 1.0594631$; 523
41. On semilog paper, the graph of $y = Ab^x$ is a line with slope $\ln b$. On log-log paper, the graph of $y = Ax^b$ is a line with slope b.
43. $f'(x) = x^{x^2+1}(1 + 2 \ln x)$, $g'(x) = x^{x^x+x}(1/x + \ln x + \ln^2 x)$

1. $y = 4e^{-5t}$ **3.** $y = 1.8835e^{0.006t}$ **5.** 89,234   **7.** 12.55 days   **9.** 4.644 million; 4.793 million; 6.166 million; 105.0 million
11. 7.736 g   **13.** 2950 years ago   **15.** 81.6°F   **17.** (a) \$449.63;   (b) \$453.13;   (c) \$453.46;   (d) \$453.47
19. (a) 69.66 months;   (b) 69.31 months   **21.** Over \$51 billion.
23. $(1 + p/100)^t = 2 \;\Rightarrow\; t \ln(1 + p/100) = \ln 2 \;\Rightarrow\; t(p/100) \approx 0.7 \;\Rightarrow\; t \approx 70/p$

25.

27. (a) e^{-1};   (b) e^3;   (c) e^2;   (d) e^{-2} **29.** 15.248 billion   **31.** About 2049

1. $\pi/3$ **3.** $-\pi/4$ **5.** $-\pi/3$ **7.** $2\pi/3$ **9.** $2\pi/3$ **11.** 0.541   **13.** 0.126235   **15.** -1.3780222 **17.** 0.30477573
19. 2.0176742   **21.** 0.60754855   **23.** -1.1598315 **25.** $\sin^{-1}(x/8)$ **27.** $\sin^{-1}(5/x)$ **29.** $\tan^{-1}(3/x) - \tan^{-1}(1/x)$
31. $\frac{1}{9}$ **33.** $\frac{56}{65}$ **35.** $x/\sqrt{1 - x^2}$ **37.** $1 - 2x^2$ **39.** $1/\sqrt{1 + x^2}$ **41.** (a) $\pi/2$;   (b) $-\pi/2$

43.

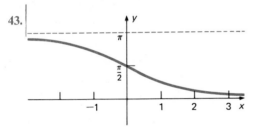

45. (a) $[0, \pi/2]; f^{-1}(x) = \frac{1}{2}\cos^{-1}(x/3)$;   (b) $[-\pi/6, \pi/6]; f^{-1}(x) = \frac{1}{3}\sin^{-1}(x/2)$;
   (c) $(-\pi/2, \pi/2); f^{-1}(x) = \tan^{-1}2x$;   (d) $(-\infty, -2/\pi] \cup [2/\pi, \infty); f^{-1}(x) = 1/\sin^{-1}x$
49. Let $\theta = 2\sin^{-1}(b/2a)$. $A = b^2(\pi/2 + \theta/2) - a^2(\theta - \sin\theta)$.

1. $-2\cos(x - 2)\sin(x - 2) = -\sin[2(x - 2)]$ **3.** $-\csc x(\cot^2 x + \csc^2 x) = \csc x - 2\csc^3 x$ **5.** $-(\csc^2 x)e^{\cot x}$
7. $(\tan x)e^{\tan x}\sec^2 x + (\sec^2 x)e^{\tan x} = (\sec^2 x)(1 + \tan x)e^{\tan x}$ **9.** $2x(1 - x^4)^{-1/2}$ **11.** $-7(2x)^{-1/2}(1 - 2x)^{-1/2}$
13. $e^x/(1 + e^{2x})$ **15.** $x(1 + x^2)^{-1} + \tan^{-1}x$ **17.** $\cos(\tan^{-1}x)/(1 + x^2) = (1 + x^2)^{-3/2}$ **19.** $2/(x\sqrt{x^4 - 1})$
21. $-1/(1 + x^2)$ **23.** $-\frac{1}{2}\cos(x^2) + C$ **25.** $\ln|\sec x| + C$ **27.** $\ln|\tan x| + C$ **29.** $\frac{1}{2}[\sin(e^2) - \sin 1] \approx 0.02619$ **31.** $\pi/4$
33. $\pi/2$ **35.** $\frac{1}{2}\tan^{-1}2x + C$ **43.** π **45.** $b = 2\sqrt{6}$ ft   **47.** $\frac{1}{13}$ rad/s   **49.** 6.283 rad/min   **51.** 3.957×10^{-4} rad/s

7. $2\sinh x \cosh x = \sinh 2x$ **9.** $2x\sinh(x^2 - 1)$ **11.** $x^2\cosh x + 2x\sinh x$
13. $2(2\cosh 2x \cosh 4x + \sinh 2x \sinh 4x)$ **15.** $3x^2/\sqrt{x^6 - 1}, x > 1$ **17.** $2x(4x^2 + 1)^{-1/2} + \sinh^{-1}2x$
19. $-\sin x \cosh(\cos x)$ **21.** $\frac{1}{2}\sinh(x^2 + 3) + C$ **23.** $\cosh(e^x) + C$ **25.** $\frac{20}{9}$ **27.** $\pi(\frac{1}{2} + \frac{1}{4}\sinh 2) \approx 4.419$
31. (b) 42,217 ft³;   (c) 5641 ft²

True-False Quiz

1. False.   2. True.   3. True.   4. False.   5. True.   6. False.   7. False.   8. True.   9. True.   10. False.   11. True.
12. True.   13. False.   14. True.   15. True.   16. True.   17. False.   18. False.   19. False.   20. False.   21. False.
22. True.   23. False.   24. True.   25. False.   26. True   27. False.   28. True.   29. False.   30. True.

Sample Test Problems

1. $4/x$ 2. $6x^2 \sin(x^3)\cos(x^3) = 3x^2 \sin(2x^3)$ 3. $(2x-4)e^{x^2-4x}$ 4. $5x^4/[(x^5-1)\ln 10]$ 5. $\sec^2 x$ 6. $-\csc^2 x$
7. $(\mathrm{sech}^2\sqrt{x})/\sqrt{x}$ 8. $\sec x$ 9. $\sec^2 x/|\sec x|$ 10. $\sqrt{3}/\sqrt{x-3x^2}$ 11. $1/\sqrt{e^{2x}-1}$ 12. $\cot(x/2)$ 13. $15e^{5x}/(e^{5x}+1)$
14. $(6x^2-4)/(2x^3-4x+5)$ 15. $-e^{\sqrt{x}}\sin(e^{\sqrt{x}})/2\sqrt{x}$ 16. $\mathrm{sech}\, x\, \mathrm{csch}\, x$ 17. $-1/\sqrt{x-x^2}$ 18. $(3\ln 4)4^{3x}+12(3x)^3$
19. $(-\csc\sqrt{x}\cot\sqrt{x})/\sqrt{x}$ 20. $2(\log_{10}2x)^{-1/3}/(3x\ln 10)$ 21. $20\sec 5x(\tan^2 5x+\sec^2 5x)$ 22. $4x^2(4+x^4)^{-1}+\tan^{-1}(x^2/2)$
23. $x^{1+x}[(1+x)x^{-1}+\ln x]$ 24. $2ex(1+x^2)^{e-1}$ 25. $\frac{1}{3}e^{3x-1}+C$ 26. $2\ln|\sin 3x|+C$ 27. $-\cos e^x+C$
28. $3\ln|x^2+x-5|+C$ 29. $e^{-1}\ln(e^{x+3}+1)+C$ 30. $2\sin x^2+C$ 31. $2\sin^{-1}2x+C$ 32. $\tan^{-1}(\sin x)+C$
33. $-\tan^{-1}(\ln x)+C$ 34. $\tanh(x-3)+C$
35. Inc. $[-\pi/2, \pi/4]$; dec. $[\pi/4, \pi/2]$; up $(-\pi/2, -\pi/4)$,   36. Inc. $[0,2]$: dec. $(-\infty, 0]$ and $[2, \infty)$;
down $(-\pi/4, \pi/2)$.   up $(-\infty, 0.59)$ and $(3.41, \infty)$; down $(0.59, 3.41)$.

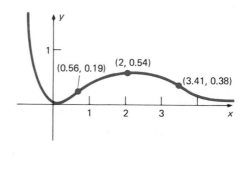

37. (a) $f'(x) = 5x^4 + 6x^2 + 4 > 0 \;\Rightarrow\; f$ increasing $\;\Rightarrow\; f^{-1}$ exists;   (b) $g(7) = 1$;   (c) $g'(7) = \frac{1}{15}$ 38. 66.4 years
39. (a) \$112;   (b) \$112.68;   (c) \$112.75;   (d) \$112.75   40. $-\frac{3}{20}$ rad/s   41. $y = 1$ 42. 27,440

PROBLEM SET 8.1   *Page 374*

1. $\frac{1}{5}(x-1)^5+C$ 3. $\frac{1}{10}(x^2+1)^5+C$ 5. $\tan^{-1}x+C$ 7. $\frac{1}{2}\ln(x^2+1)+C$ 9. $(2+t^2)^{3/2}+C$ 11. $\sec z+C$
13. $2\sin\sqrt{x}+C$ 15. $\pi/4$ 17. $x^2-x+\ln|x+1|+C$ 19. $\frac{2}{3}(\tan x)^{3/2}+C$ 21. $\frac{1}{2}\sin(\ln 4x^2)+C$ 23. $5\sin^{-1}(e^x)+C$
25. $-5\sqrt{1-e^{2x}}+C$ 27. $9/\ln 100 \approx 1.954$ 29. $x - \ln|\sin x|+C$ 31. $\frac{1}{2}\ln|\sec(z^2+4z-3)|+C$
33. $\ln|\sec e^x+\tan e^x|+C$ 35. $\tan x+e^{\sin x}+C$ 37. $-\frac{1}{3}\csc(t^3-2)+C$ 39. $-\frac{1}{3}[\cot(t^3-2)+t^3]+C$
41. $\frac{1}{2}e^{\tan^{-1}2t}+C$ 43. $\frac{1}{6}\sin^{-1}(3y^2/4)+C$ 45. $\tan^{-1}(\sec x)+C$ 47. $\frac{1}{3}\sin^{-1}(e^{3t}/2)+C$ 49. $\frac{1}{4}\tan^{-1}(\frac{1}{4}) \approx 0.0612$
51. $\frac{1}{2}\tan^{-1}[(x+1)/2]+C$ 53. $\frac{1}{3}\tan^{-1}(3x+3)+C$ 55. $\frac{1}{18}\ln|9x^2+18x+10|+C$ 57. $(\frac{1}{3})\sec^{-1}(\sqrt{2}|t|/3)+C$
59. $(\frac{2}{135})(9x-4)(3x+2)^{3/2}+C$ 61. $(\frac{1}{24})\ln|(4x+3)/(4x-3)|+C$
63. $(x/16)(4x^2-9)\sqrt{9-2x^2}+(81\sqrt{2}/32)\sin^{-1}(\sqrt{2}x/3)+C$ 65. $(\sqrt{3}/3)\ln(\sqrt{3}x+\sqrt{5+3x^2})+C$
67. $\ln|t+1+\sqrt{t^2+2t-3}|+C$ 69. $\frac{2}{27}(3\sin t-10)\sqrt{3\sin t+5}+C$ 71. $\ln(\sqrt{2}+1) \approx 0.881$ 73. π^2

PROBLEM SET 8.2   *Page 380*

1. $\frac{1}{2}x+\frac{1}{4}\sin 2x+C$ 3. $\sin x-\frac{1}{3}\sin^3 x+C$ 5. $\frac{8}{15}$ 7. $\frac{1}{2}\tan^2 y+\ln|\cos y|+C$
9. $-\frac{1}{3}\cos^3 3x[\frac{1}{3}-\frac{3}{5}\cos^2 3x+\frac{3}{7}\cos^4 3x-\frac{1}{9}\cos^6 3x]+C$ 11. $-[\csc\theta+\sin\theta]+C$ 13. $\frac{1}{64}[\frac{3}{2}t-\frac{1}{4}\sin 8t+\frac{1}{32}\sin 16t]+C$
15. $\sec^3 3y[\frac{1}{15}\sec^2 3y-\frac{1}{9}]+C$ 17. $-\frac{1}{3}\csc^3 x+C$ 19. $-\frac{1}{2}\cot^2 t+C$ 21. $-\frac{1}{18}\cos 9y+\frac{1}{2}\cos y+C$
23. $-\frac{1}{6}\cot^3 2x+\frac{1}{2}\cot 2x+x+C$ 25. $\frac{1}{21}\tan^3 7x+\frac{1}{7}\tan 7x+C$ 27. $-\frac{1}{2}\cot^2 x-\ln|\sin x|+C$ 29. $\tan x-\cot x+C$
31. $\pi^2(\frac{1}{3}\pi^2+\frac{5}{2}) \approx 57.14$

PROBLEM SET 8.3   Page 385

1. $\frac{2}{5}(x+3)^{3/2}(x-2) + C$ 3. $\frac{1}{6}(2t+7)^{3/2} - \frac{7}{2}(2t+7)^{1/2} + C$ 5. $2 + 4\ln 3 - 4\ln 4 \approx 0.8493$ 7. $\frac{1}{5}(2t+1)^{5/2}(5t-1) + C$
9. $-\ln[(1 + \sqrt{1-x^2})/|x|] + \sqrt{1-x^2} + C$ 11. $-\frac{1}{3}\ln[(\sqrt{x^2+9}+3)/|x|) + C$ 13. $\frac{1}{16}[\sqrt{3}/2 - \frac{3}{5}] \approx 0.01663$
15. $-\sqrt{4-t^2} + C$ 17. $-2\sqrt{4-x^2} - 3\sin^{-1}(x/2) + C$ 19. $(y^2+8)/\sqrt{y^2+4} + C$ 21. $\ln(\sqrt{x^2+2x+5} + x + 1) + C$
23. $3\sqrt{x^2+2x+5} - 3\ln(\sqrt{x^2+2x+5} + x + 1) + C$ 25. $\frac{9}{2}[\sin^{-1}\frac{1}{3}(x+2) + \frac{1}{9}(x+2)\sqrt{5-4x-x^2}] + C$
27. $\sin^{-1}\frac{1}{2}(x-2) + C$ 29. $\ln(x^2+2x+2) - \tan^{-1}(x+1) + C$ 31. $(\pi/16)(\pi/4 - \tan^{-1}0.5 + 0.1) \approx 0.08281$
33. $\frac{1}{2}\ln(x^2+9) + C$ 35. $\sqrt{4-x^2} - \ln[(2+\sqrt{4-x^2})^2/x^2] + C$
39. $dy/dx = -\sqrt{a^2-x^2}/x$, $y(a) = 0$; $y = a\ln[(a+\sqrt{a^2-x^2})/x] - \sqrt{a^2-x^2}$

PROBLEM SET 8.4   Page 392

1. $xe^x - e^x + C$ 3. $-\frac{1}{3}x\cos 3x + \frac{1}{9}\sin 3x + C$ 5. $x\tan^{-1}x - \frac{1}{2}\ln(1+x^2) + C$ 7. $\frac{1}{5}t\tan 5t + \frac{1}{25}\ln|\cos 5t| + C$
9. $\frac{2}{9}x^{3/2}(3\ln x - 2) + C$ 11. $\frac{1}{2}(x^2+1)\tan^{-1}x - \frac{1}{2}x + C$ 13. $\frac{1}{4}w^2(2\ln w - 1) + C$ 15. $\pi\sqrt{3}/6 + \ln 2 \approx 1.6000$
17. $a^x(x\ln a - 1)/\ln^2 a + C$ 19. $e^x(x^2 - 2x + 2) + C$ 21. $\frac{1}{2}e^t(\sin t + \cos t) + C$ 23. $\frac{1}{2}x[\sin(\ln x) - \cos(\ln x)] + C$
25. 1   27. $-108e^{-3} + 27 \approx 21.62$ 31. $\frac{16}{35} \approx 0.4571$ 33. $\int x^3 e^x\, dx = (x^3 - 3x^2 + 6x - 6)e^x + C$
37. (a) $(4n-3)\pi$;   (b) $2\pi(13\pi^2 - 4)$

PROBLEM SET 8.5   Page 399

1. $\ln|x| - \ln|x+2| + C$ 3. $2\ln|x+3| + 3\ln|x-3| + C$ 5. $3\ln|x+4| - 2\ln|x-1| + C$ 7. $2\ln|2x-1| - \ln|x+5| + C$
9. $2\ln|x| - \ln|x+1| + \ln|x-2| + C$ 11. $\frac{3}{2}x^2 - 3x + \ln|x-1| + 8\ln|x+2| + C$ 13. $\ln|x-3| - 4(x-3)^{-1} + C$
15. $2\ln|x| + \ln|x-4| + (x-4)^{-1} + C$ 17. $-2\ln|x| + 2\ln(x^2+4) + \frac{1}{2}\tan^{-1}(x/2) + C$
19. $2\ln|x+3| - \ln|x-2| - \tan^{-1}x + C$ 21. $\frac{1}{2}\ln(x^2+1) + \frac{5}{2}(x^2+1)^{-1} + C$ 23. $3\ln 10 - 2\ln 3 - 3\ln 8 \approx -1.528$
25. (a) $x = ab[e^{(b-a)kt} - 1]/[be^{(b-a)kt} - a]$;   (c) $\frac{38}{23}$;   (d) $x = a^2kt/(akt + 1)$
27. (a) $y = 10/(1 + 4e^{-0.019617t})$;   (b) 5.937 billion;   (c) 2108   29. $y = [2AB + (B^2 - AB)e^{-k(A+B)t}]/[2B + (A-B)e^{-k(A+B)t}]$

CHAPTER REVIEW 8.6   Page 400

True-False Quiz

1. True.   2. True.   3. False.   4. False.   5. True.   6. True.   7. True.   8. False.   9. True.   10. False.   11. False.
12. True.   13. True.   14. True.   15. True.   16. False.   17. False.   18. True.   19. True.   20. False.   21. True.
22. False.   23. False.   24. True.   25. True.

Sample Test Problems

1. 2   2. $-\frac{1}{2}\cot 2\theta - \theta + C$ 3. $e - 1$ 4. $\frac{1}{4}$ 5. $\frac{1}{3}y^3 - \frac{1}{2}y^2 + 2y - 2\ln|y+1| + C$ 6. $-\frac{1}{2}\cos 2t + \frac{1}{6}\cos^3 2t + C$
7. $\frac{1}{2}\ln|y^2 - 4y + 2| + C$ 8. 1   9. $e^t + 2\ln|e^t - 2| + C$ 10. $\sin x + \ln|\csc x - \cot x| + \cos x + C$
11. $(1/\sqrt{2})\sin^{-1}[(x-1)/3] + C$ 12. $x^2 e^x - 2xe^x + 2e^x + C$ 13. $(1/\sqrt{3})\ln|\sqrt{3}y + \sqrt{2 + 3y^2}| + C$
14. $-\frac{1}{2}w^2 - \frac{1}{2}\ln|w^2 - 1| + C$ 15. $-\ln|\ln|\cos x|| + C$ 16. $\ln|t-1| - \frac{1}{2}\ln(t^2 + t + 1) - \sqrt{3}\tan^{-1}[(2t+1)/\sqrt{3}] + C$
17. $\cosh x + C$ 18. $\frac{1}{6}(\ln y)^6 + C$ 19. $-x\cot x + \ln|\sin x| - \frac{1}{2}x^2 + C$ 20. $-2\cos\sqrt{x} + C$ 21. $(\ln t)^2 + C$
22. $y\ln(y^2 + 9) - 2y + 6\tan^{-1}(y/3) + C$ 23. $3e^{t/3}(\sin 3t - 9\cos 3t)/82 + C$ 24. $\ln|t| - \frac{1}{2}\ln(t^2 - 9) + \frac{1}{3}\tan^{-1}(t/3) + C$
25. $-\frac{1}{4}\cos 2x - \frac{1}{2}\cos x + C$ 26. $\frac{3}{8}x + \frac{1}{2}\sin x + \frac{1}{16}\sin 2x + C$ 27. $\frac{1}{6}\sec^3 2x - \frac{1}{2}\sec 2x + C$
28. $x - 2\sqrt{x} + 2\ln(\sqrt{x}+1) + C$ 29. $\frac{2}{9}\tan^{9/2}x + \frac{2}{5}\tan^{5/2}x + C$ 30. $\ln|t| - 6\ln(t^{1/6} + 1) + C$ 31. $-\sqrt{9 - e^{2y}} + C$
32. $\frac{2}{3}\sin^{3/2}x - \frac{4}{7}\sin^{7/2}x + \frac{2}{11}\sin^{11/2}x + C$ 33. $3\sin x + C$ 34. $\sqrt{9-y^2} + 3\ln|(3 - \sqrt{9-y^2})/y| + C$
35. $\frac{1}{4}\tan^{-1}(e^{4x}) + C$ 36. $-(a^2 + x^2)^{3/2}/(3a^2 x^3) + C$ 37. $\frac{2}{3}(w+5)^{1/2}(w - 10) + C$ 38. $-2\sqrt{1 + \cos t} + C$
39. $-\frac{1}{6}\tan^{-1}(\frac{1}{3}\cos^2 y) + C$ 40. $\sin^{-1}[(x+3)/\sqrt{10}] + C$ 41. $\ln|x/\sqrt{x^2+3}| - 2/x + (2\sqrt{3}/3)\tan^{-1}(x/\sqrt{3}) + C$
42. $x/(16\sqrt{16 + x^2}) + C$ 43. $4(3 - \sqrt{3})/3 \approx 1.691$ 44. $\ln 7 - \frac{6}{7} \approx 1.089$ 45. $3\pi(4 + 3\ln 3)/2 \approx 34.38$
46. $\ln\frac{1}{3}(2\sqrt{3} + 3) \approx 0.7677$

PROBLEM SET 9.1   Page 409

1. -1 3. -1 5. $-\frac{1}{2}$ 7. ∞ 9. 0   11. $-\frac{1}{2}$ 13. $-\frac{9}{4}$ 15. -2 17. $-\infty$ 19. $-\frac{1}{24}$ 21. $-\infty$ 23. 1
25. We did not yet know that $D_x(\sin x) = \cos x$; in fact, to show the latter, we need to know that $\lim_{x \to 0}(\sin x)/x = 1$.
27. (a) $\frac{3}{4}$;   (b) $\frac{1}{2}$ 29. $4\pi b^2$

PROBLEM SET 9.2   Page 414

1. 0   3. 0   5. 1   7. 0   9. $-\infty$ 11. 0   13. 0   15. 1   17. 0   19. e^3 21. 1   23. 1   25. 1   27. 1   29. ∞
31. ∞ 33. 1   35. Does not exist.   37. 0   39. 1   41. (a) 0;   (b) 0;   (c) $\ln a$;   (d) ∞

43.

45. $1/(k+1)$

PROBLEM SET 9.3   Page 420

1. Divergent.   3. $\frac{1}{2}e^{-16} \approx 5.63 \times 10^{-8}$ 5. Divergent.   7. 100   9. Divergent.   11. Divergent.   13. $1/\ln 2$ 15. $-\frac{1}{4}$
17. Divergent.   19. $\pi/2$ 21. π 23. $\frac{1}{2}$ 25. $(\ln 3)/2 \approx 0.5493$ 27. The work required would be $k\int_{3960}^{\infty} (1/x)\,dx = \infty$.
29. \$1,250,000   35. Region bounded by $y = 1/x$, $y = 0$, $1 \leq x < \infty$.

PROBLEM SET 9.4   Page 425

1. $\frac{3}{2}$ 3. $\pi/2$ 5. Divergent.   7. 6   9. $-\frac{3}{2}(5^{1/3} + 4^{1/3}) \approx -4.946$ 11. Divergent.   13. Divergent.   15. Divergent.
17. 0   19. $\int_0^1 1/x^p\,dx$ diverges if $p \geq 1$ (Example 4) and $\int_1^{\infty} 1/x^p\,dx$ diverges if $p \leq 1$ (Example 6, Section 9.3).   21. 6
23. (a) 3;   (b) $\pi\int_0^1 x^{-4/3}\,dx = \lim_{a \to 0^+} \pi\int_a^1 x^{-4/3}\,dx = \infty$ 25. No; $\lim_{x \to 0^+} (\sin x)/x = 1$ and so the integrand is bounded.
27. $\int_1^{\infty} e^{-x}\,dx = 1/e$. The convergence of $\int_1^{\infty} e^{-x}\,dx$ implies that of $\int_1^{\infty} e^{-x^2}\,dx$.
29. (a) $\lim_{x \to \infty} x^{n+1}/e^x = 0$. Therefore, there exists $M > 1$ such that $x^{n+1}/e^x \leq 1$ for $x \geq M$ and hence $x^{n-1}/e^x \leq 1/x^2$ for $x \geq M$.
   (b) $\int_1^{\infty} x^{n-1}e^{-x}\,dx = \int_1^M x^{n-1}e^{-x}\,dx + \int_M^{\infty} x^{n-1}e^{-x}\,dx$.
The first integral on the right is an ordinary integral. The second integral converges because $\int_M^{\infty} 1/x^2\,dx$ converges.
33. (a) $\pi/2$;   (b) π

CHAPTER REVIEW 9.5   Page 426

True-False Quiz

1. True.   2. True.   3. False.   4. False.   5. False.   6. False.   7. True.   8. True.   9. True.   10. False.   11. False.
12. True.   13. True.   14. True.   15. True.   16. True.   17. False.   18. False.   19. True.   20. False.   21. True.
22. False.   23. True.   24. True.   25. False.

Sample Test Problems

1. 4   2. $\frac{2}{3}$ 3. 0   4. ∞ 5. 2   6. 0   7. 0   8. ∞ 9. 0   10. 0   11. 1   12. e^2 13. 0   14. 1   15. 0   16. $\frac{1}{3}$
17. 1   18. -1 19. 1   20. $\pi/2$ 21. $\frac{1}{2}e^2$ 22. Divergent.   23. Divergent.   24. 0   25. $(4 - \pi)/4$ 26. 1
27. Divergent.   28. $2\sqrt{3}$ 29. $1/\ln 2$ 30. 2   31. 6   32. $\frac{1}{2}e^{-4}$ 33. Divergent.   34. 0
35. $1/\sqrt{x^6 + x} \leq 1/x^3$; convergent.   36. $(\ln x)/e^{2x} \leq 1/e^x$; convergent.   37. $(\ln x)/x \geq 1/x$ if $x \geq e$; divergent.
38. $(\ln x)/x^3 \leq 1/x^2$; convergent.

PROBLEM SET 10.1   Page 435

1. 1.72303   3. 0.119332   5. 118.113   7. $1 + 2x + 2x^2 + \frac{4}{3}x^3 + \frac{2}{3}x^4$; 1.58389   9. $2x - \frac{4}{3}x^3$; 0.443777
11. $x - \frac{1}{2}x^2 + \frac{1}{3}x^3 - \frac{1}{4}x^4$; 0.206906   13. $x - \frac{1}{3}x^3$; 0.225944   15. $e^2[1 + (x - 2) + \frac{1}{2}(x - 2)^2 + \frac{1}{6}(x - 2)^3]$
17. $1 + 2(x - \pi/4) + 2(x - \pi/4)^2 + \frac{8}{3}(x - \pi/4)^3$ 19. $\pi/4 + \frac{1}{2}(x - 1) - \frac{1}{4}(x - 1)^2 + \frac{1}{12}(x - 1)^3$
21. $7 + 2(x - 1) + (x - 1)^2 + (x - 1)^3$; expand and simplify to see that this is exact.
23. $(1 - x)^{-1} \approx 1 + x + x^2 + x^3 + \cdots + x^n$;   (a) 1.1111 compares to 1.1111;   (b) 1.9375 compares to 2;
(c) 4.0951 compares to 10;   (d) 31 compares to -1.
25. $A = \frac{1}{2}r^2(t - \sin t) \approx r^2 t^3/12$

27. (c)

r	n (Exact)	n (Approx.)	n (Rule of 70)
0.05	13.8918	13.889	14
0.10	6.9603	6.959	7
0.15	4.6498	4.649	4.67
0.20	3.4945	3.494	3.5

PROBLEM SET 10.2   Page 441

1. $\left| e^c + e^{-c} \right| = e^c + e^{-c} \leq e^2 + 1 \leq 8.39$ **3.** $\left| 2/\cos c \right| = 2/\left| \cos c \right| \leq 2/\left| \cos 2\pi/3 \right| = 4$

5. $\left| e^{-c}/(c + 4) \right| = e^{-c}/\left| c + 4 \right| \leq e^8/2 \leq 1491$ **7.** $\left| (c^2 + \cos c)/(16 \ln c) \right| \leq (c^2 + \left| \cos c \right|)/(16 \ln 2) \leq (9 + \left| \cos 3 \right|)/(16 \ln 2) \leq 1$

9. $R_6(x) = x^7/7(1 + c)^7$, where c is between 0 and x; $\left| R_6(0.5) \right| \leq 0.001117$.

11. $R_6(x) = -(\cos c)(x - 1)^7/7!$, where c is between 1 and x; $\left| R_6(0.5) \right| \leq 0.0000014$.   **13.** $n = 9$

15. $(1 + x)^{-1/2} \approx 1 - \frac{1}{2}x + \frac{3}{8}x^2 - \frac{5}{16}x^3$; $\left| R_3(x) \right| \leq 0.0000023$ **17.** $\int_0^{0.5} \sin x \, dx \approx 0.122396$ with an error less than 0.000131.

19. $-1 - (x - 1)^2 + (x - 1)^3 + (x - 1)^4$; $R_4(x) = 0$ because the fifth derivative is zero.   **21.** 0.68199832; $\left| R_3 \right| \leq 4.38 \times 10^{-8}$

PROBLEM SET 10.3   Page 448

1. 0.50227; 0.50003; 0.5   **3.** 5.2650; 5.3046; 5.3333   **5.** 1.5708; 1.9541; 1.9886   **7.** 3.1415926

9. With $\left| f''(c) \right| \leq 4$, $n = 8$; 0.8057.   **11.** With $\left| f''(c) \right| \leq 0.6$, $n = 3$; 0.9730.   **13.** With $\left| f^{(4)}(c) \right| \leq 48$, $n = 8$; -4.19745.

17. 40,825   **21.** 1,074,585,600 ft^3

PROBLEM SET 10.4   Page 454

1. 1.29   **3.** 1.29   **5.** -0.26795 **7.** 1.55715   **9.** 0.73909   **11.** 0.34711; 3.65289   **13.** 0.52658   **15.** 1.81712

17. 4.493409; -0.217234 **19.** $m = 2 = M$; $\left| x_6 - \sqrt{2} \right| \leq 1.0843 \times 10^{-19}$

21. The small value of $\left| f'(1.2) \right|$ sends the intersection of the tangent with the x-axis far to the right.

23. $i = 0.01513$, $r = 18.16\%$

PROBLEM SET 10.5   Page 460

1. 0.08446   **3.** 2.15831   **5.** (a) Graph indicates the root is near $\frac{1}{2}$.   (b) 0.49999872 after 4 steps.   (c) $\frac{1}{2}$;   (d) $g'(\frac{1}{2}) = 0$

7. (a) Graph indicates the root is near 0.9.   (b) Algorithm behaves very badly.   (c) $\left| g'(0.9) \right| \approx 6$, which is larger than 1.

9. (a) Divide by 6 and write $\frac{1}{6}x = x - \frac{5}{6}x$.   (b) 0.858736;   (c) $\left| g'(0.858736) \right| \approx 0.11$, which is much less than 1.

11. Write $x = \frac{1}{10}(1 + 11x + x^2 - x^3)$; $x = 1.8392868$.   **13.** 1.7724538; $g'(a) = 0$ **15.** (a) \$293.75;   (b) 1.5991%

19. $r = 0.6$ **21.**

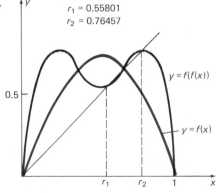

$r_1 = 0.55801$
$r_2 = 0.76457$

$y = f(f(x))$

$y = f(x)$

23. $s_1 = 0.38282$, $s_2 = 0.82694$, $s_3 = 0.50088$, $s_4 = 0.87500$; $h(x) = f(f(f(f(x)))) = 0$

True-False Quiz

1. True.   **2.** True.   **3.** True.   **4.** True.   **5.** True.   **6.** True.   **7.** True.   **8.** False.   **9.** False.   **10.** False.   **11.** True.
12. False.   **13.** True.   **14.** True.   **15.** True.   **16.** False.   **17.** False.   **18.** True.   **19.** False.   **20.** True.

Sample Test Problems

1. 47.26   **2.** (a) $x + x^2 + \frac{1}{2}x^3 + \frac{1}{6}x^4$; 0.110517;   (b) $1 + \frac{1}{2}x^2 + \frac{1}{24}x^4$; 1.0050042   **3.** $3 + 9(x - 2) + 4(x - 2)^2 + (x - 2)^3$
4. 3.941   **5.** $\frac{1}{2} - \frac{1}{4}(x - 1) + \frac{1}{8}(x - 1)^2 - \frac{1}{16}(x - 1)^3 + \frac{1}{32}(x - 1)^4$
6. $R_4(x) = -(x - 1)^5/(c + 1)^6$, where c is between 1 and x; $|R_4(1.2)| \leq 0.000005$.   **7.** $x^2 - \frac{1}{3}x^4$; $|R_5| \leq 0.000003$ **8.** $n = 5$
9. -0.0026987; 0.00002   **10.** -0.002786; 0.00014   **11.** -0.0026994; 0.0000003   **12.** -0.00269929 **13.** 0.281785
14. 0.281785   **15.** 4.49341   **16.** $D_x(\tan x) = \sec^2 x \geq 1$ **17.** -3.18306

PROBLEM SET 11.1   Page 470

1. $1, \frac{2}{3}, \frac{3}{5}, \frac{4}{7}, \frac{5}{9}$; converges; $\frac{1}{2}$.   **3.** $\frac{5}{2}, \frac{17}{3}, \frac{37}{6}, \frac{65}{11}, \frac{101}{18}$; converges; 4.   **5.** $\frac{5}{3}, \frac{2}{3}, \frac{7}{19}, \frac{8}{33}, \frac{3}{17}$; converges; 0.   **7.** $-\frac{1}{2}, \frac{2}{3}, -\frac{3}{4}, \frac{4}{5}, -\frac{5}{6}$; diverges.
9. $1, 0, -\frac{1}{3}, 0, \frac{1}{5}$; converges; 0.   **11.** $e, e^2/4, e^3/9, e^4/16, e^5/25$; diverges.
13. $-\pi/4, \pi^2/16, -\pi^3/64, \pi^4/256, -\pi^5/1024$; converges; 0   **15.** 1.9, 1.81, 1.729, 1.6561, 1.59049; converges; 1.
17. 0, $\ln 2/2$, $\ln 3/3$, $\ln 4/4$, $\ln 5/5$; converges; 0.   **19.** 2, 9/4, 64/27, 625/256, 7776/3125; converges; e.
21. $n/(n + 1)$; converges; 1.   **23.** $(-1)^n n/(2n - 1)$; diverges.   **25.** $n/[n^2 - (n - 1)^2] = n/(2n - 1)$; converges; $\frac{1}{2}$.
27. $n \sin(1/n)$; converges; 1.   **29.** $2^n/n^2$; diverges.   **31.** $\frac{1}{2}, \frac{5}{4}, \frac{9}{8}, \frac{13}{16}$; $0 < a_{n+1}/a_n = (4n + 1)/(8n - 6) < 1$ for $n \geq 2$.
33. $\frac{3}{4}, \frac{2}{3}, \frac{5}{8}, \frac{3}{5}$; $0 < a_{n+1} = [1 - 1/(n + 1)^2]a_n < a_n$ **35.** $1, \frac{3}{2}, \frac{7}{4}, \frac{15}{8}$; $0 < a_n = 2 - (\frac{1}{2})^{n-1} < a_{n+1} < 2$ **37.** 2.3028
39. $u = (1 + \sqrt{13})/2 \approx 2.302776$ **41.** 1.1118   **43.** $\int_0^1 \sin x \, dx \approx 0.4597$ **45.** Choose $N > 1/\varepsilon - 1$.
51. No; let $a_n = n + 1/n$, $b_n = -n$.   **53.** $\pi/2\sqrt{3}$

PROBLEM SET 11.2   Page 478

1. Converges; $\frac{1}{4}$.   **3.** Converges; $\frac{33}{5}$.   **5.** Diverges since $(k - 3)/k \to 1$.   **7.** Converges; 1.
9. Diverges since $k!/10^k > 1$ for large k.   **11.** Converges; $e^2/(\pi^2 - e\pi)$.   **13.** -3 **15.** $0.2 + 0.02 + 0.002 + \cdots$; $\frac{2}{9}$
17. $0.013 + 0.000013 + 0.000000013 + \cdots$; $\frac{13}{999}$ **19.** $0.4 + 0.09 + 0.009 + 0.0009 + \cdots$; $\frac{1}{2}$ **21.** 1   **23.** $S_n = -\ln(n + 1)$
25. 500 ft   **27.** \$4 billion, including the original \$1 billion.   **29.** $\frac{1}{4}$ **31.** $\frac{4}{5}$, no   **33.** 1000/9 yd
35. Try proof by contradiction.   **37.** Since $\frac{1}{2} + \frac{1}{4} + \frac{1}{6} + \cdots$ diverges, the protrusion can be made arbitrarily large.
39. Try proof by contradiction.   **41.** (a) 2;   (b) 1   **43.** (a) $A = C/(1 - e^{-kt})$;   (b) $\frac{8}{3}$ **45.** 1

PROBLEM SET 11.3   Page 485

1. Diverges.   **3.** Diverges.   **5.** Diverges.   **7.** Diverges.   **9.** Converges.   **11.** Converges.   **13.** Diverges; nth-Term Test.
15. Diverges; nth-Term Test.   **17.** Diverges; nth-Term Test.   **19.** Converges; Integral Test.   **21.** Converges; Integral Test.
23. $\int_5^\infty xe^{-x} \, dx \approx 0.0404$ **25.** $\int_5^\infty 1/(1 + x^2) \, dx \approx 0.1974$ **27.** $p > 1$ (Integral Test).   **31.** This is a famous unsolved problem.
33. 2.724×10^8

PROBLEM SET 11.4   Page 492

1. Diverges.   **3.** Converges.   **5.** Converges.   **7.** Diverges.   **9.** Converges.   **11.** Diverges; nth-Term Test.
13. Converges; Limit Comparison Test ($b_n = 1/n^{3/2}$).   **15.** Converges; Ratio Test.
17. Converges; Limit Comparison Test ($b_n = 4/n^2$).   **19.** Converges; Ordinary Comparison Test ($b_n = 1/n^2$).
21. Converges; Limit Comparison Test ($b_n = 1/n^2$).   **23.** Converges; Ratio Test.   **25.** Converges; p-Series Test ($p = \frac{3}{2}$).
27. Diverges, nth-Term Test.   **29.** Converges; Ordinary Comparison Test ($b_n = 5/n^3$).
31. Converges; Ratio Test.   **33.** Converges; Ratio Test.
35. $\sum a_n$ converges $\Rightarrow$ $0 < a_n < 1$ for large n $\Rightarrow$ $0 < a_n^2 < a_n$ for large n. Use Ordinary Comparison Test.
37. $a_n/b_n \to 0$ $\Rightarrow$ $0 \leq a_n < b_n$ for large n. Use Ordinary Comparison Test.
41. Choose r such that $R < r < 1$. Then choose N so that $n \geq N$ $\Rightarrow$ $a_n^{1/n} \leq r$—that is, $a_n \leq r^n$. Use Ordinary Comparison Test.
43. (a) Diverges;   (b) converges;   (c) converges;   (d) converges;   (e) diverges;   (f) converges.
45. Converges if $p > 1$, diverges otherwise.

PROBLEM SET 11.5   *Page 498*

1. $|S - S_9| < \frac{2}{31} \approx 0.0645$ **3.** $|S - S_9| < 1/(\ln 11) \approx 0.417$ **5.** $|S - S_9| < (\ln 10)/10 \approx 0.230$ **7.** Absolute Ratio Test.
9. Absolute Ratio Test.   **11.** Ordinary Comparison Test: $|(-1)^{n+1}/[n(n+1)]| \le 1/n^2$.   **13.** Converges conditionally.
15. Diverges.   **17.** Converges conditionally.   **19.** Converges absolutely.   **21.** Converges conditionally.
23. Converges conditionally.   **25.** Converges absolutely.   **27.** Converges conditionally.   **29.** Diverges.
31. Try proof by contradiction.   **33.** Key idea: $\sum 1/2n = \sum \frac{1}{2}(1/n)$ diverges by Problem 31 of Section 11.2   **45.** $\ln 2$

PROBLEM SET 11.6   *Page 503*

1. $[-1, 1]$ **3.** $\mathbb{R}$ **5.** $(-1, 1)$ **7.** $(-1, 1]$ **9.** $[-1, 1]$.   **11.** $(-2, 2)$ **13.** $\mathbb{R}$ **15.** $[0, 2)$ **17.** $(-3, 1)$ **19.** $[-6, -4]$
21. nth-Term Test.   **23.** $\sqrt{2}$ **25.** $1/(4 - x)$; $2 < x < 4$ **27.** (a) $[-1, \frac{1}{3})$;   (b) $(-\frac{1}{2}, \frac{7}{2}]$ **29.** $S(x) = a_0 + a_1x + a_2x^2)/(1 - x^3)$

PROBLEM SET 11.7   *Page 509*

1. $1 - x + x^2 - x^3 + \cdots = \sum_{n=0}^{\infty} (-1)^n x^n$; 1 **3.** $x^2 - x^3 + x^4 - x^5 + \cdots = \sum_{n=2}^{\infty} (-1)^n x^n$; 1

5. $\frac{1}{2}[1 + \frac{3}{2}x + (\frac{3}{2}x)^2 + \cdots] = \sum_{n=0}^{\infty} \frac{1}{2}(\frac{3}{2}x)^n$; $\frac{2}{3}$ **7.** $x^2 + x^6 + x^{10} + \cdots = \sum_{n=0}^{\infty} x^{4n+2}$; 1

9. $x^2/2 - x^3/3 \cdot 2 + x^4/4 \cdot 3 - x^5/5 \cdot 4 + \cdots = \sum_{n=2}^{\infty} (-1)^n x^n/n(n-1)$; 1 **11.** $2x + \frac{2}{3}x^3 + \frac{2}{5}x^5 + \cdots = \sum_{n=1}^{\infty} 2x^{2n-1}/(2n-1)$; 1

13. $1 - x + x^2/2! - x^3/3! + \cdots = \sum_{n=0}^{\infty} (-1)^n x^n/n!$ **15.** $2[1 + x^2/2! + x^4/4! + \cdots] = \sum_{n=0}^{\infty} 2x^{2n}/(2n)!$
17. $1 + \frac{1}{2}x^2 + \frac{1}{3}x^3 + \frac{3}{8}x^4 + \cdots$ **19.** $x - x^2 + \frac{1}{6}x^3 + \frac{1}{6}x^4 + \frac{3}{40}x^5 + \cdots$ **21.** $x + \frac{2}{3}x^3 + \frac{13}{15}x^5 - \frac{29}{105}x^7 + \cdots$
23. $x + \frac{1}{6}x^3 - \frac{1}{12}x^4 + \frac{3}{40}x^5 + \cdots$ **25.** (a) $x/(1 + x)$;   (b) $(e^x - 1 - x)/x^2$;   (c) $-\ln(1 - 2x)$
27. $x(d/dx)[1/(1 - x)] = x/(1 - x)^2$ **29.** (a) $x + \frac{1}{2}x^2 - \frac{1}{6}x^3$;   (b) $1 + x + x^2 + \frac{5}{6}x^3$ **31.** $\frac{1}{2}x + \frac{3}{4}x^2 + \frac{7}{8}x^3 + \frac{15}{16}x^4 + \cdots$
33. $F(x) = x/(1 - x - x^2)$ **35.** 3.14159

PROBLEM SET 11.8   *Page 516*

1. $x + \frac{1}{3}x^3 + \frac{2}{15}x^5$ **3.** $x + x^2 + \frac{1}{3}x^3 - \frac{1}{30}x^5$ **5.** $x - \frac{1}{2}x^2 - \frac{1}{6}x^3 + \frac{3}{40}x^5$ **7.** $1 + 3x + \frac{1}{2}x^2 + \frac{1}{24}x^4 + \frac{1}{60}x^5$
9. $1 + x + \frac{3}{2}x^2 + \frac{3}{2}x^3 + \frac{37}{24}x^4 + \frac{37}{24}x^5$ **11.** $1 - x + x^3 - x^4 + 0x^5$ **13.** $x^3 - \frac{1}{2}x^5$ **15.** $2x - \frac{1}{6}x^3 + \frac{61}{120}x^5$
17. $1 + \frac{3}{2}x + \frac{3}{8}x^2 - \frac{1}{16}x^3 + \frac{3}{128}x^4 - \frac{3}{256}x^5$ **19.** $e + e(x - 1) + \frac{1}{2}e(x - 1)^2 + \frac{1}{6}e(x - 1)^3$
21. $\frac{1}{2} - (\sqrt{3}/2)(x - \pi/3) - \frac{1}{4}(x - \pi/3)^2 + (\sqrt{3}/12)(x - \pi/3)^3$ **23.** $3 + 5(x - 1) + 4(x - 1)^2 + (x - 1)^3$
25. $\sum a_n x^n = \sum a_n(-x)^n \Rightarrow a_n = (-1)^n a_n \Rightarrow a_n = 0$ if n is odd.   **27.** $x + \frac{1}{6}x^3 + \frac{3}{40}x^5 + \frac{5}{112}x^7$ **29.** 0.9045
31. $1 - (x - 1) + (x - 1)^2 - (x - 1)^3 + \cdots$ **33.** (a) 25;   (b) -3;   (c) 0;   (d) $4e$;   (e) -4 **35.** $x - \frac{1}{3}x^3 + \frac{2}{15}x^5$

CHAPTER REVIEW 11.9   *Page 518*

True-False Quiz

1. False.   **2.** True.   **3.** True.   **4.** False.   **5.** False.   **6.** True.   **7.** False.   **8.** False.   **9.** True.   **10.** False.   **11.** True.
12. True.   **13.** False.   **14.** False.   **15.** True.   **16.** False.   **17.** False.   **18.** False.   **19.** True.   **20.** False.   **21.** True.
22. True.   **23.** True.   **24.** False.   **25.** True.   **26.** True.   **27.** True.   **28.** True.   **29.** True.   **30.** False.   **31.** True.
32. False.   **33.** False.   **34.** True.   **35.** True.

Sample Test Problems

1. Converges; 3.   **2.** Converges; 0.   **3.** Converges; e^4 **4.** Diverges.   **5.** Converges; 1.   **6.** Converges; 1.   **7.** Converges; 0.
8. Diverges.   **9.** Converges; 1.   **10.** Converges; $\frac{3}{2}$.   **11.** Diverges.   **12.** Diverges.   **13.** Converges; $e^2/(e^2 - 1)$.
14. Converges; 12.   **15.** Converges; $\frac{91}{99}$.   **16.** Diverges.   **17.** Converges; $\cos 2$.   **18.** Converges; e^{-1}.
19. Diverges; Limit Comparison Test ($b_n = 1/n$).   **20.** Converges; Limit Comparison Test ($b_n = 1/n^2$).
21. Converges; Alternating Series Test.   **22.** Diverges; nth-Term Test.   **23.** Converges; two geometric series.
24. Converges; Integral Test.   **25.** Diverges; nth-Term Test.   **26.** Converges; Limit Comparison Test ($b_n = 1/n^{3/2}$).
27. Converges; Ratio Test.   **28.** Converges; Ratio Test.   **29.** Diverges; Ratio Test.   **30.** Diverges; nth-Term Test.

31. Converges; Ratio Test.   **32.** Converges; Alternating Series Test.   **33.** Converges conditionally.
34. Converges absolutely.   **35.** Diverges.   **36.** Converges conditionally.   **37.** $[-1, 1]$ **38.** $(-\frac{1}{2}, \frac{1}{2}]$ **39.** $(3, 5]$ **40.** $\mathbb{R}$
41. $(1, 5)$ **42.** $\{-1\}$ **43.** $1 - 2x + 3x^2 - 4x^3 + \cdots; (-1, 1)$ **44.** $1 - 3x + 6x^2 - 10x^3 + \cdots$
45. $2x^2/2! - 8x^4/4! + 32x^6/6! - \cdots;$ All x.   **46.** $e^2[1 + (x - 2) + (1/2!)(x - 2)^2 + (1/3!)(x - 2)^3 + (1/4!)(x - 2)^4]$
47. $1 + x - (1/2!)x^2 - (1/3!)x^3 + (1/4!)x^4 + (1/5!)x^5 - \cdots;$ All x.
48. $1 - (1/2!)x^4 + (1/4!)x^8 - (1/6!)x^{12} + \cdots;$ 0.9045; 4 terms.   **49.** 0.21046   **50.** 999,999   **51.** 0.000004
52. (a) $1 + x^3 + x^6;$ (b) $1 + \frac{1}{2}x^2 - \frac{1}{8}x^4;$ (c) $\frac{1}{2}x^2 - \frac{1}{6}x^3 + \frac{1}{24}x^4;$ (d) $x + \frac{1}{2}x^3 + \frac{5}{24}x^5;$ (e) $x - x^2 + \frac{1}{3}x^3;$ (f) $1 - x + x^2$

PROBLEM SET 12.1   *Page 526*

1.

3.

5.

7.

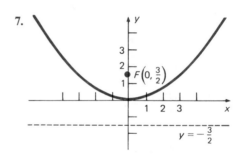

9. $y^2 = 12x$ **11.** $x^2 = 16y$ **13.** $y^2 = -20x$ **15.** $y^2 = \frac{1}{3}x$ **17.** $x^2 = -\frac{36}{5}y$

19.

21.

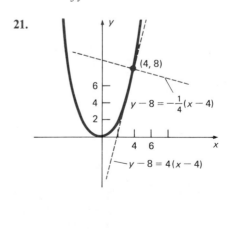

23. $y + 3\sqrt{5} = (\sqrt{5}/2)(x + 3);\; y + 3\sqrt{5} = -(2\sqrt{5}/5)(x + 3)$ **25.** $y + 3 = -\sqrt{2}(x - 3\sqrt{2});\; y + 3 = (\sqrt{2}/2)(x - 3\sqrt{2})$
27. $(4, 2\sqrt{5})$ **29.** $y + 6 = \frac{3}{2}(x + 2)$
33. Let the parabola be $y^2 = 4px$. The distance L from $(p, 0)$ to (x, y) satisfies $L^2 = (x - p)^2 + 4px = (x + p)^2$, $x \geq 0$. L^2 (and therefore L) is smallest when $x = 0$.
35. $20(1 - \cos 75°) \approx 14.8236$ million mi
37. Let $R = (x, y)$ and $G = (p, y)$, where $y^2 = 4px$. Then $|FR| = \sqrt{y^2 + (x - p)^2} = \sqrt{4px + (x - p)^2} = \sqrt{(x + p)^2} = x + p$. Also $|RG| = p - x$.   **39.** $L = 4p$ **41.** $y = \delta x^2/2H$

PROBLEM SET 12.2   *Page 532*

1. Vertical ellipse.   **3.** Vertical hyperbola.   **5.** Vertical parabola.   **7.** Horizontal ellipse.

9.

11.

13.

15.

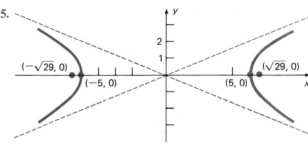

17. $x^2/36 + y^2/27 = 1$ **19.** $x^2/200 + y^2/225 = 1$ **21.** $x^2/25 + 21y^2/225 = 1$ **23.** $-x^2/9 + y^2/16 = 1$
25. $x^2/64 - y^2/16 = 1$ **27.** $x^2/16 + y^2/12 = 1$ **29.** $-x^2/20 + y^2/5 = 1$ **31.** $5\sqrt{3}$ ft   **33.** $2b^2/a$ **35.** 0.584 AU
37. 0.05175   **41.** $x^2/a^2 + y^2/b^2 = 1$

PROBLEM SET 12.3   *Page 537*

1. $x^2/9 + y^2/25 = 1$ **3.** $x^2/16 - y^2/9 = 1$ **5.** $\sqrt{6}x + y = 9$ **7.** $x + y = -2$ **9.** $3x + 4y = 25$ **11.** $y = 5$
13. $(\sqrt{3}, \frac{3}{2})$ and $(-\sqrt{3}, \frac{3}{2})$ **15.** $(7, -3)$ and $(-7, 3)$ **17.** πab **19.** $(\pi b^2/3a^2)[(b^2 - 2a^2)\sqrt{a^2 + b^2} + 2a^3]$ **21.** $\sqrt{2}a$ by $\sqrt{2}b$
23. $(6, 5\sqrt{3})$ **27.** The path will continue to cross the major axis between a focus and a neighboring vertex.
31. $(\sqrt{17}/3, 5)$

PROBLEM SET 12.4   *Page 543*

1. Circle.   **3.** Ellipse.   **5.** The point $(2, -4)$.   **7.** Parabola.   **9.** The empty set.   **11.** Intersecting lines.   **13.** Parallel lines.
15. Hyperbola.

17.

19.

21.

23.

25.

27.

29.

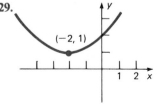

31. $(\frac{21}{20}, 1)$; $x = -\frac{29}{20}$ **33.** $(-2, -2)$ and $(4, -2)$ **35.** $(x-5)^2/25 + (y-1)^2/16 = 1$ **37.** $(x-2)^2 = 8(y-3)$
39. $-x^2/16 + (y-3)^2/9 = 1$ **41.** $(y-5)^2 = -16(x-6)$ **43.** $x^2/8 + (y-2)^2/4 = 1$
45. (a) $y = x^2 - x$; (b) $4x = y^2 - 4y$; (c) $x^2 - 5x + y^2 - 5y = 0$ **47.** $K < 0$ (ellipse), $K = 0$ (parabola), $K > 0$ (hyperbola).

ANSWERS TO ODD-NUMBERED PROBLEMS

1. $-u^2/1 + v^2/4 = 1$

$\theta = 45°$

3. $-u^2/36 + v^2/4 = 1$

$\theta = \tan^{-1} 3$

5. $v^2 = 6u$

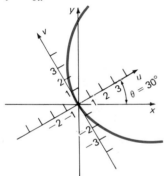

$\theta = 30°$

7. $u^2/2 + v^2/8 = 1$

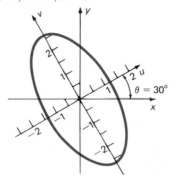

$\theta = 30°$

9. $(v + 3)(v + 1) = 0$

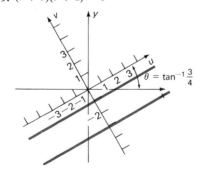

$\theta = \tan^{-1} \frac{3}{4}$

11. $(u - \sqrt{10})^2/44 + (v - 2\sqrt{10})^2/4 = 1$

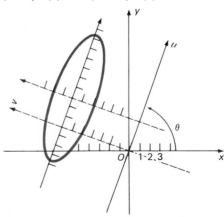

13. A rotation through θ transforms the equation to $u = d$, a line $|d|$ units from the origin.

15. $u = x \cos \theta + y \sin \theta$; $v = -x \sin \theta + y \cos \theta$ **17.** $(\frac{1}{5}, \frac{7}{5})$ and $(-\frac{1}{5}, -\frac{7}{5})$ are both the minimum distance $\sqrt{2}$ from the origin.

23. (a) $0 < |B| < 2$; (b) $B = 0$; (c) $|B| > 2$; (d) $|B| = 2$

PROBLEM SET 12.6   *Page 555*

1.

3.

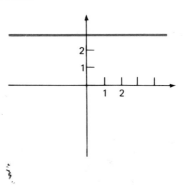

5. (a) $(4, \frac{7}{3}\pi), (4, -\frac{5}{3}\pi), (-4, \frac{4}{3}\pi), (-4, -\frac{2}{3}\pi)$;   (b) $(3, \frac{1}{4}\pi), (3, -\frac{7}{4}\pi), (-3, \frac{13}{4}\pi), (-3, -\frac{3}{4}\pi)$;

(c) $(5, \frac{7}{6}\pi), (5, -\frac{5}{6}\pi), (-5, \frac{13}{6}\pi), (-5, -\frac{11}{6}\pi)$;   (d) $(7, \frac{4}{3}\pi), (7, \frac{10}{3}\pi), (-7, \frac{1}{3}\pi), (-7, \frac{7}{3}\pi)$

7. (a) $(2, 2\sqrt{3})$; (b) $(3\sqrt{2}/2, 3\sqrt{2}/2)$; (c) $(-5\sqrt{3}/2, -\frac{5}{2})$; (d) $(-\frac{7}{2}, -7\sqrt{3}/2)$

9. (a) $(4, 7\pi/6)$; (b) $(2, \pi/3)$; (c) $(2, -\pi/4)$; (d) $(0, \pi/6)$ **11.** $r = 2/(4 \sin \theta - \cos \theta)$ **13.** $r = -5/\sin \theta$ **15.** $r = 4$

17. $y = \sqrt{3}x$ **19.** $x = -6$ **21.** $y = 4$

23. Circle.

25. Line.

27. Circle.

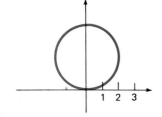

29. Parabola ($e = 1$).

31. Ellipse ($e = \frac{1}{2}$).

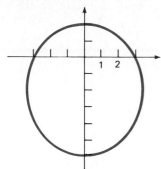

33. Parabola ($e = 1$).

35. Hyperbola ($e = 2$).

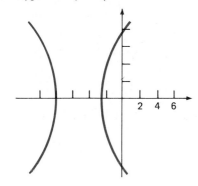

37. Use the Law of Cosines.   **39.** $2ed$ **41.** 0.83   **43.** 75 million miles.

PROBLEM SET 12.7   *Page 559*

1.

3.

5.

7.

9.

11.

13.

15.

17.

19.

21.

23.

25.

27.

29.

31.

33.

35.

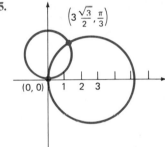

39. $r = \theta^2 + (3\pi - \theta)^2$ **41.** $r^2 = 2a^2 \cos 2\theta$

37.

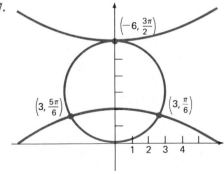

PROBLEM SET 12.8   *Page 565*

1. πa^2 **3.** $19\pi/2$ **5.** 24π **7.** $\frac{3}{2}\pi a^2$ **9.** 4   **11.** $4\pi - 6\sqrt{3} \approx 2.174$ **13.** $8\pi + 6\sqrt{3} \approx 35.525$ **15.** 4π **17.** 51π
19. $4\sqrt{3} - \frac{4}{3}\pi \approx 2.739$ **21.** $9\sqrt{2} - \frac{27}{4} \approx 5.978$ **23.** (a) $\sqrt{3}/3$;   (b) -1;   (c) $\sqrt{3}/5$;   (d) $-7\sqrt{3}/3$

25. $(-1, \pi/2)$, $(3, 3\pi/2)$, $(0.5, 0.253)$, $(0.5, 2.889)$ **27.** $8a$ **29.** $\pi a^2/4$ if n is odd, $\pi a^2/2$ if n is even.
31. (a) $\frac{1}{2}\pi b^2 - ab + (a^2 - b^2)\tan^{-1}(b/a)$ **33.** $a^2[\frac{1}{2}\pi k^2 + \frac{1}{2}k\sqrt{4 - k^2} + (k^2 - 2)\sin^{-1}(k/2)]$ **35.** $1.256569a$

CHAPTER REVIEW 12.9   *Page 566*

True-False Quiz

1. False. **2.** True. **3.** False. **4.** True. **5.** True. **6.** True. **7.** True. **8.** False. **9.** False. **10.** True. **11.** True.
12. False. **13.** True. **14.** True. **15.** True. **16.** True. **17.** True. **18.** True. **19.** True. **20.** False. **21.** False.
22. True. **23.** False. **24.** False. **25.** True. **26.** True. **27.** False. **28.** True. **29.** True. **30.** True.

Sample Test Problems

1. (a) 5; (b) 9; (c) 4; (d) 3; (e) 2; (f) 8; (g) 8; (h) 1; (i) 7; (j) 6.   **2.** Parabola; $V(0,0)$; $F(\frac{3}{2}, 0)$.
3. Ellipse; $V(0, \pm 3)$; $F(0, \pm\sqrt{5})$.   **4.** Hyperbola; $V(0, \pm 5)$; $F(0, \pm\sqrt{61})$.   **5.** Parabola; $V(0,0)$; $F(0, -\frac{9}{4})$.
6. Hyperbola; $V(\pm 4, 0)$; $F(\pm 2\sqrt{5}, 0)$.   **7.** Ellipse; $V(\pm 5, 0)$; $F(\pm 4, 0)$.   **8.** Circle.   **9.** Parabola; $V(\frac{5}{4}, \pi/2)$; $F(0,0)$.
10. Ellipse; $V(1, 0)$, $V(3, \pi)$; $F(0,0)$, $F(2, \pi)$.   **11.** $x^2/16 + y^2/12 = 1$ **12.** $x^2 = -12y$ **13.** $y^2 = -9x$
14. $-x^2/16 + y^2/9 = 1$ **15.** $x^2/4 - y^2/1 = 1$ **16.** $(x - 3)^2 = -4(y - 3)$ **17.** $(x - 1)^2/25 + (y - 2)^2/16 = 1$
18. $-(x - 2)^2/91 + (y - 3)^2/9 = 1$ **19.** $(x - 3)^2 + (y + \frac{9}{2})^2 = 9$; circle.   **20.** $(x - 3)^2/9 + (y - 2)^2/4 = 1$; ellipse.
21. $(x + 4)^2 = -6(y + 2)$; parabola.   **22.** $(x + 6)^2/10 - (y + 1)^2/3 = 1$; hyperbola.   **23.** $r = \frac{5}{2}$, $s = -\frac{1}{2}$; hyperbola; $4\sqrt{6}$.
24. $\theta = \tan^{-1}\frac{1}{2}$; $u^2/1 - v^2/9 = 1$; hyperbola.   **25.** Circle with center $(3, 0)$ and radius 3.   **26.** Horizontal line.
27. Rose with four leaves.   **28.** Vertical line.   **29.** Circle with center $(0, 0)$ and radius 4.   **30.** Cardioid.   **31.** Limaçon.
32. Limaçon.   **33.** Line.   **34.** Rose with three leaves.   **35.** Lemniscate.   **36.** Spiral.   **37.** $(x - 3)^2 + (y - 3)^2 = 9$
38. $x^2/9 - y^2/9 = 1$ **39.** -1 **40.** $(\frac{5}{2}, \pi/6)$ and $(\frac{5}{2}, 5\pi/6)$ **41.** $75\pi/2$ **42.** $\frac{8}{3}\pi + \sqrt{3}$ **43.** $k = \frac{22}{3}$

PROBLEM SET 13.1   *Page 576*

1. Simple.

3. Simple.

5. Simple.

7.

9. Simple, closed.

11. Simple.

13. t; $1/(6t)$ **15.** $(2t^2 - 3)/(2t^2 + 3)$; $24t^3/(2t^2 + 3)^3$ **17.** $\frac{5}{3}\sin t$; $\frac{5}{9}\cos^3 t$ **19.** $y - 8 = 3(x - 4)$
21. $y + \frac{2}{3}\sqrt{3} = -2(x - \frac{4}{3}\sqrt{3})$ **23.** $-\frac{44}{3}$ **25.** 8

27.

29. (a) $y = [-16/(v_0^2\cos^2\alpha)]x^2 + (\tan \alpha)x$;   (b) $(v_0 \sin \alpha)/16$;   (c) $(v_0^2 \sin 2\alpha)/32$;   (d) $\pi/4$
31. $x = (a - b)\cos t + b\cos(a/b - 1)t$; $y = (a - b)\sin t - b\sin(a/b - 1)t$
35. (a) $x = 2a \cot \theta$, $y = 2a \sin^2\theta$;   (b) $y = 8a^3/(x^2 + 4a^2)$ **37.** $8(a + b)$

PROBLEM SET 13.2   *Page 581*

1.

3.

5. $\mathbf{w} = \frac{1}{2}(\mathbf{u} + \mathbf{v})$ **7.** $|\mathbf{w}| = 1$ **9.** $|\mathbf{w}| \approx 15.87$; S 7.5° W   **11.** 125 N   **13.** N 2.68° E; 479 mi/h   **15.** 160 mi/h
23. $\alpha + \beta = 143.13°$, $\beta + \gamma = 126.87°$, $\alpha + \gamma = 90°$

PROBLEM SET 13.3   *Page 587*

1. (a) $-14\mathbf{i} + 20\mathbf{j}$;   (b) -18;   (c) -38;   (d) 425;   (e) -100;   (f) $13 - \sqrt{13}$
3. (a) $-14/\sqrt{221} \approx -0.9417$;   (b) $-5/\sqrt{29} \approx -0.9285$;   (c) $1/\sqrt{2} \approx 0.7071$;   (d) -1
5. (a) $-5\mathbf{i} + 5\mathbf{j}$;   (b) $-6\mathbf{i} - 5\mathbf{j}$;   (c) $-\sqrt{2}\mathbf{i} + e\mathbf{j}$;   (d) $3\mathbf{i} - \frac{5}{3}\mathbf{j}$ **7.** $|\mathbf{u}| = |\mathbf{v}|$
9. $(a + b)\mathbf{u} = \langle(a + b)u_1, (a + b)u_2\rangle = \langle au_1 + bu_1, au_2 + bu_2\rangle = a\langle u_1, u_2\rangle + b\langle u_1, u_2\rangle = a\mathbf{u} + b\mathbf{u}$
11. $c(\mathbf{u} \cdot \mathbf{v}) = c(u_1 v_1 + u_2 v_2) = c(u_1 v_1) + c(u_2 v_2) = (cu_1)v_1 + (cu_2)v_2 = (c\mathbf{u}) \cdot \mathbf{v}$ **13.** $9\mathbf{i} - 12\mathbf{j}$
15. $\langle 6, 3\rangle \cdot \langle -1, 2\rangle = 6(-1) + 3(2) = 0$ **17.** $c = \pm 2\sqrt{3}$ **19.** $k = 2$, $m = -1$
21. $k = (r_1 b_2 - r_2 b_1)/(a_1 b_2 - a_2 b_1)$, $m = (a_1 r_2 - a_2 r_1)/(a_1 b_2 - a_2 b_1)$ **23.** 100 ft-lb   **25.** 47 ft-lb
27. $|\mathbf{u} \cdot \mathbf{v}| = |\mathbf{u}||\mathbf{v}||\cos \theta| \leq |\mathbf{u}||\mathbf{v}|$ **29.** (a) $\mathbf{u} = \mathbf{v}$;   (b) $\theta = 60°$ **33.** (a) $\langle \frac{12}{5}, \frac{16}{5}\rangle$;   (b) $\langle -\frac{3}{25}, -\frac{4}{25}\rangle$

1. $3\mathbf{i} - \mathbf{j}$ 3. $\frac{1}{4}\mathbf{i} + 5\mathbf{j}$ 5. $\mathbf{i}$ 7. Does not exist.   9. (a) $\{t \in \mathbb{R} : t \geq -4, t \neq 2\}$;   (b) $\mathbb{R}$
11. (a) $\{t \in \mathbb{R} : t > -4, t \neq 2\}$;   (b) $\{t \in \mathbb{R} : t \text{ not an integer}\}$
13. (a) $4(2t + 3)\mathbf{i} - 2e^{2t}\mathbf{j}$; $8\mathbf{i} - 4e^{2t}\mathbf{j}$;   (b) $-2 \sin 2t\mathbf{i} - 3 \sin^2 t \cos t\mathbf{j}$; $-4 \cos 2t\mathbf{i} + (3 \sin^3 t - 6 \sin t \cos^2 t)\mathbf{j}$
15. $8e^{4t} + 9(1 - \ln t)t^{-2}$ 17. $e^{-3t}[(7 - 6t)/(2\sqrt{t - 1})\mathbf{i} + (2/t - 3 \ln 2t^2)/t\mathbf{j}]$ 19. $-6t \sin(3t^2 - 4)\mathbf{i} + 18t \exp(9t^2 - 12)\mathbf{j}$
21. $(e - 1)\mathbf{i} + (1 - e^{-1})\mathbf{j}$ 23. $\mathbf{v} = -e^{-1}\mathbf{i} + e\mathbf{j}$; $\mathbf{a} = e^{-1} + e\mathbf{j}$; $|\mathbf{v}| = \sqrt{e^{-2} + e^2} \approx 2.743$
25. $\mathbf{v} = -\sqrt{3}\mathbf{i} - \frac{3}{2}\sqrt{3}\mathbf{j}$; $\mathbf{a} = -\mathbf{i} + 3\mathbf{j}$; $|\mathbf{v}| = \sqrt{39}/2 \approx 3.122$ 27. $\mathbf{v} = 12\mathbf{i} + 12\mathbf{j}$; $\mathbf{a} = 6\mathbf{i} + 12\mathbf{j}$; $|\mathbf{v}| = 12\sqrt{2} \approx 16.97$
29. $\mathbf{v} = (\sqrt{2}/2)\mathbf{i} - 4\mathbf{j}$; $\mathbf{a} = -(\sqrt{2}/2)\mathbf{i} + 8\mathbf{j}$; $|\mathbf{v}| = \sqrt{16.5} \approx 4.062$ 31. $\mathbf{v}(t) = -32t\mathbf{j}$; $\mathbf{r}(t) = -16t^2\mathbf{j}$
33. $\mathbf{v}(t) = (t + 2)\mathbf{i} + (2 - e^{-1})\mathbf{j}$; $\mathbf{r}(t) = (\frac{1}{2}t^2 + 2t + 1)\mathbf{i} + (2t + e^{-1})\mathbf{j}$
35. $\mathbf{r}(t) = 5 \cos(6t)\mathbf{i} + 5 \sin(6t)\mathbf{j}$; $\mathbf{v}(t) = -30 \sin(6t)\mathbf{i} + 30 \cos(6t)\mathbf{j}$; $|\mathbf{v}(t)| = 30$; $\mathbf{a}(t) = -180 \cos(6t)\mathbf{i} - 180 \sin(6t)\mathbf{j}$
37. 3; 96 ft/s; $144\sqrt{3}$ ft   41. $\sqrt{9600} \approx 97.98$ ft/s   43. 4.04 rad/s   45. (a) 17,253 mi/h;   (b) 26,240 mi

1. $(\sqrt{2}/2)\mathbf{i} + (\sqrt{2}/2)\mathbf{j}$; $\sqrt{2}/8 \approx 0.177$ 3. $-\frac{4}{5}\mathbf{i} + \frac{3}{5}\mathbf{j}$; $24\sqrt{2}/125 \approx 0.272$ 5. $\kappa = 2\sqrt{5}/25 \approx 0.179$; $R = 5\sqrt{5}/2 \approx 5.59$
7. $\kappa = 2\sqrt{5}/25$; $R = 5\sqrt{5}/2$ 9. $\kappa = 1$; $R = 1$ 11. $\kappa = \frac{1}{4}$; $R = 4$ 13. $\kappa = 2\sqrt{3}/9$; $R = 3\sqrt{3}/2$ 15. $(\sqrt{2}/2, -\ln\sqrt{2})$
17. $(-\pi/2, -1)$ and $(\pi/2, 1)$ 19. $a_T = 4\sqrt{5}/5$; $a_N = 2\sqrt{5}/5$ 21. $a_T = 0$; $a_N = |a|$
25. $\mathbf{a} = \mathbf{0}$ at the origin; $\mathbf{a}$ points to the origin when $t = \pi/2$ and $t = 3\pi/2$—that is, at the points $(1, 0)$ and $(-1, 0)$.
27. The speed is constant; the curvature is 0.   31. $5\mathbf{T} + 5\mathbf{N}$, $-\mathbf{i} - 7\mathbf{j}$ 33. 72 ft/s   35. (a) $\frac{1}{2}$;   (b) $\frac{3}{4}$

True-False Quiz

1. False.   2. True.   3. False.   4. True.   5. False.   6. True.   7. True.   8. True.   9. False.   10. True.   11. True.
12. False.   13. True.   14. True.   15. True.   16. True.   17. True.   18. False.   19. True.   20. False

Sample Test Problems

1. $x - 3y = 2$ 2. $4x = y^2$ 3. $(x + 2)^2/16 + (y - 1)^2/9 = 1$ 4. $x^2/4 - y^2/1 = 1$ 5. $y = -\frac{1}{2}(x - 7)$; $y = 2(x - 7)$
6. $y - \frac{1}{2} = -\frac{1}{6}(x - 3)$; $y - \frac{1}{2} = 6(x - 3)$ 7. $2\pi^2$ 8. (a) $\mathbf{u} = \langle -3, 1\rangle$; $|\mathbf{u}| = \sqrt{10}$;   (b) $\mathbf{u} = \langle -1, 5\rangle$; $|\mathbf{u}| = \sqrt{26}$
9. (a) $\langle 4, -17\rangle$;   (b) -3;   (c) -15;   (d) -234;   (e) -36;   (f) 30 10. (a) 0.3363;   (b) -0.5369;   (c) 0.9806
11. $m = 2, k = \frac{1}{2}$ 12. $\langle -\frac{3}{5}\sqrt{5}, \frac{6}{5}\sqrt{5}\rangle$ 13. $\langle -5\sqrt{3}, 5\rangle$ 14. $-5\mathbf{i} - 9\mathbf{j}$ 15. N 12.22° W; 409.27 mi/h
16. (a) $\langle 1, 1\rangle$;   (b) $\langle 2, -1\rangle$ (c) $\langle \frac{3}{2}, \frac{1}{2}\rangle$;   (d) $\langle e^{2t}(2t + 1), e^{-t}(1 - t)\rangle$;   (e) $\langle 6e^{6t+20}, -3e^{-3t-10}\rangle$;   (f) $8e^{4t} + 2e^{-2t}$
17. (a) $t^{-1}\mathbf{i} - 6t\mathbf{j}$; $-t^{-2}\mathbf{i} - 6\mathbf{j}$;   (b) $\cos t\mathbf{i} - 2 \sin 2t\mathbf{j}$; $-\sin t\mathbf{i} - 4 \cos 2t\mathbf{j}$;   (c) $\sec^2 t\mathbf{i} - 4t^3\mathbf{j}$; $2 \sec^2 t \tan t\mathbf{i} - 12t^2\mathbf{j}$ 18. 13
19. $\mathbf{v} = -4\mathbf{i} + 4\mathbf{j}$; $\mathbf{a} = 4\mathbf{i}$; $|\mathbf{v}| = 4\sqrt{2}$ 20. $\mathbf{v} = 2\mathbf{i} + (4 - 2\sqrt{3})\mathbf{j}$; $\mathbf{a} = 2\sqrt{3}\mathbf{i} + 2\mathbf{j}$; $|\mathbf{v}| = 4\sqrt{2 - \sqrt{3}} \approx 2.07$
21. $\sqrt{2}/2$;   (b) $\frac{2}{25}$;   (c) $(1/a)\text{sech}^2 1$ 22. $\mathbf{T}(1) = (\sqrt{2}/2)\mathbf{i} + (\sqrt{2}/2)\mathbf{j}$; $\kappa(1) = \sqrt{2}/2$ 23. $a_T = \sqrt{2}$, $a_N = \sqrt{2}$,

1.

3. $y = 0$; $x = z = 0$;   **5.** (a) $3\sqrt{10}$ (b) $\sqrt{83}$; (c) 3
9. Front vertices: $(5, -2, 0)$, $(5, -2, 4)$, $(5, 3, 4)$, $(5, 3, 0)$; back vertices: $(2, 3, 4)$, $(2, 3, 0)$; $(2, -2, 0)$, $(2, -2, 4)$.
11. (a) $(x - 3)^2 + (y - 1)^2 + (z - 4)^2 = 25$; (b) $(x + 6)^2 + (y - 2)^2 + (z + 3)^2 = 4$; (c) $(x + 1)^2 + y^2 + (z - 4)^2 = 6$
13. $(6, -7, 4)$; 10   **15.** $(\frac{1}{2}, -1, -2)$; $\frac{1}{2}\sqrt{34}$

17.

19.

21.

23.

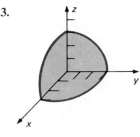

25. $(x - 1)^2 + (y - 1)^2 + (z - \frac{11}{2})^2 = \frac{53}{4}$ **27.** $(x - 6)^2 + (y - 6)^2 + (z - 6)^2 = 36$
29. (a) The horizontal plane 2 units above the xy-plane;
  (b) the vertical plane with trace $y = x$ in the xy-plane;
  (c) the union of the xz- and the yz-planes;
  (d) the union of the three coordinate planes;
  (e) the vertical cylinder of radius 2 with axis the z-axis;
  (f) the hemisphere of radius 3 centered at the origin and lying atop the xy-plane.
31. $(1, 2, 5)$, 4   **33.** 4π

PROBLEM SET 14.2   *page 617*

1. (a) $3\mathbf{i} + 3\mathbf{j} - 2\mathbf{k}$; (b) $6\mathbf{i} + 2\mathbf{j} + 4\mathbf{k}$
3. (a) $\sqrt{21}$; $\langle 4\sqrt{21}/21, \sqrt{21}/21, 2\sqrt{21}/21 \rangle$; (b) $5\sqrt{2}$; $\langle -3\sqrt{2}/10, -4\sqrt{2}/10, \sqrt{2}/10 \rangle$
5. $\langle 3/\sqrt{59}, 1/\sqrt{59}, -7/\sqrt{59} \rangle$; $\langle -15/\sqrt{59}, -5/\sqrt{59}, 35/\sqrt{59} \rangle$ **7.** 2.316 rad **9.** $\pm(-\frac{60}{7}\mathbf{i} + \frac{20}{7}\mathbf{j} + \frac{30}{7}\mathbf{k})$ **11.** 0.2288 rad
13. $5/\sqrt{19}$ **15.** $\mathbf{m} = \frac{6}{25}\mathbf{i} + \frac{12}{25}\mathbf{j} - (3\sqrt{5}/25)\mathbf{k}$; $\mathbf{n} = -\frac{81}{25}\mathbf{i} + \frac{13}{25}\mathbf{j} - (22\sqrt{5}/25)\mathbf{k}$
17. (a) $36.7°$, $122.3°$, $74.5°$; (b) $105.5°$, $57.7°$, $143.3°$ **19.** $2\sqrt{3}$ **21.** Many answers; one pair is $\langle 1, 2, 0 \rangle$ and $\langle 2, -1, 2 \rangle$.
23. (a), (b), (f)   **25.** (a) $2(x - 1) - 4(y - 2) + 3(z + 3) = 0$; (b) $3(x + 2) - 2(y + 3) - (z - 4) = 0$ **27.** 0.9933 rad
29. (a) $z = 2$; (b) $2(x + 4) - 3(y + 1) - 4(z - 2) = 0$ **31.** 0   **33.** $12/\sqrt{38} \approx 1.947$
35. $(\mathbf{u} + \mathbf{v}) \cdot (\mathbf{u} + \mathbf{v}) + (\mathbf{u} - \mathbf{v}) \cdot (\mathbf{u} - \mathbf{v}) = |\mathbf{u}|^2 + 2(\mathbf{u} \cdot \mathbf{v}) + |\mathbf{v}|^2 + |\mathbf{u}|^2 - 2(\mathbf{u} \cdot \mathbf{v}) + |\mathbf{v}|^2 = 2|\mathbf{u}|^2 + 2|\mathbf{v}|^2$
37. $35.26° = 0.6155$ rad   **39.** $37.86° = 0.6608$ rad   **41.** 15 J   **43.** $(3, 2, 1)$ **45.** 0
47. $(\frac{1}{2}(a_1 + b_1), \frac{1}{2}(a_2 + b_2), \frac{1}{2}(a_3 + b_3)), \frac{1}{2}\sqrt{(a_1 - b_1)^2 + (a_2 - b_2)^2 + (a_3 - b_3)^2}$ **49.** $(4, 2, 3)$

1. (a) $-2\mathbf{i} + 7\mathbf{j} + 10\mathbf{k}$; (b) $4\mathbf{i} + 15\mathbf{j} + 9\mathbf{k}$; (c) 29; (d) $18\mathbf{i} + 47\mathbf{j} + 20\mathbf{k}$ **3.** $c(-11\mathbf{i} - 2\mathbf{j} + 5\mathbf{k})$ **5.** $\pm(4\mathbf{i} + 9\mathbf{j} + \mathbf{k})/\sqrt{98}$
7. $3\sqrt{5}$ **9.** $6\sqrt{5}$ **11.** $20(x - 4) + 8y - 17(z - 6) = 0$ **13.** $7(x + 1) + 5(y + 2) + 4(z - 3) = 0$
15. $x - 6 - 10(y - 2) - 17(z + 1) = 0$ **17.** 69   **19.** (a) 9; (b) $\sqrt{35}$; (c) $40° = 0.6982$ rad   **27.** $\frac{1}{2}\sqrt{a^2b^2 + a^2c^2 + b^2c^2}$

1. $x = 1 + 3t, y = -2 + 7t, z = 3 + 3t$ **3.** $x = 4 + 2t, y = 2, z = 3 - 4t$
5. $x = 4 - 2t, y = -6 + t, z = 3 + 5t; (x - 4)/(-2) = (y + 6)/1 = (z - 3)/5$
7. $x = 2 - 3t, y = 5 + 4t, z = -4 + 2t; (x - 2)/(-3) = (y - 5)/4 = (z + 4)/2$ **9.** $x/4 = (y - 5)/(-5) = (z - 1)/2$
11. $x/2 = (y - \frac{23}{3})/(-2) = (z + \frac{53}{6})/3$ **13.** $(x - 4)/1 = y/(-5) = (z - 6)/2$ **15.** $x = 5 + 5t, y = -3 - 3t, z = 4$
17. $x + y + 6z = 11$ **19.** $3x - 2y = 5$ **21.** $\sqrt{6}$ **23.** $(x - 1)/(-\sqrt{3}) = (y - 3\sqrt{3})/3 = z - \pi/3$
25. $3(x + 3) - 4(y - 2) + 5(z + 1) = 0$ **27.** (b) $(3\sqrt{3}/4, 7/4, 0)$ **29.** (a) $8\sqrt{2}/3$; (b) $3\sqrt{26}/7$

1. $\mathbf{v} = 3\mathbf{i} + 8\mathbf{j} + 6\mathbf{k}$; $\mathbf{a} = 8\mathbf{j} + 12\mathbf{k}$; $|\mathbf{v}| = \sqrt{109}$ **3.** $\mathbf{v} = 3\mathbf{j} + 3\mathbf{k}$; $\mathbf{a} = -2\mathbf{i} + 6\mathbf{k}$; $|\mathbf{v}| = 3\sqrt{2}$
5. $\mathbf{v} = -2\mathbf{j} + 3\mathbf{k}$; $\mathbf{a} = -4\mathbf{i} + 2\mathbf{j}$; $|\mathbf{v}| = \sqrt{13}$ **7.** $0 = (d/dt)(\mathbf{v} \cdot \mathbf{v}) = 2\mathbf{v} \cdot \mathbf{a}$ **9.** $\int_0^4 \sqrt{9 + t^2}\, dt \approx 14.94$
11. $\int_6^{10} \sqrt{9t + 10}\, dt \approx 36.15$ **13.** $\int_{-1/2}^1 2\sqrt{2} \cosh 2t\, dt \approx 6.79$
15. $\kappa = \sqrt{30}/50$; $\mathbf{T} = (\sqrt{5}/5)(2\mathbf{i} + \mathbf{j})$; $\mathbf{N} = (\sqrt{30}/30)(\mathbf{i} - 2\mathbf{j} + 5\mathbf{k})$; $\mathbf{B} = (\sqrt{6}/6)(\mathbf{i} - 2\mathbf{j} - \mathbf{k})$
17. $\kappa = \frac{9}{10}$; $\mathbf{T} = (\sqrt{10}/20)(3\mathbf{i} - 3\sqrt{3}\mathbf{j} + 2\mathbf{k})$; $\mathbf{N} = \frac{1}{2}(-\sqrt{3}\mathbf{i} - \mathbf{j})$; $\mathbf{B} = (\sqrt{10}/20)(\mathbf{i} - \sqrt{3}\mathbf{j} - 6\mathbf{k})$
19. $\kappa = \sqrt{2}/(3e^{\pi/2})$; $\mathbf{T} = (\sqrt{3}/3)(\mathbf{i} - \mathbf{j} + \mathbf{k})$; $\mathbf{N} = -(\sqrt{2}/2)(\mathbf{i} + \mathbf{j})$; $\mathbf{B} = (\sqrt{6}/6)(\mathbf{i} - \mathbf{j} - 2\mathbf{k})$
21. $a_T = (4t + 18t^3)/(1 + 4t^2 + 9t^4)^{1/2}$; $a_N = 2(9t^4 + 9t^2 + 1)^{1/2}/(9t^4 + 4t^2 + 1)^{1/2}$
23. $a_T = 2(t^8 - 1)/t^3\sqrt{t^8 + t^4 + 1}$; $a_N = 2\sqrt{t^8 + 4t^4 + 1}/t\sqrt{t^8 + t^4 + 1}$ **25.** 207 cm   **27.** origin, $100\sqrt{3}$
29. $(2.99, 2.84, 0)$ **35.** $3x - 3y + 3z = 1$

1. Elliptic cylinder.

3. Plane.

5. Circular cylinder.

6. Ellipsoid.

886

9. Elliptic paraboloid.

11. Cylinder.

13. Hyperbolic paraboloid.   **15.** Elliptic paraboloid.   **17.** Plane.   **19.** Hemisphere.
21. (a) Replacing x by $-x$ results in an equivalent equation.   (b) Replacing x by $-x$ and y by $-y$ results in an equivalent
equation.   (c) Replacing y by $-y$ and z by $-z$ results in an equivalent equation.   (d) Replacing x, y, and z by $-x$, $-y$,
and $-z$, respectively, results in an equivalent equation.
23. $y = 2x^2 + 2z^2$ **25.** $4x^2 + 3y^2 + 4z^2 = 12$ **27.** $(0, \pm 2\sqrt{5}, 4)$ **29.** $\pi ab(c^2 - h^2)/c^2$ **31.** $4, 2\sqrt{2}$
33. $x^2 + 9y^2 - 9z^2 = 0$

PROBLEM SET 14.7   *Page 645*

1. (a) $(3\sqrt{3}, 3, -2)$;   (b) $(-2, -2\sqrt{3}, -8)$ **3.** (a) $(4\sqrt{2}, \frac{5}{3}\pi, \frac{1}{4}\pi)$;   (b) $(4, \frac{3}{4}\pi, \frac{1}{6}\pi)$

5.

7.

9.

11.

13. A sphere of radius 3 centered at the origin.   **15.** $r = 3$ **17.** $r^2 + 4z^2 = 10$ **19.** $\tan^2\phi = 2$ **21.** $\rho^2 = 4/(1 + \cos^2\phi)$
23. $r = 4/(\cos\theta + \sin\theta)$ **25.** $\rho\sin\phi = 3$ **27.** $x^2 - y^2 = z$ **29.** $z = 2r^2$ **31.** 4028 mi   **33.** 4552 mi   **35.** 2253 mi
39. (a) 3485 mi;   (b) 4552 mi;   (c) 9331 mi;   (d) 7798 mi;   (e) 3960π mi

True-False Quiz

1. True. **2.** False. **3.** True **4.** False. **5.** False. **6.** False. **7.** True. **8.** True. **9.** True. **10.** False **11.** True.
12. True. **13.** False. **14.** False. **15.** True. **16.** True. **17.** True. **18.** False. **19.** True. **20.** False. **21.** False.
22. True. **23.** False. **24.** False. **25.** True.

Sample Test Problems

1. $(x - 1)^2 + (y - 2)^2 + (z - 4)^2 = 11$ **2.** $(3, -1, 4)$; $\sqrt{26}$
3. (a) 3; $\sqrt{35}$; (b) $(\frac{2}{3}, -\frac{1}{3}, \frac{2}{3})$; $(5/\sqrt{35}, 1/\sqrt{35}, -3/\sqrt{35})$; (c) $\frac{2}{3}\mathbf{i} - \frac{1}{3}\mathbf{j} + \frac{2}{3}\mathbf{k}$; (d) $\theta \approx 80.27° \approx 1.40$ rad
4. (a) $-5\mathbf{i} - 5\mathbf{j} + 5\mathbf{k}$; (b) $-6\mathbf{i} - 2\mathbf{j} + 10\mathbf{k}$; (c) -20; (d) $4\mathbf{i} + 3\mathbf{j} - 5\mathbf{k}$ **5.** $t(10\mathbf{i} - 11\mathbf{j} - 3\mathbf{k})$, $t \in \mathbb{R}$
6. $\pm(6\mathbf{i} + 3\mathbf{j} + 5\mathbf{k})/\sqrt{70}$ **7.** (a) $y = 7$; (b) $x = -5$; (c) $z = -2$; (d) $3x - 4y + z = -45$
8. (a) $\langle x - 2, y + 4, z + 5 \rangle \cdot \langle 5, -4, 8 \rangle = 0$; (b) $5(x - 2) - 4(y + 4) + 8(z + 5) = 0$ **9.** $c = 1$ **10.** $7x - 3y + 9z = -4$
11. $x = -2 + 8t, y = 1 + t, z = 5 - 8t$ **12.** $(0, 25, 16)$ and $(-50, 0, 16)$ **13.** $x = 2t, y = 25 + t, z = 16$
14. $(x - 4)/3 = (y - 5)/5 = (z - 8)/2$ **15.** $\langle x, y, z \rangle = \langle 2, -2, 1 \rangle + t\langle -5, 4, 3 \rangle$

16.

17. $(x - 2)/1 = (y - 2)/2 = (z - \frac{8}{3})/4$; $x + 2y + 4z = \frac{50}{3}$ **18.** $\langle -\pi/2, 1, 2 \rangle$; $\langle -0.575, 0.366, 0.732 \rangle$; $\langle -2, -\pi/2, 0 \rangle$
19. $\sqrt{3}(e^5 - e) \approx 252.35$ **20.** $2\mathbf{i} - \frac{1}{2}\mathbf{j} + 2\mathbf{k}$; $2\mathbf{i} + \frac{1}{2}\mathbf{j}$; 0.1934 **21.** $a_T = 22/\sqrt{14} \approx 5.88$; $a_N = 2\sqrt{19}/\sqrt{14} \approx 2.33$
22. Circular cylinder. **23.** Sphere. **24.** Parabolic cylinder. **25.** Circular paraboloid. **26.** Plane. **27.** Plane.
28. Hyperboloid of one sheet. **29.** Ellipsoid **30.** Empty set.
31. (a) $r = 3$; (b) $r^2 = 16/(\cos^2\theta + 4\sin^2\theta)$; (c) $r^2 = 9z$; (d) $r^2 + 4z^2 = 10$
32. (a) $x^2 + y^2 + z^2 = 9$; (b) $x^2 + z^2 = 4$; (c) $x^2 - y^2 + z^2 = 1$
33. (a) $\rho = 2$; (b) $\tan^2\phi = 1$; (c) $\rho^2 = 1/(\sin^2\phi \cos 2\theta - \cos^2\phi)$; (d) $\rho = \cot\phi \csc\phi$ **34.** 9.866 **35.** $\frac{5}{4}$ **36.** 1.32 rad
37. $0 = (d/dt)(\mathbf{v} \cdot \mathbf{v}) = 2\mathbf{v} \cdot d\mathbf{v}/dt = 2\mathbf{v} \cdot \mathbf{a}$

PROBLEM SET 15.1 *Page 652*

1. (a) 5; (b) 0; (c) 6; (d) $a^6 + a^2$; (e) $2x^2$; (f) $(2, -4)$ is not in the domain of f.
3. (a) 0; (b) 2; (c) 16; (d) -4.2469; (e) 0.6311 **5.** t^2

7.

9.

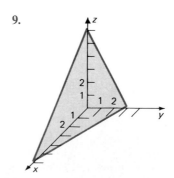

ANSWERS TO ODD-NUMBERED PROBLEMS

11.

13.

15.

17.

19.

21.

23.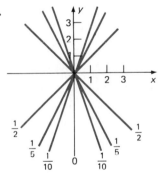

25. The set of all points on and outside the sphere $x^2 + y^2 + z^2 = 16$.

27. The set of all points on and inside the ellipsoid $x^2/9 + y^2/16 + z^2/1 = 1$.

29. The set of all spheres with centers at the origin.   **31.** A set of hyperboloids of revolution about the z-axis.

33. A set of hyperbolic cylinders parallel to the z-axis.   **35.** (a) gentle climb, steep climb;   (b) 6700 ft, 3040 ft

37. Elliptic paraboloid, $(\frac{1}{2}, -2)$, -25.25.

PROBLEM SET 15.2   *Page 660*

1. $f_x(x, y) = 8(2x - y)^3$; $f_y(x, y) = -4(2x - y)^3$ **3.** $f_x(x, y) = (x^2 + y^2)/(x^2 y)$; $f_y(x, y) = -(x^2 + y^2)/(xy^2)$

5. $f_x(x, y) = e^y \cos x$; $f_y(x, y) = e^y \sin x$ **7.** $f_x(x, y) = x(x^2 - y^2)^{-1/2}$; $f_y(x, y) = -y(x^2 - y^2)^{-1/2}$

9. $g_x(x, y) = -ye^{-xy}$; $g_y(x, y) = -xe^{-xy}$ **11.** $f_x(x, y) = 4/[1 + (4x - 7y)^2]$; $f_y(x, y) = -7/[1 + (4x - 7y)^2]$
13. $f_x(x, y) = -2xy \sin(x^2 + y^2)$; $f_y(x, y) = -2y^2 \sin(x^2 + y^2) + \cos(x^2 + y^2)$
15. $F_x(x, y) = 2 \cos x \cos y$; $F_y(x, y) = -2 \sin x \sin y$ **17.** $f_{xy}(x, y) = 12xy^2 - 15x^2y^4 = f_{yx}(x, y)$
19. $f_{xy}(x, y) = -6e^{2x} \sin y = f_{yx}(x, y)$ **21.** $F_x(3, -2) = \frac{1}{9}$; $F_y(3, -2) = -\frac{1}{2}$
23. $f_x(\sqrt{5}, -2) = -\frac{4}{21}$; $f_y(\sqrt{5}, -2) = -4\sqrt{5}/21$.   **25.** 1   **27.** 3   **29.** 120π **31.** $k/100$ **33.** $\partial^2 f/\partial x^2 = 6xy$; $\partial^2 f/\partial y^2 = -6xy$
35. $180x^4y^2 - 12x^2$ **37.** (a) $\partial^3 f/\partial y^3$;   (b) $\partial^3 f/\partial y \, \partial x^2$;   (c) $\partial^4 f/\partial y^3 \, \partial x$
39. (a) $6xy - yz$;   (b) 8;   (c) $6x - z$ **41.** $-yze^{-xyz} - y(xy - z^2)^{-1}$ **43.** $(1, 0, 29)$
45. $\{(x, y) : x < \frac{1}{2}, y > \frac{1}{2}, y < x + \frac{1}{2}\} \cup \{(x, y) : x > \frac{1}{2}, y < \frac{1}{2}, x < y + \frac{1}{2}\}$, $\{z : 0 < z \le \sqrt{3}/36\}$ **47.** (a) -4;   (b) $\frac{2}{3}$;   (c) $\frac{2}{5}$;   (d) $\frac{8}{3}$

PROBLEM SET 15.3   Page 664

1. -18 **3.** $2 - \frac{1}{2}\sqrt{3}$ **5.** $\frac{1}{3}$ **7.** Does not exist.   **9.** Entire plane.   **11.** $\{(x, y) : y \ne x^2\}$ **13.** $\{(x, y) : y \le x + 1\}$
15. $\lim_{x \to 0} f(x, 0) = \lim_{x \to 0} [0/(x^2 + 0)] = 0$; $\lim_{x \to 0} f(x, x) = \lim_{x \to 0} [x^2/(x^2 + x^2)] = \frac{1}{2}$
17. (a) $\lim_{x \to 0} f(x, mx) = \lim_{x \to 0} mx^3/(x^4 + m^2x^2) = \lim_{x \to 0} mx/(x^2 + m^2) = 0$;   (b) $\lim_{x \to 0} f(x, x^2) = \lim_{x \to 0} x^4/(x^4 + x^4) = \frac{1}{2}$,
   (c) $\lim_{(x, y) \to (0, 0)} f(x, y)$ does not exist.
19. The boundary consists of the line segments that form the outer edges of the given rectangle; the set is closed.
21. Boundary: $\{(x, y) : x^2 + y^2 = 4\} \cup \{(0, 0)\}$; the set is neither open nor closed.
23. Boundary: $\{(x, y) : y = \sin(1/x), x > 0\} \cup \{(x, y) : x = 0, y \le 1\}$; the set is open.   **25.** $g(x) = 2x$
27. (a) Continuous;   (b) discontinuous;   (c) continuous;   (d) continuous;   (e) continuous;   (f) discontinuous
29. (a) $\{(x, y, z) : x^2 + y^2 = 1, 1 \le z \le 2\}$;   (b) $\{(x, y, z) : x^2 + y^2 = 1, z = 1\}$;   (c) $\{(x, y, z) : z = 1\}$;   (d) empty set.
31. (a) $\{(x, y) : x > 0\}$;   (b) $\{(u, v, x, y) : \langle x, y \rangle = k \langle u, v \rangle, k > 0, \langle u, v \rangle \ne \langle 0, 0 \rangle\}$

PROBLEM SET 15.4   Page 671

1. $(2xy + 3y)\mathbf{i} + (x^2 + 3x)\mathbf{j}$ **3.** $e^{xy}(1 + xy)\mathbf{i} + x^2 e^{xy}\mathbf{j}$ **5.** $(x + y)^{-2}[(x^2y + 2xy^2)\mathbf{i} + x^3\mathbf{j}]$ **7.** $(x^2 + y^2 + z^2)^{-1/2}(x\mathbf{i} + y\mathbf{j} + z\mathbf{k})$
9. $xe^{x-z}[(yx + 2y)\mathbf{i} + x\mathbf{j} - xy\mathbf{k}]$ **11.** $\langle -21, 16 \rangle$, $z = -21x + 16y - 60$ **13.** $\langle 0, -2\pi \rangle$, $z = -2\pi y + \pi - 1$
15. $w = 7x - 8y - 2z + 3$ **21.** $z = -5x + 5y$

PROBLEM SET 15.5   Page 676

1. $\frac{8}{5}$ **3.** $3\sqrt{2}/2$ **5.** $(\sqrt{2} + \sqrt{6})/4$ **7.** $\frac{52}{3}$ **9.** 13   **11.** $\sqrt{21}$ **13.** $(1/\sqrt{5})(-\mathbf{i} + 2\mathbf{j})$
15. $\nabla f(\mathbf{p}) = -4\mathbf{i} + \mathbf{j}$ is perpendicular to the tangent line at $\mathbf{p}$.   **17.** $\frac{2}{3}$ **19.** (a) $(0, 0, 0)$;   (b) $-\mathbf{i} + \mathbf{j} - \mathbf{k}$;   (c) yes.
21. N 63.43° E   **23.** Descend; $-300\sqrt{2}e^{-3}$.   **25.** $x = -2y^2$ **27.** (a) $-10/\sqrt{2 + \pi^2}$ deg/m;   (b) -10 deg/s
29. (a) $(90, 125)$;   (b) $(190, 35)$;   (c) $-\frac{1}{3}, 0, \frac{2}{5}$

PROBLEM SET 15.6   Page 682

1. $12t^{11}$ **3.** $e^{3t}(3 \sin 2t + 2 \cos 2t) + e^{2t}(3 \cos 3t + 2 \sin 3t)$ **5.** $7t^6 \cos(t^7)$ **7.** $2s^3t - 3s^2t^2$
9. $(2s^2 \sin t \cos t + 2t \sin^2 s) \exp(s^2 \sin^2 t + t^2 \sin^2 s)$ **11.** $s^4t(1 + s^4t^2)^{-1/2}$ **13.** 72   **15.** $-\frac{1}{2}(\pi + 1)$
17. 244.35 board ft per year   **19.** $\sqrt{20}$ ft/s   **21.** $(3x^2 + 4xy)/(3y^2 - 2x^2)$ **23.** $(y \sin x - \sin y)/(x \cos y + \cos x)$
25. $(yz^3 - 6xz)/(3x^2 - 3xyz^2)$ **27.** $\partial T/\partial s = (\partial f/\partial x)(\partial x/\partial s) + (\partial f/\partial y)(\partial y/\partial s) + (\partial f/\partial z)(\partial z/\partial s) + (\partial f/\partial w)(\partial w/\partial s)$
31. $10\sqrt{2} - 3\pi\sqrt{2}$ **33.** 288 mi/h

PROBLEM SET 15.7   Page 687

1. $2(x - 2) + 3(y - 3) + \sqrt{3}(z - \sqrt{3}) = 0$ **3.** $(x - 1) - 3(y - 3) + \sqrt{7}(z - \sqrt{7}) = 0$ **5.** $x + y - z = 2$
7. $z + 1 = -2\sqrt{3}(x - \frac{1}{3}\pi) - 3y$ **9.** 0.08; 0.08017992   **11.** -0.03; -0.03015101 **13.** $(3, -1, -14)$
15. The common tangent plane is $y + z = 1$.   **17.** $(1, 2, -1)$ and $(-1, -2, 1)$ **19.** $x = 1 + 32t$; $y = 2 - 19t$; $z = 2 - 17t$
21. 0.004375 lb   **23.** 7%   **25.** 20 ± 0.34 **27.** $V = 9|k|/2$ **29.** (a) 4.98;   (b) 4.98196;   (c) 4.9819675

1. $(2,0)$: local minimum point.   **3.** $(0,0)$: saddle point; $(\pm\frac{1}{2},0)$: local minimum points.
5. $(0,0)$: saddle point (special argument needed).   **7.** $(1,2)$: local minimum point.   **9.** No critical points.
11. Global maximum of 7 at $(1,1)$; global minimum of -4 at $(0,-1)$.
13. Global maximum of 2 at $(\pm1,0)$; global minimum of 0 at $(0,\pm1)$.   **15.** Each of the three numbers is $N/3$.
17. A cube.   **19.** Base 8 ft by 8 ft; depth 4 ft.   **21.** $3\sqrt{3}(\mathbf{i}+\mathbf{j}+\mathbf{k})$ **23.** $2\pi/3$; 4 in.   **25.** (a) 8;   (b) -11
27. Maximum of 3 at $(1,2)$; minimum of $-\frac{12}{5}$ at $(\frac{8}{5},-\frac{2}{5})$.   **29.** $y=\frac{7}{10}x+\frac{1}{10}$
31. $(\pm\sqrt{3}/2,-\frac{1}{2})$ where $T=9/4$; $(0,1/2)$, where $T=9/4$ **33.** Equilateral triangle.

1. $f(\sqrt{3},\sqrt{3})=f(-\sqrt{3},-\sqrt{3})=6$ **3.** $f(2/\sqrt{5},-1/\sqrt{5})=f(-2/\sqrt{5},1/\sqrt{5})=5$ **5.** $f(\frac{6}{7},\frac{18}{7},-\frac{12}{7})=\frac{72}{7}$
7. Base is 4 by 4; depth is 2.   **9.** $10\sqrt{5}\ \text{ft}^3$ **11.** $8abc/3\sqrt{3}$ **13.** $x/a+y/b+z/c=3$, $V=\frac{9}{2}abc$
15. $x=\alpha d/a$, $y=\beta d/b$, $z=\gamma d/c$ **17.** $f(-1,1,0)=3$, $f(1,-1,1)=-1$ **19.** $x_i=a_i/\sqrt{a_1^2+\cdots+a_n^2}$, $w=\sqrt{a_1^2+\cdots+a_n^2}$

True-False Quiz

1. True.   **2.** False.   **3.** True.   **4.** True.   **5.** True.   **6.** True.   **7.** False.   **8.** False.   **9.** True.   **10.** False.   **11.** True.
12. True.   **13.** True.   **14.** False.   **15.** True.   **16.** True.   **17.** True.   **18.** False.   **19.** False.   **20.** False.

Sample Test Problems

1. (a) $\{(x,y):x^2+4y^2\geq100\}$ (b) $\{(x,y):2x-y\geq1\}$

2.

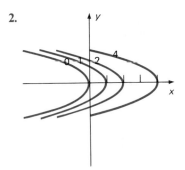

3. $12x^3y^2+14xy^7$; $36x^2y^2+14y^7$; $24x^3y+98xy^6$ **4.** $-\sin 2x$; $-2\cos 2x$; 0   **5.** $e^{-y}\sec^2 x$; $2e^{-y}\sec^2 x\tan x$; $-e^{-y}\sec^2 x$
6. $-e^{-x}\sin y$; $e^{-x}\sin y$; $-e^{-x}\cos y$ **7.** $450x^2y^4-42y^5$ **8.** 19; -14; 80   **9.** 1   **10.** $\{(x,y):y\neq x^2\}$ **11.** Does not exist.
12. (a) 0;   (b) does not exist;   (c) 0   **13.** (a) $-4\mathbf{i}-\mathbf{j}+6\mathbf{k}$; $-4(\cos 1\mathbf{i}+\sin 1\mathbf{j}-\cos 1\mathbf{k})$ **14.** $(3\sqrt{3}-6)/577\approx-0.00139$

15. $\sqrt{3}+2$ **16.** In the direction of the vector $\langle 9,4 \rangle$.   **17.** (a) $x^2 + 2y^2 = 18$;   (b) $4\mathbf{i} + 2\mathbf{j}$
18. $(v\sqrt{xy} + u\sqrt{x})/[2x(1 + u^2v^2)]$; $(v\sqrt{xy} - u\sqrt{y})/[2y(1 + u^2v^2)]$ **19.** $(x^2 + 3y - 4z)/x^2yz$; $(-x^2 - 4z)/xy^2z$; $(3y - x^2)/xyz^2$
20. 0   **21.** $15xy\sqrt{t}/z^3 + 5x^2/tz^3 - 45x^2ye^{3t}/z^4$ **22.** $\frac{7}{2} + 2\sqrt{3} \approx 6.964$ in.^{2}/s   **23.** $18\mathbf{i} + 16\mathbf{j} - 18\mathbf{k}$; $9x + 8y - 9z = 34$
24. $600\pi \pm 3.4\pi = 1884.96 \pm 10.68$ **25.** 0.7728   **26.** Maximum of 0 at $(0,0)$.   **27.** $16\sqrt{3}/3$ **28.** Maximum $\frac{1}{2}$; minimum $-\frac{1}{2}$.
29. Radius 2; height 4.

PROBLEM SET 16.1   *Page 707*

1. 14   **3.** 12   **5.** 4   **7.** 3   **9.** 168   **11.** 520   **13.** 52.57   **15.** 5.5   **19.** $c = 15.30$, $C = 30.97$ **21.** (a) -6;   (b) 6
23. Number of cubic inches of rain that fell on all of Colorado in 1980; average rainfall in Colorado during 1980.

PROBLEM SET 16.2   *Page 711*

1. $\frac{32}{3}$ **3.** $\frac{55}{4}$ **5.** 1   **7.** $\pi/2 - 1$ **9.** $\frac{4}{15}[31 - 9\sqrt{3}] \approx 4.110$ **11.** $1 - \frac{1}{2}\ln 3 \approx 0.4507$ **13.** 0   **15.** 2

17.   **19.**

21. 7   **23.** $\frac{10}{3}$ **27.** $\frac{1}{4}(e^2 - 2e + 1)$ **29.** (a) $\frac{8}{3}$;   (b) 0;   (c) $5 - \sqrt{3} - \sqrt{2}$

PROBLEM SET 16.3   *Page 717*

1. $\frac{3}{4}$ **3.** 240   **5.** $\frac{1}{2}(e^{27} - e)$ **7.** $-\sqrt{2}/(2\pi)$ **9.** $(3\ln 2 - \pi)/9$ **11.** $\frac{16}{3}$ **13.** 0   **15.** $\frac{27}{70}$ **17.** $4\tan^{-1}2 - \ln 5$
19. 6   **21.** 20   **23.** 10   **25.** $\frac{4}{15}$ **27.** $\frac{1}{2}\ln\sec 1$ **29.** 3π **31.** $\int_0^1 \int_y^1 f(x,y)\,dx\,dy$ **33.** $\int_0^1 \int_{y^4}^{\sqrt{y}} f(x,y)\,dx\,dy$
35. $\int_{-1}^0 \int_{-x}^1 f(x,y),dy\,dx + \int_0^1 \int_x^1 f(x,y)\,dy\,dx$ **37.** $\frac{256}{15}$ **39.** $\frac{1}{3}(1 - \cos 8)$ **41.** $15\pi/4$

PROBLEM SET 16.4   *Page 724*

1. $\frac{1}{12}$ **3.** $\frac{4}{9}$ **5.** $2\sqrt{3} + \frac{4}{3}\pi \approx 7.653$ **7.** $\pi a^2/8$ **9.** $8\pi + 6\sqrt{3} \approx 35.525$ **11.** $\pi(e^4 - 1) \approx 168.384$ **13.** $(\pi\ln 2)/8 \approx 0.272$
15. $\pi(2 - \sqrt{3})/2 \approx 0.421$ **17.** $\frac{1}{12}$ **19.** $81\pi/8 \approx 31.809$ **21.** $625(3\sqrt{3} + 1)/12 \approx 322.716$ **23.** $\frac{2}{3}\pi d^2(3a - d)$
25. $\frac{1}{9}a^3(6\pi - 8)$

PROBLEM SET 16.5   *Page 729*

1. $m = 30$; $\bar{x} = 2$; $\bar{y} = 1.8$ **3.** $m = \pi/4$; $\bar{x} = \pi/2$; $\bar{y} = 16/(9\pi)$ **5.** $m \approx 0.1056$; $\bar{x} \approx 0.281$; $\bar{y} \approx 0.581$
7. $m = 32/9$; $\bar{x} = 0$; $\bar{y} = 6/5$ **9.** $I_x \approx 269$; $I_y \approx 5194$; $I_z \approx 5463$ **11.** $I_x = I_y = 5a^5/12$; $I_z = 5a^5/6$
13. $\bar{r} = \sqrt{5/12}a \approx 0.6455a$ **15.** $I_x = \pi\delta a^4/4$; $\bar{r} = a/2$ **17.** $5\pi\delta a^4/4$ **19.** $\bar{x} = 0$, $\bar{y} = (15\pi + 32)a/(6\pi + 48)$
21. (a) a^3;   (b) $7a/12$;   (c) $11a^5/144$ **23.** $I_x = \pi ka^4/2$, $I_y = 17k\pi a^4/2$, $I_z = 9\pi ka^4$

PROBLEM SET 16.6   *Page 734*

1. $\sqrt{61}/3$ **3.** $\pi/3$ **5.** $9\sin^{-1}(\frac{2}{3})$ **7.** $8\sqrt{2}$ **9.** $4\pi a(a - \sqrt{a^2 - b^2})$ **11.** $2a^2(\pi - 2)$ **13.** $\frac{1}{6}\pi a^2(5\sqrt{5} - 1)$
15. $\bar{x} = \bar{y} = 0$, $\bar{z} = \frac{1}{2}(h_1 + h_2)$ **17.** $A = \pi b^2$, $B = 2\pi a^2[1 - \cos(b/a)]$, $C = \pi b^2$, $D = \pi b^2[2a/(a + \sqrt{a^2 - b^2})]$, $B < A = C < D$

1. -40 **3.** $\frac{189}{2}$ **5.** $\frac{2}{3}$ **7.** 156

9. $\int_0^1 \int_0^3 \int_0^{(1/6)(12-3x-2y)} f(x, y, z)\, dz\, dy\, dx$

11. $\int_0^4 \int_0^{(1/2)y} \int_0^2 f(x, y, z)\, dz\, dx\, dy$

13. $\int_0^{12/5} \int_{(1/3)x}^{(1/2)(4-x)} \int_0^{4-x-2z} f(x, y, z)\, dy\, dz\, dx$

15. $\int_0^3 \int_{(2/3)x}^{(1/3)(9-x)} \int_0^{(1/9)(18-2x-6y)} f(x, y, z)\, dz\, dy\, dx$

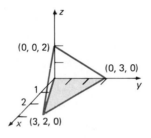

17. $\int_1^4 \int_0^1 \int_0^{\sqrt{1-z^2}} f(x, y, z)\, dy\, dz\, dx$

19. $\frac{128}{15}$ **21.** $4 \int_0^1 \int_{x^2}^1 \int_0^{\sqrt{y}} dz\, dy\, dx = 2$ **23.** $x = \bar{y} = \bar{z} = \frac{4}{15}$ **25.** $\bar{x} = \bar{y} = \bar{z} = 3a/8$ **27.** $\int_0^1 \int_0^{\sqrt{1-x^2}} \int_0^{\sqrt{1-x^2-y^2}} f(x, y, z)\, dz\, dy\, dx$
29. $\int_0^2 \int_0^{2-z} \int_0^{9-x^2} f(x, y, z)\, dy\, dx\, dz$ **31.** 4 **33.** Ave $T = 29.54$, $\bar{z} = \frac{11}{24}$

1. 8π **3.** $2\pi(5\sqrt{5} - 4)/3 \approx 15.038$ **5.** $\bar{x} = \bar{y} = 0$; $\bar{z} = \frac{16}{3}$ **7.** $k\pi(b^4 - a^4)$ **9.** $\bar{x} = \bar{y} = 0$; $\bar{z} = 2a/5$ **11.** $k\pi^2 a^6/16$
13. $\pi/9$ **15.** $\pi/32$ **17.** (a) $3a/4$;   (b) $3\pi a/16$;   (c) $6a/5$ **19.** (a) $3\pi a \sin \alpha/16\alpha$;   (b) $3\pi a/16$ **21.** $(a + b)(c - 1)/(c + 1)$

True-False Quiz

1. True.   **2.** False.   **3.** True.   **4.** True.   **5.** True.   **6.** True.   **7.** False.   **8.** False.   **9.** True.   **10.** True.   **11.** True.
12. True.   **13.** False.   **14.** False.   **15.** True.

1. $\frac{1}{24}$ **2.** 0 **3.** $\frac{2}{3}$ **4.** $-\pi/2$ **5.** $\int_0^1 \int_0^y f(x, y) \, dx \, dy$ **6.** $\int_0^{\pi/2} \int_0^{\cos x} f(x, y) \, dy \, dx$ **7.** $\int_0^{1/2} \int_0^{1-2y} \int_0^{1-2y-z} f(x, y, z) \, dx \, dz \, dy$
8. $\int_0^4 \int_0^{4-z} \int_0^{\sqrt{y}} f(x, y, z) \, dx \, dy \, dz$
9. (a) $8 \int_0^a \int_0^{\sqrt{a^2 - x^2}} \int_0^{\sqrt{a^2 - x^2 - y^2}} dz \, dy \, dx$; (b) $8 \int_0^{\pi/2} \int_0^a \int_0^{\sqrt{a^2 - r^2}} r \, dz \, dr \, d\theta$; (c) $8 \int_0^{\pi/2} \int_0^{\pi/2} \int_0^a \rho^2 \sin \phi \, d\rho \, d\phi \, d\theta$ **10.** $5\pi/4$
11. 0.8857 **12.** $2\pi \ln 1.5 \approx 2.5476$ **13.** $\bar{x} = \frac{13}{6}; \bar{y} = \frac{3}{2}$ **14.** $\frac{128}{5}$ **15.** 6 **16.** (a) 9π; (b) $8\pi/5$ **17.** $80\pi k$
18. $\bar{x} = 0; \bar{y} = \frac{10}{3}$ **19.** $ka^2bc/24$ **20.** $3\pi/2$

PROBLEM SET 17.1 Page 756

1.

3.

5.

7. $(2x - 3y)\mathbf{i} - 3x\mathbf{j} + 2\mathbf{k}$ **9.** $x^{-1}\mathbf{i} + y^{-1}\mathbf{j} + z^{-1}\mathbf{k}$ **11.** $e^y \cos z\mathbf{i} + xe^y \cos z\mathbf{j} - xe^y \sin z\mathbf{k}$ **13.** $2yz; z^2\mathbf{i} - 2y\mathbf{k}$ **15.** $0; \mathbf{0}$
17. $2e^x \cos y + 1; 2e^x \sin y \mathbf{k}$
19. (a) Meaningless; (b) vector field; (c) meaningless; (d) scalar field; (e) vector field; (f) vector field; (g) vector field;
  (h) meaningless; (i) meaningless; (j) scalar field; (k) meaningless.
25. (a) div $\mathbf{F} = 0$, div $\mathbf{G} < 0$, div $\mathbf{H} = 0$, div $\mathbf{L} > 0$; (b) clockwise for $\mathbf{H}$, not at all for others;
  (c) div $\mathbf{F} = 0$, curl $\mathbf{F} = \mathbf{0}$, div $\mathbf{G} = -2ye^{-y^2}$, curl $\mathbf{G} = \mathbf{0}$, div $\mathbf{H} = 0$, curl $\mathbf{H} = -2xe^{-x^2}\mathbf{k}$, div $\mathbf{L} = 1/\sqrt{x^2 + y^2}$, curl $\mathbf{L} = \mathbf{0}$

PROBLEM SET 17.2 Page 763

1. $14(2\sqrt{2} - 1)$ **3.** $2\sqrt{5}$ **5.** $\frac{1}{6}(14\sqrt{14} - 1)$ **7.** $\frac{100}{3}$ **9.** 144 **11.** 0 **13.** $\frac{17}{6}$ **15.** 19 **17.** $k(17\sqrt{17} - 1)/6$ **19.** $-\frac{7}{44}$
21. $-\frac{1}{2}(a^2 + b^2)$ **23.** $2 - 2/\pi$ **25.** 2.25 gal **27.** $2\pi a^2$ **29.** $4a^2$ **31.** (a) 27; (b) $-297/2$

PROBLEM SET 17.3 Page 771

1. $f(x, y) = 5x^2 - 7xy + y^2 + C$ **3.** Not conservative. **5.** $f(x, y) = \frac{2}{5}x^3y^{-2} + C$ **7.** $f(x, y) = 2xe^y - ye^x + C$
9. $f(x, y, z) = x^3 + 2y^3 + 3z^3 + C$ **11.** 14 **13.** 6 **15.** $-\pi$ **19.** $f(x, y, z) = \frac{1}{2}k(x^2 + y^2 + z^2)$
21. $\int_C \mathbf{F} \cdot d\mathbf{r} = \int_a^b m\mathbf{r}''(t) \cdot \mathbf{r}'(t) \, dt = \frac{1}{2}m \int_a^b (d/dt)[\mathbf{r}'(t) \cdot \mathbf{r}'(t)] \, dt = \frac{1}{2}m \int_a^b (d/dt)|\mathbf{r}'(t)|^2 \, dt = [\frac{1}{2}m|\mathbf{r}'(t)|^2]_a^b = \frac{1}{2}m[|\mathbf{r}'(b)|^2 - |\mathbf{r}'(a)|^2]$
23. $f(x, y, z) = -gmz$

PROBLEM SET 17.4 *Page 778*

1. $-\frac{64}{15}$ **3.** $\frac{72}{35}$ **5.** 0 **7.** $\frac{8}{3}$ **9.** (a) 0; (b) 0 **11.** (a) 0; (b) 0 **13.** 50 **15.** -2
19. (c) M and N have a discontinuity at $(0,0)$. **23.** $3\pi a^2/8$

PROBLEM SET 17.5 *Page 785*

1. $8\sqrt{3}/3$ **3.** $2 + \pi/3$ **5.** $5\pi/8$ **7.** 6 **9.** 2 **11.** 20 **13.** $\sqrt{3}ka^4/12$ **15.** $\bar{x} = \bar{y} = \bar{z} = a/3$
17. (a) 0; (b) 0; (c) $4\pi a^4$; (d) $4\pi a^4/3$; (e) $8\pi a^4/3$ **19.** (a) $4k\pi a^3$; (b) $2k\pi a^3$; (c) $hk\pi a(a+b)$

PROBLEM SET 17.6 *Page 791*

1. 0 **3.** $3a^2b^2c^2/4$ **5.** $64\pi/3$ **7.** 4π **9.** 1176π **11.** $\mathbf{V} \cdot \mathbf{F} = 3$ and so $\iint_{\delta S} \mathbf{F} \cdot \mathbf{n} \, dS = \iiint_{S} 3 \, dV = 3V(S)$.
15. (a) $20\pi/3$; (b) 4π; (c) $16\pi/3$; (d) 1; (e) 36; (f) $12\pi/5$; (g) $32\pi \ln 2$

PROBLEM SET 17.7 *Page 796*

1. 0 **3.** -2 **5.** -48π **7.** 8π **9.** 2 **11.** $\pi/4$ **15.** $1/3$ **17.** $\frac{4}{3}a^2$ J

CHAPTER REVIEW 17.8 *Page 797*

True-False Quiz

1. True. **2.** False. **3.** False. **4.** True. **5.** True. **6.** True. **7.** False. **8.** True. **9.** True. **10.** False. **11.** True.
12. True.

Sample Test Problems

2. $2yz - 6y + 2y^2$; $4yz\mathbf{i} + 2xy\mathbf{j} - 2xz\mathbf{k}$; $(2z - 6 + 4y)\mathbf{j} + 2y\mathbf{k}$; 0 **3.** $\text{curl}(f\,\nabla f) = f\,\text{curl}(\nabla f) + \nabla f \times \nabla f = \mathbf{0} + \mathbf{0} = \mathbf{0}$
4. (a) $f(x, y) = x^2y + xy + \sin y + C$; (b) $f(x, y, z) = xyz + e^{-x} + e^y + C$ **5.** (a) $\pi/4$; (b) $(3\pi - 5)/6$ **6.** 4 **7.** 47
8. $2 + e^{-1} + e$ **9.** (a) $\frac{1}{2}$; (b) $\frac{4}{3}$; (c) 0 **10.** 8 **11.** $\frac{8}{3}\pi$ **12.** $\sqrt{3}/5$ **13.** 0 **14.** 90π **15.** $9\pi(3a - 2)/\sqrt{a^2 + b^2 + 1}$

PROBLEM SET 18.1 *Page 805*

1. $y = xe^{-x} + Ce^{-x}$ **3.** $y = a + C\sqrt{1 - x^2}$ **5.** $y = xe^x + Cx$ **7.** $y = 1 + Cx^{-1}$ **9.** $y = 1 + Ce^{-\int f(x)\,dx}$
11. $y = x^4 + 2x$ **13.** $y = e^{-x}(1 - x^{-1})$ **15.** 38.506 lb **17.** $2(60 - t) - \frac{1}{1800}(60 - t)^3$
19. $I = 10^{-6}[1 - \exp(-10^6 t)]$
21. (a) $I = 0.0438 \sin 377t - 0.0578 \cos 377t + 0.0578 e^{-285.7t}$; (b) $I = 0.0438 \sin 377t - 0.0578 \cos 377t$
23. (a) 21.97; (b) 26.67; (c) $c > 7.72$; (d) $20 + 80e^{-0.05T} + 0.25T = 50$; (e) 24.10
25. (a) 200.32 ft; (b) $15200(1 - e^{-0.05T}) - 640T = 0$; (c) 7.08

PROBLEM SET 18.2 *Page 810*

1. $y = C_1 e^{2x} + C_2 e^{3x}$ **3.** $y = \frac{1}{2}e^x - \frac{1}{2}e^{-7x}$ **5.** $y = (C_1 + C_2 x)e^{2x}$ **7.** $y = e^{2x}(C_1 e^{\sqrt{3}x} + C_2 e^{-\sqrt{3}x})$
9. $y = 3\sin 2x + 2\cos 2x$ **11.** $y = e^{-x}(C_1 \cos x + C_2 \sin x)$ **13.** $y = C_1 + C_2 x + C_3 e^{-4x} + C_4 e^x$
15. $y = C_1 e^x + C_2 e^{-x} + C_3 \cos 2x + C_4 \sin 2x$ **17.** $y = D_1 \cosh 2x + D_2 \sinh 2x$
19. $y = e^{-x/2}[(C_1 + C_2 x)\cos(\sqrt{3}/2)x + (C_3 + C_4 x)\sin(\sqrt{3}/2)x]$ **21.** $y = (C_1 + C_2 \ln x)x^{-2}$

PROBLEM SET 18.3 *Page 815*

1. $y = C_1 e^{3x} + C_2 e^{-3x} - \frac{1}{9}x$ **3.** $y = (C_1 + C_2 x)e^x + x^2 + 5x + 8$ **5.** $y = C_1 e^{2x} + C_2 e^{3x} + \frac{1}{2}e^x$
7. $y = C_1 e^{-3x} + C_2 e^{-x} - \frac{1}{2}xe^{-3x}$ **9.** $y = C_1 e^{2x} + C_2 e^{-x} - \frac{3}{5}\sin x + \frac{1}{5}\cos x$ **11.** $y = C_1 \cos 2x + C_2 \sin 2x + \frac{1}{2}x \sin 2x$
13. $y = C_1 \cos 3x + C_2 \sin 3x + \frac{1}{8}\sin x + \frac{1}{13}e^{2x}$ **15.** $y = e^{2x} - e^{3x} + e^x$ **17.** $y = C_1 e^{2x} + C_2 e^x + \frac{5}{2}x + \frac{19}{4}$
19. $y = C_1 \sin x + C_2 \cos x - x \sin x - \cos x \ln|\sin x|$ **21.** $y = C_1 e^x + C_2 e^{2x} + (e^x + e^{2x})\ln(1 + e^{-x})$

1. $y = -\cos 8t$; $\pi/4$ s   **3.** 8 ft/s   **5.** $y \approx e^{-0.16t}(\cos 8t + 0.02 \sin 8t)$ **7.** 14.4 s   **9.** $Q = 10^{-6}(1 - e^{-t})$
11. (a) $Q = 2.4 \times 10^{-4} \sin 377t$;   (b) $I = 9.05 \times 10^{-2} \cos 377t$ **13.** $I \approx 12 \times 10^{-2} \sin 377t$
17. $d^2\theta/dt^2 = -(g/L) \sin \theta$

CHAPTER REVIEW 18.5   *Page 821*

True-False Quiz

1. False.   **2.** True.   **3.** True.   **4.** False.   **5.** True.   **6.** False.   **7.** True.   **8.** True.   **9.** False.   **10.** True.

Sample Test Problems

1. $y = Cx^{-1}$ **2.** $y = \frac{1}{4}x^2 + Cx^{-2}$ **3.** $y = 2e^{-x^2} + 1$ **4.** $y = xe^{ax} + Ce^{ax}$ **5.** $y = -e^x + Ce^{2x}$ **6.** $y = 2 \sin x + C \cos x$
7. $y = \frac{1}{4}e^x + C_1 e^{-3x} + C_2$ **8.** $y = C_1 e^x + C_2 e^{-x}$ **9.** $y = 3e^{2x} - 3e^x$ **10.** $y = (C_1 + C_2 x)e^{-3x/2}$
11. $y = C_1 e^x + C_2 e^{-x} - 1$ **12.** $y = (C_1 + C_2 x)e^{-2x} + \frac{1}{3}e^x$ **13.** $y = (C_1 + C_2 x + \frac{1}{2}x^2)e^{-2x}$ **14.** $y = \sin 2x$
15. $y = e^{-3x}(C_1 \cos 4x + C_2 \sin 4x)$ **16.** $y = C_1 \cos x + C_2 \sin x + x \cos x + \sin x \ln|\sec x|$ **17.** $y = C_1 + C_2 e^{-4x} + C_3 e^{2x}$
18. $y = C_1 e^{\sqrt{5}x} + C_2 e^{-\sqrt{5}x} + C_3 \cos \sqrt{2}x + C_4 \sin \sqrt{2}x$ **19.** $y = (C_1 + C_2 x)e^{\sqrt{2}x} + (C_3 + C_4 x)e^{-\sqrt{2}x}$
20. (a) $dQ/dt = 3 - 0.02Q$;   (b) $Q = 150 - 30e^{-0.02t}$;   (c) $Q \to 150$ as $t \to \infty$.   **21.** $y = -\cos 4t$; 1; $\pi/2$ **22.** 4
23. $I = e^{-t} \sin t$

Index

Interval(s):
  closed, 11
  of convergence, 501
  open, 11
Interval Additive Property, 247
Inverse, 324
  of exponential function, 337
  hyperbolic functions, 362, 363
  of logarithmic function, 330
  trigonometric functions, 348-353
Inverse Function Theorem, 327
Irrational number, 2, 6
Irrotational fluid flow, 778
Isothermal curves, 653
Isothermal surface, 676
Iteration scheme, 452

Jacobian, 746, 747

Kahoutek, 533
Kepler, Johannes, 88, 521
Kepler's equation, 460
Kepler's laws of planetary motion, 521
Kirchhoff's Law, 804, 819
Knuth, Donald E., 452
Kovalevsky, Sonya, 799

Lagrange, Joseph-Louis, 436, 649
Lagrange multipliers, 694
Lagrange's identity, 621, 624
Lamina:
  center of mass (gravity) of, 307-310,
    726
  density of, 725
  mass of, 726
  moments of inertia (second moments)
    of, 729
Laplace, P. S., 315, 607, 660
Laplace's equation, 660
Latus rectum, 526
Law of the Excluded Middle, 5
Law of Mass Action, 399
Least squares, method of, 694
Least upper bound, 10
Lebesgue, Henri, 702
Left-handed system, 608
Leibniz, Gottfried Wilhelm, 87, 120,
    240
Le Lionnais, François, 331
Lemniscate, 557

Length:
  of arc, 287, 630
  of line segment, 21
  of vector, 578, 584, 613
Level curve, 651, 674
Level surface, 653, 675
L'Hôpital, G. F. A., 368, 403
L'Hôpital's Rules, 404, 411
Limaçon, 556
Limit(s):
  of a function, 65
  of a function of two variables, 661
  of integration, 235
  of integration, change of, 256, 373
  of a sequence, 466
  trigonometric, 112
  of a vector-valued function, 590
Limit Comparison Test, 488
Limit of a function, 58-62
  definition of, 59, 65
  infinite, 187
  at infinity, 185
  left-hand, 62, 70
  one-sided, 62, 70
  right-hand, 62, 70
  theorems on, 71
Linear approximation, 430
Linear differential equations, 800
Linear equation, 28, 611
Linear operator, 103, 104, 207, 220,
    244, 476, 670, 800
Line integral(s), 758
Lines in three-space, 625
  direction numbers of, 626
  parametric equations of, 626
  symmetric equations of, 626
Lines in two-space:
  equations of, 25-29
  parallel, 28
  perpendicular, 29
  slope of, 25
Lipschitz, Rudolph, 199
Lipschitz condition, 199
Liquid pressure, 300
Logarithmic differentiation, 320
Logic, 4
Logistic model, 344, 347, 400
LORAN, 537

Machin, John, 354
Maclaurin, Colin, 432
Maclaurin polynomial, 432
Maclaurin's Formula, 437
Maclaurin's series, 511, 516

Magnetic potential, 421
Magnitude (length) of a vector, 578,
    584, 613
Major axis, 523
Marginal cost, 31, 180
Marginal price, 180
Marginal profit, 31, 180
Marginal revenue, 180
Mass, 304
  center of, 304-310, 726, 739
  moments of, 304, 306, 726, 740
Mathematical Induction, 5, 823
Maximum (value) of a function, 155,
    688
  global, 168, 688
  local, 168, 688
  tests for, 169, 170, 691
Max-Min Existence Theorem, 156, 689
Max-min problems, 158-161, 172-178,
    690-700
Mean Value Theorem for Derivatives,
    195
  Cauchy's, 407
  several-variable case, 672
Mean Value Theorem for Integrals, 251
Merz, J. T., 87
Methusaleh, 347
Midpoint of a line segment, 23, 610
Mill, John Stuart, 1
Minimum (value) of a function, 155,
    688
  global, 168, 688
  local, 168, 688
  tests for, 169, 170, 691
Minkowski's Inequality, 329
Minuit, Peter, 347
Modeling, 129, 179
Moment, first (of mass), 304
  of a homogeneous lamina, 307
  of any lamina, 726
Moment, second (of inertia), 728
Monotonic Sequence Theorem, 469
Monotonicity Theorem, 162, 197
Multiple integrals, 703-750
Multiplier effect, 479
Musical scale, 340

Napoleon, 649
Natural domain, 40
Natural exponential function, 330
Natural logarithmic function, 316-322
  graph of, 321
  inverse of, 330
Natural numbers, 2

3.14159

FORMULA CARD
to accompany
CALCULUS WITH ANALYTIC GEOMETRY
Purcell and Varberg

Fold here

DERIVATIVES

$$D_x x^r = r x^{r-1}$$

$$D_x \sin x = \cos x$$

$$D_x \tan x = \sec^2 x$$

$$D_x \sec x = \sec x \tan x$$

$$D_x \sinh x = \cosh x$$

$$D_x \cosh x = \sinh x$$

$$D_x \tanh x = \operatorname{sech}^2 x$$

$$D_x \ln x = \frac{1}{x}$$

$$D_x e^x = e^x$$

$$D_x \sin^{-1}x = \frac{1}{\sqrt{1-x^2}}$$

$$D_x \tan^{-1}x = \frac{1}{1+x^2}$$

$$D_x |x| = \frac{|x|}{x}$$

$$D_x \cos x = -\sin x$$

$$D_x \cot x = -\csc^2 x$$

$$D_x \csc x = -\csc x \cot x$$

$$D_x \coth x = -\operatorname{csch}^2 x$$

$$D_x \operatorname{sech} x = -\operatorname{sech} x \tanh x$$

$$D_x \operatorname{csch} x = -\operatorname{csch} x \coth x$$

$$D_x \log_a x = \frac{1}{x \ln a}$$

$$D_x a^x = a^x \ln a$$

$$D_x \cos^{-1}x = \frac{-1}{\sqrt{1-x^2}}$$

$$D_x \sec^{-1}x = \frac{1}{|x|\sqrt{x^2-1}}$$

INTEGRALS

1. $\int u\, dv = uv - \int v\, du$

2. $\int u^n\, du = \dfrac{1}{n+1} u^{n+1} + C,\ n \ne -1$

3. $\int \dfrac{1}{u}\, du = \ln|u| + C$

4. $\int e^u\, du = e^u + C$

5. $\int a^u\, du = \dfrac{a^u}{\ln a} + C$

6. $\int \sin u\, du = -\cos u + C$

7. $\int \cos u\, du = \sin u + C$

8. $\int \sec^2 u\, du = \tan u + C$

9. $\int \csc^2 u\, du = -\cot u + C$

10. $\int \sec u \tan u\, du = \sec u + C$

11. $\int \csc u \cot u\, du = -\csc u + C$

12. $\int \tan u\, du = \ln|\sec u| + C$

13. $\int \cot u\, du = \ln|\sin u| + C$

14. $\int \sec u\, du = \ln|\sec u + \tan u| + C$

15. $\int \csc u\, du = \ln|\csc u - \cot u| + C$

16. $\int \dfrac{1}{\sqrt{a^2-u^2}}\, du = \sin^{-1}\dfrac{u}{a} + C$

17. $\int \dfrac{1}{a^2+u^2}\, du = \dfrac{1}{a}\tan^{-1}\dfrac{u}{a} + C$

18. $\int \dfrac{1}{a^2-u^2}\, du = \dfrac{1}{2a}\ln\left|\dfrac{u+a}{u-a}\right| + C$

19. $\int \dfrac{1}{u\sqrt{u^2-a^2}}\, du = \dfrac{1}{a}\sec^{-1}\left|\dfrac{u}{a}\right| + C$

Fold here

GEOMETRY

Triangles

Pythagorean Theorem

$$a^2 + b^2 = c^2$$

Right triangle

Angles $\alpha + \beta + \gamma = 180°$
Area $A = \tfrac{1}{2}bh$

Any triangle

Circles

Circumference $\quad C = 2\pi r$
Area $\quad A = \pi r^2$

Cylinders

Surface area $\quad S = 2\pi r^2 + 2\pi rh$
Volume $\quad V = \pi r^2 h$

Cones

Surface area $\quad S = \pi r^2 + \pi r\sqrt{r^2 + h^2}$
Volume $\quad V = \dfrac{1}{3}\pi r^2 h$

Spheres

Surface area $\quad S = 4\pi r^2$
Volume $\quad V = \dfrac{4}{3}\pi r^3$

Conversions

1 inch = 2.54 centimeters
1 liter = 1000 cubic centimeters
1 kilogram = 2.20 pounds

1 kilometer = .62 miles
1 liter = 1.057 quarts
1 pound = 453.6 grams
π radians = 180 degrees

TRIGONOMETRY

Basic Identities

$$\tan t = \frac{\sin t}{\cos t} \qquad \cot t = \frac{\cos t}{\sin t} \qquad \cot t = \frac{1}{\tan t}$$

$$\sec t = \frac{1}{\cos t} \qquad \csc t = \frac{1}{\sin t}$$

$$1 + \tan^2 t = \sec^2 t \qquad \sin^2 t + \cos^2 t = 1 \qquad 1 + \cot^2 t = \csc^2 t$$

Cofunction Identities

$$\sin\left(\frac{\pi}{2} - t\right) = \cos t \qquad \cos\left(\frac{\pi}{2} - t\right) = \sin t \qquad \tan\left(\frac{\pi}{2} - t\right) = \cot t$$

Odd-even Identities

$$\sin(-t) = -\sin t \qquad \cos(-t) = \cos t \qquad \tan(-t) = -\tan t$$

Addition Formulas

$$\sin(s + t) = \sin s \cos t + \cos s \sin t \qquad \sin(s - t) = \sin s \cos t - \cos s \sin t$$

$$\cos(s + t) = \cos s \cos t - \sin s \sin t \qquad \cos(s - t) = \cos s \cos t + \sin s \sin t$$

$$\tan(s + t) = \frac{\tan s + \tan t}{1 - \tan s \tan t} \qquad \tan(s - t) = \frac{\tan s - \tan t}{1 + \tan s \tan t}$$

Double Angle Formulas

$$\sin 2t = 2 \sin t \cos t$$

$$\cos 2t = \cos^2 t - \sin^2 t = 1 - 2\sin^2 t = 2\cos^2 t - 1$$

$$\tan 2t = \frac{2 \tan t}{1 - \tan^2 t}$$

Half Angle Formulas

$$\sin \frac{t}{2} = \pm\sqrt{\frac{1 - \cos t}{2}} \qquad \cos \frac{t}{2} = \pm\sqrt{\frac{1 + \cos t}{2}} \qquad \tan \frac{t}{2} = \frac{1 - \cos t}{\sin t}$$

Product Formulas

$$2 \sin s \cos t = \sin(s + t) + \sin(s - t)$$

$$2 \cos s \cos t = \cos(s + t) + \cos(s - t)$$

$$2 \sin s \sin t = \cos(s - t) - \cos(s + t)$$

Factoring Formulas

$$\sin s + \sin t = 2 \cos \frac{s - t}{2} \sin \frac{s + t}{2}$$

$$\sin s - \sin t = 2 \cos \frac{s + t}{2} \sin \frac{s - t}{2}$$

$$\cos s + \cos t = 2 \cos \frac{s + t}{2} \cos \frac{s - t}{2}$$

$$\cos s - \cos t = -2 \sin \frac{s + t}{2} \sin \frac{s - t}{2}$$

Laws of Sines and Cosines

$$\frac{\sin \alpha}{a} = \frac{\sin \beta}{b} = \frac{\sin \gamma}{c}$$

$$a^2 = b^2 + c^2 - 2bc \cos \alpha$$

Graphs

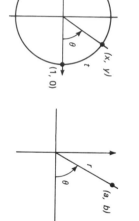

$$\sin t = \sin \theta = y = \frac{b}{r} \qquad \cos t = \cos \theta = x = \frac{a}{r}$$

$$\tan t = \tan \theta = \frac{y}{x} = \frac{b}{a} \qquad \cot t = \cot \theta = \frac{x}{y} = \frac{a}{b}$$

$y = \sin t$

$y = \cos t$

$y = \tan t$

$y = \cot t$

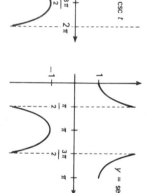

$y = \csc t$

$y = \sec t$

Inverse Trigonometric Functions

$$y = \sin^{-1}x \Leftrightarrow x = \sin y, \ -\pi/2 \le y \le \pi/2$$

$$y = \cos^{-1}x \Leftrightarrow x = \cos y, \ 0 \le y \le \pi$$

$$y = \tan^{-1}x \Leftrightarrow x = \tan y, \ -\pi/2 < y < \pi/2$$

$$y = \sec^{-1}x \Leftrightarrow x = \sec y, \ 0 \le y \le \pi, \ y \ne \pi/2$$

$$\sec^{-1}x = \cos^{-1}(1/x)$$

Hyperbolic Functions

$$\sinh x = \frac{1}{2}(e^x - e^{-x}) \qquad \cosh x = \frac{1}{2}(e^x + e^{-x})$$

$$\tanh x = \frac{\sinh x}{\cosh x} \qquad \coth x = \frac{\cosh x}{\sinh x}$$

$$\text{sech } x = \frac{1}{\cosh x} \qquad \text{csch } x = \frac{1}{\sinh x}$$

Series

$$\frac{1}{1 - x} = 1 + x + x^2 + x^3 + \cdots - 1 < x < 1$$

$$\ln(1 + x) = x - \frac{x^2}{2} + \frac{x^3}{3} - \frac{x^4}{4} + \cdots - 1 < x \le 1$$

$$e^x = 1 + x + \frac{x^2}{2!} + \frac{x^3}{3!} + \cdots$$

$$\sin x = x - \frac{x^3}{3!} + \frac{x^5}{5!} - \frac{x^7}{7!} + \cdots$$

$$\cos x = 1 - \frac{x^2}{2!} + \frac{x^4}{4!} - \frac{x^6}{6!} + \cdots$$

$$\tan^{-1}x = x - \frac{x^3}{3} + \frac{x^5}{5} - \frac{x^7}{7} + \cdots - 1 \le x \le 1$$

$$(1 + x)^p = 1 + \binom{p}{1}x + \binom{p}{2}x^2 + \binom{p}{3}x^3 + \cdots - 1 < x < 1$$

$$\binom{p}{k} = \frac{p(p - 1)(p - 2)\cdots(p - k + 1)}{k!}$$

ELEMENTARY FORMS

$1 \quad \int u\,dv = uv - \int v\,du$

$2 \quad \int u^n\,du = \frac{1}{n+1}u^{n+1} + C \quad \text{if } n \neq -1$

$3 \quad \int \frac{du}{u} = \ln|u| + C$

$4 \quad \int e^u\,du = e^u + C$

$5 \quad \int a^u\,du = \frac{a^u}{\ln a} + C$

$6 \quad \int \sin u\,du = -\cos u + C$

$7 \quad \int \cos u\,du = \sin u + C$

$8 \quad \int \sec^2 u\,du = \tan u + C$

$9 \quad \int \csc^2 u\,du = -\cot u + C$

$10 \quad \int \sec u \tan u\,du = \sec u + C$

$11 \quad \int \csc u \cot u\,du = -\csc u + C$

$12 \quad \int \tan u\,du = \ln|\sec u| + C$

$13 \quad \int \cot u\,du = \ln|\sin u| + C$

$14 \quad \int \sec u\,du = \ln|\sec u + \tan u| + C$

$15 \quad \int \csc u\,du = \ln|\csc u - \cot u| + C$

$16 \quad \int \frac{du}{\sqrt{a^2 - u^2}} = \sin^{-1}\frac{u}{a} + C$

$17 \quad \int \frac{du}{a^2 + u^2} = \frac{1}{a}\tan^{-1}\frac{u}{a} + C$

$18 \quad \int \frac{du}{a^2 - u^2} = \frac{1}{2a}\ln\left|\frac{u+a}{u-a}\right| + C$

$19 \quad \int \frac{du}{u\sqrt{u^2 - a^2}} = \frac{1}{a}\sec^{-1}\left|\frac{u}{a}\right| + C$

TRIGONOMETRIC FORMS

$20 \quad \int \sin^2 u\,du = \frac{1}{2}u - \frac{1}{4}\sin 2u + C$

$21 \quad \int \cos^2 u\,du = \frac{1}{2}u + \frac{1}{4}\sin 2u + C$

$22 \quad \int \tan^2 u\,du = \tan u - u + C$

$23 \quad \int \cot^2 u\,du = -\cot u - u + C$

$24 \quad \int \sin^3 u\,du = -\frac{1}{3}(2 + \sin^2 u)\cos u + C$

$25 \quad \int \cos^3 u\,du = \frac{1}{3}(2 + \cos^2 u)\sin u + C$

$26 \quad \int \tan^3 u\,du = \frac{1}{2}\tan^2 u + \ln|\cos u| + C$

$27 \quad \int \cot^3 u\,du = -\frac{1}{2}\cot^2 u - \ln|\sin u| + C$

$28 \quad \int \sec^3 u\,du = \frac{1}{2}\sec u \tan u + \frac{1}{2}\ln|\sec u + \tan u| + C$

$29 \quad \int \csc^3 u\,du = -\frac{1}{2}\csc u \cot u + \frac{1}{2}\ln|\csc u - \cot u| + C$

$30 \quad \int \sin au \sin bu\,du = \frac{\sin(a-b)u}{2(a-b)} - \frac{\sin(a+b)u}{2(a+b)} + C \quad \text{if } a^2 \neq b^2$

$31 \quad \int \cos au \cos bu\,du = \frac{\sin(a-b)u}{2(a-b)} + \frac{\sin(a+b)u}{2(a+b)} + C \quad \text{if } a^2 \neq b^2$

$32 \quad \int \sin au \cos bu\,du = -\frac{\cos(a-b)u}{2(a-b)} - \frac{\cos(a+b)u}{2(a+b)} + C \quad \text{if } a^2 \neq b^2$

$33 \quad \int \sin^n u\,du = -\frac{1}{n}\sin^{n-1}u\cos u + \frac{n-1}{n}\int \sin^{n-2}u\,du$

$34 \quad \int \cos^n u\,du = \frac{1}{n}\cos^{n-1}u\sin u + \frac{n-1}{n}\int \cos^{n-2}u\,du$

$35 \quad \int \tan^n u\,du = \frac{1}{n-1}\tan^{n-1}u - \int \tan^{n-2}u\,du \quad \text{if } n \neq 1$

$36 \quad \int \cot^n u\,du = \frac{-1}{n-1}\cot^{n-1}u - \int \cot^{n-2}u\,du \quad \text{if } n \neq 1$

$37 \quad \int \sec^n u\,du = \frac{1}{n-1}\sec^{n-2}u\tan u + \frac{n-2}{n-1}\int \sec^{n-2}u\,du \quad \text{if } n \neq 1$

$38 \quad \int \csc^n u\,du = \frac{-1}{n-1}\csc^{n-2}u\cot u + \frac{n-2}{n-1}\int \csc^{n-2}u\,du \quad \text{if } n \neq 1$

$39 \quad \int \sin^n u \cos^m u\,du = -\frac{\sin^{n-1}u\cos^{m+1}u}{n+m} + \frac{n-1}{n+m}\int \sin^{n-2}u\cos^m u\,du \quad \text{if } n \neq -m$

$40 \quad \int \sin^n u \cos^m u\,du = \frac{\sin^{n+1}u\cos^{m-1}u}{n+m} + \frac{m-1}{n+m}\int \sin^n u\cos^{m-2}u\,du \quad \text{if } m \neq -n$

$41 \quad \int u \sin u\,du = \sin u - u\cos u + C$

$42 \quad \int u \cos u\,du = \cos u + u\sin u + C$

$43 \quad \int u^n \sin u\,du = -u^n\cos u + n\int u^{n-1}\cos u\,du$

$44 \quad \int u^n \cos u\,du = u^n\sin u - n\int u^{n-1}\sin u\,du$

DETACH HERE

FORMS INVOLVING $\sqrt{u^2 \pm a^2}$

44. $\displaystyle\int \sqrt{u^2 \pm a^2}\, du = \frac{u}{2}\sqrt{u^2 \pm a^2} \pm \frac{a^2}{2}\ln\left|u + \sqrt{u^2 \pm a^2}\right| + C$

45. $\displaystyle\int \frac{du}{\sqrt{u^2 \pm a^2}} = \ln\left|u + \sqrt{u^2 \pm a^2}\right| + C$

46. $\displaystyle\int \frac{\sqrt{u^2 + a^2}}{u}\, du = \sqrt{u^2 + a^2} - a\ln\left(\frac{a + \sqrt{u^2 + a^2}}{u}\right) + C$

47. $\displaystyle\int \frac{\sqrt{u^2 - a^2}}{u}\, du = \sqrt{u^2 - a^2} - a\sec^{-1}\frac{u}{a} + C$

48. $\displaystyle\int u^2 \sqrt{u^2 \pm a^2}\, du = \frac{u}{8}(2u^2 \pm a^2)\sqrt{u^2 \pm a^2} - \frac{a^4}{8}\ln\left|u + \sqrt{u^2 \pm a^2}\right| + C$

49. $\displaystyle\int \frac{u^2\, du}{\sqrt{u^2 \pm a^2}} = \frac{u}{2}\sqrt{u^2 \pm a^2} \mp \frac{a^2}{2}\ln\left|u + \sqrt{u^2 \pm a^2}\right| + C$

50. $\displaystyle\int \frac{du}{u^2 \sqrt{u^2 \pm a^2}} = \mp \frac{\sqrt{u^2 \pm a^2}}{a^2 u} + C$

51. $\displaystyle\int \frac{\sqrt{u^2 \pm a^2}}{u^2}\, du = -\frac{\sqrt{u^2 \pm a^2}}{u} + \ln\left|u + \sqrt{u^2 \pm a^2}\right| + C$

52. $\displaystyle\int \frac{du}{(u^2 \pm a^2)^{3/2}} = \frac{\pm u}{a^2 \sqrt{u^2 \pm a^2}} + C$

53. $\displaystyle\int (u^2 \pm a^2)^{3/2}\, du = \frac{u}{8}(2u^2 \pm 5a^2)\sqrt{u^2 \pm a^2} + \frac{3a^4}{8}\ln\left|u + \sqrt{u^2 \pm a^2}\right| + C$

FORMS INVOLVING $\sqrt{a^2 - u^2}$

54. $\displaystyle\int \sqrt{a^2 - u^2}\, du = \frac{u}{2}\sqrt{a^2 - u^2} + \frac{a^2}{2}\sin^{-1}\frac{u}{a} + C$

55. $\displaystyle\int \frac{\sqrt{a^2 - u^2}}{u}\, du = \sqrt{a^2 - u^2} - a\ln\left|\frac{a + \sqrt{a^2 - u^2}}{u}\right| + C$

56. $\displaystyle\int \frac{u^2\, du}{\sqrt{a^2 - u^2}} = -\frac{u}{2}\sqrt{a^2 - u^2} + \frac{a^2}{2}\sin^{-1}\frac{u}{a} + C$

57. $\displaystyle\int u^2 \sqrt{a^2 - u^2}\, du = \frac{u}{8}(2u^2 - a^2)\sqrt{a^2 - u^2} + \frac{a^4}{8}\sin^{-1}\frac{u}{a} + C$

58. $\displaystyle\int \frac{du}{u^2 \sqrt{a^2 - u^2}} = -\frac{\sqrt{a^2 - u^2}}{a^2 u} + C$

59. $\displaystyle\int \frac{\sqrt{a^2 - u^2}}{u^2}\, du = -\frac{\sqrt{a^2 - u^2}}{u} - \sin^{-1}\frac{u}{a} + C$

60. $\displaystyle\int \frac{du}{u \sqrt{a^2 - u^2}} = -\frac{1}{a}\ln\left|\frac{a + \sqrt{a^2 - u^2}}{u}\right| + C$

61. $\displaystyle\int \frac{du}{(a^2 - u^2)^{3/2}} = \frac{u}{a^2 \sqrt{a^2 - u^2}} + C$

62. $\displaystyle\int (a^2 - u^2)^{3/2}\, du = \frac{u}{8}(5a^2 - 2u^2)\sqrt{a^2 - u^2} + \frac{3a^4}{8}\sin^{-1}\frac{u}{a} + C$

EXPONENTIAL AND LOGARITHMIC FORMS

63. $\displaystyle\int ue^u\, du = (u - 1)e^u + C$

64. $\displaystyle\int u^n e^u\, du = u^n e^u - n\int u^{n-1} e^u\, du$

65. $\displaystyle\int \ln u\, du = u\ln u - u + C$

66. $\displaystyle\int u^n \ln u\, du = \frac{u^{n+1}}{n+1}\ln u - \frac{u^{n+1}}{(n+1)^2} + C$

67. $\displaystyle\int e^{au}\sin bu\, du = \frac{e^{au}}{a^2 + b^2}(a\sin bu - b\cos bu) + C$

68. $\displaystyle\int e^{au}\cos bu\, du = \frac{e^{au}}{a^2 + b^2}(a\cos bu + b\sin bu) + C$

INVERSE TRIGONOMETRIC FORMS

69. $\displaystyle\int \sin^{-1} u\, du = u\sin^{-1} u + \sqrt{1 - u^2} + C$

70. $\displaystyle\int \tan^{-1} u\, du = u\tan^{-1} u - \frac{1}{2}\ln(1 + u^2) + C$

71. $\displaystyle\int \sec^{-1} u\, du = u\sec^{-1} u - \ln\left|u + \sqrt{u^2 - 1}\right| + C$

72. $\displaystyle\int u\sin^{-1} u\, du = \frac{1}{4}(2u^2 - 1)\sin^{-1} u + \frac{u}{4}\sqrt{1 - u^2} + C$

73. $\displaystyle\int u\tan^{-1} u\, du = \frac{1}{2}(u^2 + 1)\tan^{-1} u - \frac{u}{2} + C$

74. $\displaystyle\int u\sec^{-1} u\, du = \frac{u^2}{2}\sec^{-1} u - \frac{1}{2}\sqrt{u^2 - 1} + C$

75. $\displaystyle\int u^n \sin^{-1} u\, du = \frac{u^{n+1}}{n+1}\sin^{-1} u - \frac{1}{n+1}\int \frac{u^{n+1}}{\sqrt{1 - u^2}}\, du + C \quad \text{if } n \neq -1$

76. $\displaystyle\int u^n \tan^{-1} u\, du = \frac{u^{n+1}}{n+1}\tan^{-1} u - \frac{1}{n+1}\int \frac{u^{n+1}}{1 + u^2}\, du + C \quad \text{if } n \neq -1$

77. $\displaystyle\int u^n \sec^{-1} u\, du = \frac{u^{n+1}}{n+1}\sec^{-1} u - \frac{1}{n+1}\int \frac{u^n}{\sqrt{u^2 - 1}}\, du + C \quad \text{if } n \neq -1$